11TH EDITION

A Survey of

Mathematics

with Applications

ALLEN R. ANGEL
Monroe Community College

CHRISTINE D. ABBOTT
Monroe Community College

DENNIS C. RUNDE
State College of Florida

 **Pearson**

4 2020

Rental
ISBN-10:   0-13-574046-0
ISBN-13: 978-0-13-574046-0

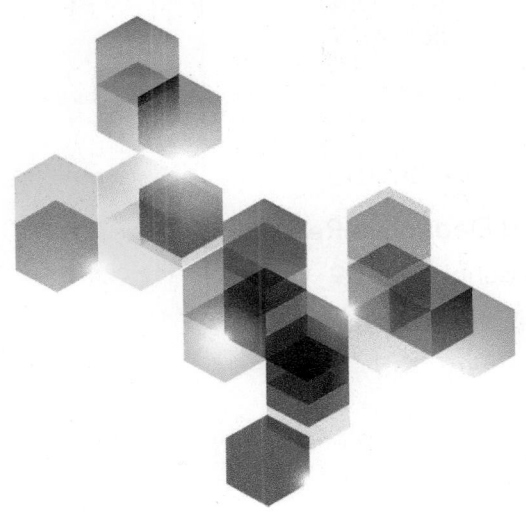

To my wife, Kathy Angel
A. R. A.

To my sons, Matthew and Jake Abbott
C. D. A.

To my mother, Tina Runde, and the memory of my father, Bud Runde
D. C. R.

Contents

Mathematics is an exciting, living study. Its applications shape the world around you and influence your everyday life. We hope that as you read this book you will realize just how important mathematics is and gain an appreciation of both its usefulness and its beauty. We also hope to teach you some practical mathematics that you can use every day and that will prepare you for further mathematics courses.

The primary purpose of this text is to provide material that you can read, understand, and enjoy. To this end, we have used straightforward language and tried to relate the mathematical concepts to everyday experiences. The concepts, definitions, and formulas that deserve special attention are in boxes or are set in boldface, italics, or color type. We have also provided many detailed examples for you to follow.

Be sure to read the chapter summary, work the review exercises, and take the chapter test at the end of each chapter. The answers to the odd-numbered exercises, all review exercises, and all chapter test exercises appear in the answer section in the back of the text. You should, however, use the answers only to check your work. The answers to all Recreational Mathematics exercises are provided either in the Recreational Mathematics boxes themselves or in the back of the book.

It is difficult to learn mathematics without becoming involved. To be successful, we suggest that you read the text carefully and *work each exercise in each assignment in detail*, for it is in doing the math that you really learn and enjoy it. If you are using this text within MyLab Math, you'll find a wealth of other learning aids available there, including tutorial videos and homework help.

We welcome your suggestions and your comments. You may contact us at math@pearson.com. (Please use the subject line "Angel Survey of Math.") Good luck with your adventure in mathematics!

Allen R. Angel

Christine D. Abbott

Dennis C. Runde

We present *A Survey of Mathematics with Applications*, Eleventh Edition, with the knowledge that we use mathematics every day. In this edition, we stress how mathematics is used in our daily lives and why it is important. Our primary goal is to give students a text they can read, understand, and enjoy while learning how mathematics affects the world around them. Numerous real-life applications are used to motivate topics. A variety of interesting and useful exercises demonstrate the real-life nature of mathematics and its importance in students' lives.

The text is intended for students who require a broad-based general overview of mathematics, especially those majoring in the liberal arts, elementary education, the social sciences, business, nursing, and allied health fields. It is particularly suitable for those courses that satisfy the minimum competency requirement in mathematics for graduation or transfer.

New to This Edition

New within the Textbook

- **Co-Requisite Course Support** – On the first page of each section in the Annotated Instructor Edition, there is a list of the prerequisite skills needed for that section. Please see the "Integrated Review" note under New to MyLab Math (below) for information on how to address gaps in prerequisite skills and use MyLab Math for co-requisite courses.
- **Data and Context Updates** – We've updated time-sensitive data to the most current available or changed the context of a problem or narrative so that it is more relevant to students.

- **Animations** – Located throughout the narrative you'll find Animation icons like the one pictured at the left. They are designed to facilitate active learning and visualization of key concepts. These Animations are housed within MyLab Math and are ideal for classroom use during lecture or by students independently. They were created in GeoGebra and are usable on any type of device. They are also editable.

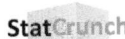

- **StatCrunch** – Located throughout the narrative you'll find StatCrunch icons like the one pictured at the left. Like the Animation features, they are designed to facilitate active learning and exploration. They are also housed within MyLab Math. StatCrunch is Pearson's online statistical software.
- **Technology Tips** – This feature of the text has been updated to go beyond graphing calculators and spreadsheets. Content now includes references to apps on smartphones and tablets.
- **Downloadable Data Sets** – For problems and examples in which students are expected to analyze a set of data, you'll see the icon **DS**, which indicates that the data is available to download in *.txt and *.csv formats. All of the data sets are housed in MyLab Math and also at **bit.ly/2Mxjj23**.
- **Learning Catalytics** – Learning Catalytics is a "bring your own device" student polling and assessment system, available in MyLab Math. Each section of the Annotated Instructor's Edition features keywords that can be entered into Learning Catalytics to bring you directly to questions for use in lecture for that section. Detailed instructions for using these keywords can be found at **bit.ly/31Z0r2p**.

New within MyLab Math

- **Integrated Review** – MyLab Math contains a chapter called "Additional Review" that features all of the necessary prerequisite objectives for this course. To help you use this content, we have also included:
  - **Skills Check Quizzes** by chapter assess the prerequisite skills students need for that chapter.

- Skills Review Homework, again by chapter, is personalized (based on the results of the Skills Check Quiz) to provide students with help on the prerequisite skills they are lacking. Students receive just the help they need—no more, no less.
    - Co-Requisite Support – In addition to the items above, for co-requisite courses (or students who just need more help), we've included videos and worksheets to provide necessary instruction for prerequisite skills. There's no need to go elsewhere for remediation.
- Tech Help for Stats Exercises – For statistics-related exercises in which students might use technology to solve, we've included a "Tech Help" button that allows them to get keystroke-by-keystroke instructions for StatCrunch, Excel, and TI Graphing Calculators. Note that this Tech Help button can be suppressed if you don't want students to use these technologies.
- Animations – Over 40 animations, referenced in the text, are designed to facilitate active learning and visualization of key concepts. They are ideal for classroom use during lecture or by students independently. They were created in GeoGebra and are usable on any type of device. They are also editable.
- StatCrunch – A dozen StatCrunch activities, referenced in the text, are designed to facilitate active learning and exploration. StatCrunch is Pearson's online statistical software.

Continuing and Revised Features

- Chapter Openers – Interesting and motivational applications introduce each chapter, which includes the Why This Is Important section, and illustrate the real-world nature of the chapter topics.
- Problem Solving – Beginning in Chapter 1, students are introduced to problem solving and critical thinking. We continue the theme of problem solving throughout the text and present special problem-solving exercises in the exercise sets.
- Critical Thinking Skills – In addition to a focus on problem solving, this book also features sections on inductive and deductive reasoning, estimation, and dimensional analysis.
- Profiles in Mathematics – Brief historical sketches and vignettes present stories of people who have advanced the discipline of mathematics. In this edition, we included more diversity among the mathematicians included.
- Did You Know? – These colorful, engaging, and lively features highlight the connections of mathematics to history, the arts and sciences, technology, and a broad variety of disciplines.
- Mathematics Today – These features discuss current real-life uses of the mathematical concepts in the chapter. Each box ends with Why This Is Important.
- Recreational Math – In these features, students are invited to apply the math in puzzles, games, and brain teasers. Answers are given in upside-down type at the bottom of the feature or (for longer answers) at the back of the book. In addition, Recreational Mathematics problems appear in the exercise sets so that they can be assigned as homework.
- Technology Tips – The material in these features explains how students can use technology including calculators, spreadsheets, and smartphone apps to explore various mathematical concepts and solve application problems.
- Timely Tips – These easy-to-identify boxes offer helpful information to make the material under discussion more understandable.
- Key-Idea Boxes – Important definitions, formulas, and procedures are boxed, making key information easy to identify for students.
- Chapter Summaries, Review Exercises, and Chapter Tests – The end-of-chapter summary charts provide an easy study experience by directing students to the location in the text where specific concepts are discussed. Review Exercises and Chapter Tests also help students review material and prepare for exams.

Acknowledgments

We thank the reviewers from all editions of the book and all the students who have offered suggestions for improving it. A list of reviewers for all editions of this book follows, with reviewers of this edition noted with an asterisk (*). Thanks to you all for helping make *A Survey of Mathematics with Applications* one of the most successful liberal arts mathematics textbooks in the country.

Reviewers for This and Previous Editions

Kate Acks, University of Hawaii Maui College

Marilyn Ahrens, Missouri Valley College

David Allen, Iona College

Mary Anne Anthony-Smith, Santa Ana College

Frank Asta, College of DuPage

Robin L. Ayers, Western Kentucky University

Hughette Bach, California State University–Sacramento

*Tammy Barker, Hillsborough Community College

Madeline Bates, Bronx Community College

Rebecca Baum, Lincoln Land Community College

Vivian Baxter, Fort Hays State University

Una Bray, Skidmore College

David H. Buckley, Polk State College

Robert C. Bueker, Western Kentucky University

Carl Carlson, Moorhead State University

Kent Carlson, St. Cloud State University

Scott Carter, Palm Beach State College

Yungchen Cheng, Missouri State University

Joseph Cleary, Massasoit Community College

*Cash Clifton, Central New Mexico Community College

Donald Cohen, SUNY College of Agriculture & Technology

*Celisa Counterman, Northampton Community College

David Dean, Santa Fe College

John Diamantopoulos, Northeastern State University

*Darlene Diaz, Santiago Canyon College

Greg Dietrich, Florida State College at Jacksonville

Charles Downey, University of Nebraska

Ryan Downie, Eastern Washington University

Jeffrey Downs, Western Nevada College

Annie Droullard, Polk State College

Patricia Dube, North Shore Community College

Ruth Ediden, Morgan State University

Lee Erker, Tri-County Community College

Nancy Eschen, Florida State College at Jacksonville

Karen Estes, St. Petersburg College

Mike Everett, Santa Ana College

Robert H. Fay, St. Petersburg College

Teklay Fessanaye, Santa Fe College

Kurtis Fink, Northwest Missouri State University

Raymond Flagg, McPherson College

*Donna Fowler, Palm Beach Atlantic University

Penelope Fowler, Tennessee Wesleyan College

Gilberto Garza, El Paso Community College

Judith L. Gersting, Indiana University–Purdue University at Indianapolis

*Rebecca Goad, Joliet Junior College

Patricia Granfield, George Mason University

Lucille Groenke, Mesa Community College

*Nick Haverhals, Avila University

Ryan Holbrook, University of Central Oklahoma

Kaylinda Holton, Tallahassee Community College

*Terri Honeycutt, Lander University

John Hornsby, University of New Orleans

*Mary Hoyt, University of North Texas at Dallas

Judith Ink, Regent University

Nancy Johnson, Broward College

Phyllis H. Jore, Valencia College

*Burcu Karabina, Florida Atlantic University

Heidi Kiley, Suffolk County Community College

Daniel Kimborowicz, Massasoit Community College

Jennifer Kimrey, Alamance Community College

Mary Lois King, Tallahassee Community College

Harriet H. Kiser, Georgia Highlands College

*Nichole Klemmer, Washtenaw Community College

Linda Kuroski, Erie Community College

Julia Ledet, Louisiana State University

David Lehmann, Southwest Missouri State University

Peter Lindstrom, North Lake College

James Magliano, Union College

Yash Manchanda, East Los Angeles College & Fullerton College

Richard Marchand, Slippery Rock University
Susan McCourt, Bristol Community College
Robert McGuigan, Westfield State College
Wallace H. Memmer, Brookdale Community College
Maurice Monahan, South Dakota State University
Julie Monte, Daytona State College
*Pedro J. Mora, Florida Gateway College
Karen Mosely, Alabama Southern Community College
*Daniel Thomas Murphree, Great Basin College
Kathleen Offenholley, Brookdale Community College
Edwin Owens, Pennsylvania College of Technology
Wing Park, College of Lake County
Bettye Parnham, Daytona State College
Joanne Peeples, El Paso Community College
Traci M. Reed, St. John's River State College
Nelson Rich, Nazareth College
Kenneth Ross, University of Oregon
*Timothy R. Ross, North Greenville University
Ronald Ruemmler, Middlesex County College
Rosa Rusinek, Queensborough Community College
Len Ruth, Sinclair Community College
John Samoylo, Delaware County Community College
Sandra Savage, Orange Coast College
Gerald Schultz, Southern Connecticut State University
Richard Schwartz, College of Staten Island
Kara Shavo, Mercer County Community College

Minnie Shuler, Chipola College
Paula R. Stickles, University of Southern Indiana
Kristin Stoley, Blinn College–Bryan
Steve Sworder, Saddleback College
*John Terrell, Troy University in Montgomery
Shirley Thompson, Moorhead College
Alvin D. Tinsley, Central Missouri State University
Sherry Tornwall, University of Florida
William Trotter, University of South Carolina
Zia Uddin, Lock Haven University of Pennsylvania
Michael Vislocky, University of Cincinnati
Thomas G. Walker II, Western Kentucky Community College
*Richard Watkins, Dabney S Lancaster Community College
*Carol Weideman, St. Petersburg College in Clearwater
Sandra Welch, Stephen F. Austin State University
Joyce Wellington, Southeastern Community College
Sue Welsch, Sierra Nevada College
Robert F. Wheeler, Northern Illinois University
Susan Wirth, Indian River State College
Judith B. Wood, College of Central Florida
Jean Woody, Tulsa Community College
James Wooland, Florida State University
*Kelly Young, Lander University
Michael A. Zwick, Monroe Community College

We thank James Lapp, Lauri Semarne, and Rick Ponticelli for their conscientious job of checking the text and answers for accuracy.

James Lapp also deserves our thanks for the excellent work he did on the Student's Solutions Manual and Instructor's Solutions Manual. We'd also like to thank Barbara Burke of Hawaii Pacific University and Deborah Doucette of Erie Community College for developing the Integrated Review worksheets to accompany the text.

Finally, we thank our family members Kathy, Robert, and Steve Angel, Mathew and Jake Abbott, and Kris, Alex, Nick, and Max Runde for their support and encouragement throughout this project. We are grateful for their wonderful support and understanding while we worked on the book. They also gave us support and encouragement and were very understanding when we could not spend as much time with them as we wished because of book deadlines. Without the support and understanding of our families, this book would not be a reality.

Allen R. Angel
Christine D. Abbott
Dennis C. Runde

Get the most out of
MyLab Math

MyLab Math for *Survey of Mathematics* 11e by Angel/Abbott/Runde *(access code required)*

MyLab Math is tightly integrated with each author team's style, offering a learning and practice experience that gives students a consistent experience from text to MyLab.

MyLab Math includes **assignable algorithmic exercises** for unlimited practice opportunities. The **complete eText** is available for learning in any environment. And a **comprehensive gradebook** gives instructors and students alike insight into how they are doing at any time.

Additionally, the following resources are available to enrich student learning.

Animations

NEW! Animations let students interact with the math in a visual, tangible way. These animations allow students to explore and manipulate the mathematical concepts, leading to more durable understanding. They can also be used by instructors in the classroom to enhance instruction.

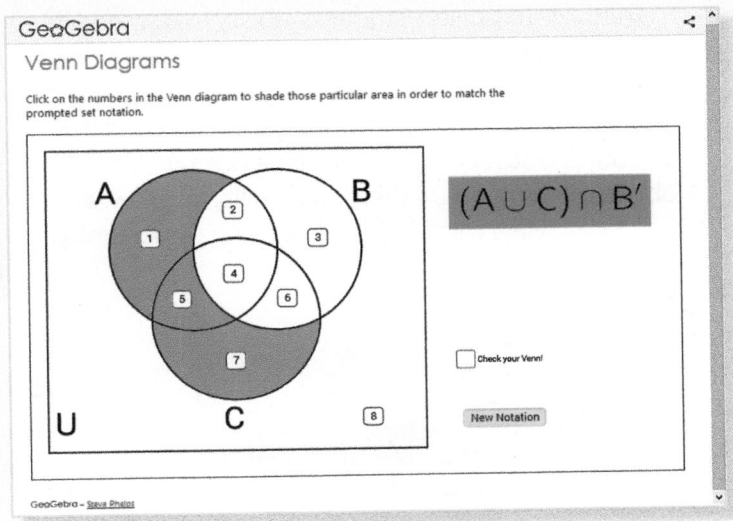

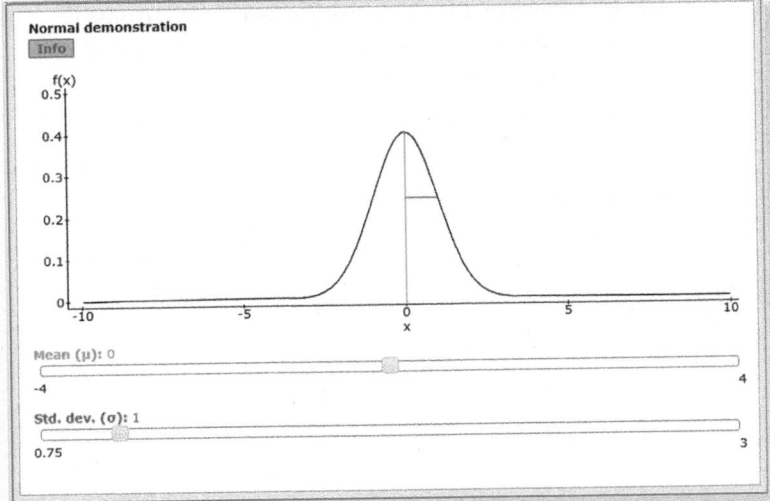

StatCrunch

NEW! StatCrunch® is a powerful web-based statistical software that allows users to collect, crunch, and communicate with data. Now integrated into this MyLab Math course, StatCrunch can be used to analyze and understand statistical concepts. **DS** icons in the text indicate that a data set is available to download online and analyze in statistical software, such as StatCrunch or Excel.

pearson.com/mylab/math

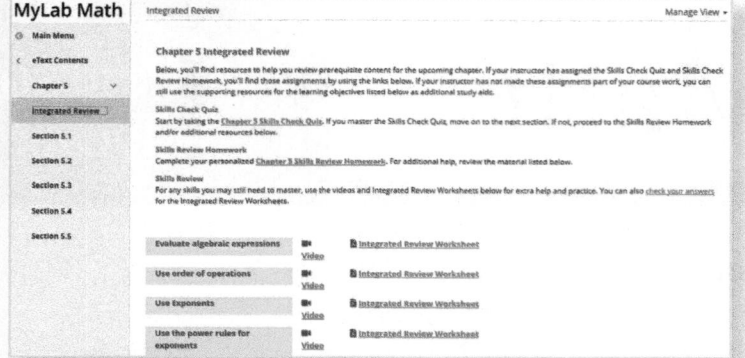

Integrated Review

Ideal for a corequisite course, or simply to get underprepared students up to speed, Integrated Review includes assessments, assignments, and resources on prerequisite topics. For each chapter, pre-made, assignable (and editable) quizzes assess students' understanding of the prerequisite skills needed for that chapter. Personalized follow-up homework assignments and re-mediation resources, in the form of videos and worksheets, are available for any gaps in skills that are identified.

Skills for Success Modules

These modules offer resources and assignments that aim to bolster students' ability to succeed in college courses and prepare for future professions. These modules now include a Mindset section with growth mindset-focused videos and exercises that encourage students to maintain a positive attitude about learning, value their own ability to grow, and view mistakes as a learning opportunity.

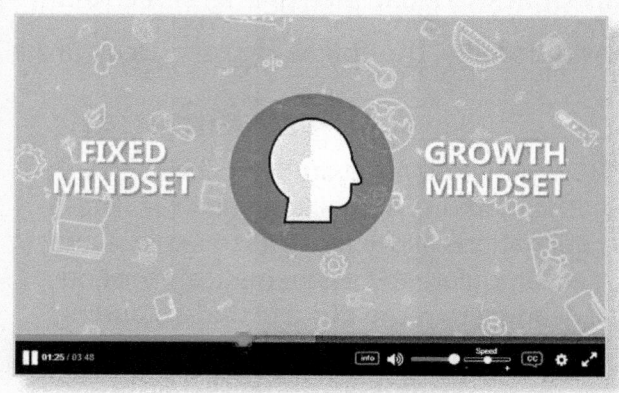

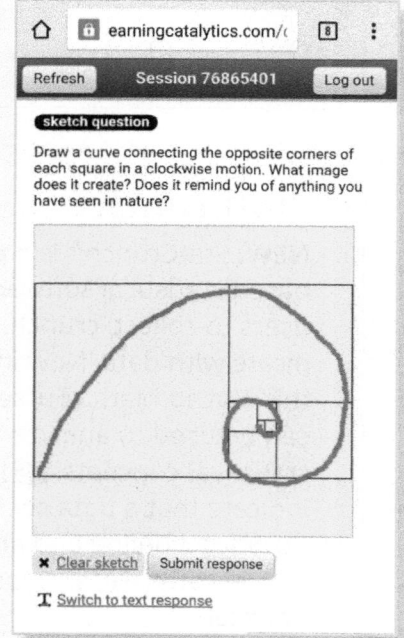

Learning Catalytics

Integrated into the MyLab course, Learning Catalytics uses students' devices in the classroom for an engagement, assessment, and classroom intelligence system that gives instructors real-time feedback on student learning. To make it easy to integrate this learning tool into the classroom, keywords are available in the Annotated Instructor's Edition at point of use. Detailed instructions for using these keywords can be found at **bit.ly/31Z0r2p**.

Resources for Success

Pearson MyLab

Instructor Resources

Annotated Instructor's Edition

All answers are included. When possible, answers are on the page with the exercises. Longer answers are in the back of the book. The AIE now includes references to Learning Catalytics to help instructors incorporate those resources. Corequisite support in the form of Integrated Review prerequisite skills are also referenced in the AIE for instructors who need to provide corequisite support in their course.

The following instructor resources are available to download from **www.pearson.com** or the MyLab Math course.

PowerPoint® Lecture Slides

Fully editable slides correlated with the textbook are available. Accessible, screen reader-friendly versions of the slides are also available.

Instructor's Solutions Manual

This manual includes fully worked solutions to all text exercises.

Instructor's Testing Manual

This manual includes tests with answer keys for each chapter of the text.

TestGen®

TestGen® (www.pearson.com/testgen) enables instructors to build, edit, print, and administer tests using a computerized bank of questions developed to cover all the objectives of the text.

Student Resources

Video Program

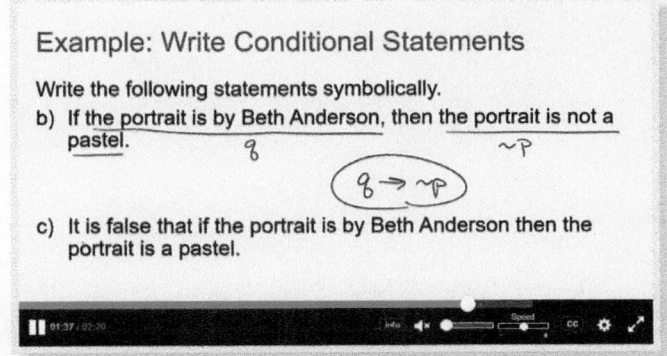

Available in MyLab Math, a modern, accessible video tutor program covers every section in the text, providing students with a video tutor at home, in a lab, or on the go. All videos have closed captioning available. Video assessment questions in the Assignment Manager check for student understanding of the video they just watched, making the videos assignable.

Student Solutions Manual

Available in MyLab Math and in print, this manual provides detailed worked-out solutions to odd-numbered exercises, and to all Chapter Review and Test exercises. (ISBN 10: 0-13-574045-2; ISBN 13: 978-0-13-574045-3)

Integrated Review Workbook

This workbook provides additional practice and study support for students. Worksheets cover prerequisite developmental-level topics necessary to move on to each chapter. This workbook is available in MyLab Math or as a printed, unbound, three-hole-punched workbook. (ISBN 10: 0-13-574028-2 ISBN 13: 978-0-13-574028-6)

pearson.com/mylab/math

◄ *Graduating from college requires the solution to many problems. Using critical thinking skills is vital when solving these problems.*

1

Critical Thinking Skills

What You Will Learn

- Inductive and deductive reasoning
- Estimation techniques
- Problem-solving procedures

Why This Is Important

Everyday life presents us with a wide range of problems that we must solve such as deciding which classes to take in the next semester of college or deciding if you can afford to buy a new car. Making good decisions can lead to desirable outcomes. For example, choosing classes that are part of your program of study may help you stay on track to graduate from college. Delaying a new car purchase may help you afford more important things in your life like college tuition.

The information you encounter in this chapter will help sharpen your critical thinking skills. These skills will help you make better decisions as you solve the problems that you encounter in your everyday life.

SECTION 1.1 Inductive and Deductive Reasoning

Upon completion of this section, you will be able to:

- Understand and use inductive reasoning to solve problems.
- Understand and use deductive reasoning to solve problems.

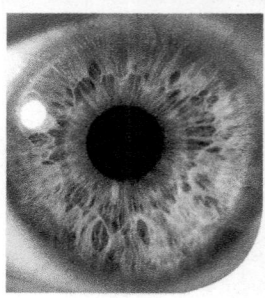

The science of biometrics involves the measurement and analysis of unique physical characteristics. Biometrics are usually used as a means of verifying personal identity. Fingerprints, iris patterns in eyes, facial recognition, DNA, and voice patterns can all be used for personal identification. Some smartphones can be unlocked by pressing a button that recognizes a unique fingerprint or by scanning an eye to recognize a unique iris pattern. Crime scene investigation often involves fingerprint and DNA evidence. Voice recognition software uses voice patterns of callers to help prevent fraud and to improve customer service.

Why This Is Important Using biometrics for personal identification involves reasoning to a general conclusion through observation of specific cases. In this section, we will discuss how inductive and deductive reasoning are essential critical thinking skills used in biometrics and in many other applications.

Inductive Reasoning

Before looking at some examples of inductive reasoning and problem solving, let us first review a few facts about certain numbers. The *natural numbers* or *counting numbers* are the numbers 1, 2, 3, 4, 5, 6, 7, 8, ... The three dots, called an *ellipsis*, mean that 8 is not the last number but that the numbers continue in the same manner. A word that we sometimes use when discussing the counting numbers is "divisible." If $a \div b$ has a remainder of zero, then *a is divisible by b*. The counting numbers that are divisible by 2 are 2, 4, 6, 8, ... These numbers are called the *even counting numbers*. The counting numbers that are not divisible by 2 are 1, 3, 5, 7, 9, ... These numbers are the *odd counting numbers*. When we refer to *odd numbers* or *even numbers*, we mean odd or even counting numbers.

Recognizing patterns is sometimes helpful in solving problems, as Examples 1 and 2 illustrate.

Example 1 *The Product of Two Odd Numbers*

If two odd numbers are multiplied together, will the product always be an odd number?

Solution To answer this question, we will examine the products of several pairs of odd numbers to see if there is a pattern.

$1 \times 5 = 5$	$3 \times 7 = 21$	$5 \times 9 = 45$
$1 \times 7 = 7$	$3 \times 9 = 27$	$5 \times 11 = 55$
$1 \times 9 = 9$	$3 \times 11 = 33$	$5 \times 13 = 65$
$1 \times 11 = 11$	$3 \times 13 = 39$	$5 \times 15 = 75$

All the products are odd numbers. Thus, we might predict from these examples that the product of any two odd numbers is an odd number. ∎

Example 2   *The Sum of an Odd Number and an Even Number*

If an odd number and an even number are added, will the sum be an odd number or an even number?

Solution   Let's look at a few examples in which one number is odd and the other number is even.

$$3 + 4 = 7 \qquad\qquad 9 + 6 = 15 \qquad\qquad 23 + 18 = 41$$

$$5 + 12 = 17 \qquad\qquad 5 + 14 = 19 \qquad\qquad 81 + 32 = 113$$

All these sums are odd numbers. Therefore, we might predict that the sum of an odd number and an even number is an odd number. ∎

In Examples 1 and 2, we cannot conclude that the results are true for all counting numbers. From the patterns developed, however, we can make predictions. This type of reasoning process, arriving at a general conclusion from specific observations or examples, is called *inductive reasoning*, or *induction*.

> Definition: **Inductive Reasoning**
> **Inductive reasoning** is the process of reasoning to a general conclusion through observations of specific cases.

Induction often involves observing a pattern and from that pattern predicting a conclusion. Imagine an endless row of dominoes. You knock down the first, which knocks down the second, which knocks down the third, and so on. Assuming the pattern will continue uninterrupted, you conclude that any one domino that you select in the row will eventually fall, even though you may not witness the event.

Inductive reasoning is often used by mathematicians and scientists to develop theories and predict answers to complicated problems. For this reason, inductive reasoning is part of the *scientific method*. When a scientist or mathematician makes a prediction based on specific observations, it is called a *hypothesis* or *conjecture*. After looking at the products in Example 1, we might conjecture that the product of two odd numbers will be an odd number. After looking at the sums in Example 2, we might conjecture that the sum of an odd number and an even number is an odd number.

Examples 3 and 4 illustrate how we arrive at a conclusion using inductive reasoning.

Example 3   *Biometrics*

As described in the opening paragraph of this section, the science of biometrics is used for personal identification. What reasoning process has led to the conclusion that no two people have the same fingerprints, iris patterns, DNA, or voice patterns?

Solution   By studying the biometrics of millions of people, scientists have never found two people who have the exact same fingerprints, iris patterns, DNA, or voice patterns. By induction, then, a conclusion can be reached that each of these biometrics provides a unique identification. A general conclusion is reached through the observation of specific cases. Therefore, the science of biometrics makes use of inductive reasoning. ∎

Example 4 *Divisibility by 4*

Consider the conjecture "If the last two digits of a number form a number that is divisible by 4, then the number itself is divisible by 4." We will test several numbers to see if the conjecture appears to be true or false.

Solution Let's look at some numbers whose last two digits form a number that is divisible by 4.

Number	Do the Last Two Digits Form a Number That is Divisible by 4?	Is the Number Divisible by 4?
324	Yes; $24 \div 4 = 6$	Yes; $324 \div 4 = 81$
4328	Yes; $28 \div 4 = 7$	Yes; $4328 \div 4 = 1082$
10,612	Yes; $12 \div 4 = 3$	Yes; $10,612 \div 4 = 2653$
21,104	Yes; $4 \div 4 = 1$	Yes; $21,104 \div 4 = 5276$

In each case, we find that if the last two digits of a number are divisible by 4, then the number itself is divisible by 4. From these examples, we might be tempted to generalize that the conjecture "If the last two digits of a number are divisible by 4, then the number itself is divisible by 4" is true.* ■

Example 5 *Pick a Number, Any Number*

Pick any number, multiply the number by 4, add 2 to the product, divide the sum by 2, and subtract 1 from the quotient. Repeat this procedure for several different numbers and then make a conjecture about the relationship between the original number and the final number.

Solution Let's go through this one together.

Pick a number:	say, 5
Multiply the number by 4:	$4 \times 5 = 20$
Add 2 to the product:	$20 + 2 = 22$
Divide the sum by 2:	$22 \div 2 = 11$
Subtract 1 from the quotient:	$11 - 1 = 10$

Note that we started with the number 5 and finished with the number 10. If you start with the number 2, you will end with the number 4. Starting with 3 would result in a final number of 6, 4 would result in 8, and so on. On the basis of these few examples, we may conjecture that when you follow the given procedure, the number you end with will always be twice the original number. ■

The result reached by inductive reasoning is often correct for the specific cases studied but not correct for all cases. History has shown that not all conclusions arrived at by inductive reasoning are correct. For example, Aristotle (384–322 B.C.) reasoned inductively that heavy objects fall at a faster rate than light objects. About 2000 years later, Galileo (1564–1642) dropped two pieces of metal—one 10 times heavier than the other—from the Leaning Tower of Pisa in Italy. He found that both hit the ground at exactly the same moment, so they must have traveled at the same rate.

---

* This statement is in fact true, as is discussed in Section 5.1.

When forming a general conclusion using inductive reasoning, you should test it with several special cases to see whether the conclusion appears correct. If a special case is found that satisfies the conditions of the conjecture but produces a different result, such a case is called a *counterexample*. A counterexample proves that the conjecture is false because only one exception is needed to show that a conjecture is not valid. Galileo's counterexample disproved Aristotle's conjecture. If a counterexample cannot be found, the conjecture is neither proven nor disproven.

Consider the statement "All birds fly." A penguin is a bird that does not fly. Therefore, a penguin is a counterexample to the statement "All birds fly."

Deductive Reasoning

A second type of reasoning process is called *deductive reasoning*, or *deduction*. Mathematicians use deductive reasoning to *prove* conjectures true or false.

> **Definition: Deductive Reasoning**
> **Deductive reasoning** is the process of reasoning to a specific conclusion from a general statement.

Timely Tip

The following diagram helps explain the difference between inductive reasoning and deductive reasoning. Inductive reasoning is the process of reasoning to a general conclusion through observations of specific cases. Deductive reasoning is the process of reasoning to a specific conclusion from a general statement.

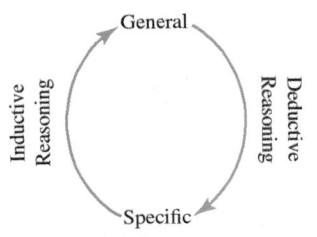

Number Tricks Using Deductive Reasoning

Example 6 illustrates deductive reasoning.

┌ Example **6** *Pick a Number, n*

Prove, using deductive reasoning, that the procedure given in Example 5 will always result in twice the original number selected.

Solution To use deductive reasoning, we begin with the *general* case rather than specific examples. In Example 5, specific cases were used. Let's select the letter n to represent *any number*.

Pick any number:	n
Multiply the number by 4:	$4n$ *4n means 4 times n.*
Add 2 to the product:	$4n + 2$
Divide the sum by 2:	$\dfrac{4n + 2}{2} = \dfrac{\overset{2}{\cancel{4}}n}{\underset{1}{\cancel{2}}} + \dfrac{\overset{1}{\cancel{2}}}{\underset{1}{\cancel{2}}} = 2n + 1$
Subtract 1 from the quotient:	$2n + 1 - 1 = 2n$

Note that for any number n selected, the result is $2n$, or twice the original number selected. Since n represented any number, we are beginning with the general case. Thus, this is deductive reasoning. ∎

In Example 5, you may have *conjectured*, using specific examples and inductive reasoning, that the result would be twice the original number selected. In Example 6, we *proved*, using deductive reasoning, that the result will always be twice the original number selected.

SECTION 1.1   *Exercises*

Warm Up Exercises

In Exercises 1–8, fill in the blank with an appropriate word, phrase, or symbol(s).

1. Another name for the counting numbers is the _____ numbers.

2. If $a \div b$ has a remainder of 0, then a is _____ by b.

3. A specific case that satisfies the conditions of a conjecture but shows the conjecture is false is called a _____.

4. A belief based on specific observations that has not been proven or disproven is called a conjecture or _____.

5. The process of reasoning to a general conclusion through observation of specific cases is called _____ reasoning.

6. The process of reasoning to a specific conclusion from a general statement is called _____ reasoning.

7. The type of reasoning used to prove a conjecture is called _____ reasoning.

8. The type of reasoning generally used to arrive at a conjecture is called _____ reasoning.

Practice the Skills

In Exercises 9–12, use inductive reasoning to predict the next line in the pattern.

9. $1 \times 3 = 3$
   $2 \times 3 = 6$
   $3 \times 3 = 9$
   $4 \times 3 = 12$

10. $15 \times 10 = 150$
    $16 \times 10 = 160$
    $17 \times 10 = 170$
    $18 \times 10 = 180$

11. 1
       $1\ \ 1$
      $1\ \ 2\ \ 1$
     $1\ \ 3\ \ 3\ \ 1$
     $\downarrow\swarrow\downarrow\swarrow\downarrow\swarrow\downarrow$
    $1\ \ 4\ \ 6\ \ 4\ \ 1$

12. $10 = 10^1$
       $100 = 10^2$
      $1000 = 10^3$
     $10{,}000 = 10^4$

In Exercises 13–16, draw the next figure in the pattern (or sequence).

13. ⬠, ▢, ⬠, . . .

14. 😊😟😔😊😟😔, . . .

15. △▢⬠⬡, . . .

16. 😊😊😊😊😊, . . .

In Exercises 17–26, use inductive reasoning to predict the next three numbers in the pattern (or sequence).

17. 1, 3, 5, 7, …

18. 4, 7, 10, 13, …

19. 5, −5, 5, −5, …

20. 1, −2, 4, −8, …

21. $1, \frac{1}{2}, \frac{1}{3}, \frac{1}{4}, \ldots$

22. $1, \frac{1}{3}, \frac{1}{5}, \frac{1}{7}, \ldots$

23. 0, 1, 3, 6, 10, 15, …

24. 1, 1, 2, 4, 7, 11, …

25. 1, 1, 2, 3, 5, 8, 13, 21, …

26. 1, 3, 4, 7, 11, 18, 29, 47, …

Problem Solving

27. Determine the letter that is the 118th entry in the following sequence. Explain how you determined your answer.

    Y, R, R, Y, R, R, Y, R, R, Y, R, R, Y, R, R, …

28. a) Select a variety of one- and two-digit numbers between 1 and 99 and multiply each by 9. Record your results.

    b) Determine the sum of the digits in each of your products in part (a). If the sum is not a one-digit number, determine the sum of the digits of the resulting sum again until you obtain a one-digit number.

    c) Make a conjecture about the sum of the digits when a one- or two-digit number is multiplied by 9.

29. *A Square Pattern* The ancient Greeks labeled certain numbers as **square numbers**. The numbers 1, 4, 9, 16, 25, and so on are square numbers.

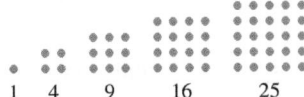

1 4 9 16 25

a) Determine the next three square numbers.

b) Describe a procedure to determine the next five square numbers without drawing the figures.

c) Is 72 a square number? Explain how you determined your answer.

30. *A Triangular Pattern* The ancient Greeks labeled certain numbers as *triangular numbers*. The numbers 1, 3, 6, 10, 15, 21, and so on are triangular numbers.

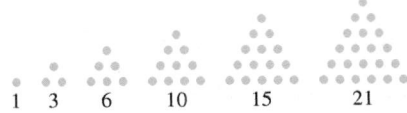

1 3 6 10 15 21

a) Can you determine the next two triangular numbers?

b) Describe a procedure to determine the next five triangular numbers without drawing the figures.

c) Is 72 a triangular number? Explain how you determined your answer.

31. *Quilt Design* The pattern shown is taken from a quilt design known as a triple Irish chain. Complete the color pattern by indicating the color assigned to each square.

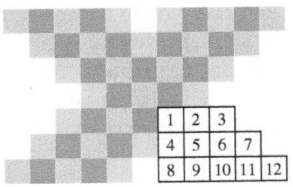

32. *Triangles in a Triangle* Four rows of a triangular figure are shown.

a) If you added six additional rows to the bottom of this triangle, using the same pattern displayed, how many triangles would appear in the 10th row?

b) If the triangles in all 10 rows were added, how many triangles would appear in the entire figure?

33. *Community College Tuition* The graph below shows the annual tuition, to the nearest hundred dollars, at Clarence Community College for the years 2015–2019.

a) Assuming this trend continues, use the graph to predict the annual tuition for the year 2020.

b) Explain how you are using inductive reasoning to determine your answer.

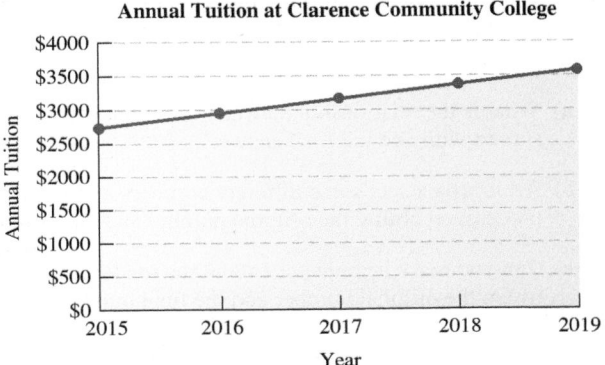

34. *Gym Membership Fees* The graph below shows the annual membership fees for Crackle Fitness for the years 2015–2019.

a) Assuming this trend continues, use the graph to predict the annual membership fee for the year 2020.

b) Explain how you are using inductive reasoning to determine your answer.

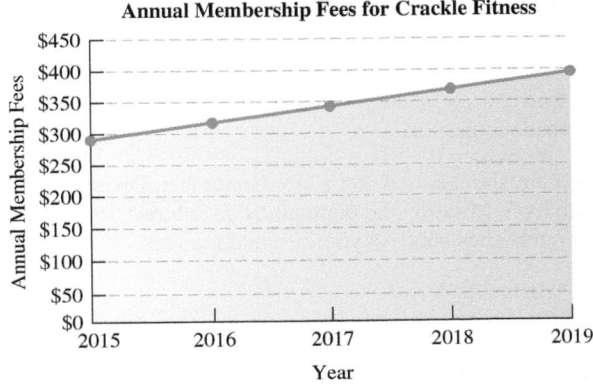

In Exercises 35 and 36, draw the next diagram in the pattern (or sequence).

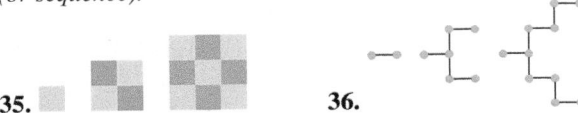

35.   **36.**

37. Pick any number, multiply the number by 3, add 6 to the product, divide the sum by 3, and subtract 2 from the quotient. See Example 5.

a) What is the relationship between the number you started with and the final number?

b) Arbitrarily select some different numbers and repeat the process, recording the original number and the result.

c) Can you make a conjecture about the relationship between the original number and the final number?

d) Prove, using deductive reasoning, the conjecture you made in part (c). See Example 6.

38. Pick any number and multiply the number by 4. Add 6 to the product. Divide the sum by 2 and subtract 3 from the quotient.

a) What is the relationship between the number you started with and the final answer?

b) Arbitrarily select some different numbers and repeat the process, recording the original number and the results.

c) Can you make a conjecture about the relationship between the original number and the final number?

d) Prove, using deductive reasoning, the conjecture you made in part (c).

39. Pick any number and add 1 to it. Determine the sum of the new number and the original number. Add 9 to the sum. Divide the new sum by 2 and subtract the original number from the quotient.

a) What is the final number?

b) Arbitrarily select some different numbers and repeat the process. Record the results.

c) Can you make a conjecture about the final number?

d) Prove, using deductive reasoning, the conjecture you made in part (c).

40. Pick any number and add 10 to the number. Divide the sum by 5. Multiply the quotient by 5. Subtract 10 from the product. Then subtract your original number.

a) What is the result?

b) Arbitrarily select some different numbers and repeat the process, recording the original number and the result.

c) Can you make a conjecture regarding the result when this process is followed?

d) Prove, using deductive reasoning, the conjecture you made in part (c).

In Exercises 41–46, determine a counterexample to show that each statement is incorrect.

41. The product of any two counting numbers is divisible by 2.

42. The sum of any three two-digit numbers is a three-digit number.

43. When a counting number is added to 3 and the sum is divided by 2, the quotient will be an even number.

44. The product of any two three-digit numbers is a five-digit number.

45. The difference of any two counting numbers will be a counting number.

46. The sum of any two odd numbers is divisible by 4.

47. *Interior Angles of a Triangle*

a) Construct a triangle and measure the three interior angles with a protractor. What is the sum of the measures?

b) Construct three other triangles, measure the angles, and record the sums. Are your answers the same?

c) Make a conjecture about the sum of the measures of the three interior angles of a triangle.

48. *Interior Angles of a Quadrilateral*

a) Construct a quadrilateral (a four-sided figure) and measure the four interior angles with a protractor. What is the sum of the measures?

b) Construct three other quadrilaterals, measure the angles, and record the sums. Are your answers the same?

c) Make a conjecture about the sum of the measures of the four interior angles of a quadrilateral.

Concept/Writing Exercises

49. *Computer Log On* While logging on to your computer, you type in your username followed by what you believe is your password. The computer indicates that a mistake has been made and asks you to try again. You retype your username and the same password. Again, the computer indicates a mistake has been made. You decide not to try again, reasoning you will get the same error message from the computer. What type of reasoning did you use? Explain.

50. *Lottery Ticket* You have purchased one lottery ticket each week for many months and have not won more than $5.00. You decide, based on your past experience, that you are not going to win the grand prize and so you stop playing the lottery. What type of reasoning did you use? Explain.

Challenge Problems/Group Activities

51. Complete the following square of numbers. Explain how you determined your answer.

1	2	3	4
2	5	10	17
3	10	25	52
4	17	52	?

52. Determine the next three numbers in the sequence.

1, 8, 11, 88, 101, 111, 181, 808, 818, 888, 1001, 1111, …

Recreational Mathematics

53.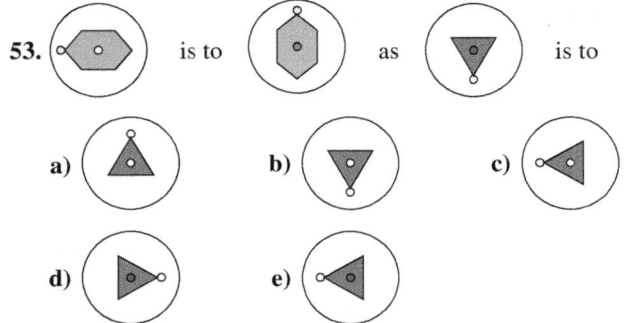

Research Activities

54. *Biometrics* Write a report on the use of biometrics as a means for verifying personal identity. Include a variety of biometrics such as fingerprints, iris patterns, DNA, voice recognition, and facial recognition. Also, include in your report a description of how inductive reasoning is used.

55. *Inductive Reasoning* Using **Stat**Crunch, the internet, magazines, or other sources, gather a set of data that involves a series of years. For two examples of such data, review Exercises 33–34.

a) Use a computer spreadsheet such as Excel or **Stat**Crunch to create a graph using the data set.

b) Use the graph of the data set to predict the outcomes for the next two years.

c) Explain how inductive reasoning was used to reach the conclusion.

56. *Jury Decision* When a jury decides whether a defendant is guilty, do the jurors collectively use primarily inductive reasoning, primarily deductive reasoning, or an equal amount of each? Write a brief report supporting your answer.

SECTION 1.2   Estimation Techniques

Upon completion of this section, you will be able to:

■ Use estimation techniques to determine an approximate answer to a question.

The cost to rent a campsite is $29 per night plus 7% sales tax. What would be the approximate cost to rent the campsite over Labor Day weekend? The cost to print T-shirts to promote a local charity is $8.99 per T-shirt. What would be the approximate cost to print 150 T-shirts? The payments on a new car would be $409 per month for 48 months. What is the approximate total amount of all these car payments? In this section, we will introduce estimation techniques that will allow us to answer these questions.

Why This Is Important Estimation techniques are helpful whenever we need to make financial decisions. By making an accurate estimate prior to making a financial decision, we often can make better financial choices and may be able to avoid unpleasant outcomes.

Estimation Techniques

An important step in solving mathematical problems—or, in fact, *any* problem—is to make sure that the answer you've arrived at makes sense. One technique for determining whether an answer is reasonable is to estimate. *Estimation* is the process of arriving at an approximate answer to a question. This section demonstrates several estimation methods.

To estimate, or approximate, an answer, we often round numbers, as illustrated in the following examples. The symbol $\approx$ means *is approximately equal to*.

Example 1 *Estimating the Cost of Cupcakes*

Malley decides to purchase cupcakes for a party. Estimate her cost if she purchases 21 cupcakes at $1.95 each.

Solution We may round the amounts as follows to obtain an estimate.

Number	Number Rounded
21	20
$\times$ $1.95	$\times$ $2.00
	$40.00

Thus, the 21 cupcakes would cost approximately $40.00, written $\approx$ $40.00.   ■

In Example 1, the true cost is $1.95 $\times$ 21, or $40.95. *Estimates are not meant to give exact values for answers but are a means of determining whether your answer is reasonable.* If you calculated an answer of $40.95 and then did a quick estimate to check it, you would know that the answer is reasonable because it is close to your estimated answer.

Example 2 *Two Ways to Estimate*

At a local discount retail store, Raj purchased a rug for $15.89, toothpaste for $1.49, diapers for $19.77, shampoo for $4.93, dog food for $12.88, and candy for $0.81. The cashier said the total bill was $69.51. Use estimation to determine whether this amount is reasonable.

Solution The most expensive item is $19.77 and the least expensive is $0.81. How should we estimate? We will estimate two different ways. First, we will round the cost of each item to the nearest 10 cents. For the second method, we will round the cost of each item to the nearest dollar. Rounding to the nearest 10 cents is more accurate. To determine whether the total bill is reasonable, however, we may need to round only to the nearest dollar.

	Round to the Nearest 10 Cents		Round to the Nearest Dollar	
Rug	$15.89 $\rightarrow$	$15.90	$15.89 $\rightarrow$	$16.00
Toothpaste	1.49 $\rightarrow$	1.50	1.49 $\rightarrow$	1.00
Diapers	19.77 $\rightarrow$	19.80	19.77 $\rightarrow$	20.00
Shampoo	4.93 $\rightarrow$	4.90	4.93 $\rightarrow$	5.00
Dog food	12.88 $\rightarrow$	12.90	12.88 $\rightarrow$	13.00
Candy	0.81 $\rightarrow$	0.80	0.81 $\rightarrow$	1.00
		$55.80		$56.00

Using either estimate, we find the bill of $69.51 is quite high. Therefore, Raj should check the bill carefully before paying it. Adding the prices of all six items gives a true cost of $55.77.   ■

Example 3 *Select the Best Estimate*

The number of bushels of grapes produced at a vineyard are 71,309 Cabernet Sauvignon, 123,879 French Colombard, 106,490 Chenin Blanc, 5960 Charbono, and 12,104 Chardonnay. Select the best estimate of the total number of bushels produced by the vineyard.

a) 500,000   b) 30,000   c) 300,000   d) 5,000,000

Solution   Following are suggested roundings. On the left, the numbers are rounded to thousands. For a less accurate estimate, round to ten thousands, as illustrated on the right.

Round to the Nearest Thousand		Round to the Nearest Ten Thousand	
71,309 →	71,000	71,309 →	70,000
123,879 →	124,000	123,879 →	120,000
106,490 →	106,000	106,490 →	110,000
5,960 →	6,000	5,960 →	10,000
12,104 →	12,000	12,104 →	10,000
	319,000		320,000

Either rounding procedure indicates that the best estimate is (c), or 300,000.   ∎

Example 4   *Using Estimation in Calculations*

The Martellos traveled 1276 miles from their home in Springfield, Missouri, to Yellowstone National Park.

a) If their trip took 19.9 hours of driving time, estimate the average speed, in miles per hour, that their car traveled.

b) If their car used 52 gallons of gasoline, estimate the average gas mileage of their car, in miles per gallon.

c) If the cost of gasoline averaged $2.05 per gallon, estimate the total cost of the gasoline used.

Solution

a) To estimate the average miles per hour, divide the number of miles by the number of hours.

$$\frac{1276}{19.9}$$

Round these numbers to obtain an estimate.

$$\frac{1300}{20} = 65$$

Therefore, the car averaged about 65 miles per hour.

b) To estimate the average gas mileage, divide the number of miles driven by the number of gallons of gasoline used.

$$\frac{1276}{52}$$

Round these numbers to obtain an estimate.

$$\frac{1300}{50} = 26$$

Therefore, the car averaged about 26 miles per gallon.

c) To estimate the total cost of gasoline, multiply the price per gallon by the number of gallons of gasoline used. Round the number of gallons of gasoline to 50 gallons and the price of the gasoline to $2.00 per gallon. The cost of the gasoline is about 50 × $2.00, or $100.   ∎

Now let's look at some different types of estimation problems.

▲ *Old Faithful geyser at Yellowstone National Park*

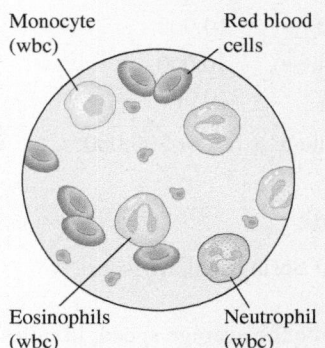
Example 5   *Estimating Distances on Trails to the West*

In the mid-1800s, thousands of settlers followed trails to the West to gain cheap, fertile land and a chance to make a fortune. One of the main trails to the West was the Oregon Trail, which ran from Independence, Missouri, to the Oregon Territory. Another trail, the California Trail, ran from Fort Hall on the Oregon Trail to Sacramento, California. A map of the Oregon Trail and the California Trail is shown below.

Trails to the West, 1850

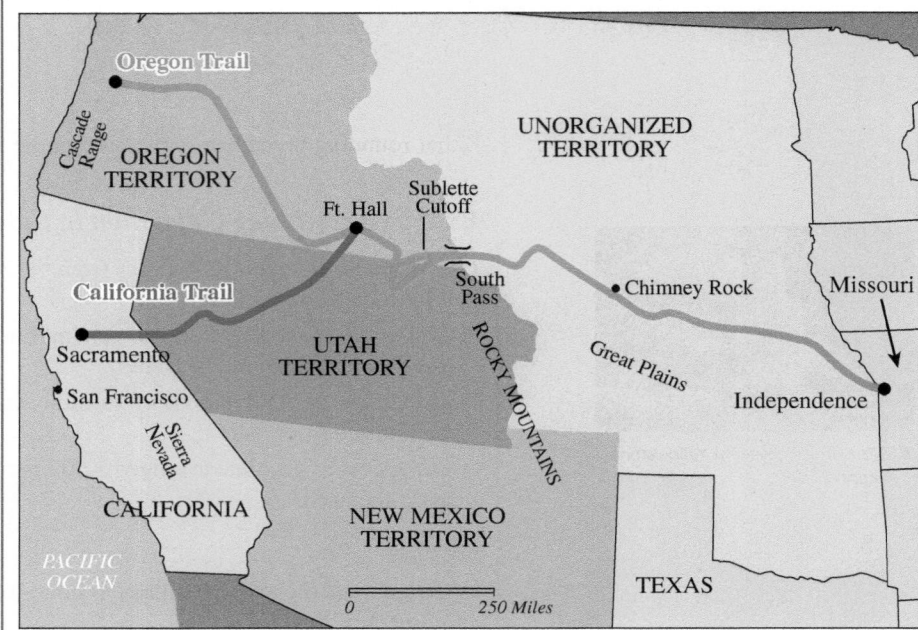

a) Using the scale shown and the Oregon Trail, estimate the distance from Independence, Missouri, to Fort Hall in the Oregon Territory.

b) Using the scale shown and the California Trail, estimate the distance from Fort Hall in the Oregon Territory to Sacramento, California.

Solution

a) Using a ruler and the scale given on the map, we can determine that approximately $\frac{3}{4}$ inch represents 250 miles (mi). One way to determine the distance from Independence, Missouri, to Fort Hall in the Oregon Territory is to mark off $\frac{3}{4}$-inch intervals along the Oregon Trail. If you do so, you should obtain approximately 4 three-quarter inch intervals. Thus, the distance is about 4×250 mi, or about 1000 mi.

   Sometimes on a map like this one, it may be difficult to get an accurate estimate because of the curves on the map. To get a more accurate estimate, you may want to use a piece of string. Place the beginning of the string at Independence, Missouri, and, using tape or pins, align the string with the road. Indicate on the string the point represented by Fort Hall. Then remove the string and make $\frac{3}{4}$-inch interval markings on the string (or measure the length of string you marked off). If, for example, your string from Independence, Missouri, to Fort Hall measures $3\frac{1}{4}$ inch, then there are $4\frac{1}{3}$ three quarter inch intervals, and the distance is about 4.33×250 or about 1083 mi.

b) Using the procedure discussed in part (a), we estimate that the distance from Fort Hall in the Oregon Territory to Sacramento, California, is about $1\frac{5}{8}$ inch, or about 2.2 three-quarter-inch intervals. Thus, the distance from Fort Hall in the Oregon Territory to Sacramento, California, is about 2.2×250 mi, or about 550 mi.   ∎

Example 6 *Estimated Energy Use*

Some utility bills contain graphs illustrating the amount of electricity and natural gas used. The following graphs show gas and electric use at a specific residence for a period of 13 months, starting in November 2019 and going through November 2020 (the month of the current bill). Also shown, in red, is the bill for the average residential customer for November 2020. Using these graphs, answer the following questions.

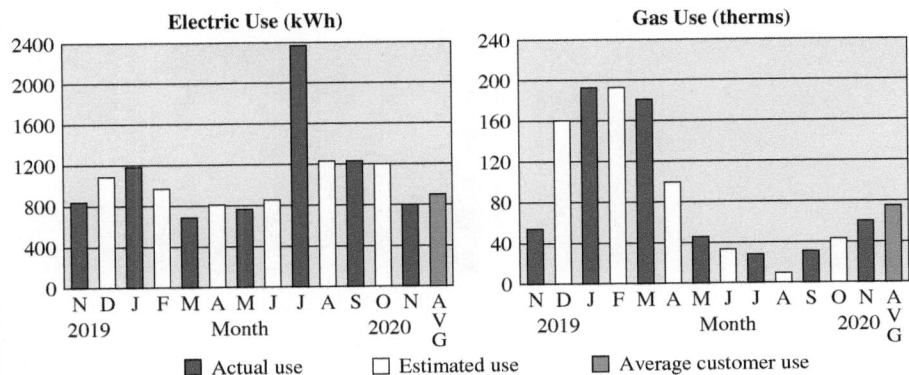

a) How often were actual gas and electric readings made?

b) Estimate the number of therms of gas used by the average residential customer in November 2020.

c) Estimate the amount of gas used by the resident in November 2020.

d) If the cost of gas is $1.12295 per therm, estimate the gas bill in March 2020.

e) In which month was the most electricity used? How many kilowatt-hours (kWh) of electricity were used in this month?

f) If the cost of electricity is 10.0127 cents per kWh, estimate the cost of the electricity in January 2020.

Solution

a) Actual readings were made every other month, in November 2019, January 2020, March 2020, May 2020, July 2020, September 2020, and November 2020. Thus, actual readings were made 7 times.

b) In November 2020, approximately 75 therms were used by the average residential customer, as shown by the height of the red bar.

c) In November 2020, approximately 60 therms were used by the resident.

d) In March 2020, about 180 therms were used. The rate is $1.12295 per therm. To get a rough approximation, round the rate to $1.12 per therm.

$$1.12 \times 180 = \$201.60$$

Thus, the cost of gas used was about $201.60.

e) The most electricity was used in July 2020. Approximately 2400 kWh of electricity were used.

f) In January 2020, about 1200 kWh were used. Write 10.0127 cents as $0.100127. Rounding the rate to $0.10 per kWh and multiplying by 1200 yields an estimate of $120.

$$0.10 \times 1200 = \$120$$

Thus, the cost of electricity in January 2020 was about $120. ∎

Example 7 *Estimating the Number of Corn Kernels in a Photo*

Estimate the number of corn kernels in the top layer in the accompanying photograph.

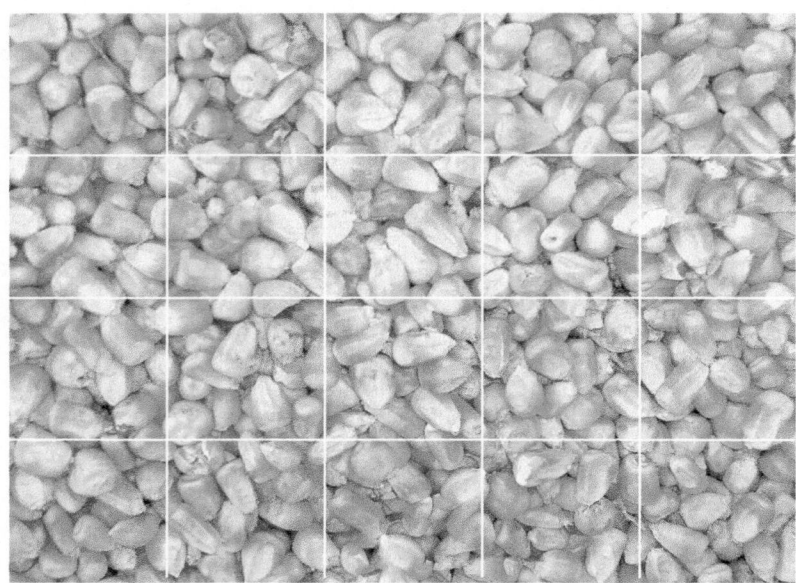

Solution To estimate the number of kernels, we can divide the photograph into rectangles with equal areas and then select one area that appears to be representative of all the areas. Estimate (or count) the number of kernels in this single area and then multiply this number by the number of equal areas.

Let's divide the photo into 20 approximately equal areas. We will select the left region in the top row as the representative region. We enlarge this region and count the kernels in it. If half a kernel is in the region, we count it (see enlargement below). There are about 18 kernels in this region. Multiplying by 20 gives $18 \times 20 = 360$. Thus, there are about 360 kernels in the photo.

In problems similar to that in Example 7, the number of regions or areas into which you choose to divide the total area is arbitrary. Generally, the more regions, the better the approximation, as long as the region selected is representative of the other regions in the map, diagram, or photo.

When you estimate an answer, the amount that your approximation differs from the actual answer will depend on how you round the numbers. For example, in estimating the product of $196{,}000 \times 0.02520$, using the rounded values $195{,}000 \times 0.025$ would yield an estimate much closer to the true answer than using the rounded values $200{,}000 \times 0.03$. Without a calculator, however, the product of $195{,}000 \times 0.025$ might be more difficult to determine than $200{,}000 \times 0.03$. When estimating, you need to determine the accuracy desired in your estimate and round the numbers accordingly.

SECTION 1.2 *Exercises*

Warm Up Exercises

In Exercises 1–2, fill in the blank with an appropriate word, phrase, or symbol(s).

1. The process of arriving at an approximate answer to a question is called _____.

2. The symbol $\approx$ means is approximately _____ to.

Practice the Skills

In Exercises 3–57, your answers may vary from the answers given in the back of the text, depending on how you round your answers.

In Exercises 3–12, estimate the answer. There is no one correct estimate. Your answer, however, should be something near the answer given.

3. $26.9 + 67.3 + 219 + 143.3$

4. $86 + 47.2 + 289.8 + 532.4 + 12.8$

5. $197{,}500 \div 4.063$

6. $\dfrac{405}{0.049}$

7. 1776×0.0098

8. 0.63×1523

9. $23.97 - 7.05$

10. $400.15 - 297.87$

11. 22% of 9116

12. 11% of 8221

Problem Solving

In Exercises 13–22, estimate the answer.

13. The cost of renting a campsite for one night if the cost for three nights was $91.35

14. The cost of dinner per person if dinner for 8 people costs $210 and they split the bill evenly

15. The amount of money Issac spends on fast food in a year if the average amount he spends per month is $47

16. The cost of 19 gallons of gas if one gallon costs $3.11

17. The cost of 5 items at a grocery store if the items cost $7.99, $4.23, $16.82, $3.51, and $20.12

18. The cost of 5 items purchased at a hardware store if the items cost $1.29, $6.86, $12.43, $25.62, and $8.99

19. The total weight of three people on an elevator if their weights are 95 lb, 127 lb, and 210 lb

20. The total weight of 5 pumpkins if the weights of the pumpkins are 16.1 pounds, 7.9 pounds, 21.4 pounds, 11.6 pounds, and 3.7 pounds

21. A 15% tip on a meal that costs $26.32

22. The number of months needed to save $400 if you save $23 each month

23. *Monthly Expenses* Li's total monthly expenses consist of room and board: $595; car payment: $289; gas: $120; insurance: $110; and miscellaneous expenses: $230. Estimate Li's total monthly expenses.

24. *Estimating Weights* In a tug of war, the weight of the members of the two three-person teams is given below. Estimate the difference in the weights of the teams.

Team A	Team B
189	183
172	229
191	167

25. *Picking Strawberries* Chuck hires 11 people to pick strawberries from his field. He agrees to pay them $1.50 for each quart they pick. Estimate the total amount of money he will have to pay if each person picks 8 quarts.

26. *Estimating Time* Donna is a long-distance runner whose average time per mile in marathons is 9 minutes, 55 seconds. Estimate the time it will take her to complete a 26-mile, 385-yard (or ≈ 26.2-mile) marathon.

27. *Exchanging Currency* Giannis visited Guanajuato, Mexico, and purchased a vase that cost 599 Mexican pesos. The current exchange rate was one U.S. dollar for 20.14 Mexican pesos. Estimate the cost of the vase in U.S. dollars.

28. *The Cost of a Vacation* The Kleins are planning a vacation in the Great Smoky Mountains National Park. Their round-trip airfare from Houston, Texas, to Knoxville, Tennessee, totals $973. Car rental is $61 per day and lodging is a total of $97 per day, and they estimate a total of $200 per day for food, gas, and other miscellaneous items. If they are planning to stay six full days and nights, estimate their total expenses.

29. *The Pacific Coast Highway* Below is a map of California's scenic Pacific Coast Highway. Using the scale on the map, estimate the distance between Carmel-by-the-Sea and Hearst Castle.

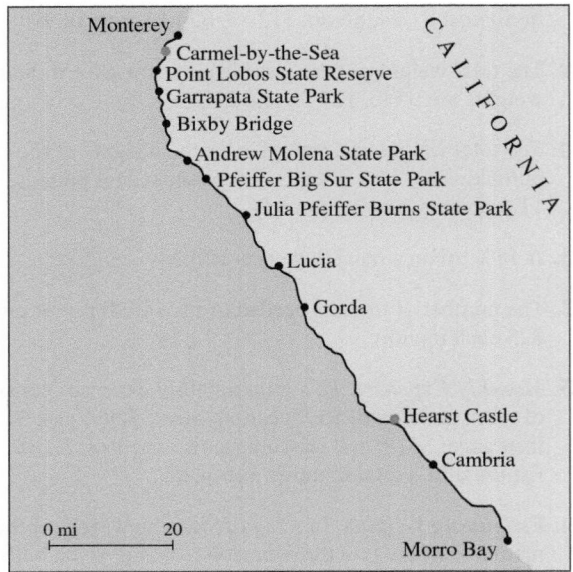

30. *The Olympic Peninsula* Following is a map of the Olympic Peninsula in the state of Washington. Using the scale on the map, estimate the distance of the route shown in red, starting at Forks and ending at Sequim, passing through Port Angeles.

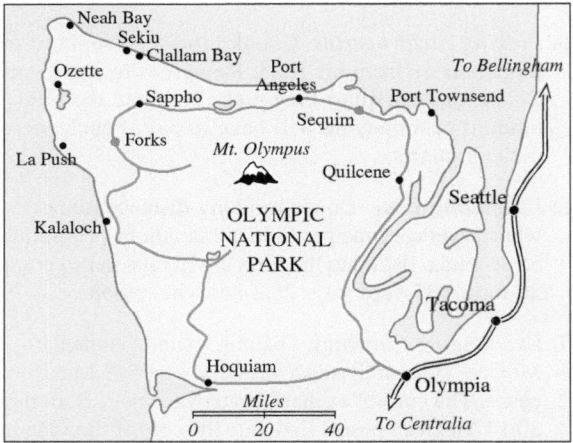

31. *Coffee Consumption* The circle graph above and to the right shows the number of cups of coffee consumed per week by people in the United States. Assume 700 people were surveyed.

a) Estimate the number of people surveyed who consume 5–7 cups of coffee per week.

b) Estimate the number of people surveyed who consume less than 1 cup of coffee per week.

c) Estimate the number of people surveyed who don't drink coffee.

U.S. Coffee Consumption

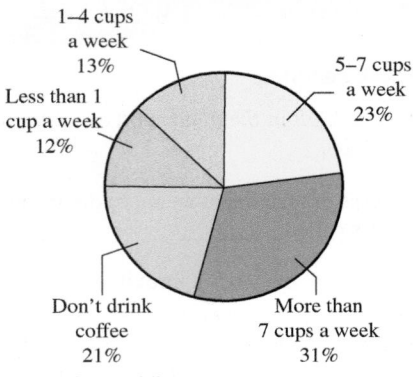

Source: Le Meridien

32. *Exercising* The circle graph below shows the biggest obstacles to sticking to an exercise program, as reported by 2987 employees at the Hiaasen Corporation.

Biggest Obstacles to Exercising

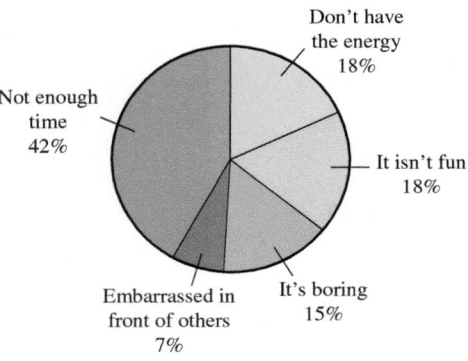

a) Estimate the number of adults surveyed who don't have the energy to exercise.

b) Estimate the number of adults surveyed who don't have enough time to exercise.

c) Estimate the number of adults surveyed who are embarrassed to exercise in front of others.

33. *An Aging Population* The bar graph on the next page shows population figures for 1900 and 2000 and estimated population figures for 2050.

a) Estimate the number of people 65 and older in 1900.

b) Estimate the number of people 65 and older in 2050.

c) Estimate the increase in the number of people 65 and older from 2000 to 2050.

d) Estimate the total U.S. population in 2000 by adding the five categories.

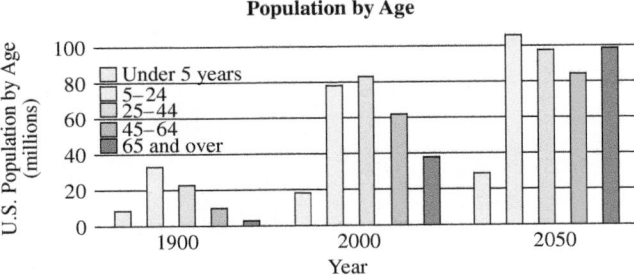

Population by Age

Source: U.S. Census Bureau

34. *Gaining Weight* The graph below shows that as a society we tend to get heavier as we grow older. Also, with age, the amount of muscle tends to drop, and fat accounts for a greater percentage of weight.

a) Estimate the average percent of body fat for a woman, age 18 to 25.

b) Estimate the average percent of body fat for a man, age 56+.

c) Greg, an average 40-year-old man, weighs 179 lb. Estimate the number of pounds of body fat he has.

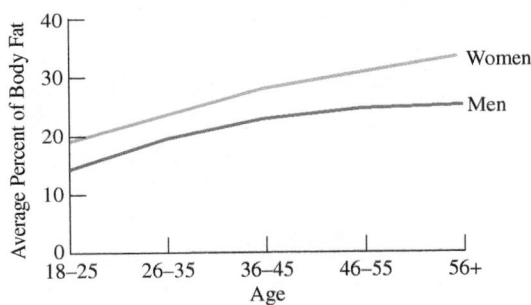

Getting Older Usually Means Getting Fatter

Source: Mayo Clinic Newsletter

35. *Land Ownership by the Federal Government* The federal government owns a great deal of land in the United States. The graph above and to the right shows the percent of land owned by the federal government in the 12 states in which the federal government owns the greatest percentage of a state's land.

a) Estimate the percent of land owned by the federal government in Nevada.

b) Estimate the difference between Alaska and Oregon in the percent of land owned by the federal government.

c) Nevada has a total area of 70,264,320 acres. Estimate the number of acres owned by the federal government in Nevada.

d) By just looking at the graph, is it possible to determine whether the federal government owns more land in Nevada or Utah? Explain.

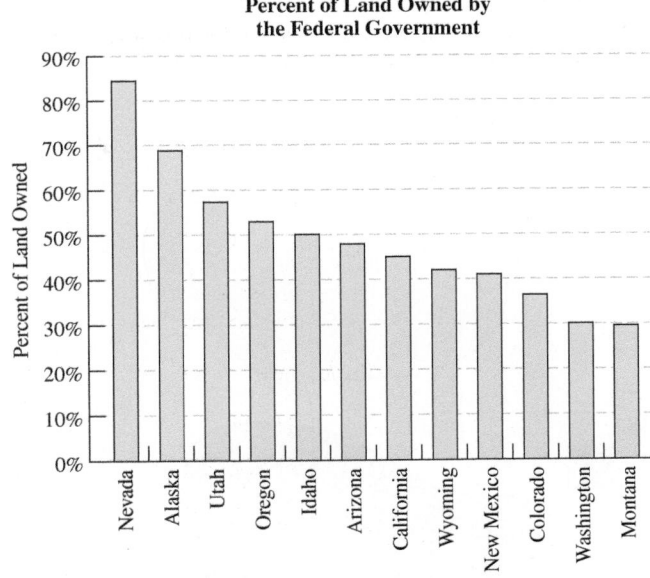

Percent of Land Owned by the Federal Government

Source: Office of Governmentwide Policy, General Services Administration

36. *Calories and Exercise* The following chart shows the calories burned per hour for an average person who weighs 150 lb.

a) Estimate the number of calories Phyllis, who weighs 150 lb, burns in a week if she lifts weight for 2 hours each week and jogs at 5 miles per hour for 4 hours each week.

b) Estimate the difference in the calories Phyllis will burn each week if she runs for 4 hours at 8 miles per hour rather than walking for 4 hours at 4 miles per hour.

c) Assume Phyllis jogs at 5 miles per hour for 3 hours and bicycles at 13 miles per hour for 3 hours each week. Estimate the number of calories she will burn in a year from these exercises.

Activity	Calories* per Hour
Running, 8 mph	920
Bicycling, 13 mph	545
Jogging, 5 mph	545
Air-walking	480
Stair-climbing	410
Weight-lifting	410
Walking, 4 mph	330
Casual bike riding	300

*For a 150-lb person.

In Exercises 37 and 38, estimate the maximum number of smaller figures (at left) that can be placed in the larger figure (at right) without the small figures overlapping.

37.

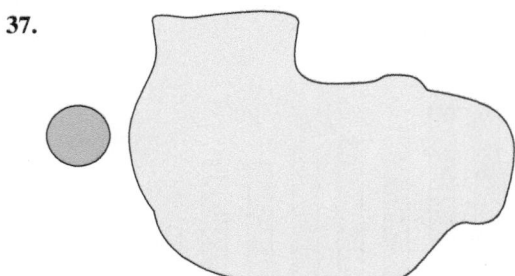

38.

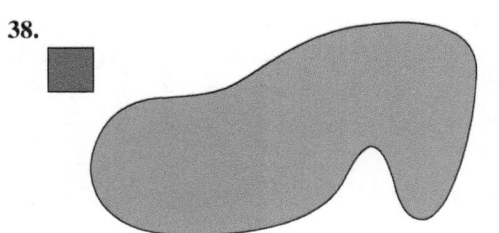

39. Estimate the number of bananas shown in the photo.

40. Estimate the number of grapes shown in the photo.

In Exercises 41 and 42, estimate, in degrees, the measure of the angles depicted. For comparison purposes a right angle, ⌐, *measures 90°.*

41. 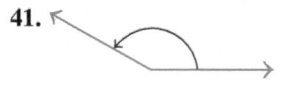   **42.**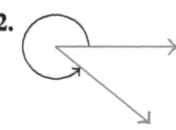

In Exercises 43 and 44, estimate the percent of area that is shaded in the following figures.

43. 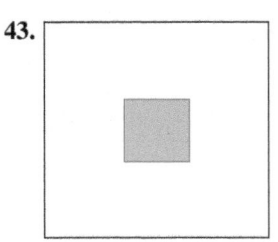   **44.**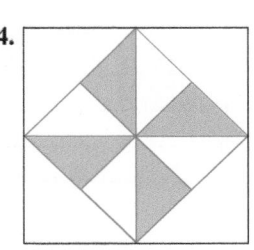

In Exercises 45 and 46, if each square represents one square unit, estimate the area of the shaded figure in square units.

45. 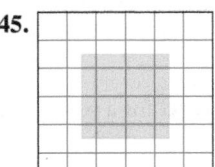   **46.**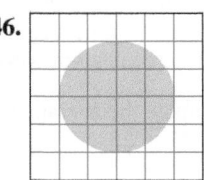

47. *Statue of Liberty* The length of the torch of the Statue of Liberty, from the tip of the flame to the bottom of the torch's baton, is 29 feet. Estimate the height of the Statue of Liberty from the top of the base to the top of the torch (the statue itself).

48. *Estimating Heights* If the height of the woman in the photo is 62 in. tall, estimate the height of the tree.

49. *Distance* Estimate, without a ruler, a distance of 12 in. Measure the distance. How good was your estimate?

50. *Weight* In a bag, place objects that you believe have a total weight of 10 lb. Weigh the bag to determine the accuracy of your estimate.

51. *Phone Call* Estimate the number of times your phone will ring in 1 minute if unanswered. Have a classmate phone you so that you can count the rings and thus test your estimate.

52. *Temperature* Fill a glass with water and estimate the water's temperature. Then use a thermometer to measure the temperature and check your estimate.

53. *Pennies* Estimate the number of pennies that will fill a 3-ounce (oz) paper cup. Then actually fill a 3-oz paper cup with pennies, counting them to determine the accuracy of your estimate.

54. *Height* Estimate the ratio of your height to your waist size. Then have a friend measure your height and waist size. Determine the stated ratio and check the accuracy of your estimate.

55. *Walking Speed* Estimate how fast you can walk 60 ft. Then mark off a distance of 60 ft and use a watch with a second hand to time yourself walking it. Determine the accuracy of your estimate.

Challenge Problems/Group Activities

56. *Shopping* Make a shopping list of 20 items you use regularly that can be purchased at a supermarket. Beside each item write down what you estimate to be its price. Add these price guesses to estimate the total cost of the 20 items. Next, make a trip to your local supermarket and record the actual price of each item. Add these prices to determine the actual total cost. How close was your estimate? (Don't forget to add tax on the taxable items.)

57. *Birthday Party* Fitz is planning a birthday party for his friend, Noelle. There will be 50 guests at the party, and Fitz must do the following

- Rent a party space
- Hire a DJ to play music for 3 hours
- Order a cake large enough so each guest gets one piece of cake
- Provide ice cream, plates, napkins, and utensils to go with the cake
- Provide two soft drinks, including ice and cups, for each of the guests
- Provide light snacks, including plates and napkins, for the guests

- Hire four waitstaff to serve the guests
- Pay the DJ and waitstaff each a 20% tip.

a) Estimate the cost of each of the items listed.

b) Estimate Fitz's total cost for the birthday party.

c) Use the Internet to determine the actual costs for this party if it were held in your location.

58. *Cost of Supplements* Each day, Roberto takes the following nutritional supplements: five tablets of vitamin C, two tablets of fish oil, one tablet of vitamin D, and one multivitamin tablet. Roberto purchases 500 vitamin C tablets for $24.99, 120 fish oil tablets for $47.99, 180 vitamin D tablets for $17.99, and 80 multivitamin tablets for $19.99. Estimate Roberto's daily cost for supplements.

Recreational Mathematics

59. *A Dime* Look at a dime. Around the edge of a dime are many lines. Estimate the number of lines there are around the edge of a dime.

60. *A Million Dollars*

a) Estimate the time it would take, in days, to spend $1 million if you spent $1 a second until the $1 million is used up.

b) Calculate the actual time it would take, in days, to spend $1 million if you spent $1 a second. How close was your estimate?

Research Activities

61. *Water Usage*

a) About how much water does your household use per day? Use the following data to estimate your household's daily water usage.

Activity	Typical Use
Running clothes washer	40 gal
Bath	35 gal
5-minute shower	25 gal
Doing dishes in sink, water running	20 gal
Running dishwasher	11 gal
Flushing toilet	4 gal
Brushing teeth, water running	2 gal

Source: U.S. Environmental Protection Agency

b) Determine from your water department (or company) your household's average daily usage by obtaining the total number of gallons used per year and dividing that amount by 365. How close was your estimate in part (a)?

c) Current records indicate that the average household uses about 300 gal of water per day (the average daily usage is 110 gal per person). Based on the number of people in your household, do you believe your household uses more or less than the average amount of water? Explain your answer.

62. *Monthly Budget* Develop a monthly budget by estimating your monthly income and your monthly output. Your monthly income should equal your monthly output.

63. *Estimation Skills* Identify three ways that you use estimation in your daily life. Discuss each of them briefly and give examples.

SECTION 1.3   Problem-Solving Procedures

Upon completion of this section, you will be able to:

■ Understand and use a general problem-solving procedure.

To maximize their profits, businesses try to keep their expenses down. We, as individuals, also try to do the same, and we often look for "bargains" or the "best deal." For example, suppose we need to purchase several pizzas and have only $87 to spend. How can we determine the maximum number of slices that we can purchase if large pizzas with 12 slices sell for $14 and medium pizzas with 8 slices sell for $10? We'll answer this question in Example 1.

Why This Is Important Problem-solving skills can help you determine solutions to problems you may encounter in everyday life.

Solving mathematical puzzles and real-life mathematical problems can be enjoyable. You should work as many exercises in this section as possible. By doing so, you will sample a variety of problem-solving techniques.

You can approach any problem by using a general procedure developed by George Polya. Before learning Polya's problem-solving procedure, let's consider an example that illustrates the procedure.

Example 1 *Saving Money When Purchasing Pizza*

Karen plans to purchase pizza for a party she is hosting. Pizza Corner sells medium pizzas (14″ diameter) for $10 and large pizzas (17″ diameter) for $14. A medium pizza has 8 slices and a large pizza has 12 slices. Only entire pizzas are sold. Assume the pizza slices are all about the same size.

a) Determine the maximum number of slices of pizza that Karen can purchase for $87 or less.

b) If the maximum number of slices determined in part (a) is purchased in the most economical way, how much will the pizzas cost?

Solution

a) The first thing to do is read the problem carefully. Read it at least twice. Be sure you understand the facts given and what you are being asked to determine. Next, make a list of the given facts and determine which facts are relevant to answering the question or questions asked.

GIVEN INFORMATION

Party host: Karen

Pizza shop: Pizza Corner

A medium pizza costs $10.

A medium pizza has a 14″ diameter.

A medium pizza contains 8 slices.

A large pizza costs $14.

A large pizza has a 17″ diameter.

A large pizza contains 12 slices.

Only entire pizzas can be purchased.

Ostriches

How many ostriches must replace the question mark to balance the fourth scale? Assume that all animals of the same kind have the same weight. That is, all giraffes weigh the same, and so forth.

We need to determine the maximum number of slices Karen can purchase for $87 or less. To determine this number, we need to know the number of slices in each pizza and the cost of each pizza. We also need to know that only entire pizzas can be purchased.

RELEVANT INFORMATION

A medium pizza costs $10.

A medium pizza contains 8 slices.

A large pizza costs $14.

A large pizza contains 12 slices.

Only entire pizzas can be purchased.

The next step is to determine the answer to the question. That is, we need to determine the maximum number of slices that can be purchased for $87 or less.

We now need a plan for solving the problem. One method is to set up a table or chart to compare costs of different combinations of pizzas. Start by using the maximum number of large pizzas that can be purchased. Then reduce the number of large pizzas and add more medium pizzas. In each case, we need to keep the cost at $87 or less.

Because one large pizza costs $14, we can determine the number of large pizzas that can be purchased by dividing $87 by $14. Because the quotient is approximately 6.21 and only entire pizzas can be purchased, only 6 large pizzas can be purchased. Six large pizzas would cost $14 × 6 = $84. Because medium pizzas cost $10, there would not be enough money left to purchase any medium pizzas. Since large pizzas have 12 slices, 6 large pizzas provide 12 × 6 = 72 slices. Thus, for $87 or less, one option is 6 large pizzas, which gives 72 slices. This option is illustrated in the first row of the table below. Also given in the table is the cost of this option, which is $84. We complete the other rows of the table in a similar manner.

Large Pizzas and Medium Pizzas	Number of Slices	Cost
6 large and 0 medium	$(6 \times 12) + (0 \times 8) = 72$	$84
5 large and 1 medium	$(5 \times 12) + (1 \times 8) = 68$	$80
4 large and 3 medium	$(4 \times 12) + (3 \times 8) = 72$	$86
3 large and 4 medium	$(3 \times 12) + (4 \times 8) = 68$	$82
2 large and 5 medium	$(2 \times 12) + (5 \times 8) = 64$	$78
1 large and 7 medium	$(1 \times 12) + (7 \times 8) = 68$	$84
0 large and 8 medium	$(0 \times 12) + (8 \times 8) = 64$	$80

The question asks us to determine the maximum number of slices of pizza that can be purchased for $87 or less. From the second column of the table, we see that the answer is 72 slices of pizza. This result can be obtained in two different ways: either 6 large pizzas and 0 medium pizzas or 4 large pizzas and 3 medium pizzas.

b) When comparing the two possibilities for purchasing the 72 slices of pizza discussed in part (a), we see that the most economical way to purchase the slices is to purchase 6 large pizzas and 0 medium pizzas. The cost is $84. ■

Problem-Solving Procedures

Following is a general procedure for problem solving as given by George Polya. Note that Example 1 demonstrates many of these guidelines.

Profile in Mathematics

George Polya

Geore Polya (1887–1985) was educated in Europe and taught at Stanford University. In his book *How to Solve It*, Polya outlines four steps in problem solving. We will use Polya's four steps as guidelines for problem solving.

PROCEDURE GUIDELINES FOR PROBLEM SOLVING

1. *Understand the problem.*
   - Read the problem *carefully* at least twice. In the first reading, get a general overview of the problem. In the second reading, determine (a) exactly what you are being asked and (b) what information the problem provides.
   - Try to make a sketch to illustrate the problem. Label the information given.
   - Make a list of the given facts that are pertinent to the problem.
   - Determine if the information you are given is sufficient to solve the problem.

2. *Devise a plan to solve the problem.*
   - Have you seen the problem or a similar problem before? Are the procedures you used to solve the similar problem applicable to the new problem?
   - Can you express the problem in terms of an algebraic equation? (We explain how to write algebraic equations in Chapter 6.)
   - Look for patterns or relationships in the problem that may help in solving it.
   - Can you express the problem more simply?
   - Can you substitute smaller or simpler numbers to make the problem more understandable?
   - Will listing the information in a table help in solving the problem?
   - Can you make an educated guess at the solution? Sometimes if you know an approximate solution, you can work backward and eventually determine the correct procedure to solve the problem.

3. *Carry out the plan.*
   - Use the plan you devised in Step 2 to solve the problem.

4. *Check the results.*
   - Ask yourself, "Does the answer make sense?" and "Is the answer reasonable?" If the answer is not reasonable, recheck your method for solving the problem and your calculations.
   - Can you check the solution using the original statement?
   - Is there an alternative method to arrive at the same conclusion?
   - Can the results of this problem be used to solve other problems?

The following examples show how to apply the guidelines for problem solving.

Example 2 *Ride-Share Service*

Latricia uses a ride-share service to travel to her local airport. Ride-share services use an app available on smartphones. Ride-share services have increasingly taken business away from traditional taxi services. The total fare that Latricia is charged includes a base fee plus a mileage fee plus a time fee. The base fee is $5.00, the mileage fee is calculated at $1.10 per mile, and the time fee is calculated at $0.21 per minute. The distance Latricia will travel to the airport is 17 miles, and it will take 25 minutes to get there. Use this information to determine Latricia's total fare for using the ride-share service to travel to the airport.

Solution We need to determine the total fare that Latricia is charged to travel to the airport. We will make a list of the information given and mark with an asterisk (*) the information that is pertinent to solving the problem.

Ride-share services use an app available on smartphones.
Ride-share services have increasingly taken business away from traditional taxi services.

* Latricia is charged a base fee plus a mileage fee plus a time fee.
* The base fee is $5.00.
* The mileage fee is calculated at $1.10 per mile.
* The time fee is calculated at $0.21 per minute.
* Latricia will travel 17 miles to the airport.
* It will take 25 minutes to get to the airport.

The first two items listed are not needed to solve the problem. To determine the total fare, we first must calculate the mileage fee and the time fee.

$$\text{Mileage fee} = \$1.10 \text{ per mile} \times 17 \text{ miles} = \$18.70$$
$$\text{Time fee} = \$0.21 \text{ per minute} \times 25 \text{ minutes} = \$5.25$$

The total fare is the sum of the base fee, the mileage fee, and the time fee.

$$\text{Total fare} = \text{Base fee} + \text{Mileage fee} + \text{Time fee}$$
$$= \$5.00 \quad + \$18.70 \quad + \$5.25$$
$$= \$28.95$$

Thus, Latricia is charged $28.95 for using the ride-share service to travel to the airport. ∎

In our next example, we will continue to use the guidelines for problem solving.

Example 3 *Retirement*

It is never too early to start planning for retirement. U.S. Census Bureau data indicate that at age 65 the average woman will live about another 21 years and the average man will live about another 19 years. The data also indicate that about 33% of the average person's retirement income will come from Social Security.

When discussing retirement planning, many investment firms and financial planners use the graph in Fig. 1.1,* which shows how long a typical retiree's assets (or "nest egg") will last based on the percentage of the assets withdrawn each year.

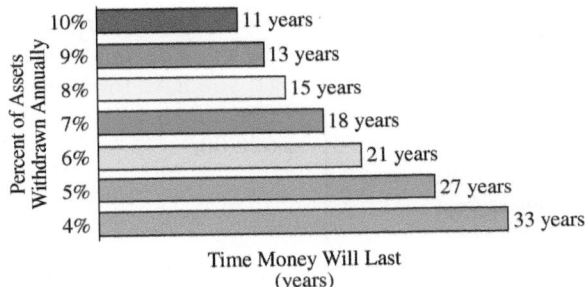

How Much You Withdraw Annually Affects How Long Your Money Will Last

Source: Ned Davis research

Figure 1.1

a) If a typical retiree has retirement assets of $500,000, how much can he or she withdraw annually if he or she wishes the assets to last 21 years?

b) How much should a retiree have in assets if he or she wishes to withdraw $25,000 annually and wishes his or her assets to last 18 years?

Solution

a) Quite a bit of information is provided in the example. We will first need to determine what information is required to answer the question. To answer the question, we need to use only the information provided in the graph. From the graph, we can see that for assets to last 21 years, about 6% of the assets can be withdrawn annually. The amount that can be withdrawn annually is determined as follows.

$$\text{Amount} = 6\% \text{ of assests}$$
$$\text{Amount} = 0.06(500,000) = 30,000$$

Thus, about $30,000 can be withdrawn annually.

*The information in this graph is based on past performance of the stock market, with 50% invested in large company stocks and 50% invested in intermediate-term bonds. Past performance is not indicative of future results.

b) Again, to solve this part of the example, we need only the information provided in the graph. From the graph, we can determine that if a retiree wishes for his or her assets to last 18 years, then 7% of the assets can be withdrawn annually. We need to determine the total assets such that 7% of the total assets is $25,000.

$$7\% \text{ of assets} = \$25{,}000 \quad \text{or} \quad 0.07 \times \text{assets} = \$25{,}000$$

Because 0.07 is multiplied by the assets to obtain $25,000, we can determine the assets by dividing the $25,000 by 0.07 as follows.

$$\text{Assets} = \frac{25{,}000}{0.07} \approx 357{,}142.86$$

Therefore, if the retiree has assets of about $357,142.86, he or she will be able to withdraw $25,000 annually and the assets will last 18 years. ∎

Example 4 *Determining a Tip*

The cost of Seana's meal before tax is $28.00.

a) If a $6\frac{1}{2}\%$ sales tax is added to her bill, determine the total cost of the meal including tax.

b) If Seana wants to leave a 15% tip on the *pretax* cost of the meal, how much should she leave?

c) If she wants to leave a 20% tip on the *pretax* cost of the meal, how much should she leave?

Solution

a) The sales tax is $6\frac{1}{2}\%$ of $28.00. To determine the sales tax, first change the $6\frac{1}{2}\%$ to a decimal number. $6\frac{1}{2}\%$ when written as a decimal number is 0.065. (If you have forgotten how to change a percent to a decimal number, read Section 10.1.) Next, multiply the decimal number, 0.065, by the amount, $28.00.

$$\text{Sales tax} = 6\tfrac{1}{2}\% \text{ of } 28.00$$
$$= 0.065(28.00) = 1.82$$

The sales tax is $1.82. The total bill is the cost of the meal + the sales tax.

$$\text{Total bill} = \text{cost of meal} + \text{sales tax}$$
$$= 28.00 + 1.82 = 29.82$$

Thus, the bill, including sales tax, is $29.82.

b) To determine 15% of any number, we can multiply the number by 0.15.

$$15\% \text{ of pretax cost} = 0.15(\$28.00)$$
$$= \$4.20$$

Thus, 15% of $28.00 is $4.20. A second method to determine a 15% tip is to determine 10% of the cost and then add to it half that amount. Following this procedure,

$$10\% \text{ of pretax cost} = 0.10(\$28.00)$$
$$= \$2.80$$

Now add half of $2.80 to $2.80.

$$\$2.80 + \frac{\$2.80}{2} = \$2.80 + \$1.40 = \$4.20$$

c) To determine 20% of any number, we can multiply the number by 0.20.

$$20\% \text{ of pretax cost} = 0.20(\$28.00)$$
$$= \$5.60$$

Thus, 20% of $28.00 is $5.60. A second method to determine a 20% tip is to determine 10% of the cost, as we showed in part (b), and then double this amount.

$$10\% \text{ of pretax cost} \times 2 = \$2.80 \times 2 = \$5.60$$

Tips are often rounded to the nearest dollar. For more information, there are many websites dedicated to the tipping process. ∎

MATHEMATICS TODAY

Decisions, Decisions

We deal with problem solving daily. Can we afford to take that vacation in the mountains? Which vehicle should I purchase? What shall I major in in college? The list of questions we ask ourselves daily goes on and on. To make decisions, we often need to consider and weigh many factors.

Often, different branches of mathematics are involved in our decision-making process. For example, we frequently use statistical data and consider the probability (or chance) of an event occurring or not occurring when we make decisions. Probability and statistics are two branches of mathematics covered in later chapters of the book.

For many people, problem solving is a recreational activity, as is evident by the great number of crossword puzzles and puzzle books sold daily.

Today, as you go about your daily business, keep a record of all the problem-solving decisions you need to make! If you do this conscientiously, you will be amazed at the outcome.

Why This Is Important There are aspects of problem solving to almost every decision we make.

TECHNOLOGY TIP

There's an App for That!
Throughout this book we will be discussing many different applications of mathematics. Many of the applications can be further explored with the use of an app on your smartphone or tablet.

For example, in Example 4, we discussed calculating a tip based on the cost of a meal. There are many apps that can assist you in calculating tips at restaurants, hair salons, golf courses, airports, and other places where tipping is customary. Furthermore, if you travel to other countries, these apps can assist you in determining an appropriate tip for these locations.

Within Technology Tip boxes, we will discuss many useful apps throughout this book. You may already be aware of some of these apps. However, always consult with your instructor before using these apps when completing work related to your course.

Example 5 *A Recipe for 6*

The chart below shows the amount of each ingredient recommended to make 2, 4, and 8 servings of Potato Buds. Determine the amount of each ingredient necessary to make 6 servings of Potato Buds by using the following procedures.

a) Multiply the amount for 2 servings by 3.*

b) Add the amounts for 2 servings to the amounts for 4 servings.

c) Determine the average of the amounts for 4 servings and for 8 servings.

d) Subtract the amounts for 2 servings from the amounts for 8 servings.

e) Compare the answers for parts (a) through (d). Are they the same? If not, explain why not.

f) Which is the correct procedure for obtaining 6 servings?

Servings	2	4	8
Water	$\frac{2}{3}$ cup	$1\frac{1}{3}$ cups	$2\frac{2}{3}$ cups
Milk	2 tbsp	$\frac{1}{3}$ cup	$\frac{2}{3}$ cup
Butter or margarine	1 tbsp	2 tbsp	4 tbsp
Salt†	$\frac{1}{4}$ tsp	$\frac{1}{2}$ tsp	1 tsp
Potato Buds	$\frac{2}{3}$ cup	$1\frac{1}{3}$ cups	$2\frac{2}{3}$ cups

†Less salt can be used if desired.

Solution

a) We multiply the amounts for 2 servings by 3.
  Water: $3\left(\frac{2}{3}\right) = 2$ cups
  Milk: $3(2) = 6$ tablespoons (tbsp)
  Butter or margarine: $3(1) = 3$ tbsp
  Salt: $3\left(\frac{1}{4}\right) = \frac{3}{4}$ teaspoon (tsp)
  Potato Buds: $3\left(\frac{2}{3}\right) = 2$ cups

---

* Addition, subtraction, multiplication, and division of fractions are discussed in detail in Section 5.3.

b) We determine the amount of each ingredient by adding the amount for 2 and 4 servings.

Water: $\frac{2}{3}$ cup $+$ $1\frac{1}{3}$ cups $=$ 2 cups

Milk: 2 tbsp $+$ $\frac{1}{3}$ cup

To add these two amounts, we must convert one of them so that both ingredients have the same units. By looking in a cookbook or a book of conversion factors, we see that 16 tbsp $=$ 1 cup. The milk in part (a) was given in tablespoons, so we will convert $\frac{1}{3}$ cup to tablespoons. One-third cup equals $\frac{1}{3}(16) = \frac{16}{3}$ or $5\frac{1}{3}$ tbsp. Therefore,

Milk: 2 tbsp $+$ $5\frac{1}{3}$ tbsp $=$ $7\frac{1}{3}$ tbsp

Let's continue with the rest of the ingredients:

Butter: 1 tbsp $+$ 2 tbsp $=$ 3 tbsp

Salt: $\frac{1}{4}$ tsp $+$ $\frac{1}{2}$ tsp $=$ $\frac{3}{4}$ tsp

Potato Buds: $\frac{2}{3}$ cup $+$ $1\frac{1}{3}$ cups $=$ 2 cups

c) We compute the amounts of the ingredients by determining the average of the amounts for 4 and 8 servings. We do so by adding the amounts for each ingredient and dividing the sum by 2.

Water: $\dfrac{1\frac{1}{3} \text{ cups} + 2\frac{2}{3} \text{ cups}}{2} = \dfrac{4 \text{ cups}}{2} = 2$ cups

Milk: $\dfrac{\frac{1}{3} \text{ cup} + \frac{2}{3} \text{ cup}}{2} = \dfrac{1 \text{ cup}}{2} = \frac{1}{2}$ cup (or 8 tbsp)

Butter: $\dfrac{2 \text{ tbsp} + 4 \text{ tbsp}}{2} = \dfrac{6 \text{ tbsp}}{2} = 3$ tbsp

Salt: $\dfrac{\frac{1}{2} \text{ tsp} + 1 \text{ tsp}}{2} = \dfrac{\frac{3}{2} \text{ tsp}}{2} = \frac{3}{4}$ tsp

Potato Buds: $\dfrac{1\frac{1}{3} \text{ cups} + 2\frac{2}{3} \text{ cups}}{2} = \dfrac{4 \text{ cups}}{2} = 2$ cups

d) We obtain the amounts of ingredients by subtracting the amounts for 2 servings from the amounts for 8 servings.

Water: $2\frac{2}{3}$ cups $-$ $\frac{2}{3}$ cup $=$ 2 cups

Milk: $\frac{2}{3}$ cup $-$ 2 tbsp $=$ $\frac{2}{3}(16)$ tbsp $-$ 2 tbsp

$= \frac{32}{3}$ tbsp $-$ $\frac{6}{3}$ tbsp

$= \frac{26}{3}$ tbsp, or $8\frac{2}{3}$ tbsp

Butter: 4 tbsp $-$ 1 tbsp $=$ 3 tbsp

Salt: 1 tsp $-$ $\frac{1}{4}$ tsp $=$ $\frac{3}{4}$ tsp

Potato Buds: $2\frac{2}{3}$ cups $-$ $\frac{2}{3}$ cup $=$ 2 cups

e) Comparing the answers in parts (a) through (d), we determine that the amounts of all ingredients, except milk, are the same. For milk, we get the following results.

Part (a): Milk $=$ 6 tbsp   Part (c): Milk $=$ 8 tbsp

Part (b): Milk $=$ $7\frac{1}{3}$ tbsp   Part (d): Milk $=$ $8\frac{2}{3}$ tbsp

Why are all these answers different? After rechecking, we determine that all our calculations are correct, so we must look deeper. Note that milk is the only ingredient that has different units for 2 servings and 4 servings. Let's check the relationship between 2 tbsp and $\frac{1}{3}$ cup. In going from 2 servings to 4 servings, we would expect that $\frac{1}{3}$ cup should be twice 2 tbsp. We know that 1 cup $=$ 16 tbsp, so

$$\frac{1}{3} \text{ cup} = \frac{1}{3}(16) = \frac{16}{3} = 5\frac{1}{3} \text{ tbsp}$$

Therefore, instead of the 4 tbsp of milk we expected for 4 servings, we get $5\frac{1}{3}$ tbsp. This change causes all our calculations for milk to be different.

f) Which is the correct answer? Because all our calculations for milk are correct, there is no single correct answer. All our answers are correct. Using $8\frac{2}{3}$ tbsp instead of 6 tbsp might make the Potato Buds a little thinner. When we cook, we generally do not add the *exact* amount recommended. We rely on experience to alter the recommended amounts according to individual taste.   ■

Many real-life problems, such as the one in Example 6, can be solved by using proportions. A proportion is a statement of equality between two ratios (or fractions).*

Example 6 *A Brine Solution*

Some recipes for smoking a turkey recommend presoaking the turkey in a brine solution. To make enough brine solution to smoke a 12-pound turkey, use 16 tablespoons (tbsp) of salt in 2 gallons (gal) of water.

a) If you wish to make 3.5 gallons of a brine solution, how much salt is needed?

b) If you wish to make enough brine solution to soak a 20-pound turkey, how much salt is needed to make the brine solution?

Solution

a) To make 3.5 gallons of a brine solution, we need to determine how much salt is needed to mix with 3.5 gallons of water. To solve this problem, we use the information that 16 tablespoons of salt is to be mixed with 2 gallons of water. We can use this ratio to set up a proportion to solve for the unknown quantity.

$$\text{Given ratio} \begin{cases} \dfrac{16 \text{ tbsp}}{2 \text{ gal}} = \dfrac{? \text{ tbsp}}{3.5 \text{ gal}} \end{cases} \begin{matrix} \leftarrow \text{Item to be found} \\ \leftarrow \text{Other information given} \end{matrix}$$

Notice in the proportion that tablespoons and gallons are placed in the same relative positions. Often, the unknown quantity is replaced with an x. The proportion may be written as follows and solved using cross-multiplication.

$$\frac{16}{2} = \frac{x}{3.5}$$

$$16(3.5) = 2x \qquad \text{Cross-multiplication}$$

$$56 = 2x$$

$$\frac{56}{2} = \frac{\cancel{2}x}{\cancel{2}} \qquad \text{Divide both sides by 2 to solve for } x.$$

$$28 = x$$

Thus, 28 tbsp of salt must be used to make 3.5 gal of a brine solution. This answer seems reasonable, since we would expect to get an answer greater than 16 tbsp.

b) To answer this question, we use the same procedure discussed in part (a). This time, we will use the information that a 12-lb turkey requires 16 tbsp of salt. The pounds may be placed in either the top or bottom of the fraction, as long as they are placed in the same relative position.

$$\text{Given ratio} \begin{cases} \dfrac{12 \text{ lb}}{16 \text{ tbsp}} = \dfrac{20 \text{ lb}}{? \text{ tbsp}} \end{cases} \begin{matrix} \leftarrow \text{Other information given} \\ \leftarrow \text{Item to be found} \end{matrix}$$

---

* Proportions are discussed in greater detail in Section 6.3.

Now replace the question mark with an x and solve the proportion.

$$\frac{12}{16} = \frac{20}{x}$$

$$12(x) = 16(20) \qquad \text{Cross-multiplication}$$

$$12x = 320$$

$$\frac{\cancel{12}x}{\cancel{12}} = \frac{320}{12} \qquad \text{Divide both sides by 12 to solve for } x.$$

$$x \approx 26.67$$

Thus, about 26.67 tbsp of salt are needed to make enough brine solution for a 20-lb turkey. This answer is reasonable because we would expect the answer to be more than the 16 tbsp of salt required for a 12-lb turkey. ∎

Most of the problems solved so far have been practical ones. Many people, however, enjoy solving brainteasers. One example of such a puzzle follows.

Example 7 *Magic Squares*

A magic square is a square array of distinct numbers such that the numbers in all rows, columns, and diagonals have the same sum. Use the digits 1, 2, 3, 4, 5, 6, 7, 8, and 9 to construct a magic square.

Solution The first step is to create a figure with nine cells, as in Fig. 1.2(a) below. We must place the nine numbers in the cells so that the same sum is obtained in each row, column, and diagonal. Common sense tells us that 7, 8, and 9 cannot be in the same row, column, or diagonal. We need some small and large numbers in the same row, column, and diagonal. To see a relationship, we list the numbers in order:

1, 2, 3, 4, 5, 6, 7, 8, 9

Note that the middle number is 5 and the smallest and largest numbers are 1 and 9, respectively. The sum of 1, 5, and 9 is 15. If the sum of 2 and 8 is added to 5, the sum is 15. Likewise 3, 5, 7, and 4, 5, 6 have sums of 15. We see that in each group of three numbers the sum is 15 and 5 is a member of the group.

9	5	1

(a)

		8
9	5	1
2		

(b)

4		8
9	5	1
2		6

(c)

4	3	8
9	5	1
2	7	6

(d)

Figure 1.2

Because 5 is the middle number in the list of numbers, place 5 in the center square. Place 9 and 1 to the left and right of 5, as in Fig. 1.2(a). Now we place the 2 and the 8. The 8 cannot be placed next to 9 because $8 + 9 = 17$, which is greater than 15. Place the smaller number 2 next to the larger number 9. We elected to place the 2 in the lower left-hand cell and the 8 in the upper right-hand cell, as in Fig. 1.2(b). The sum of 8 and 1 is 9. To arrive at a sum of 15, we place 6 in the lower right-hand cell, as in Fig. 1.2(c). The sum of 9 and 2 is 11. To arrive at a sum of 15, we place 4 in the upper left-hand cell as in Fig. 1.2(c). Now the diagonals 2, 5, 8, and 4, 5, 6 have sums of 15. The numbers that remain to be placed in the empty cells are 3 and 7. Using arithmetic, we can see that 3 goes in the top middle cell and 7 goes in the bottom middle cell, as in Fig. 1.2(d). A check shows that the sum of the numbers in all the rows, columns, and diagonals is 15. ∎

Magic Squares

The solution to Example 7 is not unique. Other arrangements of the nine numbers in the cells will produce a magic square. Also, other techniques of arriving at a solution for a magic square may be used. In fact, the process described will not work if the number of squares is even, for example, 16 instead of 9. Magic squares are not limited to the operation of addition or to the set of counting numbers.

SECTION 1.3 *Exercises*

Practice the Skills/Problem Solving

1. **Reading a Map** The scale on a map is 1 inch = 12 miles. How long a distance is a route on the map if it measures 4.25 in.?

2. **Blueprints** Tony, an architect, is designing a new enclosure for the giraffes at a zoo. The scale of his plan is 1 in. = 2.5 yd. He draws a 22.4-in. line on the blueprint to represent the northern boundary of the enclosure. What actual distance does this boundary line represent?

3. **Height of a Tree** At a given time of day, the ratio of the height of an object to the length of its shadow is the same for all objects. If a 3-ft stick in the ground casts a shadow of 1.2 ft, determine the height of a tree that casts a shadow that is 15.36 ft.

4. **Grass Seed** A bag of Turf Builder Sun and Shade Mix grass seed covers an area of 4000 ft². How many bags are needed to cover an area of 35,000 ft²?

5. **Sales Tax** Quan purchased four tickets to see Luke Bryan at the Ryman Auditorium in Nashville. The tickets cost $113 each plus tax. The sales tax rate in Nashville is 9.25%. Determine the total price Quan paid for the four tickets.

6. **Sales Tax** Michelle purchased two tickets to see Kendrick Lamar at the Hollywood Bowl in Los Angeles. The tickets cost $187 each plus tax. The sales tax rate in Los Angeles is 9.5%. Determine the total price Michelle paid for the two tickets.

7. **How Americans Spend Their Money** The circle graph below shows the percent of Americans' net monthly income spent on housing, transportation, food, insurance/retirement, healthcare, and entertainment/miscellaneous. Sammy has a net monthly income of $1950. Assuming this graph applies to Sammy's spending,

How Americans Spend Their Money

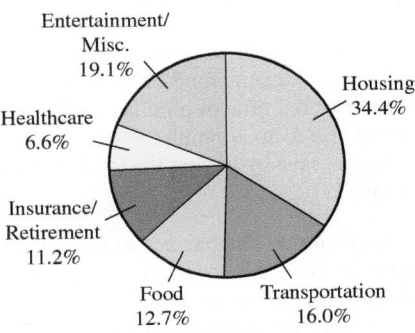

Source:U.S. Bureau of Labor Statistics

a) how much more money per month does Sammy spend on entertainment/miscellaneous than on food?

b) how much more money per month does Sammy spend on housing than on transportation?

8. **Bachelor's Degrees** The circle graph below shows the percent of bachelor's degrees awarded in 2016 in business, health professions, social sciences and history, psychology, engineering, and other fields in the United States. In 2016, the number of bachelor's degrees awarded was 2,895,000.

Bachelor's Degrees Awarded in 2016

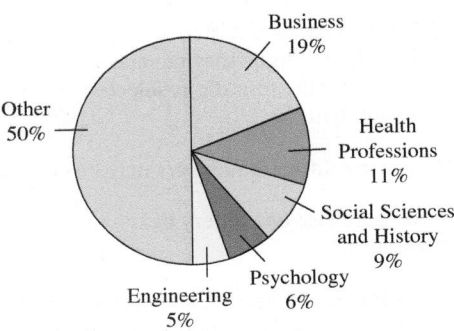

Source:U.S. Department of Education

a) How many more bachelor's degrees were awarded in business than in engineering?

b) How many more bachelor's degrees were awarded in health professions than in psychology?

9. ***Boston Public Transportation*** Marcelo buys a monthly CharlieCard, which allows him unlimited bus rides in Boston, for $84.50. Without the CharlieCard, each bus ride costs $2.00. How many bus rides per month would Marcelo have to take so that the cost of the rides with the CharlieCard is less than the cost of the rides without the CharlieCard? Source: www.mbta.com

10. ***New York Public Transportation*** Karissa buys a monthly MetroCard, which allows her unlimited subway rides in New York City, for $121.50. Without the MetroCard, each subway ride costs $2.75. How many subway rides per month would Karissa have to take so that the cost of the rides with the MetroCard is less than the cost of the rides without the MetroCard? Source: www.mta.info

11. ***Home Theater System*** Jackson wants to purchase a home theater system that sells for $2500. Either he can pay the total amount at the time of purchase or he can agree to pay $250 down and $130 a month for 18 months. How much money can he save by paying the total amount at the time of purchase?

12. ***Checking Account*** The balance in Gabriela's checking account is $349.72. She purchased four video games at $32.39 each, including tax. If she pays by check, what is the new balance in her checking account?

13. ***Buying a House*** The Browns want to purchase a house that costs $150,000. They plan to take out a $120,000 mortgage on the house and put $30,000 as a down payment. The bank informs them that with a 15-year mortgage their monthly payment would be $887.63 and with a 30-year mortgage their monthly payment would be $572.90. Determine the amount they would save on the cost of the house if they selected the 15-year mortgage rather than the 30-year mortgage.

14. ***Getting an 80 Average*** On four exams, Wallace's grades were 79, 93, 91, and 68. What grade must he obtain on his fifth exam to have an 80 average?

15. ***Playing a Lottery*** In one state lottery game, you must select four digits (digits may be repeated). If your number matches exactly the four digits selected by the lottery commission, you win.

a) How many different numbers may be chosen?

b) If you purchase one lottery ticket, what is your chance of winning?

16. ***Energy Value and Energy Consumption*** The table above and to the right gives the approximate energy values of some foods, in kilojoules (kJ), and the energy requirements of some activities.

Food	Energy Value (kJ)	Activity	Energy Consumption (kJ/min)
Chocolate milkshake	2200	Walking	25
Fried egg	460	Cycling	35
Hamburger	1550	Swimming	50
Strawberry shortcake	1400	Running	80
Glass of skim milk	350		

a) How soon would you use up the energy from a fried egg by walking?

b) How soon would you use up the energy from a hamburger by swimming?

c) How soon would you use up the energy from a piece of strawberry shortcake by cycling?

d) How soon would you use up the energy from a hamburger and a chocolate milkshake by running?

17. ***Gas Mileage*** Wendy fills her gas tank completely and makes a note that the odometer reads 38,451.4 miles. The next time she put gas in her car, filling the tank took 12.6 gal and the odometer read 38,687.0 miles. Determine the number of miles per gallon that Wendy's car got on this tank of gas.

18. ***Saving for a Stereo*** Fernando works 40 hours per week and makes $8.50 per hour.

a) How much money can he expect to earn in 1 year (52 weeks)?

b) If he saves all the money he earns, how long will he have to work to save for a stereo receiver that costs $1275?

19. ***Mail-Order Purchase*** Fatima purchased 4 tires by mail order. She paid $52.80 per tire plus $5.60 per tire for shipping and handling. There is no sales tax on this purchase because the tires were purchased out of state. She also had to pay $8.56, including tax, per tire for mounting and balancing. At a local tire store, her total for the 4 tires with mounting and balancing would be $324 plus an 8% sales tax. How much did Mary save by purchasing the tires through the mail?

20. ***Furnace Technician*** Shawn, a furnace technician, charges a $50 flat rate plus $40 per hour to repair furnaces. During one month, his total income from repairing 15 furnaces was $1750. How many hours did he spend repairing furnaces?

21. ***Rain or Snow*** A plane is flying at an altitude of 15,000 ft, where the temperature is −6°F. The nearby airport, at an altitude of 3000 ft, is reporting precipitation. If the temperature increases 2.4°F for every 1000-ft decrease in altitude, will the precipitation at the airport be rain or snow? Assume that rain changes to snow at 32°F.

22. *Profit Margins* The following chart shows retail stores' average percent profit margin on certain items.

Product Category	Average Profit Margin (%)
Video equipment	12
Audio components	14
Stereo speakers	20–25
Extended warranties	50–60

Source: *Consumer Reports*

a) Determine the average profit of a store that has the list price of $620 on a camcorder.

b) Determine the average profit of a store that has a list price of $1200 on a pair of speakers (use a 22% profit margin).

c) If you negotiate and get the salesperson to sell the speakers for $1000, determine the store's profit (use a 22% profit margin).

23. *Income Taxes* The federal Income tax rate schedule for a person filing a single return in 2018 is shown below.

Taxable Income	Taxes
$0–$9,525	10% of taxable income
$9,526–$38,700	$952.50 + 12% of the amount over $9,525
$38,701–$82,500	$4,453.50 + 22% of the amount over $38,700
$82,501–$157,500	$14,089.50 + 24% of the amount over $82,500
$157,501–$200,000	$32,089.50 + 32% of the amount over $157,500
$200,001–$500,000	$45,689.50 + 35% of the amount over $200,000
$500,001 or more	$150,689.50 + 37% of the amount over $500,000

Source: www.irs.gov

a) Shenile works as a student tutor and paid $876 in federal taxes in 2018. Determine Shenile's taxable income.

b) Logan works as a caddy at a golf course and paid $2017.50 in federal taxes in 2018. Determine Logan's taxable income.

24. *Income Taxes* Use the federal income tax rate schedule for a person filing a single return in 2018 shown in Exercise 23 to answer the following questions.

a) Lewis works as a waiter and paid $4541.50 in federal taxes in 2018. Determine Lewis's taxable income.

b) Leticia works as an occupational therapist and paid $14,593.50 in federal taxes in 2018. Determine Leticia's taxable income.

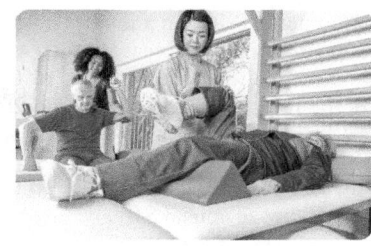

▲ See Exercise 24

25. *Wasted Water* A faucet leaks 1 oz of water per minute.

a) How many gallons of water are wasted in a year? (A gallon contains 128 oz.)

b) If water costs $11.20 per 1000 gal, how much additional money is being spent on the water bill?

26. *Leaking Faucet* A faucet is leaking at a rate of one drop of water per second. Assume that the volume of one drop of water is 0.1 cubic centimeter (0.1 cm^3).

a) Determine the volume of water in cubic centimeters lost in 1 year.

b) How many days would it take to fill a rectangular basin 30 cm by 20 cm by 20 cm?

27. *Tire Pressure* When a car's tire pressure is 30 pounds per square inch (psi), it averages 20.8 mpg of gasoline. If the tire pressure is increased to 35 psi, the car averages 21.6 mpg of gasoline.

a) If Mr. Levy drives an average of 20,000 mi per year, how many gallons of gasoline will he save in a year by increasing his tire pressure from 30 to 35 psi?

b) If gasoline costs $3.00 per gallon, how much will he save in a year?

c) If we assume that there are about 140 million cars in the United States and that these changes are typical of each car, how many gallons of gasoline would be saved if all drivers increased their cars' tire pressure?

28. *Parking at LAX* Ivana is flying out of Los Angeles International Airport (LAX) for business. She can park her car in long-term parking at LAX for $30 per day, or she can park her car at an off-site location, Park N Fly, for $17.95 per day. The tax rate for parking at LAX is 9.5%, and the tax rate for parking at Park N Fly is 8.75%. Ivana plans to be gone for 6 days.

a) Determine Ivana's total cost, including tax, to park her car at LAX for the 6 days.

b) Determine Ivana's total cost, including tax, to park her car at Park N Fly for the 6 days. Round your answer to the nearest cent.

c) How much money can Ivana save by parking at Park N Fly instead of LAX?

29. *Adjusting for Inflation* Assume that the rate of inflation is 2% per year for the next 2 years. What will the price of a sofa be 2 years from now if the sofa costs $999 today?

30. Investing You place $1000 in a mutual fund. The first year, the value of the fund increases by 10%. The second year, the value of the fund decreases by 10%. Determine the value of the fund at the end of the second year. Is it greater than, less than, or equal to your initial investment?

31. X-rays With a certain medical insurance policy, the customer must first pay an annual $100 deductible, and then the policy covers 80% of the cost of x-rays remaining after the deductible. The first insurance claims for a specific year submitted by Yungchen are for two x-rays. The first x-ray cost $720, and the second x-ray cost $980. How much, in total, will Yungchen need to pay for these x-rays?

32. Buying a Boat Four partners decide to share the cost of a boat equally. By bringing in an additional partner, they can reduce the cost to each of the five partners by $3000. What is the total cost of the boat?

33. Making Cream of Wheat The following amounts of ingredients are recommended to make various servings of Nabisco Instant Cream of Wheat. *Note:* 16 tbsp = 1 cup.

Ingredient	1 Serving	2 Servings	4 Servings
Mix water or milk	1 cup	2 cups	$3\frac{3}{4}$ cups
With salt (optional)	$\frac{1}{8}$ tsp	$\frac{1}{4}$ tsp	$\frac{1}{2}$ tsp
Add Cream of Wheat	3 tbsp	$\frac{1}{2}$ cup	$\frac{3}{4}$ cups

Determine the amount of each ingredient needed to make 3 servings using the following procedures.

a) Multiply the amounts for 1 serving by 3.

b) Determine the average of the amounts for 2 and 4 servings.

c) Subtract the amounts for 1 serving from the amounts for 4 servings.

d) Compare the answers obtained in parts (a) through (c) and explain any differences.

34. Making Rice Following are the amounts of ingredients recommended to make various servings of Uncle Ben's Original Converted Rice. *Note:* 1 tbsp = 3 tsp.

Ingredient	2 Servings	4 Servings	6 Servings	12 Servings
Rice (cups)	$\frac{1}{2}$	1	$1\frac{1}{2}$	3
Water (cups)	$1\frac{1}{3}$	$2\frac{1}{4}$	$3\frac{1}{3}$	6
Salt (teaspoons)	$\frac{1}{4}$	$\frac{1}{2}$	$\frac{3}{4}$	$1\frac{1}{2}$
Butter	1 tsp	2 tsp	1 tbsp	2 tbsp

Determine the amount of each ingredient needed to make 8 servings using the following procedures.

a) Multiply the amount for 2 servings by 4.

b) Multiply the amount for 4 servings by 2.

c) Add the amounts for 2 and 6 servings.

d) Subtract the amount for 4 servings from the amount for 12 servings.

e) Compare the answers obtained in parts (a) through (d) and explain any differences.

35. Purchasing DVDs The manager of the video department at Target plans to purchase a large number of DVDs of *Black Panther*. One supplier is selling boxes of 20 DVD movies for $240, and a second supplier is selling boxes of 12 DVD movies for $180. Only complete boxes of DVD movies can be purchased.

a) If the manager can purchase boxes of DVD movies from either or both suppliers, determine the maximum number of DVD movies that can be purchased for $425. Indicate how many boxes of 20 and how many boxes of 12 will be purchased.

b) How much will the DVD movies cost?

36. Who Will Win? Steve and Mark ran a 100-yard race. Mark won by 5 yards, which means Steve had run only 95 yards when Mark crossed the finish line. They decided to race again, with Mark starting 5 yards behind the starting line. Assuming both runners run the second race at the same pace as the first race, who will win?

37. One Square Foot How many square inches, 1 in. by 1 in., fit in an area of 1 square foot, 1 ft by 1 ft?

38. Cubic Inches How many cubic inches fit in 1 cubic foot?

39. Rectangle If the length and width of a rectangle each double, what happens to the area of the rectangle?

40. A Square Partition A 20 ft by 20 ft carpet is partitioned into 5 ft by 5 ft squares. How many squares will there be?

41. Cube If the length, width, and height of a cube all double, what happens to the volume of the cube?

42. Pole in a Lake A pole is in the middle of a small lake. One half of the pole is in the ground. One third of the pole is covered by water. Eleven feet, or one sixth, of the pole are out of water. What is the length of the pole?

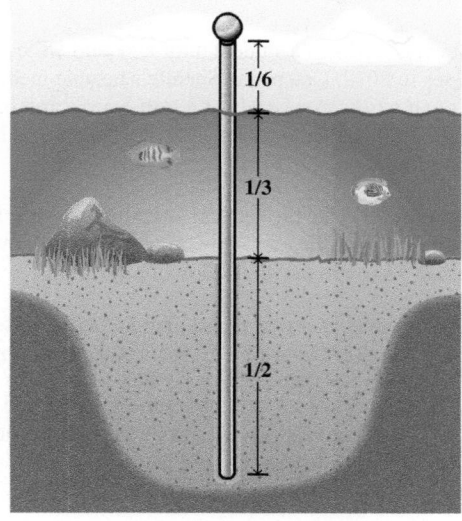

43. Buying Candy How much do 10 pieces of candy cost if 1000 pieces cost $10?

44. A Balance On the following balance, where should the one missing block ▪ be placed so that the balance would balance on the triangle (the fulcrum)? Assume that each block has the same weight.

45. Ties, Ties, Ties I have at least three ties. All my ties are red except two. All my ties are blue except two. All my ties are brown except two. How many ties do I have?

46. Palindromes A *palindrome* is a number (or word) that reads the same forward and backward. The numbers 1991 and 43234 are examples of palindromes. How many palindromes are there between the numbers 2000 and 3000? List them.

47. Telephone Number Words Below is a photo of the keyboard of a phone.

Certain businesses like to use a word to help people remember their phone number. For example, a doctor's office may use the word *medical* to represent its phone number, which would be 633-4225, or a shoe company may use the word *running* to represent its phone number, which would be 786-6464. Following are two numbers. Each number can be represented as a word that signifies something related to that business. Determine the words.

a) 733-7374 (*Hint:* could be used by a soft drink company)

b) 967-5688 (*Hint:* could be used by a gym/fitness center)

48. Tracing a Figure Try to retrace the figure without lifting your pencil from the paper or retracing a line. It can be done, but you need to begin at certain locations. We will discuss problems of this type in Chapter 13.

49. Numbers in Circles Place the numbers 1 through 6 in the circles below so that the sum along each of the three straight lines is the same. Each number must be used exactly once. (*Note:* There is more than one correct answer.)

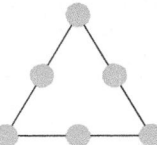

50. Cuts in Cheese If you make the three complete cuts in the cheese as shown, how many pieces of cheese will you have?

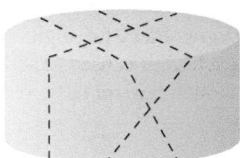

51. Magic Square Create a magic square by using the numbers 2, 4, 6, 8, 10, 12, 14, 16, and 18. The sum of the numbers in every column, row, and diagonal must be 30.

52. Magic Square Create a magic square by using the numbers 3, 4, 5, 6, 7, 8, 9, 10, and 11. The sum of the numbers in every column, row, and diagonal must be 21.

In Exercises 53–55, use the three magic squares illustrated to obtain the answers.

8	1	6
3	5	7
4	9	2

3	2	7
8	4	0
1	6	5

10	9	14
15	11	7
8	13	12

53. Magic Square Examine the 3 by 3 magic squares and determine the sum of the four corner entries of each magic square. How can you determine the sum by using a key number in the magic square?

54. Magic Square For a 3 by 3 magic square, how can you determine the sum of the numbers in any particular row, column, or diagonal by using a key value in the magic square?

55. Magic Square For a 3 by 3 magic square, how can you determine the sum of all the numbers in the square by using a key value in the magic square?

56. Stack of Cubes Identical cubes are stacked in the corner of a room as shown. How many of the cubes are not visible?

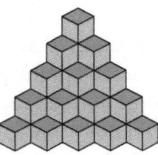

57. Dominos Consider a domino with six dots as shown. Two ways of connecting the three dots on the left with the three dots on the right are illustrated. Using three lines, in how many ways can the three dots on the left be connected with the three dots on the right?

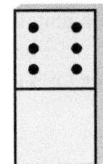 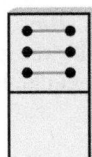 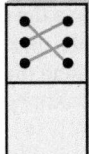

58. *Handshakes All Around* Five salespeople gather for a sales meeting. How many handshakes will each person make if each must shake hands with each of the four others?

59. *Consecutive Digits* Place the digits 1 through 8 in the eight boxes so that each digit is used exactly once and no two consecutive digits touch horizontally, vertically, or diagonally.

60. *A Digital Clock* Digital clocks display numerals by lighting all or some of the seven parts of the pattern shown. If each digit 0 through 9 is displayed once, which of the seven parts is used least often? Which part is used most often?

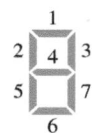

61. *A Grid* Place five 1's, five 2's, five 3's, five 4's, and five 5's in a 5 × 5 grid so that each digit—that is, 1, 2, 3, 4, 5—appears exactly once in each row and exactly once in each column.

Challenge Problems/Group Activities

62. *Insurance Policies* Ray owns two cars (a Ford Mustang and a Ford Focus), a house, and a rental apartment. He has auto insurance for both cars, a homeowner's policy, and a policy for the rental property. The costs of the policies are
    Mustang: $1648 per year
    Focus: $1530 per year
    Homeowner's: $640 per year
    Rental property: $750 per year

Ray is considering taking out a $1 million personal umbrella liability policy. The annual cost of the umbrella policy would be $450. If he has the umbrella policy, he can lower the limits on parts of his auto policies and still have equal or better protection. If Ray purchases the umbrella policy, he can reduce his premium on the Mustang by $90 per year and his premium on the Focus by 12%. If he purchases the umbrella policy and reduces the amount he pays for auto insurance, what is the net amount he is actually paying for the umbrella policy?

63. *Musicians* Jaquan, Cindy, and Mark are musicians. One plays the guitar, one plays the saxophone, and one plays the drums. They live in three adjacent houses on Lake View Drive. From the following information, determine who plays the drums. (*Hint:* A table may be helpful.)
    Mark does not play the guitar.
    Jaquan plays the guitar, but does not play the saxophone.

    The saxophone player and drummer live next to each other.
    Three years ago, Cindy broke her wrist playing the drums and has not played them since.
    Mark lives in the last house.
    The saxophone player and the guitar player share a common backyard swimming pool.

64. *Counting Triangles* How many triangles are in the figure?

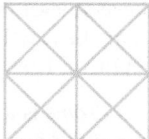

65. *Area of a Rectangle* Rectangle ABCD is made up entirely of squares. The black square has a side of 1 unit. Determine the area of rectangle ABCD.

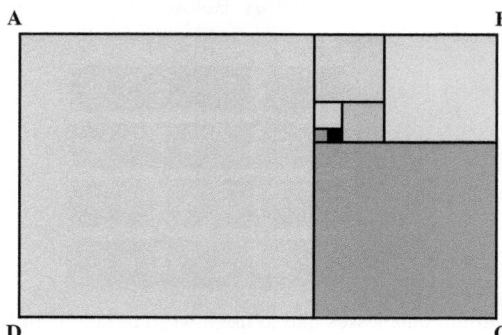

66. *Spending Money* Samantha went into a store and spent half her money and then spent $20 more. Samantha then went into a second store and spent half her remaining money and then spent $20 more. After spending money in the second store, Samantha had no money left. How much money did she have when she went into the first store?

67. *Boxes of Fruit* There are three boxes on a table, each with a label. Thomas knows that one box contains grapes, one box contains cherries, and the third box contains both grapes and cherries. He also knows that the three labels used—grapes, cherries, and grapes and cherries—were mixed up and that none of the boxes received the correct label. He opens just one box and, without looking into the box, takes out one piece of fruit. He looks at the fruit and immediately labels all the boxes correctly. Which box did Thomas open? How did he know how to correctly label the boxes?

Research Activities

68. *Puzzles* Many fun and interesting puzzle books and magazines are available. Using this chapter and puzzle books as a guide, construct five of your own puzzles and present them to your instructor.

CHAPTER 1 *Summary*

Important Facts and Concepts	Examples and Discussion
Section 1.1	
The **natural numbers** or **counting numbers** are 1, 2, 3, 4,….	Examples 1, 2, 4, 5, pages 2–4
A **conjecture** is a prediction based on specific observations.	Example 4, page 4
A **counterexample** is a special case that satisfies all the conditions of a conjecture but proves the conjecture false.	Discussion page 5
Inductive reasoning is the process of reasoning to a general conclusion through observations of specific cases.	Examples 1 and 2, pages 2–3
Deductive reasoning is the process of reasoning to a specific conclusion from a general statement.	Example 6, page 5
Section 1.2	
Estimation is the process of arriving at an approximate answer to a question.	Examples 1–7, pages 10–14
Section 1.3	
Guidelines for Problem Solving	
1. Understand the problem.	Discussion page 22
2. Devise a plan to solve the problem.	Examples 1–7, pages 20–28
3. Carry out the plan.	
4. Check the results.	

CHAPTER 1 *Review Exercises*

1.1†

In Exercises 1–8, use inductive reasoning to predict the next three numbers or figures in the pattern.

1. 3, 8, 13, 18, …

2. 28, 25, 22, 19, …

3. 4, −8, 16, −32, …

4. 5, 7, 10, 14, 19, …

5. 25, 24, 22, 19, 15, …

6. 6, 3, $\frac{3}{2}$, $\frac{3}{4}$, …

7. ◐, ⊟, ◑, ⊟, …

8. ⊡ , ⊡ , △ , ⊡ , ⊡ , …

9. *Pattern* Examine the grid above and to the right for a pattern and then select the answer which completes the pattern. (*Hint:* Think about rotating groups of four squares at a time.)

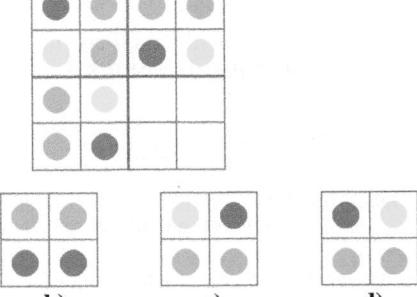

10. Pick any number and multiply the number by 10. Add 5 to the product. Divide the sum by 5. Subtract 1 from the quotient.

a) What is the relationship between the number you started with and the final number?

b) Arbitrarily select some different numbers and repeat the process. Record the original number and the results.

---

c) Make a conjecture about the original number and the final number.

d) Prove, using deductive reasoning, the conjecture you made in part (c).

11. Pick any number between 1 and 20. Add 5 to the number. Multiply the sum by 6. Subtract 12 from the product. Divide the difference by 2. Divide the quotient by 3. Subtract the number you started with from the quotient. What is your answer? Try this process with a different number. Make a conjecture as to what your final answer will always be.

12. *Counterexample* Determine a counterexample to the statement "The sum of two squares is an even number."

1.2

In Exercises 13–25, estimate the answer. Your answers may vary from those given in the back of the book, depending on how you round to arrive at the answer, but your answers should be something near the answers given.

13. $205{,}123 \times 4002$

14. $215.9 + 128.752 + 3.6 + 861 + 792$

15. 21% of 2095

16. *Distance* Estimate the distance from your wrist to your elbow and estimate the length of your foot. Which do you think is greater? With the help of a friend, measure both lengths to determine which is longer.

17. *Cost* Estimate the cost of 48 decorative bricks if one brick costs $3.97.

18. *Sales Tax* Estimate the sales tax on a boat that costs $21,000 if the sales tax rate is 8%.

19. *Walking Speed* Estimate your average walking speed in miles per hour if you walked 1.1 mi in 22 min.

20. *Groceries* Estimate the total cost of six grocery items that cost $2.49, $0.79, $1.89, $0.10, $2.19, and $6.75.

21. *A Walking Path* The scale of the map is $\frac{1}{4}$ in. = 0.1 mi. Estimate the distance of the walking path indicated in red.

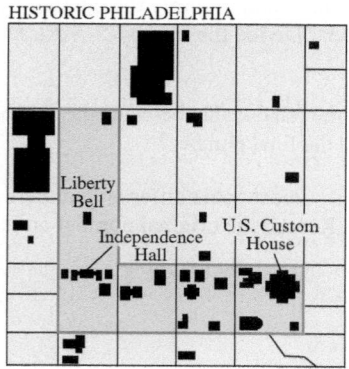

HISTORIC PHILADELPHIA

In Exercises 22 and 23, refer to the following graph, which illustrates the average daily high temperatures by month for the cities of Tampa and Chicago.

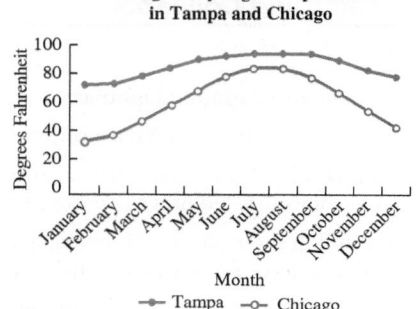

Average Daily High Temperatures in Tampa and Chicago

Source: www.noaa.gov

22. *January Temperatures in Tampa and Chicago* Estimate the difference in the average daily high temperatures for Tampa and Chicago in January.

23. *July Temperatures in Tampa and Chicago* Estimate the difference in the average daily high temperatures for Tampa and Chicago in July.

24. *Estimating an Area* If each square represents one square unit, estimate the size of the shaded area.

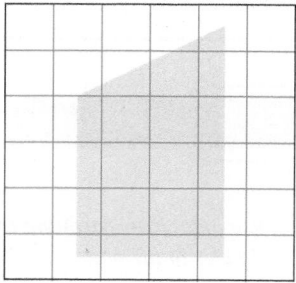

25. *Railroad Car Estimation* The scale of a model railroad is 1 in. = 12.5 ft. Estimate the length and height of an actual box car if this drawing is the same size as the model box car.

1.3

Solve the following problems.

26. *Buying a Copy Machine* David can buy a copier from Staples for $500 that prints, copies, and scans. He can either pay the total amount at the time of purchase, or he can agree to pay the store $50 down and $40 a month for 12 months. How much money can he save by paying the total amount at the time of purchase?

27. Buying in Quantity A six-pack of bottled water costs $1.99. A case of 4 six-packs costs $4.99. How much will be saved by purchasing the case rather than 4 individual six-packs?

28. Jet Ski Rental The rental cost of a jet ski from Freemac Marina is $15 for 15 minutes and the cost from Silvan Rental is $25 per half-hour. If you plan to rent the jet ski for 2 hours, which is the better deal and by how much?

29. Fishing Charter Sam and four of his friends want to schedule a 6-hour fishing trip on Lake Ontario with Reel Easy Sport Fishing. The cost for five people is $445. The cost for six people is $510. How much money per person will Sam and each of his friends save if they can find one more friend to go on their fishing trip? Assume that they will split the total cost equally.

30. Applying Fertilizer Ronaldo needs to apply fertilizer to his lawn on his farm. A 30-pound bag of fertilizer will cover an area of 2500 square feet.

a) How many pounds of fertilizer are needed to cover an area of 24,000 square feet?

b) If Ronaldo has only 150 pounds of fertilizer, how many square feet can he fertilize?

31. Auto Insurance Most insurance companies reduce premiums by 10% for people under the age of 25 who successfully pass a driver's education course. A particular driver's education course costs $60. Patrick, who just turned 18, has auto insurance that costs $1030 per year. By taking the driver's education course, how much would he save in auto insurance, including the cost of the course, from the age of 18 until the age of 25?

32. Pediatric Dosage If 1.5 milligrams of a medicine is to be given for 10 lb of body weight, how many milligrams should be given to a child who weighs 52 lb?

33. Qualifying for a Mortgage Serena's bank will grant her a mortgage provided the monthly mortgage payments are not greater than 28% of her monthly take-home pay. What is the maximum monthly mortgage payment the bank will grant to Serena if her gross monthly salary is $5500 and her payroll deductions are 30% of her gross monthly salary?

34. Flying West New York City is on eastern standard time, St. Louis is on central standard time (1 hour earlier than eastern standard time), and Las Vegas is on Pacific standard time (3 hours earlier than eastern standard time). A flight leaves New York City at 9 A.M. eastern standard time, stops for 50 min in St. Louis, and arrives in Las Vegas at 1:35 P.M. Pacific time. How long is the plane actually flying?

35. Crossing Time Zones The international date line is an imaginary line of longitude (from the North Pole to the South Pole) on Earth's surface between Japan and Hawaii in the Pacific Ocean. Crossing the line east to west adds a day to the present date. Crossing the line west to east subtracts a day. At 3:00 P.M. on July 25 in Hawaii, what is the time and date in Tokyo, Japan, which is four time zones to the west?

36. Conversions If 1 mi = 1.6 km,

a) convert 65 miles per hour to kilometers per hour.

b) convert 90 kilometers per hour to miles per hour.

37. Dot Pattern If the following pattern is continued, how many dots will be in the hundredth figure?

38. Magic Square The following magic square uses each number from 6 to 21 exactly once. Complete the magic square by using the unused numbers from 6 through 21 exactly once.

21	7		18
10		15	
14	12	11	17
9	19		

39. Magic Square Create a magic square by using the numbers 13, 15, 17, 19, 21, 23, 25, 27, and 29. The sum of the numbers in every row, column, and diagonal must be 63.

40. Microbes in a Jar A colony of microbes doubles in number every second. A single microbe is placed in a jar, and in an hour the jar is full. When was the jar half full?

41. Brothers and Sisters Jim has four more brothers than sisters. How many more brothers than sisters does his sister Mary have?

42. A Missing Dollar Three friends check into a single room in a motel and pay $10 apiece. The room costs $25 instead of $30, so a clerk is sent to the room to give $5 back. The friends each take back $1, and the clerk is given $2 for his trouble. Now each of the friends paid $9, a total of $27, and the clerk received $2. What happened to the missing dollar?

43. Average Grade An average grade of 80 on all exams in a course is needed to earn a B in the course. On her first four exams, Jacqueline's grades are 93, 88, 81, and 86. What is the minimum grade that Jacqueline can receive on the fifth exam to earn a B in the course?

44. Change for a Dollar Could a person have $1.15 worth of change in his pocket and still not be able to give someone change for a dollar bill? If so, what coins might he have?

45. *Volume of a Cube* Here is a flat pattern for a cube to be formed by folding. The sides of each square are 6 cm. Find the volume of the cube.

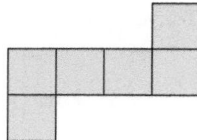

46. *The Heavier Coin* You have 13 coins, which all look alike. Twelve coins weigh exactly the same, but the other one is heavier. You have a pan balance. Tell how to determine the heavier coin in just three weighings.

47. *The Sum of Numbers* Determine the sum of the first 500 counting numbers. (*Hint:* Group in pairs.)

48. *Balancing a Scale* On a balance scale, three green balls balance six blue balls, two yellow balls balance five blue balls, and six blue balls balance four white balls. How many blue balls are needed to balance four green, two yellow, and two white balls?

49. *Palindromes* How many three-digit numbers greater than 100 are palindromes?

50. *Figures* Draw the next figure in the pattern.

51. *Patterns* How many orange tiles will be required to build the sixth figure in the pattern above and to the right?

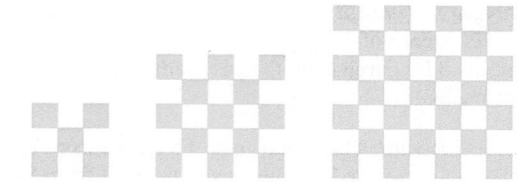

52. *Sum of Numbers* Place the numbers 1 through 12 in the 12 circles so that the sum of the numbers in each of the six rows is 26. Use each number from 1 through 12 exactly once.

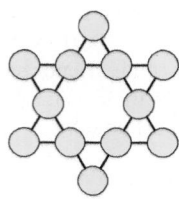

53. *People in a Line* People stand in a line at movies, the grocery store, and many other places.

a) In how many ways can two people stand in a line?

b) In how many ways can three people stand in a line?

c) In how many ways can four people stand in a line?

d) In how many ways can five people stand in a line?

e) Using the results from parts (a) through (d), make a conjecture about the number of ways in which n people can stand in a line.

CHAPTER 1   Test

In Exercises 1 and 2, use inductive reasoning to determine the next three numbers in the pattern.

1. 2, 8, 14, 20, …

2. $1, \frac{1}{2}, \frac{1}{3}, \frac{1}{4}, \ldots$

3. Pick any number, multiply the number by 5, and add 10 to the number. Divide the sum by 5. Subtract 1 from the quotient.

a) What is the relationship between the number you started with and the final answer?

b) Arbitrarily select some different numbers and repeat the process. Record the original number and the results.

c) Make a conjecture about the relationship between the original number and the final answer.

d) Prove, using deductive reasoning, the conjecture made in part (c).

In Exercises 4 and 5, estimate the answers.

4. $0.51 \times 96{,}000$

5. $\dfrac{188{,}000}{0.11}$

6. *Estimating Area* If each square represents one square unit, estimate the area of the shaded figure.

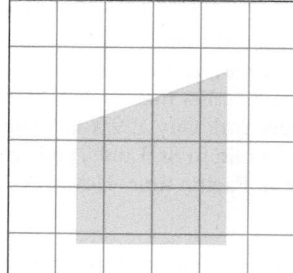

7. *Body Mass Index* The federal government gives a procedure to determine if a child is overweight. To make this decision, first determine the child's body mass index

(BMI). Then compare the BMI with one of the two charts, one for boys and one for girls, provided by the government. Below, we give the chart for boys up to age 20. To determine a child's BMI:

1) Divide the child's weight (in pounds) by the child's height (in inches).

2) Divide the results from part 1 by the child's height again.

3) Multiply the result from part 2 by 703.

Richard is a 14-year-old boy who weighs 130 lb and is 63 in. tall.

a) Determine his BMI.

b) Does he appear to be at risk for being overweight, or is he overweight? Explain.

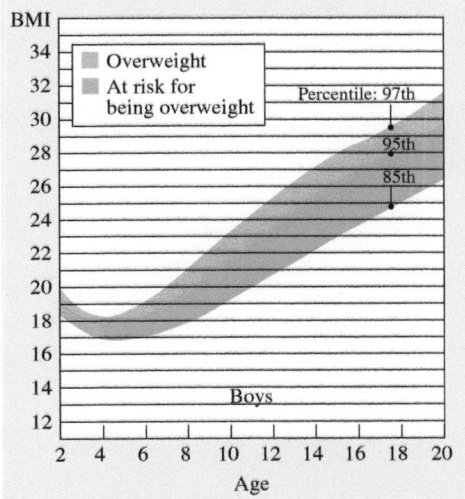

Source: Rod Little-USN&WR

8. Visiting Washington, DC The following graph shows the number of visitors, in thousands, from China, United Kingdom, Germany, South Korea, and France to Washington, DC, for the year 2017.

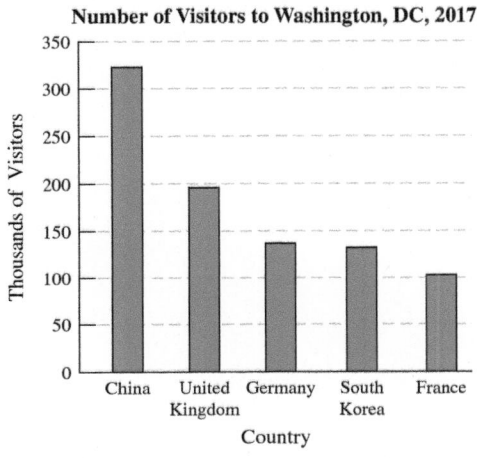

Source: U.S. Department of Commerce

a) Estimate the number of visitors to Washington, DC, from China in 2017.

b) Estimate the number of visitors to Washington, DC, from France in 2017

9. Truck Rental A truck costs $35 a day plus 20 cents a mile to rent. If the total cost for a one-day rental is $50.40, how many miles were driven?

10. Cans of Soda At a local store a six-pack of soda costs $2.59 and individual cans cost $0.80. What is the maximum number of cans of soda that can be purchased for $15?

11. Cutting Wood How much time does it take Carla, a carpenter, to cut a 10-ft length of wood into four equal pieces, if each cut takes $2\frac{1}{2}$ min?

12. Rocky Mountain National Park The scale on a map of Rocky Mountain National Park indicates that 1 inch on the map equals about $\frac{3}{4}$ mile in the park. If the distance on the map from Grand Lake Entrance to Holzwarth Historic Site is about 10 inches, how far is this distance in the park?

▲ Rocky Mountain National Park

13. Payment Shortfall Kono gets $12.75 per hour with time and a half for any time over 40 hours per week. If she works a 50-hr week and gets paid $652.25, by how much was she underpaid?

14. Magic Square Create a magic square by using the numbers 5, 10, 15, 20, 25, 30, 35, 40, and 45. The sum of the numbers in every row, column, and diagonal must be 75.

15. A Drive to the Beach Mary drove from her home to the beach that is 30 mi from her house. The first 15 mi she drove at 60 mph, and the next 15 mi she drove at 30 mph. Would the trip take more, less, or the same time if she traveled the entire 30 mi at a steady 45 mph?

16. Chili Recipe Three teaspoons equal one tablespoon. A recipe for chili calls for $\frac{1}{2}$ teaspoon of salt for 2 pounds of ground beef. How many tablespoons of salt are needed for 6 pounds of ground beef?

17. Area of a Concrete Walkway A 10 meter by 12 meter rectangular lawn has a concrete walkway 1 meter wide around the outside of the lawn. What is the area of the walkway?

18. *Jelly Bean Guess* One guess is off by 9, another guess is off by 17, and yet another guess is off by 31. How many jelly beans are in the jar below?

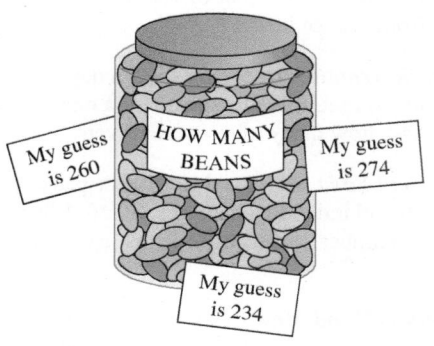

19. *Buying Plants* David wants to purchase nine herb plants. Countryside Nursery has herbs that are on sale at three for $3.99. David has a coupon for 25% off an unlimited number of herb plants at the original price of $1.75 per plant.

a) Determine the cost of purchasing nine plants at the sale price.

b) Determine the cost of purchasing nine plants if the coupon is used.

c) Which is the least expensive way to purchase the nine plants, and by how much?

20. *Arranging Letters* In how many different ways can four letters, A, B, C, D, be arranged?

◄ When you create a shopping list of items to purchase, you are placing items in a set.

2

Sets

What You Will Learn

- Set concepts including methods to indicate sets, equal sets, and equivalent sets
- Subsets
- Venn diagrams and set operations such as complement, intersection, union, difference, and Cartesian product
- Equality of sets
- Applications of sets
- Infinite sets

Why This Is Important

A basic human impulse is to sort or classify things. As you will see in this chapter, putting elements into sets helps you order and arrange your world. It allows you to deal with large quantities of information. Computer programs such as Python and technological tools such as Excel and StatCrunch that allow us to analyze large sets of data rely on set concepts. Set building is a learning tool that helps answer the question "What are the characteristics of this group?" For example, when you are creating a shopping list of items to purchase for a meal, you are placing items in a particular set. You may want to create a set of items needed from the produce section, a set of items needed from the meat department, and a set of items needed from the snack aisle. Organizing items into sets helps ensure that you buy all the needed items. Studying sets is also important because sets underlie other mathematical topics such as logic, probability, statistics, and abstract algebra.

SECTION 2.1   Set Concepts

Upon completion of this section, you will be able to:

- Understand methods to indicate a set including description, roster form, and set-builder notation.
- Understand fundamental set concepts including finite sets, infinite sets, equal sets, equivalent sets, the cardinal number of a set, the null or empty set, and universal sets.

Throughout your day you may have many different tasks to complete. Going to class, going to work, making appointments, shopping, spending time with family and friends, responding to email, and checking Instagram are just some of the things you may do on a daily basis. In this section, we will discuss ways to sort or classify items, such as tasks we need to complete, into sets. We will also discuss different methods that can be used to indicate sets.

Why This Is Important   Set classifications are important in a range of applications that affect our daily lives. Organizing your day into different sets of tasks to complete can help you with time management while limiting your amount of stress.

We encounter sets in many different ways every day of our lives. A *set* is a collection of objects, which are called *elements* or *members* of the set. For example, the United States is a collection, or set, of 50 states plus the District of Columbia. The 50 individual states plus the District of Columbia are the members or elements of the set that is called the United States.

A set is *well defined* if its contents can be clearly determined. The set of U.S. presidents is a well-defined set because its contents, the presidents, can be named. The set of the three best movies is not a well-defined set because the word *best* is interpreted differently by different people. In this text, we use only well-defined sets.

Methods to Indicate a Set

Three methods are commonly used to indicate a set: (1) description, (2) roster form, and (3) set-builder notation.

The method of indicating a set by *description* is illustrated in Example 1.

Example 1   *Description of Sets*

Write a description of the set containing the elements Sunday, Monday, Tuesday, Wednesday, Thursday, Friday, and Saturday.

Solution   The set is the days of the week.   ∎

Listing the elements of a set inside a pair of *braces*, $\{\ \ \}$, is called *roster form*. The braces are an essential part of the notation because they identify the contents as a set. For example, $\{1, 2, 3\}$ is notation for the set whose elements are 1, 2, and 3, but $(1, 2, 3)$ and $[1, 2, 3]$ are not sets because parentheses and brackets do not indicate a set. For a set written in roster form, commas separate the elements of the set. The order in which the elements are listed is not important. Additionally, we do not list an element more than once. For example, the set $\{1, 1, 2\}$ would be written as set $\{1, 2\}$.

Sets are generally named with capital letters. For example, the name commonly selected for the set of *natural numbers* or *counting numbers* is N.

Definition: **Natural Numbers**
$$N = \{1, 2, 3, 4, 5, \ldots\}$$

The three dots after the 5, called an *ellipsis,* indicate that the elements in the set continue in the same manner. An ellipsis followed by a last element indicates that the elements continue in the same manner up to and including the last element. This notation is illustrated in Example 2(b).

Example 2 *Roster Form of Sets*

Express the following in roster form.
a) Set A is the set of natural numbers less than 6.
b) Set B is the set of natural numbers less than or equal to 45.
c) Set P is the set of planets in Earth's solar system.

Solution

a) The natural numbers less than 6 are 1, 2, 3, 4, and 5. Thus, set A in roster form is
$A = \{1, 2, 3, 4, 5\}$.
b) $B = \{1, 2, 3, 4, \ldots, 45\}$. The 45 after the ellipsis indicates that the elements continue in the same manner up to and including the number 45.
c) $P = \{$Mercury, Venus, Earth, Mars, Jupiter, Saturn, Uranus, Neptune$\}$* ■

Example 3 *The Word* Inclusive

Express the following in roster form.
a) Set A is the set of natural numbers between 5 and 9.
b) Set B is the set of natural numbers between 5 and 9, inclusive.

Solution

a) $A = \{6, 7, 8\}$
b) $B = \{5, 6, 7, 8, 9\}$. Note that the word *inclusive* indicates that the values of 5 and 9 are included in the set. ■

The symbol $\in$, read "is an element of," is used to indicate membership in a set. In Example 3, because 8 is an element of set A, we write $8 \in A$. This may also be written $8 \in \{6, 7, 8\}$. We may also write $5 \notin A$, meaning that 5 is not an element of set A.

Set-builder notation (sometimes called *set-generator notation*) may be used to symbolize a set. Set-builder notation is frequently used in algebra. The following example illustrates its form and how it is read.

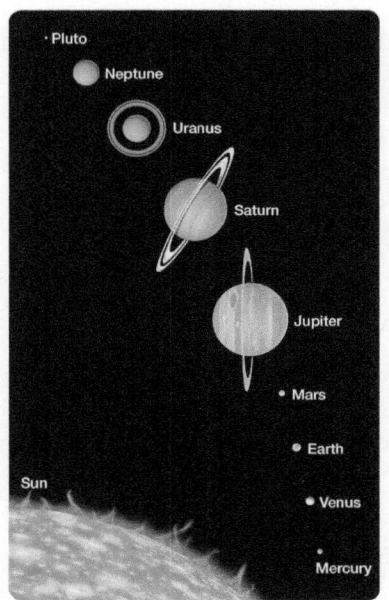

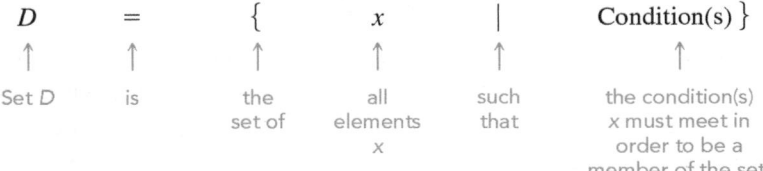

D	$=$	$\{$	x	$\mid$	Condition(s) $\}$
↑	↑	↑	↑	↑	↑
Set D	is	the set of	all elements x	such that	the condition(s) x must meet in order to be a member of the set.

Consider $E = \{x \mid x \in N \text{ and } x > 10\}$. The statement is read: "Set E is the set of all the elements x such that x is a natural number and x is greater than 10." The conditions that x must meet to be a member of the set are $x \in N$, which means that x must be a natural number, and $x > 10$, which means that x must be greater than 10. The numbers that meet both conditions are the set of natural numbers greater than 10. Set E in roster form is

$$E = \{11, 12, 13, 14, \ldots\}$$

The planets of Earth's solar system

* In August 2006, Pluto was reclassified as a dwarf planet.

Example 4   *Using Set-Builder Notation*

a) Write set $B = \{1, 2, 3, 4, 5\}$ in set-builder notation.
b) Write, in words, how you would read set B in set-builder notation.

Solution

a) Because set B consists of the natural numbers less than 6, we write

$$B = \{x \mid x \in N \text{ and } x < 6\}.$$

Another acceptable answer is $B = \{x \mid x \in N \text{ and } x \leq 5\}$.

b) Set B is the set of all elements x such that x is a natural number and x is less than 6. Another acceptable answer is set B is the set of all elements x such that x is a natural number and x is less than or equal to 5.   ■

Example 5   *Roster Form to Set-Builder Notation*

a) Write set $C = \{$North America, South America, Europe, Asia, Australia, Africa, Antarctica$\}$ in set-builder notation.
b) Write in words how you would read set C in set-builder notation.

Solution

a) $C = \{x \mid x \text{ is a continent}\}$.
b) Set C is the set of all elements x such that x is a continent.   ■

Example 6   *Set-Builder Notation to Roster Form*

Write set $A = \{x \mid x \in N \text{ and } 2 \leq x < 8\}$ in roster form.

Solution $A = \{2, 3, 4, 5, 6, 7\}$ ■

Example 7   *Elements of a Set*

Determine whether each statement is true or false. If false, give the reason.
a) blue $\in \{$orange, yellow, blue, green$\}$
b) peppermint $\in \{$spearmint, cinnamon, licorice$\}$
c) $\{$sapphire$\} \in \{$gold, diamond, ruby, sapphire$\}$

Solution

a) blue $\in \{$orange, yellow, blue, green$\}$ is a true statement since blue is an element of the set $\{$orange, yellow, blue, green$\}$.
b) peppermint $\in \{$spearmint, cinnamon, licorice$\}$ is a false statement since peppermint is not an element of the set $\{$spearmint, cinnamon, licorice$\}$
c) $\{$sapphire$\} \in \{$gold, diamond, ruby, sapphire$\}$ is a false statement because $\{$sapphire$\}$ is a set and the set $\{$sapphire$\}$ is not an element of the set $\{$gold, diamond, ruby, sapphire$\}$.   ■

Example 8   *Most Expensive Colleges in the United States*

The table on the next page shows the total costs for one year of college for the 2019 academic year at the 10 most expensive colleges in the United States. Let set C be the set of colleges located in California that are among the 10 most expensive colleges in the United States. Write set C in roster form.

▲ *University of Southern California*

Ten Most Expensive Colleges in the United States	State	Total Cost
Columbia University	NY	$59,430
University of Chicago	IL	$57,006
Vassar College	NY	$56,960
Trinity College	CT	$56,910
Harvey Mudd College	CA	$56,876
Franklin and Marshall College	PA	$56,550
Amherst College	MA	$56,426
Tufts University	MA	$56,382
University of Southern California	CA	$56,225
Bucknell University	PA	$56,092

Source: *US News*

Solution By examining the table, we determine that two colleges located in California appear in the table. They are Harvey Mudd College and University of Southern California. Thus, set $C = \{$Harvey Mudd College, University of Southern California$\}$. ∎

Fundamental Set Concepts

A set is said to be *finite* if it either contains no elements or the number of elements in the set is a natural number. The set $B = \{2, 4, 6, 8, 10\}$ is a finite set because the number of elements in the set is 5, and 5 is a natural number. A set that is not finite is said to be *infinite*. The set of counting numbers is one example of an infinite set. Infinite sets are discussed in more detail in Section 2.6.

Another important concept is equality of sets.

> **Definition: Equal Sets**
> Set A is **equal** to set B, symbolized by $A = B$, if and only if set A and set B contain exactly the same elements.

For example, if set $A = \{1, 2, 3\}$ and set $B = \{3, 1, 2\}$, then $A = B$ because they contain exactly the same elements. The order of the elements in the set is not important. If two sets are equal, both must contain the same number of elements. The number of elements in a set is called its *cardinal number*.

> **Definition: Cardinal Number**
> The **cardinal number** of set A, symbolized by $n(A)$, is the number of elements in set A.

Both set $A = \{1, 2, 3\}$ and set $B = \{$England, Brazil, Japan$\}$ have a cardinal number of 3; that is, $n(A) = 3$, and $n(B) = 3$. We can say that set A and set B both have a cardinality of 3.

Two sets are said to be *equivalent* if they contain the same number of elements.

> **Definition: Equivalent Sets**
> Set A is **equivalent** to set B if and only if $n(A) = n(B)$.

Any sets that are equal must also be equivalent. Not all sets that are equivalent are equal, however. The sets $D = \{a, b, c\}$ and $E = \{\text{apple, orange, pear}\}$ are equivalent because both have the same cardinal number, 3. Because the elements differ, however, the sets are not equal.

Two sets that are equivalent or have the same cardinality can be placed in *one-to-one correspondence*. Set A and set B can be placed in one-to-one correspondence if every element of set A can be matched with exactly one element of set B and every element of set B can be matched with exactly one element of set A. For example, there is a one-to-one correspondence between the student names on a class list and the student identification numbers because we can match each student with a student identification number.

Consider set S, states, and set C, state capitals.

$$S = \{\text{North Carolina, Georgia, South Carolina, Florida}\}$$

$$C = \{\text{Columbia, Raleigh, Tallahassee, Atlanta}\}$$

Two different one-to-one correspondences for sets S and C follow.

$$S = \{\text{North Carolina, Georgia, South Carolina, Florida}\}$$

$$C = \{\text{Columbia, Raleigh, Tallahassee, Atlanta}\}$$

$$S = \{\text{North Carolina, Georgia, South Carolina, Florida}\}$$

$$C = \{\text{Columbia, Raleigh, Tallahassee, Atlanta}\}$$

Other one-to-one correspondences between sets S and C are possible. Do you know which capital goes with which state?

Some sets do not contain any elements, such as the set of zebras that live in your house.

Definition: Empty Set

The set that contains no elements is called the **empty set** or **null set** and is symbolized by $\{\ \}$ or $\varnothing$.

Note that $\{\varnothing\}$ is not the empty set. This set contains the element $\varnothing$ and has a cardinality of 1. The set $\{0\}$ is also not the empty set because it contains the element 0. It also has a cardinality of 1.

Example 9 *Natural Number Solutions*

Indicate the set of natural numbers that satisfies the equation $x + 2 = 0$.

Solution The values that satisfy the equation are those natural numbers that make the equation a true statement. Only the number -2 satisfies this equation. Because -2 is not a natural number, the solution set of this equation is $\{\ \}$ or $\varnothing$. ∎

Another important set is a *universal set*.

Definition: Universal Set

A **universal set**, symbolized by U, is a set that contains all the elements for any specific discussion.

When a universal set is given, only the elements in the universal set may be considered when working the problem. If, for example, the universal set for a particular problem is defined as $U = \{1, 2, 3, 4, \ldots, 10\}$, then only the natural numbers 1 through 10 may be used in that problem.

SECTION 2.1   *Exercises*

Warm Up Exercises

In Exercises 1–10, fill in the blank with an appropriate word, phrase, or symbol(s).

1. A collection of objects is called a(n) _____.

2. Three dots placed in a set to show that the set continues in the same manner is called a(n) _____.

3. The three ways a set can be written are _____, _____, and _____.

4. A set that contains no elements or the number of elements in the set is a natural number is called a(n) _____ set.

5. A set that is not finite is called a(n) _____ set.

6. Two sets that contain the same number of elements are called _____ sets.

7. Two sets that contain the same elements are called _____ sets.

8. The number of elements in a set is called the _____ number of the set.

9. The set that contains no elements is called the _____ set.

10. The two ways to indicate an empty set are _____ and _____.

Practice the Skills

In Exercises 11–16, determine whether each set is well defined or not well defined.

11. The set of best artists

12. The set of the most interesting movies in 2019

13. The set of countries with a population of more than 1 trillion people

14. The set of Academy Awards winners in 2019

15. The set of astronauts who walked on the moon

▲ *Eugene A. Cernan on the moon*

16. The set of the most interesting teachers at your school

In Exercises 17–22, determine whether each set is finite or infinite.

17. $\{1, 2, 3, 4, \ldots\}$

18. The set of multiples of 3 between 0 and 40

19. The set of odd numbers greater than 25

20. The set of fractions between 1 and 2

21. The set of odd numbers greater than 15

22. The set of apple trees in Gro-More Farms Orchards

In Exercises 23–32, express each set in roster form. You may need to use the Internet or some other reference source.

23. The set of states in the United States whose names begin with the letter H

24. The set of months whose names begin with the letter J

25. The set of natural numbers between 10 and 178

26. $C = \{x \mid x + 6 = 10\}$

27. $B = \{x \mid x \in N \text{ and } x \text{ is even}\}$

28. The set of states in the United States that have a common border with Alaska

29. The set of football players over the age of 70 who are still playing in the National Football League

30. The set of states in the United States that have a common border with the state of Washington

31. $E = \{x \mid x \in N \text{ and } 14 \leq x < 85\}$

32. The set of states in the United States that are not in the contiguous 48 states

In Exercises 33–36, use the following table, which shows the number of monthly users, in millions, for the 10 most used mobile gaming apps in the United States in July 2018. Let the 10 mobile gaming apps be the universal set.

Mobile Gaming App	Mobile Users (in millions)
Google Play Games	24
Pokémon Go	13
IGN	12
GameSpot	11
Candy Crush Saga	11
Rave	11
Words with Friends	9
Solitaire by MobilityWare	7
Candy Crush Soda Saga	7
ApplsLoading	6

Source: CNET

Use the list to determine each set in roster form.

33. The set of mobile gaming apps in which the number of monthly users was greater than 12 million.

34. The set of mobile gaming apps in which the number of monthly users was less than 7 million.

35. The set of mobile gaming apps in which the number of monthly users was between 5 million and 8 million.

36. The set of mobile gaming apps in which the number of monthly users was between 15 million and 20 million.

In Exercises 37–40, use the following graph, created using **StatCrunch***, which shows the worldwide sales of the Apple iPhone, in millions of units, for the years 2011–2018. Let the 8 years be the universal set.*

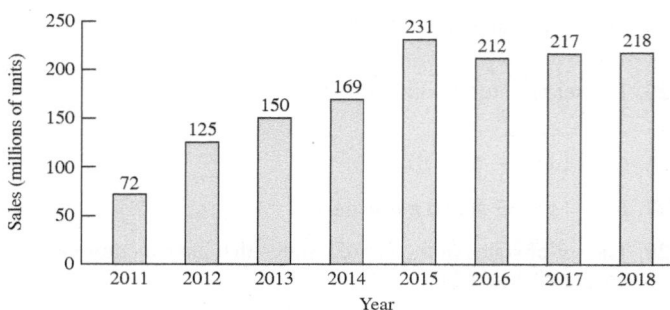

Unit Sales of the Apple iPhone Worldwide

Source: ZDNet

Use the graph to determine each set in roster form.

37. The set of years in which Apple iPhone unit sales were greater than 200 million

38. The set of years in which Apple iPhone unit sales were less than 200 million

39. The set of years in which Apple iPhone unit sales were less than 20 million

40. The set of years in which Apple iPhone unit sales were between 140 million and 200 million

In Exercises 41–48, express each set in set-builder notation.

41. $B = \{7, 8, 9, 10, 11, 12, 13, 14\}$

42. $A = \{1, 2, 3, 4, 5, 6, 7, 8, 9\}$

43. $C = \{3, 6, 9, 12, \dots\}$

44. $D = \{5, 10, 15, 20, \dots\}$

45. E is the set of odd natural numbers.

46. A is the set of national holidays in the United States in July.

47. C is the set of months that contain less than 30 days.

48. $F = \{15, 16, 17, \dots, 100\}$

In Exercises 49–56, write a description of each set.

49. $A = \{1, 2, 3, 4, 5, 6, 7\}$

50. $D = \{3, 6, 9, 12, 15, 18, \dots\}$

51. $V = \{a, e, i, o, u\}$

52. $D = \{$January, February, March, April, May, June, July, August, September, October, November, December$\}$

53. $T = \{$oak, maple, elm, pine, $\dots\}$

54. $E = \{x \mid x \in N$ and $4 \le x < 11\}$

55. $S = \{$spring, summer, fall, winter$\}$

56. $R = \{$Washington, Jefferson, Lincoln, Roosevelt$\}$

▲ Mount Rushmore

In Exercises 57–60, use the following list, which shows the 10 social media apps in the United States with the most monthly mobile users, in millions, in July 2018. Let the 10 social media apps in the list represent the universal set.

Social Media App	Number of Monthly Mobile Users (in millions)
Facebook	169
Instagram	117
Facebook Messenger	111
Twitter	70
Pinterest	58
Snapchat	52
Reddit	33
Tumblr	23
WhatsApp	21
Google Hangouts	15

Source: Digital Information World

Use the list to determine each set in roster form.

57. $\{x \mid x$ is a social media app with at least 100 million mobile users$\}$

58. $\{x \mid x$ is a social media app with fewer than 25 million mobile users$\}$

59. $\{x \mid x$ is a social media app with between 50 million and 75 million mobile users$\}$

60. $\{x \mid x$ is a social media app with between 115 million and 150 million mobile users$\}$

In Exercises 61–64, use the graph, which shows the cost of a 30-second commercial, in millions of dollars, during the Super Bowl from 2010 through 2019. Let the 10 years represent the universal set.

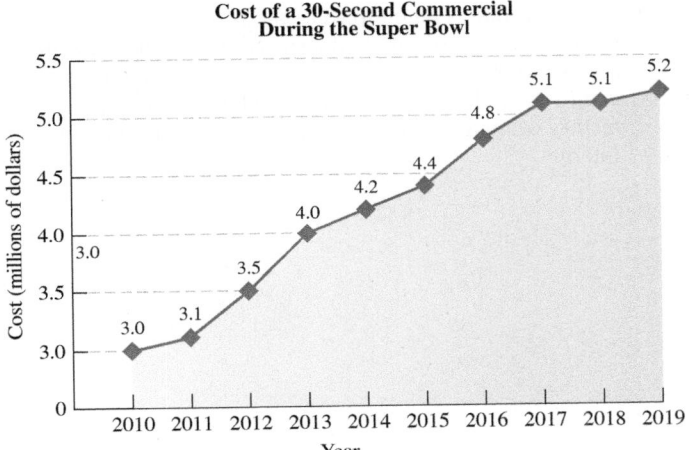

Source: Kantar Media, Nielsen, Ad Age

Use the graph to determine the following sets in roster form.

61. The set of years in which the cost of a Super Bowl commercial was more than $4.5 million

62. The set of years in which the cost of a Super Bowl commercial was less than $3.5 million

63. The set of years in which the cost of a Super Bowl commercial was between $3.6 million and $4.5 million

64. The set of years in which the cost of a Super Bowl commercial was greater than $4.0 million and less than $4.5 million

In Exercises 65–72, state whether each statement is true or false. If false, give the reason.

65. $\{e\} \in \{a, e, i, o, u\}$

66. $b \in \{a, b, c, d, e, f\}$

67. $h \in \{a, b, c, d, e, f\}$

68. February $\in \{$months of the year$\}$

69. $3 \notin \{x \mid x \in N$ and x is odd$\}$

70. Tokyo $\in \{$cities in the United States$\}$

71. $9 \in \{1, 3, 5, 7, \dots\}$

72. $2 \in \{x \mid x$ is an odd natural number$\}$

In Exercises 73–76, use the sets $A = \{2, 4, 6, 8\}$, $B = \{1, 3, 7, 9, 13, 21\}, C = \{\ \ \}, and D = \{\#, \&, \%, \square, \$\}$.

73. Determine $n(A)$.   **75.** Determine $n(C)$.

74. Determine $n(B)$.   **76.** Determine $n(D)$.

In Exercises 77–82, determine whether the pairs of sets are equal, equivalent, both, or neither.

77. $A = \{$lion, tiger, monkey$\}$, $B = \{$tiger, monkey, lion$\}$

78. $A = \{$purple, green, yellow$\}$, $B = \{q, r, s\}$

79. $A = \{$grapes, apples, oranges$\}$, $B = \{$grapes, peaches, apples, oranges$\}$

80. A is the set of Siamese cats. B is the set of cats.

81. A is the set of letters in the word *bank*. B is the set of letters in the word *post*.

82. A is the set of states. B is the set of state capitals.

Problem Solving

83. Set-builder notation is often more versatile and efficient than listing a set in roster form. This versatility is illustrated with the following two sets.

$A = \{x \mid x \in N$ and $x > 2\}$
$B = \{x \mid x > 2\}$

a) Write a description of set A and set B.

b) Explain the difference between set A and set B. (*Hint:* Is $4\frac{1}{2} \in A$? Is $4\frac{1}{2} \in B$?)

c) Write set A in roster form.

d) Can set B be written in roster form? Explain your answer.

84. Consider sets A and B below

$$A = \{x \mid 2 < x \le 5 \text{ and } x \in N\}$$

and

$$B = \{x \mid 2 < x \le 5\}$$

a) Write a description of set A and set B.

b) Explain the difference between set A and set B.

c) Write set A in roster form.

d) Can set B be written in roster form? Explain your answer.

*A cardinal number answers the question "How many?" An **ordinal number** describes the relative position that an element occupies. For example, Molly's desk is the third desk from the aisle.*

In Exercises 85–88, determine whether the number used is a cardinal number or an ordinal number.

85. J. K. Rowling has written 7 Harry Potter books.

86. Study the chart on page 25 in the book.

87. Lincoln was the sixteenth president of the United States.

88. Emily paid $35 for her new blouse.

▲ *Abraham Lincoln*

89. Describe three sets of which you are a member.

90. Describe three sets that have no members.

91. Write a short paragraph explaining why the universal set and the empty set are necessary in the study of sets.

Challenge Problem/Group Activity

92. a) In a given exercise, a universal set is not specified, but we know that actor Bradley Cooper is a member of the universal set. Describe five different possible universal sets of which Bradley Cooper is a member.

b) Write a description of one set that includes all the universal sets in part (a).

Research Activity

93. *Georg Cantor* is recognized as the founder and a leader in the development of set theory. Do research and write a paper on his life and his contributions to set theory and to the field of mathematics. References include history of mathematics books, encyclopedias, and the Internet.

SECTION 2.2   Subsets

Upon completion of this section, you will be able to:

■ Determine and recognize subsets.

■ Determine and recognize proper subsets.

■ Determine the number of subsets and proper subsets of a given set.

Consider the following sets. Set $A =$ {baseball, basketball, hockey}. Set $B =$ {baseball, football, basketball, hockey, softball}. Note that each element of set A is also an element of set B. In this section, we will discuss how to illustrate the relationship between two sets, A and B, when each element of set A is also an element of set B.

Why This Is Important   The relationship between sets is important throughout life. For example, to gain a promotion at work, you may need to fulfill different sets of criteria or sets of goals. Organizing your goals into different sets may make it easier for you to meet these goals and fulfill the requirements for a promotion.

Subsets

In our complex world, we often break larger sets into smaller, more manageable sets, called *subsets*. For example, consider the set of people in your class. Suppose we categorize the set of people in your class according to the first letter of their last name (the A's, B's, C's, etc.). When we do so, each of these sets may be considered a subset of the original set. Each of these subsets can be separated further. For example, the set of people whose last name begins with the letter A can be categorized as either male or female or by their age. Each of these collections of people is also a subset. A given set may have many different subsets.

> **Definition: Subset**
> Set A is a **subset** of set B, symbolized by $A \subseteq B$, if and only if all the elements of set A are also elements of set B.

The symbol $A \subseteq B$ indicates that "set A is a subset of set B." The symbol $\not\subseteq$ is used to indicate "is not a subset." Thus, $A \not\subseteq B$ indicates that set A is not a subset of set B. *To show that set A is not a subset of set B, we must find at least one element of set A that is not an element of set B.*

Example 1 A Subset?

Determine whether set A is a subset of set B.

a) $A = \{\text{Brazil, Argentina, Peru}\}$
   $B = \{\text{Brazil, Argentina, Colombia, Peru}\}$
b) $A = \{1, 3, 5, 7\}$
   $B = \{1, 3\}$
c) $A = \{x \mid x \text{ is a yellow fruit}\}$
   $B = \{x \mid x \text{ is a red fruit}\}$
d) $A = \{\text{pink, purple, blue}\}$
   $B = \{\text{purple, blue, pink}\}$

Solution

a) All the elements of set A are contained in set B; therefore $A \subseteq B$.
b) The elements 5 and 7 are in set A but not in set B; therefore $A \not\subseteq B$ (A is not a subset of B). In this example, however, all the elements of set B are contained in set A; therefore, $B \subseteq A$.
c) There are fruits, such as bananas, that are in set A that are not in set B, so $A \not\subseteq B$.
d) All the elements of set A are contained in set B, so $A \subseteq B$. Note also that $B \subseteq A$. In fact, set A = set B. ∎

Proper Subsets

> **Definition: Proper Subset**
> Set A is a **proper subset** of set B, symbolized by $A \subset B$, if and only if all the elements of set A are elements of set B and set $A \neq$ set B (that is, set B must contain at least one element not in set A).

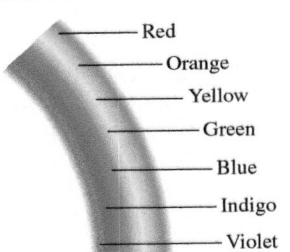
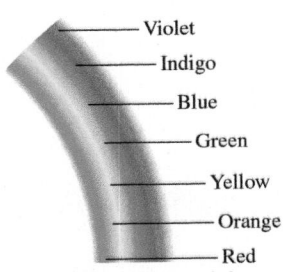

MATHEMATICS TODAY

The Ladder of Life

In biology, the science of classifying all living things is called *taxonomy*. More than 2000 years ago, Aristotle formalized animal classification with his "ladder of life": higher animals, lower animals, higher plants, lower plants. Today, living organisms are classified into six kingdoms (or sets) called animalia, plantae, archaea, eubacteria, fungi, and protista. Even more specific groupings of living things are made according to shared characteristics. The groupings, from most general to most specific, are kingdom, phylum, class, order, family, genus, and species. For example, a zebra, *Equus burchelli*, is a member of the genus *Equus*, as is the horse, *Equus caballus*. Both the zebra and the horse are members of the universal set called the kingdom of animals and the same family, Equidae; they are members of different species (*E. burchelli* and *E. caballus*), however.

Why This Is Important Scientists use sets to classify and categorize animals, plants, and all forms of life. These sets make it easier to understand the behaviors and characteristics of living organisms.

Consider the sets $A = \{$red, blue, yellow$\}$ and $B = \{$red, orange, yellow, green, blue, violet$\}$. Set A is a *subset* of set B, $A \subseteq B$, because every element of set A is also an element of set B. Set A is also a *proper subset* of set B, $A \subset B$, because set A and set B are not equal. Now consider $C = \{$car, bus, train$\}$ and $D = \{$train, car, bus$\}$. Set C is a subset of set D, $C \subseteq D$, because every element of set C is also an element of set D. Set C, however, is not a proper subset of set D, $C \not\subset D$, because set C and set D are equal sets.

Example 2 A Proper Subset?

Determine whether set A is a proper subset of set B.
a) $A = \{$jazz, pop, hip hop$\}$
$\quad B = \{$classical, jazz, pop, rap, hip hop$\}$
b) $A = \{a, b, c, d\}$ $B = \{a, c, b, d\}$

Solution

a) All the elements of set A are contained in set B, and sets A and B are not equal; thus, $A \subset B$.

b) Set $A = $ set B, so $A \not\subset B$. (However, $A \subseteq B$.)   ∎

Every set is a subset of itself, but no set is a proper subset of itself. For all sets A, $A \subseteq A$, but $A \not\subset A$. For example, if $A = \{1, 2, 3\}$, then $A \subseteq A$ because every element of set A is contained in set A, but $A \not\subset A$ because set $A = $ set A.

Let $A = \{\ \}$ and $B = \{1, 2, 3, 4\}$. Is $A \subseteq B$? To show $A \not\subseteq B$, you must determine at least one element of set A that is not an element of set B. Because this cannot be done, $A \subseteq B$ must be true. Using the same reasoning, we can show that *the empty set is a subset of every set, including itself.*

Example 3 Element or Subset?

Determine whether the following are true or false.
a) $3 \in \{3, 4, 5\}$
b) $\{3\} \in \{3, 4, 5\}$
c) $\{3\} \in \{\{3\}, \{4\}, \{5\}\}$
d) $\{3\} \subseteq \{3, 4, 5\}$
e) $3 \subseteq \{3, 4, 5\}$
f) $\{\ \} \subseteq \{3, 4, 5\}$

Solution

a) $3 \in \{3, 4, 5\}$ is a true statement because 3 is an element of the set $\{3, 4, 5\}$.

b) $\{3\} \in \{3, 4, 5\}$ is a false statement because $\{3\}$ is a set, and the set $\{3\}$ is not an element of the set $\{3, 4, 5\}$.

c) $\{3\} \in \{\{3\}, \{4\}, \{5\}\}$ is a true statement because $\{3\}$ is an element in the set. The elements of the set $\{\{3\}, \{4\}, \{5\}\}$ are themselves sets.

d) $\{3\} \subseteq \{3, 4, 5\}$ is a true statement because every element of the first set is an element of the second set.

e) $3 \subseteq \{3, 4, 5\}$ is a false statement because the 3 is not in braces, so it is not a set and thus cannot be a subset. The 3 is an element of the set as indicated in part (a).

f) $\{\ \} \subseteq \{3, 4, 5\}$ is a true statement because the empty set is a subset of every set.   ∎

Number of Subsets

How many distinct subsets can be made from a given set? The empty set has no elements and has exactly one subset, the empty set. A set with one element has two subsets. A set with two elements has four subsets. A set with three elements has eight subsets. This information is illustrated in Table 2.1.

Table 2.1 Number of Subsets

Set	Subsets	Number of Subsets
$\{\ \}$	$\{\ \}$	$1 = 2^0$
$\{a\}$	$\{a\}$	
	$\{\ \}$	$2 = 2^1$
$\{a, b\}$	$\{a, b\}$	
	$\{a\}, \{b\}$	
	$\{\ \}$	$4 = 2 \times 2 = 2^2$
$\{a, b, c\}$	$\{a, b, c\}$	
	$\{a, b\}, \{a, c\}, \{b, c\}$	
	$\{a\}, \{b\}, \{c\}$	
	$\{\ \}$	$8 = 2 \times 2 \times 2 = 2^3$

By continuing this table with larger and larger sets, we can develop a general expression for determining the number of distinct subsets that can be made from any given set.

> ### Number of Distinct Subsets
> The **number of distinct subsets** of a finite set A is 2^n, where n is the number of elements in set A.

Every set is a subset of itself, but no set is a proper subset of itself. Thus, the number of proper subsets will always be one less than the number of subsets that can be made from any given set. We summarize this concept in the following expression.

> ### Number of Distinct Proper Subsets
> The **number of distinct proper subsets** of a finite set A is $2^n - 1$, where n is the number of elements in set A.

Example 4 Distinct Subsets

a) Determine the number of distinct subsets for the set $\{S, L, E, D\}$.

b) List all the distinct subsets for the set $\{S, L, E, D\}$.

c) How many of the distinct subsets are proper subsets?

Solution

a) Since the number of elements in the set is 4, the number of distinct subsets is
$2^4 = 2 \times 2 \times 2 \times 2 = 16$.

b)
$\{S, L, E, D\}$	$\{S, L, E\}$	$\{S, L\}$	$\{S\}$	$\{\ \}$
	$\{S, L, D\}$	$\{S, E\}$	$\{L\}$	
	$\{S, E, D\}$	$\{S, D\}$	$\{E\}$	
	$\{L, E, D\}$	$\{L, E\}$	$\{D\}$	
		$\{L, D\}$		
		$\{E, D\}$		

c) There are 15 proper subsets. Every subset except $\{S, L, E, D\}$ is a proper subset. ∎

Example **5** *Car Options*

Janelle is ordering a new car. She can order some, all, or none of the following options: leather interior, moon roof, navigation system, power windows, power seats, alarm system, and premium sound system. How many different variations of the set of options are possible?

Solution Janelle can order the car with no options, any one option, any two options, any three options, and so on, up to seven options. One technique used in problem solving is to consider similar problems that you have solved previously. If you think about this problem, you will realize it is the same as asking how many distinct subsets can be made from a set with seven elements. The number of different variations of the set of options is the same as the number of possible subsets of a set with seven elements. There are 2^7, or 128, possible subsets of a set with seven elements. Thus, there are 128 possible variations of the set of options for the car. ∎

SECTION 2.2   *Exercises*

Warm Up Exercises

In Exercises 1–4, fill in the blank with an appropriate word, phrase, or symbol(s).

1. If all the elements of set A are also elements of set B, then set A is a(n) _____ of set B.

2. If all the elements of set A are also elements of set B, and set $A \neq$ set B, then set A is a(n) _____ subset of set B.

3. The expression for determining the number of distinct subsets for a set with n distinct elements is _____.

4. The expression for determining the number of distinct proper subsets for a set with n distinct elements is _____.

Practice the Skills

In Exercises 5–26, answer true or false. If false, give the reason.

5. $\{\text{table}\} \subseteq \{\text{sofa, chair, table}\}$

6. $\{\text{Pacific}\} \subseteq \{\text{Atlantic, Pacific, Indian, Artic}\}$

7. $\{\text{apple, pear}\} \subseteq \{\text{pear, peach, grape, orange}\}$

8. $\{\text{dog, cat}\} \subseteq \{\text{dog, bird, hamster}\}$

9. $\{\text{AT\&T, Verizon}\} \subset \{\text{Verizon, Sprint, T-Mobile, AT\&T}\}$

10. $\{\text{American, Southwest}\} \subset \{\text{Delta, American, Southwest}\}$

11. $\{\text{rock, paper, scissors}\} \subset \{\text{rock, paper, scissors}\}$

12. $\{\text{engineer, social worker, teacher, architect}\} \subset \{\text{social worker, teacher, engineer, architect}\}$

13. book $\in \{\text{book, magazine, newspaper}\}$

14. necklace $\in \{\text{necklace, ring, bracelet, earring}\}$

15. $\{\text{cookie}\} \in \{\text{brownie, ice cream, cake, cookie}\}$

16. $\{\ \} \in \{\text{a, b, c, d}\}$

17. tiger $\notin \{\text{zebra, giraffe, polar bear}\}$

18. $\{\ \} \subseteq \{\text{vanilla, chocolate, strawberry}\}$

19. $\{\text{chair}\} \subset \{\text{sofa, table, chair}\}$

20. $\{a, b, c\} \not\subset \{c, b, a\}$

21. $\{\ \} = \{\varnothing\}$

22. $\varnothing = \{\ \}$

23. $\{0\} = \varnothing$

24. $\{\ \} \subseteq \{\ \}$

25. $0 = \{\ \}$

26. $\{1\} \in \{\{1\}, \{2\}, \{3\}\}$

In Exercises 27–34, determine whether $A = B, A \subseteq B, B \subseteq A$, $A \subset B, B \subset A$, or if none of these applies. (There may be more than one answer.)

27. $A = \{\text{Coca-Cola, Dr Pepper, Sprite}\}$
    $B = \{\text{Coca-Cola, Sprite}\}$

28. $A = \{2, 4, 6, 8\}$
    $B = \{2, 3, 4, 5, 6, 7, 8\}$

29. $A = \{x \mid x$ is the set of boats sold in the United States in 2019$\}$
$B = \{x \mid x$ is the set of fishing boats sold in the United States in 2019$\}$

30. $A = \{x \mid x$ is the set of science fiction books written in 2019$\}$
$B = \{x \mid x$ the set of books written in 2019$\}$

31. $A = \{a, b, c, d, e\}$
$B = \{a, c, d, e, f\}$

32. $A = \{x \mid x$ is a sport that uses a ball$\}$
$B = \{$basketball, soccer, tennis$\}$

33. Set A is the set of natural numbers between 2 and 7. Set B is the set of natural numbers greater than 2 and less than 7.

34. Set A is the set of all cars manufactured by General Motors. Set B is the set of sports cars manufactured by General Motors.

In Exercises 35–38, list all the subsets of the sets given.

35. $D = \varnothing$

36. $A = \{\bigcirc\}$

37. $B = \{$cow, horse$\}$

38. $C = \{$steak, pork, chicken$\}$

Problem Solving

39. For set $A = \{a, b, c, d\}$,

   a) list all the subsets of set A.

   b) state which of the subsets in part (a) are not proper subsets of set A.

40. A set contains nine elements.

   a) How many subsets does it have?

   b) How many proper subsets does it have?

In Exercises 41–52, if the statement is true for all sets A and B, write "true." If it is not true for all sets A and B, write "false." Assume that $A \neq \varnothing$, $U \neq \varnothing$, and $A \subset U$.

41. If $A \subseteq B$, then $A \subset B$.

42. If $A \subset B$, then $A \subseteq B$.

43. $A \subseteq A$

44. $A \subset A$

45. $\varnothing \subset A$

46. $\varnothing \subseteq A$

47. $A \subseteq U$

48. $\varnothing \subset \varnothing$

49. $\varnothing \subset U$

50. $U \subseteq \varnothing$

51. $\varnothing \subseteq \varnothing$

52. $U \subset \varnothing$

53. **Building a House** The Li family is planning to build a house in a new development. They can either build the base model offered by the builder or add any of the following options: security system, finished basement, deck, hardwood flooring. How many different variations of the house are possible?

54. **Installing an In-ground Pool** The Fitzgeralds are installing an in-ground pool in their backyard. They can either install the base model offered by the pool company or add any of the following options: automatic pool cleaner, solar cover, waterfall, hot tub, fountain, slide, diving board. How many different variations of the pool are possible?

55. **Salad Toppings** Donald is making a salad at Sweet Tomatoes. His salad can consist of just lettuce, or he can add any of the following items: cucumbers, onions, tomatoes, carrots, green peppers, olives, mushrooms. How many different variations of a salad are possible?

56. **Telephone Features** A customer with Verizon can order telephone service with some, all, or none of the following features: call waiting, call forwarding, caller identification, three-way calling, voice mail, fax line. How many different variations of the set of features are possible?

57. If $E \subseteq F$ and $F \subseteq E$, what other relationship exists between E and F?

58. How can you determine whether the set of boys is equivalent to the set of girls at a roller-skating rink?

59. For the set $D = \{a, b, c\}$

   a) is a an element of set D?

   b) is c a subset of set D?

   c) is $\{a, b\}$ a subset of set D?

Challenge Problem/Group Activity

60. **Hospital Expansion** A hospital has four members on the board of directors: Arnold, Benitez, Cathy, and Dominique.

   a) When the members vote on whether to add a wing to the hospital, how many different ways can they vote (abstentions are not allowed)? For example, Arnold—yes, Benitez—no, Cathy—no, and Dominique—yes is one of the many possibilities.

b) Make a listing of all the possible outcomes of the vote. For example, the vote described in part (a) could be represented as (YNNY).

c) How many of the outcomes given in part (b) would result in a majority supporting the addition of a wing to the hospital? That is, how many of the outcomes have three or more Y's?

Recreational Mathematics

61. How many elements must a set have if the number of proper subsets of the set is $\frac{1}{2}$ of the total number of subsets of the set?

62. If $A \subset B$ and $B \subset C$, must $A \subset C$?

63. If $A \subset B$ and $B \subseteq C$, must $A \subset C$?

64. If $A \subseteq B$ and $B \subseteq C$, must $A \subset C$?

Research Activity

65. On page 52, we discussed the ladder of life. Do research and indicate all the different classifications in the Linnaean system, from most general to the most specific, in which a koala belongs.

SECTION 2.3   Venn Diagrams and Set Operations

Upon completion of this section, you will be able to:

- Construct a Venn diagram with two sets.
- Determine the complement of a set.
- Determine the intersection of two sets.
- Determine the union of two sets.
- Understand the relationship between $n(A \cup B)$, $n(A)$, $n(B)$, and $n(A \cap B)$.
- Determine the difference of two sets.
- Determine the Cartesian product of two sets.

To be eligible to vote in the United States, a person must be at least 18 years old and must be a U.S. citizen. We could define set A as the set of people who are at least 18 years old and set B as the set of people who are U.S. citizens. Then the set of eligible voters would be the set of people who belong to both set A *and* set B. Note that this is a very different set than the set of people who belong to set A *or* set B. In this section, we will use a tool that will allow us to better understand the difference between statements involving the word *and* and statements involving the word *or*.

Why This Is Important   The words *and* and *or* play an important role in many everyday applications. In addition to determining voter eligibility, these applications may include the wording of employment offers, real estate contracts, and other legal documents. Knowing the proper interpretation of statements involving *and* and *or* may help you make wise legal and financial decisions.

Construct Venn Diagrams with Two Sets

A useful technique for illustrating set relationships is the Venn diagram, named for English mathematician John Venn (1834–1923). Venn invented the diagrams and used them to illustrate ideas in his text on symbolic logic, published in 1881.

In a Venn diagram, a rectangle usually represents the universal set, U. The items inside the rectangle may be divided into subsets of the universal set. The subsets are usually represented by circles. In Fig. 2.1, the circle labeled A represents set A, which is a subset of the universal set.

Two sets may be represented in a Venn diagram in any of four different ways, as shown in Fig. 2.2 on page 57. Two sets A and B are *disjoint* when they have no elements in common. Two disjoint sets A and B are illustrated in Fig. 2.2(a). If set A is a

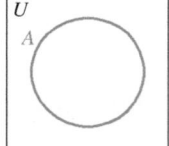

Figure 2.1

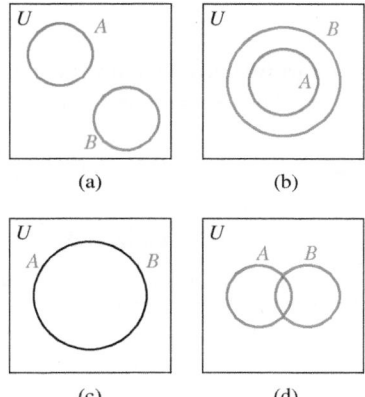

(a)   (b)

(c)   (d)

Figure 2.2

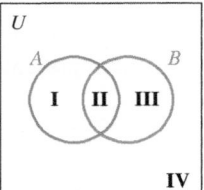

Figure 2.3

proper subset of set B, $A \subset B$, the two sets may be illustrated as in Fig. 2.2(b). If set A contains exactly the same elements as set B, that is, $A = B$, the two sets may be illustrated as in Fig. 2.2(c). Two sets A and B with some elements in common are shown in Fig. 2.2(d), which is regarded as the most general form of a Venn diagram.

If we label the regions of the diagram in Fig. 2.2(d) using I, II, III, and IV, we can illustrate the four possible cases with this one diagram, Fig. 2.3.

CASE 1: DISJOINT SETS When sets A and B are disjoint, they have no elements in common. Therefore, region II of Fig. 2.3 is empty.

CASE 2: SUBSETS When $A \subseteq B$, every element of set A is also an element of set B. Thus, there can be no elements in region I of Fig. 2.3. If $B \subseteq A$, however, then region III of Fig. 2.3 is empty.

CASE 3: EQUAL SETS When set $A = $ set B, all the elements of set A are elements of set B and all the elements of set B are elements of set A. Thus, regions I and III of Fig. 2.3 are empty.

CASE 4: OVERLAPPING SETS When sets A and B have elements in common, those elements are in region II of Fig. 2.3. The elements that belong to set A but not to set B are in region I. The elements that belong to set B but not to set A are in region III.

In each of the four cases, any element belonging to the universal set but not belonging to set A or set B is placed in region IV.

Next we introduce set operations. Venn diagrams will be helpful in understanding set operations. The basic operations of arithmetic are $+$, $-$, $\times$, and $\div$. When we see these symbols, we know what procedure to follow to determine the answer. Some of the operations in set theory are $'$, $\cap$, $\cup$, $-$, and $\times$. They represent complement, intersection, union, difference, and Cartesian product, respectively.

Complement

> **Definition: Complement**
> The **complement** of set A, symbolized by A', is the set of all the elements in the universal set that are not in set A.

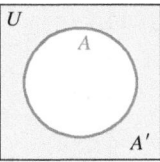

Figure 2.4

Using set-builder notation, the complement of set A is indicated by $A' = \{x \mid x \in U$ and $x \notin A\}$. In Fig. 2.4, the shaded region outside set A within the universal set represents the complement of set A, or A'.

Example 1 *A Set and Its Complement*

Given

$$U = \{a, b, c, d, e, f, g\} \text{ and } A = \{a, c, e\}$$

determine A' and illustrate the relationship among sets U, A, and A' in a Venn diagram.

Solution The elements in U that are not in set A are b, d, f, g. Thus, $A' = \{b, d, f, g\}$. The Venn diagram is illustrated in Fig. 2.5. The elements b, d, f, g can be placed anywhere in the shaded area. ∎

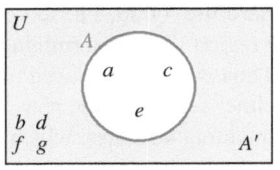

Figure 2.5

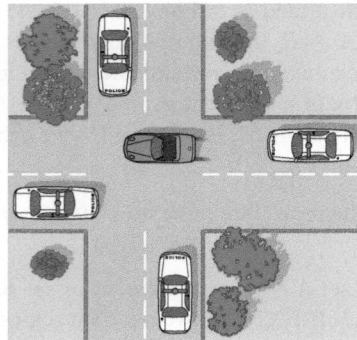

Intersection

The word *intersection* brings to mind the area common to two crossing streets. The red car in the figure is in the intersection of the two streets. The set operation intersection is defined as follows.

> **Definition: Intersection**
> The **intersection** of sets A and B, symbolized by $A \cap B$, is the set containing all the elements that are common to both set A and set B.

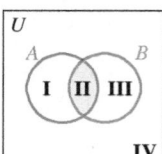

Figure 2.6

Using set-builder notation, the intersection of set A and set B is indicated by $A \cap B = \{x \,|\, x \in A \text{ and } x \in B\}$. The shaded region, region II, in Fig. 2.6 represents the intersection of sets A and B.

Example 2 *Sets with Overlapping Regions*

Let the universal set, U, represent the 50 states in the United States. Let set A represent the set of states with a population of more than 10 million people as of 2019. Let set B represent the set of states that have at least one city with a population of more than 1 million people, as of 2019 (see the table). Draw a Venn diagram illustrating the relationship between set A and set B.

Set A States with a Population of More Than 10 Million People	Set B States with at Least One City with a Population of More Than 1 Million People
California	California
Texas	Texas
Florida	New York
New York	Illinois
Pennsylvania	Pennsylvania
Illinois	Arizona
Ohio	
Georgia	
North Carolina	

Source: Bureau of the U.S. Census

Solution First determine the intersection of sets A and B. The states common to both sets are California, Texas, New York, Illinois, and Pennsylvania. Therefore,

$$A \cap B = \{\text{California, Texas, New York, Illinois, Pennsylvania}\}$$

Place these elements in region II of Fig. 2.7. Complete region I by determining the elements in set A that have not been placed in region II. Therefore, Ohio, Florida, Georgia, and North Carolina are placed in region I. Complete region III by determining the elements in set B that have not been placed in region II. Thus, Arizona is placed in region III. Finally, place those elements in U that are not in either set within the rectangle but are outside both circles. This group includes the remaining 40 states, which are placed in region IV.   ∎

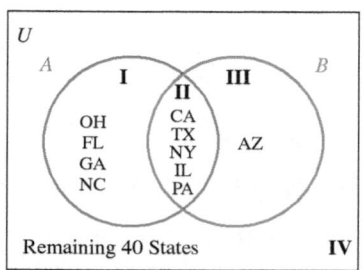

Figure 2.7

Example **3** *The Intersection of Sets*

Given

$$U = \{1, 2, 3, 4, 5, 6, 7, 8, 9, 10\}$$
$$A = \{1, 2, 3, 8\}$$
$$B = \{1, 3, 6, 7, 8\}$$
$$C = \{\ \}$$

determine

a) $A \cap B$.   b) $A \cap C$.   c) $A' \cap B$.   d) $(A \cap B)'$.

Solution

a) $A \cap B = \{1, 2, 3, 8\} \cap \{1, 3, 6, 7, 8\} = \{1, 3, 8\}$. The elements common to both set A and set B are 1, 3, and 8.

b) $A \cap C = \{1, 2, 3, 8\} \cap \{\ \} = \{\ \}$. There are no elements common to both set A and set C.

c) To determine $A' \cap B$, we must first determine A'.

$$A' = \{4, 5, 6, 7, 9, 10\}$$
$$A' \cap B = \{4, 5, 6, 7, 9, 10\} \cap \{1, 3, 6, 7, 8\}$$
$$= \{6, 7\}$$

d) To determine $(A \cap B)'$, first determine $A \cap B$.

$$A \cap B = \{1, 3, 8\} \text{ from part (a)}$$
$$(A \cap B)' = \{1, 3, 8\}' = \{2, 4, 5, 6, 7, 9, 10\}$$

∎

Union

The word *union* means to unite or join together, as in marriage, and that is exactly what is done when we perform the operation of union.

> **Definition: Union**
> The **union** of set A and set B, symbolized by $A \cup B$, is the set containing all the elements that are members of set A or of set B (or of both sets).

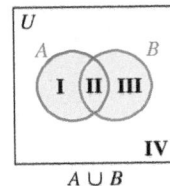

A ∪ B

Figure 2.8

Using set-builder notation, the union of set A and set B is indicated by $A \cup B = \{x \,|\, x \in A \text{ or } x \in B\}$. The three shaded regions of Fig. 2.8, regions I, II, and III, together represent the union of sets A and B. If an element is common to both sets, it is listed only once in the union of the sets.

Example **4** *Determining Sets from a Venn Diagram*

Use the Venn diagram in Fig. 2.9 to determine the following sets.

a) U b) A c) B' d) $A \cap B$

e) $A \cup B$ f) $(A \cup B)'$ g) $n(A \cup B)$

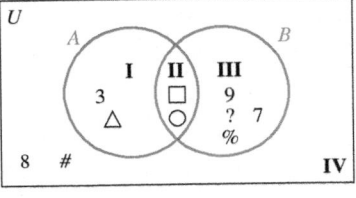

Figure 2.9

Solution

a) The universal set consists of all the elements within the rectangle, that is, the elements in regions I, II, III, and IV. Thus, $U = \{3, \triangle, \square, \bigcirc, 9, 7, \%, ?, \#, 8\}$.

b) Set A consists of the elements in regions I and II. Thus, $A = \{3, \triangle, \square, \bigcirc\}$.

c) B' consists of the elements outside set B, or the elements in regions I and IV. Thus, $B' = \{3, \triangle, \#, 8\}$.

d) $A \cap B$ consists of the elements that belong to both set A and set B (region II). Thus, $A \cap B = \{\square, \bigcirc\}$.

e) $A \cup B$ consists of the elements that belong to set A or set B (regions I, II, or III). Thus, $A \cup B = \{3, \triangle, \square, \bigcirc, 9, 7, ?, \%\}$.

f) $(A \cup B)'$ consists of the elements in U that are not in $A \cup B$. Thus, $(A \cup B)' = \{\#, 8\}$.

g) $n(A \cup B)$ represents the *number of elements* in the union of sets A and B. Thus, $n(A \cup B) = 8$, as there are eight elements in the union of sets A and B. ∎

Union of Two Sets

Example 5 *The Union of Sets*

Given

$$U = \{1, 2, 3, 4, 5, 6, 7, 8, 9, 10\}$$
$$A = \{1, 2, 4, 6\}$$
$$B = \{1, 3, 6, 7, 9\}$$
$$C = \{\ \}$$

determine each of the following.

a) $A \cup B$ b) $A \cup C$ c) $A' \cup B$ d) $(A \cup B)'$

Solution

a) $A \cup B = \{1, 2, 4, 6\} \cup \{1, 3, 6, 7, 9\} = \{1, 2, 3, 4, 6, 7, 9\}$

b) $A \cup C = \{1, 2, 4, 6\} \cup \{\ \} = \{1, 2, 4, 6\}$. Note that $A \cup C = A$.

c) To determine $A' \cup B$, we must determine A'.

$$A' = \{3, 5, 7, 8, 9, 10\}$$
$$A' \cup B = \{3, 5, 7, 8, 9, 10\} \cup \{1, 3, 6, 7, 9\}$$
$$= \{1, 3, 5, 6, 7, 8, 9, 10\}$$

d) Determine $(A \cup B)'$ by first determining $A \cup B$, and then determine the complement of $A \cup B$.

$$A \cup B = \{1, 2, 3, 4, 6, 7, 9\} \text{ from part (a)}$$
$$(A \cup B)' = \{1, 2, 3, 4, 6, 7, 9\}' = \{5, 8, 10\}$$ ∎

Example 6 *Union and Intersection*

Given

$$U = \{a, b, c, d, e, f, g\}$$
$$A = \{a, b, e, g\}$$
$$B = \{a, c, d, e\}$$
$$C = \{b, e, f\}$$

determine each of the following.

a) $(A \cup B) \cap (A \cup C)$ b) $(A \cup B) \cap C'$ c) $A' \cap B'$

Solution

a) $(A \cup B) \cap (A \cup C) = \{a, b, c, d, e, g\} \cap \{a, b, e, f, g\}$
$$= \{a, b, e, g\}$$

b) $(A \cup B) \cap C' = \{a, b, c, d, e, g\} \cap \{a, c, d, g\}$
$$= \{a, c, d, g\}$$

c) $A' \cap B' = \{c, d, f\} \cap \{b, f, g\}$
$$= \{f\}$$

The Meaning of *and* and *or*

The words *and* and *or* are very important in many areas of mathematics. We use these words in several chapters in this book, including Chapter 11, Probability. The word *and* is generally interpreted to mean *intersection*, whereas *or* is generally interpreted to mean *union*. Suppose $A = \{1, 2, 3, 5, 6, 8\}$ and $B = \{1, 3, 4, 7, 9, 10\}$. The elements that belong to set A *and* set B are 1 and 3. These are the elements in the intersection of the sets. The elements that belong to set A *or* set B are 1, 2, 3, 4, 5, 6, 7, 8, 9, and 10. These are the elements in the union of the sets.

The Relationship Between $n(A \cup B)$, $n(A)$, $n(B)$, and $n(A \cap B)$

Having looked at unions and intersections, we can now determine a relationship between $n(A \cup B)$, $n(A)$, $n(B)$, and $n(A \cap B)$. Suppose set A has eight elements, set B has five elements, and $A \cap B$ has two elements. How many elements are in $A \cup B$? Let's make up some arbitrary sets that meet the criteria specified and draw a Venn diagram. If we let set $A = \{a, b, c, d, e, f, g, h\}$, then set B must contain five elements, two of which are also in set A. Let set $B = \{g, h, i, j, k\}$. We construct a Venn diagram by filling in the intersection first, as shown in Fig. 2.10. The number of elements in $A \cup B$ is 11. The elements g and h are in both sets, and if we add $n(A) + n(B)$, we are counting these elements twice.

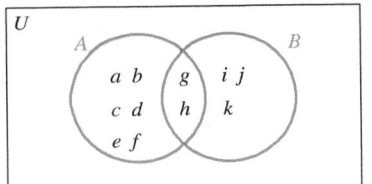

Figure 2.10

To determine the number of elements in the union of sets A and B, we can add the number of elements in sets A and B and then subtract the number of elements common to both sets.

> ### The Number of Elements in $A \cup B$
> For any finite sets A and B,
> $$n(A \cup B) = n(A) + n(B) - n(A \cap B)$$

Example 7 *How Many Visitors Speak Spanish or French?*

The results of a survey of visitors at the Grand Canyon showed that 25 speak Spanish, 14 speak French, and 4 speak both Spanish and French. How many speak Spanish or French?

Solution If we let set A be the set of visitors who speak Spanish and let set B be the set of visitors who speak French, then we need to determine $n(A \cup B)$. We can use the above formula to determine $n(A \cup B)$.

$$n(A \cup B) = n(A) + n(B) - n(A \cap B)$$
$$n(A \cup B) = 25 + 14 - 4$$
$$= 35$$

Thus, 35 of the visitors surveyed speak either Spanish or French.

⌐ Example **8** *The Number of Elements in Set A*

The results of a survey of customers at a McDonald's restaurant showed that 28 purchased either a hamburger or french fries, 20 purchased french fries, and 17 purchased both a hamburger and french fries. How many customers purchased only a hamburger?

Solution If we let set A be the set of customers who purchased a hamburger and set B be the set of customers who purchased french fries, we need to determine $n(A)$. We are given the number of customers who purchased either a hamburger or french fries, which is $n(A \cup B)$. We are also given the number of customers who purchased french fries, $n(B)$, and the number of customers who purchased both a hamburger and french fries, $n(A \cap B)$. We can use the formula $n(A \cup B) = n(A) + n(B) - n(A \cap B)$ to solve for $n(A)$.

$$n(A \cup B) = n(A) + n(B) - n(A \cap B)$$
$$28 = n(A) + 20 - 17$$
$$28 = n(A) + 3$$
$$28 - 3 = n(A) + 3 - 3$$
$$25 = n(A)$$

Thus, the number of customers who purchased only a hamburger is 25.   ∎

Two other set operations are the difference of two sets and the Cartesian product. We will first discuss the difference of two sets.

Difference of Two Sets

> Definition: **Difference of Two Sets**
> The **difference of two sets** A and B, symbolized $A - B$, is the set of elements that belong to set A but not to set B.

Using set-builder notation, the difference of two sets A and B is indicated by $A - B = \{x \mid x \in A \text{ and } x \notin B\}$. The shaded region, region I, in Fig. 2.11 represents the difference of two sets A and B, or $A - B$. Notice in Fig. 2.11 that $A - B = A \cap B'$.

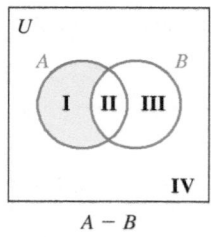

$A - B$

Figure 2.11

⌐ Example **9** *The Difference of Two Sets*

Given

$$U = \{a, b, c, d, e, f, g, h, i, j, k\}$$
$$A = \{b, d, e, f, g, h\}$$
$$B = \{a, b, d, h, i\}$$
$$C = \{b, e, g\}$$

determine

a) $A - B$ b) $A - C$ c) $A' - B$ d) $A - C'$

Solution

a) $A - B$ is the set of elements that are in set A but not set B. The elements that are in set A but not set B are $e, f,$ and g. Therefore, $A - B = \{e, f, g\}$.

b) $A - C$ is the set of elements that are in set A but not set C. The elements that are in set A but not set C are $d, f,$ and h. Therefore, $A - C = \{d, f, h\}$.

c) To determine $A' - B$, we must first determine A'.

$$A' = \{a, c, i, j, k\}$$

$A' - B$ is the set of elements that are in set A' but not set B. The elements that are in set A' but not set B are c, j, and k. Therefore, $A' - B = \{c, j, k\}$.

d) To determine $A - C'$, we must first determine C'.

$$C' = \{a, c, d, f, h, i, j, k\}$$

$A - C'$ is the set of elements that are in set A but not set C'. The elements that are in set A but not set C' are b, e, and g. Therefore, $A - C' = \{b, e, g\}$.   ∎

Next we discuss the Cartesian product.

Cartesian Product

> **Definition: Cartesian Product**
> The **Cartesian product** of set A and set B, symbolized by $A \times B$ and read "A cross B," is the set of all possible *ordered pairs* of the form (a, b), where $a \in A$ and $b \in B$.

To determine the ordered pairs in a Cartesian product, select the first element of set A and form an ordered pair with each element of set B. Then select the second element of set A and form an ordered pair with each element of set B. Continue in this manner until you have used each element of set A.

Example 10 *The Cartesian Product of Two Sets*

Given $A = \{\text{orange, banana, apple}\}$ and $B = \{1, 2\}$, determine the following.

a) $A \times B$ b) $B \times A$ c) $A \times A$ d) $B \times B$

Solution

a) $A \times B = \{(\text{orange}, 1), (\text{orange}, 2), (\text{banana}, 1), (\text{banana}, 2), (\text{apple}, 1), (\text{apple}, 2)\}$

b) $B \times A = \{(1, \text{orange}), (1, \text{banana}), (1, \text{apple}), (2, \text{orange}), (2, \text{banana}), (2, \text{apple})\}$

c) $A \times A = \{(\text{orange}, \text{orange}), (\text{orange}, \text{banana}), (\text{orange}, \text{apple}), (\text{banana}, \text{orange}), (\text{banana}, \text{banana}), (\text{banana}, \text{apple}), (\text{apple}, \text{orange}), (\text{apple}, \text{banana}), (\text{apple}, \text{apple})\}$

d) $B \times B = \{(1, 1), (1, 2), (2, 1), (2, 2)\}$ ∎

We can see from Example 10 that, in general, $A \times B \neq B \times A$. The ordered pairs in $A \times B$ are not the same as the ordered pairs in $B \times A$. For example $(\text{orange}, 1) \neq (1, \text{orange})$.

In general, if a set A has m elements and a set B has n elements, then the number of ordered pairs in $A \times B$ will be $m \times n$. In Example 10, set A contains 3 elements and set B contains 2 elements. Notice that $A \times B$ contains 3×2, or 6, ordered pairs.

SECTION 2.3
Exercises

Warm Up Exercises

In exercises 1– 8, fill in the blank with an appropriate word, phrase, or symbol(s).

1. The set of all the elements in the universal set that are not in set A is called the _____ of set A.

2. The set containing all the elements that are members of set A or of set B or of both sets is called the _____ of set A and set B.

3. The set containing all the elements that are common to both set A and set B is called the _____ of set A and set B.

4. The set of elements that belong to set A, but not to set B, is called the _____ of two sets A and B.

5. The set of all possible ordered pairs of the form (a, b), where $a \in A$ and $b \in B$, is called the _____ product of set A and set B.

6. Two sets with no elements in common are called _____ sets.

7. If set A has m elements and set B has n elements, the Cartesian product $A \times B$ has _____ elements.

8. In a Venn diagram with two overlapping sets there are _____ regions.

Practice the Skills

In Exercises 9–13, use Fig. 2.2 on page 57 as a guide to draw a Venn diagram that illustrates the situation described.

9. Set A and set B are disjoint sets.

10. $A \subset B$

11. $B \subset A$

12. $A = B$

13. Set A and set B are overlapping sets.

14. For the sets U, A, and B, construct a Venn diagram and place the elements in the proper regions.
$$U = \{a, b, c, d, e, f\}$$
$$A = \{a, c\}$$
$$B = \{b, e, f\}$$

15. For the sets U, A, and B, construct a Venn diagram and place the elements in the proper regions.
$$U = \{1, 2, 3, 4, 5, 6, 7, 8\}$$
$$A = \{1, 6, 8\}$$
$$B = \{2, 4, 6, 7, 8\}$$

16. For the sets U, A, and B, construct a Venn diagram and place the elements in the proper regions.
$$U = \{a, b, c, d, e, f, g, h, i, j, k\}$$
$$A = \{a, b, e, f, h, j\}$$
$$B = \{b, c, d, f, j\}$$

Problem Solving

17. *Restaurants* For the sets U, A, and B, construct a Venn diagram and place the elements in the proper regions.

$U = \{$Burger King, Chick-fil-A, Chipotle, Domino's, McDonald's, Panera Bread, Pizza Hut, Subway$\}$

$A = \{$Chick-fil-A, Domino's, Panera Bread, Subway$\}$

$B = \{$Burger King, Chipotle, Domino's, Pizza Hut, Subway$\}$

18. *National Parks* For the sets U, A, and B, construct a Venn diagram and place the elements in the proper regions.

$U = \{$Acadia, Badlands, Death Valley, Glacier, Mammoth Cave, Mount Rainier, North Cascades, Shenandoah, Yellowstone, Zion$\}$

$A = \{$Acadia, Badlands, Glacier, Mount Rainier, Yellowstone$\}$

$B = \{$Death Valley, Glacier, Mammoth Cave, Mount Rainier, Zion$\}$

▲ *Mount Rainier*

19. *Occupations* The following table shows the fastest-growing occupations that require a bachelor's degree, based on the estimated employment in 2024 and the predicted percent increase in employment from 2014 to 2024. Let the occupations in the table represent the universal set.

Fastest-Growing Occupations Requiring a Bachelor's Degree, 2014–2024		
Occupation	**Estimated Employment in 2024 (in thousands of jobs)**	**Percent increase**
Operations research analysts (ORA)	119	30
Personal financial advisors (PFA)	323	30
Cartographers and photogrammetrists (CP)	16	29
Translators and interpreters (TI)	79	29
Forensic science technicians (FST)	18	27
Biomedical engineers (BE)	27	23
Substance abuse counselors (SAC)	116	22
Athletic trainers (AT)	31	21
Computer systems analysts (CSA)	686	21
Mental health social workers (MH)	140	19

Source: U.S. Bureau of Labor Statistics

Let A = the set of fastest-growing occupations that require a bachelor's degree whose estimated employment in 2024 is at least 100,000.

Let B = the set of fastest-growing occupations that require a bachelor's degree whose predicted percent increase in employment from 2014 to 2024 is at least 25%.

Using the abbreviations listed in the table for each occupation and sets A and B as described above, construct a Venn diagram illustrating the sets.

20. *NASCAR Standings* The table below shows the 2018 Monster Energy NASCAR Cup Series Standings. The table shows the 12 drivers having the highest point total and the number of Monster Energy races they won. Let the drivers in the table represent the universal set.

2018 Monster Energy NASCAR Cup Series Final Standings		
Driver	**Points**	**Wins**
Joey Logano (JL)	5040	3
Martin Truex Jr. (MT)	5035	4
Kevin Harvick (KH)	5034	8
Kyle Busch (KB)	5033	8
Aric Almirola (AA)	2354	1
Chase Elliott (CE)	2350	3
Kurt Busch (KuB)	2350	1
Brad Keselowski (BK)	2343	3
Kyle Larson (KL)	2299	0
Ryan Blaney (RB)	2298	1
Denny Hamlin (DH)	2285	0
Clint Bowyer (CB)	2272	2

Source: NASCAR

Let A = the set of drivers with at least 2350 points.

Let B = the set of drivers with more than 1 win.

Construct a Venn diagram illustrating the sets. Use the driver's initials in the Venn diagram.

21. Let U represent the set of retail stores in the United States. Let A represent the set of retail stores that sell children's clothing. Describe A'.

22. Let U represent the set of animals in U.S. zoos. Let A represent the set of animals in the San Diego Zoo. Describe A'.

In Exercises 23–28,

   U is the set of cities in the United States.

   A is the set of cities that have a professional sports team.

   B is the set of cities that have a symphony.

Describe each of the following sets in words.

23. A'

24. B'

25. $A \cup B$

26. $A \cap B$

27. $A \cap B'$

28. $A \cup B'$

In Exercises 29–34,

   U is the set of furniture stores.

   A is the set of furniture stores that sell mattresses.

   B is the set of furniture stores that sell outdoor furniture.

   C is the set of furniture stores that sell leather furniture.

Describe the following sets.

29. $A \cup C$

30. $A \cap B$

31. $B' \cap C$

32. $A \cap B \cap C$

33. $A \cup B \cup C$

34. $A' \cup C'$

In Exercises 35–44, use the Venn diagram in Fig. 2.12 to list the set of elements in roster form.

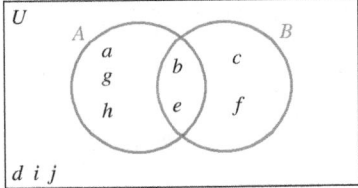

Figure 2.12

35. A	36. B
37. $A \cap B$	38. U
39. $A \cup B$	40. $(A \cup B)'$
41. $A' \cap B'$	42. $(A \cap B)'$
43. $A - B$	44. $A - B'$

In Exercises 45–54, use the Venn diagram in Fig. 2.13 to list the set of elements in roster form.

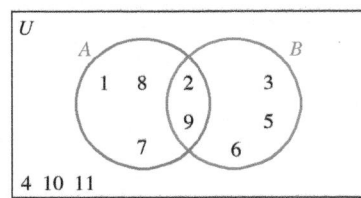

Figure 2.13

45. A

46. B

47. U

48. $A \cap B$

49. $A' \cup B$

50. $A \cup B'$

51. $A' \cap B$

52. $(A \cup B)'$

53. $A' - B$

54. $(A - B)'$

In Exercises 55–64, let

$$U = \{1, 2, 3, 4, 5, 6, 7, 8\}$$
$$A = \{1, 2, 4, 5, 7\}$$
$$B = \{2, 3, 5, 6\}$$

Determine the following.

55. $A \cup B$

56. $A \cap B$

57. $(A \cup B)'$

58. $A' \cap B'$

59. $(A \cup B)' \cap B$

60. $(A \cup B) \cap (A \cup B)'$

61. $(B \cup A)' \cap (B' \cup A')$

62. $A' \cup (A \cap B)$

63. $(A - B)'$

64. $A' - B'$

In Exercises 65–74, let

$$U = \{a, b, c, d, e, f, g, h, i, j, k\}$$
$$A = \{a, c, d, f, g, i\}$$
$$B = \{b, c, d, f, g\}$$
$$C = \{a, b, f, i, j\}$$

Determine the following.

65. $B \cup C$

66. $A \cap C$

67. $A' \cup B'$

68. $(A \cap C)'$

69. $(A \cap B) \cup C$

70. $A \cup (C \cap B)'$

71. $(A' \cup C) \cup (A \cap B)$

72. $(C \cap B) \cap (A' \cap B)$

73. $(A - B)' - C$

74. $(C - A)' - B$

In Exercises 75–80, let

$$A = \{1, 2, 3\}$$
$$B = \{a, b\}$$

75. Determine $A \times B$.

76. Determine $B \times A$.

77. Does $A \times B = B \times A$?

78. Determine $n(A \times B)$.   **79.** Determine $n(B \times A)$.

80. Does $n(A \times B) = n(B \times A)$?

In Exercises 81–90, let

$$U = \{x \mid x \in N \text{ and } x < 10\}$$
$$A = \{x \mid x \in N \text{ and } x \text{ is odd and } x < 10\}$$
$$B = \{x \mid x \in N \text{ and } x \text{ is even and } x < 10\}$$
$$C = \{x \mid x \in N \text{ and } x < 6\}$$

Determine the following.

81. $A \cap B$

82. $A \cup B$

83. $(B \cup C)'$

84. $A \cap C'$

85. $A \cap B'$

86. $(B \cap C)'$

87. $(A \cup C) \cap B$

88. $(A \cap B)' \cup C$

89. $(A' \cup B') \cap C$

90. $(A' \cap C) \cup (A \cap B)$

91. When will a set and its complement be disjoint? Explain and give an example.

92. When will $n(A \cap B) = 0$? Explain and give an example.

93. *Pet Ownership* The results of a survey of customers at PetSmart showed that 27 owned dogs, 38 owned cats, and 16 owned both dogs and cats. How many people owned either a dog or a cat?

94. *Student Government and Clubs* At St. Petersburg College, 46 students were members of student government or clubs, 30 were members of student government, and 4 were members of student government and clubs. How many students were members only of clubs?

95. Consider the formula

$$n(A \cup B) = n(A) + n(B) - n(A \cap B)$$

a) Show that this relation holds for $A = \{a, b, c, d\}$ and $B = \{b, d, e, f, g, h\}$.

b) Make up your own sets A and B, each consisting of at least six elements. Using these sets, show that the relation holds.

c) Use a Venn diagram and explain why the relation holds for any two sets A and B.

96. The Venn diagram in Fig. 2.14 shows a technique of labeling the regions to indicate membership of elements in a particular region. Define each of the four regions with a set statement. (*Hint:* $A \cap B'$ defines region I.)

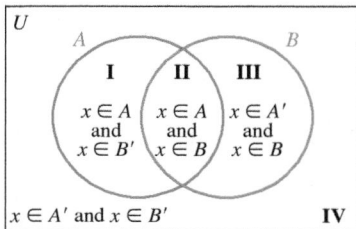

Figure 2.14

In Exercises 97–104, let $U = \{0, 1, 2, 3, 4, 5, \ldots\}$, $A = \{1, 2, 3, 4, \ldots\}$, $B = \{4, 8, 12, 16, \ldots\}$, and $C = \{2, 4, 6, 8, \ldots\}$. Determine the following.

97. $A \cup B$

98. $A \cap B$

99. $B \cup C$

100. $B \cap C$

101. $A \cap C$

102. $A' \cap C$

103. $(B \cup C)' \cup C$

104. $(A \cap C) \cap B'$

Challenge Problems/Group Activities

In Exercises 105–110, determine whether the answer is $\varnothing$, A, or U. (Assume $A \neq \varnothing$, $A \neq U$.)

105. $A \cap A'$

106. $A \cup A'$

107. $A \cup \varnothing$

108. $A \cap \varnothing$

109. $A' \cup U$

110. $A \cap U$

In Exercises 111–114, determine the relationship between set A and set B if

111. $A \cap B = B$.

112. $A \cup B = B$.

113. $A \cap B = \varnothing$.

114. $A \cup B = \varnothing$.

Research Activity

115. *Fastest-Growing Occupations* Use the U.S. Bureau of Labor Statistics website to create a Venn diagram illustrating the 10 fastest-growing occupations for college graduates with an Associate's degree based on employment in 2014 and the estimated employment for those same occupations in 2024.

Venn Diagrams with Three Sets and Verification of Equality of Sets

Upon completion of this section, you will be able to:

- Construct a Venn diagram with three sets.
- Use Venn diagrams to verify equality of sets.
- Understand De Morgan's laws for sets.

You are looking for a career that best matches your strengths, which include adaptability, dependability, and honesty. Is there a way that you can determine which careers match all three of these strengths? How about only two of these strengths? The answer is yes! In this section, we will use Venn diagrams to answer similar questions. We will learn that a Venn diagram can be used to display important information quickly and efficiently.

Why This Is Important Classifying sets using Venn diagrams often helps us to understand the relationship among various sets such as the sets of careers that require similar strengths. We will be happiest in a career that matches our strengths. For example, if one of your strengths is mathematics then you might want to work in science, engineering, or accounting. Using Venn diagrams is a valuable way to categorize your strengths and weaknesses when choosing a career or place to work.

Construct a Venn Diagram with Three Sets

In Section 2.3, we learned how to use Venn diagrams to illustrate two sets. Venn diagrams can also be used to illustrate three sets.

For three sets, *A*, *B*, and *C*, the diagram is drawn so the three sets overlap (Fig. 2.15), creating eight regions. The diagrams in Fig. 2.16 emphasize selected regions of three intersecting sets. *When constructing Venn diagrams with three sets, we generally start with region V and work outward,* as explained in the procedure given below.

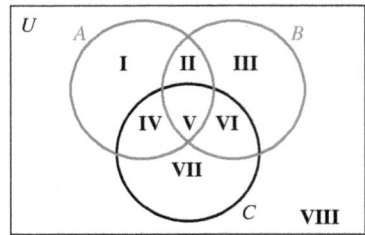

Figure 2.15

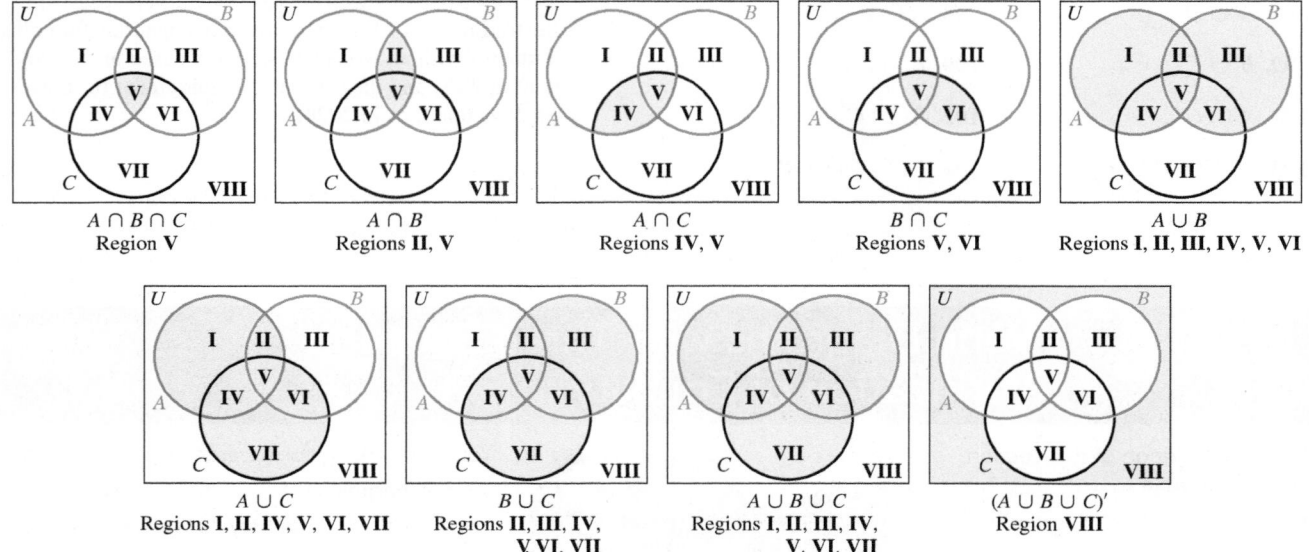

Figure 2.16

PROCEDURE   GENERAL PROCEDURE FOR CONSTRUCTING VENN
DIAGRAMS WITH THREE SETS, *A*, *B*, AND *C*

1. Determine the elements to be placed in region V by determining the elements that are common to all three sets, $A \cap B \cap C$.

2. Determine the elements to be placed in region II. Determine the elements in $A \cap B$. The elements in this set belong in regions II and V. Place the elements in the set $A \cap B$ that are not listed in region V in region II. The elements in regions IV and VI are determined in a similar manner.

3. Determine the elements to be placed in region I by determining the elements in set *A* that are not in regions II, IV, and V. The elements in regions III and VII are determined in a similar manner.

4. Determine the elements to be placed in region VIII by determining the elements in the universal set that are not in regions I through VII.

Example 1 illustrates the general procedure.

Example 1 *Constructing a Venn Diagram for Three Sets*

Construct a Venn diagram illustrating the following sets.

$$U = \{1, 2, 3, 4, 5, 6, 7, 8, 9, 10, 11, 12\}$$
$$A = \{5, 9, 10, 12\}$$
$$B = \{1, 2, 4, 5, 8, 9, 10\}$$
$$C = \{1, 3, 5, 8, 9, 11\}$$

Solution First determine the intersection of all three sets. Because the elements 5 and 9 are in all three sets, $A \cap B \cap C = \{5, 9\}$. The elements 5 and 9 are placed in region V in Fig. 2.17. Next complete region II by determining the intersection of sets A and B.

$$A \cap B = \{5, 9, 10\}$$

$A \cap B$ consists of regions II and V. The elements 5 and 9 have already been placed in region V, so 10 must be placed in region II.

Now determine what elements go in region IV.

$$A \cap C = \{5, 9\}$$

Since 5 and 9 have already been placed in region V, there are no elements to be placed in region IV. Now determine the elements to go in region VI.

$$B \cap C = \{1, 5, 8, 9\}$$

Since both the 5 and 9 have already been placed in region V, place the 1 and 8 in region VI. Now complete set A. The only element of set A that has not previously been placed in regions II, IV, or V is 12. Therefore, place the element 12 in region I. The element 12 that is placed in region I is only in set A and not in set B or set C. Using set B, complete region III using the same general procedure used to determine the elements in region I. Using set C, complete region VII by using the same procedure used to complete regions I and III. To determine the elements in region VIII, determine the elements in U that have not been placed in regions I–VII. The elements 6 and 7 have not been placed in regions I–VII, so place them in region VIII. ∎

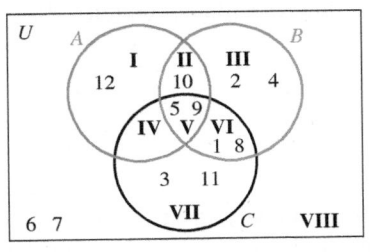

Figure 2.17

Venn diagrams can be used to illustrate and analyze many everyday problems. One example follows.

Example 2 *Blood Types*

Human blood is classified (typed) according to the presence or absence of the specific antigens A, B, and Rh in the red blood cells. Antigens are highly specified proteins and carbohydrates that will trigger the production of antibodies in the blood to fight infection. Blood containing the Rh antigen is labeled positive, +, while blood lacking the Rh antigen is labeled negative, −. Blood lacking both A and B antigens is called type O. Sketch a Venn diagram with three sets A, B, and Rh and place each type of blood listed in the proper region. A person has only one type of blood.

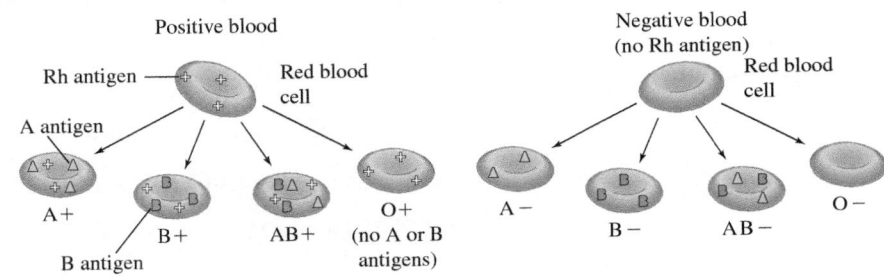

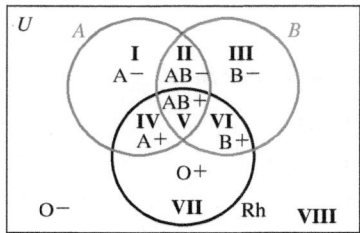

Figure 2.18

As illustrated in Chapter 1, the first thing to do is to read the question carefully and make sure you understand what is given and what you are asked to determine. There are three antigens A, B, and Rh. Therefore, begin by naming the three circles in a Venn diagram with the three antigens; see Fig. 2.18.

Any blood containing the Rh antigen is positive, and any blood not containing the Rh antigen is negative. Therefore, all blood in the Rh circle is positive, and all blood outside the Rh circle is negative. The intersection of all three sets, region V, is AB+. Region II contains only antigens A and B and is therefore AB−. Region I is A− because it contains only antigen A. Region III is B−, region IV is A+, and region VI is B+. Region VII is O+, containing only the Rh antigen. Region VIII, which lacks all three antigens, is O−. ∎

Verification of Equality of Sets

In this chapter, for clarity we may refer to operations on sets, such as $A \cup B'$ or $A \cap B \cap C$, as *statements involving sets* or simply as *statements*. Now we discuss how to determine if two statements involving sets are equal.

Consider the question: Is $A' \cup B = A' \cap B$ *for all sets A and B*? For the specific sets $U = \{1, 2, 3, 4, 5\}$, $A = \{1, 3\}$, and $B = \{2, 4, 5\}$, is $A' \cup B = A' \cap B$? To answer the question, we do the following.

Determine $A' \cup B$	**Determine $A' \cap B$**
$A' = \{2, 4, 5\}$	$A' = \{2, 4, 5\}$
$B = \{2, 4, 5\}$	$B = \{2, 4, 5\}$
$A' \cup B = \{2, 4, 5\}$	$A' \cap B = \{2, 4, 5\}$

For these sets, $A' \cup B = A' \cap B$, because both set statements are equal to $\{2, 4, 5\}$. At this point you may believe that $A' \cup B = A' \cap B$ for all sets A and B.

If we select the sets $U = \{1, 2, 3, 4, 5\}$, $A = \{1, 3, 5\}$, and $B = \{2, 3\}$, we see that $A' \cup B = \{2, 3, 4\}$ and $A' \cap B = \{2\}$. For this case, $A' \cup B \neq A' \cap B$. Thus, we have proved that $A' \cup B \neq A' \cap B$ for all sets A and B by using a *counter example*. A counterexample, as explained in Chapter 1, is an example that shows a statement is not true.

In Chapter 1, we explained that proofs involve the use of *deductive reasoning*. Recall that deductive reasoning begins with a general statement and works to a specific conclusion. To verify, or determine whether set statements are equal for any two sets selected, we use deductive reasoning with Venn diagrams. Venn diagrams are used because they can illustrate general cases. To determine if statements that contain sets, such as $(A \cap B)'$ and $A' \cup B'$, are equal for all sets A and B, we use the regions of Venn diagrams. If both statements represent the same regions of the Venn diagram, then the statements are equal for all sets A and B. See Example 3.

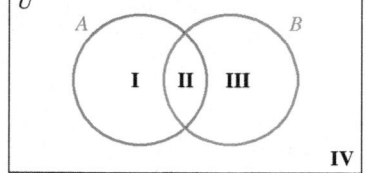

Figure 2.19

Example 3 *Equality of Sets*

Determine whether $(A \cap B)' = A' \cup B'$ for all sets A and B.

Draw a Venn diagram with two sets A and B, as in Fig. 2.19. Label the regions as indicated.

Determine $(A \cap B)'$

Set	Corresponding Regions
A	I, II
B	II, III
$A \cap B$	II
$(A \cap B)'$	I, III, IV

Determine $A' \cup B'$

Set	Corresponding Regions
A'	III, IV
B'	I, IV
$A' \cup B'$	I, III, IV

Both statements are represented by the same regions, I, III, IV, of the Venn diagram. Thus, $(A \cap B)' = A' \cup B'$ for all sets A and B. ∎

In Example 3, when we proved that $(A \cap B)' = A' \cup B'$, we started with two general sets and worked to the specific conclusion that both statements represented the same regions of the Venn diagram. We showed that $(A \cap B)' = A' \cup B'$ *for all sets A and B*. No matter what sets we choose for A and B, this statement will be true. For example, let $U = \{1, 2, 3, 4, 5, 6, 7, 8, 9, 10\}$, $A = \{3, 4, 6, 10\}$, and $B = \{1, 2, 4, 5, 6, 8\}$.

$$(A \cap B)' = A' \cup B'$$
$$\{4, 6\}' = \{3, 4, 6, 10\}' \cup \{1, 2, 4, 5, 6, 8\}'$$
$$\{1, 2, 3, 5, 7, 8, 9, 10\} = \{1, 2, 5, 7, 8, 9\} \cup \{3, 7, 9, 10\}$$
$$\{1, 2, 3, 5, 7, 8, 9, 10\} = \{1, 2, 3, 5, 7, 8, 9, 10\}$$

We can also use Venn diagrams to prove statements involving three sets.

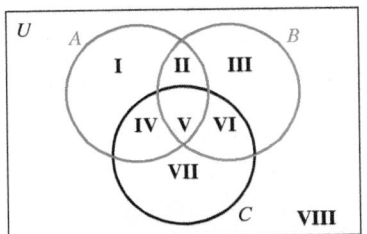

Figure 2.20

Animation
→

Venn Diagrams

Example 4 *Equality of Sets*

Determine whether $A \cup (B \cap C) = (A \cup B) \cap (A \cup C)$ for all sets, A, B, and C.

Solution Because the statements include three sets, A, B, and C, three circles must be used. The Venn diagram illustrating the eight regions is shown in Fig. 2.20.

First we will determine the regions that correspond to $A \cup (B \cap C)$, and then we will determine the regions that correspond to $(A \cup B) \cap (A \cup C)$. If both answers are the same, the statements are equal.

Determine $A \cup (B \cap C)$

Set	Corresponding Regions
A	I, II, IV, V
$B \cap C$	V, VI
$A \cup (B \cap C)$	I, II, IV, V, VI

Determine $(A \cup B) \cap (A \cup C)$

Set	Corresponding Regions
$A \cup B$	I, II, III, IV, V, VI
$A \cup C$	I, II, IV, V, VI, VII
$(A \cup B) \cap (A \cup C)$	I, II, IV, V, VI

The regions that correspond to $A \cup (B \cap C)$ are I, II, IV, V, and VI, and the regions that correspond to $(A \cup B) \cap (A \cup C)$ are also I, II, IV, V, and VI. The results show that both statements are represented by the same regions, namely, I, II, IV, V, and VI, and therefore $A \cup (B \cap C) = (A \cup B) \cap (A \cup C)$ for all sets A, B, and C. ∎

In Example 4, we proved that $A \cup (B \cap C) = (A \cup B) \cap (A \cup C)$ for all sets A, B, and C. Show that this statement is true for the specific sets $U = \{1, 2, 3, 4, 5, 6, 7, 8, 9, 10\}$, $A = \{1, 2, 3, 7\}$, $B = \{2, 3, 4, 5, 7, 9\}$, and $C = \{1, 4, 7, 8, 10\}$.

De Morgan's Laws for Sets

In set theory, logic, and other branches of mathematics, a pair of related theorems known as De Morgan's laws make it possible to transform statements and formulas into alternative and often more convenient forms. In set theory, *De Morgan's laws* are symbolized as follows.

De Morgan's Laws for Sets

1. $(A \cap B)' = A' \cup B'$
2. $(A \cup B)' = A' \cap B'$

Law 1 was verified in Example 3. We suggest that you verify law 2 at this time. The laws were expressed verbally by William of Ockham in the fourteenth century. In the nineteenth century, Augustus De Morgan expressed them mathematically. De Morgan's laws will be discussed more thoroughly in Chapter 3, Logic.

SECTION 2.4 *Exercises*

Warm Up Exercises

In Exercises 1–4, fill in the blank with an appropriate word, phrase, or symbol(s).

1. The number of regions created when constructing a Venn diagram with three overlapping sets is _____.

2. a) When constructing a Venn diagram with three overlapping sets, region _____ is generally completed first.

   b) When constructing a Venn diagram with three overlapping sets, after completing region V, the next regions generally completed are II, IV, and _____.

3. Complete De Morgan's laws:

   a) $(A \cup B)' =$ _____

   b) $(A \cap B)' =$ _____

4. When using Venn diagrams to verify or determine whether set statements are equal we use _____ reasoning.

Practice the Skills/Problem Solving

5. A Venn diagram contains three sets, *A*, *B*, and *C*, as in Fig. 2.15 on page 68. If region V contains 4 elements and there are 12 elements in $B \cap C$, how many elements belong in region VI? Explain.

6. A Venn diagram contains three sets, *A*, *B*, and *C*, as in Fig. 2.15 on page 68. If region V contains 4 elements and there are 9 elements in $A \cap B$, how many elements belong in region II? Explain.

7. Construct a Venn diagram illustrating the following sets.

$$U = \{1, 2, 3, 4, 5, 6, 7, 8, 9, 10\}$$
$$A = \{2, 3, 4, 7, 8\}$$
$$B = \{1, 2, 3, 6, 8\}$$
$$C = \{2, 4, 5, 6, 8, 10\}$$

8. Construct a Venn diagram illustrating the following sets.

$$U = \{a, b, c, d, e, f, g, h, i, j\}$$
$$A = \{c, d, e, g, h, i\}$$
$$B = \{a, c, d, g\}$$
$$C = \{c, f, i, j\}$$

9. Construct a Venn diagram illustrating the following sets.

$$U = \{Bambi, Pinocchio, Moana, Up, Zootopia,$$
$$\quad Ratatouille, Frozen, Cinderella, WALL\text{-}E,$$
$$\quad Aladdin\}$$
$$A = \{Bambi, Moana, Up, Frozen\}$$
$$B = \{Bambi, Moana, Cinderella, Aladdin\}$$
$$C = \{Bambi, Up, Zootopia, Aladdin\}$$

▲ *Olaf from the movie Frozen*

10. Construct a Venn diagram illustrating the following sets.

$$U = \{microwave\ oven, freezer, dishwasher, refrigerator,$$
$$\quad washer, dryer, toaster, blender, food\ processor, iron\}$$
$$A = \{toaster, blender, iron, dishwasher, washer, dryer\}$$
$$B = \{dishwasher, iron, freezer\}$$
$$C = \{washer, dryer, iron, freezer, microwave\ oven\}$$

11. Construct a Venn diagram illustrating the following sets.

$$U = \{Costco, Dollar\ General, Marshall's, Nordstrom,$$
$$\quad H\ \&\ M, Lowe's, Sprouts, Ross, Dollar\ Tree,$$
$$\quad Aldi, Urban\ Outfitters, Five\ Below, Ulta\ Beauty,$$
$$\quad Burlington\}$$

$A = \{$Costco, Marshall's, H & M, Sprouts, Ulta Beauty$\}$
$B = \{$Dollar General, H & M, Ross, Ulta Beauty, Burlington$\}$
$C = \{$H & M, Lowe's, Aldi, Dollar Tree, Sprouts$\}$

12. Construct a Venn diagram illustrating the following sets.

$U = \{$Louis Armstrong, Glenn Miller, Stan Kenton, Charlie Parker, Duke Ellington, Benny Goodman, Count Basie, John Coltrane, Dizzy Gillespie, Miles Davis, Thelonius Monk$\}$
$A = \{$Stan Kenton, Count Basie, Dizzy Gillespie, Duke Ellington, Thelonius Monk$\}$
$B = \{$Louis Armstrong, Glenn Miller, Count Basie, Duke Ellington, Miles Davis$\}$
$C = \{$Count Basie, Miles Davis, Stan Kenton, Charlie Parker, Duke Ellington$\}$

13. ***Olympic Medals*** Consider the following table, which shows countries that won at least 15 medals in the 2018 Winter Olympics. Let the countries in the table represent the universal set.

Country	Gold Medals	Silver Medals	Bronze Medals	Total Medals
Norway	14	14	11	39
Germany	14	10	7	31
Canada	11	8	10	29
United States	9	8	6	23
Netherlands	8	6	6	20
South Korea	5	8	4	17
OAR*	2	6	9	17
Switzerland	5	6	4	15
France	5	4	6	15

Source: United States Olympic Committee

Let $A =$ the set of countries that won at least 10 gold medals.

Let $B =$ the set of countries that won at least 8 silver medals.

Let $C =$ the set of countries that won at least 6 bronze medals.

Construct a Venn diagram that illustrates the set A, B, and C.

14. ***Popular TV Shows*** Construct a Venn diagram illustrating the following sets.

$U = \{$*The Voice* (*V*), *Dancing with the Stars* (*DWS*), *Game of Thrones* (*GT*), *The Handmaid's Tale* (*HT*), *Monday Night Football* (*MNF*), *The Big Bang Theory* (*BBT*), *This Is Us* (*TIU*), *Young Sheldon* (*YS*), *The Good Doctor* (*GD*), *NCIS* (*N*)$\}$
$A = \{V, GT, MNF, BBT, GD, N\}$
$B = \{V, GT, HT, YS\}$
$C = \{GT, HT, MNF, TIU\}$

* Olympic Athletes from Russia (OAR).

Grain Production In Exercises 15–20, use the following table, which shows the top 10 countries for production of corn, rice, and wheat in 2016. The universal set is the set of countries in the world.

Corn	Rice	Wheat
1. United States	1. China	1. China
2. China	2. India	2. India
3. Brazil	3. Indonesia	3. Russia
4. Argentina	4. Bangladesh	4. United States
5. Mexico	5. Vietnam	5. Canada
6. Ukraine	6. Myanmar	6. France
7. India	7. Thailand	7. Ukraine
8. Indonesia	8. Philippines	8. Pakistan
9. Russia	9. Brazil	9. Germany
10. Canada	10. Pakistan	10. Australia

Rankings of Corn, Rice, and Wheat Production in 2016

Source: UN Food and Agriculture Organization

Indicate in which regions, I–VIII in Fig. 2.21, each of the following countries belongs.

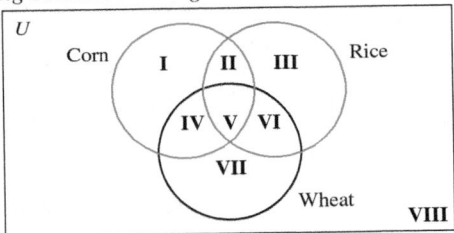

Figure 2.21

15. Canada

16. Pakistan

17. China

18. Germany

19. Spain

20. Brazil

Figures In Exercises 21–32, indicate in Fig. 2.22 the region in which each of the figures should be placed.

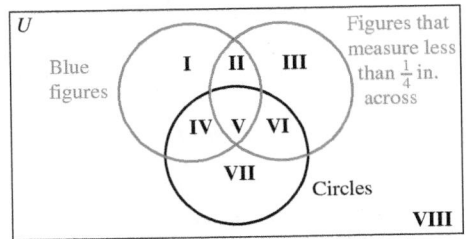

Figure 2.22

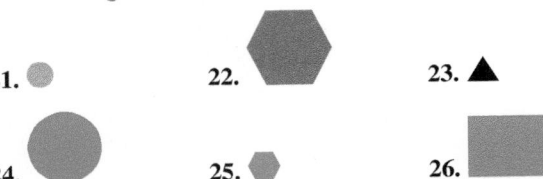

21.

22.

23.

24.

25.

26.

27. 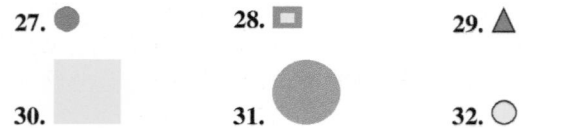   **28.** ☐   **29.** ▲

30. ☐   **31.** ●   **32.** ○

Senate Bills In Exercises 33–38, use Fig. 2.23. During a session of the U.S. Senate, three bills were voted on. The votes of six senators are shown. Determine in which region of the figure each senator should be placed. The set labeled Bill 1 represents the set of senators who voted yes on Bill 1, and so on.

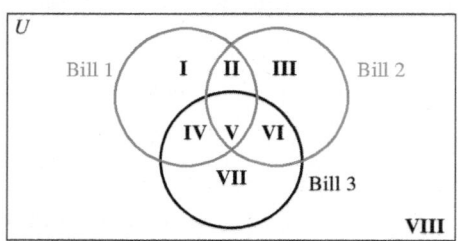

Figure 2.23

SENATOR	BILL 1	BILL 2	BILL 3
33. Cruz	yes	no	no
34. Blackburn	no	no	yes
35. Feinstein	no	no	no
36. Schumer	yes	yes	yes
37. Hirono	no	yes	yes
38. Rubio	no	yes	no

In Exercises 39–50, use the Venn diagram in Fig. 2.24 to list the sets in roster form.

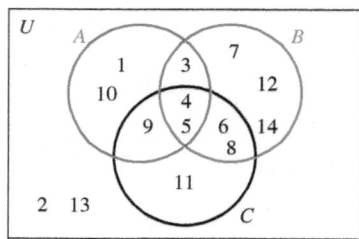

Figure 2.24

39. A **40.** U

41. C **42.** B'

43. $A \cap B$ **44.** $A \cap C$

45. $(B \cap C)'$ **46.** $A \cap B \cap C$

47. $(A \cup C)'$ **48.** $A \cap (B \cup C)$

49. $(A - B)'$

50. $A' - B$

In Exercises 51–56, use Venn diagrams to determine whether the following statements are equal for all sets A and B.

51. $(A \cap B)'$, $A' \cup B'$

52. $(A \cap B)'$, $A' \cup B$

53. $A' \cup B'$, $A \cap B$

54. $A' \cup B'$, $(A \cup B)'$

55. $A' \cap B'$, $A \cup B'$

56. $(A' \cap B)'$, $A \cup B'$

In Exercises 57–64, use Venn diagrams to determine whether the following statements are equal for all sets A, B, and C.

57. $A \cap (B \cup C)$, $(A \cap B) \cup C$

58. $A \cup (B \cap C)$, $(B \cap C) \cup A$

59. $A \cap (B \cup C)$, $(B \cup C) \cap A$

60. $A \cup (B \cap C)'$, $A' \cap (B' \cup C)$

61. $A \cap (B \cup C)$, $(A \cap B) \cup (A \cap C)$

62. $A \cup (B \cap C)$, $(A \cup B) \cap (A \cup C)$

63. $(A \cup B) \cap (B \cup C)$, $B \cup (A \cap C)$

64. $(A \cup B)' \cap C$, $(A' \cup C') \cap (B' \cup C)$

In Exercises 65–68, use a set statement to write a description of the shaded area. Use union, intersection, and complement as necessary. More than one answer may be possible.

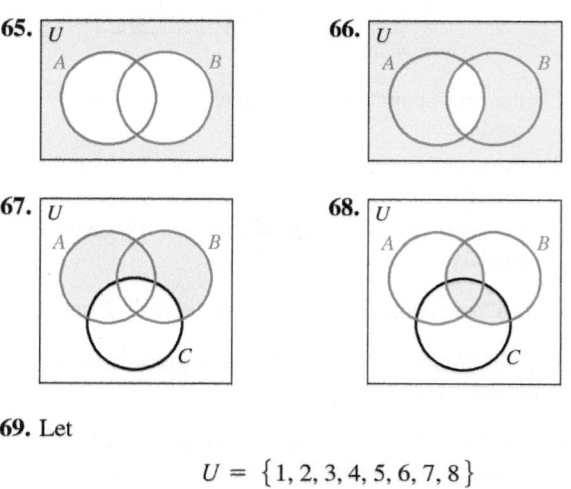

69. Let
$$U = \{1, 2, 3, 4, 5, 6, 7, 8\}$$
$$A = \{1, 2, 3, 6\}$$
$$B = \{3, 6, 7\}$$

a) Apply De Morgan's Law 1 on p. 72 to show that $(A \cap B)' = A' \cup B'$ for these sets.

b) Make up your own sets A and B. Verify that $(A \cap B)' = A' \cup B'$ for your sets A and B.

70. Let

$$U = \{a, b, c, d, e, f, g, h, i\}$$
$$A = \{a, e, g, h, i\}$$
$$B = \{a, g, i\}$$

a) Apply De Morgan's Law 2 on p. 72 to show that $(A \cup B)' = A' \cap B'$ for these sets.

b) Make up your own sets A and B. Verify that $(A \cap B)' = A' \cup B'$ for your sets A and B.

71. *Blood Types* A hematology text gives the following information on percentages of the different types of blood worldwide.

Type	Positive Blood, %	Negative Blood, %
A	37	6
O	32	6.5
B	11	2
AB	5	0.5

Construct a Venn diagram similar to the one in Example 2 and place the correct percentage in each of the eight regions.

72. Define each of the eight regions in Fig. 2.25 using sets A, B, and C and a set operation. (*Hint:* $A \cap B' \cap C'$ defines region I.)

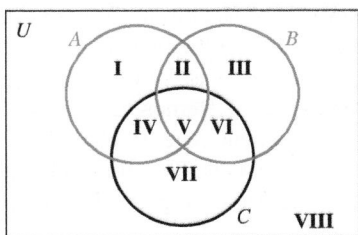

Figure 2.25

73. *Categorizing Contracts* J & C Mechanical Contractors wants to classify its projects. The contractors categorize set A as construction projects, set B as plumbing projects, and set C as projects with a budget greater than $300,000.

a) Draw a Venn diagram that can be used to categorize the company projects according to the listed criteria.

b) Determine the region of the diagram that contains construction projects and plumbing projects with a budget greater than $300,000. Describe the region using sets A, B, and C with set operations. Use union, intersection, and complement as necessary.

c) Determine the region of the diagram that contains plumbing projects with a budget greater than $300,000 that are not construction projects. Describe the region using sets A, B, and C with set operations. Use union, intersection, and complement as necessary.

d) Determine the region of the diagram that contains construction projects and nonplumbing projects whose budget is less than or equal to $300,000. Describe the region using sets A, B, and C with set operations. Use union, intersection, and complement as necessary.

Challenge Problem/Group Activity

74. We were able to determine the number of elements in the union of two sets with the formula

$$n(A \cup B) = n(A) + n(B) - n(A \cap B).$$

Can you determine a formula for determining the number of elements in the union of three sets? In other words, write a formula to determine $n(A \cup B \cup C)$. [*Hint:* The formula will contain each of the following: $n(A), n(B), n(C)$, $n(A \cap B \cap C'), n(A \cap B' \cap C), n(A' \cap B \cap C)$, and $2n(A \cap B \cap C)$.

Recreational Mathematics

75. a) Construct a Venn diagram illustrating four sets, A, B, C, and D. (*Hint:* Four circles cannot be used, and you should end up with 16 *distinct* regions.) Have fun!

b) Label each region with a set statement (see Exercise 72). Check all 16 regions to make sure that *each is distinct*.

Research Activity

76. *Combining Colors* The two Venn diagrams below illustrate what happens when colors are added or subtracted. Do research in an art text, the Internet, or another source, and write a report explaining the creation of the colors in the Venn diagrams, using such terms as union of colors and subtraction (or difference) of colors.

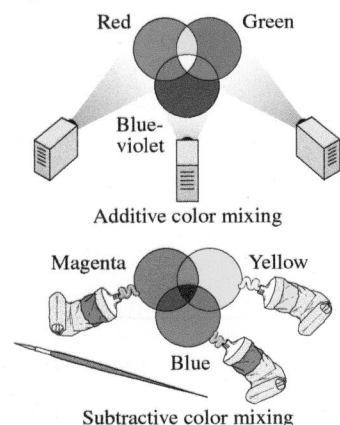

SECTION 2.5   Applications of Sets

Upon completion of this section, you will be able to:

- Use Venn diagrams and set concepts to solve application problems.

The members of a health club were surveyed about taking fitness classes at the club. Suppose the results of the survey show how many members took a yoga class, how many members took a spinning class, and how many members took a class in yoga and a class in spinning. How can the manager of the club use this information to determine how many members took only a yoga class? In this section, we will learn how to use Venn diagrams to answer this type of question.

Why This Is Important As you read through this section, you will see many real-life applications of set theory. These applications can help us understand how to determine the number of elements in different sets. In addition to the health club application described above, we will also work with applications involving homeowners' insurance policies, bookstore sales, restaurant surveys, and many other financial and legal applications. Knowing how to analyze such sets is important for making decisions affecting businesses and our personal lives.

We can solve practical problems involving sets by using the problem-solving process discussed in Chapter 1: Understand the problem, devise a plan, carry out the plan, and then examine and check the results. First determine: What is the problem? or What am I looking for? To devise the plan, list all the facts that are given and how they are related. *Look for key words or phrases* such as "only set *A*," "set *A* and set *B*," "set *A* or set *B*," "set *A* and set *B* and not set *C*." Remember that *and* means intersection, *or* means union, and *not* means complement. The problems we solve in this section contain two or three sets of elements, which can be represented in a Venn diagram. Our plan will generally include drawing a Venn diagram, labeling the diagram, and filling in the regions of the diagram.

Whenever possible, follow the procedure in Section 2.4 for completing the Venn diagram and then answer the questions. *Remember: When drawing Venn diagrams, we generally start with the intersection of the sets and work outward.*

Timely Tip

When drawing Venn diagrams, we generally start with the intersection of the sets and work outward.

Example 1 *Pizza King Purchases*

Pizza King surveyed 250 of its customers and collected the following information regarding purchases.

175 purchased pizza.

141 purchased chicken wings.

80 purchased both pizza and chicken wings.

Of those surveyed, how many customers

a) did not purchase either pizza or chicken wings?

b) purchased chicken wings but not pizza?

c) purchased pizza but not chicken wings?

d) purchased pizza or chicken wings?

Solution We will use P to represent customers who purchased pizza and C to represent customers who purchased chicken wings. The example provides the following information.

The number of customers surveyed is 250: $n(U) = 250$.

The number of customers surveyed who purchased pizza is 175: $n(P) = 175$.

The number of customers surveyed who purchased chicken wings is 141: $n(C) = 141$.

The number of customers surveyed who purchased both pizza and chicken wings is 80: $n(P \cap C)$ is 80.

We illustrate this information in the Venn diagram shown in Fig. 2.26. We already know that $P \cap C$ corresponds to region II. Because $n(P \cap C) = 80$, we write 80 in region II. Set P consists of regions I and II. We know that set P, the number of customers who purchased pizza, contains 175 customers. The total of these two regions must be 175. Region II contains 80 customers. Therefore, region I contains $175 - 80 = 95$ customers. We write the number 95 in region I. Set C consists of regions II and III. Because $n(C) = 141$, the total in these two regions must be 141. Region II contains 80, leaving $141 - 80 = 61$ customers for region III. We write 61 in region III.

The total number of customers surveyed who purchased pizza or chicken wings is determined by adding the numbers in regions I, II, and III. Therefore $n(P \cup C) = 95 + 80 + 61 = 236$. The number of customers in region IV is the difference between $n(U)$ and $n(P \cup C)$. There are $250 - 236$, or 14, customers in region IV.

a) The customers surveyed who did not purchase either pizza or chicken wings are those customers in the universal set who are not contained in set P or set C. The 14 customers in region IV did not purchase pizza or chicken wings.

b) The 61 customers in region III are those customers surveyed who purchased chicken wings but not pizza.

c) The 95 customers in region I are those customers surveyed who purchased pizza but not chicken wings.

d) The customers in regions I, II, and III are those customers surveyed who purchased either pizza or chicken wings. Thus, $95 + 80 + 61 = 236$ customers surveyed purchased either pizza or chicken wings. Notice that the 80 customers in region II who purchased both pizza and chicken wings are included in those customers surveyed who purchased either pizza or chicken wings. ∎

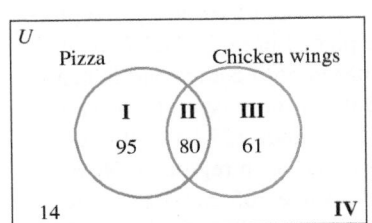

Figure 2.26

Application of Venn Diagrams

Similar problems involving three sets can be solved, as illustrated in Example 2.

Example 2 *Software Purchases*

Best Buy has recorded recent sales for three types of computer software: games, educational software, and utility programs. The following information regarding software purchases was obtained from a survey of 893 customers.

545 purchased games.

497 purchased educational software.

290 purchased utility programs.

297 purchased games and educational software.

196 purchased educational software and utility programs.

205 purchased games and utility programs.

157 purchased all three types of software.

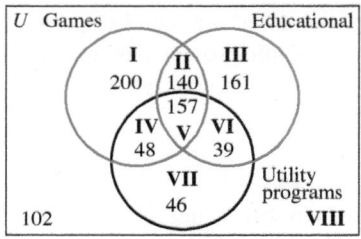

Figure 2.27

Use a Venn diagram to answer the following questions. How many customers purchased

a) none of these types of software?

b) only games?

c) at least one of these types of software?

d) exactly two of these types of software?

Solution   Begin by constructing a Venn diagram with three overlapping circles. One circle represents games, another educational software, and the third utility programs, See Fig. 2.27. Label the eight regions.

Whenever possible, work from the center of the diagram outward. First fill in region V. Since 157 customers purchased all three types of software, we place 157 in region V. Next determine the number to be placed in region II. Regions II and V together represent the customers who purchased both games and educational software. Since 297 customers purchased both of these types of software, the sum of the numbers in these regions must be 297. Since 157 have already been placed in region V, $297 - 157 = 140$ must be placed in region II. Now we determine the number to be placed in region IV. Since 205 customers purchased both games and utility programs, the sum of the numbers in regions IV and V must be 205. Since 157 customers have already been placed in region V, $205 - 157 = 48$ must be placed in region IV. Now determine the number to be placed in region VI. A total of 196 customers purchased educational software and utility programs. The numbers in regions V and VI must total 196. Since 157 customers have already been placed in region V, the number to be placed in region VI is $196 - 157 = 39$.

Now that we have determined the numbers for regions V, II, IV, and VI, we can determine the numbers to be placed in regions I, III, and VII. We are given that 545 customers purchased games. The sum of the numbers in regions I, II, IV, and V must be 545. To determine the number to be placed in region I, subtract the amounts in regions II, IV, and V from 545. There must be $545 - 140 - 48 - 157 = 200$ in region I. Determine the numbers to be placed in regions III and VII in a similar manner.

$$\text{Region III } = 497 - 140 - 157 - 39 = 161$$
$$\text{Region VII} = 290 - 48 - 157 - 39 = 46$$

Now that we have determined the numbers in regions I through VII, we can determine the number to be placed in region VIII. Adding the numbers in regions I through VII yields a sum of 791. The difference between the total number of customers surveyed, 893, and the sum of the numbers in regions I through VII must be placed in region VIII.

$$\text{Region VIII} = 893 - 791 = 102$$

Now that we have completed the Venn diagram, we can answer the questions.

a) One hundred two customers did not purchase any of these types of software. These customers are indicated in region VIII.

b) Region I represents those customers who purchased only games. Thus, 200 customers purchased only games.

c) The words *at least one* mean "one or more." All those in regions I through VII purchased at least one of the types of software. The sum of the numbers in regions I through VII is 791, so 791 customers purchased at least one of the types of software.

d) The customers in regions II, IV, and VI purchased exactly two of the types of software. Summing the numbers in these regions $140 + 48 + 39$ we determine that 227 customers purchased exactly two of these types of software. Notice that we did not include the customers in region V. Those customers purchased all three types of software.

Timely Tip

When constructing a Venn diagram, remember to subtract the number in region V from the respective values in determining the numbers to be placed in regions II, IV, and VI.

The procedure to work problems like those given in Example 2 is generally the same. Start by completing region V. Next complete regions II, IV, and VI. Then complete regions I, III, and VII. Finally, complete region VIII. When you are constructing Venn diagrams, be sure to check your work carefully.

Example 3 *Travel Packages*

Liberty Travel surveyed 125 potential customers. The following information was obtained.

> 68 wished to travel to Hawaii.
>
> 53 wished to travel to Las Vegas.
>
> 47 wished to travel to Disney World.
>
> 34 wished to travel to Hawaii and Las Vegas.
>
> 26 wished to travel to Las Vegas and Disney World.
>
> 23 wished to travel to Hawaii and Disney World.
>
> 18 wished to travel to all three destinations.

Use a Venn diagram to answer the following questions. How many of those surveyed

a) did not wish to travel to any of these destinations?

b) wished to travel only to Hawaii?

c) wished to travel to Disney World *and* Las Vegas, but not to Hawaii?

d) wished to travel to Disney World *or* Las Vegas, but not to Hawaii?

e) wished to travel to exactly one of these destinations?

Solution The Venn diagram is constructed using the procedures we outlined in Example 2. The diagram is illustrated in Fig. 2.28. We suggest you construct the diagram by yourself now and check your diagram with Fig. 2.28.

a) Twenty-two potential customers did not wish to travel to any of these destinations (see region VIII).

b) Twenty-nine potential customers wished to travel only to Hawaii (see region I).

c) Those potential customers in region VI wished to travel to Disney World *and* Las Vegas, but not to Hawaii. Therefore, eight customers satisfied the criteria.

d) The word *or* in this type of problem means one or the other or both. All the potential customers in regions II, III, IV, V, VI, and VII wished to travel to Disney World or Las Vegas. Those in regions II, IV, and V also wished to travel to Hawaii. The potential customers who wished to travel to Disney World or Las Vegas, but not to Hawaii, are determined by adding the numbers in regions III, VI, and VII. There are 16 + 8 + 11 = 35 potential customers who satisfy the criteria.

e) Those potential customers in regions I, III, and VII wished to travel to exactly one of the destinations. Therefore, 29 + 16 + 11 = 56 customers wished to travel to exactly one of these destinations. ∎

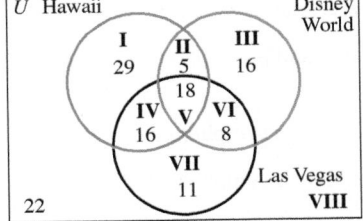

Figure 2.28

Exercises

Practice the Skills/Problem Solving

In Exercises 1–15, draw a Venn diagram to obtain the answers.

1. **Study Locations** At Tallahassee Community College, a survey was taken to determine where students studied on campus. Of the 250 students surveyed, it was determined that

   > 177 studied in the library.
   >
   > 138 studied in the cafeteria.
   >
   > 80 studied in both the library and cafeteria.

Of those students surveyed,

a) how many studied only in the library?

b) how many studied only in the cafeteria?

c) how many did not study in either the library or cafeteria?

2. Food Delivery Services In a survey of 130 people who used food delivery services, it was determined that

   74 used Grubhub.
   70 used Uber Eats.
   41 used both Grubhub and Uber Eats.

Of those surveyed,

a) how many used only Grubhub?

b) how many used only Uber Eats?

c) how many did not use either of these services?

3. Real Estate The Greens are moving to Athens, Georgia. Their real estate agent located 77 houses listed for sale in the Athens area in their price range. Of these houses listed for sale,

   47 had a finished basement.
   52 had a three-car garage.
   35 had a finished basement and a three-car garage.

How many had

a) a finished basement but not a three-car garage?

b) a three-car garage but not a finished basement?

c) either a finished basement or a three-car garage?

4. Racing Fleet Foot Racing interviewed 150 long-distance runners to determine the type of races in which they participated. The following information was determined.

   102 participated in a marathon.
    93 participated in a triathlon.
    55 participated in both a marathon and a triathlon.

How many

a) participated in only a marathon?

b) participated in only a triathlon?

c) participated in either a marathon or a triathlon?

d) had not participated in either a marathon or a triathlon?

5. Social Media A survey of 108 high school students determined whether they used Instagram, Twitter, or Facebook. The following information was determined.

   68 used Instagram.
   58 used Twitter.
   75 used Facebook.

   34 used Instagram and Twitter.
   38 used Twitter and Facebook.
   45 used Instagram and Facebook.
   22 used all three.

How many of the students surveyed used

a) only Instagram?

b) Instagram and Twitter, but not Facebook?

c) Instagram or Twitter?

d) Instagram or Twitter, but not Facebook?

e) exactly two of the social media?

6. Cultural Activities Thirty-three U.S. cities were researched to determine whether they had a professional sports team, a symphony, or a children's museum. The following information was determined.

   16 had a professional sports team.
   17 had a symphony.
   15 had a children's museum.
   11 had a professional sports team and a symphony.
    7 had a professional sports team and a children's museum.
    9 had a symphony and children's museum.
    5 had all three activities.

How many of the cities surveyed had

a) only a professional sports team?

b) a professional sports team and a symphony, but not a children's museum?

c) a professional sports team or a symphony?

d) a professional sports team or a symphony, but not a children's museum?

e) exactly two of the activities?

7. Movies A survey of 350 customers was taken at Regal Cinemas in Austin, Texas, regarding the type of movies customers liked. The following information was determined.

   196 liked dramas.
   153 liked comedies.
    88 liked science fiction.
    59 liked dramas and comedies.

37 liked dramas and science fiction.

32 liked comedies and science fiction.

21 liked all three types of movies.

Of the customers surveyed, how many liked

a) none of these types of movies?

b) only dramas?

c) exactly one of these types of movies?

d) exactly two of these types of movies?

e) dramas or comedies?

8. ***Book Purchases*** A survey of 85 customers was taken at Barnes & Noble regarding the types of books purchased. The survey determined that

44 purchased mysteries.

33 purchased science fiction.

29 purchased romance novels.

13 purchased mysteries and science fiction.

5 purchased science fiction and romance novels.

11 purchased mysteries and romance novels.

2 purchased all three types of books.

How many of the customers surveyed purchased

a) only mysteries?

b) mysteries and science fiction, but not romance novels?

c) mysteries or science fiction?

d) mysteries or science fiction, but not romance novels?

e) exactly two types?

9. ***TV Choices*** *TV Guide* surveyed subscribers, asking which of the following reality shows they watched on a regular basis: *Survivor, The Voice, America's Got Talent.* The results of the questionnaires that were returned showed that

253 watched *Survivor.*

181 watched *The Voice.*

202 watched *America's Got Talent.*

98 watched *Survivor* and *The Voice.*

85 watched *Survivor* and *America's Got Talent.*

90 watched *The Voice* and *America's Got Talent.*

58 watched all three shows.

29 watched none of these shows.

a) How many subscribers returned the survey?

Of the subscribers who returned the survey, how many watched

b) *Survivor* and *The voice,* but not *America's Got Talent?*

c) *America's Got Talent,* but neither *Survivor* nor *The Voice?*

d) exactly two of these shows?

e) at least one of these shows?

10. ***Jobs at a Restaurant*** In a survey of employees at Qdoba Mexican Eats, it was determined that

8 cooked food.

9 washed dishes.

18 operated the cash register.

4 cooked food and washed dishes.

5 washed dishes and operated the cash register.

3 cooked food and operated the cash register.

2 did all three jobs.

5 did none of these jobs.

a) How many Qdoba Mexican Eats employees were surveyed?

Of the Qdoba Mexican Eats employees surveyed, how many

b) only cooked food?

c) only operated the cash register?

d) washed dishes and operated the cash register, but did not cook food?

e) washed dishes or operated the cash register, but did not cook food?

f) did at least two of these jobs?

11. ***Homeowners' Insurance Policies*** A committee of the Florida legislature decided to analyze 350 homeowners' insurance policies to determine if the consumers' homes were covered for damage due to sinkholes, mold, and floods. The following results were determined.

170 homes were covered for damage due to sinkholes.

172 homes were covered for damage due to mold.

234 homes were covered for damage due to floods.

105 homes were covered for damage due to sinkholes and mold.

115 homes were covered for damage due to mold and floods.

109 homes were covered for damage due to sinkholes and floods.

78 homes were covered for damage due to all three conditions.

How many of the homes

a) were covered for damage due to mold but were not covered for damage due to sinkholes?

b) were covered for damage due to sinkholes or mold?

c) were covered for damage due to mold and floods, but were not covered for damage due to sinkholes?

d) were not covered for damage due to any of the three conditions?

12. *Appetizers Survey* Da Tulio's Restaurant hired Dennis to determine what kind of appetizers customers liked. He surveyed 100 people, with the following results: 78 liked shrimp cocktail, 56 liked mozzarella sticks, and 35 liked both shrimp cocktail and mozzarella sticks. Every person interviewed liked one or the other or both kinds of appetizers. Does this result seem correct? Explain your answer.

13. *Discovering an Error* A New York State Thruway agent sampled cars going from New York into Pennsylvania. In his report, he indicated that of the 85 cars sampled,

    35 cars were driven by women.

    53 cars were sport-utility vehicles.

    43 cars had two or more passengers.

    27 cars were driven by women in sport-utility vehicles.

    25 cars were cars driven by women and had two or more passengers.

    20 cars were sport-utility vehicles with two or more passengers.

    15 cars were driven by women in sport-utility vehicles and had two or more passengers.

After his supervisor reads the report, she explains to the agent that he made a mistake. Explain how his supervisor knew that the agent's report contained an error.

Challenge Problems/Group Activities

14. *Parks* A survey of 300 parks showed the following.

    15 had only camping.

    20 had only hiking trails.

    35 had only picnicking.

    185 had camping.

    140 had camping and hiking trails.

    125 had camping and picnicking.

    210 had hiking trails.

Determine the number of parks that

a) had at least one of these features.

b) had all three features.

c) did not have any of these features.

d) had exactly two of these features.

15. *Surveying Farmers* A survey of 500 farmers in a midwestern state showed the following.

    125 grew only wheat.

    110 grew only corn.

    90 grew only oats.

    200 grew wheat.

    60 grew wheat and corn.

    50 grew wheat and oats.

    180 grew corn.

Determine the number of farmers who

a) grew at least one of the three.

b) grew all three.

c) did not grow any of the three.

d) grew exactly two of the three.

16. *Family Reunion* When the Montesano family members discussed where their annual reunion should take place, they determined that of all the family members,

    8 would not go to a park.

    7 would not go to a beach.

    11 would not go to the family cottage.

    3 would go to neither a park nor a beach.

    4 would go to neither a beach nor the family cottage.

    6 would go to neither a park nor the family cottage.

    2 would not go to a park or a beach or to the family cottage.

    1 would go to all three places.

What is the total number of family members?

Recreational Mathematics

17. *Number of Elements* A universal set U consists of 12 elements. If sets A, B, and C are proper subsets of U and $n(U) = 12, n(A \cap B) = n(A \cap C) = n(B \cap C) = 6$, $n(A \cap B \cap C) = 4$, and $n(A \cup B \cup C) = 10$, determine

a) $n(A \cup B)$

b) $n(A' \cup C)$

c) $n(A \cap B)'$

SECTION 2.6 Infinite Sets

Upon completion of this section, you will be able to:

- Understand infinite sets.
- Understand countable sets.

▲ *Georg Cantor*

Which set is larger, the set of integers or the set of even integers? One might argue that because the set of even integers is a subset of the set of integers, the set of integers must be larger than the set of even integers. Yet both sets are infinite sets, so how can we determine which set is larger? This question puzzled mathematicians for centuries until 1874, when Georg Cantor developed a method of determining the cardinal number of an infinite set. In this section, we will discuss infinite sets and how to determine the number of elements in an infinite set.

Why This Is Important The concept of infinity plays an important role in many applications of mathematics and science. For example, calculus relies on the concept of infinity, and calculus allows doctors to determine the amount of blood flow in a human heart. Calculus also is used by businesses to make important decisions that can maximize profits and minimize costs.

Infinite Sets

On page 45, we state that a finite set is a set in which the number of elements is zero or the number of elements can be expressed as a natural number. On page 46, we define a one-to-one correspondence. To determine the number of elements in a finite set, we can place the set in a one-to-one correspondence with a subset of the set of counting numbers. For example, the set $A = \{\#, ?, \$\}$ can be placed in one-to-one correspondence with set $B = \{1, 2, 3\}$, a subset of the set of counting numbers.

$$A = \{\#, ?, \$\}$$
$$\downarrow \downarrow \downarrow$$
$$B = \{1, 2, 3\}$$

Because the cardinal number of set B is 3, the cardinal number of set A is also 3. Any two sets, such as set A and set B, that can be placed in a one-to-one correspondence must have the same number of elements (therefore the same cardinality) and must be equivalent sets. Note that $n(A)$ and $n(B)$ both equal 3.

German mathematician Georg Cantor (1845–1918), known as the father of set theory, thought about sets that were not bounded. He called an unbounded set an *infinite set* and provided the following definition.

> **Definition: Infinite Set**
> An **infinite set** is a set that can be placed in a one-to-one correspondence with a proper subset of itself.

In Example 1, we use Cantor's definition of an infinite set to show that the set of counting numbers is infinite.

Example 1 *The Set of Natural Numbers*

Show that $N = \{1, 2, 3, 4, 5, \ldots, n, \ldots\}$ is an infinite set.

Solution To show that the set N is infinite, we establish a one-to-one correspondence between the counting numbers and a proper subset of itself. By removing the first element from the set of counting numbers, we get the set $\{2, 3, 4, 5, 6, \dots \}$, which is a proper subset of the set of counting numbers. Now we establish the one-to-one correspondence.

$$\text{Counting numbers} = \{1, 2, 3, 4, 5, \dots, \quad n, \dots \}$$
$$\downarrow \downarrow \downarrow \downarrow \downarrow \qquad\quad \downarrow$$
$$\text{Proper subset} \quad = \{2, 3, 4, 5, 6, \dots, n + 1, \dots \}$$

Note that for any number, n, in the set of counting numbers, its corresponding num-ber in the proper subset is one greater, or $n + 1$. We have now shown the desired one-to-one correspondence, and thus the set of counting numbers is infinite.   ∎

Note in Example 1 that we showed the pairing of the general terms $n \rightarrow (n + 1)$. Showing a one-to-one correspondence of infinite sets requires showing the pairing of the general terms in the two infinite sets.

In the set of counting numbers, n represents the general term. For any other set of numbers, the general term will be different. The general term in any set should be written in terms of n such that when 1 is substituted for n in the general term, we get the first number in the set; when 2 is substituted for n in the general term, we get the second number in the set; when 6 is substituted for n in the general term, we get the sixth number in the set; and so on.

Consider the set $\{4, 9, 14, 19, \dots \}$. Suppose we want to write the general term for this set (or sequence) of numbers. What would the general term be? The numbers differ by 5, so the general term will be of the form $5n$ plus or minus some number. Substituting 1 for n yields $5(1)$, or 5. Because the first number in the set is 4, we need to subtract 1 from the 5. Thus, the general term is $5n - 1$. Note that when $n = 1$, the value is $5(1) - 1$ or 4; when $n = 2$, the value is $5(2) - 1$ or 9; when $n = 3$, the value is $5(3) - 1$ or 14; and so on. Therefore, we write the set of numbers with the general term as

$$\{4, 9, 14, 19, \dots, 5n - 1, \dots \}$$

Now that you are aware of how to determine the general term of a set of numbers, we can do some more problems involving sets.

Example 2   *The Set of Even Numbers*

Show that the set of even counting numbers $\{2, 4, 6, 8, \dots, 2n, \dots \}$ is an infinite set.

Solution   First create a proper subset of the set of even counting numbers by re-moving the first number from the set. Then establish a one-to-one correspondence.

$$\text{Even counting numbers:} \quad \{2, 4, 6, 8, \dots, \quad 2n, \dots \}$$
$$\downarrow \downarrow \downarrow \downarrow \qquad\quad \downarrow$$
$$\text{Proper subset:} \qquad\qquad \{4, 6, 8, 10, \dots, 2n + 2, \dots \}$$

A one-to-one correspondence exists between the two sets, so the set of even counting numbers is infinite.   ∎

Example 3   *The Set of Multiples of Five*

Show that the set $\{5, 10, 15, 20, \dots, 5n, \dots \}$ is an infinite set.

Solution

$$\text{Given set:} \qquad \{5, 10, 15, 20, \dots, \quad 5n, \dots \}$$
$$\downarrow \quad \downarrow \quad \downarrow \quad \downarrow \qquad\quad \downarrow$$
$$\text{Proper subset:} \quad \{10, 15, 20, 25, \dots, 5n + 5, \dots \}$$

Therefore, the given set is an infinite set.   ∎

Countable Sets

In his work with infinite sets, Cantor developed ideas on how to determine the cardinal number of an infinite set. He called the cardinal number of infinite sets "transfinite cardinal numbers" or "transfinite powers." He defined a set as *countable* if it is finite or if it can be placed in a one-to-one correspondence with the set of counting numbers. All infinite sets that can be placed in a one-to-one correspondence with the set of counting numbers have cardinal number *aleph-null*, symbolized $\aleph_0$ (the first Hebrew letter, aleph, with a zero subscript, read "null").

Example 4 *The Cardinal Number of the Set of Even Numbers*

Show that the set of even counting numbers has cardinal number $\aleph_0$.

Solution In Example 2, we showed that the set of even counting numbers is infinite by setting up a one-to-one correspondence between the set and a proper subset of itself.

Now we will show that it is countable and has cardinality $\aleph_0$ by setting up a one-to-one correspondence between the set of counting numbers and the set of even counting numbers.

$$\text{Counting numbers:} \quad N = \{1, 2, 3, 4, \ldots, n, \ldots\}$$
$$\downarrow \downarrow \downarrow \downarrow \qquad \downarrow$$
$$\text{Even counting numbers:} \quad E = \{2, 4, 6, 8, \ldots, 2n, \ldots\}$$

For each number n in the set of counting numbers, its corresponding number is $2n$. Since we determined a one-to-one correspondence between the set of counting numbers and the set of even counting numbers, the set of even counting numbers is countable. Thus, the cardinal number of the set of even counting numbers is $\aleph_0$; that is, $n(E) = \aleph_0$. As we mentioned earlier, the set of even counting numbers is an infinite set, since it can be placed in a one-to-one correspondence with a proper subset of itself. Therefore, the set of even counting numbers is both infinite and countable. ∎

> **Definition: Cardinal Number of Infinite Sets**
> Any set that can be placed in a one-to-one correspondence with the set of counting numbers has **cardinal number** (or cardinality) $\aleph_0$ and is infinite and is countable.

Example 5 *The Cardinal Number of the Set of Odd Numbers*

Show that the set of odd counting numbers has cardinality $\aleph_0$.

Solution To show that the set of odd counting numbers has cardinality $\aleph_0$, we need to show a one-to-one correspondence between the set of counting numbers and the set of odd counting numbers.

$$\text{Counting numbers:} \quad N = \{1, 2, 3, 4, 5, \ldots, \quad n, \ldots\}$$
$$\downarrow \downarrow \downarrow \downarrow \downarrow \qquad \downarrow$$
$$\text{Odd counting numbers:} \quad O = \{1, 3, 5, 7, 9, \ldots, 2n - 1, \ldots\}$$

Since there is a one-to-one correspondence, the odd counting numbers have cardinality $\aleph_0$; that is, $n(O) = \aleph_0$. ∎

We have shown that both the odd and the even counting numbers have cardinality $\aleph_0$. Merging the odd counting numbers with the even counting numbers gives the set of counting numbers, and we may reason that

$$\aleph_0 + \aleph_0 = \aleph_0$$

This result may seem strange, but it is true. What could such a statement mean? Well, consider a hotel with infinitely many rooms. If all the rooms are occupied, the hotel is, of course, full. If more guests appear, wanting accommodations, will they be turned away?

Welcome to

H⬤TEL
INFINITY

▲ *... where there's always room for one more...*

The answer is *no,* for if the room clerk were to reassign each guest to a new room with a room number twice that of the present room, all the odd-numbered rooms would become unoccupied and there would be space for infinitely many more guests!

Cantor showed that there are different orders of infinity. Sets that are countable and have cardinal number $\aleph_0$ are the lowest order of infinity. Cantor showed that the set of integers and the set of rational numbers (fractions of the form p/q, where $q \neq 0$) are infinite sets with cardinality $\aleph_0$. He also showed that the set of real numbers (discussed in Chapter 5) could not be placed in a one-to-one correspondence with the set of counting numbers and that they have a higher order of infinity.

SECTION 2.6   *Exercises*

Warm Up Exercises

In Exercises 1–2, fill in the blank with an appropriate word, phrase, or symbol(s).

1. A set that can be placed in a one-to-one correspondence with a proper subset of itself is called a(n) _____ set.

2. A set that is finite or can be placed in a one-to-one correspondence with the set of counting numbers is called a(n) _____ set.

Practice the Skills

In Exercises 3–12, show that the set is infinite by placing it in a one-to-one correspondence with a proper subset of itself. Be sure to show the pairing of the general terms in the sets.

3. $\{5, 6, 7, 8, 9, \dots\}$

4. $\{30, 31, 32, 33, 34, \dots\}$

5. $\{6, 8, 10, 12, 14, \dots\}$

6. $\{5, 8, 11, 14, 17, \dots\}$

7. $\{5, 7, 9, 11, 13, \dots\}$

8. $\{5, 9, 13, 17, 21, \dots\}$

9. $\{\frac{1}{2}, \frac{1}{4}, \frac{1}{6}, \frac{1}{8}, \frac{1}{10}, \dots\}$

10. $\{1, \frac{1}{2}, \frac{1}{3}, \frac{1}{4}, \frac{1}{5}, \dots\}$

11. $\{\frac{4}{11}, \frac{5}{11}, \frac{6}{11}, \frac{7}{11}, \frac{8}{11}, \dots\}$

12. $\{\frac{6}{13}, \frac{7}{13}, \frac{8}{13}, \frac{9}{13}, \frac{10}{13}, \dots\}$

In Exercises 13–22, show that the set has cardinal number $\aleph_0$ by establishing a one-to-one correspondence between the set of counting numbers and the given set. Be sure to show the pairing of the general terms in the sets.

13. $\{3, 6, 9, 12, 15, \dots\}$

14. $\{20, 21, 22, 23, 24, \dots\}$

15. $\{4, 6, 8, 10, 12, \dots\}$

16. $\{0, 2, 4, 6, 8, \dots\}$

17. $\{2, 5, 8, 11, 14, \dots\}$

18. $\{7, 11, 15, 19, 23, \dots\}$

19. $\{\frac{1}{3}, \frac{1}{6}, \frac{1}{9}, \frac{1}{12}, \frac{1}{15}, \dots\}$

20. $\{\frac{1}{2}, \frac{1}{4}, \frac{1}{6}, \frac{1}{8}, \dots\}$

21. $\{\frac{1}{3}, \frac{1}{4}, \frac{1}{5}, \frac{1}{6}, \frac{1}{7}, \dots\}$

22. $\{\frac{1}{2}, \frac{2}{3}, \frac{3}{4}, \frac{4}{5}, \frac{5}{6}, \dots\}$

Challenge Problems/Group Activities

In Exercises 23–26, show that the set has cardinality $\aleph_0$ by establishing a one-to-one correspondence between the set of counting numbers and the given set.

23. $\{1, 4, 9, 16, 25, \dots\}$

24. $\{2, 4, 8, 16, 32, \dots\}$

25. $\{3, 9, 27, 81, 243, \dots\}$

26. $\{\frac{1}{3}, \frac{1}{6}, \frac{1}{12}, \frac{1}{24}, \frac{1}{48}, \dots\}$

Recreational Mathematics

In Exercises 27–31, insert the symbol $<$, $>$, or $=$ in the shaded area to make a true statement.

27. $\aleph_0$ ▨ $\aleph_0 + \aleph_0$

28. $2\aleph_0$ ▨ $\aleph_0 + \aleph_0$

29. $2\aleph_0$ ▨ $\aleph_0$

30. $\aleph_0 + 5$ ▨ $\aleph_0 + 3$

31. $n(N)$ ▨ $\aleph_0$

32. *Zeno's Paradox* There are a number of paradoxes (a statement that appears to be true and false at the same time) associated with infinite sets and the concept of infinity. One of these, called *Zeno's paradox*, is named after the mathematician Zeno, born about 496 B.C. in Italy. According to Zeno's paradox, suppose Achilles starts out 1 meter behind a tortoise. Also, suppose Achilles walks 10 times as fast as the tortoise crawls. When Achilles reaches the point where the tortoise started, the tortoise is 1/10 of a meter ahead of Achilles; when Achilles reaches the point where the tortoise was 1/10 of a meter ahead, the tortoise is now 1/100 of a meter ahead; and so on. According to Zeno's paradox, Achilles gets closer and closer to the tortoise but never catches up to the tortoise.

a) Do you believe the reasoning process is sound? If not, explain why not.

b) In actuality, if this situation were real, would Achilles ever pass the tortoise?

Research Activities

33. *Rational Numbers* Do research to explain how Cantor proved that the set of rational numbers has cardinal number $\aleph_0$.

34. *Real Numbers* Do research to explain how it can be shown that the set of real numbers do not have cardinal number $\aleph_0$.

CHAPTER 2 Summary

Important Facts and Concepts

Examples and Discussion

Section 2.1

Methods Used to Indicate a Set

Description — Example 1, page 42

Roster form — Examples 2–3, 5, 6, 8, pages 43–46

Set-builder notation — Examples 4–6, page 44

Symbol	Meaning
$\in$	is an element of
$\notin$	is not an element of
$n(A)$	number of elements in set A
$\varnothing$ or $\{\ \}$	the empty set
U	the universal set

Examples 4–7, 9, page 44, 46

Section 2.2

Symbol	Meaning
$\subseteq$	is a subset of
$\nsubseteq$	is not a subset of
$\subset$	is a proper subset of
$\not\subset$	is not a proper subset of

Examples 1 and 3, pages 51-52

Number of distinct subsets of a finite set with n elements is 2^n.

Number of distinct proper subsets of a finite set with n elements is $2^n - 1$.

Examples 4 and 5, pages 53–54

Example 4, page 53

Section 2.3

Symbol	Meaning
$'$	complement
$\cap$	intersection
$\cup$	union
$-$	difference of two sets
$\times$	Cartesian product

Examples 1, 3–6, pages 57–61

Examples 2, 3, 6, pages 58–61

Examples 4–6, pages 59–61

Example 9, pages 62–63

Example 10, page 63

And is generally interpreted to mean *intersection*.

Or is generally interpreted to mean *union*.

For any sets A and B,

$$n(A \cup B) = n(A) + n(B) - n(A \cap B)$$

Examples 7 and 8, pages 61–62

Examples 7 and 8, pages 61–62

Section 2.4

De Morgan's Laws

$$(A \cap B)' = A' \cup B'$$
$$(A \cup B)' = A' \cap B'$$

Example 3, pages 70–71

Section 2.6

An **infinite set** is a set that can be placed in a one-to-one correspondence with a proper subset of itself.

Examples 1–5, pages 83–85, Discussion page 83

$\aleph_0$ is read aleph-null

Examples 4–5, page 85

Countable sets

Examples 4–5, page 85

CHAPTER 2   Review Exercises

2.1, 2.2, 2.3, 2.4, 2.6

In Exercises 1–14, state whether each statement is true or false. If false, give a reason.

1. The set of stores located in the state of Wyoming is a well-defined set.

2. The set of the three best songs is a well-defined set.

3. maple $\in$ {oak, elm, maple, sycamore}

4. { } $\subset \varnothing$

5. {3, 6, 9, 12, ...} and {2, 4, 6, 8, ...} are disjoint sets.

6. {Mercury, Venus, Earth, Mars} is an example of a set in roster form.

7. {candle, picture, lamp} = {picture, chair, lamp}

8. {apple, orange, banana, pear} is equivalent to {tomato, corn, spinach, radish}.

9. If $A =$ {pen, pencil, book, calculator}, then $n(A) = 4$.

10. $A =$ {1, 3, 5, 7, ...} is a countable set.

11. $A =$ {1, 4, 7, 10, ..., 31} is a finite set.

12. {2, 5, 7} $\subseteq$ {2, 5, 7, 10}.

13. $\{x \mid x \in N$ and $3 < x \leq 9\}$ is a set in set-builder notation.

14. $\{x \mid x \in N$ and $2 < x \leq 12\} \subseteq \{1, 2, 3, 4, 5, ..., 20\}$

In Exercises 15–18, express each set in roster form.

15. Set A is the set of odd natural numbers between 5 and 16.

16. Set B is the set of states that border Kansas.

17. $C = \{x \mid x \in N$ and $x < 175\}$

18. $D = \{x \mid x \in N$ and $8 < x \leq 80\}$

In Exercises 19–22, express each set in set-builder notation.

19. Set A is the set of natural numbers between 50 and 150.

20. Set B is the set of natural numbers greater than 42.

21. Set C is the set of natural numbers less than 7.

22. Set D is the set of natural numbers between 27 and 51, inclusive.

In Exercises 23–26, express each set with a written description.

23. $A = \{x \mid x$ is a capital letter of the English alphabet from E through M inclusive}

24. $B =$ {penny, nickel, dime, quarter, half-dollar}

25. $C = \{a, b, c\}$

26. $D = \{x \mid 3 \leq x < 9\}$

In Exercises 27–36, let

$$U = \{1, 2, 3, 4, ..., 10\}$$
$$A = \{1, 3, 5, 7\}$$
$$B = \{3, 7, 9, 10\}$$
$$C = \{1, 7, 10\}$$

Determine the following.

27. $A \cap B$

28. $A \cup B'$

29. $A' \cap B$

30. $(A \cup B)' \cup C$

31. $A - B$

32. $A - C'$

33. $A \times C$

34. $B \times A$

35. The number of subsets of set B

36. The number of proper subsets of set A

37. For the following sets, construct a Venn diagram and place the elements in the proper region.

$U =$ {lion, tiger, leopard, cheetah, puma, lynx, panther, jaguar}
$A =$ {tiger, puma, lynx}
$B =$ {lion, tiger, jaguar, panther}
$C =$ {tiger, lynx, cheetah, panther}

In Exercises 38–43, use Fig. 2.29 to determine the sets.

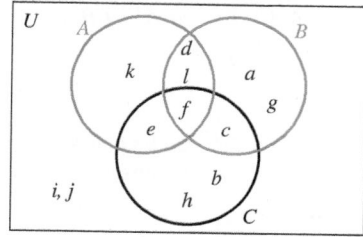

Figure 2.29

38. $A \cup C$

39. $A \cap B'$

40. $A \cup B \cup C$

41. $A \cap B \cap C$

42. $(A \cup B) \cap C$

43. $A - B'$

Construct a Venn diagram to determine whether the following statements are true for all sets A, B, and C.

44. $(A' \cup B')' = A \cap B$

45. $(A \cup B') \cup (A \cup C') = A \cup (B \cap C)'$

In Exercises 46–52, use the following table, which shows the amount of sugar, in grams (g), and caffeine, in milligrams (mg), in an 8-oz serving of selected beverages. Let the beverages listed represent the universal set.

Beverage	Sugar (grams, g)	Caffeine (milligrams, mg)
Mountain Dew	31	37
Coca-Cola	27	23
Pepsi	27	25
Sprite	26	0
Brewed coffee	0	108
Brewed tea	0	47
Orange juice	24	0
Grape juice	40	0
Gatorade	14	0
Red Bull	26	76
Arizona Lemon Iced Tea	24	15
Vitamin Water	13	17
Water	0	0

Source: International Food Information Council

Let A be the set of beverages that contain at least 20 g of sugar.

Let B be the set of beverages that contain at least 20 mg of caffeine.

Indicate in Fig. 2.30 in which region, I–IV, each of the following beverages belongs.

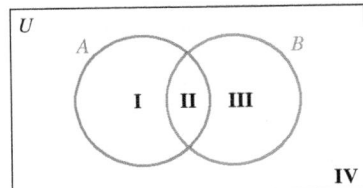

Figure 2.30

46. Pepsi

47. Brewed coffee

48. Orange juice

49. Vitamin Water

50. Gatorade

51. Mountain Dew

52. Red Bull

2.5

53. *Pizza Survey* A pizza chain was willing to pay $1 to each person interviewed about his or her likes and dislikes of types of pizza crust. Of the people interviewed, 200 liked thin crust, 270 liked thick crust, 70 liked both, and 50 did not like either type of crust. What was the total cost of the survey?

54. *Shopping Preferences* Visitors to a shopping mall in Atlanta, Georgia, were surveyed to determine their preference for shopping in wholesale warehouse stores. The following information was determined.

 58 shopped in BJ's Wholesale Club.

 49 shopped in Sam's Club.

 45 shopped in Costco.

 15 shopped in BJ's Wholesale Club and Sam's Club.

 16 shopped in BJ's Wholesale Club and Costco.

 12 shopped in Sam's Club and Costco.

  5 shopped in all three stores.

 17 did not shop in any of the three stores.

Construct a Venn diagram and then determine how many people

a) completed the survey.

b) shopped only in BJ's Wholesale Club.

c) shopped in BJ's Wholesale Club and Sam's Club, but not Costco.

d) shopped in BJ's Wholesale Club or Costco, but not Sam's Club.

55. *Electronic Devices* In a survey of 496 college students, it was determined that

 356 owned an iPad.

 293 owned a laptop.

 285 owned a gaming system.

 193 owned an iPad and a laptop.

 200 owned an iPad and a gaming system.

 139 owned a laptop and a gaming system.

  68 owned an iPad, a laptop, and a gaming system.

Construct a Venn diagram and determine how many students surveyed owned

a) only an iPad.

b) exactly one of these devices.

c) a laptop and a gaming system, but not an iPad.

d) an iPad or a laptop, but not a gaming system.

e) exactly two of these devices.

2.6

In Exercises 56 and 57, show that the sets are infinite by placing each set in a one-to-one correspondence with a proper subset of itself.

56. $\{2, 4, 6, 8, 10, \dots\}$

57. $\{3, 5, 7, 9, 11, \dots\}$

In Exercises 58 and 59, show that each set has cardinal number $\aleph_0$ by setting up a one-to-one correspondence between the set of counting numbers and the given set.

58. $\{5, 8, 11, 14, 17, \dots\}$

59. $\{4, 9, 14, 19, 24, \dots\}$

CHAPTER 2   *Test*

In Exercises 1–8, state whether each statement is true or false. If the statement is false, explain why.

1. $\{2, y, \triangle, \$\}$ is equivalent to $\{t, \#, 6, \square\}$.

2. $\{\text{apple, peach, orange, pear}\} = \{\text{bean, carrot, broccoli, cauliflower}\}$

3. $\{\text{star, moon, sun}\} \subset \{\text{star, moon, sun, planet}\}$

4. $7 \in \{x \mid x \in N \text{ and } x < 7\}$

5. $\{p, q, r, s\}$ has 15 subsets.

6. If $A \cap B = \{\ \}$, then A and B are disjoint sets.

7. For any set A, $A \cup A' = \{\ \}$.

8. For any set A, $A \cap U = A$.

In Exercises 9 and 10, use set

$$A = \{x \mid x \in N \text{ and } x < 12\}$$

9. Write set A in roster form.

10. Write a description of set A.

In Exercises 11–16, use the following information.

$$U = \{2, 4, 6, 8, 10, 12, 14\}$$
$$A = \{2, 4, 6, 8\}$$
$$B = \{6, 8, 10, 12\}$$
$$C = \{2, 10, 14\}$$

Determine the following.

11. $A \cap B$

12. $A \cup C'$

13. $A \cap (B \cap C')$

14. $n(A \cap B')$

15. $A - B$

16. $A \times C$

17. Using the sets provided for Exercises 11–16, draw a Venn diagram illustrating the relationship among the sets.

18. Use a Venn diagram to determine whether

$$A \cap (B \cup C') = (A \cap B) \cup (A \cap C')$$

for all sets A, B, and C. Show your work.

19. *Water Activities* A survey of 155 residents of Lake Placid were asked what kind of activities they participated in on a daily basis during the summer months. The following information was determined.

107 swam.
90 fished.
76 walked.
57 swam and fished.
54 swam and walked.
52 fished and walked.
35 swam, fished, and walked.

Construct a Venn diagram and then determine the number of residents who participated in

a) exactly two of these activities.

b) none of these activities.

c) at least one of these activities.

d) swimming and fishing, but not walking.

e) swimming or fishing, but not walking.

f) only fishing.

20. Show that the following set is infinite by setting up a one-to-one correspondence between the set and a proper subset of itself.

$$\{7, 8, 9, 10, \dots\}$$

◄ *Logic is used in the programming of the apps we use on our smartphones.*

3

Logic

What You Will Learn

- Statements, logical connectives, quantifiers, and compound statements that involve the words *not, and, or, if ... then,* and *if and only if*
- Truth tables for negations, conjunctions, and disjunctions
- Truth tables for conditionals and biconditionals and identifying self-contradictions, tautologies, and implications
- Equivalent statements, De Morgan's laws, and variations of conditional statements
- Symbolic arguments and standard forms of arguments
- Euler diagrams and syllogistic arguments
- Using logic to analyze switching circuits

Throughout this chapter we will see how logic can be used to analyze and understand statements we encounter in our everyday lives. Such statements may appear in advertisers' claims and political statements. Failure to completely understand such statements could cause us to be misled by advertisers or to be uninformed when voting for political candidates. We can also use logic to analyze and understand statements that appear in legal documents. Failure to understand legal documents such as car loans, home rental agreements, or credit card statements could lead to making poor financial decisions. In addition to helping us understand statements, studying logic helps us communicate effectively, make more convincing arguments, and develop patterns of reasoning for decision making. Logic is also used in the programming of most electronic devices including smartphones whenever we use an app.

SECTION 3.1   Statements and Logical Connectives

- Identify statements and logical connectives.
- Understand quantifiers and identify the negations of statements that contain quantifiers.
- Identify compound statements.
- Identify statements using the words *not, and, or, if–then,* and *if and only if.*

Advertisements often rely on spoken or written statements that are used to favorably portray the advertised product and form a convincing argument that will persuade us to purchase the product. Some familiar advertising statements are: *It keeps going and going and going; Once you pop, you can't stop; Impossible is nothing;* and *What happens here, stays here.* In this section, we will learn how to represent statements using logic symbols that may help us better understand the nature of the statement. We will use these symbols throughout the chapter to analyze more complicated statements.

Why This Is Important   Statements appear everywhere in our lives. In addition to statements in advertising, we see statements in legal documents, product instructions, and game rules. By using the symbols introduced in this section, we can represent such statements, and in turn we can better understand these statements.

We begin this chapter with a discussion of the history of logic. We also discuss the use of logic in language.

History

The ancient Greeks were the first people to systematically analyze the way humans think and arrive at conclusions. Aristotle (384–322 B.C.) organized the study of logic for the first time in a work called *Organon.* As a result of his work, Aristotle is called the father of logic. The logic from this period, called *Aristotelian logic,* has been taught and studied for more than 2000 years.

Since Aristotle's time, the study of logic has been continued by other great philosophers and mathematicians. Gottfried Wilhelm Leibniz (1646–1716) had a deep conviction that all mathematical and scientific concepts could be derived from logic. As a result, he became the first serious student of *symbolic logic.* A self-educated English mathematician, George Boole (1815–1864), is considered to be the founder of symbolic logic because of his impressive work in this area. Mathematician Charles Dodgson, better known as Lewis Carroll, incorporated many interesting ideas from logic into his books *Alice's Adventures in Wonderland* and *Through the Looking Glass* and his other children's stories.

The study of logic is also good preparation for other areas of mathematics. If you preview Chapter 11, on probability, you will see formulas for the probability of *A or B* and the probability of *A and B*, symbolized as $P(A \text{ or } B)$ and $P(A \text{ and } B)$, respectively. Special meanings of common words such as *or* and *and* apply to all areas of mathematics. The meanings of these and other special words are discussed in this chapter.

Logic in Language

In reading, writing, and speaking, we use many words such as *and, or,* and *if … then …* to connect thoughts. In logic we call these words *connectives.* How are these words interpreted in daily communication? A judge announces to a convicted offender, "I hereby sentence you to five months of community service *and* a fine of $100." In this case, we normally interpret the word *and* to indicate that *both* events will take place. That is, the person must perform community service and must also pay a fine.

Now suppose a judge states, "I sentence you to six months in prison *or* 10 months of community service." In this case, we interpret the connective *or* as meaning the convicted person must either spend the time in jail or perform community service, but not both. The word *or* in this case is the *exclusive or*. When the *exclusive or* is used, one or the other of the events can take place, but *not both*.

In a restaurant, a waiter asks, "May I interest you in a cup of soup or a sandwich?" This question offers three possibilities: You may order soup, you may order a sandwich, or you may order both soup and a sandwich. The *or* in this case is the *inclusive or*. When the *inclusive or* is used, one or the other, *or both* events can take place. *In this chapter, when we use the word* or *in a logic statement, it will mean the inclusive* or *unless stated otherwise.*

If–then statements are often used to relate two ideas, as in the bank policy statement "If the average daily balance is greater than $500, then there will be no service charge." If–then statements are also used to emphasize a point or add humor, as in the statement "If the Cubs win, then I will be a monkey's uncle."

Now let's look at logic from a mathematical point of view.

Statements, Logical Connectives, and Quantifiers

A sentence that can be judged either true or false is called a *statement*. Labeling a statement true or false is called *assigning a truth value* to the statement. Here are some examples of statements.

1. There are 100 inches in a yard.
2. The Empire State Building is in the state of New York.
3. The National Basketball Association (NBA) team from Portland, Oregon, is the Lakers.

In each case, we may need to consult a resource such as the Internet, but we can determine that each statement is either true or false. Statement 1 is false because there are 36 inches in a yard, not 100 inches. Statement 2 is true because the Empire State Building is in New York City, which is in the state of New York. Statement 3 is false because the NBA team from Portland, Oregon, is the Trailblazers, not the Lakers. The Lakers are from Los Angeles, California.

The three sentences discussed above are examples of *simple statements* because they convey one idea. Sentences combining two or more ideas that can be assigned a truth value are called *compound statements*. Compound statements are discussed shortly.

Sometimes it is necessary to change a statement to its opposite meaning. To do so, we use the *negation* of a statement. For example, the negation of the statement "Emily is at home" is "Emily is not at home." The negation of a true statement is always a false statement, and the negation of a false statement is always a true statement. We must use special caution when negating statements containing the words *all*, *none* (or *no*), and *some*. These words are referred to as *quantifiers*.

Consider the statement "All lakes contain fresh water." We know this statement is false because the Great Salt Lake in Utah contains salt water. Its negation must therefore be true. We may be tempted to write its negation as "No lake contains fresh water," but this statement is also false because Lake Superior contains fresh water. Therefore, "No lakes contain fresh water" is not the negation of "All lakes contain fresh water." The correct negation of "All lakes contain fresh water" is "Not all lakes contain fresh water" or "At least one lake does not contain fresh water" or "Some lakes do not contain fresh water." These statements all imply that at least one lake does not contain fresh water, which is a true statement.

Now consider the statement "No birds can swim." This statement is false because at least one bird, the penguin, can swim. Therefore, the negation of this statement must be true. We may be tempted to write the negation as "All birds can swim," but because this statement is also false it cannot be the negation. The correct negation

▲ *The Empire State Building in New York*

of the statement is "Some birds can swim" or "At least one bird can swim," each of which is a true statement.

Now let's consider statements involving the quantifier *some,* as in "Some students have a driver's license." This statement is true, meaning that at least one student has a driver's license. The negation of this statement must therefore be false. The negation is "No student has a driver's license," which is a false statement.

Consider the statement "Some students do not ride motorcycles." This statement is true because it means "At least one student does not ride a motorcycle." The negation of this statement must therefore be false. The negation is "All students ride motorcycles," which is a false statement.

The negation of quantified statements is summarized as follows:

Form of statement	Form of negation
All are.	Some are not.
None are.	Some are.
Some are.	None are.
Some are not.	All are.

The following diagram might help you to remember the statements and their negations:

The quantifiers diagonally opposite each other are the negations of each other.

Example 1   *Write Negations of Statements Involving Quantifiers*

Write the negation of each statement.

a) All cocker spaniels are dogs.

b) Some insects are mammals.

c) Some days are not Mondays.

d) No coins have the image of Abraham Lincoln.

Solution

a) The statement "*All* cocker spaniels *are* dogs" is a true statement and is of the form "All are." Referring to the text above, to negate a statement of the form "All are," we write a statement of the form "Some are not." Thus, the negation of the statement "*All* cocker spaniels *are* dogs" is "*Some* cocker spaniels *are not* dogs," which is a false statement.

b) The statement "*Some* insects *are* mammals" is a false statement and is of the form "Some are." To negate a statement of the form "Some are," we write a statement of the form "None are." Thus, the negation of the statement "*Some* insects *are* mammals" is "*None* of the insects *are* mammals" or "No insects are mammals," which are true statements.

c) The statement "*Some* days *are not* Mondays" is a true statement and is of the form "Some are not." To negate a statement of the form "Some are not," we write a statement of the form "All are." Thus, the negation of the statement "*Some* days *are not* Mondays" is "*All* days *are* Mondays," which is a false statement.

d) The statement "No coins have the image of Abraham Lincoln" is false because pennies have the image of Abraham Lincoln. Although the statement "No coins have the image of Abraham Lincoln" does not exactly fit any of the forms discussed in the text above, it can be considered to be of the form "None are." To negate a statement of the form "None are," we write a statement of the form "Some are." Thus, the negation of the statement "No coins have the image of Abraham Lincoln" is "Some coins have the image of Abraham Lincoln," which is a true statement.   ∎

Sudoku

	1		4	8		5	6	
5					9	8		
	3				1	4		7
8	2			9		1		
6			1		4			9
		3		6			4	5
9		1	5				2	
		7	2					4
	5	2		7	8		3	

Solving puzzles requires us to use logic. Sudoku is a puzzle that originated in Japan and continues to gain popularity worldwide. To solve the puzzle, you need to place every digit from 1 to 9 exactly one time in each row, in each column, and in each of the nine 3 by 3 boxes. For more information and a daily puzzle see www.websudoku.com. Furthermore, there are many apps that can be used to play Sudoku on your tablet or smartphone. The solution to the puzzle above can be found in the Answers section in the back of this book. For an additional puzzle see Exercise 82 on page 102.

Compound Statements

Statements consisting of two or more simple statements are called **compound statements**. The connectives often used to join two simple statements are

and,   or,   if … then …,   if and only if

In addition, we consider a simple statement that has been negated to be a compound statement. The word *not* is generally used to negate a statement.

To reduce the amount of writing in logic, it is common to represent each simple statement with a lowercase letter. For example, suppose we are discussing the simple statement "Leland is a farmer." Instead of writing "Leland is a farmer" over and over again, we can let p represent the statement "Leland is a farmer." Thereafter we can simply refer to the statement with the letter p. It is customary to use the letters p, q, r, and s to represent simple statements, but other letters may be used instead. Let's now look at the connectives used to make compound statements.

Not Statements

The negation is symbolized by $\sim$ and read "not." For example, the negation of the statement "Steve is a college student" is "Steve is not a college student." If p represents the simple statement "Steve is a college student," then $\sim p$ represents the compound statement "Steve is not a college student." For any statement p, $\sim(\sim p) = p$. For example, the negation of the statement "Steve is not a college student" is "Steve is a college student."

Consider the statement "Inga is not at home." This statement contains the word *not*, which indicates that it is a negation. To write this statement symbolically, we let p represent "Inga *is* at home." Then $\sim p$ would be "Inga is not at home." *We will use this convention of letting letters such as p, q, or r represent statements that are not negated. We will indicate that a statement is negated with the negation symbol, $\sim$.*

And Statements

The *conjunction* is symbolized by $\wedge$ and read "and." The $\wedge$ looks like an A (for And) with the bar missing. Let p and q represent the simple statements.

p: You will perform 5 months of community service.

q: You will pay a $100 fine.

Then the following is the conjunction written in symbolic form.

$\underbrace{\text{You will perform 5 months of community service}}_{p}$ $\underbrace{\text{and}}_{\wedge}$ $\underbrace{\text{you will pay a \$100 fine.}}_{q.}$

The conjunction is generally expressed as *and*. Other words sometimes used to express a conjunction are *but, however,* and *nevertheless*.

Example 2 *Write a Conjunction*

The following statement involves Jon Batiste, an American jazz musician, band leader, and television personality. Write the following conjunction in symbolic form.

Jon Batiste is the leader of the band Stay Human, but Jon Batiste is not playing the Chicago Jazz Festival.

Solution Let l and p represent the following simple statements.

l: Jon Batiste is the leader of the band Stay Human.

p: Jon Batiste is playing the Chicago Jazz Festival.

In symbolic form, the compound statement is $l \wedge \sim p$. ∎

In Example 2, the given statement "Jon Batiste is the leader of the band Stay Human, but Jon Batiste is not playing the Chicago Jazz Festival" could also have been given as "Jon Batiste is the leader of the band Stay Human, but *he* is not playing the Chicago Jazz Festival." It should be clear that the word *he* refers to Jon Batiste. Therefore, the statement "Jon Batiste is the leader of the band Stay Human, but *he* is not playing the Chicago Jazz Festival" can also be symbolized as $l \wedge \sim p$.

Or Statements

The *disjunction* is symbolized by $\vee$ and read "or." The *or* we use in this book (except where indicated in the exercise sets) is the *inclusive or* described on page 93.

Example 3 *Write a Disjunction*

Let

$$p: \quad \text{Camila will take chemistry.}$$
$$q: \quad \text{Camila will take French.}$$

Write the following statements in symbolic form.

a) Camila will take chemistry or Camila will take French.

b) Camila will not take chemistry or Camila will not take French.

c) Camila will take French or Camila will not take chemistry.

Solution

a) $p \vee q$ b) $\sim p \vee \sim q$ c) $q \vee \sim p$ ∎

Because *or* represents the *inclusive or*, the statement "Camila will take chemistry or Camila will take French" in Example 3(a) may mean that Camila will take chemistry, or that Camila will take French, or that Camila will take both chemistry *and* French. The statement in Example 3(a) could also be written as "Camila will take chemistry or French."

When a compound statement contains more than one connective, a comma can be used to indicate which simple statements are to be grouped together. When we write the compound statement symbolically, *the simple statements on the same side of the comma are to be grouped together within parentheses.*

For example, "Pink is a singer (p) or Jennifer Aniston is an actress (j), and Dallas is in Texas (d)" is written $(p \vee j) \wedge d$. Note that the p and j are both on the same side of the comma in the written statement. They are therefore grouped together within parentheses. The statement "Pink is a singer, or Jennifer Aniston is an actress and Dallas is in Texas" is written $p \vee (j \wedge d)$. In this case, j and d are on the same side of the comma and are therefore grouped together within parentheses.

Example 4 *Understand How Commas Are Used to Group Statements*

Let

$$p: \quad \text{Dinner includes soup.}$$
$$q: \quad \text{Dinner includes salad.}$$
$$r: \quad \text{Dinner includes the vegetable of the day.}$$

Write the following statements in symbolic form and indicate whether the statement is a negation, conjunction, or disjunction.

a) Dinner includes soup, and salad or the vegetable of the day.

b) Dinner includes soup and salad, or the vegetable of the day.

Solution

a) The comma tells us to group the statement "Dinner includes salad" with the statement "Dinner includes the vegetable of the day." Note that both statements are on the same side of the comma. The statement in symbolic form is $p \wedge (q \vee r)$.

   In mathematics, we always evaluate the information within the parentheses first. Since the conjunction, $\wedge$, is outside the parentheses and is evaluated *last,* this statement is considered a *conjunction*.

b) The comma tells us to group the statement "Dinner includes soup" with the statement "Dinner includes salad." Note that both statements are on the same side of the comma. The statement in symbolic form is $(p \wedge q) \vee r$. Since the disjunction, $\vee$, is outside the parentheses and is evaluated *last,* this statement is considered a *disjunction*. ∎

The information provided in Example 4 is summarized below.

Statement	Symbolic representation	Type of statement
Dinner includes soup, and salad or the vegetable of the day.	$p \wedge (q \vee r)$	conjunction
Dinner includes soup and salad, or the vegetable of the day.	$(p \wedge q) \vee r$	disjunction

A negation symbol has the effect of negating only the statement that directly follows it. To negate a compound statement, we must use parentheses. When a negation symbol is placed in front of a statement in parentheses, it negates the entire statement in parentheses. The negation symbol in this case is read, "It is not true that …" or "It is false that …"

Example 5 *Change Symbolic Statements into Words*

Let

> p: The house is for sale.
>
> q: We can afford to buy the house.

Write the following symbolic statements in words.

a) $p \wedge {\sim}q$ b) ${\sim}p \vee {\sim}q$ c) ${\sim}(p \wedge q)$

Solution

a) The house is for sale and we cannot afford to buy the house.
b) The house is not for sale or we cannot afford to buy the house.
c) It is false that the house is for sale and we can afford to buy the house. ∎

Recall that the word *but* may also be used in a conjunction. Therefore, Example 5(a) could also be written "The house is for sale, *but* we cannot afford to buy the house."

Part (b) of Example 5 is a disjunction, since it can be written $({\sim}p) \vee ({\sim}q)$. Part (c), which is ${\sim}(p \wedge q)$, is a negation, since the negation symbol negates the entire statement within parentheses. The similarity of these two statements is discussed in Section 3.4.

Occasionally, we come across a *neither–nor* statement, such as "John is neither handsome nor rich." This statement means that John is not handsome *and* John is not rich. If p represents "John is handsome" and q represents "John is rich," this statement is symbolized by ${\sim}p \wedge {\sim}q$.

If–Then Statements

The *conditional* is symbolized by $\rightarrow$ and is read "if–then." The statement $p \rightarrow q$ is read "If p, then q."* The conditional statement consists of two parts: The part that precedes the arrow is the *antecedent*, and the part that follows the arrow is the *consequent*.† In the conditional statement $p \rightarrow q$, the p is the antecedent and the q is the consequent.

In the conditional statement $\sim(p \vee q) \rightarrow (p \wedge q)$, the antecedent is $\sim(p \vee q)$ and the consequent is $(p \wedge q)$. An example of a conditional statement is "If you drink your milk, then you will grow up to be healthy." A conditional symbol may be placed between any two statements even if the statements are not related.

Sometimes the word *then* in a conditional statement is not explicitly stated. For example, the statement "If you get an A, I will buy you a car" is a conditional statement because it actually means "If you get an A, then I will buy you a car."

Example 6 *Write Conditional Statements*

Let

    s: The shoes were purchased at the thrift store.

    e: The outfit was expensive.

Write the following statements symbolically.

a) If the shoes were purchased at the thrift store, then the outfit was not expensive.

b) If the outfit was expensive, then the shoes were not purchased at the thrift store.

c) It is false that if the shoes were not purchased at the thrift store, then the outfit was not expensive.

Solution

a) $s \rightarrow \sim e$ b) $e \rightarrow \sim s$ c) $\sim(\sim s \rightarrow \sim e)$

Example 7 *Use Commas When Writing a Symbolic Statement in Words*

Let

    p: Jorge is enrolled in calculus.

    q: Jorge's major is criminal justice.

    r: Jorge's major is engineering.

Write the following symbolic statements in words and indicate whether the statement is a negation, conjunction, disjunction, or conditional.

a) $(q \rightarrow \sim p) \vee r$ b) $q \rightarrow (\sim p \vee r)$

Solution The parentheses indicate where to place the commas in the sentences.

a) "If Jorge's major is criminal justice then Jorge is not enrolled in calculus, or Jorge's major is engineering." This statement is a disjunction because $\vee$ is outside the parentheses.

b) "If Jorge's major is criminal justice, then Jorge is not enrolled in calculus or Jorge's major is engineering." This statement is a conditional because $\rightarrow$ is outside the parentheses.

---

*Some books indicate that $p \rightarrow q$ may also be read "p implies q." Many higher-level mathematics books, however, indicate that $p \rightarrow q$ may be read "p implies q" only under certain conditions. Implications are discussed in Section 3.3.

†Some books refer to the antecedent as the hypothesis or premise and the consequent as the conclusion.

If and Only If Statements

The *biconditional* is symbolized by ↔ and is read "if and only if." The phrase *if and only if* is sometimes abbreviated as "iff." The statement $p \leftrightarrow q$ is read "p if and only if q."

Example 8 Write Statements Using the Biconditional

Let

> p: Alex plays goalie on the lacrosse team.
>
> q: The Huskies win the Champion's Cup.

Write the following symbolic statements in words.

a) $p \leftrightarrow q$ b) $q \leftrightarrow \sim p$ c) $\sim (p \leftrightarrow \sim q)$

Solution

a) Alex plays goalie on the lacrosse team if and only if the Huskies win the Champion's Cup.

b) The Huskies win the Champion's Cup if and only if Alex does not play goalie on the lacrosse team.

c) It is false that Alex plays goalie on the lacrosse team if and only if the Huskies do not win the Champion's Cup. ∎

You will learn later that $p \leftrightarrow q$ means the same as $(p \rightarrow q) \wedge (q \rightarrow p)$. Therefore, the statement "I will go to college if and only if I can pay the tuition" has the same logical meaning as "If I go to college then I can pay the tuition, and if I can pay the tuition then I will go to college."

A summary of the connectives discussed in this section is given in Table 3.1.

Table 3.1 Logical Connectives

Formal Name	Symbol	Read	Symbolic Form
Negation	~	"Not"	$\sim p$
Conjunction	∧	"And"	$p \wedge q$
Disjunction	∨	"Or"	$p \vee q$
Conditional	→	"If–then"	$p \rightarrow q$
Biconditional	↔	"If and only if"	$p \leftrightarrow q$

SECTION 3.1   Exercises

Warm Up Exercises

In Exercises 1–6, fill in the blanks with an appropriate word, phrase, or symbol(s).

1. A sentence that can be judged either true or false is called a(n) _____.

2. A statement that conveys only one idea is called a(n) _____ statement.

3. A statement that consists of two or more simple statements is called a(n) _____ statement.

4. Words such as *all, none* (or *no*), and *some* are examples of _____.

5. a) The negation is symbolized by ~ and is read "_____."

   b) The conjunction is symbolized by ∧ and is read "_____."

   c) The disjunction is symbolized by ∨ and is read "_____."

6. a) The conditional is symbolized by → and is read "_____."

   b) The biconditional is symbolized by ↔ and is read "_____."

Practice the Skills/Problem Solving

In Exercises 7–16, indicate whether the statement is a simple statement or a compound statement. If it is a compound statement, indicate whether it is a negation, conjunction, disjunction, conditional, or biconditional by using both the word and its appropriate symbol (for example, "a negation," ~).

7. John is taking Liberal Arts Mathematics during the fall semester.

8. If you write a history report, then the life of Benjamin Banneker would be a good topic.

9. Time will go backward if and only if you travel faster than the speed of light.

10. Miles Davis did not play rap music.

11. Tony Stark makes an appearance in *Spider Man* and *The Incredible Hulk*.

12. The book was neither a novel nor an autobiography.

13. If Andres plays in the World Cup, then he will play for Spain.

14. The Thunder will make the playoffs if and only if they win their last five games.

15. It is false that Jeffery is a high school teacher and a grade school teacher.

16. The hurricane did $400,000 worth of damage to DeSoto County.

In Exercises 17–26, write the negation of the statement.

17. All Eco Sun scooters are made by Amigo.

18. No freshmen can participate in the graduation ceremonies.

19. Some turtles do not have claws.

20. All Apple smartphones run on an IOS operating system.

21. No bicycles have three wheels.

22. All horses have manes.

23. Some pedestrians are in the crosswalk.

24. Some teachers are not professors.

25. No mountain climbers are teachers.

26. Some vitamins contain sugar.

In Exercises 27–32, write the statement in symbolic form.

Let
    p: A panther has a long tail.
    q: A bobcat can purr.

27. A panther does not have a long tail.

28. A panther has a long tail and a bobcat can purr.

29. A bobcat cannot purr or a panther does not have a long tail.

30. A bobcat cannot purr if and only if a panther does not have a long tail.

31. If a panther does not have a long tail, then a bobcat cannot purr.

32. A bobcat cannot purr, but a panther has a long tail.

In Exercises 33–38, write the statement in symbolic form.

Let
    p: The burrito contains sriracha.
    q: The tacos are vegan.

33. The burrito contains sriracha and the tacos are not vegan.

34. If the burrito does not contain sriracha, then the tacos are vegan.

35. The tacos are not vegan if and only if the burrito contains sriracha.

36. The burrito does not contain sriracha or the tacos are not vegan.

37. It is false that the burrito contains sriracha or the tacos are vegan.

38. It is false that if the tacos are not vegan then the burrito contains sriracha.

In Exercises 39–48, write the compound statement in words.

Let
    p: Joe has an iPad.
    q: Brie has a MacBook.

57. $(p \lor q) \land \sim r$ **58.** $(p \land q) \lor r$

59. $\sim p \land (q \lor r)$ **60.** $(q \rightarrow p) \lor r$

61. $\sim r \rightarrow (q \land p)$ **62.** $(q \land r) \rightarrow p$

63. $(q \rightarrow r) \land p$ **64.** $\sim p \rightarrow (q \lor r)$

65. $(q \leftrightarrow p) \land r$ **66.** $q \rightarrow (p \leftrightarrow r)$

39. $\sim q$ **40.** $\sim p$

41. $p \land q$ **42.** $q \lor p$

43. $\sim p \rightarrow q$ **44.** $\sim p \leftrightarrow \sim q$

45. $\sim p \lor \sim q$ **46.** $\sim (q \lor p)$

47. $\sim (p \land q)$ **48.** $\sim p \land \sim q$

In Exercises 49–56, write the statements in symbolic form.

Let

    p: The temperature is 90°.
    q: The air conditioner is working.
    r: The apartment is vacant.

49. The temperature is 90° and the air conditioner is not working, and the apartment is vacant.

50. The temperature is not 90° and the air conditioner is working, but the apartment is vacant.

51. The temperature is 90° and the air conditioner is working, or the apartment is vacant.

52. If the temperature is 90°, then the air conditioner is working or the apartment is not vacant.

53. If the apartment is vacant and the air conditioner is working, then the temperature is 90°.

54. The temperature is not 90° if and only if the air conditioner is not working, or the apartment is not vacant.

55. The apartment is vacant if and only if the air conditioner is working, and the temperature is 90°.

56. If the air conditioner is working, then the temperature is 90° if and only if the apartment is vacant.

In Exercises 57–66, write each symbolic statement in words.

Let

    p: The water is 70°.
    q: The sun is shining.
    r: We go swimming.

Dinner Menu *In Exercises 67–70, use the following information to arrive at your answers. Many restaurant dinner menus include statements such as the following: All dinners are served with a choice of: soup or salad, and potatoes or pasta, and carrots or peas. Which of the following selections are permissible? If a selection is not permissible, explain why. See the discussion of the* exclusive or *on page 93.*

Welcome to
ANTONIO'S RESTAURANTE
"All dinners are served
with a choice of:
soup or salad, potatoes
or pasta, and carrots or peas"

67. Soup, salad, and peas

68. Salad, pasta, and carrots

69. Soup, potatoes, pasta, and peas

70. Soup, pasta, and potatoes

In Exercises 71–79, (a) select letters to represent the simple statements and write each statement symbolically by using parentheses if necessary and (b) indicate whether the statement is a negation, conjunction, disjunction, conditional, or biconditional.

71. Johnny started the bonfire and Jaci did not forget the marshmallows.

72. If the meeting is in Las Vegas, then we can see a show or we can go to the casino.

▲ Las Vegas

73. It is false that if you work out then you will not gain weight.

74. If dinner is ready then we can eat, or we cannot go to the restaurant.

75. If the food has fiber or the food has vitamins, then you will be healthy.

76. If Corliss is teaching then Faye is in the math lab, if and only if it is not a weekend.

77. You may take this course, if and only if you did not fail the previous course or you passed the placement test.

78. If the car has gas and the battery is charged, then the car will start.

79. The classroom is empty if and only if it is the weekend, or it is 7 A.M.

Challenge Problems/Group Activities

80. *An Ancient Question* If Zeus could do anything, could he build a wall that he could not jump over? Explain your answer.

81. a) Make up three simple statements and label them p, q, and r. Then write compound statements to represent $(p \vee q) \wedge r$ and $p \vee (q \wedge r)$.

    b) Do you think that the statements for $(p \vee q) \wedge r$ and $p \vee (q \wedge r)$ mean the same thing? Explain.

Recreational Mathematics

82. *Sudoku* Refer to the *Recreational Mathematics* on page 95. Complete the following Sudoku puzzle.

	8			3				
	9		6	1			4	
	5	1	9			8	6	3
		6				4	8	
5								1
	2	4				5		
4	3	5			8	1	9	
2			5	1			7	
			2			6		

Research Activities

83. *Legal Documents* Obtain a legal document such as a will or rental agreement and copy one page of the document. Circle every connective used. Then list the number of times each connective appeared. Be sure to include conditional statements from which the word *then* was omitted from the sentence.

84. *George Boole* Write a report on the life and accomplishments of George Boole, who was an important contributor to the development of logic. In your report, indicate how his work eventually led to the development of the computer.

85. *Mind Your P's and Q's* The study of logic often involves the use of the letters p and q to represent statements. Research the use of the letters p and q in the study of logic. Also include a brief description of the various possible sources of the phrase "mind your p's and q's."

SECTION 3.2 — Truth Tables for Negation, Conjunction, and Disjunction

Upon completion of this section, you will be able to:

- Construct truth tables for statements involving negations, conjunctions, and disjunctions.
- Understand an alternate method for constructing truth tables.
- Determine truth values of statements without constructing truth tables.

Consider the following statement: Yogurt contains probiotics and whole wheat bread contains antioxidants. Under what conditions can the statement be considered true? Under what conditions can the statement be considered false? In this section, we will introduce a tool used to help us analyze such statements.

Why This Is Important We often encounter statements that include the words *not*, *and*, and *or*. These statements are called negations, conjunctions, and disjunctions, respectively. Understanding when such statements are true and when they are false can affect our daily lives. Making informed decisions when encountering these statements can have long-term effects on many aspects of our future.

A *truth table* is a device used to determine when a compound statement is true or false. Five basic truth tables are used in constructing other truth tables. Three are discussed in this section (Tables 3.2, 3.4, and 3.5), and two are discussed in the next section. Section 3.5 uses truth tables in determining whether a logical argument is valid or invalid.

Negation

The first truth table is for *negation*. If p is a true statement, then the negation of p, "not p," is a false statement. If p is a false statement, then "not p" is a true statement. For example, if the statement "The shirt is blue" is true, then the statement "The shirt is not blue" is false. These relationships are summarized in Table 3.2. For a simple statement, there are exactly two true–false cases, as shown.

Table 3.2 Negation

	p	$\sim p$
Case 1	T	F
Case 2	F	T

Conjunction

If a compound statement consists of two simple statements p and q, there are four possible cases, as illustrated in Table 3.3. Consider the statement "The test is today and the test covers Chapter 5." The simple statement "The test is today" has two possible truth values, true or false. The simple statement "The test covers Chapter 5" also has two possible truth values, true or false. Thus, for these two simple statements there are four distinct possible true–false arrangements. Whenever we construct a truth table for a compound statement that consists of two simple statements, we begin by listing the four true–false cases shown in Table 3.3.

Table 3.3

	p	q
Case 1	T	T
Case 2	T	F
Case 3	F	T
Case 4	F	F

To illustrate the conjunction, consider the following situation. You have recently purchased a new house. To decorate it, you ordered a new carpet and new furniture from the same store. You explain to the salesperson that the carpet must be delivered before the furniture. He promises that the carpet will be delivered on Thursday and that the furniture will be delivered on Friday.

To help determine whether the salesperson kept his promise, we assign letters to each simple statement. Let p be "The carpet will be delivered on Thursday" and q be "The furniture will be delivered on Friday." The salesperson's statement written in symbolic form is $p \land q$. There are four possible true–false situations to be considered (Table 3.4).

Table 3.4 Conjunction

	p	q	$p \land q$
Case 1	T	T	T
Case 2	T	F	F
Case 3	F	T	F
Case 4	F	F	F

CASE 1: p is true and q is true. The carpet is delivered on Thursday and the furniture is delivered on Friday. The salesperson has kept his promise and the compound statement is true. Thus, we put a T in the $p \land q$ column.

CASE 2: p is true and q is false. The carpet is delivered on Thursday but the furniture is not delivered on Friday. Since the furniture was not delivered as promised, the compound statement is false. Thus, we put an F in the $p \land q$ column.

CASE 3: p is false and q is true. The carpet is not delivered on Thursday but the furniture is delivered on Friday. Since the carpet was not delivered on Thursday as promised, the compound statement is false. Thus, we put an F in the $p \land q$ column.

CASE 4: p is false and q is false. The carpet is not delivered on Thursday and the furniture is not delivered on Friday. Since the carpet and furniture were not delivered as promised, the compound statement is false. Thus, we put an F in the $p \land q$ column.

Examining the four cases, we see that in only one case did the salesperson keep his promise: in case 1. Therefore, case 1 (TT) is true. In cases 2, 3, and 4, the salesperson did not keep his promise and the compound statement is false. As this example illustrates, an *and* statement is true only when both simple statements are true. The results are summarized in Table 3.4, the truth table for the conjunction.

> The **conjunction** $p \wedge q$ is true only when both p and q are true.

Disjunction

Consider the job description in the margin that describes two job requirements. Who qualifies for the job? To help analyze the statement, translate it into symbolic form. Let p be "A requirement for the job is a two-year college degree in civil technology" and q be "A requirement for the job is five years of related experience." The statement in symbolic form is $p \vee q$. For the two simple statements, there are four distinct cases (see Table 3.5).

CASE 1: p is true and q is true. A candidate has a two-year college degree in civil technology and five years of related experience. The candidate has both requirements and qualifies for the job. Consider qualifying for the job as a true statement and not qualifying as a false statement. Since the candidate qualifies for the job, we put a T in the $p \vee q$ column.

CASE 2: p is true and q is false. A candidate has a two-year college degree in civil technology but does not have five years of related experience. The candidate still qualifies for the job with only the two-year college degree. Thus, we put a T in the $p \vee q$ column.

CASE 3: p is false and q is true. The candidate does not have a two-year college degree in civil technology but does have five years of related experience. The candidate still qualifies for the job with only the five years of related experience. Thus, we put a T in the $p \vee q$ column.

CASE 4: p is false and q is false. The candidate does not have a two-year college degree in civil technology and does not have five years of related experience. The candidate does not meet either of the two requirements and therefore does not qualify for the job. Thus, we put an F in the $p \vee q$ column.

In examining the four cases, we see that there is only one case in which the candidate does not qualify for the job: case 4 (FF). As this example indicates, an *or* statement will be true in every case, except when both simple statements are false. The results are summarized in Table 3.5, the truth table for the disjunction.

Table 3.5   Disjunction

p	q	$p \vee q$
T	T	T
T	F	T
F	T	T
F	F	F

> The **disjunction** $p \vee q$ is true when either p is true, q is true, or both p and q are true.

The disjunction $p \vee q$ is false only when p and q are both false.

Constructing Truth Tables for Statements Involving Negations, Conjunctions, and Disjunctions

We will now construct additional truth tables for statements involving the negation, conjunction, and disjunction. We summarize these compound statements below.

Negation, Conjunction, and Disjunction

- **Negation,** $\sim p$, is read "not p." If p is true, then $\sim p$ is false; if p is false, then $\sim p$ is true. In other words, $\sim p$ will always have the *opposite* truth value of p.
- **Conjunction,** $p \wedge q$, is read "p and q." $p \wedge q$ is true only when both p and q are true.
- **Disjunction,** $p \vee q$, is read "p or q." $p \vee q$ is true when either p is true, q is true, or both p and q are true. In other words, $p \vee q$ is false only when both p and q are false.

We will discuss two methods for constructing truth tables. Although the two methods produce tables that will look different, the answer columns will be the same regardless of which method you use.

Example 1 *Construct a Truth Table*

Construct a truth table for $p \wedge \sim q$.

Solution Because there are two statements, p and q, construct a truth table with four cases; see Table 3.6(a). Then write the truth values under the p in the compound statement and label this column 1, as in Table 3.6(b). Copy these truth values directly from the p column on the left. Write the corresponding truth values under the q in the compound statement and call this column 2, as in Table 3.6(c). Copy the truth values for column 2 directly from the q column on the left. Now find the truth values of $\sim q$ by negating the truth values in column 2 and call this

Table 3.6

(a)

	p	q	$p \wedge \sim q$
Case 1	T	T	
Case 2	T	F	
Case 3	F	T	
Case 4	F	F	

(b)

p	q	$p \wedge \sim q$
T	T	T
T	F	T
F	T	F
F	F	F
		1

(c)

p	q	p	$\wedge$	$\sim q$
T	T	T		T
T	F	T		F
F	T	F		T
F	F	F		F
		1		2

(d)

p	q	p	$\wedge$	$\sim$	q
T	T	T		F	T
T	F	T		T	F
F	T	F		F	T
F	F	F		T	F
		1		3	2

(e)

p	q	p	$\wedge$	$\sim$	q
T	T	T	F	F	T
T	F	T	T	T	F
F	T	F	F	F	T
F	F	F	F	T	F
		1	4	3	2

Answer column

column 3, as in Table 3.6(d). Use the conjunction table, Table 3.4, and the entries in the columns labeled 1 and 3 to complete the column labeled 4, as in Table 3.6(e). The results in column 4 are obtained as follows:

Row 1: T ∧ F is F.   Row 2: T ∧ T is T.
Row 3: F ∧ F is F.   Row 4: F ∧ T is F.

The answer is always the last column completed. The columns labeled 1, 2, and 3 assist in arriving at the answer labeled column 4.   ■

The statement $p \land \sim q$ in Example 1 actually means $p \land (\sim q)$. In the future, instead of listing a column for q and a separate column for its negation, we will make one column for $\sim q$, which will have the opposite values of those in the q column on the left. Similarly, when we evaluate $\sim p$, we will use the opposite values of those in the p column on the left. This procedure is illustrated in Example 2.

In Example 1, we spoke about *cases* and also *columns*. Consider Table 3.6(e) on page 105. This table has four cases indicated by the four different rows of the two left-hand (unnumbered) columns. The four *cases* are TT, TF, FT, and FF. In every truth table with two letters, we list the four cases (the first two columns) first. Then we complete the remaining columns in the truth table. In Table 3.6(e), after completing the two left-hand columns, we complete the remaining columns in the order indicated by the numbers below the columns. We will continue to place numbers below the columns to show the order in which the columns are completed.

In discussion of the truth table in Example 2, and all following truth tables, if we say column 1, it means the column labeled 1. Column 2 will mean the column labeled 2, and so on.

Timely Tip

When constructing truth tables it is very important to keep your entries in neat columns and rows. If you are using lined paper, put only one row of the table on each line. If you are not using lined paper, using a straightedge may help you correctly enter the information into the truth table's rows and columns.

▲ *An Amethyst Crystal*

Example 2 *Construct and Interpret a Truth Table*

a) Construct a truth table for the following statement: The lamp is not a Himalayan salt lamp and the crystal is not amethyst.

b) Under what conditions will the compound statement be true?

c) Suppose "The lamp is a Himalayan salt lamp" is a false statement and "The crystal is amethyst" is a true statement. Is the compound statement given in part (a) true or false?

Solution

a) First, write the simple statements in symbolic form by using simple statements that are not negations.

    Let

p: The lamp is a Himalayan salt lamp.

q: The crystal is amethyst.

Therefore, the compound statement may be written $\sim p \land \sim q$. Now construct a truth table with four cases, as shown in Table 3.7.

Table 3.7

p	q	$\sim p$	$\land$	$\sim q$
T	T	F	F	F
T	F	F	F	T
F	T	T	Ⓕ	F
F	F	T	Ⓣ	T
		1	3	2

Fill in the column labeled 1 by negating the truth values under p on the far left. Fill in the column labeled 2 by negating the values under q in the second column from the left. Fill in the column labeled 3 by using the columns labeled 1 and 2 and the definition of conjunction.

In the first row, to determine the entry for column 3, we use false for $\sim p$ and false for $\sim q$. Since false $\wedge$ false is false (see case 4 of Table 3.4 on page 103), we place an F in column 3, row 1. In the second row, we use false for $\sim p$ and true for $\sim q$. Since false $\wedge$ true is false (see case 3 of Table 3.4), we place an F in column 3, row 2. In the third row, we use true for $\sim p$ and false for $\sim q$. Since true $\wedge$ false is false (see case 2 of Table 3.4), we place an F in column 3, row 3. In the fourth row, we use true for $\sim p$ and true for $\sim q$. Since true $\wedge$ true is true (see case 1 of Table 3.4), we place a T in column 3, row 4.

b) The compound statement in part (a) will be true only in case 4 from Table 3.7 (circled in blue) when both simple statements, p and q, are false—that is, when the lamp is not a Himalayan salt lamp and the crystal is not an amethyst.

c) We are told that p, "The lamp is a Himalayan salt lamp," is a false statement and that q, "The crystal is amethyst," is a true statement. From Table 3.7, we can determine that when p is false and q is true, the compound statement, case 3 (circled in red), is false. ∎

Example 3   *Truth Table with a Negation*

Construct a truth table for $\sim(\sim q \vee p)$.

Solution First construct the standard truth table listing the four cases. Then work within parentheses. The order to be followed is indicated by the numbers below the columns (see Table 3.8). Under $\sim q$, column 1, write the negation of the q column. Then, in column 2, copy the values from the p column. Next, complete the *or* column, column 3, using columns 1 and 2 and the truth table for the disjunction (see Table 3.5 on page 104). The *or* column is false only when both statements are false, as in case 3. Finally, negate the values in the *or* column, column 3, and place these negated values in column 4. By examining the truth table you can see that the compound statement $\sim(\sim q \vee p)$ is true only in case 3, that is, when p is false and q is true. ∎

Table 3.8

p	q	$\sim$	($\sim q$	$\vee$	p)
T	T	F	F	T	T
T	F	F	T	T	T
F	T	T	F	F	F
F	F	F	T	T	F
		4	1	3	2

PROCEDURE   CONSTRUCTING TRUTH TABLES

1. Study the compound statement and determine whether it is a negation, conjunction, disjunction, conditional, or biconditional statement, as was done in Section 3.1. The answer to the truth table will appear under $\sim$ if the statement is a negation, under $\wedge$ if the statement is a conjunction, under $\vee$ if the statement is a disjunction, under $\rightarrow$ if the statement is a conditional, and under $\leftrightarrow$ if the statement is a biconditional.

2. Complete the columns under the simple statements, p, q, r, and their negations, $\sim p$, $\sim q$, $\sim r$. If there are nested parentheses (one pair of parentheses within another pair), work with the innermost pair first.

3. Complete the column under the connective within the parentheses, if present. You will use the truth values of the connective in determining the final answer in Step 5.

4. Complete the column under any remaining statements and their negations.

5. Complete the column under any remaining connectives. Recall that the answer will appear under the column determined in Step 1. If the statement is a conjunction, disjunction, conditional, or biconditional, you will obtain the truth values for the connective by using the last column completed on the left side and on the right side of the connective. If the statement is a negation, you will obtain the truth values by negating the truth values of the last column completed within the grouping symbols on the right side of the negation. Be sure to circle or highlight your answer column or number the columns in the order they were completed.

Table 3.9

p	q	$(p$	$\wedge$	$\sim q)$	$\vee$	$\sim p$
T	T	T	F	F	F	F
T	F	T	T	T	T	F
F	T	F	F	F	T	T
F	F	F	F	T	T	T
		1	3	2	5	4

Example 4   *Use the General Procedure to Construct a Truth Table*

Construct a truth table for the statement $(p \wedge \sim q) \vee \sim p$.

Solution This statement is a disjunction, so the answer will be under the disjunction symbol, $\vee$. To begin, we complete the columns under p and $\sim q$ within the parentheses and call these columns 1 and 2, respectively (see Table 3.9). Next, complete the column under the conjunction symbol, $\wedge$, using the truth values in columns 1 and 2, and call this column 3. Then, complete the column under $\sim p$ and call this column 4. To complete the answer, column 5, use the definition of disjunction and the truth values in columns 3 and 4. ∎

So far, all the truth tables we have constructed have contained at most two simple statements. Now we will explain how to construct a truth table that consists of three simple statements, such as $(p \wedge q) \wedge r$. When a compound statement consists of three simple statements, there are eight different true–false possibilities, as illustrated in Table 3.10. To begin such a truth table, write four T's and four F's in the column under p. Under the second statement, q, pairs of T's alternate with pairs of F's. Under the third statement, r, T alternates with F. This technique is not the only way of listing the cases, but it ensures that each case is unique and that no cases are omitted.

Table 3.10

	p	q	r
Case 1	T	T	T
Case 2	T	T	F
Case 3	T	F	T
Case 4	T	F	F
Case 5	F	T	T
Case 6	F	T	F
Case 7	F	F	T
Case 8	F	F	F

Example 5   *Construct a Truth Table with Eight Cases*

a) Construct a truth table for the statement "Santana is home and he is not at his desk, or he is sleeping."

b) Suppose that "Santana is home" is a false statement, that "Santana is at his desk" is a true statement, and that "Santana is sleeping" is a true statement. Is the compound statement in part (a) true or false?

Solution

a) First we will translate the statement into symbolic form.

Let

> p: Santana is home.
> q: Santana is at his desk.
> r: Santana is sleeping.

In symbolic form, the statement is $(p \wedge \sim q) \vee r$.

Since the statement is composed of three simple statements, there are eight cases. Begin by listing the eight cases in the three left-hand columns; see Table 3.11. By examining the statement, you can see that it is a disjunction. Therefore, the answer will be in the $\vee$ column. Fill out the truth table by working in parentheses first. Place values under p, column 1, and $\sim q$, column 2. Then determine the conjunction

Table 3.11

p	q	r	$(p$	$\wedge$	$\sim q)$	$\vee$	r
T	T	T	T	F	F	T	T
T	T	F	T	F	F	F	F
T	F	T	T	T	T	T	T
T	F	F	T	T	T	T	F
F	T	T	F	F	F	Ⓣ	T
F	T	F	F	F	F	F	F
F	F	T	F	F	T	T	T
F	F	F	F	F	T	F	F
			1	3	2	5	4

of columns 1 and 2 to obtain column 3. Place the values of r in column 4. To obtain the answer, column 5, use columns 3 and 4 and the information for the disjunction contained in Table 3.5 on page 104.

b) We are given the following:

p: Santana is home—false.

q: Santana is at his desk—true.

r: Santana is sleeping—true.

We need to determine the truth value of the following case: false, true, true. In case 5 of the truth table, p, q, and r are F, T, and T, respectively. Therefore, under these conditions, the original compound statement is true (as circled in the table). ∎

We have learned that a truth table with one simple statement has two cases, a truth table with two simple statements has four cases, and a truth table with three simple statements has eight cases. In general, *the number of distinct cases in a truth table with n distinct simple statements is 2^n.* The compound statement $(p \lor q) \lor (r \land \sim s)$ has four simple statements, p, q, r, s. Thus, a truth table for this compound statement would have 2^4, or 16, distinct cases.

TECHNOLOGY TIP

Truth Table Builder Apps

There are several apps available for use on most smartphones and tablets that can assist in the building of truth tables. Most of these apps can also be used to test your ability to complete truth tables accurately. Although these apps can be helpful as you learn how to build truth tables, be careful to completely understand the instructions and notation used before utilizing such an app. For example, the plus sign, +, is often used to represent "and"; a vertical bar (found on most computer keyboards above the backward slash), I, is often used to represent "or"; and the apostrophe,', is often used to represent "not." Finally, as always, check with your instructor prior to using any apps to complete required work for your course.

Alternate Method for Constructing Truth Tables

We now present an alternate method for constructing truth tables. We will use the alternate method to construct truth tables for the same statements we analyzed in Examples 1, 2, and 3.

Example 6   *Use the Alternate Method to Construct a Truth Table*

Construct a truth table for $p \land \sim q$.

Solution   We begin by constructing the first two columns of a truth table with four cases, as shown in Table 3.12(a) on page 110. We will add additional columns to Table 3.12(a) to develop our answer column. Since we wish to determine the truth table for the compound statement $p \land \sim q$, we need to be able to compare the truth values for p with the truth values for $\sim q$. Table 3.12(a) already has a column showing the truth values for p. We next add a column showing the truth values for $\sim q$, as shown in Table 3.12(b). Recall that the values of $\sim q$ are the opposite of those for q.

Finally, we add the answer column for the compound statement $p \land \sim q$, as shown in Table 3.12(c). To determine the truth values for the $p \land \sim q$ column, use the p column and the $\sim q$ column, and the conjunction table, Table 3.4, on page 103.

Table 3.12

(a)		
	p	**q**
Case 1	T	T
Case 2	T	F
Case 3	F	T
Case 4	F	F

(b)		
p	**q**	**~q**
T	T	F
T	F	T
F	T	F
F	F	T

(c)			
p	**q**	**~q**	**p ∧ ~q**
T	T	F	F
T	F	T	T
F	T	F	F
F	F	T	F

↑ ↑ ↑
Use these columns
to determine the   Answer
answer column.   column

Note that the answer column of Table 3.12(c) is the same as the answer column of Table 3.6(e) on page 105. ∎

Example 7   *Use the Alternate Method to Construct a Truth Table*

Construct a truth table for $\sim p \wedge \sim q$.

Solution Begin by constructing a truth table with four cases, as shown in Table 3.13(a). Since we wish to determine the truth table for the compound statement $\sim p \wedge \sim q$, we will add a column for $\sim p$ and a column for $\sim q$, as shown in Table 3.13(b).

Table 3.13

(a)	
p	**q**
T	T
T	F
F	T
F	F

(b)			
p	**q**	**~p**	**~q**
T	T	F	F
T	F	F	T
F	T	T	F
F	F	T	T

(c)				
p	**q**	**~p**	**~q**	**~p ∧ ~q**
T	T	F	F	F
T	F	F	T	F
F	T	T	F	F
F	F	T	T	T

↑ ↑ ↑
Use these columns
to determine the   Answer
answer column.   column

Finally, we add our answer column for the compound statement $\sim p \wedge \sim q$. To determine the truth values for the $\sim p \wedge \sim q$ column, use the $\sim p$ column and the $\sim q$ column, and the conjunction table, Table 3.4, on page 103. Note that the answer column of Table 3.13c is the same as the answer column of Table 3.7 on page 106. ∎

Example 8   *Use the Alternate Method to Construct a Truth Table*

Construct a truth table for $\sim (\sim q \vee p)$.

Solution Begin by constructing a truth table with four cases, as shown in Table 3.14(a). To complete the truth table, we will work within parentheses first. Thus, we next add a column for $\sim q$, as shown in Table 3.14(b). We then will construct a column for the expression within parentheses $\sim q \vee p$ by using the $\sim q$ column and the p column, and the disjunction table, Table 3.5, on page 104.

Table 3.14

(a)	
p	q
T	T
T	F
F	T
F	F

(b)		
p	q	$\sim q$
T	T	F
T	F	T
F	T	F
F	F	T

(c)			
p	q	$\sim q$	$\sim q \vee p$
T	T	F	T
T	F	T	T
F	T	F	F
T	F	T	T

Use these columns to
determine the
$\sim q \vee p$ column.

$\sim q \vee p$
column

Table 3.15

p	q	$\sim q$	$\sim q \vee p$	$\sim (\sim q \vee p)$
T	T	F	T	F
T	F	T	T	F
F	T	F	F	T
F	F	T	T	F

Take the opposite
of this column to
get the answer
column.

Answer
column

Finally, we add the answer column for the compound statement, $\sim (\sim q \vee p)$, as shown in Table 3.15. To determine the truth values for the $\sim (\sim q \vee p)$ column, take the opposite values of those shown in the $\sim q \vee p$ column in Table 3.14(c). Note that the answer column of Table 3.15 is the same as the answer column of Table 3.8 on page 107. ∎

We have demonstrated two methods for constructing truth tables. Unless your instructor indicates otherwise, you may use either method. Both methods, if done correctly, will lead to the correct answer. *In the remainder of this chapter, we will demonstrate the construction of truth tables using only the first method.*

Determine Truth Values Without Constructing a Truth Table

When we construct a truth table, we determine the truth values of a compound statement for every possible case. If we want to determine the truth value of the compound statement for any specific case when we know the truth values of the simple statements, we do not have to develop the entire table. For example, to determine the truth value for the statement

$$2 + 3 = 5 \qquad \text{and} \qquad 1 + 1 = 3$$

we let p be $2 + 3 = 5$ and q be $1 + 1 = 3$. Now we can write the compound statement as $p \wedge q$. We know that p is a true statement and q is a false statement. Thus, we can substitute T for p and F for q and evaluate the statement:

$$p \wedge q$$
$$T \wedge F$$
$$F$$

Therefore, the compound statement $2 + 3 = 5$ and $1 + 1 = 3$ is a false statement.

In the remaining examples in this section, we will determine the truth values of compound statements without constructing a truth table.

Exploring Truth Values
for Compound Statements

Example 9 *Determine the Truth Value of Compound Statements*

Determine the truth value of each simple statement. Then, using these truth values, determine the truth value of the compound statement.

a) 1 million is greater than or equal to 1 billion.

b) Montreal is in Canada or Beijing is in Mexico, but Santiago is not in Japan.

Solution

a) Let

p: 1 million is greater than 1 billion.

q: 1 million is equal to 1 billion.

The statement "1 million is greater than or equal to 1 billion" means that 1 million is greater than 1 billion or 1 million is equal to 1 billion. The compound statement can be expressed as $p \vee q$. We know that both p and q are false statements, since 1 million is less than 1 billion and 1 million is not equal to 1 billion. We substitute F for p and F for q and evaluate the statement:

$$p \vee q$$
$$F \vee F$$
$$F$$

Therefore, the compound statement "1 million is greater than or equal to 1 billion" is a false statement.

b) Let

p: Montreal is in Canada.

q: Beijing is in Mexico.

r: Santiago is in Japan.

Recall that the word *but* is used to express a conjunction. Thus, the compound statement, "Montreal is in Canada or Beijing is in Mexico, but Santiago is not in Japan" can be written in symbolic form as $(p \vee q) \wedge \sim r$. If necessary, we can consult the Internet to determine that since Montreal is in Canada, statement p is a true statement. Since Beijing is in China, not in Mexico, statement q is a false statement. And since Santiago is in Chile, not in Japan, statement r is also a false statement. To evaluate the compound statement, $(p \vee q) \wedge \sim r$, we substitute T for p, F for q, and F for r and then simplify.

$$(p \vee q) \wedge \sim r$$
$$(T \vee F) \wedge \sim F$$
$$T \quad \wedge \quad T$$
$$T$$

Therefore, the original statement is a true statement.   ■

Example 10   *Pet Ownership in the United States*

The number of pets owned in the United States in 2018 is shown in Figure 3.1. Use this graph to determine the truth value of the following statement: There are more dogs owned than cats and there are fewer reptiles owned than birds, or the most numerous pets owned are not fish.

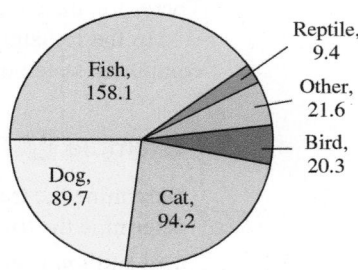

Pets Owned in the United States (millions)

Fish, 158.1
Reptile, 9.4
Other, 21.6
Bird, 20.3
Dog, 89.7
Cat, 94.2

Source: American Pet Products Manufacturing Association

Figure 3.1

Solution Let

p:	There are more dogs owned than cats.
q:	There are fewer reptiles owned than birds.
r:	The most numerous pets owned are fish.

The given compound statement can be written in symbolic form as $(p \wedge q) \vee \sim r$. From Fig. 3.1, we see that statement p is false: There are actually more cats owned than dogs. We also see that statement q is true: There are fewer reptiles owned than birds. We also see that statement r is true: The most numerous pets owned are fish. Since r is true, its negation, $\sim r$, is false. Therefore, we substitute F for p, T for q, and F for $\sim r$, which gives

$$(p \wedge q) \vee \sim r$$
$$(F \wedge T) \vee F$$
$$F \qquad \vee F$$
$$F$$

Thus, the original compound statement is a false statement. ∎

SECTION 3.2 *Exercises*

Warm Up Exercises

In Exercises 1–4, fill in the blanks with an appropriate word, phrase, or symbol(s).

1. The negation $\sim p$ will always have the _____ truth value of p.

2. The conjunction $p \wedge q$ is true only when both p and q are _____.

3. The disjunction $p \vee q$ is false only when both p and q are _____.

4. A truth table for a compound statement with

   a) one distinct simple statement will have _____ cases.

   b) two distinct simple statements will have _____ cases.

   c) three distinct simple statements will have _____ cases.

Practice the Skills/Problem Solving

In Exercises 5–18, construct a truth table for the statement.

5. $p \wedge \sim p$

6. $p \vee \sim p$

7. $\sim p \vee q$

8. $\sim p \vee \sim q$

9. $p \wedge \sim q$

10. $\sim p \wedge \sim q$

11. $\sim(p \wedge \sim q)$

12. $\sim(p \vee \sim q)$

13. $p \vee (\sim q \vee r)$

14. $\sim p \wedge (q \vee r)$

15. $(p \wedge \sim q) \vee r$

16. $(p \vee \sim q) \wedge r$

17. $(\sim p \wedge q) \vee \sim r$

18. $\sim p \vee (\sim q \wedge r)$

In Exercises 19–26, write the statement in symbolic form and construct a truth table.

19. Apples are a good source of fiber and oranges are a good source of vitamin C.

20. Sophia did not work today, but she volunteered at the shelter.

21. Joaquin will pitch or he will not play first base.

22. Katie will not go to the zoo or she will go to the library.

23. It is false that the car is not a Chevrolet, but the car is a Corvette.

24. It is false that Charlie is not a tiger or Patty is a dolphin.

25. School is not in session and the kids are home, or I am working.

26. The drier is loud or I am drying shoes, and I cannot fix dinner.

In Exercises 27–36, determine the truth value of the statement if

    a) *p* is true, *q* is false, and *r* is true.

    b) *p* is false, *q* is true, and *r* is true.

27. $(p \land \sim q) \lor r$

28. $p \land (\sim q \lor r)$

29. $(\sim p \land \sim q) \lor \sim r$

30. $\sim p \land (\sim q \lor \sim r)$

31. $(p \lor \sim q) \land [\sim(p \land \sim r)]$

32. $(p \land \sim q) \lor r$

33. $(\sim r \land p) \lor q$

34. $\sim q \lor (r \land p)$

35. $(\sim p \lor \sim q) \lor (\sim r \lor q)$

36. $(\sim r \land \sim q) \land (\sim r \lor \sim p)$

In Exercises 37– 44, determine the truth value for each simple statement. Then use these truth values to determine the truth value of the compound statement. (You may have to use a reference source such as the Internet.)

37. $8 + 7 = 20 - 5$ and $63 \div 7 = 3 \cdot 3$

38. $0 < -3$ or $5 \geq 10$

39. Chevrolet makes trucks or Toyota makes shoes.

40. Dell makes blue jeans and Apple makes computers.

41. George Washington was the first U.S. president or Abraham Lincoln was the second U.S. president, but Harry Truman was not the third U.S. president.

42. Paris is in England or London is in France, and Berlin is in Germany.

▲ *Paris*

43. Chicago is in Mexico or Los Angeles is in California, or Dallas is in Canada.

44. Holstein is a breed of cattle and collie is a breed of dogs, or beagle is not a breed of cats.

Five Largest Countries In Exercises 45–48, use the table to determine the truth value of each simple statement. Then determine the truth value of the compound statement.

Countries with the Largest Land Areas	
Country	**Area (in millions of square km)**
Russia	17.08
Canada	9.98
United States	9.83
China	9.60
Brazil	8.51
Source: CIA World Fact Book	

45. Russia is larger than Canada and China is larger than Brazil.

46. The United States is over twice as large as Brazil or Canada is over twice as large as China.

47. The area of Brazil and China combined is larger than the area of Russia, and the area of the United States and Canada combined is not larger than the area of Russia.

48. The area of the United States is not larger than the area of Russia but the area of Canada is larger than the area of China.

Sleep Time In Exercises 49–52, use the graph, which shows the number of hours Americans sleep, to determine the truth value of each simple statement. Then determine the truth value of the compound statement.

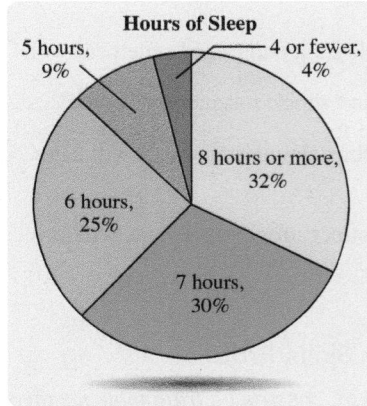

49. It is false that 30% of Americans get 6 hours of sleep each night and 9% get 5 hours of sleep each night.

50. Twenty-five percent of Americans get 6 hours of sleep each night, and 30% get 7 hours of sleep each night or 9% do not get 5 hours of sleep each night.

51. Thirteen percent of Americans get 5 or fewer hours of sleep each night or 32% get 6 or more hours of sleep each night, and 30% get 8 or more hours of sleep each night.

52. Over one-half of Americans get 7 or fewer hours of sleep each night, and over one-quarter get 6 or fewer hours of sleep each night.

In Exercises 53–56, let

  p: Tanisha owns a convertible.
  q: Joan owns a Volvo.

Translate each statement into symbols. Then construct a truth table for each compound statement and indicate under what conditions the compound statement is true.

53. Tanisha owns a convertible and Joan does not own a Volvo.

54. Tanisha does not own a convertible, but Joan owns a Volvo.

55. Tanisha owns a convertible or Joan does not own a Volvo.

56. Tanisha does not own a convertible or Joan does not own a Volvo.

In Exercises 57–60, let

  p: The house is owned by an engineer.
  q: The heat is solar generated.
  r: The car is run by electric power.

Translate each statement into symbols. Then construct a truth table for each compound statement and indicate under what conditions the compound statement is true.

57. The car is run by electric power or the heat is solar generated, but the house is owned by an engineer.

58. The house is owned by an engineer and the heat is solar generated, or the car is run by electric power.

59. The heat is solar generated, or the house is owned by an engineer and the car is not run by electric power.

60. The house is not owned by an engineer, and the car is not run by electric power and the heat is solar generated.

Obtaining a Loan *In Exercises 61 and 62, read the requirements and each applicant's qualifications for obtaining a loan.*

a) Identify which of the applicants would qualify for the loan.

b) For the applicants who do not qualify for the loan, explain why.

61. To qualify for a loan of $40,000, an applicant must have a gross income of $28,000 if single, $46,000 combined income if married, and assets of at least $6,000.

Mrs. Rusinek, married with three children, earns $42,000. Mr. Rusinek does not have an income. The Rusineks have assets of $42,000.

Mr. Duncan is not married, works in sales, and earns $31,000. He has assets of $9000.

Mrs. Tuttle and her husband have total assets of $43,000. One earns $35,000, and the other earns $23,500.

62. To qualify for a loan of $45,000, an applicant must have a gross income of $30,000 if single, $50,000 combined income if married, and assets of at least $10,000.

Mr. Argento, married with two children, earns $37,000. Mrs. Argento earns $15,000 at a part-time job. The Argentos have assets of $25,000.

Ms. McVey, single, has assets of $19,000. She works in a store and earns $25,000.

Mr. Siewert earns $24,000 and Ms. Fox, his wife, earns $28,000. Their assets total $8000.

63. ***Airline Special Fares*** An airline advertisement states, "To get the special fare you must purchase your tickets between January 1 and February 15 and fly round trip between March 1 and April 1. You must depart on a Monday, Tuesday, or Wednesday, and return on a Tuesday, Wednesday, or Thursday, and stay over at least one Saturday."

a) Determine which of the following individuals will qualify for the special fare.

b) If the person does not qualify for the special fare, explain why.

Xavier plans to purchase his ticket on January 15, depart on Monday, March 3, and return on Tuesday, March 18.

Gina plans to purchase her ticket on February 1, depart on Wednesday, March 12, and return on Thursday, April 3.

Kara plans to purchase her ticket on February 14, depart on Tuesday, March 4, and return on Monday, March 19.

Christos plans to purchase his ticket on January 4, depart on Monday, March 10, and return on Thursday, March 13.

Chang plans to purchase his ticket on January 1, depart on Monday, March 3, and return on Monday, March 10.

Problem Solving/Group Activities

In Exercises 64 and 65, construct a truth table for the symbolic statement.

64. $\sim[(\sim(p \lor q)) \lor (q \land r)]$

65. $[(q \land \sim r) \land (\sim p \lor \sim q)] \lor (p \lor \sim r)$

66. On page 109, we indicated that a compound statement consisting of n simple statements had 2^n distinct true–false cases.

a) How many distinct true–false cases does a truth table containing simple statements p, q, r, and s have?

b) List all possible true–false cases for a truth table containing the simple statements p, q, r, and s.

c) Use the list in part (b) to construct a truth table for $(q \land p) \lor (\sim r \land s)$.

d) Construct a truth table for $(\sim r \land \sim s) \land (\sim p \lor q)$.

67. Must $(p \land \sim q) \lor r$ and $(q \land \sim r) \lor p$ have the same number of trues in their answer columns? Explain.

Research Activity

68. *Logic and Set Theory* Do research and write a report on each of the following.

a) The relationship between *negation* in logic and *complement* in set theory

b) The relationship between *conjunction* in logic and *intersection* in set theory

c) The relationship between *disjunction* in logic and *union* in set theory

---

SECTION 3.3   Truth Tables for the Conditional and Biconditional

Upon completion of this section, you will be able to:

- Construct truth tables involving conditional statements.
- Construct truth tables involving biconditional statements.
- Understand self-contradictions, tautologies, and implications.

Suppose I said to you, "If you get an A, then I will buy you a car." As we discussed in Section 3.1, this statement is called a *conditional* statement. In this section, we will discuss under what conditions a conditional statement is true and under what conditions a conditional statement is false.

Why This Is Important We often encounter statements of the form *if ... then* and *if and only if*. These statements are called conditional statements and biconditional statements, respectively. Such statements can be found in advertising claims, political speeches, and many legal documents. Understanding when these statements are true and when these statements are false can have a large impact on our lives. For example, when purchasing a home, it is very important that you understand the statements contained in the real estate contract.

Conditional

In Section 3.1, we mentioned that the statement preceding the conditional symbol is called the *antecedent* and that the statement following the conditional symbol is called the *consequent*. For example, consider $(p \lor q) \rightarrow [\sim(q \land r)]$. In this statement, $(p \lor q)$ is the antecedent and $[\sim(q \land r)]$ is the consequent.

To develop a truth table for the conditional statement, consider the statement "If you get an A, then I will buy you a car." Assume this statement is true except when I have actually broken my promise to you.

Let

p: You get an A.

q: I buy you a car.

Translated into symbolic form, the statement becomes $p \rightarrow q$. Let's examine the four cases shown in Table 3.16.

Table 3.16 Conditional

p	q	$p \rightarrow q$
T	T	T
T	F	F
F	T	T
F	F	T

CASE 1: (T, T) You get an A, and I buy a car for you. I have met my commitment, and the statement is true.

CASE 2: (T, F) You get an A, and I do not buy a car for you. I have broken my promise, and the statement is false.

What happens if you don't get an A? If you don't get an A, I no longer have a commitment to you, and therefore I cannot break my promise.

CASE 3: (F, T) You do not get an A, and I buy you a car. I have not broken my promise, and therefore the statement is true.

CASE 4: (F, F) You do not get an A, and I don't buy you a car. I have not broken my promise, and therefore the statement is true.

The conditional statement is false when the antecedent is true and the consequent is false. In every other case the conditional statement is true.

> The **conditional statement** $p \rightarrow q$ is true in every case except when p is a true statement and q is a false statement.

Table 3.17

p	q	$\sim p$	$\rightarrow$	$\sim q$
T	T	F	T	F
T	F	F	T	T
F	T	T	F	F
F	F	T	T	T
		1	3	2

Table 3.18

p	q	r	p	$\rightarrow$	$(\sim q$	$\wedge$	$r)$
T	T	T	T	F	F	F	T
T	T	F	T	F	F	F	F
T	F	T	T	T	T	T	T
T	F	F	T	F	T	F	F
F	T	T	F	T	F	F	T
F	T	F	F	T	F	F	F
F	F	T	F	T	T	T	T
F	F	F	F	T	T	F	F
			4	5	1	3	2

Example 1 *A Truth Table with a Conditional*

Construct a truth table for the statement $\sim p \rightarrow \sim q$.

Solution Because this statement is a conditional, the answer will lie under the $\rightarrow$. Fill out the truth table by placing the appropriate truth values under $\sim p$, column 1, and under $\sim q$, column 2 (see Table 3.17). Then, using the information given in the truth table for the conditional (Table 3.16 above) and the truth values in columns 1 and 2, determine the solution, column 3. In row 1, the antecedent, $\sim p$, is false and the consequent, $\sim q$, is also false. Row 1 is F $\rightarrow$ F, which according to row 4 of Table 3.16, is T. Likewise, row 2 of Table 3.17 is F $\rightarrow$ T, which is T. Row 3 is T $\rightarrow$ F, which is F. Row 4 is T $\rightarrow$ T, which is T. ∎

Example 2 *A Conditional Truth Table with Three Simple Statements*

Construct a truth table for the statement $p \rightarrow (\sim q \wedge r)$.

Solution Because this statement is a conditional, the answer will lie under the $\rightarrow$. Work within the parentheses first. Place the truth values under $\sim q$, column 1, and r, column 2 (Table 3.18). Then take the conjunction of columns 1 and 2 to obtain column 3. Next, place the truth values under p in column 4. To determine the answer, column 5, use columns 3 and 4 and the information of the conditional

statement given in Table 3.16. Column 4 represents the truth values of the antecedent, and column 3 represents the truth values of the consequent. Remember that the conditional is false only when the antecedent is true and the consequent is false, as in cases (rows) 1, 2, and 4 of column 5. ∎

Example 3 *Examining an Advertisement*

An advertisement for Top Power nutritional supplements makes the following claim: "If you use Top Power, then you will not feel tired and you will get a solid workout." Translate the statement into symbolic form and construct a truth table.

Solution Let

p: You use Top Power.

q: You will feel tired.

r: You will get a solid workout.

In symbolic form, the claim is

$$p \rightarrow (\sim q \wedge r).$$

This symbolic statement is identical to the statement in Table 3.18 on page 117. Thus, the truth tables are the same. Column 3 represents the truth value of $(\sim q \wedge r)$, which corresponds to the statement "You will not feel tired and you will get a solid workout." Note that column 3 is true in cases (rows) 3 and 7. In case 3, since p is true, you used Top Power. In case 7, however, since p is false, you did not use Top Power. From case 7 we can conclude that it is possible for you to not feel tired and for you to get a solid workout without using Top Power. ∎

A truth table cannot by itself determine whether a compound statement is true or false. However, a truth table does allow us to examine all possible cases for compound statements.

Biconditional

The *biconditional statement* $p \leftrightarrow q$ means that $p \rightarrow q$ and $q \rightarrow p$, or, symbolically, $(p \rightarrow q) \wedge (q \rightarrow p)$. To determine the truth table for $p \leftrightarrow q$, we will construct the truth table for $(p \rightarrow q) \wedge (q \rightarrow p)$. In Table 3.19 below, we use column 3 and column 6 to obtain the answer in column 7. Table 3.20 below shows the truth values for the biconditional statement.

Table 3.19

p	q	$(p$	$\rightarrow$	$q)$	$\wedge$	$(q$	$\rightarrow$	$p)$
T	T	T	T	T	T	T	T	T
T	F	T	F	F	F	F	T	T
F	T	F	T	T	F	T	F	F
F	F	F	T	F	T	F	T	F
		1	3	2	7	4	6	5

Table 3.20 Biconditional

p	q	$p \leftrightarrow q$
T	T	T
T	F	F
F	T	F
F	F	T

From Table 3.20 we see that the biconditional statement is true when the antecedent and the consequent have the same truth value and false when the antecedent and consequent have different truth values.

> The **biconditional statement** $p \leftrightarrow q$ is true only when p and q have the same truth value, that is, when both are true or both are false.

Example **4** *A Truth Table Using a Biconditional*

Construct a truth table for the statement $\sim p \leftrightarrow (\sim q \rightarrow r)$.

Solution Since there are three letters, there must be eight cases. The parentheses indicate that the answer must be under the biconditional, as shown in Table 3.21. Use columns 3 and 4 to obtain the answer in column 5. When columns 3 and 4 have the same truth values, place a T in column 5. When columns 3 and 4 have different truth values, place an F in column 5.

Table 3.21

p	q	r	$\sim p$	$\leftrightarrow$	$(\sim q$	$\rightarrow$	$r)$
T	T	T	F	F	F	T	T
T	T	F	F	F	F	T	F
T	F	T	F	F	T	T	T
T	F	F	F	T	T	F	F
F	T	T	T	T	F	T	T
F	T	F	T	T	F	T	F
F	F	T	T	T	T	T	T
F	F	F	T	F	T	F	F
			4	5	1	3	2

In Section 3.2, we showed that determining the truth value of a compound statement for a specific case does not require constructing an entire truth table. Examples 5 and 6 illustrate this technique for the conditional and the biconditional.

Animation

→

Exploring Truth Values
for Conditional Statements

Example **5** *Determine the Truth Value of a Compound Statement*

Determine the truth value of the biconditional statement $(\sim p \rightarrow q) \leftrightarrow (q \rightarrow \sim r)$ when p is true, q is false, and r is false.

Solution Substitute the truth value for each simple statement and simplify.

$$(\sim p \rightarrow q) \leftrightarrow (q \rightarrow \sim r)$$
$$(\sim T \rightarrow F) \leftrightarrow (F \rightarrow \sim F)$$
$$(F \rightarrow F) \leftrightarrow (F \rightarrow T)$$
$$\quad T \quad \leftrightarrow \quad T$$
$$T$$

Thus, for this specific case, the biconditional statement is true.

Example **6** *Determine the Truth Value of a Compound Statement*

Determine the truth value for each simple statement. Then use the truth values to determine the truth value of the compound statement.

a) If 15 is an even number, then 29 is an even number.
b) Vanderbilt University is in Tennessee and Wake Forest University is in Alaska, if and only if Syracuse University is in Alabama.

Solution

a) Let

p: 15 is an even number.

q: 29 is an even number.

Then the statement "If 15 is an even number, then 29 is an even number" can be written $p \rightarrow q$. Since 15 is not an even number, p is a false statement. Also, since 29 is not an even number, q is a false statement. We substitute F for p and F for q and evaluate the statement:

$$p \rightarrow q$$
$$\text{F} \rightarrow \text{F}$$
$$\text{T}$$

Therefore, "If 15 is an even number, then 29 is an even number" is a true statement.

b) Let

p: Vanderbilt University is in Tennessee.

q: Wake Forest University is in Alaska.

r: Syracuse University is in Alabama.

The original compound statement can be written $(p \wedge q) \leftrightarrow r$. By checking the Internet we can determine that Vanderbilt University is in Tennesee, Wake Forest University is in North Carolina, and Syracuse University is in New York. Therefore, p is a true statement, but q and r are false statements. We will substitute T for p, F for q, and F for r and evaluate the compound statement:

$$(p \wedge q) \leftrightarrow r$$
$$(\text{T} \wedge \text{F}) \leftrightarrow \text{F}$$
$$\text{F} \quad \leftrightarrow \text{F}$$
$$\text{T}$$

Therefore, the original compound statement is true. ∎

▲ *Wake Forest University*

Example 7 *Using Real Data in Compound Statements*

The graph in Fig. 3.2 was created using **Stat**Crunch with data obtained from the website www.counterpointresearch.com. The graph shows the United States market share for various brands of cell phones in 2018. Use the data presented in the graph to determine the truth value of the following compound statements.

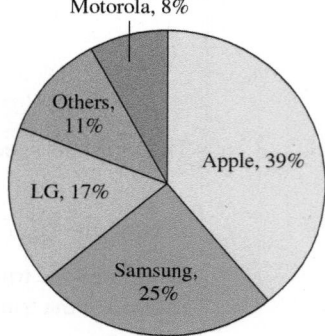

U.S. Cell Phone Market Share 2018

Source: www.counterpointresearch.com

Figure 3.2

a) If Apple has 39% of the U.S. market share and Samsung has 36% of the U.S. market share, then LG has 17% of the U.S. market share.

b) Motorola has 18% of the U.S. market share or others have 20% of the U.S. market share, if and only if Samsung has 25% of the U.S. market share.

RECREATIONAL MATH

Satisfiability Problems

Suppose you are hosting a dinner party for seven people: Yasumasa, Marie, Albert, Stephen, Leonhard, Karl, and Emmy. You need to develop a seating plan around your circular dining room table that would satisfy all your guests. Albert and Emmy are great friends and must sit together. Yasumasa and Karl haven't spoken to each other in years and cannot sit by each other. Leonhard must sit by Marie or by Albert, but he cannot sit by Karl. Stephen insists he sit by Albert. Can you come up with a plan that would satisfy all your guests? Now imagine the difficulty of such a problem as the list of guests, and their demands, grows.

Problems such as this are known as *satisfiability* problems. The symbolic logic you are studying in this chapter allows computer scientists to represent these problems with symbols and solve the problems using computers. Even with the fastest computers, some satisfiability problems take an enormous amount of time to solve.

One solution to the problem posed above is shown upside down below. Exercises 74 and 75 on page 126 have other satisfiability problems.

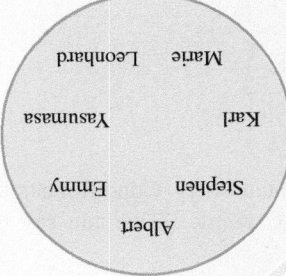

Solution

a) Let

p: Apple has 39% of the U.S. market share.

q: Samsung has 36% of the U.S. market share.

r: LG has 17% of the U.S. market share.

Then the original compound statement can be written with symbols as $(p \land q) \rightarrow r$. From Fig. 3.2 we can determine that both p and r are true statements and that q is a false statement. We substitute T for p, F for q, and T for r and evaluate the compound statement.

$$(p \land q) \rightarrow r$$
$$(T \land F) \rightarrow T$$
$$F \quad \rightarrow T$$
$$T$$

Therefore, the original compound statement "If Apple has 39% of the U.S. market share and Samsung has 36% of the U.S. market share, then LG has 17% of the U.S. market share" is true.

b) Let

p: Motorola has 18% of the U.S. market share.

q: Others have 20% of the U.S. market share.

r: Samsung has 25% of the U.S. market share.

Then the original compound statement can be written with symbols as $(p \lor q) \leftrightarrow r$. From Fig. 3.2 we can determine that both p and q are false statements and that r is a true statement. We substitute F for p, F for q, and T for r and evaluate the compound statement.

$$(p \lor q) \leftrightarrow r$$
$$(F \lor F) \leftrightarrow T$$
$$F \quad \leftrightarrow T$$
$$F$$

Therefore, the original compound statement "Motorola has 18% of the U.S. market share or others have 20% of the U.S. market share, if and only if Samsung has 25% of the U.S. market share" is a false statement. ∎

Self-Contradictions, Tautologies, and Implications

Two special situations can occur in the answer column of a truth table of a compound statement: The statement may always be false, or the statement may always be true. We give such statements special names.

> Definition: **Self-contradiction**
>
> A **self-contradiction** is a compound statement that is always false.

When every truth value in the answer column of the truth table is false, then the statement is a self-contradiction.

Example 8 *All Falses, a Self-Contradiction*

Construct a truth table for the statement $(p \leftrightarrow q) \land (p \leftrightarrow \sim q)$.

Solution See Table 3.22 on page 122. In this example, the truth values are false in each case of column 5. This statement is an example of a self-contradiction or a *logically false statement*.

Table 3.22

p	q	$(p \leftrightarrow q)$	$\wedge$	$(p$	$\leftrightarrow$	$\sim q)$
T	T	T	F	T	F	F
T	F	F	F	T	T	T
F	T	F	F	F	T	F
F	F	T	F	F	F	T
		1	5	2	4	3

Definition Tautology

A **tautology** is a compound statement that is always true.

When every truth value in the answer column of the truth table is true, the statement is a tautology.

Example 9 *All Trues, a Tautology*

Construct a truth table for the statement $(p \wedge q) \rightarrow (p \vee r)$.

Solution The answer is given in column 3 of Table 3.23. The truth values are true in every case. Thus, the statement is an example of a tautology or a *logically true statement*.

Table 3.23

p	q	r	$(p \wedge q)$	$\rightarrow$	$(p \vee r)$
T	T	T	T	T	T
T	T	F	T	T	T
T	F	T	F	T	T
T	F	F	F	T	T
F	T	T	F	T	T
F	T	F	F	T	F
F	F	T	F	T	T
F	F	F	F	T	F
			1	3	2

The conditional statement $(p \wedge q) \rightarrow (p \vee r)$ is a tautology. Conditional statements that are tautologies are called *implications*. In Example 9, we can say that *$p \wedge q$ implies $p \vee r$*.

Definition: Implication

An **implication** is a conditional statement that is a tautology.

In any implication the antecedent of the conditional statement implies the consequent. In other words, if the antecedent is true, then the consequent must also be true. That is, the consequent will be true whenever the antecedent is true. We will use implications when we study symbolic arguments in Section 3.5.

Example **10** *An Implication?*

Determine whether the conditional statement $[(p \lor q) \land \sim q] \to p$ is an implication.

Solution   If the conditional statement is a tautology, the conditional statement is an implication. Because the conditional statement is a tautology (see Table 3.24), the conditional statement is an implication. The antecedent $[(p \lor q) \land \sim q]$ implies the consequent p. Note that the antecedent is true only in case 2 and the consequent is also true in case 2.

Table 3.24

p	q	$[(p \lor q)$	$\land$	$\sim q]$	$\to$	p
T	T	T	F	F	T	T
T	F	T	T	T	T	T
F	T	T	F	F	T	F
F	F	F	F	T	T	F
		1	3	2	5	4

SECTION 3.3 *Exercises*

Warm Up Exercises

In Exercises 1–16, fill in the blanks with an appropriate word, phrase, or symbol(s).

1. The conditional statement $p \to q$ is _____ only when p is true and q is false.

2. In the conditional statement $p \to q$,

   a) The lower-case letter p represents the _____.

   b) The lower-case letter q represents the _____.

3. The biconditional statement $p \leftrightarrow q$ is _____ only when p and q have the same truth value.

4. A compound statement that is always true is known as a(n) _____.

5. A compound statement that is always false is known as a(n) _____.

6. A conditional statement that is a tautology is known as a(n) _____.

Practice the Skills

In Exercises 7–16, construct a truth table for the statement.

7. $\sim p \to q$

8. $p \to \sim q$

9. $\sim p \leftrightarrow \sim q$

10. $\sim (p \leftrightarrow q)$

11. $(p \to q) \leftrightarrow p$

12. $(p \leftrightarrow q) \to p$

13. $p \leftrightarrow (q \lor p)$

14. $(p \leftrightarrow q) \lor p$

15. $(\sim p \to q) \leftrightarrow (p \to \sim q)$

16. $(\sim p \leftrightarrow q) \to (\sim p \leftrightarrow \sim q)$

In Exercises 17–24, construct a truth table for the statement.

17. $\sim p \to (q \land r)$

18. $(\sim p \to q) \land r$

19. $\sim p \leftrightarrow (q \lor \sim r)$

20. $(\sim p \leftrightarrow q) \lor \sim r$

21. $(\sim p \land q) \to \sim r$

22. $(p \lor \sim q) \leftrightarrow r$

23. $(p \to q) \leftrightarrow (\sim q \to \sim r)$

24. $(\sim p \leftrightarrow \sim q) \to (\sim q \leftrightarrow r)$

In Exercises 25–30, write the statement in symbolic form. Then construct a truth table for the symbolic statement.

25. If today is Monday, then the library is open and we can study together.

26. Jana is the CEO if and only if Allia is the CFO, or Camila is not the COO.

27. Blackberries are high in vitamin K if and only if mangos are high in B vitamins, or cherries are high in vitamin C.

28. If the dam holds then we can go fishing, if and only if the pole is not broken.

29. If it is not too cold then we can take a walk, or we can go to the gym.

30. It is false that if you play spider solitaire, then you do not play free cell and you play minesweeper.

In Exercises 31–36, use a truth table to determine whether the statement is a tautology, self-contradiction, or neither.

31. $\sim p \vee (p \vee q)$

32. $\sim p \wedge (p \wedge q)$

33. $(\sim p \wedge q) \wedge (p \vee \sim q)$

34. $(p \vee \sim q) \vee (\sim p \vee q)$

35. $\sim [(p \vee q) \leftrightarrow q]$

36. $\sim [(p \wedge q) \rightarrow p]$

In Exercises 37–42, use a truth table to determine whether the statement is an implication.

37. $\sim p \rightarrow p$

38. $p \rightarrow (p \vee q)$

39. $\sim p \rightarrow \sim (p \wedge q)$

40. $(p \vee q) \rightarrow (p \vee \sim r)$

41. $[(p \rightarrow q) \wedge (q \rightarrow p)] \rightarrow (p \leftrightarrow q)$

42. $[(p \vee q) \wedge r] \rightarrow (p \vee q)$

In Exercises 43–50, if p is true, q is false, and r is false, determine the truth value of the statement.

43. $p \rightarrow (q \rightarrow r)$

44. $(p \wedge \sim q) \rightarrow \sim r$

45. $(p \wedge q) \leftrightarrow (q \vee \sim r)$

46. $r \rightarrow (\sim p \leftrightarrow \sim q)$

47. $(\sim p \wedge \sim q) \vee \sim r$

48. $\sim [p \rightarrow (q \wedge r)]$

49. $(\sim p \leftrightarrow r) \vee (\sim q \leftrightarrow r)$

50. $\sim [(p \vee q) \leftrightarrow (p \rightarrow \sim r)]$

Problem Solving

In Exercises 51–58, determine the truth value for each simple statement. Then, using the truth values, determine the truth value of the compound statement.

51. If $1 + 1 = 3$, then $5 - 2 = 3$.

52. If $7^2 = 49$, then $\sqrt{49} = 7$.

53. If the United States has 60 states or the United States capital is Washington, D.C., then the capital of Mexico is Philadelphia.

▲ The U.S. Capitol Building

54. If the moon circles the earth and the earth circles the sun, then the moon is made of cheese.

55. Snickers is a brand of watches and Hershey's is a brand of cell phones, if and only if Reese's is a brand of trucks.

56. Quentin Tarantino is a movie director, or if Halle Berry is a school teacher then Denzel Washington is a circus clown.

57. Independence Day is in July and Labor Day is in September, if and only if Thanksgiving is in April.

58. Honda makes automobiles or Honda makes motorcycles, if and only if Mazda makes cereal.

In Exercises 59–62, use the information provided about the moons for the planets Jupiter and Saturn to determine the truth values of the simple statements. Then determine the truth value of the compound statement.

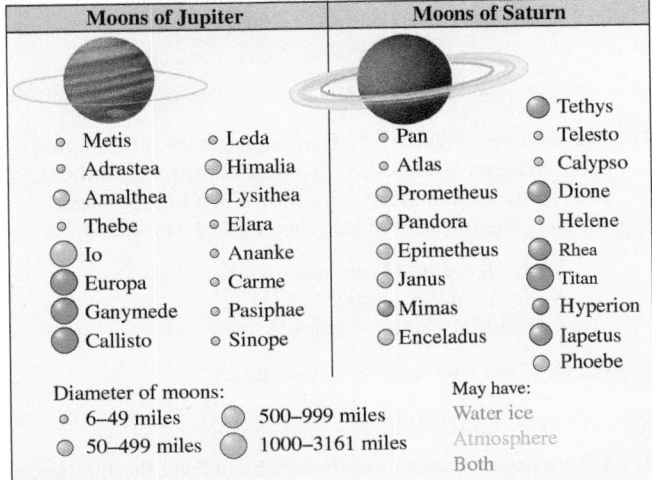

Source: Data from *Time* magazine

59. *Jupiter's Moons* Io has a diameter of 1000–3161 miles or Thebe may have water, and Io may have atmosphere.

60. *Moons of Saturn* Titan may have water and Titan may have atmosphere, if and only if Janus may have water.

61. *Moon Comparisons* Phoebe has a larger diameter than Rhea if and only if Callisto may have water ice, and Calypso has a diameter of 6–49 miles.

62. *Moon Comparisons* If Jupiter has 16 moons or Saturn does not have 18 moons, then Saturn has 7 moons that may have water ice.

College Credits *In Exercises 63 and 64, use the graph to determine the truth value of each simple statement. Then determine the truth value of the compound statement.*

The following graph shows the number of credits in various categories needed by Arturo to earn his Associate in Arts degree from Van Pelt University.

Credits Needed for an A.A. Degree

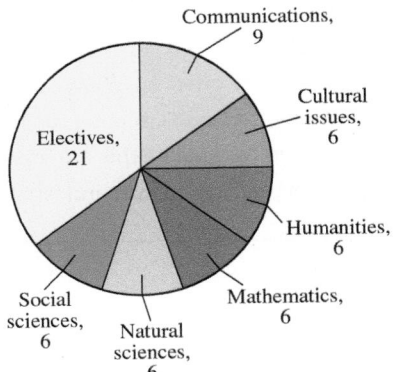

63. The number of communications credits needed is more than the number of mathematics credits needed and the number of cultural issues credits needed is equal to the number of humanities credits needed, if and only if the number of social sciences credits needed is more than the number of natural sciences credits needed.

64. If the number of elective credits needed is 21 or the number of communications credits needed is 15, then the number of humanities credits needed is equal to the number of mathematics credits needed.

In Exercises 65–68, suppose both of the following statements are false.

p: Muhundan spoke at the teachers' conference.

q: Muhundan received the outstanding teacher award.

Determine the truth values of each compound statement.

65. If Muhundan received the outstanding teacher award, then Muhundan spoke at the teachers' conference.

66. If Muhundan did not receive the outstanding teacher award, then Muhundan spoke at the teachers' conference.

67. Muhundan did not receive the outstanding teacher award if and only if Muhundan spoke at the teachers' conference.

68. If Muhundan did not receive the outstanding teacher award, then Muhundan did not speak at the teachers' conference.

Concept/Writing Exercises

69. *A New Computer* Your parents make the following statement to your sister, "If you get straight A's this semester, then we will buy you a new computer." At the end of the semester your parents buy your sister a new computer. Can you conclude that your sister got straight A's? Explain.

70. *Job Interview* Consider the statement "If your interview goes well, then you will be offered the job." If you are interviewed and then offered the job, can you conclude that your interview went well? Explain.

Challenge Problems/Group Activities

In Exercises 71 and 72, construct truth tables for the symbolic statement.

71. $[p \vee (q \rightarrow \sim r)] \leftrightarrow (p \wedge \sim q)$

72. $[(r \rightarrow \sim q) \rightarrow \sim p] \vee (q \leftrightarrow \sim r)$

73. Construct a truth table for

  a) $(p \vee q) \rightarrow (r \wedge s)$.

  b) $(q \rightarrow \sim p) \vee (r \leftrightarrow s)$.

Recreational Mathematics

74. Satisfiability Problem Refer to the Recreational Mathematics box on page 121 and then solve the following satisfiability problem. Allen, Booker, Chris, and Dennis all were born in the same year—one in January, one in February, one in March, and one in April. Chris was born before Dennis. Dennis was born two months after Booker. Booker was born after Allen, but before Chris. Determine who was born in each month.

75. Cat Puzzle Solve the following puzzle. Pablo has four cats. The parents are Tiger and Boots, and the kittens are Sam and Sue. Each cat insists on eating out of its own bowl. To complicate matters, each cat will eat only its own brand of cat food. The colors of the bowls are yellow, pink, blue, and green. The different types of cat food are Whiskas, Friskies, Nine Lives, and Meow Mix. Tiger will eat Meow Mix if and only if it is in a yellow bowl. If Boots is to eat her food, then it must be in a yellow bowl. Pablo knows that the label on the can containing Sam's food is the same color as his bowl. Boots eats Whiskas. Meow Mix and Nine Lives are packaged in a brown paper bag. The color of Sue's bowl is green if and only if she eats Meow Mix. The label on the Friskies can is pink. Match each cat with its food and the bowl of the correct color.

▲ *See Exercise 75*

76. The Youngest Triplet The Barr triplets have an annoying habit: Whenever a question is asked of the three of them, two tell the truth and the third lies. When I asked them which of them was born last, they replied as follows.

> *Mary:* Katie was born last.
> *Katie:* I am the youngest.
> *Annie:* Mary is the youngest.

Which of the Barr triplets was born last?

Research Activity

77. Advertisement Select an advertisement from the Internet, a newspaper, or a magazine that makes or implies a conditional statement. Analyze the advertisement to determine whether the consequent necessarily follows from the antecedent. Explain your answer. (See Example 3.)

SECTION 3.4 Equivalent Statements

Upon completion of this section, you will be able to:

- Determine if two statements are equivalent.
- Use De Morgan's laws for logic.
- Write conditional statements as disjunctions.
- Write the negation of conditional statements.
- Write variations of conditional statements, including inverses, converses, and contrapositives.

Consider the following statements:

If you got straight A's, then you are on the president's list.

If you are on the president's list, then you got straight A's.

If you did not get straight A's, then you are not on the president's list.

If you are not on the president's list, then you did not get straight A's.

Are these statements all saying the same thing, or does each one say something completely different from the others? In this section, we will use logic symbols and truth tables to compare each of these statements. We will also learn several forms of equivalent statements.

Why This Is Important In our everyday lives we often encounter statements that are initially difficult to understand. The tools presented in this section can help us better understand such statements. Understanding when two statements are equivalent is important to understanding advertisers' claims, political statements, and legal documents.

Equivalent Statements

Equivalent statements are an important concept in the study of logic.

3.4 Equivalent Statements

> **Definition: Equivalent**
> Two statements are **equivalent**, symbolized ⇔, if both statements have exactly the same truth values in the answer columns of the truth tables.*

Sometimes the words *logically equivalent* are used in place of the word *equivalent*.

To determine whether two statements are equivalent, construct a truth table for each statement and compare the answer columns of the truth tables. If the answer columns are identical, the statements are equivalent. If the answer columns are not identical, the statements are not equivalent.

Example 1 *Equivalent Statements*

Determine whether the following two statements are equivalent.

$$p \wedge (q \vee r)$$
$$(p \wedge q) \vee (p \wedge r)$$

Solution Construct a truth table for each statement (see Table 3.25).

Table 3.25

p	q	r	p	$\wedge$	$(q \vee r)$	$(p \wedge q)$	$\vee$	$(p \wedge r)$
T	T	T	T	T	T	T	T	T
T	T	F	T	T	T	T	T	F
T	F	T	T	T	T	F	T	T
T	F	F	T	F	F	F	F	F
F	T	T	F	F	T	F	F	F
F	T	F	F	F	T	F	F	F
F	F	T	F	F	T	F	F	F
F	F	F	F	F	F	F	F	F
			1	3	2	1	3	2

Because the truth tables have the same answer (column 3 for both tables), the statements are equivalent. Therefore, we can write

$$p \wedge (q \vee r) \Leftrightarrow (p \wedge q) \vee (p \wedge r)$$ ∎

Example 2 *Are the Following Equivalent Statements?*

Determine whether the following statements are equivalent.

a) If you win your fantasy football league and you are here Friday, then we will celebrate.

b) If you do not win your fantasy football league or you are not here Friday, then we will not celebrate.

Solution First write each statement in symbolic form, then construct a truth table for each statement. If the answer columns of both truth tables are identical, then the statements are equivalent. If the answer columns are not identical, then the statements are not equivalent.

Let:

p: You win your fantasy football league.

q: You are here Friday.

r: We will celebrate.

*The symbol ≡ is also used to indicate equivalent statements.

RECREATIONAL MATH

Kakuro

Like Sudoku (see the *Recreational Mathematics* box on Sudoku on page 95), Kakuro is a puzzle that requires logic. The aim of the game is to fill all the blank squares in the grid with only the digits 1–9 so that the numbers you enter add up to the corresponding clues. For example, in the puzzle shown below, the 17 near the upper left corner of the puzzle is a clue that indicates that the two numbers that you place in the two squares to the right of the 17 must add up to equal 17. The 9 near the upper left corner of the puzzle is a clue that indicates that the two numbers that you place in the two squares below the 9 must add up to equal 9. In addition, you may not repeat any digits in any given "word" or string of numbers.

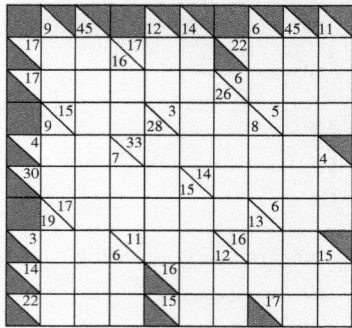

For more information and more puzzles, see www.kakuro.com. Furthermore, there are many apps that can be used to play Kakuro on your tablet or smartphone. The solution to the puzzle can be found in the answer section in the back of this book. For an additional puzzle, see Exercise 76 on page 139.

In symbolic form, the statements are

a) $(p \wedge q) \rightarrow r$.   b) $(\sim p \vee \sim q) \rightarrow \sim r$.

The truth tables for these statements are given in Tables 3.26 and 3.27, respectively. The answers in the columns labeled 5 are not identical, so the statements are not equivalent.

Table 3.26

p	q	r	(p	∧	q)	→	r
T	T	T	T	T	T	T	T
T	T	F	T	T	T	F	F
T	F	T	T	F	F	T	T
T	F	F	T	F	F	T	F
F	T	T	F	F	T	T	T
F	T	F	F	F	T	T	F
F	F	T	F	F	F	T	T
F	F	F	F	F	F	T	F
			1	3	2	5	4

Table 3.27

p	q	r	(~p	∨	~q)	→	~r
T	T	T	F	F	F	T	F
T	T	F	F	F	F	T	T
T	F	T	F	T	T	F	F
T	F	F	F	T	T	T	T
F	T	T	T	T	F	F	F
F	T	F	T	T	F	T	T
F	F	T	T	T	T	F	F
F	F	F	T	T	T	T	T
			1	3	2	5	4

Example 3 *Which Statements Are Logically Equivalent?*

Determine which, if any, of the following statements are equivalent to "It is not true that the story is on social media and the story is real."

a) If the story is not real, then the story is not on social media.

b) The story is not on social media or the story is not real.

c) The story is not real and the story is not on social media.

d) If the story is not on social media, then the story is not real.

Solution To determine whether any of the choices are equivalent to the given statement, first write the given statement and the choices in symbolic form. Then construct truth tables and compare the answer columns of the truth tables.

Let

p: The story is on social media.

q: The story is real.

The given statement is expressed in symbolic form as $\sim (p \wedge q)$. Using p and q as indicated, choices (a) through (d) may be expressed symbolically as

a) $\sim q \rightarrow \sim p$.   b) $\sim p \vee \sim q$.   c) $\sim q \wedge \sim p$.   d) $\sim p \rightarrow \sim q$.

Table 3.28 shows the truth table for the given statement and Table 3.29 shows the truth table for each statement (a) through (d). By examining the truth tables, we see that the given statement, $\sim (p \wedge q)$, is logically equivalent to choice (b), $\sim p \vee \sim q$. Therefore, the correct answer is "The story is not on social media or the story is not real." This statement is logically equivalent to the original statement "It is not true that the story is on social media and the story is real."

Table 3.28

p	q	~	(p	∧	q)
T	T	F	T	T	T
T	F	T	T	F	F
F	T	T	F	F	T
F	F	T	F	F	F
		4	1	3	2

Table 3.29

		(a)			(b)			(c)			(d)		
p	q	~q	→	~p	~p	∨	~q	~q	∧	~p	~p	→	~q
T	T	F	T	F	F	F	F	F	F	F	F	T	F
T	F	T	F	F	F	T	T	T	F	F	F	T	T
F	T	F	T	T	T	T	F	F	F	T	T	F	F
F	F	T	T	T	T	T	T	T	T	T	T	T	T

De Morgan's Laws for Logic

Example 3 showed that a statement of the form $\sim(p \wedge q)$ is equivalent to a statement of the form $\sim p \vee \sim q$. Thus, we may write $\sim(p \wedge q) \Leftrightarrow \sim p \vee \sim q$. This equivalent statement is the first of two special laws called De Morgan's laws. The laws, named after Augustus De Morgan, an English mathematician, were first introduced in Section 2.4, where they applied to sets.

De Morgan's Laws for Logic

1. $\sim(p \wedge q) \Leftrightarrow \sim p \vee \sim q$
2. $\sim(p \vee q) \Leftrightarrow \sim p \wedge \sim q$

You can demonstrate that De Morgan's second law is true by constructing and comparing truth tables for $\sim(p \vee q)$ and $\sim p \wedge \sim q$. Do so now.

When using De Morgan's laws, if it becomes necessary to negate an already negated statement, use the fact that $\sim(\sim p)$ is equivalent to p. For example, the negation of the statement "Today is not Monday" is "Today is Monday."

Example 4 Use De Morgan's Laws

Select the statement that is logically equivalent to "I do not have investments, but I do not have debts."

a) I do not have investments or I do not have debts.

b) It is false that I have investments and I have debts.

c) It is false that I have investments or I have debts.

d) I have investments or I have debts.

Solution To determine which statement is equivalent, write each statement in symbolic form.

Let

p: I have investments.

q: I have debts.

The statement "I do not have investments, but I do not have debts" written symbolically is $\sim p \wedge \sim q$. Recall that the word *but* means the same thing as *and*. Now, write parts (a) through (d) symbolically.

a) $\sim p \vee \sim q$ b) $\sim(p \wedge q)$ c) $\sim(p \vee q)$ d) $p \vee q$

De Morgan's law shows that $\sim p \wedge \sim q$ is equivalent to $\sim(p \vee q)$. Therefore, the answer is (c): "It is false that I have investments or I have debts." ∎

Example 5 Use De Morgan's Laws

Write a statement that is logically equivalent to "It is false that you can take Trigonometry to meet your mathematics requirement or you can take Biology to meet your science requirement."

Solution Let

p: You can take Trigonometry to meet your mathematics requirement.

q: You can take Biology to meet your science requirement.

The given statement is of the form $\sim(p \vee q)$. Using the second of De Morgan's laws, we see that an equivalent statement in symbols is $\sim p \wedge \sim q$. Therefore, an equivalent statement is "You cannot take Trigonometry to meet your mathematics requirement and you cannot take Biology to meet your science requirement." ∎

Consider $\sim(p \wedge q) \Leftrightarrow \sim p \vee \sim q$, one of De Morgan's laws. To go from $\sim(p \wedge q)$ to $\sim p \vee \sim q$, we negate both the p and the q within parentheses; change the conjunction, $\wedge$, to a disjunction, $\vee$; and remove the negation symbol preceding the left parentheses and the parentheses themselves. We can use a similar procedure to obtain equivalent statements. For example,

$$\sim(\sim p \wedge q) \Leftrightarrow p \vee \sim q$$
$$\sim(p \wedge \sim q) \Leftrightarrow \sim p \vee q$$

We can use a similar procedure to obtain equivalent statements when a disjunction is within parentheses. Note that

$$\sim(\sim p \vee q) \Leftrightarrow p \wedge \sim q$$
$$\sim(p \vee \sim q) \Leftrightarrow \sim p \wedge q$$

Example 6   *Using De Morgan's Laws*

Use De Morgan's laws to write a statement logically equivalent to "We cannot go to the fireworks, but we can go to the ice cream shop."

Solution Let

p: We can go to the fireworks.

q: We can go to the ice cream shop.

The statement written symbolically is $\sim p \wedge q$. Earlier we showed that

$$\sim p \wedge q \Leftrightarrow \sim(p \vee \sim q)$$

Therefore, the statement "It is false that we can go to the fireworks or we cannot go to the ice cream shop" is logically equivalent to the given statement. ∎

There are strong similarities between the topics of set theory and logic. We can see them by examining De Morgan's laws for sets and logic.

De Morgan's laws: set theory	De Morgan's laws: logic
$(A \cap B)' = A' \cup B'$	$\sim(p \wedge q) \Leftrightarrow \sim p \vee \sim q$
$(A \cup B)' = A' \cap B'$	$\sim(p \vee q) \Leftrightarrow \sim p \wedge \sim q$

The complement in set theory, ′, is similar to the negation, $\sim$, in logic. The intersection, $\cap$, is similar to the conjunction, $\wedge$; and the union, $\cup$, is similar to the disjunction, $\vee$. If we were to interchange the set symbols with the logic symbols, De Morgan's laws would remain, but in a different form.

Both ′ and $\sim$ can be interpreted as *not*.
Both $\cap$ and $\wedge$ can be interpreted as *and*.
Both $\cup$ and $\vee$ can be interpreted as *or*.

For example, the set statement $A' \cup B$ can be written as a statement in logic as $\sim a \vee b$.

Conditional Statements as Disjunctions

Statements containing connectives other than *and* and *or* may have equivalent statements. To illustrate this point, construct truth tables for $p \rightarrow q$ and for $\sim p \lor q$. The truth tables will have the same answer columns and therefore the statements are equivalent. We summarize this as follows.

The Conditional Statement Written as a Disjunction

$$p \rightarrow q \Leftrightarrow \sim p \lor q$$

With these equivalent statements, we can write a conditional statement as a disjunction or a disjunction as a conditional statement. For example, the statement "If the game is polo, then you ride a horse" can be equivalently stated as "The game is not polo or you ride a horse."

To change a conditional statement to a disjunction, negate the antecedent, change the conditional symbol to a disjunction symbol, and keep the consequent the same. To change a disjunction statement to a conditional statement, negate the first statement, change the disjunction symbol to a conditional symbol, and keep the second statement the same.

Example 7 *Rewriting a Disjunction as a Conditional Statement*

Write a conditional statement that is logically equivalent to "The cows are in the pasture or the horses are not in the barn."

Solution Let

p: The cows are in the pasture.

q: The horses are in the barn.

The original statement may be written symbolically as $p \lor \sim q$. To write an equivalent conditional statement, negate the first statement, p; replace the disjunction symbol, $\lor$, with a conditional symbol, $\rightarrow$; and keep the second statement the same. Symbolically, the equivalent statement is $\sim p \rightarrow \sim q$. The equivalent statement in words is "If the cows are not in the pasture, then the horses are not in the barn." ∎

Negation of the Conditional Statement

Now we will discuss how to negate a conditional statement. To negate a conditional statement we use the fact that $p \rightarrow q \Leftrightarrow \sim p \lor q$ and De Morgan's laws. Examples 8 and 9 show this process.

Example 8 *The Negation of a Conditional Statement*

Determine a statement equivalent to $\sim (p \rightarrow q)$.

Solution Begin with $p \rightarrow q \Leftrightarrow \sim p \lor q$, negate both statements, and use De Morgan's laws.

$$p \rightarrow q \Leftrightarrow \sim p \lor q$$
$$\sim (p \rightarrow q) \Leftrightarrow \sim (\sim p \lor q) \quad \text{Negate both statements.}$$
$$\Leftrightarrow p \land \sim q \quad \text{De Morgan's laws}$$

Therefore, $\sim (p \rightarrow q)$ is equivalent to $p \land \sim q$. ∎

We summarize the result of Example 8 as follows.

> ### The Negation of the Conditional Statement Written as a Conjunction
> $$\sim(p \to q) \Leftrightarrow p \wedge \sim q$$

▲ Spikeball

Example 9 Write an Equivalent Statement

Write a statement that is equivalent to "It is not true that if I play spikeball then I sprain my ankle."

Solution Let

p: I play spikeball.

q: I sprain my ankle.

The given statement can be represented symbolically as $\sim(p \to q)$. We showed in Example 8 that $\sim(p \to q)$ is logically equivalent to $p \wedge \sim q$. Therefore, an equivalent statement is "I play spikeball and I do not sprain my ankle."   ■

Using the fact that $\sim(p \to q) \Leftrightarrow p \wedge \sim q$, can you determine what $\sim(p \to \sim q)$ is equivalent to as a conjunction? If you answered $p \wedge q$ you answered correctly.

Variations of the Conditional Statement

We know that $p \to q$ is equivalent to $\sim p \vee q$. Are any other statements equivalent to $p \to q$? Yes, there are many. Now let's look at the variations of the conditional statement to determine whether any are equivalent to the conditional statement. *The variations of the conditional statement are made by switching and/or negating the antecedent and the consequent of a conditional statement.* The variations of the conditional statement are the *converse* of the conditional, the *inverse* of the conditional, and the *contrapositive* of the conditional.

Listed here are the variations of the conditional with their symbolic form and the words we say to read each one.

Variations of the
Conditional Statement

> ### Variations of the Conditional Statement
>
> | Name | Symbolic form | Read |
> |------|---------------|------|
> | Conditional | $p \to q$ | "If p, then q" |
> | Converse of the conditional | $q \to p$ | "If q, then p" |
> | Inverse of the conditional | $\sim p \to \sim q$ | "If not p, then not q" |
> | Contrapositive of the conditional | $\sim q \to \sim p$ | "If not q, then not p" |

To write the converse of the conditional statement, switch the order of the antecedent and the consequent. To write the inverse, negate both the antecedent and the consequent. To write the contrapositive, switch the order of the antecedent and the consequent and then negate both of them.

Are any of the variations of the conditional statement equivalent? To determine the answer, we can construct a truth table for each variation, as shown in Table 3.30 on page 133. It reveals that *the conditional statement is equivalent to the contrapositive statement and that the converse statement is equivalent to the inverse statement.*

Timely Tip

From Table 3.30 we can see that a conditional statement and the contrapositive statement are always equivalent to each other. We can also see that the converse statement and the inverse statement are always equivalent to each other, but they are not equivalent to the original conditional statement.

Table 3.30

p	q	Conditional $p \rightarrow q$	Contrapositive $\sim q \rightarrow \sim p$	Converse $q \rightarrow p$	Inverse $\sim p \rightarrow \sim q$
T	T	T	T	T	T
T	F	F	F	T	T
F	T	T	T	F	F
F	F	T	T	T	T

▲ Andy Warhol

Example 10   The Converse, Inverse, and Contrapositive

For the conditional statement "If the painting is by Andy Warhol, then the painting is valuable," write the
a) converse.   b) inverse.   c) contrapositive.

Solution

a) Let

   p: The painting is by Andy Warhol.

   q: The painting is valuable.

The conditional statement is of the form $p \rightarrow q$, so the converse must be of the form $q \rightarrow p$. Therefore, the converse is "If the painting is valuable, then the painting is by Andy Warhol."

b) The inverse is of the form $\sim p \rightarrow \sim q$. Therefore, the inverse is "If the painting is not by Andy Warhol, then the painting is not valuable."

c) The contrapositive is of the form $\sim q \rightarrow \sim p$. Therefore, the contrapositive is "If the painting is not valuable, then the painting is not by Andy Warhol." ∎

Example 11   Determine the Truth Values

Let

   p: The number is divisible by 10.

   q: The number is divisible by 5.

Write the following statements and determine which are true.
a) The conditional statement, $p \rightarrow q$
b) The converse of $p \rightarrow q$
c) The inverse of $p \rightarrow q$
d) The contrapositive of $p \rightarrow q$

Solution

a) The conditional statement in symbols is $p \rightarrow q$. Therefore, in words the conditional statement is *If the number is divisible by 10, then the number is divisible by 5.* This statement is true. A number divisible by 10 must also be divisible by 5, since 5 is a divisor of 10.

b) The converse of the conditional statement in symbols is $q \rightarrow p$. Therefore, in words the converse is *If the number is divisible by 5, then the number is divisible by 10.* This statement is false. For example, 15 is divisible by 5, but 15 is not divisible by 10.

c) The inverse of the conditional statement in symbols is $\sim p \rightarrow \sim q$. Therefore, in words the inverse is *If the number is not divisible by 10, then the number is not divisible by 5.* This statement is false. For example, 25 is not divisible by 10, but 25 is divisible by 5.

d) The contrapositive of the conditional statement in symbols is $\sim q \rightarrow \sim p$. Therefore, in words the contrapositive is *If the number is not divisible by 5, then the number is not divisible by 10.* This statement is true. Any number that is not divisible by 5 cannot be divisible by 10, since 5 is a divisor of 10.   ■

Since the contrapositive statement is always equivalent to the original conditional statement, in Example 11(d) we should have expected the answer to be a true statement because the original conditional statement was also a true statement.

Example 12   *Use the Contrapositive*

Use the contrapositive to write a statement logically equivalent to "If you don't eat your meat, then you can't have any pudding."

Solution Let

$$p: \quad \text{You do eat your meat.}$$
$$q: \quad \text{You can have any pudding.}$$

The given statement written symbolically is

$$\sim p \rightarrow \sim q$$

The contrapositive of the statement is

$$q \rightarrow p$$

Therefore, an equivalent statement is "If you can have any pudding, then you do eat your meat."   ■

The contrapositive of the conditional is very important in mathematics. Consider the statement "If a^2 is not a whole number, then a is not a whole number." Is this statement true? You may find this question difficult to answer. Writing the statement's contrapositive may enable you to answer the question. The contrapositive is "If a is a whole number, then a^2 is a whole number." Since the contrapositive is a true statement, the original statement must also be true.

Example 13   *Which Are Equivalent?*

Determine which, if any, of the following statements are equivalent. You may use De Morgan's laws, the fact that $p \rightarrow q \Leftrightarrow \sim p \vee q$, information from the variations of the conditional, or truth tables.

a) If you leave by 9 A.M., then you will get to your destination on time.

b) You do not leave by 9 A.M. or you will get to your destination on time.

c) It is false that you will get to your destination on time or you did not leave by 9 A.M.

d) If you do not get to your destination on time, then you did not leave by 9 A.M.

Solution Let

$$p: \quad \text{You leave by 9 A.M.}$$
$$q: \quad \text{You will get to your destination on time.}$$

In symbolic form, the four statements are

a) $p \rightarrow q$.   b) $\sim p \vee q$.   c) $\sim (q \vee \sim p)$.   d) $\sim q \rightarrow \sim p$.

Which of these statements are equivalent? Earlier in this section, you learned that $p \rightarrow q$ is equivalent to $\sim p \lor q$. Therefore, statements (a) and (b) are equivalent. Statement (d) is the contrapositive of statement (a). Therefore, statement (d) is also equivalent to statement (a) and statement (b). Statements (a), (b), and (d) all have the same truth table (Table 3.31).

Table 3.31

		(a)	(b)	(d)
p	q	$p \rightarrow q$	$\sim p \lor q$	$\sim q \rightarrow \sim p$
T	T	T	T	T
T	F	F	F	F
F	T	T	T	T
F	F	T	T	T

Now let's look at statement (c). To determine whether $\sim(q \lor \sim p)$ is equivalent to the other statements, we will construct its truth table (Table 3.32) and compare the answer column with the answer columns in Table 3.31.

Table 3.32

			(c)		
p	q	$\sim$	$(q$	$\lor$	$\sim p)$
T	T	F	T	T	F
T	F	T	F	F	F
F	T	F	T	T	T
F	F	F	F	T	T
		4	1	3	2

None of the three answer columns of the truth table in Table 3.31 is the same as the answer column of the truth table in Table 3.32. Therefore $\sim(q \lor \sim p)$ is not equivalent to any of the other statements. Therefore, only statements (a), (b), and (d) are equivalent to each other. ∎

SECTION 3.4 *Exercises*

Warm Up Exercises

In Exercises 1–8, fill in the blanks with an appropriate word, phrase, or symbol(s).

1. Statements that have exactly the same truth values in the answer columns of their truth tables are called _____ statements.

2. De Morgan's laws state that

   a) $\sim(p \land q)$ is equivalent to _____, and

   b) $\sim(p \lor q)$ is equivalent to _____.

3. The conditional statement $p \rightarrow q$ is equivalent to the following disjunction statement: _____.

4. The negation of the conditional statement $\sim(p \rightarrow q)$ is equivalent to the following conjunction statement: _____.

5. Given the conditional statement $p \rightarrow q$, the converse of the conditional statement in symbolic form is _____.

6. Given the conditional statement $p \rightarrow q$, the inverse of the conditional statement in symbolic form is _____.

7. Given the conditional statement $p \rightarrow q$, the contrapositive of the conditional statement in symbolic form is _____.

8. Of the converse, inverse, and contrapositive, only the contrapositive of the conditional statement is _____ to the conditional statement.

Practice the Skills

In Exercises 9–16, use a truth table to determine whether the two statements are equivalent.

9. $p \rightarrow q, \sim p \vee q$

10. $\sim(p \rightarrow q), p \wedge \sim q$

11. $\sim q \rightarrow \sim p, p \rightarrow q$

12. $q \rightarrow p, \sim p \rightarrow \sim q$

13. $(p \vee q) \wedge r, p \vee (q \wedge r)$

14. $p \wedge (q \vee r), (p \wedge q) \vee r$

15. $(p \rightarrow q) \wedge (q \rightarrow p), (p \leftrightarrow q)$

16. $(p \rightarrow q) \rightarrow r, p \rightarrow (q \rightarrow r)$

In Exercises 17–24, use De Morgan's laws to determine whether the two statements are equivalent.

17. $\sim(p \wedge q), \sim p \wedge \sim q$

18. $\sim(p \wedge q), \sim p \vee \sim q$

19. $\sim(p \vee q), \sim p \wedge \sim q$

20. $\sim(p \vee q), \sim p \vee \sim q$

21. $\sim(\sim p \wedge q), p \wedge \sim q$

22. $\sim(\sim p \wedge q), p \vee \sim q$

23. $(\sim p \vee \sim q) \rightarrow r, \sim(p \wedge q) \rightarrow r$

24. $q \rightarrow \sim(p \wedge \sim r), q \rightarrow \sim p \vee r$

Problem Solving

In Exercises 25–30, use De Morgan's laws to write an equivalent statement for the given sentence.

25. It is false that Jay-Z sings opera and Beyoncé sings country.

26. It is false that the Camaro is a Dodge or the Challenger is a Chevy.

27. The dog was not a bulldog and the dog was not a boxer.

28. Franklin is wealthy, but he is not famous.

29. If I am late for class, then I will not get bonus points or I will fail the test.

30. If Phil buys us dinner, then we will not go to the top of the CN Tower but we will be able to walk to the Red Bistro Restaurant.

In Exercises 31–36, use the fact that $p \rightarrow q$ is equivalent to $\sim p \vee q$ to write an equivalent form of the given statement.

31. If I see a movie, then I buy popcorn.

32. Byron did not walk to the meeting or we started late.

33. Chase is not hiding or the pitcher is broken.

34. If Joanne goes to the Lightning game, then she will not go to the Rays game.

35. Opal exercises daily or she is not healthy.

36. If Weezer is not on the radio, then Oly is working.

In Exercises 37–42, use the fact that $\sim(p \rightarrow q)$ is equivalent to $p \wedge \sim q$ to write the statement in an equivalent form.

37. It is false that if we go to Chicago, then we will go to Navy Pier.

38. It is not true that if Ghana is in the United Nations then Ghana is in the World Trade Organization.

39. I am cold and the heater is not working.

40. The Badgers beat the Nittany Lions and the Bucks beat the 76ers.

41. It is not true that if Amazon has a sale then we will buy $100 worth of books.

42. Dexter works in Miami, but Debra does not work for him.

In Exercises 43–48, write the converse, inverse, and contrapositive of the statement in sentence form.

43. If you got straight A's, then you are on the president's list.

44. If the water is running, then Linus is getting a drink.

45. If I go to Mexico, then I buy silver jewelry.

46. If I write a haiku, then you write a sonnet.

47. If the menu includes calzones, then I do not stay on my diet.

48. If Grandma is not available to babysit, then we cannot go to the movie.

In Exercises 49–54, write the contrapositive of the statement. Use the contrapositive to determine whether the conditional statement is true or false.

49. If a natural number is divisible by 14, then the natural number is divisible by 7.

50. If the quadrilateral is a parallelogram, then the opposite sides are parallel.

51. If a natural number is divisible by 3, then the natural number is divisible by 6.

52. If the number is not a natural number, then the number is not a positive number.

53. If two lines do not intersect in at least one point, then the two lines are parallel.

54. If the two lines are perpendicular, then the two lines are not parallel.

In Exercises 55–68, determine which, if any, of the three statements are equivalent (see Example 13).

55. a) If the player gets a red card, then the player sits out.

   b) If the player sits out, then the player gets a red card.

   c) The player does not sit out or the player gets a red card.

56. a) If the dress is a Prada, then it is not a Gucci.

   b) It is not true that the dress is a Prada and it is not a Gucci.

   c) The dress is not a Prada or it is a Gucci.

57. a) The shoes are not on sale but the purse is on sale.

   b) If the shoes are not on sale then the purse is not on sale.

   c) It is false that the shoes are on sale or the purse is not on sale.

58. a) LeBron did not play for the Bucks or Giannis did not play for the Cavaliers.

   b) If LeBron played for the Bucks, then Giannis played for the Cavaliers.

   c) It is not true that Giannis played for the Cavaliers and LeBron did not play for the Bucks.

59. a) We do not go hiking or we go fishing.

   b) If we go hiking, then we do not go fishing.

   c) If we go fishing, then we do not go hiking.

60. a) If Tim is the director then Johnny is the actor.

   b) Tim is not the director or Johnny is the actor.

   c) It is false that Tim is the director and Johnny is not the actor.

61. a) The grass grows and the trees are blooming.

   b) If the trees are blooming, then the grass does not grow.

   c) The trees are not blooming or the grass does not grow.

62. a) Akemi plays *Star Stable* if and only if she plays *Just Dance*.

   b) If Akemi plays *Star Stable*, then she plays *Just Dance*, and if Akemi plays *Just Dance*, then she plays *Star Stable*.

   c) Akemi does not play *Star Stable* and she does not play *Just Dance*.

63. a) If the corn bag goes in the hole, then you are awarded three points.

   b) It is false that the corn bag goes in the hole and you are awarded three points.

   c) The corn bag does not go in the hole and you are not awarded three points.

64. a) *Fitz and the Tantrums* will not go on tour if and only if James King does not play the saxophone.

b) It is false that *Fitz and the Tantrums* will go on tour if and only if James King does not play the saxophone.

c) If *Fitz and the Tantrums* go on tour, then James King plays saxophone.

65. a) If the pay is good and today is Monday, then I will take the job.

b) If I do not take the job, then it is false that the pay is good or today is Monday.

c) The pay is good and today is Monday, or I will take the job.

66. a) If you are 18 years old and a citizen of the United States, then you can vote in the presidential election.

b) You can vote in the presidential election, if and only if you are a citizen of the United States and you are 18 years old.

c) You cannot vote in the presidential election, or you are 18 years old and you are not a citizen of the United States.

67. a) The spacecraft was made by SpaceX, or it was not made by Orbital and it was made by Boeing.

b) The spacecraft was made by Boeing if and only if, it was made by SpaceX or it was not made by Orbital.

c) If the spacecraft was not made by SpaceX, then it was not made by Orbital and it was made by Boeing.

68. a) The mortgage rate went down, if and only if Tim purchased the house and the down payment was 10%.

b) The down payment was 10%, and if Tim purchased the house then the mortgage rate went down.

c) If Tim purchased the house, then the mortgage rate went down and the down payment was not 10%.

Concept/Writing Exercises

69. If p and q represent two simple statements, and if $p \rightarrow q$ is a false statement, what must be the truth value of the converse, $q \rightarrow p$? Explain.

70. If p and q represent two simple statements, and if $p \rightarrow q$ is a false statement, what must be the truth value of the inverse, $\sim p \rightarrow \sim q$? Explain.

71. If p and q represent two simple statements, and if $p \rightarrow q$ is a false statement, what must be the truth value of the contrapositive, $\sim q \rightarrow \sim p$? Explain.

72. If p and q represent two simple statements, and if $p \rightarrow q$ is a true statement, what must be the truth value of the contrapositive, $\sim q \rightarrow \sim p$? Explain.

Challenge Problems/Group Activities

73. We learned that $p \rightarrow q \Leftrightarrow \sim p \ \vee \ q$. Determine a conjunction that is equivalent to $p \rightarrow q$. (*Hint:* There are many answers.)

74. Determine whether $\sim [\sim (p \vee \sim q)] \Leftrightarrow p \vee \sim q$. Explain the method(s) you used to determine your answer.

75. *Fuzzy Logic* Read the *Mathematics Today* box on fuzzy logic on page 134. In symbolic logic, a statement is either true or false (consider true to have a value of 1 and false a value of 0). In fuzzy logic, nothing is true or false, but everything is a matter of degree. For example, consider the statement "The sun is shining." In fuzzy logic, this statement may have a value between 0 and 1 and may be constantly changing. For example, if the sun is partially blocked by clouds, the value of this statement may be 0.25. In fuzzy logic, the values of connective statements are found as follows for statements p and q.

Not p has a truth value of $1 - p$.
$p \ \wedge \ q$ has a truth value equal to the minimum of p and q.
$p \ \vee \ q$ has a truth value equal to the maximum of p and q.
$p \rightarrow q$ has a truth value equal to the minimum of 1 and $1 - p + q$.
$p \leftrightarrow q$ has a truth value equal to $1 - |p - q|$, that is, 1 minus the absolute value[†] of p minus q.

Suppose the statement "*p*: The sun is shining" has a truth value of 0.25 and the statement "*q*: Mary is getting a tan" has a truth value of 0.20. Determine the truth value of

a) $\sim p$ **b)** $\sim q$

c) $p \wedge q$ **d)** $p \vee q$

e) $p \rightarrow q$ **f)** $p \leftrightarrow q$

[†]Absolute values are discussed in Section 12.6.

Recreational Exercises

76. *Kakuro* Refer to the *Recreational Mathematics* box on page 128. Complete the following Kakuro puzzle.

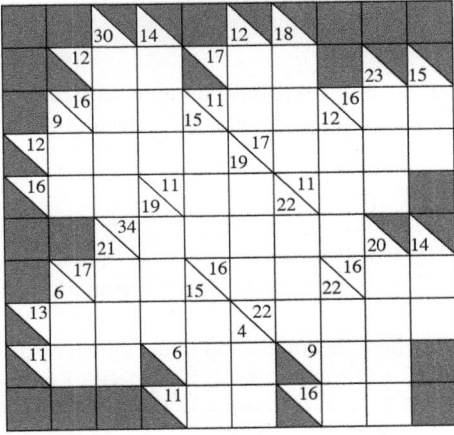

Research Activities

77. *Fuzzy Logic* Review the *Mathematics Today* box regarding fuzzy logic on page 134 and Exercise 75. Do additional research and write a report on how fuzzy logic is used in many devices we encounter in our everyday lives.

78. *Lewis Carroll* Read one of Lewis Carroll's books and write a report on how he used logic in the book. Give at least five specific examples.

79. *Augustus De Morgan* Do research and write a report on the life and achievements of Augustus De Morgan. Indicate in your report his contributions to sets and logic.

SECTION 3.5 Symbolic Arguments

Upon completion of this section, you will be able to:

■ Determine the validity of a symbolic argument by using truth tables or by using standard arguments.

■ Determine a logical conclusion from a given set of premises.

Consider the following statements.

If we go to the Bahamas, then we will go snorkeling.

We go to the Bahamas.

What logical conclusion can you draw from these two statements? Do you agree that you can logically conclude that *We will go snorkeling*? In this section, we will use our knowledge of logic to study the structure of such statements to draw logical conclusions.

Why This Is Important We often learn information that allows us to draw conclusions from this information. In this section we will study how we can use logic to draw such conclusions using what are called *standard arguments*. We will also learn how to use truth tables to determine the accuracy of conclusions drawn from a set of given logical statements. The information presented in this section can help us make wise decisions in our everyday lives.

Determine the Validity of Arguments

Previously in this chapter, we used symbolic logic to determine the truth value of a compound statement. We now extend those basic ideas to determine whether we can draw logical conclusions from a set of given statements. Consider once again the two statements.

If we go to the Bahamas, then we go snorkeling.

We go to the Bahamas.

These statements in the following form constitute what we will call a *symbolic argument*.

Premise 1:	If we go to the Bahamas, then we will go snorkeling.
Premise 2:	We go to the Bahamas.
Conclusion:	We will go snorkeling.

A *symbolic argument* consists of a set of *premises* and a *conclusion*. The argument is called a symbolic argument because we generally write the argument in symbolic form to determine its validity.

> **Definition: Valid and Invalid Arguments**
> An **argument is valid** when its conclusion necessarily follows from a given set of premises.
> An **argument is invalid** or a **fallacy** when the conclusion does not necessarily follow from the given set of premises.

An argument that is not valid is invalid. The argument just presented is an example of a valid argument, as the conclusion necessarily follows from the premises. Now we will discuss a procedure to determine whether an argument is valid or invalid. We begin by writing the argument in symbolic form. To write the argument in symbolic form, we let p and q be

p:	We go to the Bahamas.
q:	We will go snorkeling.

Symbolically, the argument is written

Premise 1: $p \rightarrow q$
Premise 2: p
Conclusion: $\therefore q$ (The three-dot-triangle is read "therefore.")

Write the argument in the following form.

If   [*premise 1*   **and**   *premise 2*]   **then**   *conclusion*
      $[(p \rightarrow q)$ $\wedge$ $p]$ $\rightarrow$ q

Then construct a truth table for the conditional statement $[(p \rightarrow q) \wedge p] \rightarrow q$ (Table 3.33). *If the truth table answer column is true in every case, then the statement is a tautology, and the argument is valid. If the conditional statement is not a tautology, then the argument is invalid. Recall conditional statements that are tautologies are called implications.* Since the conditional statement $[(p \rightarrow q) \wedge p] \rightarrow q$ is a tautology (see column 5), the conditional statement is also an implication. Thus, the conclusion necessarily follows from the premises and the argument is valid.

Table 3.33

p	q	$[(p \rightarrow q)$	$\wedge$	$p]$	$\rightarrow$	q
T	T	T	T	T	T	T
T	F	F	F	T	T	F
F	T	T	F	F	T	T
F	F	T	F	F	T	F
		1	3	2	5	4

Once we have demonstrated that an argument in a particular form is valid, all arguments with exactly the same form will also be valid. In fact, many of these forms have been assigned names. The argument form just discussed,

$$p \rightarrow q$$
$$\underline{p}$$
$$\therefore q$$

is called the *law of detachment*, or *modus ponens*.

Example 1 *Determining the Validity of an Argument Without a Truth Table*

Determine whether the following argument is valid or invalid.

If the marsupial is taller than one meter, then the marsupial is a kangaroo.

The marsupial is taller than one meter.

∴ The marsupial is a kangaroo.

Solution Translate the argument into symbolic form.
Let

p: The marsupial is taller than one meter.
q: The marsupial is a kangaroo.

In symbolic form, the argument is

$$p \rightarrow q$$
$$\underline{p}$$
$$\therefore q$$

This argument is also the law of detachment. Therefore, it is a valid argument.   ∎

Next, we summarize the procedure for determining whether an argument is valid or not valid.

PROCEDURE   TO DETERMINE WHETHER AN ARGUMENT IS VALID

1. Write the argument in symbolic form.
2. Compare the form of the argument with forms that are known to be valid or invalid. If there are no known forms to compare the argument to, or you do not remember the forms, go to Step 3.
3. If the argument contains two premises, write a conditional statement of the form
$$\left[(\text{premise 1}) \wedge (\text{premise 2}) \right] \rightarrow \text{conclusion}$$
4. Construct a truth table for the statement in Step 3.
5. If the answer column of the truth table has all trues, the statement is a tautology, and the argument is valid. If the answer column does not have all trues, the argument is invalid.

Examples 1 through 4 contain two premises. When an argument contains more than two premises, Step 3 of the procedure will change slightly, as will be explained shortly.

Example 2 *Determining the Validity of an Argument with a Truth Table*

Determine whether the following argument is valid or invalid.

If Detroit is in Michigan, then Dallas is in California.

Dallas is not in California.

∴ Detroit is not in Michigan.

Solution Let

p: Detroit is in Michigan.

q: Dallas is in California.

In symbolic form, the argument is

$$p \rightarrow q$$
$$\sim q$$
$$\overline{\therefore \sim p}$$

As we have not tested an argument in this form, we will construct a truth table to determine whether the argument is valid or invalid. We write the argument in the form $\left[(p \rightarrow q) \wedge \sim q\right] \rightarrow \sim p$, and construct a truth table (Table 3.34). Since the answer, column 5, has all T's, the argument is valid.

Table 3.34

p	q	$[(p \rightarrow q)$	$\wedge$	$\sim q]$	$\rightarrow$	$\sim p$
T	T	T	F	F	T	F
T	F	F	F	T	T	F
F	T	T	F	F	T	T
F	F	T	T	T	T	T
		1	3	2	5	4

The argument form in Example 2 is an example of the *law of contraposition*, or *modus tollens*.

Note that the argument in Example 2 is valid even though the conclusion, "Detroit is not in Michigan," is a false statement. It is also possible to have an invalid argument in which the conclusion is a true statement. *When an argument is valid, the conclusion necessarily follows from the premises. It is not necessary for the premises or the conclusion to be true statements in the argument.*

Example 3   *Another Symbolic Argument*

Determine whether the following argument is valid or invalid.

The grass is green or the grass is full of weeds.
The grass is not green.

$\therefore$ The grass is full of weeds.

Solution Let

p: The grass is green.

q: The grass is full of weeds.

In symbolic form, the argument is

$$p \vee q$$
$$\sim p$$
$$\overline{\therefore q}$$

As this form is not one of those we are familiar with, we will construct a truth table. We write the argument in the form $\left[(p \vee q) \wedge \sim p\right] \rightarrow q$. Next we construct a truth table, as shown in Table 3.35 on page 143. The answer to the truth table, column 5, is true in *every case*. Therefore, the statement is a tautology, and the argument is valid.

Table 3.35

p	q	$[(p \lor q)$	$\land$	$\sim p]$	$\rightarrow$	q
T	T	T	F	F	T	T
T	F	T	F	F	T	F
F	T	T	T	T	T	T
F	F	F	F	T	T	F
		1	3	2	5	4

The argument form in Example 3 is an example of a *disjunctive syllogism*. Other standard forms of arguments are given in the following chart.

Standard Forms of Arguments

Valid Arguments	*Law of Detachment*	*Law of Contraposition*	*Law of Syllogism*	*Disjunctive Syllogism*
	$p \rightarrow q$	$p \rightarrow q$	$p \rightarrow q$	$p \lor q$
	p	$\sim q$	$q \rightarrow r$	$\sim p$
	$\therefore q$	$\therefore \sim p$	$\therefore p \rightarrow r$	$\therefore q$
Invalid Arguments	*Fallacy of the Converse*	*Fallacy of the Inverse*		
	$p \rightarrow q$	$p \rightarrow q$		
	q	$\sim p$		
	$\therefore p$	$\therefore \sim q$		

As we saw in Example 1, it is not always necessary to construct a truth table to determine whether or not an argument is valid. The next two examples will show how we can identify an argument as one of the standard arguments given in the chart above.

Example 4 *Identifying a Standard Argument*

Determine whether the following argument is valid or invalid.

> If you are on Instagram, then you see my pictures.
> If you see my pictures, then you know I have a dog.
> ∴ If you are on Instagram, then you know I have a dog.

Solution Let

p: You are on Instagram.
q: You see my pictures.
r: You know I have a dog.

In symbolic form, the argument is

$$p \rightarrow q$$
$$q \rightarrow r$$
$$\therefore p \rightarrow r$$

The argument is in the form of the law of syllogism. Therefore, the argument is valid, and there is no need to construct a truth table.

Example 5   *Identifying Common Fallacies in Arguments*

Determine whether the following arguments are valid or invalid.

a)   If it is snowing, then we put salt on the driveway.
      We put salt on the driveway.
      _____
      ∴ It is snowing.

b)   If it is snowing, then we put salt on the driveway.
      It is not snowing.
      _____
      ∴ We do not put salt on the driveway.

Solution

a) Let

$$p: \quad \text{It is snowing.}$$
$$q: \quad \text{We put salt on the driveway.}$$

In symbolic form, the argument is written as follows.

$$p \rightarrow q$$
$$\underline{q}$$
$$\therefore p$$

This argument is in the form of the fallacy of the converse. Therefore, the argument is a fallacy, or invalid.

b) Using the same symbols defined in the solution to part (a), in symbolic form, the argument is written as follows.

$$p \rightarrow q$$
$$\underline{\sim p}$$
$$\therefore \sim q$$

This argument is in the form of the fallacy of the inverse. Therefore, the argument is a fallacy, or invalid.   ∎

Now we consider an argument that has more than two premises. When an argument contains more than two premises, the statement we test, using a truth table, is formed by taking the conjunction of all the premises as the antecedent of a conditional statement and the conclusion as the consequent of the conditional statement. One example is an argument of the following form

$$p_1$$
$$p_2$$
$$\underline{p_3}$$
$$\therefore c$$

To determine whether this argument is valid, we determine the truth table for $[p_1 \wedge p_2 \wedge p_3] \rightarrow c$. When we evaluate $[p_1 \wedge p_2 \wedge p_3]$, it makes no difference whether we evaluate $[(p_1 \wedge p_2) \wedge p_3]$ or $[p_1 \wedge (p_2 \wedge p_3)]$ because both give the same answer. In Example 6, we evaluate $[p_1 \wedge p_2 \wedge p_3]$ from left to right; that is, $[(p_1 \wedge p_2) \wedge p_3]$.

Example 6   *An Argument with Three Premises*

Use a truth table to determine whether the following argument is valid or invalid.

If my cell phone company is Verizon, then I can call you free of charge.
I can call you free of charge or I can send you a text message.
I can send you a text message or my cell phone company is Verizon.
_____
∴ My cell phone company is Verizon.

Timely Tip

If you are not sure whether an argument with two premises is one of the standard forms or if you do not remember the standard forms, you can always determine whether a given argument is valid or invalid by using a truth table. To do so, follow the boxed procedure on page 141.

In Example 5(b), if you did not recognize that this argument was of the same form as the fallacy of the inverse you could construct the truth table for the conditional statement

$$[(p \rightarrow q) \wedge \sim p] \rightarrow \sim q$$

The true–false values under the conditional column, $\rightarrow$, would be T, T, F, T. Since the statement is not a tautology, the argument is invalid.

Solution This argument contains three simple statements.

Let

p: My cell phone company is Verizon.

q: I can call you free of charge.

r: I can send you a text message.

In symbolic form, the argument is written as follows.

$$p \rightarrow q$$
$$q \lor r$$
$$\underline{r \lor p}$$
$$\therefore p$$

Write the argument in the form

$$\left[(p \rightarrow q) \land (q \lor r) \land (r \lor p)\right] \rightarrow p.$$

Now construct the truth table (Table 3.36). The answer, column 7 of the truth table, is not true in every case. Thus, the argument is a fallacy, or invalid.

Table 3.36

p	q	r	$[(p \rightarrow q)$	$\land$	$(q \lor r)$	$\land$	$(r \lor p)]$	$\rightarrow$	p
T	T	T	T	T	T	T	T	T	T
T	T	F	T	T	T	T	T	T	T
T	F	T	F	F	T	F	T	T	T
T	F	F	F	F	F	F	T	T	T
F	T	T	T	T	T	T	T	F	F
F	T	F	T	T	T	F	F	T	F
F	F	T	T	T	T	T	T	F	F
F	F	F	T	F	F	F	F	T	F
			1	3	2	5	4	7	6

Let's now investigate how we can arrive at a valid conclusion from a given set of premises.

Determine Logical Conclusions

Example 7 *Determine a Logical Conclusion*

Determine a logical conclusion that follows from the given statements. "I will attend the concert or I will attend the volleyball game. I will not attend the concert. Therefore …"

Solution If we recognize a specific form of an argument, we can use our knowledge of that form to draw a logical conclusion.

Let

p: I will attend the concert.

q: I will attend the volleyball game.

The argument is of the following form:

$$p \lor q$$
$$\underline{\sim p}$$
$$\therefore ?$$

If the question mark is replaced with q, this argument is of the form disjunctive syllogism. Thus, a logical conclusion is "Therefore, I will attend the volleyball game."

SECTION 3.5
Exercises

Warm Up Exercises

In Exercises 1–6, fill in the blanks with an appropriate word, phrase, or symbol(s).

1. When the conclusion of an argument necessarily follows from the given set of premises it is a(n) _____ argument.

2. When the conclusion of an argument does not necessarily follow from the given set of premises it is a(n) _____ argument.

3. An argument that is invalid is also known as a(n) _____ .

4. To determine the validity of an argument with two premises, construct a truth table of the form $[(\text{premise 1}) \wedge (\text{premise 2})] \rightarrow$ _____ .

5. If the conditional statement referred to in Exercise 4 is a tautology, then the argument is a(n) _____ argument.

6. If the conditional statement referred to in Exercise 4 is not a tautology, then the argument is a(n) _____ argument.

Practice the Skills

For Exercises 7–12, fill in the blank to identify the standard form of the argument.

7. $p \vee q$ Disjunctive _____ .
$\dfrac{\sim p}{\therefore q}$

8. $p \rightarrow q$ Law of _____ .
$\dfrac{p}{\therefore q}$

9. $p \rightarrow q$ Fallacy of the _____ .
$\dfrac{\sim p}{\therefore \sim q}$

10. $p \rightarrow q$ Fallacy of the _____ .
$\dfrac{q}{\therefore p}$

11. $p \rightarrow q$ Law of _____ .
$\dfrac{q \rightarrow r}{\therefore p \rightarrow r}$

12. $p \rightarrow q$ Law of _____ .
$\dfrac{\sim q}{\therefore \sim p}$

In Exercises 13–32, determine whether the argument is valid or invalid. You may compare the argument to a standard form, given on page 143, or use a truth table.

13. $a \rightarrow b$
$\dfrac{\sim a}{\therefore \sim b}$

14. $a \vee b$
$\dfrac{\sim a}{\therefore \sim b}$

15. $e \rightarrow f$
$\dfrac{e}{\therefore f}$

16. $g \rightarrow h$
$\dfrac{h \rightarrow i}{\therefore g \rightarrow i}$

17. $j \vee k$
$\dfrac{\sim k}{\therefore j}$

18. $l \rightarrow m$
$\dfrac{\sim m}{\therefore \sim l}$

19. $n \rightarrow o$
$\dfrac{o}{\therefore n}$

20. $r \rightarrow s$
$\dfrac{r}{\therefore s}$

21. $t \rightarrow u$
$\dfrac{\sim u}{\therefore \sim t}$

22. $v \rightarrow w$
$\dfrac{w}{\therefore v}$

23. $x \rightarrow y$
$\dfrac{y \rightarrow z}{\therefore x \rightarrow z}$

24. $x \rightarrow y$
$\dfrac{\sim x}{\therefore \sim y}$

25. $p \leftrightarrow q$
$\dfrac{q \wedge r}{\therefore p \vee r}$

26. $p \leftrightarrow q$
$\dfrac{q \rightarrow r}{\therefore \sim r \rightarrow \sim p}$

27. $p \wedge q$
$\dfrac{q \vee r}{\therefore r}$

28. $p \rightarrow \sim q$
$\dfrac{q \leftrightarrow \sim r}{\therefore p \rightarrow \sim r}$

29. $p \rightarrow q$
$q \vee r$
$\dfrac{r \vee p}{\therefore p}$

30. $p \rightarrow q$
$q \rightarrow r$
$\dfrac{r \rightarrow p}{\therefore p \leftrightarrow r}$

31. $p \rightarrow q$
$r \rightarrow \sim p$
$\dfrac{p \vee r}{\therefore q \vee \sim p}$

32. $p \vee \sim q$
$q \vee \sim r$
$\dfrac{r \vee \sim p}{\therefore p \vee r}$

Problem Solving

In Exercises 33–48, (a) translate the argument into symbolic form and (b) determine if the argument is valid or invalid. You may compare the argument to a standard form or use a truth table.

33. If the Rays win the division, then the Rays go to the playoffs.

The Rays do not win the division.

∴ The Rays do not go to the playoffs.

34. If the car is a Road Runner, then the car is fast.

The car is fast.

∴ The car is a Road Runner.

35. If we visit the zoo, then we see a zebra.

We visit the zoo.

∴ We see a zebra.

36. If we visit Harlem, then we go to the Apollo Theater.

If we go to the Apollo Theater, then we see a concert.

∴ If we visit Harlem, then we see a concert.

37. If the guitar is a Les Paul model, then the guitar is made by Gibson.

The guitar is not made by Gibson.

∴ The guitar is not a Les Paul model.

38. We will go for a bike ride or we will go shopping.

We will not go shopping.

∴ We will go for a bike ride.

39. If we take the kayaks on the river, then we see alligators.

We see alligators.

∴ We take the kayaks on the river.

40. If you pass general chemistry, then you can take organic chemistry.

You pass general chemistry.

∴ You can take organic chemistry.

41. Lucious will give the company to Hakeem or Lucious will give the company to Cookie.

Lucious will not give the company to Hakeem.

∴ Lucious will give the company to Cookie.

42. If you solve the *New York Times* crossword puzzle, then you are a genius.

You are not a genius.

∴ You did not solve the *New York Times* crossword puzzle.

43. If it is cold, then graduation will be held indoors.

If graduation is held indoors, then the fireworks will be postponed.

∴ If it is cold, then the fireworks will be postponed.

44. If Lucinda is singing in the band, then Steve is playing the guitar.

Lucinda is not singing in the band.

∴ Steve is not playing the guitar.

45. Jevon is a bowler and Michael is a golfer.

If Michael is a golfer, then Alisha is a curler.

∴ If Alisha is a curler, then Jevon is a bowler.

46. Vitamin C helps your immune system or niacin helps reduce cholesterol.

If niacin helps reduce cholesterol, then vitamin E enhances your skin.

∴ Vitamin C helps your immune system and vitamin E enhances your skin.

47. Javier is a police officer or Javier is a little league coach.

If Javier is a police officer, then Javier is a community leader.

∴ Javier is a little league coach if and only if Javier is a community leader.

48. The garden has vegetables or the garden has flowers.

If the garden does not have flowers, then the garden has vegetables.

∴ The garden has flowers or the garden has vegetables.

In Exercises 49–58, translate the argument into symbolic form. Then determine whether the argument is valid or invalid.

49. If you read *The Riptide Ultra-Glide* then you can understand *Tiger Shrimp Tango*. You cannot understand *Tiger Shrimp Tango*. Therefore, you did not read *The Riptide Ultra-Glide*.

50. The printer has a clogged nozzle or the printer does not have toner. The printer has toner. Therefore, the printer has a clogged nozzle.

51. If it rains on Monday, then we will go shopping. It does not rain on Monday. Therefore, we will not go shopping.

52. If Xiomara runs a dance school, then Jane becomes an author. Jane becomes an author. Therefore, Xiomara runs a dance school.

53. If you did not submit your application, then you will not be accepted to school. You did not submit your application. Therefore, you will not be accepted to school.

54. If Bonnie passes the bar exam, then she will practice law. Bonnie will not practice law. Therefore, Bonnie did not pass the bar exam.

55. The baby is crying but the baby is not hungry. If the baby is hungry then the baby is crying. Therefore, the baby is hungry.

56. If the car is new, then the car has air conditioning. The car is not new and the car has air conditioning. Therefore, the car is not new.

57. If you liked *This Is Spinal Tap* then you liked *Best in Show*. If you liked *Best in Show* then you did not like *A Mighty Wind*. Therefore, if you liked *This Is Spinal Tap* then you liked *A Mighty Wind*.

58. The engineering courses are difficult and the chemistry labs are long. If the chemistry labs are long, then the art tests are easy. Therefore, the engineering courses are difficult and the art tests are not easy.

In Exercises 59–64, using the standard forms of arguments and other information you have learned, supply what you believe is a logical conclusion to the argument. Verify that the argument is valid for the conclusion you supplied.

59. If the radio is a Superadio, then the radio is made by RCA. The radio is a Superadio. Therefore, . . .

60. If the temperature hits 100°, then we will go swimming. We did not go swimming.
Therefore, . . .

61. I am stressed out or I have the flu.
I do not have the flu.
Therefore, . . .

62. If I can get Nick to his piano lesson by 3:30 P.M., then I can do my shopping.
If I can do my shopping, then we do not need to order pizza again.
Therefore, . . .

63. If you close the deal, then you will get a commission.
You did not get a commission.
Therefore, . . .

64. If Katherine is at her ballet lesson, then Allyson will take a nap.
Katherine is at her ballet lesson.
Therefore, . . .

Concept/Writing Exercises

65. Is it possible for an argument to be valid if its conclusion is false? Explain your answer.

66. Is it possible for an argument to be invalid if its conclusion is true? Explain your answer.

67. Is it possible for an argument to be invalid if the premises are all true? Explain your answer.

68. Is it possible for an argument to be valid if the premises are all false? Explain your answer.

Challenge Problems/Group Activities

69. Determine whether the argument is valid or invalid.

If Lynn wins the contest or strikes oil, then she will be rich.

If Lynn is rich, then she will stop working.

∴ If Lynn does not stop working, then she did not win the contest.

70. Is it possible for an argument to be invalid if the conjunction of the premises is false in every case of the truth table? Explain your answer.

Recreational Mathematics

71. René Descartes was a seventeenth-century French mathematician and philosopher. One of his most memorable statements is, "I think, therefore, I am." This statement is the basis for the following joke.

Descartes walks into an inn. The innkeeper asks Descartes if he would like something to drink. Descartes replies, "I think not," and promptly vanishes into thin air!

This joke can be summarized in the following argument: If I think, then I am. I think not. Therefore, I am not.

a) Represent this argument symbolically.

b) Is it a valid argument?

c) Explain your answer using either a standard form of argument or using a truth table.

Research Activities

72. Show how logic is used in advertising. Discuss several advertisements and show how logic is used to persuade the reader.

73. Find examples of valid (or invalid) arguments in printed matter such as newspaper or magazine articles. Explain why the arguments are valid (or invalid).

Upon completion of this section, you will be able to:

■ Determine the validity of syllogistic arguments using Euler diagrams.

Deandre is studying to be a nurse. While reading the program literature, he learns that "All art classes are electives." He also learns that "Introduction to Theater is an elective." From these two statements can Deandre conclude that Introduction to Theater is considered an art class? In this section we will examine arguments that consist of statements that contain quantifiers. We learned in Section 3.1 that quantifiers include words such as *all*, *some*, or *none*.

Why This Is Important   When several sentences containing the words *all*, *some*, or *none* are put together into an argument, we need to be able to determine if the conclusion necessarily follows from the premises. In this section we will study such arguments, and we will be able to see how we can analyze similar arguments that we encounter in our everyday lives.

In Section 3.5, we showed how to determine the validity of *symbolic arguments* using truth tables and comparing the arguments to standard forms. This section presents another form of argument called a *syllogistic argument*, better known by the shorter name *syllogism*. The validity of a syllogistic argument is determined by using Euler (pronounced "oiler") diagrams, as is explained shortly.

Syllogistic logic, a deductive process of arriving at a conclusion, was developed by Aristotle in about 350 B.C. Aristotle considered the relationships among the four types of statements that follow.

$$\text{All} \underline{\hspace{1.5cm}} \text{are} \underline{\hspace{1.5cm}}.$$
$$\text{No} \underline{\hspace{1.5cm}} \text{are} \underline{\hspace{1.5cm}}.$$
$$\text{Some} \underline{\hspace{1.5cm}} \text{are} \underline{\hspace{1.5cm}}.$$
$$\text{Some} \underline{\hspace{1.5cm}} \text{are not} \underline{\hspace{1.5cm}}.$$

Examples of these statements are as follows: *All doctors are tall. No doctors are tall. Some doctors are tall. Some doctors are not tall.* Since Aristotle's time, other types of statements have been added to the study of syllogistic logic, two of which are

$$\underline{\hspace{1.5cm}} \text{is a} \underline{\hspace{1.5cm}}.$$
$$\underline{\hspace{1.5cm}} \text{is not a} \underline{\hspace{1.5cm}}.$$

Examples of these statements are as follows: *Maria is a doctor. Maria is not a doctor.*

The difference between a symbolic argument and a syllogistic argument can be seen in the following chart. Symbolic arguments use the connectives *and, or, not, if–then,* and *if and only if*. Syllogistic arguments use the quantifiers *all, some,* and *none,* which were discussed in Section 3.1.

Symbolic Arguments Versus Syllogistic Arguments

	Words or phrases used	*Method of determining validity*
Symbolic argument	and, or, not, if–then, if and only if	Truth tables or by comparison with standard forms of arguments
Syllogistic argument	all are, some are, none are, some are not	Euler diagrams

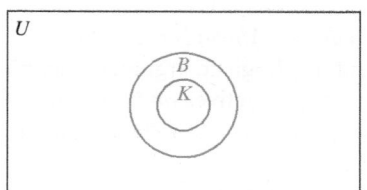

Figure 3.3

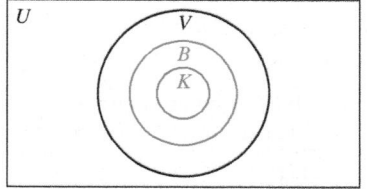

Figure 3.4

As with symbolic logic, the premises and the conclusion together form an argument. An example of a syllogistic argument is

All beagles are dogs.

All dogs bark.

∴ All beagles bark.

This is an example of a valid argument. Recall from Section 3.5 that an argument is *valid* when its conclusion necessarily follows from a given set of premises. Recall that an argument in which the conclusion does not necessarily follow from the given premises is said to be an *invalid argument* or a *fallacy*.

Before we give another example of a syllogism, let's review the Venn diagrams discussed in Section 2.3 in relationship with Aristotle's four statements.

All A's are B's	No A's are B's	Some A's are B's	Some A's are not B's
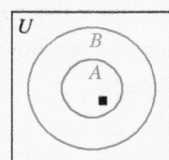	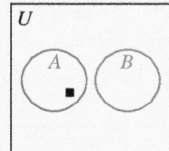	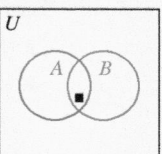	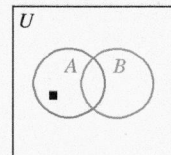
If an element is in set A, then it is in set B.	If an element is in set A, then it is not in set B.	There is at least one element that is in both set A and set B.	There is at least one element that is in set A that is not in set B.

One method used to determine whether an argument is valid or is a fallacy is by means of an *Euler diagram*, named after Leonhard Euler, who used circles to represent sets in syllogistic arguments. The technique of using Euler diagrams is illustrated in Example 1.

Example 1 *Using an Euler Diagram*

Determine whether the following syllogism is valid or invalid.

All keys are made of brass.

All things made of brass are valuable.

∴ All keys are valuable.

Solution To determine whether this syllogism is valid or invalid, we will construct an Euler diagram. We begin with the first premise, "All keys are made of brass." As shown in Fig. 3.3, the inner blue circle labeled K represents the set of all keys and the outer red circle labeled B represents the set of all brass objects. The first premise requires that the inner blue circle must be entirely contained within the outer red circle. Next, we will represent the second premise, "All things made of brass are valuable." As shown in Fig. 3.4, the outermost black circle labeled V represents the set of all valuable objects. The second premise dictates that the red circle, representing the set of brass objects, must be entirely contained within the black circle, representing the set of valuable objects. Now, examine the completed Euler diagram in Fig. 3.4. Note that the premises force the set of keys to be within the set of valuable objects. Therefore, the argument is valid, since the conclusion, "All keys are valuable," necessarily follows from the set of premises. ∎

The syllogism in Example 1 is valid even though the conclusion, "All keys are valuable," is not a true statement. Similarly, a syllogism can be invalid, or a fallacy, even if the conclusion is a true statement.

When we determine the validity of an argument, we are determining whether the conclusion necessarily follows from the premises. When we say that an argument is valid, we are saying that if all the premises are true statements, then the conclusion must also be a true statement.

The form of the argument determines its validity, not the particular statements. For example, consider the syllogism

All Earth people have two heads.

All people with two heads can fly.

∴ All Earth people can fly.

The form of this argument is the same as that of the previous valid argument in Example 1. Therefore, this argument is also valid.

Example 2 *Analyzing a Syllogism*

Determine whether the following syllogism is valid or invalid.

All Packers are athletes.

Brett is a Packer.

∴ Brett is an athlete.

Solution The statement "All Packers are athletes" is illustrated in Fig. 3.5. Note that the *P* circle must be completely inside the *A* circle. The second premise, "Brett is a Packer," tell us that Brett must be placed in the inner circle labeled *P* (Fig. 3.6). The Euler diagram illustrates that by placing Brett inside the *P* circle, Brett must also be inside the *A* circle. Therefore, the conclusion "Brett is an athlete," necessarily follows from the premises and the argument is valid. ∎

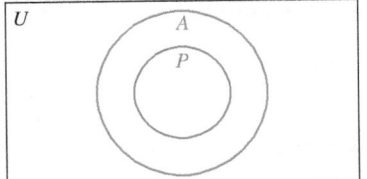

Figure 3.5

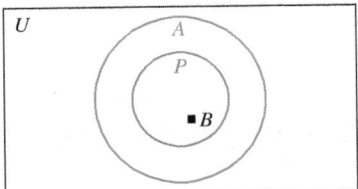

Figure 3.6

In both Example 1 and Example 2, we had no choice as to where the second premise was to be placed in the Euler diagram. In Example 1, the set of brass objects had to be placed inside the set of valuable objects. In Example 2, Brett had to be placed inside the set of Packers. Often when determining the truth value of a syllogism, a premise can be placed in more than one area in the diagram. *We always try to draw the Euler diagram so that the conclusion **does not necessarily** follow from the premises. If that can be done, then the conclusion **does not necessarily** follow from the premises and the argument is invalid.* If we cannot show that the argument is invalid, only then do we accept the argument as valid. We illustrate this process in Example 3.

Example 3 *Harmonicas and Trumpets*

Determine whether the following syllogism is valid or invalid.

All harmonicas are musical instruments.

A trumpet is a musical instrument.

∴ A trumpet is a harmonica.

Solution The premise "All harmonicas are musical instruments" is illustrated in Fig. 3.7(a) on page 152. The next premise, "A trumpet is a musical instrument," tells us that a trumpet must be placed in the set of musical instruments. Two diagrams in which both premises are satisfied are shown in Fig. 3.7(b) and (c). By examining 3.7(b), however, we see that a trumpet is not a harmonica. Therefore, the conclusion, "A trumpet is a harmonica," does not necessarily follow from the set of premises. The argument is thus invalid, or the argument is a fallacy.

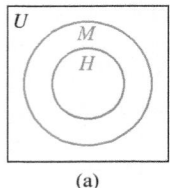 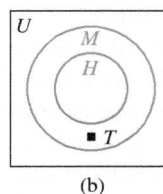 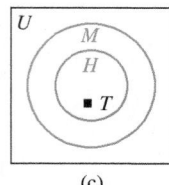

(a)   (b)   (c)

Figure 3.7

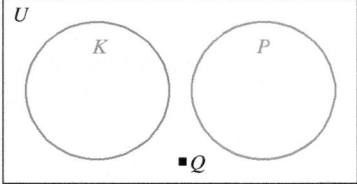

Figure 3.8

Timely Tip

Note that in Example 4 if we placed Quentin in the *P* circle, the argument would appear to be valid. Remember that *whenever testing the validity of an argument, always try to show that the argument is invalid.* If there is any way of showing that the conclusion does not necessarily follow from the premises, then the argument is invalid.

Example 4 *Team Mascots*

Determine whether the following syllogism is valid or invalid.

> No Knights are Pioneers.
> Quentin is not a Knight.
> ─────────────────
> ∴ Quentin is a Pioneer.

Solution The first premise tells us that Knights and Pioneers are disjoint sets. This means that the two sets do not intersect, and the *K* circle and the *P* circle should be drawn so they do not overlap. The second premise tells us that Quentin is not a Knight. Thus, Quentin should be placed outside the *K* circle. Notice from Fig. 3.8 that it is possible to place Quentin outside the *K* circle without placing him inside the *P* circle. The Euler diagram shows that the two given premises are satisfied and shows that the conclusion does not necessarily follow from the given premises. Therefore, the argument is invalid. ∎

Example 5 *A Syllogism Involving the Word* Some

Determine whether the following syllogism is valid or invalid.

> All *A*'s are *B*'s.
> Some *B*'s are *C*'s.
> ─────────────────
> ∴ Some *A*'s are *C*'s.

Solution The premise "All *A*'s are *B*'s" is illustrated in Fig. 3.9. The premise "Some *B*'s are *C*'s" means that there is at least one *B* that is a *C*. We can illustrate this set of premises in four ways, as illustrated in Fig. 3.10.

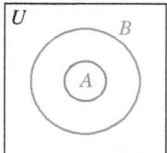

Figure 3.9

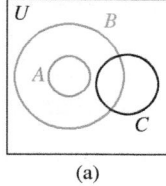 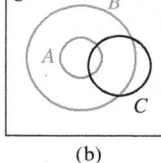 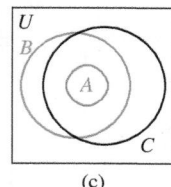 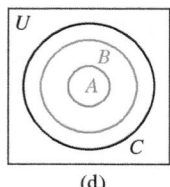

(a)   (b)   (c)   (d)

Figure 3.10

In all four illustrations, we see that (1) all *A*'s are *B*'s and (2) some *B*'s are *C*'s. The conclusion is "Some *A*'s are *C*'s." Since at least one of the illustrations, Fig. 3.10(a), shows that the conclusion does not necessarily follow from the given premises, the argument is invalid. ∎

Example 6 *Fish and Cows*

Determine whether the following syllogism is valid or invalid.

> No fish are mammals.
> All cows are mammals.
> ─────────────────
> ∴ No fish are cows.

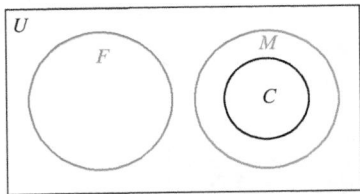

Figure 3.11

Solution The first premise tells us that fish and mammals are disjoint sets, as shown in Fig. 3.11. The second premise tells us that the set of cows is a subset of the set of mammals. Therefore, the circle representing the set of cows must go within the circle representing the set of mammals.

Note that the set of fish and the set of cows cannot be made to intersect without violating a premise. Thus, the conclusion "No fish are cows" necessarily follows from the premises and the argument is valid. Note that we did not say that this conclusion is true, only that the argument is valid. ■

SECTION 3.6 Exercises

Warm Up Exercises

In Exercises 1–6, fill in the blanks with an appropriate word, phrase, or symbol(s).

1. The validity of a syllogistic argument can be determined using a(n) _____ diagram.

2. If an Euler diagram can be drawn only in a way in which the conclusion necessarily follows from the premises, the syllogistic argument is a(n) _____ argument.

3. If an Euler diagram can be drawn in a way in which the conclusion does not necessarily follow from the premises, the syllogistic argument is a(n) _____ argument.

4. The following Euler diagram represents the statement _____ *A*'s are *B*'s.

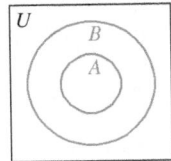

5. The following Euler diagram represents the statement _____ *A*'s are *B*'s.

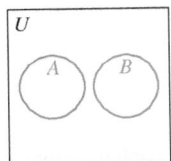

6. The following Euler diagram represents the statement _____ *A*'s are *B*'s.

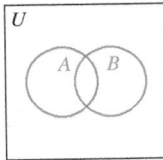

Practice the Skills/Problem Solving

In Exercises 7–26, use an Euler diagram to determine whether the syllogism is valid or invalid.

7. All kangaroos are marsupials.
   Bounder is a kangaroo.
   ∴ Bounder is a marsupial.

8. All dogs are canines.
   Boomer is a canine.
   ∴ Boomer is a dog.

9. All yo-yos are toys.
   All Slinkies are toys.
   ∴ All Slinkies are yo-yos.

10. All dolphins are mammals.
    All mammals are vertebrates.
    ∴ All dolphins are vertebrates.

11. No PCs are Macs.
    All Dells are PCs.
    ∴ No Dells are Macs.

12. All American Beauties are roses.
    No roses are daffodils.
    ∴ No American Beauties are daffodils.

13. No lawn weeds are flowers.
Sedge is not a flower.

∴ Sedge is a lawn weed.

14. No Cyclones are Longhorns.
Zendaya is not a Longhorn.

∴ Zendaya is a Cyclone.

15. Some mushrooms are poisonous.
A morel is a mushroom.

∴ A morel is poisonous.

16. Some skunks are tame.
Pepé is a skunk.

∴ Pepé is not tame.

17. Some stamps are collectors' items.
Some collectors' items are valuable.

∴ Some stamps are valuable.

18. Some chemicals are poisonous.
Some poisonous things are beautiful.

∴ Some chemicals are beautiful.

19. Some flowers love sunlight.
All things that love sunlight love water.

∴ Some flowers love water.

20. Some caterpillars are furry.
All furry things are mammals.

∴ Some caterpillars are mammals.

21. No scarecrows are tin men.
No tin men are lions.

∴ No scarecrows are lions.

22. No squirrels are reptiles.
No reptiles are birds.

∴ No squirrels are birds.

23. Some nurses work in pediatrics.
Seth works in pediatrics.

∴ Seth is a nurse.

24. All rainy days are cloudy.
Today it is cloudy.

∴ Today is a rainy day.

25. All Rolexes are watches.
A Pulsar is not a Rolex.

∴ A Pulsar is not a watch.

26. All Zebras are pens.
A Ticonderoga is not a pen.

∴ A Ticonderoga is not a Zebra.

27. All sweet things taste good.
All things that taste good are fattening.
All things that are fattening put on pounds.

∴ All sweet things put on pounds.

28. All cats are dogs.
All dogs are cows.
All cows are pigs.

∴ All cats are pigs.

Concept/Writing Exercises

29. Can an argument be valid if the conclusion is a false statement? Explain your answer.

30. Can an argument be invalid if the conclusion is a true statement? Explain.

Challenge Problem/Group Activity

31. **Sets and Logic** Statements in logic can be translated into set statements: For example, $p \wedge q$ is similar to $P \cap Q$; $p \vee q$ is similar to $P \cup Q$; and $p \rightarrow q$ is equivalent to $\sim p \vee q$, which is similar to $P' \cup Q$. Euler diagrams can also be used to show that arguments similar to those discussed in Section 3.5 are valid or invalid. Use Euler diagrams to show that the following symbolic argument is invalid.

$$p \rightarrow q$$
$$\underline{p \vee q}$$
$$\therefore \sim p$$

Research Activity

32. **Leonhard Euler** Leonhard Euler is considered one of the greatest mathematicians of all time. Do research and write a report on Euler's life. Include information on his contributions to sets and to logic. Also indicate other areas of mathematics in which he made important contributions.

Switching Circuits

Upon completion of this section, you will be able to:

- Use symbolic statements to represent switching circuits.
- Draw switching circuits that represent symbolic statements.
- Determine whether two switching circuits are equivalent.

Suppose you are sitting at your desk and want to turn on a lamp so you can read a book. If a wall switch controls the power to the outlet that the lamp is plugged into and the lamp has an on/off switch, what conditions must be true for the lamp's bulb to light? The electrical wiring in our homes can be explained and described using logic.

Why This Is Important In this section we will show how electrical circuits can be represented using logic. Understanding these basic circuits can help us understand how logic is used in all modern electronic devices such as smartphones, computers, and appliances.

Using Symbolic Statements to Represent Switching Circuits

A common application of logic is switching circuits. To understand the basic concepts of switching circuits, let us examine a few simple circuits that are common in most homes. The typical lamp has a cord, which is plugged into a wall outlet. Somewhere between the bulb in the lamp and the outlet is a switch to turn the lamp on and off. A switch is often referred to as being on or off. When the switch is in the *on* position, the current flows through the switch and the bulb lights up. When the switch is in the *on* position, we can say that the switch is *closed* and that current will flow through the switch. When the switch is in the *off* position, the current does not flow through the switch and the bulb does not light. When the switch is in the *off* position, we can say that the switch is *open*, and the current does not flow through the switch. The basic configuration of a switch is shown in Fig. 3.12.

Electric circuits can be expressed as logical statements. We represent switches as letters, using T to represent a closed switch (or current flow) and F to represent an open switch (or no current flow). This relationship is indicated in Table 3.37.

Occasionally, we have a wall switch connected to a wall outlet and a lamp plugged into the wall outlet (Fig. 3.13). We then say that the wall switch and the switch on the lamp are in *series*, meaning that for the bulb in the lamp to light, both switches must be on at the same time (Fig. 3.14). On the other hand, the bulb will not light if either switch is off or if both switches are off. In either of these conditions, the electricity will not flow through the circuit. In a *series circuit*, the current can take only one path. If any switch in the path is open, the current cannot flow.

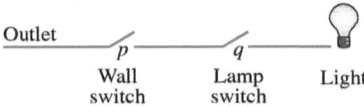

Outlet — *p* — *q* — Light

Wall switch   Lamp switch   Light

Figure 3.14

_____/_____

Figure 3.12

Table 3.37

Switch	Lightbulb
T	on (switch closed)
F	off (switch open)

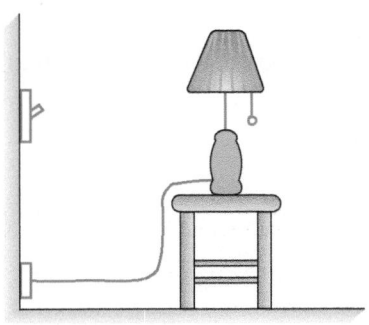

Figure 3.13

To illustrate this situation symbolically, let *p* represent the wall switch and *q* the lamp switch. The letter T will be used to represent both a closed switch and the bulb lighting. The letter F will represent an open switch and the bulb not lighting. Thus, we have four possible cases as shown in Table 3.38 on page 156.

CASE 1: Both switches are closed; that is, p is T and q is T. The light is on, T.

CASE 2: Switch p is closed and switch q is open; that is, p is T and q is F. The light is off, F.

CASE 3: Switch p is open and switch q is closed; that is, p is F and q is T. The light is off, F.

CASE 4: Both switches are open; that is, p is F and q is F. The light is off, F.

Table 3.38 summarizes the results. The on–off results are the same as the truth table for the conjunction $p \wedge q$ if we think of "on" as true and "off" as false.

Table 3.38

p	q	Light	$p \wedge q$
T	T	on (T)	T
T	F	off (F)	F
F	T	off (F)	F
F	F	off (F)	F

Series Circuits

Switches in **series** will always be represented with a conjunction, $\wedge$.

Another type of electric circuit used in the home is the *parallel circuit*, in which there are two or more paths that the current can take. If the current can pass through either path or both (see Fig. 3.15), the light will go on. The letter T will be used to represent both a closed switch and the bulb lighting. The letter F will represent an open switch and the bulb not lighting. Thus, we have four possible cases as shown in Table 3.39.

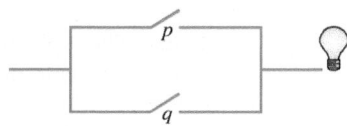

Figure 3.15

CASE 1: Both switches are closed; that is, p is T and q is T. The light is on, T.

CASE 2: Switch p is closed and switch q is open; that is, p is T and q is F. The light is on, T.

CASE 3: Switch p is open and switch q is closed; that is, p is F and q is T. The light is on, T.

CASE 4: Both switches are open; that is, p is F and q is F. The light is off, F.

Table 3.39 summarizes the results. The on–off results are the same as the $p \vee q$ truth table if we think of "on" as true and "off" as false.

Table 3.39

p	q	Light	$p \vee q$
T	T	on (T)	T
T	F	on (T)	T
F	T	on (T)	T
F	F	off (F)	F

Parallel Circuits

Switches in **parallel** will always be represented with a disjunction, $\vee$.

Sometimes it is necessary to have two or more switches in the same circuit that will both be open at the same time and both be closed at the same time. In such circuits, we will use the same letter to represent both switches. For example, in the circuit shown in Fig. 3.16, there are two switches labeled p. Therefore, both of these switches must be open at the same time and both must be closed at the same time. One of the p switches cannot be open at the same time the other p switch is closed.

We can now combine some of these basic concepts to analyze more circuits.

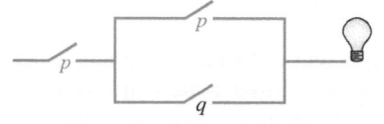

Figure 3.16

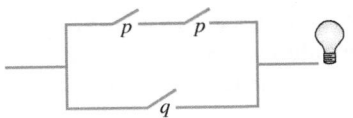

Figure 3.17

Table 3.40

p	q	p	∧	(p ∨ q)
T	T	T	T	T
T	F	T	T	T
F	T	F	F	T
F	F	F	F	F
		1	3	2

Example 1 *Representing a Switching Circuit with Symbolic Statements*

a) Write a symbolic statement that represents the circuit shown in Fig. 3.16 on page 156.

b) Construct a truth table to determine when the light will be on.

Solution

a) In Fig. 3.16, there is a blue switch *p* on the left, and to the right there is a branch containing a red switch *p* and a switch *q*. The current must flow through the blue switch *p* into the branch on its right. Therefore, the blue switch *p* is in series with the branch containing the red switch *p* and switch *q*. After the current flows through the blue switch *p* it reaches the branch on its right. At this point, the current has the option of flowing into the red switch *p*, switch *q*, or both of these switches. Therefore, the branch containing the red switch *p* and switch *q* is a parallel branch. We say these two switches are in parallel. The entire circuit in symbolic form is *p* ∧ (*p* ∨ *q*). Note that the parentheses are very important. Without parentheses, the symbolic statement could be interpreted as (*p* ∧ *p*) ∨ *q*. The diagram for (*p* ∧ *p*) ∨ *q* is illustrated in Fig. 3.17. We shall see later in this section that the circuits in Fig. 3.16 and Fig. 3.17 are not the same.

b) The truth table for the statement (Table 3.40) indicates that the light will be on only in the cases in which *p* is true or when switch *p* is closed. ∎

Example 2 *Representing a Switching Circuit with Symbolic Statements*

a) Write a symbolic statement that represents the circuit in Fig. 3.18.

b) Construct a truth table to determine when the light will be on.

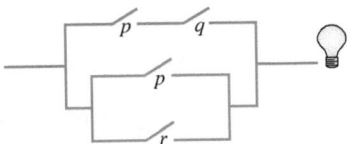

Figure 3.18

Solution

a) The upper branch of the circuit contains two switches *p* and *q* in series. We represent this branch with the statement *p* ∧ *q*. The lower branch of the circuit contains two switches *p* and *r* in parallel. We represent this branch with the statement *p* ∨ *r*. The upper branch is in parallel with the lower branch. Putting the two branches together, we get the statement (*p* ∧ *q*) ∨ (*p* ∨ *r*).

b) The truth table for the statement (Table 3.41) shows that cases (rows) 6 and 8 are false. Thus, the light will be off in these two cases and on in all other cases.

Table 3.41

p	q	r	(p ∧ q)	∨	(p ∨ r)
T	T	T	T	T	T
T	T	F	T	T	T
T	F	T	F	T	T
T	F	F	F	T	T
F	T	T	F	T	T
F	T	F	F	F	F
F	F	T	F	T	T
F	F	F	F	F	F
			1	3	2

Drawing Switching Circuits That Represent Symbolic Statements

We will next study how to draw a switching circuit that represents a given symbolic statement. Suppose we are given the statement $(p \land q) \lor r$ and are asked to construct a circuit corresponding to it. Remember that $\land$ indicates a series branch and $\lor$ indicates a parallel branch. Working first within parentheses, we see that switches p and q are in series. This series branch is in parallel with switch r, as indicated in Fig. 3.19.

Occasionally, it is necessary to have two switches in the same circuit such that when one switch is open, the other switch is closed; and when one switch is closed, the other switch is open. Therefore, the two switches will never both be open together and never be closed together. This situation can be represented by using p for one of the switches and $\bar{p}$ for the other switch. The switch labeled $\bar{p}$ corresponds to $\sim p$ in a logic statement. For example, in a series circuit, $p \land \sim p$ would be represented by Fig. 3.20. In this case, the light would never go on; the switches would counteract each other. When switch p is closed, switch $\bar{p}$ is open; and when p is open, $\bar{p}$ is closed.

Figure 3.19   Figure 3.20

Example 3 *Representing a Symbolic Statement as a Switching Circuit*

Draw a switching circuit that represents $\big[(p \land \sim q) \lor (r \lor q)\big] \land s$.

Solution In the statement, p and $\sim q$ have $\land$ between them, so switches p and $\bar{q}$ are in series, as represented in Fig. 3.21. Also in the statement, r and q have $\lor$ between them, so switches r and q are in parallel, as represented in Fig. 3.22.

Figure 3.21   Figure 3.22

Because $(p \land \sim q)$ and $(r \lor q)$ are connected with $\lor$, the two branches are in parallel with each other. The parallel branches are represented in Fig. 3.23. Finally, s in the statement is connected to the rest of the statement with $\land$. Therefore, switch s is in series with the entire rest of the circuit, as illustrated in Fig. 3.24.

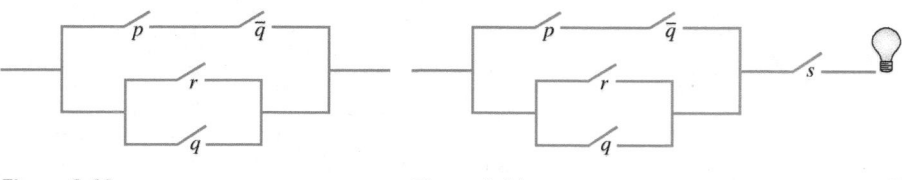

Figure 3.23   Figure 3.24

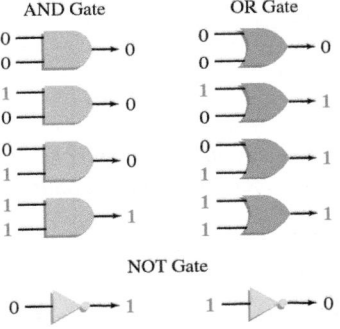

Equivalent Circuits

Sometimes two circuits that look very different will actually have exactly the same conditions under which the light will be on. If we were to analyze the truth tables for the corresponding symbolic statements for such circuits, we would determine that they have identical answer columns. In other words, the corresponding symbolic statements are equivalent.

> **Definition: Equivalent Circuits**
> **Equivalent circuits** are two circuits that have equivalent corresponding symbolic statements.

To determine whether two circuits are equivalent, we will analyze the answer columns of the truth tables of their corresponding symbolic statements. If the answer columns from their corresponding symbolic statements are identical, then the circuits are equivalent.

Example 4 *Are the Circuits Equivalent?*

Determine whether the two circuits are equivalent.

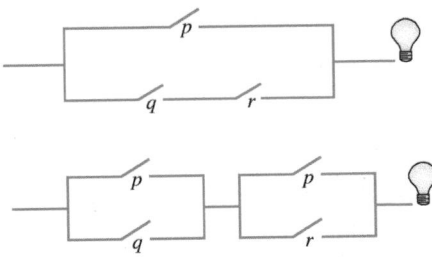

Solution The symbolic statement that represents the top circuit is $p \vee (q \wedge r)$. The symbolic statement that represents the bottom circuit is $(p \vee q) \wedge (p \vee r)$. The truth tables for these statements are shown in Table 3.42.

Table 3.42

p	q	r	p	$\vee$	$(q \wedge r)$	$(p \vee q)$	$\wedge$	$(p \vee r)$
T	T	T	T	T	T	T	T	T
T	T	F	T	T	F	T	T	T
T	F	T	T	T	F	T	T	T
T	F	F	T	T	F	T	T	T
F	T	T	F	T	T	T	T	T
F	T	F	F	F	F	T	F	F
F	F	T	F	F	F	F	F	T
F	F	F	F	F	F	F	F	F

Note that the answer columns for the two statements are identical. Therefore, $p \vee (q \wedge r)$ is equivalent to $(p \vee q) \wedge (p \vee r)$ and the two circuits are equivalent.

SECTION 3.7   *Exercises*

Warm Up Exercises

In Exercises 1–4, fill in the blanks with an appropriate word, phrase, or symbol(s).

1. A conjunction, ∧, is used to represent switches in a(n) _____ circuit.

2. A disjunction, ∨, is used to represent switches in a(n) _____ circuit.

3. When a switch labeled as p is open, a switch labeled as $\bar{p}$ is _____ .

4. Two circuits are equivalent if they have equivalent _____ statements.

Practice the Skills/Problem Solving

In Exercises 5–12, (a) write a symbolic statement that represents the circuit and (b) construct a truth table to determine when the lightbulb will be on. That is, determine which switches must be open and which switches must be closed for the lightbulb to be on.

5.
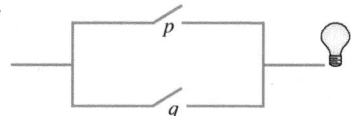

6.
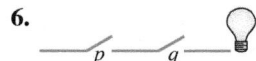

7.
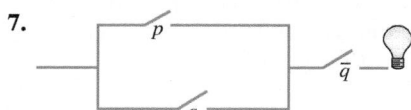

8.
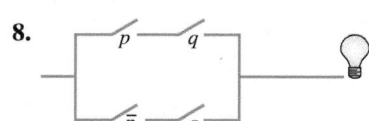

9.
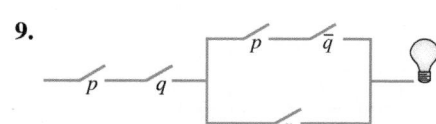

10.
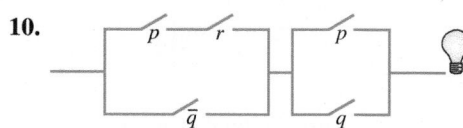

11.
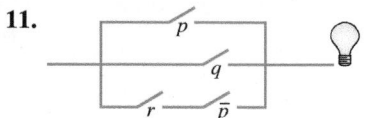

12.
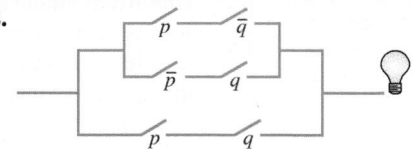

In Exercises 13–20, draw a switching circuit that represents the symbolic statement.

13. $(p \lor q) \land r$

14. $p \lor (q \land r)$

15. $p \lor (\sim q \land \sim r)$

16. $\sim p \land (q \lor r)$

17. $(p \lor q) \land (r \lor s)$

18. $(p \land q) \lor (r \land s)$

19. $\big[(p \lor q) \lor (r \land q) \big] \land (\sim p)$

20. $\big[(p \lor q) \land r \big] \lor (\sim p \land q)$

In Exercises 21–26, represent each circuit with a symbolic statement. Then use a truth table to determine if the circuits are equivalent.

21.
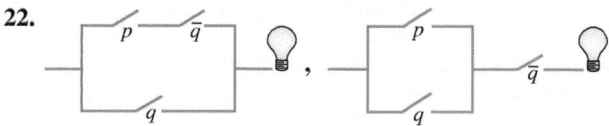

22.
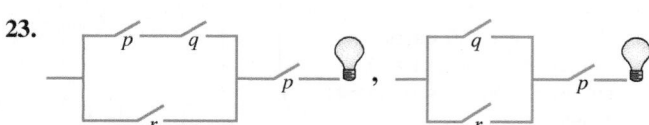

23.
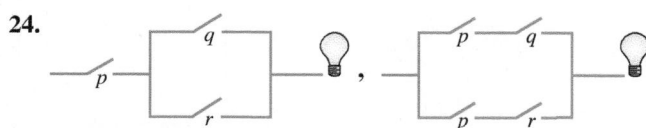

24.

25.

26.

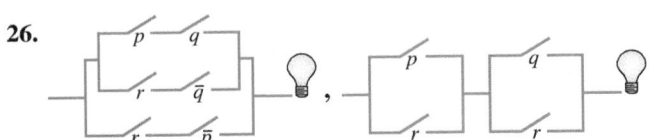

Concept/Writing Exercises

27. Explain why the lightbulb will never go on in the following circuit.

28. Explain why the lightbulb will always be on in the following circuit.

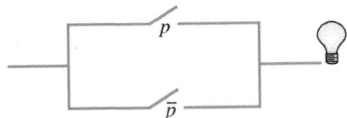

Challenge Problems/Group Activities

29. Design a circuit that can be represented by

   a) $p \to q$

   b) $\sim(p \to q)$

   (*Hint*: See Section 3.4.)

30. Design two circuits that each involve switches labeled p, q, r, and s that appear to be different but are actually equivalent circuits.

Research Activities

31. *Electronic Gates* Digital computers use gates that work like switches to perform calculations. (See *Did You Know?* box on page 159.) Information is fed into the gates and information leaves the gates, according to the type of gate. The three basic gates used in computers are the NOT gate, the AND gate, and the OR gate. Do research on the three types of gates.

   a) Explain how each gate works.

   b) Explain the relationship between each gate and the corresponding logic connectives *not*, *and*, and *or*.

   c) Illustrate how two or more gates can be combined to form a more complex gate.

CHAPTER 3 *Summary*

Important Facts and Concepts	Examples
Section 3.1	
Quantifiers	Example 1, page 94

Form of Statement	Form of Negation
All are.	Some are not.
None are.	Some are.
Some are.	None are.
Some are not.	All are.

Summary of Connectives — Examples 2–8, pages 95–99

Formal Name	Symbol	Read	Symbolic Form
Negation	$\sim$	not	$\sim p$
Conjunction	$\wedge$	and	$p \wedge q$
Disjunction	$\vee$	or	$p \vee q$
Conditional	$\to$	if–then	$p \to q$
Biconditional	$\leftrightarrow$	if and only if	$p \leftrightarrow q$

Section 3.2 & Section 3.3

Basic Truth Tables

Examples 1–10, pages 105–113
Examples 1–10, pages 117–123

Negation				Conjunction	Disjunction	Conditional	Biconditional
p	$\sim p$	p	q	$p \wedge q$	$p \vee q$	$p \rightarrow q$	$p \leftrightarrow q$
T	F	T	T	T	T	T	T
F	T	T	F	F	T	F	F
		F	T	F	T	T	F
		F	F	F	F	T	T

Section 3.4

De Morgan's Laws

Examples 4–6, pages 129–130

$$\sim(p \wedge q) \Leftrightarrow \sim p \vee \sim q$$
$$\sim(p \vee q) \Leftrightarrow \sim p \wedge \sim q$$

Other Equivalent Forms

Examples 7–9, pages 131–132

$$p \rightarrow q \Leftrightarrow \sim p \vee q$$
$$\sim(p \rightarrow q) \Leftrightarrow p \wedge \sim q$$
$$p \leftrightarrow q \Leftrightarrow \left[(p \rightarrow q) \wedge (q \rightarrow p)\right]$$

Variations of the Conditional Statement

Examples 10–12, pages 133–144

Name	Symbolic Form	Read
Conditional	$p \rightarrow q$	If p, then q.
Converse of the conditional	$q \rightarrow p$	If q, then p.
Inverse of the conditional	$\sim p \rightarrow \sim q$	If not p, then not q.
Contrapositive of the conditional	$\sim q \rightarrow \sim p$	If not q, then not p.

Section 3.5

Standard Forms of Valid Arguments

Examples 1–4, pages 141–143

Law of Detachment	Law of Contraposition	Law of Syllogism	Disjunctive Syllogism
$p \rightarrow q$	$p \rightarrow q$	$p \rightarrow q$	$p \vee q$
p	$\sim q$	$q \rightarrow r$	$\sim p$
$\therefore q$	$\therefore \sim p$	$\therefore p \rightarrow r$	$\therefore q$

Standard Forms of Invalid Arguments

Example 5, page 144

Fallacy of the Converse	Fallacy of the Inverse
$p \rightarrow q$	$p \rightarrow q$
q	$\sim p$
$\therefore p$	$\therefore \sim q$

Section 3.5 & Section 3.6

Symbolic Argument Versus Syllogistic Argument

Examples 1–5, pages 141–144
Examples 1–6, pages 150–153

	Words or Phrases Used	Method of Determining Validity
Symbolic argument	and, or, not, if–then, if and only if	Truth tables or by comparison with standard forms of arguments
Syllogistic argument	all are, some are, none are, some are not	Euler diagrams

Section 3.7 *Switching Circuits as Symbolic Statements* Switches in series will always be represented with a conjunction, $\land$. Switches in parallel will always be represented with a disjunction, $\lor$.	Examples 1–4, pages 157–159

CHAPTER 3   *Review Exercises*

3.1

In Exercises 1–4, write the negation of the statement.

1. All diamonds are made of carbon.

2. No pets are allowed in this park.

3. Some women are presidents.

4. Some pine trees are not green.

In Exercises 5–10, write each compound statement in words.

> p: The coffee is Maxwell House.
> q: The coffee is hot.
> r: The coffee is strong.

5. $q \lor r$

6. $\sim q \land r$

7. $q \rightarrow (r \land \sim p)$

8. $p \leftrightarrow \sim r$

9. $\sim p \leftrightarrow (r \land \sim q)$

10. $(p \lor \sim q) \land \sim r$

3.2

In Exercises 11–16, use the statements for p, q, and r as in Exercises 5–10 to write the statement in symbolic form.

11. The coffee is hot, but the coffee is not strong.

12. If the coffee is strong, then the coffee is not Maxwell House.

13. If the coffee is strong then the coffee is hot, or the coffee is not Maxwell House.

14. The coffee is hot if and only if the coffee is Maxwell House, and the coffee is not strong.

15. The coffee is strong and the coffee is hot, or the coffee is not Maxwell House.

16. It is false that the coffee is strong and the coffee is hot.

In Exercises 17–22, construct a truth table for the statement.

17. $(p \lor q) \land \sim p$

18. $q \leftrightarrow (p \lor \sim q)$

19. $(p \lor q) \leftrightarrow (p \lor r)$

20. $p \land (\sim q \lor r)$

21. $p \rightarrow (q \land \sim r)$

22. $(p \land q) \rightarrow \sim r$

3.2, 3.3

In Exercises 23–26, determine the truth value of the statement. You may need to use the Internet as a reference.

23. Apple makes iPhones and Dell makes canoes, or Hewlett Packard makes laser printers.

24. If a minute has 60 seconds, then an hour has 60 minutes if and only if a day has 20 hours.

25. If Oregon borders the Pacific Ocean or California borders the Atlantic Ocean, then Minnesota is south of Texas.

26. President's Day is in February, or Memorial Day is in May and Labor Day is in December.

3.3

In Exercises 27–30, determine the truth value of the statement when p is T, q is F, and r is F.

27. $(\sim p \lor q) \rightarrow \sim (p \land \sim q)$

28. $(p \leftrightarrow q) \rightarrow (\sim p \lor r)$

29. $\sim r \leftrightarrow \big[(p \lor q) \leftrightarrow \sim p\big]$

30. $\sim \big[(q \land r) \rightarrow (\sim p \lor r)\big]$

3.4

In Exercises 31–34, determine whether the pairs of statements are equivalent. You may use De Morgan's laws, the fact that $(p \rightarrow q) \Leftrightarrow (\sim p \lor q)$, the fact that $\sim (p \rightarrow q) \Leftrightarrow (p \land \sim q)$, truth tables, or equivalent forms of the conditional statement.

31. $\sim (p \land \sim q),\ \sim p \land q$

32. $p \lor q,\ \sim p \rightarrow q$

33. $\sim p \lor (q \land r),\ \ (\sim p \lor q) \land (\sim p \lor r)$

34. $(\sim q \rightarrow p) \land p,\ \ \sim (\sim p \leftrightarrow q) \lor p$

In Exercises 35–39, use De Morgan's laws, the fact that $(p \rightarrow q) \Leftrightarrow (\sim p \vee q)$, *or the fact that* $\sim (p \rightarrow q) \Leftrightarrow (p \wedge \sim q)$, *to write an equivalent statement for the given statement.*

35. A grasshopper is an insect and a spider is not an insect.

36. Lynn Swann played for the Steelers or Jack Tatum played for the Raiders.

37. It is not true that Altec Lansing only produces speakers or Harman Kardon only produces stereo receivers.

38. We did not go to the beach and we did not find sharks' teeth.

39. If the temperature is not above 32°, then we will go ice fishing at O'Leary's Lake.

In Exercises 40–44, write the (a) converse, (b) inverse, and (c) contrapositive for the given statement.

40. If Maya plays basketball, then she sits on the bench.

41. If we take the table to *Antiques Roadshow*, then we will learn the table's value.

42. If you do not advertise, then you do not sell more doughnuts.

43. If you do not study for the math test, then you fail the course.

44. If you get straight A's on your report card, then I will let you attend the prom.

In Exercises 45–48, determine which, if any, of the three statements are equivalent.

45. a) If Cheap Trick plays at the White House, then Jacque is the president.

  b) Cheap Trick does not play at the White House or Jacque is the president.

  c) It is false that Cheap Trick plays at the White House and Jacque is not the president.

46. a) The screwdriver is on the workbench if and only if the screwdriver is not on the counter.

  b) If the screwdriver is not on the counter, then the screwdriver is not on the workbench.

  c) It is false that the screwdriver is on the counter and the screwdriver is not on the workbench.

47. a) If $2 + 3 = 6$, then $3 + 1 = 5$.

  b) $2 + 3 = 6$ if and only if $3 + 1 \neq 5$.

  c) If $3 + 1 \neq 5$, then $2 + 3 \neq 6$.

48. a) If the sale is on Tuesday and I have money, then I will go to the sale.

  b) If I go to the sale, then the sale is on Tuesday and I have money.

  c) I go to the sale, or the sale is on Tuesday and I have money.

3.5

In Exercises 49–52, determine whether the argument is valid or invalid. You may compare the argument to a standard form or use a truth table.

49. $p \rightarrow q$
   $\sim p$
   $\overline{\therefore \sim q}$

50. $p \wedge q$
   $q \rightarrow r$
   $\overline{\therefore p \rightarrow r}$

51. If Jose is the manager, then Kevin is the coach.
   If Kevin is the coach, then Tim is the umpire.
   $\overline{\therefore \text{ If Jose is the manager, then Tim is the umpire.}}$

52. If the truck is a diesel, then the truck is too cold to start. The truck is too cold to start or the car has a flat tire. Therefore, the truck is not a diesel.

3.6

In Exercises 53–56, use an Euler diagram to determine whether the argument is valid or invalid.

53. All marigolds are annuals.
   Some tulips are annuals.
   $\overline{\therefore \text{ Some tulips are marigolds.}}$

54. All hackysack players are students.
   Joel is a hackysack player.
   $\overline{\therefore \text{ Joel is a student.}}$

55. No Beatles are Rolling Stones.
   Some Rolling Stones are Small Faces.
   $\overline{\therefore \text{ No Beatles are Small Faces.}}$

56. All bears are furry.
   Teddy is furry.
   $\overline{\therefore \text{ Teddy is a bear.}}$

3.7

57. a) Write the corresponding symbolic statement of the circuit shown.

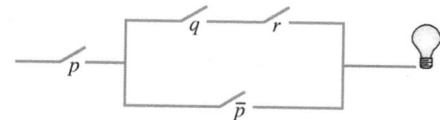

  b) Construct a truth table to determine when the lightbulb will be on.

58. Construct a diagram of a circuit that corresponds to the symbolic statement $(p \vee q) \vee (p \wedge q)$.

59. Determine whether the circuits shown on the right are equivalent.

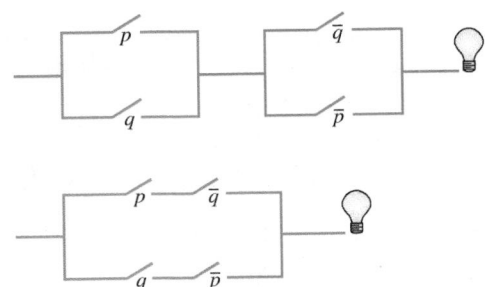

In Exercises 1–3, write the statement in symbolic form

    p: Phobos is a moon of Mars.

    q: Callisto is a moon of Jupiter.

    r: Rosalind is a moon of Uranus.

1. Phobos is not a moon of Mars or Callisto is a moon of Jupiter, and Rosalind is not a moon of Uranus.

2. If Rosalind is a moon of Uranus then Callisto is a moon of Jupiter, or Phobos is not a moon of Mars.

3. It is false that Rosalind is a moon of Uranus if and only if Callisto is not a moon of Jupiter.

In Exercises 4 and 5, use p, q, and r as above to write each symbolic statement in words.

4. $(\sim p \wedge r) \leftrightarrow \sim q$ **5.** $(p \vee \sim q) \rightarrow r$

In Exercises 6 and 7, construct a truth table for the given statement.

6. $[\sim(p \rightarrow r)] \wedge q$ **7.** $(q \leftrightarrow \sim r) \vee p$

In Exercises 8 and 9, determine the truth value of the statement.

8. $2 + 6 = 8$ or $7 - 12 = 5$.

9. A leap year has 366 days and a week has eight days if and only if an hour has 24 minutes.

In Exercises 10 and 11, given that p is true, q is false, and r is true, determine the truth value of the statement.

10. $(\sim p \wedge q) \leftrightarrow (q \vee \sim r)$

11. $[\sim(r \rightarrow \sim p)] \wedge (q \rightarrow p)$

12. Determine whether the pair of statements are equivalent.

    $\sim p \vee q$, $\sim(p \wedge \sim q)$

In Exercises 13 and 14, determine which, if any, of the three statements are equivalent.

13. a) If the bird is red, then it is a cardinal.

    b) The bird is not red or it is a cardinal.

    c) If the bird is not red, then it is not a cardinal.

14. a) It is not true that the test is today or the concert is tonight.

    b) The test is not today and the concert is not tonight.

    c) If the test is not today, then the concert is not tonight.

15. *Translate the following argument into symbolic form. Determine whether the argument is valid or invalid by comparing the argument to a recognized form or by using a truth table.*

    If the soccer team wins the game, then Sue played fullback. If Sue played fullback, then the team is in second place. Therefore, if the soccer team wins the game, then the team is in second place.

16. *Use an Euler diagram to determine whether the syllogism is valid or is a fallacy.*

    All living things contain carbon.

    Roger contains carbon.
    ―――――――――――――――――

    ∴ Therefore, Roger is a living thing.

In Exercises 17 and 18, write the negation of the statement.

17. All coffee beans contain caffeine.

18. Nick played football and Max played baseball.

19. Write the converse, inverse, and contrapositive of the conditional statement "If the garbage truck comes, then today is Saturday."

20. Construct a diagram of a circuit that corresponds to $(p \wedge q) \vee (\sim p \vee \sim q)$

◄ *Appliances use a computer chip that operates with numbers in bases other than base 10.*

4

Systems of Numeration

What You Will Learn

- Additive, multiplicative, and ciphered systems of numeration
- Place-value systems of numeration
- Use bases other than base 10
- Perform computations in other bases
- Understand early computational methods

Why This Is Important

The number system most of the world uses today, called the Hindu–Arabic system, is only one way to communicate numerically. In this chapter, we will study other systems of numeration. We will also study how to perform basic arithmetic in other bases and with methods other than those most of us were taught when we were children. Computers, appliances, GPS applications, gaming systems, smartphones, and other electronic devices all use a computer chip that operates with numbers in bases other than base 10. Understanding how to work with numbers in other bases can provide us with insight into how these devices work. These devices certainly play an increasing role in our lives today!

Additive, Multiplicative, and Ciphered Systems of Numeration

Upon completion of this section, you will be able to:

- Write numbers using additive systems of numeration.
- Write numbers using multiplicative systems of numeration.
- Write numbers using ciphered systems of numeration.

We are all familiar with the numerals we grew up with: 1, 2, 3, and so forth. Now think of another set of numerals. You may have thought of Roman numerals: I, II, III, and so forth. Some common modern-day uses of Roman numerals include clock and watch numbers, the copyright year on movie and television credits, outlines, and page numbers at the beginning of books. In this section, we will study several sets of numerals besides our familiar 1, 2, 3, ...

Why This Is Important Seeing numerals represented many different ways helps us to increase our number sense and gain a better understanding of how numbers work. Thinking about numerals in different ways can also help us to form connections or see patterns. Understanding patterns is important in many scientific areas such as predicting weather so we are prepared in an emergency situation. By studying other types of numerals, we can better understand the underlying concepts used in many applications of mathematics.

Numbers and Numerals

Just as the first attempts to write were made long after the development of speech, the first representation of numbers by symbols came long after people had learned to count. A tally system using physical objects, such as scratch marks in the soil or on a stone, notches on a stick, pebbles, or knots on a vine, was probably the earliest method of recording numbers.

In primitive societies, such a tally system adequately served the limited need for recording livestock, agriculture, or whatever was counted. As civilization developed, however, more efficient and accurate methods of calculating and keeping records were needed. Because tally systems are impractical and inefficient, societies developed symbols to replace them. The Egyptians, for example, used the symbol ∩, and the Babylonians used the symbol ◁ to represent the number we symbolize by 10.

A *number* is a quantity, and it answers the question "How many?" A *numeral* is a symbol such as ∩, ◁, or 10 used to represent the number. We think a number but write a numeral. The distinction between number and numeral will be made here only if it is helpful to the discussion.

In language, relatively few letters of the alphabet are used to construct a large number of words. Similarly, in arithmetic, a small variety of numerals can be used to represent all numbers. In general, when representing a number, we use as few numerals as possible. One of the greatest accomplishments of humankind has been the development of systems of numeration, whereby all numbers are "created" from a few symbols. Without such systems, mathematics would not have developed to its present level.

> **Definition: System of Numeration**
> A **system of numeration** consists of a set of numerals and a scheme or rule for combining the numerals to represent numbers.

Did You Know?

The Rhind Papyrus

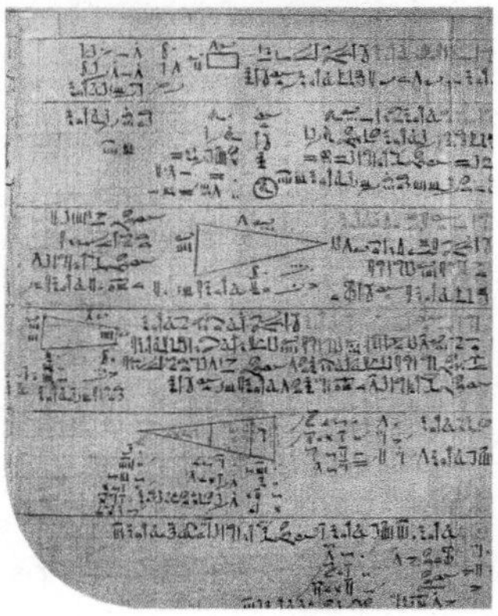

Much of our knowledge of Egyptian mathematics and numeration comes from a roll of papyrus measuring 18 ft by 13 in. Written around 1650 B.C., it was discovered in a shop in Luxor, Egypt, in 1858 by Henry Rhind, a Scottish lawyer turned archaeologist. Unlike the straightforward accounting of property and events common to Egyptian tombs, the Rhind Papyrus has inscribed on it 85 mathematical problems and solutions involving addition, subtraction, multiplication, division, and geometry.

The key to translation of the papyrus was the Rosetta Stone, which had been discovered some 60 years earlier by one of Napoleon's officers. Both of these treasures now reside at the British Museum in London.

Four types of numeration systems used by different cultures are the topic of this chapter. They are additive (or repetitive), multiplicative, ciphered, and place-value systems. You do not need to memorize all the symbols, but you should understand the principles behind each system. By the end of this chapter, we hope that you better understand the system we use, the *Hindu–Arabic system*, and its relationship to other types of systems.

Additive Systems

An additive system is one in which the number represented by a particular set of numerals is simply the sum of the values of the numerals. The additive system of numeration is one of the oldest and most primitive types of numeration systems. One of the first additive systems, the Egyptian hieroglyphic system, dates back to about 3000 B.C. The Egyptians used symbols for the powers of 10: 10^0 or 1, 10^1 or 10, 10^2 or $10 \cdot 10$, 10^3 or $10 \cdot 10 \cdot 10$, and so on. Table 4.1 lists the Egyptian hieroglyphic numerals with the equivalent Hindu–Arabic numerals.

To write the number 600 in Egyptian hieroglyphics, we write the numeral for 100 six times: 𓍢𓍢𓍢𓍢𓍢𓍢.

Table 4.1 Egyptian Hieroglyphics

Hindu–Arabic Numerals	Egyptian Numerals	Description
1	𓏺	Staff (vertical stroke)
10	𓎆	Heel bone (arch)
100	𓍢	Scroll (coiled rope)
1000	𓆼	Lotus flower
10,000	𓂭	Pointing finger
100,000	𓆐	Tadpole (or whale)
1,000,000	𓁨	Astonished person

Example 1 *From an Egyptian to a Hindu–Arabic Numeral*

Write the following numeral as a Hindu–Arabic numeral.

𓂭𓂭𓍢𓍢𓍢𓍢𓎆𓏺𓏺𓏺𓏺𓏺

Solution

$$10{,}000 + 10{,}000 + 100 + 100 + 100 + 100 + 10 + 1 + 1 + 1 + 1 + 1$$
$$= 20{,}415$$

■

Example 2 *From a Hindu–Arabic to an Egyptian Numeral*

Write 1,203,462 as an Egyptian numeral.

Solution

$$1{,}203{,}462 = 1{,}000{,}000 + 200{,}000 + 3000 + 400 + 60 + 2$$

𓁨𓆐𓆐𓆼𓆼𓆼𓍢𓍢𓍢𓍢𓎆𓎆𓎆𓎆𓎆𓎆𓏺𓏺

■

In the Egyptian hieroglyphic system, the order of the symbols is not important. For example, 𓆐𓏺𓏺𓍢𓍢𓎆 and 𓏺𓏺𓍢𓍢𓆐𓎆 both represent 100,212.

Users of additive systems easily accomplished addition and subtraction by combining or removing symbols. Multiplication and division were more difficult; they were performed by a process called *duplation and mediation* (see Section 4.5). The Egyptians had no symbol for zero, but they did have an understanding of fractions. The symbol ⌣ was used to take the reciprocal of a number; thus, ⅏ meant $\frac{1}{3}$ and ⅏ was $\frac{1}{11}$. Writing large numbers in the Egyptian hieroglyphic system takes longer than in other systems because so many symbols have to be listed. For example, 45 symbols are needed to represent 99,999.

The Roman numeration system, a second example of an additive system, was developed later than the Egyptian system. Roman numerals (Table 4.2) were used in most European countries until the eighteenth century. They are still commonly seen on buildings, on clocks, and in books. Roman numerals are selected letters of the Roman alphabet.

Table 4.2 Roman Numerals

Roman numerals	I	V	X	L	C	D	M
Hindu–Arabic numerals	1	5	10	50	100	500	1000

Timely Tip

When working with Roman numerals, we work from *left to right*. We add each numeral unless its value is *smaller* than the value of the numeral to its right. In that case, we *subtract* its value from the value of the numeral to its right.

The Roman system has two advantages over the Egyptian system. The first is that it uses the subtraction principle as well as the addition principle. Starting from the left, we add each numeral unless its value is smaller than the value of the numeral to its right. In that case, we subtract its value from the value of the numeral to its right. Only the numbers 1, 10, 100, 1000, ... can be subtracted, and they can only be subtracted from the next two higher numbers. For example, C (100) can be subtracted only from D (500) or M (1000). The symbol DC represents $500 + 100$, or 600, and CD represents $500 - 100$, or 400. Similarly, MC represents $1000 + 100$, or 1100, and CM represents $1000 - 100$, or 900. In addition, CX represents $100 + 10$, or 110, and XC represents $100 - 10$, or 90. Also, XI represents $10 + 1$, or 11, and IX represents $10 - 1$, or 9.

Write A Roman
numeral As
A Hindu-Arabic numeral

Example 3 *From a Roman Numeral to a Hindu–Arabic Numeral*

Write MDCCCLXVII as a Hindu–Arabic numeral.

Solution Since each numeral is larger than the one to its right, no subtraction is necessary.

$$\text{MDCCCLXVII} = 1000 + 500 + 100 + 100 + 100 + 50 + 10 + 5 + 1 + 1$$
$$= 1867$$ ∎

Example 4 *From a Roman Numeral to a Hindu–Arabic Numeral*

Write CMLXIV as a Hindu–Arabic numeral.

Solution As you read the numerals from left to right, you will note that C (100) has a smaller value than M (1000) and that C is to the left of M. Therefore, CM represents $1000 - 100$, or 900. Also note that I (1) has a smaller value than V (5) and that I is to the left of V. Therefore, IV represents $5 - 1$, or 4. The rest of the numerals can be added from left to right.

$$\text{CMLXIV} = (1000 - 100) + 50 + 10 + (5 - 1) = 964$$ ∎

Example 5 *From a Hindu–Arabic Numeral to a Roman Numeral*

Write 439 as a Roman numeral.

Solution

$$439 = 400 + 30 + 9$$

To represent 400, we will subtract 100 (C) from 500 (D). We will place C to the left of D to indicate the subtraction ($500 - 100$). Likewise, to represent 9, we will subtract 1 (I) from 10 (X). We will place I to the left of X to indicate the subtraction ($10 - 1$). To represent 30 we will simply write X three times. Therefore,

$$439 = 400 + 30 + 9 = (500 - 100) + 10 + 10 + 10 + (10 - 1)$$

$$= \text{CDXXXIX}$$

In the Roman numeration system, a symbol is not repeated more than three consecutive times. For example, the number 646 would be written DCXLVI instead of DCXXXXVI.

The second advantage of the Roman numeration system over the Egyptian numeration system is that it makes use of the multiplication principle for numerals greater than 1000. A bar above a symbol or group of symbols indicates that the symbol or symbols are to be multiplied by 1000. Thus, $\overline{V} = 5 \times 1000 = 5000$, $\overline{X} = 10 \times 1000 = 10,000$, and $\overline{CD} = 400 \times 1000 = 400,000$. Other examples are $\overline{VI} = 6 \times 1000 = 6000$, $\overline{XIX} = 19 \times 1000 = 19,000$, and $\overline{XCIV} = 94 \times 1000 = 94,000$. This practice greatly reduces the number of symbols needed to write large numbers.

Example 6 *Writing a Large Roman Numeral*

Write 12,345 as a Roman numeral.

Solution

$$12,345 = (12 \times 1000) + 300 + 40 + 5$$
$$= \left[(10 + 1 + 1) \times 1000\right] + (100 + 100 + 100) + (50 - 10) + 5$$
$$= \overline{XII}\text{CCCXLV}$$

Write a Hindu-Arabic Numeral as a Roman Numeral

Multiplicative Systems

Multiplicative numeration systems are more similar to our Hindu–Arabic system than are additive systems. In a multiplicative system, 642 might be written (6) (100) (4) (10) (2) or

6
100
4
10
2

Note that no addition signs are needed in the representation. From this illustration, try to formulate a rule explaining how multiplicative systems work.

The principal example of a multiplicative system is the traditional Chinese system. The numerals used in this system are given in Table 4.3. The numeration system

Table 4.3 Traditional Chinese Numerals

Traditional Chinese numerals	零	一	二	三	四	五	六	七	八	九	十	百	千
Hindu–Arabic numerals	0	1	2	3	4	5	6	7	8	9	10	100	1000

used in China today is different from the traditional system discussed in this chapter. The present-day numeration system in China is a positional-value system rather than a multiplicative system, and in some areas of China, 0 is used as the numeral for zero.

Chinese numerals are always written vertically. The numeral on top usually will be from 1 to 9 inclusive. This numeral is to be multiplied by the power of 10 below it. Thus, 20 is written

$$二\atop十 \Big\} \; 2 \times 10 = 20$$

and 400 is written

$$四\atop百 \Big\} \; 4 \times 100 = 400$$

Example 7 *A Traditional Chinese Numeral*

Write 538 as a Chinese numeral.

Solution

$$538 = \begin{cases} 500 = \begin{cases} 5 & 五 \\ 100 & 百 \end{cases} \\ 30 = \begin{cases} 3 & 三 \\ 10 & 十 \end{cases} \\ 8 = \quad\;\; 8 \quad 八 \end{cases}$$

When writing traditional Chinese numerals, the units digit is never multiplied by a power of the base. Note that in Example 7 the units digit, the 8, was not multiplied by a power of the base.

When writing Chinese numerals, there are some special cases that need to be considered. When writing a numeral between 11 and 19 it is <u>not</u> necessary to include the 1 before the 10. Thus, 18 would be written $十\atop八$ rather than $十\atop八$. Another special case involves the use of zero.

When more than one consecutive zero occurs (except at the end of a numeral) you need to write a zero, but only once for two or more consecutive zeros. Zeros are not included at the end of numerals. The top two illustrations that follow show how zeros are used within a numeral, and the bottom two show that zeros are not used at the end of a numeral.

$$406 = \begin{matrix} 四\atop百 \big\} \; 4 \times 100 = 400 \\ 零 \big\} \; 0 \times 10 = \;\;\; 0 \\ 六 \big\} \; 6 \qquad\quad = \;\;\; 6 \end{matrix}$$

$$4006 = \begin{matrix} 四\atop千 \big\} \; 4 \times 1000 = 4000 \\ 零 \big\} \begin{matrix} 0 \times 100 = \;\;\; 0 \\ 0 \times 10 = \;\;\; 0 \end{matrix} \\ 六 \big\} \qquad\quad 6 = \;\;\; 6 \end{matrix}$$

$$460 = \begin{matrix} 四\atop百 \big\} \; 4 \times 100 = 400 \\ 六\atop十 \big\} \; 6 \times 10 = 60 \end{matrix}$$

$$4600 = \begin{matrix} 四\atop千 \big\} \; 4 \times 1000 = 4000 \\ 六\atop百 \big\} \; 6 \times 100 = 600 \end{matrix}$$

Example 8 *Traditional Chinese Numerals*

Write the following as traditional Chinese numerals.

a) 7080   b) 7008

Solution In part (a), there is one zero between the 7 and the 8. In part (b), there are two zeros between the 7 and the 8. As mentioned previously, the symbol for zero is used only once in each of these numerals.

a) $7080 =$
$$\left.\begin{array}{c}七\\千\end{array}\right\} 7 \times 1000$$
$$零 \} 0 \times 100$$
$$\left.\begin{array}{c}八\\十\end{array}\right\} 8 \times 10$$

b) $7008 =$
$$\left.\begin{array}{c}七\\千\end{array}\right\} 7 \times 1000$$
$$零 \left.\begin{array}{c}\\\end{array}\right\} \begin{array}{l}0 \times 100,\\0 \times 10\end{array}$$
$$八 \} 8$$

Timely Tip

Notice the difference between our Hindu–Arabic numeration system, which is a positional numeration system, and the Chinese system, which is a multiplicative numeration system. Below we show how to write 5678 as a Chinese numeral if the Chinese system were a positional value system similar to ours.

Multiplicative		Positional Value	
五	5	五 5	
千	1000	六 6	
六	6	七 7	
百	100	八 8	
七	7		
十	10		
八	8		

Note that the multiples of base 10 are removed when writing positional value numerals. We will discuss positional value systems in more detail shortly.

Ciphered Systems

A ciphered numeration system is one in which there are numerals for numbers up to and including the base and for multiples of the base. The numbers represented by a particular set of numerals is the sum of the values of the numerals.

Ciphered numeration systems require the memorization of many different symbols but have the advantage that numbers can be written in a compact form. The ciphered numeration system that we discuss is the Ionic Greek system (Table 4.4 on page 173). The Ionic Greek system was developed in about 3000 B.C., and it used letters of the Greek alphabet for numerals. Other ciphered systems include the Hebrew, Coptic, Hindu, Brahmin, Syrian, Egyptian Hieratic, and early Arabic systems.

The classic Greek alphabet contains only 24 letters; however, 27 symbols were needed. Thus, the Greeks used three obsolete Greek letters—ϝ, ϟ, and ϡ—that are not part of the classic Greek alphabet. To distinguish words from numerals, the Greeks would place a mark similar to our apostrophe to the right and above each letter that was used as a numeral. In this text, we will not use this mark because we will not be using Greek words and numerals together.

We can write 45 as $40 + 5$. When 45 is written as a Greek numeral, the plus sign is omitted:

$$45 = \mu\epsilon$$

Similarly, 768 written as a Greek numeral is $\psi\xi\eta$.

Table 4.4 Ionic Greek Numerals

1	α	alpha	60	ξ	xi
2	β	beta	70	o	omicron
3	γ	gamma	80	π	pi
4	δ	delta	90	φ^*	koppa
5	ϵ	epsilon	100	ρ	rho
6	f^*	digamma	200	σ	sigma
7	ζ	zeta	300	τ	tau
8	η	eta	400	υ	upsilon
9	θ	theta	500	ϕ	phi
10	ι	iota	600	χ	chi
20	κ	kappa	700	ψ	psi
30	λ	lambda	800	ω	omega
40	μ	mu	900	$\gimel^*$	sampi
50	ν	nu			

*Ancient Greek letters not used in the classic or modern Greek language

When the Greek letter iota, ι, is placed to the left and above a numeral, it represents the numeral multiplied by 1000. For example,

$$'\theta = 9 \times 1000 = 9000$$
$$'\zeta = 7 \times 1000 = 7000$$

Example 9 *From an Ionic Greek Numeral to a Hindu–Arabic Numeral*

Write $\tau\pi\epsilon$ as a Hindu–Arabic numeral.

Solution $\tau = 300$, $\pi = 80$, and $\epsilon = 5$. The sum is 385. ∎

Example 10 *Writing an Ionic Greek Numeral*

Write 1654 as an Ionic Greek numeral.

Solution

$$
\begin{aligned}
1654 &= 1000 + 600 + 50 + 4 \\
&= (1 \times 1000) + 600 + 50 + 4 \\
&= '\alpha \qquad\qquad \chi \quad \nu \quad \delta \\
&= '\alpha\, \chi\, \nu\, \delta
\end{aligned}
$$
∎

SECTION 4.1 *Exercises*

Warm Up Exercises

In Exercises 1–8, fill in the blanks with an appropriate word, phrase, or symbol(s).

1. A number is a quantity, and it answers the question "_____?"

2. A symbol used to represent a number is called a(n) _____.

3. The system of numeration that we presently use is the _____ system.

4. The Egyptian hieroglyphic system and the Roman numeration system are examples of _____ numeration systems.

5. As you read a Roman numeral from left to right, if a smaller number comes before a larger number, you should _____ the smaller number from the larger number.

6. A bar above a Roman numeral means to multiply that numeral by _____ .

7. The Chinese numeration system is an example of a _____ numeration system.

8. The Ionic Greek numeration system is an example of a _____ numeration system.

Practice the Skills

In Exercises 9–14, write the Egyptian numeral as a Hindu–Arabic numeral.

9. 𓏺𓏺𓎋𓎋𓏭𓏭𓏭𓏭𓏭

10. 𓂋𓏺𓏺𓏺𓎋𓎋𓎋𓏭

11. 𓋃𓋃𓋃𓏺𓏺𓏺𓎋𓎋𓎋𓎋𓏭

12. 𓆼𓆼𓆼𓆼𓆼𓂋𓏺𓏺𓎋

13. 𓍢𓍢𓆼𓆼𓆼𓋃𓋃𓋃𓋃𓏺𓏺𓎋𓏭𓏭𓏭𓏭

14. 𓆐𓆐𓆐𓍢𓍢𓏺𓏺𓎋𓎋𓎋

In Exercises 15–20, write the numeral as an Egyptian numeral.

15. 22   16. 134

17. 2045   18. 20,321

19. 173,845   20. 3,235,614

In Exercises 21–32, write the Roman numeral as a Hindu–Arabic numeral.

21. XXXVI   22. XXII

23. CXCIV   24. CDLXXXVIII

25. MMDCXLII   26. MCMLXIV

27. MMCMXLVI   28. MDCCXLVI

29. $\overline{\text{IV}}$CDXCIX   30. $\overline{\text{L}}$MCMXLIV

31. $\overline{\text{X}}$MMDCLXVI   32. $\overline{\text{XCIX}}$CMXCIX

In Exercises 33–44, write the numeral as a Roman numeral.

33. 62   34. 76

35. 349   36. 692

37. 1914   38. 4285

39. 4793   40. 6274

41. 9999   42. 14,315

43. 46,281   44. 89,119

In Exercises 45–52, write the Chinese numeral as a Hindu–Arabic numeral.

45. 七十四   46. 六十二

47. 四千零八十一   48. 三千零二十九

49. 八千五百五十   50. 三千四百八十七

51. 四千零三   52. 五千六百零二

In Exercises 53–60, write the numeral as a traditional Chinese numeral.

53. 53   54. 178

55. 378   56. 2001

57. 4260   58. 6905

59. 7056   60. 3009

In Exercises 61–66, write the numeral as a Hindu–Arabic numeral.

61. $\mu\varepsilon$

62. $\sigma\pi\beta$

63. $\omega o \eta$

64. $\partial\epsilon$

65. $'\beta\omega\pi\gamma$

66. $'\zeta\upsilon\xi\zeta$

In Exercises 67–72, write the numeral as an Ionic Greek numeral.

67. 543

68. 139

69. 784

70. 2008

71. 5005

72. 9999

In Exercises 73–76, write the numeral as numerals in the indicated systems of numeration.

73. $\overset{s}{\,}\cap\cap\mathbb{I}$ in Hindu–Arabic, Roman, traditional Chinese, and Greek

74. MCMXXXVI in Hindu–Arabic, Egyptian, Greek, and traditional Chinese

75. 五百二十七 in Hindu–Arabic, Egyptian, Roman, and Greek

76. $\upsilon\kappa\beta$ in Hindu–Arabic, Egyptian, Roman, and traditional Chinese

Concept/Writing Exercises

77. What is the difference between a number and a numeral?

78. What is a system of numeration?

Challenge Problems/Group Activities

79. Write the Roman numeral for 999,999.

80. Make up your own additive system of numeration and indicate the symbols and rules used to represent numbers. Using your system of numeration, write

a) your age.

b) the year you were born.

c) the current year.

Recreational Mathematics

81. *A Roman Numeral Equation* Without using any type of writing instrument, what can you do to make the following incorrect equation a correct equation?

$$XI + I = X$$

82. *Palindromes* Words and numerals that read the same both backward and forward are called *palindromes*. Some examples are the words CIVIC and RACECAR, and the numerals 121 and 32523. Using Roman numerals, list the last year that was a palindrome.

83. *Roman Numeral Years* Which year in the past 2000 years required the most Roman numerals to write? Write out the year in Roman numerals.

Research Activity

84. *Egypt, China, and Greece* In this section, we discussed the numeration systems of Egypt, China, and Greece. Choose one of these countries and write a paper that includes the following.

a) State the current numerals used in your chosen country.

b) Explain how that country's current system of numeration works. If more than one system is used, discuss the system used most commonly.

▲ Great Pyramid of Giza in Egypt

85. *Rhind Papyrus* Write a report on the Rhind Papyrus (see the *Did You Know* on page 168). Include in your report some of the mathematics problems inscribed on the papyrus.

SECTION 4.2 Place-Value or Positional-Value Numeration Systems

Upon completion of this section, you will be able to:

- Understand place-value numeration-systems.
- Use and understand Babylonian numerals.
- Use and understand Mayan numerals

Chances are you have seen one of the many games that require contestants to place digits in the correct order to guess the price of a car, appliance, vacation to Paris or elsewhere, or other fabulous prizes. Such games require the contestant to have an understanding of the concept of place value that is used in our modern number system. In this section, we will study two other number systems that also rely on place-value.

Why This Is Important Understanding place-value helps you understand the difference between a paycheck for $500 and the $5000 price tag on a car you may like to purchase. By studying other place-value numeration systems, we will gain a better understanding of our own numeration system.

Eighteenth-century mathematician Pierre Simon, Marquis de Laplace, speaking of the positional principle, said: "The idea is so simple that this very simplicity is the reason for our not being sufficiently aware of how much attention it deserves."

Did You Know?

Babylonian Numerals

The form Babylonian numerals took is directly related to their writing materials. Babylonians used a reed (later a stylus) to make their marks in wet clay. The end could be used to make a thin wedge, $\mathbf{I}$, which represents a unit, or a wider wedge, $\blacktriangleleft$, which represents 10 units. The clay dried quickly, so the writings tended to be short but extremely durable.

Place-Value Numeration Systems

Today the most common type of numeration system is the place-value system. The Hindu–Arabic numeration system, used in the United States and many other countries, is an example of a place-value system. In a *place-value system*, which is also called a *positional-value system*, the value of the symbol depends on its position in the representation of the number. For example, the 2 in 20 represents 2 tens, and the 2 in 200 represents 2 hundreds. A true positional-value system requires a *base* and a set of symbols, including a symbol for zero and one for each counting number less than the base. Although any number can be written in any base, the most common positional-value system is the base 10 system, which is called the *decimal number system*.

The Hindus in India are credited with the invention of zero and the other symbols used in our system. The Arabs, who traded regularly with the Hindus, also adopted the system, thus the name Hindu–Arabic. Not until the middle of the fifteenth century, however, did the Hindu–Arabic numerals take the form we know today.

The Hindu–Arabic numerals and the positional-value system of numeration revolutionized mathematics by making addition, subtraction, multiplication, and division much easier to learn and very practical to use. Merchants and traders no longer had to depend on the counting board or abacus. The first group of mathematicians who computed with the Hindu–Arabic system, rather than with pebbles or beads on a wire, were known as the "algorists."

In the Hindu–Arabic system, the symbols 0, 1, 2, 3, 4, 5, 6, 7, 8, and 9 are called *digits*. The base 10 system was developed from counting on fingers, and the word *digit* comes from the Latin word for fingers.

The positional values in the Hindu–Arabic system are

$$\ldots, (10)^5, (10)^4, (10)^3, (10)^2, 10, 1$$

To evaluate a numeral in the Hindu–Arabic system, we multiply the digit on the right by 1. We multiply the second digit from the right by the base, 10. We multiply the third digit from the right by the base squared, 10^2 or 100. We multiply the fourth digit from the right by the base cubed, 10^3 or 1000, and so on. In general, we multiply the digit n places from the right by 10^{n-1}. Therefore, we multiply the digit eight places from the right by 10^7. Using the place-value rule, we can write a number in *expanded form*.

Example 1 *Writing a Hindu–Arabic Numeral in Expanded Form*

Write 1234 in expanded form.

Solution

$$1234 = (1 \times 10^3) + (2 \times 10^2) + (3 \times 10) + (4 \times 1)$$
$$= (1 \times 1000) + (2 \times 100) + (3 \times 10) + (4 \times 1) \quad \blacksquare$$

Babylonian Numerals

Table 4.5 Babylonian Numerals

Babylonian numerals	!	<
Hindu–Arabic numerals	1	10

The oldest known numeration system that resembled a place-value system was developed by the Babylonians in about 2500 B.C. Their system resembled a place-value system with a base of 60, a *sexagesimal* system. It was not a true place-value system because it lacked a symbol for zero. The lack of a symbol for zero led to a great deal of ambiguity and confusion. Table 4.5 gives the Babylonian numerals.

The positional values in the Babylonian system are

$$\ldots, (60)^3, (60)^2, 60, 1$$

In a Babylonian numeral, a gap is left between the characters to distinguish between the various place values. From right to left, the sum of the first group of numerals is multiplied by 1. The sum of the second group is multiplied by 60. The sum of the third group is multiplied by $(60)^2$, and so on.

Example 2 *The Babylonian System: A Place-Value System*

Write ⟨⟨ ! ! !   ⟨⟨ ! ! as a Hindu–Arabic numeral.

Solution

⟨⟨ ! ! !	⟨⟨ ! !
60's	units
10 + 10 + 1 + 1 + 1	10 + 10 + 1 + 1
60's	units
↓	↓
(23 × 60)	+ (22 × 1)

$$1380 + 22 = 1402 \quad \blacksquare$$

The Babylonians used the symbol 𝈖 to indicate subtraction. The numeral ⟨𝈖 ! ! represents $10 - 2$, or 8. The numeral ⟨⟨⟨⟨ 𝈖 ! ! ! ! represents $40 - 3$, or 37 in base 10 or decimal notation.

Example 3 *From a Babylonian to a Hindu–Arabic Numeral*

Write ! !   ⟨ !   ⟨⟨ 𝈖 ! ! ! as a Hindu–Arabic numeral.

Solution The place value of these three groups of numerals from left to right is

	$(60)^2$,	60,	1
or	3600,	60,	1

The numeral in the group on the right has a value of $20 - 2$, or 18. The numeral in the center group has a value of $10 + 1$, or 11. The numeral on the left represents $1 + 1$, or 2. Multiplying each group by its positional value gives

$$(2 \times 60^2) + (11 \times 60) + (18 \times 1)$$
$$= (2 \times 3600) + (11 \times 60) + (18 \times 1)$$
$$= 7200 + 660 + 18$$
$$= 7878$$

To explain the procedure used to convert from a Hindu–Arabic numeral to a Babylonian numeral, we will consider a length of time. How can we change 9820 seconds into hours, minutes, and seconds? Since there are 3600 seconds in an hour (60 seconds to a minute and 60 minutes to an hour), we can determine the number of hours in 9820 seconds by dividing 9820 by 60^2, or 3600.

$$\begin{array}{r} 2 \quad \leftarrow \text{Hours} \\ 3600\overline{)9820} \\ \underline{7200} \\ 2620 \quad \leftarrow \text{Remaining seconds} \end{array}$$

Now we can determine the number of minutes by dividing the remaining seconds by 60, the number of seconds in a minute.

$$\begin{array}{r} 43 \quad \leftarrow \text{Minutes} \\ 60\overline{)2620} \\ \underline{2400} \\ 220 \\ \underline{180} \\ 40 \quad \leftarrow \text{Remaining seconds} \end{array}$$

Since the remaining number of seconds, 40, is less than the number of seconds in a minute, our task is complete.

$$9820 \text{ sec} = 2 \text{ hr}, 43 \text{ min}, 40 \text{ sec}$$

The same procedure is used to convert a decimal (base 10) numeral to a Babylonian numeral or any numeral in a different base.

Example 4 *From a Hindu–Arabic to a Babylonian Numeral*

Write 2519 as a Babylonian numeral.

Solution The Babylonian numeration system has positional values of

$$\dots, (60)^3, (60)^2, 60, 1$$

which can be expressed as

$$\dots, 216000, 3600, 60, 1$$

The largest positional value less than or equal to 2519 is 60. To determine how many groups of 60 are in 2519, divide 2519 by 60.

$$\begin{array}{r} 41 \quad \leftarrow \text{Groups of 60} \\ 60\overline{)2519} \\ \underline{240} \\ 119 \\ \underline{60} \\ 59 \quad \leftarrow \text{Units remaining} \end{array}$$

Thus, $2519 \div 60 = 41$ with remainder 59. There are 41 groups of 60 and 59 units remaining. Because the remainder, 59, is less than the base, 60, no further division is necessary. The remainder represents the number of units when the numeral is written in expanded form. Therefore, $2519 = (41 \times 60) + (59 \times 1)$. When written as a Babylonian numeral, 2519 is

$$\text{<<<<I} \quad \text{<<<<<<<ĩI}$$

Example 5 *Using Division to Determine a Babylonian Numeral*

Write 5363 as a Babylonian numeral.

Solution To begin, divide 5363 by the largest positional value less than or equal to 5363. That value is 3600.

$$5363 \div 3600 = 1 \text{ with remainder } 1763$$

There is one group of 3600 in 5363. Next divide the remainder, 1763, by 60 to determine the number of groups of 60 in 1763.

$$1763 \div 60 = 29 \text{ with remainder } 23$$

There are 29 groups of 60 with 23 units remaining.

$$5363 = (1 \times 3600) + (29 \times 60) + (23 \times 1)$$

Thus, 5363 written as a Babylonian numeral is

$$\text{I} \quad \text{<<<ĩI} \quad \text{<<III}$$

Example 6 *A Babylonian Numeral with a Blank Space*

Write 7223 as a Babylonian numeral.

Solution As we did in the last example, we begin by dividing 7223 by 3600.

$$7223 \div 3600 = 2 \text{ with remainder } 23$$

Therefore, there are two groups of 3600 in 7223. The next positional value in the Babylonian numeration system is 60. However, the remainder from the above division, 23, is less than 60. So there are zero groups of 60 in 7223 and 23 units remaining. Therefore, we can write 7223 as follows:

$$7223 = (2 \times 60^2) + (0 \times 60) + (23 \times 1)$$

Recall that the Babylonian numeration system does not contain a symbol for 0. Therefore, we will need to leave a larger blank space when writing the numeral. This larger blank space will indicate that there are no groups of 60s present in 7223.

$$7223 = \underline{(2 \times 60^2)} + (0 + 60) + \underline{(23 \times 1)}$$
$$\text{II} \qquad\qquad\qquad\quad \text{<<III}$$
$$\uparrow$$

The larger blank space here represents 0 groups of 60.

Thus, the answer is **II**   **<<III**.

Mayan Numerals

Another place-value system is the Mayan numeration system. The Mayans, who lived on the Yucatan Peninsula in present-day Mexico, developed a sophisticated numeration system based on their religious and agricultural calendar. The numerals in this system are written vertically rather than horizontally, with the units position on the bottom. In the Mayan system, the numeral in the bottom row is to be multiplied by 1. The numeral in the second row from the bottom is to be multiplied by 20. The numeral in the third row is to be multiplied by 18×20, or 360. You probably expected the numeral in the third row to be multiplied by 20^2 rather than 18×20. It is believed that the Mayans used 18×20 so that their numeration system would conform to their calendar of 360 days. The positional values greater than 18×20 are 18×20^2, 18×20^3, and so on.

Positional Values in the Mayan System

$\dots 18 \times (20)^3,$	$18 \times (20)^2,$	$18 \times 20,$	$20,$	1
or $\dots 144{,}000,$	$7200,$	$360,$	$20,$	1

The digits 0, 1, 2, 3, $\dots$, 19 of the Mayan system are formed by a simple grouping of dots and lines, as shown in Table 4.6.

Table 4.6 Mayan Numerals

0	1	2	3	4	5	6	7	8	9
10	11	12	13	14	15	16	17	18	19

Example 7 *The Mayan System: A Positional-Value System*

Write ⠿ as a Hindu–Arabic numeral.

Solution In the Mayan numeration system, the first three positional values are

$$18 \times 20$$
$$20$$
$$1$$

$$= 11 \times (18 \times 20) = 3960$$
$$= 3 \times 20 \qquad = 60$$
$$= 16 \times 1 \qquad = \underline{16}$$
$$4036$$

Example 8 *From a Mayan to a Hindu–Arabic Numeral*

Write ⠿ as a Hindu–Arabic numeral.

Solution

$$= 2 \times \left[18 \times (20)^2\right] = 14{,}400$$
$$= 8 \times (18 \times 20) \qquad = 2880$$
$$= 0 \times 20 \qquad\qquad = 0$$
$$= 4 \times 1 \qquad\qquad = \underline{4}$$
$$17{,}284$$

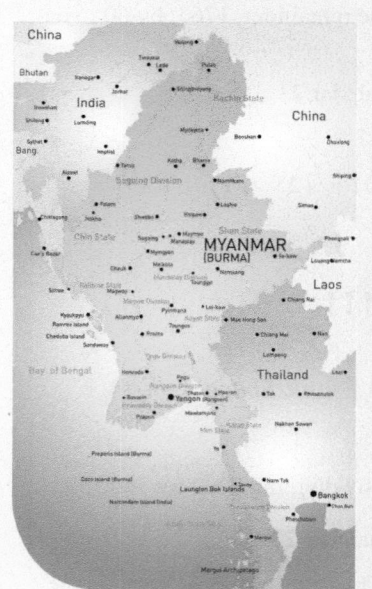

Timely Tip

Notice that changing a numeral from the Babylonian or Mayan numeration system *to the Hindu–Arabic* (or decimal or base 10) system involves *multiplication*. Changing a numeral from *the Hindu–Arabic system* to the Babylonian or Mayan numeration system involves *division*.

Example 9 *From a Hindu–Arabic to a Mayan Numeral*

Write 4025 as a Mayan numeral.

Solution To convert from a Hindu–Arabic to a Mayan numeral, we use a procedure similar to the one used to convert to a Babylonian numeral. The Mayan numeration system positional values are ..., 7200, 360, 20, 1. The greatest positional value less than or equal to 4025 is 360. Divide 4025 by 360.

$$4025 \div 360 = 11 \text{ with remainder } 65$$

There are 11 groups of 360 in 4025. Next, divide the remainder, 65, by 20.

$$65 \div 20 = 3 \text{ with remainder } 5$$

There are 3 groups of 20 with five units remaining.

$$4025 = (11 \times 360) + (3 \times 20) + (5 \times 1)$$

4025 written as a Mayan numeral is

$$\left. \begin{array}{c} 11 \times 360 \\ 3 \times 20 \\ 5 \times 1 \end{array} \right\} =$$

Example 10 *From a Hindu–Arabic to Mayan Numeral*

Write 21,846 as a Mayan numeral.

Solution To begin, divide 21,846 by the largest positional value less than or equal to 21,846. That value is 7200.

$$21,846 \div 7200 = 3 \text{ with remainder } 246$$

There are 3 groups of 7200 in 21,846. The next positional value in the Mayan numeration system is 360. However, the remainder from the above division, 246, is less than 360. So there are zero groups of 360 in 21,846 with a remainder of 246. Next we divide 246 by 20.

$$246 \div 20 = 12 \text{ with remainder } 6.$$

There are 12 groups of 20 with 6 units remaining.

$$21,846 = (3 \times 7200) + (0 \times 360) + (12 \times 20) + (6 \times 1)$$

21,846 written as a Mayan numeral is

$$\left. \begin{array}{c} 3 \times 7200 \\ 0 \times 360 \\ 12 \times 20 \\ 6 \times 1 \end{array} \right\} =$$

SECTION 4.2 *Exercises*

Warm Up Exercises

In Exercises 1–10, fill in the blanks with an appropriate word, phrase, or symbol(s).

1. The most common place-value system is the decimal number system that uses a base of _____ .

2. When we write 789 as $(7 \times 10^2) + (8 \times 10) + (9 \times 1)$ the number is written in _____ form.

3. In the numeral 234, the 2 represents 2 _____ .

4. In the numeral 234, the 3 represents 3 _____ .

5. In the numeral 234, the 4 represents 4 _____ .

6. The Babylonian system is not considered a true place-value system because it lacked a symbol for _____ .

7. In the Babylonian numeration system,

   a) the symbol ❚ represents the numeral _____ .

   b) the symbol ❮ represents the numeral _____ .

   c) the symbol ❚ indicates _____ .

8. The base used in the Babylonian numeration system is _____ .

9. In the Mayan numeration system,

   a) the symbol ⬯ represents the numeral _____ .

   b) the symbol • represents the numeral _____ .

   c) the symbol — represents the numeral _____ .

10. It is believed that the Mayans used 18×20 instead of 20^2 so that their numeration system would conform to their _____ , which had 360 days.

Practice the Skills

In Exercises 11–20, write the Hindu–Arabic numeral in expanded form.

11. 25   **12.** 57

13. 712   **14.** 562

15. 4387   **16.** 3769

17. 16,402   **18.** 23,468

19. 346,861   **20.** 3,765,934

In Exercises 21–26, write the Babylonian numeral as a Hindu–Arabic numeral.

21. ❮❚❚❚❚   **22.** ❮❮❚❚

23. ❮❚❚❚ ❚❚❚❚   **24.** ❮❚ ❮❮❚❚❚

25. ❚ ❮❮❚ ❮❚❚❚   **26.** ❮   ❚❚

In Exercises 27–32, write the numeral as a Babylonian numeral.

27. 23   **28.** 129   **29.** 471

30. 512   **31.** 3605   **32.** 12,435

In Exercises 33–38, write the Mayan numeral as a Hindu–Arabic numeral.

33. (Mayan)   **34.** (Mayan)

35. (Mayan)   **36.** (Mayan)

37. (Mayan)   **38.** (Mayan)

In Exercises 39–44, write the numeral as a Mayan numeral.

39. 16   **40.** 33

41. 297   **42.** 406

43. 2163   **44.** 30,967

In Exercises 45 and 46, write the numeral in the indicated systems of numeration.

45. (Mayan) in Hindu–Arabic and Babylonian

46. ❮❮❮❚❚❚ in Hindu–Arabic and Mayan

In Exercises 47 and 48, suppose a place-value numeration system has base ⬯, with digits represented by the symbols △, ♢, □, and ⊗. Write each expression in expanded form.

47. △ □ ♢   **48.** ⊗ △ ♢ □

Concept/Writing Exercises

49. Describe two ways that the Mayan place-value system differs from the Hindu–Arabic place-value system.

50. a) The Babylonian system did not have a symbol for zero. Why did this lead to some confusion?

   b) Write 133 and 7980 as Babylonian numerals.

Challenge Problems/Group Activities

51. a) Write the Mayan numeral for 999,999.

   b) Write the Babylonian numeral for 999,999.

In Exercises 52–55, first convert each numeral to a Hindu–Arabic numeral and then perform the indicated operation. Finally, convert the answer back to a numeral in the original numeration system.

52. ❚❚ ❮❮❚❚❚ + ❮❮❚❚❚

53. ❚❚❚ ❮❮❮❚❚❚ − ❮❮❮❚❚

54. (Mayan) + (Mayan)   **55.** (Mayan) − (Mayan)

56. *Your Own System* Create your own place-value system. Write 2021 in your system.

Research Activities

57. *Hindu-Arabic System* Investigate and write a report on the development of the Hindu–Arabic system of numeration. Start with the earliest records of this system in India.

58. *Eastern Arabic Numerals* The Arabic numeration system currently in use in Iran, Afghanistan, Egypt, and Sudan, is a base 10 positional-value system, which uses different symbols than the Hindu–Arabic numeration system. Write the symbols used in the Eastern Arabic system of numeration and their equivalent symbols in the Hindu–Arabic numeration system. Write 54,607, and 2021 in Eastern Arabic numerals.

SECTION 4.3   Other Bases

Upon completion of this section, you will be able to:

- Convert between Hindu–Arabic numerals and numerals with bases less than ten.
- Convert between Hindu–Arabic numerals and numerals with bases greater than ten.

Have you done any of the following today: spent time on Facebook, Instagram, or Twitter, shopped online, used your smartphone to send a text message or to listen to music, or received directions from a GPS device? These activities all involve using a device with a computer chip that operates with numerals other than our familiar base 10 (decimal) numerals. In this section, we will study different bases, including base 2 (binary), base 8 (octal), and base 16 (hexadecimal) numerals, all of which are used in most modern computers and electronic devices.

Why This Is Important In addition to learning about the mathematics used in designing and programming electronic devices, studying other bases helps us better understand how our own decimal system works.

The positional values in the Hindu–Arabic numeration system are

$$\ldots, (10)^4, \ (10)^3, \ (10)^2, 10, 1$$

The positional values in the Babylonian numeration system are

$$\ldots, (60)^4, (60)^3, (60)^2, 60, 1$$

Thus, 10 and 60 are called the *bases* of the Hindu–Arabic and Babylonian systems, respectively.

Any counting number greater than 1 may be used as a base for a positional-value numeration system. If a positional-value system has a base b, then its positional values will be

$$\ldots, b^4, b^3, b^2, b, 1$$

For example, the positional values in a base 8 system are

$$\ldots, 8^4, 8^3, 8^2, 8, 1$$

and the positional values in a base 2 system are

$$\ldots, 2^4, 2^3, 2^2, 2, 1$$

As we indicated in Section 4.2, the Mayan numeration system is based on the number 20. It is not, however, a true base 20 positional-value system. Why not?

The reason for the almost universal acceptance of the base 10 numeration system is that most human beings have 10 fingers. However, throughout history, some societies have used numeration systems that use a base other than 10. Two such systems are the base 12, or *duodecimal*, system and the base 60, or *sexagesimal* system that we studied in Section 4.2. Our present-day society still contains remnants of these other base systems. For example, there are 12 inches in a foot, 12 months in a year, 12 items

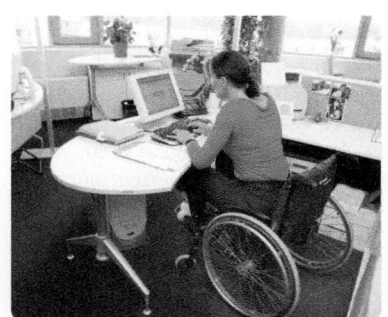

▲ *Computers make use of the binary, octal, and hexadecimal number systems.*

in a dozen, and 12 numbers on a clock. Furthermore, each hour contains 60 minutes and each minute contains 60 seconds. When measuring angles, 1 degree contains 60 minutes and each minute contains 60 seconds.

Computers and many other electronic devices make use of three numeration systems: the *binary* (base 2), the *octal* (base 8), and the *hexadecimal* (base 16) numeration systems. One familiar example of the binary numeration system is the bar codes found on most items purchased in stores today. Computers can use the binary numeration system because the binary number system consists of only the digits 0 and 1. These digits can be represented with electronic switches that are either off (0) or on (1). All data that we enter into a computer can be converted into a series of single binary digits. Each binary digit is known as a *bit*. The octal numeration system is used when eight bits of data are grouped together to form a *byte*. In the American Standard Code for Information Interchange (ASCII) code, the byte 01000001 represents the character A, and 01100001 represents the character a. Other characters along with their decimal, binary, octal, and hexadecimal representation can be found at the website www.asciitable.com. The hexadecimal numeration system is used to create computer languages. Examples of computer languages that rely on the hexadecimal system are HTML, JavaScript, and CSS, all of which are used heavily in creating Internet web pages. Computers can easily convert between binary (base 2), octal (base 8), and hexadecimal (base 16) numbers.

Bases Less Than 10

A place-value system with base b must have b distinct symbols, a symbol for zero and a symbol for each numeral less than the base. For example, a base 6 system must have symbols for 0, 1, 2, 3, 4, and 5. All numerals in base 6 are constructed from these 6 symbols. A base 8 system must have symbols for 0, 1, 2, 3, 4, 5, 6, and 7. All numerals in base 8 are constructed from these 8 symbols, and so on.

A numeral in a base other than base 10 will be indicated by a subscript to the right of the numeral. Thus, 123_5 represents a base 5 numeral. The numeral 123_6 represents a base 6 numeral. The value of 123_5 is not the same as the value of 123_{10}, and the value of 123_6 is not the same as the value of 123_{10}. A base 10 numeral may be written without a subscript. For example, 123 means 123_{10} and 456 means 456_{10}. For clarity in certain problems, we will use the subscript 10 to indicate a numeral in base 10.

The symbols that represent the base itself, in any base b, are 10_b. For example, in base 5, the symbols 10_5 represent the number five. Note that $10_5 = 1 \times 5 + 0 \times 1 = 5 + 0 = 5_{10}$, or the number five in base 10. The numeral 10_5 means one group of 5 and no units. In base 6, the symbols 10_6 represent the number six. The symbols 10_6 represent one group of 6 and no units, and so on.

To change a numeral in a base other than 10 to a base 10 numeral, we follow the same procedure we used in Section 4.2 to change the Babylonian and Mayan numerals to base 10 numerals. Multiply each digit by its respective positional value. Then determine the sum of the products.

Example 1 *Converting from Base 4 to Base 10*

Convert 223_4 to base 10.

Solution In base 4, the positional values are ... , $4^3, 4^2, 4, 1$. In expanded form,

$$223_4 = (2 \times 4^2) + (2 \times 4) + (3 \times 1)$$
$$= (2 \times 16) + (2 \times 4) + (3 \times 1)$$
$$= 32 + 8 + 3 = 43$$ ∎

In Example 1, the units digit in 223_4 is 3. Notice that 3_4 has the same value as 3_{10}, since both are equal to the number 3. That is, $3_4 = 3_{10}$. If n is a digit less than the base b, and the base b is less than or equal to 10, then $n_b = n_{10}$.

Example 2 *Converting from Base 6 to Base 10*

Convert 5124_6 to base 10.

Solution In base 6, the positional values are ... , $6^3, 6^2, 6, 1$. In expanded form,

$$
\begin{aligned}
5124_6 &= (5 \times 6^3) + (1 \times 6^2) + (2 \times 6) + (4 \times 1) \\
&= (5 \times 216) + (1 \times 36) + (2 \times 6) + (4 \times 1) \\
&= 1080 + 36 + 12 + 4 = 1132
\end{aligned}
$$
∎

Example 3 *Converting from Base 2 to Base 10*

Convert 110010_2 to base 10.

Solution

$$
\begin{aligned}
110010_2 &= (1 \times 2^5) + (1 \times 2^4) + (0 \times 2^3) + (0 \times 2^2) + (1 \times 2) + (0 \times 1) \\
&= (1 \times 32) + (1 \times 16) + (0 \times 8) + (0 \times 4) + (1 \times 2) + (0 \times 1) \\
&= 32 + 16 + 0 + 0 + 2 + 0 \\
&= 50
\end{aligned}
$$
∎

To change a base 10 numeral to a different base, we will use a procedure similar to the one we used to convert base 10 numerals to Babylonian and Mayan numerals, as was explained in Section 4.2. Divide the base 10 numeral by the highest power of the new base that is less than or equal to the given base 10 numeral. Record this quotient. Then divide the remainder by the next smaller power of the new base and record this quotient. Repeat this procedure until the remainder is less than the new base. The answer is the set of quotients listed from left to right, with the remainder on the far right. This procedure is illustrated in Examples 4 and 5.

Example 4 *Converting from Base 10 to Base 8*

Convert 486 to base 8.

Solution We are converting a numeral in base 10 to a numeral in base 8. The positional values in the base 8 system are ... , $8^3, 8^2, 8, 1$, or ... , 512, 64, 8, 1. The highest power of 8 that is less than or equal to 486 is 8^2, or 64. Divide 486 by 64.

First digit in answer
↓

$$486 \div 64 = 7 \text{ with remainder } 38$$

Therefore, there are seven groups of 8^2 in 486. Next divide the remainder, 38, by 8.

Second digit in answer
↓

$$38 \div 8 = 4 \text{ with remainder } 6$$

↑
Third digit in answer

There are four groups of 8 in 38 and 6 units remaining. Since the remainder, 6, is less than the base, 8, no further division is required.

$$
\begin{aligned}
&= (7 \times 64) + (4 \times 8) + (6 \times 1) \\
&= (7 \times 8^2) + (4 \times 8) + (6 \times 1) \\
&= 746_8
\end{aligned}
$$

Notice that we placed the subscript 8 to the right of 746 to show that it is a base 8 numeral.
∎

The same procedure is used in Example 5, but it is given in a simplified manner.

Example 5 *Converting from Base 10 to Base 3*

Convert 273 to base 3.

Solution The place values in the base 3 system are $\ldots, 3^6, 3^5, 3^4, 3^3, 3^2, 3, 1$, or $\ldots, 729, 243, 81, 27, 9, 3, 1$. The highest power of the base that is less than or equal to 273 is 3^5, or 243. Successive divisions by the powers of the base give the following result.

$$273 \div 243 = 1 \text{ with remainder } 30$$
$$30 \div 81 = 0 \text{ with remainder } 30$$
$$30 \div 27 = 1 \text{ with remainder } 3$$
$$3 \div 9 = 0 \text{ with remainder } 3$$
$$3 \div 3 = 1 \text{ with remainder } 0$$

The remainder, 0, is less than the base, 3, so no further division is necessary. To obtain the answer, list the quotients from top to bottom followed by the remainder in the last division.

We can represent 273 as one group of 243, no groups of 81, one group of 27, no groups of 9, one group of 3, and no units.

$$273 = (1 \times 243) + (0 \times 81) + (1 \times 27) + (0 \times 9) + (1 \times 3) + (0 \times 1)$$
$$= (1 \times 3^5) + (0 \times 3^4) + (1 \times 3^3) + (0 \times 3^2) + (1 \times 3) + (0 \times 1)$$
$$= 101010_3 \qquad \blacksquare$$

Bases Greater Than 10

Recall that a place-value system with base b must have symbols for the digits from 0 up to one less than the base. For example, base 6 uses the digits 0, 1, 2, 3, 4, and 5; and base 8 uses the digits 0, 1, 2, 3, 4, 5, 6, and 7. What happens if the base is larger than ten? We will need single-digit symbols to represent the numbers ten, eleven, twelve, ... up to one less than the base. *In this textbook, whenever a base larger than ten is used we will use the capital letter A to represent ten, the capital letter B to represent eleven, the capital letter C to represent twelve, and so on.* For example, for base 12, known as the *duodecimal system*, we use the symbols 0, 1, 2, 3, 4, 5, 6, 7, 8, 9, A, and B, where A represents ten and B represents eleven. For base 16, known as the *hexadecimal system*, we use the symbols 0, 1, 2, 3, 4, 5, 6, 7, 8, 9, A, B, C, D, E, and F.

Example 6 *Converting to and from Base 12*

a) Convert $39BA_{12}$ to base 10.
b) Convert 6893 to base 12.

Solution

a) In base 12, the positional values are $\ldots, 12^3, 12^2, 12, 1$ or $\ldots, 1728, 144, 12, 1$. Since B has the value of eleven and A has the value of ten, we perform the following calculation.

$$39BA_{12} = (3 \times 12^3) + (9 \times 12^2) + (B \times 12) + (A \times 1)$$
$$= (3 \times 1728) + (9 \times 144) + (11 \times 12) + (10 \times 1)$$
$$= 5184 + 1296 + 132 + 10 = 6622$$

...nues until the quotient is zero.
...the bottom number to the top
... column. Thus, $328 = 2303_5$.

...e 5 by this method.

...ase 8 by this method.

Exercises

...ual to in base 10?

...qual to in base 10?

...$_4$ equal to in base 10?

...$_4$ equal to in base 10?

...l, if n is a digit less than the base b, and the
... less than or equal to 10, then describe what n_b
...to in base 10.

...is 10_2 equal to in base 10?

...t is 10_8 equal to in base 10?

...at is 10_{16} equal to in base 10?

...hat is 10_{32} equal to in base 10?

...a general, for any base b, what is 10_b equal
...o in base 10?

...What is 11_2 equal to in base 10?

What is 11_8 equal to in base 10?

What is 11_{16} equal to in base 10?

What is 11_{32} equal to in base 10?

In general, for any base b, what is 11_b equal
to in base 10?

...enge Problems/Group Activities

...termine b if $111_b = 43$.

...ermine d if $ddd_5 = 124$.

...Use the numerals 0, 1, and 2 to write the first
...20 numbers in the base 3 numeration system.

...What is the next numeral after 222_3?

68. ...puter Code The ASCII code used by most computers
...the last seven positions of an eight-bit byte to repre-
...all the characters on a standard keyboard. How many
...rent orderings of 0's and 1's (or how many different
...acters) can be made by using the last seven positions
...eight-bit byte?

69. a) *Your Own Numeration System* Make up your
own base 20 positional-value numeration system.
Indicate the 20 numerals you will use to represent
the 20 numbers less than the base.

b) Write 523 and 5293 in your base 20 numeration
system.

Recreational Mathematics

70. *The Price Is Right* Refer to the *Recreational Mathematics*
on page 185. Determine the correct order in which to place
the digits 1, 5, 5, 2 to match the price of the 2018 Sunfish
Laser Performance sailboat.

▲ *Place the digits 1, 5, 5, 2 in the correct
order to match the price of the sailboat.*

$ ___ ___ ___ ___

71. Suppose a base 4 place-value system has its digits represented by colors as follows:

● = 0   ● = 1   ○ = 2   ● = 3

a) Determine the value of ● ● ○ ● ●$_4$ in base
10.

b) Write 177 in the base 4 system using only the four
colors given in the exercise.

Research Activities

72. *Duodecimal System* Write a report on the use of the duodecimal (base 12) system of numeration. You may wish to
contact the Dozenal Society (see the *Did You Know?* on
page 186) for more information.

73. *Numeration Systems* Throughout history, various societies have used different bases for their numeration systems.
Groups of people in North and South America, Africa, and
Australia all used a variety of numeration systems. Choose
several such groups and write a report on their numeration
systems.

74. *Papua New Guinea* Today, the various groups of people
in Papua New Guinea still use a variety of numeration systems. Write a report on the different systems used by these
groups.

Perform Computations in Other Bases

Upon completion of this section, you will be able to:

- Add numerals with bases other than base 10.
- Subtract numerals with bases other than base 10.
- Multiply numerals with bases other than base 10.
- Divide numerals with bases other than base 10.

You may recall learning how to add, subtract, multiply, and divide when you were in grade school. You may recall "carrying" when adding and "borrowing" when subtracting. For example, when performing 75 + 18, you would add 5 and 8 to get 13. You would write down the 3 and "carry the 1." When performing 43 − 19 you would "borrow from the 4" so you could perform 13 − 9 to get 4.

In this section, we will use similar procedures when working with numerals that have bases other than base 10.

Why This Is Important By working in other bases, we will gain a better understanding of our own base 10 place-value system. We also gain an understanding of bases such as base 2, base 8, and base 16, which are used in computers and electronic devices.

Addition

When computers perform calculations, they do so in base 2, the binary system. In this section, we will explain how to perform calculations in base 2 and other bases.

In a base 2 system, the only digits are 0 and 1, and the place values are

$$\dots, 2^4, 2^3, 2^2, 2, 1$$

$$\text{or} \quad \dots, 16, 8, 4, 2, 1$$

Suppose we want to add $1_2 + 1_2$. The subscript 2 indicates that we are adding in base 2. Remember that the answer to $1_2 + 1_2$ must be written using only the digits 0 and 1. The sum of $1_2 + 1_2$ is 10_2, which represents 1 group of two and 0 units in base 2. Recall that 10_2 means $(1 \times 2) + (0 \times 1)$.

If we wanted to determine the sum of $10_2 + 1_2$, we would add the digits in the right-hand, or units, column. Since $0_2 + 1_2 = 1_2$, the sum of $10_2 + 1_2 = 11_2$.

We are going to work additional examples and exercises in base 2, so rather than performing individual calculations in every problem, we can construct and use an addition table for base 2, Table 4.7 (just as we used an addition table in base 10 when we first learned to add in base 10).

Table 4.7   Base 2 Addition Table

+	0	1
0	0	1
1	1	10

Example 1 *Adding in Base 2*

Add 1101_2
       $+ \ 111_2$

Solution Begin by adding the digits in the right-hand, or units, column. From previous discussion, and as can be seen in Table 4.7, $1_2 + 1_2 = 10_2$. Place the 0 under the units column and carry the 1 to the 2's column, the second column from the right.

$$
\begin{array}{cccc}
2^3 & 2^2 & 2 & 1 \\
\downarrow & \downarrow & \downarrow & \downarrow \\
1 & 1 & {}^{1}0 & 1 \\
+ & 1 & 1 & 1 \\
\hline
 & & & 0_2
\end{array}
$$

Place value of columns

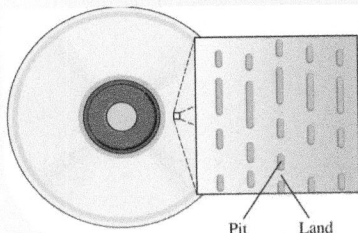
Now add the three digits in the 2's column, $1_2 + 0_2 + 1_2$. Treat it as $(1_2 + 0_2) + 1_2$. Therefore, add $1_2 + 0_2$ to get 1_2, then add $1_2 + 1_2$ to get 10_2. Place the 0 under the 2's column and carry the 1 to the 2^2 column (the third column from the right).

$$\begin{array}{r} 1\ {}^1 1\ {}^1 0\ 1 \\ +\ \ 1\ 1\ 1 \\ \hline 0\ 0_2 \end{array}$$

Now add the three 1's in the 2^2 column to get $(1_2 + 1_2) + 1_2 = 10_2 + 1_2 = 11_2$. Place the 1 under the 2^2 column and carry the 1 to the 2^3 column (the fourth column from the right).

$$\begin{array}{r} {}^1 1\ {}^1 1\ {}^1 0\ 1 \\ +\ \ 1\ 1\ 1 \\ \hline 1\ 0\ 0_2 \end{array}$$

Now add the two 1's in the 2^3 column, $1_2 + 1_2 = 10_2$. Place the 10 as follows.

$$\begin{array}{r} {}^1 1\ {}^1 1\ {}^1 0\ 1 \\ +\ \ \ \ \ 1\ 1\ 1 \\ \hline 1\ 0\ 1\ 0\ 0_2 \end{array}$$

Therefore, the sum is 10100_2. ∎

Let's now look at addition in a base 5 system. In base 5, the only digits are 0, 1, 2, 3, and 4, and the positional values are

$$\ldots,\ 5^4,\ \ 5^3,\ 5^2, 5, 1$$
$$\text{or}\quad \ldots, 625, 125,\ \ 25, 5, 1$$

What is the sum of $4_5 + 3_5$? We can consider this to mean $(1 + 1 + 1 + 1) + (1 + 1 + 1)$. We can regroup the seven 1's into one group of five and two units as $(1 + 1 + 1 + 1 + 1) + (1 + 1)$. Thus, the sum of $4_5 + 3_5 = 12_5$ (circled in Table 4.8, the base 5 addition table). Recall that 12_5 means $(1 \times 5) + (2 \times 1)$. We can use this same procedure in obtaining the remaining values in the base 5 addition table.

Example 2 *Use the Base 5 Addition Table*

Add 34_5
$+\ 23_5$

Solution First determine from Table 4.8 that $4_5 + 3_5$ is 12_5. Record the 2 and carry the 1 to the 5's column.

$$\begin{array}{r} {}^1 3\ 4_5 \\ +\ 2\ 3_5 \\ \hline 2_5 \end{array}$$

Add the numbers in the second column, $(1_5 + 3_5) + 2_5 = 4_5 + 2_5 = 11_5$. Record the 11.

$$\begin{array}{r} {}^1 3\ 4_5 \\ +\ \ 2\ 3_5 \\ \hline 1\ 1\ 2_5 \end{array}$$

The sum is 112_5. ∎

Table 4.8 Base 5 Addition Table

+	0	1	2	3	4
0	0	1	2	3	4
1	1	2	3	4	10
2	2	3	4	10	11
3	3	4	10	11	12
4	4	10	11	(12)	13

Example **3** *Add in Base 5*

Add
$$3214_5$$
$$+ \ 2011_5$$

Solution

$$3 \ 2 \ {}^11 \ 4_5$$
$$+ \ 2 \ 0 \ 1 \ 1_5$$
$$\overline{1 \ 0 \ 2 \ 3 \ 0_5}$$

The sum is 10230_5.   ∎

You can develop an addition table for any base and use it to add in that base. As you get more comfortable with addition in other bases, however, you may prefer to add in other bases by using mental arithmetic. To do so, convert the sum from the given base to base 10 and then convert the base 10 numeral back into the given base. You must clearly understand how to convert from base 10 to the given base, as discussed in Section 4.3. For example, to add $7_9 + 8_9$, add $7 + 8$ in base 10 to get 15_{10} and then mentally convert 15_{10} to 16_9 using the procedure given earlier. Remember, 16_9 when converted to base 10 becomes $(1 \times 9) + (6 \times 1)$, or 15. Addition using this procedure is illustrated in Example 4.

Example **4** *Adding in Base 10; Converting to Base 4*

Add
$$1321_4$$
$$+ \ 1133_4$$

Solution To solve this problem, make the necessary conversions by using mental arithmetic.

$1 + 3 = 4_{10} = 10_4$. Record the 0 and carry the 1.

$$1 \ 3 \ {}^12 \ 1_4$$
$$+ \ 1 \ 1 \ 3 \ 3_4$$
$$\overline{0_4}$$

$1 + 2 + 3 = 6_{10} = 12_4$. Record the 2 and carry the 1.

$$1 \ {}^13 \ {}^12 \ 1_4$$
$$+ \ 1 \ 1 \ 3 \ 3_4$$
$$\overline{2 \ 0_4}$$

$1 + 3 + 1 = 5_{10} = 11_4$. Record the 1 and carry the 1.

$$^11 \ {}^13 \ {}^12 \ 1_4$$
$$+ \ 1 \ 1 \ 3 \ 3_4$$
$$\overline{1 \ 2 \ 0_4}$$

$1 + 1 + 1 = 3_{10} = 3_4$. Record the 3.

$$1 \ 3 \ 2 \ 1_4$$
$$+ \ 1 \ 1 \ 3 \ 3_4$$
$$\overline{3 \ 1 \ 2 \ 0_4}$$

The sum is 3120_4.   ∎

Subtraction

Subtraction can also be performed in bases other than base 10. Always remember that when you "borrow," you borrow the amount of the base given in the subtraction problem. For example, if subtracting in base 5, when you borrow, you borrow 5. If subtracting in base 12, when you borrow, you borrow 12.

Example 5 *Subtracting in Base 5*

Subtract 3032_5
          -1004_5

Solution We will perform the subtraction in base 10 and convert the results to base 5. Since 4 is greater than 2, we must borrow one group of 5 from the preceding column. This action gives a sum of $5 + 2$, or 7, in base 10. Now we subtract 4 from 7; the difference is 3. We complete the problem in the usual manner. The 3 in the second column becomes a 2, $2 - 0 = 2$. In the third column, $0 - 0 = 0$. Finally, in the fourth column, $3 - 1 = 2$.

$$\begin{array}{r} 3032_5 \\ -1004_5 \\ \hline 2023_5 \end{array}$$

In the next example, we will subtract base 12 numerals. Recall from Section 4.3 that we use the capital letter A to represent the number ten and the capital letter B to represent the number eleven in base 12.

Example 6 *Subtracting in Base 12*

Subtract $97A_{12}$
          $-4B8_{12}$

Solution In base 12, A represents 10. Therefore, in the units column we have $10 - 8 = 2$. Next, in base 12, B represents 11. Therefore, in the next column we must subtract 11 from 7. Since 11 is greater than 7, borrowing is necessary. We must borrow one group of 12 from the preceding column. We then have a sum of $12 + 7$, or 19. We can now subtract 11 from 19 and the difference is 8. Since we borrowed 1 from the far-left column, the 9 becomes 8, and $8 - 4 = 4$.

$$\begin{array}{r} 97A_{12} \\ -4B8_{12} \\ \hline 482_{12} \end{array}$$

The difference is 482_{12}.

Multiplication

Multiplication can also be performed in bases other than base 10. Doing so is helped by forming a multiplication table for the base desired. Suppose we want to determine the product of $4_5 \times 3_5$. In base 10, 4×3 means there are four groups of three units. Similarly, in a base 5 system, $4_5 \times 3_5$ means there are four groups of three units, or

$$(1 + 1 + 1) + (1 + 1 + 1) + (1 + 1 + 1) + (1 + 1 + 1)$$

Regrouping the 12 units above into groups of five gives

$$(1 + 1 + 1 + 1 + 1) + (1 + 1 + 1 + 1 + 1) + (1 + 1)$$

or two groups of five, and two units. Thus, $4_5 \times 3_5 = 22_5$.

Table 4.9 Base 5 Multiplication Table

×	0	1	2	3	4
0	0	0	0	0	0
1	0	1	2	3	4
2	0	2	4	11	13
3	0	3	11	14	22
4	0	4	13	(22)	31

Addition and Multiplication
in Base 5

The product of $4_5 \times 3_5$, or 22_5, is circled in Table 4.9, the base 5 multiplication table. We can construct other values in the base 5 multiplication table in the same way. You may, however, find it easier to multiply the values in the base 10 system and then change the product to base 5 by using the procedure discussed in Section 4.3. Multiplying 4×3 in base 10 gives 12, and converting 12 from base 10 to base 5 gives 22_5. The other values in the table may be determined by either method discussed.

Example 7 *Using the Base 5 Multiplication Table*

Multiply 13_5
              $\times\ 3_5$

Solution Use the base 5 multiplication table to determine the products. When the product consists of two digits, record the right digit and carry the left digit. Multiplying gives $3_5 \times 3_5 = 14_5$. Record the 4 and carry the 1.

$$\begin{array}{r} {}^{1}13_5 \\ \times\ 3_5 \\ \hline 4 \end{array}$$

$(3_5 \times 1_5) + 1_5 = 4_5$. Record the 4.

$$\begin{array}{r} {}^{1}13_5 \\ \times\ 3_5 \\ \hline 44_5 \end{array}$$

The product is 44_5.   ∎

Constructing a multiplication table is often tedious, especially when the base is large. To multiply in a given base without the use of a table, multiply in base 10 and convert the products to the appropriate base before recording them. This procedure is illustrated in Example 8.

Example 8 *Multiplying in Base 7*

Multiply 43_7
            $\times\ 25_7$

Solution $5 \times 3 = 15_{10} = (2 \times 7) + (1 \times 1) = 21_7$. Record the 1 and carry the 2.

$$\begin{array}{r} {}^{2}43_7 \\ \times\ 25_7 \\ \hline 1 \end{array}$$

$(5 \times 4) + 2 = 20 + 2 = 22_{10} = (3 \times 7) + (1 \times 1) = 31_7$. Record the 31.

$$\begin{array}{r} {}^{2}43_7 \\ \times\ 25_7 \\ \hline 311 \end{array}$$

$2 \times 3 = 6_{10} = 6_7$. Record the 6.

$$\begin{array}{r} {}^{2}43_7 \\ \times\ 25_7 \\ \hline 311 \\ 6 \end{array}$$

$2 \times 4 = 8_{10} = (1 \times 7) + (1 \times 1) = 11_7$. Record the 11. Now add in base 7 to determine the answer. Remember, in base 7, there are no digits greater than 6.

$$\begin{array}{r} {}^2 43_7 \\ \times\ 25_7 \\ \hline 311 \\ 116 \\ \hline 1501_7 \end{array}$$

Therefore, the product is 1501_7. ∎

Division

Division is performed in much the same manner as long division in base 10. A detailed example of a division in base 5 is illustrated in Example 9. The same procedure is used for division in any other base.

Example 9 *Dividing in Base 5*

Divide $2_5\overline{)143_5}$.

Solution Using the multiplication table for base 5, Table 4.9 on page 194, we list the multiples of the divisor, 2.

$$\begin{aligned} 2_5 \times 1_5 &= 2_5 \\ 2_5 \times 2_5 &= 4_5 \\ 2_5 \times 3_5 &= 11_5 \\ 2_5 \times 4_5 &= 13_5 \end{aligned}$$

Since $2_5 \times 4_5 = 13_5$, which is the largest product less than 14_5, 2_5 divides into 14_5 four times. Record the 13. Subtract 13_5 from 14_5. The difference is 1_5. Record the 1.

$$\begin{array}{r} 4 \\ 2_5\overline{)143_5} \\ 13 \\ \hline 1 \end{array}$$

Now bring down the 3 as when dividing in base 10.

$$\begin{array}{r} 4 \\ 2_5\overline{)143_5} \\ 13 \\ \hline 13 \end{array}$$

We see that $2_5 \times 4_5 = 13_5$. Use this information to complete the problem.

$$\begin{array}{r} 44_5 \\ 2_5\overline{)143_5} \\ 13 \\ \hline 13 \\ 13 \\ \hline 0 \end{array}$$

Therefore, $143_5 \div 2_5 = 44_5$ with remainder 0_5. ∎

A division problem can be checked by multiplication. If the division was performed correctly, (quotient $\times$ divisor) + remainder = dividend. We can check Example 9 as follows.

$$(44_5 \times 2_5) + 0_5 = 143_5$$

$$\begin{array}{r} 44_5 \\ \times\ \ 2_5 \\ \hline 143_5 \quad \text{Check.} \end{array}$$

Example 10   Dividing in Base 6

Divide $5_6 \overline{)3410_6}$

Solution   The multiples of 5 in base 6 are

$$5_6 \times 1_6 = 5_6 \qquad 5_6 \times 2_6 = 14_6 \qquad 5_6 \times 3_6 = 23_6$$
$$5_6 \times 4_6 = 32_6 \qquad 5_6 \times 5_6 = 41_6$$

$$
\begin{array}{r}
423_6 \\
5_6 \overline{)3410_6} \\
\underline{32} \\
21 \\
\underline{14} \\
30 \\
\underline{23} \\
3
\end{array}
$$

Thus, the quotient is 423_6, with remainder 3_6.

CHECK: Does $(423_6 \times 5_6) + 3_6 = 3410_6$?

$$
\begin{array}{r}
423_6 \\
\times\ \ 5_6 \\
\hline
3403_6
\end{array}
$$

$$3403_6 + 3_6 \stackrel{?}{=} 3410_6 \quad \text{True} \qquad \blacksquare$$

Timely Tip

Be careful when subtracting in bases other than base 10! In the Example 10 division problem, we had to borrow when subtracting two different times. In each case, since we are working in base 6, we must borrow a group of 6 from the preceding column.

SECTION 4.4   Exercises

Warm Up Exercises

In Exercises 1–6, fill in the blanks with an appropriate word, phrase, or symbol(s).

1. In base 5, $3_5 + 4_5 = $ _____.

2. In base 6, $2_6 + 5_6 = $ _____.

3. When subtracting in base 5, when we borrow, we must borrow one group of _____ from the preceding column.

4. When subtracting in base 8, when we borrow, we must borrow one group of _____ from the preceding column.

5. In base 5, $3_5 \times 4_5 = $ _____.

6. In base 9, $3_9 \times 7_9 = $ _____.

Practice the Skills

In Exercises 7–18, add in the indicated base.

7. $\begin{array}{r} 21_3 \\ + 20_3 \\ \hline \end{array}$

8. $\begin{array}{r} 23_4 \\ + 13_4 \\ \hline \end{array}$

9. $\begin{array}{r} 132_5 \\ + 34_5 \\ \hline \end{array}$

10. $\begin{array}{r} 254_6 \\ + 15_6 \\ \hline \end{array}$

11. $\begin{array}{r} 654_7 \\ + 463_7 \\ \hline \end{array}$

12. $\begin{array}{r} 727_8 \\ + 564_8 \\ \hline \end{array}$

13. $\begin{array}{r} 1011_2 \\ + 1110_2 \\ \hline \end{array}$

14. $\begin{array}{r} 1110_2 \\ + 1111_2 \\ \hline \end{array}$

15. $\begin{array}{r} A734_{12} \\ + 128B_{12} \\ \hline \end{array}$

16. $\begin{array}{r} B293_{12} \\ + 486A_{12} \\ \hline \end{array}$

17. $\begin{array}{r} A734_{16} \\ + 128B_{16} \\ \hline \end{array}$

18. $\begin{array}{r} B293_{16} \\ + 486A_{16} \\ \hline \end{array}$

In Exercises 19–30, subtract in the indicated base.

19. $\begin{array}{r} 201_3 \\ - 120_3 \\ \hline \end{array}$

20. $\begin{array}{r} 512_6 \\ - 420_6 \\ \hline \end{array}$

21. $\begin{array}{r} 338_9 \\ - 274_9 \\ \hline \end{array}$

22. $\begin{array}{r} 1221_3 \\ - 202_3 \\ \hline \end{array}$

23. 1101_2
$- \quad 111_2$

24. 1001_2
$- \quad 110_2$

25. 4223_7
$- \quad 304_7$

26. 2173_8
-1654_8

27. 4232_5
-2341_5

28. 2100_3
-1012_3

29. $A3B3_{12}$
$- \quad 21B4_{12}$

30. $4E7_{16}$
-189_{16}

In Exercises 31–42, multiply in the indicated base.

31. 22_3
$\times \ 2_3$

32. 32_4
$\times \ 3_4$

33. 647_8
$\times \ 5_8$

34. 124_{12}
$\times \ 6_{12}$

35. 37_8
$\times 21_8$

36. 512_6
$\times 23_6$

37. $B12_{12}$
$\times 83_{12}$

38. $6A3_{12}$
$\times 24_{12}$

39. 110_2
$\times 11_2$

40. 111_2
$\times 101_2$

41. 316_7
$\times 16_7$

42. $7A9_{12}$
$\times 12_{12}$

In Exercises 43–54, divide in the indicated base.

43. $2_4 \overline{)312_4}$

44. $2_8 \overline{)730_8}$

45. $3_7 \overline{)506_7}$

46. $4_5 \overline{)302_5}$

47. $3_8 \overline{)200_8}$

48. $4_8 \overline{)252_8}$

49. $2_4 \overline{)213_4}$

50. $5_6 \overline{)214_6}$

51. $3_5 \overline{)224_5}$

52. $4_6 \overline{)210_6}$

53. $6_7 \overline{)404_7}$

54. $3_7 \overline{)2101_7}$

Problem Solving

In Exercises 55–58, suppose the numerals in a base 5 numeration system are as illustrated with their equivalent Hindu–Arabic numerals.

$\bigcirc = 0 \quad \ominus = 1 \quad \oplus = 2 \quad \ominus = 3 \quad \oplus = 4$

Add the following base 5 numerals.

55.
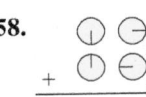

56.

57.

58.

In Exercises 59–66, assume the numerals given are in a base 4 numeration system. In this system, suppose colors are used as numerals, as indicated below.

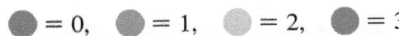

Add the following base 4 numerals. Your answers will contain a variety of the colors indicated.

59.
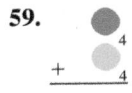

60.
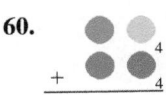

61.

62.
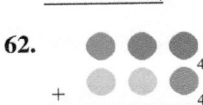

Subtract the following in base 4. Your answer will contain a variety of the colors indicated.

63.
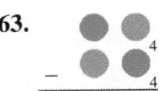

64.

65.
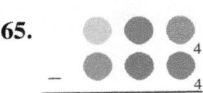

66.
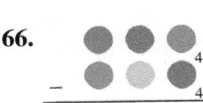

Challenge Problems/Group Activities

In Exercises 67–68, perform the indicated operation.

67. FAB_{16}
$\times \quad 4_{16}$

68. $D_{16} \overline{)FACE_{16}}$

69. Consider the multiplication
$$462_8$$
$$\times \ 35_8$$

a) Multiply the numerals in base 8.

b) Convert 462_8 and 35_8 to base 10.

c) Multiply the base 10 numerals determined in part (b).

d) Convert the answer obtained in base 8 in part (a) to base 10.

e) Are the answers obtained in parts (c) and (d) the same? Why or why not?

70. Determine b, by trial and error, if $1304_b = 204$.

71. In a base 4 system, suppose each of the four numerals is represented by one of the following colors:

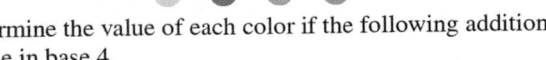

Determine the value of each color if the following addition is true in base 4.

Research Activities

72. Computer Numeration Systems Write a report on how computers use the binary (base 2), octal (base 8), and hexadecimal (base 16) numeration systems.

73. Computer Subtraction One method used by computers to perform subtraction is the "end around carry method." Do research and write a report explaining, with specific examples, how a computer performs subtraction by using the end around carry method.

SECTION 4.5 Early Computational Methods

Upon completion of this section, you will be able to:

- Multiply whole numbers using duplation and mediation.
- Multiply whole numbers using lattice multiplication.
- Multiply whole numbers using Napier's rods.

Suppose that last week you worked at your job in a movie theater for 27 hours at a rate of $8 per hour. How much money did you make? You may determine the answer by multiplying 27×8 "by hand."

$$\begin{array}{r} {}^5 27 \\ \times\ 8 \\ \hline 216 \end{array}$$

Although most of us would use this method, it is not the only method for multiplying two numbers together. Early civilizations used various other methods. In this section, we will study three other methods of multiplication.

Why This Is Important Studying other methods of multiplication can help us better understand how traditional multiplication works. In addition, there is a link between multiplication methods used by early civilizations and the way microprocessors used in computers and electronic devices multiply. Understanding other multiplication methods helps us gain a better understanding of computing in other bases.

Duplation and Mediation

The first method of multiplication we will study is *duplation and mediation*. This method, also known as *Russian peasant multiplication*, is still used in parts of Russia today. Duplation refers to doubling one number, and mediation refers to halving one number. The duplation and mediation method is similar to a method used by the ancient Egyptians as described on the Rhind Papyrus. Example 1 illustrates this method of multiplication.

Example 1 *Using Duplation and Mediation*

Multiply 39×23 using duplation and mediation.

Solution Write 39 and 23 with a dash between the two numbers. Divide the number on the left, 39, by 2, drop the remainder and place the quotient, 19, under the 39. Double the number on the right, 23, to obtain 46 and place it under the 23. You will then have the following number pairs.

$$39—23$$
$$19—46$$

Continue this process, dividing the number in the left column by 2, disregarding the remainder, and doubling the number in the right-hand column, as shown below. When a 1 appears in the left-hand column, stop.

$$39—23$$
$$19—46$$
$$9—92$$
$$4—184$$
$$2—368$$
$$1—736$$

Next, cross out all the even numbers in the left-hand column and the corresponding numbers in the right-hand column.

$$39—23$$
$$19—46$$
$$9—92$$
$$4—\cancel{184}$$
$$2—\cancel{368}$$
$$1—736$$

Now add the remaining numbers in the right-hand column to obtain $23 + 46 + 92 + 736 = 897$. The number obtained, 897, is the product of 39×23. Thus, $39 \times 23 = 897$.   ▪

Lattice Multiplication

Another method of multiplication is *lattice multiplication*. This method's name comes from the use of a grid, or lattice, when multiplying two numbers. It is also known as the *gelosia* method. The lattice method was brought to Europe in the 1200s by Fibonacci, who likely learned it from the Egyptians, Arabs, or Hindus (see the *Profile in Mathematics* on page 276).

Example 2   *Using Lattice Multiplication*

Multiply 312×75 using lattice multiplication.

Solution   To multiply 312×75 using lattice multiplication, first construct a rectangle consisting of three columns (one for each digit of 312) and two rows (one for each digit of 75).

Place the digits 3, 1, 2 above the boxes and the digits 7, 5 on the right of the boxes. Then place a diagonal in each box, as shown in Fig. 4.1 below.

Complete each box by multiplying the number on top of the box by the number to the right of the box (Fig. 4.2). Place the units digit of the product below the diagonal and the tens digit of the product above the diagonal.

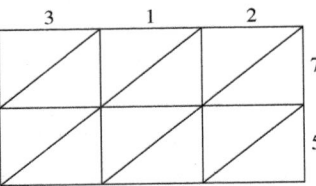

Figure 4.1

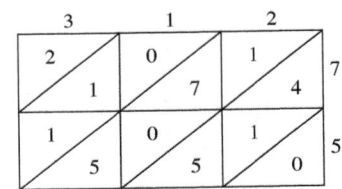

Figure 4.2

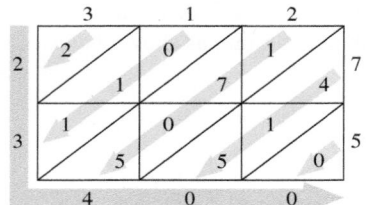

Figure 4.3

Add the numbers along the diagonals, as shown with the blue shaded arrows in Fig. 4.3, starting with the bottom right diagonal. If the sum in a diagonal is 10 or greater, record the units digit below the rectangle and carry the tens digit to the next diagonal to the left.

For example, when adding 4, 1, and 5 (along the second blue diagonal from the right), the sum is 10. Record the 0 below the rectangle and carry the 1 to the next blue diagonal. The sum of $1 + 1 + 7 + 0 + 5$ is 14. Record the 4 and carry the 1 to the next blue diagonal. The sum of the numbers in the next blue diagonal is $1 + 0 + 1 + 1$ or 3.

The answer is read down the left-hand column and along the bottom, as shown by the purple arrow in Fig. 4.3. Therefore, $312 \times 75 = 23,400$. ∎

Napier's Rods

The third method used to multiply numbers was developed from the lattice method by John Napier in the early 1600s and is known as *Napier's rods* or *Napier's bones*. This method is similar to that used in modern computers. Napier developed a system of separate rods (which were often made out of bones) numbered 0 through 9 and an additional rod for an index, numbered vertically 1 through 9 (Fig. 4.4). Each rod is divided into 10 blocks. Each block below the first block contains a multiple of the number in the first block, with a diagonal separating the digits. The units digits are placed to the right of the diagonals and the tens digits to the left. Example 3 explains how Napier's rods are used to multiply numbers.

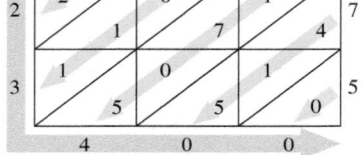

▲ Napier's Rods

Figure 4.4

Figure 4.5

Example 3 *Using Napier's Rods*

Multiply 8×365, using Napier's rods.

Solution To multiply 8×365, line up the rods 3, 6, and 5 to the right of the index 8, as shown in Fig. 4.5. Below the 3, 6, and 5 place the blocks that contain the products of 8×3, 8×6, and 8×5, respectively. To obtain the answer, add along the diagonals as is done with the lattice method.

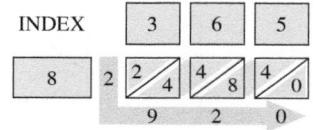

Thus, $8 \times 365 = 2920$. ∎

Profile in Mathematics

During the seventeenth century, the growth of scientific fields such as astronomy required the ability to perform often unwieldy calculations. Scottish mathematician John Napier (1550–1617) made great contributions toward solving the problem of computing these numbers. His inventions include simple calculating machines and a device for performing multiplication and division known as Napier's rods. Napier also developed the theory of logarithms.

Example 4 illustrates the procedure for multiplying numbers containing more than one digit, using Napier's rods.

Example 4 *Using Napier's Rods to Multiply Two- and Three-Digit Numbers*

Multiply 48 × 365, using Napier's rods.

Solution 48 × 365 = (40 + 8) × 365

Write (40 + 8) × 365 = (40 × 365) + (8 × 365). To find 40 × 365, determine 4 × 365 and multiply the product by 10. To evaluate 4 × 365, set up **Napier's rods** for 3, 6, and 5 with index 4, and then evaluate along the diagonals, as indicated.

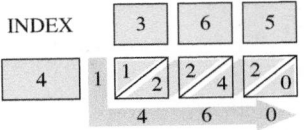

Therefore, 4 × 365 = 1460. Then 40 × 365 = 1460 × 10 = 14,600.

$$48 \times 365 = (40 \times 365) + (8 \times 365) \quad \begin{array}{l} 8 \times 365 = 2920 \\ \text{from Example 3} \end{array}$$
$$= 14{,}600 + 2920$$
$$= 17{,}520$$

SECTION 4.5   *Exercises*

Warm Up Exercises

In Exercises 1–4, fill in the blanks with an appropriate word, phrase, or symbol(s).

1. When determining the product 39 × 73, using duplation and mediation,

   a) the 39 is first _____ by 2, and then

   b) the 73 is _____ .

2. When using duplation and mediation, when you divide the number on the left by two you disregard the _____ .

3. When determining the product 327 × 45, using lattice multiplication, first construct a rectangle consisting of _____ columns and _____ rows.

4. When using lattice multiplication, after using multiplication to complete the boxes in the rectangle, add the numbers along the _____ .

Practice the Skills

In Exercises 5–12, multiply using duplation and mediation.

5. 11 × 19   6. 15 × 21   7. 29 × 35

8. 138 × 41   9. 35 × 236   10. 96 × 53

11. 93 × 93   12. 49 × 124

In Exercises 13–20, use lattice multiplication to determine the product.

13. 3 × 229   14. 7 × 212   15. 6 × 425

16. 9 × 509   17. 75 × 12   18. 47 × 259

19. 314 × 652   20. 634 × 832

In Exercises 21–28, multiply using Napier's rods.

21. 2 × 52   22. 4 × 68   23. 5 × 81

24. 6 × 171   25. 5 × 125   26. 75 × 125

27. 9 × 6742   28. 7 × 3456

Problem Solving

In Exercises 29 and 30, we show lattice multiplications.
(a) Determine the numbers being multiplied. (b) Determine the product.

29.
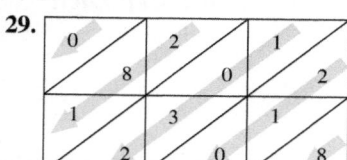

30.
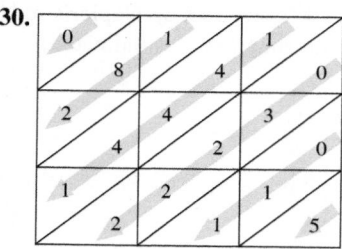

In Exercises 31 and 32, we solve a multiplication problem using Napier's rods. (a) Determine the numbers being multiplied. Each empty box contains a single digit. (b) Determine the product.

31.
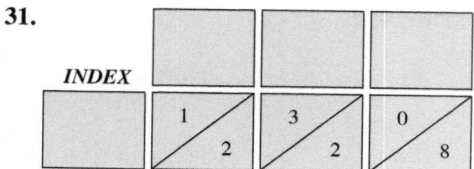

32.
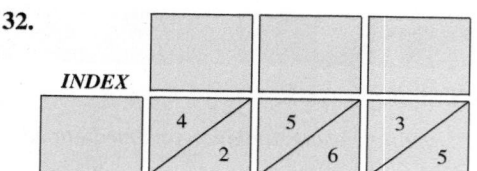

Challenge Problems/Group Activities

In Exercises 33 and 34, use the method of duplation and mediation to perform the multiplication. See Section 4.1 for Egyptian and Roman numerals. Write the answer in the numeration system in which the exercise is given.

33. $(\cap|||) \cdot (\cap\cap||)$

34. $(XXVI) \cdot (LXVII)$

In Exercises 35 and 36, (a) use lattice multiplication to perform the multiplication. (Hint: Be sure not to list any number greater than or equal to the base within the box.) Write the answer in the base in which the exercise is given. (b) Multiply the numbers as explained in Section 4.4. If you do not obtain the results obtained in part (a), explain why.

35. $21_3 \times 21_3$

36. $24_5 \times 234_5$

Research Activities

37. *John Napier* In addition to Napier's rods, John Napier is credited with making other important contributions to mathematics. Write a report on John Napier and his contributions to mathematics.

38. *Duplation and Mediation* Write a paper explaining why the duplation and mediation method works.

CHAPTER 4   *Summary*

Important Facts and Concepts

	Examples/Discussion
Section 4.1	
Additive Numeration Systems	
Egyptian Hieroglyphics	Discussion page 168, Examples 1–2, page 168
Roman Numerals	Discussion page 169, Examples 3–6, pages 169–170
Multiplicative Numeration Systems	
Traditional Chinese Numerals	Discussion pages 170–171, Examples 7–8, page 171–172
Ciphered Numeration Systems	
Ionic Greek Numerals	Discussion page 172, Examples 9–10, page 172

The following symbols were used in Section 4.1 and Section 4.2.

Egyptian Hieroglyphics

Egyptian numerals	∣	∩	◎	𝄃	∿	⌒	𓂀
Hindu–Arabic numerals	1	10	100	1000	10,000	100,000	1,000,000

Roman Numerals

Roman numerals	I	V	X	L	C	D	M
Hindu–Arabic numerals	1	5	10	50	100	500	1000

Traditional Chinese Numerals

Traditional Chinese numerals	零	一	二	三	四	五	六	七	八	九	十	百	千
Hindu–Arabic numerals	0	1	2	3	4	5	6	7	8	9	10	100	1000

Ionic Greek Numerals

1	α	alpha	60	ξ	xi
2	β	beta	70	o	omicron
3	γ	gamma	80	π	pi
4	δ	delta	90	$\koppa$	koppa
5	ϵ	epsilon	100	ρ	rho
6	$\digamma$	digamma	200	σ	sigma
7	ζ	zeta	300	τ	tau
8	η	eta	400	υ	upsilon
9	θ	theta	500	ϕ	phi
10	ι	iota	600	χ	chi
20	κ	kappa	700	ψ	psi
30	λ	lambda	800	ω	omega
40	μ	mu	900	$\sampi$	sampi
50	ν	nu			

Babylonian Numerals

Babylonian numerals	❙	❮
Hindu–Arabic numerals	1	10

Mayan Numerals

0	1	2	3	4	5	6	7	8	9
⬭	•	••	•••	••••	▬	$\overset{\bullet}{\underline{}}$	$\overset{\bullet\bullet}{\underline{}}$	$\overset{\bullet\bullet\bullet}{\underline{}}$	$\overset{\bullet\bullet\bullet\bullet}{\underline{}}$

10	11	12	13	14	15	16	17	18	19
═	$\overset{\bullet}{═}$	$\overset{\bullet\bullet}{═}$	$\overset{\bullet\bullet\bullet}{═}$	$\overset{\bullet\bullet\bullet\bullet}{═}$	≡	$\overset{\bullet}{≡}$	$\overset{\bullet\bullet}{≡}$	$\overset{\bullet\bullet\bullet}{≡}$	$\overset{\bullet\bullet\bullet\bullet}{≡}$

CHAPTER 4   *Review Exercises*

4.1, 4.2

In Exercises 1–4, write the Egyptian numeral as a Hindu–Arabic numeral.

1. 99∩∩II

2.)))))𝄢𝄢∩∩∩∩∩III

3. ∝◁∝◁9999∩∩I

4. 𝑋∝◁∝◁\𝄢𝄢𝄢𝄢9999∩∩∩

In Exercises 5–8, write the Hindu–Arabic numeral as an Egyptian numeral.

5. 3213

6. 10,200

7. 124,321

8. 1,003,042

In Exercises 9–12, write the Roman numeral as a Hindu–Arabic numeral.

9. XIV

10. CXLII

11. MMCDXXXVII

12. $\overline{\text{V}}$DCCLIX

In Exercises 13–16, write the Hindu–Arabic numeral as a Roman numeral.

13. 29

14. 543

15. 1964

16. 6491

In Exercises 17–20, write the Chinese numeral as a Hindu–Arabic numeral.

17. 三十二

18. 四十五

19. 二百六十七

20. 三千四百二十九

In Exercises 21–24, write the Hindu–Arabic numeral as a Chinese numeral.

21. 23

22. 54

23. 492

24. 2652

In Exercises 25–28, write the Ionic Greek numeral as a Hindu–Arabic numeral.

25. $\pi\alpha$ **26.** $\phi\mu\eta$

27. $\chi\epsilon$ **28.** $'\gamma\tau\lambda\delta$

In Exercises 29–32, write the Hindu–Arabic numeral as an Ionic Greek numeral.

29. 22   **30.** 773

31. 867   **32.** 4219

In Exercises 33–36, write the Babylonian numeral as a Hindu–Arabic numeral.

33. ❚❚❚ ＜❚   **34.** ❚❚❚ ＜T❚

35. ❚ ❚ ＜❚❚❚   **36.** ❚❚ ❚❚❚ ＜T❚❚

In Exercises 37–40, write the Hindu–Arabic numeral as a Babylonian numeral.

37. 181   **38.** 3679

39. 7280   **40.** 36,184

In Exercises 41–44, write the Mayan numeral as a Hindu–Arabic numeral.

41. ••
•••

42. •••
••••
••

43. ••
◯
•

44. ••
••••
••••
•••

In Exercises 45–48, write the Hindu–Arabic numeral as a Mayan numeral.

45. 69   **46.** 812

47. 1571   **48.** 17,913

4.3

In Exercises 49–54, convert the numeral to a Hindu–Arabic numeral.

49. 47_8 **50.** 111_2 **51.** 130_4

52. 3425_7 **53.** $A94_{12}$ **54.** 20220_3

In Exercises 55–60, convert 463 to a numeral in the base indicated.

55. base 2   **56.** base 4

57. base 5   **58.** base 8

59. base 12   **60.** base 16

4.4

In Exercises 61–66, add in the base indicated.

61. 121_4
$+ 322_4$

62. 10110_2
$+ 11001_2$

63. $9B_{12}$
$+ 87_{12}$

64. $2B9_{16}$
$+ 456_{16}$

65. 3024_5
$+ 4023_5$

66. 3407_8
$+ 7014_8$

In Exercises 67–72, subtract in the base indicated.

67. 321_4
$- 133_4$

68. 1001_2
$- 101_2$

69. $A7B_{12}$
$- 95_{12}$

70. 4321_5
$- 442_5$

71. 1713_8
$- 1243_8$

72. $F64_{16}$
$- 2A3_{16}$

In Exercises 73–78, multiply in the base indicated.

73. 431_6
$\times 3_6$

74. 2321_4
$\times 3_4$

75. 34_5
$\times 21_5$

76. 476_8
$\times 23_8$

77. 126_{12}
$\times 47_{12}$

78. $1A3_{16}$
$\times 12_{16}$

In Exercises 79–84, divide in the base indicated.

79. $2_3 \overline{)120_3}$ **80.** $2_4 \overline{)320_4}$

81. $3_5 \overline{)130_5}$ **82.** $4_6 \overline{)3020_6}$

83. $3_6 \overline{)2034_6}$ **84.** $6_8 \overline{)5072_8}$

4.5

85. Multiply 125×23, using the duplation and mediation method.

86. Multiply 125×23, using lattice multiplication.

87. Multiply 125×23, using Napier's rods.

CHAPTER 4 — *Test*

In Exercises 1–6, convert the numeral to a Hindu–Arabic numeral.

1. MMCDLXXIX

2. <<<ꔶ <ꔶꔶꔶ

3. 八
   千
   零
   九
   十

5. ∝𓂝𓎛𓎛𓏤9∩∩∩∩‖

6. ʹβψμϵ

In Exercises 7–11, convert the base 10 numeral to a numeral in the numeration system indicated.

7. 2124 to Egyptian

8. 2476 to Ionic Greek

9. 1434 to Mayan

10. 1596 to Babylonian

11. 2749 to Roman

In Exercises 12–13, convert the given numeral to a numeral in base 10.

12. 23_4

13. $B92_{12}$

In Exercises 14–15, convert the given numeral to a numeral in the base indicated.

14. 36 to base 2

15. 2938 to base 16

In Exercises 16–18, perform the indicated operations.

16. $\begin{array}{r} 1101_2 \\ +\ 1011_2 \end{array}$

17. $\begin{array}{r} 45_6 \\ \times\ 23_6 \end{array}$

18. $3_5\overline{)1210_5}$

19. Multiply 35×28, using duplation and mediation.

20. Multiply 43×196, using lattice multiplication.

◄ *Number theory helps scientists and researchers model problems.*

5

Number Theory and the Real Number System

What You Will Learn

- An introduction to number theory
- Integers
- Rational numbers
- Irrational numbers
- Real numbers and their properties
- Rules of exponents and scientific notation
- Arithmetic and geometric sequences
- The Fibonacci sequence

Why This Is Important

Every time we use a computer, make a telephone call, or watch television we use a device that relies on numbers to operate. In addition to playing many roles in our everyday lives, numbers are also used to describe the natural world, communicate information, and describe problems facing scientists and researchers. In this chapter, we will see how number theory, the study of numbers and their properties, makes all these roles possible.

Number theory has many applications that help scientists and researchers solve problems in many areas including communications, biology, medicine, computers, and other modern technical devices. Number theory is also used by architects to design concert hall ceilings for optimal acoustics.

Number Theory

Upon completion of this section, you will be able to:

- Classify natural numbers as prime or composite.
- Understand the rules of divisibility for the numbers 2, 3, 4, 5, 6, 8, 9, and 10.
- Write the prime factorization of a natural number.
- Determine the greatest common divisor of a group of natural numbers.
- Determine the least common multiple of a group of natural numbers.
- Understand the nature of prime numbers.

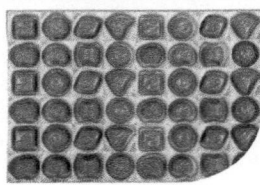

Suppose you decide to start your own candy company. You would like to sell boxes containing 48 pieces of candy. You must decide how to arrange the pieces within the box. You could have six rows with eight pieces in each row, or four rows with twelve pieces in each row, or three layers each with four rows with four pieces in each row. Each of these possibilities involves writing the number 48 as the product of two or more smaller numbers. In this section, we will discuss many other similar problems related to the writing of numbers as the product of smaller numbers.

Why This Is Important Writing numbers as the product of smaller numbers has many applications to the branch of mathematics we know as number theory. For example, banks and retail companies use number theory to secure online transactions.

This chapter introduces *number theory*, the study of numbers and their properties. The numbers we use to count are called the *counting numbers* or *natural numbers*. Because we begin counting with the number 1, the set of natural numbers begins with 1. The set of natural numbers is frequently denoted by N:

$$N = \{1, 2, 3, 4, 5, \dots\}$$

Any natural number can be expressed as a product of two or more natural numbers. For example, $8 = 2 \times 4$, $16 = 4 \times 4$, and $19 = 1 \times 19$. The natural numbers that are multiplied together are called *factors* of the product. For example,

$$2 \times 4 = 8$$
$$\uparrow \quad \uparrow$$
Factors

A natural number may have many factors. For example, what pairs of numbers have a product of 18?

$$1 \cdot 18 = 18$$
$$2 \cdot 9 = 18$$
$$3 \cdot 6 = 18$$

The numbers 1, 2, 3, 6, 9, and 18 are all factors of 18. Each of these numbers divides 18 without a remainder.

If a and b are natural numbers, we say that a is a *divisor* of b or a *divides* b, symbolized $a \mid b$, if the quotient of b divided by a has a remainder of 0. If a divides b, then b is *divisible* by a. For example, 4 divides 12, symbolized $4 \mid 12$, since the quotient of 12 divided by 4 has a remainder of 0. Note that 12 is divisible by 4. The notation $7 \nmid 12$ means that 7 does not divide 12. Note that every factor of a natural number is also a divisor of the natural number. *Caution:* Do not confuse the symbols $a \mid b$ and a/b; $a \mid b$ means "a divides b" and a/b means "a divided by b" ($a \div b$). The symbols a/b and $a \div b$ indicate that the operation of division is to be performed, and b may or may not be a divisor of a.

Prime and Composite Numbers

Every natural number greater than 1 can be classified as either a prime number or a composite number.

Profile in Mathematics

Eratosthenes of Cyrene

Eratosthenes of Cyrene (275–195 B.C.) was born in northern Africa near the present-day city of Shahhat, Libya. Eratosthenes is best known for being the first to estimate accurately the diameter of Earth. He is also credited for developing a method of determining prime numbers known as the **sieve of Eratosthenes**. Although he is most known for his work in mathematics, Eratosthenes also was influential in the fields of history, geography, and astronomy. In addition, Eratosthenes served for many years as the director of the famous library in Alexandria, Egypt. Although Eratosthenes was a highly regarded scholar throughout the ancient world, only fragments of his writing remain today. Eratosthenes was near 80 years old when, after going blind, he died from voluntary starvation.

Definition: Prime Number

A **prime number** is a natural number greater than 1 that has exactly two factors (or divisors), itself and 1.

The number 5 is a prime number because it is divisible only by the factors 1 and 5. The first eight prime numbers are 2, 3, 5, 7, 11, 13, 17, and 19. The number 2 is the only even prime number. All other even numbers have at least three divisors: 1, 2, and the number itself.

Definition: Composite Number

A **composite number** is a natural number that is divisible by a number other than itself and 1.

Any natural number greater than 1 that is not prime is composite. The first eight composite numbers are 4, 6, 8, 9, 10, 12, 14, and 15.

The number 1 is neither prime nor composite; it is called a *unit*. The number 38 has at least three divisors, 1, 2, and 38, and hence is a composite number. In contrast, the number 23 is a prime number, since its only divisors are 1 and 23.

More than 2000 years ago, the ancient Greeks developed a technique for determining which numbers are prime numbers and which are not. This technique is known as the *sieve of Eratosthenes*, for the Greek mathematician Eratosthenes of Cyrene, who first used it.

To determine the prime numbers less than or equal to any natural number, say, 50, using this method, list the first 50 counting numbers (Fig. 5.1). Cross out 1, since it is not a prime number. Circle 2, the first prime number. Then cross out all the multiples of 2: 4, 6, 8, . . . , 50. Circle the next prime number, 3. Cross out all multiples of 3 that are not already crossed out. Continue this process of crossing out multiples of prime numbers until you reach the prime number p, such that $p \cdot p$, or p^2, is greater than the last number listed, in this case 50. Therefore, we next circle 5 and cross out its multiples. Then circle 7 and cross out its multiples. The next prime number is 11, and $11 \cdot 11$, or 121, is greater than 50, so you are done. At this point, circle all the remaining numbers to obtain the prime numbers less than or equal to 50. The prime numbers less than or equal to 50 are 2, 3, 5, 7, 11, 13, 17, 19, 23, 29, 31, 37, 41, 43, and 47.

1	2	3	4	5	6	7	8	9	10
11	12	13	14	15	16	17	18	19	20
21	22	23	24	25	26	27	28	29	30
31	32	33	34	35	36	37	38	39	40
41	42	43	44	45	46	47	48	49	50

Figure 5.1

Rules of Divisibility

Now we turn our attention to composite numbers and their factors. The rules of divisibility given in the chart on page 210 are helpful in determining divisors (or factors) of composite numbers.

▲ *Factoring Machine*

In 1989, Jeffrey Shallit of the University of Waterloo came across an article in a 1920 French journal regarding a machine that was built in 1914 for factoring numbers. Eugene Oliver Carissan, an amateur mathematician who had invented the machine, wrote the article. After reading the article, Shallit wondered whatever became of the factoring machine.

After considerable searching, Shallit found the machine in a drawer of an astronomical observatory in Floirac, France. The machine was in good condition, and it still worked. By rotating the machine by a hand crank at two revolutions per minute, an operator could process 35 to 40 numbers per second. Carissan needed just 10 minutes to prove that 708,158,977 is a prime number, an amazing feat in pre-computer times. The machine is now housed at the Conservatoire National des Arts et Métiers in Paris.

Rules of Divisibility

Divisible by	Test	Example
2	The number is even.	924 is divisible by 2, since 924 is even.
3	The sum of the digits of the number is divisible by 3.	924 is divisible by 3, since the sum of the digits is $9 + 2 + 4 = 15$, and 15 is divisible by 3.
4	The number formed by the last two digits of the number is divisible by 4.	924 is divisible by 4, since the number formed by the last two digits, 24, is divisible by 4.
5	The number ends in 0 or 5.	265 is divisible by 5, since the number ends in 5.
6	The number is divisible by both 2 and 3.	924 is divisible by 6, since it is divisible by both 2 and 3.
8	The number formed by the last three digits of the number is divisible by 8.	5824 is divisible by 8, since the number formed by the last three digits, 824, is divisible by 8.
9	The sum of the digits of the number is divisible by 9.	837 is divisible by 9, since the sum of the digits, 18, is divisible by 9.
10	The number ends in 0.	290 is divisible by 10, since the number ends in 0.

Note that the chart does not list rules for the divisibility for the number 7. The rule for 7 is given in Exercise 90 on page 218. You may find after using the rule that the easiest way to check divisibility by 7 is simply to perform the division.

The test for divisibility by 6 is a particular case of the general statement that the product of two prime divisors of a number is a divisor of the number. Thus, for example, if both 3 and 5 divide a number, then 15 will also divide the number.

Example 1 *Using the Divisibility Rules*

Determine whether 563,244 is divisible by

a) 2   b) 3   c) 4   d) 5   e) 6   f) 8   g) 9   h) 10

Solution

a) 2: Since 563,244 is even, it is divisible by 2.

b) 3: The sum of the digits of 563,244 is $5 + 6 + 3 + 2 + 4 + 4 = 24$. Since 24 is divisible by 3, the number 563,244 is divisible by 3.

c) 4: The number formed by the last two digits is 44. Since 44 is divisible by 4, the number 563,244 is divisible by 4.

d) 5: Since 563,244 does not end in 0 or 5, the number 563,244 is not divisible by 5.

e) 6: Since 563,244 is divisible by both 2 and 3, the number 563,244 is divisible by 6.

f) 8: The number formed by the last 3 digits is 244. Since 244 is not divisible by 8, the number 563,244 is not divisible by 8.

g) 9: The sum of the digits of 563,244 is 24. Since 24 is not divisible by 9, the number 563,244 is not divisible by 9.

h) 10: Since 563,244 does not end in 0, the number 563,244 is not divisible by 10. ∎

Prime Factorization of a Number

Every composite number can be expressed as a product of two or more prime numbers. The process of breaking a given composite number down into a product of prime numbers is called *prime factorization*. The prime factorization of 18 is $3 \times 3 \times 2$.

No other natural number listed as a product of primes will have the same prime factorization as 18. The *fundamental theorem of arithmetic* states this concept formally. (A *theorem* is a statement or proposition that can be proven true.)

> **The Fundamental Theorem of Arithmetic**
> Every composite number can be expressed as a *unique* product of prime numbers.

In writing the prime factorization of a number, the order of the factors does not matter. However, for consistency, we will write the prime factorization with the factors from smallest to largest and we will use exponents to represent repeated factors. For example, we will write the prime factorization of 18 as $2 \cdot 3^2$. We will illustrate two methods to determine the prime factorization of a number: *branching* and *division.*

To determine the prime factorization of a number by branching, write the number as the product of two factors. If one or both of the factors are not prime numbers, continue factoring each composite number until all the factors are prime. We illustrate branching in Example 2.

Example 2 *Prime Factorization by Branching*

Write 1500 as a product of primes.

Solution Select any two numbers whose product is 1500. Among the many choices, two possibilities are $15 \cdot 100$ and $30 \cdot 50$. First consider $15 \cdot 100$. Since neither 15 nor 100 is a prime number, determine any two numbers whose product is 15 and any two numbers whose product is 100. Continue branching as shown in Fig. 5.2 until the numbers in the last row are all prime numbers. To determine the answer, write the product of all the prime factors. The branching diagram is sometimes called a *factor tree.*

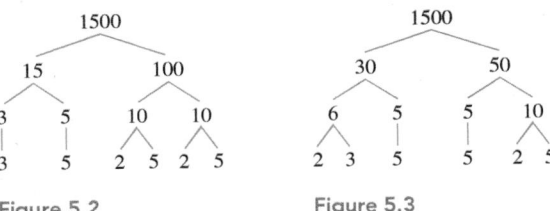

Figure 5.2   Figure 5.3

We see that the numbers in the last row of factors in Fig. 5.2 are all prime numbers. Thus, the prime factorization of 1500 is $3 \cdot 5 \cdot 2 \cdot 5 \cdot 2 \cdot 5 = 2 \cdot 2 \cdot 3 \cdot 5 \cdot 5 \cdot 5 = 2^2 \cdot 3 \cdot 5^3$. Note from Fig. 5.3 that choosing 30 and 50 as the first pair of factors also leads to the same prime factorization of $2^2 \cdot 3 \cdot 5^3$. ∎

To obtain the prime factorization of a number by the division method, divide the given number by the smallest prime number by which it is divisible. Place the quotient under the given number. Then divide the quotient by the smallest prime number by which it is divisible and again record the quotient. Repeat this process until the quotient is a prime number. The prime factorization is the product of all the prime divisors and the prime (or last) quotient. This procedure is illustrated in Example 3.

Example 3 *Prime Factorization by Division*

Write 1500 as a product of prime numbers.

Solution Because 1500 is an even number, the smallest prime number that divides it is 2. Divide 1500 by 2. Place the quotient, 750, below the 1500. Continue to divide each quotient by the smallest prime number that divides it.

2	1500
2	750
3	375
5	125
5	25
	5

The final quotient, 5, is a prime number, so we stop. The prime factorization is determined by multiplying all divisors on the left side and the final prime number quotient as shown by the shaded arrow. Thus, the prime factorization of 1500 is

$$2 \cdot 2 \cdot 3 \cdot 5 \cdot 5 \cdot 5 = 2^2 \cdot 3 \cdot 5^3.$$

Note that despite the different methods used in Examples 2 and 3, the answer is the same.

Greatest Common Divisor

The discussion in Section 5.3 of how to reduce fractions makes use of the greatest common divisor (GCD). We will now discuss how to determine the GCD of a set of numbers. One technique of determining the GCD is to use prime factorization.

> **Definition: Greatest Common Divisor**
> The **greatest common divisor (GCD)** of a set of natural numbers is the largest natural number that divides (without remainder) every number in that set.

What is the GCD of 12 and 18? One way to determine the GCD is to list the divisors (or factors) of 12 and 18:

Divisors of 12 $\{\mathbf{1}, \mathbf{2}, \mathbf{3}, 4, \mathbf{6}, 12\}$
Divisors of 18 $\{\mathbf{1}, \mathbf{2}, \mathbf{3}, \mathbf{6}, 9, 18\}$

The common divisors are 1, 2, 3, and 6, in bold above. Therefore, the greatest common divisor is 6.

If the numbers are large, this method of determining the GCD is not practical. The GCD can be determined more efficiently by using prime factorization.

> **PROCEDURE** TO DETERMINE THE GREATEST COMMON DIVISOR OF TWO OR MORE NUMBERS
>
> 1. Determine the prime factorization of each number.
> 2. List each prime factor with the *smallest* exponent that appears in each of the prime factorizations.
> 3. Determine the product of the factors determined in Step 2.

Example 4 illustrates this procedure.

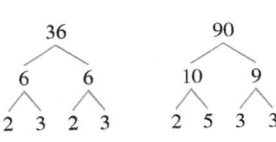

Figure 5.4

Example 4 *Using Prime Factorization to Determine the GCD*

Determine the GCD of 36 and 90.

Solution The branching method of determining the prime factors of 36 and 90 is illustrated in Fig. 5.4.

1. The prime factorization of 36 is $2^2 \cdot 3^2$, and the prime factorization of 90 is $2 \cdot 3^2 \cdot 5$.

2. The prime factors with the smallest exponents that appear in each of the factorizations of 36 and 90 are 2 and 3^2. Note that because 5 is not in the prime factorization of 36, it is not included when determining the GCD.

3. The product of the factors determined in Step 2 is $2 \cdot 3^2 = 2 \cdot 9 = 18$. The GCD of 36 and 90 is 18. Eighteen is the largest natural number that divides both 36 and 90. ∎

Example 5 *Determining the GCD*

Determine the GCD of 315 and 450.

Solution

1. The prime factorization of 315 is $3^2 \cdot 5 \cdot 7$, and the prime factorization of 450 is $2 \cdot 3^2 \cdot 5^2$. You should verify these answers using either the branching method or the division method.

2. The prime factors with the smallest exponents that appear in each of the factorizations of 315 and 450 are 3^2 and 5.

3. The product of the factors determined in Step 2 is $3^2 \cdot 5 = 9 \cdot 5 = 45$. The greatest common divisor of 315 and 450 is 45. It is the largest natural number that will evenly divide both 315 and 450. ∎

Did You Know?

Friendly Numbers

The ancient Greeks often thought of numbers as having human qualities. For example, the numbers 220 and 284 were considered "friendly" or "amicable" numbers because each number was the sum of the other number's *proper factors*. (A proper factor is any factor of a number other than the number itself.) If you sum all the proper factors of 284 $(1 + 2 + 4 + 71 + 142)$, you get the number 220, and if you sum all the proper factors of 220 $(1 + 2 + 4 + 5 + 10 + 11 + 20 + 22 + 44 + 55 + 110)$, you get 284.

Least Common Multiple

To perform addition and subtraction of fractions (Section 5.3), we use the least common multiple (LCM). We will now discuss how to determine the LCM of a set of numbers.

> **Definition: Least Common Multiple**
> The **least common multiple (LCM)** of a set of natural numbers is the smallest natural number that is divisible (without remainder) by each element of the set.

Least Common Multiple and Greatest Common Divisor

What is the least common multiple of 12 and 18? One way to determine the LCM is to list the multiples of each number:

Multiples of 12 $\{12, 24, \mathbf{36}, 48, 60, \mathbf{72}, 84, 96, \mathbf{108}, 120, 132, \mathbf{144}, \dots \}$

Multiples of 18 $\{18, \mathbf{36}, 54, \mathbf{72}, 90, \mathbf{108}, 126, \mathbf{144}, 162, \dots \}$

Some common multiples of 12 and 18 are 36, 72, 108, and 144, in bold above. The least common multiple, 36, is the smallest number that is divisible by both 12 and 18. Usually, the most efficient method of determining the LCM is to use prime factorization.

> **PROCEDURE** TO DETERMINE THE LEAST COMMON MULTIPLE
> OF TWO OR MORE NUMBERS
>
> 1. Determine the prime factorization of each number.
> 2. List each prime factor with the *greatest* exponent that appears in any of the prime factorizations.
> 3. Determine the product of the factors determined in Step 2.

MATHEMATICS TODAY

Online Credit Card Number Safety

Prime numbers play an essential role in protecting the credit card numbers of consumers who are making online purchases. The encryption or coding process, known as the *RSA algorithm*, relies on a key number, *n*, which is the product of two very large prime numbers, *p* and *q*. The prime numbers *p* and *q* are each over 150 digits long. The product, *n*, is very difficult to factor—even with the use of the world's fastest computers. The number *n* is publicly available and is used by merchants to encrypt the credit card number after it is entered by the consumer. The prime numbers *p* and *q* are known only by the credit card company and are used to decrypt the credit card number. This same algorithm was used by the U.S. government for many years to protect key databases.

Why This Is Important Encryption systems involve many areas of mathematics, including many that are discussed in this book, such as prime numbers, permutations, modular arithmetic, polynomials, and group theory. Encryption systems, relying on number theory, enable secure online transactions.

Example 6 illustrates this procedure.

Example 6 *Using Prime Factorization to Determine the LCM*

Determine the LCM of 36 and 90.

Solution

1. Determine the prime factors of each number. In Example 4, we determined that

$$36 = 2^2 \cdot 3^2 \quad \text{and} \quad 90 = 2 \cdot 3^2 \cdot 5$$

2. List each prime factor with the greatest exponent that appears in either of the prime factorizations: $2^2, 3^2, 5$.

3. Determine the product of the factors determined in Step 2:

$$2^2 \cdot 3^2 \cdot 5 = 4 \cdot 9 \cdot 5 = 180$$

Thus, 180 is the LCM of 36 and 90. It is the smallest natural number that is divisible by both 36 and 90. ∎

Example 7 *Determining the LCM*

Determine the LCM of 315 and 450.

Solution

1. Determine the prime factorization of each number. In Example 5, we determined that

$$315 = 3^2 \cdot 5 \cdot 7 \quad \text{and} \quad 450 = 2 \cdot 3^2 \cdot 5^2$$

2. List each prime factor with the greatest exponent that appears in either of the prime factorizations: $2, 3^2, 5^2, 7$.

3. Determine the product of the factors from Step (2):

$$2 \cdot 3^2 \cdot 5^2 \cdot 7 = 2 \cdot 9 \cdot 25 \cdot 7 = 3150$$

Thus, 3150 is the least common multiple of 315 and 450. It is the smallest natural number that is evenly divisible by both 315 and 450. ∎

More About Prime Numbers

More than 2000 years ago, the Greek mathematician Euclid proved that there is no largest prime number. Mathematicians, however, continue to strive to discover larger and larger prime numbers.

Marin Mersenne (1588–1648), a seventeenth-century monk, discovered that numbers of the form $2^n - 1$ are often prime numbers when n is a prime number. For example,

$$2^2 - 1 = 4 - 1 = 3 \qquad 2^3 - 1 = 8 - 1 = 7$$
$$2^5 - 1 = 32 - 1 = 31 \qquad 2^7 - 1 = 128 - 1 = 127$$

Numbers of the form $2^n - 1$ that are prime are referred to as *Mersenne primes*. The first 10 Mersenne primes occur when $n = 2, 3, 5, 7, 13, 17, 19, 31, 61, 89$. The first time the expression $2^n - 1$ does not generate a prime number, for prime number n, is when n is 11. The number $2^{11} - 1$ is a composite number (see Exercise 88 on page 218).

Scientists frequently use Mersenne primes in their search for larger and larger primes. The largest prime number found to date was discovered on December 7, 2018, by Patrick Laroche of Ocala, Florida. Laroche worked in conjunction with the Great Internet Mersenne Prime Search (GIMPS; see *Mathematics Today*, at left) to conduct the search. The number is Mersenne prime $2^{82,589,933} - 1$. This record prime number is the fifty-first known Mersenne prime. When written out, it is 24,862,048 digits long. When written out using a standard 12-point font, the number is almost 33 miles long!

Another mathematician who studied prime numbers was Pierre de Fermat (1601–1665). A lawyer by profession, Fermat became interested in mathematics as a hobby. He became one of the finest mathematicians of the seventeenth century. Fermat conjectured that each number of the form $2^{2^n} + 1$, now referred to as a *Fermat number*, was prime for each natural number n. Recall that a *conjecture* is a hypothesis that has not been proved or disproved. In 1732, Leonhard Euler proved that for $n = 5$, $2^{2^5} + 1 = 2^{32} + 1$ was a composite number, thus disproving Fermat's conjecture.

Since Euler's time, mathematicians have been able to evaluate only the sixth, seventh, eighth, ninth, tenth, and eleventh Fermat numbers to determine whether they are prime or composite. Each of these numbers has been shown to be composite. The eleventh Fermat number was factored by Richard Brent and François Morain in 1988. The sheer magnitude of the numbers involved makes it difficult to test these numbers, even with supercomputers.

In 1742, German mathematician Christian Goldbach conjectured in a letter to Euler that every even number greater than or equal to 4 can be represented as the sum of two (not necessarily distinct) prime numbers (for example, $4 = 2 + 2$, $6 = 3 + 3, 8 = 3 + 5, 10 = 5 + 5, 12 = 5 + 7$). This conjecture became known as *Goldbach's conjecture*, and it remains unproven to this day. The *twin prime conjecture* is another famous long-standing conjecture. *Twin primes* are primes of the form p and $p + 2$ (for example, 3 and 5, 5 and 7, 11 and 13). This conjecture states that there are an infinite number of pairs of twin primes. At the time of this writing, the largest known twin primes are of the form $2{,}996{,}863{,}034{,}895 \cdot 2^{1,290,000} \pm 1$, which were discovered by a collaborative effort of two research groups, *Twin Prime Search* and *PrimeGrid*, on September 15, 2016, and contains 388,342 digits.

SECTION 5.1 *Exercises*

Warm Up Exercises

In Exercises 1–10, fill in the blanks with an appropriate word, phrase, or symbol(s).

1. The study of numbers and their properties is known as number _____.

2. If a is a factor of b, then $b \div a$ is a(n) _____.

3. If a is divisible by b, then $a \div b$ has a remainder of _____.

4. A natural number greater than 1 that has exactly two factors, itself and 1, is known as a(n) _____ number.

5. A natural number that is divisible by a number other than itself and 1 is known as a(n) _____ number.

6. The smallest natural number that is divisible (without remainder) by each number in a set of numbers is known as the least common _____ of the set of numbers.

7. The largest natural number that divides (without remainder) each number in a set of numbers is known as the greatest common _____ of the set of numbers.

8. Prime numbers of the form $2^n - 1$, where n is a prime number, are known as _____ prime numbers.

9. A hypothesis that has not been proved or disproved is known as a(n) _____.

10. Goldbach's conjecture states that every even number greater than or equal to 4 can be written as the sum of two (not necessarily) distinct _____ numbers.

Practice the Skills

In Exercises 11 and 12, use the sieve of Eratosthenes to determine the prime numbers less than or equal to the given number.

11. 100   **12.** 150

In Exercises 13–26, determine whether the statement is true or false. Modify each false statement to make it a true statement.

13. 64 is a factor of 8.   **14.** 6 is a factor of 36.

15. 7 is a multiple of 35.   **16.** 54 is a multiple of 9.

17. 4 is a divisor of 32.   **18.** 49 is a divisor of 7.

19. 5|40.   **20.** 36|9.

21. 72 is a divisor of 8.   **22.** 6 is a divisor of 48.

23. If a number is not divisible by 5, then it is not divisible by 10.

24. If a number is not divisible by 10, then it is not divisible by 5.

25. If a number is divisible by 2 and 3, then the number is divisible by 6.

26. If a number is divisible by 3 and 5, then the number is divisible by 15.

In Exercises 27–32, determine whether the number is divisible by each of the following numbers: 2, 3, 4, 5, 6, 8, 9, and 10.

27. 3225   **28.** 25,620

29. 72,216   **30.** 170,820

31. 1,882,320   **32.** 3,941,221

33. Determine a number that is divisible by 2, 3, 4, 5, and 6.

34. Determine a number that is divisible by 3, 4, 5, 9, and 10.

In Exercises 35–44, determine the prime factorization of the number.

35. 12   **36.** 50

37. 180   **38.** 315

39. 332   **40.** 399

41. 513   **42.** 663

43. 1336   **44.** 3190

In Exercises 45–54, determine (a) the greatest common divisor (GCD) and (b) the least common multiple (LCM).

45. 6 and 21   **46.** 18 and 24

47. 20 and 35   **48.** 32 and 224

49. 40 and 900   **50.** 120 and 240

51. 96 and 212   **52.** 240 and 285

53. 24, 48, and 128   **54.** 18, 78, and 198

Problem Solving

55. *Landscaping* Sin-Liu weeds her flower beds every four days and trims her shrubs every 14 days. If Sin-Liu weeds her flower beds and trims her shrubs on August 1, how many days will it be before she weeds her flower beds and trims her shrubs on the same day again?

56. *U.S. Senate Committees* The U.S. Senate consists of 100 members. Senate committees are to be formed so that each of the committees contains the same number of senators and each senator is a member of exactly one committee. The committees are to have more than 2 members but fewer than 50 members. There are various ways that these committees can be formed.

▲ *U.S. Capitol Building*

a) What size committees are possible?

b) How many committees are there for each size?

57. *Setting Up Chairs* Jerrett is setting up chairs for his school band concert. He needs to put 120 chairs on the gymnasium floor in rows of equal size, and there must be at least 2 chairs in each row. List the number of rows and the number of chairs in each row that are possible.

58. *Barbie and Ken* Mary collects Barbie dolls and Ken dolls. She has 390 Barbie dolls and 468 Ken dolls. Mary wishes to display the dolls in groups so that the same number of dolls are in each group and that each doll belongs to one group. If each group is to consist only of Barbie dolls or only of Ken dolls, what is the largest number of dolls Mary can have in each group?

59. *Toy Car Collection* Martha collects Matchbox and

HotWheels toy cars. She has 70 red cars and 175 blue cars. She wants to line up her cars in groups so that each group has the same number of cars and each group contains only red cars or only blue cars. What is the largest number of cars she can have in a group?

60. *Stacking Trading Cards* Desmond collects trading cards. He has 432 baseball cards and 360 football cards. He wants to make stacks of cards on a table so that each stack contains the same number of cards and each card belongs to one stack. If the baseball and football cards must not be mixed in the stacks, what is the largest number of cards that he can have in a stack?

61. *Tree Rows* Elizabeth is the manager at Queen Palm Nursery and is in charge of displaying potted trees in rows. Elizabeth has 150 citrus trees and 180 palm trees. She wants to make rows of trees so that each row has the same number of trees and each tree is in a row. If the citrus trees and the palm trees must not be mixed in the rows, what is the largest number of trees that she can have in a row?

62. *Car Maintenance* For many sport utility vehicles, it is recommended that the oil be changed every 3500 miles and that the tires be rotated every 6000 miles. If Carmella just had the oil changed and tires rotated on her SUV during the same visit to her mechanic, how many miles will she drive before she has the oil changed and tires rotated again during the same visit?

63. *Work Schedules* Sara and Harry both work the same night shift. Sara has every fifth night off and Harry has every sixth

night off. If they both have tonight off, how many days will pass before they have the same night off again?

64. *Taking Medicine* Elijah takes the medicine bisphosphonate once every 30 days and the medicine pegaspargase once every 14 days. If Elijah took both medicines on February 1, how many days would it be before he has to take both medicines on the same day again?

65. *Relatively Prime* *Two numbers with a greatest common divisor of 1 are said to be* relatively prime. *For example, the numbers 9 and 14 are relatively prime, since their GCD is 1. Determine whether the following pairs of numbers are relatively prime. Write* yes *or* no *as your answer.*

a) 8, 9   **b)** 15, 24   **c)** 39, 52   **d)** 177, 178

66. The primes 2 and 3 are consecutive natural numbers. Is there another pair of consecutive natural numbers both of which are prime? Explain.

67. Determine the next two sets of twin primes that follow the set 11, 13.

68. Determine the first five Mersenne prime numbers.

69. Determine the first three Fermat numbers and determine whether they are prime or composite.

70. Show that Goldbach's conjecture is true for the even numbers 4 through 20.

71. State a procedure that defines a divisibility test for 14.

72. State a procedure that defines a divisibility test for 22.

Euclidean Algorithm *Another method that can be used to determine the greatest common divisor is known as the* Euclidean algorithm. *We illustrate this procedure by determining the GCD of 60 and 220.*

First divide 220 by 60 as shown below. Disregard the quotient 3 and then divide 60 by the remainder 40. Continue this process of dividing the divisors by the remainders until you obtain a remainder of 0. The divisor in the last division, in which the remainder is 0, is the GCD.

$$
\begin{array}{ccc}
3 & 1 & 2 \\
60\overline{)220} & 40\overline{)60} & 20\overline{)40} \\
\underline{180} & \underline{40} & \underline{40} \\
40 & 20 & \mathbf{0}
\end{array}
$$

Since 40/20 had a remainder of 0, the GCD is 20.

In Exercises 73–78, use the Euclidean algorithm to determine the GCD.

73. 15, 40   **74.** 12, 28

75. 35, 105   **76.** 78, 104

77. 150, 180   **78.** 210, 560

Perfect Numbers *A number whose proper factors (factors other than the number itself) add up to the number is called a perfect number. For example, 6 is a perfect number because its proper factors are 1, 2, and 3, and 1 + 2 + 3 = 6. Determine which, if any, of the following numbers are perfect numbers.*

79. 28   **80.** 36   **81.** 72   **82.** 496

Challenge Problems/Group Activities

83. ***Number of Positive Factors*** The following procedure can be used to determine the *number of positive factors* (or *divisors*) of a composite number. Write the number in prime factorization form. Examine the exponents on the prime numbers in the prime factorization. Add 1 to each exponent and then determine the product of these numbers. This product gives the number of positive factors of the composite number. For example, the prime factorization of 75 is $3 \cdot 5^2$. The exponents on the prime factors are 1 and 2. Therefore, 75 has $(1 + 1) \cdot (2 + 1) = 2 \cdot 3$ or 6 positive factors (the factors are 1, 3, 5, 15, 25, and 75).

    a) Use this procedure to determine the number of positive factors of 60.

    b) To check your answer, list all the positive factors of 60. You should obtain the same number of factors determined in part (a).

84. Recall that if a number is divisible by both 2 and 3, then the number is divisible by 6. If a number is divisible by both 2 and 4, is the number necessarily divisible by 8? Explain your answer.

85. The product of any three consecutive natural numbers is divisible by 6. Explain why.

86. A number in which each digit except 0 appears exactly three times is divisible by 3. For example, 888,444,555 and 714,714,714 are both divisible by 3. Explain why this outcome must be true.

87. Use the fact that if $a \mid b$ and $a \mid c$, then $a \mid (b + c)$ to determine whether 54,036 is divisible by 18. (*Hint:* Write 54,036 as 54,000 + 36.)

88. ***Mersenne Primes*** Show that $2^n - 1$ is a (Mersenne) prime for $n = 2, 3, 5,$ and 7 but composite for $n = 11$.

89. ***Another Conjecture*** Goldbach also conjectured in his letter to Euler that *every* integer greater than 5 is the sum of three prime numbers. For example, $6 = 2 + 2 + 2$ and $7 = 2 + 2 + 3$. Show that this conjecture is true for integers 8 through 20.

90. ***Divisibility by Seven*** The following describes a procedure to determine whether a number is divisible by 7. We will demonstrate the procedure with the number 203.

    i) Remove the units digit from the number, double the units digit, and subtract it from the remaining number. For 203, we will remove the 3, double it to get 6, and then subtract $20 - 6$ to get 14.

    ii) If this new number is divisible by 7, then so is the original number. In our case, since 14 is divisible by 7, then 203 is also divisible by 7.

    iii) If you are not sure if the new number obtained in Step ii is divisible by 7, you can repeat the process described in the first step.

Use the procedure described above to determine whether the following numbers are divisible by 7.

    a) 329   **b)** 553   **c)** 583   **d)** 4823

91. ***Prime Numbers*** Consider the first eight prime numbers greater than 3. The numbers are 5, 7, 11, 13, 17, 19, 23, and 29.

    a) Determine which of these prime numbers differs by 1 from a multiple of the number 6.

    b) Use inductive reasoning and the results obtained in part (a) to make a conjecture regarding prime numbers.

    c) Select a few more prime numbers and determine whether your conjecture appears to be correct.

Research Activities

92. ***GIMPS*** (See *Mathematics Today* on page 215.) Write a report on the GIMPS project. Describe the history and development of the project. Include a current update of the project's findings.

93. ***Deficient and Abundant Numbers*** Explain what *deficient numbers* and *abundant numbers* are. Give an example of each type of number.

SECTION 5.2   The Integers

Upon completion of this section, you will be able to:

- Add integers.
- Subtract integers.
- Multiply integers.
- Divide integers.
- Simplify expressions involving exponents on integers.
- Simplify expressions using the order of operations.

Answer the question, What is the temperature, to the nearest degree Fahrenheit, outside right now? Depending on where you are and what time of year it is, you may have answered with a positive number, a negative number, or zero. These numbers are examples of the group of numbers called the *integers* that we will study in this section. We will also study the operations of addition, subtraction, multiplication, and division using the integers.

Why This Is Important   We all use integers daily. An understanding of the integers and their properties is important to successfully complete our everyday tasks. Applications we study in this section include elevation differences, temperature changes, and time zone calculations.

In Sections 1.1 and 5.1, we defined the natural or counting numbers:

$$N = \{1, 2, 3, 4, \dots\}$$

Another important set of numbers, the *whole numbers*, helps to answer the question, How many?

$$\text{Whole numbers} = \{0, 1, 2, 3, 4, \dots\}$$

Note that the set of whole numbers contains the number 0 but that the set of counting numbers does not. Although we use the number 0 daily and take it for granted, the number 0 as we know it was not used and accepted until the sixteenth century.

If the temperature is 12°F and drops 20°F, the resulting temperature is −8°F. This type of problem shows the need for negative numbers. The set of *integers* consists of the negative integers, 0, and the positive integers.

$$\text{Integers} = \{\underbrace{\dots, -4, -3, -2, -1}_{\text{Negative integers}}, 0, \underbrace{1, 2, 3, \dots}_{\text{Positive integers}}\}$$

The term *positive integers* is yet another name for the natural numbers or counting numbers.

An understanding of addition, subtraction, multiplication, and division of the integers is essential in understanding algebra (Chapter 6). To aid in our explanation of addition and subtraction of integers, we introduce the real number line (Fig. 5.5). The real number line contains the integers and all the other real numbers that are not integers. Some examples of real numbers that are not integers are indicated in Fig. 5.5, namely, $-\frac{5}{2}, \frac{1}{2}, \sqrt{2}$, and π. We discuss real numbers that are not integers in the next two sections.

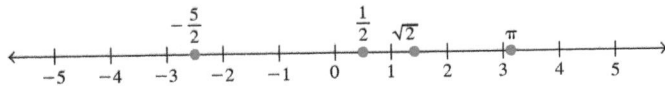

Figure 5.5

The arrows at the ends of the real number line indicate that the line continues indefinitely in both directions. Note that for any natural number, n, on the number line, the *opposite of* that number, $-n$, is also on the number line. This real number line was drawn horizontally, but it could just as well have been drawn vertically. In fact, in the next chapter, we show that the axes of a graph are the union of two number lines, one horizontal and the other vertical.

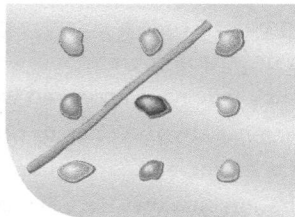
The number line can be used to determine the greater (or lesser) of two integers. Two *inequality symbols* that we will use in this chapter are $>$ and $<$. The symbol $>$ is read "is greater than," and the symbol $<$ is read "is less than." Expressions that contain an inequality symbol are called *inequalities*. On the number line, the numbers increase from left to right. The number 3 is greater than 2, written $3 > 2$. Observe that 3 is to the right of 2 on the number line. Similarly, we can see that $0 > -1$ by observing that 0 is to the right of -1 on the number line.

Instead of stating that 3 is greater than 2, we could state that 2 is less than 3, written $2 < 3$. Note that 2 is to the left of 3 on the number line. We can also see that $-1 < 0$ by observing that -1 is to the left of 0. The inequality symbol always points to the smaller of the two numbers when the inequality is true.

Example 1 *Writing an Inequality*

Insert either $>$ or $<$ in the shaded area between the paired numbers to make the statement correct.
a) $-7 \;\; 8$ b) $-7 \;\; -8$ c) $-7 \;\; 0$ d) $-7 \;\; -4$

Solution

a) $-7 < 8$, since -7 is to the left of 8 on the number line.
b) $-7 > -8$, since -7 is to the right of -8 on the number line.
c) $-7 < 0$, since -7 is to the left of 0 on the number line.
d) $-7 < -4$, since -7 is to the left of -4 on the number line. ∎

Addition of Integers

Addition of integers can be represented geometrically using a number line. To do so, begin at 0 on the number line. Represent the first addend (the first number to be added) by an arrow starting at 0. Draw the arrow to the right if the addend is positive. If the addend is negative, draw the arrow to the left. From the tip of the first arrow, draw a second arrow to represent the second addend. Draw the second arrow to the right or left, as just explained. The sum of the two integers is found at the tip of the second arrow.

Example 2 *Adding Integers*

Evaluate the following using the number line.
a) $3 + (-5)$ b) $-1 + (-4)$ c) $-6 + 4$ d) $3 + (-3)$

Solution

a)
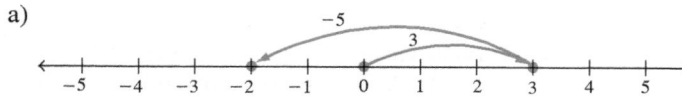
Thus, $3 + (-5) = -2$.

b)
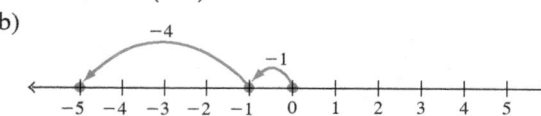
Thus, $-1 + (-4) = -5$.

c)
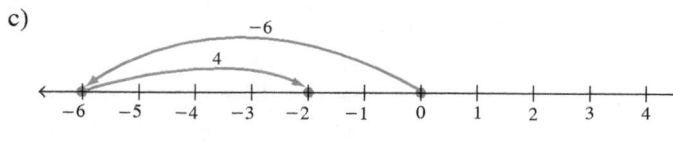
Thus, $-6 + 4 = -2$.

d)

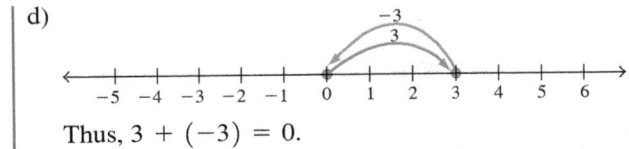

Thus, $3 + (-3) = 0$.

In Example 2(d), the number -3 is said to be the additive inverse of 3 and 3 is said to be the additive inverse of -3 because their sum is 0. In general, the *additive inverse* of the number n is $-n$, since $n + (-n) = 0$. Inverses are discussed more formally in Chapter 9.

Notice in the paragraph above we used the letter n to make a general statement about additive inverses. In mathematics, and throughout this textbook, we will often use letters called **variables** to represent numbers. Variables are usually shown in italics.

Subtraction of Integers

Any subtraction problem can be rewritten as an addition problem. To do so, we use the following rule of subtraction.

> **Subtraction**
> $$a - b = a + (-b)$$

The rule for subtraction indicates that to subtract b from a, *add* the additive inverse of b to a. For example,

$$3 - 5 = 3 + (-5)$$

    ↑   ↑ ↑

Subtraction Addition   Additive inverse of 5

Now we can determine the value of $3 + (-5)$.

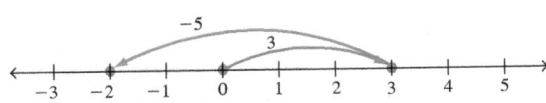

Thus, $3 - 5 = 3 + (-5) = -2$.

Adding and
Subtracting Integers

Example 3 *Subtracting Integers*

Evaluate $-4 - (-3)$ using the number line.

Solution We are subtracting -3 from -4. The additive inverse of -3 is 3; therefore, we add 3 to -4. We now add $-4 + 3$ on the number line to obtain the answer -1.

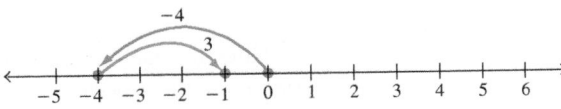

Thus, $-4 - (-3) = -4 + 3 = -1$.

In Example 3, we determined that $-4 - (-3) = -4 + 3$. In general, $a - (-b) = a + b$. As you get more proficient in working with integers, you should be able to answer questions involving them without drawing a number line.

Example 4 *Subtracting: Adding the Inverse*

Evaluate:

a) $-5 - 3$ b) $-5 - (-3)$ c) $5 - (-3)$ d) $5 - 3$

Solution

a) $-5 - 3 = -5 + (-3) = -8$ b) $-5 - (-3) = -5 + 3 = -2$
c) $5 - (-3) = 5 + 3 = 8$ d) $5 - 3 = 5 + (-3) = 2$ ∎

Example 5 *Elevation Difference*

The highest point in the United States is Denali in Alaska at a height of 20,237 ft above sea level. The lowest point in the United States is Death Valley in California and Nevada at a depth of 282 ft below sea level (-282 ft). Determine the vertical height difference between Denali and Death Valley.

Solution We obtain the vertical difference by subtracting the lower elevation from the higher elevation.

$$20{,}237 - (-282) = 20{,}237 + 282 = 20{,}519$$

The vertical difference is 20,519 ft.   ∎

Multiplication of Integers

The multiplication property of zero is important in our discussion of multiplication of integers. It indicates that the product of 0 and any number is 0.

> **Multiplication Property of Zero**
> $$a \cdot 0 = 0 \cdot a = 0$$

We will develop the rules for multiplication of integers using number patterns. The four possible cases are

1. positive integer $\times$ positive integer,
2. positive integer $\times$ negative integer,
3. negative integer $\times$ positive integer, and
4. negative integer $\times$ negative integer.

CASE 1: *POSITIVE INTEGER $\times$ POSITIVE INTEGER* The product of two positive integers can be defined as repeated addition of a positive integer. Thus, $3 \cdot 2$ means $2 + 2 + 2$. This sum will always be positive. Thus, *a positive integer times a positive integer is a positive integer.*

CASE 2: *POSITIVE INTEGER $\times$ NEGATIVE INTEGER* Consider the following patterns:

$$3(3) = 9$$
$$3(2) = 6$$
$$3(1) = 3$$

Note that each time the second factor is reduced by 1, the product is reduced by 3. Continuing the process gives

$$3(0) = 0$$

What comes next?

$$3(-1) = -3$$
$$3(-2) = -6$$

The pattern indicates that *a positive integer times a negative integer is a negative integer.*

Four 4's

The game of Four 4's is a challenging way to learn about the operations on integers. In this game, you must use exactly four 4's, and no other digits, along with one or more of the operations of addition, subtraction, multiplication, and division* to write expressions.

You may also use as many grouping symbols (that is, parentheses and brackets) as you wish. The object of the game is to first write an expression that is equal to 0. Then write a second expression that is equal to 1, then write a third expression that is equal to 2, and so on up to 9. For example, one way to represent the number 2 is as follows: $\frac{4 \cdot 4}{4 + 4} = \frac{16}{8} = 2$. There are several other acceptable ways as well. For more information see www.mathsisfun.com/puzzles/four-fours.html. Solutions for the whole numbers 0 –9 can be found in the back of this book. See also Exercise 82 on page 228 for an expansion of the rules needed to represent larger whole numbers.

*This game will be expanded to include other operations such as exponents and square roots later in the book.

We can confirm this result by using the number line. The expression $3(-2)$ means $(-2) + (-2) + (-2)$. Adding $(-2) + (-2) + (-2)$ on the number line, we obtain a sum of -6.

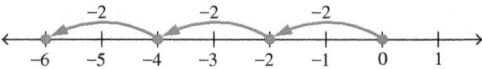

CASE 3: ***NEGATIVE INTEGER × POSITIVE INTEGER*** A procedure similar to that used in case 2 will indicate that *a negative integer times a positive integer is a negative integer.*

CASE 4: ***NEGATIVE INTEGER × NEGATIVE INTEGER*** We have illustrated that a positive integer times a negative integer is a negative integer. We make use of this fact in the following pattern:

$$4(-4) = -16$$
$$3(-4) = -12$$
$$2(-4) = -8$$
$$1(-4) = -4$$

In this pattern, each time the first factor is decreased by 1, the product is increased by 4. Continuing this process gives

$$0(-4) = 0$$
$$(-1)(-4) = 4$$
$$(-2)(-4) = 8$$

This pattern illustrates that *a negative integer times a negative integer is a positive integer.*

The examples were restricted to integers. The rules for multiplication, however, can be used for any numbers. We summarize them as follows.

> **Rules for Multiplication**
> 1. The product of two numbers with *like signs* (positive × positive or negative × negative) is a *positive number.*
> 2. The product of two numbers with *unlike signs* (positive × negative or negative × positive) is a *negative number.*

Example 6 *Multiplying Integers*

Evaluate
a) $5 \cdot 6$ b) $5 \cdot (-6)$ c) $(-5) \cdot 6$ d) $(-5) \cdot (-6)$

Solution

a) $5 \cdot 6 = 30$ b) $5 \cdot (-6) = -30$
c) $(-5) \cdot 6 = -30$ d) $(-5) \cdot (-6) = 30$ ∎

Division of Integers

You may already realize that a relationship exists between multiplication and division.

$$6 \div 2 = 3 \quad \text{means that} \quad 3 \cdot 2 = 6$$
$$\frac{20}{10} = 2 \quad \text{means that} \quad 2 \cdot 10 = 20$$

Timely Tip

Why can't we divide by 0? Suppose we are trying to determine the quotient $\frac{5}{0}$. If such an expression were defined, then we could write $\frac{5}{0} = x$. If true, this would mean that $5 = 0 \cdot x$, or $5 = 0$, which is false. Thus, there is no number, x, that makes the equation $\frac{5}{0} = x$ true. Therefore, we say a quotient of any nonzero* number divided by zero is *undefined*.

These examples demonstrate that division is the reverse process of multiplication.

Division

For any a, b, and c where $b \neq 0$, $\frac{a}{b} = c$ means that $c \cdot b = a$.

In the definition of division, note that we cannot divide by 0 (see the *Timely Tip* at left). The quotient of a nonzero number divided by 0 is said to be *undefined*. For example, $\frac{5}{0}$ is undefined.

Next, we discuss the four possible cases for division of integers, which are similar to those for multiplication.

CASE 1: *POSITIVE INTEGER ÷ POSITIVE INTEGER* *A positive integer divided by a positive integer is positive.*

$$\frac{6}{2} = 3, \quad \text{since} \quad 3(2) = 6$$

CASE 2: *POSITIVE INTEGER ÷ NEGATIVE INTEGER* *A positive integer divided by a negative integer is negative.*

$$\frac{6}{-2} = -3, \quad \text{since} \quad (-3)(-2) = 6$$

CASE 3: *NEGATIVE INTEGER ÷ POSITIVE INTEGER* *A negative integer divided by a positive integer is negative.*

$$\frac{-6}{2} = -3, \quad \text{since} \quad (-3)(2) = -6$$

CASE 4: *NEGATIVE INTEGER ÷ NEGATIVE INTEGER* *A negative integer divided by a negative integer is positive.*

$$\frac{-6}{-2} = 3, \quad \text{since} \quad 3(-2) = -6$$

The examples were restricted to integers. The rules for division, however, can be used for any numbers. You should realize that division of integers does not always result in an integer. The rules for division are summarized as follows.

Rules for Division

1. The quotient of two numbers with *like signs* (positive ÷ positive or negative ÷ negative) is a *positive number*.
2. The quotient of two numbers with *unlike signs* (positive ÷ negative or negative ÷ positive) is a *negative number*.

Example 7 *Dividing Integers*

Evaluate.

a) $\frac{63}{9}$ b) $\frac{-63}{9}$ c) $\frac{63}{-9}$ d) $\frac{-63}{-9}$

Solution

a) $\frac{63}{9} = 7$ b) $\frac{-63}{9} = -7$ c) $\frac{63}{-9} = -7$ d) $\frac{-63}{-9} = 7$ ∎

*The expression $\frac{0}{0}$ is considered *indeterminate*.

Exponents

An understanding of exponents is important in mathematics. In the expression 5^2, the 2 is referred to as the *exponent* and the 5 is referred to as the *base*. We read 5^2 as 5 to the second power, or 5 squared, which means

$$5^2 = \underbrace{5 \cdot 5}_{2 \text{ factors of } 5}$$

The number 5 to the third power, or 5 cubed, written 5^3, means

$$5^3 = \underbrace{5 \cdot 5 \cdot 5}_{3 \text{ factors of } 5}$$

In general, the number b to the nth power, written b^n, means

$$b^n = \underbrace{b \cdot b \cdot b \cdot \cdots \cdot b}_{n \text{ factors of } b}$$

Example 8 *Evaluating the Power of a Number*

Evaluate.

a) 5^2 b) $(-3)^2$ c) 3^4 d) 1^{1000} e) 1000^1

Solution

a) $5^2 = 5 \cdot 5 = 25$

b) $(-3)^2 = (-3) \cdot (-3) = 9$

c) $3^4 = 3 \cdot 3 \cdot 3 \cdot 3 = 81$

d) $1^{1000} = 1$

  The number 1 multiplied by itself any number of times equals 1.

e) $1000^1 = 1000$

  Any number with an exponent of 1 equals the number itself.   ∎

In general, $-x^n$ means take the opposite of x^n, or $-1 \cdot x^n$. For example, -5^2 means take the opposite of 5^2 or $-1 \cdot 5^2$ to obtain an answer of -25

Example 9 *The Importance of Parentheses*

Evaluate.

a) $(-2)^4$ b) -2^4 c) $(-2)^5$ d) -2^5

Solution

a) $(-2)^4 = (-2)(-2)(-2)(-2) = 4(-2)(-2) = -8(-2) = 16$

b) -2^4 means take the opposite of 2^4 or $-1 \cdot 2^4$.
  $-1 \cdot 2^4 = -1 \cdot 2 \cdot 2 \cdot 2 \cdot 2 = -1 \cdot 16 = -16$

c) $(-2)^5 = (-2)(-2)(-2)(-2)(-2) = 4(-2)(-2)(-2) = -8(-2)(-2)$
  $= 16(-2) = -32$

d) $-2^5 = -1 \cdot 2^5 = -1 \cdot 2 \cdot 2 \cdot 2 \cdot 2 \cdot 2 = -1 \cdot 32 = -32$ ∎

From Example 9, we can see that $(-x)^n \neq -x^n$ when n is an even natural number and that $(-x)^n = -x^n$ when n is an odd natural number.

Order of Operations

Now that we have introduced exponents, we can present the *order of operations*. Can you evaluate the expression $2 + 5 \cdot 4$? Is it 28? Or is it 22? To answer this, you must know the order in which to perform the indicated operations.

Timely Tip

When an expression involves a negative number and an exponent, we have to be careful when evaluating the expression. Let's evaluate the two expressions $(-7)^2$ and -7^2. The location of the negative sign is very important. In general,

$$(-x)^n = \underbrace{(-x)(-x)(-x)\cdots(-x)}_{n \text{ factors of } (-x)},$$

and

$$-x^n = -1 \cdot x^n.$$

Therefore,

$$(-7)^2 = (-7)(-7) = 49$$
$$-7^2 = -1 \cdot 7^2 = -1 \cdot 49 = -49.$$

> **PROCEDURE   ORDER OF OPERATIONS**
>
> 1. First, perform all operations within parentheses or other grouping symbols.
> 2. Next, evaluate all exponents.
> 3. Next, perform all multiplications and divisions going from left to right.
> 4. Finally, perform all additions and subtractions going from left to right.

In the expression $2 + 5 \cdot 4$, since multiplication is performed before addition:

$$2 + 5 \cdot 4 = 2 + 20 = 22$$

Example 10   *Using the Order of Operations*

Evaluate $100 - 3 \cdot 5^2 + 8$.

Solution   We will evaluate this expression using the order of operations.

$$100 - 3 \cdot 5^2 + 8$$
$$= 100 - 3 \cdot 25 + 8$$
$$= 100 - 75 + 8$$
$$= 25 + 8$$
$$= 33$$

Example 11   *Using the Order of Operations*

Evaluate $(41 + 23) \div 8 \cdot (13 - 15)^2$.

Solution   We begin by evaluating the expressions within the parentheses.

$$(41 + 23) \div 8 \cdot (13 - 15)^2$$
$$= 64 \div 8 \cdot (-2)^2$$
$$= 64 \div 8 \cdot 4$$
$$= 8 \cdot 4$$
$$= 32$$

SECTION 5.2   *Exercises*

Warm Up Exercises

In Exercises 1–4, fill in the blanks with an appropriate word, phrase, or symbol(s).

1. The set of numbers $\{0, 1, 2, 3, 4, \ldots\}$ is known as the _____ numbers.

2. The rule for subtraction indicates that to subtract b from a, _____ the additive inverse of b to a.

3. a) The product of two numbers with *like* signs is a(n) _____ number.

b) The product of two numbers with *unlike* signs is a(n) _____ number.

4. a) The quotient of two numbers with *unlike* signs is a(n) _____ number.

b) The quotient of two numbers with *like* signs is a(n) _____ number.

In Exercises 5 and 6, insert either $>$ or $<$ in the shaded area between the paired numbers to make the statement correct.

5. a) $-3 \quad 2$

b) $-3 \quad -2$

c) $-3 \quad 0$

d) $-3 \quad -4$

6. a) $-5 \quad -8$

b) $-5 \quad 8$

c) $-5 \quad -1$

d) $-5 \quad 0$

Practice the Skills

In Exercises 7–10, write the numbers in increasing order from left to right.

7. $0, -3, 3, -6, 6, -9$

8. $100, -10, 1, 0, -1, 10$

9. $-5, -2, -3, -1, -4, -6$

10. $106, 33, -47, -108, 72, -76$

In Exercises 11–18, evaluate the expression.

11. $-4 + 7$ **12.** $10 + (-3)$

13. $-9 + 6$ **14.** $-4 + (-4)$

15. $6 + (-11) + 0$ **16.** $2 + 5 + (-4)$

17. $(-3) + (-4) + 9$ **18.** $8 + (-3) + (-2)$

In Exercises 19–26, evaluate the expression.

19. $4 - 6$ **20.** $-9 - 4$

21. $-3 - 5$ **22.** $8 - (-2)$

23. $-5 - (-3)$ **24.** $-4 - 4$

25. $14 - 20$ **26.** $8 - (-3)$

In Exercises 27–34, evaluate the expression.

27. $-2 \cdot 5$ **28.** $7(-3)$

29. $(-4)(-4)$ **30.** $-10(2)$

31. $(-8)(-2) \cdot 6$ **32.** $4(-5)(-6)$

33. $(5 \cdot 6)(-2)$ **34.** $(-9)(-1)(-2)$

In Exercises 35–42, evaluate the expression.

35. $-12 \div (-6)$ **36.** $-72 \div 9$

37. $\dfrac{20}{-20}$ **38.** $\dfrac{-90}{9}$

39. $56/-8$ **40.** $-75/15$

41. $-210/14$ **42.** $186/-6$

In Exercises 43–50, evaluate the expression.

43. a) 3^2 **b)** 2^3

44. a) 2^5 **b)** 5^2

45. a) $(-5)^2$ **b)** -5^2

46. a) -3^2 **b)** $(-3)^2$

47. a) -2^4 **b)** $(-2)^4$

48. a) $(-3)^4$ **b)** -3^4

49. a) -4^3 **b)** $(-4)^3$

50. a) 2020^1 **b)** 1^{2020}

In Excercises 51–60, evaluate the expression using the order of operations.

51. $15 - 8 \cdot 2$

52. $-15 + 25 \div 5$

53. $(24 \div 4 \cdot 2) - (4 - 7 + 5)$

54. $(50 \div 5 \cdot 2) + (6 \div 3 \cdot 2 - 1 \cdot 2)$

55. $(-17 + 21)^2 \cdot 3 \div (13 - 18 \div 2)$

56. $(13 - 17)^2 \cdot 3 \div (-4 \cdot 8 \div 2)$

57. $-3[(-5^2 + 5^3) \div (-2^2 + 2^3)]$

58. $-5[(-2^3 - 2^2) \cdot 2^4 \div 4^2]$

59. $\dfrac{-[4 - (6 - 12)^2]}{[9 \div 3 + 4]^2 - 34/2}$

60. $\dfrac{2[2^3 + (10 - 4/2)^2]}{24 + 3^2 \cdot 5 - [3^3 \div (-10 + 1)]}$

Problem Solving

61. *Submarine Depth* The *USS Virginia* submarine is at an underwater depth of 600 feet. It then rises 200 feet, dives 400 feet, and finally dives another 300 feet. What is the *USS Virginia's* final depth?

62. *Dow Jones Industrial Average* On February 20, 2019, the Dow Jones Industrial Average (DJIA) opened at 25,872 points. During the day, it gained 82 points. On February 21 it lost 103 points. On February 22 it gained 181 points. Determine the DJIA at closing on February 22, 2019.

▲ The New York Stock Exchange

63. *Elevation Difference* Mount Everest in the Himalayas is the highest point on Earth. It is 29,035 ft above sea level. The Mariana Trench in the Pacific Ocean is the lowest point on Earth. It is 36,198 ft below sea level. Determine the vertical height difference between Mount Everest and the Mariana Trench.

64. *Extreme Temperatures* The hottest temperature ever recorded in the United States was 134°F, which occurred at Greenland Ranch, California, in Death Valley on July 10, 1913. The coldest temperature ever recorded in the United States was −80°F, which occurred at Prospect Creek Camp, Alaska, in the Endicott Mountains on January 23, 1971. Determine the difference between these two temperatures.

65. *Time Zone Calculations* Part of a World Standard Time Zones chart used by airlines and the United States Navy is shown. The scale along the bottom is just like a number line with the integers −12, −11, . . . , 11, 12 on it. Note that only part of the scale from −12 to 12 is shown.

a) Determine the difference in time between Nairobi (zone +3) and Los Angeles (zone −8).

b) Determine the difference in time between Casablanca (zone −1) and Rio de Janeiro (zone −3).

c) Determine the difference in time between Boston (zone −5) and Tokyo (zone +9).

d) Determine the difference in time between Puerto Vallarta (zone −7) and Seoul (zone +8).

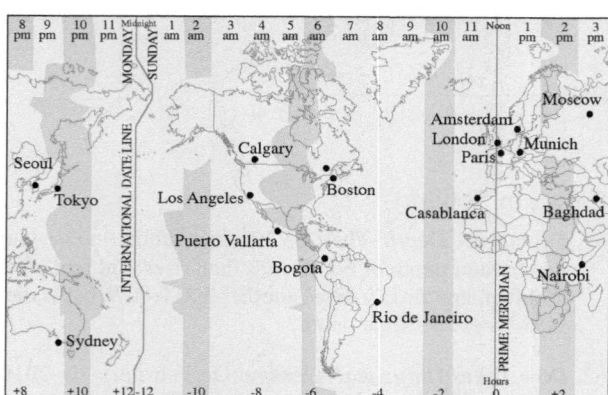

66. *Bus Stops* A city bus currently has 23 people on it. At the next stop, 10 people get off the bus, and 8 people get on the bus. At the next stop, 15 people get off the bus, and 6 people get on the bus. At the next stop, 2 people get off the bus, and 17 people get on the bus. At this point, how many people are on the bus?

Concept/Writing Exercises

In Exercises 67–76, determine whether the statement is true or false. Modify each false statement to make it a true statement.

67. Every whole number is an integer.

68. Every integer is a whole number.

69. The difference of any two negative integers is a negative integer.

70. The sum of any two negative integers is a negative integer.

71. The product of any two positive integers is a positive integer.

72. The difference of a positive integer and a negative integer is always a negative integer.

73. The quotient of a negative integer and a positive integer is always a negative integer.

74. The quotient of any two negative integers is a negative integer.

75. The sum of a positive integer and a negative integer is always a positive integer.

76. The product of a positive integer and a negative integer is always a positive integer.

77. Explain why the quotient of a nonzero number divided by 0 is undefined.

78. Explain why $\dfrac{a}{b} = \dfrac{-a}{-b}$.

Challenge Problems/Group Activities

79. Determine the quotient:
$$\frac{-1 + 2 - 3 + 4 - 5 + \cdots - 99 + 100}{1 - 2 + 3 - 4 + 5 - \cdots + 99 - 100}$$

80. *Pentagonal Numbers* Triangular numbers and square numbers were introduced in the Section 1.1 Exercises. There are also **pentagonal numbers**, which were also studied by the Greeks. Four pentagonal numbers are 1, 5, 12, and 22.

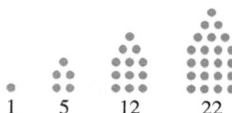

a) Determine the next three pentagonal numbers.

b) Describe a procedure to determine the next five pentagonal numbers without drawing the figures.

c) Is 72 a pentagonal number? Explain how you determined your answer.

81. Place the appropriate plus or minus signs between each digit so that the total will equal 1.
$$0 \quad 1 \quad 2 \quad 3 \quad 4 \quad 5 \quad 6 \quad 7 \quad 8 \quad 9 = 1$$

Recreational Mathematics

82. *Four 4's* Refer to the *Recreational Mathematics* box on page 223.

a) Use the rules given on page 223 to represent the following whole numbers: 12, 15, 16, 17, 20.

b) We will now change our rules to allow the number 44 to count as two of the four 4's. Use the number 44 and two other fours to represent the whole number 10.

Research Activity

83. Write a report on the history of the number 0 in the Hindu–Arabic numeration system.

The Rational Numbers

Upon completion of this section, you will be able to:

- Reduce fractions to lowest terms.
- Convert mixed numbers to improper fractions and vice versa.
- Express rational numbers as terminating or repeating decimal numbers.
- Convert decimal numbers to fractions.
- Multiply and divide fractions.
- Add and subtract fractions.

Among the ingredients called for in a recipe for chocolate chip cookies are $2\frac{1}{4}$ cups flour, $\frac{1}{2}$ teaspoon salt, $1\frac{1}{3}$ teaspoons baking soda, $\frac{3}{4}$ cup sugar, and $\frac{3}{4}$ cup brown sugar. How much of each of these ingredients would you need to use if you wish to double the recipe? In this section, we will review the use of fractions such as those in this chocolate chip cookie recipe. We will also learn about the operations of addition, subtraction, multiplication, and division using fractions.

Why This Is Important Fractions are everywhere in our everyday lives. Kitchen measurements, tool sizes, and many other real-life applications all involve fractions.

We introduced the number line and discussed the integers in Section 5.2. The numbers that fall between the integers on the number line are either rational or irrational numbers. In this section, we discuss the rational numbers, and in Section 5.4, we discuss the irrational numbers.

Any number that can be expressed as a quotient of two integers (denominator not 0) is a rational number.

> **Definition: Rational Numbers**
> The set of **rational numbers**, denoted by Q, is the set of all numbers of the form $\frac{p}{q}$, where p and q are integers and $q \neq 0$.

Recall from Section 5.2 that division by zero is *undefined*. For this reason, the denominator, q, of a rational number, $\frac{p}{q}$, cannot be zero.

The following numbers are examples of rational numbers:

$$\frac{1}{3}, \quad \frac{3}{4}, \quad -\frac{7}{8}, \quad 1\frac{2}{3}, \quad 2, \quad 0, \quad \frac{15}{7}$$

The integers 2 and 0 are rational numbers because each can be expressed as the quotient of two integers: $2 = \frac{2}{1}$ and $0 = \frac{0}{1}$. In fact, every integer n is a rational number because it can be written in the form of $\frac{n}{1}$.

Numbers such as $\frac{1}{3}$ and $-\frac{7}{8}$ are also called *fractions*. The number above the fraction line is called the *numerator*, and the number below the fraction line is called the *denominator*.

Reducing Fractions

Sometimes the numerator and denominator in a fraction have a common divisor (or common factor). For example, both the numerator and denominator of the fraction $\frac{6}{10}$ have the common divisor 2. When a numerator and denominator have a common divisor, we can *reduce the fraction to its lowest terms*.

A fraction is said to be in its *lowest terms* (or reduced) when the numerator and denominator are relatively prime (that is, have no common divisors other than 1). To reduce a fraction to its lowest terms, divide both the numerator and the denominator by the greatest common divisor. Recall that a procedure for determining the greatest common divisor was discussed in Section 5.1.

The fraction $\frac{6}{10}$ is reduced to its lowest terms as follows.

$$\frac{6}{10} = \frac{6 \div 2}{10 \div 2} = \frac{3}{5}$$

Example 1 *Reducing a Fraction to Lowest Terms*

Reduce $\frac{36}{90}$ to its lowest terms.

Solution On page 213 in Example 4 of Section 5.1, we determined that the GCD of 36 and 90 is 18. Divide the numerator and the denominator by GCD, 18.

$$\frac{36}{90} = \frac{36 \div 18}{90 \div 18} = \frac{2}{5}$$

Since there are no common divisors of 2 and 5 other than 1, the fraction $\frac{2}{5}$ is in its lowest terms. ∎

Mixed Numbers and Improper Fractions

Consider the number $2\frac{3}{4}$. It is an example of a *mixed number*. It is called a mixed number because it consists of an integer, 2, and a fraction, $\frac{3}{4}$. The mixed number $2\frac{3}{4}$ means $2 + \frac{3}{4}$. The mixed number $-4\frac{1}{4}$ means $-(4 + \frac{1}{4})$. Rational numbers greater than 1 or less than -1 that are not integers may be represented as mixed numbers, or as *improper fractions*. An improper fraction is a fraction whose numerator is greater than its denominator. An example of an improper fraction is $\frac{8}{5}$. Figure 5.6 shows both mixed numbers and the corresponding improper fractions indicated on a number line. In this section, we show how to convert mixed numbers to improper fractions and vice versa.

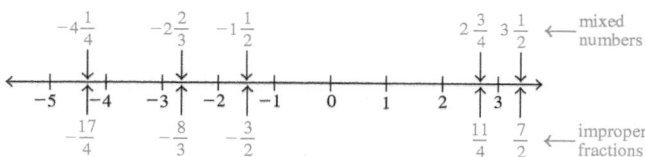

Figure 5.6

We begin by limiting our discussion to positive mixed numbers and positive improper fractions.

PROCEDURE CONVERTING A POSITIVE MIXED NUMBER TO AN IMPROPER FRACTION

1. Multiply the denominator of the fraction in the mixed number by the integer preceding it.

2. Add the product obtained in Step 1 to the numerator of the fraction in the mixed number. This sum is the numerator of the improper fraction we are seeking. The denominator of the improper fraction we are seeking is the same as the denominator of the fraction in the mixed number.

Example **2** *Converting Mixed Numbers to Improper Fractions*

Convert the following mixed numbers to improper fractions.

a) $3\dfrac{5}{6}$ b) $5\dfrac{2}{7}$

Solution

a) $3\dfrac{5}{6} = \dfrac{6 \cdot 3 + 5}{6} = \dfrac{18 + 5}{6} = \dfrac{23}{6}$

b) $5\dfrac{2}{7} = \dfrac{7 \cdot 5 + 2}{7} = \dfrac{35 + 2}{7} = \dfrac{37}{7}$

Notice that both $\dfrac{23}{6}$ and $\dfrac{37}{7}$ have numerators that are larger than their denominators. Therefore, both numbers are improper fractions. ■

Now let's discuss converting a positive improper fraction to a mixed number.

PROCEDURE **CONVERTING A POSITIVE IMPROPER FRACTION TO A MIXED NUMBER**

1. Divide the numerator by the denominator. Identify the quotient and the remainder.
2. The quotient obtained in Step 1 is the integer part of the mixed number. The remainder is the numerator of the fraction in the mixed number. The denominator in the fraction of the mixed number will be the same as the denominator in the original fraction.

Example **3** *From Improper Fraction to Mixed Number*

Convert the following improper fractions to mixed numbers.

a) $\dfrac{8}{5}$ b) $\dfrac{225}{8}$

Solution

a) Divide the numerator, 8, by the denominator, 5.

$$
\begin{array}{r}
1 \quad \leftarrow \text{Quotient} \\
\text{Divisor} \rightarrow \ 5\overline{)8} \quad \leftarrow \text{Dividend} \\
5 \\
\overline{3} \quad \leftarrow \text{Remainder}
\end{array}
$$

Therefore,

$$\dfrac{8}{5} = 1\dfrac{3}{5} \quad \begin{array}{l} \leftarrow \text{Remainder} \\ \leftarrow \text{Divisor} \end{array}$$

Quotient ↓ (above the 1)

The mixed number is $1\dfrac{3}{5}$.

b) Divide the numerator, 225, by the denominator, 8.

$$
\begin{array}{r}
28 \quad \leftarrow \text{Quotient} \\
\text{Divisor} \rightarrow \; 8\overline{)225} \quad \leftarrow \text{Dividend} \\
\underline{16} \quad\quad \\
65 \quad\quad \\
\underline{64} \quad\quad \\
1 \quad \leftarrow \text{Remainder}
\end{array}
$$

Therefore,

$$
\overset{\text{Quotient}}{\underset{\downarrow}{}}
$$

$$
\frac{225}{8} = 28\frac{1}{8} \quad \begin{array}{l} \leftarrow \text{Remainder} \\ \leftarrow \text{Divisor} \end{array}
$$

The mixed number is $28\frac{1}{8}$. ∎

Up to this point, we have only worked with positive mixed numbers and positive improper fractions. When converting a negative mixed number to an improper fraction, or a negative improper fraction to a mixed number, it is best to ignore the negative sign temporarily. Perform the calculation as described earlier and then reattach the negative sign.

Example 4 *Negative Mixed Numbers and Improper Fractions*

a) Convert $-3\frac{5}{6}$ to an improper fraction.

b) Convert $-\frac{8}{5}$ to a mixed number.

Solution

a) First ignore the negative sign and examine $3\frac{5}{6}$. We learned in Example 2(a) that $3\frac{5}{6} = \frac{23}{6}$. To convert $-3\frac{5}{6}$ we reattach the negative sign. Thus, $-3\frac{5}{6} = -\frac{23}{6}$

b) We learned in Example 3(a) that $\frac{8}{5} = 1\frac{3}{5}$. Therefore, $-\frac{8}{5} = -1\frac{3}{5}$. ∎

Terminating or Repeating Decimal Numbers

Note the following important property of the rational numbers. *Every rational number when expressed as a decimal number will be either a terminating or a repeating decimal number.*

Examples of terminating decimal numbers are 0.5, 0.75, and 4.65. Examples of repeating decimal numbers are $0.333\ldots$, $0.2323\ldots$, and $8.13456456\ldots$. One way to indicate that a number or group of numbers repeats is to place a bar above the number or group of numbers that repeats. Thus, $0.333\ldots$ may be written $0.\overline{3}$, $0.2323\ldots$ may be written $0.\overline{23}$, and $8.13456456\ldots$ may be written $8.13\overline{456}$.

Example 5 *Terminating Decimal Numbers*

Show that the following rational numbers can be expressed as terminating decimal numbers.

a) $\dfrac{3}{4}$ b) $-\dfrac{7}{20}$ c) $\dfrac{125}{40}$

Solution To express the rational number in decimal form, divide the numerator by the denominator. If you use a calculator or long division, you will see that each fraction results in a terminating decimal number.

a) $\dfrac{3}{4} = 0.75$ b) $-\dfrac{7}{20} = -0.35$ c) $\dfrac{125}{40} = 3.125$ ■

Example 6 *Repeating Decimal Numbers*

Show that the following rational numbers can be expressed as repeating decimal numbers.

a) $\dfrac{2}{3}$ b) $\dfrac{14}{99}$ c) $1\dfrac{5}{36}$

Solution If you use a calculator or long division, you will see that each fraction results in a repeating decimal number.

a) $2 \div 3 = 0.6666\ldots$ or $0.\overline{6}$

b) $14 \div 99 = 0.141414\ldots$ or $0.\overline{14}$

c) $1\frac{5}{36} = 1 + \frac{5}{36} = 1 + 0.138888\ldots = 1.138888\ldots$ or $1.13\overline{8}$ ■

Note that in each part of Example 6, the quotient when expressed as a decimal number has no final digit and continues indefinitely. Each number is a repeating decimal number.

When a fraction is converted to a decimal number, the maximum number of digits that can repeat is $n - 1$, where n is the denominator of the fraction. For example, when $\frac{2}{7}$ is converted to a decimal number, the maximum number of digits that can repeat is $7 - 1$, or 6.

Converting Decimal Numbers to Fractions

We can convert a terminating or repeating decimal number into a quotient of integers. The explanation of the procedure will refer to the positional values to the right of the decimal point, as illustrated here:

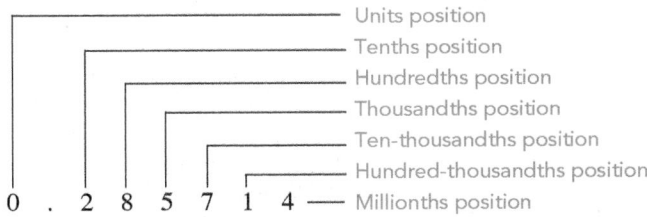

Example 7 demonstrates how to convert from a decimal number to a fraction.

Example 7 *Converting Decimal Numbers to Fractions*

Convert the following terminating decimal numbers to a quotient of integers. Reduce the quotient to lowest terms.

a) 0.7   b) 0.35   c) 0.016   d) 3.41

Solution When converting a terminating decimal number to a quotient of integers, we observe the last digit to the right of the decimal point. The position of this digit will indicate the denominator of the quotient of integers.

a) $0.7 = \frac{7}{10}$ because the 7 is in the tenths position. The fraction $\frac{7}{10}$ is in its lowest term.

b) $0.35 = \frac{35}{100}$ because the right-most digit, 5, is in the hundredths position. We finish by reducing the fraction to lowest terms:

$$\frac{35}{100} = \frac{35 \div 5}{100 \div 5} = \frac{7}{20}$$

c) $0.016 = \frac{16}{1000}$ because the right-most digit, 6, is in the thousandths position. We finish by reducing the fraction to lowest terms:

$$\frac{16}{1000} = \frac{16 \div 8}{1000 \div 8} = \frac{2}{125}$$

d) We begin by writing 3.41 as the mixed number $3\frac{41}{100}$ because the right-most digit, 1, is in the hundredths position. To write the mixed number as a ratio of two integers, we follow the procedure shown in Example 2, on page 231:

$$3\frac{41}{100} = \frac{100 \cdot 3 + 41}{100} = \frac{341}{100}$$

The fraction $\frac{341}{100}$ cannot be reduced further. ∎

Converting a repeating decimal number to a quotient of integers is more difficult than converting a terminating decimal number to a quotient of integers. To do so, we must "create" another repeating decimal number with the same repeating digits so that when one repeating decimal number is subtracted from the other repeating decimal number, the difference will be a whole number. To create a number with the same repeating digits, multiply the original repeating decimal number by an appropriate power of 10. Examples 8 through 10 demonstrate this procedure.

Example 8   *Converting a Repeating Decimal Number to a Fraction*

Convert $0.\overline{3}$ to a quotient of integers.

Solution $0.\overline{3} = 0.3\overline{3} = 0.33\overline{3}$, and so on.

Let the original repeating decimal number be n; thus, $n = 0.\overline{3}$. Because one digit repeats, we multiply both sides of the equation by 10, which gives $10n = 3.\overline{3}$. Then we subtract.

$$10n = 3.\overline{3}$$
$$-\quad n = 0.\overline{3}$$
$$\overline{9n = 3.0}$$

Note that $10n - n = 9n$ and $3.\overline{3} - 0.\overline{3} = 3.0$.

Next, we solve for n by dividing both sides of the equation by 9.

$$\frac{9n}{9} = \frac{3.0}{9}$$

$$n = \frac{3}{9} = \frac{1}{3}$$

Therefore, $0.\overline{3} = \frac{1}{3}$. Evaluate $1 \div 3$ on a calculator now and see what value you get. ∎

Example 9 *Converting a Repeating Decimal Number to a Fraction*

Convert $0.\overline{35}$ to a quotient of integers.

Solution Let $n = 0.\overline{35}$. Since two digits repeat, multiply both sides of the equation by 100. Thus, $100n = 35.\overline{35}$. Now we subtract n from $100n$.

$$
\begin{array}{r}
100n = 35.\overline{35} \\
-\quad n = 0.\overline{35} \\
\hline
99n = 35
\end{array}
$$

Finally, we divide both sides of the equation by 99.

$$\frac{99n}{99} = \frac{35}{99}$$

$$n = \frac{35}{99}$$

Therefore, $0.\overline{35} = \frac{35}{99}$. Evaluate $35 \div 99$ on a calculator now and see what value you get. ∎

Example 10 *Converting a Repeating Decimal Number to a Fraction*

Convert $12.14\overline{2}$ to a quotient of integers.

Solution This example is different from the two preceding examples in that the repeating digit, 2, is not directly to the right of the decimal point. When this situation arises, use multiplication, as shown below, to move the decimal point to the right until the repeating terms are directly to its right. If the decimal point needs to be moved one place to the right, multiply the number by 10. If the decimal point needs to be moved two places to the right, multiply the number by 100, and so on. In this example, the decimal point must be moved two places to the right. Thus, the number must be multiplied by 100.

$$n = 12.14\overline{2}$$
$$100n = 100 \times 12.14\overline{2} = 1214.\overline{2}$$

Now proceed as in the previous two examples. Since one digit repeats, multiply both sides by 10.

$$100n = 1214.\overline{2}$$
$$10 \times 100n = 10 \times 1214.\overline{2}$$
$$1000n = 12142.\overline{2}$$

Now subtract $100n$ from $1000n$ so that the repeating part will drop out.

$$
\begin{array}{r}
1000n = 12{,}142.\overline{2} \\
-\quad 100n = \phantom{12{,}}1214.\overline{2} \\
\hline
900n = 10{,}928
\end{array}
$$

$$n = \frac{10{,}928}{900} = \frac{2732}{225}$$

Therefore, $12.14\overline{2} = \dfrac{2732}{225}$. Evaluate $2732 \div 225$ on a calculator now and see what value you get. ∎

Multiplication and Division of Fractions

The product of two fractions is determined by multiplying the numerators together and multiplying the denominators together.

Multiplication of Fractions

$$\frac{a}{b} \cdot \frac{c}{d} = \frac{a \cdot c}{b \cdot d} = \frac{ac}{bd}, \quad b \neq 0, \quad d \neq 0$$

Example 11 *Multiplying Fractions*

Evaluate.

a) $\dfrac{2}{5} \cdot \dfrac{9}{11}$ b) $\left(\dfrac{-2}{3}\right)\left(\dfrac{-4}{9}\right)$ c) $\left(1\dfrac{7}{8}\right)\left(2\dfrac{1}{4}\right)$

Solution

a) $\dfrac{2}{5} \cdot \dfrac{9}{11} = \dfrac{2 \cdot 9}{5 \cdot 11} = \dfrac{18}{55}$ b) $\left(\dfrac{-2}{3}\right)\left(\dfrac{-4}{9}\right) = \dfrac{(-2)(-4)}{(3)(9)} = \dfrac{8}{27}$

c) $\left(1\dfrac{7}{8}\right)\left(2\dfrac{1}{4}\right) = \dfrac{15}{8} \cdot \dfrac{9}{4} = \dfrac{15 \cdot 9}{8 \cdot 4} = \dfrac{135}{32} = 4\dfrac{7}{32}$ ∎

To divide fractions we make use of the *reciprocal* of a number. The *reciprocal* of any number is 1 divided by that number. The product of a number and its reciprocal must equal 1. Examples of some numbers and their reciprocals follow.

Number		Reciprocal		Product
3	·	$\dfrac{1}{3}$	=	1
$\dfrac{3}{5}$	·	$\dfrac{5}{3}$	=	1
-6	·	$-\dfrac{1}{6}$	=	1

To determine the quotient of two fractions, multiply the first fraction by the reciprocal of the second fraction.

Division of Fractions

$$\frac{a}{b} \div \frac{c}{d} = \frac{a}{b} \cdot \frac{d}{c} = \frac{ad}{bc}, \quad b \neq 0, \quad d \neq 0, \quad c \neq 0$$

Example 12 *Dividing Fractions*

Evaluate.

a) $\dfrac{2}{7} \div \dfrac{3}{5}$ b) $\left(-\dfrac{2}{9}\right) \div \dfrac{3}{13}$

Solution

a) $\dfrac{2}{7} \div \dfrac{3}{5} = \dfrac{2}{7} \cdot \dfrac{5}{3} = \dfrac{2 \cdot 5}{7 \cdot 3} = \dfrac{10}{21}$

b) $\left(-\dfrac{2}{9}\right) \div \dfrac{3}{13} = \dfrac{-2}{9} \cdot \dfrac{13}{3} = \dfrac{-2 \cdot 13}{9 \cdot 3} = \dfrac{-26}{27} = -\dfrac{26}{27}$ ∎

Addition and Subtraction of Fractions

Before we can add or subtract fractions, the fractions must have a common denominator. A common denominator is another name for a common multiple of the denominators. The *lowest common denominator (LCD)* is the least common multiple of the denominators.

To add or subtract two fractions with a common denominator, we add or subtract their numerators and retain the common denominator.

> ### Addition and Subtraction of Fractions
> $$\frac{a}{c} + \frac{b}{c} = \frac{a+b}{c}, \quad c \neq 0; \qquad \frac{a}{c} - \frac{b}{c} = \frac{a-b}{c}, \quad c \neq 0$$

Example 13 *Adding and Subtracting Fractions*

Evaluate.

a) $\dfrac{1}{8} + \dfrac{3}{8}$
b) $\dfrac{19}{24} - \dfrac{5}{24}$

Solution

a) $\dfrac{1}{8} + \dfrac{3}{8} = \dfrac{1+3}{8} = \dfrac{4}{8} = \dfrac{1}{2}$

b) $\dfrac{19}{24} - \dfrac{5}{24} = \dfrac{19-5}{24} = \dfrac{14}{24} = \dfrac{7}{12}$ ∎

Note that in Example 13, the denominators of the fractions being added or subtracted were the same; that is, they have a common denominator. *When adding or subtracting two fractions with unlike denominators, first rewrite each fraction with a common denominator. Then add or subtract the fractions.*

Writing fractions with a common denominator is accomplished with the *fundamental law of rational numbers.*

> ### Fundamental Law of Rational Numbers
> If a, b, and c are integers, with $b \neq 0$ and $c \neq 0$, then
> $$\frac{a}{b} = \frac{a}{b} \cdot \frac{c}{c} = \frac{a \cdot c}{b \cdot c}$$

The terms $\dfrac{a}{b}$ and $\dfrac{a \cdot c}{b \cdot c}$ are called *equivalent fractions*. For example, since $\dfrac{7}{12} = \dfrac{7 \cdot 5}{12 \cdot 5} = \dfrac{35}{60}$, the fractions $\dfrac{7}{12}$ and $\dfrac{35}{60}$ are equivalent fractions. We will see the importance of equivalent fractions in the next two examples.

Example 14 *Adding and Subtracting Fractions with Unlike Denominators*

Evaluate.

a) $\dfrac{13}{15} - \dfrac{5}{6}$
b) $\dfrac{1}{36} + \dfrac{1}{90}$

Solution

a) Using prime factorization (Section 5.1), we determine that the LCM of 15 and 6 is 30. We will therefore express each fraction as an equivalent fraction with a denominator of 30. First, consider $\dfrac{13}{15}$. To obtain a denominator of 30, we must multiply the denominator, 15, by 2. If the denominator is multiplied by 2, the numerator must also be multiplied by 2 to obtain an equivalent fraction. Next, consider $\frac{5}{6}$. To obtain a denominator of 30, we must multiply the denominator, 6, by 5. Therefore, we must multiply both numerator and denominator by 5 to obtain an equivalent fraction.

$$\frac{13}{15} - \frac{5}{6} = \left(\frac{13}{15} \cdot \frac{2}{2} \right) - \left(\frac{5}{6} \cdot \frac{5}{5} \right)$$

$$= \frac{26}{30} - \frac{25}{30}$$

$$= \frac{1}{30}$$

b) On page 214, in Example 6 of Section 5.1, we determined that the LCM of 36 and 90 is 180. Rewrite each fraction as an equivalent fraction using the LCM as the common denominator.

$$\frac{1}{36} + \frac{1}{90} = \left(\frac{1}{36} \cdot \frac{5}{5} \right) + \left(\frac{1}{90} \cdot \frac{2}{2} \right)$$

$$= \frac{5}{180} + \frac{2}{180}$$

$$= \frac{7}{180}$$ ∎

In Section 5.2 we introduced the order of operations, which is used to simplify expressions. We will use the order of operations in the next example.

Example 15 *Using the Order of Operations*

Evaluate the expression using the order of operations: $\dfrac{13}{15} - \dfrac{14}{45} \div \dfrac{7}{9}$

Solution Since division is performed before subtraction, we begin by performing the division. To divide $\frac{14}{45}$ by $\frac{7}{9}$, multiply by the reciprocal, $\frac{9}{7}$.

$$= \frac{13}{15} - \frac{14}{45} \div \frac{7}{9}$$

$$= \frac{13}{15} - \frac{14}{45} \cdot \frac{9}{7}$$

$$= \frac{13}{15} - \frac{\overset{2 \cdot 7}{\cancel{14}}}{\underset{5 \cdot 9}{\cancel{45}}} \cdot \frac{\overset{9}{\cancel{9}}}{\cancel{7}}$$ Write 14 as 2 · 7 and 45 as 5 · 9 and cross out common factors

$$= \frac{13}{15} - \frac{2}{5}$$

$$= \frac{13}{15} - \left(\frac{2}{5} \cdot \frac{3}{3} \right)$$ Rewrite $\frac{2}{5}$ with the LCD, 15.

$$= \frac{13}{15} - \frac{6}{15} = \frac{7}{15}$$ ∎

Example 16 *Rice Preparation*

Following are the instructions given on a box of instant rice. Determine the amount of (a) rice and water, (b) salt, and (c) butter or margarine needed to make 3 servings of rice.

Directions
1. Bring water, salt, and butter (or margarine) to a boil.
2. Stir in rice. Cover; remove from heat. Let stand 5 minutes. Fluff with fork.

To Make	Rice and Water (Equal Measures)	Salt	Butter or Margarine
2 servings	$\frac{2}{3}$ cup	$\frac{1}{4}$ tsp	1 tsp
4 servings	$1\frac{1}{3}$ cups	$\frac{1}{2}$ tsp	2 tsp

Solution Since 3 is halfway between 2 and 4, we can determine the amount of each ingredient by determining the average of the amount for 2 and 4 servings. To do so, we add the amounts for 2 servings and 4 servings and divide the sum by 2.

a) Rice and water: $\dfrac{\frac{2}{3} + 1\frac{1}{3}}{2} = \dfrac{\frac{2}{3} + \frac{4}{3}}{2} = \dfrac{\frac{6}{3}}{2} = \dfrac{2}{2} = 1$ cup

b) Salt: $\dfrac{\frac{1}{4} + \frac{1}{2}}{2} = \dfrac{\frac{1}{4} + \frac{2}{4}}{2} = \dfrac{\frac{3}{4}}{2} = \dfrac{3}{4} \cdot \dfrac{1}{2} = \dfrac{3}{8}$ tsp

c) Butter or margarine: $\dfrac{1 + 2}{2} = \dfrac{3}{2}$, or $1\frac{1}{2}$ tsp   ∎

The solution to Example 16 can be determined in other ways. Suggest two other procedures for solving the same problem.

SECTION 5.3   *Exercises*

Warm Up Exercises

In Exercises 1–10, fill in the blanks with an appropriate word, phrase, or symbol(s).

1. The set of rational numbers is the set of numbers of the form $\dfrac{p}{q}$, where p and q are _____ and $q \neq 0$.

2. For the rational number $\dfrac{p}{q}$, p is called the _____.

3. For the rational number $\dfrac{p}{q}$, q is called the _____.

4. Rational numbers such as $2\frac{3}{4}$ and $-1\frac{1}{2}$ are examples of _____ numbers.

5. Rational numbers such as $\frac{11}{4}$ and $-\frac{3}{2}$ are examples of _____ fractions.

6. The number $\frac{1}{2}$ can be represented as a(n) _____ decimal number.

7. The number $\frac{1}{3}$ can be represented as a(n) _____ decimal number.

8. In the decimal number 0.285714, the 2 is in the _____ position.

9. In the decimal number 0.285714, the 7 is in the _____ position.

10. The rational numbers $\frac{1}{2}$ and $\frac{5}{10}$ are examples of _____ fractions.

Practice the Skills

In Exercises 11–16, reduce each fraction to lowest terms.

11. $\dfrac{6}{8}$

12. $\dfrac{12}{30}$

13. $\dfrac{24}{42}$

14. $\dfrac{36}{56}$

15. $\dfrac{95}{125}$

16. $\dfrac{13}{221}$

In Exercises 17–22, convert each mixed number to an improper fraction.

17. $2\dfrac{5}{6}$

18. $4\dfrac{1}{3}$

19. $-5\dfrac{3}{4}$

20. $-7\dfrac{1}{5}$

21. $-4\dfrac{15}{16}$

22. $11\dfrac{9}{16}$

In Exercises 23–26, write the number of inches indicated by the arrows as an improper fraction.

23.

24.

25.

26.

In Exercises 27–32, convert each improper fraction to a mixed number.

27. $\dfrac{13}{8}$

28. $\dfrac{23}{6}$

29. $-\dfrac{46}{5}$

30. $-\dfrac{157}{12}$

31. $-\dfrac{878}{15}$

32. $\dfrac{1028}{21}$

In Exercises 33–40, express each rational number as terminating or repeating decimal number.

33. $\dfrac{2}{5}$

34. $\dfrac{5}{8}$

35. $\dfrac{1}{3}$

36. $\dfrac{1}{6}$

37. $\dfrac{3}{8}$

38. $\dfrac{23}{7}$

39. $\dfrac{13}{6}$

40. $\dfrac{115}{15}$

In Exercises 41–48, express each terminating decimal number as a quotient of two integers. Reduce the quotient to lowest terms.

41. 0.6

42. 0.13

43. 0.175

44. 0.375

45. 0.295

46. 0.251

47. 0.0131

48. 0.2345

In Exercises 49–56, express each repeating decimal number as a quotient of two integers. Reduce the quotient to lowest terms.

49. $0.\overline{1}$

50. $0.\overline{8}$

51. $0.\overline{9}$

52. $0.\overline{23}$

53. $1.\overline{36}$

54. $0.\overline{135}$

55. $2.0\overline{5}$

56. $2.4\overline{9}$

In Exercises 57–66, perform the indicated operation and reduce your answer to lowest terms.

57. $\dfrac{2}{7} \cdot \dfrac{14}{15}$

58. $\dfrac{4}{9} \cdot \dfrac{18}{21}$

59. $\left(\dfrac{-3}{8}\right)\left(\dfrac{-16}{15}\right)$

60. $\left(-\dfrac{3}{5}\right) \div \dfrac{10}{21}$

61. $\dfrac{7}{8} \div \dfrac{8}{7}$

62. $\dfrac{3}{7} \div \dfrac{3}{7}$

63. $\left(\dfrac{3}{5} \cdot \dfrac{4}{7}\right) \div \dfrac{1}{3}$

64. $\left(\dfrac{4}{7} \div \dfrac{4}{5}\right) \cdot \dfrac{1}{7}$

65. $\left[\left(-\dfrac{2}{3}\right)\left(\dfrac{5}{8}\right)\right] \div \left(-\dfrac{7}{16}\right)$

66. $\left(\dfrac{7}{15} \cdot \dfrac{5}{8}\right) \div \left(\dfrac{7}{9} \cdot \dfrac{5}{2}\right)$

In Exercises 67–76, perform the indicated operation and reduce your answer to lowest terms.

67. $\dfrac{1}{5} + \dfrac{2}{3}$

68. $\dfrac{7}{8} - \dfrac{1}{6}$

69. $\dfrac{2}{11} + \dfrac{5}{22}$

70. $\dfrac{8}{9} - \dfrac{5}{6}$

71. $\dfrac{3}{14} + \dfrac{8}{21}$

72. $\dfrac{2}{15} + \dfrac{6}{35}$

73. $\dfrac{1}{12} + \dfrac{1}{48} + \dfrac{1}{72}$

74. $\dfrac{3}{5} + \dfrac{7}{15} + \dfrac{9}{75}$

75. $\dfrac{1}{30} - \dfrac{3}{40} - \dfrac{7}{50}$

76. $\dfrac{4}{25} - \dfrac{9}{100} - \dfrac{7}{40}$

Problem Solving

Alternative methods for adding and subtracting two fractions are shown below. These methods may not result in a solution in its lowest terms.

$$\frac{a}{b} + \frac{c}{d} = \frac{ad + bc}{bd} \quad \text{and} \quad \frac{a}{b} - \frac{c}{d} = \frac{ad - bc}{bd}$$

In Exercises 77–80, use the appropriate formula above to evaluate the expression.

77. $\dfrac{2}{7} + \dfrac{1}{2}$

78. $\dfrac{4}{7} + \dfrac{2}{5}$

79. $\dfrac{5}{6} - \dfrac{7}{8}$

80. $\dfrac{7}{3} - \dfrac{5}{12}$

In Exercises 81–90, evaluate each expression using the order of operations.

81. $\dfrac{1}{3} + \dfrac{1}{4} \cdot \dfrac{2}{5}$

82. $\dfrac{4}{9} + \dfrac{5}{6} \div \dfrac{2}{3}$

83. $\dfrac{7}{8} - \dfrac{3}{4} \div \dfrac{5}{6}$

84. $\dfrac{13}{15} - \dfrac{9}{10} \cdot \dfrac{5}{6}$

85. $\dfrac{7}{8} \div \dfrac{3}{4} \cdot \dfrac{3}{16}$

86. $-\dfrac{8}{15} \div \dfrac{2}{3} \cdot \dfrac{7}{9}$

87. $\dfrac{1}{3} \cdot \dfrac{3}{7} + \dfrac{3}{5} \cdot \dfrac{10}{11}$

88. $\dfrac{3}{5} \cdot \dfrac{10}{21} + \dfrac{6}{11} \cdot \dfrac{33}{32}$

89. $\left(\dfrac{3}{4} + \dfrac{1}{6}\right) \div \left(2 - \dfrac{7}{6}\right)$

90. $\left(3 - \dfrac{4}{9}\right) \div \left(4 + \dfrac{2}{3}\right)$

In Exercises 91–102, write an expression that will solve the problem and then evaluate the expression.

91. *Running* Duy started a running program in January. At that time, he could run a mile in $10\frac{1}{5}$ minutes. By July, Duy could run a mile in $7\frac{3}{5}$ minutes. Determine the difference in Duy's times from January to July.

92. *Thistles* Rachel has four different varieties of thistles invading her pasture. She estimates that of these thistles, $\frac{1}{2}$ are Canada thistles, $\frac{1}{4}$ are bull thistles, $\frac{1}{6}$ are plumeless thistles, and the rest are musk thistles. What fraction of the thistles are musk thistles?

93. *Stairway Height* A stairway consists of 14 stairs, each $8\frac{5}{8}$ inches high. What is the vertical height of the stairway?

94. *Alphabet Soup* Margaret's recipe for alphabet soup calls for (among other items) $\frac{1}{4}$ cup snipped parsley, $\frac{1}{8}$ teaspoon pepper, and $\frac{1}{2}$ cup sliced carrots. Margaret is expecting company and needs to multiply the amounts

of the ingredients by $1\frac{1}{2}$ times. Determine the amount of (a) snipped parsley, (b) pepper, and (c) sliced carrots she needs for the soup.

95. *Sprinkler System* To repair his sprinkler system, Tony needs a total of $20\frac{5}{16}$ inches of PVC pipe. He has on hand pieces that measure $2\frac{1}{4}$ inches, $3\frac{7}{8}$ inches, and $4\frac{1}{4}$ inches in length. If he can combine these pieces and use them in the repair, how long a piece of PVC pipe will Tony need to purchase to repair his sprinkler system?

96. *Crop Storage* Todd has a silo on his farm in which he stores silage made from various crops. His silo is currently $\frac{1}{4}$ full of corn silage, $\frac{2}{5}$ full of hay silage, and $\frac{1}{3}$ full of oats silage. What fraction of Todd's silo is currently in use?

97. *Department Budget* Jaime is chair of the humanities department at Santa Fe College. Jaime has a budget in which $\frac{1}{2}$ of the money is for photocopying, $\frac{2}{5}$ of the money is for computer-related expenses, and the rest of the money is for student tutors in the foreign languages lab. What fraction of Jaime's budget is for student tutors?

98. *Art Supplies* Denise teaches kindergarten and is buying supplies for her class to make papier-mâché piggy banks. Each piggy bank to be made requires $1\frac{1}{4}$ cups of flour. If Denise has 15 students who are going to make piggy banks, how much flour does Denise need to purchase?

99. *Height of a Computer Stand* The instructions for assembling a computer stand include a diagram illustrating its dimensions. Determine the total height of the stand.

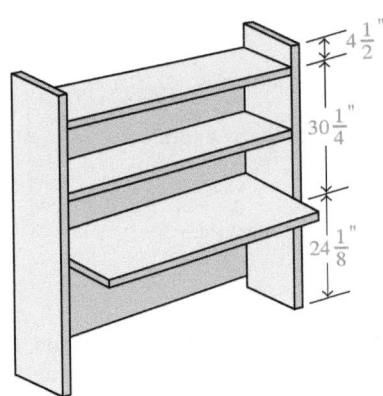

100. ***Width of a Picture*** The width of a picture is $24\frac{7}{8}$ in., as shown in the diagram. Determine x, the distance from the edge of the frame to the center.

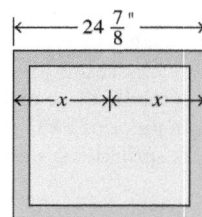

101. ***Traveling Interstate 5*** While on vacation, Janet traveled the full length of Interstate Highway 5 from the U.S.–Canada border to the U.S.–Mexico border (see the map below). She recorded the following travel times between cities: Blaine, Washington, to Seattle: 1 hour 49 minutes; Seattle to Portland: 2 hours 48 minutes; Portland to Sacramento: 9 hours 6 minutes; Sacramento to Los Angeles: 6 hours 3 minutes; Los Angeles to San Diego: 2 hours 9 minutes; San Diego to San Ysidro, California: 22 minutes.

a) Write each of these times as a fraction or as a mixed number with minutes represented as a fraction with a denominator of 60. Do not reduce the fractional part of the mixed number.

b) What is Janet's total driving time from Blaine, Washington, to San Ysidro, California? Give your answer as a mixed number and in terms of hours and minutes.

▲ *Interstate Highway 5*

102. ***Traveling Across Texas*** While on vacation, Omar traveled the entire length of Interstate Highway 10 in Texas (see the map that follows). Omar recorded the following travel times between cities on his trip. Anthony to El Paso: 25 minutes; El Paso to San Antonio: 7 hours 54 minutes; San Antonio to Houston: 3 hours 1 minute; Houston to Beaumont: 1 hour 23 minutes; Beaumont to Orange: 28 minutes.

a) Write each of these times as a fraction or as a mixed number with minutes represented as a fraction with a denominator of 60. Do not reduce the fractional part of the mixed number.

b) What is Omar's total driving time from Anthony to Orange? Give your answer as a mixed number and in terms of hours and minutes.

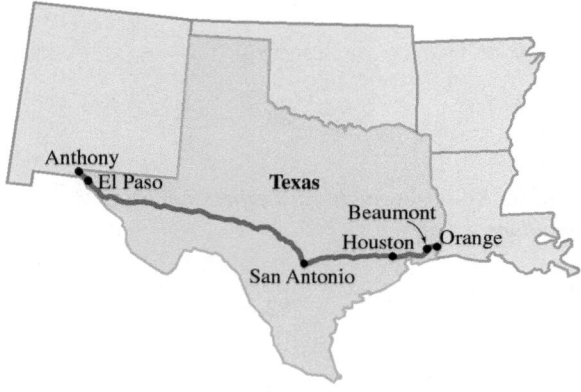

▲ *Interstate Highway 10*

Challenge Problems/Group Activities

103. ***Cutting Lumber*** If a piece of wood $8\frac{3}{4}$ ft long was cut into four equal pieces, determine the length of each piece. (Allow $\frac{1}{8}$ in. for each saw cut.)

104. ***Dimensions of a Room*** A rectangular room measures 8 ft 3 in. wide by 10 ft 8 in. long by 9 ft 2 in. high.

a) Determine the perimeter of the room in feet. Write your answer as a mixed number.

b) Calculate the area of the floor of the room in square feet.

c) Calculate the volume of the room in cubic feet. Use volume = length × width × height. Write your answer as a mixed number.

105. ***Hanging a Picture*** The back of a framed picture that is to be hung is shown at the top of the next page. A nail is to be hammered into the wall, and the picture will be hung by the wire on the nail.

a) If the center of the wire is to rest on the nail and a side of the picture is to be 20 in. from the window, how far from the window should the nail be placed?

b) If the top of the frame is to be $26\frac{1}{4}$ in. from the ceiling, how far from the ceiling should the nail be placed? (Assume the wire will not stretch.)

c) Repeat part (b) if the wire will stretch $\frac{1}{4}$ in. when the picture is hung.

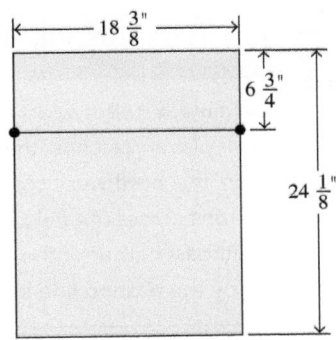

106. *Cooking Oatmeal* Following are the instructions given on a box of oatmeal. Determine the amount of water (or milk) and oats needed to make $1\frac{1}{2}$ servings by:

a) Adding the amount of each ingredient needed for 1 serving to the amount needed for 2 servings and dividing by 2.

b) Adding the amount of each ingredient needed for 1 serving to half the amount needed for 1 serving.

Directions

1. Boil water or milk and salt (if desired).
2. Stir in oats.
3. Stirring occasionally, cook over medium heat for 5 minutes.

Servings	1	2
Water (or milk)	1 cup	$1\frac{3}{4}$ cup
Oats	$\frac{1}{2}$ cup	1 cup
Salt (optional)	dash	$\frac{1}{8}$ tsp

Dense Set of Numbers A set of numbers is said to be a *dense set* if between any two distinct members of the set there exists a third distinct member of the set. The set of integers is not dense, since between any two consecutive integers there is not another integer. For example, between 1 and 2 there are no other integers. The set of rational numbers is dense because between any two distinct rational numbers there exists a third distinct rational number. For example, we can determine a rational number between 0.243 and 0.244. The number 0.243 can be written as 0.2430, and 0.244 can be written as 0.2440. There are many numbers between these two numbers. Some of them are 0.2431, 0.2435, and 0.243912. In Exercises 107–110, determine a rational number between the two numbers in each pair. Many answers are possible.

107. 0.21 and 0.22

108. 4.005 and 4.05

109. −2.176 and −2.175

110. 1.3457 and 1.34571

Halfway Between Two Numbers To determine a rational number halfway between any two rational numbers given in fraction form, add the two numbers together and divide their sum by 2. In Exercises 111–114, determine a rational number halfway between the two fractions in each pair.

111. $\frac{1}{4}$ and $\frac{3}{4}$

112. $\frac{1}{5}$ and $\frac{2}{5}$

113. $\frac{1}{100}$ and $\frac{1}{10}$

114. $\frac{7}{13}$ and $\frac{8}{13}$

115. Consider the rational number $0.\overline{9}$.

a) Use the method from Example 8 on page 234 to convert $0.\overline{9}$ to a quotient of integers.

b) Determine a number halfway between $0.\overline{9}$ and 1 by adding the two numbers and dividing by 2.

c) Determine $\frac{1}{3} + \frac{2}{3}$ by adding the fractions. Now express $\frac{1}{3}$ and $\frac{2}{3}$ as repeating decimals and determine the sum using the decimal representation of $\frac{1}{3}$ and $\frac{2}{3}$.

d) What conclusion can you draw from parts (a), (b), and (c)?

Research Activity

116. *Ancient Greek Mathematicians* The ancient Greeks are often considered the first true mathematicians. Write a report summarizing the ancient Greeks' contributions to rational numbers. Include in your report what they learned and believed about the rational numbers.

SECTION 5.4   The Irrational Numbers

Upon completion of this section, you will be able to:

- Simplify radicals.
- Add and subtract radicals.
- Multiply radicals.
- Divide radicals.
- Rationalize the denominator.
- Estimate square roots without a calculator.

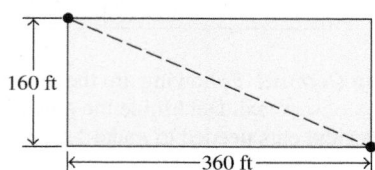

A football field is 160 ft wide and 360 ft long. A football player catches the football from a kickoff in the northwest corner of one end zone and runs across the field in a straight line to the southeast corner of the other end zone, as shown by the dashed line in the diagram.

How far did the football player run? The answer to this question is an example of an *irrational* number. In this section, we will introduce irrational numbers along with the operations of addition, subtraction, multiplication, and division of irrational numbers. The answer to the above question is $40\sqrt{97}$ feet, or about 394 feet.

Why This Is Important   Irrational numbers play an important role in many areas of mathematics, science, business, architecture, banking, and other fields.

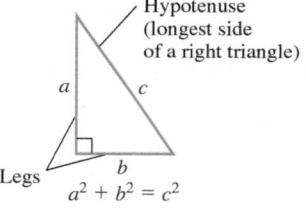

Figure 5.7

Pythagoras (ca. 585–500 B.C.), a Greek mathematician, is credited with providing a written proof that in any *right triangle* (a triangle with a 90° angle; see Fig. 5.7), the square of the length of one side (a^2) added to the square of the length of the other side (b^2) equals the square of the length of the hypotenuse (c^2). The formula $a^2 + b^2 = c^2$ is now known as the *Pythagorean theorem*.* Pythagoras discovered that when using the formula, with $a = 1$ and $b = 1$, the value of c was not a rational number.

$$a^2 + b^2 = c^2$$
$$1^2 + 1^2 = c^2$$
$$1 + 1 = c^2$$
$$2 = c^2$$

There is no rational number that when squared will equal 2. This fact prompted a need for a new set of numbers, the irrational numbers.

In Section 5.2, we introduced the real number line. The points on the number line that are not rational numbers are referred to as *irrational numbers*. Recall that every rational number can be expressed as a ratio of two integers. This leads us to the definition of irrational number.

> **Definition: Irrational Number**
> An **irrational number** is a real number that cannot be written as the ratio of two integers.

Also recall that every rational number can be written as either a terminating or a repeating decimal number. Therefore, an irrational number has a decimal representation that is a nonterminating, nonrepeating decimal number. For example, a

_____
* The Pythagorean theorem is discussed in more detail in Section 8.3.

Pythagoras of Samos founded a philosophical and religious school in southern Italy in the sixth century B.C. The scholars at the school, known as Pythagoreans, produced important works of mathematics, astronomy, and the theory of music. Although the Pythagoreans are credited with proving the Pythagorean theorem, it was known to the ancient Babylonians 1000 years earlier. The Pythagoreans were a secret society that formed a model for many secret societies in existence today. One practice was that students were to spend their first three years of study in silence, while their master, Pythagoras, spoke to them from behind a curtain. Among other philosophical beliefs of the Pythagoreans was "that at its deepest level, reality is mathematical in nature."

Simplifying Radicals

nonrepeating decimal number such as 5.12639537… can be used to indicate an irrational number. Notice that no number or set of numbers repeat on a continuous basis, and the three dots at the end of the number indicate that the number continues indefinitely. Nonrepeating number patterns can be used to indicate irrational numbers. For example, 6.1011011101111… and 0.525225222… are both irrational numbers.

The expression $\sqrt{2}$ is read "the square root of 2" or "radical 2." The symbol $\sqrt{}$ is called the *radical sign*, and the number or expression inside the radical sign is called the *radicand*. In $\sqrt{2}$, 2 is the radicand.

The square roots of some numbers are rational, whereas the square roots of other numbers are irrational. The *principal* (or *positive*) *square root* of a number n, written $\sqrt{n}$, is the positive number that, when multiplied by itself, gives n. Whenever we mention the term "square root" in this text, we mean the principal square root. For example,

$$\sqrt{9} = 3, \quad \text{since} \quad 3 \cdot 3 = 9$$

$$\sqrt{36} = 6, \quad \text{since} \quad 6 \cdot 6 = 36$$

Both $\sqrt{9}$ and $\sqrt{36}$ are examples of numbers that are rational numbers because 3 and 6, respectively, are terminating decimal numbers.

We return to the problem faced by Pythagoras: If $c^2 = 2$, then it can be shown that c has a value of $\sqrt{2}$, but what is $\sqrt{2}$ equal to? The $\sqrt{2}$ is an irrational number, and it cannot be expressed as a terminating or repeating decimal number. It can only be approximated by a decimal number: $\sqrt{2}$ is approximately 1.4142136 (to seven decimal places). Later in this section, we will discuss using a calculator to approximate irrational numbers.

Other irrational numbers include $\sqrt{3}$, $\sqrt{5}$, and $\sqrt{37}$. Another important irrational number used to represent the ratio of a circle's circumference to its diameter is pi, symbolized π. Pi is approximately 3.1415927.

We have discussed procedures for performing the arithmetic operations of addition, subtraction, multiplication, and division with rational numbers. We can perform the same operations with the irrational numbers. Before we can proceed, however, we must understand the numbers called perfect squares. Any number that is the square of a natural number is said to be a *perfect square*. Some perfect squares are shown in the following chart.

Natural numbers	1,	2,	3,	4,	5,	6,…
Squares of the natural numbers	1^2,	2^2,	3^2,	4^2,	5^2,	6^2,…
Perfect squares	1,	4,	9,	16,	25,	36,…

The numbers 1, 4, 9, 16, 25, and 36 are some of the perfect square numbers. Can you determine the next two perfect square numbers? How many perfect square numbers are there? The square root of a perfect square number will be a natural number. For example, $\sqrt{1} = 1$, $\sqrt{4} = 2$, $\sqrt{9} = 3$, $\sqrt{16} = 4$, $\sqrt{25} = 5$, and so on.

The number that multiplies a radical is called the radical's *coefficient*. For example, in $3\sqrt{5}$, the 3 is the coefficient of the radical.

Simplify Radicals

Some irrational numbers can be simplified by determining whether there are any perfect square factors in the radicand. If there are, the following rule can be used to simplify the radical.

Product Rule for Radicals
$$\sqrt{a \cdot b} = \sqrt{a} \cdot \sqrt{b}, \quad a \geq 0, \quad b \geq 0$$

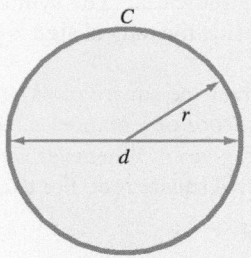

$$C = \pi d$$
$$\text{or } \pi = \frac{C}{d}$$

To simplify a radical, write the radical as a product of two radicals. One of the radicals should contain the greatest perfect square that is a factor of the radicand in the original expression. Then simplify the radical containing the perfect square factor. For example,

$$\sqrt{12} = \sqrt{4 \cdot 3} = \sqrt{4} \cdot \sqrt{3} = 2 \cdot \sqrt{3} = 2\sqrt{3}$$

and

$$\sqrt{75} = \sqrt{25 \cdot 3} = \sqrt{25} \cdot \sqrt{3} = 5 \cdot \sqrt{3} = 5\sqrt{3}$$

Example 1 Simplifying Radicals

Simplify.

a) $\sqrt{28}$ b) $\sqrt{72}$ c) $\sqrt{54}$

Solution

a) Since 4 is a perfect square factor of 28, we write

$$\sqrt{28} = \sqrt{4 \cdot 7} = \sqrt{4} \cdot \sqrt{7} = 2 \cdot \sqrt{7} = 2\sqrt{7}$$

Since 7 has no perfect square factors, $\sqrt{7}$ cannot be simplified.

b) Since 36 is a perfect square factor of 72, we write

$$\sqrt{72} = \sqrt{36 \cdot 2} = \sqrt{36} \cdot \sqrt{2} = 6\sqrt{2}$$

c) Since 9 is a perfect square factor of 54, we write

$$\sqrt{54} = \sqrt{9 \cdot 6} = \sqrt{9} \cdot \sqrt{6} = 3\sqrt{6}$$ ∎

In Example 1(b), you can obtain the correct answer if you start out factoring differently:

$$\sqrt{72} = \sqrt{9 \cdot 8} = \sqrt{9} \cdot \sqrt{8} = 3 \cdot \sqrt{8} = 3\sqrt{8}$$

Note that 8 has 4 as a perfect square factor.

$$3\sqrt{8} = 3\sqrt{4 \cdot 2} = 3 \cdot \sqrt{4} \cdot \sqrt{2} = 3 \cdot 2 \cdot \sqrt{2} = 6\sqrt{2}$$

The second method will eventually give the same answer, but it requires more work. It is best to try to factor out the *largest* perfect square factor from the radicand.

Addition and Subtraction of Radicals

To add or subtract two or more square roots with the same radicand, add or subtract their coefficients while keeping the common radicand. The answer is the sum or difference of the coefficients multiplied by the common radical.

Example 2 Adding and Subtracting Radicals with the Same Radicand

Simplify.

a) $5\sqrt{3} + 4\sqrt{3}$ b) $2\sqrt{5} - 6\sqrt{5} + \sqrt{5}$

Solution

a) $5\sqrt{3} + 4\sqrt{3} = (5 + 4)\sqrt{3} = 9\sqrt{3}$
b) $2\sqrt{5} - 6\sqrt{5} + \sqrt{5} = (2 - 6 + 1)\sqrt{5} = -3\sqrt{5}$

Note that $\sqrt{5} = 1\sqrt{5}$.   ∎

┌─

Example 3 *Subtracting Radicals with Different Radicands*

Simplify $6\sqrt{2} - \sqrt{18}$.

Solution   When two square roots have different radicands, they cannot be added or subtracted. When this occurs, we should try to simplify one or both square roots. If the simplified square roots have the same radicand, then we can perform addition or subtraction.

$$6\sqrt{2} - \sqrt{18} = 6\sqrt{2} - \sqrt{9 \cdot 2}$$
$$= 6\sqrt{2} - \sqrt{9} \cdot \sqrt{2}$$
$$= 6\sqrt{2} - 3\sqrt{2}$$
$$= (6 - 3)\sqrt{2}$$
$$= 3\sqrt{2} \qquad \blacksquare$$

└─

Multiplication of Radicals

When multiplying radicals, we again make use of the product rule for radicals, as introduced on page 245. After the radicands are multiplied, simplify the remaining radical when possible.

┌─

Example 4 *Multiplying Radicals*

Simplify.

a) $\sqrt{3} \cdot \sqrt{27}$ b) $\sqrt{3} \cdot \sqrt{13}$ c) $\sqrt{6} \cdot \sqrt{10}$

Solution

a) $\sqrt{3} \cdot \sqrt{27} = \sqrt{3 \cdot 27} = \sqrt{81} = 9$
b) $\sqrt{3} \cdot \sqrt{13} = \sqrt{3 \cdot 13} = \sqrt{39}$
c) $\sqrt{6} \cdot \sqrt{10} = \sqrt{6 \cdot 10} = \sqrt{60} = \sqrt{4 \cdot 15} = \sqrt{4} \cdot \sqrt{15} = 2\sqrt{15}$ $\blacksquare$

└─

Division of Radicals

To divide radicals, use the following rule. After performing the division, simplify when possible.

╭───╮

Quotient Rule for Radicals

$$\frac{\sqrt{a}}{\sqrt{b}} = \sqrt{\frac{a}{b}}, \quad a \geq 0, \quad b > 0$$

╰───╯

┌─

Example 5 *Dividing Radicals*

Divide.

a) $\dfrac{\sqrt{50}}{\sqrt{2}}$ b) $\dfrac{\sqrt{90}}{\sqrt{5}}$

Solution

a) $\dfrac{\sqrt{50}}{\sqrt{2}} = \sqrt{\dfrac{50}{2}} = \sqrt{25} = 5$

b) $\dfrac{\sqrt{90}}{\sqrt{5}} = \sqrt{\dfrac{90}{5}} = \sqrt{18} = \sqrt{9 \cdot 2} = \sqrt{9} \cdot \sqrt{2} = 3\sqrt{2}$ $\blacksquare$

└─

Rationalizing the Denominator

A denominator of a fraction is *rationalized* when it contains no radical expressions. To rationalize a denominator that contains only a square root, multiply both the numerator and denominator of the fraction by the square root of a number that will result in the radicand in the denominator becoming a perfect square. This action is the equivalent of multiplying the fraction by 1 because the value of the fraction does not change. Then simplify the fractions when possible.

> ### Example 6 *Rationalizing the Denominator*
>
> Rationalize the denominator of the following.
>
> a) $\dfrac{5}{\sqrt{2}}$ b) $\dfrac{5}{\sqrt{12}}$ c) $\dfrac{\sqrt{5}}{\sqrt{10}}$
>
> **Solution**
>
> a) Multiply the numerator and denominator by the square root of a number that will make the radicand in the denominator a perfect square.
>
> $$\frac{5}{\sqrt{2}} = \frac{5}{\sqrt{2}} \cdot \frac{\sqrt{2}}{\sqrt{2}} = \frac{5\sqrt{2}}{\sqrt{4}} = \frac{5\sqrt{2}}{2}$$
>
> Note that the 2's in the answer cannot be divided out because one 2 is a radicand and the other is not.
>
> b) $\dfrac{5}{\sqrt{12}} = \dfrac{5}{\sqrt{12}} \cdot \dfrac{\sqrt{3}}{\sqrt{3}} = \dfrac{5\sqrt{3}}{\sqrt{36}} = \dfrac{5\sqrt{3}}{6}$
>
> You could have also obtained the same answer to this problem by multiplying both the numerator and denominator by $\sqrt{12}$ and then simplifying. Try to do so now.
>
> c) Write $\dfrac{\sqrt{5}}{\sqrt{10}}$ as $\sqrt{\dfrac{5}{10}}$ and reduce the fraction to obtain $\sqrt{\dfrac{1}{2}}$. By the quotient rule
>
> for radicals, $\sqrt{\dfrac{1}{2}} = \dfrac{\sqrt{1}}{\sqrt{2}}$ or $\dfrac{1}{\sqrt{2}}$. Now rationalize the denominator of $\dfrac{1}{\sqrt{2}}$.
>
> $$\frac{1}{\sqrt{2}} = \frac{1}{\sqrt{2}} \cdot \frac{\sqrt{2}}{\sqrt{2}} = \frac{\sqrt{2}}{2}$$
>
> The answer could also be obtained by multiplying the numerator and denominator by $\sqrt{10}$ and then simplifying. Try this now.   ∎

Estimating Square Roots Without a Calculator

Consider the irrational number $\sqrt{17}$. The radicand 17 is between the two perfect squares 16 and 25. Since $\sqrt{16} = 4$ and $\sqrt{25} = 5$, we can reason that $\sqrt{17}$ is between 4 and 5. We summarize this discussion as follows.

$$16 < \quad 17 \quad < 25$$
$$\sqrt{16} < \sqrt{17} < \sqrt{25}$$
$$4 < \sqrt{17} < 5$$

Furthermore, since 17 is closer to 16 than it is to 25, we can further estimate that $\sqrt{17}$ is closer to 4 than it is to 5. In other words, $\sqrt{17}$ is between 4 and 4.5.

Example 7 *Estimating Square Roots*

The following diagram is a sketch of a 16-in. ruler marked using $\frac{1}{2}$ inches.

```
  1   2   3   4   5   6   7   8   9  10  11  12  13  14  15
|..|..|..|..|..|..|..|..|..|..|..|..|..|..|..|..|
```

Indicate between which two adjacent ruler marks each of the following irrational numbers will fall.

a) $\sqrt{7}$ b) $\sqrt{89}$

Solution

a) $4 < 7 < 9$

   $\sqrt{4} < \sqrt{7} < \sqrt{9}$

   $2 < \sqrt{7} < 3$

Therefore, $\sqrt{7}$ is between 2 and 3. Furthermore, since 7 is closer to 9 than it is to 4, $\sqrt{7}$ is closer to 3 than it is to 2. Thus, $\sqrt{7}$ is between 2.5 and 3.

b) $81 < 89 < 100$

   $\sqrt{81} < \sqrt{89} < \sqrt{100}$

   $9 < \sqrt{89} < 10$

Therefore, $\sqrt{89}$ is between 9 and 10. Furthermore, since 89 is closer to 81 than it is to 100, $\sqrt{89}$ is closer to 9 than it is to 10. Thus, $\sqrt{89}$ is between 9 and 9.5. ∎

TECHNOLOGY TIP

Approximating Square Roots on a Scientific Calculator

Consider the irrational number the square root of two. We use the symbol $\sqrt{2}$ to represent the *exact value* of this number. Although exact values are important, approximations are also important, especially when working with application problems. We can use a scientific calculator to obtain approximations for square roots. Scientific calculators generally have one of the following square root keys:

$$\boxed{\sqrt{}} \quad \text{or} \quad \boxed{\sqrt{x}}$$

For simplicity, we will refer to the square root key with the $\boxed{\sqrt{}}$ symbol.
   To approximate $\sqrt{2}$, perform the following keystrokes:

$$\boxed{\sqrt{}} \quad \boxed{2} \quad \boxed{\text{ENTER}}$$

or, depending on your model of calculator, you may have to just enter the following:

$$\boxed{2} \quad \boxed{\sqrt{}}$$

The display on your calculator may read 1.414213562. Your calculator may display more or fewer digits. It is important to realize that 1.414213562 is a rational number *approximation* for the irrational number $\sqrt{2}$. The symbol $\approx$ means *is approximately equal to*, and we write

$$\sqrt{2} \approx 1.414213562$$

⬆ Exact value (irrational number)   ⬆ Approximation (rational number)

Note: If your calculator has the $\sqrt{}$ symbol printed *above* the key instead of on the face of the key, you can access the square root function by first pressing the "2nd" or the "SHIFT" key.

Example **8** *Approximating Square Roots*

Use a scientific calculator to approximate the following square roots. Round each answer to two decimal places.

a) $\sqrt{7}$ b) $\sqrt{89}$

Solution

a) $\sqrt{7} \approx 2.65$ b) $\sqrt{89} \approx 9.43$

Note that these approximations confirm the estimates that were made in Example 7.

SECTION 5.4   Exercises

Warm Up Exercises

In Exercises 1–8, fill in the blanks with an appropriate word, phrase, or symbol(s).

1. A real number that cannot be written as a ratio of two integers is known as a(n) _____ number.

2. The symbol $\sqrt{\ }$ is called the _____ sign.

3. In the radical expression $3\sqrt{5}$, the 5 is called the _____.

4. In the radical expression $3\sqrt{5}$, the 3 is called the _____ of the radical expression.

5. The principal (or positive) square root of a number n, written $\sqrt{n}$, is the positive number that, when multiplied by _____, gives n.

6. Any number that is the square of a natural number is called a(n) _____ square.

7. When the denominator of a fraction contains no radical expressions, it is said to be _____.

8. Consider the irrational number the square root of two.

   a) We use the symbol $\sqrt{2}$ to represent the _____ value of the number.

   b) The number 1.414213562 is a(n) _____ of the number $\sqrt{2}$.

Practice the Skills

In Exercises 9–18, determine whether the number is rational or irrational.

9. $\sqrt{36}$

10. $\dfrac{5}{8}$

11. $\sqrt{10}$

12. π

13. $3.575775777\ldots$

14. $0.212112111\ldots$

15. $\dfrac{22}{7}$

16. 3.14159

17. $\dfrac{\sqrt{75}}{\sqrt{3}}$

18. $\dfrac{\sqrt{5}}{\sqrt{5}}$

In Exercises 19–26, evaluate the expression.

19. $\sqrt{0}$

20. $\sqrt{1}$

21. $\sqrt{25}$

22. $\sqrt{36}$

23. $-\sqrt{36}$

24. $-\sqrt{81}$

25. $-\sqrt{100}$

26. $-\sqrt{225}$

In Exercises 27–36, classify the number as a member of one or more of the following sets: the rational numbers, the integers, the natural numbers, the irrational numbers.

27. 0

28. -5

29. $\sqrt{13}$

30. $2\sqrt{7}$

31. 0.040040004

32. 2.718

33. $0.\overline{123}$

34. 0.123123123

35. $\sqrt{123}$

36. $0.123112311123\ldots$

In Exercises 37–44, simplify the radical.

37. $\sqrt{20}$

38. $\sqrt{27}$

39. $\sqrt{40}$

40. $\sqrt{54}$

41. $\sqrt{63}$

42. $\sqrt{75}$

43. $\sqrt{84}$

44. $\sqrt{90}$

In Exercises 45–52, perform the indicated operation.

45. $4\sqrt{6} + 3\sqrt{6}$

46. $8\sqrt{3} - 10\sqrt{3}$

47. $5\sqrt{18} - 7\sqrt{8}$

48. $2\sqrt{20} - 3\sqrt{45}$

49. $4\sqrt{12} - 7\sqrt{27}$

50. $2\sqrt{7} + 5\sqrt{28}$

51. $5\sqrt{3} + 7\sqrt{12} - 3\sqrt{75}$

52. $\sqrt{63} + 13\sqrt{98} - 5\sqrt{112}$

In Exercises 53–60, perform the indicated operation. Simplify the answer when possible.

53. $\sqrt{2}\sqrt{18}$

54. $\sqrt{5}\sqrt{15}$

55. $\sqrt{6}\sqrt{10}$

56. $\sqrt{10}\sqrt{35}$

57. $\dfrac{\sqrt{20}}{\sqrt{5}}$

58. $\dfrac{\sqrt{125}}{\sqrt{5}}$

59. $\dfrac{\sqrt{72}}{\sqrt{8}}$

60. $\dfrac{\sqrt{145}}{\sqrt{5}}$

In Exercises 61–68, rationalize the denominator.

61. $\dfrac{1}{\sqrt{5}}$

62. $\dfrac{3}{\sqrt{3}}$

63. $\dfrac{\sqrt{3}}{\sqrt{7}}$

64. $\dfrac{\sqrt{2}}{\sqrt{7}}$

65. $\dfrac{\sqrt{20}}{\sqrt{3}}$

66. $\dfrac{\sqrt{50}}{\sqrt{14}}$

67. $\dfrac{\sqrt{10}}{\sqrt{6}}$

68. $\dfrac{\sqrt{2}}{\sqrt{27}}$

Problem Solving

Approximating Radicals The following diagram is a sketch of a 16-in. ruler marked using $\frac{1}{2}$ inches.

```
1  2  3  4  5  6  7  8  9  10 11 12 13 14 15
```

In Exercises 69–74, without using a calculator, indicate between which two adjacent ruler marks each of the following irrational numbers will fall. Support your answer by obtaining an approximation with a calculator.

69. $\sqrt{37}$ in.

70. $\sqrt{61}$ in.

71. $\sqrt{97}$ in.

72. $\sqrt{123}$ in.

73. $\sqrt{170}$ in.

74. $\sqrt{200}$ in.

75. *Height of Boys* The median height of boys less than 5 years old can be estimated using the formula $H = 2.9\sqrt{x} + 20.1$, where H is the height in inches and x is the boys' age in months. Estimate the median height of boys who are 30 months old. Round your answer to the nearest tenth of an inch.

76. *A Swinging Pendulum* The time T required for a pendulum to swing back and forth may be determined by the formula

$$T = 2\pi\sqrt{\dfrac{l}{g}}$$

where l is the length of the pendulum and g is the acceleration of gravity. Determine the time in seconds for the pendulum to swing back and forth if $l = 35$ cm and $g = 980$ cm/sec². Round your answer to the nearest tenth of a second.

77. *Dropping an Object* The formula $t = \dfrac{\sqrt{d}}{4}$ can be used to estimate the time, t, in seconds it takes for an object dropped to travel d feet.

a) Estimate the time it takes for an object to drop 100 ft.

b) Estimate the time it takes for an object to drop 400 ft.

c) Estimate the time it takes for an object to drop 900 ft.

d) Estimate the time it takes for an object to drop 1600 ft.

78. *Estimating Speed of a Vehicle* The speed that a vehicle was traveling, s, in miles per hour, when the brakes were first applied, can be estimated using the formula $s = \sqrt{\dfrac{d}{0.04}}$ where d is the length of the vehicle's skid marks, in feet.

a) Estimate the speed of a car that made skid marks 4 ft long.

b) Estimate the speed of a car that made skid marks 16 ft long.

c) Estimate the speed of a car that made skid marks 64 ft long.

d) Estimate the speed of a car that made skid marks 256 ft long.

Concept/Writing Exercises

In Exercises 79–84, determine whether the statement is true or false. Rewrite each false statement to make it a true statement. A false statement can be modified in more than one way to be made a true statement.

79. $\sqrt{c}$ is a rational number for any composite number c.

80. $\sqrt{p}$ is a rational number for any prime number p.

81. The sum of any two rational numbers is always a rational number.

82. The product of any two rational numbers is always a rational number.

83. The product of an irrational and a rational number is always an irrational number.

84. The product of any two irrational numbers is always an irrational number.

In Exercises 85–88, give an example to show that the stated case can occur.

85. The sum of two irrational numbers may be a rational number.

86. The sum of two irrational numbers may be an irrational number.

87. The product of two irrational numbers may be an irrational number.

88. The product of two irrational numbers may be a rational number.

89. Without doing any calculations, determine whether $\sqrt{2} = 1.414$.

90. Without doing any calculations, determine whether $\sqrt{17} = 4.123$.

91. The number π is an irrational number. Often the values 3.14 or $\frac{22}{7}$ are used for π. Does π equal either 3.14 or $\frac{22}{7}$?

92. Give an example to show that $\sqrt{a + b} \neq \sqrt{a} + \sqrt{b}$.

93. Give an example to show that $\sqrt{a \cdot b} = \sqrt{a} \cdot \sqrt{b}$.

94. Give an example to show that for $b \neq 0$, $\sqrt{\dfrac{a}{b}} = \dfrac{\sqrt{a}}{\sqrt{b}}$.

Challenge Problems/Group Activities

95. a) Is $\sqrt{0.04}$ rational or irrational? Explain.

  b) Is $\sqrt{0.7}$ rational or irrational? Explain.

96. One way to determine a rational number between two distinct rational numbers is to add the two distinct rational numbers and divide by 2. Do you think that this method will always work for finding an irrational number between two distinct irrational numbers? Explain.

Recreational Mathematics

97. *More Four 4's* In the *Recreational Mathematics* box on page 223 and in Exercise 82 on page 228, we introduced some of the basic rules of the game Four 4's. We now expand our operations to include square roots. For example, one way to represent the whole number 8 is $\sqrt{4} + \sqrt{4} + \sqrt{4} + \sqrt{4} = 2 + 2 + 2 + 2 = 8$. Using at least one square root of 4, $\sqrt{4}$, play Four 4's to represent the following whole numbers:

  a) 11

  b) 13

  c) 14

  d) 18

Research Activities

98. Write a report on the history of the development of the irrational numbers.

99. Write a report on the history of pi. In your report, indicate when the symbol π was first used and list the first 100 digits of π.

SECTION 5.5   Real Numbers and Their Properties

Upon completion of this section, you will be able to:

■ Understand the properties of the real number system.

Evaluate $10 + 5$ and then evaluate $5 + 10$. Evaluate 10×5 and then evaluate 5×10. Evaluate $10 - 5$ and then evaluate $5 - 10$. Finally, evaluate $10 \div 5$ and then evaluate $5 \div 10$. You will notice that $10 + 5 = 5 + 10$ and that $10 \times 5 = 5 \times 10$. On the other hand, $10 - 5 \neq 5 - 10$ and $10 \div 5 \neq 5 \div 10$. These simple examples demonstrate a certain property that holds for addition and multiplication, but not for subtraction and division. In this section, we will study the set of *real* numbers and their properties.

Why This Is Important Understanding the properties of real numbers is vital to learning algebra and many other areas of mathematics. Many applications that we will see in this section rely on our understanding of real numbers. For example, building a bookcase, creating a budget, determining price per unit, and determining the percent discount of an item are just some of the ways we use the properties of real numbers.

Now that we have discussed both the rational and irrational numbers, we can discuss the real numbers and the properties of the real number system. The union of the rational numbers and the irrational numbers is the *set of real numbers*, symbolized by $\mathbb{R}$.

Figure 5.8 illustrates the relationship among various sets of numbers. It shows that the natural numbers are a subset of the whole numbers, the integers, the rational numbers, and the real numbers. For example, since the number 3 is a natural or counting number, it is also a whole number, an integer, a rational number, and a real number. Since the rational number $\frac{1}{4}$ is outside the set of integers, it is not an integer, a whole number, or a natural number. The number $\frac{1}{4}$ is a real number, however, as is the irrational number $\sqrt{2}$. Note that the real numbers are the union of the rational numbers and the irrational numbers.

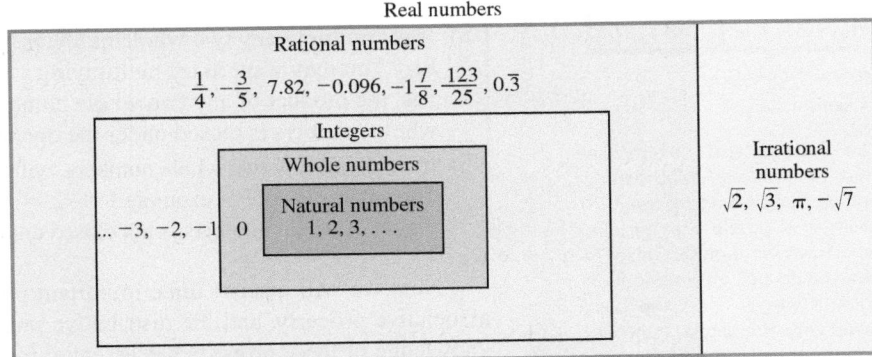

Figure 5.8

The relationship between the various sets of numbers in the real number system can also be illustrated with a tree diagram, as in Fig. 5.9.

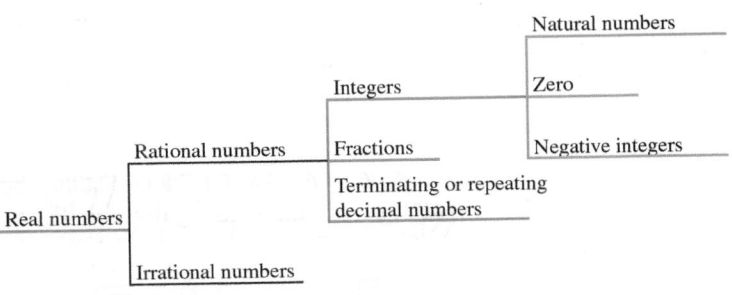

Figure 5.9

Figure 5.8 and Figure 5.9 show that, for example, the natural numbers are a subset of the integers, the rational numbers, and the real numbers. We can also see, for example, the natural numbers, zero, and the negative integers together form the integers.

Properties of the Real Number System

We are now prepared to consider the properties of the real number system. The first property we will discuss is *closure*.

Definition: Closure

If an operation is performed on any two elements of a set and the result is always an element of the set, we say that the set has **closure** or is **closed** under that given operation.

KenKen

16×		7+	
2−			4
	12×	2÷	
		2÷	

© KenKen Puzzle LLC.
www.kenken.com

Like Sudoku and Kakuro (see the *Recreational Mathematics* boxes on page 95 and page 128), KenKen is a puzzle that requires logic. However, KenKen also requires you to use arithmetic facts. To play KenKen, fill in the blank squares with the digits 1 through the number of rows or columns contained in the puzzle. For example, the puzzle above has 4 rows and 4 columns; therefore, you fill in the blank squares with the digits 1 through 4. Like Sudoku, you are not allowed to repeat a digit in any row or column. The heavily outlined sets of squares are called cages. The numbers placed in each cage must combine, *in any order*, to produce the given number using the given mathematical operation. For example, the cage in the lower right corner of the puzzle above must contain two digits whose quotient is 2. Cages with just one box should be filled in with the given number in the top corner. A number can be repeated within a cage as long as it is not in the same row or column. For more information and more puzzles, see www.kenken.com. Complete the above puzzle.

The solution to the above puzzle can be found in the answer section in the back of this book. For an additional puzzle, see Exercise 78 on page 258.

Note: KenKen® is a registered trademark of Nextoy, LLC. Puzzle content © 2014 KenKen Puzzle LLC. All rights reserved.

Is the sum of any two natural numbers a natural number? The answer is yes. Thus, we say that the natural numbers are closed under the operation of addition.

Are the natural numbers closed under the operation of subtraction? If we subtract one natural number from another natural number, must the difference always be a natural number? The answer is no. For example, $3 - 5 = -2$, and -2 is not a natural number. Therefore, the natural numbers are not closed under the operation of subtraction.

Example 1 *Closure of Sets*

Determine whether the set of whole numbers is closed under the operations of (a) multiplication and (b) division.

Solution

a) If we multiply any two whole numbers, will the product always be a whole number? You may want to try multiplying several whole numbers. You will see that yes, the product of any two whole numbers is a whole number. Thus, the set of whole numbers is closed under the operation of multiplication.

b) If we divide any two whole numbers, will the quotient always be a whole number? The answer is no. For example $1 \div 2 = \frac{1}{2}$, and $\frac{1}{2}$ is not a whole number. Therefore, the set of whole numbers is not closed under the operation of division. ∎

Next we will discuss three important properties: the commutative property, the associative property, and the distributive property of multiplication over addition. A knowledge of these properties is essential for the understanding of algebra. We begin with the commutative property.

Commutative Property

Addition	Multiplication
$a + b = b + a$	$a \cdot b = b \cdot a$

for any real numbers a and b.

The commutative property states that the *order* in which two numbers are added or multiplied is not important. For example, $4 + 5 = 5 + 4 = 9$ and $3 \cdot 6 = 6 \cdot 3 = 18$. Note that the commutative property does not hold for the operations of subtraction or division. For example,

$$4 - 7 \neq 7 - 4 \quad \text{and} \quad 9 \div 3 \neq 3 \div 9$$

Now we introduce the associative property.

Associative Property

Addition	Multiplication
$(a + b) + c = a + (b + c)$	$(a \cdot b) \cdot c = a \cdot (b \cdot c)$

for any real numbers a, b, and c.

The associative property states that when adding or multiplying three real numbers, we may place parentheses around any two adjacent numbers. The result is the same regardless of the placement of parentheses. For example,

$$\begin{array}{cc} (3 + 4) + 5 = 3 + (4 + 5) & (3 \cdot 4) \cdot 5 = 3 \cdot (4 \cdot 5) \\ 7 + 5 = 3 + 9 & 12 \cdot 5 = 3 \cdot 20 \\ 12 = 12 & 60 = 60 \end{array}$$

The associative property does not hold for the operations of subtraction or division. For example,

$$(10 - 6) - 2 \neq 10 - (6 - 2) \quad \text{and} \quad (27 \div 9) \div 3 \neq 27 \div (9 \div 3)$$

Note the difference between the commutative property and the associative property. The commutative property involves a change in *order*, whereas the associative property involves a change in *grouping* (or the *association* of numbers that are grouped together).

Another property of the real numbers is the distributive property of multiplication over addition.

> ### Distributive Property of Multiplication over Addition
> $$a \cdot (b + c) = a \cdot b + a \cdot c$$
> for any real numbers a, b, and c.

For example, if $a = 3$, $b = 4$, and $c = 5$, then

$$3 \cdot (4 + 5) = (3 \cdot 4) + (3 \cdot 5)$$
$$3 \cdot 9 = 12 + 15$$
$$27 = 27$$

This result indicates that, when using the distributive property, you may either add the numbers within parentheses first and then multiply or multiply first and then add. Note that the distributive property involves two operations, addition and multiplication. Although positive integers were used in the example, any real numbers could have been used.

We frequently use the commutative, associative, and distributive properties without realizing that we are doing so. To add $13 + 4 + 6$, we may add the $4 + 6$ first to get 10. To this sum we then add 13 to get 23. Here we have done the equivalent of placing parentheses around the $4 + 6$. We can do so because of the associative property of addition.

To multiply 102×11 in our heads, we might multiply $100 \times 11 = 1100$ and $2 \times 11 = 22$ and add these two products to get 1122. We are permitted to do so because of the distributive property.

$$102 \times 11 = (100 + 2) \times 11 = (100 \times 11) + (2 \times 11)$$
$$= 1100 + 22 = 1122$$

Earlier we stated that variables are letters used to represent numbers. In algebra, we will often use the properties of real numbers to simplify expressions involving variables. When a number is placed directly to the left of a variable, it indicates multiplication. For example, $5x$ means $5 \cdot x$. We will use variables in the next example.

Example 2   *Identifying Properties of Real Numbers*

Name the property illustrated.
a) $x \cdot 9 = 9x$
b) $(x + 3) + 7 = x + (3 + 7)$
c) $3(x + 5) = 3 \cdot x + 3 \cdot 5$
d) $5(3y) = (5 \cdot 3)y$
e) $(2w) \cdot 6 = 6 \cdot (2w)$
f) $7 + (z + 4) = 7 + (4 + z)$

Solution

a) Commutative property of multiplication
b) Associative property of addition
c) Distributive property of multiplication over addition

d) Associative property of multiplication

e) The order of $2w$ and 6 is changed using the commutative property of multiplication.

f) The only change between the left and right sides of the equal sign is the order of the z and 4 within the parentheses. The order is changed from $z + 4$ to $4 + z$ using the commutative property of addition. ∎

Example 3 *Simplifying by Using the Distributive Property*

Simplify.

a) $4(5 + \sqrt{6})$ b) $\sqrt{2}(3 + \sqrt{5})$

Solution

a) $4(5 + \sqrt{6}) = (4 \cdot 5) + (4 \cdot \sqrt{6})$

$= 20 + 4\sqrt{6}$

b) $\sqrt{2}(3 + \sqrt{5}) = (\sqrt{2} \cdot 3) + (\sqrt{2} \cdot \sqrt{5})$

$= 3\sqrt{2} + \sqrt{10}$

Note that $\sqrt{2} \cdot 3$ is written as $3\sqrt{2}$. ∎

Example 4 *Distributive Property*

Use the distributive property to multiply $2(x + 5)$. Then simplify the result.

Solution $2(x + 5) = 2 \cdot x + 2 \cdot 5$

$= 2x + 10$ ∎

We summarize the properties mentioned in this section as follows, where a, b, and c are any real numbers.

The Properties of Real Numbers

Commutative property of addition	$a + b = b + a$
Commutative property of multiplication	$a \cdot b = b \cdot a$
Associative property of addition	$(a + b) + c = a + (b + c)$
Associative property of multiplication	$(a \cdot b) \cdot c = a \cdot (b \cdot c)$
Distributive property of multiplication over addition	$a \cdot (b + c) = a \cdot b + a \cdot c$

SECTION 5.5   *Exercises*

Warm Up Exercises

In Exercises 1–8, fill in the blanks with an appropriate word, phrase, or symbol(s).

1. The union of the rational numbers and the irrational numbers is the set of _____ numbers.

2. The symbol used to represent the set of real numbers is _____.

3. If an operation is performed on any two elements of a set and the result is always an element of the set, we say that the set is _____ under that given operation.

4. The equation $2 + 3 = 3 + 2$ demonstrates the _____ property of addition.

5. The equation $4 \cdot 5 = 5 \cdot 4$ demonstrates the _____ property of multiplication.

6. The equation $(2 + 3) + 4 = 2 + (3 + 4)$ demonstrates the _____ property of addition.

7. The equation $(5 \cdot 6) \cdot 7 = 5 \cdot (6 \cdot 7)$ demonstrates the _____ property of multiplication.

8. The equation $2 \cdot (3 + 4) = 2 \cdot 3 + 2 \cdot 4$ demonstrates the _____ property of multiplication over addition.

Practice the Skills

In Exercises 9–12, determine whether the natural numbers are closed under the given operation.

9. Addition

10. Multiplication

11. Subtraction

12. Division

In Exercises 13–16, determine whether the integers are closed under the given operation.

13. Multiplication

14. Division

15. Subtraction

16. Addition

In Exercises 17–20, determine whether the rational numbers are closed under the given operation.

17. Addition

18. Division

19. Subtraction

20. Multiplication

In Exercises 21–24, determine whether the irrational numbers are closed under the given operation.

21. Division

22. Addition

23. Multiplication

24. Subtraction

In Exercises 25–28, determine whether the real numbers are closed under the given operation.

25. Addition

26. Multiplication

27. Division

28. Subtraction

29. Does $(5 + x) + 6 = (x + 5) + 6$ illustrate the commutative property or the associative property? Explain your answer.

30. Does $(x + 5) + 6 = x + (5 + 6)$ illustrate the commutative property or the associative property? Explain your answer.

31. Does the commutative property hold for the rational numbers under the operation of division? Give an example to support your answer.

32. Does the commutative property hold for the integers under the operation of subtraction? Give an example to support your answer.

33. Give an example to show that the commutative property of multiplication may be true for the negative integers.

34. Give an example to show that the commutative property of addition may be true for the negative integers.

35. Give an example to show that the associative property of addition may be true for the negative integers.

36. Give an example to show that the associative property of multiplication may be true for the negative integers.

37. Does the associative property hold for the integers under the operation of division? Give an example to support your answer.

38. Does the associative property hold for the integers under the operation of subtraction? Give an example to support your answer.

39. Does the associative property hold for the real numbers under the operation of division? Give an example to support your answer.

40. Does $a + (b \cdot c) = (a + b) \cdot (a + c)$? Give an example to support your answer.

In Exercises 41–52, state the name of the property illustrated.

41. $4(x + 3) = 4 \cdot x + 4 \cdot 3$

42. $8 + 7 = 7 + 8$

43. $(7 \cdot 8) \cdot 9 = 7 \cdot (8 \cdot 9)$

44. $c + d = d + c$

45. $(24 + 7) + 3 = 24 + (7 + 3)$

46. $4 \cdot (11 \cdot x) = (4 \cdot 11) \cdot x$

47. $\sqrt{3} \cdot 7 = 7 \cdot \sqrt{3}$

48. $\frac{3}{8} + \left(\frac{1}{8} + \frac{3}{2}\right) = \left(\frac{3}{8} + \frac{1}{8}\right) + \frac{3}{2}$

49. $-1(x + 4) = (-1) \cdot x + (-1) \cdot 4$

50. $(r + s) \cdot t = (r \cdot t) + (s \cdot t)$

51. $(r + s) + t = t + (r + s)$

52. $g \cdot (h + i) = (h + i) \cdot g$

In Exercises 53–64, use the distributive property to multiply. Then, if possible, simplify the resulting expression.

53. $3(b + 7)$

54. $9(a + 2)$

55. $6(c - 2)$

56. $4(d - 5)$

57. $-2(4x - 1)$

58. $-3(6x - 5)$

59. $32\left(\frac{1}{16}x - \frac{1}{32}\right)$

60. $15\left(\frac{2}{3}x - \frac{4}{5}\right)$

61. $\sqrt{2}(\sqrt{8} - \sqrt{2})$

62. $-3(2 - \sqrt{3})$

63. $5(\sqrt{2} + \sqrt{3})$

64. $\sqrt{5}(\sqrt{15} - \sqrt{20})$

Problem Solving

In Exercises 65–68, determine whether the activity can be used to illustrate the commutative property. For the property to hold, the end result must be identical, regardless of the order in which the actions are performed.

65. Feeding your cats and giving your cats water

66. Putting your wallet in your back pocket and putting your keys in your front pocket

67. Washing clothes and drying clothes

68. Turning on a computer and sending an email on the computer

In Exercises 69–74, determine whether the activity can be used to illustrate the associative property. For the property to hold, doing the first two actions followed by the third would produce the same end result as doing the second and third actions followed by the first.

69. Putting batteries in your calculator, pressing the $\boxed{\text{ON}}$ key, and pressing the $\boxed{7}$ key

70. Filling your car with gas, washing the windshield, and checking the tire pressure

71. Sending a text message to your grandmother, sending one to your parents, and sending one to your friend

72. Mowing the lawn, trimming the bushes, and removing dead limbs from trees

73. Brushing your teeth, washing your face, and combing your hair

74. Cracking an egg, pouring out the egg, and cooking the egg

Challenge Problems/Group Activities

75. Describe two other activities that can be used to illustrate the commutative property (see Exercises 65–68).

76. Describe three other activities that can be used to illustrate the associative property (see Exercises 69–74).

77. Does $0 \div a = a \div 0$ (assume $a \neq 0$)? Explain.

Recreational Math

78. *KenKen* Refer to the *Recreational Mathematics* box on page 254. Complete the following KenKen puzzle.

2	2−	2÷	24×
3−			
	7+		
1−		5+	

79. a) Consider the three words *man eating tiger*. Does (*man eating*) *tiger* mean the same as *man* (*eating tiger*)?

   b) Does (*horse riding*) *monkey* mean the same as *horse* (*riding monkey*)?

   c) Can you find three other nonassociative word triples?

Research Activity

80. *Complex Numbers* A set of numbers that was not discussed in this chapter is the set of *complex numbers*. Write a report on complex numbers. Include their relationship to the real numbers.

SECTION 5.6   Rules of Exponents and Scientific Notation

Upon completion of this section, you will be able to:

- Evaluate expressions using the rules of exponents.
- Write numbers using scientific notation.
- Use a scientific calculator to solve problems involving scientific notation.

What is the current world population? What is the diameter of an atom? What is the diameter of a galaxy? What is the wavelength of an x-ray? What is the current national federal debt? All these questions have answers that may involve either very large numbers or very small numbers. One way to accurately represent such numbers is with *scientific notation*. In this section, we will study how to use scientific notation when answering questions such as those posed above.

Why This Is Important Scientific notation is used in many fields of study, including economics, physics, chemistry, and astronomy. Items such as a computer hard drive's capacity, measured in gigabytes, can also be represented using scientific notation.

In Section 5.2 we introduced exponents. Here we discuss the rules of exponents, which are very important to the study of mathematics.

Rules of Exponents

We begin with the product rule for exponents. Consider

$$2^2 \cdot 2^3 = \underbrace{2 \cdot 2}_{\text{2 factors}} \cdot \underbrace{2 \cdot 2 \cdot 2}_{\text{3 factors}} = 2^5$$

This example illustrates the product rule for exponents.

> ### Product Rule for Exponents
> $$a^m \cdot a^n = a^{m+n}$$

Therefore, by using the product rule, $2^2 \cdot 2^3 = 2^{2+3} = 2^5$.

Example 1 Using the Product Rule for Exponents

Use the product rule to simplify.
a) $4^3 \cdot 4^2$ b) $6 \cdot 6^3$

Solution

a) $4^3 \cdot 4^2 = 4^{3+2} = 4^5 = 1024$ b) $6 \cdot 6^3 = 6^1 \cdot 6^3 = 6^4 = 1296$ ∎

Consider

$$\frac{2^5}{2^2} = \frac{2 \cdot 2 \cdot 2 \cdot \cancel{2} \cdot \cancel{2}}{\cancel{2} \cdot \cancel{2}} = 2 \cdot 2 \cdot 2 = 2^3$$

This example illustrates the quotient rule for exponents.

> ### Quotient Rule for Exponents
> $$\frac{a^m}{a^n} = a^{m-n}, \quad a \neq 0$$

Therefore, $\dfrac{2^5}{2^2} = 2^{5-2} = 2^3$.

Did You Know?

Catalan's Conjecture

Consider all the squares of positive integers:

$1^2 = 1$, $2^2 = 4$, $3^2 = 9$,
$4^2 = 16$,   and so on

Now consider all the cubes of positive integers:

$1^3 = 1$, $2^3 = 8$, $3^3 = 27$,
$4^3 = 64$,   and so on

Next, consider all the fourth powers of positive integers:

$1^4 = 1$, $2^4 = 16$, $3^4 = 81$,
$4^4 = 256$,   and so on

If we were to put *all* powers of the positive integers greater than one, into one set, the set would begin as follows: $\{1, 4, 8, 9, 16, 25, 27, 32, 36, 49, 64, \ldots\}$. In 1844, Belgian mathematician Eugène Catalan proposed that of all the numbers in this infinite set, only 8 and 9 are consecutive integers. Although this conjecture was easily stated, the formal proof of it eluded mathematicians for more than 150 years. In April 2002, Preda Mihailescu, a Romanian-born mathematician, proved Catalan's Conjecture. The study of mathematics involves many conjectures that are often easy to pose, but difficult to prove. For another example of a now-proven longstanding conjecture, see the *Did You Know* box on page 472 regarding Fermat's Last Theorem. For two more still unproven conjectures see the discussion of Goldbach's conjecture and the twin-prime conjecture in Section 5.1.

Example 2 *Using the Quotient Rule for Exponents*

Use the quotient rule to simplify.

a) $\dfrac{3^7}{3^5}$

b) $\dfrac{5^9}{5^5}$

Solution

a) $\dfrac{3^7}{3^5} = 3^{7-5} = 3^2 = 9$

b) $\dfrac{5^9}{5^5} = 5^{9-5} = 5^4 = 625$

Consider $2^3 \div 2^3$. The quotient rule gives

$$\frac{2^3}{2^3} = 2^{3-3} = 2^0$$

But $\dfrac{2^3}{2^3} = \dfrac{8}{8} = 1$. Therefore, 2^0 must equal 1. This example illustrates the zero exponent rule.

Zero Exponent Rule

$$a^0 = 1, \quad a \neq 0$$

Note that 0^0 is not defined by the zero exponent rule.

Example 3 *Using the Zero Exponent Rule*

Use the zero exponent rule to simplify. Assume $x \neq 0$.
a) 2^0 b) $(-2)^0$ c) -2^0 d) $(5x)^0$ e) $5x^0$

Solution

a) $2^0 = 1$ b) $(-2)^0 = 1$ c) $-2^0 = -1 \cdot 2^0 = -1 \cdot 1 = -1$
d) $(5x)^0 = 1$ e) $5x^0 = 5 \cdot x^0 = 5 \cdot 1 = 5$

Consider $2^3 \div 2^5$. The quotient rule yields

$$\frac{2^3}{2^5} = 2^{3-5} = 2^{-2}$$

But $\dfrac{2^3}{2^5} = \dfrac{\not{2} \cdot \not{2} \cdot \not{2}}{\not{2} \cdot \not{2} \cdot \not{2} \cdot 2 \cdot 2} = \dfrac{1}{2^2}$. Since $\dfrac{2^3}{2^5}$ equals both 2^{-2} and $\dfrac{1}{2^2}$, then 2^{-2} must equal $\dfrac{1}{2^2}$. This example illustrates the negative exponent rule.

Negative Exponent Rule

$$a^{-m} = \frac{1}{a^m}, \quad a \neq 0$$

Example 4 *Using the Negative Exponent Rule*

Use the negative exponent rule to simplify.
a) 3^{-2} b) 7^{-1}

Solution

a) $3^{-2} = \dfrac{1}{3^2} = \dfrac{1}{9}$

b) $7^{-1} = \dfrac{1}{7^1} = \dfrac{1}{7}$

Consider $(2^3)^2$.

$$(2^3)^2 = (2^3)(2^3) = 2^{3+3} = 2^6$$

This example illustrates the power rule for exponents.

Power Rule for Exponents

$$(a^m)^n = a^{m \cdot n}$$

Thus, $(2^3)^2 = 2^{3 \cdot 2} = 2^6$.

Example 5 Using the Power Rule for Exponents

Use the power rule for exponents to simplify.

a) $(2^4)^5$ 　　　　　　　　　b) $(3^7)^2$

Solution

a) $(2^4)^5 = 2^{4 \cdot 5} = 2^{20}$ 　　b) $(3^7)^2 = 3^{7 \cdot 2} = 3^{14}$ ■

Summary of the Rules of Exponents

$a^m \cdot a^n = a^{m+n}$	Product rule for exponents
$\dfrac{a^m}{a^n} = a^{m-n}, \quad a \neq 0$	Quotient rule for exponents
$a^0 = 1, \quad a \neq 0$	Zero exponent rule
$a^{-m} = \dfrac{1}{a^m}, \quad a \neq 0$	Negative exponent rule
$(a^m)^n = a^{m \cdot n}$	Power rule for exponents

Scientific Notation

Often scientific problems deal with very large and very small numbers. For example, the distance from Earth to the sun is about 93,000,000 miles. The wavelength of a yellow color of light is about 0.0000006 meter. Because working with many zeros is difficult, scientists developed a notation that expresses such numbers with exponents. For example, consider the distance from Earth to the sun, 93,000,000 miles.

$$93,000,000 = 9.3 \times 10,000,000$$
$$= 9.3 \times 10^7$$

The wavelength of a yellow color of light is about 0.0000006 meter.

$$0.0000006 = 6.0 \times 0.0000001$$
$$= 6.0 \times 10^{-7}$$

The numbers 9.3×10^7 and 6.0×10^{-7} are written in a form called *scientific notation*. Each number written in scientific notation is written as a number greater than or equal to 1 and less than 10 multiplied by some power of 10.

Some examples of numbers in scientific notation are

$$3.7 \times 10^3, \quad 2.05 \times 10^{-3}, \quad 5.6 \times 10^8, \quad \text{and} \quad 1.00 \times 10^{-5}$$

The following is a procedure for writing a number in scientific notation.

> **PROCEDURE** **TO WRITE A NUMBER IN SCIENTIFIC NOTATION**
>
> 1. Move the decimal point in the original number to the right or left until you obtain a number greater than or equal to 1 and less than 10.
> 2. Count the number of places you have moved the decimal point to obtain the number in Step 1. If the decimal point was moved to the left, the count is to be considered positive. If the decimal point was moved to the right, the count is to be considered negative.
> 3. Multiply the number obtained in Step 1 by 10 raised to the count determined in Step 2. (Note that the count determined in Step 2 is the exponent on the base 10.)

Example 6 *Converting from Decimal Notation to Scientific Notation*

Write each number in scientific notation.

a) One drop of blood contains about 5,800,000 blood cells.
b) In 2018, there were about 276,100,000 million cars registered in the United States.
c) In 2018, the population of China was about 1,415,000,000.
d) The diameter of a hydrogen atom nucleus is about 0.0000000000011 millimeter.
e) The wavelength of an x-ray is about 0.000000000492 meter.

> **Solution**
>
> a) $5,800,000 = 5.8 \times 10^6$
> b) $276,100,000 = 2.761 \times 10^8$
> c) $1,415,000,000 = 1.415 \times 10^9$
> d) $0.0000000000011 = 1.1 \times 10^{-12}$
> e) $0.000000000492 = 4.92 \times 10^{-10}$ ∎

To convert from a number given in scientific notation to decimal notation, we reverse the procedure.

> **PROCEDURE** **TO CHANGE A NUMBER IN SCIENTIFIC NOTATION TO DECIMAL NOTATION**
>
> 1. Observe the exponent on the 10.
> 2. a) If the exponent is positive, move the decimal point in the number to the right the same number of places as the exponent. Adding zeros to the number might be necessary.
>   b) If the exponent is negative, move the decimal point in the number to the left the same number of places as the exponent. Adding zeros might be necessary.

Scientific
Notation

Example 7 *Converting from Scientific Notation to Decimal Notation*

Write each number in decimal notation.

a) The average distance from Mars to the sun is about 1.4×10^8 miles.
b) The half-life of uranium-235 is about 4.5×10^9 years.
c) The average grain size in siltstone is about 1.35×10^{-3} inch.
d) A *millimicron* is a unit of measure used for very small distances. One millimicron is about 3.94×10^{-8} inch.

Solution

a) $1.4 \times 10^8 = 140{,}000{,}000$ b) $4.5 \times 10^9 = 4{,}500{,}000{,}000$

c) $1.35 \times 10^{-3} = 0.00135$ d) $3.94 \times 10^{-8} = 0.0000000394$ ■

In scientific journals and books, we occasionally see numbers like 10^{15} and 10^{-6}. We interpret these numbers as 1×10^{15} and 1×10^{-6}, respectively, when converting the numbers to decimal form.

Example 8 *Multiplying Numbers in Scientific Notation*

Multiply $(2.1 \times 10^5)(9 \times 10^{-3})$. Write the answer in scientific notation and in decimal notation.

Solution We begin by rewriting the product so that 2.1 and 9 are multiplied together and the powers of 10 are multiplied together.

$$
\begin{aligned}
(2.1 \times 10^5)(9 \times 10^{-3}) &= (2.1 \times 9)(10^5 \times 10^{-3}) \\
&= 18.9 \times 10^{5+(-3)} \\
&= 18.9 \times 10^2 \\
&= 1.89 \times 10^3 \qquad \text{Scientific notation} \\
&= 1890 \qquad\qquad \text{Decimal notation} \qquad ■
\end{aligned}
$$

Example 9 *Dividing Numbers Using Scientific Notation*

Divide $\dfrac{0.000000000048}{24{,}000{,}000{,}000}$. Write the answer in scientific notation.

Solution First write each number in scientific notation.

$$
\begin{aligned}
\frac{0.000000000048}{24{,}000{,}000{,}000} = \frac{4.8 \times 10^{-11}}{2.4 \times 10^{10}} &= \left(\frac{4.8}{2.4}\right)\left(\frac{10^{-11}}{10^{10}}\right) \\
&= 2.0 \times 10^{-11-10} \\
&= 2.0 \times 10^{-21} \qquad ■
\end{aligned}
$$

Scientific Notation on a Scientific Calculator

Performing operations on numbers written in scientific notation can be done with most scientific calculators.

TECHNOLOGY TIP

Scientific Notation on a Scientific Calculator

Most scientific calculators have a scientific notation key labeled "Exp," "EXP," or "EE." We will refer to the scientific notation key as ⬚EXP⬚ . The following keystrokes can be used to enter the number 3.7×10^6.

Keystroke	Calculator Display
3.7	3.7
EXP	3.7E
6	3.7E6

Your calculator may have some variations to the display shown here. The display 3.7E6 means 3.7×10^6.

We will now use a scientific calculator to perform some computations using scientific notation.

MATHEMATICS TODAY

What's the Difference Between Debt and Deficit?

▲ *A World War I–era poster encouraging investment in U.S. government bonds*

We often hear economists and politicians talk about such things as revenue, expenditures, deficit, surplus, and national debt. Revenue is the money a government collects annually, mostly through taxes. Expenditures are the money a government spends annually. If revenue exceeds expenditures, a surplus occurs; if expenditures exceed revenue, a deficit occurs for that year. The national debt is the total of all the budget deficits and (the few) surpluses encountered by the U.S. federal government for more than 200 years. In 2018, the national debt was about \$21.97 trillion, or about $\$2.197 \times 10^{13}$. This amount is owed to investors worldwide—perhaps including you—who own U.S. government bonds.

Why This Is Important Understanding the financial structure of our government is an important part of understanding our democratic system. The same principles discussed here also apply to our own finances.

Example 10 *Use Scientific Notation on a Calculator to Determine a Product*

Multiply $(3.7 \times 10^6)(2.1 \times 10^{-2})$ using a scientific calculator. Write the answer in decimal notation.

Solution Our sequence of keystrokes is as follows:

Keystroke	Display
3.7	3.7
EXP	3.7E
6	3.7E6
×	3.7E6∗
2.1	3.7E6∗2.1
EXP	3.7E6∗2.1E
(−) ∗	3.7E6∗2.1E−
2	3.7E6∗2.1E−2
=	77700[†]

Thus the product is 77,700. ∎

Example 11 *Use Scientific Notation on a Calculator to Determine a Quotient*

Divide $\dfrac{0.000000000048}{24{,}000{,}000{,}000}$ using a scientific calculator. Write the answer in scientific notation.

Solution We first rewrite the numerator and denominator using scientific notation. See Example 9.

$$\frac{0.000000000048}{24{,}000{,}000{,}000} = \frac{4.8 \times 10^{-11}}{2.4 \times 10^{10}}$$

Next we use a scientific calculator to perform the computation. The keystrokes are as follows:

4.8 EXP (−) 11 ÷ 2.4 EXP 10 =

The display on the calculator is $2.^{-21}$, which means 2.0×10^{-21}. ∎

Example 12 *U.S. Debt per Person*

In 2018, the U.S. Department of the Treasury estimated the U.S. national debt to be about \$21.97 trillion. In 2018, the population of the United States was estimated to be about 323 million. Determine the average debt per person by dividing the U.S. national debt by the U.S. population.

Solution We first write the numbers involved using decimal notation. We then convert them to scientific notation.

$$21.97 \text{ trillion} = 21{,}970{,}000{,}000{,}000 = 2.197 \times 10^{13}$$
$$323 \text{ million} = 323{,}000{,}000 = 3.23 \times 10^8$$

[∗]Some calculators will require you to enter the negative sign using a +/− key *after* entering the exponent.
[†]Some calculators will display the answer in scientific notation: 7.77E4, which means 7.77×10^4 and equals 77,700.

Next we divide 2.197×10^{13} by 3.23×10^8 using a scientific calculator. The keystrokes are

$$2.197 \boxed{\text{EXP}} \ 13 \ \boxed{\div} \ 3.23 \boxed{\text{EXP}} \ 8 \ \boxed{=}$$

The display shows 68,018.57585. This number indicates that in 2018, the U.S. government owed about \$68,018.58 per person living in the United States. ∎

SECTION 5.6 Exercises

Warm Up Exercises

In Exercises 1–6, fill in the blanks with an appropriate word, phrase, or symbol(s).

1. With the product rule for exponents, the expression $x^2 \cdot x^3$ can be simplified to _____.

2. With the quotient rule for exponents, the expression $\dfrac{x^7}{x^4}$ can be simplified to _____.

3. With the zero exponent rule, the expression 7^0 can be simplified to _____.

4. With the negative exponent rule, for $x \neq 0$, the expression x^{-5} can be simplified to _____.

5. With the power rule for exponents, the expression $(x^2)^3$ can be simplified to _____.

6. A number written in scientific notation is a number greater than or equal to 1 and less than 10 multiplied by a power of _____.

Practice the Skills

In Exercises 7–22, evaluate the expression. Assume $x \neq 0$.

7. a) $2^2 \cdot 2^3$ b) $(-2)^2 \cdot (-2)^3$

8. a) $3 \cdot 3^3$ b) $(-3) \cdot (-3)^3$

9. a) $\dfrac{5^7}{5^5}$ b) $\dfrac{(-5)^7}{(-5)^5}$

10. a) $\dfrac{4^5}{4^2}$ b) $\dfrac{(-4)^5}{(-4)^2}$

11. a) 6^0 b) -6^0

12. a) $(-6)^0$ b) $-(-6)^0$

13. a) $(6x)^0$ b) $6x^0$

14. a) $-6x^0$ b) $(-6x)^0$

15. a) 3^{-3} b) 7^{-2}

16. a) 5^{-1} b) 2^{-4}

17. a) -9^{-2} b) $(-9)^{-2}$

18. a) -5^{-2} b) $(-5)^{-2}$

19. a) $(2^3)^2$ b) $(3^2)^3$

20. a) $\left[\left(\dfrac{1}{2}\right)^2\right]^3$ b) $\left[\left(-\dfrac{1}{2}\right)^2\right]^3$

21. a) $4^3 \cdot 4^{-2}$ b) $2^{-2} \cdot 2^{-2}$

22. a) $(-5)^{-2}(-5)^{-2}$ b) $(-1)^{-5}(-1)^{-5}$

In Exercises 23–32, express the number in scientific notation.

23. 503,000

24. 81,000,000

25. 0.00042

26. 0.0000063

27. 0.56

28. 0.00467

29. 19,000

30. 1,260,000,000

31. 0.000186

32. 0.0003

In Exercises 33–40, express the number in decimal notation.

33. 2.3×10^3

34. 9.6×10^5

35. 1.68×10^{-3}

36. 4.42×10^{-4}

37. 8.62×10^{-5}

38. 2.19×10^{-4}

39. 2.01×10^0

40. 4.6×10^1

In Exercises 41–48, (a) perform the indicated operation without the use of a calculator and express each answer in decimal notation. (b) Confirm your answer from part (a) by using a scientific calculator to perform the operations. If the calculator displays the answer in scientific notation, convert the answer to decimal notation.

41. $(4.1 \times 10^2)(2 \times 10^3)$

42. $(2.2 \times 10^3)(3 \times 10^3)$

43. $(5.1 \times 10^1)(3 \times 10^{-4})$

44. $(1.6 \times 10^{-2})(4 \times 10^{-3})$

45. $\dfrac{7.5 \times 10^6}{3 \times 10^4}$

46. $\dfrac{2.7 \times 10^3}{9 \times 10^{-2}}$

47. $\dfrac{8.4 \times 10^{-6}}{4 \times 10^{-3}}$

48. $\dfrac{5.2 \times 10^7}{4 \times 10^9}$

In Exercises 49–56, (a) perform the indicated operation without the use of a calculator and express each answer in scientific notation. (b) Confirm your answer from part (a) by using a scientific calculator to perform the operations. If the calculator displays the answer in decimal notation, convert the answer to scientific notation.

49. $(200{,}000)(3600)$

50. $(230{,}000)(3000)$

51. $(0.003)(0.00015)$

52. $(50{,}000)(0.00011)$

53. $\dfrac{5{,}600{,}000}{80{,}000}$

54. $\dfrac{28{,}000}{0.004}$

55. $\dfrac{0.00004}{200}$

56. $\dfrac{0.0012}{0.000006}$

Problem Solving

In Exercises 57–60, list the numbers from smallest to largest.

57. 1.03×10^4; 1.7; 3.6×10^{-3}; 9.8×10^2

58. 8.5×10^{-5}; 8.2×10^3; 1.3×10^{-1}; 6.2×10^4

59. $40{,}000$; 4.1×10^3; 0.00079; 8.3×10^{-5}

60. $267{,}000{,}000$; 3.14×10^7; $1{,}962{,}000$; 4.79×10^6

In Exercises 61–78, use a scientific calculator to perform the necessary operations.

61. U.S. Debt per Person in 2014 Versus 2018 In Example 12 on page 264, it was determined that the U.S. national debt in 2018 was about \$68,018.58 per person. In 2014, the U.S. national debt was about \$17.82 trillion, and the U.S. population was about 318.6 million people.

a) Determine the amount of debt per person in 2014.

b) How much more per person did the U.S. government owe in 2018 than in 2014?

62. Spain's Debt per Person In 2018, the Spanish national debt was about 1.171 trillion euros (the euro is the unit of currency in Spain). Spain's population in 2018 was about 46.4 million people.

a) Determine the amount of debt per person for Spain in 2014.

b) On March 9, 2019, 1 euro was worth about \$1.12. Convert the amount determined in part (a) to dollars.

63. U.S. Population On March 9, 2019, the population of the United States was about 328 million people, and the population of the world was about 7.69 billion people. Determine the ratio of the U.S. population to the world's population by dividing the U.S. population by the world's population. Write your answer as a decimal number to the nearest thousandth.

64. China's Population. On March 9, 2019, the population of China was about 1.42 billion people, and the population of the world was about 7.69 billion people. Determine the ratio of China's population to the world's population by dividing China's population by the world's population. Write your answer as a decimal number to the nearest thousandth.

65. Gross Domestic Product The gross domestic product (GDP) of a country is the total output of goods and services produced within that country. In 2018, the GDP of the United States was about \$20.66 trillion, and the U.S. population was about 328 million people. Determine the U.S. GDP per person by dividing the GDP by the population.

66. China's GDP In 2018, the GDP (see Exercise 65) of China was about \$14.2 trillion, and the population of China was about 1.42 billion people. Determine China's GDP per person by dividing the GDP by the population.

67. Traveling to the Moon The distance from Earth to the moon is approximately 239,000 mi. If a spacecraft travels at a speed of 20,000 mph, how many hours would the spacecraft need to travel from Earth to the moon? Use distance = rate × time.

68. Traveling to Proxima Centauri The star closest in distance to our own sun is Proxima Centauri. The distance from Earth to Proxima Centauri is approximately 2.5×10^{13} miles. If a spacecraft travels at a speed of 20,000 mph, how many hours would the spacecraft need to travel to Proxima Centauri? Use distance = rate × time and write your answer in scientific notation.

69. 1950 Niagara Treaty The 1950 Niagara Treaty between the United States and Canada requires that during the tourist season a minimum of 100,000 cubic feet of water per second (ft^3/sec) flow over Niagara Falls (another 130,000–160,000 ft^3/sec are diverted for power generation). Determine the minimum amount of water that will flow over the falls in a 24-hour period during the tourist season. Write your answer in scientific notation.

70. ***Computer Speed*** As of March 13, 2019, the fastest computer in the world was the IBM *Summit* located at Oak Ridge National Laboratory in Tennessee. This computer is capable of performing about 200,000 trillion (2.0×10^{17}) calculations per second. At this rate, how long with it take for this computer to perform a task requiring 7.6×10^{65} calculations? Write your answer in scientific notation.

71. ***Blood Cells in a Cubic Millimeter*** If a cubic millimeter of blood contains 5,800,000 red blood cells, how many red blood cells are contained in 50 cubic millimeters of blood? Write your answer in scientific notation.

72. ***Bucket Full of Molecules*** A drop of water contains about 40 billion molecules. If a bucket has half a million drops of water in it, how many molecules of water are in the bucket? Write your answer in scientific notation.

73. ***Earth to Sun Comparison*** The mass of the sun is approximately 2×10^{30} kilograms, and the mass of Earth is approximately 6×10^{24} kilograms. How many times greater is the mass of the sun than the mass of Earth? Write your answer in decimal notation.

74. ***Radioactive Isotopes*** The half-life of a radioactive isotope is the time required for half the quantity of the isotope to decompose. The half-life of uranium-238 is 4.5×10^{9} years, and the half-life of uranium-234 is 2.5×10^{5} years. How many times greater is the half-life of uranium-238 than uranium-234?

75. ***Metric System Comparison*** In the metric system, 1 meter $= 10^{3}$ millimeters. How many times greater is a meter than a millimeter?

76. ***Milligrams and Kilograms*** In the metric system, 1 gram $= 10^{3}$ milligrams and 1 gram $= 10^{-3}$ kilogram. What is the relationship between milligrams and kilograms?

77. ***Internet Usage*** In 2018, the Falkland Islands was the country with the largest percentage of its population using the Internet. The population of the Falkland Islands is about 2.92×10^{3} people. If 97% of the Falklands Islands population used the Internet in 2018, approximately how many people used the Internet?

▲ Falkland Islands

78. ***Smartphone Usage*** In 2019, there were approximately 2.50×10^{9} smartphone users worldwide. If the world's population in 2019 was about 7.71×10^{9}, determine the approximate percentage of the worldwide population that used a smartphone.

Challenge Problems/Group Activities

79. ***Comparing a Million to a Billion*** Many people have no idea of the difference in size between a million (1,000,000), a billion (1,000,000,000), and a trillion (1,000,000,000,000).

    a) Write a million, a billion, and a trillion in scientific notation.

    b) Determine how long it would take to spend one million dollars if you spent $1000 a day.

    c) Repeat part (b) for one billion dollars.

    d) Repeat part (b) for one trillion dollars.

    e) How many times greater is one billion dollars than one million dollars?

80. ***Speed of Light***

    a) Light travels at a speed of 1.86×10^{5} mi/sec. A *lightyear* is the distance that light travels in 1 year. Determine the number of miles in a light-year.

    b) Earth is approximately 93,000,000 mi from the sun. How long does it take light from the sun to reach Earth?

81. ***Bacteria in a Culture*** The exponential function $E(t) = 2^{10} \cdot 2^{t}$ approximates the number of bacteria in a certain culture after t hours.

    a) The initial number of bacteria is determined when $t = 0$. What is the initial number of bacteria?

    b) How many bacteria are there after $\frac{1}{2}$ hour?

Research Activities

82. ***U.S. Government Debt*** Obtain data from the U.S. Department of the Treasury and from the U.S. Bureau of the Census Internet websites to calculate the current U.S. national debt per person. Write a report in which you compare your figure with those obtained in Exercise 61 and Example 12. Include in your report definitions of the following terms: revenues, expenditures, deficit, and surplus.

83. ***John Allen Paulos*** Read the *Profiles in Mathematics* box on page 262. Read one of Paulos's books and write a 500-word report on it.

84. ***Scientific Notation*** Find an article in a newspaper or magazine or on the Internet that contains scientific notation. Write a paragraph explaining how scientific notation was used.

Arithmetic and Geometric Sequences

Upon completion of this section, you will be able to:

■ Identify and solve problems involving arithmetic sequences.

■ Identify and solve problems involving geometric sequences.

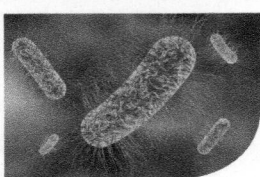

Under optimal laboratory conditions, the number of *Escherichia coli* (*E. coli*) bacteria will double every 17 minutes. If an experiment conducted under optimal conditions begins with 100 bacteria, how many bacteria will be present after 24 hours? Questions such as these are important to the modern study of diseases. In this section, we will study such questions by using lists of numbers known as *sequences*.

Why This Is Important Sequences are used throughout many branches of mathematics to model real-life phenomena, such as population growth, disease control, and the inflation rates of currencies.

Now we will introduce sequences. A *sequence* is a list of numbers that are related to each other by a rule. The numbers that form the sequence are called its *terms*. If your salary increases or decreases by a fixed amount over a period of time, the listing of the amounts, over time, would form an arithmetic sequence. When interest in a savings account is compounded at regular intervals, the listing of the amounts in the account over time will be a geometric sequence.

Arithmetic Sequences

A sequence in which each term after the first term differs from the preceding term by a constant amount is called an *arithmetic sequence*. The amount by which each pair of successive terms differs is called the *common difference*, *d*. The common difference can be determined by subtracting any term from the term that directly follows it.

Examples of arithmetic sequences	**Common differences**
$1, 5, 9, 13, 17, \ldots$	$d = 5 - 1 = 4$
$-7, -5, -3, -1, 1, \ldots$	$d = -5 - (-7) = -5 + 7 = 2$
$\dfrac{5}{2}, \dfrac{3}{2}, \dfrac{1}{2}, -\dfrac{1}{2}, \ldots$	$d = \dfrac{3}{2} - \dfrac{5}{2} = -\dfrac{2}{2} = -1$

Example 1 *The First Five Terms of an Arithmetic Sequence*

Write the first five terms of an arithmetic sequence with first term 4 and common difference of 6.

Solution The first term is 4. The second term is $4 + 6$, or 10. The third term is $10 + 6$, or 16. The fourth term is $16 + 6$, or 22. The fifth term is $22 + 6$, or 28. Thus, the first five terms of the sequence are 4, 10, 16, 22, 28. ∎

Example 2 *An Arithmetic Sequence with a Negative Difference*

Write the first five terms of the arithmetic sequence with first term 9 and a common difference of -4.

Solution The first five terms of the sequence are

$$9, 5, 1, -3, -7$$

∎

When discussing a sequence, we often represent the first term as a_1 (read "a sub 1"), the second term as a_2, the fifteenth term as a_{15}, and so on. We use the notation a_n to represent the general or nth term of a sequence. Thus, a sequence may be symbolized as

$$a_1, a_2, a_3, a_4, \ldots, a_n, \ldots$$

For example, in the sequence 2, 5, 8, 11, 14, $\ldots$, we have

$$a_1 = 2, a_2 = 5, a_3 = 8, a_4 = 11, a_5 = 14, \ldots.$$

The following formula can be used to determine the nth term of an arithmetic sequence.

General or nth Term of an Arithmetic Sequence

For an arithmetic sequence with first term a_1 and common difference d, the general or nth term can be determined using the following formula.

$$a_n = a_1 + (n - 1)d$$

Example 3 *Determining the Twelfth Term of an Arithmetic Sequence*

Determine the twelfth term of the arithmetic sequence whose first term is -2 and whose common difference is 5.

Solution To determine the twelfth term, or a_{12}, replace n in the formula with 12, a_1 with -2, and d with 5.

$$a_n = a_1 + (n - 1)d$$
$$a_{12} = -2 + (12 - 1)5$$
$$= -2 + (11)5$$
$$= -2 + 55$$
$$= 53$$

The twelfth term is 53. As a check, we have listed the first 12 terms of the sequence:

$$-2, 3, 8, 13, 18, 23, 28, 33, 38, 43, 48, 53.$$ ∎

Example 4 *Determining an Expression for the nth Term*

Write an expression for the general or nth term, a_n, for the sequence 1, 6, 11, 16, $\ldots$.

Solution In this sequence, the first term, a_1, is 1 and the common difference, d, is 5. We substitute these values into $a_n = a_1 + (n - 1)d$ to obtain an expression for the nth term, a_n.

$$a_n = a_1 + (n - 1)d$$
$$= 1 + (n - 1)5$$
$$= 1 + 5n - 5$$
$$= 5n - 4$$

The expression is $a_n = 5n - 4$. Note that when $n = 1$, the first term is $5(1) - 4 = 1$. When $n = 2$, the second term is $5(2) - 4 = 6$, and so on.   ∎

The Martingale system is a strategy used by gamblers to determine the amount to bet while playing multiple rounds of a game of chance. The strategy requires the player to consistently bet a standard amount following a win and to double the previous bet following a loss. The underlying principle is that the first win will recover all the previous losses, thus ensuring that the player does not lose money in the long run. Let's use a standard bet of $1 as an example. In the first round, the player bets $1. If the player wins, the bet in the second round remains $1. If the player loses, the bet in the second round doubles to $2. Next, if the player wins in the second round, the bet in the third round returns to $1; however, if the player loses in the second round, the bet in the third round doubles to $4. The system continues with the player betting $1 following a win, and doubling the previous bet following a loss. Although the strategy may work in the short term, a player may have difficulty doubling the bet after several consecutive losses. Casinos generally place a maximum on the amount a person can bet. Exercise 74 on page 275 demonstrates why the Martingale system is a risky wagering strategy.

We can use the following formula to determine the sum of the first n terms in an arithmetic sequence.

Sum of the First n Terms in an Arithmetic Sequence
The sum of the first n terms of an arithmetic sequence can be determined with the following formula where a_1 represents the first term and a_n represents the nth term.

$$s_n = \frac{n(a_1 + a_n)}{2}$$

In this formula, s_n represents the sum of the first n terms, a_1 is the first term, a_n is the nth term, and n is the number of terms in the sequence from a_1 to a_n.

Example 5 *Determining the Sum of an Arithmetic Sequence*

Determine the sum of the first 25 even natural numbers.

Solution The sequence we are discussing is

$$2, 4, 6, 8, 10, \ldots, 50$$

In this arithmetic sequence the first term is 2, the 25th term is 50, and there are 25 terms; therefore, $a_1 = 2$, $a_{25} = 50$, and $n = 25$. Thus, the sum of the first 25 terms is

$$s_n = \frac{n(a_1 + a_n)}{2}$$

$$s_{25} = \frac{25(2 + 50)}{2}$$

$$= \frac{25(52)}{2}$$

$$= \frac{1300}{2}$$

$$= 650$$

The sum of the first 25 even natural numbers, $2 + 4 + 6 + 8 + \cdots + 50$, is 650. ∎

Geometric Sequences

The next type of sequence we will discuss is the geometric sequence. A *geometric sequence* is one in which the ratio of any term to the term that directly precedes it is a constant. This constant is called the *common ratio*. The common ratio, r, can be determined by taking any term except the first and dividing that term by the preceding term.

Examples of geometric sequences	Common ratios
$2, 4, 8, 16, 32, \ldots$	$r = 4 \div 2 = 2$; $r = 8 \div 4 = 2$
$-3, 6, -12, 24, -48, \ldots$	$r = 6 \div (-3) = -2$; $r = (-12) \div 6 = -2$
$\dfrac{2}{3}, \dfrac{2}{9}, \dfrac{2}{27}, \dfrac{2}{81}, \ldots$	$r = \dfrac{2}{9} \div \dfrac{2}{3} = \left(\dfrac{2}{9}\right)\left(\dfrac{3}{2}\right) = \dfrac{1}{3}$; $r = \dfrac{2}{27} \div \dfrac{2}{9} = \left(\dfrac{2}{27}\right)\left(\dfrac{9}{2}\right) = \dfrac{1}{3}$

To construct a geometric sequence when the first term, a_1, and common ratio, r, are known, multiply the first term by the common ratio to get the second term. Then multiply the second term by the common ratio to get the third term, and so on.

Example 6   *The First Five Terms of a Geometric Sequence*

Write the first five terms of the geometric sequence whose first term, a_1, is 3 and whose common ratio, r, is 4.

Solution   The first term is 3. The second term, determined by multiplying the first term by 4, is $3 \cdot 4$, or 12. The third term is $12 \cdot 4$, or 48. The fourth term is $48 \cdot 4$ or 192. The fifth term is $192 \cdot 4$, or 768. Thus, the first five terms of the sequence are 3, 12, 48, 192, 768. ∎

The following formula can be used to determine the nth term of a geometric sequence.

> ### General or nth Term of a Geometric Sequence
> For a geometric sequence with first term a_1 and common ratio r, the general or nth term can be determined using the following formula.
> $$a_n = a_1 r^{n-1}$$

Example 7   *Determining the Twelfth Term of a Geometric Sequence*

Determine the twelfth term of the geometric sequence whose first term is -4 and whose common ratio is 2.

Solution   In this sequence, $a_1 = -4$, $r = 2$, and $n = 12$. Substituting the values into the formula, we obtain

$$a_n = a_1 r^{n-1}$$
$$a_{12} = -4 \cdot 2^{12-1}$$
$$= -4 \cdot 2^{11}$$
$$= -4 \cdot 2048$$
$$= -8192$$

As a check, we have listed the first twelve terms of the sequence:

$-4, -8, -16, -32, -64, -128, -256, -512, -1024, -2048, -4096, -8192$ ∎

Example 8   *Determining an Expression for the nth Term*

Write an expression for the general or nth term, a_n, of the sequence 2, 6, 18, 54, ...

Solution   In this sequence, the first term is 2, so $a_1 = 2$. Note that $\frac{6}{2} = 3$, so $r = 3$. We substitute $a_1 = 2$ and $r = 3$ into $a_n = a_1 r^{n-1}$ to obtain an expression for the nth term, a_n.

$$a_n = a_1 r^{n-1}$$
$$= 2(3)^{n-1}$$

The expression is $a_n = 2(3)^{n-1}$. Note that when $n = 1$, $a_1 = 2(3)^0 = 2(1) = 2$. When $n = 2$, $a_2 = 2(3)^1 = 6$, and so on. ∎

The following formula can be used to determine the sum of the first n terms of a geometric sequence.

> ### Sum of the First n Terms of a Geometric Sequence
> The sum of the first n terms of a geometric sequence can be determined with the following formula where a_1 represents the first term and r represents the common ratio.
> $$s_n = \frac{a_1(1 - r^n)}{1 - r}, r \neq 1$$

Example **9** *Determining the Sum of a Geometric Sequence*

Determine the sum of the first five terms of the geometric sequence whose first term is 4 and whose common ratio is 2.

Solution In this sequence, $a_1 = 4$, $r = 2$, and $n = 5$. Substituting these values into the formula, we get

$$s_n = \frac{a_1(1 - r^n)}{1 - r}$$

$$s_5 = \frac{4\left[1 - (2)^5\right]}{1 - 2}$$

$$= \frac{4(1 - 32)}{-1}$$

$$= \frac{4(-31)}{-1} = \frac{-124}{-1} = 124$$

The sum of the first five terms of the sequence is 124. The first five terms of the sequence are 4, 8, 16, 32, 64. If you add these five numbers, you will obtain the sum 124. ■

Example **10** *Pounds and Pounds of Silver*

As a reward for saving his kingdom from a band of thieves, a king offered a knight one of two options. The knight's first option was to be paid 100,000 pounds of silver all at once. The second option was to be paid over the course of a month. On the first day, he would receive one pound of silver. On the second day, he would receive two pounds of silver. On the third day, he would receive four pounds of silver, and so on, each day receiving double the amount given on the previous day. Assuming the month has 30 days, which option would provide the knight with more silver?

Solution The first option pays the knight 100,000 pounds of silver. The second option pays according to the geometric sequence 1, 2, 4, 8, 16, ... In this sequence, $a_1 = 1$, $r = 2$, and $n = 30$. The sum of this sequence can be determined by substituting these values into the formula to obtain

$$s_n = \frac{a_1(1 - r^n)}{1 - r}$$

$$s_{30} = \frac{1(1 - 2^{30})}{1 - 2}$$

$$= \frac{1 - 1,073,741,824}{-1}$$

$$= \frac{-1,073,741,823}{-1}$$

$$= 1,073,741,823$$

Thus, the knight would get 1,073,741,823 pounds of silver with the second option. Compared with the first option, the second would result in an additional 1,073,641,823 pounds of silver. ■

SECTION 5.7   *Exercises*

Warm Up Exercises

In Exercises 1–6, fill in the blanks with an appropriate word, phrase, or symbol(s).

1. A list of numbers that are related to each other by a rule is called a(n) _____.

2. The numbers that form a sequence are called its _____.

3. A sequence in which each term differs from the preceding term by a constant amount is called a(n) _____ sequence.

4. The amount by which each pair of successive terms differ in an arithmetic sequence is called the common _____.

5. A sequence in which the ratio of any term to the term that directly precedes it is a constant is called a(n) _____ sequence.

6. The constant determined by dividing any term in a geometric sequence by the term that directly precedes it is called the common _____.

Practice the Skills

In Exercises 7–12, write the first five terms of the arithmetic sequence with the first term, a_1, and common difference, d.

7. $a_1 = 3, d = 4$

8. $a_1 = 8, d = 9$

9. $a_1 = 25, d = -5$

10. $a_1 = -11, d = 5$

11. $a_1 = 5, d = -2$

12. $a_1 = -3, d = -4$

In Exercises 13–18, determine the indicated term for the arithmetic sequence with the first term, a_1, and common difference, d.

13. Determine a_{20} when $a_1 = 4, d = 3$.

14. Determine a_{22} when $a_1 = 9, d = -3$.

15. Determine a_{30} when $a_1 = -20, d = 5$.

16. Determine a_{27} when $a_1 = 16, d = -2$.

17. Determine a_{20} when $a_1 = \frac{4}{5}, d = -1$.

18. Determine a_{25} when $a_1 = \frac{1}{3}, d = \frac{2}{3}$.

In Exercises 19–24, write an expression for the general or nth term, a_n, of the arithmetic sequence.

19. $1, 2, 3, 4, \ldots$

20. $2, 4, 6, 8, \ldots$

21. $1, 3, 5, 7, \ldots$

22. $-1, 1, 3, 5$

23. $5, 8, 11, 14$

24. $-7, -2, 3, 8, \ldots$

In Exercises 25–30, determine the sum of the terms of the arithmetic sequence. The number of terms, n, is given.

25. $1, 2, 3, 4, \ldots, 50; n = 50$

26. $2, 4, 6, 8, \ldots, 100; n = 50$

27. $1, 3, 5, 7, \ldots, 99; n = 50$

28. $-4, -7, -10, -13, \ldots, -28; n = 9$

29. $11, 6, 1, -4, \ldots, -24; n = 8$

30. $-4, -11, -18, -25, \ldots, -193; n = 28$

In Exercises 31–36, write the first five terms of the geometric sequence with the first term, a_1, and common ratio, r.

31. $a_1 = 2, r = 5$

32. $a_1 = 3, r = 2$

33. $a_1 = 5, r = 2$

34. $a_1 = 8, r = \frac{1}{2}$

35. $a_1 = -3, r = -1$

36. $a_1 = -6, r = -2$

In Exercises 37–42, determine the indicated term for the geometric sequence with the first term, a_1, and common ratio, r.

37. Determine a_6 when $a_1 = 5, r = 2$.

38. Determine a_6 when $a_1 = 1, r = 3$.

39. Determine a_7 when $a_1 = 64, r = \frac{1}{2}$.

40. Determine a_8 when $a_1 = \frac{1}{2}, r = -2$.

41. Determine a_7 when $a_1 = -5, r = 3$.

42. Determine a_7 when $a_1 = -3, r = -3$.

In Exercises 43–48, write an expression for the general or nth term, a_n, for the geometric sequence.

43. $3, 9, 27, 81, \ldots$

44. $3, 12, 48, 192, \ldots$

45. $2, 1, \frac{1}{2}, \frac{1}{4}, \ldots$

46. $72, 12, 2, \frac{1}{3}, \ldots$

47. $-16, -8, -4, -2, \ldots$

48. $-3, 6, -12, 24, \ldots$

In Exercises 49–54, determine the sum of the first n terms of the geometric sequence for the values of a_1 and r.

49. $n = 6, a_1 = 3, r = 2$

50. $n = 7, a_1 = 1, r = 3$

51. $n = 6, a_1 = -3, r = 4$

52. $n = 9, a_1 = -3, r = 5$

53. $n = 15, a_1 = 2, r = 3$

54. $n = 20, a_1 = 4, r = 2$

Problem Solving

55. Determine the sum of the first 100 natural numbers.

56. Determine the sum of the first 100 odd natural numbers.

57. Determine the sum of the first 100 even natural numbers.

58. Determine the sum of the first 50 multiples of 3.

59. *Clock Strikes* A clock strikes once at 1 o'clock, twice at 2 o'clock, three times at 3 o'clock, and so on. How many times does it strike over a 12-hr period?

60. *A Bouncing Ball* Each time a ball bounces, the height attained by the ball is 6 in. less than the previous height attained. If on the first bounce the ball reaches a height of 6 ft, determine the height attained on the eleventh bounce.

61. *Pendulum Movement* Each swing of a pendulum (from far left to far right) is 3 in. shorter than the preceding swing. The first swing is 8 ft.

a) Determine the length of the twelfth swing.

b) Determine the total distance traveled by the pendulum during the first 12 swings.

62. *Annual Pay Raises* Rita is given a starting salary of $35,000 and promised a $1400 raise per year after each of the next 8 years.

a) Determine her salary during her eighth year of work.

b) Determine the total salary she received over the 8 years.

63. *Enrollment Increase* The current enrollment at Loras College is 8000 students. If the enrollment increases by 8% per year, determine the enrollment 10 years later.

64. *Samurai Sword Construction* While making a traditional Japanese samurai sword, the master sword maker prepares the blade by heating a bar of iron until it is white hot. He then folds it over and pounds it smooth. Therefore, after each folding, the number of layers of steel is doubled. Assuming the sword maker starts with a bar of one layer and folds it 15 times, how many layers of steel will the finished sword contain?

65. *Salary Increase* If your salary were to increase at a rate of 3% per year, determine your salary during your 10th year if your original salary was $35,000. (*Hint*: Use $a_1 = 35,000, r = 1.03$, and $n = 10$.)

66. *A Bouncing Ball* When dropped, a ball rebounds to four-fifths of its original height. How high will the ball rebound after the fourth bounce if it is dropped from a height of 30 ft?

67. *China's Population Growth* In 2018 China's population was about 1.4 billion people. If China's population is growing by about 0.6% per year, estimate China's population in the year 2030. (*Hint:* Use $a_1 = 1.4, r = 1.006$, and $n = 12$.) Round your answer to the nearest tenth of a billion people.

68. *India's Population Growth* In 2018 India's population was about 1.4 billion people. If India's population is growing by about 1.1% per year, estimate India's population in the year 2030. Round your answer to the nearest tenth of a billion people.

Challenge Problems/Group Activities

69. A geometric sequence has $a_1 = 82$ and $r = \frac{1}{2}$; determine s_6.

70. *Sums of Interior Angles* The sums of the interior angles of a triangle, a quadrilateral, a pentagon, and a hexagon are 180°, 360°, 540°, and 720°, respectively. Use this pattern to determine a formula for the general term, a_n, where a_n represents the sum of the interior angles of an n-sided polygon.

71. *Divisibility by 6* Determine how many numbers between 7 and 1610 are divisible by 6.

72. Determine r and a_1 for the geometric sequence with $a_2 = 24$ and $a_5 = 648$.

73. *Total Distance Traveled by a Bouncing Ball* A ball is dropped from a height of 30 ft. On each bounce, it attains a height four-fifths of its original height (or of the previous bounce). Determine the total vertical distance traveled by the ball after it has completed its fifth bounce (has therefore hit the ground six times).

Recreational Mathematics

74. ***Martingale System*** See the *Recreational Mathematics* box on page 270. Use the Martingale system to answer the following questions.

a) Assume a player is using a $10 standard bet and loses five times in a row. How much money should the player bet in the sixth round? How much money has the player lost at the end of the fifth round?

b) Assume a player is using a $10 standard bet and loses 10 times in a row. How much money should the player bet in the eleventh round? How much money has the player lost at the end of the tenth round?

c) Why is this a risky strategy?

Research Activity

75. ***Series*** A topic generally associated with sequences is *series*.

a) Research *series* and explain what a series is and how it differs from a sequence. Also write a formal definition of series. Give examples of different kinds of series.

b) Write the arithmetic series associated with the arithmetic sequence 1, 4, 7, 10, 13, ….

c) Write the geometric series associated with the geometric sequence 3, 6, 12, 24, 48, ….

d) What is an infinite geometric series?

e) Determine the sum of the terms of the infinite geometric series $1 + \dfrac{1}{2} + \dfrac{1}{4} + \dfrac{1}{8} + \dfrac{1}{16} + \cdots$.

SECTION 5.8   Fibonacci Sequence

Upon completion of this section, you will be able to:

- Identify the Fibonacci sequence.
- Understand the relationship between Fibonacci numbers and the golden ratio.

Consider the following numbers found in nature: the number of petals on a daisy, the number of sunflower seeds in a sunflower, the number of scales on a pineapple, the number of ancestors of a male bee. In this section, we will determine that all these numbers are part of a special sequence of numbers called the Fibonacci sequence.

Why This Is Important The Fibonacci sequence and the related topic of the golden ratio appear in many places in nature and play an important role in geometry, art, and music.

The Fibonacci Sequence

Our discussion of sequences would not be complete without mentioning a sequence known as the *Fibonacci sequence*. The sequence is named after Leonardo of Pisa, also known as Fibonacci. He was one of the most distinguished mathematicians of the Middle Ages. This sequence is first mentioned in his book *Liber Abacci* (Book of the Abacus), which contained many interesting problems, such as: "A certain man put a pair of rabbits in a place surrounded on all sides by a wall. How many pairs of rabbits can be produced from that pair in a year if it is assumed that every month each pair begets a new pair which from the second month becomes productive?"

The solution to this problem (Fig. 5.10) led to the development of the sequence that bears its author's name: the Fibonacci sequence. The sequence is shown in Table 5.1 on page 276. The numbers in the columns titled *Pairs of Adults* form the Fibonacci sequence.

▲ *The head of a sunflower*

Table 5.1

Month	Pairs of Adults	Pairs of Babies	Total Pairs
1	1	0	1
2	1	1	2
3	2	1	3
4	3	2	5
5	5	3	8
6	8	5	13
7	13	8	21
8	21	13	34
9	34	21	55
10	55	34	89
11	89	55	144
12	144	89	233

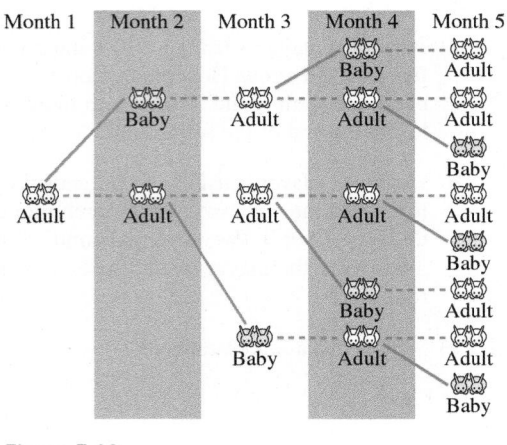

Figure 5.10

Following is the Fibonacci sequence

Fibonacci Sequence

$$1, 1, 2, 3, 5, 8, 13, 21, \ldots$$

In the Fibonacci sequence, the first and second terms are 1. The sum of these two terms is the third term. The sum of the second and third terms is the fourth term, and so on.

In the middle of the nineteenth century, mathematicians made a serious study of this sequence and found strong similarities between it and many natural phenomena. Fibonacci numbers appear in the seed arrangement of many species of plants and in the petal counts of various flowers. For example, when the flowering head of the sunflower matures to seed, the seeds' spiral arrangement becomes clearly visible. A typical count of these spirals may give 89 steeply curving to the right, 55 curving more shallowly to the left, and 34 again shallowly to the right. The largest known specimen to be examined had spiral counts of 144 right, 89 left, and 55 right. These numbers, like the other three mentioned, are consecutive terms of the Fibonacci sequence.

On the heads of many flowers, petals surrounding the central disk generally yield a Fibonacci number. For example, some daisies contain 21 petals, and others contain 34, 55, or 89 petals. (People who use a daisy to play the "love me, love me not" game will likely pluck 21, 34, 55, or 89 petals before arriving at an answer.)

Fibonacci numbers are also observed in the structure of pinecones and pineapples. The tablike or scalelike structures called bracts that make up the main body of the pinecone form a set of spirals that start from the cone's attachment to the branch. Two sets of oppositely directed spirals can be observed, one steep and the other more gradual. A count on the steep spiral will reveal a Fibonacci number, and a count on the gradual one will be the adjacent smaller Fibonacci number, or if not, the next smaller Fibonacci number. One investigation of 4290 pinecones from 10 species of pine trees found in California revealed that only 74 cones, or 1.7%, deviated from this Fibonacci pattern.

Like pinecone bracts, pineapple scales are patterned into spirals, and because they are roughly hexagonal in shape, three distinct sets of spirals can be counted. Generally, the number of pineapple scales in each spiral are Fibonacci numbers.

Table 5.2

Numbers	Ratio
1, 1	$\dfrac{1}{1} = 1$
1, 2	$\dfrac{2}{1} = 2$
2, 3	$\dfrac{3}{2} = 1.5$
3, 5	$\dfrac{5}{3} = 1.666\ldots$
5, 8	$\dfrac{8}{5} = 1.6$
8, 13	$\dfrac{13}{8} = 1.625$
13, 21	$\dfrac{21}{13} \approx 1.615$
21, 34	$\dfrac{34}{21} \approx 1.619$
34, 55	$\dfrac{55}{34} \approx 1.618$
55, 89	$\dfrac{89}{55} \approx 1.618$

Example 1 *Determining a Fibonacci-Type Sequence*

Determine if the sequence is a Fibonacci-type sequence. If it is, determine the next two terms of the sequence.

a) 2, 3, 6, 18, 108, 1944, …   b) 1, 3, 4, 7, 11, 18, …

Solution

a) A Fibonacci-type sequence is a sequence in which each term, after the first two terms, is the sum of the two preceding terms. The third term, 6, is not the sum of the two preceding terms, 2 and 3. Therefore, this sequence is not a Fibonacci-type sequence.

b) Each term is the sum of the two preceding terms. Thus, this sequence is a Fibonacci-type sequence. The next term is $11 + 18$, or 29. The following term is $18 + 29$, or 47. ∎

Fibonacci Numbers and the Golden Ratio

In 1753, while studying the Fibonacci sequence, Robert Simson, a mathematician at the University of Glasgow, noticed that when he took the ratio of any term to the term that immediately preceded it, the value he obtained remained in the vicinity of one specific number. To illustrate this, we indicate in Table 5.2 the ratio of various pairs of sequential Fibonacci numbers.

The ratio of the 50th term to the 49th term is 1.6180. Simson proved that the ratio of the $(n + 1)$ term to the nth term as n gets larger and larger is the irrational number $(\sqrt{5} + 1)/2$, which begins 1.61803…. This number was already well known to mathematicians at that time as the *golden number*.

Many years earlier, Bavarian astronomer and mathematician Johannes Kepler wrote that for him the golden number symbolized the Creator's intention "to create like from like." The golden number $(\sqrt{5} + 1)/2$ is frequently referred to as "phi," symbolized by the Greek letter Φ.

The ancient Greeks, in about the sixth century B.C., sought unifying principles of beauty and perfection, which they believed could be described by using mathematics. In their study of beauty, the Greeks used the term *golden ratio*. To understand the golden ratio, let's consider the line segment AB in Fig. 5.11.

Figure 5.11

When this line segment is divided at a point C such that the ratio of the whole, AB, to the larger part, AC, is equal to the ratio of the larger part, AC, to the smaller part, CB, each ratio AB/AC and AC/CB is referred to as a *golden ratio*. The proportion these ratios form, $AB/AC = AC/CB$, is called the *golden proportion*. Furthermore, each ratio in the proportion will have a value equal to the golden number, $(\sqrt{5} + 1)/2$.

$$\frac{AB}{AC} = \frac{AC}{CB} = \frac{\sqrt{5} + 1}{2} \approx 1.618$$

The Great Pyramid of Giza in Egypt, built about 2600 B.C., is the earliest known example of use of the golden ratio in architecture. The ratio of any of its sides of the square base (775.75 ft) to its altitude (481.4 ft) is about 1.611. In medieval times, people referred to the golden proportion as the *divine proportion*, reflecting their belief in its relationship to the will of God.

Twentieth-century architect Le Corbusier developed a scale of proportions for the human body that he called the Modulor (Fig. 5.12). Note that the navel separates the entire body into golden proportions, as do the neck and knee.

▲ The Great Pyramid of Giza

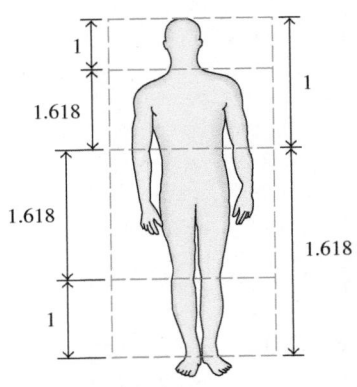

Figure 5.12

From the golden proportion, the *golden rectangle* can be formed, as shown in Fig. 5.13.

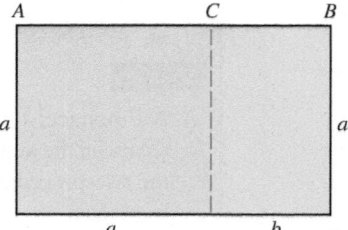

Figure 5.13

$$\frac{\text{Length}}{\text{Width}} = \frac{a + b}{a} = \frac{a}{b} = \frac{\sqrt{5} + 1}{2}$$

Note that when a square is cut off one end of a golden rectangle, as in Fig. 5.13, the remaining rectangle has the same properties as the original golden rectangle (creating "like from like") and is therefore itself a golden rectangle. Interestingly, the curve derived from a succession of diminishing golden rectangles, as shown in Fig. 5.14, is the same as the spiral curve of the chambered nautilus. The same curve appears on the horns of rams and some other animals. It is the same curve that is observed in the plant structures mentioned earlier: sunflowers, other flower heads, pinecones, and pineapples. The curve shown in Fig. 5.14 closely approximates what mathematicians call a *logarithmic spiral.*

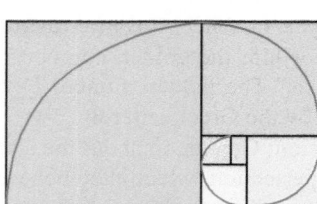 

Figure 5.14

Ancient Greek civilization used the golden rectangle in art and architecture. The main measurements of many buildings of antiquity, including the Parthenon in Athens, are governed by golden ratios and rectangles. It is for Phidas, considered the greatest

▲ *The Parthenon*

▲ *Circus Sideshow (La Parade de Cirque),* 1887, by Georges Seurat

Fibonacci and the Male Bee's Ancestors

An excellent example of the Fibonacci sequence in nature comes from the breeding practices of bees. Female or worker bees are produced when the queen bee mates with a male bee. Male bees are produced from the queen's unfertilized eggs. In essence, then, female bees have two parents, whereas male bees only have one parent. The family tree of a male bee would look like this:

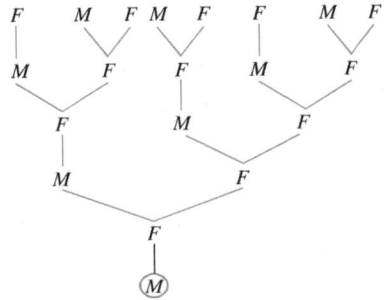

From this tree, we can see that the 1 male bee (circled) has 1 parent, 2 grandparents, 3 great-grandparents, 5 great-great-grandparents, 8 great-great-great-grandparents, and so on. We see the Fibonacci sequence as we move back through the male bees' generations.

▲ *Fibonacci's Garden by Caryl Bryer Fallert*

of Greek sculptors, that the golden ratio was named "phi." The proportions can be found abundantly in his work. The proportions of the golden rectangle can be found in the work of many artists, from the old masters to the moderns. For example, the golden rectangle can be seen on page 278 in the painting *Circus Sideshow (La Parade de Cirque),* 1887, by Georges Seurat, a French neoimpressionist artist.

In addition to using the golden rectangle in art, several artists have used Fibonacci numbers in art. One contemporary example is the 1995 work by Caryl Bryer Fallert called *Fibonacci's Garden,* which is shown in the margin. This artwork is a quilt constructed from two separate fabrics that are put together in a pattern based on the Fibonacci sequence.

Fibonacci numbers are also found in another form of art, namely, music. Perhaps the most obvious link between Fibonacci numbers and music can be found on the piano keyboard. An octave (Fig. 5.15) on a keyboard has 13 keys: 8 white keys and 5 black keys (the 5 black keys are in one group of 2 and one group of 3).

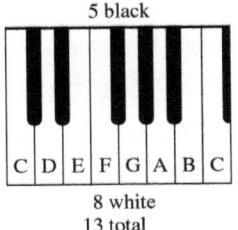

Figure 5.15

In Western music, the most complete scale, the chromatic scale, consists of 13 notes (from C to the next higher C). Its predecessor, the diatonic scale, contains 8 notes (an octave). The diatonic scale was preceded by a 5-note pentatonic scale (*penta* is Greek for "five"). Each number is a Fibonacci number. In popular music, the song "Lateralas" by the band Tool uses the Fibonacci sequence in both the time signature and the lyric arrangement. The song also contains several references to the golden ratio and to the logarithmic spiral.

The visual arts deal with what is pleasing to the eye, and musical composition deals with what is pleasing to the ear. Whereas art achieves some of its goals by using division of planes and area, music achieves some of its goals by a similar division of time, using notes of various duration and spacing. The musical intervals considered by many to be the most pleasing to the ear are the major sixth and minor sixth. A major sixth, for example, consists of the note C, vibrating at about 264[*] vibrations per second, and note A, vibrating at about 440 vibrations per second. The ratio of 440 to 264 reduces to 5 to 3, or $\frac{5}{3}$, a ratio of two consecutive Fibonacci numbers. An example of a minor sixth is E (about 330 vibrations per second) and C (about 528 vibrations per second). The ratio 528 to 330 reduces to 8 to 5, or $\frac{8}{5}$, the next ratio of two consecutive Fibonacci numbers. The vibrations of any sixth interval reduce to a similar ratio.

Patterns that can be expressed mathematically in terms of Fibonacci relationships have been found in Gregorian chants and works of many composers, including Bach, Beethoven, and Bartók. A number of twentieth-century musical works, including Ernst Krenek's *Fibonacci Mobile,* were deliberately structured by using Fibonacci proportions.

A number of studies have tried to explain why the Fibonacci sequence and the golden ratio are linked to so many real-life situations. It appears that the Fibonacci numbers are a part of a natural harmony that is pleasing to both the eye and the ear. In the nineteenth century, German physicist and psychologist Gustav Fechner tried to determine which dimensions were most pleasing to the eye. Fechner, along with psychologist Wilhelm Wundt, determined that most people do unconsciously favor golden dimensions when purchasing greeting cards, mirrors, and other rectangular objects. This discovery has been widely used by commercial manufacturers in their packaging and labeling designs, by retailers in their store displays, and in other areas of business and advertising.

[*]Frequencies of notes vary in different parts of the world and change over time.

SECTION 5.8 Exercises

Warm Up Exercises

In Exercises 1–6, fill in the blanks with an appropriate word, phrase, or symbol(s).

1. The sequence of numbers 1, 1, 2, 3, 5, 8, 13, 21, ... is known as the _____ sequence.

2. The irrational number $\frac{\sqrt{5}+1}{2}$ is known as the _____ number.

3. In medieval times, people referred to the golden proportion as the _____ proportion.

4. A rectangle whose ratio of its length to its width is equal to the golden number is known as a golden _____.

In Exercises 5 and 6, use the following diagram and assume that the ratio $\dfrac{AB}{AC}$ is equal to the ratio $\dfrac{AC}{CB}$

5. Each ratio $\dfrac{AB}{AC}$ and $\dfrac{AC}{CB}$ is referred to as a golden _____.

6. The proportion $\dfrac{AB}{AC} = \dfrac{AC}{CB}$ is called the golden _____.

Practice the Skills/Problem Solving

In Exercises 7–14, determine whether the sequence is a Fibonacci-type sequence (each term is the sum of the two preceding terms). If it is, determine the next two terms of the sequence.

7. 3, 5, 8, 13, 21, 34, ...

8. 2, 4, 6, 10, 16, 26, ...

9. 1, 2, 2, 4, 8, 32, ...

10. 1, 4, 9, 16, 25, 36, ...

11. −1, 1, 0, 1, 1, 2, ...

12. 0, π, π, 2π, 3π, 5π, ...

13. 5, 10, 15, 25, 40, 65, ...

14. $\frac{1}{4}, \frac{1}{4}, \frac{1}{2}, \frac{3}{4}, 1\frac{1}{4}, 2, \ldots$

15. Write out the first 15 terms of the Fibonacci sequence.

16. The sum of any six consecutive Fibonacci numbers is always divisible by 4. Select any six consecutive Fibonacci numbers and show that for your selection this statement is true.

17. The sum of any 10 consecutive Fibonacci numbers is always divisible by 11. Select any 10 consecutive Fibonacci numbers and show that for your selection this statement is true.

18. The greatest common factor of any two consecutive Fibonacci numbers is 1. Show that this statement is true for the first 15 Fibonacci numbers.

19. Twice any Fibonacci number minus the next Fibonacci number equals the second number preceding the original number. Select a number in the Fibonacci sequence and show that this pattern holds for the number selected.

20. For any four consecutive Fibonacci numbers, the difference of the squares of the middle two numbers equals the product of the smallest and largest numbers. Select four consecutive Fibonacci numbers and show that this pattern holds for the numbers you selected.

21. a) Select any two one-digit (nonzero) numbers and add them to obtain a third number. Continue adding the two previous terms to get a Fibonacci-type sequence.

b) Form ratios of successive terms to show how they will eventually approach the golden number.

22. a) Select any three consecutive terms of a Fibonacci sequence. Subtract the product of the terms on each side of the middle term from the square of the middle term. What is the difference?

b) Repeat part (a) with three different consecutive terms of the sequence.

c) Make a conjecture about what will happen when you repeat this process for any three consecutive terms of a Fibonacci sequence.

23. *Pascal's Triangle* One of the most famous number patterns involves *Pascal's triangle*. The Fibonacci sequence can be determined by using Pascal's triangle. Can you explain how that can be done? A hint is shown.

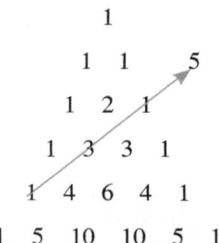

24. *Lucas Sequence* **a)** A sequence related to the Fibonacci sequence is the *Lucas sequence*. The Lucas sequence is formed in a manner similar to the Fibonacci sequence. The first two numbers of the Lucas sequence are 1 and 3. Write the first eight terms of the Lucas sequence.

b) Complete the next two lines of the following chart.

$$1 + 2 = 3$$
$$1 + 3 = 4$$
$$2 + 5 = 7$$
$$3 + 8 = 11$$
$$5 + 13 = 18$$

c) What do you observe about the first column in the chart in part (b)?

25. a) To what decimal value is $(\sqrt{5} + 1)/2$ approximately equal?

   b) To what decimal value is $(\sqrt{5} - 1)/2$ approximately equal?

   c) By how much do the results in parts (a) and (b) differ?

26. Determine the ratio of the length to the width of your cell phone screen and compare this ratio to Φ.

27. Determine the ratio of the length to width of various photographs and compare these ratios to Φ.

28. Find three physical objects whose dimensions are very close to a golden rectangle.

   a) List the objects and record the dimensions.

   b) Compute the ratios of their lengths to their widths.

   c) Determine the difference between the golden ratio and the ratio you obtain in part (b)—to the nearest tenth—for each object.

Concept/Writing Exercises

29. The eleventh Fibonacci number is 89. Examine the first six digits in the decimal expression of its reciprocal, $\frac{1}{89}$. What do you determine?

30. Determine the ratio of the second to the first term of the Fibonacci sequence. Then determine the ratio of the third to the second term of the sequence and determine whether this ratio was an increase or decrease from the first ratio. Continue this process for 10 ratios and then make a conjecture regarding the increasing or decreasing values in consecutive ratios.

31. A musical composition is described as follows. Explain why this piece is based on the golden ratio.

Entire Composition

34 measures	55 measures	21 measures	34 measures
Theme	Fast, Loud	Slow	Repeat of theme

Challenge Problems/Group Activities

32. Draw a line of length 5 in. Determine and mark the point on the line that will create the golden ratio. Explain how you determined your answer.

33. The divine proportion is $(a + b)/a = a/b$ (see Fig. 5.13 on page 278), which can be written $1 + (b/a) = a/b$. Now let $x = a/b$, which gives $1 + (1/x) = x$. Multiply both sides of this equation by x to get a quadratic equation and then use the quadratic formula (Section 6.9) to show that one answer is $x = (1 + \sqrt{5})/2$ (the golden ratio).

34. *Pythagorean Triples* A Pythagorean triple is a set of three whole numbers, $\{a, b, c\}$, such that $a^2 + b^2 = c^2$. For example, since $6^2 + 8^2 = (10)^2$, $\{6, 8, 10\}$ is a Pythagorean triple. The following steps show how to determine Pythagorean triples using any four consecutive Fibonacci numbers. Here we will demonstrate the process with the Fibonacci numbers 3, 5, 8, and 13.

   1) Determine the product of 2 and the two inner Fibonacci numbers. We have $2(5)(8) = 80$, which is the first number in the Pythagorean triple. So $a = 80$.

   2) Determine the product of the two outer numbers. We have $3(13) = 39$, which is the second number in the Pythagorean triple. So $b = 39$.

   3) Determine the sum of the squares of the inner two numbers. We have $5^2 + 8^2 = 25 + 64 = 89$, which is the third number in the Pythagorean triple. So $c = 89$.

   This process has produced the Pythagorean triple, $\{80, 39, 89\}$. To verify,

   $$(80)^2 + (39)^2 = (89)^2$$
   $$6400 + 1521 = 7921$$
   $$7921 = 7921$$

   Use this process to produce four other Pythagorean triples.

35. *Reflections* When two panes of glass are placed face to face, four interior reflective surfaces exist, labeled 1, 2, 3, and 4. If light is not reflected, it has just one path through the glass (see the figure below). If it has one reflection, it can be reflected in two ways. If it has two reflections, it can be reflected in three ways. Use this information to answer parts (a) through (c).

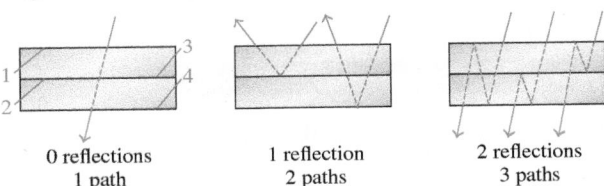

0 reflections	1 reflection	2 reflections
1 path	2 paths	3 paths

   a) If a ray is reflected three times, there are five paths it can follow. Show the paths.

   b) If a ray is reflected four times, there are eight paths it can follow. Show the paths.

   c) How many paths can a ray follow if it is reflected five times? Explain how you determined your answer.

Research Activities

36. *Fibonacci* Write a report on the history and mathematical contributions of Fibonacci.

37. *Digits* The digits 1 through 9 have evolved considerably since they appeared in Fibonacci's book *Liber Abacci*. Write a report tracing the history of the evolution of the digits 1 through 9 since Fibonacci's time.

38. *Golden Ratio and Golden Rectangle* Write a report indicating where the golden ratio and golden rectangle have been used in art and architecture. You may wish to include information on art and architecture related to the golden ratio and Fibonacci sequences.

CHAPTER 5   *Summary*

Important Facts and Concepts

	Examples and Discussion

Section 5.1

Fundamental Theorem of Arithmetic

Every composite number can be expressed as a unique product of prime numbers.

Examples 2–7, pages 211–214

Sections 5.1–5.5

Sets of Numbers

Natural or counting numbers: $\{1, 2, 3, 4, \dots\}$

Whole numbers: $\{0, 1, 2, 3, 4, \dots\}$

Integers: $\{\dots, -3, -2, -1, 0, 1, 2, 3, \dots\}$

Rational numbers: Numbers of the form p/q, where p and q are integers, $q \neq 0$. Every rational number when expressed as a decimal number will be either a terminating or repeating decimal number.

Irrational number: A real number that cannot be written as the ratio of two integers (not a rational number)

Real numbers: The union of the rational numbers and the irrational numbers

Discussion pages 208, 219, 229, 244, and 253

Section 5.2

Order of Operations

1. First, perform all operations within parentheses or other grouping symbols.
2. Next, evaluate all exponents.
3. Next, perform all multiplications and divisions from left to right.
4. Finally, perform all additions and subtractions from left to right.

Discussion pages 225–226, Examples 10–11, page 226

Section 5.3

Fundamental Law of Rational Numbers

$$\frac{a}{b} = \frac{a}{b} \cdot \frac{c}{c} = \frac{ac}{bc} \quad b \neq 0, \quad c \neq 0$$

Discussion page 237, Examples 14–16, pages 237–239

Section 5.4

Rules of Radicals

Product rule for radicals:

$$\sqrt{a \cdot b} = \sqrt{a} \cdot \sqrt{b}, \quad a \geq 0, \quad b \geq 0$$

Quotient rule for radicals:

$$\frac{\sqrt{a}}{\sqrt{b}} = \sqrt{\frac{a}{b}}, \quad a \geq 0, \quad b > 0$$

Discussion pages 245 and 247, Examples 1–6, pages 246–248

Section 5.5

Properties of Real Numbers

Commutative property of addition: $a + b = b + a$

Commutative property of multiplication: $a \cdot b = b \cdot a$

Associative property of addition:

$(a + b) + c = a + (b + c)$

Associative property of multiplication:

$(a \cdot b) \cdot c = a \cdot (b \cdot c)$

Distributive property: $a \cdot (b + c) = ab + ac$

Discussion pages 253–256, Examples 1–4, pages 254–256

Section 5.6

Rules of Exponents

Product rule for exponents: $a^m \cdot a^n = a^{m+n}$

Quotient rule for exponents: $\dfrac{a^m}{a^n} = a^{m-n}, \quad a \neq 0$

Zero exponent rule: $a^0 = 1, \quad a \neq 0$

Negative exponent rule: $a^{-m} = \dfrac{1}{a^m}, \quad a \neq 0$

Power rule: $(a^m)^n = a^{m \cdot n}$

Examples 1–5, pages 259–261

Scientific Notation

Scientific notation is used to write very small or very large numbers. Each number in scientific notation is written as a number greater than or equal to 1 and less than 10 multiplied by an integer power of 10.

Discussion pages 261–262,
Examples 6–12, pages 262–265

Section 5.7

Arithmetic Sequence

General or nth term: $a_n = a_1 + (n-1)d$

Sum of the first n terms: $s_n = \dfrac{n(a_1 + a_n)}{2}$

Discussion page 268,
Examples 1–5, pages 268–270

Geometric Sequence

General or nth term: $a_n = a_1 r^{n-1}$

Sum of the first n terms: $s_n = \dfrac{a_1(1 - r^n)}{1 - r}, \quad r \neq 1$

Discussion pages 270–271,
Examples 6–10, pages 271–272

Section 5.8

Fibonacci Sequence

$1, 1, 2, 3, 5, 8, 13, 21, \ldots$

Discussion pages 275–279,
Example 1, page 277

Golden Number

$\dfrac{\sqrt{5} + 1}{2} \approx 1.618$

Discussion pages 277–279

Golden Proportion

$\dfrac{a + b}{a} = \dfrac{a}{b}$

Discussion pages 277–279

CHAPTER 5	*Review Exercises*

5.1

In Exercises 1 and 2, determine whether the number is divisible by each of the following numbers: 2, 3, 4, 5, 6, 8, 9, and 10.

1. 56,340

2. 400,644

In Exercises 3 and 4, determine the prime factorization of the number.

3. 840

4. 1452

In Exercises 5 and 6, determine the GCD and LCM of the numbers.

5. 24, 36

6. 63, 108

7. *Airport Activity* O'Hare International Airport in Chicago has a flight leaving for New York City every 45 minutes and a flight leaving for Washington, DC, every 60 minutes. If a flight to New York City and a flight to Washington, DC, leave at the same time, how many minutes will it be before a flight to New York City and a flight to Washington, DC, again leave at the same time?

5.2

In Exercises 8–13, use a number line to evaluate the expression.

8. $2 + (-6)$ **9.** $-2 + 5$

10. $-2 + (-4)$ **11.** $4 - 8$

12. $-5 - 4$ **13.** $-3 - (-6)$

In Exercises 14–17, evaluate the expression.

14. $6 \cdot (-4)$ **15.** $(-2)(-12)$

16. $\dfrac{-35}{-7}$ **17.** $\dfrac{12}{-6}$

In Exercises 18–21, evaluate the expression using the order of operations.

18. $8 + 5 \cdot 3$ **19.** $4 \cdot 3^2 - 2^3 \cdot 3$

20. $(64 \div 4 \cdot 2) - (25 - 7 \cdot 4)$

21. $-2(5 \cdot 3^2 - 4^3 \div 8)^2$

5.3

In Exercises 22–25, express the fraction as a terminating or repeating decimal number.

22. $\dfrac{11}{25}$ **23.** $\dfrac{13}{4}$

24. $\dfrac{6}{7}$ **25.** $\dfrac{7}{12}$

In Exercises 26–29, express the decimal number as a quotient of two integers.

26. 4.2 **27.** $0.\overline{6}$

28. $0.\overline{51}$ **29.** 0.083

In Exercises 30 and 31, express each mixed number as an improper fraction.

30. $2\dfrac{4}{7}$ **31.** $-3\dfrac{1}{4}$

In Exercises 32 and 33, express each improper fraction as a mixed number.

32. $\dfrac{19}{5}$ **33.** $-\dfrac{136}{5}$

In Exercises 34–37, perform the indicated operation and reduce your answer to lowest terms.

34. $\dfrac{4}{15} - \dfrac{2}{5}$ **35.** $\dfrac{1}{6} + \dfrac{5}{4}$

36. $\dfrac{7}{16} \cdot \dfrac{12}{21}$ **37.** $\dfrac{5}{9} \div \dfrac{6}{7}$

In Exercises 38–41, evaluate the expression using the order of operations.

38. $\dfrac{1}{3} + \dfrac{2}{5} \cdot \dfrac{5}{11}$ **39.** $\dfrac{3}{4} - \dfrac{1}{8} \div \dfrac{5}{12}$

40. $\dfrac{13}{16} \div \dfrac{7}{8} \cdot \dfrac{14}{15}$ **41.** $\left(\dfrac{1}{4} + \dfrac{5}{6}\right) \div \left(3 - \dfrac{1}{6}\right)$

42. *Cajun Turkey* A recipe for roasted cajun turkey calls for $\frac{1}{8}$ teaspoon of cayenne pepper per pound of turkey. If Jennifer is preparing a turkey that weighs $17\frac{3}{4}$ pounds, how much cayenne pepper does she need?

5.4

In Exercises 43–54, simplify the expression. Rationalize the denominator when necessary.

43. $\sqrt{60}$ **44.** $\sqrt{2} - 4\sqrt{2}$

45. $\sqrt{8} + 6\sqrt{2}$ **46.** $\sqrt{3} - 7\sqrt{27}$

47. $\sqrt{28} + \sqrt{63}$ **48.** $\sqrt{3} \cdot \sqrt{6}$

49. $\sqrt{8} \cdot \sqrt{6}$ **50.** $\dfrac{\sqrt{300}}{\sqrt{3}}$

51. $\dfrac{4}{\sqrt{3}}$ **52.** $\dfrac{\sqrt{7}}{\sqrt{5}}$

53. $3(2 + \sqrt{7})$ **54.** $\sqrt{3}(4 + \sqrt{6})$

5.5

In Exercises 55–59, state the name of the property illustrated.

55. $7 + 9 = 9 + 7$

56. $5 \cdot m = m \cdot 5$

57. $(1 + 2) + 3 = 1 + (2 + 3)$

58. $9(x + 1) = 9 \cdot x + 9 \cdot 1$

59. $(3 \cdot a) \cdot b = 3 \cdot (a \cdot b)$

In Exercises 60–64, determine whether the set of numbers is closed under the given operation.

60. Natural numbers, subtraction

61. Whole numbers, multiplication

62. Integers, division

63. Real numbers, subtraction

64. Irrational numbers, multiplication

5.6

In Exercises 65–72, evaluate each expression.

65. $3^2 \cdot 3^3$ **66.** $\dfrac{6^4}{6^2}$

67. 9^0

68. 5^{-3}

69. $(2^3)^4$

70. -8^0

71. -7^{-2}

72. $3 \cdot 2^{-3}$

In Exercises 73–74, write each number in scientific notation.

73. 36,200,000

74. 0.0000158

In Exercises 75–76, express each number in decimal notation.

75. 2.8×10^5

76. 1.39×10^{-4}

In Exercises 77–78, (a) perform the indicated operation and write your answer in scientific notation. (b) Confirm the result determined in part (a) by performing the calculation on a scientific calculator.

77. $(3 \times 10^4)(2 \times 10^{-9})$

78. $\dfrac{1.5 \times 10^{-3}}{5 \times 10^{-4}}$

In Exercises 79–81, (a) perform the indicated calculation by first converting each number to scientific notation. Write your answer in decimal notation. (b) Confirm the result determined in part (a) by performing the calculation on a scientific calculator.

79. (550,000)(2,000,000)

80. $\dfrac{8,400,000}{70,000}$

81. $\dfrac{0.000002}{0.0000004}$

82. *Space Distances* The distance from Earth to the sun is about 1.49×10^{11} meters. The distance from Earth to the moon is about 3.84×10^8 meters. The distance from Earth to the sun is about how many times larger than the distance from Earth to the moon? Use a scientific calculator and round your answer to the nearest whole number.

83. *Outstanding Debt* As a result of a recent water and sewer system improvement, the city of Galena, Illinois, has an outstanding debt of $20,000,000. If the population of Galena is 3600 people, how much would each person have to contribute to pay off the outstanding debt?

5.7

In Exercises 84–85, determine whether the sequence is arithmetic or geometric. Then determine the next two terms of the sequence.

84. $3, 9, 15, 21, \ldots$

85. $\frac{1}{2}, 1, 2, 4, \ldots$

In Exercises 86–89, determine the indicated term of the sequence with the given first term, a_1, and common difference, d, or common ratio, r.

86. Determine a_9 when $a_1 = -6$, $d = 2$.

87. Determine a_{10} when $a_1 = -20$, $d = 5$.

88. Determine a_5 when $a_1 = 3$, $r = 2$.

89. Determine a_{10} when $a_1 = -1$, $r = 3$.

In Exercises 90 and 91, determine the sum of the arithmetic sequence. The number of terms, n, is given.

90. $3, 6, 9, 12, \ldots, 150$; $n = 50$

91. $0.5, 0.75, 1.00, 1.25, \ldots, 5.25$; $n = 20$

In Exercises 92 and 93, determine the sum of the first n terms of the geometric sequence for the values of a_1 and r.

92. $n = 4$, $a_1 = 3$, $r = 2$

93. $n = 6$, $a_1 = 1$, $r = -2$

In Exercises 94 and 95, first determine whether the sequence is arithmetic or geometric; then write an expression for the general or nth term, a_n.

94. $1, 4, 7, 10, \ldots$

95. $2, -2, 2, -2, \ldots$

5.8

96. Write down the first 15 terms of the Fibonacci sequence.

In Exercises 97 and 98, determine whether the sequence is a Fibonacci-type sequence. If so, determine the next two terms.

97. $0, 1, 1, 2, 2, 3, 3, 4, 4, \ldots$

98. $-1, 0, -1, -1, -2, -3, -5, \ldots$

CHAPTER 5   Test

1. Which of the numbers 2, 3, 4, 5, 6, 8, 9, and 10 divide 20,270?

2. Determine the prime factorization of 315.

3. Evaluate $[(-3) + 7] - (-4)$.

4. Evaluate $[(-70)(-5)] \div (8 - 10)$.

5. Convert $3\frac{2}{11}$ to an improper fraction.

6. Write $\frac{13}{25}$ as a terminating or repeating decimal.

7. Express 6.45 as a quotient of two integers.

8. Evaluate the expression using the order of operations:
$$\frac{7}{20} - \frac{12}{25} \div \frac{9}{10}.$$

9. Simplify $\sqrt{63} + \sqrt{28}$.

10. Rationalize the denominator $\dfrac{\sqrt{2}}{\sqrt{7}}$.

11. Determine whether the integers are closed under the operation of multiplication. Explain your answer.

12. Name the property illustrated: $3(x + 7) = 3x + 21$.

Evaluate the expression.

13. $\dfrac{4^5}{4^2}$

14. $4^3 \cdot 4^2$

15. 3^{-4}

16. Perform the operation by first converting the numerator and denominator to scientific notation. Write the answer in scientific notation.
$$\frac{7,200,000}{0.000009}$$

17. Write an expression for the general or nth term, a_n, of the sequence $-2, -6, -10, -14, \dots$.

18. Determine the sum of the terms of the arithmetic sequence. The number of terms, n, is given.
$$-2, -5, -8, -11, \dots, -32; n = 11$$

19. Write an expression for the general or nth term, a_n, of the sequence 3, 6, 12, 24, ….

20. Write the first 10 terms of the Fibonacci sequence.

◀ Algebra is used in many aspects of the financial decisions we make including the interest we pay on a car loan.

6

Algebra, Graphs, and Functions

What You Will Learn

- Using the order of operations and solving linear equations
- Working with formulas
- Solving applications of algebra
- Solving variation problems
- Solving linear inequalities
- Graphing linear equations
- Solving systems of linear equations
- Graphing linear inequalities and systems of linear inequalities in two variables
- Solving quadratic equations
- Working with functions and their graphs

Why This Is Important

Algebra is one of the most practical tools available to us for solving everyday problems. For example, we use algebra when we calculate the sales tax on purchases. We use a formula to calculate the compound interest that we earn when investing our money. We also use a formula to calculate the interest that we pay when we borrow money for a student loan, a car loan, or a home mortgage. The coordinate system we use when graphing is similar to the coordinate system we use on maps. The distance we travel can be determined by a formula relating our rate of speed and our time traveled. The symbolic nature of algebra allows us to represent problems in a simple and compact manner. This makes algebra an excellent tool for solving a wide variety of problems that we face in our everyday lives.

SECTION 6.1   Order of Operations and Solving Linear Equations

Upon completion of this section, you will be able to:

- Understand algebraic concepts.
- Use the order of operations.
- Solve linear equations.

Alia works in a surf shop. She earns a weekly salary plus a commission on all items that she sells. In this section, we will learn that Alia's weekly earnings can be represented with a linear equation. Furthermore, we will learn that many real-world applications can be represented with linear equations.

Why This Is Important   Representing problems with equations and then solving these equations is a powerful problem-solving tool that can be used in many areas of our lives. In addition to representing weekly earnings, linear equations can also be used to calculate the price we pay, including sales tax, when we purchase items. Furthermore, price markups and discounts that retailers apply to prices of items can also be represented with linear equations. Having an understanding of how linear equations can be used to solve these and other everyday financial problems can lead to us making better financial decisions.

Algebraic Concepts

English philosopher Alfred North Whitehead explained the power of algebra when he stated, "By relieving the brain of unnecessary work, a good notation sets the mind free to concentrate on more advanced problems."

Algebra is a generalized form of arithmetic. The word *algebra* is derived from the Arabic word *al-jabr* (meaning "reunion of broken parts"), which was the title of a book written by the mathematician Muhammed ibn-Musa al Khwarizmi in about A.D. 825.

Algebra uses letters of the alphabet called *variables* to represent numbers. Often the letters x and y are used to represent variables. However, any letter may be used as a variable. A symbol that represents a specific quantity is called a *constant*.

Multiplication of numbers and variables may be represented in several different ways in algebra. Because the "times" sign might be confused with the variable x, a dot between two numbers or variables indicates multiplication. Thus, $3 \cdot 4$ means 3 times 4, and $x \cdot y$ means x times y. Placing two letters or a number and a letter next to one another, with or without parentheses, also indicates multiplication. Thus, $3x$ means 3 times x, xy means x times y, and $(x)(y)$ means x times y.

An *algebraic expression* (or simply an *expression*) is a collection of variables, numbers, grouping symbols, and operation symbols. Some examples of algebraic expressions are

$$x, \qquad x + 2, \qquad 3(2x + 3), \qquad \frac{3x + 1}{2x - 3}, \qquad \text{and} \qquad x^2 + 7x + 3$$

The parts that are added or subtracted in an algebraic expression are called *terms*. The expression $0.51x + 40.1$ contains two terms: $0.51x$ and $+ 40.1$. The expression $4x - 3y - 5$ contains three terms: $4x$, $-3y$, and -5. The $+$ and $-$ signs that break the expression into terms are part of the terms. When listing the terms of an expression, however, it is not necessary to include the $+$ sign at the beginning of the term.

The numerical part of a term is called its *numerical coefficient* or, simply, its *coefficient*. In the term $4x$, the 4 is the numerical coefficient. In the term $-4y$, the -4 is the numerical coefficient.

Like terms are terms that have the same variables with the same exponents on the variables. *Unlike terms* have different variables or different exponents on the variables.

Like Terms	Unlike Terms
$2x$, $7x$ (same variable, x)	$2x$, 9 (only first term has a variable)
$-8y$, $3y$ (same variable, y)	$5x$, $6y$ (different variables)
-4, 10 (both constants)	x, 8 (only first term has a variable)
$-5x^2$, $6x^2$ (same variable with same exponent)	$2x^3$, $3x^2$ (different exponents)

To *simplify an expression* means to combine like terms by using the commutative, associative, and distributive properties discussed in Chapter 5. For convenience, we list these properties below.

Properties of the Real Numbers

$a(b + c) = ab + ac$	Distributive property
$a + b = b + a$	Commutative property of addition
$ab = ba$	Commutative property of multiplication
$(a + b) + c = a + (b + c)$	Associative property of addition
$(ab)c = a(bc)$	Associative property of multiplication

Example 1 *Combining Like Terms*

Combine like terms in each expression.

a) $3x + 6x$
b) $5y - 2y$
c) $x + 3y - 7 + 3x - 5y - 1$
d) $2x^2 - 3x + 7 - 5x^2 + 4x - 9$

Solution

a) We use the distributive property (in reverse) to combine like terms.

$$3x + 6x = (3 + 6)x \qquad \text{Distributive property}$$
$$= 9x$$

b) $5y - 2y = (5 - 2)y = 3y$

c)
$$x + 3y - 7 + 3x - 5y - 1$$
$$= x + 3x + 3y - 5y - 7 - 1 \qquad \text{Rearrange terms; place like terms together.}$$
$$= 4x - 2y - 8 \qquad \text{Combine like terms.}$$

d)
$$2x^2 - 3x + 7 - 5x^2 + 4x - 9 \qquad \text{Rearrange terms; place like terms together.}$$
$$= 2x^2 - 5x^2 - 3x + 4x + 7 - 9$$
$$= -3x^2 + x - 2 \qquad \text{Combine like terms.} \qquad ■$$

   In Example 1(c) and 1(d), we can rearrange the terms of the expressions by using the commutative and associative properties that were discussed in Section 5.5. In Example 1(c), we generally list the terms in alphabetical order with the constant term, the term without a variable, last. In Example 1(d), we list the terms in *descending order* of the variable x. This means the exponents on the variable x get lower as the terms go from left to right with the constant term last.

Order of Operations

To *evaluate an expression* means to determine the value of the expression for a given value(s) of the variable(s). To evaluate an expression we need to know the order of operations to follow. In Section 5.2 we introduced the order of operations using only numbers. Now we will use the order of operations with expressions that also contain variables.

For example, suppose we want to evaluate the expression $2 + 3x$ when $x = 4$. Substituting 4 for x, we obtain $2 + 3 \cdot 4$. What is the value of $2 + 3 \cdot 4$? Does it equal 20, or does it equal 14? In mathematics, unless parentheses indicate otherwise, always perform multiplication before addition. Thus, the correct answer is 14.

$$2 + 3 \cdot 4 = 2 + (3 \cdot 4) = 2 + 12 = 14$$

The order of operations for evaluating an expression is repeated here for your convenience.

> **PROCEDURE** ORDER OF OPERATIONS
>
> 1. First, perform all operations within parentheses or other grouping symbols (according to the following order).
> 2. Next, perform all exponential operations (that is, raising to powers or determining roots).
> 3. Next, perform all multiplications and divisions from left to right.
> 4. Finally, perform all additions and subtractions from left to right.

Example 2 *Evaluating an Expression*

Evaluate the expression $-x^2 + 3x + 10$ for $x = 2$.

Solution Substitute 2 for each x and use the order of operations to evaluate the expression.

$$
\begin{aligned}
&-x^2 + 3x + 10 \\
&= -(2)^2 + 3(2) + 10 \\
&= -4 + 6 + 10 \\
&= 2 + 10 \\
&= 12
\end{aligned}
$$

■

Solving Equations

Two algebraic expressions joined by an equal sign form an *equation*. Some examples of equations are

$$x + 2 = 4, \qquad 3x + 4 = 1, \qquad \text{and} \qquad x + 3 = 2x$$

The *solution to an equation* is the number or numbers that replace the variable to make the equation a true statement. For example, the solution to the equation $x + 3 = 4$ is 1. When we determine the solution to an equation, we *solve the equation*.

We can determine if any number is a solution to an equation by *checking the solution*. To check the solution, we substitute the number for the variable in the equation. If the resulting statement is a true statement, that number is a solution to the equation. If the resulting statement is a false statement, the number is not a solution to the equation. To check to see if 1 is a solution to the equation $x + 3 = 4$, we do the following. We use the symbol $\overset{?}{=}$ to indicate that we are in the process of checking to see if we have a true statement.

$$
\begin{aligned}
x + 3 &= 4 \\
1 + 3 &\overset{?}{=} 4 \qquad \text{Substitute 1 for } x. \\
4 &= 4 \qquad \text{True}
\end{aligned}
$$

The same number is obtained on both sides of the equal sign, so 1 is the solution. For the equation $x + 3 = 4$, the only solution is 1. Any other value of x would result in the check being a false statement.

Example 3 *Checking a Solution*

Determine whether -1 is a solution to the equation

$$7(x + 2) - 5x - 28 = 4(x - 3) - 3(x + 1).$$

Solution To determine whether -1 is a solution to the equation, we substitute -1 for each x in the equation. We then use the order of operations to evaluate the expression on each side of the equation. If the two sides of the equation are equal, then -1 is a solution to the equation.

$$7(x + 2) - 5x - 28 = 4(x - 3) - 3(x + 1)$$

$7(-1 + 2) - 5(-1) - 28 \stackrel{?}{=} 4(-1 - 3) - 3(-1 + 1)$ Substituted -1 for x.

$7(1) - 5(-1) - 28 \stackrel{?}{=} 4(-4) - 3(0)$ Evaluated expressions within parentheses.

$7 + 5 - 28 \stackrel{?}{=} -16 - 0$ Performed multiplications.

$-16 = -16$ True

Because -1 makes the equation a true statement, -1 is a solution to the equation. ∎

In this section, we discuss solving *linear (or first-degree) equations*. A linear equation in one variable is one in which the largest exponent on the variable is 1. Examples of linear equations are $5x - 1 = 3$ and $2x + 4 = 6x - 5$.

Equivalent equations are equations that have the same solution. The equations $2x - 5 = 1$, $2x = 6$, and $x = 3$ are all equivalent equations, since they all have the same solution, 3. When we solve an equation, we write the given equation as a series of simpler equivalent equations until we obtain an equation of the form $x = c$, where c is some real number.

To solve any equation, we have to *isolate the variable*. That means getting the variable by itself on one side of the equal sign. The four properties of equality that we are about to discuss are used to isolate the variable. The first is the addition property.

Addition Property of Equality
If $a = b$, then $a + c = b + c$ for all real numbers a, b, and c.

The addition property of equality indicates that the same number can be added to both sides of an equation without changing the solution.

Example 4 *Using the Addition Property of Equality*

Determine the solution to the equation $x - 8 = 12$.

Solution To isolate the variable, add 8 to both sides of the equation.

$$x - 8 = 12$$
$$x - 8 + 8 = 12 + 8$$
$$x + 0 = 20$$
$$x = 20$$

CHECK:

$$x - 8 = 12$$
$20 - 8 \stackrel{?}{=} 12$ Substitute 20 for x.
$12 = 12$ True   ∎

In Example 4, we showed the step $x + 0 = 20$. This step is usually done mentally, and the step is usually not listed.

> ## Subtraction Property of Equality
> If $a = b$, then $a - c = b - c$ for all real numbers a, b, and c.

The subtraction property of equality indicates that the same number can be subtracted from both sides of an equation without changing the solution.

Example 5 *Using the Subtraction Property of Equality*

Determine the solution to the equation $x + 11 = 19$.

Solution To isolate the variable, subtract 11 from both sides of the equation.

$$x + 11 = 19$$
$$x + 11 - 11 = 19 - 11$$
$$x = 8$$

Note in Example 5 that we did not subtract 19 from both sides of the equation, because doing so would not result in getting x by itself on one side of the equation.

Now we discuss the multiplication property.

> ## Multiplication Property of Equality
> If $a = b$, then $a \cdot c = b \cdot c$ for all real numbers a, b, and c, where $c \neq 0$.

The multiplication property of equality indicates that both sides of the equation can be multiplied by the same nonzero number without changing the solution.

Example 6 *Using the Multiplication Property of Equality*

Determine the solution to $\dfrac{x}{4} = 8$.

Solution To solve this equation, we isolate the variable by multiplying both sides of the equation by 4.

$$\frac{x}{4} = 8$$

$$4\left(\frac{x}{4}\right) = 4(8)$$

$$\frac{\overset{1}{\cancel{4}}x}{\underset{1}{\cancel{4}}} = 32$$

$$1x = 32$$

$$x = 32$$

In Example 6, we showed the steps $\dfrac{4x}{4} = 32$ and $1x = 32$. Usually, we will not illustrate these steps.

> **Division Property of Equality**
>
> If $a = b$, then $\dfrac{a}{c} = \dfrac{b}{c}$ for all real numbers a, b, and c, $c \neq 0$.

The division property of equality indicates that both sides of an equation can be divided by the same nonzero number without changing the solution. Note that the divisor, c, cannot be 0 because division by 0 is undefined.

Example 7 *Using the Division Property of Equality*

Solve the equation $-6x = 42$.

Solution To solve this equation, we isolate the variable by dividing both sides of the equation by -6.

$$-6x = 42$$
$$\frac{-6x}{-6} = \frac{42}{-6}$$
$$x = -7$$

An *algorithm* is a general procedure for accomplishing a task. The following general procedure is an algorithm for solving linear (or first-degree) equations. Sometimes the solution to an equation may be determined more easily by using a variation of this general procedure. Remember that the primary objective in solving any equation is to isolate the variable.

> **PROCEDURE**   A GENERAL PROCEDURE FOR SOLVING LINEAR EQUATIONS
>
> 1. If the equation contains fractions, multiply both sides of the equation by the lowest common denominator (or least common multiple). This step will eliminate all fractions from the equation.
> 2. Use the distributive property to remove parentheses when necessary.
> 3. Combine like terms on the same side of the equal sign when possible.
> 4. Use the addition or subtraction property to collect all terms with a variable on one side of the equation and all constants on the other side of the equation. It may be necessary to use the addition or subtraction property more than once. This process will eventually result in an equation of the form $ax = b$, where a and b are real numbers.
> 5. Solve for the variable using the division or multiplication property. The result will be an answer in the form $x = c$, where c is a real number.
> 6. Check your answer by substituting the value obtained in Step 5 back into the original equation.

Example 8 *Solving a Linear Equation*

Solve the equation $3x - 5 = 28$.

Solution Our goal is to isolate the variable; therefore we start by getting the term $3x$ by itself on one side of the equation.

$$3x - 5 = 28$$
$$3x - 5 + 5 = 28 + 5 \qquad \text{Add 5 to both sides of the equation (addition property) (Step 4).}$$
$$3x = 33$$

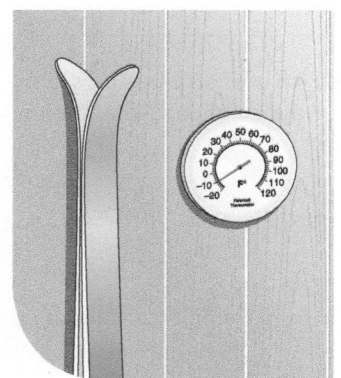
$$\frac{3x}{3} = \frac{33}{3}$$ Divide both sides of the equation by 3 (division property) (Step 5).

$$x = 11$$

A check will show that 11 is the solution to $3x - 5 = 28$.   ∎

We did not show the checking of the solution to Example 8. To save space, we will not show all checks. However, you should check all your answers when solving equations.

Example 9   *Solving a Linear Equation*

Solve the equation $6t + 11 = 2(t + 6) - 5$.

Solution   Our goal is to isolate the variable t. To do so, follow the general procedure for solving equations.

$$6t + 11 = 2(t + 6) - 5$$

$$6t + 11 = 2t + 12 - 5$$ Distributive property (Step 2)

$$6t + 11 = 2t + 7$$ Combine like terms (Step 3).

$$6t - 2t + 11 = 2t - 2t + 7$$ Subtraction property (Step 4)

$$4t + 11 = 7$$

$$4t + 11 - 11 = 7 - 11$$ Subtraction property (Step 4)

$$4t = -4$$

$$\frac{4t}{4} = \frac{-4}{4}$$ Division property (Step 5)

$$t = -1$$ ∎

Example 10   *Solving an Equation Containing Fractions*

Solve the equation $\dfrac{2x}{3} + \dfrac{1}{3} = \dfrac{3}{4}$.

Solution   When an equation contains fractions, we generally begin by multiplying both sides of the equation by the lowest common denominator, LCD (see page 237 in Chapter 5). In this example, the LCD is 12 because 12 is the smallest number that is divisible by both 3 and 4.

$$12\left(\frac{2x}{3} + \frac{1}{3}\right) = 12\left(\frac{3}{4}\right)$$ Multiply both sides of the equation by the LCD (Step 1).

$$12\left(\frac{2x}{3}\right) + 12\left(\frac{1}{3}\right) = 12\left(\frac{3}{4}\right)$$ Distributive property (Step 2)

$$\overset{4}{\cancel{12}}\left(\frac{2x}{\underset{1}{\cancel{3}}}\right) + \overset{4}{\cancel{12}}\left(\frac{1}{\underset{1}{\cancel{3}}}\right) = \overset{3}{\cancel{12}}\left(\frac{3}{\underset{1}{\cancel{4}}}\right)$$ Divide out common factors.

$$8x + 4 = 9$$

$$8x + 4 - 4 = 9 - 4$$ Subtraction property (Step 4)

$$8x = 5$$

$$\frac{8x}{8} = \frac{5}{8}$$ Division property (Step 5)

$$x = \frac{5}{8}$$

A check will show that $\frac{5}{8}$ is the solution to the equation.   ∎

Timely Tip

Remember that the goal in solving an equation is to get the variable alone on one side of the equal sign by using the general procedure for solving equations.

Example 11   *Solving an Equation Containing Decimals*

Solve the equation $4x - 0.48 = 0.8x + 4$ and check your solution.

Solution   This equation may be solved with the decimals, or you may multiply each term by 100 and eliminate the decimals. We will solve the equation with the decimals.

$$4x - 0.48 = 0.8x + 4$$
$$4x - 0.48 + 0.48 = 0.8x + 4 + 0.48 \qquad \text{Addition property}$$
$$4x = 0.8x + 4.48$$
$$4x - 0.8x = 0.8x - 0.8x + 4.48 \qquad \text{Subtraction Property}$$
$$3.2x = 4.48$$
$$\frac{3.2x}{3.2} = \frac{4.48}{3.2} \qquad \text{Division property}$$
$$x = 1.4$$

CHECK:
$$4x - 0.48 = 0.8x + 4$$
$$4(1.4) - 0.48 \stackrel{?}{=} 0.8(1.4) + 4 \qquad \text{Substitute 1.4 for each } x \text{ in the equation.}$$
$$5.6 - 0.48 \stackrel{?}{=} 1.12 + 4$$
$$5.12 = 5.12 \qquad \text{True} \qquad ■$$

So far, every equation has had exactly one solution. Some equations, however, have no solution, and others have more than one solution. Example 12 illustrates an equation that has no solution, and Example 13 illustrates an equation that has an infinite number of solutions.

Example 12   *An Equation with No Solution*

Solve $3(x - 5) + x + 12 = 6x - 2(x + 2)$.

Solution

$$3(x - 5) + x + 12 = 6x - 2(x + 2)$$
$$3x - 15 + x + 12 = 6x - 2x - 4 \qquad \text{Distributive property}$$
$$4x - 3 = 4x - 4 \qquad \text{Combine like terms.}$$
$$4x - 4x - 3 = 4x - 4x - 4 \qquad \text{Subtraction property}$$
$$-3 = -4 \qquad \text{False}$$

During the process of solving an equation, if you obtain a false statement like $-3 = -4$, or $1 = 0$, the equation has no solution.   ■

An equation that has no solution is called a *contradiction*. The equation in Example 12, $3(x - 5) + x + 12 = 6x - 2(x + 2)$, is a contradiction and thus has no solution.

Example 13   *An Equation with Infinitely Many Solutions*

Solve $2(x + 4) - 3(x - 5) = -x + 23$.

Solution

$$2(x + 4) - 3(x - 5) = -x + 23$$
$$2x + 8 - 3x + 15 = -x + 23 \qquad \text{Distributive property}$$
$$-x + 23 = -x + 23 \qquad \text{Combine like terms.}$$

Note that at this point both sides of the equation are the same. Every real number will satisfy this equation. The solution to this equation is *all real numbers*. This equation has an infinite number of solutions.   ■

An equation that is true for all real numbers, like the equation in Example 13, is called an *identity*. An identity has the solution *all real numbers*.

The following table summarizes information regarding contradictions and identities.

Type of Equation	When Equation Is True	How to Write Answer
Contradiction	Never	No solution
Identity	Always	All real numbers

SECTION 6.1   *Exercises*

Warm Up Exercises

In Exercises 1–10, fill in the blank with an appropriate word, phrase, or symbol(s).

1. A letter of the alphabet used to represent numbers is called a(n) _____.

2. A symbol that represents a specific quantity is called a(n) _____.

3. A collection of variables, numbers, grouping symbols, and operation symbols is called an algebraic _____.

4. The number or numbers that replace the variable to make an equation a true statement is called the _____ to the equation.

5. The parts that are added or subtracted in an algebraic expression are called _____.

6. Terms that have the same variables with the same exponents on the variables are called _____ terms.

7. The numerical part of a term is called its numerical _____.

8. An equation in which the highest exponent on the variable is a 1 is called a(n) _____ equation.

9. An equation that has an infinite number of solutions is called a(n) _____.

10. An equation that has no solution is called a(n) _____.

Practice the Skills

In Exercises 11–30, combine like terms.

11. $4x + 9x$

12. $8x - 5x$

13. $-7x + 3x - 8$

14. $-9x - 2x + 15$

15. $7x + 3y - 4x + 8y$

16. $-5y + 4z + 2y - 9z$

17. $5x - 7y + 8 - 3x - 11y - 13$

18. $-4x + 2y - 17 + x - 9y + 12$

19. $10x^2 + x - 21 - 11x^2 + 7x + 16$

20. $-13x^2 - 3x + 23 + 4x^2 - x - 29$

21. $3(t + 3) + 5(t - 2) + 1$

22. $6(r - 3) - 2(r + 5) + 10$

23. $6.2x - 8.3 + 7.1x$

24. $4.7x - 6.1 + 8.2x$

25. $\dfrac{3}{5}x + \dfrac{3}{10}x - 8$

26. $\dfrac{1}{5}x - \dfrac{1}{3}x - 4$

27. $0.2(x + 4) + 1.2(x - 3)$

28. $0.9(2.3x - 2) + 1.7(3.2x - 5)$

29. $\dfrac{2}{3}(x + 6) + \dfrac{1}{6}(x + 6)$

30. $\dfrac{2}{3}(3x + 9) - \dfrac{1}{4}(2x + 5)$

In Exercises 31–44, evaluate the expression for the given value(s) of the variable(s).

31. $x - 7$, $x = 4$

32. $4x - 3$, $x = -1$

33. $-3x + 7$, $x = -2$

34. $4x - 7$, $x = \frac{5}{2}$

35. x^2, $x = 8$

36. x^2, $x = -3$

37. $-x^2$, $x = -5$

38. $-x^2$, $x = 1$

39. $x^2 - 5x + 12$, $x = 3$

40. $x^2 + 7x - 3$, $x = -4$

41. $-2x^2 + 5x - 9$, $x = 3$

42. $-x^2 + 3x - 10$, $x = -2$

43. $x^3 - 3x^2 + 7x - 5$, $x = 2$

44. $2x^3 + x^2 - 3x - 4$, $x = -1$

In Exercises 45–50, determine whether the given value is a solution to the equation.

45. $4x + 5 = 17$, $x = 3$

46. $7x - 1 = -29$, $x = -4$

47. $8x - 7 = 5x + 1$, $x = -2$

48. $4x + 3 = 7x - 6$, $x = 3$

49. $3(2x - 7) + 1 = -4(3x + 1) + 20$, $x = 2$

50. $-2(5x + 1) + 3 = 7(4x + 7) + 4$, $x = -3$

In Exercises 51–76, solve the equation.

51. $y + 7 = 9$ **52.** $y - 4 = 13$

53. $\dfrac{x}{9} = -5$ **54.** $-\dfrac{x}{7} = -8$

55. $7x = -63$ **56.** $-8x = 32$

57. $16 = -3t - 2$ **58.** $25 = 4x + 5$

59. $6t - 8 = 4t - 2$ **60.** $6t - 7 = 8t + 9$

61. $3x + 2 - 6x = -x - 15 + 8 - 5x$

62. $6x + 8 - 22x = 28 + 14x - 10 + 12x$

63. $4(3n + 1) = 5(4n - 6) + 9n$

64. $4(t - 3) + 8 = 4(2t - 6)$

65. $\dfrac{1}{2}x + \dfrac{1}{3} = \dfrac{2}{3}$ **66.** $\dfrac{1}{2}y + \dfrac{1}{3} = \dfrac{1}{4}$

67. $\dfrac{x}{4} + 2x = \dfrac{1}{3}$ **68.** $\dfrac{r}{3} + 2r = 7$

69. $0.9x - 1.2 = 2.4$

70. $5x + 0.050 = -0.732$

71. $0.2x + 1.3 = 0.4x - 4.5$

72. $3.01x - 5.2 = 2.91x + 1.57$

73. $2(x + 3) - 4 = 2(x - 4)$

74. $3(x + 2) + 2(x - 1) = 5x - 7$

75. $4(x - 4) + 12 = 4(x - 1)$

76. $6(t + 2) - 14 = 6t - 2$

Problem Solving

77. *Sales Tax* If the sales tax, s, on an item is 8%, the sales tax on an item costing d dollars can be determined by the equation $s = 0.08d$. Determine the sales tax on a pair of sneakers that cost $79.

78. *Cost of a Tour* The cost, c, in dollars, for Crescent City Tours to provide a tour for x people can be determined by the equation $c = 220 + 2.75x$. Determine the cost for Crescent City Tours to provide a tour for 75 people.

79. *Temperature Conversion* Temperatures measured in degrees Celsius, C, can be converted to degrees Fahrenheit, F, using the equation $F = \dfrac{9}{5}C + 32$. Use this formula to convert the following temperatures to degrees Fahrenheit.

a) 0°C (freezing temperature of water)

b) 100°C (boiling temperature of water)

c) 35°C (temperature of a summer day in Rio de Janeiro, Brazil)

▲ Rio de Janeiro, Brazil

d) −5°C (temperature of a winter day in Toronto, Canada)

80. *Salary Plus Commission* Alia works in a surf shop. She earns a weekly salary of $475 plus a commission of 15% of her weekly sales total. Alia's weekly earnings, E, can be represented with the linear equation $E = 475 + 0.15x$, where x represents her weekly sales.

a) Determine Alia's weekly earnings if her weekly sales were $1500.

b) For one week, Alia's weekly earnings were $649. Determine her weekly sales.

Concept/Writing Exercises

81. Explain why $(-1)^n = 1$ for any even number n.

82. Does $(x + y)^2 = x^2 + y^2$? Complete the table and state your conclusion.

x	y	$(x + y)^2$	$x^2 + y^2$
2	3		
−2	−3		
−2	3		
2	−3		

83. Suppose n represents any natural number. Explain why 1^n equals 1.

84. a) Explain why $-x^2$ will always be a negative number for any nonzero real number selected for x.
   b) Explain why $(-x)^2$ will always be a positive number for any nonzero real number selected for x.

Challenge Problem/Group Activitiy

85. ***Depth of a Submarine*** The pressure, P, in pounds per square inch (psi) exerted on an object x ft below the sea is given by the equation $P = 14.70 + 0.43x$. The 14.70 represents the weight in pounds of the column of air (from sea level to the top of the atmosphere) standing over a 1 in. by 1 in. square of seawater. The $0.43x$ represents the weight in pounds of a column of water 1 in. by 1 in. by x ft (see Fig. 6.1).

a) A submarine is built to withstand a pressure of 148 psi. How deep can that submarine go?

b) If the pressure gauge in the submarine registers a pressure of 128.65 psi, how deep is the submarine?

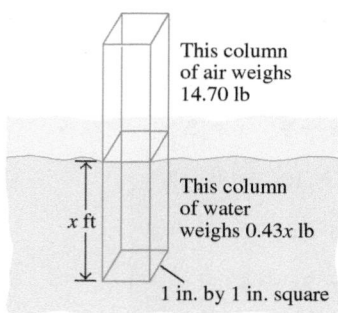

Figure 6.1

Research Activities

86. ***Exponents*** When were exponents first used? Write a paper explaining how exponents were first used and when mathematicians began writing them in the present form.

87. ***Ancient Equations*** Write a report explaining how the ancient Egyptians used equations. Include in your discussion the forms of the equations used.

SECTION 6.2   Formulas

Upon completion of this section, you will be able to:

- Evaluate a formula.
- Solve for a variable in a formula or equation.

Breno received a holiday bonus of $1500 at work. He wishes to invest his money in a certificate of deposit to help save for a new boat. In this section, we will use a special kind of equation, called a *formula*, to determine the amount of interest that Breno will earn from the certificate of deposit.

Why This Is Important Formulas are used in many applications, especially financial applications, that affect our lives. Included in these applications are interest earned on investments as well as interest charged on loans. Understanding how to use formulas may enable us to make better financial decisions.

Evaluating Formulas

A *formula* is an equation that typically has a real-life application. To *evaluate* a formula, substitute the given values for their respective variables and then evaluate using the order of operations given in Section 6.1. Many of the formulas given in this section are discussed in greater detail in other parts of the book.

Example 1 *Certificate of Deposit*

Breno invests $1500 in a certificate of deposit (CD) that pays 1.5% (or 0.015) simple interest for 2 years. The simple interest formula, which will be discussed again in Section 10.2, is used to calculate the simple interest earned on an investment. The simple interest formula is interest = principal × rate × time, or $i = prt$.

a) Determine the amount of simple interest that Breno earns on the CD.

b) What is the total value of the CD, including interest, at the end of 2 years?

Solution

a) The principal, p, is the amount of money invested, or $1500. The rate, r, is 0.015, and the time, t, is 2 years. We substitute these values into the simple interest formula and evaluate.

$$i = prt$$
$$= 1500(0.015)(2)$$
$$= 45$$

Thus, Breno earns $45 simple interest.

b) The total value of the CD is the original principal, $1500, plus the simple interest earned, $45. Thus, the total value of the CD is $1500 + $45, or $1545. ■

Example 2 *Ice-Cream Box*

The volume of a rectangular solid, which will be discussed in Section 8.4, can be determined using the formula volume = length × width × height, or $V = lwh$. Use this formula to determine the width of a rectangular box of ice cream if $l = 7$ in., $h = 3.5$ in., and $V = 122.5 \text{ in.}^3$.

Solution We substitute the appropriate values into the volume formula and solve for the desired quantity, w.

$$V = lwh$$
$$122.5 = (7)w(3.5)$$
$$122.5 = 24.5w$$
$$\frac{122.5}{24.5} = \frac{24.5w}{24.5}$$
$$5 = w$$

Therefore, the width of the ice-cream box is 5 in. ■

In Examples 1 and 2, we used an equation to represent real-life phenomena. By doing so, we say we have created a *mathematical model* or simply a *model* to represent the situation. A model may be a formula, a single equation, or several equations considered simultaneously. In some exercises, when we are asking you to write a model, we will mark the exercise with the word **MODELING**.

Some formulas contain *subscripts*. Subscripts are numbers (or letters) placed below and to the right of variables. They are used to help clarify a formula. For example, if two different amounts are used in a problem, they may be symbolized as A and A_0, or A_1 and A_2. Subscripts are read using the word *sub*; for example, A_0 is read "A sub zero" and A_1 is read "A sub one." Our next example involves variables with subscripts.

Example 3 *Calculating Slope*

The slope, m, of a line, which will be discussed in Section 6.6, can be determined using the formula $m = \dfrac{y_2 - y_1}{x_2 - x_1}$, where (x_1, y_1) and (x_2, y_2) are the coordinates of two points on the line. Use this formula to determine the slope of a line containing the points $(3, -2)$ and $(-1, 6)$.

Solution Here $x_1 = 3, y_1 = -2, x_2 = -1$, and $y_2 = 6$. Substituting these values into the formula we have the following.

$$m = \frac{y_2 - y_1}{x_2 - x_1} = \frac{6 - (-2)}{-1 - 3} = \frac{8}{-4} = -2$$

Thus, the slope of the line containing the points $(3, -2)$ and $(-1, 6)$ is -2.   ∎

Solving for a Variable in a Formula or Equation

Often in mathematics and science courses, you are given a formula or an equation expressed in terms of one variable and asked to express it in terms of a different variable. For example, you may be given the formula $P = i^2 r$ and asked to solve the formula for r. To do so, treat each of the variables, except the one you are solving for, as if it were a constant. Then solve for the variable desired, using the properties previously discussed. Examples 4 through 6 show how to do this task.

When graphing equations in Section 6.6, you will sometimes have to solve the equation for the variable y, as is done in Example 4.

Example 4   *Solving for a Variable in an Equation*

Solve the equation $2x + 5y - 10 = 0$ for y.

Solution We need to isolate the term containing the variable y. Begin by adding 10 and subtracting $2x$ from both sides of the equation.

$$2x + 5y - 10 = 0$$
$$2x + 5y - 10 + 10 = 0 + 10 \qquad \text{Addition property}$$
$$2x + 5y = 10$$
$$-2x + 2x + 5y = -2x + 10 \qquad \text{Subtraction property}$$
$$5y = -2x + 10$$
$$\frac{5y}{5} = \frac{-2x + 10}{5} \qquad \text{Division property}$$
$$y = \frac{-2x + 10}{5}$$
$$y = -\frac{2x}{5} + \frac{10}{5}$$
$$y = -\frac{2}{5}x + 2$$ ∎

Note that once you have found $y = \frac{-2x + 10}{5}$, you have solved the equation for y. The solution can also be expressed in the form $y = -\frac{2}{5}x + 2$. This form of the equation is convenient for graphing equations, as will be explained in Section 6.6.

Many formulas contain Greek letters, such as μ (mu), σ (sigma), Σ (capital sigma), δ (delta), Δ (capital delta), ε (epsilon), π (pi), θ (theta), and λ (lambda). Our next example involves Greek letter variables.

Example 5   *Standard Score Formula*

This example involves an important formula from statistics used to determine the standard score, or z-score. This formula will be discussed in Section 12.5.

$$z = \frac{x - \mu}{\sigma}$$

Solve this formula for x.

Solution To isolate the term x, use the general procedure for solving linear equations given in Section 6.1. Treat each variable, except x, as if it were a constant.

$$z = \frac{x - \mu}{\sigma}$$

$$z \cdot \sigma = \frac{x - \mu}{\sigma} \cdot \sigma \qquad \text{Multiply both sides of the equation by } \sigma.$$

$$z\sigma = x - \mu$$

$$z\sigma + \mu = x - \mu + \mu \qquad \text{Add } \mu \text{ to both sides of the equation.}$$

$$z\sigma + \mu = x$$

$$\text{or} \qquad x = z\sigma + \mu \qquad \blacksquare$$

Example 6 The Slope–Intercept Form of the Equation of a Line

A formula used to represent a linear equation, the slope–intercept form of the equation of a line, is

$$y = mx + b$$

Solve this equation for m.

Solution To isolate the variable m, use the general procedure for solving linear equations given in Section 6.1. Treat each variable, except m, as if it were a constant.

$$y = mx + b$$

$$y - b = mx + b - b \qquad \text{Subtract } b \text{ from both sides of the equation.}$$

$$y - b = mx$$

$$\frac{y - b}{x} = \frac{mx}{x} \qquad \text{Divide both sides of the equation by } x.$$

$$\frac{y - b}{x} = m \qquad \blacksquare$$

We will discuss the slope–intercept form of a line in detail in Section 6.6.

SECTION 6.2 Exercises

Warm Up Exercises

In Exercises 1 and 2, fill in the blank with an appropriate word, phrase, or symbol(s).

1. Numbers or letters that are placed below and to the right of variables are called _____ .

2. An equation that typically has a real-life application is called a(n) _____ .

Practice the Skills

In Exercises 3–30, use the formula to determine the value of the indicated variable for the values given. When appropriate, use the $\boxed{\pi}$ key on your calculator and round your answer to the nearest hundredth.

3. $P = 4s$; determine P when $s = 10$. (geometry)

4. $A = lw$; determine A when $l = 9$ and $w = 4$. (geometry)

5. $P = 2l + 2w$; determine P when $l = 15$ and $w = 8$ (geometry).

6. $A = 2lw + 2wh + 2lh$; determine A when $l = 2$, $w = 3$, and $h = 4$ (geometry).

7. $F = ma$; determine m when $F = 40$ and $a = 5$ (physics).

8. $p = i^2r$; determine r when $p = 62{,}500$ and $i = 5$ (electronics).

9. $A = 2\pi rh + 2\pi r^2$; determine A when $r = 2$, and $h = 3$ (geometry).

10. $A = \pi(R^2 - r^2)$; determine A when $R = 5$ and $r = 4$ (geometry).

11. $K = \frac{1}{2}mv^2$; determine m when $K = 4500$ and $v = 30$ (physics).

12. $\bar{v} = \frac{v + v_0}{2}$; determine v when $\bar{v} = 8$ and $v_0 = 3$ (physics).

13. $T = \frac{PV}{k}$; determine P when $T = 80$, $V = 20$, and $k = 0.5$ (physics).

14. $P = \frac{kAT}{l}$; determine k when $P = 35$, $A = 14$, $T = 2$, and $l = 4$ (physics).

15. $V = -\frac{1}{2}at^2$; determine a when $V = 2304$ and $t = 12$ (physics).

16. $A = \frac{1}{2}h(b_1 + b_2)$; determine h when $A = 36$, $b_1 = 4$, and $b_2 = 8$ (geometry).

17. $V = \pi r^2 h$; determine h when $V = 942$ and $r = 5$ (geometry).

18. $V = \frac{1}{3}\pi r^2 h$; determine h when $V = 47.10$ and $r = 3$ (geometry).

19. $F = \frac{9}{5}C + 32$; determine F when $C = 7$ (temperature conversion).

20. $C = \frac{5}{9}(F - 32)$; determine C when $F = 77$ (temperature conversion).

21. $m = \frac{y_2 - y_1}{x_2 - x_1}$; determine m when $y_2 = 8$, $y_1 = -4$, $x_2 = -3$, and $x_1 = -5$ (mathematics).

22. $m = \frac{y_2 - y_1}{x_2 - x_1}$; determine m when $x_1 = 7, y_1 = -3, x_2 = -4$, and $y_2 = 3$ (mathematics).

23. $x = \frac{-b + \sqrt{b^2 - 4ac}}{2a}$; determine x when $a = 2$, $b = -5$, and $c = -12$ (mathematics).

24. $x = \frac{-b - \sqrt{b^2 - 4ac}}{2a}$; determine x when $a = 2, b = -5$, and $c = -12$ (mathematics).

25. $s = -16t^2 + v_0 t + s_0$; determine s when $t = 4$, $v_0 = 30$, and $s_0 = 150$ (physics).

26. $c = \sqrt{a^2 + b^2}$; determine c when $a = 5$ and $b = 12$ (geometry).

27. $P = \frac{nRT}{V}$; determine R when $P = 63, n = 27, T = 2$, and $V = 6$ (chemistry).

28. $P = \frac{nRT}{V}$; determine T when $P = 81, n = 15, R = 18$, and $V = 10$ (chemistry).

29. $A = P\left(1 + \frac{r}{n}\right)^{nt}$; determine A when $P = 1200, r = 0.04, n = 2$, and $t = 5$ (banking).

30. $A = P\left(1 + \frac{r}{n}\right)^{nt}$; determine A when $P = 100, r = 0.06$ $n = 1$, and $t = 3$ (banking).

In Exercises 31–36, solve the equation for y.

31. $6x + 3y = 9$

32. $4x - 2y = 10$

33. $x - 6y = 12$

34. $3x + 10y = 20$

35. $2x - 3y + 6 = 0$

36. $10x - 3y = 0$

In Exercises 37–52, solve for the variable indicated.

37. $d = rt$ for r

38. $V = lwh$ for w

39. $p = a + b + c$ for a

40. $v - e + f = 2$ for v

41. $C = 2\pi r$ for r

42. $V = \pi r^2 h$ for h

43. $A = \frac{1}{2}bh$ for b

44. $V = \frac{1}{3}\pi r^2 h$ for h

45. $y = a + bx$ for x

46. $y = mx + b$ for x

47. $P = 2l + 2w$ for w

48. $P = 2l + 2w$ for l

49. $F = \frac{9}{5}C + 32$ for C

50. $C = \frac{5}{9}(F - 32)$ for F

51. $A = P(1 + rt)$ for r

52. $A = P(1 + rt)$ for t

Problem Solving

In Exercises 53–57, When appropriate, use the $\boxed{\pi}$ key on your calculator and round your answer to the nearest hundredth.

53. *Driving Time* A family is planning to drive 403 miles from their home in Richmond, Virginia, to the Great Smoky Mountains National Park. If they drive at an average rate of 62 miles per hour, how long will their trip take? Use the formula $d = rt$.

54. *Savings Account*   Christine borrowed \$5200 from a bank at a simple interest rate of 2.5% (or 0.025) for one year.

   a) Determine how much interest Christine paid at the end of 1 year. Use the formula $i = prt$.

   b) Determine the total amount Christine will repay the bank at the end of 1 year.

55. *Interest on a Loan*   Jeff borrowed \$800 from his brother for 2 years. At the end of 2 years, Jeff repaid the \$800 plus \$128 in interest. What simple interest rate did Jeff pay?

56. *Volume of an Ice-Cream Cone*   An ice-cream cone is filled with ice cream to the top of the cone. Determine the volume, in cubic inches, of ice cream in the cone if the cone's radius is 1.5 in. and the height is 7 in. (See the figure.) The formula for the volume of a cone is $V = \dfrac{1}{3}\pi r^2 h$.

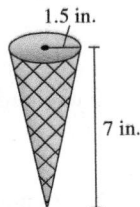

1.5 in.

7 in.

57. *Body Mass Index*   A person's body mass index (BMI) is determined by the formula $B = \dfrac{703w}{h^2}$, where w is the person's weight, in pounds, and h is the person's height, in inches. John is 6 ft tall and weighs 200 lb.

   a) Determine his BMI.

   b) If John would like to have a BMI of 26, how much weight would he need to gain or lose?

Challenge Problem/Group Activity

58. Determine the volume of the block shown in Fig. 6.2, excluding the hole. The formula for the volume of a rectangular solid is $V = lwh$. The formula for the volume of a cylinder is $V = \pi r^2 h$.

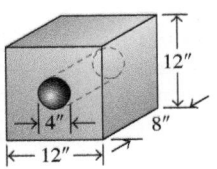

12″

4″   8″

12″

Figure 6.2

Applications of Algebra

Upon completion of this section, you will be able to:

- Solve applications of linear equations in one variable.
- Solve applications using proportions.

Brittany purchased a new battery for her SUV. Her total cost will include the price of the battery, 7% sales tax, and a battery recycling fee. The total cost that Brittany pays can be represented with an equation. When we use an equation to represent an application, we say that we are *modeling* the application using algebra. Throughout this section we will use algebra to model many real-life applications.

Why This Is Important   Being able to model real-life problems with algebra allows us to determine solutions to these problems. Some of the problems we will encounter in this section involve other financial applications as well as applications involving geometry, medicine, cooking, and traveling. Learning how to use algebra to solve application problems provides us with a powerful tool that can be used in many aspects of our lives.

Applications of Linear Equations in One Variable

One reason to study algebra is that it can be used to solve everyday problems. In this section, we will do two things: (1) show how to translate a written problem into a mathematical equation and (2) show how linear equations can be used in solving everyday problems. We begin by illustrating how English phrases can be written as mathematical expressions. When writing a mathematical expression, we may use any letter to represent the variable. In the following illustrations, we use the letter x.

Phrase	Mathematical expression
Six more than a number	$x + 6$
A number increased by 3	$x + 3$
Four less than a number	$x - 4$
A number decreased by 9	$x - 9$
Twice a number	$2x$
The product of 7 and a number	$7x$
The quotient of a number and 8	$\dfrac{x}{8}$
Four times a number	$4x$
3 decreased by a number	$3 - x$
The difference between a number and 5	$x - 5$

Sometimes the phrase that must be converted to a mathematical expression involves more than one operation.

Phrase	Mathematical expression
Four less than 3 times a number	$3x - 4$
Ten more than twice a number	$2x + 10$
The sum of 5 times a number, and 3	$5x + 3$
Eight times a number, decreased by 7	$8x - 7$

The word *is* often represents the equal sign.

Phrase	Mathematical equation
Six more than a number is 10.	$x + 6 = 10$
Five less than a number is 20.	$x - 5 = 20$
Twice a number, decreased by 6 is 12.	$2x - 6 = 12$
A number decreased by 13 is 6 times the number.	$x - 13 = 6x$

Matching Verbal Statements to Equations

The following is a general procedure for solving word problems.

> **PROCEDURE**   SOLVING WORD PROBLEMS
>
> 1. Read the problem carefully at least twice to be sure that you understand it.
> 2. If possible, draw a sketch to help visualize the problem.
> 3. Identify which quantity you are being asked to determine. Choose a letter to represent this unknown quantity. Write down exactly what this letter represents.
> 4. Write the word problem as an equation.
> 5. Solve the equation for the unknown quantity.
> 6. Answer the question or questions asked.
> 7. Check the solution.

This general procedure for solving word problems is illustrated in Examples 1 through 4. In these examples, the equations we obtain are mathematical models of the given situations.

Example 1 *Copy Machine*

Jayna purchased a copier for $2300. The cost to make each copy is $0.02. If Jayna spends a total of $2626 in a year, which includes the cost of the copier and the copies made, determine the number of copies she made.

Solution In this problem, the unknown quantity is the number of copies Jayna made. Let's select n to represent the number of copies made. Then we construct an equation using the given information that will allow us to solve for n.

Let

$$n = \text{number of copies}$$

Then

$$\$0.02n = \text{the cost for making } n \text{ copies at \$0.02 per copy}$$

$$\text{Purchase price} + \text{copy cost} = \text{total spent}$$

$$\$2300 + \$0.02n = \$2626$$

Now solve the equation.

$$2300 + 0.02n = 2626$$
$$2300 - 2300 + 0.02n = 2626 - 2300$$
$$0.02n = 326$$
$$\frac{0.02n}{0.02} = \frac{326}{0.02}$$
$$n = 16{,}300$$

Therefore, Jayna made 16,300 copies.

CHECK: The check is made with the information given in the original problem.

$$\text{Total spent} = \text{purchase price} + \text{copy cost}$$
$$= 2300 + 0.02(16{,}300)$$
$$= 2300 + 326$$
$$= 2626$$

Therefore, this answer checks. ∎

Example 2 *Party Punch*

Casey needs to make 2 gallons, or 256 ounces, of fruit punch for a birthday party. Her recipe requires mixing pineapple juice, ginger ale, and orange juice. The amount of ginger ale required is twice the amount of pineapple juice, and the amount of orange juice required is five times the amount of pineapple juice. How many ounces of pineapple juice, ginger ale, and orange juice should Casey mix together to make the punch?

Solution The amount of ginger ale and the amount of orange juice are both described in terms of the amount of pineapple juice. The amount of ginger ale is twice the amount of pineapple juice. The amount of orange juice is five times the amount of pineapple juice. Therefore, we will let x represent the number of ounces of pineapple juice. We then can write expressions for the number of ounces of ginger ale and the number of ounces of orange juice in terms of the variable x.

Let

$$x = \text{number of ounces of pineapple juice}$$
$$2x = \text{number of ounces of ginger ale}$$
$$5x = \text{number of ounces of orange juice}$$

Next, the sum of these amounts must equal 256 ounces.

$$x + 2x + 5x = 256$$
$$8x = 256$$
$$x = 32$$

Thus, Casey must mix 32 ounces of pineapple juice, $2(32)$, or 64 ounces of ginger ale, and $5(32)$, or 160 ounces of orange juice. To check this answer, we note that $32 + 64 + 160 = 256$. ∎

Example 3 *Dimensions of a Sandbox*

Javier is building a rectangular sandbox for his children. He has 34 ft of wood to use for the perimeter of the sandbox. What should be the dimensions of the sandbox if he wants the length to be 3 ft greater than the width?

Solution The formula for determining the perimeter of a rectangle is $P = 2l + 2w$, where P is the perimeter, l is the length, and w is the width. The length is 3 ft more than the width; therefore, $l = w + 3$. A diagram, such as the one below, is often helpful in solving problems of this type.

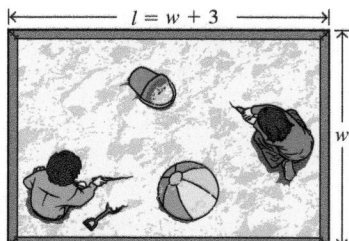

The perimeter of the sandbox, P, is 34 ft.
  Substitute the known quantities in the formula.

$$P = 2l + 2w$$
$$34 = 2(w + 3) + 2w$$
$$34 = 2w + 6 + 2w$$
$$34 = 4w + 6$$
$$28 = 4w$$
$$7 = w$$

The width of the sandbox is 7 ft and the length of the sandbox is $7 + 3 = 10$ ft. ∎

In everyday life we often encounter percents. The word *percent* means "per hundred." Thus, for example, 7% means 7 per hundred, or $\frac{7}{100}$. When $\frac{7}{100}$ is converted to a decimal number, we obtain 0.07. Thus, 7% = 0.07.

Let's look at one example involving percent. (See Section 10.1 for a more detailed discussion of percent.)

Example 4 *Craft Show*

Kameko is selling her homemade jewelry. Determine the cost of a necklace before tax if the total cost of a necklace, including an 8% sales tax, is $34.56.

Solution We are asked to determine the cost of a necklace before sales tax.
Let
$$x = \text{cost of the necklace before sales tax}$$

Then
$$0.08x = 8\% \text{ of the cost of the necklace (the sales tax)}$$

$$\text{Cost of the necklace before tax } + \text{ tax on the necklace} = 34.56$$

$$x + 0.08x = 34.56$$

$$1.08x = 34.56$$

$$\frac{1.08x}{1.08} = \frac{34.56}{1.08}$$

$$x = 32$$

Thus, the cost of the necklace before tax is $32. ∎

Proportions

A *ratio* is a quotient of two quantities. An example is the ratio of 2 to 5, which can be written 2:5 or $\frac{2}{5}$ or 2/5. Ratios are used in proportions.

> **Definition: Proportion**
> A **proportion** is a statement of equality between two ratios.

An example of a proportion is $\frac{a}{b} = \frac{c}{d}$, where $b \neq 0$ and $d \neq 0$. Consider the proportion

$$\frac{x+2}{5} = \frac{x+5}{8}$$

We can solve this proportion by first multiplying both sides of the equation by the LCD, 40.

$$\frac{x+2}{5} = \frac{x+5}{8}$$

$$\overset{8}{\cancel{40}}\left(\frac{x+2}{\cancel{5}}\right) = \overset{5}{\cancel{40}}\left(\frac{x+5}{\cancel{8}}\right) \qquad \text{Multiplication property}$$

$$8(x+2) = 5(x+5)$$

$$8x + 16 = 5x + 25$$

$$3x + 16 = 25$$

$$3x = 9$$

$$x = 3$$

Did You Know?

Survival of the Fittest

Symbols come and symbols go; the ones that find the greatest acceptance are the ones that survive. The Egyptians used pictorial symbols: a pair of legs walking forward for addition or backward for subtraction. Robert Recorde (1510–1558) used two parallel lines, =, to represent "equals" because "no 2 thynges can be moore equalle." Some symbols evolved from abbreviations, such as the "+" sign, which comes from the Latin *et* meaning "and." The evolution of others is less clear. The invention of the printing press in the fifteenth century led to a greater standardization of symbols already in use.

A check will show that 3 is the solution.

Proportions can often be solved more easily by using cross-multiplication.

> ### Cross-Multiplication
>
> If $\dfrac{a}{b} = \dfrac{c}{d}$, then $ad = bc$,   where $b \neq 0, d \neq 0$.

Let's use cross-multiplication to solve the proportion $\dfrac{x + 2}{5} = \dfrac{x + 5}{8}$.

$$\frac{x + 2}{5} = \frac{x + 5}{8}$$
$$8(x + 2) = 5(x + 5) \quad \text{Cross-multiplication}$$
$$8x + 16 = 5x + 25$$
$$3x + 16 = 25$$
$$3x = 9$$
$$x = 3$$

Notice we obtained the same answer as we did by multiplying both sides of the equation by the LCD, 40.

Many practical application problems can be solved using proportions.

> ### PROCEDURE   SOLVING APPLICATION PROBLEMS USING PROPORTIONS
>
> 1. Represent the unknown quantity by a variable.
> 2. Set up the proportion by listing the given ratio on the left-hand side of the equation and the unknown quantity and other given quantity on the right-hand side of the equation. When setting up the right-hand side of the proportion, the same respective quantities should occupy the same respective positions on the left and right. For example, an acceptable proportion might be
>
> $$\frac{\text{miles}}{\text{hour}} = \frac{\text{miles}}{\text{hour}}$$
>
> 3. Once the proportion is properly written, use cross multiplication to solve the equation.
> 4. Answer the question or questions asked using the appropriate units.

▲ *Hendersonville, North Carolina*

Example 5   *Water Usage*

The cost for water in the city of Hendersonville, North Carolina, is $2.14 per 1000 gallons (gal) of water used. What is the water bill if 25,000 gallons of water are used?

Solution This problem may be solved by setting up a proportion. One proportion that can be used is

$$\frac{\text{cost of 1000 gal}}{1000 \text{ gal}} = \frac{\text{cost of 25,000 gal}}{25,000 \text{ gal}}$$

The unknown quantity is the cost for 25,000 gallons of water, so we will call this quantity x. The proportion then becomes

$$\text{Given ratio} \left\{ \frac{2.14}{1000} = \frac{x}{25,000} \right.$$

Now we solve for x by using cross-multiplication.

$$(2.14)(25{,}000) = 1000x$$

$$53{,}500 = 1000x$$

$$\frac{53{,}500}{1000} = \frac{1000x}{1000}$$

$$53.50 = x$$

The cost of 25,000 gallons of water is $53.50.

Example 6 *Determining the Amount of Insulin*

Insulin comes in vials labeled with the number of insulin units per cubic centimeter (cc) of fluid. An insulin vial labeled U40 has 40 insulin units per cc of fluid. Tyler's insulin dosage is 30 insulin units. How many cubic centimeters are needed for Tyler's dosage?

Solution The unknown quantity, x, is the number of cubic centimeters of fluid needed for Tyler's dosage. Below is one proportion that can be used to determine that quantity.

$$\text{Given ratio} \left\{ \frac{40 \text{ units}}{1 \text{ cc}} = \frac{30 \text{ units}}{x \text{ cc}} \right.$$

$$40x = 30(1)$$

$$40x = 30$$

$$x = \frac{30}{40} = 0.75$$

Tyler's insulin dosage requires 0.75 cc of the fluid.

SECTION 6.3 *Exercises*

Warm Up Exercises

In Exercises 1 and 2, fill in the blank with an appropriate word, phrase, or symbol(s).

1. A quotient of two quantities is call a(n) _____.

2. A statement of equality between two ratios is called a(n) _____.

Practice the Skills

In Exercises 3–18, write the phrase as a mathematical expression.

3. 3 more than x

4. 8 less than y

5. The product of 11 and x

6. The quotient of y and 3

7. 6, decreased by 4 times y

8. 7 times x, increased by 4

9. The product of 8 and w, increased by 9

10. 8, increased by 5 times x

11. 6 more than 4 times x

12. 5 less than twice a number

13. The quotient of z and 13, decreased by 4

14. 9 less than the product of 7 and w

15. 12 decreased by s, divided by 4

16. The sum of 8 and t, divided by 2

17. Three times the sum of a number and 7

18. -4 times the difference between x and 1.

In Exercises 19–30, write an equation and solve.

19. The sum of a number and 8 is 23

20. The difference between a number and 11 is 2

21. The product of 9 and a number is 54

22. The quotient of a number and 3 is 10

23. Ten less than four times a number is 42.

24. Fourteen increased by 6 times a number is 32.

25. Twelve more than 4 times a number is 32.

26. Eight less than the product of 3 and a number is 6 times the number, increased by 10.

27. A number increased by 6 is 3 less than twice the number.

28. The quotient of a number and 4 is 12 less than the number.

29. A number increased by 10 is 2 times the sum of the number and 3.

30. Three times the difference of a number and 4 is 6 more than the number.

Problem Solving

In Exercises 31–42, set up an equation that can be used to solve the problem. Solve the equation and determine the desired value(s). When necessary, round answers to the nearest hundredth.

31. MODELING—*Reimbursed Expenses* When sales representatives for PT Pharmaceuticals drive to out-of-town meetings that require an overnight stay, they receive $150 for lodging plus $0.50 per mile driven. How many miles did Joe drive if PT reimbursed him $255 for an overnight trip?

32. MODELING—*Custom Printing* Kobe has written a book and wishes to use a custom printer to print copies of his book. There is a one-time setup fee of $250 plus a fee of $9.50 per book printed. If the total cost to Kobe was $3575, how many copies of the book were printed?

33. MODELING—*Jet Skiing* At Action Water Sports of Ocean City, Maryland, the cost of renting a Jet Ski is $42 per

half hour, which includes a 5% sales tax. Determine the cost of a half hour Jet Ski rental before tax.

34. MODELING—*SUV Battery* When Brittany purchased a new battery for her SUV, she paid 7% sales tax plus a $5 battery recycling fee. If the total cost Brittany paid was $208.30, determine the cost of the battery before the sales tax and the recycling fee.

35. MODELING—*Sales Commission* Tito receives a weekly salary of $400 at Anderson's Appliances. He also receives a 6% commission on the total dollar amount of all sales he makes. What must his total sales be in a week if he is to make a total of $790?

36. MODELING—*Furniture* Pier 1 Imports has a sale offering 10% off of all furniture. If Amanda spent $378.99 on furniture before tax, what was the price of the furniture she purchased before the discount and before tax?

37. MODELING—*Investment* Marty received an inheritance of $20,000. If he wants to invest three times as much money in stocks as in mutual funds, how much money should Marty invest in mutual funds and in stocks?

38. MODELING—*Scholarship Donation* Each year, Andrea donates a total of $1000 for scholarships at Nassau Community College. This year, she wants the amount she donates for scholarships for liberal arts students to be three times the amount she donates for scholarships for business students. Determine the amount she should donate for each type of scholarship.

39. MODELING—*Dimensions of a Deck* Jim is building a rectangular deck and wants the length to be 2 ft greater than the width. What will be the dimensions of the deck if the perimeter is to be 52 ft?

40. MODELING—*Picture Frame* Marty is building a rectangular picture frame and has 42 inches of framing material to work with. What should be the dimensions of the frame if Marty wants the width to be 5 inches less than the length?

41. MODELING—*Enclosing Two Pens* Chuck has 140 ft of fencing with which he wants to fence in two connecting, adjacent square pens with fencing between the two pens (see the figure). What will be the dimensions if the length of the entire enclosed region is to be twice the width?

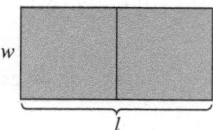

42. MODELING—*Chicken Yard* Rodney has 84 feet of fencing with which to build 3 separate pens for his chickens. The pens will be adjacent to the chicken coop as shown in the diagram on page 311.

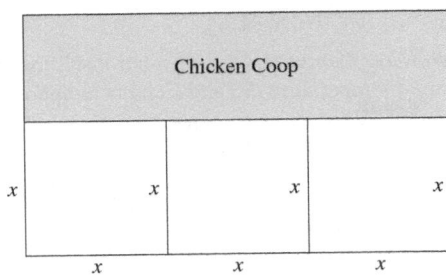

Chicken Coop

x x x x

x x x

The square pens are all the same size and will consist of fencing on only 3 sides with the coop on the remaining side. Determine the dimensions of each pen.

In Exercises 43–51, use proportions to solve the problem.

43. *Water Bill* The water rate in Aurora, Colorado, is $5.87 per 1000 gallons of water used.

   a) During May, the Hindermans used 4327 gallons of water. Determine their water bill for May.

   b) During June, the Hindermans had a water bill of $28.71. Determine the amount of water they used in June.

44. *Paint* One gallon of paint covers a surface area of 825 ft². Assuming paint can only be purchased in whole gallons, how many gallons are needed to paint a house with a surface area of 6600 ft²?

45. *Nielsen Ratings* Nielsen Media Research estimates the number of people who watch a television show. One rating point means that about 1,800,000 households watched the show. The top-rated television show for the week of February 18, 2019, was *The Oscars* with a rating of 16.4. About how many households watched *The Oscars*?

▲ Lady Gaga with her Academy Award

46. *Lasagna* A recipe for 6 servings of lasagna uses 16 ounces of Italian sausage.

   a) If the recipe were to be made for 15 servings, how many ounces of Italian sausage would be needed?

   b) How many servings of lasagna can be made with 24 ounces of Italian sausage?

47. *Topsoil* A 40-pound bag of topsoil will cover a surface area of 12 ft².

   a) How many pounds of topsoil are needed to cover a surface area of 480 ft²?

   b) How many bags of topsoil must be purchased to cover a surface area of 480 ft²?

48. *Speed Limit* When Jacob crossed over from Niagara Falls, New York, to Niagara Falls, Canada, he saw a sign that said 50 miles per hour (mph) is equal to 80 kilometers per hour (kph).

▲ Niagara Falls

   a) How many kilometers per hour are equal to 1 mph?

   b) On a stretch of the Queen Elizabeth Way, the speed limit is 90 kph. What is the speed limit in miles per hour?

49. *The Proper Dosage* A doctor asks a nurse to give a patient 250 milligrams (mg) of the drug simethicone. The drug is available only in a solution whose concentration is 40 mg simethicone per 0.6 milliliter (mℓ) of solution. How many milliliters of solution should the nurse give the patient?

Amount of Insulin In Exercises 50 and 51, how much insulin (in cc) would be given for the following doses? (Refer to Example 6 on page 309.)

50. 15 units of insulin from a vial marked U40

51. 35 units of insulin from a vial marked U40

Challenge Problems/Group Activities

52. *Income Tax* Some states allow spouses to file individual tax returns (on a single form) even though they have filed a joint federal tax return. It is usually to the taxpayers' advantage to do so when both spouses work. The smallest amount of tax owed (or the largest refund) will occur when the spouses taxable incomes are the same.

Mr. McAdams's 2020 taxable income was $34,200, and Mrs. McAdams's taxable income for that year was $36,400. The McAdams' total tax deduction for the year was $3640. This deduction can be divided between Mr. and Mrs. McAdams any way they wish. How should the $3640 be divided between them to result in each individual having the same taxable income and therefore the greatest tax refund?

53. *Auto Insurance* A driver education course at the East Lake School of Driving costs $45 but saves those under 25 years of age 10% of their annual insurance premiums until they are 25. Dan has just turned 18, and his insurance costs $100.00 per month.

a) When will the amount saved from insurance equal the price of the course?

b) Including the cost of the course, when Dan turns 25, how much will he have saved?

Recreational Mathematics

54. *Thermometer Comparison* The relationship between Fahrenheit temperature (F) and Celsius temperature (C) is shown by the formula $F = \dfrac{9}{5}C + 32$. At what temperature will a Fahrenheit thermometer read the same as a Celsius thermometer?

Research Activity

55. *Ratio and Proportion* Ratio and proportion are used in many different ways in everyday life. Submit two articles from newspapers, magazines, or the Internet in which ratios and/or proportions are used. Write a brief summary of each article explaining how ratio and/or proportion were used.

SECTION 6.4 Variation

Upon completion of this section, you will be able to:

- Understand direct variation.
- Understand inverse variation.
- Understand joint variation.
- Understand combined variation.

When you set up your new surround sound stereo system, you place a microphone at the place you typically sit when watching television. The computer within the receiver then generates a series of sounds, and automatically adjusts the volume of all the speakers in the system to give the best balance of sounds from the speakers. Because the sound you hear from each speaker is dependent on (or varies with) your distance from the speaker, the computer is using a concept called variation when activating your sound system.

Why This Is Important Throughout this section we will use the concept of variation to help us model a wide variety of real-life applications. These applications involve medicine, geometry, economics, physics, chemistry, and music. Having knowledge of variation is a powerful problem-solving tool that we can use in many aspects of our lives.

Direct Variation

Many scientific formulas are expressed as variation equations. A *variation equation* is an equation that relates one variable to one or more other variables through the operations of multiplication or division (or both operations). There are essentially four types of variation problems: direct, inverse, joint, and combined variation.

With *direct variation*, the values of the two related variables increase together or decrease together; that is, as one increases so does the other, and as one decreases so does the other.

Consider a car traveling at 40 miles an hour. The car travels 40 miles in 1 hour, 80 miles in 2 hours, and 120 miles in 3 hours. Note that, as the time increases, the distance traveled increases, and, as the time decreases, the distance traveled decreases.

The formula used to calculate distance traveled is

$$\text{Distance} = \text{rate} \cdot \text{time}$$

Since the rate is a constant 40 miles per hour, the formula can be written

$$d = 40t$$

We say that distance *varies directly* as time or that distance is *directly proportional* to time.

The preceding equation is an example of direct variation.

> **Direct Variation**
> If a variable y varies directly with a variable x, then
> $$y = kx$$
> where k is the **constant of proportionality** (or the variation constant).

Examples 1 through 4 illustrate direct variation.

Example 1 *Direct Variation in Physics*

The length that a spring will stretch, S, varies directly with the force (or weight), F, attached to the spring. Write the equation for the length that a spring will stretch, S, if the constant of proportionality is 0.07.

Solution

$$S = kF \qquad \text{\small S varies directly as F.}$$
$$S = 0.07F \qquad \text{\small Constant of proportionality, k, is 0.07.}$$ ∎

Example 2 *Direct Variation in Medicine*

The recommended dosage, d, of the antibiotic drug vancomycin is directly proportional to a person's weight, w.
a) Write this variation equation.
b) Determine the recommended dosage, in milligrams, for Jayce, who weighs 128 lb. Assume the constant of proportionality for the dosage is 18.

Solution

a) $d = kw$
b) $d = 18(128) = 2304$
The recommended dosage for Jayce is 2304 mg. ∎

In certain variation problems, the constant of proportionality, k, may not be known. In such cases, we can often determine it by substituting the given values in the variation formula and solving for k.

Example 3 *Constant of Proportionality*

For a certain physics application, E varies directly as the square of v. If E is 196 when v is 7, determine the constant of proportionality.

Solution Since E varies directly as the *square of v*, we begin with the formula $E = kv^2$. Since the constant of proportionality, k, is not given, we must determine k using the given information. Substitute 196 for E and 7 for v.

$$E = kv^2$$
$$196 = k(7)^2$$
$$196 = 49k$$
$$\frac{196}{49} = \frac{49k}{49}$$
$$4 = k$$

Thus, the constant of proportionality is 4. ∎

Example 4   *Using the Constant of Proportionality*

The area, a, of a picture projected on a movie screen varies directly as the square of the distance, d, from the projector to the screen. If a projector at a distance of 25 feet projects a picture with an area of 100 square feet, what is the area of the projected picture when the projector is at a distance of 40 feet?

Solution We begin with the formula $a = kd^2$. Since the constant of proportionality is not given, we must determine k, using the given information.

$$a = kd^2$$
$$100 = k(25)^2$$
$$100 = k(625)$$
$$\frac{100}{625} = k$$
$$0.16 = k$$

We now use $k = 0.16$ to determine a when $d = 40$.

$$a = kd^2$$
$$a = 0.16d^2$$
$$a = 0.16(40)^2$$
$$a = 0.16(1600)$$
$$a = 256$$

Thus, the area of a projected picture is 256 ft^2 when the projector is at a distance of 40 ft. ∎

Inverse Variation

A second type of variation is *inverse variation*. When two quantities vary inversely, as one quantity increases the other quantity decreases, and vice versa.

To explain inverse variation, we use the formula distance = rate · time. If we solve for time, we get time = distance/rate. Assume the distance is fixed at 100 miles; then

$$\text{time} = \frac{100}{\text{rate}}$$

At 100 miles per hour, it would take 1 hour to cover this 100 mile distance. At a rate of 50 miles an hour, it would take 2 hours. At a rate of 25 miles an hour, it would take 4 hours. Note that as the rate (or speed) decreases, the time increases and vice versa.

The preceding equation can be written

$$t = \frac{100}{r}$$

This equation is an example of an inverse variation equation. The time and rate are inversely proportional. The constant of proportionality in this case is 100.

Inverse Variation
If a variable y varies inversely with a variable x, then

$$y = \frac{k}{x}$$

where k is the constant of proportionality.

Two quantities *vary inversely*, or are *inversely proportional*, when as one quantity increases the other quantity decreases and vice versa. Examples 5 and 6 illustrate inverse variation.

Example 5 *Inverse Variation in Speaker Loudness*

The loudness, l, of a stereo speaker, measured in decibels (dB), varies inversely as the square of the distance, d, of the listener from the speaker. If the loudness is 20 dB when the listener is 6 ft from the speaker, determine the equation that expresses the relationship between the loudness and the distance.

Solution   Since the loudness varies inversely as the square of the distance, the general form of the equation is

$$l = \frac{k}{d^2}$$

To determine k, we substitute the given values for l and d and solve for k.

$$20 = \frac{k}{6^2}$$

$$20 = \frac{k}{36}$$

$$(20)(36) = k$$

$$720 = k$$

Thus, the equation is $l = \frac{720}{d^2}$.   ■

Example 6 *Using the Constant of Proportionality*

For a certain chemistry application, V varies inversely as P. If $V = 2$ when $P = 26$, determine V when $P = 4$.

Solution   First, write the inverse variation equation. Then substitute $V = 2$ and $P = 26$ into the variation equation and solve for k.

$$V = \frac{k}{P}$$

$$2 = \frac{k}{26}$$

$$52 = k$$

Thus, the constant of proportionality, k, is 52. Now we substitute 52 for k and 4 for P into the original inverse variation equation to determine V.

$$V = \frac{k}{P}$$

$$V = \frac{52}{4}$$

$$V = 13$$

Thus, when P is 4 V is 13.   ■

Joint Variation

One quantity may vary directly as a product of two or more other quantities. This type of variation is called *joint variation*.

> ### Joint Variation
> If a variable y varies jointly with variables x and z, then
> $$y = kxz$$
> where k is the constant of proportionality.

Example 7 *Joint Variation in Geometry*

The area, A, of a triangle varies jointly as its base, b, and height, h. If the area of a triangle is 48 in.^{2} when its base is 12 in. and its height is 8 in., determine the area of a triangle whose base is 15 in. and whose height is 20 in.

Solution First write the joint variation, then substitute the known values and solve for k.

$$A = kbh$$
$$48 = k(12)(8)$$
$$48 = k(96)$$
$$\frac{48}{96} = k$$
$$\frac{1}{2} = k$$

Now solve for the area of the given triangle.

$$A = kbh$$
$$= \frac{1}{2}(15)(20)$$
$$= 150$$

Thus, the area of a triangle with a base of 15 in. and a height of 20 in. is 150 in.^{2} ■

> ### Summary of Variation Equations
>
> | Direct | Inverse | Joint |
> |--------|---------|-------|
> | $y = kx$ | $y = \dfrac{k}{x}$ | $y = kxz$ |

Combined Variation

Often in real-life situations, one variable varies as a combination of variables. The following examples illustrate the use of *combined variation*.

Example 8 *Combined Variation in Engineering*

The load, L, that a horizontal beam can safely support varies jointly as the width, w, and the square of the depth, d, and inversely as the length, l. Express L in terms of w, d, l, and the constant of proportionality, k.

Solution

$$L = \frac{kwd^2}{l}$$

■

Example 9 *Hot Dog Price, Combined Variation*

The owners of Henrietta Hots determined their weekly sales of hot dogs, S, vary directly with their advertising budget, A, and inversely with their hot dog price, P. When their advertising budget is $600 and the price of a hot dog is $1.50, they sell 5600 hot dogs a week.

a) Write a variation equation expressing S in terms of A and P. Include the value of the constant of proportionality.

b) Determine the expected sales if the advertising budget is $800 and the hot dog price is $1.75.

Solution

a) Since S varies directly as A and inversely as P, we begin with the equation

$$S = \frac{kA}{P}$$

We now determine k using the known values.

$$5600 = \frac{k(600)}{1.50}$$

$$5600 = 400k$$

$$14 = k$$

Therefore, the equation for the weekly sales of hot dogs is $S = \dfrac{14A}{P}$.

b)
$$S = \frac{14A}{P}$$

$$= \frac{14(800)}{1.75}$$

$$= 6400$$

Henrietta Hots can expect to sell 6400 hot dogs a week if the advertising budget is $800 and the hot dog price is $1.75. ∎

Example 10 *Combined Variation*

A varies jointly as B and C and inversely as the square of D. If $A = 1$ when $B = 9$, $C = 4$, and $D = 6$, determine A when $B = 8$, $C = 12$, and $D = 5$.

Solution We begin with the equation

$$A = \frac{kBC}{D^2}$$

We must first determine the constant of proportionality, k, by substituting the known values for A, B, C, and D and solving for k.

$$1 = \frac{k(9)(4)}{6^2}$$

$$1 = \frac{36k}{36}$$

$$1 = k$$

Thus, the constant of proportionality equals 1. Now we determine A for the corresponding values of B, C, and D.

$$A = \frac{kBC}{D^2}$$

$$A = \frac{(1)(8)(12)}{5^2} = \frac{96}{25} = 3.84$$

SECTION 6.4 Exercises

Warm Up Exercises

In Exercises 1–4, fill in the blank with an appropriate word, phrase, or symbol(s).

1. When the values of two related variables increase together or decrease together, it is an example of _____ variation.

2. When one quantity increases while the other quantity decreases, it is an example of _____ variation.

3. A variation equation that relates one variable to two or more other variables using the operation of multiplication is called a(n) _____ variation equation.

4. In the equation $y = kx$, k is called the constant of _____.

Practice the Skills

In Exercises 5–20, use your intuition to determine whether the variation between the indicated quantities is direct or inverse.

5. The distance between two cities on a map and the actual distance between them

6. The time required to fill a pool with a hose and the volume of water coming from the hose

7. The time required to boil water on a burner and the temperature of the burner

8. The radius of a balloon and its volume

9. The interest earned on an investment and the interest rate

10. The diameter of a hose and the volume of water coming out of the hose

11. The speed of a runner and the time it takes the runner to complete a 10-kilometer race

12. The number of volunteers present and the time it takes to clean up the city park

13. The number of calories eaten and the amount of exercise required to burn off those calories

14. The number of miles that a tire has been driven and the depth of the tread remaining on the tire

15. The width of the lawn mower and the time it takes to mow the lawn

16. On Earth, the weight and mass of an object

17. The number of people in line at a bank and the amount of time required for the last person to reach the teller

18. The time it takes an ice cube to melt in water and the temperature of the water

19. The number of ceramic tiles needed to cover a floor and the area of the floor

20. The percent of light that filters through water and the depth of the water

In Exercises 21 and 22, use Exercises 5–20 as a guide.

21. Name two items that have not been mentioned in this section that vary directly.

22. Name two items that have not been mentioned in this section that vary inversely.

In Exercises 23–38, (a) write the variation equation and (b) determine the quantity indicated.

23. y varies directly as x. Determine y when $x = 12$ and $k = 9$.

24. x varies inversely as y. Determine x when $y = 15$ and $k = 45$.

25. m varies inversely as the square of n. Determine m when $n = 10$ and $k = 20$.

26. r varies directly as the square of s. Determine r when $s = 2$ and $k = 13$.

27. A varies directly as B and inversely as C. Determine A when $B = 5$, $C = 10$, and $k = 5$.

28. M varies directly as J and inversely as C. Determine M when $J = 12$, $C = 16$, and $k = 4$.

29. F varies jointly as D and E. Determine F when $D = 3$, $E = 10$, and $k = 7$.

30. A varies jointly as R_1 and R_2 and inversely as the square of L. Determine A when $R_1 = 120$, $R_2 = 8$, $L = 5$, and $k = \frac{3}{2}$.

31. H varies directly as L. If $H = 12$ when $L = 40$, determine H when $L = 10$.

32. C varies inversely as J. If $C = 7$ when $J = 0.7$, determine C when $J = 14$.

33. q varies inversely as the square of w. If $q = 9$ when $w = 5$, determine q when $w = 3$.

34. A varies directly as the square of B. If $A = 245$ when $B = 7$, determine A when $B = 12$.

35. J varies directly as the square root of b. If $J = 2$ when $b = 36$, determine J when $b = 81$.

36. C varies inversely as the square root of m. If $C = 8$ when $m = 9$, determine C when $m = 100$.

37. Z varies jointly as W and Y. If $Z = 12$ when $W = 9$ and $Y = 4$, determine Z when $W = 50$ and $Y = 6$.

38. S varies jointly as I and the square of T. If $S = 4$ when $I = 10$ and $T = 4$, determine S when $I = 4$ and $T = 6$.

39. F varies jointly as M_1 and M_2 and inversely as the square of d. If $F = 20$ when $M_1 = 5$, $M_2 = 10$, and $d = 0.2$, determine F when $M_1 = 10$, $M_2 = 20$, and $d = 0.4$.

40. F varies jointly as q_1 and q_2 and inversely as the square of d. If $F = 80$ when $q_1 = 4$, $q_2 = 16$, and $d = 0.4$, determine F when $q_1 = 12$, $q_2 = 20$, and $d = 0.2$.

Problem Solving

In Exercises 41–50, (a) write the variation equation and (b) determine the quantity indicated.

41. *Profit* The profit, p, from selling bicycles is directly proportional to n, the number of bicycles sold. If the profit from selling 8 bicycles is $450, determine the profit from selling 18 bicycles.

42. *Property Tax* The property tax, t, on a home is directly proportional to the assessed value, v, of the home. If the

property tax on a home with an assessed value of $140,000 is $2100, what is the property tax on a home with an assessed value of $180,000?

43. *Melting an Ice Cube* The time, t, for an ice cube to melt is inversely proportional to the temperature, T, of the water in which the ice cube is placed. If it takes an ice cube 2 minutes to melt in 75°F water, how long will it take an ice cube of the same size to melt in 80°F water?

44. *Pressure and Volume* The volume of a gas, V, varies inversely as its pressure, P. If the volume of a gas is 6400 cubic centimeters (cc) when the pressure is 25 millimeters (mm) of mercury, determine the volume when the pressure is 200 mm of mercury.

45. *Distance* The distance, d, that a rock falls when dropped from a cliff varies directly as the square of time, t, the rock is falling. If a rock falls 64 feet in 2 seconds, how far will it fall in 3 seconds?

46. *Stopping Distance of a Car* The stopping distance, d, of a car after the brakes are applied varies directly as the square of the speed, s, of the car. If a car traveling at a speed of 40 miles per hour can stop in 80 feet, what is the stopping distance of a car traveling at 55 miles per hour?

47. *Electrical Resistance* The electrical resistance of a wire, R, varies directly as its length, L, and inversely as its cross-sectional area, A. If the resistance of a wire is 0.2 ohm when the length is 200 ft and its cross-sectional area is 0.05 in.^{2}, what is the resistance of a wire whose length is 5000 ft with a cross-sectional area of 0.01 in.^{2}?

48. *Guitar Strings* The number of vibrations per second, v, of a guitar string varies directly as the square root of the tension, t, and inversely as the length of the string, l. If the number of vibrations per second is 5 when the tension is 225 kg and the length of the string is 0.60 m, determine the number of vibrations per second when the tension is 196 kg and the length of the string is 0.70 m.

49. *Wattage Rating* The wattage rating of an appliance, W, varies jointly as the square of the current, I, and the resistance, R. If the wattage is 6 watts when the current is 0.2 ampere and the resistance is 150 ohms, determine the wattage when the current is 0.3 ampere and the resistance is 100 ohms.

50. *Phone Calls* The number of phone calls between two cities during a given time period, N, varies directly as the populations p_1 and p_2 of the two cities and inversely to the distance, d, between them. If 100,000 calls are made between two cities 300 mi apart and the populations of the cities are 60,000 and 200,000, how many calls are made between two cities with populations of 125,000 and 175,000 that are 450 mi apart?

Concept/Writing Exercises

51. a) If y varies directly as x and the constant of proportionality is 2, does x vary directly or inversely as y?

b) Give the new constant of proportionality as x varies directly as y.

52. a) If y varies inversely as x and the constant of proportionality is 0.3, does x vary directly or inversely as y?

b) Give the new constant of proportionality as x varies inversely as y.

53. a) Assume y varies directly as x. If x is doubled, how will it affect y?

b) Assume y varies inversely as x. If x is doubled, how will it affect y?

54. a) Assume a varies directly as b^2. If b is doubled, how will it affect a?

b) Assume a varies inversely as b^2. If b is doubled, how will it affect a?

Challenge Problems/Group Activities

55. *Photography* An article in the magazine *Outdoor and Travel Photography* states, "If a surface is illuminated by a point-source of light, the intensity of illumination produced is inversely proportional to the square of the distance separating them. In practical terms, this means that foreground objects will be grossly overexposed if your background subject is properly exposed with a flash. Thus direct flash will not offer pleasing results if there are any intervening objects between the foreground and the subject."

If the subject you are photographing is 4 ft from the flash and the illumination on this subject is $\frac{1}{16}$ of the light of the flash, what is the intensity of illumination on an intervening object that is 3 ft from the flash?

56. *Water Cost* In a specific region of the country, the amount of a customer's water bill, W, is directly proportional to the average daily temperature for the month, T, the lawn area, A, and the square root of F, where F is the family size, and inversely proportional to the number of inches of rain, R.

In one month, the average daily temperature is 78°F and the number of inches of rain is 5.6. If the average family of four who has a thousand square feet of lawn pays $72.00 for water for that month, estimate the water bill in the same month for the average family of six who has 1500 ft^2 of lawn.

SECTION 6.5   Solving Linear Inequalities

Upon completion of this section, you will be able to:

- Solve a linear inequality.
- Solve compound inequalities.

Devon is currently taking Sociology and needs to earn a course grade of B to maintain his scholarship. Devon has taken 4 tests and is about to take his fifth and final test. Devon would like to know what range of grades on this last test would give him a grade of B in the course. In this section, we will see how we can use a linear inequality to represent this and many other everyday problems.

Why This Is Important Linear inequalities can be used to represent many problems that we face in our everyday lives. The problems we will see in this section involve employee salaries, college employment, parking fees, business profit, and many other financial applications. Understanding how to model and solve such problems using algebra is a very powerful tool.

The symbols of inequality are as follows.

> **Symbols of Inequality**
> $a < b$ means that a is less than b.
> $a \le b$ means that a is less than or equal to b.
> $a > b$ means that a is greater than b.
> $a \ge b$ means that a is greater than or equal to b.

An *inequality* consists of two (or more) expressions joined by an inequality symbol.

Examples of inequalities

$$3 < 5, \qquad x < 2, \qquad 3x - 2 \ge 5$$

A statement of inequality can be used to indicate a set of real numbers. For example, $x < 2$ represents the set of all real numbers less than 2. Listing all these numbers is impossible, but some are $-2, -1.234, -1, -\frac{1}{2}, 0, \frac{97}{163}, 1$.

Solving Linear Inequalities

To indicate all real numbers less than 2, we can use the number line. The number line was discussed in Chapter 5.

To indicate the solution set of $x < 2$ on the number line, we draw an open circle at 2 and a line to the left of 2 with an arrow at its end. This technique indicates that all points to the left of 2 are part of the solution set. The open circle indicates that the solution set does not include the number 2.

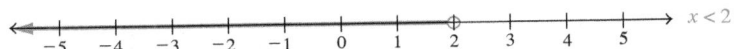

To indicate the solution set of $x \le 2$ on the number line, we draw a closed (or darkened) circle at 2 and a line to the left of 2 with an arrow at its end. The closed circle indicates that the 2 is part of the solution.

⌐ **Example 1** *Graphing an Inequality Involving $\le$*

Graph the solution set of $x \le 1$, where x is a real number, on the number line.

Solution   The numbers less than or equal to 1 are all the points on the number line to the left of 1 and 1 itself. The closed circle at 1 shows that 1 is included in the solution set.

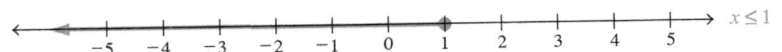

■

The inequality statements $x < 2$ and $2 > x$ have the same meaning. Note that the inequality symbol points to the x in both cases. Thus, one inequality may be written in place of the other. Likewise, $x > 2$ and $2 < x$ have the same meaning. Note that the inequality symbol points to the 2 in both cases. We make use of this fact in Example 2.

Example 2 *Graphing an Inequality Involving <*

Graph the solution set of $3 < x$, where x is a real number, on the number line.

Solution We can restate $3 < x$ as $x > 3$. Both statements have identical solutions. Any number that is greater than 3 satisfies the inequality $x > 3$. The graph includes all the points to the right of 3 on the number line. To indicate that 3 is not part of the solution set, we place an open circle at 3.

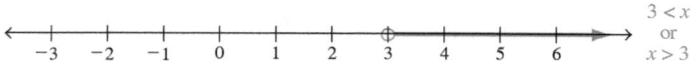

$3 < x$
or
$x > 3$

We can determine the solution to an inequality by adding, subtracting, multiplying, or dividing both sides of the inequality by the same number or expression. We use the procedure discussed in Section 6.1 to isolate the variable, with one important exception: *When both sides of an inequality are multiplied or divided by a negative number, the direction of the inequality symbol is reversed.*

> **Timely Tip**
>
> Remember to change the direction of the inequality symbol when multiplying or dividing both sides of an inequality by a *negative* number.

Example 3 *Multiplying by a Negative Number*

Solve the inequality $-x > 3$, where x is a real number, and graph the solution set on the number line.

Solution To solve this inequality, we must eliminate the negative sign in front of the x. To do so, we multiply both sides of the inequality by -1 and change the direction of the inequality symbol.

$$-x > 3$$
$$-1(-x) < -1(3)$$
$$x < -3$$

Multiply both sides of the inequality by -1 and change the direction of the inequality symbol.

The solution set is graphed on the number line as follows.

$x < -3$

Example 4 *Dividing by a Negative Number*

Solve the inequality $-6x < 18$, where x is a real number, and graph the solution set on the number line.

Solution To solve the inequality, we need to divide both sides of the inequality by the coefficient of the x-term, -6. When dividing both sides of the inequality by -6, change the direction of the inequality symbol because we are dividing by a negative number.

$$-6x < 18$$
$$\frac{-6x}{-6} > \frac{18}{-6}$$
$$x > -3$$

Divide both sides of the inequality by -6 and change the direction of the inequality symbol.

The solution set is graphed on the number line as follows.

$x > -3$

Example 5 *Solving an Inequality*

Solve the inequality $3x - 5 > 7$, where x is a real number, and graph the solution set on the number line.

Solution To determine the solution set, isolate x on one side of the inequality.

$$3x - 5 > 7$$

$$3x - 5 + 5 > 7 + 5 \qquad \text{Add 5 to both sides of the inequality.}$$

$$3x > 12$$

$$\frac{3x}{3} > \frac{12}{3} \qquad \text{Divide both sides of the inequality by 3.}$$

$$x > 4$$

Thus, the solution set of $3x - 5 > 7$ is all real numbers greater than 4.

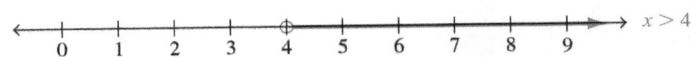 $x > 4$

     Note that in Example 5, the *direction of the inequality symbol* did not change when both sides of the inequality were divided by the positive number 3.

Example 6 *A Solution of Only Integers*

Solve the inequality $x - 1 < 3$, where x is an integer, and graph the solution set on the number line.

Solution To determine the solution set, isolate x on one side of the inequality symbol.

$$x - 1 < 3$$

$$x - 1 + 1 < 3 + 1 \qquad \text{Add 1 to both sides of the inequality.}$$

$$x < 4$$

Since x is an integer and is less than 4, the solution set is the set of integers less than 4, or $\{\,\dots,\,-2,\,-1,\,0,\,1,\,2,\,3\,\}$. To graph the solution set, we make solid dots at the corresponding points on the number line. The ellipsis (the three dots) to the left of -3 on the number line indicate that all the integers to the left of -3 are included.

 $x < 4$, x an integer

     Some linear inequalities have no solution. When solving an inequality, if you obtain a statement with no variables that is *always false*, such as $3 > 5$, the inequality has *no solution*. There is no real number that makes the statement true.

     Some linear inequalities have all real numbers as their solution. When solving an inequality, if you obtain a statement with no variables that is *always true*, such as $2 > 1$, the solution is *all real numbers*.

Timely Tip

If, while solving an inequality, you get an inequality with no variables that is always

- false, such as $3 > 5$, then the inequality has *no solution*.
- true, such as $2 > 1$, then the solution is *all real numbers*.

Solving Compound Inequalities

An inequality of the form $a < x < b$ is called a *compound inequality*. Consider the compound inequality $-3 < x \leq 2$, which means that $-3 < x$ *and* $x \leq 2$.

Example 7 *A Compound Inequality*

Graph the solution set of the inequality $-3 < x \leq 2$

a) where x is an integer.   b) where x is a real number.

Solution

a) The solution set is all the integers greater than -3 and less than or equal to 2, or $\{-2, -1, 0, 1, 2\}$.

$-3 < x \leq 2$, x an integer

b) The solution set consists of all the real numbers greater than -3 and less than or equal to 2.

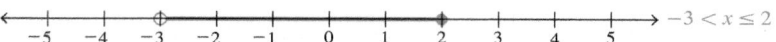

$-3 < x \leq 2$

Example 8 *Solving a Compound Inequality*

Solve the compound inequality for x, where x is a real number, and graph the solution set.

$$-4 < \frac{x+3}{2} \leq 5$$

Solution To solve this compound inequality, we must isolate the x. To do so, we use the same principles used to solve inequalities.

$$-4 < \frac{x+3}{2} \leq 5$$

$$2(-4) < 2\left(\frac{x+3}{2}\right) \leq 2(5) \qquad \text{Multiply each part of the inequality by 2.}$$

$$-8 < x+3 \leq 10$$

$$-8 - 3 < x+3-3 \leq 10-3 \qquad \text{Subtract 3 from each part of the inequality.}$$

$$-11 < x \leq 7$$

The solution set is graphed on the number line as follows.

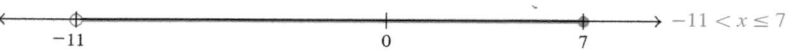

$-11 < x \leq 7$

Example 9 *Average Grade*

A student must have an average, or mean, on five tests that is greater than or equal to 80% but less than 90% to receive a final grade of B. Devon's grades on the first four tests were 98%, 76%, 86%, and 92%. What range of grades on the fifth test would give him a B in the course?

Solution The unknown quantity is the range of grades on the fifth test. First construct an inequality that can be used to determine the range of grades on the fifth test. The average (mean) is determined by adding the grades and dividing the sum by the number of tests.

Let $x =$ the fifth grade. Then

$$\text{Average} = \frac{98 + 76 + 86 + 92 + x}{5}$$

For Devon to obtain a B in this course, his average must be greater than or equal to 80 but less than 90.

$$80 \leq \frac{98 + 76 + 86 + 92 + x}{5} < 90$$

$$80 \leq \frac{352 + x}{5} < 90$$

$$5(80) \leq \cancel{5}\left(\frac{352 + x}{\cancel{5}}\right) < 5(90)$$ Multiply each part of the inequality by 5.

$$400 \leq 352 + x < 450$$

$$400 - 352 \leq 352 - 352 + x < 450 - 352$$ Subtract 352 from each part of the inequality

$$48 \leq x < 98$$

Thus, a grade greater than or equal to 48% and less than 98% on the fifth test will result in Devon getting a grade of B in this course. ■

"In mathematics the art of posing problems is easier than that of solving them."

Georg Cantor

SECTION 6.5

Exercises

Warm Up Exercises

In Exercises 1–8, fill in the blank with an appropriate word, phrase, or symbol(s).

1. Two or more expressions joined by an inequality symbol is called a(n) _____.

2. a) $a < b$ means that a is _____ than b.

   b) $a > b$ means that a is _____ than b.

3. An inequality of the form $a < x < b$ is called a(n) _____ inequality.

4. When solving an inequality, if you obtain a statement that is always false, such as $3 > 5$, the answer is _____ solution.

5. When solving an inequality, if you obtain a statement that is always true, such as $2 > 1$, the solution is _____ real numbers.

6. When both sides of an inequality are multiplied or divided by a(n) _____ number, the direction of the inequality symbol is reversed.

7. To indicate the solution set of $x \leq 5$ on the number line, we draw a(n) _____ circle at 5 and a line to the left of the circle.

8. To indicate the solution set of $x < 5$ on the number line, we draw a(n) _____ circle at 5 and a line to the left of the circle.

Practice the Skills

In Exercises 9–30, graph the solution set of the inequality, where x is a real number, on the number line.

9. $x \geq 4$

10. $x < 4$

11. $x + 9 \geq 4$

12. $x - 8 > -9$

13. $-6x \leq 36$

14. $-4x < 12$

15. $\frac{-x}{3} \geq 3$

16. $\frac{-x}{2} \leq -4$

17. $2x + 6 \geq 14$

18. $-3x + 7 > 22$

19. $-2x + 5 \leq 4x - 25$

20. $3x + 12 < 5x + 14$

21. $4(2x - 1) < 2(4x - 3)$

22. $-3x + 7 \geq -6(x - 1) + 3(x + 1)$

23. $5(2x - 7) \leq 2(5x + 4)$

24. $3x + 11 \geq 6(x + 2) - 3(x + 4)$

25. $-1 \leq x \leq 3$

26. $-4 < x \leq 1$

27. $2 < x - 4 \leq 6$

28. $-3 \leq 2x + 5 < 11$

29. $\frac{1}{2} < \frac{x + 4}{2} \leq 4$

30. $-1 \leq \frac{2x + 3}{5} \leq 1$

In Exercises 31–46, graph the solution set of the inequality, where x is an integer, on the number line.

31. $x > 1$

32. $-2 < x$

33. $-4x \leq 36$

34. $\frac{x}{3} \leq -2$

35. $-\frac{x}{4} \geq 2$

36. $\frac{3x}{5} \leq 3$

37. $-15 < -4x - 3$

38. $-5x - 1 < 17 - 2x$

39. $3(x + 4) \geq 4x + 13$

40. $-2(x - 1) < 3(x - 4) + 5$

41. $5(x + 4) - 6 \leq 2x + 8$

42. $-2 \leq x < 6$

43. $1 > -x > -6$

44. $-2 < 2x + 3 < 6$

45. $0.3 \leq \frac{x + 2}{10} \leq 0.5$

46. $-\frac{1}{3} < \frac{x - 2}{12} \leq \frac{1}{4}$

Problem Solving

47. *U.S. Motor Vehicle Production* The graph below was created using **Stat**Crunch with data obtained from the U.S. Bureau of Transportation website (www.bts .gov). The graph shows the total vehicle production, in millions of vehicles, in the United States for the years 2010–2017.

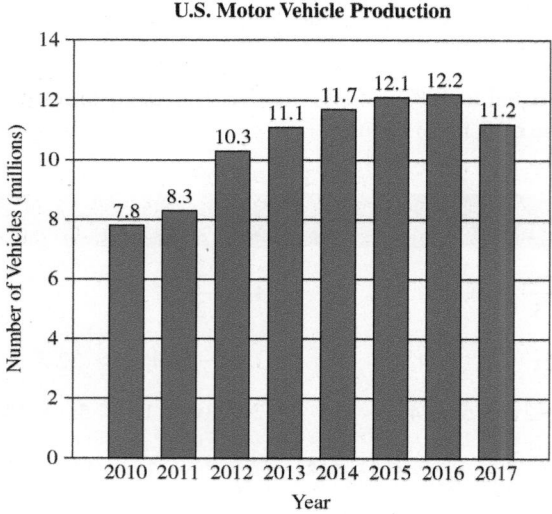

U.S. Motor Vehicle Production

Number of Vehicles (millions) vs. *Year*

Values: 2010: 7.8, 2011: 8.3, 2012: 10.3, 2013: 11.1, 2014: 11.7, 2015: 12.1, 2016: 12.2, 2017: 11.2

Source: U.S. Bureau of Transportation

In which years was the number of vehicles produced

a) > 12 million?

b) < 11 million?

c) ≤ 10.3 million?

d) ≥ 11.7 million?

48. *U.S. Presidential Salary* The graph below was created using **Stat**Crunch with data obtained from the website www.salary.com. The graph shows the presidential salary, in thousands of dollars, for the years 1789 to the present*.

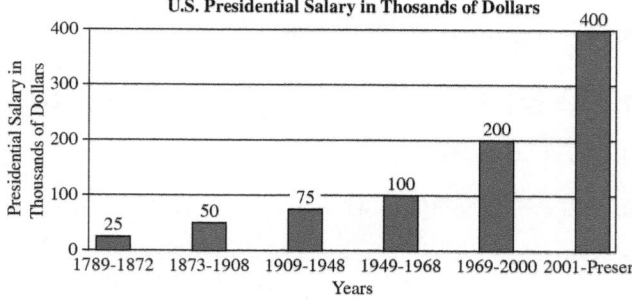

U.S. Presidential Salary in Thosands of Dollars

Presidential Salary in Thousands of Dollars vs. *Years*

Values: 1789-1872: 25, 1873-1908: 50, 1909-1948: 75, 1949-1968: 100, 1969-2000: 200, 2001-Present: 400

Source: www.salary.com

---

*As of March 17, 2019, the presidential salary was still $400,000 per year.

In which years was the presidential salary

a) < $100,000?

b) ≤ $100,000?

c) > $200,000?

d) ≥ $200,000?

49. *Gym Membership* FitForLife gym offers two membership options. Option A requires an initial membership fee along with a monthly payment. Option B requires only a monthly payment. The two membership options are shown in the table below. Determine the number of months it will take for Option A to cost less than Option B.

Membership Plan	Initial Membership Fee	Monthly Fee
Option A	$150	$20
Option B	None	$35

50. *Salary Plans* Bobby recently accepted a sales position in Portland, Oregon. He can select between the two salary plans shown in the table. Determine the dollar amount of weekly sales that would result in Bobby earning more with Plan B than with Plan A.

Salary Plan	Weekly Salary	Commission on Sales
Plan A	$500	6%
Plan B	$400	8%

51. *College Employment* To be eligible for financial assistance for college, Julie can earn no more than $3200 during her summer employment. She will earn $662.50 cutting lawns. She is also earning $7.25 per hour at an evening job as a cashier at a grocery store. What is the maximum number of hours Julie can work at the grocery store without jeopardizing her financial assistance?

52. *Anniversary* The Kubys are celebrating their 75th wedding anniversary by having a party at the Diplomat Party House. They have budgeted $4000 for the party. If the party house charges a $100 booking fee, plus $30 per person, determine the maximum number of guests the Kubys can invite and remain within their budget.

53. *Parking Costs* A parking garage in Rockford, Illinois, charges $1.75 for the first hour and $0.50 for each additional hour or part of an hour. What is the maximum length of time Tom can park in the garage if he wishes to pay no more than $4.25?

54. *Ride-Share Service* While on vacation, Tamara is using a ride-share service to visit attractions. The ride-share fees include a pickup fee of $2.40 plus a mileage fee of $1.80 per mile traveled. If Tamara would like to spend at most $24 for the ride-share service, how many miles can she travel?

55. *Finding Velocity* The velocity, v, in feet per second, t sec after a tennis ball is projected directly upward is given by the formula $v = 84 - 32t$. How many seconds after being projected upward will the velocity be between 36 ft/sec and 68 ft/sec?

56. *Speed Limit* The minimum speed for vehicles on a highway is 40 mph, and the maximum speed is 55 mph. If Philip has been driving nonstop along the highway for 4 hr, what range in miles could he have legally traveled?

57. *Grade of A* To earn a grade of A in Philosophy, a student must have a 93% or greater average on four tests. Hyun currently has grades of 87%, 91%, and 98%. What range of grades on the fourth test would earn Hyun a grade of A in the class?

58. *Passing Grade* To earn a passing grade in Topology at Old College of Alabama, a student must have a 60% or greater average on 5 tests. Justice currently has grades of 61%, 75%, 57%, and 51%. What range of grades on the fifth test would earn Justice a passing grade?

Challenge Problems/Group Activities

59. *Painting a House* Donovan is painting the exterior of his house. The instructions on the paint can indicate that 1 gal covers from 250 ft^2 to 400 ft^2. The total surface of the house to be painted is 2750 ft^2. Determine the number of gallons of paint he could use and express the answer as an inequality.

60. *Final Exam* Teresa's five test grades for the semester are 86%, 78%, 68%, 92%, and 72%. Her final exam counts one-third of her final grade. What range of grades on her final exam would result in Teresa receiving a final grade of B in the course? (See Example 9.)

61. A student multiplied both sides of the inequality $-\frac{1}{3}x \leq 4$ by -3 and forgot to reverse the direction of the inequality symbol. What is the relation between the student's incorrect solution set and the correct solution set? Is there any number in both the correct solution set and the student's incorrect solution set? If so, what is it?

Research Activity

62. *Inequalities* Find a newspaper, magazine, or Internet article that contains the mathematical concept of inequality.

a) From the information in the article write a statement of inequality.

b) Summarize the article and explain how you arrived at the inequality statement in part (a).

SECTION 6.6   Graphing Linear Equations

Upon completion of this section, you will be able to:

- Understand the Cartesian coordinate system.
- Graph linear equations by plotting points.
- Graph linear equations by using the x- and y-intercepts.
- Determine the slope of a line.
- Graph linear equations by using the slope and y-intercept.

The profit, p, of a company may depend on the amount of sales, s. The cost, c, of mailing a package may depend on the weight, w, of the package. Real-life problems, such as these two examples, often involve two or more variables. In this section, we will discuss how to graph equations with two variables.

Why This Is Important Many applications, including those described above, can be modeled using two variables. The relationship between two variables can be represented with a graph, and many times this graph is a straight line. In this section, we will encounter many real-life applications including those involving a company's profits, simple interest, test grades, and retail spending on electronics. An understanding of how to graph lines is an important skill in many occupations.

Cartesian Coordinate System

To be able to work with equations with two variables requires understanding the *Cartesian* (or *rectangular*) *coordinate system*, named after the French mathematician René Descartes. The rectangular coordinate system consists of two perpendicular number lines (Fig. 6.3). The horizontal line is the *x-axis*, and the vertical line is the *y-axis*. The point of intersection of the x-axis and y-axis is called the *origin*. The numbers on the axes to the right and above the origin are positive. The numbers on the axes on the left and below the origin are negative. The axes divide the plane into four parts: the first, second, third, and fourth *quadrants*.

We indicate the location of a point in the rectangular coordinate system by means of an *ordered pair* of the form (x, y). The x-coordinate is always placed first, and the y-coordinate is always placed second in the ordered pair. Consider the point illustrated in Fig. 6.4. Since the x-coordinate of the point is 5 and the y-coordinate is 3, the ordered pair that represents this point is $(5, 3)$.

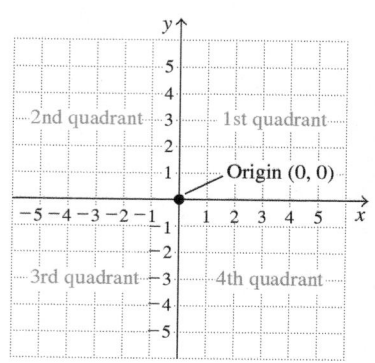

Figure 6.3   Figure 6.4

The origin is represented by the ordered pair $(0, 0)$. Every point on the plane can be represented by one and only one ordered pair (x, y), and every ordered pair (x, y) represents one and only one point on the plane.

Example 1   *Plotting Points*

Plot the points $A(-2, 4)$, $B(3, -2)$, $C(-1, -3)$, $D(6, 0)$, and $E(0, -4)$.

Solution Point A has an x-coordinate of -2 and a y-coordinate of 4. Project a vertical line up from -2 on the x-axis and a horizontal line to the left from 4 on the y-axis. The two lines intersect at the point denoted A (Fig. 6.5). The other points are plotted in a similar manner.

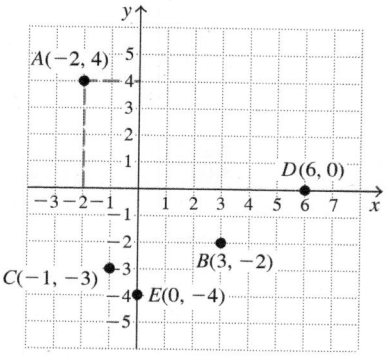

Figure 6.5

According to legend, French mathematician and philosopher René Descartes (1596–1650) did some of his best thinking in bed. He was a sickly child, and so the Jesuits who undertook his education allowed him to stay in bed each morning as long as he liked. He carried this practice into adulthood, seldom getting up before noon. One morning as he watched a fly crawl about the ceiling, near the corner of his room, he was struck with the idea that the fly's position could best be described by the connecting distances from it to the two adjacent walls. These became the coordinates of his rectangular coordinate system and were appropriately named after him (Cartesian coordinates) and not the fly.

Example 2 *A Parallelogram*

The points, A, B, and C are three vertices of a parallelogram with two sides parallel to the x-axis. Plot the three ordered pairs below and determine the coordinates of the fourth vertex, D, of the parallelogram.

$$A(1, 2) \quad B(2, 4) \quad C(7, 4)$$

Solution A parallelogram is a figure that has opposite sides that are of equal length and are parallel. Parallel lines are two lines in the same plane that do not intersect. The horizontal distance between points B and C is 5 units (see Fig. 6.6). Therefore, the horizontal distance between points A and D must also be 5 units. This problem has two possible solutions, as illustrated in Fig. 6.6. In each figure, we have indicated the given points in red.

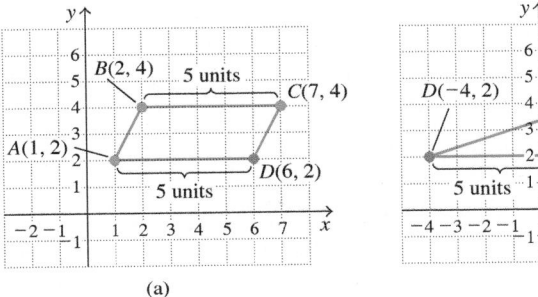

(a)   (b)

Figure 6.6

The solutions are the points $(6, 2)$ and $(-4, 2)$. ∎

Graphing Linear Equations by Plotting Points

Consider the following equation in two variables: $y = x + 1$. Every ordered pair that makes the equation a true statement is a solution to, or satisfies, the equation. We can mentally determine some ordered pairs that satisfy the equation $y = x + 1$ by picking some values of x and determining the corresponding values of y. For example, suppose we let $x = 1$; then $y = 1 + 1 = 2$. The ordered pair $(1, 2)$ is a solution to the equation $y = x + 1$. We can make a chart of other ordered pairs that are solutions to the equation.

x	y	Ordered Pair
1	2	$(1, 2)$
2	3	$(2, 3)$
3	4	$(3, 4)$
4.5	5.5	$(4.5, 5.5)$
-3	-2	$(-3, -2)$

How many other ordered pairs satisfy the equation? Infinitely many ordered pairs satisfy the equation. Since we cannot list all the solutions, we show them by means of a graph. A *graph* is an illustration of all the points whose coordinates satisfy an equation.

The points $(1, 2)$, $(2, 3)$, $(3, 4)$, $(4.5, 5.5)$, and $(-3, -2)$ are plotted in Fig. 6.7. With a straightedge we can draw one line that contains all these points. This line, when extended indefinitely in both directions, passes through all the points in the plane that satisfy the equation $y = x + 1$. The arrows on the ends of the line indicate that the line extends indefinitely.

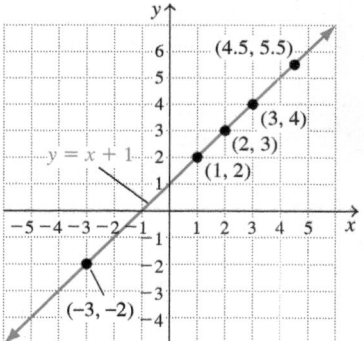

Figure 6.7

All equations of the form $ax + by = c$, $a \neq 0$ and $b \neq 0$, will be straight lines when graphed. Thus, such equations are called *linear equations in two variables*. The exponents on the variables x and y must be 1 for the equation to be linear. Since only two points are needed to draw a line, only two points are needed to graph a linear equation. However, it is always a good idea to plot a third point as a checkpoint. If no error has been made, all three points will be in a line, or *collinear*. One method that can be used to obtain points is to solve the equation for y, substitute values for x, and determine the corresponding values of y.

Example 3 *Graphing an Equation by Plotting Points*

Graph $y = 2x + 2$.

Solution Since the equation is already solved for y, select values for x and determine the corresponding values for y. The table below indicates values arbitrarily selected for x and their corresponding values for y. The ordered pairs are $(0, 2)$, $(1, 4)$, and $(-1, 0)$. The graph is shown in Fig. 6.8.

x	y
0	2
1	4
-1	0

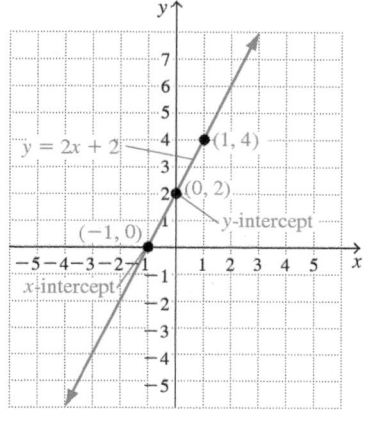

Figure 6.8

> **PROCEDURE**   GRAPHING LINEAR EQUATIONS BY PLOTTING POINTS
>
> 1. Solve the equation for y.
> 2. Select at least three values for x and determine their corresponding values of y.
> 3. Plot the points.
> 4. The points should be in a straight line. Draw a line through the set of points and place arrow tips at both ends of the line.

In Step 4 of the procedure, if the points are not in a straight line, recheck your calculations and determine your error.

Graphing by Using the x- and y-Intercepts

Example 3 contained two special points on the graph, $(-1, 0)$ and $(0, 2)$. At these points, the line crosses the x-axis and the y-axis, respectively. The ordered pairs $(-1, 0)$ and $(0, 2)$ represent the *x-intercept* and the *y-intercept*, respectively. Another method that can be used to graph linear equations is to determine the x- and y-intercepts of the graph.

> **PROCEDURE**   DETERMINING THE X- AND Y-INTERCEPTS
>
> To determine the x-intercept, set $y = 0$ and solve the equation for x.
> To determine the y-intercept, set $x = 0$ and solve the equation for y.

A linear equation may be graphed by determining the x- and y-intercepts, plotting the intercepts, and drawing a straight line through the intercepts. When graphing by this method, you should always plot a checkpoint before drawing your graph. To obtain a checkpoint, select a nonzero value for x and determine the corresponding value of y. The checkpoint should be collinear with the x- and y-intercepts.

Example 4   *Graphing an Equation by Using Intercepts*

Graph $2x + 3y = 6$ by using the x- and y-intercepts.

Solution To determine the x-intercept, set $y = 0$ and solve for x.

$$2x + 3y = 6$$
$$2x + 3(0) = 6$$
$$2x = 6$$
$$x = 3$$

The x-intercept is $(3, 0)$. To determine the y-intercept, set $x = 0$ and solve for y.

$$2x + 3y = 6$$
$$2(0) + 3y = 6$$
$$3y = 6$$
$$y = 2$$

The y-intercept is $(0, 2)$.

As a checkpoint, substitute $x = -3$ and determine the corresponding value for y.

$$2x + 3y = 6$$
$$2(-3) + 3y = 6$$
$$-6 + 3y = 6$$
$$3y = 12$$
$$y = 4$$

The checkpoint is the ordered pair $(-3, 4)$. The three points are plotted in Figure 6.9.

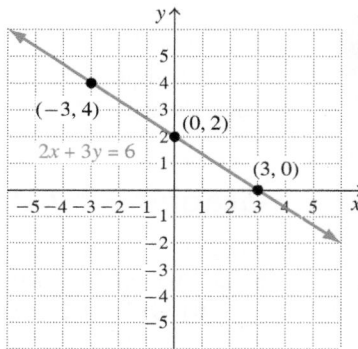

Figure 6.9

Since all three points are collinear, draw a line through the three points to obtain the graph. ∎

TECHNOLOGY TIP

Graphing Calculators and Graphing Apps

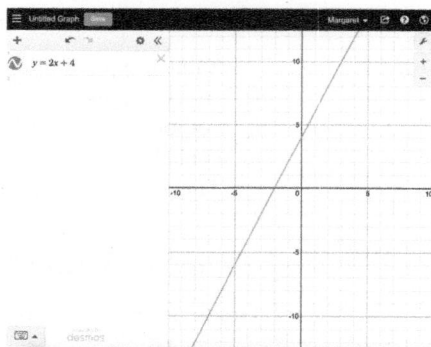

Figure 6.10

Figure 6.10 shows the graph of the equation $y = 2x + 4$, as illustrated on the screen of a graphing app that is available on smartphones, tablets, and other computers. Graphing calculators and graphing apps are often used in teaching mathematics courses. Makers of graphing calculators include Texas Instruments, Casio, Sharp, and Hewett-Packard. Popular graphing apps include Desmos, Free Graphing Calculator, Quick Graph, Graph +, and many more.

In addition to graphing equations, these calculators and apps have many features such as zoom, trace, and table that allow you to explore many aspects of the equations and their graphs. Other features allow you to determine the x- and y-intercepts of a graph, the intersection points of two graphs, and other concepts of graphs that are discussed in calculus. Although most standardized testing allows use of specified models of graphing calculators, the use of smartphones and tablets is usually not allowed. However, some states are incorporating graphing calculator apps into computerized testing.

Check with your instructor prior to using a graphing calculator or a graphing app to complete required work for your course.

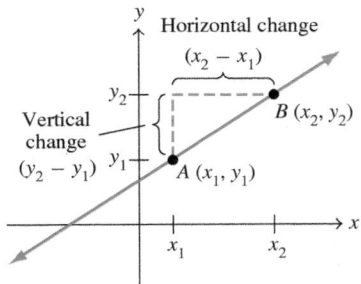

Figure 6.11

Slope

Another useful concept when you are working with straight lines is slope, which is a measure of the "steepness" of a line. The *slope of a line* is the ratio of the vertical change to the horizontal change for any two points on the line. Consider Fig. 6.11.

Point A has coordinates (x_1, y_1), and point B has coordinates (x_2, y_2). The vertical change between points A and B is $y_2 - y_1$. The horizontal change between points A and B is $x_2 - x_1$. Thus, the slope, which is symbolized with the letter m, can be determined as follows.

> **Slope of a Line**
>
> $$\textbf{Slope} = \frac{\text{vertical change}}{\text{horizontal change}}$$
>
> $$m = \frac{y_2 - y_1}{x_2 - x_1}$$

The Greek capital letter delta, Δ, is often used to represent the words "the change in." Therefore, slope may be defined as

$$m = \frac{\Delta y}{\Delta x}$$

A line may have a positive slope, a negative slope, or a zero slope, or the slope may be undefined, as indicated in Fig. 6.12. A line with a positive slope rises from left to right, as shown in Fig. 6.12(a). A line with a negative slope falls from left to right, as shown in Fig. 6.12(b). A horizontal line, which neither rises nor falls, has a slope of zero, as shown in Fig. 6.12(c). Since a vertical line does not have any horizontal change (the x-value remains constant) and since we cannot divide by 0, the slope of a vertical line is undefined, as shown in Fig. 6.12(d).

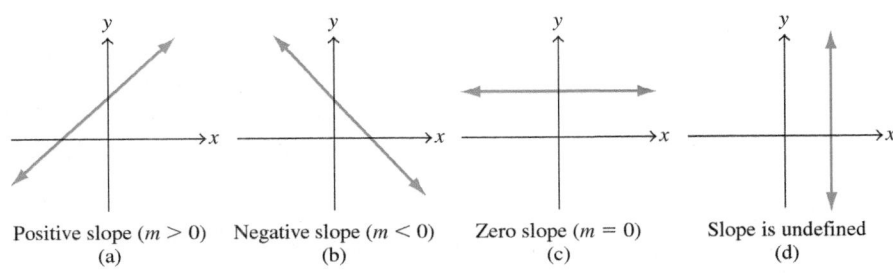

Positive slope ($m > 0$)	Negative slope ($m < 0$)	Zero slope ($m = 0$)	Slope is undefined
(a)	(b)	(c)	(d)

Figure 6.12

Example 5 *Determining the Slope of a Line*

Determine the slope of the line that passes through the points $(-1, -3)$ and $(1, 5)$.

Solution Let's begin by drawing a sketch, illustrating the points and the line. See Fig. 6.13(a) on page 334.

We will let (x_1, y_1) be $(-1, -3)$ and (x_2, y_2) be $(1, 5)$. Then

$$\text{Slope} = \frac{y_2 - y_1}{x_2 - x_1} = \frac{5 - (-3)}{1 - (-1)} = \frac{5 + 3}{1 + 1} = \frac{8}{2} = \frac{4}{1} = 4$$

The slope of 4 means that there is a vertical change of 4 units for each horizontal change of 1 unit; see Fig. 6.13(b). The slope is positive, and the line rises from left to right. Note that we would have obtained the same result if we let (x_1, y_1) be $(1, 5)$ and (x_2, y_2) be $(-1, -3)$. Try this method now and see.

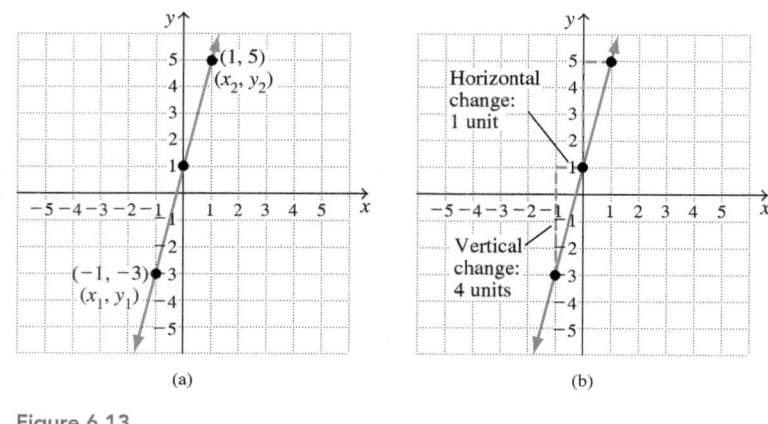

(a)   (b)

Figure 6.13

Exploring Slope
and y-intercept

Graphing Linear Equations by Using the Slope and y-Intercept

A linear equation given in the form $y = mx + b$ is said to be in slope–intercept form.

> ### Slope–Intercept Form of the Equation of a Line
>
> $$y = mx + b$$
>
> where m is the slope of the line and $(0, b)$ is the y-intercept of the line.

In the equation $y = mx + b$, b represents the value of y where the graph of the equation $y = mx + b$ crosses the y-axis.

Consider the graph of the equation $y = 3x + 4$, which appears in Fig. 6.14. By examining the graph, we can see that the y-intercept is $(0, 4)$. We can also see that the graph has a positive slope because it rises from left to right. Because the vertical change is 3 units for every 1 unit of horizontal change, the slope must be $\frac{3}{1}$, or 3.

We could graph this equation by marking the y-intercept at $(0, 4)$ and then moving *up* 3 units and to the *right* 1 unit to get another point. If the slope were -3, which means $\frac{-3}{1}$, we could start at the y-intercept and move *down* 3 units and to the *right* 1 unit. Thus, if we know the slope and y-intercept of a line, we can graph the line.

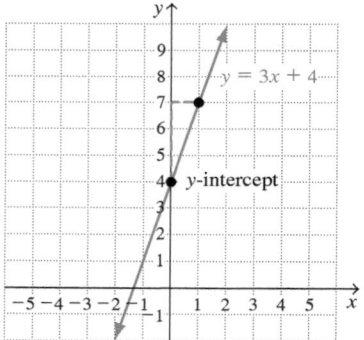

Figure 6.14

> **PROCEDURE**   GRAPHING LINEAR EQUATIONS BY USING THE SLOPE AND Y-INTERCEPT
>
> 1. Solve the equation for y to place the equation in slope–intercept form.
> 2. Determine the slope and y-intercept from the equation.
> 3. Plot the y-intercept.
> 4. Obtain a second point using the slope.
> 5. Draw a straight line through the points.

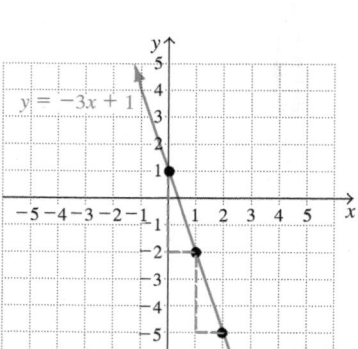

Figure 6.15

Example 6   *Graphing an Equation by Using the Slope and y-Intercept*

Graph $y = -3x + 1$ by using the slope and y-intercept.

Solution   The slope is -3, or $\frac{-3}{1}$, and the y-intercept is $(0, 1)$ (see Fig. 6.15). Plot $(0, 1)$ on the y-axis. Then plot the next point by moving *down* 3 units and to the *right* 1 unit. A third point has been plotted in the same way. The graph of $y = -3x + 1$ is the line drawn through these three points.

Example 7 Writing an Equation in Slope–Intercept Form

a) Write $4x - 3y = 9$ in slope–intercept form.

b) Graph the equation.

Solution

a) To write $4x - 3y = 9$ in slope–intercept form, we solve the given equation for y.

$$4x - 3y = 9$$
$$4x - 4x - 3y = -4x + 9$$
$$-3y = -4x + 9$$
$$\frac{-3y}{-3} = \frac{-4x + 9}{-3}$$
$$y = \frac{-4x}{-3} + \frac{9}{-3} \quad \text{or} \quad y = \frac{4}{3}x - 3$$

Thus, in slope–intercept form, the equation is $y = \frac{4}{3}x - 3$.

b) The y-intercept is $(0, -3)$, and the slope is $\frac{4}{3}$. Plot a point at $(0, -3)$ on the y-axis, then move *up* 4 units and to the *right* 3 units to obtain the second point (see Fig. 6.16). Draw a line through the two points. ■

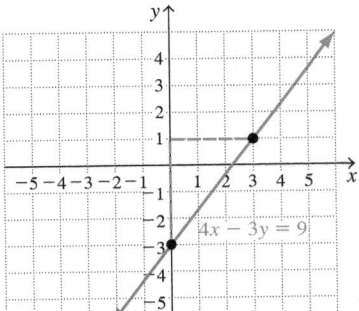

Figure 6.16

Example 8 Determining the Equation of a Line from Its Graph

Determine the equation of the line in Fig. 6.17.

Solution If we determine the slope and the y-intercept of the line, we can write the equation using slope–intercept form, $y = mx + b$. We see from the graph that the y-intercept is $(0, 1)$; thus, $b = 1$. The slope of the line is negative because the graph falls from left to right. The change in y is 1 unit for every 3-unit change in x. Thus, m, the slope of the line, is $-\frac{1}{3}$.

$$y = mx + b$$

$$y = -\frac{1}{3}x + 1$$

The equation of the line is $y = -\frac{1}{3}x + 1$. ■

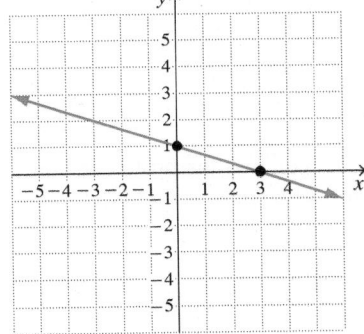

Figure 6.17

Example 9 Horizontal and Vertical Lines

In the Cartesian coordinate system, graph (a) $y = 2$ and (b) $x = -3$.

Solution

a) For any value of x, the value of y is 2. Therefore, the graph will be a horizontal line through $y = 2$ (Fig. 6.18).

b) For any value of y, the value of x is -3. Therefore, the graph will be a vertical line through $x = -3$ (Fig. 6.19).

Note that the graph of $y = 2$ has a slope of 0. The slope of the graph of $x = -3$ is undefined. ■

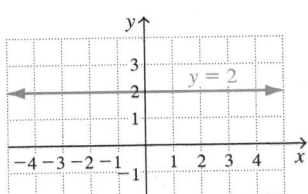

Figure 6.18

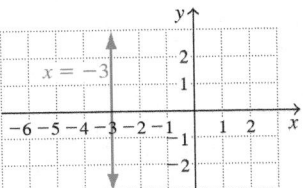

Figure 6.19

In graphing the equations in this section, we labeled the horizontal axis the x-axis and the vertical axis the y-axis. For each equation, we can determine values for y by substituting values for x. Since the value of y depends on the value of x, we refer to y as the *dependent variable* and x as the *independent variable*. We label the *vertical axis* with the *dependent variable* and the *horizontal axis* with the *independent variable*. For the equation $C = 3n + 5$, the C is the dependent variable and the n is the independent variable. Thus, to graph this equation, we label the vertical axis C and the horizontal axis n.

In many graphs, the values to be plotted on one axis are much greater than the values to be plotted on the other axis. When that occurs, we can use different scales on the horizontal and the vertical axes, as illustrated in Examples 10 and 11. The next two examples illustrate applications of graphing.

Example 10 *Using a Graph to Determine Distance*

The distance in miles, d, a car travels in t hours, traveling at a constant rate of 40 miles per hour, can be determined by the equation $d = 40t$.

a) Graph $d = 40t$, for $t \le 6$.

b) Use the graph to estimate the distance the car travels in 4 hours.

Solution

a) Since $d = 40t$ is a linear equation, the graph will be a straight line. Select three values for t, determine the corresponding values for d, and then draw the graph (Fig. 6.20).

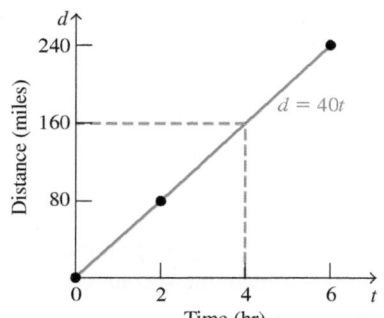

Figure 6.20

$$d = 40t$$

Let $t = 0$, $d = 40(0) = 0$

Let $t = 2$, $d = 40(2) = 80$

Let $t = 6$, $d = 40(6) = 240$

t	d
0	0
2	80
6	240

b) By drawing a vertical line from $t = 4$ on the time axis up to the graph and then drawing a horizontal line across to the distance axis, we can estimate that the distance the car travels in 4 hours is about 160 miles. ■

Example 11 *Using a Graph to Determine Profits*

Javier owns a business that sells mountain bicycles. He believes the profit (or loss) from the bicycles sold can be estimated by the formula $P = 60S - 90{,}000$, where S is the number of bicycles sold.

a) Graph $P = 60S - 90{,}000$, for $S \le 5000$ bicycles.

b) From the graph, estimate the number of bicycles that must be sold for the business to break even.

c) From the graph, estimate the number of bicycles sold if the profit from selling bicycles is about \$150,000.

Solution

a) Select values for S and determine the corresponding values of P. The graph is illustrated in Fig. 6.21.

S	P
0	$-90{,}000$
2000	30,000
5000	210,000

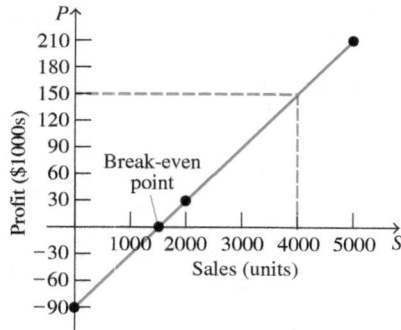

Figure 6.21

b) The break-even point is the number of bicycles that must be sold for the company to have neither a profit nor a loss. The break-even point is where the graph intercepts the horizontal, or *S*-axis, for that is where the profit, *P*, is 0. To break even, about 1500 bicycles must be sold. The actual value can be obtained by substituting 0 for *P* in the equation and solving the equation for *S*. Do this now.

c) We can estimate the answer by drawing a horizontal line from 150 on the profit axis. Since the horizontal line cuts the graph at about 4000 on the *S*-axis, about 4000 bicycles were sold. ∎

SECTION 6.6 *Exercises*

Warm Up Exercises

In Exercises 1–8, fill in the blank with an appropriate word, phrase, or symbols(s).

1. An illustration of all the points whose coordinates satisfy an equation is called a(n) _____.

2. To determine the *x*-intercept of the graph of a linear equation, set _____ equal to zero and solve the equation for *x*.

3. To determine the *y*-intercept of the graph of a linear equation, set _____ equal to zero and solve the equation for *y*.

4. The ratio of the vertical change to the horizontal change for any two points on a line is called the _____ of the line.

5. The three methods used to graph a linear equation given in this section are _____, _____, and _____.

6. The minimum number of points needed to graph a linear equation is _____.

7. In the equation $y = mx + b$, the slope of the line is represented by _____.

8. a) The point whose coordinates are (2, 7) is located in the _____ quadrant.

    b) The point whose coordinates are $(-3, -5)$ is located in the _____ quadrant.

Practice the Skills

In Exercises 9–16, plot the given points on the same axes.

9. (3, 1)

10. $(-2, -4)$

11. $(-5, -1)$

12. (4, 0)

13. (0, 2)

14. (0, 0)

15. $(0, -5)$

16. $\left(2\frac{1}{2}, 5\frac{1}{2}\right)$

In Exercises 17–24, plot the given points on the same axes.

17. $(4, -1)$

18. (3, 2)

19. $(-1, -5)$

20. $(0, -1)$

21. (0, 2)

22. $(-3, 0)$

23. (5, 1)

24. (4.5, 3.5)

Exercises 25–34 are indicated on the graph in Fig. 6.22. Write the coordinates of the indicated point.

25.

26.

27.

28.

29.

30.

31.

32.

33.

34.

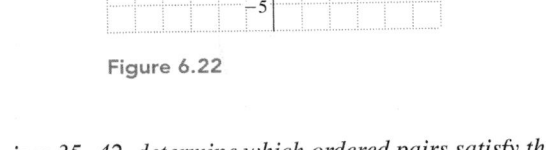

Figure 6.22

In Exercises 35–42, determine which ordered pairs satisfy the given equation.

35. $2x + 3y = 6$ (0, 2) (3, 0) (2, 3)

36. $3x - 5y = 15$ $(0, -3)$ (5, 0) $(-3, 5)$

37. $3x - 2y = 10$ (8, 7) $(-1, 4)$ $\left(\frac{10}{3}, 0\right)$

38. $2x - 3y = 9$ $(0, -3)$ (2, 1) $(-3, -5)$

39. $6y = 3x + 6$ $(1, -1)$ (4, 3) (2, 5)

40. $3y = 4x + 2$ (2, 1) (1, 2) $\left(0, \frac{2}{3}\right)$

41. $y = -\frac{3}{8}x + 3$ (8, 0) $(-8, 9)$ (0, 3)

42. $y = -\frac{5}{7}x + 10$ (14, 0) (0, 10) (7, 15)

In Exercises 43–52, graph the equation by plotting points, as in Example 3.

43. $y = x + 1$

44. $y = x - 2$

45. $y = 2x - 1$

46. $y = -2x + 3$

47. $y + 3x = 6$

48. $y - 4x = 8$

49. $y = \frac{1}{2}x + 4$

50. $3y = 2x - 3$

51. $y = \frac{1}{3}x$

52. $y = -\frac{2}{3}x$

In Exercises 53–62, graph the equation, using the x- and y-intercepts, as in Example 4.

53. $x + y = 2$ **54.** $x - y = 4$

55. $x + y = 1$ **56.** $x - y = 3$

57. $2x + y = 6$ **58.** $3x - y = -6$

59. $x + 2y = 6$ **60.** $2x - y = 4$

61. $2x = -4y - 8$ **62.** $6x = 3y - 9$

In Exercises 63–70, determine the slope of the line through the given points. If the slope is undefined, so state.

63. $(3, 5)$ and $(5, 9)$ **64.** $(4, 3)$ and $(3, 4)$

65. $(5, 1)$ and $(8, 8)$ **66.** $(-2, 6)$ and $(-4, 9)$

67. $(1, 5)$ and $(4, 5)$ **68.** $(-3, -4)$ and $(5, -4)$

69. $(8, -3)$ and $(8, 3)$ **70.** $(2, 6)$ and $(2, -3)$

In Exercises 71–78, graph the equation using the slope and y-intercept, as in Examples 6 and 7.

71. $y = x + 2$ **72.** $y = -x - 2$

73. $y = -2x + 3$ **74.** $y = 3x - 1$

75. $y = -\frac{1}{2}x + 1$ **76.** $y = \frac{1}{2}x - 2$

77. $3x - 2y + 6 = 0$ **78.** $3x + 4y - 8 = 0$

In Exercises 79–82, determine the equation of the graph.

79.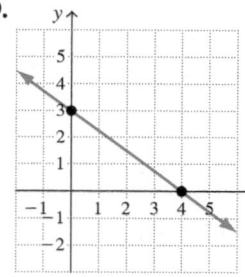

80.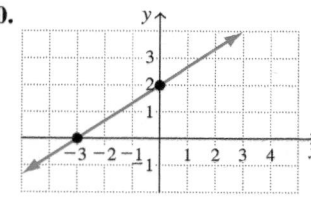

81.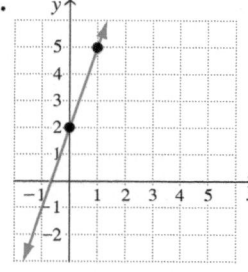

82.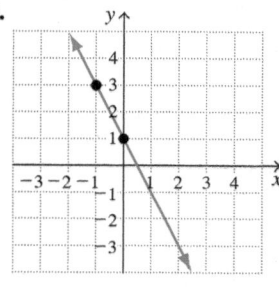

In Exercises 83–86, graph the equation and state the slope of the line if the slope exists (see Example 9).

83. $x = 3$ **84.** $x = -2$

85. $y = 3$ **86.** $y = -4$

In Exercises 87–90, determine the equation of the graph.

87. 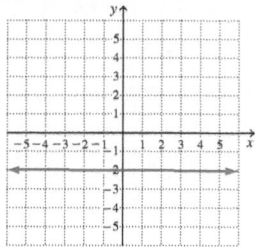   **88.**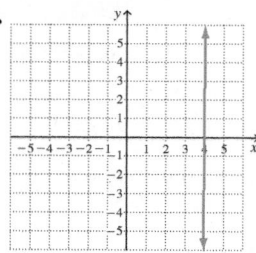

89. 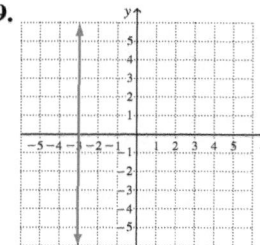   **90.**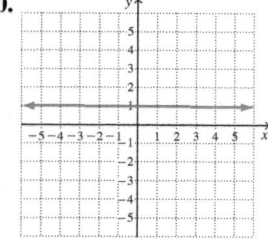

Problem Solving

In Exercises 91 and 92, points A, B, and C are three vertices of a rectangle. Plot the three points. (a) Determine the coordinates of the fourth point, D, to complete the rectangle. (b) Determine the area of the rectangle; use $A = lw$.

91. $A(-1, 4)$, $B(4, 4)$, $C(-1, 2)$

92. $A(-4, 2)$, $B(7, 2)$, $C(7, 8)$

In Exercises 93 and 94, points A, B, and C are three vertices of a parallelogram (see Example 2) with sides parallel to the x-axis. Plot the three points. Determine the coordinates of the fourth point, D, to complete the parallelogram. Note: There are two possible answers for point D.

93. $A(3, 2)$, $B(5, 5)$, $C(9, 5)$

94. $A(-2, 2)$, $B(3, 2)$, $C(6, -1)$

In Exercises 95–98, for what value of b will the line joining the points P and Q be parallel to the indicated axis?

95. $P(2, -3)$, $Q(6, b)$;   x-axis

96. $P(4, 7)$, $Q(b, -2)$;   y-axis

97. $P(3b - 1, 5)$, $Q(8, 4)$;   y-axis

98. $P(-6, 2b + 3)$, $Q(7, -1)$;   x-axis

99. *Hanging Wallpaper* Tanisha owns an interior design business. Her charge, C, for hanging wallpaper is $40 plus $0.30 per square foot of wallpaper she hangs, or $C = 40 + 0.30s$, where s is the number of square feet of wallpaper she hangs.

a) Graph $C = 40 + 0.30s$, for $s \leq 500$.

b) From the graph, estimate her charge if she hangs 300 square feet of wallpaper.

c) If her charge is $70, use the equation for C to determine how many square feet of wallpaper she hung.

100. *Selling Chocolates* Ryan sells chocolate on the Internet. His monthly profit, p, in dollars, can be estimated by $p = 15n - 300$, where n is the number of dozens of chocolates he sells in a month.

a) Graph $p = 15n - 300$, for $n \leq 60$.

b) From the graph, estimate his profit if he sells 40 dozen chocolates in a month.

c) How many dozens of chocolates must he sell in a month to break even?

101. *Security System* The Kamdars installed a home security system that includes an installation fee of $75 plus a monthly service fee of $25. The total cost of the system is given by $C = 75 + 25x$, where x is the number of months of service.

a) Graph $C = 75 + 25x$, for $x \leq 50$.

b) From the graph, estimate the Kamdars' total cost after 30 months.

c) If the Kamdars' total cost is $975, estimate the number of months of service.

102. *Earning Simple Interest* When $1000 is invested in a savings account paying simple interest for a year, the interest, i, in dollars, earned can be determined by the formula $i = 1000r$, where r is the interest rate in decimal form.

a) Graph $i = 1000r$, for r up to and including an interest rate of 15%.

b) If the interest rate is 4%, what is the simple interest?

c) If the interest rate is 6%, what is the simple interest?

103. *Determining the Number of Defects* The top graph on the right shows the daily number of workers absent from

the assembly line at J. B. Davis Corporation and the number of defects coming off the assembly line for each of 10 days. The blue line on the graph can be used to approximate the number of defects coming off the assembly line per day for a given number of workers absent.

a) Determine the slope of the blue line using the two points indicated with ordered pairs.

b) Using the slope determined in part (a) and the y-intercept, $(0, 9)$, determine the equation of the blue line.

c) Using the equation you determined in part (b), determine the approximate number of defects for a day if 3 workers are absent.

d) Using the equation you determined in part (b), approximate the number of workers absent for a day if there are 17 defects that day.

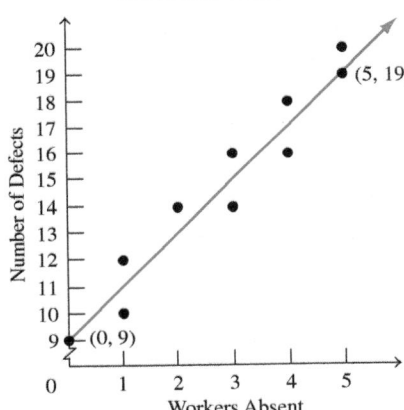

104. *Determining a Test Grade* The graph below shows the hours studied and the test grades on a biology test for 8 students. The red line on the graph can be used to approximate the test grade the student receives for the number of hours he or she studies.

a) Determine the slope of the red line using the two points indicated with ordered pairs.

b) Using the slope determined in part (a) and the y-intercept, $(0, 53)$, determine the equation of the red line.

c) Using the equation you determined in part (b), determine the approximate test grade for a student who studied for 3 hours.

d) Using the equation you determined in part (b), approximate the amount of time a student would need to study to receive a grade of 80 on the biology test.

105. ***Online Electronics Sales*** The graph below was created with **StatCrunch** using data from the U. S. Department of Commerce website: (www.commerce.gov). The blue dots on the graph show the percentage of all spending on electronics in the United States that was done online for the years 2010–2017. The red line can be used to approximate this same spending. On the graph, $x = 0$ represents the year 2010, $x = 1$ represents the year 2011, $x = 2$ represents the year 2012, and so on.

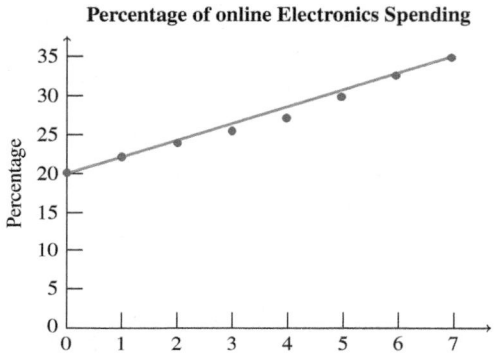

Percentage of online Electronics Spending

Year After 2010

Source: U.S. Department of Commerce

a) Use the points $(0, 20)$ and $(7, 34.7)$ to calculate the slope of the red line.

b) Determine the equation of the red line.

c) Using the equation you determined in part (b), estimate the percentage in 2016.

d) Using the equation you determined in part (b), determine the year in which the percentage was 24.

106. ***Landlines*** The graph above and to the right was created with **StatCrunch** using data from the Federal Communication Commission website: (www.fcc.gov). The blue dots on the graph show the percentage of all households in the United States that had a landline for

non-cellular telephone service for the years 2010–2017. The red line can be used to approximate this same percentage. On the graph, $x = 0$ represents the year 2010, $x = 1$ represents the year 2011, $x = 2$ represents the year 2012, and so on.

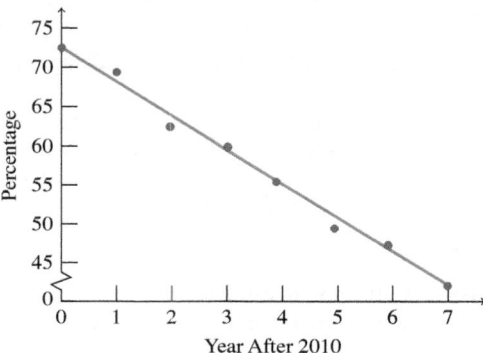

Percentage of U.S. Households with Landlines

Year After 2010

Source: Federal Communications Commission

a) Use the points $(0, 72.5)$ and $(7, 43.8)$ to calculate the slope of the red line.

b) Determine the equation of the red line.

c) Using the equation you determined in part (b), estimate the percentage in 2016.

d) Using the equation you determined in part (b), determine the year in which the percentage was 60.2.

Challenge Problems/Group Activities

107. a) Two lines are parallel when they do not intersect no matter how far they are extended. Explain how you can determine, without graphing the equations, whether the graphs of two different equations will be parallel lines when graphed.

b) Determine whether the graphs of the equations $2x - 3y = 6$ and $4x = 6y + 6$ are parallel lines.

108. In which quadrants will the set of points that satisfy the equation $x + y = 1$ lie? Explain.

Research Activity

109. ***René Descartes*** René Descartes is known for his contributions to algebra. Write a paper on his life and his contributions to algebra.

SECTION 6.7   Solving Systems of Linear Equations

Upon completion of this section, you will be able to:

- Determine if an ordered pair is a solution to a system of linear equations.
- Solve a system of linear equations by graphing.
- Solve a system of linear equations by the substitution method.
- Solve a system of linear equations by the addition method.
- Solve applications of systems of linear equations.

Delbert is choosing between two membership options at the local movie society. A *director* membership has a lower initial cost to join, but the cost per movie is higher. A *producer* membership has a higher initial cost to join, but the cost per movie is lower. Which membership would be a better financial choice for Delbert? In this section, we will learn how to use two equations, considered simultaneously, to help answer this question.

Why This Is Important   As we will see in this section, we will use a system of equations to model many real-life problems including business profit or loss, medicine mixtures, and a variety of financial applications. These applications have relevance to a wide variety of occupations and to many aspects of our daily lives.

In algebra, it is often necessary to determine the common solution to two or more linear equations. When two or more linear equations are considered simultaneously, the equations are called a *system of linear equations*. The solution to a system of linear equations may be determined by a number of techniques. In this section, we illustrate three different methods for solving systems of linear equations.

Solutions to Systems of Linear Equations

A *solution to a system of equations* is the ordered pair or ordered pairs that satisfy *all* equations in the system. A system of linear equations may have exactly one solution, no solution, or infinitely many solutions.

Example 1 *Is the Ordered Pair a Solution?*

Determine which of the ordered pairs is a solution to the system of equations.

$$y = \frac{1}{2}x - 4$$
$$3x + 5y = -9$$

a) $(0, -4)$ b) $(12, -9)$ c) $(2, -3)$

Solution To be a solution, the ordered pair must satisfy each equation in the system.

a) For $(0, -4)$ we substitute $x = 0$ and $y = -4$ into each equation in the system.

$$y = \frac{1}{2}x - 4 \qquad\qquad 3x + 5y = -9$$
$$-4 = \frac{1}{2}(0) - 4 \qquad\qquad 3(0) + 5(-4) = -9$$
$$-4 = 0 - 4 \qquad\qquad\qquad 0 - 20 = -9$$
$$-4 = -4 \quad \text{True} \qquad\qquad -20 = -9 \quad \text{False}$$

Since $(0, -4)$ does not satisfy *both* equations, it is not a solution to the system.

b) For $(12, -9)$ we substitute $x = 12$ and $y = -9$ into each equation in the system.

$$y = \frac{1}{2}x - 4 \qquad\qquad\qquad 3x + 5y = -9$$

$$-9 = \frac{1}{2}(12) - 4 \qquad\qquad 3(12) + 5(-9) = -9$$

$$-9 = 6 - 4 \qquad\qquad\qquad 36 - 45 = -9$$

$$-9 = 2 \quad \text{False} \qquad\qquad\quad -9 = -9 \quad \text{True}$$

Since $(12, -9)$ does not satisfy *both* equations, it is not a solution to the system.

c) For $(2, -3)$ we substitute $x = 2$ and $y = -3$ into each equation in the system.

$$y = \frac{1}{2}x - 4 \qquad\qquad\qquad 3x + 5y = -9$$

$$-3 = \frac{1}{2}(2) - 4 \qquad\qquad 3(2) + 5(-3) = -9$$

$$-3 = 1 - 4 \qquad\qquad\qquad 6 - 15 = -9$$

$$-3 = -3 \quad \text{True} \qquad\qquad\quad -9 = -9 \quad \text{True}$$

Since $(2, -3)$ satisfies *both* equations, it is a solution to the system. ∎

Solving a Linear
System by Graphing

Solving a System of Linear Equations by Graphing

To solve a system of linear equations graphically, we graph both equations on the same axes. If the system has a unique solution, then the solution will be the ordered pair that represents the point of intersection of the lines in the system.

When two lines are graphed, three situations are possible, as illustrated in Figure 6.23.

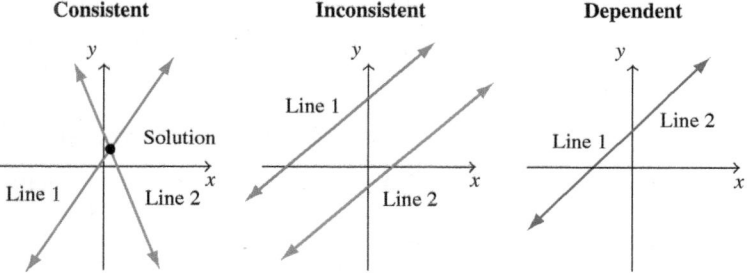

Consistent	Inconsistent	Dependent
Line 1 *Intersects* Line 2	Line 1 Is *Parallel* to Line 2	Line 1 Is the *Same Line* as Line 2
• Exactly one solution	• No Solution	• Infinite number of solutions
• The solution is at the point of intersection.	• Since Parallel lines do not intersect, there is no solution.	• Every point on the common line is a solution.
• **Consistent system***	• **Inconsistent system**	• **Dependent system***
• Lines have different slopes	• Lines have same slopes and different y-intercepts	• Lines have same slope and same y-intercept

Figure 6.23

Examples 2 and 3, on pages 343 and 344, respectively, illustrate consistent systems. Example 4, on page 345, illustrates an inconsistent system. And Example 5, on page 345, illustrates a dependent system. We next provide a procedure for solving a system of linear equations by graphing.

---

*A consistent system has *at least one* solution. Since a dependent system has at least one solution, a dependent system is also a consistent system.

> **PROCEDURE**   SOLVING A SYSTEM OF LINEAR EQUATIONS BY GRAPHING
>
> 1. Determine three ordered pairs that satisfy each equation.
> 2. Plot the points that correspond to the ordered pairs and graph *both* equations on the same axes.
> 3. The coordinates of the point or points of intersection of the graphs are the solution or solutions to the system of equations.

Example 2 *A Unique Solution by Graphing*

Determine the solution to the following system of equations graphically.

$$x + y = 4$$
$$2x - y = -1$$

Solution For each equation, substitute values for x and determine the corresponding values for y. Three ordered pairs that satisfy $x + y = 4$ are $(0, 4)$, $(1, 3)$, and $(4, 0)$, as shown in the left hand table below. Three ordered pairs that satisfy $2x - y = -1$ are $(-2, -3)$, $(0, 1)$, and $(1, 3)$, as shown in the right-hand table below. Next, graph both $x + y = 4$ and $2x - y = -1$ on the same axes, Fig. 6.24.

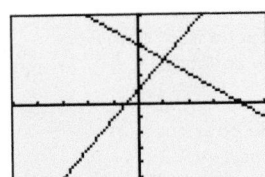

Figure 6.24

$x + y = 4$	
x	y
0	4
1	3
4	0

$2x - y = -1$	
x	y
-2	-3
0	1
1	3

The graphs intersect at $(1, 3)$, which is the solution to the system of equations. This point is the only point that satisfies *both* equations.

CHECK:

$x + y = 4$	$2x - y = -1$
$1 + 3 = 4$	$2(1) - 3 = -1$
$4 = 4$ True	$2 - 3 = -1$
	$-1 = -1$ True ∎

Fig. 6.25 shows the system of equations in Example 2, $x + y = 4$ and $2x - y = -1$, graphed on a Texas Instrument TI-84 Plus calculator.

In Example 2 we were able to estimate the solution to problems by determining what appeared to be the point of intersection of the graphs in the systems of linear equations. Since the solution to a system of equations may not be integer values, you may not be able to obtain the exact solution by graphing. If an exact answer to a system of linear equations is needed, we can solve the system with algebraic methods. Two such methods are the substitution method and the addition method.

We first discuss the substitution method.

Figure 6.25

Substitution Method

> **PROCEDURE** SOLVING A SYSTEM OF EQUATIONS BY THE SUBSTITUTION METHOD
>
> 1. Solve one of the equations for one of the variables. If possible, solve for a variable with a numerical coefficient of 1. By doing so, you may avoid working with fractions.
> 2. Substitute the expression determined in Step 1 into the other equation. This step yields an equation in terms of a single variable.
> 3. Solve the equation determined in Step 2 for the variable.
> 4. Substitute the value determined in Step 3 into the equation you rewrote in Step 1 and solve for the remaining variable.

Examples 3, 4, and 5 illustrate the *substitution method*.

Example 3 *A Unique Solution by the Substitution Method*

Solve the following system of equations by substitution.

$$x + y = 4$$
$$2x - y = -1$$

Solution The numerical coefficients of the x- and y-terms in the equation $x + y = 4$ are both 1. Thus, we can solve this equation for either x or y. Let's solve for x in the first equation.

STEP 1.

$$x + y = 4$$
$$x + y - y = 4 - y \qquad \text{Subtract } y \text{ from both sides of the equations.}$$
$$x = 4 - y$$

STEP 2. Substitute $4 - y$ for x in the second equation.

$$2x - y = -1$$
$$2(4 - y) - y = -1$$

STEP 3. Now solve the equation for y.

$$8 - 2y - y = -1 \qquad \text{Distributive property}$$
$$8 - 3y = -1$$
$$8 - 8 - 3y = -1 - 8 \qquad \text{Subtract 8 from both sides of the equation.}$$
$$-3y = -9$$
$$\frac{-3y}{-3} = \frac{-9}{-3} \qquad \text{Divide both sides of the equation by } -3.$$
$$y = 3$$

STEP 4. Substitute $y = 3$ in the equation solved for x and determine the value of x.

$$x = 4 - y$$
$$x = 4 - 3$$
$$x = 1$$

Thus, the solution is the ordered pair $(1, 3)$. ∎

Timely Tip

When solving a system of equations, once you successfully solve for one of the variables, make sure you solve for the other variable. Remember that a solution to a system of equations must contain a numerical value for each variable in the system.

In Example 2 on page 343, we graphed the same system of equations that we solved algebraically in Example 3. Notice we get the same answer graphically as we do when solving algebraically.

Example 4 No Solution by the Substitution Method

Solve the following system of equations by substitution.

$$x + y = 3$$
$$2x + 2y = -4$$

Solution The numerical coefficients of the x- and y-terms in the equation $x + y = 3$ are both 1. Thus, we can solve this equation for either x or y. Let us solve for y in the first equation.

$$x + y = 3$$
$$x - x + y = 3 - x \qquad \text{Subtract } x \text{ from both sides of the equation.}$$
$$y = 3 - x$$

Now substitute $3 - x$ for y in the second equation.

$$2x + 2y = -4$$
$$2x + 2(3 - x) = -4$$
$$2x + 6 - 2x = -4 \qquad \text{Distributive property}$$
$$6 = -4 \qquad \text{False}$$

Since 6 cannot be equal to -4, there is no solution to the system of equations. Thus, the system of equations is inconsistent. ∎

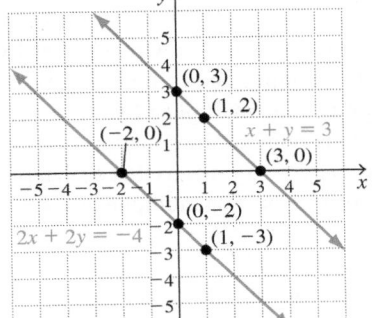

Figure 6.26

Timely Tip

When solving a system of equations, if you obtain a statement that is always
- false, such as $6 = -4$, the system is inconsistent and has no solution. The equations in this system represent parallel lines.
- true, such as $6 = 6$, the system is dependent and has an infinite number of solutions. The equations in this system both represent the same line.

The system of equations from Example 4 is graphed in Fig. 6.26. Notice that the lines do not intersect, and therefore, there is no solution to the system. This agrees with the answer we obtained algebraically in Example 4.

When solving the system in Example 4, we obtained $6 = -4$ and indicated that the system was inconsistent and that there was no solution. When solving a system of equations, if you obtain a false statement, such as $4 = 0$ or $-2 = 0$, the system is *inconsistent* and has *no solution*.

Example 5 An Infinite Number of Solutions by the Substitution Method

Solve the following system of equations by substitution.

$$y = -2x + 3$$
$$2y = 6 - 4x$$

Solution The first equation $y = -2x + 3$ is already solved for y, so we will substitute $-2x + 3$ for y in the second equation.

$$2y = 6 - 4x$$
$$2(-2x + 3) = 6 - 4x$$
$$-4x + 6 = 6 - 4x$$
$$4x - 4x + 6 = 6 - 4x + 4x \qquad \text{Add } 4x \text{ to both sides of the equation.}$$
$$6 = 6 \qquad \text{True}$$

Since $6 = 6$, the system has an infinite number of solutions. Thus, the system of equations is dependent. ∎

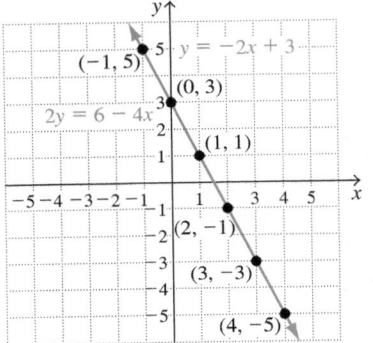

Figure 6.27

The system of equations from Example 5 is graphed in Fig. 6.27. Notice that the graph of both equations is the same line. Every point on the line is a solution to both equations and so there are an infinite number of solutions to the system. This agrees with the answer we obtained algebraically in Example 5.

When solving Example 5, we obtained $6 = 6$ and indicated that the system was dependent and had an infinite number of solutions. When solving a system of equations, if you obtain a true statement, such as $0 = 0$ or $6 = 6$, the system is *dependent* and has an *infinite number of solutions*.

Addition Method

If neither of the equations in a system of linear equations has a variable with a coefficient of 1, it is generally easier to solve the system by using the *addition* (or *elimination*) *method*.

To solve a system of linear equations by the addition method, it is necessary to obtain two equations whose sum will be a single equation containing only one variable. To achieve this goal, we rewrite the system of equations as two equations where the coefficients of one of the variables are opposites of each other. For example, if one equation has a term of $2x$, we might rewrite the other equation so that its x term will be $-2x$. To obtain the desired equations, it might be necessary to multiply one or both equations in the original system by a number. When an equation is to be multiplied by a number, we will place brackets around the equation and place the number the equation is to be multiplied by before the brackets. For example, $4\lbrack 2x + 3y = 6 \rbrack$ means that each term on both sides of the equal sign in the equation $2x + 3y = 6$ is to be multiplied by 4:

$$4\lbrack 2x + 3y = 6 \rbrack \quad \text{gives} \quad 8x + 12y = 24$$

This notation will make our explanations much more efficient and easier for you to follow.

PROCEDURE SOLVING A SYSTEM OF EQUATIONS BY THE ADDITION METHOD

1. If necessary, rewrite the equations so that the terms containing the variables appear on one side of the equal sign and the constants appear on the other side of the equal sign.

2. If necessary, multiply one or both equations by a constant(s) so that when you add the equations, the sum will be an equation containing only one variable.

3. Add the equations to obtain a single equation in one variable.

4. Solve for the variable in the equation obtained in Step 3.

5. Substitute the value determined in Step 4 into either of the original equations and solve for the other variable.

Example **6** *Eliminating a Variable by the Addition Method*

Solve the following system of equations by the addition method.

$$4x + y = 13$$
$$2x - y = 5$$

Solution Since the coefficients of the y-terms, 1 and -1, are opposites, the sum of the y-terms will be zero when the equations are added. Thus, the sum of the two equations will contain only one variable, x. Add the two equations to obtain one equation in one variable. Then solve for the remaining variable.

$$\begin{array}{r} 4x + y = 13 \\ \underline{2x - y = 5} \\ 6x \quad\;\; = 18 \\ x = 3 \end{array}$$

Now substitute 3 for x in either of the original equations to determine the value of y.

$$4x + y = 13$$
$$4(3) + y = 13$$
$$12 + y = 13$$
$$y = 1$$

The solution to the system is $(3, 1)$.   ∎

Example 7 *Multiplying One Equation While Using the Addition Method*

Solve the following system of equations by the addition method.

$$4x - y = 2$$
$$3x + 2y = 7$$

Solution We can multiply the top equation by 2 and then add the two equations to eliminate the variable y.

$$2[4x - y = 2] \quad \text{gives} \quad 8x - 2y = 4$$
$$3x + 2y = 7 \qquad\qquad\qquad 3x + 2y = 7$$

$$8x - 2y = 4$$
$$\underline{3x + 2y = 7}$$
$$11x \qquad = 11$$
$$x = 1$$

Now we determine y by substituting 1 for x in either of the original equations.

$$4x - y = 2$$
$$4(1) - y = 2$$
$$4 - y = 2$$
$$-y = -2$$
$$y = 2$$

The solution to the system is $(1, 2)$.   ∎

Example 8 *Multiplying Both Equations While Using the Addition Method*

Solve the following system of equations by the addition method.

$$3x - 4y = 8$$
$$2x + 3y = 9$$

Solution In this system, we cannot eliminate a variable by multiplying only one equation by an integer value and then adding. To eliminate a variable, we can multiply each equation by a different number. To eliminate the variable x, we can multiply the top equation by 2 and the bottom equation by -3 (or the top by -2 and the bottom by 3) and then add the two equations.

$$2[3x - 4y = 8] \quad \text{gives} \quad 6x - 8y = 16$$
$$-3[2x + 3y = 9] \quad \text{gives} \quad -6x - 9y = -27$$

$$6x - 8y = \quad 16$$
$$\underline{-6x - 9y = -27}$$
$$-17y = -11$$
$$y = \frac{11}{17}$$

We could now determine x by substituting $\frac{11}{17}$ for y in either of the original equations. Although it can be done, it gets messy. Instead, let's solve for x by eliminating the

Timely Tip

Following is a summary of the different types of systems of linear equations.
- A *consistent system of equations* is one that has at least one solution.
- An *inconsistent system of equations* is one that has no solution.
- A *dependent system of equations* is one that has an infinite number of solutions.

variable y from the two original equations. To do so, we multiply the first equation by 3 and the second equation by 4.

$$3[3x - 4y = 8] \quad \text{gives} \quad 9x - 12y = 24$$

$$4[2x + 3y = 9] \quad \text{gives} \quad 8x + 12y = 36$$

$$9x - 12y = 24$$
$$\underline{8x + 12y = 36}$$
$$17x \qquad = 60$$
$$x = \frac{60}{17}$$

The solution to the system is $\left(\frac{60}{17}, \frac{11}{17}\right)$. ■

Applications of Systems of Linear Equations

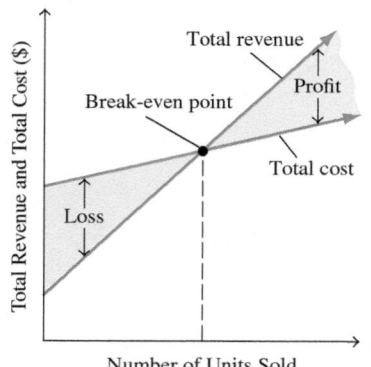

Figure 6.28

Our first application involves a business application known as *break-even analysis*. Manufacturers use *break-even analysis* to determine how many units of an item must be sold for the business to "break even," that is, for its total revenue to equal its total cost. Suppose we let the horizontal axis of a graph represent the number of units manufactured and sold and the vertical axis represent dollars. Then linear equations for cost, C, and revenue, R, can both be graphed on the same axes (Fig. 6.28). Both C and R are expressed in dollars, and both are a function of the number of units. Profit, P, is the difference between revenue, R, and cost, C. Thus, $P = R - C$. If revenue is greater than cost, the company makes a profit. If cost is greater than revenue, the company has a loss.

Initially, the cost graph is higher than the revenue graph because of fixed overhead costs such as rent and utilities. During low levels of sales, the manufacturer suffers a loss (the cost graph is greater). During higher levels of sales, the manufacturer realizes a profit (the revenue graph is greater). The point at which the two graphs intersect is called the *break-even point*. At that number of units sold, revenue equals cost and the manufacturer breaks even.

In Section 6.2, we introduced *modeling*. Recall that a *mathematical model* is an equation or system of equations that represents a real-life situation. In Examples 9 and 10, we develop equations that model real-life situations.

Example 9 *Modeling—Profit and Loss in Business*

At a collectibles show, Richard can sell model trains for $35. The costs for making the trains are a fixed cost of $200 and a production cost of $15 apiece.

a) Write an equation that represents Richard's revenue. Write an equation that represents Richard's cost.

b) How many model trains must Richard sell to break even?

c) Write an equation for the profit formula. Use the formula to determine Richard's profit if he sells 15 model trains.

d) How many model trains must Richard sell to make a profit of $600?

Solution

a) Let x denote the number of model trains made and sold. The revenue is given by the equation

$$R = 35x \quad (\$35 \text{ times the number of trains})$$

and the cost is given by the equation

$$C = 200 + 15x \quad (\$200 \text{ plus } \$15 \text{ times the number of trains})$$

Did You Know?

How to Succeed in Business

Economics, a science dependent on mathematics, dates back to just before the Industrial Revolution of the eighteenth century. Technologies were being invented and applied to the manufacture of cloth, iron, transportation, and agriculture. These new technologies led to the development of mathematically based economic models that often included systems of equations. French economist Jules Dupuit (1804–1866) suggested a method to calculate the value of railroad bridges; Irish economist Dionysis Larder (1793–1859) showed railroad companies how to structure their rates so as to increase their profits.

b) The break-even point is the point at which the revenue and cost graphs intersect. In Fig. 6.29, the graphs intersect at the point $(10, 350)$, which is the break-even point. Thus, for Richard to break even, he must sell 10 model trains. When 10 model trains are made and sold, the cost and revenue are both $350.

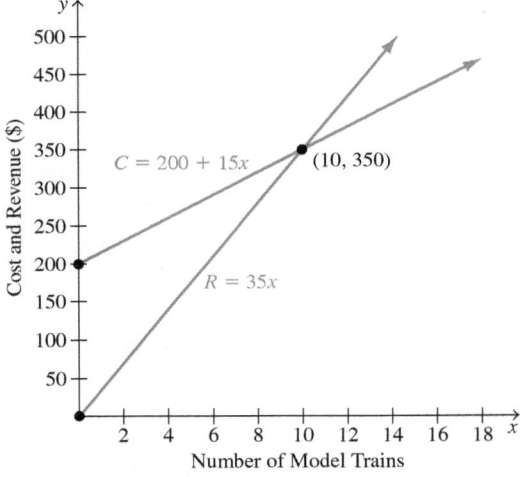

Figure 6.29

c) Profit is equal to the revenue minus the cost. Therefore, the profit formula is

$$P = R - C$$
$$= 35x - (200 + 15x)$$
$$= 35x - 200 - 15x$$
$$= 20x - 200$$

For 15 trains, the profit is determined as follows.

$$P = 20x - 200$$
$$= 20(15) - 200 = 100$$

Richard has a profit of $100 if he sells 15 model trains. By observing the graph, we can see that if Richard sells 15 model trains, he will have a profit since at 15 trains the revenue line is above the cost line.

d) We can determine the number of model trains that Richard must sell to have a profit of $600 by using the profit formula. Substituting 600 for P we have

$$P = 20x - 200$$
$$600 = 20x - 200$$
$$800 = 20x$$
$$40 = x$$

Thus, Richard must sell 40 model trains to make a profit of $600. ∎

Example 10 MODELING—*A Mixture Problem*

Shandra, a pharmacist, needs 500 milliliters $(m\ell)$ of a 10% phenobarbital solution. She has only a 5% phenobarbital solution and a 25% phenobarbital solution available. How many milliliters of each solution should she mix to obtain the desired solution?

Solution First we set up a system of equations. The unknown quantities are the amount of the 5% solution and the amount of the 25% solution that must be used. Let

$$x = \text{number of m}\ell \text{ of 5\% solution}$$
$$y = \text{number of m}\ell \text{ of 25\% solution}$$

We know that 500 mℓ of solution are needed. Thus,

$$x + y = 500$$

The total amount of phenobarbital in a solution is determined by multiplying the percent of phenobarbital by the number of milliliters of solution. The second equation comes from the fact that

$$\begin{pmatrix} \text{Total amount of} \\ \text{phenobarbital in} \\ \text{5\% solution} \end{pmatrix} + \begin{pmatrix} \text{total amount of} \\ \text{phenobarbital in} \\ \text{25\% solution} \end{pmatrix} = \begin{pmatrix} \text{total amount of} \\ \text{phenobarbital} \\ \text{in 10\% mixture} \end{pmatrix}$$

$$0.05x \qquad + \qquad 0.25y \qquad = \qquad 0.10(500)$$

$$\text{or} \qquad 0.05x + 0.25y = 50$$

The system of equations is

$$x + y = 500$$
$$0.05x + 0.25y = 50$$

Let's solve this system of equations by using the addition method. There are various ways of eliminating one variable. To obtain integer values in the second equation, we can multiply both sides of the equation by 100. The result will be an x-term of $5x$. If we multiply both sides of the first equation by -5, that will result in an x-term of $-5x$. By following this process, we can eliminate the x-terms from the system.

$$-5\left[x + y = 500\right] \qquad \text{gives} \qquad -5x - 5y = -2500$$
$$100\left[0.05x + 0.25y = 50\right] \qquad \text{gives} \qquad 5x + 25y = 5000$$

$$\begin{aligned} -5x - 5y &= -2500 \\ \underline{5x + 25y} &= \underline{5000} \\ 20y &= 2500 \end{aligned}$$

$$\frac{20y}{20} = \frac{2500}{20}$$

$$y = 125$$

Now we determine x.

$$x + y = 500$$
$$x + 125 = 500$$
$$x = 375$$

Therefore, 375 mℓ of a 5% phenobarbital solution must be mixed with 125 mℓ of a 25% phenobarbital solution to obtain 500 mℓ of a 10% phenobarbital solution. ■

Example 10 can also be solved by using substitution.

SECTION 6.7
Exercises

Warm Up Exercises

In Exercises 1–10, fill in the blank with an appropriate word, phrase, or symbol(s).

1. When two or more linear equations are considered simultaneously, the equations are called a(n) _____ of linear equations.

2. The ordered pair or ordered pairs that satisfy all equations in a system of equations is called the _____ to the system of equations.

3. A system of equations that has no solution is called a(n) _____ system of equations.

4. A system of equations that has at least one solution is called a(n) _____ system of equations.

5. A system of equations that has an infinite number of solutions is called a(n) _____ system of equations.

6. If the graphs of the equations in a system of linear equations are parallel lines, then the system has _____ solution(s).

7. If the graphs of the equations in a system of linear equations intersect at only one point, then the system has _____ solution(s).

8. If the graphs of the equations in a system of linear equations are the same line, then the system has a(n) _____ number of solutions.

9. When solving a system of linear equations, if you obtain a true statement such as $0 = 0$ or $6 = 6$, the system is dependent and has a(n) _____ number of solutions.

10. When solving a system of linear equations, if you obtain a false statement such as $4 = 0$ or $-2 = 0$, the system is inconsistent and has _____ solution.

Practice the Skills

In Exercises 11 and 12, determine which ordered pairs are solutions to the given system.

11. $y = 4x - 2$ $(2, 6)$ $(3, -10)$ $(1, 7)$
     $y = -x + 8$

12. $x + 2y = 6$ $(-2, 4)$ $(2, 2)$ $(3, -9)$
     $x - y = -6$

In Exercises 13–22, solve the system of equations graphically. If the system does not have a single ordered pair as a solution, state whether the system is inconsistent or dependent.

13. $y = \dfrac{2}{3}x + 3$

     $y = -5x + 3$

14. $y = 6x - 2$

     $y = -\dfrac{1}{3}x - 2$

15. $y = -x + 3$

     $y = x - 1$

16. $y = -x + 2$

     $y = -3x + 4$

17. $x + 2y = 4$

     $3x + y = -3$

18. $x + 3y = -3$

     $2x + y = 4$

19. $y = -\dfrac{1}{2}x + 1$

     $x + 2y = 4$

20. $y = \dfrac{1}{2}x + 2$

     $x - 2y = 6$

21. $y = \dfrac{2}{3}x + 1$

     $2x - 3y = -3$

22. $y = -\dfrac{3}{2}x + 2$

     $3x + 2y = 4$

In Exercises 23–34, solve the system of equations by the substitution method. If the system does not have a single ordered pair as a solution, state whether the system is inconsistent or dependent.

23. $y = x + 9$
     $y = -x + 11$

24. $y = 2x - 1$
     $y = 3x - 4$

25. $6x + 3y = 3$
     $x - 3y = 4$

26. $3x + 4y = -2$
     $-5x + y = 11$

27. $y - x = 4$
     $x - y = 3$

28. $x + y = 3$
     $y + x = 5$

29. $y = -\dfrac{2}{3}x + 3$

     $x + 3y = 6$

30. $5x + 3y = -5$

     $x = \dfrac{3}{5}y + 5$

31. $x - y = -2$
     $y = 3x - 4$

32. $x - y = -1$
     $y = 2x + 4$

33. $3x + y = 15$
     $x = -\dfrac{1}{3}y + 5$

34. $2x + y = 12$
     $x = -\dfrac{1}{2}y + 6$

In Exercises 35–48, solve the system of equations by the addition method. If the system does not have a unique ordered pair as a solution, state whether the system is inconsistent or dependent.

35. $x + y = 4$
     $3x - y = 8$

36. $x - 3y = -1$
     $5x + 3y = 13$

37. $4x + 3y = -1$
     $2x - y = -13$

38. $x + 2y = 15$
     $3x + y = 5$

39. $2x + 5y = 6$
     $4x + 3y = -2$

40. $3x - 2y = 9$
     $9x - 5y = 24$

41. $5x - 2y = 16$
     $4x + 3y = -1$

42. $2x + 3y = 6$
     $5x - 4y = -8$

43. $3x - 2y = 7$
     $4y - 6x = -14$

44. $2x - 7y = 17$
     $21y - 6x = -51$

45. $2x - 7y = 1$
     $4x - 14y = 3$

46. $4x + y = 6$
     $-8x - 2y = 13$

47. $3x + 2y = 8$
     $5x - 3y = 2$

48. $6x + 6y = 1$
     $4x + 9y = 4$

Problem Solving

In Exercises 49–54, part of the question involves determining a system of equations that models the situation.

49. MODELING—*Truck Rentals* The cost of renting a medium-sized truck at U-Haul Rental is $30 per day plus $0.79 a mile. The cost of renting a similar truck at Discount Rentals is $24 per day plus $0.85 a mile.

  a) Write a system of equations, with one equation representing the total cost of renting a truck from U-Haul for a day and the other equation representing the total cost of renting a truck from Discount Rentals for a day.

  b) Graph both equations for up to and including 200 miles on the same axes.

  c) Use the graph to estimate the number of miles that would need to be driven in a day for the cost of renting a truck from U-Haul to equal the cost of renting a truck from Discount Rentals.

50. MODELING—*Landscaping Costs* Tom's Tree and Landscape Service charges $200 for a consultation fee plus $60 per hour for labor, and Lawn Perfect Landscape Service charges $305 for a consultation fee plus $25 per hour for labor.

  a) Write the system of equations to represent the cost of the two landscaping services.

  b) Graph both equations for up to and including 10 hours on the same axes.

  c) Use the graph to estimate the number of hours of landscaping that must be used for both services to have the same cost.

51. MODELING—*Selling Backpacks* Benjamin's Backpacks can sell backpacks for $25 per backpack. The costs for making the backpacks are a fixed cost of $400 and a production cost of $15 per backpack (see Example 9 for an example of cost and revenue equations).

  a) Write the cost and revenue equations.

  b) Graph both equations, for up to and including 50 backpacks, on the same axes.

  c) Use the graph to estimate the number of backpacks Benjamin's Backpacks must sell to break even.

  d) Write the profit formula.

  e) Use the profit formula to determine whether Benjamin's Backpacks makes a profit or loss if it sells 30 backpacks. What is the profit or loss?

  f) How many backpacks must Benjamin's Backpacks sell to make a profit of $1000?

52. MODELING—*Manufacturing Blu-ray Disc Players* A manufacturer sells Blu-ray Disc players for $150 per unit. Manufacturing costs consist of a fixed cost of $3850 and a production cost of $115 per unit.

  a) Write the cost and revenue equations.

  b) Graph both equations for up to and including 150 units on the same axes.

  c) Use the graph to estimate the number of units the manufacturer must sell to break even.

  d) Write the profit formula.

  e) Use the profit formula to determine the manufacturer's profit or loss if 100 units are sold.

  f) How many units must the manufacturer sell to make a profit of $875?

53. MODELING—*Selling Earbuds* SoundsGreat manufactures and sells earbuds. They sell their earbuds for $40 per unit. The manufacturing costs consist of a fixed cost of $4050 and a production cost of $15 per unit.

  a) Write the cost and revenue equations.

  b) Graph both equations for up to and including 200 units on the same axes.

  c) Use the graph to estimate the number of units SoundsGreat must sell to break even.

  d) Write the profit formula.

  e) Use the profit formula to determine whether SoundsGreat makes a profit or loss if 155 units are sold.

  f) How many units must SoundsGreat sell in order to make a profit of $575?

54. MODELING—*Manufacturing Television Stands* A manufacturer sells television stands for $125 per stand. The manufacturing costs consist of a fixed cost of $6750 and a production cost of $50 per stand.

a) Write the cost and revenue equations.

b) Graph both equations for up to and including 150 stands on the same axes.

c) Use the graph to estimate the number of stands the manufacturer must sell to break even.

d) Write the profit formula.

e) Use the profit formula to determine the manufacturer's profit or loss if 80 units are sold.

f) How many stands must the manufacturer sell in order to make a profit of $3750?

In Exercises 55–62, write a system of equations that can be used to solve the problem. Then solve the system and determine the answer.

55. MODELING—*Shares of Stock* Jaafar owns a total of 50 shares of stock, some in Under Armour and some in Hershey Company. On a particular day, Under Armour stock closed at $73 per share and Hershey Company stock closed at $85 per share. If Jaafar's stock for these companies was worth $4070, how many shares of each type of stock does he own?

56. MODELING—*Landscaping* The Garden Factory purchased 4 tons of topsoil and 3 tons of mulch for $1529. The next week the company purchased 2 tons of topsoil and 5 tons of mulch for $1405. Determine the price per ton for topsoil and the price per ton for mulch.

57. MODELING—*Chemical Mixture* Antonio is a chemist and needs 15 liters (ℓ) of a 25% hydrochloric acid solution. He discovers he is out of the 25% hydrochloric acid solution. He checks his supply shelf and finds he has a large supply of both 40% and 10% hydrochloric acid solutions. He decides to use the 40% and 10% solutions to make 15 ℓ of a 25% solution. How many liters of the 40% solution and of the 10% solution should he mix?

58. MODELING—*Nut and Pretzel Mix* Dave wants to purchase 20 pounds of party mix for a total of $30. To obtain the

mixture, he will mix nuts that cost $3 per pound with pretzels that cost $1 per pound. How many pounds of each should he use?

59. MODELING—*Sales Position* Nathan, a luggage salesperson, is considering two job offers. At Harbor Sales, Nathan's salary would be $249 per week plus a 15% commission of sales. At The Pampered Traveler, his salary would be $300 per week plus a 12% commission of sales.

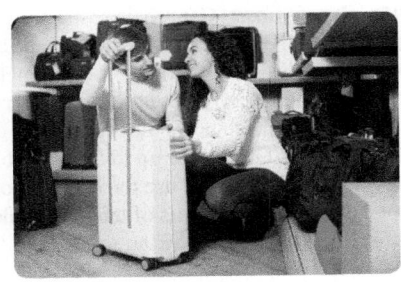

a) What weekly dollar volume of sales would Nathan need to make for the total income from both companies to be the same?

b) If he expects to makes weekly sales of $2500, which company would give the greater salary?

60. MODELING—*Copy Service* Jayne recently purchased a high-speed copier for her home office and wants to purchase a service contract on the copier. She is considering two sources for the service contract. Ace Office charges $18 a month plus 2 cents per copy. Copy Haven charges $24 a month plus 1.5 cents per copy.

a) How many copies would Jayne need to make for the monthly costs of both plans to be the same?

b) If Jayne plans to make 2000 copies a month, which plan would be the least expensive?

61. MODELING—*Hardwood Floor Installation* The cost to purchase a particular type of hardwood flooring at Home Depot is $2.65 per square foot. In addition, the installation cost is $468.75. The cost to purchase the same flooring at Hardwood Guys is $3.10 per square foot. In addition, the installation cost is $412.50.

a) Determine the number of square feet of this flooring that Roberto must purchase for the total cost of the flooring and installation to be the same from both stores.

b) If Roberto needs to purchase and have installed 196 square feet of this flooring, which store would be less expensive?

62. MODELING—*Movie Society Membership* Delbert is choosing between two membership options at the local movie society. A *Director* membership costs $250 per year, and it costs $3.75 each time Delbert sees a movie. A *Producer* membership costs $325 per year, and it costs $2.50 each time Delbert sees a movie.

a) How many movies would Delbert have to see in one year for the two options to cost the same amount?

b) If Delbert plans to see 100 movies in a year, which membership would be the least expensive?

63. MODELING—*Repair Costs* Car repair costs at Steve's Garage can be modeled by the equation $C = 200 + 50x$, and the repair costs at Greg's Garage can be modeled by the equation $C = 375 + 25x$, where x represents the number of hours of labor. The graph below shows the repair costs at Steve's Garage and at Greg's Garage for up to 5 hours of labor. Assuming this trend continues, use the substitution method to determine the number of hours of labor it would take for the total cost at each garage to be the same.

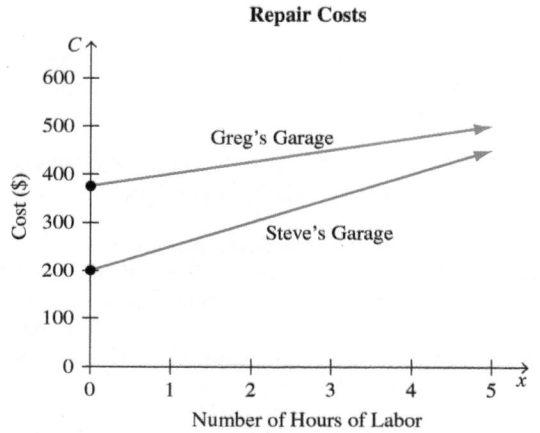

Repair Costs

64. *Salaries* Jose's salary can be modeled by the equation $y = 39,000 + 1200t$, and Charlie's salary can be modeled by the equation $y = 45,000 + 700t$, where t represents the number of years since 2020. The graph below shows Jose's salary and Charlie's salary for up to 10 years after 2020. Assuming this trend continues, use the substitution method to determine when Jose's salary will equal Charlie's salary.

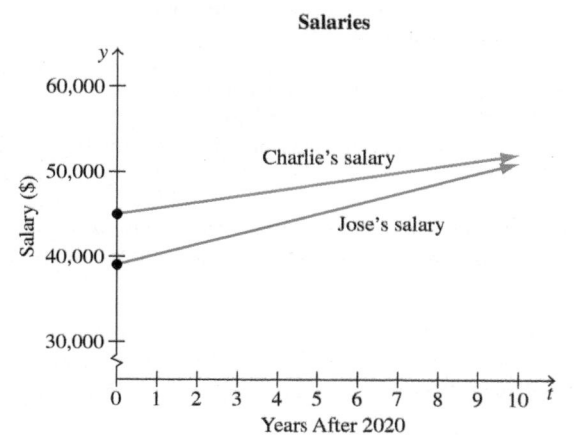

Salaries

Concept/Writing Exercises

65. a) If the two lines in a system of equations have different slopes, how many solutions will the system have? Explain.

b) If the two lines in a system of equations have the same slope but different y-intercepts, how many solutions will the system have? Explain.

c) If the two lines in a system of equations have the same slope and the same y-intercept, how many solutions will the system have? Explain.

66. Indicate whether the graph shown represents a consistent, inconsistent, or dependent system.

a)

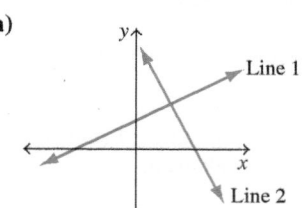

b)

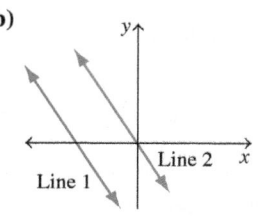

c)

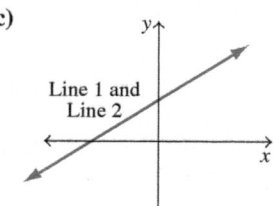

Challenge Problems/Group Activities

67. Solve the following system of equations for u and v by first substituting x for $\frac{1}{u}$ and y for $\frac{1}{v}$.

$$\frac{1}{u} + \frac{2}{v} = 8$$

$$\frac{3}{u} - \frac{1}{v} = 3$$

68. Develop a system of linear equations that has $(6, 5)$ as its solution. Explain how you developed your system of equations.

69. The substitution or addition methods can also be used to solve a system of three linear equations in three variables. Consider the following system.

$$x + y + z = 7$$
$$x - y + 2z = 9$$
$$-x + 2y + z = 4$$

The *ordered triple* (x, y, z) is the solution to the system if it satisfies all three equations.

a) Show that the ordered triple $(2, 1, 4)$ is a solution to the system.

b) Use the substitution or addition method to determine the solution to the system. (*Hint:* Eliminate one variable by using two equations. Then eliminate the same variable by using two different equations.)

70. Construct a system of two linear equations that has no solution. Explain how you know the system has no solution.

71. Construct a system of two linear equations that has an infinite number of solutions. Explain how you know the system has an infinite number of solutions.

72. When solving a system of linear equations by the substitution method, a student obtained the equation $0 = 0$ and gave the solution as $(0, 0)$. What is the student's error?

73. In parts (a)–(d), make up a system of linear equations whose solution will be the ordered pair given. (*Hint:* It may be helpful to visualize possible graphs that have the given solution. There are many possible answers for each part.)

 a) $(0, 0)$ b) $(1, 0)$ c) $(0, 1)$ d) $(1, 1)$

Research Activity

74. ***The Rhind Papyrus*** The Rhind Papyrus indicates that the early Egyptians used linear equations. Do research and write a paper on the symbols used in linear equations and the use of the linear equations by the early Egyptians.

SECTION 6.8

Linear Inequalities in Two Variables and Systems of Linear Inequalities

Upon completion of this section, you will be able to:

- Graph linear inequalities in two variables.
- Graph systems of linear inequalities.
- Solve linear programming problems.

A manufacturer of gas grills needs to determine how many grills to ship to each of two stores. The manufacturer can ship at most 300 grills in total. A linear inequality in two variables can be used to represent the number of grills that can be shipped to each store. In this section, we will discuss how to set up and solve linear inequalities in two variables.

Why This Is Important Linear inequalities can be used to model problems similar to the one described above and many others that we encounter daily. Furthermore, inequalities are used in a branch of mathematics called linear programming, which can be used in maximizing a company's revenue and profits or in minimizing the company's costs. Linear programming is used in many occupations.

Graphing Linear Inequalities in Two Variables

In Section 6.5, we introduced linear inequalities in one variable. Now we will introduce linear inequalities in two variables. Some examples of linear inequalities in two variables are $2x + 3y \leq 7$, $x + 7y \geq 5$, and $x - 3y < 6$.

The solution set of a linear inequality in one variable may be indicated on a number line. The solution set of a linear inequality in two variables is indicated on a coordinate plane.

An inequality that is strictly less than $(<)$ or greater than $(>)$ will have as its solution set a *half-plane*. A half-plane is the set of all the points on one side of a line. An inequality that is less than or equal to $(\leq)$ or greater than or equal to $(\geq)$ will have as its solution set the set of points that consists of a half-plane and a line. To indicate that the line is part of the solution set, we draw a solid line. To indicate that the line is not part of the solution set, we draw a dashed line.

Graphing Linear Inequalities
with 2 Variables

> ## PROCEDURE   GRAPHING LINEAR INEQUALITIES IN TWO VARIABLES
>
> 1. Mentally substitute the equal sign for the inequality sign and plot points as if you were graphing the equation.
> 2. If the inequality is $<$ or $>$, draw a dashed line through the points. If the inequality is $\leq$ or $\geq$, draw a solid line through the points.
> 3. Select a test point not on the line and substitute the x- and y-coordinates into the inequality. If the substitution results in a true statement, shade in the area on the same side of the line as the test point. If the test point results in a false statement, shade in the area on the opposite side of the line as the test point.

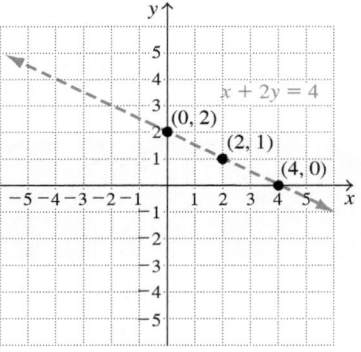

Figure 6.30

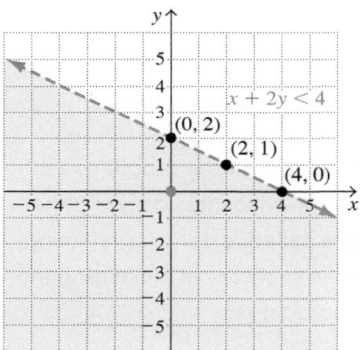

Figure 6.31

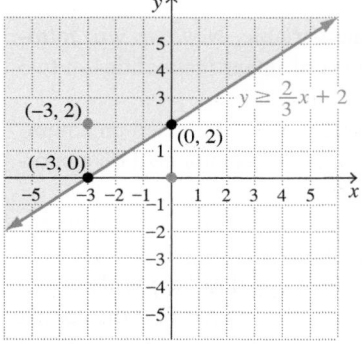

Figure 6.32

Example 1 *Graphing an Inequality*

Graph the inequality $x + 2y < 4$.

Solution   To obtain the solution set, start by graphing $x + 2y = 4$. Since the original inequality is strictly "less than," draw a dashed line (Fig. 6.30). The dashed line indicates that the points on the line are not part of the solution set.

The line $x + 2y = 4$ divides the plane into three parts, the line itself and two *half-planes*. The line is the boundary between the two half-planes. The points in only one of the two half-planes will satisfy the inequality $x + 2y < 4$. The points in the other half-plane will satisfy the inequality $x + 2y > 4$.

To determine the solution set of the inequality $x + 2y < 4$, pick any point on the plane that is not on the line. The simplest point to work with is the origin, $(0, 0)$. Substitute $x = 0$ and $y = 0$ into $x + 2y < 4$.

$$x + 2y < 4$$
$$\text{Is } 0 + 2(0) < 4?$$
$$0 + 0 < 4$$
$$0 < 4 \quad \text{True}$$

Since 0 is less than 4, the point $(0, 0)$ is part of the solution set. All the points on the same side of the graph of $x + 2y = 4$ as the point $(0, 0)$ are members of the solution set. We indicate that by shading the half-plane that contains $(0, 0)$. The graph is shown in Fig. 6.31.   ∎

Example 2 *Graphing an Inequality*

Graph the inequality $y \geq \dfrac{2}{3}x + 2$.

Solution   First draw the graph of the equation $y = \frac{2}{3}x + 2$. Use a solid line because the inequality is greater than or equal to and the points on the boundary line are included in the solution set (see Fig. 6.32). Now pick a point that is not on the line. Take $(0, 0)$ as the test point.

$$y \geq \frac{2}{3}x + 2$$
$$\text{Is } 0 \geq \frac{2}{3}(0) + 2?$$
$$0 \geq 2 \quad \text{False}$$

Since 0 is not greater than or equal to 2 $(0 \not\geq 2)$, the solution set is the line and the half-plane that does not contain the point $(0, 0)$. This area is shaded in Fig. 6.32.

If you had arbitrarily selected the test point $(-3, 2)$ from the other half-plane, you would have determined that the inequality would be true: $2 \geq \frac{2}{3}(-3) + 2$, or $2 \geq 0$. Because 2 is greater than or equal to 0, the point $(-3, 2)$ would be in the half-plane containing the solution set. ∎

Example 3 *Graphing an Inequality*

Graph the inequality $y < x$.

Solution The inequality is strictly "less than," so the boundary line is not part of the solution set. In graphing the equation $y = x$, draw a dashed line (Fig. 6.33). Since $(0, 0)$ is *on* the line, it cannot serve as a test point. Let's pick the point $(1, -1)$.

$$y < x$$
$$-1 < 1 \quad \text{True}$$

Since $-1 < 1$ is true, the solution set is the half-plane containing the point $(1, -1)$. The solution set is indicated in Fig. 6.33. ∎

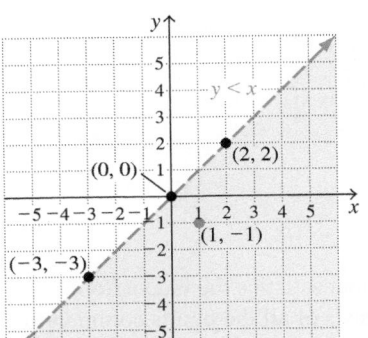

Figure 6.33

Graphing Systems of Linear Inequalities

When two or more linear inequalities are considered simultaneously, the inequalities are called a *system of linear inequalities*. The solution set of a system of linear inequalities is the set of points that satisfy all inequalities in the system. The solution set of a system of linear inequalities may consist of infinitely many ordered pairs. To determine the solution set to a system of linear inequalities, graph each inequality on the same axes. The ordered pair solutions common to all the inequalities are the solution set to the system.

PROCEDURE   GRAPHING A SYSTEM OF LINEAR INEQUALITIES

1. Select one of the inequalities. Replace the inequality symbol with an equal sign and draw the graph of the equation. Draw the graph with a dashed line if the inequality is $<$ or $>$ and with a solid line if the inequality is $\leq$ or $\geq$.

2. Select a test point on one side of the line and determine whether the point is a solution to the inequality. If so, shade the area on the side of the line containing the point. If the point is not a solution, shade the area on the other side of the line.

3. Repeat Steps 1 and 2 for the other inequality.

4. The intersection of the two shaded areas and any solid line common to both inequalities form the solution set to the system of inequalities.

Example 4 *Graphing a System of Inequalities*

Graph the following system of inequalities and indicate the solution set.

$$y < -x + 1$$
$$x - y < 5$$

Solution Graph both inequalities on the same axes. First draw the graph of $y < -x + 1$. When drawing the graph, remember to use a dashed line, since the inequality is "less than" (see Fig. 6.34a on page 358). Shade the half-plane that satisfies the inequality $y < -x + 1$.

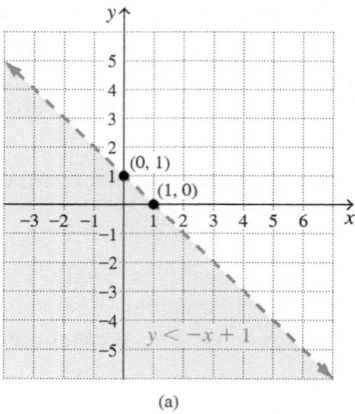

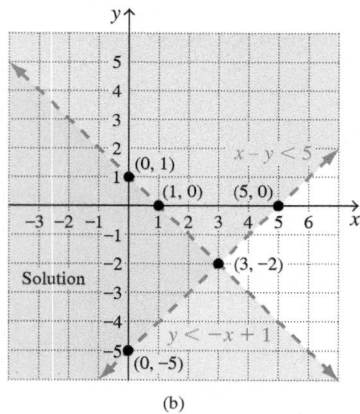

(a)   (b)

Figure 6.34

Now, on the same axes, shade the half-plane that satisfies the inequality $x - y < 5$ (see Fig. 6.34b). The solution set consists of all the points common to the two shaded half-planes. These are the points in the region on the graph containing both color shadings. In Figure 6.34(b), we have indicated this region in green. Figure 6.34(b) shows that the two lines intersect at $(3, -2)$. This ordered pair can also be determined by any of the algebraic methods discussed in Section 6.7. The ordered pair $(3, -2)$ is not part of the solution set.   ∎

Example 5 *Graphing a System of Linear Inequalities*

Graph the following system of inequalities and indicate the solution set.

$$4x - 2y \geq 8$$
$$2x + 3y < 6$$

Solution   Graph the inequality $4x - 2y \geq 8$. Remember to use a solid line because the inequality is "greater than or equal to"; see Fig. 6.35(a).

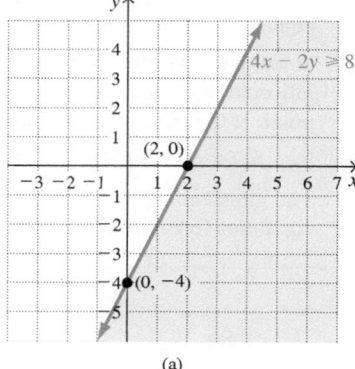

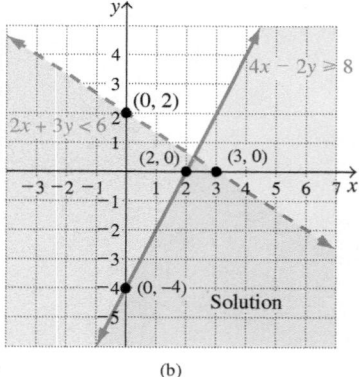

(a)   (b)

Figure 6.35

On the same set of axes, draw the graph of $2x + 3y < 6$. Use a dashed line since the inequality is "less than"; see Fig. 6.35(b). The solution is the region of the graph that contains both color shadings and the part of the solid blue line that satisfies the inequality $2x + 3y < 6$. Note that the point of intersection of the two lines is not a part of the solution set.   ∎

Example 6 *Another System of Inequalities*

Graph the following system of inequalities and indicate the solution set.

$$x \geq -2$$
$$y < 3$$

Solution Graph the inequality $x \geq -2$; see Fig. 6.36(a). On the same axes, graph the inequality $y < 3$; see Fig. 6.36(b). The solution set is that region of the graph that is shaded in both colors and the part of the solid blue line that satisfies the inequality $y < 3$. The point of intersection of the two lines, $(-2, 3)$, is not part of the solution because it does not satisfy the inequality $y < 3$.

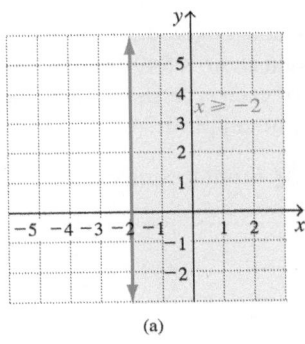
(a)

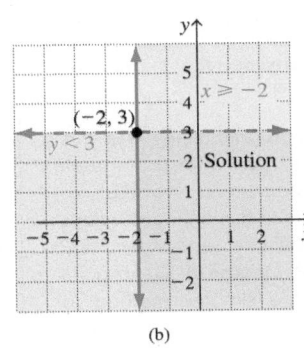
(b)

Figure 6.36

Linear Programming

Linear programming provides businesses and governments with a mathematical tool for determining the most efficient use of resources. Linear programming is used to solve problems in many areas of business, science, medicine, and the military by using a technique called the *simplex method*, which was developed in the 1940's by George Dantzig.

In a linear programming problem, restrictions called *constraints* are represented with a system of linear inequalities. When the system is graphed, a shaded region (Fig. 6.37), called the *feasible region*, is obtained.

For each linear programming problem, we will obtain a formula of the form $K = Ax + By$, called the *objective function*. The objective function represents a model for the quantity, in this case K, that we want to maximize or minimize. For example, a typical objective function that might be used to maximize a company's profit, P, is $P = 3x + 7y$. We determine the maximum profit by substituting the ordered pairs (x, y) of the vertices of the feasible region into the objective function to see which ordered pair yields the maximum profit.

The fundamental principle of linear programming provides a rule for determining the maximum or minimum value of an objective function used in a linear programming problem.

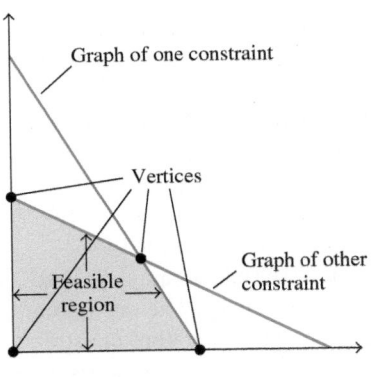

Figure 6.37

> ### Fundamental Principle of Linear Programming
> If the objective function, $K = Ax + By$, is evaluated at each point in a feasible region, the maximum and minimum values of the equation occur at vertices of the region.

Linear programming is a powerful tool for determining the maximum and minimum values of an objective function. Using the fundamental principle of linear

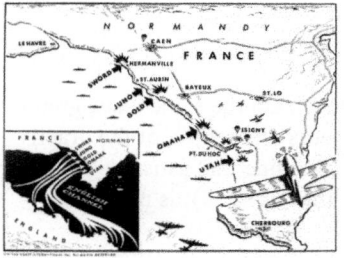
programming, we are quickly able to determine the maximum and minimum values of an objective function by using just a few of the infinitely many points in the feasible region.

Example 7 illustrates how the fundamental principle of linear programming is used to solve a linear programming problem.

Example 7 MODELING—*Using the Fundamental Principle of Linear Programming*

The Down Home Chair company makes two types of rocking chairs, a plain chair and a fancy chair. Each rocking chair must be assembled and then finished. The plain chair takes 4 hours to assemble and 4 hours to finish. The fancy chair takes 8 hours to assemble and 12 hours to finish. The company can provide at most 160 worker-hours of assembling and 180 worker-hours of finishing a day. If the profit on a plain chair is $40 and the profit on a fancy chair is $65, how many rocking chairs of each type should the company make per day to maximize profits? What is the maximum profit?

Solution From the information given, we know the following facts.

	Assembly Time (hr)	Finishing Time (hr)	Profit ($)
Plain chair	4	4	40.00
Fancy chair	8	12	65.00

Let

$$x = \text{the number of plain chairs made per day}$$
$$y = \text{the number of fancy chairs made per day}$$
$$40x = \text{profit on the plain chairs}$$
$$65y = \text{profit on the fancy chairs}$$
$$P = \text{the total profit}$$

The total profit is the sum of the profit on the plain chairs and the profit on the fancy chairs. Since $40x$ is the profit on the plain chairs and $65y$ is the profit on the fancy chairs, the profit formula is $P = 40x + 65y$.

The maximum profit, P, is dependent on several conditions, or *constraints*. The number of chairs manufactured each day cannot be a negative amount. This condition gives us the constraints $x \geq 0$ and $y \geq 0$. Another constraint is determined by the total number of hours allocated for assembling. Four hours are required to assemble the plain chair, so the total number of hours needed to assemble x plain chairs is $4x$. Eight hours are required to assemble a fancy chair, so the total number of hours needed to assemble y fancy chairs is $8y$. The maximum number of hours allocated for assembling is 160 per day. Thus, the third constraint is $4x + 8y \leq 160$. The final constraint is determined by the number of hours allotted for finishing. Finishing a plain chair takes 4 hours, or $4x$ hours to finish x plain chairs. Finishing a fancy chair takes 12 hours, or $12y$ hours to finish y fancy chairs. The total number of hours allotted for finishing is 180 per day. Therefore, the fourth constraint is $4x + 12y \leq 180$. Thus, the four constraints are

$$x \geq 0$$
$$y \geq 0$$
$$4x + 8y \leq 160$$
$$4x + 12y \leq 180$$

The list of constraints is a system of linear inequalities in two variables. The solution to the system of inequalities is the set of ordered pairs that satisfy all the constraints. The constraints and the vertices of the feasible region are indicated in Fig. 6.38. Note that the solution to the system consists of the shaded region and the solid boundaries. The vertices at $(0, 0)$, $(0, 15)$, $(30, 5)$, and $(40, 0)$ are the points at which the boundaries intersect. These points can also be determined by the addition or substitution method described in Section 6.7.

The goal in this example is to maximize the profit. The objective function is given by the profit formula $P = 40x + 65y$. According to the fundamental principle, the maximum profit will be determined at one of the vertices of the feasible region. Calculate P for each one of the vertices.

$$P = 40x + 65y$$

$$\text{At } (0, 0), \quad P = 40(0) + 65(0) = 0$$

$$\text{At } (0, 15), \quad P = 40(0) + 65(15) = 975$$

$$\text{At } (30, 5), \quad P = 40(30) + 65(5) = 1525$$

$$\text{At } (40, 0), \quad P = 40(40) + 65(0) = 1600$$

The maximum profit is at $(40, 0)$, which means that the company should manufacture 40 plain rocking chairs and no fancy rocking chairs. The maximum profit would be $1600. The minimum profit would be at $(0, 0)$, when no rocking chairs of either style were manufactured. ∎

Example 8   MODELING—*Washers and Dryers, Maximizing Profit*

The Admiral Appliance Company makes washers and dryers. The company must manufacture at least one washer per day to ship to one of its customers. No more than 6 washers can be manufactured due to production restrictions. The number of dryers manufactured cannot exceed 7 per day. Also, the number of washers manufactured cannot exceed the number of dryers manufactured per day. If the profit on each washer is $20 and the profit on each dryer is $30, how many of each appliance should the company make per day to maximize profits? What is the maximum profit?

Solution Let

$$x = \text{the number of washers manufactured per day}$$

$$y = \text{the number of dryers manufactured per day}$$

$$20x = \text{the profit on washers}$$

$$30y = \text{the profit on dryers}$$

$$P = \text{the total profit}$$

The maximum profit is dependent on several constraints. The number of appliances manufactured each day cannot be a negative amount. This condition gives us the constraints $x \geq 0$ and $y \geq 0$. The company must manufacture at least one washer per day; therefore, $x \geq 1$. No more than 6 washers can be manufactured per day; therefore, $x \leq 6$. No more than 7 dryers can be manufactured per day; therefore, $y \leq 7$. The number of washers cannot exceed the number of dryers manufactured per day; therefore, $x \leq y$. Thus, the six constraints are

$$x \geq 0, \; y \geq 0, \; x \geq 1, \; x \leq 6, \; y \leq 7, \; x \leq y$$

In this example, the objective function is the profit function. Since $20x$ is the profit on x washers and $30y$ is the profit on y dryers, the profit function is

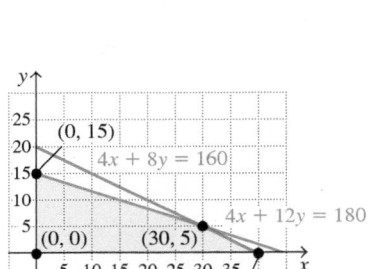

Figure 6.38

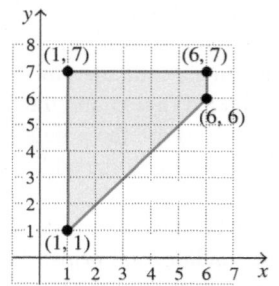

Figure 6.39

$P = 20x + 30y$. Figure 6.39 shows the feasible region. The feasible region consists of the shaded region and the boundaries. The vertices of the feasible region are at $(1, 1)$, $(1, 7)$, $(6, 7)$, and $(6, 6)$.

Next we calculate the value of the objective function, P, at each one of the vertices.

$$P = 20x + 30y$$

At $(1, 1)$, $P = 20(1) + 30(1) = 50$

At $(1, 7)$, $P = 20(1) + 30(7) = 230$

At $(6, 7)$, $P = 20(6) + 30(7) = 330$

At $(6, 6)$, $P = 20(6) + 30(6) = 300$

The maximum profit is at $(6, 7)$. Therefore, the company should manufacture 6 washers and 7 dryers to maximize its profit. The maximum profit is $330.   ∎

Use the following steps to solve a linear programming problem.

PROCEDURE   **SOLVING A LINEAR PROGRAMMING PROBLEM**

1. Determine all necessary constraints.

2. Determine the objective function.

3. Graph the constraints and determine the feasible region.

4. Determine the vertices of the feasible region.

5. Determine the value of the objective function at each vertex.

The solution is the maximum or minimum value obtained when the coordinates of the vertices are substituted into the objective function.

SECTION 6.8
Exercises

Warm Up Exercises

In Exercises 1–10, fill in the blank with an appropriate word, phrase, or symbol(s).

1. The set of all points in a plane on one side of a line is called a(n) _____.

2. When graphing an inequality in two variables, if the inequality contains either $<$ or $>$ use a(n) _____ line.

3. When graphing an inequality in two variables, if the inequality contains either $\leq$ or $\geq$ use a(n) _____ line.

4. The set of points that satisfy all inequalities in a system of linear inequalities is called the _____ set.

5. The solution set to a system of linear inequalities is the set of all _____ pairs that are common to all the inequalities in the system.

6. In a linear programming problem the restrictions that are represented by linear inequalities are called the _____ .

7. In a linear programming problem, the shaded region formed by graphing the system of inequalities is called the _____ region.

8. The points of intersection of the boundaries of the feasible region are called _____ .

9. In a linear programming problem, the function used for the quantity that we want to maximize or minimize is called the _____ function.

10. The maximum and minimum values of the objective function occur at the _____ of the feasible region.

Practice the Skills

11. Determine whether $(0, 0)$ is a solution to the given inequality.

   a) $3x - 4y > 12$ b) $2x + 3y \leq 6$

   c) $y \geq \dfrac{2}{3}x - 5$ d) $y < -\dfrac{5}{4}x - 3$

12. Determine whether $(0, 1)$ is a solution to the given inequality.

   a) $x + y \geq 0$ b) $3x - 2y < 0$

   c) $y < \dfrac{2}{3}x$ d) $y \leq -\dfrac{5}{7}x$

In Exercises 13–28, graph the inequality.

13. $y > x + 1$

14. $y < -x - 2$

15. $y \geq 2x - 6$

16. $y < -2x + 2$

17. $2x - 3y > 6$

18. $x + 2y > 4$

19. $y \geq \dfrac{3}{4}x - 2$

20. $y \leq -\dfrac{2}{3}x + 1$

21. $3x + 2y < 6$

22. $-x + 2y < 2$

23. $5x + 2y \geq 10$

24. $3x - 2y < 12$

25. $y \geq -2x + 1$

26. $y \leq 3x - 4$

27. $x + y > 0$

28. $x + 2y \leq 0$

In Exercises 29–38, graph the system of linear inequalities and indicate the solution set.

29. $y > x - 4$
   $x + y < 5$

30. $3x - y \leq 6$
   $y < x - 4$

31. $y \leq -x + 5$
   $x - y < 3$

32. $4x + 2y > 8$
   $y \leq x + 1$

33. $x - 3y \leq 3$
   $x + 2y \geq 4$

34. $x + y < 4$
   $3x + 2y \geq 6$

35. $x + 2y \geq 4$
   $3x - y \geq -6$

36. $3x - 2y \geq 6$
   $x - y \leq 3$

37. $x \geq 1$
   $y \leq 1$

38. $x \leq 0$
   $y \leq 0$

In Exercises 39–42, a feasible region and its vertices are shown. Determine the maximum and minimum values of the given objective function.

39. $K = 6x + 4y$

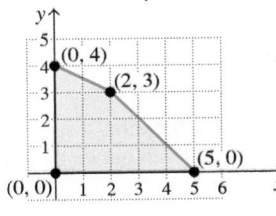

40. $K = 10x + 8y$

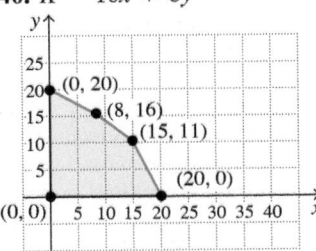

41. $K = 2x + 3y$

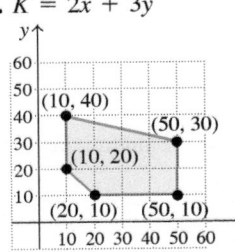

42. $K = 40x + 50y$

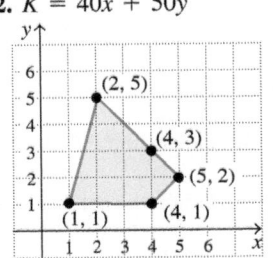

In Exercises 43–48, a set of constraints and a profit function are given.

a) *Graph the constraints and determine the vertices of the feasible region.*

b) *Use the vertices obtained in part (a) to determine the maximum and minimum profit.*

43. $x + y \leq 5$
   $2x + y \leq 8$
   $x \geq 0$
   $y \geq 0$
   $P = 4x + 3y$

44. $x + 2y \leq 6$
   $3x + 2y \leq 12$
   $x \geq 0$
   $y \geq 0$
   $P = 2x + 6y$

45. $x + y \leq 4$
   $x + 3y \leq 6$
   $x \geq 0$
   $y \geq 0$
   $P = 7x + 6y$

46. $x + y \leq 50$
   $x + 3y \leq 90$
   $x \geq 0$
   $y \geq 0$
   $P = 20x + 40y$

47. $2x + 3y \geq 18$
   $4x + 2y \leq 20$
   $x \geq 1$
   $y \geq 4$
   $P = 2.20x + 1.65y$

48. $x + 2y \leq 14$
   $7x + 4y \geq 28$
   $x \geq 2$
   $x \leq 10$
   $y \geq 1$
   $P = 15.13x + 9.35y$

Problem Solving

In Exercises 49 and 50, use a system of linear inequalities.

49. **MODELING—*Special Diet*** Ruben is on a special diet. He must consume fewer than 500 calories at a meal that consists of one serving of chicken and one serving of rice. The meal must contain at least 150 calories from each source.

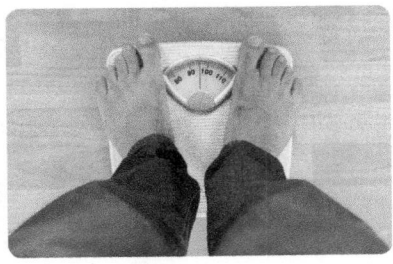

a) Using x to represent the number of calories from chicken and y to represent the number of calories from rice, translate the problem into a system of linear inequalities.

b) Graph the system of linear inequalities and indicate the solution set. Graph calories from chicken on the horizontal axis and calories from rice on the vertical axis.

c) There are about 180 calories in 3 oz of chicken and about 200 calories in 8 oz of rice. Select a point in the solution set. For the point selected, determine the number of ounces of chicken and the number of ounces of rice to be served.

50. MODELING— *Tablet Sales* The bookstore at St. Petersburg College sells Apple tablets and Samsung tablets. Based on demand, it is necessary to stock at least twice as many Apple tablets as Samsung tablets. The costs to the store are $300 for an Apple tablet and $200 for a Samsung tablet. Management wants at least 10 Apple tablets and at least 5 Samsung tablets in inventory at all times and does not want more than $12,000 in inventory at any one time.

a) Using A to represent Apple tablets and S to represent Samsung tablets, translate the problem into a system of linear inequalities.

b) Graph the system of linear inequalities and indicate the solution set. Graph the number of Apple tablets on the horizontal axis and the number of Samsung tablets on the vertical axis

c) Select a point in the solution set and determine the inventory cost for the two models that corresponds to the point.

In Exercises 51–56, solve the linear programming problems.

51. MODELING— *On Wheels* The Boards and Blades Company manufactures skateboards and in-line skates. The company can produce a maximum of 20 skateboards and pairs of in-line skates per day. It makes a profit of $25 on a skateboard and a profit of $20 on a pair of in-line skates. The company's planners want to make at least 3 skateboards but not more than 6 skateboards per day. To keep customers happy, they must make at least 2 pairs of in-line skates per day.

a) List the constraints.

b) Determine the objective function.

c) Graph the set of constraints.

d) Determine the vertices of the feasible region.

e) How many skateboards and pairs of in-line skates should be made to maximize the profit?

f) Determine the maximum profit.

52. MODELING— *Washing Machine Production* A company manufactures two types of washers, top load and front

load. The company can manufacture a maximum of 18 washers per day. It makes a profit of $20 on top load machines and $25 on front load machines. No more than 5 front load machines can be manufactured due to production restrictions. To meet consumer demand, the company must manufacture at least 2 front load machines and 2 top load machines per day.

a) List the constraints.

b) Determine the objective function.

c) Graph the set of constraints.

d) Determine the vertices of the feasible region.

e) How many washing machines of each type should be made to maximize profit?

f) Determine the maximum profit.

53. MODELING— *Paint Production* A paint supplier has two machines that produce both indoor paint and outdoor paint. To meet one of its contractual obligations, the company must produce at least 60 gal of indoor paint and 100 gal of outdoor paint. Machine I makes 3 gal of indoor paint and 10 gal of outdoor paint per hour. Machine II makes 4 gal of indoor paint and 5 gal of outdoor paint per hour. It costs $28 per hour to run machine I and $33 per hour to run machine II.

a) List the constraints.

b) Determine the objective function.

c) Graph the set of constraints.

d) Determine the vertices of the feasible region.

e) How many hours should each machine be operated to fulfill the contract at a minimum cost?

f) Determine the minimum cost.

54. MODELING— *Special Diet* A dietitian prepares a special diet using two food groups, A and B. Each ounce of food group A contains 3 units of vitamin C and 1 unit of vitamin D. Each ounce of food group B contains 1 unit of vitamin C and 2 units of vitamin D. The minimum daily requirements with this diet are at least 9 units of vitamin C and at least 8 units of vitamin D. Each ounce of food group A costs 50 cents, and each ounce of food group B costs 30 cents.

a) List the constraints.

b) Determine the objective function for minimizing cost.

c) Graph the set of constraints.

d) Determine the vertices of the feasible region.

e) How many ounces of each food group should be used to meet the daily requirements and minimize the cost?

f) Determine the minimum cost.

Challenge Problems/Group Activities

55. MODELING—*Hot Dog Profits* To make one pack of all-beef hot dogs, a manufacturer uses 1 lb of beef; to make one pack of regular hot dogs, the manufacturer uses $\frac{1}{2}$ lb each of beef and pork. The profit on the all-beef hot dogs is 40 cents per pack, and the profit on regular hot dogs is 30 cents per pack. If there are 200 lb of beef and 150 lb of pork available, how many packs of all-beef and how many packs of regular hot dogs should the manufacturer make to maximize the profit? What is the profit?

56. MODELING—*Car Seats and Strollers* A company makes car seats and strollers. Each car seat and stroller passes through three processes: assembly, safety testing, and packaging. A car seat requires 1 hr in assembly, 2 hr in safety testing, and 1 hr in packaging. A stroller requires 3 hr in assembly, 1 hr in safety testing, and 1 hr in packaging. Employee work schedules allow for 24 hr per day for assembly, 16 hr per day for safety testing, and 10 hr per day for packaging. The profit for each car seat is $25, and the profit for each stroller is $35. How many units of each type should the company make per day to maximize the profit? What is the maximum profit?

Concept/Writing Exercises

57. a) Do all systems of linear inequalities have solutions? Explain.

b) Write a system of linear inequalities that has no solution.

58. Can a system of linear inequalities have a solution set consisting of a single point? Explain.

59. Which of the following inequalities have the same graph? Explain how you determined your answer.

a) $2x - y < 8$ **b)** $-2x + y > -8$

c) $2x - 4y < 16$ **d)** $y > 2x - 8$

Research Activity

60. *Operations Research* The study of operations research draws on several disciplines, including mathematics, probability theory, statistics, and economics. George B. Dantzig (see the *Profile in Mathematics* on page 359) was one of the key people in developing operations research. Write a paper on Dantzig and his contributions to operations research and linear programming.

SECTION 6.9

Solving Quadratic Equations by Using Factoring and by Using the Quadratic Formula

Upon completion of this section, you will be able to:

- Multiply binomials.
- Factor trinomials of the form $x^2 + bx + c$.
- Factor trinomials of the form $ax^2 + bx + c, a \neq 1$.
- Solve quadratic equations by factoring.
- Solve quadratic equations by using the quadratic formula.
- Use quadratic equations to model application problems.

Pablo and Maya recently installed an in-ground rectangular swimming pool in their backyard. They have enough money budgeted to pay for the installation of a brick border around the pool of uniform width that has an area of 296 square feet. They would like to know the width of this border. This problem can be modeled using a quadratic equation. In this section, we will learn two methods to solve quadratic equations.

Why This Is Important Quadratic equations play an important role in modeling many applications from the fields of engineering, physics, chemistry, statistics, business, building construction, and scientific research. As our society becomes more technologically advanced, many occupations rely on basic mathematical understanding, which includes knowledge of quadratic equations.

Multiply Binomials

Before discussing quadratic equations, we begin by discussing binomials and multiplying binomials.

A *binomial* is an expression that contains two terms in which each exponent that appears on a variable is a whole number.

Examples of binomials

$x + 3$	$x - 5$
$3x + 5$	$4x - 2$

To multiply two binomials, we use the distributive property, which was discussed in Section 5.5. Each term in the first binomial must be multiplied by each term in the second binomial. A convenient way to carry out this process is the *FOIL method*. The acronym FOIL indicates that you multiply the *F*irst terms, *O*uter terms, *I*nner terms, and *L*ast terms of the binomials, in that order.

$$\overset{\text{First}\quad\text{Outer}\quad\text{Inner}\quad\text{Last}}{(a + b)(c + d) = a \cdot c + a \cdot d + b \cdot c + b \cdot d}$$

After multiplying the first, outer, inner, and last terms, combine all like terms.

Timely Tip

When we use FOIL to multiply two binomials, we are really using the distributive property twice.

$(a + b)(c + d)$
 $= a(c + d) + b(c + d)$
 $= ac + ad + bc + bd$

Example 1 *Multiplying Binomials*

Multiply $(x + 4)(x + 7)$.

Solution The FOIL method of multiplication yields

$$\begin{aligned}(x + 4)(x + 7) &= \overset{F}{x \cdot x} + \overset{O}{x \cdot 7} + \overset{I}{4 \cdot x} + \overset{L}{4 \cdot 7} \\ &= x^2 + 7x + 4x + 28 \\ &= x^2 + 11x + 28\end{aligned}$$

Note that the like terms $7x$ and $4x$ were combined to get $11x$. ∎

Example 2 *Multiplying Binomials*

Multiply $(3x - 4)(2x + 7)$.

Solution

$$\begin{aligned}(3x - 4)(2x + 7) &= \overset{F}{3x \cdot 2x} + \overset{O}{3x \cdot 7} + \overset{I}{(-4) \cdot 2x} + \overset{L}{(-4) \cdot 7} \\ &= 6x^2 + 21x - 8x - 28 \\ &= 6x^2 + 13x - 28\end{aligned}$$

∎

Factoring Trinomials of the Form $x^2 + bx + c$

The expression $x^2 + 8x + 15$ is an example of a trinomial. A *trinomial* is an expression containing three terms in which each exponent that appears on a variable is a whole number.

Example 1 showed that

$$(x + 4)(x + 7) = x^2 + 11x + 28$$

Since the product of $x + 4$ and $x + 7$ is $x^2 + 11x + 28$, we say that $x + 4$ and $x + 7$ are *factors* of $x^2 + 11x + 28$. *To factor* an expression means to write the expression as a product of its factors. For example, to factor $x^2 + 11x + 28$ we write

$$x^2 + 11x + 28 = (x + 4)(x + 7)$$

Let's look at the factors more closely.

$$
\begin{array}{c}
\overline{}\; 4 + 7 = 11 \\
\; 4 \cdot 7 = 28 \\
x^2 + 11x + 28 = (x + 4)(x + 7)
\end{array}
$$

Note that the sum of the two numbers in the factors is $4 + 7$, or 11. The 11 is the coefficient of the x-term. Also note that the product of the numbers in the two factors is $4 \cdot 7$, or 28. The 28 is the constant in the trinomial. In general, when factoring an expression of the form $x^2 + bx + c$, we need to determine two numbers whose product is c and whose sum is b. When we determine the two numbers, the factors will be of the form

$$(x + \ \square\ )\, (x + \ \square\ )$$

$$
\begin{array}{cc}
\uparrow & \uparrow \\
\text{One} & \text{Other} \\
\text{number} & \text{number}
\end{array}
$$

Example 3 *Factoring a Trinomial*

Factor $x^2 + 6x + 8$.

Solution We need to determine two numbers whose product is 8 and whose sum is 6. Since the product is $+8$, the two numbers must both be positive or both be negative. Because the coefficient of the x-term is positive, only the positive factors of 8 need to be considered. Can you explain why? We begin by listing the positive numbers whose product is 8.

Factors of 8	Sum of Factors
1(8)	$1 + 8 = 9$
2(4)	$2 + 4 = 6$

Since $2 \cdot 4 = 8$ and $2 + 4 = 6$, the numbers we are seeking are 2 and 4. Thus, we write

$$x^2 + 6x + 8 = (x + 2)(x + 4)$$

Note that $(x + 4)(x + 2)$ is also an acceptable answer. ∎

PROCEDURE FACTORING TRINOMIALS OF THE FORM $x^2 + bx + c$

1. Determine two numbers whose product is c and whose sum is b.
2. Write factors in the form

$$(x + \ \square\ )\, (x + \ \square\ )$$

$$
\begin{array}{cc}
\uparrow & \uparrow \\
\text{One number} & \text{Other number} \\
\text{from Step 1} & \text{from Step 1}
\end{array}
$$

3. Check your answer by multiplying the factors using the FOIL method.

If, for example, the numbers determined in Step 1 of the procedure were 6 and -4, the factors would be written $(x + 6)(x - 4)$.

Example 4 *Factoring a Trinomial*

Factor $x^2 - 6x - 16$.

Solution We must determine two numbers whose product is -16 and whose sum is -6. Begin by listing the factors of -16.

Factors of -16	Sum of Factors
$-16(1)$	$-16 + 1 = -15$
$-8(2)$	$-8 + 2 = -6$
$-4(4)$	$-4 + 4 = 0$
$-2(8)$	$-2 + 8 = 6$
$-1(16)$	$-1 + 16 = 15$

The table lists all the factors of -16. The only factors listed whose product is -16 and whose sum is -6 are -8 and 2. We listed all factors in this example so that you could see, for example, that $-8(2)$ is a different set of factors than $-2(8)$. Once you determine the factors you are looking for, there is no need to go any further. The trinomial can be written in factored form as

$$x^2 - 6x - 16 = (x - 8)(x + 2)$$ ∎

Factoring Trinomials of the Form $ax^2 + bx + c, a \neq 1$

Now we discuss how to factor an expression of the form $ax^2 + bx + c$, where a, the coefficient of the squared term, is not equal to 1.

Consider the multiplication problem $(2x + 1)(x + 3)$.

$$(2x + 1)(x + 3) = 2x \cdot x + 2x \cdot 3 + 1 \cdot x + 1 \cdot 3$$
$$= 2x^2 + 6x + x + 3$$
$$= 2x^2 + 7x + 3$$

Since $(2x + 1)(x + 3) = 2x^2 + 7x + 3$, the factors of $2x^2 + 7x + 3$ are $2x + 1$ and $x + 3$.

Let's study the coefficients more closely.

$$\mathbf{F} = 2 \cdot 1 = 2 \quad \mathbf{O} + \mathbf{I} = (2 \cdot 3) + (1 \cdot 1) = 7 \quad \mathbf{L} = 1 \cdot 3 = 3$$

Note that the product of the coefficients of the first terms in the multiplication of the binomials equals 2, the coefficient of the squared term. The sum of the products of the coefficients of the outer and inner terms equals 7, the coefficient of the x-term. The product of the last terms equals 3, the constant.

A procedure to factor expressions of the form $ax^2 + bx + c, a \neq 1$, follows.

> **PROCEDURE**   FACTORING TRINOMIALS OF THE FORM
> $ax^2 + bx + c, a \neq 1$
>
> 1. Write all pairs of factors of the coefficient of the squared term, a.
> 2. Write all pairs of factors of the constant, c.
> 3. Try various combinations of these factors until the sum of the products of the outer and inner terms is bx.
> 4. Check your answer by multiplying the factors using the FOIL method.

Example 5 *Factoring a Trinomial, $a \neq 1$*

Factor $3x^2 + 17x + 10$.

Solution   The only positive factors of 3 are 3 and 1. Therefore, we write

$$3x^2 + 17x + 10 = (3x \quad)(x \quad)$$

The number 10 has both positive and negative factors. However, since both the constant, 10, and the sum of the products of the outer and inner terms, 17, are positive, the two factors must be positive. The positive factors of 10 are $1(10)$ and $2(5)$. The following is a list of the possible factors.

Possible Factors	Sum of Products of Outer and Inner Terms
$(3x + 1)(x + 10)$	$31x$
$(3x + 10)(x + 1)$	$13x$
$(3x + 2)(x + 5)$	$17x$ ← Correct middle term
$(3x + 5)(x + 2)$	$11x$

Thus, $3x^2 + 17x + 10 = (3x + 2)(x + 5)$.   ■

Note that factoring problems of this type may be checked by using the FOIL method of multiplication. We will check the results to Example 5:

$$(3x + 2)(x + 5) = 3x \cdot x + 3x \cdot 5 + 2 \cdot x + 2 \cdot 5$$
$$= 3x^2 + 15x + 2x + 10$$
$$= 3x^2 + 17x + 10$$

Since we obtained the expression we started with, our factoring is correct.

Example 6 *Factoring a Trinomial, $a \neq 1$*

Factor $6x^2 - 11x - 10$.

Solution   The factors of 6 will be either $6 \cdot 1$ or $2 \cdot 3$. Therefore, the factors may be of the form $(6x \quad)(x \quad)$ or $(2x \quad)(3x \quad)$. When there is more than one set of factors for the first term, we generally try the medium-sized factors first. If that does not work, we try the other factors. Thus, we write

$$6x^2 - 11x - 10 = (2x \quad)(3x \quad)$$

The factors of -10 are $(-1)(10)$, $(1)(-10)$, $(-2)(5)$, and $(2)(-5)$.
    The correct factoring is $6x^2 - 11x - 10 = (2x - 5)(3x + 2)$.   ■

Timely Tip

To factor a trinomial, $ax^2 + bx + c$, first observe the sign of the constant term, c.

$$x^2 + 11x + 28 \rightarrow (x + 7)(x + 4)$$

$$x^2 - 11x + 28 \rightarrow (x - 7)(x - 4)$$

If c is positive, numbers have the same sign

$$x^2 + 3x - 28 \rightarrow (x + 7)(x - 4)$$

$$x^2 - 3x - 28 \rightarrow (x - 7)(x + 4)$$

If c is negative, numbers have different signs

Note that in Example 6 we first tried factors of the form $(2x\quad)(3x\quad)$. If we had not determined the correct factors using them, we would have tried $(6x\quad)(x\quad)$.

Solving Quadratic Equations by Factoring

In Section 6.1, we solved linear, or first-degree, equations. In those equations, the exponent on all variables was 1. Now we deal with the *quadratic equation*. The standard form of a quadratic equation in one variable is shown in the box.

Standard Form of a Quadratic Equation

$$ax^2 + bx + c = 0, \quad a \neq 0$$

Note that in the standard form of a quadratic equation, the greatest exponent on x is 2 and the right side of the equation is equal to zero. In this section, we will solve quadratic equations by factoring and by the quadratic formula.

To solve a quadratic equation by factoring, set one side of the equation equal to 0 and then use the *zero-factor* property.

Zero-Factor Property

If $a \cdot b = 0$, then $a = 0$ or $b = 0$.

The zero-factor property indicates that if the product of two factors is 0, then one (or both) of the factors must have a value of 0.

Example 7   *Using the Zero-Factor Property*

Solve the equation $(x + 4)(x - 3) = 0$.

Solution When we use the zero-factor property, either $(x + 4)$ or $(x - 3)$ must equal 0 for the product to equal 0. Thus, we set each individual factor equal to 0 and solve each resulting equation for x.

$$(x + 4)(x - 3) = 0$$

$$x + 4 = 0 \qquad \text{or} \qquad x - 3 = 0$$

$$x = -4 \qquad\qquad\qquad x = 3$$

Thus, the solutions are -4 and 3.

CHECK: $x = -4$ $x = 3$

$$(x + 4)(x - 3) = 0 \qquad\qquad (x + 4)(x - 3) = 0$$

$$(-4 + 4)(-4 - 3) = 0 \qquad\qquad (3 + 4)(3 - 3) = 0$$

$$0(-7) = 0 \qquad\qquad\qquad (7)(0) = 0$$

$$0 = 0 \text{ True} \qquad\qquad\qquad 0 = 0 \text{ True} \quad \blacksquare$$

> **PROCEDURE** SOLVING QUADRATIC EQUATIONS BY FACTORING
>
> 1. Use the addition or subtraction property to make one side of the equation equal to 0.
> 2. Factor the side of the equation not equal to 0.
> 3. Use the zero-factor property to solve the equation.

Examples 8 and 9 illustrate this procedure.

Example 8 *Solving a Quadratic Equation by Factoring*

Solve the equation $x^2 - 8x = -15$.

Solution First add 15 to both sides of the equation to make the right-hand side of the equation equal to 0.

$$x^2 - 8x = -15$$
$$x^2 - 8x + 15 = -15 + 15$$
$$x^2 - 8x + 15 = 0$$

Factor the left-hand side of the equation. The goal is to determine two numbers whose product is 15 and whose sum is -8. Since the product of the numbers is positive and the sum of the numbers is negative, the two numbers must both be negative. The numbers are -3 and -5. Note that $(-3)(-5) = 15$ and $-3 + (-5) = -8$.

$$x^2 - 8x + 15 = 0$$
$$(x - 3)(x - 5) = 0$$

Now use the zero-factor property to determine the solution.

$$x - 3 = 0 \quad \text{or} \quad x - 5 = 0$$
$$x = 3 \qquad\qquad x = 5$$

The solutions are 3 and 5. ∎

Timely Tip

Recall that every factoring problem can be checked by multiplying the factors. The product of the factors should be identical to the original expression that was factored. If we wished to check the factoring of Example 9, we would multiply $(3x + 2)(x - 3)$. Since the product of the factors is $3x^2 - 7x - 6$, which is the expression we started with, our factoring is correct.

Example 9 *Solving a Quadratic Equation by Factoring*

Solve the equation $3x^2 - 7x - 6 = 0$.

Solution $3x^2 - 7x - 6$ factors into $(3x + 2)(x - 3)$. Thus, we write

$$3x^2 - 7x - 6 = 0$$
$$(3x + 2)(x - 3) = 0$$
$$3x + 2 = 0 \quad \text{or} \quad x - 3 = 0$$
$$3x = -2 \qquad\qquad x = 3$$
$$x = -\frac{2}{3}$$

The solutions are $-\frac{2}{3}$ and 3. ∎

Solving Quadratic Equations by Using the Quadratic Formula

Not all quadratic equations can be solved by factoring. When a quadratic equation cannot be easily solved by factoring, we can solve the equation with the *quadratic formula*. The quadratic formula can be used to solve any quadratic equation.

> **Quadratic Formula**
> For a quadratic equation in standard form, $ax^2 + bx + c = 0, a \neq 0$, the quadratic formula is
> $$x = \frac{-b \pm \sqrt{b^2 - 4ac}}{2a}$$

In the quadratic formula, the plus or minus symbol, $\pm$, is used. If, for example, $x = 2 \pm 3$, then $x = 2 + 3 = 5$ or $x = 2 - 3 = -1$.

It is possible for a quadratic equation to have no real solution. In solving an equation, if the radicand (the expression inside the square root) is a negative number, the quadratic equation has *no real solution*.

To use the quadratic formula, first write the quadratic equation in standard form. Then determine the values of a (the coefficient of the squared term), b (the coefficient of the x term), and c (the constant). Finally, substitute the values of a, b, and c into the quadratic formula and evaluate.

Example 10 Solving a Quadratic Equation by Using the Quadratic Formula

Solve the equation $x^2 - 2x - 15 = 0$ by using the quadratic formula.

Solution In this equation, $a = 1, b = -2$, and $c = -15$.

$$x = \frac{-b \pm \sqrt{b^2 - 4ac}}{2a} = \frac{-(-2) \pm \sqrt{(-2)^2 - 4(1)(-15)}}{2(1)}$$

$$= \frac{2 \pm \sqrt{4 + 60}}{2}$$

$$= \frac{2 \pm \sqrt{64}}{2}$$

$$= \frac{2 \pm 8}{2}$$

$$\frac{2 + 8}{2} = \frac{10}{2} = 5 \quad \text{or} \quad \frac{2 - 8}{2} = \frac{-6}{2} = -3$$

The solutions are 5 and -3.   ■

Note that Example 10 can also be solved by factoring.

Example 11 Irrational Solutions to a Quadratic Equation

Solve $4x^2 - 8x = -1$ by using the quadratic formula.

Solution Begin by writing the equation in standard form by adding 1 to both sides of the equation, which gives the following.

$$4x^2 - 8x + 1 = 0$$

$$a = 4, \qquad b = -8, \qquad c = 1$$

$$x = \frac{-b \pm \sqrt{b^2 - 4ac}}{2a} = \frac{-(-8) \pm \sqrt{(-8)^2 - 4(4)(1)}}{2(4)}$$

Did You Know?

The Mathematics of Motion

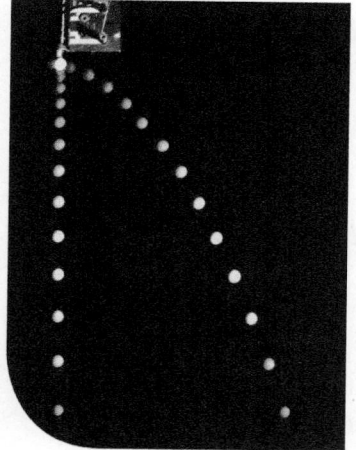

The free fall of an object is something that has interested scientists and mathematicians for centuries. It is described by a quadratic equation. Shown here is a time-lapse photo that shows the free fall of a ball in equal time intervals. What you see can be described verbally this way: The rate of change in velocity in each interval is the same; therefore, velocity is continuously increasing and acceleration is constant.

$$= \frac{8 \pm \sqrt{64 - 16}}{8}$$

$$= \frac{8 \pm \sqrt{48}}{8}$$

Since $\sqrt{48} = \sqrt{16}\sqrt{3} = 4\sqrt{3}$ (see Section 5.4), we write

$$\frac{8 \pm \sqrt{48}}{8} = \frac{8 \pm 4\sqrt{3}}{8} = \frac{\overset{1}{\cancel{4}}(2 \pm \sqrt{3})}{\underset{2}{\cancel{8}}} = \frac{2 \pm \sqrt{3}}{2}$$

The solutions are $\dfrac{2 + \sqrt{3}}{2}$ and $\dfrac{2 - \sqrt{3}}{2}$. ∎

Note that the solutions to Example 11 are irrational numbers.

Applications of Quadratic Equations

Example 12 *Brick Border*

Pablo and Maya recently installed an inground rectangular swimming pool measuring 40 ft by 30 ft. They want to add a brick border of uniform width around all sides of the pool. How wide can they make the brick border if they purchased enough brick to cover an area of 296 ft^2?

Solution Let's make a diagram of the pool and the brick border (Fig. 6.40). Let $x =$ the uniform width of the brick border. Then the total length of the larger rectangular area, the pool plus the border, is $2x + 40$. The total width of the larger rectangular area is $2x + 30$.

The area of the brick border can be determined by subtracting the area of the pool from the area of the pool plus the brick border.

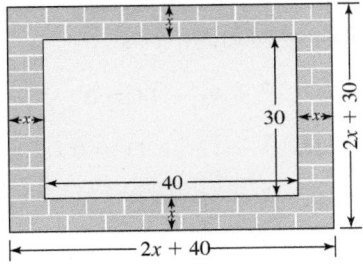

Figure 6.40

$$\text{Area of pool} = l \cdot w = (40)(30) = 1200 \text{ ft}^2$$

$$\text{Area of pool plus brick border} = l \cdot w = (2x + 40)(2x + 30)$$

$$= 4x^2 + 140x + 1200$$

$$\text{Area of the brick border} = \text{area of pool plus brick border} - \text{area of pool}$$

$$= (4x^2 + 140x + 1200) - 1200$$

$$= 4x^2 + 140x$$

The area of the brick border must be 296 ft^2. Therefore,

$$296 = 4x^2 + 140x$$

or

$$4x^2 + 140x - 296 = 0$$

$$4(x^2 + 35x - 74) = 0 \qquad \text{Factor out 4 from each term.}$$

$$\frac{\cancel{4}}{\cancel{4}}(x^2 + 35x - 74) = \frac{0}{4} \qquad \begin{array}{l}\text{Divide both sides of the} \\ \text{equation by 4.}\end{array}$$

$$x^2 + 35x - 74 = 0$$

$$(x + 37)(x - 2) = 0 \qquad \text{Factor trinomial.}$$

$$x + 37 = 0 \qquad \text{or} \qquad x - 2 = 0$$

$$x = -37 \qquad\qquad\quad x = 2$$

Since lengths are positive, the only possible answer is $x = 2$. Thus, they can make a brick border 2 ft wide all around the pool. ∎

SECTION 6.9
Exercises

Warm Up Exercises

In Exercises 1–6, fill in the blank with an appropriate word, phrase, or symbol(s).

1. An expression that contains two terms in which each exponent that appears on a variable is a whole number is called a(n) _____ .

2. An expression that contains three terms in which each exponent that appears on a variable is a whole number is called a(n) _____ .

3. When multiplying two binomials, a method that obtains the products of the first, outer, inner, and last terms is called the _____ method.

4. The property that indicates that if the product of two factors is 0, then one or both of the factors must have a value of 0 is called the _____ property.

5. When a quadratic equation cannot be easily solved by factoring, the equation can be solved using the _____ formula.

6. For a quadratic equation in standard form $ax^2 + bx + c = 0$, $a \neq 0$, the quadratic formula is

   _____ .

Practice the Skills

In Exercises 7–12, multiply the binomials.

7. $(x + 3)(x - 11)$ 8. $(x - 7)(x + 5)$

9. $(2x + 3)(3x - 1)$ 10. $(3x - 2)(2x + 1)$

11. $(6x - 7)(8x + 9)$ 12. $(7x + 9)(6x - 5)$

In Exercises 13–24, factor the trinomial.

13. $x^2 + 5x + 6$ 14. $x^2 + 7x + 10$

15. $x^2 - x - 6$ 16. $x^2 - 3x - 10$

17. $x^2 + 3x - 10$ 18. $x^2 + 5x - 14$

19. $x^2 - 2x - 3$ 20. $x^2 - 4x - 5$

21. $x^2 - 9x + 18$ 22. $x^2 - 6x + 8$

23. $x^2 + 3x - 28$ 24. $x^2 + 4x - 32$

In Exercises 25–36, factor the trinomial.

25. $2x^2 + 7x + 3$ 26. $3x^2 + 7x + 2$

27. $3x^2 + x - 2$ 28. $2x^2 - x - 3$

29. $3x^2 - 17x + 10$ 30. $2x^2 - 9x + 10$

31. $4x^2 + 11x + 6$ 32. $4x^2 + 20x + 21$

33. $4x^2 - 11x + 6$ 34. $6x^2 - 11x + 4$

35. $8x^2 - 2x - 3$ 36. $6x^2 - 13x + 2$

In Exercises 37–40, solve each equation, using the zero-factor property.

37. $(x - 6)(5x - 4) = 0$

38. $(2x + 3)(x - 1) = 0$

39. $(3x + 4)(2x - 1) = 0$

40. $(5x - 4)(3x + 2) = 0$

In Exercises 41–58, solve each equation by factoring.

41. $x^2 + 8x + 15 = 0$ 42. $x^2 + 9x + 14 = 0$

43. $x^2 - 8x + 7 = 0$ 44. $x^2 - 12x + 11 = 0$

45. $x^2 - 15 = 2x$ 46. $x^2 - 21 = 4x$

47. $x^2 = 4x - 3$ 48. $x^2 = 8x - 7$

49. $6x^2 + 19x + 15 = 0$

50. $6x^2 + 25x + 14 = 0$

51. $3x^2 + 10x = 8$ 52. $2x^2 + 15x = 27$

53. $3x^2 = -5x + 2$ 54. $2x^2 = -5x + 3$

55. $3x^2 = -5x - 2$ 56. $2x^2 = -5x - 3$

57. $20x^2 = 15 - 13x$ 58. $15x^2 = 16 - 8x$

In Exercises 59–78, solve the equation, using the quadratic formula. If the equation has no real solution, so state.

59. $x^2 + 2x - 3 = 0$ 60. $x^2 - 11x + 24 = 0$

61. $x^2 - 3x - 18 = 0$ 62. $x^2 - 6x - 16 = 0$

63. $x^2 - 4x + 2 = 0$ 64. $x^2 - 2x - 2 = 0$

65. $x^2 - 2x + 3 = 0$ 66. $2x^2 - 4x + 5 = 0$

67. $3x^2 + 9x + 5 = 0$ 68. $2x^2 - 5x - 2 = 0$

69. $3x^2 - 8x + 1 = 0$ 70. $2x^2 + 4x + 1 = 0$

71. $4x^2 - 5x - 3 = 0$

72. $3x^2 - 7x - 5 = 0$

73. $2x^2 + 7x + 5 = 0$

74. $3x^2 - 10x + 7 = 0$

75. $3x^2 = 9x - 5$

76. $4x^2 = -7x + 1$

77. $4x^2 + 6x = -5$

78. $5x^2 - 4x = -2$

Problem Solving

79. *Drying Time* The time, t, in minutes, needed for clothes hanging on a line outdoors to dry, at a specific temperature, depends on the humidity, h. The time can be approximated by the equation $t = 2h^2 + 80h + 40$, where h is the humidity expressed as a decimal number. Determine the length of time required for clothing to dry if there is 60% humidity.

80. *Height of a Diver* A diver jumps from a platform diving board that is 32 feet above water. The height, h, of the diver above water, in feet, t seconds after jumping from the platform, can be determined by the equation $h = -16t^2 + 20t + 32$. Determine the diver's height above water 2 seconds after jumping from the platform.

81. *Flower Garden* Karen's back yard has a width of 20 meters and a length of 30 meters. Karen wants to put a rectangular flower garden in the middle of the back yard, leaving a strip of grass of uniform width around all sides of the flower garden. If she wants to have 336 square meters of grass, what will be the width and length of the garden?

82. *Garden and Walkway* CJ's garden is surrounded by a uniform-width walkway. The garden and walkway together cover an area of 320 square feet. If the dimensions of the garden are 12 feet by 16 feet, determine the width of the walkway.

83. *Height of a Ball* A ball is projected upward from the ground. Its height h, above ground, in feet, can be determined by the equation

$$h = -16t^2 + 128t$$

where t is the number of seconds after the ball is projected. After how many seconds will the ball be 256 ft above ground?

84. *Height of a Rocket* A model rocket is launched from a hill 80 feet above sea level. The launch site is next to the ocean, and the rocket will fall into the ocean. The rocket's height, h, above sea level can be determined by the equation $h = -16t^2 + 64t + 80$ where t is the number of

seconds after the rocket is launched. Determine the time it takes the rocket to strike the ocean.

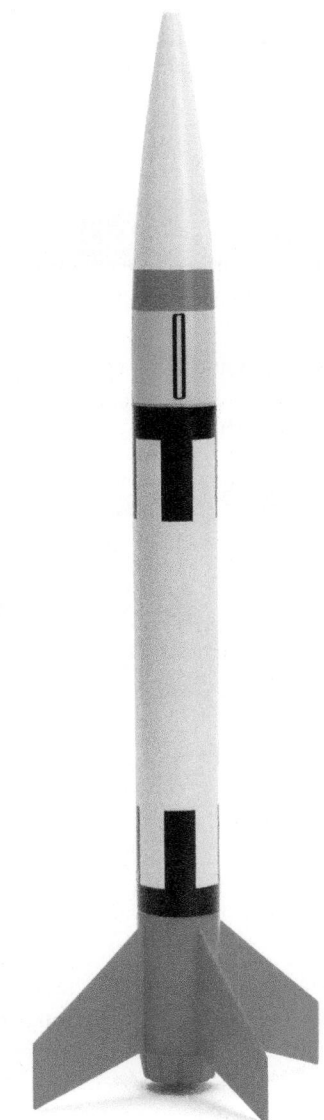

Challenge Problems/Group Activities

85. **a)** Explain why solving the equation $(x - 4)(x - 7) = 6$ by setting each factor equal to 6 is not correct.

b) Determine the correct solution to $(x - 4)(x - 7) = 6$.

86. The radicand in the quadratic formula, $b^2 - 4ac$, is called the *discriminant*. How many real number solutions will the quadratic equation have if the discriminant is (a) greater than 0, (b) equal to 0, or (c) less than zero? Explain your answer.

87. Write an equation that has solutions -1 and 3.

Research Activities

88. *Girolamo Cordano* Italian mathematician Girolamo Cardano (1501–1576) is recognized for his skill in solving equations. Write a paper about his life and his contributions to mathematics, in particular his contribution to solving equations.

89. *Foo Ling Awong* Chinese mathematician Foo Ling Awong, who lived during the Pong dynasty, developed a technique, other than trial and error, to factor trinomials of the form $ax^2 + bx + c$, $a \neq 1$. Write a paper about his life and his contributions to mathematics, in particular his technique for factoring trinomials in the form $ax^2 + bx + c$, $a \neq 1$.

SECTION 6.10   Functions and Their Graphs

Upon completion of this section, you will be able to:

- Understand relations and functions.
- Understand and graph linear functions.
- Understand and graph quadratic functions.
- Understand and graph exponential functions.
- Solve problems involving natural exponential functions.

Nikita wishes to download songs from iTunes and each song costs $1.29. Thus, one song costs 1 × $1.29, or $1.29. Two songs cost 2 × $1.29, or $2.58. Three songs cost 3 × $1.29, or $3.87, and so on. The relationship between two quantities—in this case, the number of songs downloaded and the cost of the songs—will be discussed in this section.

Why This Is Important We will introduce the concept of *function*, an extremely important mathematical concept. Examples of functions in this section involve weekly salary for sales representatives, estimating the height of a spacecraft above the surface of the moon, estimating the age of fossils, estimating the growth of bacteria, and estimating the population of earth in the future. Knowledge of functions is a crucial aspect of mathematical understanding and is useful in many areas of modern society.

Relations and Functions

We begin with a discussion of *relations*. A *relation* is any set of ordered pairs. Every graph is a set of ordered pairs. Therefore, any graph will be a relation. The following table indicates the relation between the number of songs purchased and the cost of the songs discussed in the section opening paragraph.

Number of Songs	Cost ($)
0	0.00
1	1.29
2	2.58
3	3.87
⋮	⋮
10	12.90
⋮	⋮

In general, the cost for purchasing n songs will be $1.29 times the number of songs, or $1.29n$ dollars. We can represent the cost, c, of n songs by the equation $c = 1.29n$. Since the value of c depends on the value of n, we refer to c as the *dependent variable* and n as the *independent variable*. *Note that for each value of the independent variable, n, there is one and only one value of the dependent variable, c.* An equation such as $c = 1.29n$ is called a ***function***. In the equation $c = 1.29n$, the value of c depends on the value of n, so we say that "c is a function of n."

> **Definition: Function**
> A **function** is a special type of relation where each value of the independent variable corresponds to exactly one value of the dependent variable.

The set of values that can be used for the independent variable is called the *domain* of the function, and the resulting set of values obtained for the dependent variable is called the *range*. The domain and range for the function $c = 1.29n$ are illustrated in Fig. 6.41 below.

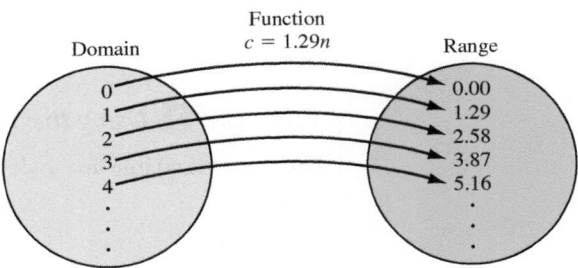

Figure 6.41

Example 1 *Is the Relation Also a Function?*

Determine whether the given relation is also a function.

a) $\{(1,4),(2,3),(3,5),(-1,3),(0,6)\}$
b) $\{(-1,2),(4,2),(3,1),(2,6),(3,5)\}$

Solution In both parts (a) and (b), each of the ordered pairs is of the form (x, y). The x represents the independent variable and the y represents the dependent variable.

a) Each x corresponds with a unique y. For example, the value $x = 1$ corresponds only with the value $y = 4$. The value $x = 2$ corresponds only with the value $y = 3$, and so on. Since each value of the independent variable corresponds to a unique value of the dependent variable, this relation is also a function.

b) Notice that the value $x = 3$ corresponds to both $y = 1$ and $y = 5$. Therefore, each value of the independent variable does not correspond to a unique value of the dependent variable. Thus, this relation is not a function. ∎

When we graphed equations of the form $ax + by = c$ in Section 6.6, we determined that they were straight lines. For example, the graph of $y = 2x - 1$ is illustrated in Fig. 6.42.

Is the equation $y = 2x - 1$ a function? To answer this question, we must ask, Does each value of x correspond to a unique value of y? The answer is yes; therefore, this equation is a function.

For the equation $y = 2x - 1$, we say that "y is a function of x" and write $y = f(x)$. The notation $f(x)$ is read "f of x." When we are given an equation that is a function, we may replace the y in the equation with $f(x)$ because $f(x)$ represents y. Thus, $y = 2x - 1$ may be written $f(x) = 2x - 1$.

To evaluate a function for a specific value of x, replace each x in the function with the given value, then evaluate. For example, to evaluate $f(x) = 2x - 1$ when $x = 8$, we do the following.

$$f(x) = 2x - 1$$
$$f(8) = 2(8) - 1 = 16 - 1 = 15$$

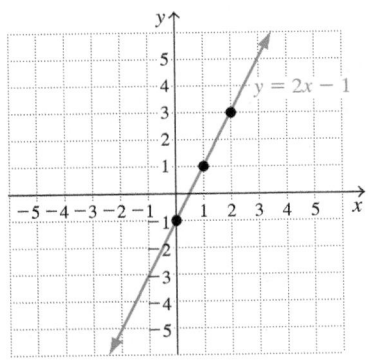

Figure 6.42

Thus, $f(8) = 15$. Since $f(x) = y$, when $x = 8$, $y = 15$. What is the domain and range of $f(x) = 2x - 1$? Because x can be any real number, the domain is the set of real numbers, symbolized $\mathbb{R}$. The range is also $\mathbb{R}$.

When looking at a graph, the following test can be used to quickly determine whether the graph represents a function.

> ### Definition: Vertical Line Test
> - If a vertical line can be drawn so that it intersects a graph at more than one point, then *the graph does not represent a function.*
> - If a vertical line cannot be drawn so that it intersects a graph at more than one point, then *the graph represents a function.*

Example 2 *Using the Vertical Line Test*

Use the vertical line test to determine which of the graphs in Figure 6.43 represent functions.

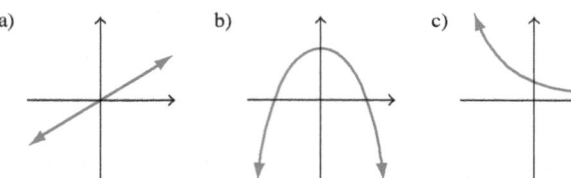

Figure 6.43

Solution The graphs in parts (a), (b), and (c) represent functions, but the graph in part (d) does not.

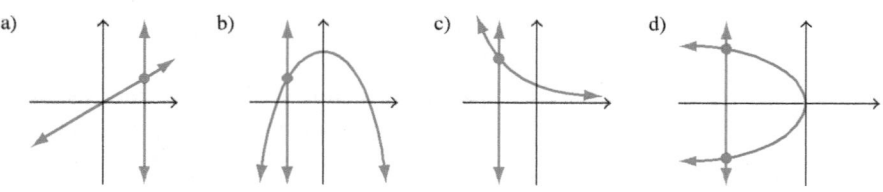

There are many real-life applications of functions. In fact, all the applications illustrated in Sections 6.1 through 6.3 are functions.

In this section, we will discuss three types of functions: linear functions, quadratic functions, and exponential functions.

Linear Functions and Their Graphs

In Section 6.6, we graphed linear equations. The graph of any linear equation of the form $y = ax + b$ will pass the vertical line test, and so equations of the form $y = ax + b$ are *linear functions*. If we wished, we could write the linear function as $f(x) = ax + b$.

Example 3 *Salary as a Linear Function*

Claudia works at a new vehicle dealership as a sales representative. Claudia's weekly salary, s, is given by the function $s(x) = 200 + 0.06x$, where x represents Claudia's weekly sales, in dollars.

a) What is Claudia's weekly salary if during that week she sells one truck for $68,000?

b) If for one week, Claudia's weekly salary was $1580, what were her weekly sales?

Solution

a) We substitute $68,000$ for x in the function.

$$s(x) = 200 + 0.06x$$
$$s(68,000) = 200 + 0.06(68,000)$$
$$= 200 + 4080 = 4280$$

Thus, if Claudia sells one truck for $68,000 in a given week, her salary for that week will be $4280.

b) In the function, replace $s(x)$ with the weekly salary, 1580, and solve for x.

$$s(x) = 200 + 0.06x$$
$$1580 = 200 + 0.06x$$
$$1380 = 0.06x$$
$$23,000 = x$$

Thus, if Claudia's weekly salary was $1580, her weekly sales were $23,000. ■

The graphs of linear functions are straight lines that will pass the vertical line test. In Section 6.6, we discussed how to graph linear equations. Linear functions can be graphed by plotting points, by using intercepts, or by using the slope and y-intercept.

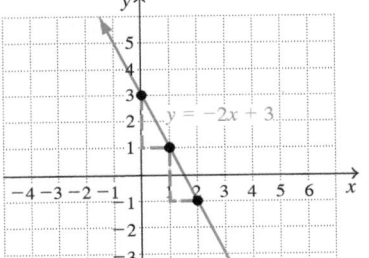

Figure 6.44

Example 4 *Graphing a Linear Function*

Graph $f(x) = -2x + 3$ by using the slope and y-intercept.

Solution We can rewrite this function as $y = -2x + 3$. From Section 6.6, we know that the slope is -2 and the y-intercept is $(0, 3)$. Plot $(0, 3)$ on the y-axis. Then plot the next point by moving *down* 2 units and to the *right* 1 unit (see Fig. 6.44). A third point has been plotted in the same way. The graph of $f(x) = -2x + 3$ is the line drawn through these three points. ■

Quadratic Functions and Their Graphs

The standard form of a quadratic equation is $y = ax^2 + bx + c, a \neq 0$. We will learn shortly that graphs of equations of this form always pass the vertical line test and are functions. Therefore, equations of the form $y = ax^2 + bx + c, a \neq 0$, may be referred to as *quadratic functions*. We may express quadratic functions using function notation as $f(x) = ax^2 + bx + c$. Two examples of quadratic functions are $y = 2x^2 + 5x - 7$ and $f(x) = -\frac{1}{2}x^2 + 4$.

▲ *Neil Armstrong on the moon*

Example 5 *Landing on the Moon*

On July 20, 1969, Neil Armstrong became the first person to walk on the moon. The velocity, v, of his spacecraft, the *Eagle*, in meters per second, was a function of time before touchdown, t, given by

$$v = f(t) = 3.2t + 0.45$$

The height of the spacecraft, h, above the moon's surface, in meters, was also a function of time before touchdown, given by

$$h = g(t) = 1.6t^2 + 0.45t$$

What was the velocity of the spacecraft and its height above the moon's surface

a) at 3 seconds before touchdown?   b) at touchdown (0 seconds)?

Solution

a) $v = f(t) = 3.2t + 0.45$
$\quad f(3) = 3.2(3) + 0.45$
$\qquad\quad = 9.6 + 0.45$
$\qquad\quad = 10.05$

$h = g(t) = 1.6t^2 + 0.45t$
$\quad g(3) = 1.6(3)^2 + 0.45(3)$
$\qquad\quad = 1.6(9) + 1.35$
$\qquad\quad = 14.4 + 1.35$
$\qquad\quad = 15.75$

The velocity 3 seconds before touchdown was 10.05 meters per second, and the height above the moon's surface 3 seconds before touchdown was 15.75 meters.

b) $\quad v = f(t) = 3.2t + 0.45,$
$\qquad f(0) = 3.2(0) + 0.45$
$\qquad\quad = 0 + 0.45$
$\qquad\quad = 0.45$

$h = g(t) = 1.6t^2 + 0.45t$
$\quad g(0) = 1.6(0)^2 + 0.45(0)$
$\qquad\quad = 0 + 0$
$\qquad\quad = 0$

The touchdown velocity was 0.45 meter per second. At touchdown, the *Eagle* is on the moon, and the height above the moon's surface is therefore 0 meter. ∎

The graph of every quadratic function is a *parabola*. Two parabolas are illustrated in Fig. 6.45. Note that both graphs represent functions, since they pass the vertical line test. A parabola opens upward when the coefficient of the squared term, a, is greater than 0, as shown in Figure 6.45(a). A parabola opens downward when the coefficient of the squared term, a, is less than 0, as shown in Fig. 6.45(b).

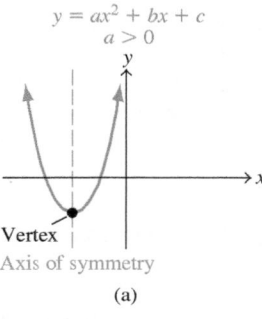

$y = ax^2 + bx + c$
$a > 0$

Vertex
Axis of symmetry
(a)

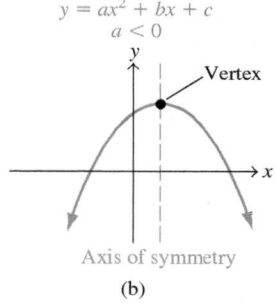

$y = ax^2 + bx + c$
$a < 0$

Vertex
Axis of symmetry
(b)

Figure 6.45

The *vertex* of a parabola is the lowest point on a parabola that opens upward and the highest point on a parabola that opens downward. Every parabola is *symmetric* with respect to a vertical line through its vertex. This line is called the *axis of symmetry* of the parabola.

Axis of Symmetry of a Parabola

The axis of symmetry of the graph of an equation of the form $y = ax^2 + bx + c$ can be determined by the following formula.

$$x = \frac{-b}{2a}$$

This formula also gives the x-coordinate of the vertex of a parabola.

Once the x-coordinate of the vertex has been determined, the y-coordinate can be determined by substituting the value determined for the x-coordinate into the quadratic equation and evaluating the equation. This procedure is illustrated in Example 6.

MATHEMATICS TODAY

Computing Reality with Mathematical Models

It is a relatively easy matter for sci-entists and mathematicians to de-scribe and predict a simple motion like that of a falling object. When the phenomenon is complicated, such as making an accurate predic-tion of the weather, the mathemat-ics becomes much more difficult. The National Weather Service has devised an algorithm that takes the temperature, pressure, mois-ture content, and wind velocity of more than 250,000 points in Earth's atmosphere and applies a set of equations that, it believes, will rea-sonably predict what will happen at each point over time. Researchers at the National Center for Super-computing Applications are working on a computer model that simulates thunderstorms. They want to know why some thunderstorms can turn severe, even deadly.

Why This Is Important Com-puters are used to model real-life situations, including weather fore-casting, travel time estimations, and stock market predictions.

Example 6 Describing the Graph of a Quadratic Equation

Consider the equation $y = -3x^2 + 12x - 5$.

a) Determine whether the graph of the equation will be a parabola that opens upward or downward.

b) Determine the equation of the axis of symmetry of the parabola.

c) Determine the vertex of the parabola.

Solution

a) Since $a = -3$, which is less than 0, the parabola opens downward.

b) To determine the equation of the axis of symmetry, we use the equation $x = \frac{-b}{2a}$. In the equation

$$y = -3x^2 + 12x - 5, a = -3, b = 12, \text{ and } c = -5, \text{ so}$$

$$x = \frac{-b}{2a} = \frac{-(12)}{2(-3)} = \frac{-12}{-6} = 2$$

The equation of the axis of symmetry is $x = 2$.

c) The x-coordinate of the vertex is 2 from part (b). To determine the y-coordinate, we substitute 2 for x in the equation $y = -3x^2 + 12x - 5$ and then evaluate.

$$\begin{aligned} y &= -3x^2 + 12x - 5 \\ &= -3(2)^2 + 12(2) - 5 \\ &= -3(4) + 24 - 5 \\ &= -12 + 24 - 5 \\ &= 7 \end{aligned}$$

Therefore, the vertex of the parabola is located at the point $(2, 7)$ on the graph. ∎

PROCEDURE   GRAPHING A QUADRATIC EQUATION

1. Determine whether the parabola opens upward or downward.
2. Determine the equation of the axis of symmetry.
3. Determine the vertex of the parabola.
4. Determine the y-intercept by substituting $x = 0$ into the equation.
5. Determine the x-intercepts (if they exist) by substituting $y = 0$ into the equation and solving for x.
6. Draw the graph, making use of the information gained in Steps 1 through 5. Remember that the parabola will be symmetric with respect to the axis of symmetry.

In Step 5, you may use either factoring or the quadratic formula to determine the x-intercepts.

Example 7 Graphing a Quadratic Equation

Graph the equation $y = x^2 - 6x + 8$.

Solution We follow the steps outlined in the general procedure.

1. Since $a = 1$, which is greater than 0, the parabola opens upward.

2. Axis of symmetry: $x = \frac{-b}{2a} = \frac{-(-6)}{2(1)} = \frac{6}{2} = 3$

Thus, the axis of symmetry is $x = 3$.

3. y-coordinate of vertex:

$$y = x^2 - 6x + 8$$
$$= (3)^2 - 6(3) + 8 = 9 - 18 + 8 = -1$$

Thus, the vertex is at $(3, -1)$.

4. y-intercept: Substitute $x = 0$ and solve for y.

$$y = x^2 - 6x + 8$$
$$y = (0)^2 - 6(0) + 8 = 8$$

Thus, the y-intercept is at $(0, 8)$.

5. x-intercepts: Substitute $y = 0$ and solve for x.

$$0 = x^2 - 6x + 8, \text{ or } x^2 - 6x + 8 = 0$$

We can solve this equation by factoring.

$$x^2 - 6x + 8 = 0$$
$$(x - 4)(x - 2) = 0$$
$$x - 4 = 0 \quad \text{or} \quad x - 2 = 0$$
$$x = 4 \qquad\qquad x = 2$$

Thus, the x-intercepts are $(4, 0)$ and $(2, 0)$.

6. Plot the vertex $(3, -1)$, the axis of symmetry $x = 3$, the y-intercept $(0, 8)$, and the x-intercepts $(4, 0)$ and $(2, 0)$. Then graph the equation (Fig. 6.46). ∎

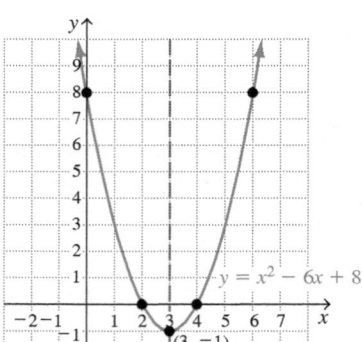

Figure 6.46

Note that the domain of the graph in Example 7, the possible x-values, is the set of all real numbers, $\mathbb{R}$. The range, the possible y-values, is the set of all real numbers greater than or equal to -1. When graphing parabolas, if you think you need additional points to graph the equation, you can always substitute values for x and determine the corresponding values of y and plot those points. For example, if you substituted 1 for x, the corresponding value of y is 3. Thus, you could plot the point $(1, 3)$.

Example 8 *Domain and Range of a Quadratic Function*

a) Graph the function $f(x) = -2x^2 + 3x + 4$.

b) Determine the domain and range of the function.

Solution

a) Since $f(x)$ means y, we can replace $f(x)$ with y to obtain $y = -2x^2 + 3x + 4$. Now graph $y = -2x^2 + 3x + 4$ using the steps outlined in the general procedure.

1. Since $a = -2$, which is less than 0, the parabola opens downward.

2. Axis of symmetry:

$$x = \frac{-b}{2a} = \frac{-(3)}{2(-2)} = \frac{-3}{-4} = \frac{3}{4}$$

Thus, the axis of symmetry is $x = \frac{3}{4}$.

3. y-coordinate of vertex:

$$y = -2x^2 + 3x + 4$$

$$= -2\left(\frac{3}{4}\right)^2 + 3\left(\frac{3}{4}\right) + 4$$

$$= -2\left(\frac{9}{16}\right) + \frac{9}{4} + 4$$

$$= -\frac{9}{8} + \frac{9}{4} + 4$$

$$= -\frac{9}{8} + \frac{18}{8} + \frac{32}{8} = \frac{41}{8} \quad \text{or} \quad 5\frac{1}{8}$$

Thus, the vertex is at $\left(\frac{3}{4}, 5\frac{1}{8}\right)$.

4. y-intercept: $y = -2x^2 + 3x + 4$
$$= -2(0)^2 + 3(0) + 4 = 4$$

Thus, the y-intercept is $(0, 4)$.

5. x-intercepts: $y = -2x^2 + 3x + 4$
$$0 = -2x^2 + 3x + 4 \quad \text{or} \quad -2x^2 + 3x + 4 = 0$$

This equation cannot be factored, so we will use the quadratic formula to solve it.

$$a = -2, \qquad b = 3, \qquad c = 4$$

$$x = \frac{-b \pm \sqrt{b^2 - 4ac}}{2a}$$

$$= \frac{-3 \pm \sqrt{3^2 - 4(-2)(4)}}{2(-2)}$$

$$= \frac{-3 \pm \sqrt{9 + 32}}{-4}$$

$$= \frac{-3 \pm \sqrt{41}}{-4}$$

Since $\sqrt{41} \approx 6.4$,

$$x \approx \frac{-3 + 6.4}{-4} \approx \frac{3.4}{-4} \approx -0.85 \quad \text{or} \quad x \approx \frac{-3 - 6.4}{-4} \approx \frac{-9.4}{-4} \approx 2.35$$

Thus, the x-intercepts are about $(-0.85, 0)$ and $(2.35, 0)$.

6. Plot the vertex $\left(\frac{3}{4}, 5\frac{1}{8}\right)$, the y-intercept $(0, 4)$, and the x-intercepts $(-0.85, 0)$ and $(2.35, 0)$. Then graph the function (Fig. 6.47).

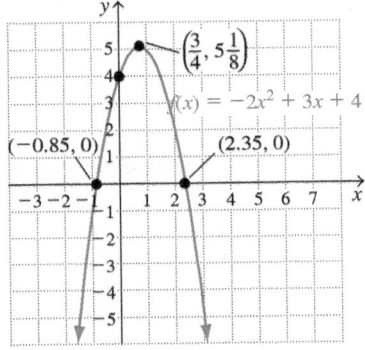

Figure 6.47

b) The domain, the values that can be used for x, is the set of all real numbers, $\mathbb{R}$. The range, the values of y, is $y \le 5\frac{1}{8}$.

When we use the quadratic formula to determine the x-intercepts of a graph, if the radicand, $b^2 - 4ac$, is a negative number, the graph has no x-intercepts. The graph will lie totally above or below the x-axis.

Exponential Functions and Their Graphs

Many real-life problems, including population growth, growth of bacteria, and decay of radioactive substances, increase or decrease at a very rapid rate. For example, the graph in Fig. 6.48 shows the growth of the world population. Notice how the graph is increasing rapidly. This is an example in which the graph is increasing *exponentially*. The equation of a graph that increases or decreases exponentially is called an *exponential equation* or an *exponential function*. An exponential equation or exponential function is of the form $y = a^x$ or $f(x) = a^x$, where in both cases $a > 0, a \neq 1$. When graphing exponential functions we often use the form $y = a^x$, and when working with applications we often use the form $f(x) = a^x$. However, since $y = f(x)$, we can use either form to represent exponential functions.

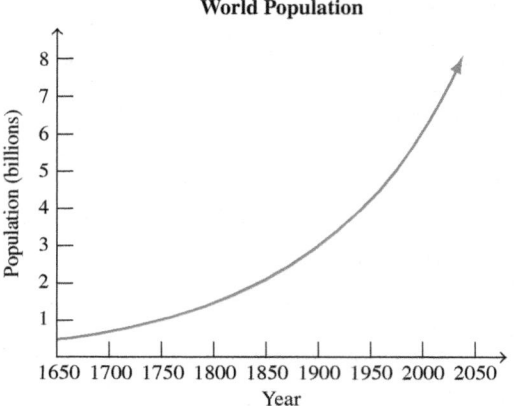

World Population

Source: U.S. Census Bureau

Figure 6.48

In an exponential function, letters other than x and y may be used for the variables. The following are examples of exponential functions: $f(x) = 2^x$, $A(r) = \left(\frac{1}{2}\right)^r$ and $P(t) = 2.3^t$. Note in exponential functions that the *variable is the exponent* of some positive constant that is not equal to 1. In many real-life applications, the variable t will be used to represent time. Problems involving exponential functions can be evaluated much more easily if you use a calculator containing a $\boxed{y^x}$, $\boxed{x^y}$, or $\boxed{\wedge}$ key.

The following function, referred to as the *exponential growth* or *decay formula*, is used to solve many real-life problems.

Exponential Growth or Decay Formula

$$P(t) = P_0 a^{kt}, \qquad a > 0, \quad a \neq 1$$

In the exponential growth or decay formula, P_0 represents the original amount present, $P(t)$ represents the amount present after t years, and a and k are constants.

When $k > 0$, and $a > 1$, $P(t)$ increases as t increases and we have exponential growth. When $k < 0$, $P(t)$ decreases as t increases and we have exponential decay.

Example 9 *Using an Exponential Decay Formula*

Carbon dating is used by scientists to determine the age of fossils, bones, and other items. The formula used in carbon dating is

$$P(t) = P_0 2^{-t/5600}$$

where P_0 represents the original amount of carbon-14 (C_{14}) present and $P(t)$ represents the amount of C_{14} present after t years. If 10 mg of C_{14} is present in an animal bone recently excavated, how many milligrams will be present in 3000 years?

Solution In this example, $P_0 = 10$ and $t = 3000$. Substituting these values into the formula gives

$$P(t) = P_0 2^{-t/5600}$$
$$P(3000) = 10(2)^{-3000/5600}$$
$$\approx 10(2)^{-0.54}$$
$$\approx 10(0.69)$$
$$\approx 6.9 \text{ mg}$$

(Recall that $\approx$ means "is approximately equal to.")

Thus, in 3000 years, approximately 6.9 mg of the original 10 mg of C_{14} will remain. ■

In Example 9, we used a scientific calculator to evaluate $(2)^{-3000/5600}$. The steps need to do this are described in the following Technology Tip.

TECHNOLOGY TIP

Evaluating an Exponential Function

The steps to evaluate $(2)^{-3000/5600}$ on a calculator with a ⌐∧⌐ key are given below.

2 ⌐∧⌐ ⌐(⌐ ⌐(-)⌐ 3000 ⌐÷⌐ 5600 ⌐)⌐ ⌐=⌐ .6898170602

Notice that before the 3000, we used the negative key ⌐(-)⌐, and not the ⌐–⌐ subtraction key.
After the ⌐=⌐ key is pressed, the calculator displays the answer 0.6898170602.
To evaluate $10(2)^{-3000/5600}$ on a scientific calculator, we can press the following keys.

10 ⌐×⌐ 2 ⌐∧⌐ ⌐(⌐ ⌐(-)⌐ 3000 ⌐÷⌐ 5600 ⌐)⌐ ⌐=⌐ 6.898170602.

Notice that the answer obtained using the calculator steps shown above is a little more accurate than the answer we gave when we rounded the values before the final answer in Example 9.

What does the graph of an *exponential function* of the form $y = a^x$, $a > 0$, $a \neq 1$, look like? Examples 10 and 11 illustrate graphs of exponential functions.

Example 10 *Graphing an Exponential Function with a Base Greater Than 1*

a) Graph $y = 2^x$.
b) Determine the domain and range of the function.

Solution

a) Substitute values for x and determine the corresponding values of y. The graph is shown in Fig. 6.49 on page 386.

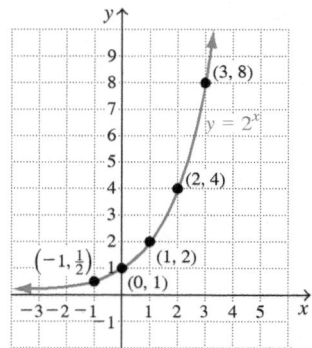

Figure 6.49

$$y = 2^x$$

$x = -3,$	$y = 2^{-3} = \dfrac{1}{2^3} = \dfrac{1}{8}$
$x = -2,$	$y = 2^{-2} = \dfrac{1}{2^2} = \dfrac{1}{4}$
$x = -1,$	$y = 2^{-1} = \dfrac{1}{2^1} = \dfrac{1}{2}$
$x = 0,$	$y = 2^0 = 1$
$x = 1,$	$y = 2^1 = 2$
$x = 2,$	$y = 2^2 = 4$
$x = 3,$	$y = 2^3 = 8$

x	y
-3	$\dfrac{1}{8}$
-2	$\dfrac{1}{4}$
-1	$\dfrac{1}{2}$
0	1
1	2
2	4
3	8

b) The domain is all real numbers, $\mathbb{R}$. The range is $y > 0$. Note that y can never have a value of 0.   ∎

All exponential functions of the form $y = a^x$, or $f(x) = a^x$ where $a > 1$, will have the general shape of the graph illustrated in Fig. 6.49.

Example 11 *Graphing an Exponential Function with a Base Between 0 and 1*

a) Graph $y = \left(\frac{1}{2}\right)^x$.

b) Determine the domain and range of the function.

Solution

a) We begin by substituting values for x and calculating values for y. We then plot the ordered pairs and use these points to graph the function. To evaluate a fraction with a negative exponent, we use the fact that

$$\left(\frac{a}{b}\right)^{-x} = \left(\frac{b}{a}\right)^{x}$$

For example,

$$\left(\frac{1}{2}\right)^{-3} = \left(\frac{2}{1}\right)^{3} = 8$$

Then

$$y = \left(\frac{1}{2}\right)^x$$

$x = -3,$	$y = \left(\dfrac{1}{2}\right)^{-3} = 2^3 = 8$
$x = -2,$	$y = \left(\dfrac{1}{2}\right)^{-2} = 2^2 = 4$
$x = -1,$	$y = \left(\dfrac{1}{2}\right)^{-1} = 2^1 = 2$
$x = 0,$	$y = \left(\dfrac{1}{2}\right)^{0} = 1$
$x = 1,$	$y = \left(\dfrac{1}{2}\right)^{1} = \dfrac{1}{2}$
$x = 2,$	$y = \left(\dfrac{1}{2}\right)^{2} = \dfrac{1}{4}$
$x = 3,$	$y = \left(\dfrac{1}{2}\right)^{3} = \dfrac{1}{8}$

x	y
-3	8
-2	4
-1	2
0	1
1	$\dfrac{1}{2}$
2	$\dfrac{1}{4}$
3	$\dfrac{1}{8}$

The graph is illustrated in Fig. 6.50.

b) The domain is the set of all real numbers, $\mathbb{R}$. The range is $y > 0$.   ∎

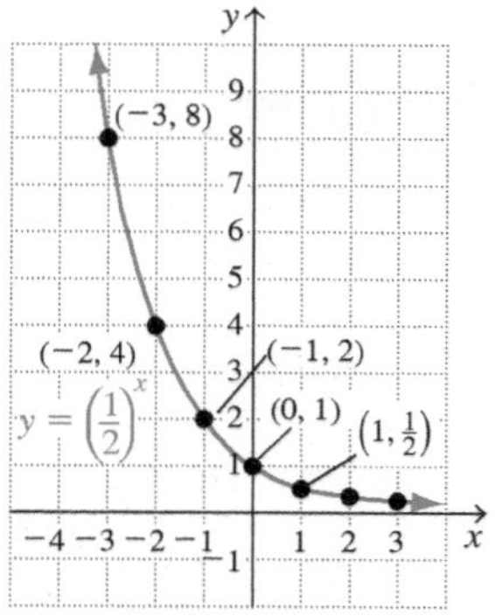

Figure 6.50

All exponential functions of the form $y = a^x$ or $f(x) = a^x, 0 < a < 1$, will have the general shape of the graph illustrated in Fig. 6.50.

Linear, Quadratic and Exponential Growth

Example 12 *Is the Growth Exponential?*

The graph below shows the population, in millions, of people age 85 and older in the United States for selected years from 1960 to 2010 and projected to 2050.

a) Does the graph approximate the graph of an exponential function?

b) Estimate the number of U.S. people age 85 and older in 2000.

Solution

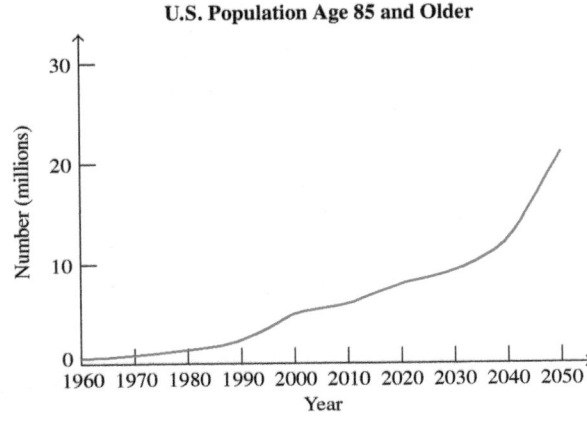

U.S. Population Age 85 and Older

Source: U.S. Census Bureau

a) Yes, the graph has the approximate shape of an exponential function. A function that increases rapidly with this general shape is, or approximates, an exponential function.

b) From the graph, we see that there were about 5 million people in the United States who were age 85 or older in 2000. ∎

Natural Exponential Functions

When the a in the exponential function $f(x) = a^x$ is replaced with a very special letter e, we get the *natural exponential function* $f(x) = e^x$. The exponential growth or decay formula is $P(t) = P_0 a^{kt}$. Replacing the letter a with the letter e in the exponential growth or decay formula leads us to the *natural exponential growth or decay formula*,

Natural Exponential Growth or Decay Formula

$$P(t) = P_0 e^{kt}$$

The letter e represents an irrational number called the natural base whose value is approximately 2.7183. The number e plays an important role in mathematics and is used in determining the solution to many application problems.

To evaluate $e^{0.012(6)}$ on a calculator, as will be needed in Example 13, press*

| 2nd | ln | (0.012 | × | 6 |) | = | 1.074655344

Notice that there is not a box around the left-hand parentheses. After the | ln | key is pressed the calculator automatically generates the left-hand parentheses.

---

*Keys to press may vary on some calculators.

After the $\boxed{=}$ key is pressed, the calculator displays the answer 1.074655344. To evaluate $7.53\,e^{0.012(6)}$ on a calculator, press

7.53 $\boxed{\times}$ $\boxed{2^{nd}}$ $\boxed{\ln}$ (0.012 $\boxed{\times}$ 6 $\boxed{)}$ $\boxed{=}$ 8.092154741

In this calculation, after the $\boxed{=}$ key is pressed, the calculator displays the answer 8.092154741. This calculation will be used in Example 13.

Example 13 *Evaluating an Exponential Growth Function*

The world population in 2017 was about 7.53 billion people. Assume that the world population continues to grow exponentially at the growth rate of 1.2% per year. The expected world population, in billions of people, t years after 2017, is given by the formula $P(t) = 7.53e^{0.012t}$. Determine the expected world population in 2023.

Source: U.S. Census Bureanu

Solution Since 2023 is 6 years after 2017, $t = 6$.

$$P(t) = 7.53e^{0.012t}$$
$$P(6) = 7.53e^{0.012(6)}$$
$$= 7.53e^{0.072}$$
$$\approx 7.53(1.074655344)$$
$$\approx 8.092154741$$

Thus, in 2023, the world population is expected to be about 8.09 billion people. ∎

Example 14 *Evaluating an Exponential Decay Function*

Plutonium, a radioactive material used in most nuclear reactors, decays exponentially at a rate of 0.003% per year. If there are originally 2000 grams of plutonium, the amount of plutonium, P, remaining after t years is $P(t) = 2000e^{-0.00003t}$. How much plutonium will remain after 50 years?

Solution Substitute 50 years for t in the function, then evaluate using a calculator as described earlier.

$$P(t) = 2000e^{-0.00003t}$$
$$P(50) = 2000e^{-0.00003(50)}$$
$$= 2000e^{-0.0015}$$
$$\approx 2000(0.9985011244)$$
$$\approx 1997.0 \text{ grams}$$

Thus, after 50 years, the amount of plutonium remaining will be about 1997 grams. ∎

Recall that all exponential functions of the form $y = a^x$ or $f(x) = a^x$, where $a > 1$ will have the general shape of the graph illustrated in Figure 6.49 on page 386. Since the natural base, $e \approx 2.7183$, is greater than 1, the shape of the graph of the exponential function, $y = e^x$ or $f(x) = e^x$, will also be similar to the graph in Figure 6.49.

SECTION 6.10
Exercises

Warm Up Exercises

In Exercises 1–8, fill in the blank with an appropriate word, phrase, or symbol(s).

1. Any set of ordered pairs is called a(n) _____ .

2. A special type of relation where each value of the independent variable corresponds to exactly one value of the dependent variable is called a(n) _____ .

3. The set of values that can be used for the independent variable is called the _____ of the function.

4. The set of values obtained for the dependent variable is called the _____ of the function.

5. In a quadratic function of the form $f(x) = ax^2 + bx + c$, if a is greater than 0, then the graph of $f(x)$ is a parabola that opens _____.

6. In a quadratic function of the form $f(x) = ax^2 + bx + c$, if a is less than 0, then the graph of $f(x)$ is a parabola that opens _____.

7. The formula used to determine the equation for the axis of symmetry of a parabola is _____.

8. A test that is used to determine whether a graph represents a function is called the _____ line test.

In Exercises 9–16, determine whether the set of ordered pairs is a function.

9. $\{(2, 5), (4, 9), (1, 6), (-1, 3)\}$

10. $\{(-4, 2), (1, 5), (3, -7), (5, 6)\}$

11. $\{(2, 5), (2, 7), (2, -1), (2, 2)\}$

12. $\{(-1, 2), (-1, 4), (-1, 6), (-1, 9)\}$

13. $\{(-1, -1), (0, -1), (3, -1)\}$

14. $\{(1, 1), (2, 1), (3, 1), (4, 1)\}$

15. $\{(1, 1), (1, 2), (1, 3), (1, 4)\}$

16. $\{(4, 2), (4, -3), (4, 5)\}$

Practice the Skills

In Exercises 17–30, determine whether the graph represents a function. If it does represent a function, give its domain and range.

17.
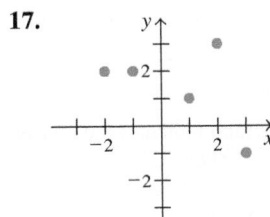

18.
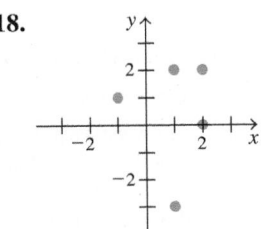

19.
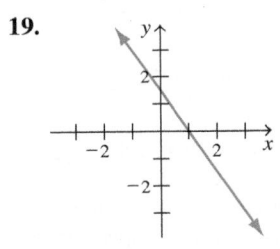

20.
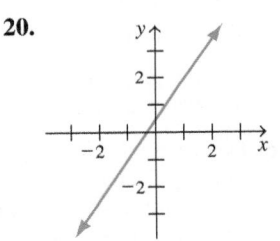

21.
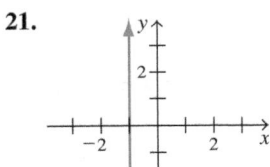

22.
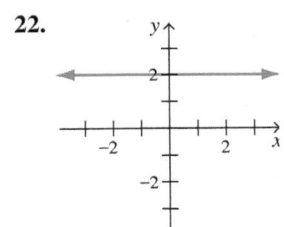

23.
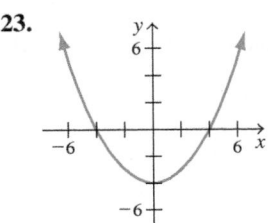

24.
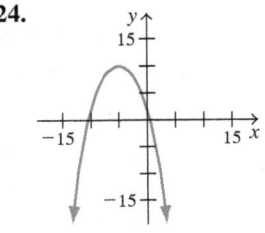

25.
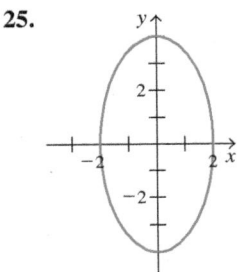

26.
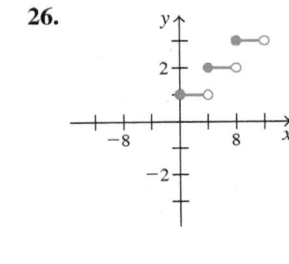

27.
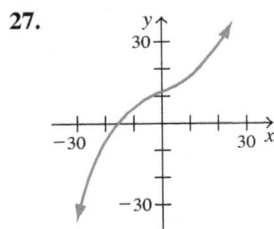

28.
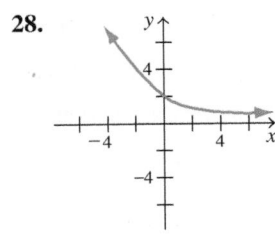

29.
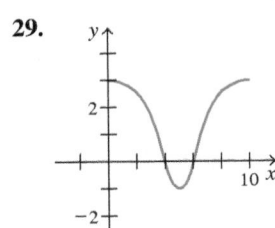

30.
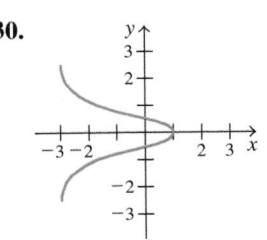

In Exercises 31–40, evaluate the function for the given value of x.

31. $f(x) = x + 3, \quad x = 5$

32. $f(x) = 4x - 1, \quad x = 2$

33. $f(x) = -3x + 3, \quad x = -1$

34. $f(x) = 8x + 2, \quad x = -2$

35. $f(x) = x^2 - 3x + 1, \quad x = 4$

36. $f(x) = x^2 - 5, \quad x = 7$

37. $f(x) = -x^2 - 2x + 1, \quad x = 4$

38. $f(x) = -2x^2 + 3x - 1, \quad x = 3$

39. $f(x) = 4x^2 - 6x - 9, \quad x = -3$

40. $f(x) = 3x^2 + 2x - 5, \quad x = -4$

In Exercises 41–46, graph the function by using the slope and y-intercept.

41. $f(x) = -x + 3$

42. $f(x) = 2x + 1$

43. $f(x) = -3x + 2$

44. $f(x) = 2x + 5$

45. $f(x) = \frac{3}{2}x - 1$

46. $f(x) = -\frac{1}{2}x + 3$

In Exercises 47–60,

a) *determine whether the parabola will open upward or downward.*

b) *determine the equation of the axis of symmetry.*

c) *determine the vertex.*

d) *determine the y-intercept.*

e) *determine the x-intercepts if they exist.*

f) *graph the function.*

g) *determine the domain and range of the function.*

47. $y = x^2 - 1$

48. $y = x^2 - 9$

49. $y = -x^2 + 4$

50. $y = -x^2 + 16$

51. $y = -2x^2 - 8$

52. $y = 2x^2 - 3$

53. $f(x) = x^2 + 4x - 4$

54. $f(x) = x^2 - 8x + 1$

55. $y = x^2 + 5x + 6$

56. $y = x^2 - 7x - 8$

57. $f(x) = -x^2 + 4x - 6$

58. $f(x) = -x^2 + 8x - 8$

59. $f(x) = -2x^2 + 3x - 2$

60. $f(x) = -4x^2 - 6x + 4$

In Exercises 61–70, graph the function and state the domain and range.

61. $y = 3^x$

62. $f(x) = 4^x$

63. $y = \left(\frac{1}{3}\right)^x$

64. $y = \left(\frac{1}{4}\right)^x$

65. $f(x) = 2^x + 1$

66. $y = 3^x - 1$

67. $y = 4^x + 1$

68. $y = 2^x - 1$

69. $y = 3^{x-1}$

70. $f(x) = 4^{x+1}$

Problem Solving

71. *Yearly Profit* Marta is a part owner of a newly opened candy company. Marta's yearly profit, in dollars, is given by the function $p(x) = 3.5x - 15{,}000$, where x is the number of pounds of candy sold per year. If Marta sells 20,000 pounds of candy a year, determine her yearly profit.

72. *Distances Traveled* The distance a car travels, $d(t)$, at a constant 60 mph is given by the function $d(t) = 60t$, where t is the time in hours. Determine the distance traveled in

a) 3 hours.

b) 7 hours.

73. *Lakewood Ranch Hourly Temperature* The graph on the top of page 391 was created using **Stat**Crunch with data obtained from AccuWeather.com. The graph indicates the hourly temperature, in degrees Fahrenheit, in Lakewood Ranch, Florida, from 10 A.M. to 6 P.M. on Sunday, March 17, 2019. The quadratic function $t(x) = -0.65x^2 + 5.16x + 64.42$ can be used to estimate the hourly temperature, where x is the number of hours since 10 A.M. and $0 \le x \le 8$.

a) Use the function $t(x)$ to estimate the temperature in Lakewood Ranch at 1 P.M. Round your answer to the nearest degree.

b) Use the graph to determine the hour that the temperature was a maximum.

c) Determine the x-coordinate of the vertex of the graph of $t(x)$. Then use this value in $t(x)$ to estimate the maximum temperature. Round your value of x to the nearest hour. Round your answer for the temperature to the nearest degree.

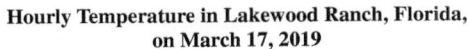

Hourly Temperature in Lakewood Ranch, Florida, on March 17, 2019

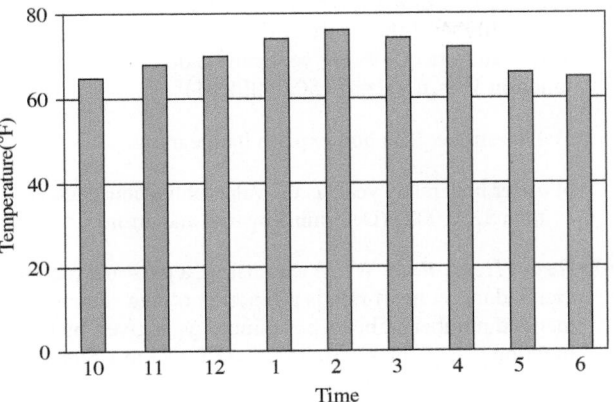

Source: AccuWeather.com

74. *Dubuque Hourly Temperature* The following graph was created using **StatCrunch** with data obtained from AccuWeather.com. The graph indicates the hourly temperature, in degrees Fahrenheit, in Dubuque, Iowa, from 9 A.M. to 5 P.M. on Sunday, March 17, 2019. The quadratic function $t(x) = -0.59x^2 + 4.71x + 44.39$ can be used to estimate the hourly temperature, where x is the number of hours since 9 A.M. and $0 \le x \le 8$.

Hourly Temperature in Dubuque, Iowa, on March 17, 2019

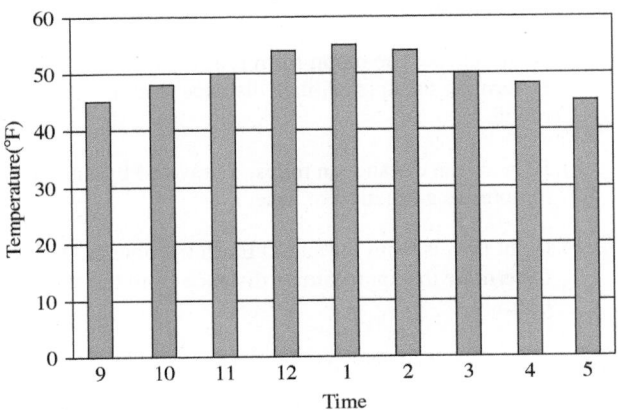

Source: AccuWeather.com

a) Use the function $t(x)$ to estimate the temperature in Dubuque at 12 P.M. Round your answer to the nearest degree.

b) Use the graph to determine the hour that the temperature was a maximum.

c) Determine the x-coordinate of the vertex of the graph of $t(x)$. Then use this value in $t(x)$ to estimate the maximum temperature. Round your value of x to the nearest hour. Round your answer for the temperature to the nearest degree.

75. *Bacteria* The number of a certain type of bacteria, y, present in a culture is determined by the equation $y = 2000(3)^x$, where x is the number of days the culture has been growing. Determine the number of bacteria present after 5 days.

76. *Adjusting for Inflation* If P is the price of an item today, the price of the same item n years from today, P_n, is $P_n = P(1 + r)^n$, where r is the constant rate of inflation. Determine the price of a movie ticket 10 years from today if the price today is $8.00 and the annual rate of inflation is constant at 3%.

77. *Expected Growth* The town of Lockport currently has 4000 residents. The expected future population can be approximated by the function $P(x) = 4000(1.3)^{0.1x}$, where x is the number of years in the future. Determine the expected population of Lockport in

a) 10 years.

b) 50 years.

78. *Value of a Car* The cost of a new car is $30,000. If it depreciates at a rate of 18% per year, the value of the car in t years can be approximated by the function $V(t) = 30,000(0.82)^t$. Determine the value of the car in 4 years. Round your answer to the nearest cent.

79. *Population of China* In 2017, the population of China was about 1.39 billion people, and the population growth rate was about 0.6% per year. Assuming the population of China continues to grow exponentially at this rate, the expected population of China, in billions of people, t years after 2017, is given by the exponential function $P(t) = 1.39e^{0.006t}$. Use this function to predict the population of China in the year 2025. Source: U.S. Census Bureau

80. *Population of India* In 2017, the population of India was about 1.34 billion people, and the population growth rate was about 1.1% per year. Assuming the population of India continues to grow exponentially at this rate, the expected population of India, in billions of people, t years after 2017, is given by the exponential function $P(t) = 1.34e^{0.011t}$. Use this function to predict the population of India in the year 2025. Source: U.S. Census Bureau

81. *Value of New York City* Assume the value of the island of Manhattan has grown at an exponential rate of 8% per year since 1626 when Peter Minuit of the Dutch West India Company purchased the island for $24. The value of the island, V, at any time, t, in years after 1626, can be determined by the formula $V(t) = 24e^{0.08t}$. What is the value of the island in 2020, 394 years after Minuit purchased it? Use a scientific calculator and give your answer in scientific notation as provided by your calculator.

82. *Louisiana Purchase* The map on page 392 shows the land purchased by the United States from France in 1803 as part of the Louisiana Purchase. Assume the value of the land has grown at a rate of 8% per year since the land was purchased for about 4 cents, or $0.04, per acre. The value of the land per acre can be approximated by the exponential function $V(t) = 0.04e^{0.08t}$. Approximate the value of the land per acre in 2020, 217 years after the purchase.

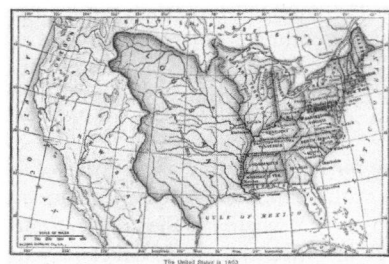

▲ *See Exercise 82.*

83. ***Radioactive Decay*** Radium-226 is a radioactive isotope that decays exponentially at a rate of 0.0428% per year. The amount of radium-226, R, remaining after t years can be determined by the formula $R(t) = R_0 e^{-0.000428t}$, where R_0 is the original amount present. If there are originally 10 grams of radium-226, determine the amount of radium-226 remaining after 1000 years.

84. ***Radioactive Decay*** Strontium-90 is a radioactive isotope that decays exponentially at a rate of 2.8% per year. The amount of strontium-90, S, remaining after t years can be determined by the function $S(t) = S_0 e^{-0.028t}$, where S_0 is the original amount present. If there are originally 1000 grams of strontium-90, determine the amount of strontium-90 remaining after 40 years. Round your answer to the nearest gram.

85. The spacing of the frets on the neck of a classical guitar is determined from the equation $d = (21.9)(2)^{(20-x)/12}$, where $x = $ the fret number and $d = $ the distance in centimeters of the xth fret from the bridge.

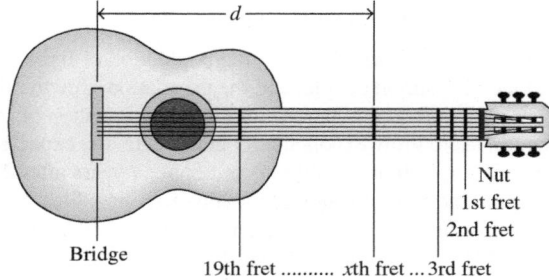

a) Determine how far the 19th fret should be from the bridge (rounded to one decimal place).

b) Determine how far the 4th fret should be from the bridge (rounded to one decimal place).

c) The distance of the nut from the bridge can be determined by letting $x = 0$ in the given exponential equation. Determine the distance from the nut to the bridge (rounded to one decimal place).

Challenge Problems/Group Activities

86. ***Appreciation of a House*** A house initially cost $150,000. The value, V, of the house after n years if it appreciates at a constant rate of 4% per year can be determined by the function $V = f(n) = \$150{,}000(1.04)^n$.

a) Determine $f(8)$ and explain its meaning.

b) After how many years is the value of the house greater than $250,000? (Determine by trial and error.)

87. ***Target Heart Rate*** While exercising, a person's recommended target heart rate is a function of age. The recommended number of beats per minute, y, is given by the function $y = f(x) = -0.85x + 187$, where x represents a person's age in years. Determine the number of recommended heart beats per minute for the following ages.

a) 20

b) 30

c) 50

d) 60

e) What is the age of a person with a recommended target heart rate of 85?

88. ***Speed of Light*** Light travels at about 186,000 miles per second through space. The distance, d, in miles that light travels in t seconds can be determined by the function $d(t) = 186{,}000t$.

a) Light reaches the moon from Earth in about 1.3 sec. Determine the approximate distance from Earth to the moon.

b) Express the distance in miles, d, traveled by light in t minutes as a function of time, t.

c) Light travels from the sun to Earth in about 8.3 min. Determine the approximate distance from the sun to Earth.

Research Activity

89. ***François Viète*** The idea of using variables in algebraic equations was introduced by the French mathematician François Viète (1540–1603). Write a paper about his life and his contributions to mathematics. In particular, discuss his work with algebraic equations.

CHAPTER 6   *Summary*

Important Facts and Concepts	Examples and Discussion
Section 6.1	
Order of Operations	Examples 1–2, pages 289–290
Properties Used to Solve Equations	Examples 4–13, pages 291–296
Addition property of equality	
If $a = b$, then $a + c = b + c$.	
Subtraction property of equality	
If $a = b$, then $a - c = b - c$.	
Multiplication property of equality	
If $a = b$, then $ac = bc, c \neq 0$.	
Division property of equality	
If $a = b$, then $\dfrac{a}{c} = \dfrac{b}{c}, c \neq 0$.	
Section 6.2	
Evaluating a formula	Examples 1–3, pages 298–300
Solving for a variable in a formula	Examples 4–6, pages 300–301
Section 6.3	
Applications of Linear Equations	
Proportions	Examples 1–4, pages 305–307
If $\dfrac{a}{b} = \dfrac{c}{d}$, then $ad = bc$.	Examples 5–6, pages 308–309
Section 6.4	
Variation	
Direct: $y = kx$ Inverse: $y = \dfrac{k}{x}$ Joint: $y = kxz$	Examples 1–10, pages 313–318
Section 6.5	
Inequality Symbols	Examples 1–9, pages 321–325
$a < b$ means that a is less than b.	
$a \leq b$ means that a is less than or equal to b.	
$a > b$ means that a is greater than b.	
$a \geq b$ means that a is greater than or equal to b.	
Section 6.6	
Intercepts	
To determine the x-intercept, set $y = 0$ and solve the resulting equation for x.	Example 4, pages 331–332
To determine the y-intercept, set $x = 0$ and solve the resulting equation for y.	
Slope	
Slope (m): $m = \dfrac{y_2 - y_1}{x_2 - x_1}$	Example 5, page 333–334
Linear equation in two variables:	Examples 3, 4, 6, 7, 10, 11, pages 330–336
$ax + by = c, \quad a \neq 0 \quad \text{and} \quad b \neq 0$	
Slope–intercept form of a line:	Examples 6–8, pages 334–335
$y = mx + b$	

Section 6.7

Solving Systems of Equations by Graphing

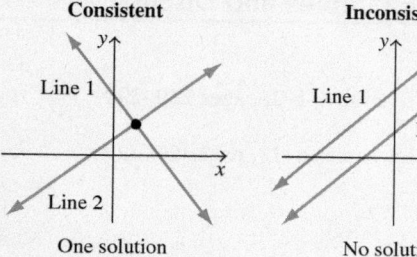

Consistent

Line 1

Line 2

One solution

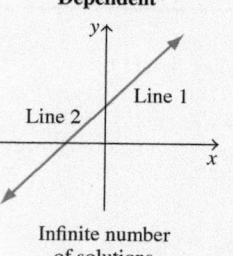

Inconsistent

Line 1

Line 2

No solution

Dependent

Line 2

Line 1

Infinite number
of solutions

Discussion page 342

Example 2, page 343

Example 9, pages 348–349

Solving Systems of Equations by Algebraic Methods

Substitution

Addition (or elimination) method

Examples 3–5, pages 344–345

Examples 6–8, 10, pages 346–348, 349–350

Section 6.8

Linear Inequalities in Two Variables

Examples 1–3, pages 356–357

System of Linear Inequalities

Graph both inequalities on the same axes. The solution is the set of points that satisfy both inequalities.

Examples 4–6, pages 357–359

Fundamental Principle of Linear Programming

If the objective function $K = Ax + By$ is evaluated at each point in a feasible region, the maximum and minimum values of the equation occur at vertices of the region.

Discussion pages 359–360

Examples 7–8, pages 360–362

Section 6.9

Quadratic equation in one variable:
$$ax^2 + bx + c = 0, \quad a \neq 0$$

Examples 8, 9, page 371–372

Quadratic formula:

$$x = \frac{-b \pm \sqrt{b^2 - 4ac}}{2a}$$

Examples 10,11, pages 372–373

Zero-Factor Property

If $a \cdot b = 0$, then $a = 0$ or $b = 0$.

Example 7–8, pages 370–371

Section 6.10

Quadratic equation (or function) in two variables:
$$y = ax^2 + bx + c, \quad a \neq 0$$

Examples 5–8, pages 379–383

Axis of symmetry of a parabola:
$$x = \frac{-b}{2a}$$

Examples 5–8, pages 379–383

Exponential function:
$$y = a^x \quad \text{or} \quad f(x) = a^x, \quad a \neq 1, \quad a > 0$$

Examples 9–14, pages 385–388

Exponential growth or decay formula:
$$P(t) = P_0 a^{kt}, a > 0, a \neq 1$$

Example 9, page 385

Natural exponential function:
$$f(x) = e^x, e \approx 2.7183$$

Examples 13, 14, page 388

Natural exponential growth or decay function:
$$P(t) = P_0 e^{kt}$$

Examples 13, 14, page 388

CHAPTER 6   *Review Exercises*

6.1

In Exercises 1–4, evaluate the expression for the given value(s) of the variable.

1. $x^2 + 10x$, $x = 3$

2. $-x^2 - 5$, $x = -2$

3. $3x^2 - 2x + 7$, $x = 2$

4. $4x^3 + 7x^2 - 3x + 1$, $x = -1$

In Exercises 5 and 6, combine like terms.

5. $3x - 9 + x + 6$

6. $2x + 3(x - 2) + 7x$

In Exercises 7–10, solve the equation for the given variable.

7. $4t + 8 = -12$

8. $4(x - 2) = 3 + 5(x + 4)$

9. $\dfrac{x + 5}{2} = \dfrac{x - 3}{4}$

10. $\dfrac{x + 2}{5} = \dfrac{x + 1}{4}$

6.2

In Exercises 11–14, use the formula to determine the value of the indicated variable for the values given.

11. $F = ma$
    Determine F when $m = 40$ and $a = 5$ (physics).

12. $v = lwh$
    Determine v when $l = 10$, $w = 7$, and $h = 3$ (geometry).

13. $z = \dfrac{\bar{x} - \mu}{\dfrac{\sigma}{\sqrt{n}}}$
    Determine $\bar{x}$ when $z = 2$, $\mu = 100$, $\sigma = 3$, and $n = 16$ (statistics).

14. $E = mc^2$
    Determine m when $E = 400$ and $c = 4$ (physics).

In Exercises 15 and 16, solve for y.

15. $8x - 4y = 16$

16. $2x + 9y = 17$

In Exercises 17 and 18, solve for the variable indicated.

17. $A = lw$,   for w

18. $L = 2(wh + lh)$,   for l

6.3

In Exercises 19 and 20, write the phrase in mathematical terms.

19. 5, increased by 3 times x

20. The difference between 9 divided by q and 15

In Exercises 21–24, write an equation that can be used to solve the problems. Solve the equation and determine the desired value(s).

21. Three, increased by 7 times a number is 17.

22. The product of 3 and a number, increased by 8, is 6 less than the number.

23. Five times the difference of a number and 4 is 45.

24. Fourteen more than 10 times a number is 8 times the sum of the number and 12.

In Exercises 25–28, write the equation and then determine the solution.

25. MODELING— *Investing* Jim received an inheritance of $15,000. If he wants to invest twice as much money in mutual funds as in bonds, how much should he invest in mutual funds?

26. MODELING— *Restaurant Profit* John owns two restaurants. His profit for a year at restaurant A is $15,000 greater than his profit at restaurant B. The total profit from both restaurants is $75,000. Determine the profit at each restaurant.

27. *Making Oatmeal* A recipe for Hot Oats Cereal calls for 2 cups of water and for $\frac{1}{3}$ cup of dry oats. How many cups of dry oats would be used with 3 cups of water?

28. *Laying Blocks* A mason lays 120 blocks in 1 hr 40 min. How long will it take her to lay 300 blocks?

6.4

In Exercises 29–32, determine the quantity indicated.

29. s is directly proportional to t. If $s = 60$ when $t = 10$, determine s when $t = 12$.

30. J is inversely proportional to the square of A. If $J = 25$ when $A = 2$, determine J when $A = 5$.

31. W is directly proportional to L and inversely proportional to A. If $W = 80$ when $L = 100$ and $A = 20$, determine W when $L = 50$ and $A = 40$.

32. z is jointly proportional to x and y and inversely proportional to the square of r. If $z = 12$ when $x = 20$, $y = 8$, and $r = 8$, determine z when $x = 10$, $y = 80$, and $r = 3$.

33. *Property Tax* The property tax, t, on a house is directly proportional to the assessed value, v, of the house. If the property tax on a house with an assessed value of $155,000 is $2325, what is the property tax on a house with an assessed value of $210,000?

34. *Electric Bill* An electric company charges $0.162 per kilowatt-hour (kWh). What is the electric bill if 740 kWh are used in a month?

6.5

In Exercises 35 and 36, graph the solution set for the set of real numbers on the real number line.

35. $8 + 9x \le 6x - 4$

36. $3(x + 9) \le 4x + 11$

In Exercises 37 and 38, graph the solution set for the set of integers on the real number line.

37. $5x + 13 \ge -22$

38. $-8 \le x + 2 \le 7$

6.6

In Exercises 39–42, graph the ordered pair in the Cartesian coordinate system.

39. $(-2, 1)$ **40.** $(2, -2)$

41. $(-4, 3)$ **42.** $(5, 3)$

In Exercises 43 and 44, graph the equation by plotting points.

43. $x - y = 2$ **44.** $2x + 3y = 12$

In Exercises 45 and 46, graph the equation, using the x- and y-intercepts.

45. $x - 2y = 6$ **46.** $4x - 3y = 12$

In Exercises 47–50, determine the slope of the line through the given points.

47. $(1, 3), (4, 7)$ **48.** $(3, -1), (5, -4)$

49. $(-1, -4), (2, 3)$ **50.** $(6, 2), (6, -2)$

In Exercises 51 and 52, graph the equation by plotting the y-intercept and then plotting a second point by making use of the slope.

51. $y = 2x - 5$ **52.** $2y - 4 = 3x$

In Exercises 53 and 54, determine the equation of the graph.

53. 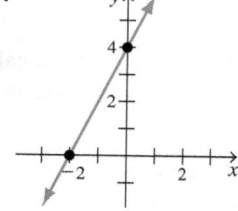   **54.**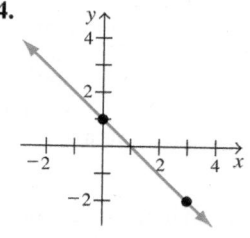

55. MODELING— *Taxi Business Cost* Tony's weekly cost, C, in dollars, of operating a taxi business can be approximated by the equation $C = 0.20m + 80$, where m is the number of miles driven in a week.

a) Draw a graph of weekly cost versus the number of miles driven for up to and including 500 miles driven per week.

b) If, during one week, Tony drove the taxi 150 miles, what would be the cost?

c) How many miles would Tony have to drive for the weekly cost to be $104?

6.7

In Exercises 56 and 57, solve the system of equations graphically. If the system does not have a single ordered pair as a solution, state whether the system is inconsistent or dependent.

56. $x + 3y = 5$
$\quad\; 3x - 3y = 3$
57. $x + 2y = 5$
$\quad\; 2x + 4y = 4$

In Exercises 58–61, solve the system of equations by the substitution method. If the system does not have a single ordered pair as a solution, state whether the system is inconsistent or dependent.

58. $-x + y = -2$
$\quad\;\; x + 2y = 5$
59. $x - 2y = 9$
$\quad\;\; y = 2x - 3$

60. $2x - y = 4$
$\quad\; 3x - y = 2$
61. $3x + y = 1$
$\quad\; 3y = -9x - 4$

In Exercises 62–65, solve the system of equations by the addition method. If the system does not have a single ordered pair as a solution, state whether the system is inconsistent or dependent.

62. $x + y = 2$
$\quad\; x + 3y = -2$
63. $4x - 8y = 16$
$\quad\;\; x - 2y = 4$

64. $3x - 4y = 10$
$\quad\; 5x + 3y = 7$
65. $3x + 4y = 6$
$\quad\; 2x - 3y = 4$

66. MODELING— *Borrowing Money* A company borrows $400,000 for 1 year to expand its product line. Some of the money was borrowed at a 3% simple interest rate, and the rest of the money was borrowed at a 6% simple interest rate. How much money was borrowed at each rate if the total annual interest was $16,500?

67. MODELING— *Chemistry* In chemistry class, Tom has an 80% acid solution and a 50% acid solution. How much of each solution should he mix to get 100 liters of a 75% acid solution?

68. MODELING — *Cool Air* Emily needs to purchase a new air conditioner for the office. Model 1600A costs $950 to purchase and $32 per month to operate. Model 6070B, a more efficient unit, costs $1275 to purchase and $22 per month to operate.

a) After how many months will the total cost of both units be equal?

b) Which model will be the more cost effective if the life of both units is guaranteed for 10 years?

69. MODELING— *Minimizing Parking Costs* The cost of parking in All-Day parking lot is $5 for the first hour and $0.50 for each additional hour. Sav-a-Lot parking lot costs $4.25 for the first hour and $0.75 for each additional hour.

    a) In how many hours after the first hour would the total cost of parking at All-Day and Sav-a-Lot be the same?

    b) If Mark needed to park his car for 5 hr, which parking lot would be less expensive?

6.8

In Exercises 70 and 71, graph the inequality.

70. $2x + 3y \le 12$ **71.** $4x + 2y \ge 12$

In Exercises 72 and 73, graph the system of linear inequalities and indicate the solution set.

72. $2x + y < 8$ **73.** $x - y > 5$
     $y \ge 2x - 1$ $6x + 5y \le 30$

74. The set of constraints and profit formula for a linear programming problem are

$$2x + 3y \le 12$$
$$2x + y \le 8$$
$$x \ge 0$$
$$y \ge 0$$
$$P = 5x + 3y$$

    a) Draw the graph of the constraints and determine the vertices of the feasible region.

    b) Use the vertices to determine the maximum and minimum profit.

6.9

In Exercises 75–78, factor the trinomial.

75. $x^2 + 6x + 9$

76. $x^2 + 2x - 15$

77. $x^2 - 10x + 24$

78. $6x^2 + 7x - 3$

In Exercises 79–82, solve the equation by factoring.

79. $x^2 + 9x + 20 = 0$

80. $x^2 + 3x = 10$

81. $3x^2 - 17x + 10 = 0$

82. $3x^2 = -7x - 2$

In Exercises 83–86, solve the equation using the quadratic formula. If the equation has no real solution, so state.

83. $x^2 - 6x - 16 = 0$

84. $2x^2 - x - 3 = 0$

85. $2x^2 - 3x + 4 = 0$

86. $x^2 - 3x - 2 = 0$

6.10

In Exercises 87–90, determine whether the graph represents a function. If it does represent a function, give its domain and range.

87.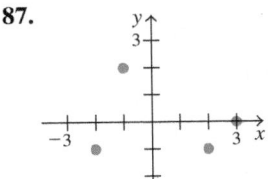

88.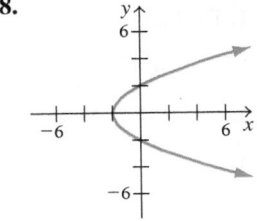

89.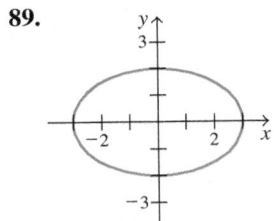

90.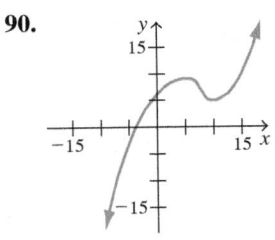

In Exercises 91–94, evaluate f(x) for the given value of x.

91. $f(x) = 4x + 3, \quad x = 4$

92. $f(x) = -2x + 5, \quad x = -3$

93. $f(x) = 3x^2 - 2x + 1, \quad x = 5$

94. $f(x) = -4x^2 + 7x + 9, \quad x = -1$

In Exercises 95 and 96, for each function

    a) *determine whether the parabola will open upward or downward.*

    b) *determine the equation of the axis of symmetry.*

    c) *determine the vertex.*

    d) *determine the y-intercept.*

    e) *determine the x-intercepts if they exist.*

    f) *graph the function.*

    g) *determine the domain and range.*

95. $y = -x^2 - 4x + 21$

96. $f(x) = 2x^2 + 8x + 6$

In Exercises 97 and 98, graph the function and state the domain and range.

97. $y = 4^x$

98. $y = \left(\dfrac{1}{2}\right)^x$

6.1, 6.2, 6.10

99. *Gas Mileage* The gas mileage, m, of a specific car can be estimated by the equation (or function)

$$m = 30 - 0.002n^2, \quad 20 \le n \le 80$$

where n is the speed of the car in miles per hour. Estimate the gas mileage when the car travels at 60 mph.

100. *Filtered Light* The percent of light filtering through Swan Lake, P, can be approximated by the function $P(x) = 100(0.92)^x$, where x is the depth in feet. Determine the percent of light filtering through at a depth of 4.5 ft.

CHAPTER 6   *Test*

1. Evaluate $3x^2 + 6x - 1$, when $x = -2$.

In Exercises 2 and 3, solve the equation.

2. $4x + 6 = 2(3x - 7)$

3. $-2(x - 3) + 6x = 2x + 3(x - 4)$

In Exercise 4, write an equation to represent the problem. Then solve the equation.

4. *Salary* Mary's salary is $350 per week plus a 6% commission of sales. How much in sales must Mary make to earn a total of $710 per week?

5. Evaluate $L = ah + bh + ch$ when $a = 2, b = 5, c = 4$, and $h = 7$.

6. Solve $3x + 5y = 11$ for y.

7. For a constant area, the length, l, of a rectangle varies inversely as the width, w. If $l = 15$ ft when $w = 9$ ft, determine the length of a rectangle with the same area if the width is 20 ft.

8. Graph the solution set of $-5x + 14 \le 2x + 35$ on the real number line.

9. Determine the slope of the line through the points $(-2, 8)$ and $(1, 14)$.

10. Graph the equation $2x - 3y = 15$

11. Solve the system of equations graphically.

$$y = 2x - 12$$
$$2x + 2y = -6$$

In Exercises 12 and 13, solve the system of equations by the method indicated.

12. $x + y = -1$
$\quad\;\; 2x + 3y = -5$
$\quad\;\;$ (substitution)

13. $4x + 3y = 5$
$\quad\;\; 2x + 4y = 10$
$\quad\;\;$ (addition)

14. MODELING—*Truck Rental* U-Haul charges a daily fee plus a mileage charge to rent a truck. Dorothy rented

a truck from U-Haul and was charged $132 for 3 days' rental and 150 miles driven. Elena rented the same truck and was charged $142 for 2 days' rental and 400 miles driven. Determine the daily fee and the mileage charge for renting this truck.

15. Graph the inequality $3y \ge 5x - 12$.

16. The set of constraints and profit formula for a linear programming problem are

$$x + 3y \le 6$$
$$4x + 3y \le 15$$
$$x \ge 0$$
$$y \ge 0$$
$$P = 6x + 4y$$

a) Graph the constraints and determine the vertices of the feasible region.

b) Use the vertices to determine the maximum and minimum profit.

17. Solve the equation $x^2 + 7x = -6$ by factoring.

18. Solve the equation $3x^2 + 2x = 8$ by using the quadratic formula.

19. Evaluate $f(x) = -2x^2 - 8x + 7$ when $x = -2$.

20. For the graph of the equation $y = x^2 - 2x + 4$,

a) determine whether the parabola will open upward or downward.

b) determine the equation of the axis of symmetry.

c) determine the vertex.

d) determine the y-intercept.

e) determine the x-intercepts if they exist.

f) graph the function.

g) determine the domain and range of the function.

◄ *Everywhere outside the United States you may see traffic and other signs given in metric units.*

7

The Metric System

What You Will Learn

- Basic terms and conversions within the metric system
- Length, area, and volume in the metric system
- Mass and temperature in the metric system
- Dimensional analysis and conversions to and from the metric system

The United States is the only major country not presently using the metric system as its primary system of measurement. For example, if you travel to Canada, Mexico, or overseas, you will find that when you purchase gasoline, the amount is measured in liters, not gallons.

In the United States, because of the growth in technology and the increase in international trade, more and more of the products we use daily are measured in metric units. For example, soda is sold in liter bottles, medicines are measured in milligrams, and tire sizes are given in centimeters. When you purchase a computer, the size of the hard drive, such as 12 terabytes (TB), is given using a metric measurement. When you purchase a storage device to store your music or videos, the size of the flash drive or other storage device may be 128 gigabytes (GB), also a metric measurement. Throughout this chapter we will discuss the importance of the metric system more thoroughly.

SECTION 7.1   Basic Terms and Conversions Within the Metric System

Upon completion of this section, you will be able to:

■ Perform conversions within the metric system.

Recall from your science courses that most of the measurements were in metric units. For example, in chemistry you may have worked with liter containers, and in physics you may have worked with a mass in kilograms rather than a weight in pounds. Have you ever asked yourself why the sciences use metric measurement? In this section, we will explain some benefits of using the metric system.

Why This Is Important In addition to being used in science courses, most countries use the metric system for measurement. A knowledge of the metric system will help you not only at home but also when you travel to other countries. It will also help you when you shop for electronic and other products.

Most countries of the world use the *Système international d'unités* or *SI system*. The SI system is generally referred to as the *metric system* in the United States. The metric system was named for the Greek word *metron*, meaning "measure." The metric system was first developed in France during the French Revolution. Two systems of weights and measures exist side by side in the United States today, the *U.S. customary system* and the metric system. The metric system is used predominantly in the automotive, construction, farm equipment, computer, bottling industries, and in health-related professions. Furthermore, almost every industry that ships internationally uses at least some metric measures.

In this chapter, we will discuss the metric measurements of length, area, volume, mass, and temperature. Using the metric system has many advantages. Some of them are summarized here.

1. The metric system is the worldwide accepted standard measurement system. All industrial nations that trade internationally, except the United States, use the metric system as the official system of measurement.

2. In the metric system there is only one basic unit of measurement for each physical quantity. In the U.S. customary system, many units are often used to represent the same physical quantity. For example, when discussing length, we use inches, feet, yards, miles, and so on. Converting from one of these units to the other is often a tedious task (consider changing 12 miles to inches). In the metric system, we can make many conversions by simply moving the decimal point.

3. The SI system is based on the number 10, and there is less need for fractions because most quantities can be expressed as decimals.

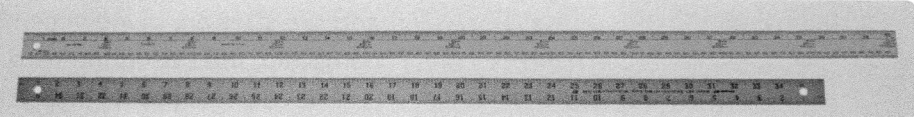

▲ *A meter (top figure) is a little longer than a yard (bottom figure).*

Write Small and Large Numbers in the Metric System

In some countries that use the metric system, a comma is used as we use a decimal point. For example, we write one-tenth as 0.1. Some countries, however, write one-tenth as 0,1. The photo below on the left, taken in Italy, shows the comma being used in the price in euros. Notice the comma being used as we use a decimal point.

When writing large numbers, some countries that use the metric system use a decimal point as we use a comma. The photo below on the right shows a decimal point being used as we use a comma. In this photo HUF stands for Hungarian forint, the Hungarian unit of currency. The blanket shown in the photo costs 25 thousand HUF.

When writing large or very small numbers in the metric system, instead of using either commas or decimal points, groups of three digits are often separated by spaces. Thus, one million is written 1 000 000 in some countries. The number we would write as 0.00003 would be written as 0.000 03. When there are only four digits in a number, the space is generally omitted. Thus, four ten-thousandths would be written 0.0004, and six thousand four hundred would be written 6400. The photo below, taken in Europe, shows the number one million written using spaces between groups of three digits. In this chapter, when a number has 5 or more digits, we will use spaces to separate groups of 3 digits.

Did You Know?

Lost in Space

▲ *The missing Mars Climate Orbiter*

In September 1999, the United States lost a $125 million space-craft, the *Mars Climate Orbiter*, as it approached Mars. Two space-craft teams, one at NASA's Jet Propulsion Laboratory (JPL) and the other at a Lockheed Martin facility, were unknowingly ex-changing some vital information in different measurement units. The spacecraft team at Lockheed sent some measurements to the spacecraft team at JPL using U.S. customary units. The JPL team as-sumed the information it received was in metric units. The mix-up in units led to the JPL scientists giving the spacecraft's computer the wrong information, which led to the spacecraft entering the Martian atmosphere, where it burned up. NASA has taken steps to prevent this error from ever happening again.

Basic Terms in the Metric System

Because the official definitions of many metric terms are quite technical, we present them informally.

The *meter* (m) is commonly used to measure *length* in the metric system. One meter is a little more than a yard. A door is about 2 meters high.

The *kilogram* (kg) is commonly used to measure *mass*. The difference between mass and weight is discussed in Section 7.3. One kilogram is about 2.2 pounds. A newborn baby may have a mass of about 3 kilograms. The *gram* (g), a unit of mass derived from the kilogram, is used to measure small amounts. A nickel has a mass of about 5 grams.

The *liter* (ℓ) is commonly used to measure *volume*. One liter is a little more than a quart. The gas tank of a compact car may hold 50 liters of gasoline.

Thus,

$$1 \text{ m } \approx 1 \text{ yd}$$
$$1 \text{ kg } \approx 2.2 \text{ lb}$$
$$1 \, \ell \approx 1 \text{ qt}$$

The term *degree Celsius* (°C)* is used to measure *temperature*. The freezing point of water is 0°C, and the boiling point of water is 100°C. The temperature on a warm day may be 30°C.

0°C = 32°F	Water freezes
23°C = 73.4°F	Comfortable room temperature
37°C = 98.6°F	Average body temperature
100°C = 212°F	Water boils

Prefixes Used in the Metric System

The metric system is based on the number 10 and therefore is a decimal system. Prefixes are used to denote a multiple or part of a basic unit. Table 7.1 summarizes the more commonly used prefixes and their meanings. In the table, where we mention "basic units" we mean metric units without prefixes, such as meter, gram, or liter. From Table 7.1, we can determine that a *deka*meter represents 10 meters and a *centi*meter represents $\frac{1}{100}$ of a meter. Also, 1 kiloliter = 1000 liters, 1 kilogram = 1000 grams, and 1 milliliter = $\frac{1}{1000}$ liter, and so on.

Table 7.1   Metric Prefixes

Prefix	Symbol	Meaning
kilo	k	1000 × basic unit
hecto	h	100 × basic unit
deka	da	10 × basic unit
—	—	basic unit (meter, gram, liter, . . .)
deci	d	$\frac{1}{10}$ of basic unit
centi	c	$\frac{1}{100}$ of basic unit
milli	m	$\frac{1}{1000}$ of basic unit

*The metric system uses the Kelvin temperature scale. However, many of the sciences use the Celsius temperature scale.

For scientific work that involves very large and very small quantities, the following prefixes are also used: *mega* (M) is one million times the basic unit, *giga* (G) is one billion times the basic unit, *tera* (T) is one trillion times the basic unit, *micro* (μ, the Greek letter mu) is one millionth of the basic unit, *nano* (n) is one billionth of the basic unit, and *pico* (p) is one trillionth of the basic unit.

In this book, the abbreviations or symbols for units of measure are not pluralized, but full names are. For example, 5 milliliters is symbolized as 5 mℓ, not 5 mℓs. Some countries that use the metric system do not use an "s" in their abbreviations, whereas others do.

MATHEMATICS TODAY   Additional Metric Prefaces

Information about the metric system is available through the National Institute of Standards and Technology website, www.nist.gov. Two documents included on the website are the *Metric Style Guide* and *A Brief History of Measurement Systems*. The chart on the right was obtained from the latter.

Why This Is Important We often deal with very large and very small quantities that may be stated using prefixes like giga, mega, micro, and nano. For example, a camera's memory card may contain 128 gigabytes of memory.

METRIC PREFIXES		
Multiples and Submultiples	**Prefixes**	**Symbols**
1 000 000 000 000 000 000 000 000 = 10^{24}	yotta	Y
1 000 000 000 000 000 000 000 = 10^{21}	zetta	Z
1 000 000 000 000 000 000 = 10^{18}	exa	E
1 000 000 000 000 000 = 10^{15}	peta	P
1 000 000 000 000 = 10^{12}	tera	T
1 000 000 000 = 10^{9}	giga	G
1 000 000 = 10^{6}	mega	M
1000 = 10^{3}	kilo	k
100 = 10^{2}	hecto	h
10 = 10^{1}	deka	da*
1 = 10^{0}		
0.1 = 10^{-1}	deci	d
0.01 = 10^{-2}	centi	c
0.001 = 10^{-3}	milli	m
0.000 001 = 10^{-6}	micro	μ
0.000 000 001 = 10^{-9}	nano	n
0.000 000 000 001 = 10^{-12}	pico	p
0.000 000 000 000 001 = 10^{-15}	femto	f
0.000 000 000 000 000 001 = 10^{-18}	atto	a
0.000 000 000 000 000 000 001 = 10^{-21}	zepto	z
0.000 000 000 000 000 000 000 001 = 10^{-24}	yocto	y

*Some countries use D for deka.

Perform Conversions Within the Metric System

We will use Table 7.2 to help demonstrate how to change from one metric unit to another metric unit using the meter as our basic unit (for example, meters to kilometers, and so on).

The meters in Table 7.2 can be replaced by grams, liters, or any other basic unit of the metric system. Regardless of which unit we choose, the procedure is the same. For purposes of explanation, we have used the meter.

Table 7.2 Changing Metric Units

Measure of length	kilometer	hectometer	dekameter	meter	decimeter	centimeter	millimeter
Symbol	km	hm	dam	m	dm	cm	mm
Number of meters	1000 m	100 m	10 m	1 m	0.1 m	0.01 m	0.001 m

▲ Our neighbors in Canada (and also Mexico) use the metric system.

Table 7.2 shows that 1 hectometer equals 100 meters and 1 millimeter is 0.001 (or $\frac{1}{1000}$) meter. The millimeter is the smallest unit in the table. A centimeter is 10 times as large as a millimeter, a decimeter is 10 times as large as a centimeter, a meter is 10 times as large as a decimeter, and so on. Because each unit is 10 times as large as the unit on its right, converting from one unit to another is simply a matter of multiplying or dividing by powers of 10.

PROCEDURE   CHANGING UNITS WITHIN THE METRIC SYSTEM

1. To change from a smaller unit to a larger unit (for example, from meters to kilometers), move the decimal point in the original quantity one place to the left for each larger unit of measurement until you obtain the desired unit of measurement.

2. To change from a larger unit to a smaller unit (for example, from kilometers to meters), move the decimal point in the original quantity one place to the right for each smaller unit of measurement until you obtain the desired unit of measurement.

Metric System Conversions

Example 1 *Converting from Basic Metric Units*

a) Convert 457.3 m to km.
b) Convert 14 g to cg.
c) Convert 0.53 ℓ to mℓ.
d) Convert 2450 ℓ to kℓ.

Solution

a) Table 7.2 above shows that kilometers are larger units of measurement than meters. Kilometers appear three places to the left of meters in the table. Therefore, to change a measure from meters to kilometers, we must move the decimal point in the given number three places to the left, or

$$457.3 \text{ m} = 0.4573 \text{ km}$$

Note that since we are changing from a smaller unit of measurement (meter) to a larger unit of measurement (kilometer), the answer will be a smaller number of units.

b) Grams are a larger unit of measurement than centigrams. To convert grams to centigrams, we move the decimal point two places to the right, or

$$14 \text{ g} = 1400 \text{ cg}$$

Note that since we are changing from a larger unit of measurement (gram) to a smaller unit of measurement (centigram), the answer will be a larger number of units.

c) $0.53 \ \ell = 530 \ m\ell$

d) $2450 \ \ell = 2.45 \ k\ell$

Example 2 *Converting Metric Units*

a) Convert 704 mm to hectometers.

b) Convert 6.34 dam to decimeters.

Solution

a) Table 7.2 shows that hectometers are five places to the left of millimeters. Therefore, to make the conversion, we must move the decimal point in the given number five places to the left, or

$$704 \ mm = 0.007 \ 04 \ hm$$

b) Table 7.2 shows that decimeters are two places to the right of dekameters. Therefore, to make the conversion, we must move the decimal point in the given number two places to the right, or

$$6.34 \ dam = 634 \ dm$$

Example 3 *Metric Road Signs*

The signs in the photo, from Beijing, China, show that the distance to Xizhimen Inner Street is 500 meters, and the speed limit is 80 kilometers per hour.

a) Determine the distance to Xizhimen Inner Street in kilometers.

b) Determine the speed limit in centimeters per hour.

Solution

a) We must move the decimal point three places to the left to change from meters to kilometers. Therefore,

$$500 \ m = 0.5 \ km$$

b) To change from kilometers per hour to centimeters per hour, we must change from kilometers to centimeters. To change from kilometers to centimeters we must move the decimal point five places to the right. Therefore,

$$80 \ km = 8 \ 000 \ 000 \ cm$$

Thus, 80 kilometers per hour is equal to 8 000 000 centimeters per hour.

Example **4** *Comparing Lengths*

Arrange in order from smallest to largest length: 2.3 m, 3421 mm, and 104 cm.

Solution To be compared, these lengths should all be in the same units of measure. Let's convert all the measures to millimeters, the smallest units of the lengths being compared.

$$2.3 \text{ m} = 2300 \text{ mm} \qquad 3421 \text{ mm} \qquad 104 \text{ cm} = 1040 \text{ mm}$$

Since the lengths, in millimeters, from smallest to largest are 1040, 2300, 3421, the lengths arranged in order from smallest to largest are 104 cm, 2.3 m, and 3421 mm. ∎

TECHNOLOGY TIP

Metric System Apps

There are several apps available for use on most smartphones and tablets that can assist in the conversion between the U.S. customary system and the metric system. These apps are particularly helpful when traveling outside of the United States to countries that use the metric system. These apps can also be used to check your work throughout this chapter. In addition to these apps, several websites including www.us-metric.org and www.nist.gov provide useful information about the metric system. As always, check with your instructor prior to using any apps or websites to complete required work for your course.

SECTION 7.1
Exercises

Warm Up Exercises

In Exercises 1–10, fill in the blanks with an appropriate word, phrase, or symbol(s).

1. The name commonly used for the Système international d'unités in the United States is the _____ system.

2. The name of the system of measurement primarily used in the United States today is the U.S. _____ system.

3. In the metric system,

   a) the commonly used unit to measure length is the _____,

   b) the commonly used units to measure mass are the gram and the _____,

   c) the commonly used unit to measure liquid volume is the _____, and

   d) the commonly used unit to measure temperature is degrees _____.

4. Three advantages of the metric system are that

   a) it is the worldwide accepted system of ____,

   b) there is only one base unit of measurement for each physical _____, and

   c) it is based on the number _____.

5. In the metric system, the prefix used to indicate

   a) ten times the basic unit is _____.

   b) one tenth of the basic unit is _____.

6. In the metric system, the prefix used to indicate

   a) one hundred times the basic unit is _____.

   b) one hundredth of the basic unit is _____.

7. In the metric system, the prefix used to indicate

   a) one thousand times the basic unit is _____.

   b) one thousandth of the basic unit is _____.

8. In the metric system, the prefix used to indicate

   a) one million times the basic unit is _____.

   b) one millionth of the basic unit is ____.

9. a) The freezing point of water in the metric system is
   _____.

   b) The boiling point of water in the metric system is
   _____.

   c) Normal body temperature in the metric system is
   _____.

10. a) A meter is a little longer than a _____.

   b) A liter is a little more than a _____.

   c) A kilogram is about _____ pounds.

Practice the Skills

In Exercises 11–16, match the given prefix with the one letter,
a)–f), that gives the meaning of the prefix.

11. Milli   **a)** $\dfrac{1}{100}$ of basic unit

12. Kilo   **b)** $\dfrac{1}{1000}$ of basic unit

13. Hecto   **c)** 100 times basic unit

14. Deka   **d)** 1000 times basic unit

15. Deci   **e)** 10 times basic unit

16. Centi   **f)** $\dfrac{1}{10}$ of basic unit

17. Complete the following.

   a) 1 milligram = _____ gram

   b) 1 hectogram = _____ grams

   c) 1 kilogram = _____ grams

   d) 1 centigram = _____ gram

   e) 1 dekagram = _____ grams

   f) 1 decigram = _____ gram

18. Complete the following.

   a) 1 dekaliter = _____ liters

   b) 1 centiliter = _____ liter

   c) 1 milliliter = _____ liter

   d) 1 deciliter = _____ liter

   e) 1 kiloliter = _____ liters

   f) 1 hectoliter = _____ liters

In Exercises 19–23, without referring to any of the tables or your
notes, give the symbol and the equivalent in grams for the unit.

19. Centigram   **20.** Milligram

21. Decigram   **22.** Dekagram

23. Kilogram   **24.** Hectogram

World's Fastest Train   The world's fastest train is the Japanese
Maglev train. The Maglev train can travel up to 603 kilometers
per hour (km/h).

25. What is this speed in m/h?

26. What is this speed in mm/h?

In Exercises 27–32, fill in the missing values.

27. 9.2 m = _____ mm

28. 17.3 kg = _____ hg

29. 0.057 m = _____ km

30. 8 dam = _____ m

31. 186.2 cm = _____ m

32. 9.32 mℓ = _____ ℓ

In Exercises 33–38, convert the given unit to the unit indicated.

33. 6.5 km to millimeters

34. 69.3 kg to hectograms

35. 24 hm to kilometers

36. 895 kℓ to milliliters

37. 40 302 mℓ to dekaliters

38. 0.034 mℓ to deciliters

Giraffes *In Exercises 39–46, use the following information about the giraffe to answer the questions. The Masai giraffe is the largest species of giraffe. The male giraffe can reach a height of about 5.8 meters and have a mass of about 1270 kilograms. Its tongue can be about 50 centimeters long. It can run for short distances at a speed of about 56.3 kilometers per hour.*

39. What is the height of the giraffe in centimeters?

40. What is the height of the giraffe in millimeters?

41. What is the length of its tongue in meters?

42. What is the length of its tongue in kilometers?

43. What is the mass of the giraffe in grams?

44. What is the mass of the giraffe in decigrams?

45. What is the speed of the giraffe in meters per hour?

46. What is the speed of the giraffe in dekameters per hour?

In Exercises 47–52, arrange the quantities in order from smallest to largest.

47. 4.4 dam, 0.52 km, 620 cm

48. 514 hm, 62 km, 680 m

49. 1.4 kg, 1600 g, 16 300 dg

50. 4.3 ℓ, 420 cℓ, 0.045 kℓ

51. 2.6 km, 105 000 mm, 52.6 hm

52. 0.032 kℓ, 460 dℓ, 48 000 cℓ

Problem Solving

53. *Walking* Would you be walking faster if you walked 1 dam in 10 min or 1 hm in 10 min? Explain.

54. *Who Ran Faster* Jim ran 100 m, and Bob ran 100 yd in the same length of time. Who ran faster? Explain.

55. *Water Removal* One pump removes 1 daℓ of water in 1 min, and another pump removes 1 dℓ of water in 1 min. Which pump removes water faster? Explain.

56. *Balance* If 5 kg are placed on one side of a balance and a 15 lb weight is placed on the other side, which side is heavier? Explain.

57. *Mount Fuji and Pikes Peak* The most visited mountain in the world is Mount Fuji in Japan, which has a height of 3776 meters. The second most visited mountain in the world is Pikes Peak in Colorado, which has a height of 4302 meters.

a) How many meters higher is Pikes Peak than Mount Fuji?

b) How many centimeters higher is Pikes Peak than Mount Fuji?

c) How many hectometers higher is Pikes Peak than Mount Fuji?

▲ *Mount Fuji*

58. *2018 Ford Mustang* The 2018 Ford Mustang gets about 13.60 km/ℓ on the highway. The 2018 Ford Mustang GT gets about 10.63 km/ℓ.

a) What is the highway gas mileage of the Ford Mustang in meters per liter?

b) What is the highway gas mileage of the Ford Mustang GT in dekameters per liter?

c) How much better mileage, in kilometers per liter, does the Mustang get than the Mustang GT?

59. *Calcium Tablets* Deandra takes two 250-mg chewable calcium tablets each day.

a) How many milligrams of calcium will Deandra take in a week?

b) How many grams of calcium will Deandra take in a week?

60. *Framing a Painting* A painting, including the frame, measures 74 cm by 99 cm.

a) How many centimeters of framing were needed to frame the painting?

b) How many millimeters of framing were needed to frame the painting?

61. *Liters of Soda* A bottle of soda contains 360 mℓ.

  a) How many milliliters are contained in a six-bottle carton?

  b) How many liters does the amount in part (a) equal?

  c) At $2.45 for the carton of soda, what is its cost per liter?

62. *Track and Field* A high school has a 400-m oval track. If Patty runs around the track eight times, how many kilometers has she traveled?

63. *A Home Run* A baseball diamond is a square whose sides are about 27 m in length.

  a) How many meters does a batter run if he hits a home run?

  b) How many kilometers?

  c) How many millimeters?

64. *Tennis Stadium* This photo taken at the Roland Garros Tennis Stadium in Paris shows the distances from the Roland Garros Stadium to tennis stadiums where the other three grand slams of tennis are played.

  a) How much farther from the Roland Garros Tennis Stadium is Melbourne Park (in Australia) than Flushing Meadows (in New York)?

  b) What is the distance determined in part (a) in meters?

Challenge Exercises/Group Activities

In Exercises 65–68, fill in the blank to make a true statement.

65. 1 gigameter = _____ megameters

66. 1 nanogram = _____ micrograms

67. 1 teraliter = _____ picoliters

68. 1 megagram = _____ nanograms

Large and Small Numbers *One advantage of the metric system is that by using the proper prefix, you can write large and small numbers without large groups of zeros. In Exercises 69–74, write an equivalent metric measurement without using any zeros. For example, you can write 3000 m without zeros as 3 km and 0.0003 hm as 3 cm.*

69. 9000 cm

70. 2000 mm

71. 0.000 06 hg

72. 3000 dm

73. 0.02 kℓ

74. 500 cm

Research Activity

75. *Development of the Metric System* Write a report on the development of the metric system in Europe. Indicate which individual people had the most influence in its development.

SECTION 7.2 **Length, Area, and Volume**

Upon completion of this section, you will be able to:

- Understand the units of length in the metric system.
- Understand the units of area in the metric system.
- Understand the units of volume in the metric system.

When U.S. automobile manufacturers ship vehicles overseas, the stated gas mileage on the window sticker is usually given in liters per hundred kilometers rather than miles per gallon. For example, a Jeep Wrangler's gas mileage may be stated as $\frac{13\,\ell}{100\,\text{km}}$. Other measurements, such as the vehicle's length and weight, are also given in metric measurements. When clothing manufacturers ship clothing overseas, the clothing sizes will be given in metric units, such as centimeters. What would a pants waist size of 38 inches be equal to in centimeters? In this chapter, you will learn how to make conversions to and from the metric system.

Why This Is Important If we export goods to other countries, our exports need to be sized in metric units. Imports from other countries, such as tires, are often sized in metric units. More and more items in the United States are being sized using metric or both metric and U.S. customary units.

Tire Size

On the side of every tire you will see a code that provides information about the tire. The P at the beginning of the code given below indicates that the tire is intended for a passenger vehicle.

The three digits after the P, 205, represent the tire's width in millimeters; see Figure 7.2. The two-digit number after the slash, 65, represents the tire's aspect ratio. This is the ratio of the tire's height (from tread to rim) to its width. The height of this tire would be 65% of the tire's width of 205 mm, or 0.65 × 205 = 133.25 mm. The R that follows the 65 indicates that the tire is a radial tire. The last two digits, 16, indicate that the tire is meant to be mounted on a wheel with a diameter of 16 inches.

Figure 7.2

Why This Is Important We see metric units used in many aspects of our lives today—from the medicines we take to tires on vehicles we drive. Having a knowledge of metric units allows us to be better consumers.

This section and the next section are designed to help you *think metric*, that is, to become acquainted with day-to-day usage of metric units. In this section, we consider length, area, and volume.

Units of Length in the Metric System

The basic unit of length in the metric system is the meter. In all English-speaking countries except the United States, *meter* is spelled "metre." Until 1960, the meter was officially defined by the length of a platinum bar kept in a vault in France. The modern definition of the meter is based on the speed of light, a constant that has been determined with great precision. Other commonly used units of length are the kilometer, centimeter, and millimeter. The meter, which is a little longer than 1 yard, is used to measure things that we normally measure in yards and feet. A man whose height is about 2 meters is a tall man. A tractor trailer unit (an 18-wheeler) is about 18 meters long.

The kilometer is used to measure what we normally measure in miles. For example, the distance from New York to Seattle is about 5120 kilometers. One kilometer is about 0.6 mile, and 1 mile is about 1.6 kilometers.

Centimeters and millimeters are used to measure what we normally measure in inches. The centimeter is a little less than $\frac{1}{2}$ inch (see Fig. 7.1), and the millimeter is a little less than $\frac{1}{20}$ inch. A millimeter is about the thickness of a dime. A book may measure 20 cm by 25 cm with a thickness of about 3 cm. Millimeters are often used in scientific work and other areas in which small quantities must be measured. The length of a small insect may be measured in millimeters.

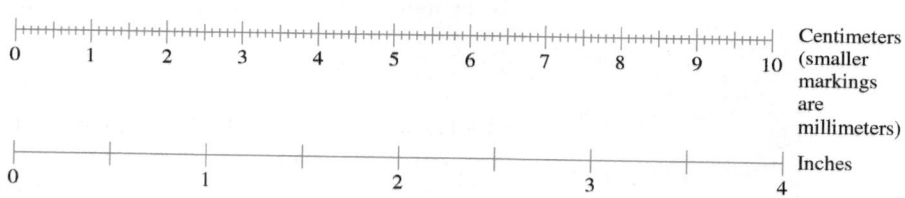

Figure 7.1

From Figure 7.1 we can see that 1 inch is about 2.5 centimeters, or about 25 millimeters.

Example 1 *Choosing an Appropriate Unit of Length*

Determine which metric unit of length you would use to express the following.
a) The diameter of a basketball rim
b) The length of a necklace
c) The height of a five-story apartment building
d) The distance between New Orleans and San Diego
e) Your height
f) Your waist size

Solution

a) Centimeters   b) Millimeters or centimeters
c) Meters   d) Kilometers
e) Meters or centimeters   f) Centimeters

In some parts of this solution, more than one possible answer is listed. Measurements can often be made using more than one unit. For example, if someone asks your height, you might answer $5\frac{1}{2}$ feet, or 66 inches. Both answers are correct. ■

Did You Know?

1 Yard = 3 Feet = 36 Inches = . . .

In the U.S. customary system, 27 different units of length are used. How many of them can you name? Don't forget rod, mil, paris line, toise, cubit, and light-year. The different units can be found in the *CRC Handbook of Chemistry and Physics* or at various websites.

Units of Area in the Metric System

In Chapter 8, we provide formulas and discuss procedures for determining the area and volume of many geometric figures. The procedures and formulas for determining area and volume are the same regardless of whether the units are metric units or customary units. When determining areas and volumes, each side of the figure must be given in (or converted to) the same unit.

The area enclosed in a square with 1-centimeter sides (Fig. 7.3) is $1 \text{ cm} \times 1 \text{ cm} = 1 \text{ cm}^2$. A square whose sides are 2 cm (Fig. 7.4) has an area of $2 \text{ cm} \times 2 \text{ cm} = 2^2 \text{ cm}^2 = 4 \text{ cm}^2$.

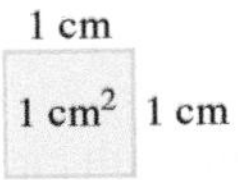

Figure 7.3

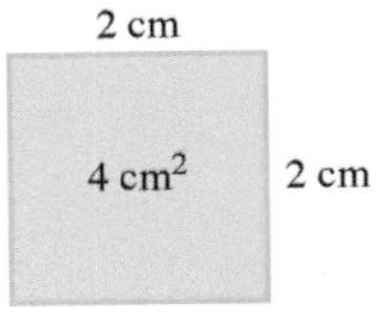

Figure 7.4

Areas are always expressed in square units, such as square centimeters, square kilometers, or square meters. When determining areas, be careful that all the numbers being multiplied are expressed in the same units.

In the metric system, the square centimeter replaces the square inch. The square meter replaces the square foot and square yard. In the future, you might purchase carpet or other floor covering by the square meter instead of by the square foot.

For measuring large land areas, the metric system uses a square unit 100 meters on each side (a square hectometer). This unit is called a *hectare* (pronounced "hectair" and symbolized ha). A hectare is about 2.5 acres. One square mile of land contains about 260 hectares. Very large units of area are measured in square kilometers. One square kilometer is about $\frac{4}{10}$ square mile.

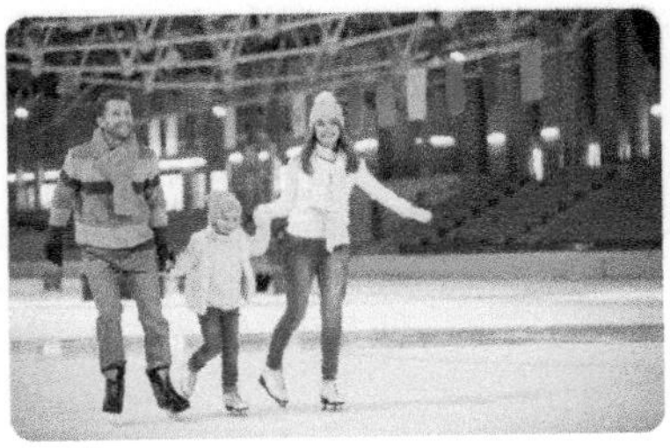

Example 2 *Choosing an Appropriate Unit of Area*

Determine which metric unit of area you would use to measure the area of the following.

a) Bryce Canyon National Park
b) The top of a kitchen table
c) A person's property with an average-sized lot
d) The screen of a tablet, such as an iPad
e) A football field
f) An ice-skating rink (see photo)
g) A nickel
h) A DVD

Solution

a) Square kilometers or hectares
b) Square meters
c) Square meters or hectares
d) Square centimeters
e) Hectares or square meters
f) Square meters
g) Square millimeters or square centimeters
h) Square centimeters ■

To determine the area, A, of a square, we use the formula $A = s^2$, where s is the length of a side of the square. The area can also be determined by the formula area $=$ side $\times$ side. We use this information in Example 3.

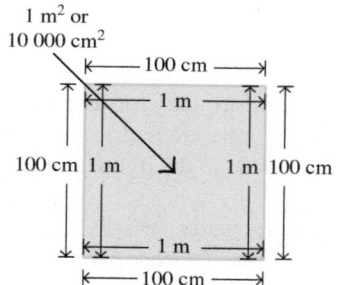

1 m² or 10 000 cm²

Figure 7.5

Example 3 *Converting Square Meters to Square Centimeters*

A square meter is how many times larger than a square centimeter?

Solution A square meter is a square whose sides are 1 meter long. Since 1 m equals 100 cm, we can replace 1 m with 100 cm (see Fig. 7.5). The area of $1\ m^2 = 1\ m \times 1\ m = 100\ cm \times 100\ cm = 10\ 000\ cm^2$. Thus, the area of one square meter is 10,000 times larger than the area of one square centimeter. This technique can be used to convert from any square unit to a different square unit. ∎

Example 4 *Tabletop*

Determine the area of a rectangular tabletop if its length is 1.5 m and its width is 1.1 m (see Fig. 7.6).

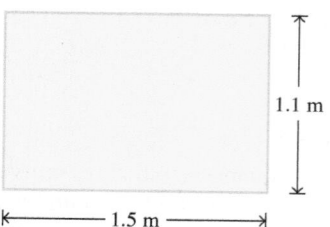

1.1 m

1.5 m

Figure 7.6

Solution To determine the area, we use the formula

$$\text{Area} = \text{length} \times \text{width}$$

or

$$A = l \times w$$

Substituting values for l and w, we have

$$A = 1.5\ m \times 1.1\ m$$
$$= 1.65\ m^2$$

Notice that the area is measured in square meters. ∎

Timely Tip

The number π was introduced in Chapter 5. Recall that π is approximately 3.14. However, when solving problems we will use the $\boxed{\pi}$ key on a scientific calculator to get a more accurate answer.

Example 5 *Face of a Clock*

The face of a circular clock has a diameter of 36 cm. Determine the surface area of the face of the clock.

Solution The formula for determining the area of a circle is $A = \pi r^2$, where π is *approximately* 3.14. The radius, r, is one-half the diameter. Since the diameter is 36 cm, the radius is 18 cm. Using the $\boxed{\pi}$ key on a calculator we get the following.

$$A = \pi r^2$$
$$A = \pi(18)^2$$
$$A \approx 1017.88\ cm^2$$

Thus, the area of the face of the clock is approximately 1017.88 cm². ∎

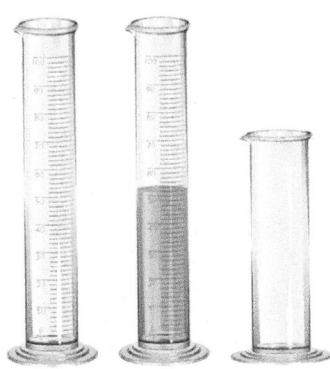

Figure 7.7

Units of Volume in the Metric System

When a figure has only two dimensions—length and width—we can determine its area. When a figure has three dimensions—length, width, and height—we can determine its volume. The volume of an item can be considered the space occupied by the item.

In the metric system, volume may be expressed in terms of liters or cubic meters, depending on what is being measured. In all English-speaking countries except the United States, *liter* is spelled "litre."

The volume of liquids is expressed in liters. A liter is a little larger than a quart. Liters are used in place of pints, quarts, and gallons. A liter can be divided into 1000 equal parts, each of which is called a milliliter. Figure 7.7 illustrates a graduated cylinder. In chemistry, 100 mℓ and other metric graduated cylinders are often used. Milliliters are used to express the volume of very small amounts of liquid. Drug dosages are often expressed in milliliters. An 8-oz cup will hold about 240 mℓ of liquid.

The kiloliter, 1000 liters, is used to represent the volume of large amounts of liquid. Tank trucks carrying gasoline to service stations hold about 10.5 kℓ of gasoline.

Cubic meters are used to express the volume of large amounts of solid, gaseous, and liquid material. The volume of a dump truck's load of topsoil is measured in cubic meters. The volume of natural gas used to heat a house may soon be measured in cubic meters instead of cubic feet.

The liquid in a liter container will fit exactly in a cubic decimeter (1ℓ = 1 dm^3, see Fig. 7.8). Note that 1 ℓ = 1000 mℓ and that 1 dm^3 = 1000 cm^3. Because 1 ℓ = 1 dm^3, *1 mℓ must equal 1 cm^3*. Other useful facts are illustrated in Table 7.3. Thus, within the metric system, conversions are much simpler than in the U.S. customary system. For example, how would you convert cubic feet of water into gallons of water?

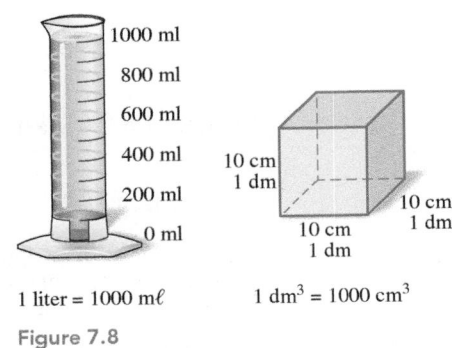

1 liter = 1000 mℓ 1 dm^3 = 1000 cm^3

Figure 7.8

Table 7.3

Volume in Cubic Units		Volume in Liters
1 cm^3	=	1 mℓ
1 dm^3	=	1 ℓ
1 m^3	=	1 kℓ

Example 6 *Choosing an Appropriate Unit of Volume*

Determine which metric unit of volume you would use to measure the volume of the following.

a) The water in Lake Tahoe

b) A carton of milk

c) A truckload of topsoil

d) A liquid drug dosage

e) Sand in a paper cup

f) A dime

g) Water in a drinking glass

h) Water in a swimming pool

i) A kitchen

j) Concrete used to lay the foundation for a basement

Solution

a) Kiloliters or cubic meters

b) Liters

c) Cubic meters

d) Milliliters

e) Cubic centimeters

f) Cubic millimeters

g) Milliliters

h) Kiloliters or liters or cubic meters

i) Cubic meters

j) Cubic meters

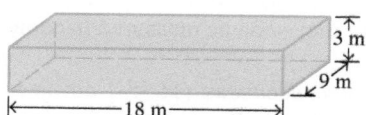

Figure 7.9

Example 7 *Swimming Pool Volume*

A rectangular swimming pool is 18 m long and 9 m wide, and it has a uniform depth of 3 m (Fig. 7.9). Determine (a) the volume of the pool in cubic meters and (b) the volume of the pool in kiloliters.

Solution

a) To determine the volume in cubic meters, we use the formula

$$V = l \times w \times h$$

Substituting values for l, w, and h we have

$$V = 18 \text{ m} \times 9 \text{ m} \times 3 \text{ m}$$
$$= 486 \text{ m}^3$$

b) Since $1 \text{ m}^3 = 1 \text{ k}\ell$, the pool will hold 486 kℓ of water.

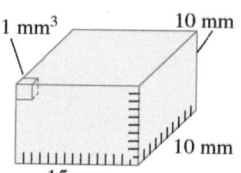

Figure 7.10

Example 8 *Choose an Appropriate Unit*

Select the most appropriate answer. The volume of a shoe box is approximately

a) 1500 mm^3.   b) 6500 ℓ.   c) 6500 cm^3.

Solution A shoe box is not a liquid, so its volume is not expressed in liters. Thus, (b) is not the answer. The volume of the rectangular solid in Fig. 7.10 is approximately 1500 mm^3, so (a) is not an appropriate answer. A shoe box may measure about 33 cm × 18 cm × 11 cm, or 6534 cm^3. Therefore, 6500 cm^3 or (c) is the most appropriate answer.

When the volume of a liquid is measured, the abbreviation cc is often used instead of cm^3 to represent cubic centimeters. For example, a nurse may give a patient an injection of 3 cc or 3 mℓ of the drug ampicillin.

Example 9 *Measuring Medicine*

A nurse must give a patient 3 cc of the drug gentamicin mixed in 100 cc of a normal saline solution.

a) How many milliliters of the drug will the nurse administer?

b) What is the total volume of the drug and saline solution in milliliters?

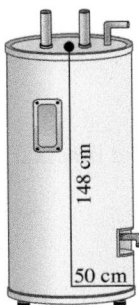

148 cm

50 cm

Figure 7.11

Solution

a) Because 1 cc is equal in volume to 1 mℓ, the nurse will administer 3 mℓ of the drug.
b) The total volume is $3 + 100$, or 103 cc, which is equal to 103 mℓ.   ∎

Example 10   *A Hot-Water Heater*

A hot-water heater, in the shape of a right circular cylinder, has a radius of 50 cm and a height of 148 cm. What is the capacity, in liters, of the hot-water heater?

Solution   The hot-water heater is illustrated in Fig. 7.11. The formula for the volume of a right circular cylinder is $V = \pi r^2 h$. Since we want the capacity in liters, we will express all the measurements in meters. The volume will then be given in cubic meters, which can be easily converted to liters. Thus, 50 cm = 0.5 m, and 148 cm = 1.48 m.

$$V = \pi r^2 h$$
$$= \pi(0.5 \text{ m})^2(1.48 \text{ m})$$
$$= \pi(0.25)(1.48) \approx 1.1624 \text{ m}^3$$

We want the volume in liters, so we must change the answer from cubic meters to liters.

$$1 \text{ m}^3 = 1000 \text{ }\ell$$

So,

$$1.1624 \text{ m}^3 = 1.1624 \times 1000 = 1162.4 \text{ }\ell$$

Therefore, the hot-water heater's capacity is about 1162.4 ℓ.   ∎

Example 11   *Comparing Volume Units*

a) How many times larger is a cubic meter than a cubic centimeter?
b) How many times larger is a cubic dekameter than a cubic meter?

Solution

a) The procedure used to determine the answer is similar to that used in Example 3 in this section. First we draw a cubic meter, which is a cube 1 m long by 1 m wide by 1 m high. In Fig. 7.12, we represent each meter as 100 centimeters. The volume of the cube is its length times its width times its height, or

$$V = l \times w \times h$$
$$= 100 \text{ cm} \times 100 \text{ cm} \times 100 \text{ cm} = 1\,000\,000 \text{ cm}^3$$

Since $1 \text{ m}^3 = 1\,000\,000 \text{ cm}^3$, a cubic meter is one million times larger than a cubic centimeter.

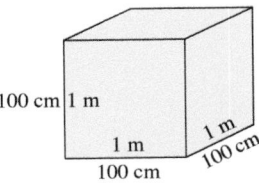

100 cm|1 m
1 m
1 m
100 cm
100 cm

Figure 7.12

b) Work part (b) in a similar manner (Fig. 7.13). A dekameter is 10 meters. Thus,

$$V = l \times w \times h$$
$$= 10 \text{ m} \times 10 \text{ m} \times 10 \text{ m} = 1000 \text{ m}^3$$

Since $1 \text{ dam}^3 = 1000 \text{ m}^3$, a cubic dekameter is one thousand times larger than a cubic meter.

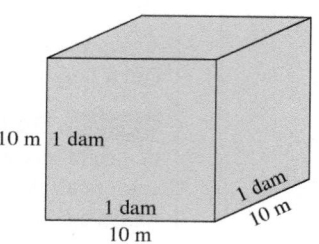

10 m | 1 dam

1 dam
10 m

1 dam
10 m

Figure 7.13

SECTION 7.2
Exercises

Warm Up Exercises

In Exercises 1–10, fill in the blanks with an appropriate word, phrase, or symbol(s).

1. Millimeters, mm, are a measure of _____.

2. Cubic meters, m^3, are a measure of _____.

3. Hectares, ha, are a measure of _____.

4. Centimeters, cm, are a measure of _____.

5. Kilometers, km, are a measure of _____.

6. Kiloliters, $k\ell$, are a measure of _____.

7. Cubic centimeters, cm^3, are a measure of _____.

8. Square meters, m^2, are a measure of _____.

9. Cubic kilometers, km^3, are a measure of _____.

10. Meters, m, are a measure of _____.

Practice the Skills

In Exercises 11–18, indicate the metric unit of measurement that you would use to express the following.

11. The length of a pencil

12. The distance between freeway exits

13. The height of the U.S. capitol building

14. The diameter of a quarter

15. The length of a newborn infant

16. The length of an adult whale

17. The distance to the moon

18. The length of a butterfly

In Exercises 19–24, choose the best answer.

19. The distance traveled by a home run ball in a major league baseball stadium is about how far?

    a) 120 hm   **b)** 120 km   **c)** 120 m

20. A U.S. postage stamp is about how wide and how long?

    a) 2 cm × 3 cm   **b)** 2 mm × 3 mm   **c)** 2 hm × 3 hm

21. The distance between Orlando, Florida, and Boston, Massachusetts, is about how far?

    a) 2070 cm   **b)** 2070 m   **c)** 2070 km

22. A ruler is about how wide?

    a) 4 mm   **b)** 4 cm   **c)** 4 dm

23. The world's tallest Ferris wheel, the High Roller, in Las Vegas, Nevada is about how high?

    a) 168 cm   **b)** 168 m   **c)** 168 km

24. The diameter of a coffee cup is about which of the following?

    a) 8 mm   b) 8 cm   **c)** 8 dm

In Exercises 25–30, (a) estimate the item in metric units and (b) measure it with a metric ruler. Record your result.

25. The height of a kitchen table

26. The width of a classroom door

27. The length of a sofa

28. The diameter of a soda can

29. The height of a container of milk

30. The thickness of a quarter

In Exercises 31–36, indicate the metric unit of measurement you would use to express the area of the following.

31. A house

32. A computer monitor's screen

33. A national park

34. A building lot for a house

35. A postage stamp

36. Disney World

In Exercises 37–42, choose the best answer.

37. A credit card has an area of about

   a) 50 cm². **b)** 50 mm². **c)** 50 dm².

38. The area of a city lot is about

   a) 800 m². **b)** 800 hm². **c)** 800 cm².

39. The area of a city lot is about

   a) $\frac{1}{8}$ m². **b)** $\frac{1}{8}$ ha. **c)** $\frac{1}{8}$ dam².

40. The area of one side of a nickel is about

   a) 3 cm². **b)** 3 m². **c)** 3 mm².

41. The area of a slice of bread is about

   a) 100 cm². **b)** 100 mm². **c)** 100 m².

42. The area of the Great Smoky Mountains National Park is about

   a) 2100 m². **b)** 2100 cm². **c)** 2100 km².

In Exercises 43–48, (a) estimate the area of the item in metric units and (b) measure its dimensions in metric units and compute its area.

43. The cover of a hardbound book

44. A typical photograph

45. A $20 bill

46. The top of your kitchen table

47. A DVD

48. The face of a penny

In Exercises 49–56, determine the metric unit that would best express the volume of the following.

49. Liquid in an eye dropper

50. Water in a hot-water heater

51. Water flowing over Niagara Falls per minute

52. A child's pail filled with beach sand

53. A clothing closet

54. Air in a hot air balloon

55. A truckload of top soil

56. Air in a soccer ball

In Exercises 57–64, choose the best answer to indicate the volume of the following.

57. A desk drawer

   a) 7780 mm³ **b)** 7780 dm³ **c)** 7780 cm³

58. Water that could fill an Olympic-size swimming pool

   a) 2500 kℓ **b)** 2500 ℓ **c)** 2500 mℓ

59. The volume of a typical living room

    a) 45 km^3 **b)** 45 m^3 **c)** 45 cm^3

60. A carry-on suitcase

    a) 0.04 cm^3 **b)** 0.04 mm^3 **c)** 0.04 m^3

61. Juice that can be squeezed out of an orange

    a) $120 \text{ k}\ell$ **b)** $120 \text{ m}\ell$ **c)** $120 \ \ell$

62. The air in a basketball

    a) 6370 m^3 **b)** 6370 mm^3 **c)** 6370 cm^3

63. Soda in a full soda can

    a) $355 \ \ell$ **b)** $355 \text{ m}\ell$ **c)** 355 m^3

64. Water in a 24-ft-diameter above-ground circular swimming pool

    a) $55 \ \ell$ **b)** $55 \text{ m}\ell$ **c)** $55 \text{ k}\ell$

In Exercises 65–68, (a) estimate the volume in metric units and (b) compute the actual volume of the item. When appropriate, use the 🔲 key on your calculator. Round answers to the nearest hundredth.

65. Air in a pizza box that is 46 cm long, 46 cm wide, and 5 cm tall (Use $V = lwh$.)

66. Water in a water bed that is 2 m long, 1.5 m wide, and 25 cm deep (Use $V = lwh$.)

67. Oil in a barrel that has a height of 1 m and a diameter of 0.5 m (Use $V = \pi r^2 h$.)

68. Water in a cylindrical tank that is 40 cm in diameter and 2 m high (Use $V = \pi r^2 h$.)

In Exercises 69–72, select the best answer.

69. One liter of liquid has the equivalent volume of which of the following: a cubic centimeter, a cubic decimeter, or a cubic meter?

70. One cubic meter has the equivalent volume of which of the following liquid measures: a liter, a milliliter, or a kiloliter?

71. One milliliter of liquid has the equivalent volume of which of the following: a cubic centimeter, a cubic decimeter, or a cubic meter?

72. A hectare has an area of about how many acres: 2.5, 25, or 250?

Problem Solving

In Exercises 73–86, when appropriate, use the 🔲 key on your calculator and round your answer to the nearest hundredth.

73. *Area* Use a metric ruler to measure the length and width of the sides of the rectangle. Then compute the area of the rectangle. Give your answers in metric units. (Use the formula $A = lw$.)

74. *Area* Use a metric ruler to determine the radius of the circle. Then compute the area of the circle. Give your answers in metric units. (Use the formula $A = \pi r^2$.)

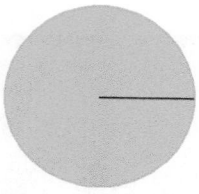

75. *Helicopter's Rotor* The world's largest helicopter that went into production is the Russian-built MI-26. Its single top rotor blade is 32.0 meters long. What is the area of the circle formed by the spinning rotor blade?

76. *Flower Garden* A rectangular flower garden 60 m by 80 m is surrounded by a dirt walkway 1.6 m wide.

    a) Determine the area of the region covering the flower garden and the walkway.

    b) Determine the area of the walkway.

77. *Central Park* Central Park in New York City is approximately rectangular in shape with a length of about 4 km and a width about 0.8 km.

    a) Determine the approximate area of the park in square kilometers.

    b) If 1 km^2 equals 100 ha, determine the area of the park in hectares.

78. *Real Estate Sign* In Costa Rica, a real estate sign indicated that a piece of property was for sale. It listed the total area of the property as 1 ha, 1531 m^2 with frontal area of 67 m^2.

    a) If 1 ha equals $10\ 000 \text{ m}^2$, what is the total area of the property in square meters?

b) What is the total area in square kilometers?

c) What is the frontal area in hectares?

79. *Children's Playground* The dimensions of a large rectangular playground are 42.4 m by 32.5 m.

a) Determine the area of the playground in square meters.

b) If 1 m² equals 0.0001 ha, determine the area of the playground in hectares.

80. *Carry-On Suitcase* At most airports in countries that use the metric system, the maximum allowable dimensions of a carry-on suitcase are length 55 cm, width 25 cm, and height 35 cm.

a) What is the maximum volume, in cubic centimeters, of the carry-on suitcase?

b) What is the maximum volume in cubic meters?

c) What is the maximum volume in cubic millimeters?

81. *Painting* The total area of a framed painting including the matting is 2540 cm². If the length and width of the actual painting are 37 cm and 28 cm, respectively, determine the area of the matting.

82. *Volume of a Saucepan* A cylindrical saucepan has a diameter of 18.0 cm and a height of 9.0 cm. Determine the volume of the saucepan in cubic centimeters.

83. *Fish Tank Volume* A rectangular fish tank is 70 cm long, 40 cm wide, and 20 cm high.

a) How many cubic centimeters of water will the tank hold?

b) How many milliliters of water will the tank hold?

c) How many liters of water will the tank hold?

84. *Hot-Water Heater* A cylindrical hot-water heater has a diameter of 0.56 m and a height of 1.17 m. If the hot-water heater is filled, determine the volume of water in the hot-water heater in

a) cubic meters.   **b)** liters.

85. How many times larger is a square kilometer than a square hectometer?

86. How many times larger is a square kilometer than a square dekameter?

87. How many times larger is a cubic meter than a cubic decimeter?

88. How many times larger is a cubic centimeter than a cubic millimeter?

In Exercises 89–94, replace the question mark with the appropriate value.

89. $1 \text{ cm}^2 = ? \text{ mm}^2$

90. $1 \text{ km}^2 = ? \text{ dam}^2$

91. $1 \text{ cm}^2 = ? \text{ m}^2$

92. $1 \text{ mm}^3 = ? \text{ dm}^3$

93. $1 \text{ m}^3 = ? \text{ cm}^3$

94. $1 \text{ hm}^3 = ? \text{ km}^3$

In Exercises 95–98, fill in the blank.

95. $724 \text{ cm}^3 = $ _____ $m\ell$

96. $435 \text{ cm}^3 = $ _____ ℓ

97. $189 \text{ k}\ell = $ _____ m^3

98. $4.2 \text{ } \ell = $ _____ cm^3

Glacier In Exercises 99 and 100, assume that a part of a glacier that contains 60 cubic meters of ice breaks off and falls into the ocean.

99. When the ice that has fallen into the ocean melts, determine the approximate amount of water, in kiloliters, obtained from the ice.

100. When the ice melts, determine the approximate amount of water, in cubic centimeters, obtained from the ice.

Challenge Problems/Group Activities

In Exercises 101 and 102, fill in the blank to make a true statement.

101. $5.3 \text{ k}\ell = $ _____ dm^3

102. $1.4 \text{ ha} = $ _____ cm^2

103. _Conversions_ In Example 3, we illustrated how to change an area in a metric unit to an area measured with a different metric unit.

a) Using Example 3 as a guide, change 1 square mile to square inches.

b) Is converting from one unit of area to a different unit of area generally easier in the metric system or the U.S. customary system? Explain.

Recreational Mathematics

104. _Water Usage_ How much water do we use daily? On the average, people in the United States use more water than people anywhere else in the world.

a) Take a guess at the number of liters of water used per day per person in the United States.

b) Now take a guess at the number of liters used per day per person in China.

c) The country where the least amount of water is used per day per person is Mozambique. Take a guess at the number of liters used per day per person in Mozambique.

Research Activity

105. _The Meter_ The definition of the meter has changed several times throughout history. Write a report on the history of the meter, from when it was first named to the present.

SECTION 7.3   Mass and Temperature

Upon completion of this section, you will be able to:

- Understand mass in the metric system.
- Understand temperature in the metric system.
- Convert temperature from Celsius to Fahrenheit or from Fahrenheit to Celsius.

It is January, and you and your friends go for a visit to Niagara Falls, Canada. The thermostat in your hotel room is set on 20° Celsius. Will you need to adjust the thermostat? If the temperature outside is 15°C, how should you dress?

While watching television, you see an astronaut floating while working on the International Space Station. Does this floating astronaut have any weight? Does the astronaut have any mass? In this section, we will discuss these two important metric measurements, temperature and mass.

Why This Is Important Temperatures outside the United States are almost always given in degrees Celsius. If you stay in a hotel or motel in Canada or Mexico, your room temperature and the outdoor temperature will be given in degrees Celsius. Also, if you take a cruise to the Caribbean or elsewhere, the room thermostat will most likely show degrees Celsius. When planning a trip to a different country you may want to review the 10-day forecast so you know what type of clothing to bring so that you will be comfortable. If the website you are looking at says the low temperature for a particular day will be 15°C, will you need a heavy coat? Also, temperatures used in the sciences are measured in degrees Celsius. The concept of mass is important to all sciences.

Identify a Suitable
Metric Mass Unit

Mass in the Metric System

Weight and mass are not the same. _Mass_ is a measure of the amount of matter in an object. It is determined by the molecular structure of the object, and it will not change from place to place. Weight is a measure of the gravitational pull on an object. For example, the gravitational pull of Earth is about six times as great as the gravitational pull of the moon. Thus, a person on the moon weighs about $\frac{1}{6}$ as much as on Earth, even though the person's mass remains the same. In space, where there is no gravity, a person has no weight but does have mass.

Even on Earth, the gravitational pull varies from point to point. The closer you are to Earth's center, the greater the gravitational pull. Thus, a person weighs very slightly less on a mountain than in a nearby valley. Because the mass of an object does not vary with location, scientists generally use mass rather than weight.

Although weight and mass are not the same, on Earth they are proportional to each other (the greater the weight, the greater the mass). Therefore, for our purposes, we can treat weight and mass as the same.

The *kilogram* is the base unit of mass in the metric system. It is about 2.2 pounds. The official kilogram is a cylinder of platinum–iridium alloy kept by the International Bureau of Weights and Measures, located in Sèvres, France, near Paris. (See the *Did You Know?* in the margin.)

Items that we normally measure in pounds are usually measured in kilograms in other parts of the world. For example, an average-sized man has a mass of about 75 kg.

The *gram* (a unit that is 0.001 kg) is relatively small and is used for items normally measured in ounces. A nickel has a mass of about 5 g, a cube of sugar has a mass of about 2 g, and a large paper clip has a mass of about 1 g.

The *milligram* is used extensively in the medical and scientific fields as well as in the pharmaceutical industry. Nearly all bottles of tablets are now labeled in either milligrams or grams.

The *metric tonne* (t) is used to express the mass of heavy items. One metric tonne equals 1000 kg. A metric tonne is a little larger than our customary ton of 2000 lb. The mass of a large truck may be expressed in metric tonnes.

Example 1 *Choosing the Appropriate Unit*

Determine which metric unit you would use to express the mass of the following.

a) An 8-year-old child   b) A locomotive   c) A dinner plate

d) A box of cereal   e) A laptop computer   f) A spider

g) A frog   h) A refrigerator

Solution

a) Kilograms   b) Metric tonnes   c) Grams

d) Grams   e) Grams or kilograms   f) Milligrams

g) Grams   h) Kilograms   ■

One kilogram of water has a volume of exactly 1 liter. In fact, 1 liter is defined to be the volume of 1 kilogram of water at a specified temperature and pressure. Thus, mass and volume are easily interchangeable in the metric system. Converting from weight to volume is not nearly as convenient in the U.S. customary system. For example, how would you change pounds of water to cubic feet or gallons of water in our customary system?

Figure 7.14 shows that for water, the volume of 1 cubic decimeter is equivalent to a volume of 1 liter, and the water has a mass of 1 kilogram.

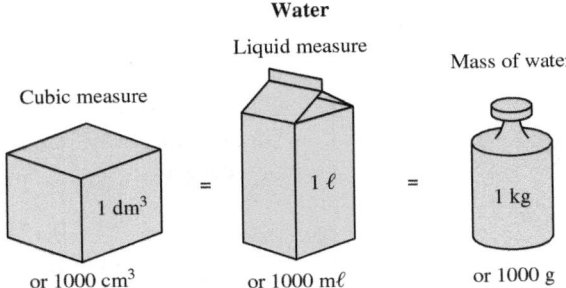

Figure 7.14 One cubic decimeter of water has the volume of one liter and the mass of one kilogram.

Since $1\,dm^3 = 1000\,cm^3$, $1\,\ell = 1000\,m\ell$, and $1\,kg = 1000\,g$, we can reason that for water

$$1\,cm^3 \;=\; 1\,m\ell \;=\; 1\,g$$

Table 7.4 gives some important information regarding measurements involving water.

Table 7.4   Volume and Mass of Water

Volume in Cubic Units		Volume in Liters		Mass of Water
$1\,cm^3$	=	$1\,m\ell$	=	$1\,g$
$1\,dm^3$	=	$1\,\ell$	=	$1\,kg$
$1\,m^3$	=	$1\,k\ell$	=	$1\,t$ (1000 kg)

Example 2 *Volume of a Fish Tank*

A fish tank is 1 m long, 50 cm high, and 250 mm wide (Fig. 7.15).

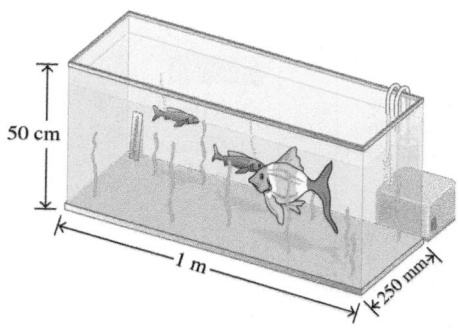

Figure 7.15

a) Determine the number of liters of water the tank holds.
b) What is the mass of the water in kilograms?

Solution

a) We must convert all the measurements to the same units. Let's convert them all to meters: 50 cm is 0.5 m, and 250 mm is 0.25 m.

$$V = l \times w \times h$$
$$= 1 \times 0.25 \times 0.5$$
$$= 0.125\,m^3$$

Since $1\,m^3$ of water $= 1\,k\ell$ of water,

$$0.125\,m^3 = 0.125\,k\ell, \text{ or } 125\,\ell \text{ of water}$$

Thus the tank holds $125\,\ell$ of water.

b) Since $1\,\ell$ of water has a mass of 1 kg, $125\,\ell$ of water has a mass of 125 kg. Thus the water in the fish tank has a mass of 125 kg. ∎

To convince yourself of the advantages of the metric system, do a similar problem involving the U.S. customary system of measurement, such as Challenge Problems/Group Activities Exercise 70 at the end of this section.

Temperature in the Metric System

The Celsius scale is used to measure temperatures in the metric system. Figure 7.17 shows a thermometer with the Fahrenheit scale on the left and the Celsius scale on the right.

The Celsius scale was named for Swedish astronomer Anders Celsius (1701–1744), who first devised it in 1742. On the Celsius scale, water freezes at 0°C and boils at 100°C. In the past, the Celsius thermometer was called a "centigrade thermometer." Recall that *centi* means $\frac{1}{100}$, and there are 100 degrees between the freezing point of water and the boiling point of water. Thus, 1°C is $\frac{1}{100}$ of this interval. Table 7.5 gives some common temperatures in both degrees Celsius (°C) and degrees Fahrenheit (°F).

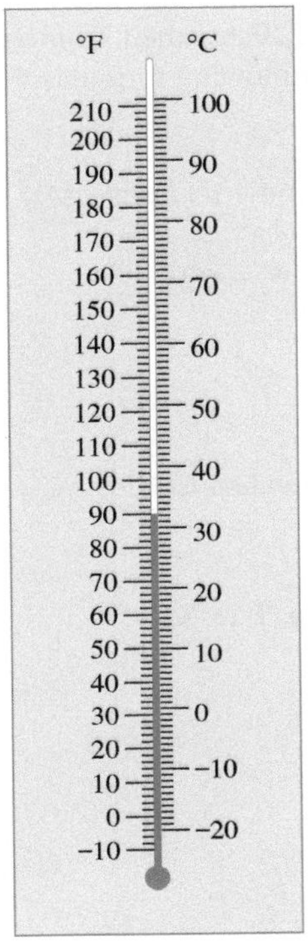

Figure 7.17

Table 7.5

Celsius Temperature		Fahrenheit Temperature
−18°C	A very cold day	0°F
0°C	Freezing point of water	32°F
10°C	A warm winter day	50°F
20°C	A mild spring day	68°F
30°C	A warm summer day	86°F
37°C	Body temperature	98.6°F
100°C	Boiling point of water	212°F
177°C	Oven temperature for baking	351°F

At a temperature of −40° the Celsius and Fahrenheit temperatures are the same. That is, −40°C = −40°F.

Example *Metric Temperatures*

Choose the best answer. (Refer to the dual-scale thermometer in Fig. 7.17.)

a) Minneapolis, Minnesota, on New Year's Day might have a temperature of
   i) −5°C.   ii) 20°C.   iii) 40°C.

b) Baltimore, Maryland, on July 4 might have a temperature of
   i) 15°C.   ii) 30°C.   iii) 45°C.

c) The oven temperature for baking a cake might be
   i) 60°C.   ii) 100°C.   iii) 175°C.

Solution

a) A temperature of 20°C is possible if it is a very mild winter, but 40°C is much too hot. The best answer for a normal winter is −5°C.

b) The best estimate is 30°C. A temperature of 15°C is too chilly, and 45°C is too hot for July 4.

c) A cake bakes at temperatures well above boiling, so the only reasonable answer is 175°C.   ■

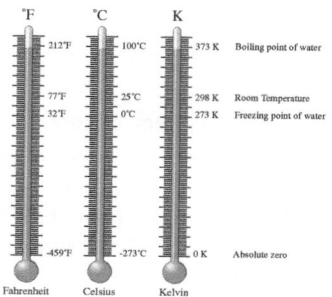
Convert Temperature from Celsius to Fahrenheit or from Fahrenheit to Celsius

Comparing the temperatures in Table 7.5, we see that the Celsius scale has 100° from the boiling point of water to the freezing point of water and the Fahrenheit scale has 180° from the boiling point of water to the freezing point of water. Therefore, one Celsius degree represents a greater change in temperature than one Fahrenheit degree does. In fact, one Celsius degree is the same as $\frac{180}{100}$, or $\frac{9}{5}$, Fahrenheit degrees. When converting from one system to the other system, use the following formulas.

From Celsius to Fahrenheit

$$F = \frac{9}{5}C + 32$$

From Fahrenheit to Celsius

$$C = \frac{5}{9}(F - 32)$$

Example 4 *Convert to °C*

An ideal temperature for a wine cellar is 55°F. What is the equivalent Celsius temperature?

Solution We use the formula $C = \frac{5}{9}(F - 32)$ to convert from °F to °C. Substituting $F = 55$ gives

$$C = \frac{5}{9}(F - 32)$$
$$= \frac{5}{9}(55 - 32)$$
$$= \frac{5}{9}(23)$$
$$\approx 12.8$$

Thus, the equivalent temperature is about 12.8°C.   ∎

Example 5 *Old Faithful*

The water shooting out of the vent of the Old Faithful geyser at Yellowstone National Park is often about 204 °F. What is the equivalent Celsius temperature?

Solution We use the formula $C = \frac{5}{9}(F - 32)$ to convert from °F to °C. Substituting F = 204 gives the following.

$$C = \frac{5}{9}(F - 32)$$
$$= \frac{5}{9}(204 - 32)$$
$$= \frac{5}{9}(172)$$
$$\approx 95.56$$

Thus, the equivalent temperature is about 95.56 °C.   ∎

Did You Know?

Year of Metrification

Almost every country in the world, except for the United States, Liberia, and Myanmar, presently use the metric system. The chart below shows the official year of metrification for various countries. Space prohibits us from listing all the countries. For a more complete chart see http://lamar.colostate.edu/~hillger/internat.htm.

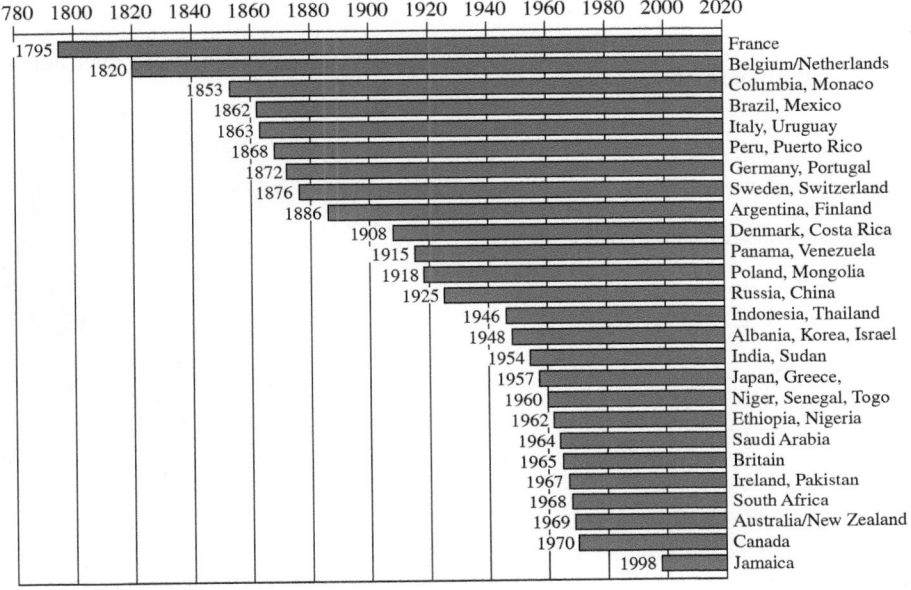

Year of Official Metrification

Advance of Metric Usage in the World

Courtesy of U.S. Metric Association

MATHEMATICS TODAY   The Metric System: Why It Is Important

Although Americans primarily use the U.S. Customary System, we see and use the metric system daily. The active ingredients in medicines are measured in milligrams or milliliters. Nutritional food labels report nutritional quantities in grams or milligrams.

Many items we see at a store are measured using both metric units and U.S. Customary units. For example, soda is measured in both liters and fluid ounces, protein bars are measured in both grams and dry ounces, wine is measured in both milliliters and liquid ounces, and dental floss is measured in both meters and yards.

Metric measurements are also used in many of the electronics we use every day. A new smartphone may have 256 gigabytes of memory, a computer hard drive may be able to hold 500 gigabytes of data, the focal length of a camera lens may be 50 millimeters, and the speed of a new laptop may be 1.6 gigahertz. Many parts of an automobile are measured with metric units including the engine and the tires (see *Did You Know?* on page 410). Exercise 63 on page 439 states that the engine displacement of a 2019 Chevrolet Camaro SS is 6.2 liters.

Other units of metric measurement include the use of kilowatts to measure electrical usage, millimeters of mercury to measure blood pressure (see *Mathematics Today* on page 435), decibels to measure sound intensity, and meters and kilometers to measure distances at sporting events. Government agencies often include metric measurements in their reports, especially when the reports involve importing and exporting goods to other countries.

Why This Is Important   The metric system is used in many aspects of our daily lives. Furthermore, with the globalization of our modern world, an understanding of the metric system is vital to an increasing number of occupations.

SECTION 7.3
Exercises

Warm Up Exercises

In Exercises 1–8, fill in the blanks with an appropriate word, phrase, or symbol(s).

1. In the metric system the kilogram measures _____.

2. The mass of a nickel is about 5 _____.

3. The metric unit that is a little more than 2.2 pounds is the _____.

4. A metric tonne is the unit used to express the _____ of a very heavy item, such as a dump truck.

5. Temperatures in the metric system are measured in degrees _____.

6. The boiling point of water is _____ °C.

7. The freezing point of water is _____ °C.

8. Normal body temperature is _____ °C.

Practice the Skills

In Exercises 9–18, indicate the metric unit of measurement that would best express the mass of the following.

9. A horse

10. A hummingbird

11. A book

12. A box of cereal

13. A new pencil

14. A sports utility vehicle

15. A refrigerator

16. A mosquito

17. A full-grown elephant

18. A calculator

In Exercises 19–24, select the best answer.

19. The mass of a large coconut is about how much?

   a) 1.4 mg   b) 1.4 g   c) 1.4 kg

20. The mass of a doughnut is about how much?

   a) 20 g   b) 20 kg   c) 20 mg

21. The mass of a pencil is about how much?

   a) 5.3 g   b) 5.3 mg   c) 5.3 hg

22. The mass of a bathroom scale is about how much?

   a) 1.8 g   b) 1.8 mg   c) 1.8 kg

23. The mass of a full-grown male lion is about how much?

   a) 190 hg   b) 190 kg   c) 190 dag

24. The mass of a full-size car is about how much?

   a) 1800 hg   b) 1800 g   c) 1800 kg

In Exercises 25–28, estimate the mass of the item. If a scale with metric measure is available, determine the mass.

25. Your body

26. A gallon of water

27. A telephone book

28. A watermelon

In Exercises 29–36, choose the best answer. Use Table 7.5 and Fig. 7.17 to help select your answers.

29. Freezing rain is most likely to occur at a temperature of

   a) −25°C.   b) 32°C.   c) 0°C.

30. The thermostat for an air conditioner was set for 70°F. This setting is closest to

   a) 8°C.   b) 21°C.   c) 40°C.

31. The temperature of the water in a certain lake is 12°C. You could

   a) ice fish.

   b) dress warmly and walk along the lake.

   c) swim in the lake.

32. The temperature of the water of a hot shower might be

   a) 25°C.   b) 40°C.   c) 60°C.

33. The temperature of flowing volcanic lava might be

a) 50°C. **b)** 100°C. **c)** 1200°C.

34. The temperature of an apple pie baking in the oven might be

a) 90°C **b)** 100°C **c)** 177°C

35. The temperature of the snow on top of Mount Rainier, Washington, might be

a) −15°C. **b)** −5°C. **c)** 5°C.

▲ Mount Rainier, Washington

36. The temperature of water in a hot tub might be

a) 30°C. **b)** 50°C. **c)** 40°C.

In Exercises 37–50, convert each temperature as indicated. When appropriate, give your answer to the nearest tenth of a degree.

37. 25°C = _____ °F

38. 80°F = _____ °C

39. −25°F = _____ °C

40. −10°F = _____ °C

41. 0°F = _____ °C

42. 100°F = _____ °C

43. 37°C = _____ °F

44. −4°C = _____ °F

45. 10°F = _____ °C

46. 70°C = _____ °F

47. 0°C = _____ °F

48. 50°C = _____ °F

49. −40°F = _____ °C

50. 425°F = _____ °C

In Exercises 51–56, use the graph above and to the right, which shows the daily low and high temperatures, in degrees Celsius, for the week in the outback in Australia. Determine the following temperatures in degrees Fahrenheit.

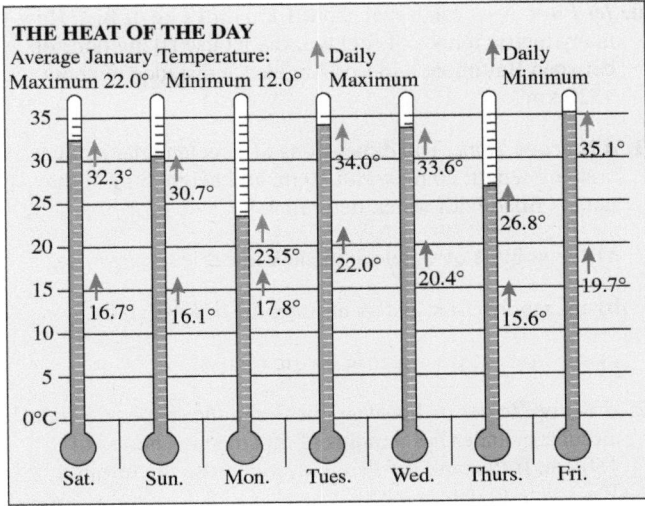

51. The average January maximum temperature

52. The maximum temperature for the week

53. The minimum temperature on Friday

54. The maximum temperature on Saturday

55. The range of temperatures on Monday

56. The range of temperatures on Tuesday

Problem Solving

57. *Sand Pail* A pail has a mass of 560 grams. If 62 grams of sand and 130 milliliters of water are added to the pail, what is the mass of the pail and its contents in grams?

58. *Bridge Weight Restriction* A bridge has a weight restriction of 3.5 metric tonnes (t). If the weight of an empty truck is 2.92 t and the driver and passengers together weigh 180 kilograms, how much cargo, in metric tonnes, can be added to the truck without exceeding the 3.5-t weight limit?

59. *Curry* As shown in the photo, curry being sold at an outdoor market in Paris costs 7 euros per 100 grams.

a) Determine the cost, in euros, for 1 kilogram of the curry.

b) Determine the cost, in euros, for 500 grams of the curry.

c) If the total cost for the curry is 21 euros, how much curry was purchased?

60. Jet Fuel A jet can travel about 1 km on 17 kg of fuel. How many metric tonnes of fuel will the jet use flying nonstop between Baltimore and Los Angeles, a distance of about 4320 km?

61. A Storage Tank The dimensions of a rectangular storage tank are length 20 m, width 10 m, and height 8 m. If the tank is filled with water, determine

   a) the volume of water in the tank in cubic meters.

   b) the number of kiloliters of water the tank will hold.

   c) the mass of the water in metric tonnes.

62. A Water Heater A hot-water heater in the shape of a right circular cylinder has a radius of 50 cm and a height of 150 cm. If the tank is filled with water, use the formula, $V = \pi r^2 h$ to determine

   a) the volume of water in the tank in cubic meters.

   b) the number of liters of water the tank will hold.

   c) the mass of the water in kilograms.

In Exercises 63–66, convert as indicated.

63. 3.5 kg = _____ t

64. 17.4 t = _____ g

65. 1 460 000 mg = _____ t

66. 9.52 t = _____ kg

Writing Exercises

67. Earth or Space
   a) Is a person's mass the same in space as on Earth? Explain.

   b) Is a person's weight the same in space as on Earth? Explain.

68. Is There a Problem? A temperature display at a bank flashes the temperature in degrees Fahrenheit and then flashes the temperature in degrees Celsius. If it flashes 78°F, then 20°C, is there a problem with the display? Explain.

Challenge Problems/Group Activities

69. Gatorade Gatorade is poured into a plastic bottle that holds 1.2 ℓ of liquid. The bottle is then placed in a freezer. When the bottle is removed from the freezer, the plastic is cut away, leaving just the frozen Gatorade.

   a) What is the approximate mass of the frozen Gatorade in grams?

   b) What is the approximate volume of the frozen Gatorade in cubic centimeters?

In Example 2, we showed how to determine the volume and mass of water in a fish tank. Exercise 70 demonstrates how much more complicated solving a similar problem is in the U.S. customary system.

70. Fish Tank A rectangular fish tank is 1 yd long by 1.5 ft high by 15 in. wide.

   a) Determine the volume of water in the fish tank in cubic feet.

   b) Determine the weight of the water in pounds. One cubic foot of water weighs about 62.5 lb.

   c) If 1 gal of water weighs about 8.3 lb, how many gallons will the tank hold?

Recreational Mathematics

71. Balance the Scale Determine the quantity needed to replace the question mark to make the scale balance. The weight times the distance on both sides of the fulcrum (the triangle) must be the same to make the scale balance.

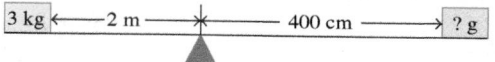

72. Interesting Facts The 2018 *Guinness Book of World Records* provides some interesting facts.

   a) The lowest temperature ever recorded in the United States was −62.11°C on January 23, 1971, in Prospect Creek, Alaska. What is this temperature in degrees Fahrenheit?

   b) International Falls, Minnesota, has the lowest annual mean (average) temperature in the United States (including Alaska). Its mean annual temperature is about 2.5°C. What is this temperature in degrees Fahrenheit?

Research Activity

73. Exporting Goods Do industries located near your school export goods? If so, are they training employees to use and understand the metric system? Contact local industries that export goods and write a report on your findings.

Dimensional Analysis and Conversions to and from the Metric System

Upon completion of this section, you will be able to:

- Understand dimensional analysis.
- Make conversions to and from the metric system.

In this section we will use dimensional analysis to convert measurements from our U.S. customary system to the metric system, and vice versa. Dimensional analysis can be useful in many of your science courses, as well as in real-life situations. For example, if you drive to Mexico for a vacation, you could use dimensional analysis to determine if your 6 foot 2 inch high SUV will fit in an underground garage that accepts cars with a maximum height of 2.3 meters. Or you can use dimensional analysis to figure out how many pesos you should get when you convert your U.S. currency into Mexican currency.

Why This Is Important As you will see by the examples, the conversion techniques discussed in this section can be used in everyday life. If driving in Canada or Mexico, for example, you can use dimensional analysis to convert from kilometers per hour to miles per hour so you can avoid getting a speeding ticket. If you need to convert currency, dimensional analysis can be used to help you determine how much foreign currency you will get in return for your U.S. dollars.

Understand Unit Fractions

You may sometimes need to change units of measurement in the metric system to equivalent units in the U.S. customary system or vice versa. To do so, use *dimensional analysis*, which is a procedure used to convert from one unit of measurement to a different unit of measurement. To perform dimensional analysis, you must first understand what is meant by a unit fraction. A *unit fraction* is any fraction in which the numerator and denominator contain different units and the value of the fraction is 1. From Table 7.6, we can obtain many unit fractions involving U.S. customary units.

Table 7.6 **U.S. Customary Units**

1 foot = 12 inches
1 yard = 3 feet
1 mile = 5280 feet
1 pound = 16 ounces
1 ton = 2000 pounds
1 cup (liquid) = 8 fluid ounces
1 pint = 2 cups
1 quart = 2 pints
1 gallon = 4 quarts
1 minute = 60 seconds
1 hour = 60 minutes
1 day = 24 hours
1 year = 365 days

Examples of Unit Fractions

$$\frac{12 \text{ in.}}{1 \text{ ft}} \qquad \frac{1 \text{ ft}}{12 \text{ in.}} \qquad \frac{16 \text{ oz}}{1 \text{ lb}} \qquad \frac{1 \text{ lb}}{16 \text{ oz}} \qquad \frac{60 \text{ min}}{1 \text{ hr}} \qquad \frac{1 \text{ hr}}{60 \text{ min}}$$

In each of these examples, the numerator equals the denominator, so the value of the fraction is 1.

Dimensional Analysis

To convert an expression from one unit of measurement to a different unit, multiply the given expression by the unit fraction (or fractions) that will result in the answer having the units you are seeking. When two fractions are being multiplied and the same unit appears in the numerator of one fraction and in the denominator of the other fraction, that common unit may be divided out and eliminated. For example, suppose we want to convert 30 inches to feet. We consider the following:

$$30 \text{ in.} = ? \text{ ft}$$

Since inches are given, we will need to eliminate them. Thus, inches will need to appear in the denominator of the unit fraction. We need to convert to feet, so feet will need to appear in the numerator of the unit fraction. If we multiply a quantity in inches by a unit fraction containing feet/inches, the inches will divide out as follows, leaving feet. In the following illustration we have omitted the numbers in the unit fraction so we can concentrate on the units.

$$(\cancel{\text{in.}})\left(\frac{\text{ft}}{\cancel{\text{in.}}}\right) = \text{ft}$$

Thus, to convert 30 inches to feet, we do the following.

$$30 \text{ in.} = (30 \cancel{\text{in.}})\left(\frac{1 \text{ ft}}{12 \cancel{\text{in.}}}\right) = \frac{30}{12} \text{ ft} = 2.5 \text{ ft}$$

In Examples 1 through 3, we will give examples that do not involve the metric system. After that, we will use dimensional analysis to make conversions to and from the metric system.

Example 1 *Using Dimensional Analysis*

A billboard on the side of a road indicates that a restaurant is 420 feet ahead. Convert 420 feet to miles.

> *Solution* From Table 7.6 we see that 1 mile is 5280 feet. Therefore, we solve the problem as follows.
>
> $$420 \text{ ft} = (420 \cancel{\text{ft}})\left(\frac{1 \text{ mi}}{5280 \cancel{\text{ft}}}\right) = \frac{420}{5280} \text{ mi} \approx 0.08 \text{ mi}$$
>
> Thus, 420 feet is approximately 0.08 mile. ∎

Example 2 *Mexican Pesos*

On January 28, 2019, $1 U.S. could be exchanged for about 19.03 Mexican pesos. What was the amount in U.S. dollars of 2500 pesos?

> *Solution*
>
> $$2500 \text{ pesos} = 2500 \cancel{\text{pesos}}\left(\frac{\$1.00}{19.03 \cancel{\text{pesos}}}\right) = \frac{\$2500}{19.03} \approx \$131.37$$
>
> Thus, 2500 Mexican pesos had a value of about $131.37. ∎

If more than one unit needs to be changed, more than one multiplication may be needed, as illustrated in Example 3.

Example 3 *Using Several Unit Fractions*

Convert 60 miles per hour to feet per second.

Solution Let's consider the units given and where we want to end up. We are given $\frac{mi}{hr}$ and wish to end with $\frac{ft}{sec}$. Thus, we need to change miles into feet and hours into seconds. Because two units need to be changed, we will need to multiply the given quantity by two unit fractions, one for each conversion. First we show how to convert the units of measurement from miles per hour to feet per second:

$$\left(\frac{\cancel{mi}}{\cancel{hr}}\right)\left(\frac{ft}{\cancel{mi}}\right)\left(\frac{\cancel{hr}}{sec}\right) \qquad \text{gives an answer in} \qquad \frac{ft}{sec}$$

Since 1 hour contains 60 minutes and 1 minute contains 60 seconds, there are $60 \cdot 60 = 3600$ seconds in one hour. Now we multiply the given quantity by the appropriate unit fractions to obtain the answer:

$$60\,\frac{mi}{hr} = \left(60\,\frac{\cancel{mi}}{\cancel{hr}}\right)\left(\frac{5280\,ft}{1\,\cancel{mi}}\right)\left(\frac{1\,\cancel{hr}}{3600\,sec}\right) = \frac{(60)(5280)}{(1)(3600)}\,\frac{ft}{sec}$$

$$= 88\,\frac{ft}{sec}$$

Note that $\left(60\,\frac{mi}{hr}\right)\left(\frac{1\,hr}{3600\,sec}\right)\left(\frac{5280\,ft}{1\,mi}\right)$ will give the same answer. ∎

Conversions to and from the Metric System

Now we will apply dimensional analysis to the metric system.

Table 7.7 is used in making conversions between the U.S. customary system and the metric system. The values given in Table 7.7 are often approximations. A more exact table of conversion factors may be found on the Internet. However, we can use this table to obtain many unit fractions.

Table 7.7 Conversions Table

Length	
1 inch (in.)	= 2.54 centimeters (cm)
1 foot (ft)	= 30.5 centimeters (cm)
1 foot (ft)	= 0.305 meter (m)
1 yard (yd)	= 0.9 meter (m)
1 mile (mi)	= 1.6 kilometers (km)
Area	
1 square inch (in.^{2})	= 6.5 square centimeters (cm^2)
1 square foot (ft^2)	= 0.09 square meter (m^2)
1 square yard (yd^2)	= 0.8 square meter (m^2)
1 square mile (mi^2)	= 2.6 square kilometers (km^2)
1 acre (ac)	= 0.4 hectare (ha)

(Continued)

Table 7.7　Conversions Table (*Continued*)

Volume	
1 teaspoon (tsp)	= 5 milliliters (mℓ)
1 tablespoon (tbsp)	= 15 milliliters (mℓ)
1 fluid ounce (fl oz)	= 30 milliliters (mℓ)
1 cup (c)	= 0.24 liter (ℓ)
1 pint (pt)	= 0.47 liter (ℓ)
1 quart (qt)	= 0.95 liter (ℓ)
1 gallon (gal)	= 3.8 liters (ℓ)
1 cubic foot (ft^3)	= 0.03 cubic meter (m^3)
1 cubic yard (yd^3)	= 0.76 cubic meter (m^3)
Weight (Mass)	
1 ounce (oz)	= 28 grams (g)
1 pound (lb)	= 0.45 kilogram (kg)
1 ton (T)	= 0.9 tonne (t)

Table 7.7 shows that 1 in. = 2.54 cm. From this equality, we can write the two unit fractions

$$\frac{1 \text{ in.}}{2.54 \text{ cm}} \quad \text{or} \quad \frac{2.54 \text{ cm}}{1 \text{ in.}}$$

Examples of other unit fractions from Table 7.7 are

$$\frac{1 \text{ yd}}{0.9 \text{ m}}, \quad \frac{0.9 \text{ m}}{1 \text{ yd}}, \quad \frac{1 \text{ gal}}{3.8 \ \ell}, \quad \frac{3.8 \ \ell}{1 \text{ gal}}, \quad \frac{1 \text{ lb}}{0.45 \text{ kg}}, \quad \text{and} \quad \frac{0.45 \text{ kg}}{1 \text{ lb}}$$

To change from a metric unit to a customary unit or vice versa, multiply the given quantity by the unit fraction whose product will result in the units you are seeking. For example, to convert 5 in. to centimeters, multiply 5 in. by a unit fraction with centimeters in the numerator and inches in the denominator.

$$5 \text{ in.} = (5 \text{ in.})\left(\frac{2.54 \text{ cm}}{1 \text{ in.}}\right)$$
$$= 5(2.54) \text{ cm}$$
$$= 12.7 \text{ cm}$$

Example 4　*Volume, Area, and Length Conversions*

The tallest flagpole with the largest flag is in Sheboygan, Wisconsin. The steel pole weighs 353,000 pounds. The foundation of the pole is made from 700 cubic yards of concrete. The flagpole stands 400 feet high. The 4-story-high flag on top of the flagpole has an area of 648 square meters. It took 1900 liters of paint to paint the pole.

a) What is the mass, in kilograms, of the flagpole?

b) What is the volume, in cubic meters, of the concrete used for the foundation?

c) What is the height, in meters, of the flagpole?

d) What is the area, in square feet, of the flag?

e) What is the volume, in gallons, of the paint used to paint the flagpole?

Solution

a) In Table 7.7, under the heading of mass, we see that 1 pound = 0.45 kilogram. Thus, the unit fractions we need to consider to change from pounds to kilograms are

$$\frac{1 \text{ lb}}{0.45 \text{ kg}} \quad \text{or} \quad \frac{0.45 \text{ kg}}{1 \text{ lb}}$$

We need to convert from pounds to kilograms. Thus, we need to use the unit fraction with pounds in the denominator so that the pounds will divide out. The solution follows.

$$353{,}000 \text{ lb} = 353{,}000 \text{ lb} \left(\frac{0.45 \text{ kg}}{1 \text{ lb}} \right) = 353{,}000(0.45 \text{ kg}) = 158\ 850 \text{ kg}$$

Thus, the mass of the flagpole is 158 850 kilograms.

b) In Table 7.7, under the heading of volume, we see that 1 cubic yard = 0.76 cubic meter. Thus, the unit fractions we need to consider to change from cubic yards to cubic meters are

$$\frac{1 \text{ yd}^3}{0.76 \text{ m}^3} \quad \text{or} \quad \frac{0.76 \text{ m}^3}{1 \text{ yd}^3}$$

We need to convert from cubic yards to cubic meters. Thus, we need to use the unit fraction with cubic yards in the denominator so that cubic yards will divide out.

$$700 \text{ yd}^3 = 700 \text{ yd}^3 \left(\frac{0.76 \text{ m}^3}{1 \text{ yd}^3} \right) = 700 \ (0.76 \text{ m}^3) = 532 \text{ m}^3$$

Thus, the volume of the concrete used is 532 cubic meters.

c) We need to change from feet to meters. Table 7.7 shows that 1 foot = 0.305 meter. We convert 400 feet to meters as follows.

$$400 \text{ ft} = 400 \text{ ft} \left(\frac{0.305 \text{ m}}{1 \text{ ft}} \right) = 400(0.305 \text{ m}) = 122 \text{ m}$$

Thus, the height of the flagpole is 122 meters.

d) To convert from square meters to square feet we need to consider the unit fractions below.

$$\frac{1 \text{ ft}^2}{0.09 \text{ m}^2} \quad \text{or} \quad \frac{0.09 \text{ m}^2}{1 \text{ ft}^2}$$

Since we are starting with square meters, the unit fraction must have square meters in the denominator. The solution follows.

$$648 \text{ m}^2 = 648 \text{ m}^2 \left(\frac{1 \text{ ft}^2}{0.09 \text{ m}^2} \right) = \frac{648}{0.09} \text{ ft}^2 = 7200 \text{ ft}^2$$

Thus, the area of the flag is 7200 square feet.

e) To change from 1900 liters to gallons we do the following.

$$1900 \ \ell = 1900 \ \ell \left(\frac{1 \text{ gal}}{3.8 \ \ell} \right) = \frac{1900}{3.8} \text{ gal} = 500 \text{ gal}$$

Thus, the volume of paint used to paint the flagpole was 500 gallons. ∎

RECREATIONAL MATH

What Metric Unit Am I?

(a) I am a length greater than your foot, but less than the length of a large kitchen table top.

(b) I am a weight greater than a calculator, but less than a rocking chair.

(c) I am a liquid volume greater than a glass of water, but less than a large jug of milk or a gallon of gasoline.

(d) I am an area greater than a football field, but less than the Magic Kingdom at Disney World.

Example **5** *Administering a Medicine*

A nurse must administer 4 cc of codeine medicine to a patient.

a) How many milliliters of the medicine will be administered?

b) How many ounces is this dosage equivalent to?

Solution

a) Since 1 cc = 1 mℓ, the nurse will administer 4 mℓ of the medicine.

b) Since 1 fl oz = 30 mℓ,

$$4 \text{ mℓ} = (4 \text{ mℓ})\left(\frac{1 \text{ fl oz}}{30 \text{ mℓ}}\right) = \frac{4}{30} \text{ fl oz} \approx 0.13 \text{ fl oz} \quad \blacksquare$$

Suppose we want to convert 150 millimeters to inches. Table 7.7 does not have a conversion factor from millimeters to inches, but it does have one for inches to centimeters. Because 1 inch = 2.54 centimeters and 1 centimeter = 10 millimeters, we can reason that 1 inch = 25.4 millimeters. Therefore, unit fractions we may use are as follows.

$$\frac{1 \text{ in.}}{25.4 \text{ mm}} \quad \text{or} \quad \frac{25.4 \text{ mm}}{1 \text{ in.}}$$

We can solve the problem as follows.

$$150 \text{ mm} = (150 \text{ mm})\left(\frac{1 \text{ in.}}{25.4 \text{ mm}}\right) = \frac{150}{25.4} \text{ in.}$$

$$\approx 5.91 \text{ in.}$$

If we wish, we can use dimensional analysis using two unit fractions to make the conversion. The procedure follows:

$$150 \text{ mm} = (150 \text{ mm})\left(\frac{1 \text{ cm}}{10 \text{ mm}}\right)\left(\frac{1 \text{ in.}}{2.54 \text{ cm}}\right) = \frac{150}{(10)(2.54)} \text{ in.}$$

$$\approx 5.91 \text{ in.}$$

Example **6** *Blueberries*

As shown in the photo, blueberries at an outdoor market in Italy cost 9.00 euros per kilogram. If 1 euro can be converted to $1.14, determine the cost of the blueberries in U.S. dollars per pound.

Solution In Table 7.7 we see that 1 pound = 0.45 kilogram. The two unit fractions involving pounds and kilograms are

$$\frac{1 \text{ lb}}{0.45 \text{ kg}} \quad \text{or} \quad \frac{0.45 \text{ kg}}{1 \text{ lb}}$$

The two unit fractions involving euros and dollars are

$$\frac{1 \text{ euro}}{\$1.14} \quad \text{or} \quad \frac{\$1.14}{1 \text{ euro}}$$

The conversion from euros per kilogram to U.S. dollars per pound follows. Note that 9 euros per kilogram can be expressed as 9 euros/1 kg.

$$\frac{9 \text{ euros}}{1 \text{ kg}} = \left(\frac{9 \text{ euros}}{1 \text{ kg}}\right)\left(\frac{0.45 \text{ kg}}{1 \text{ lb}}\right)\left(\frac{\$1.14}{1 \text{ euro}}\right) = \frac{\$(9)(0.45)(1.14)}{1 \text{ lb}} \approx \frac{\$4.62}{\text{lb}}$$

Therefore, 9.00 euros per kilogram is equivalent to about $4.62 per pound.   ∎

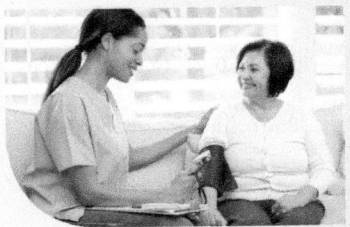
Example 7 *Active Ingredient in Cough Syrup*

The label on a bottle of Vicks Formula 44D cough syrup indicates that the active ingredient is dextromethorphan hydrobromide and that 5 mℓ (or 1 teaspoon) of cough syrup contains 10 mg of this ingredient. The recommended adult dosage is 3 teaspoons.

a) How many milliliters of cough syrup are in the recommended adult dosage?
b) If the bottle contains 8 fluid ounces of cough syrup, how many milligrams of the active ingredient are in the bottle?

Solution

a) The recommended adult dosage is 3 teaspoons, and each teaspoon contains 5 mℓ of cough syrup. Therefore, there are 3 × 5, or 15, mℓ of cough syrup in the recommended adult dosage. Using dimensional analysis, we have the following

$$3 \text{ tsp} = (3 \text{ tsp})\frac{5 \text{ m}\ell}{1 \text{ tsp}} = 15 \text{ m}\ell$$

b) We will work this problem in two parts. First we will convert 8 fluid ounces into milliliters, and then we will determine the amount of active ingredient in the bottle of cough syrup. Table 7.7 shows that 1 fluid ounce is equivalent to 30 milliliters. To change from 8 fluid ounces to milliliters we do the following:

$$8 \text{ fl oz} = (8 \text{ fl oz})\left(\frac{30 \text{ m}\ell}{1 \text{ fl oz}}\right) = 240 \text{ m}\ell$$

Next we determine the amount of active ingredient in the bottle by determining the amount of active ingredient in 240 milliliters of cough syrup. We are told that 5 milliliters of cough syrup contains 10 milligrams of the active ingredient. We calculate the amount of active ingredient as follows.

$$\text{Amount of activate ingredient} = 240 \text{ m}\ell \left(\frac{10 \text{ mg}}{5 \text{ m}\ell}\right) = 480 \text{ mg}$$

Therefore, there are 480 mg (or 0.48 g) of the active ingredient in the 8-ounce bottle. ∎

Example 8 *Determining Dosage by Weight*

Drug dosage is often administered according to a patient's weight. For example, 30 mg of the drug vancomycin is to be given for each kilogram of a person's weight. If Martha, who weighs 136 lb, is to be administered the drug, what dosage should she be given?

Solution First we need to convert Martha's weight into kilograms. From Table 7.7, we see that 1 lb = 0.45 kg. We obtain our unit fraction from this information.

$$136 \text{ lb} = 136 \text{ lb}\left(\frac{0.45 \text{ kg}}{1 \text{ lb}}\right) = 61.2 \text{ kg}$$

Next, since 30 mg of the drug is to be given for each kilogram of a person's weight, we multiply 61.2 kg by 30 to determine the dosage.

$$\text{Amount of drug} = 61.2 \text{ kg}\left(\frac{30 \text{ mg}}{1 \text{ kg}}\right) = 1836 \text{ mg}$$

Thus, 1836 mg, or 1.836 g, of the drug should be given. ∎

SECTION 7.4   Exercises

Warm Up Exercises

In Exercises 1–10, fill in the blanks with an appropriate word, phrase, or symbol(s).

1. A procedure used to convert from one unit of measurement to a different unit of measurement is called _____ analysis.

2. A fraction in which the numerator and denominator contain different units, and the value of the fraction is 1, is called a(n) _____ fraction.

3. The two unit fractions that relate seconds and minutes are _____ and _____.

4. The two unit fractions that relate feet and inches are _____ and _____.

5. The two unit fractions that relate meters and centimeters are _____ and _____.

6. The two unit fractions that relate liters and milliliters are _____ and _____.

7. a) When converting from kilograms to pounds, you would use the unit fraction _____.

   b) When converting from pounds to kilograms, you would use the unit fraction _____.

8. a) When converting from centimeters to feet, you would use the unit fraction _____.

   b) When converting from feet to centimeters, you would use the unit fraction _____.

9. a) When converting from gallons to liters, you would use the unit fraction _____.

   b) When converting from liters to gallons, you would use the unit fraction _____.

10. a) When converting from square yards to square meters, you would use the unit fraction _____.

    b) When converting from square meters to square yards, you would use the unit fraction _____.

Practice the Skills

In Exercises 11–26, convert the quantity to the indicated units. When appropriate, round your answer to the nearest hundredth.

11. 18 lb to kilograms
12. 90 in. to centimeters
13. 425 g to ounces
14. 9.6 m to yards
15. 175 kg to pounds
16. 20 yd² to square meters
17. 39 mi to kilometers
18. 765 mm to inches
19. 675 ha to acres
20. 253 oz to grams
21. 25.2 ℓ to pints
22. 4 T to tonnes
23. 3.8 km² to square miles
24. 25.6 mℓ to fluid ounces
25. 120 lb to kilograms
26. 6.2 acres to hectares

In Exercises 27–30, use the part of the scorecard that shows the distance in meters for the first four holes of the Millbrook Resort Golf Course in Queenstown, New Zealand. Determine the distances indicated. When appropriate round your answer to the nearest hundredth.

Hole	Black Tees	Blue Tees	Handicap	Par			White Tees	Red Tees
1	505	505	3	5			466	414
2	185	175	15	3			137	91
3	366	357	11	4			344	287
4	396	376	7	4			376	303

27. Hole 1, red tees, in yards
28. Hole 2, white tees, in yards
29. Hole 3, blue tees, in feet
30. Hole 4, black tees, in feet

Problem Solving

In Exercises 31–59, when necessary round your answers to the nearest hundredth.

31. **Speed Limit** A speed limit sign in Canada shows 80 km/h. Determine this speed in miles per hour.

32. **Distance Traveled** Rosa's new car travels 94 miles on 3 gallons of gasoline. How many kilometers can Rosa's car travel with the same amount of gasoline?

33. **Garden** Isaac is planting a rectangular vegetable garden that is 15 yards long by 10 yards wide. What area, in square meters, will his garden be?

34. **Pitcher of Water** How many milliliters of water will a 32 fl oz pitcher hold?

35. **Poison Dart Frog** A full-grown poison dart frog has a weight of about 6 g. What is its weight in ounces?

36. Swimming Pool A swimming pool holds 12,500 gal of water. What is this volume in kiloliters?

37. Area of Tribal Park The Monument Valley Navajo Tribal Park has an area of 91,696 acres. What is its area in hectares?

38. Building a Basement A basement is to be 50 ft long, 30 ft wide, and 8 ft high. How much dirt will have to be removed when this basement is built? Answer in cubic meters.

39. Yellowstone National Park The plastic bottles recycled annually at the general stores at Yellowstone National Park would reach a height of about 1,000,000 feet if stacked end to end. What is this distance in

a) meters?

b) kilometers?

40. Weight Restriction The weight restriction on a road in Mexico is 7.0 t.

a) How many tons does this weight equal?

b) How many pounds does this weight equal?

41. Flying In the United States, the maximum volume of liquid travelers can pack in their carry-on suitcase is 3.4 fluid ounces. When travelers are flying elsewhere in the world, they can generally pack no more than 100 milliliters of liquid in their carry-on.

a) Convert 3.4 ounces to milliliters.

b) Which is greater, 3.4 ounces or 100 milliliters?

42. Aquarium A large tank at an aquarium holds 250 kℓ of salt water. How many gallons does it hold?

43. Earth The artwork that follows shows the dimensions of the various parts of Earth, from its crust to its inner core. Use a thickness of 15 miles for the crust.

a) Determine the radius of Earth in miles.

b) Determine the radius of Earth in kilometers.

c) The deepest hole dug in Earth, called the Kola Well, was in the Kola Peninsula in northwest Russia. The Russians dug the well for 27 years before stopping because the high internal temperature of Earth was melting their drills. The deepest the hole reached was about 12.2 km. How far down, in miles, did the hole reach?

d) At the location of the hole, the crust was about 24 km. How thick was the crust at that location in miles?

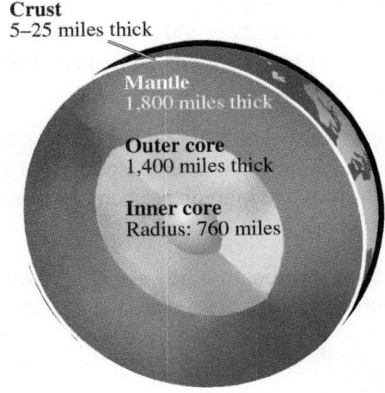

Crust
5–25 miles thick

Mantle
1,800 miles thick

Outer core
1,400 miles thick

Inner core
Radius: 760 miles

44. Square Meters to Square Feet One meter is about 3.3 ft. Use this information to determine

a) the equivalent of one square meter in square feet.

b) the equivalent of one cubic meter in cubic feet.

45. Acres to Hectares One acre is about 0.4 hectare.

a) Use this information to determine the equivalent of 18.5 acres in hectares.

b) If 1 hectare is 10 000 m^2, determine the area of the 18.5 acres in square meters.

46. Dosage for a Child The recommended dosage of the drug codeine for pediatric patients is 1 mg per kilogram of a child's weight. What dosage of codeine should be given to April, who weighs 56 lb?

47. Medicine Dosage For each kilogram of a person's weight, 1.5 mg of the antibiotic drug gentamicin is to be administered. If Li weighs 120 lb, how much of the drug should she receive?

48. Medicine for a Dog For each kilogram of weight of a dog, 5 mg of the drug bretylium is to be given. If Blaster, a golden retriever, weighs 82 lb, how much of the drug should be given?

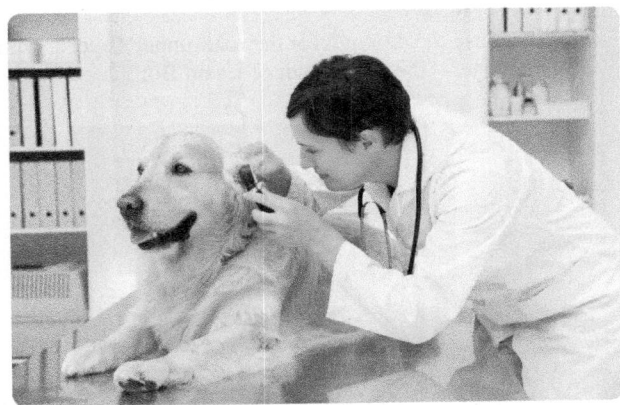

49. *Ampicillin* The recommended dosage of the drug ampicillin for pediatric patients is 200 mg per kilogram of a patient's weight. If Imani weighs 76 lb, how much ampicillin should she receive?

50. *Active Ingredients* The label on the bottle of Triaminic expectorant indicates that each teaspoon (5 mℓ) contains 12.5 mg of the active ingredient phenylpropanolamine hydrochloride.

  a) Determine the amount of the active ingredient in the recommended adult dosage of 2 teaspoons.

  b) Determine the quantity of the active ingredient in a 12-oz bottle.

51. *Stomach Ache Remedy* The label on the bottle of Maximum Strength Pepto-Bismol indicates that each tablespoon contains 236 mg of the active ingredient bismuth subsalicylate.

  a) Determine the amount of the active ingredient in the recommended dosage of 2 tablespoons.

  b) If the bottle contains 8 fl oz, determine the quantity of the active ingredient in the bottle.

52. *Gibraltar* Gibraltar is a British territory that shares a 1.2-kilometer border with Spain. The territory covers an area of about 6.84 square kilometers and the Rock of Gibraltar has a maximum height of 426 meters.

  a) Determine the length of the border with Spain in miles.

  b) Determine the area of the territory in square miles.

  c) Determine the height of the Rock of Gibraltar in feet.

53. *Roadrunner* The great roadrunner, found in parts of the Southwest, is a very fast bird on the ground. It has been recorded as reaching speeds as high as 42 km/h, although it normally runs about 32 km/h. Usain Bolt has a record speed in the Olympics of running about 23.35 mph.

  a) If we use the 32 km/h for the roadrunner's speed, who ran faster—the roadrunner or Usain Bolt?

  b) If we use the 42 km/h for the roadrunner's speed, who ran faster—the roadrunner of Usain Bolt?

54. *Two Charlestons* Charleston, West Virginia, is the capital of West Virginia. It has an area of 8458 hectares.

A well-known vacation spot is Charleston, South Carolina. It has an area of 81,600 acres.

  a) Convert the area of Charleston, South Carolina, to hectares. Then determine which Charleston has the largest area.

  b) If 1 km^2 = 100 ha, determine the area of Charleston, South Carolina, in square kilometers.

  c) Determine the area of Charleston, South Carolina, in square miles.

55. *Overseas Flight* The photo of the monitor taken during an overseas flight shows information about the location and speed of the airplane.

  a) Determine the altitude in miles.

  b) Determine the ground speed in miles per hour.

  c) Determine the headwind in miles per hour.

  d) Determine the temperature in degrees Fahrenheit using the formula $F = \dfrac{9}{5}C + 32$ (see Section 7.3).

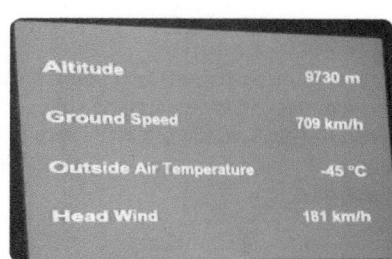

56. *Peppers* Peppers in Rome, Italy, cost 2 euros (€) per kilogram.

  a) Determine the cost per pound, in euros, for the peppers.

  b) If 1 € can be converted to $1.14 U.S., determine the cost, in U.S. dollars, of 1 pound of peppers.

57. *Eggplant* The photo shown, taken in Italy, shows melanzane (eggplant) selling for 2.40 euros per kilogram.

  a) Determine the price of 1 pound of melanzane in euros.

  b) If 1 euro can be converted to $1.14 U.S, determine the cost in U.S. dollars for 1 pound of melanzane.

58. *Unit Fraction* On February 1, 2019, the conversion rates for U.S. dollars were as follows: One U.S. dollar could buy 0.87 euro, 1.33 Canadian dollars, or 19.05 Mexican pesos. Using this information, determine the unit fractions to convert

a) Canadian dollars to euros.

b) euros to Mexican pesos.

c) Mexican pesos to Canadian dollars.

59. *Unit Fractions* An advertisement for Piaggio airplanes stated that their Avanti II had a speed of 402 knots, or 745 kilometers per hour, or 463 miles per hour. Using this information, determine a unit fraction to convert

a) knots to kilometers per hour.

b) miles per hour to knots.

c) kilometers per hour to miles per hour.

60. *Making Cookies* Change all the measurements in the cookie recipe below to metric units. Do not forget pan size, temperature, and size of cookies.

Magic Cookie Bars

$\frac{1}{2}$ C graham cracker crumbs

12 oz nuts

8 oz chocolate pieces

$1\frac{1}{3}$ C flaked coconut

$1\frac{1}{3}$ C condensed milk

Coat the bottom of a 9 in. × 13 in. pan with melted margarine. Add rest of ingredients one by one: crumbs, nuts, chocolate, and coconut. Pour condensed milk over all. Bake at 350°F for 25 minutes. Allow to cool 15 minutes before cutting. Makes about two dozen $1\frac{1}{2}$ in. by 3 in. bars.

Challenge Problems/Group Activities

61. *Nursing Question* The following question was selected from a nursing exam. Can you answer it?

In caring for a patient after delivery, you are to give 0.2 mg Ergotrate Maleate. The ampule is labeled $\frac{1}{300}$ grain/mℓ. How much would you draw and give? (60 mg = 1 grain)

a) 15 cc   b) 1.0 cc   c) 0.5 cc   d) 0.01 cc

62. *How Much Beef* Paul is organizing a picnic and plans to purchase 0.18 kg of ground beef for each 100 lb of weight of guests who will be in attendance. If he expects 15 people whose average weight is 130 lb, how many pounds of beef should he purchase?

63. *Auto Engine* The displacement of an automobile engine is measured in liters. A 2019 Chevrolet Camaro SS has a 6.2 ℓ engine.

a) Determine the displacement of the engine in cubic centimeters.

b) If 1 cubic inch is approximately equal to 16.39 cubic centimeters, determine the displacement of the engine in cubic inches.

Research Activity

64. *U.S. Customary Units of Length* The website www.onlineconversion.com gives 16 units of length used in the U.S. customary system. It also gives their metric equivalent. Determine the metric equivalent of each of the following units of length to the nearest hundredth.

a) 1 pica   b) 1 link

c) 1 rod   d) 1 league

e) 1 fathom   f) 1 nautical mile

g) 1 furlong

CHAPTER 7   *Summary*

Important Facts and Concepts	Examples and Discussion

Section 7.1
Metric Units and Conversions Within the Metric System

Metric Units

Prefix	Symbol	Meaning
kilo	k	$1000 \times$ basic unit
hecto	h	$100 \times$ basic unit
deka	da	$10 \times$ basic unit
—	—	basic unit (meter, gram, liter, . . .)
deci	d	$\frac{1}{10}$ of basic unit
centi	c	$\frac{1}{100}$ of basic unit
milli	m	$\frac{1}{1000}$ of basic unit

Discussion pages 402–404

Examples 1–4, pages 404–406

Section 7.2
Length, Area, Volume

Length

Discussion page 410

Example 1, pages 410–411

Area

Discussion page 411.
Examples 2–5, pages 411–412

Volume

Discussion page 413.
Examples 6–11, pages 413–416

Section 7.3
Mass

Discussion pages 420–422.
Examples 1–2, pages 421–422

Water Volume in Cubic Units		Volume in Liters		Mass of Water
$1\ cm^3$	=	$1\ m\ell$	=	$1\ g$
$1\ dm^3$	=	$1\ \ell$	=	$1\ kg$
$1\ m^3$	=	$1\ k\ell$	=	$1\ t\ (1000\ kg)$

Temperature

$$C = \frac{5}{9}(F - 32)$$

$$F = \frac{9}{5}C + 32$$

Discussion pages 423–424.
Examples 3–5, pages 423–424

Section 7.4
Dimensional Analysis

Using dimensional analysis

Discussion pages 429–430.
Examples 1–3, pages 430–431

Using dimensional analysis with the metric system

Discussion pages 431–432.
Examples 4–8, pages 432–435

CHAPTER 7   Review Exercises

7.1

In Exercises 1–6, indicate the meaning of the prefix.

1. Centi

2. Kilo

3. Hecto

4. Milli

5. Deka

6. Deci

In Exercises 7–12, change the given quantity to the quantity indicated.

7. 80 mg to grams

8. 0.52 km to decimeters

9. 5700 cm to hectometers

10. 1 000 000 mg to kilograms

11. 4.62 kℓ to liters

12. 192.6 dag to decigrams

In Exercises 13 and 14, arrange the quantities from smallest to largest.

13. 2.67 kℓ, 3000 mℓ, 14 630 cℓ

14. 0.047 km, 4700 m, 47 000 cm

7.2, 7.3

In Exercises 15–20, indicate the metric unit of measurement that would best express the following.

15. The temperature of the water in an ocean

16. The length of a toothpick

17. The area of a typical garden

18. The volume of a bottle of water

19. The mass of a car

20. The distance from Earth to the moon

In Exercises 21 and 22, (a) first estimate the following in metric units and then (b) measure with a metric ruler. Record your results.

21. Your waist

22. The length of a new pencil

In Exercises 23–28, select the best answer.

23. The distance between Buffalo, New York, and Tampa, Florida, is about

  a) 20 000 cm.   **b)** 2000 km.   **c)** 20 000 m.

24. The mass of a sports utility vehicle is about

  a) 1300 kg.   **b)** 13 000 g.   **c)** 130 000 mg.

25. The volume of a jug of water is about

  a) 0.1 kℓ.   **b)** 0.5 ℓ.   **c)** 4 ℓ.

26. The area of a large classroom floor may be

  a) 200 m².   **b)** 0.5 ha.   **c)** 0.02 km².

27. The temperature on a hot summer day in Florida may be

  a) 34°C.   **b)** 55°C.   **c)** 25°C.

28. The height of a cell phone tower is about

  a) 300 cm.   **b)** 30 m.   **c)** 3 km.

29. Convert 1800 kilograms to tonnes.

30. Convert 9.2 t to grams.

31. The temperature on a thermometer is 24°C. What is the Fahrenheit temperature?

32. If the room temperature is 68°F, what is the Celsius temperature?

33. If your outdoor thermometer shows a temperature of −6°F, what is the Celsius temperature?

34. If Joshua's body temperature is 39°C, what is his Fahrenheit temperature?

35. Measure, in centimeters, each of the line segments, then compute the area of the figure. Use the formula $A = lw$.

36. Measure, in centimeters, the radius of the circle, then compute the area of the circle. Use the formula $A = \pi r^2$.

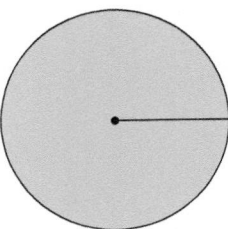

37. a) *Hot Tub Volume* What is the volume of water in a full circular hot tub that has a diameter of 2 m and has a depth of 1 m? Answer in cubic meters.

b) What is the mass of the water in kilograms?

38. *Computer Monitor* A rectangular computer screen measures 33.7 cm by 26.7 cm. Determine the screen's area in

a) square centimeters.

b) square meters.

39. A fish tank is shown below.

a) What is its volume in cubic centimeters?

b) What is its volume in cubic meters?

c) How many milliliters of water will the tank hold?

d) How many kiloliters of water will the tank hold?
0.18

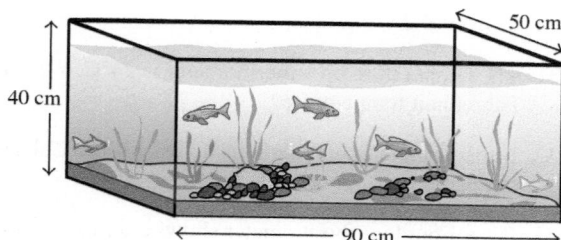

40. A square kilometer is a square with length and width both 1 km. How many times larger is a square kilometer than a square dekameter?

7.4

In Exercises 41–50, change the given quantity to the indicated quantity. When appropriate, round answers to the nearest hundredth. Use the conversion chart on pages 430–431.

41. 41 kg = _____ lb

42. 18 cm = _____ in.

43. 37 yd = _____ m

44. 100 m = _____ yd

45. 52 mi = _____ km

46. 27 ℓ = _____ qt

47. 20 gal = _____ ℓ

48. 60 m^3 = _____ yd^3

49. 96 cm^2 = _____ in.^{2}

50. 4 qt = _____ ℓ

51. *Pediatrician's Office* At a pediatrician's office, 9-year-old Reinhardt measured 137 cm and weighed 44.2 kilograms.

a) What is his height in inches?

b) What is his weight in pounds?

52. *Distance Traveled* The distance from Dallas, Texas, to Houston, Texas, is 241 miles. Determine this distance in kilometers.

53. *Carpeting a Room* Irene is buying new carpet for her rectangular family room. The room is 15 ft wide and 24 ft long. The carpeting is sold only in square meters. How many square meters of carpeting will she need? Round your answer to the nearest tenth of a square meter.

54. *Milk Tank* A cylindrical milk tank can store 50,000 gal of milk.

a) Determine the volume in kiloliters.

b) Estimate the weight of the milk in kilograms. Assume that milk has the same weight as water.

55. *Speed Limit* The speed limit on a road in Canada is 70 kilometers per hour. What is the speed limit in

a) miles per hour?

b) meters per hour?

56. *A Water Tank* A rectangular tank used to test leaks in tires is 90 cm long by 70 cm wide by 40 cm deep.

a) Determine the number of liters of water the tank holds.

b) What is the mass of the water in kilograms?

57. *Oranges* If the cost of oranges is $3.50 per kilogram, determine the cost of 1 lb of oranges.

CHAPTER 7   *Test*

1. Convert 2400 ℓ to kℓ.

2. Convert 46.2 cm to hm.

3. How many times greater is a kilometer than a dekameter?

4. *Jogging* A high school track is an oval that measures 400 m around. If Savannah jogs around the track six times, how many kilometers has she gone?

In Exercises 5–9, choose the best answer.

5. The width of a page in a hardbound book is about

    a) 5 cm.   **b)** 20 cm.   **c)** 60 cm.

6. The surface area of the top of a kitchen table is about

    a) 2 m^2.   **b)** 200 cm^2.   **c)** 2000 cm^2.

7. The amount of gasoline that an automobile's gas tank can hold is about

    a) 200 ℓ.   **b)** 20 ℓ.   **c)** 75 ℓ.

8. The distance that a typical car can go on a full tank of gas is about

    a) 50 m.   **b)** 500 km.   **c)** 5000 km.

9. The outside temperature on a warm day is about

    a) 8°C.   **b)** −2°C.   **c)** 24°C.

10. How many times greater is a square meter than a square centimeter?

11. How many times greater is a cubic meter than a cubic centimeter?

12. Convert 88 lb to kilograms.

13. *Washington Monument* The Washington Monument is the world's tallest stone structure and the world's tallest obelisk. It stands about 169.29 meters tall. Determine its height in feet.

14. *National Park* Wrangell–St. Elias National Park and Preserve in southern Alaska is the largest national park in the United States. Its area is 53 321 square kilometers. Determine the park's area in square miles.

15. Convert 40°F to degrees Celsius.

16. Convert −5°C to degrees Fahrenheit.

17. *Elevator* A sign in an elevator in France indicates that its maximum capacity is 300 kg. Determine the maximum capacity in

    a) grams.   **b)** pounds.

18. *At the Aquarium* A rectangular fish tank at an aquarium is 20 m long by 20 m wide by 8 m deep.

    a) Determine the volume of the tank in cubic meters.

    b) Determine the number of liters of water the tank holds.

    c) Determine the weight of the water in kilograms.

19. *Cost of Paint* The first coat of paint for the outside walls of a building requires 1 ℓ of paint for each 10 m^2 of wall surface. The second coat requires 1 ℓ for every 15 m^2. If the paint costs $3.50 per liter, what will be the cost of two coats of paint for the four outside walls of a rectangular building 20 m long, 15 m wide, and 6 m high?

20. *Regular Gasoline* Regular gasoline at a gas station in Europe costs 1.61 euros per liter.

    a) Determine the cost, in euros per gallon, of regular gasoline.

    b) If one dollar could be exchanged for 0.87 euro, determine the cost, in U.S. dollars per gallon, of regular gasoline.

◀ Basketballs and basketball courts are examples of geometric objects.

8

Geometry

What You Will Learn

- Points, lines, planes, and angles
- Polygons, similar figures, and congruent figures
- Perimeter and area of polygons, circumference and area of circles, and the Pythagorean theorem
- Volume and surface area
- Transformational geometry, symmetry, and tessellations
- Topology including Möbius strips, Klein bottles, and maps
- Non-Euclidean geometry and fractal geometry

Many objects we encounter daily can be described using geometry. Spherical basketballs bounced on a rectangular court, cylindrical cans of vegetables, rectangular tablet computer screens, and ice cream cones are all examples of geometric objects we frequently encounter. Geometry also plays a key role in scientific research. The shape of DNA molecules, animal and plant cells, bacteria, and viruses are described using geometry. The position of atoms within a molecule or of molecules within a crystal are described using geometry. Furthermore, engineers use geometry extensively when they create items that are used in the products we use every day. For example, the antennae of most cell phones are in the geometric shape described as fractal. In this chapter, we will study some of these and many other applications of geometry that affect our everyday lives.

Points, Lines, Planes, and Angles

Upon completion of this section, you will be able to:

- Recognize and solve problems involving points and lines.
- Recognize and solve problems involving planes.
- Recognize and solve problems involving angles.

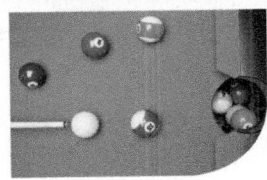

Playing billiards involves many geometric concepts. In this section, we will introduce the geometric terms *point*, *line*, *plane*, and *angle*. In a typical billiard game, each of these terms is evident. A billiard ball rests on a specific point on the table, which is part of a plane. After being struck with a cue, the ball travels along a path that is part of a line. Angles are involved in the path the ball takes when the ball hits a table bumper or another ball.

Why This Is Important The concepts introduced in this section form a basis for the study of geometry. These concepts allow us to better apply geometry to solving a variety of problems that we encounter daily. These concepts are also used by scientists to describe the universe and by engineers to design and improve many of the products we rely on, including the digital camera on your smartphone.

The History of Geometry

Geometry as a science is said to have begun in the Nile Valley of ancient Egypt. The Egyptians used geometry to measure land and to build pyramids and other structures. The word *geometry* is derived from two Greek words: *ge*, meaning earth, and *metron*, meaning measure. Thus, geometry means "earth measure."

Unlike the Egyptians, the Greeks were interested in more than just the applied aspects of geometry. The Greeks attempted to apply their knowledge of logic to geometry. In about 600 B.C., Thales of Miletus was the first to be credited with using deductive methods to develop geometric concepts. Later, Pythagoras continued the systematic development of geometry that Thales had begun. In about 300 B.C., Euclid (see *Profile in Mathematics* in the margin) collected and summarized much of the Greek mathematics of his time. In a set of 13 books called *Elements*, Euclid laid the foundation for plane geometry, which is also called *Euclidean geometry*.

Euclid is credited with being the first mathematician to use the *axiomatic method* in developing a branch of mathematics. First, Euclid introduced *undefined terms* such as point, line, plane, and angle. He related these to physical space by such statements as "A line is length without breadth" so that we may intuitively understand them. Because such statements play no further role in his system, they constitute primitive or undefined terms.

Second, Euclid introduced certain *definitions*. The definitions are introduced when needed and are often based on the undefined terms. Some terms that Euclid introduced and defined include triangle, right angle, and hypotenuse.

Third, Euclid stated certain primitive propositions called *postulates* (now called *axioms**) about the undefined terms and definitions. The reader is asked to accept these statements as true on the basis of their "obviousness" and their relationship with the physical world. For example, the Greeks accepted all right angles as being equal, which is Euclid's fourth postulate.

Profile in Mathematics

Euclid

Euclid (320–275 B.C.) lived in Alexandria, Egypt, and was a teacher and scholar at Alexandria's school called the *Museum*. Here Euclid collected and arranged many of the mathematical results known at the time. This collection of works became his 13-volume masterpiece known as *Elements*. Beginning with a list of definitions, postulates, and axioms, Euclid proved one theorem after another, using only previously proven results. This method of proof became a model of mathematical and scientific investigation that survives today. Remarkably, the geometry in *Elements* does not rely on making exact geometric measurements using a ruler or protractor. Rather, the work is developed using only an unmarked straightedge and a drawing compass. Next to the Bible, Euclid's *Elements* may be the most translated, published, and studied of all the books produced in the Western world.

---

*The concept of the axiom has changed significantly since Euclid's time. Now any statement may be designated as an axiom, whether it is self-evident or not. All axioms are *accepted* as true. A set of axioms forms the foundation for a mathematical system.

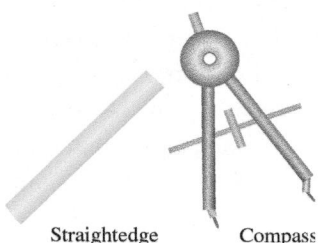
Fourth, Euclid proved, using deductive reasoning (see Section 1.1), other propositions called *theorems*. One theorem that Euclid proved is the Pythagorean theorem (see Section 8.3). He also proved that the sum of the angles of a triangle is 180°. Using only 10 axioms, Euclid deduced 465 theorems in plane and solid geometry, number theory, and Greek geometric algebra.

Point and Line

Three basic terms in geometry are *point*, *line*, and *plane*. These three terms are not given a formal definition, but we recognize points, lines, and planes when we see them.

Let's consider some properties of a line. Assume that a line means a straight line unless otherwise stated.

1. A line is a set of points. Each point is on the line and the line passes through each point. When we wish to refer to a specific point, we will label it with a single capital letter. For example, in Fig. 8.1(a) three points are labeled A, B, and C, respectively.

2. Any two distinct points determine a unique line. Figure 8.1(a) illustrates a line. The arrows at both ends of the line indicate that the line continues in each direction. The line in Fig. 8.1(a) may be symbolized with any two points on the line by placing a line with a double-sided arrow above the letters that correspond to the points, such as $\overleftrightarrow{AB}$, $\overleftrightarrow{BA}$, $\overleftrightarrow{AC}$, $\overleftrightarrow{CA}$, $\overleftrightarrow{BC}$, or $\overleftrightarrow{CB}$.

3. Any point on a line separates the line into three parts: the point itself and two *half lines* (neither of which includes the point). For example, in Fig. 8.1(a) point B separates the line into the point B and two half lines. Half line BA, symbolized $\overset{\circ}{BA}\!\!\rightarrow$, is illustrated in Fig. 8.1(b). The open circle above the B indicates that point B is not included in the half line. Figure 8.1(c) illustrates half line BC, symbolized $\overset{\circ}{BC}\!\!\rightarrow$.

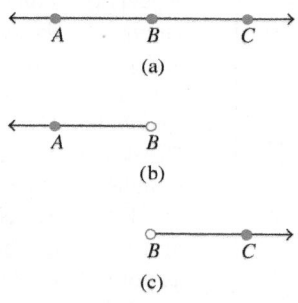

Figure 8.1

Look at the half line $\overset{\circ}{AB}\!\!\rightarrow$ in Fig. 8.2(b) on page 447. If the *end point*, A, is included with the set of points on the half line, the result is called a *ray*. Ray AB, symbolized $\overrightarrow{AB}$, is illustrated in Fig. 8.2(c). Ray BA, symbolized $\overrightarrow{BA}$, is illustrated in Fig. 8.2(d). A *line segment* is that part of a line between two points, including the end points. Line segment AB, symbolized $\overline{AB}$, is illustrated in Fig. 8.2(e). An open line segment is the set of points on a line between two points, excluding the end points. Open line segment AB, symbolized $\overset{\circ}{AB}\overset{\circ}{}$, is illustrated in Fig. 8.2(f). Figure 8.2(g) illustrates two half open line segments, symbolized $\overline{AB}\overset{\circ}{}$ and $\overset{\circ}{AB}$.

Recall from Chapter 2 that the intersection (symbolized $\cap$) of two sets is the set of elements (points in this case) common to both sets. Consider the rays $\overrightarrow{AB}$ and $\overrightarrow{BA}$ in Fig. 8.3(a) on page 447. The intersection of $\overrightarrow{AB}$ and $\overrightarrow{BA}$ is the set of points that are common to both rays, or line segment AB, $\overline{AB}$. Thus, $\overrightarrow{AB} \cap \overrightarrow{BA} = \overline{AB}$.

Description	Diagram	Symbol
(a) Line AB		$\overleftrightarrow{AB}$
(b) Half line AB		$\overrightarrow{AB}$
(c) Ray AB		$\overrightarrow{AB}$
(d) Ray BA		$\overleftarrow{BA}$
(e) Line segment AB		$\overline{AB}$
(f) Open line segment AB		$\overset{\circ\quad\circ}{AB}$
(g) Half open line segments AB		$\overset{\quad\circ}{AB}$ $\overset{\circ}{AB}$

Figure 8.2

Also recall from Chapter 2 that the union (symbolized $\cup$) of two sets is the set of elements (points in this case) that belong to either of the sets or both sets. The union of $\overrightarrow{AB}$ and $\overrightarrow{BA}$ is $\overleftrightarrow{AB}$ (Fig. 8.3b). Thus, $\overrightarrow{AB} \cup \overrightarrow{BA} = \overleftrightarrow{AB}$.

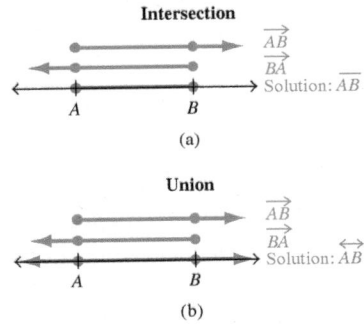

Figure 8.3

Example 1 · *Unions and Intersections of Parts of a Line*

Refer to line AD below. Determine the following.

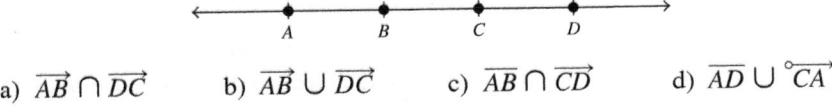

a) $\overrightarrow{AB} \cap \overrightarrow{DC}$ b) $\overrightarrow{AB} \cup \overrightarrow{DC}$ c) $\overrightarrow{AB} \cap \overrightarrow{CD}$ d) $\overleftrightarrow{AD} \cup \overset{\circ}{CA}$

Solution

a) $\overrightarrow{AB} \cap \overrightarrow{DC}$

Ray AB and ray DC are shown below. The intersection of these two rays is that part of line AD that is a part of *both* ray AB *and* ray DC. The intersection of ray AB and ray DC is line segment AD, $\overline{AD}$.

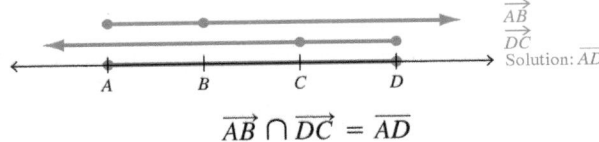

$$\overrightarrow{AB} \cap \overrightarrow{DC} = \overline{AD}$$

b) $\overrightarrow{AB} \cup \overrightarrow{DC}$

Once again ray AB and ray DC are shown below. The union of these two rays is that part of line AD that is part of *either* ray AB *or* ray DC. The union of ray AB and ray DC is the entire line AD, $\overleftrightarrow{AD}$.

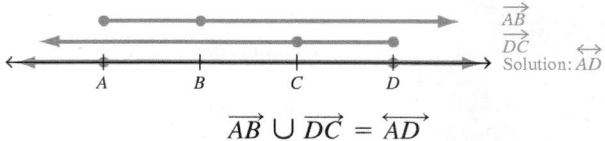

$$\overrightarrow{AB} \cup \overrightarrow{DC} = \overleftrightarrow{AD}$$

c) $\overline{AB} \cap \overrightarrow{CD}$

Line segment AB and ray CD have no points in common, so their intersection is empty.

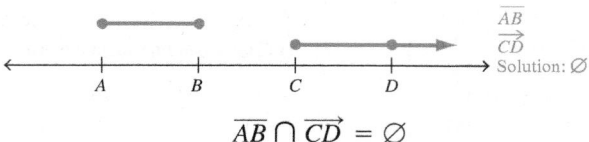

$$\overline{AB} \cap \overrightarrow{CD} = \varnothing$$

d) $\overline{AD} \cup \overset{\circ}{\overrightarrow{CA}}$

The union of line segment AD and half line CA is ray DA, $\overrightarrow{DA}$ (or, equivalently, $\overrightarrow{DB}$ or $\overrightarrow{DC}$).

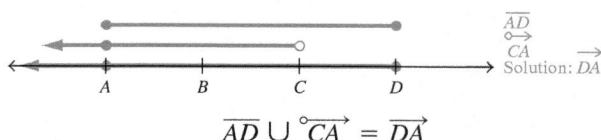

$$\overline{AD} \cup \overset{\circ}{\overrightarrow{CA}} = \overrightarrow{DA}$$ ∎

Plane

The term *plane* is one of Euclid's undefined terms. For our purposes, we can think of a plane as a two-dimensional surface that extends infinitely in both directions, like an infinitely large white board. Euclidean geometry is called *plane geometry* because it is the study of two-dimensional figures in a plane.

Two lines in the same plane that do not intersect are called *parallel lines*. Figure 8.4(a) illustrates two parallel lines in a plane ($\overleftrightarrow{AB}$ is parallel to $\overleftrightarrow{CD}$).

Properties of planes include the following:

1. Any three points that are not on the same line (noncollinear points) determine a unique plane (Fig. 8.4b).

2. A line in a plane divides the plane into three parts, the line itself and two half planes (Fig. 8.4c).

3. Any line and a point not on the line determine a unique plane.

4. The intersection of two distinct planes is a line (Fig. 8.4d).

Two planes that do not intersect are said to be *parallel planes*. For example, in Fig. 8.5 plane ABE is parallel to plane GHF.

Two lines that do not lie in the same plane and do not intersect are called *skew lines*. Figure 8.5 illustrates many skew lines (for example, $\overleftrightarrow{AB}$ and $\overleftrightarrow{CD}$).

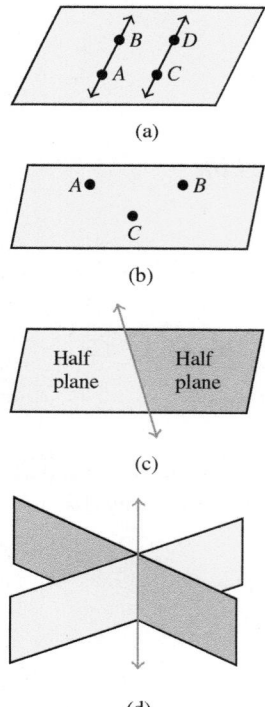

(a)

(b)

(c)

(d)

Figure 8.4

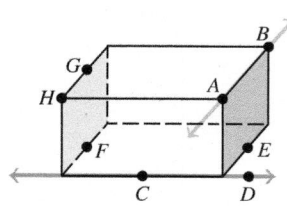

Figure 8.5

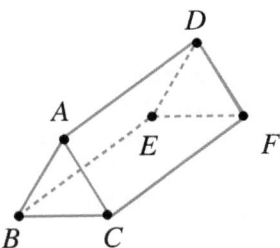

Figure 8.6

Example 2 *Working with Lines and Planes*

Figure 8.6 suggests several lines and planes. The lines may be described by naming two points, and the planes may be described by naming three points. Determine the following.

a) plane $ABC \cap$ plane BCF
b) plane $ABC \cap \overleftrightarrow{CF}$
c) plane $ABC \cap \overleftrightarrow{DF}$
d) two parallel lines
e) two parallel planes
f) two skew lines

Solution

a) Plane ABC can be described as containing the "front" of Figure 8.6. Plane BCF can be described as containing the "bottom" of Figure 8.6. The intersection of these two planes is the line that contains point B and point C. Thus, plane $ABC \cap$ plane $BCF = \overleftrightarrow{BC}$.

b) From Figure 8.6, we see that $\overleftrightarrow{CF}$, the line that contains the points C and F, intersects plane ABC at the single point C. Thus, plane $ABC \cap \overleftrightarrow{CF} = \{C\}$.

c) From Figure 8.6, we see that $\overleftrightarrow{DF}$, the line that contains the points D and F, has no points in common with plane ABC. Thus, plane $ABC \cap \overleftrightarrow{DF} = \varnothing$, the empty set.

d) Parallel lines are lines that are in the same plane but do not intersect. From Figure 8.6, we see that $\overleftrightarrow{AD}$ and $\overleftrightarrow{CF}$ are parallel, $\overleftrightarrow{AD}$ and $\overleftrightarrow{BE}$ are parallel, $\overleftrightarrow{AC}$ and $\overleftrightarrow{DF}$ are parallel, and $\overleftrightarrow{BC}$ and $\overleftrightarrow{EF}$ are parallel. There are several other parallel lines in Fig. 8.6. Can you determine two more parallel lines?

e) Parallel planes are planes that do not intersect. From Fig. 8.6, we see that plane ABC and plane DEF do not intersect. Thus, plane ABC and plane DEF are parallel planes.

f) Skew lines are two lines that do not lie in the same plane and do not intersect. From Fig. 8.6, we see that $\overleftrightarrow{AB}$ and $\overleftrightarrow{DF}$ are not in the same plane and do not intersect. Thus, $\overleftrightarrow{AB}$ and $\overleftrightarrow{DF}$ are skew lines. Other examples of skew lines are $\overleftrightarrow{AC}$ and $\overleftrightarrow{DE}$, $\overleftrightarrow{AD}$ and $\overleftrightarrow{BC}$, and $\overleftrightarrow{AB}$ and $\overleftrightarrow{CF}$. There are several other skew lines in Fig. 8.6. Can you determine two more skew lines? ∎

Angles

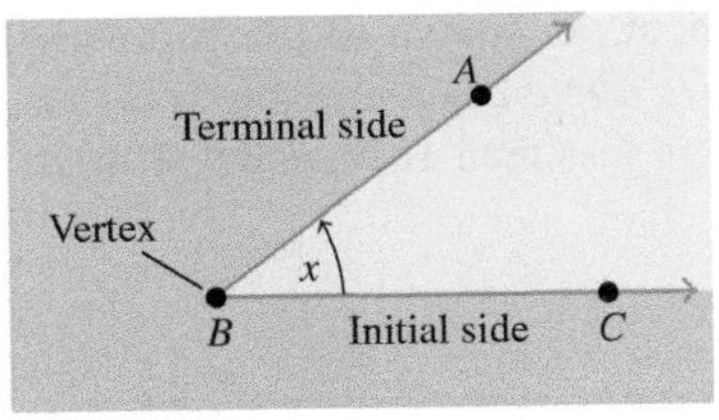

An *angle*, denoted $\angle$, is the union of two rays with a common end point (Fig. 8.7):

$$\overrightarrow{BA} \cup \overrightarrow{BC} = \angle ABC \text{ (or } \angle CBA)$$

An angle can be formed by the rotation of a ray about its end point. An angle has an initial side and a terminal side. The initial side indicates the position of the ray prior to rotation; the terminal side indicates the position of the ray after rotation. The point common to both rays is called the *vertex* of the angle. The letter designating the vertex is always the middle one of the three letters designating an angle. The rays that make up the angle are called its *sides*.

There are several ways to name an angle. The angle in Fig. 8.7 may be denoted

$$\angle ABC, \qquad \angle CBA, \qquad \text{or} \qquad \angle B$$

An angle divides a plane into three distinct parts: the angle itself, its interior, and its exterior. In Fig. 8.7, the angle is represented by the blue lines, the interior of the angle is shaded pink, and the exterior is shaded tan.

The *measure of an angle*, symbolized m, is the amount of rotation from its initial side to its terminal side. In Fig. 8.7, the letter x represents the measure of $\angle ABC$; therefore, we may write $m\angle ABC = x$.

Figure 8.7

Angles can be measured in *degrees*, radians, or gradients. In this text, we will discuss only the degree unit of measurement. Consider a circle whose circumference is divided into 360 equal parts. If we draw a line from each mark on the circumference to the center of the circle, we get 360 wedge-shaped pieces. The measure of an angle formed by the straight sides of each wedge-shaped piece is defined to be 1 degree, written 1°. An angle of 45 degrees is written 45°. A *protractor* is used to measure angles. The angle shown being measured by the protractor in Fig. 8.8 is 50°.

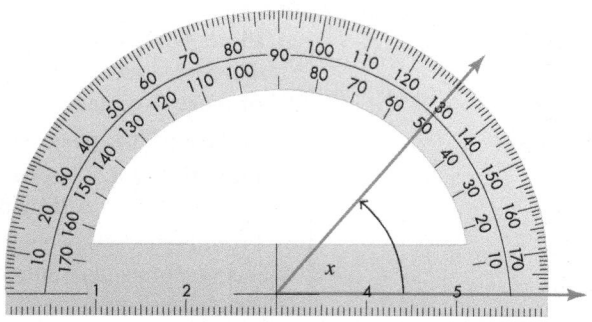

Figure 8.8

Example 3 *Working with Angles*

Refer to Fig. 8.9. Determine the following.

a) $\overrightarrow{BE} \cup \overrightarrow{BC}$ b) $\overleftrightarrow{AC} \cap \angle CBG$ c) $\angle CBE \cap \angle CBG$

d) $\angle EBF \cap \angle DBG$

> **Solution**
>
> a) $\overrightarrow{BE} \cup \overrightarrow{BC} = \angle CBE$ or $\angle EBC$ b) $\overleftrightarrow{AC} \cap \angle CBG = \overrightarrow{BC}$
>
> c) $\angle CBE \cap \angle CBG = \overrightarrow{BC}$ d) $\angle EBF \cap \angle DBG = \{B\}$ ∎

Angles are classified by their degree measurement, as shown in the following summary. A *right angle* has a measure of 90°, an *acute angle* has a measure less than 90°, an *obtuse angle* has a measure greater than 90° but less than 180°, and a *straight angle* has a measure of 180°.

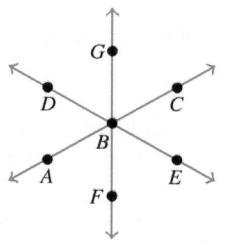

Figure 8.9

Right Angle	Acute Angle	Obtuse Angle	Straight Angle
$x = 90°$	$0° < x < 90°$	$90° < x < 180°$	$x = 180°$
The symbol ∟ is used to indicate right angles.			

Two angles in the same plane are *adjacent angles* when they have a common vertex and a common side but no common interior points. In Fig. 8.10, ∡*DBC* and ∡*CBA* are adjacent angles, but ∡*DBA* and ∡*CBA* are not adjacent angles.

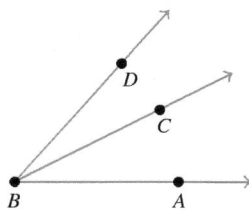

Figure 8.10

Two angles are called *complementary angles* if the sum of their measures is 90°. Two angles are called *supplementary angles* if the sum of their measures is 180°.

Example 4 *Determining Complementary and Supplementary Angles*

In Fig. 8.11, we see that $m∡ABC = 31°$.

a) ∡*ABC* and ∡*CBD* are complementary angles. Determine $m∡CBD$.

b) ∡*ABC* and ∡*CBE* are supplementary angles. Determine $m∡CBE$.

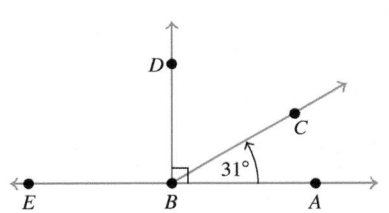

Figure 8.11

Solution

a) The sum of two complementary angles must be 90°, so

$$m∡ABC + m∡CBD = 90°$$
$$31° + m∡CBD = 90°$$
$$m∡CBD = 90° - 31° = 59°$$

Subtract 31° from each side of the equation.

b) The sum of two supplementary angles must be 180°, so

$$m∡ABC + m∡CBE = 180°$$
$$31° + m∡CBE = 180°$$
$$m∡CBE = 180° - 31° = 149°$$

Subtract 31° from each side of the equation. ∎

Example 5 *Determining Complementary Angles*

If ∡*ABC* and ∡*CBD* are complementary angles and $m∡ABC$ is 26° less than $m∡CBD$, determine the measure of each angle (Fig. 8.12).

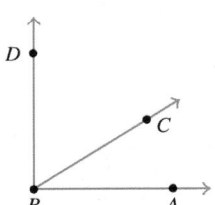

Figure 8.12

Solution Let $m∡CBD = x$. Then $m∡ABC = x - 26$, since it is 26° less than $m∡CBD$. Because these angles are complementary, we have

$$m∡CBD + m∡ABC = 90°$$
$$x + (x - 26°) = 90°$$
$$2x - 26° = 90°$$
$$2x = 116°$$
$$x = 58°$$

Therefore, $m∡CBD = 58°$ and $m∡ABC = 58° - 26°$, or 32°. Note that $58° + 32° = 90°$, which is what we expected. ∎

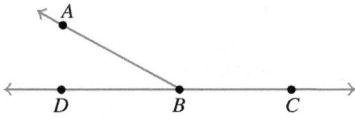

Figure 8.13

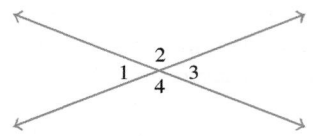

Figure 8.14

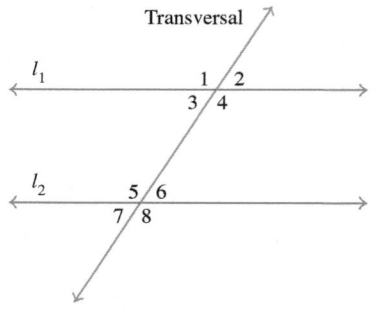

Figure 8.15

Example **6** *Determining Supplementary Angles*

If $\angle ABC$ and $\angle ABD$ are supplementary angles and $m\angle ABC$ is five times greater than $m\angle ABD$, determine $m\angle ABC$ and $m\angle ABD$ (Fig. 8.13).

Solution Let $m\angle ABD = x$, then $m\angle ABC = 5x$. Since these angles are supplementary, we have

$$m\angle ABD + m\angle ABC = 180°$$
$$x + 5x = 180°$$
$$6x = 180°$$
$$x = 30°$$

Thus, $m\angle ABD = 30°$ and $m\angle ABC = 5(30°) = 150°$. Note that $30° + 150° = 180°$, which is what we expected. ∎

When two straight lines intersect, the nonadjacent angles formed are called *vertical angles*. In Fig. 8.14, $\angle 1$ and $\angle 3$ are vertical angles, and $\angle 2$ and $\angle 4$ are vertical angles. We can show that vertical angles have the same measure; that is, they are equal. For example, Fig. 8.14 shows that

$$m\angle 1 + m\angle 2 = 180°.$$
$$m\angle 2 + m\angle 3 = 180°.$$

Since $\angle 2$ has the same measure in both cases, $m\angle 1$ must equal $m\angle 3$.

> **Vertical Angles**
> Vertical angles have the same measure.

A line that intersects two different lines, l_1 and l_2, at two different points is called a *transversal*. Figure 8.15 illustrates that when two parallel lines are cut by a transversal, eight angles are formed. Angles 3, 4, 5, and 6 are called *interior angles*, and angles 1, 2, 7, and 8 are called *exterior angles*. Eight pairs of supplementary angles are formed. Can you list them?

Special names are given to the angles formed by a transversal crossing two parallel lines. We describe these angles below.

Name	Description	Illustration	Pairs of Angles Meeting Criteria
Alternate interior angles	Interior angles on opposite sides of the transversal		$\angle 3$ and $\angle 6$ $\angle 4$ and $\angle 5$
Alternate exterior angles	Exterior angles on opposite sides of the transversal		$\angle 1$ and $\angle 8$ $\angle 2$ and $\angle 7$
Corresponding angles	One interior and one exterior angle on the same side of the transversal		$\angle 1$ and $\angle 5$ $\angle 2$ and $\angle 6$ $\angle 3$ and $\angle 7$ $\angle 4$ and $\angle 8$

> **Parallel Lines Cut by a Transversal**
> When two parallel lines are cut by a transversal,
> 1. alternate interior angles have the same measure.
> 2. alternate exterior angles have the same measure.
> 3. corresponding angles have the same measure.

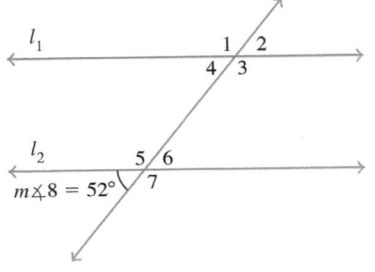

Figure 8.16

Example 7 *Determining Angle Measures*

Figure 8.16 shows two parallel lines cut by a transversal. If $m \angle 8 = 52°$, determine the measure of $\angle 1$ through $\angle 7$.

Solution

$m\angle 6 = 52°$	$\angle 8$ and $\angle 6$ are vertical angles.
$m\angle 5 = 128°$	$\angle 8$ and $\angle 5$ are supplementary angles.
$m\angle 7 = 128°$	$\angle 5$ and $\angle 7$ are vertical angles.
$m\angle 1 = 128°$	$\angle 1$ and $\angle 7$ are alternate exterior angles.
$m\angle 4 = 52°$	$\angle 4$ and $\angle 6$ are alternate interior angles.
$m\angle 2 = 52°$	$\angle 6$ and $\angle 2$ are corresponding angles.
$m\angle 3 = 128°$	$\angle 3$ and $\angle 1$ are vertical angles.

In Example 7, the angles could have been determined in alternate ways. For example, we mentioned $m\angle 1 = 128°$ because $\angle 1$ and $\angle 7$ are alternate exterior angles. We could have also stated that $m\angle 1 = 128°$ because $\angle 1$ and $\angle 5$ are corresponding angles.

SECTION 8.1 *Exercises*

Warm Up Exercises

In Exercises 1–10, fill in the blanks with an appropriate word, phrase, or symbol(s).

1. Two lines in the same plane that do not intersect are called _____ lines.

2. Two lines that do not lie in the same plane and do not intersect are called _____ lines.

3. The union of two rays with a common endpoint is called a(n) _____.

4. Two angles, the sum of whose measures is 90°, are called _____ angles.

5. Two angles, the sum of whose measures is 180°, are called _____ angles.

6. An angle whose measure is 180° is called a(n) _____ angle.

7. An angle whose measure is 90° is called a(n) _____ angle.

8. An angle whose measure is greater than 90° but less than 180° is called a(n) _____ angle.

9. An angle whose measure is less than 90° is called a(n) _____ angle.

10. When two straight lines intersect, the nonadjacent angles formed are called _____ angles.

Practice the Skills

In Exercises 11–18, identify the figure as a line, half line, ray, line segment, open line segment, or half open line segment. Denote the figure by its appropriate symbol.

11.

12.

13.

14.

15.

16.

17.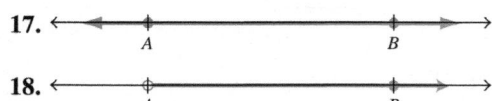

18.

In Exercises 19–30, use the figure to determine the following:

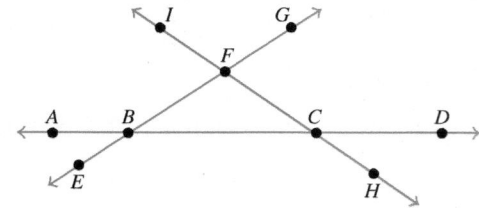

19. $\overleftrightarrow{AB} \cap \overrightarrow{DC}$

20. $\overset{\circ}{BD} \cap \overset{\circ}{CB}$

21. $\overline{BC} \cup \overset{\circ}{CD}$

22. $\overset{\circ}{FE} \cup \overrightarrow{FG}$

23. $\overset{\circ}{BD} \cup \overset{\circ}{CB}$

24. $\overrightarrow{BG} \cup \overrightarrow{FE}$

25. $\overrightarrow{FC} \cup \overrightarrow{FI}$

26. $\overrightarrow{BC} \cup \overrightarrow{BA}$

27. $\overrightarrow{BC} \cup \overrightarrow{CF} \cup \overrightarrow{FB}$

28. $\overset{\circ}{BC} \cup \overset{\circ}{CF} \cup \overset{\circ}{FB}$

29. $\overrightarrow{BG} \cap \overrightarrow{CI}$

30. $\{C\} \cap \overset{\circ}{CH}$

In Exercises 31–42, use the figure to determine the following.

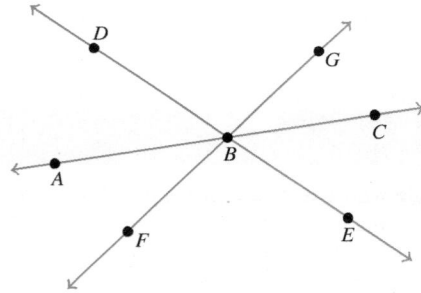

31. $\overrightarrow{BC} \cup \overrightarrow{BG}$

32. $\overrightarrow{BC} \cup \overrightarrow{BD}$

33. $\angle GBC \cap \angle CBE$

34. $\angle ABD \cap \angle DBG$

35. $\angle ABG \cap \angle CBD$

36. $\angle ABD \cap \angle CBG$

37. $\angle DBF \cap \overrightarrow{FG}$

38. $\angle EBF \cap \overrightarrow{ED}$

39. $\angle ABE \cup \overset{\circ}{AB}$

40. $\angle CBF \cup \overrightarrow{BC}$

41. $\angle CBG \cup \overrightarrow{BG}$

42. $\overset{\circ}{BF} \cup \overrightarrow{BE}$

In Exercises 43–48, classify the angle as acute, right, straight, obtuse, or none of these angles.

43.

44.

45.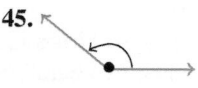

46.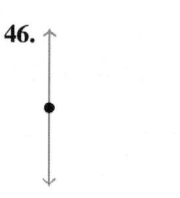

47.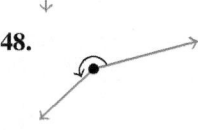

48.

In Exercises 49–54, determine the complementary angle of the given angle.

49. $13°$

50. $62°$

51. $32\frac{3}{4}°$

52. $31\frac{2}{5}°$

53. $64.7°$

54. $0.01°$

In Exercises 55–60, determine the supplementary angle of the given angle.

55. $29°$

56. $167°$

57. $20.5°$

58. $148.7°$

59. $43\frac{5}{7}°$

60. $64\frac{7}{16}°$

In Exercises 61–66, match the descriptions of the angles with the corresponding figure in parts (a)–(f).

61. Supplementary angles

62. Complementary angles

63. Vertical angles

64. Corresponding angles

65. Alternate exterior angles

66. Alternate interior angles

a)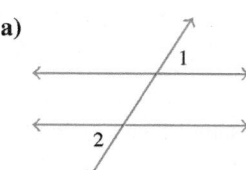

b)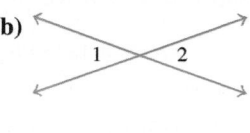

c)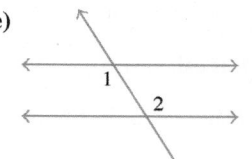

d)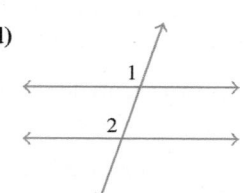

e)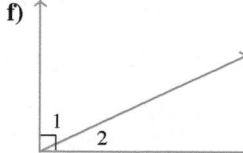

f)

Problem Solving

67. MODELING—*Complementary Angles* If $\angle 1$ and $\angle 2$ are complementary angles and if the measure of $\angle 2$ is 5 times the measure of $\angle 1$, determine the measures of $\angle 1$ and $\angle 2$.

68. MODELING—*Complementary Angles* The difference between the measures of two complementary angles is 8°. Determine the measures of the two angles.

69. MODELING—*Supplementary Angles* The difference between the measures of two supplementary angles is 102°. Determine the measures of the two angles.

70. MODELING—*Supplementary Angles* If $\angle 1$ and $\angle 2$ are supplementary angles and if the measure of $\angle 2$ is 17 times the measure of $\angle 1$, determine the measures of the two angles.

In Exercises 71–74, parallel lines are cut by the transversal shown. Determine the measures of $\angle 1$ through $\angle 7$.

71.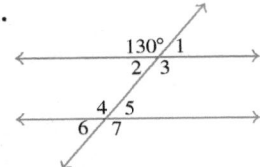

72.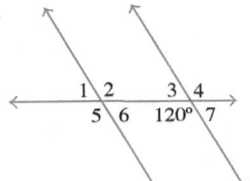

73.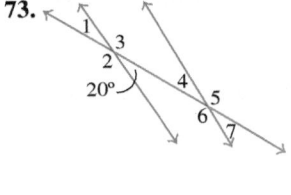

74.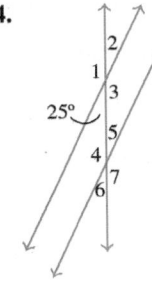

In Exercises 75–78, the angles are complementary angles. Determine the measures of $\angle 1$ and $\angle 2$.

75.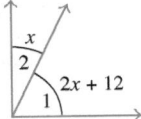

76.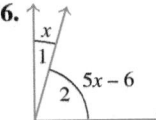

77.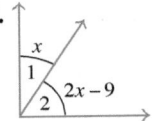

78.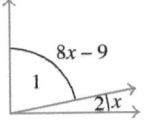

In Exercises 79–82, the angles are supplementary angles. Determine the measures of $\angle 1$ and $\angle 2$.

79.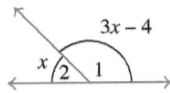

80.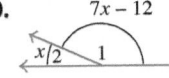

81.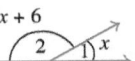

82.

In Exercises 83–90, the figure below suggests several lines and planes. The lines may be described by naming two points, and the planes may be described by naming three points. Use the figure to name the following. Many answers are possible.

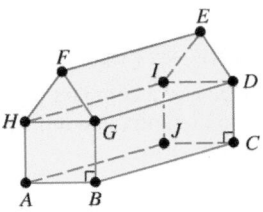

83. Two parallel lines

84. Two skew lines

85. Two parallel planes

86. Two planes that intersect at right angles

87. Three planes whose intersection is a single point

88. Three planes whose intersection is a line

89. A line and a plane whose intersection is a point

90. A line and a plane whose intersection is a line

Concept/Writing Exercises

In Exercises 91–96, determine whether the statement is always true, sometimes true, or never true. Explain your answer.

91. Two lines that are both parallel to a third line must be parallel to each other.

92. A triangle contains exactly two acute angles.

93. Vertical angles are complementary angles.

94. Alternate exterior angles are supplementary angles.

95. Alternate interior angles are complementary angles.

96. A triangle contains two obtuse angles.

97. a) How many lines can be drawn through a given point?

   b) How many planes can be drawn through a given point?

98. What is the intersection of two distinct nonparallel planes?

99. How many planes can be drawn through a given line?

100. a) Will three noncollinear points *A*, *B*, and *C* always determine a plane? Explain.

   b) Is it possible to determine more than one plane with three noncollinear points? Explain.

   c) How many planes can be constructed through three collinear points?

Challenge Problems/Group Activities

101. Use a straightedge and a compass to construct a triangle with sides of equal length (an equilateral triangle) by doing the following:

   a) Use the straightedge to draw a line segment of any length and label the end points *A* and *B* (Fig. a).

   b) Place one end of the compass at point *A* and the other end at point *B* and draw an arc as shown (Fig. b).

   c) Now turn the compass around and draw another arc as shown. Label the point of intersection of the two arcs *C* (Fig. c).

   d) Draw line segments *AC* and *BC*. This completes the construction of equilateral triangle *ABC* (Fig. d).

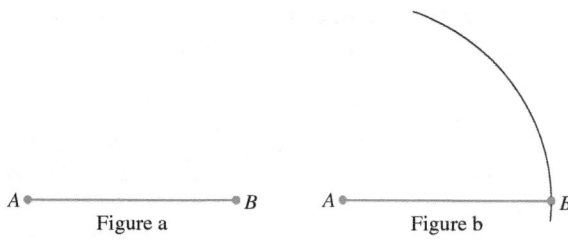

Figure a   Figure b

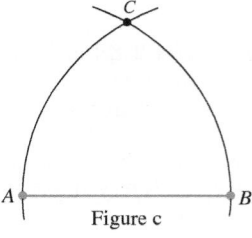

Figure c

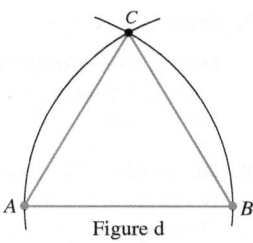

Figure d

102. If lines *l* and *m* are parallel lines and if lines *l* and *n* are skew lines, is it true that lines *m* and *n* must also be skew? (*Hint*: Look at Fig. 8.5 on page 448.) Explain your answer and include a sketch to support your answer.

103. Two lines are *perpendicular* if they intersect at right angles. If lines *l* and *m* are perpendicular and if lines *m* and *n* are perpendicular, is it true that lines *l* and *n* must also be perpendicular? Explain your answer and include a sketch to support your answer.

104. Suppose you have three distinct lines, all lying in the same plane. Determine all the possible ways in which the three lines can be related. There are four cases. Sketch each case.

105. $\angle ABC$ and $\angle CBD$ are complementary and $m\angle CBD$ is twice the $m\angle ABC$. $\angle ABD$ and $\angle DBE$ are supplementary angles.

   a) Draw a sketch illustrating $\angle ABC$, $\angle CBD$, and $\angle DBE$.

   b) Determine $m\angle ABC$.

   c) Determine $m\angle CBD$.

   d) Determine $m\angle DBE$.

Research Activities

106. *Euclid* Write a research paper on Euclid's contributions to geometry.

107. *Classic Geometry* Write a research paper on the three classic geometry problems of Greek antiquity (see the *Did You Know?* on page 450).

108. *Geometric Constructions* Research the geometric constructions that use a straightedge and a compass only. Prepare a poster demonstrating five of these basic constructions.

Upon completion of this section, you will be able to:

- Solve problems involving the sides and angles of polygons.
- Solve problems involving similar figures.
- Solve problems involving congruent figures.

What shape would you use to best describe each of the following road signs: a stop sign, a yield sign, a speed limit sign? In this section, we will study the shapes of these and other geometric figures that can be classified as *polygons*.

Why This Is Important In addition to road signs, polygons and their properties are used in many applications we see in our daily lives. Polygons play an important role in the architecture of buildings, in floor tile patterns, in map making, and in many engineering applications. A basic understanding of polygons can be helpful in solving a variety of problems.

Sides and Angles of Polygons

A *polygon* is a closed figure in a plane determined by three or more straight line segments. Examples of polygons are given in Fig. 8.17.

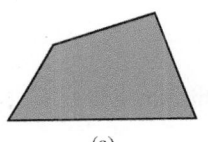 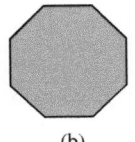 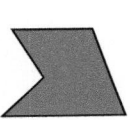 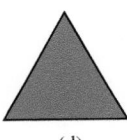

(a)   (b)   (c)   (d)

Figure 8.17

The straight line segments that form the polygon are called its *sides*, and a point where two sides meet is called a *vertex* (plural, *vertices*). The union of the sides of a polygon and its interior is called a *polygonal region*. A *regular polygon* is one whose sides are all the same length and whose interior angles all have the same measure. Figures 8.17(b) and (d) are regular polygons.

Polygons are named according to their number of sides. The names of some polygons are given in Table 8.1.

Table 8.1

Number of Sides	Name	Number of Sides	Name
3	Triangle	8	Octagon
4	Quadrilateral	9	Nonagon
5	Pentagon	10	Decagon
6	Hexagon	12	Dodecagon
7	Heptagon	20	Icosagon

One of the most important polygons is the triangle. The sum of the measures of the interior angles of a triangle is 180°. To illustrate, consider triangle ABC given in Fig. 8.18. The triangle is formed by drawing two transversals through two parallel lines l_1 and l_2 with the two transversals intersecting at a point on l_1.

In Fig. 8.18, notice that $\angle A$ and $\angle A'$ are corresponding angles. Recall from Section 8.1 that corresponding angles are equal, so $m\angle A = m\angle A'$. Also, $\angle C$ and $\angle C'$ are corresponding angles; therefore, $m\angle C = m\angle C'$. Next, we notice that $\angle B$ and $\angle B'$ are vertical angles. In Section 8.1, we learned that vertical angles are equal; therefore,

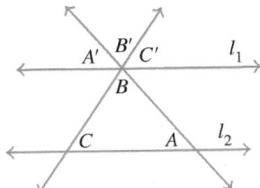

Figure 8.18

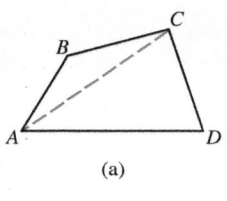

(a)

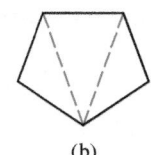

(b)

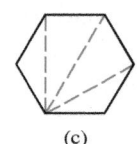

(c)

Figure 8.19

Sum of Interior
Angles of Polygons

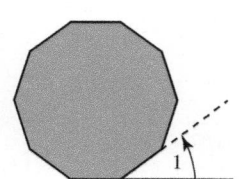

$m \angle B = m \angle B'$. Figure 8.18 shows that $\angle A'$, $\angle B'$, and $\angle C'$ form a straight angle; therefore, $m \angle A' + m \angle B' + m \angle C' = 180°$. Since $m \angle A = m \angle A'$, $m \angle B = m \angle B'$, and $m \angle C = m \angle C'$, we can reason that $m \angle A + m \angle B + m \angle C = 180°$. This discussion illustrates that the sum of the interior angles of a triangle is 180°.

Consider the quadrilateral *ABCD* (Fig. 8.19a). Drawing a straight line segment between any two vertices forms two triangles. Since the sum of the measures of the angles of a triangle is 180°, the sum of the measures of the interior angles of a quadrilateral is 2 · 180°, or 360°.

Now let's examine a pentagon (Fig. 8.19b). We can draw two straight line segments to form three triangles. Thus, the sum of the measures of the interior angles of a five-sided figure is 3 · 180°, or 540°. Figure 8.19(c) shows that four triangles can be drawn in a six-sided figure. Table 8.2 summarizes this information.

Table 8.2

Sides	Triangles	Sum of the Measures of the Interior Angles
3	1	$1(180°) = 180°$
4	2	$2(180°) = 360°$
5	3	$3(180°) = 540°$
6	4	$4(180°) = 720°$

If we continue this procedure, we can see that for an *n*-sided polygon the sum of the measures of the interior angles is $(n - 2)180°$.

> **Sum of the Measures of Interior Angles**
> The **sum** of the measures of the interior angles of an *n*-sided polygon is $(n - 2)180°$.

Example 1 *Angles of a Regular Decagon*

A regular decagon (see the figure in the margin) is a 10-sided figure with all sides having the same length and all interior angles having the same measure. Determine

a) the measure of an interior angle.

b) the measure of exterior $\angle 1$.

Solution

a) Using the formula $(n - 2)180°$, we can determine the sum of the measures of the interior angles of the regular decagon.

$$\text{Sum} = (10 - 2)180°$$
$$= 8 \cdot 180°$$
$$= 1440°$$

The measure of an interior angle of a regular polygon can be determined by dividing the sum of the interior angles by the number of angles. For a regular decagon,

$$\text{Measure of one interior angle} = \frac{1440°}{10} = 144°$$

b) Since $\angle 1$ is the supplement of an interior angle,

$$\angle 1 = 180° - 144° = 36°$$

■

To discuss area in the next section, we must be able to identify various types of triangles and quadrilaterals. The following is a summary of certain types of triangles and their characteristics.

Triangles

Acute Triangle

All angles are acute angles.

Obtuse Triangle

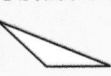

One angle is an obtuse angle.

Right Triangle

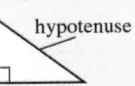
hypotenuse

One angle is a right angle.

Isosceles Triangle

Two equal sides Two equal angles

Equilateral Triangle

Three equal sides Three equal angles (60° each)

Scalene Triangle

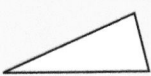

No two sides are equal in length. No two angles are equal in measure.

Similar Figures

In everyday living, we often have to deal with geometric figures that have the "same shape" but are of different sizes. For example, an architect will make a small-scale drawing of a floor plan or a photographer will make an enlargement of a photograph. Figures that have the same shape but may be of different sizes are called *similar figures*. Two similar figures are illustrated in Fig. 8.20.

Similar figures have *corresponding angles* and *corresponding sides*. In Fig. 8.20, triangle *ABC* has angles *A*, *B*, and *C*. Their respective corresponding angles in triangle *DEF* are angles *D*, *E*, and *F*. Sides $\overline{AB}$, $\overline{BC}$, and $\overline{AC}$ in triangle *ABC* have corresponding sides $\overline{DE}$, $\overline{EF}$, and $\overline{DF}$, respectively, in triangle *DEF*.

> **Definition: Similar Figures**
> Two figures are **similar** if their corresponding angles have the same measure and the lengths of their corresponding sides are in proportion.

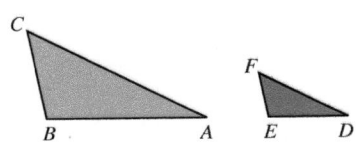

Figure 8.20

In Fig. 8.20, $\angle A$ and $\angle D$ have the same measure, $\angle B$ and $\angle E$ have the same measure, and $\angle C$ and $\angle F$ have the same measure. Also, the lengths of corresponding sides of similar triangles are in proportion.

When we refer to the line segment *AB*, we place a bar over the *AB* and write $\overline{AB}$. *When we refer to the length of a line segment, we do not place a bar above the two letters.* For example, if we write *AB* = 12, we are indicating the length of line segment $\overline{AB}$ is 12. The proportion below shows that the lengths of the corresponding sides of the similar triangles in Fig. 8.20 are in proportion.

$$\frac{AB}{DE} = \frac{BC}{EF} = \frac{AC}{DF}$$

Example 2 *Similar Figures*

Consider the similar figures in Fig. 8.21.

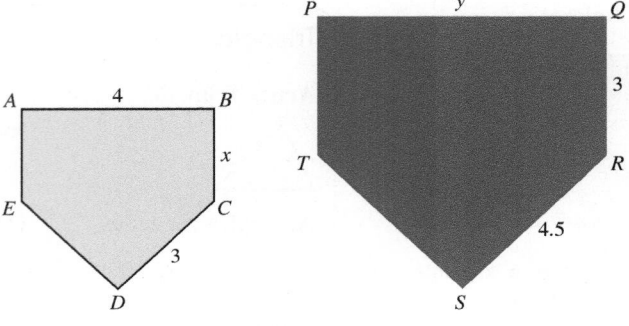

Figure 8.21

Determine

a) the length of side $\overline{BC}$.   b) the length of side $\overline{PQ}$.

Solution

a) We will represent the length of side $\overline{BC}$ with the variable x. Because the corresponding sides of similar figures must be in proportion, we can write a proportion (as explained in Section 6.3) to determine the length of side $\overline{BC}$. The lengths of corresponding sides $\overline{CD}$ and $\overline{RS}$ are known, so we use them as one ratio in the proportion. The side corresponding to $\overline{BC}$ is $\overline{QR}$.

$$\frac{BC}{QR} = \frac{CD}{RS}$$

$$\frac{x}{3} = \frac{3}{4.5}$$

Now we solve for x.

$$x \cdot 4.5 = 3 \cdot 3 \quad \text{Cross multiply}$$
$$4.5x = 9$$
$$x = 2$$

Thus, the length of side $\overline{BC}$ is 2 units.

b) We will represent the length of side $\overline{PQ}$ with the variable y. The side corresponding to $\overline{PQ}$ is $\overline{AB}$. We will work part (b) in a manner similar to part (a).

$$\frac{PQ}{AB} = \frac{RS}{CD}$$

$$\frac{y}{4} = \frac{4.5}{3}$$

$$y \cdot 3 = 4 \cdot 4.5$$
$$3y = 18$$
$$y = 6$$

Thus, the length of side $\overline{PQ}$ is 6 units.   ■

Example 3 *Determine the Height of a Tree*

Monique plans to remove a tree from her backyard. She needs to know the height of the tree. Monique is 6 ft tall and determines that when her shadow is 9 ft long, the shadow of the tree is 45 ft long (see Fig. 8.22 on page 461). How tall is the tree?

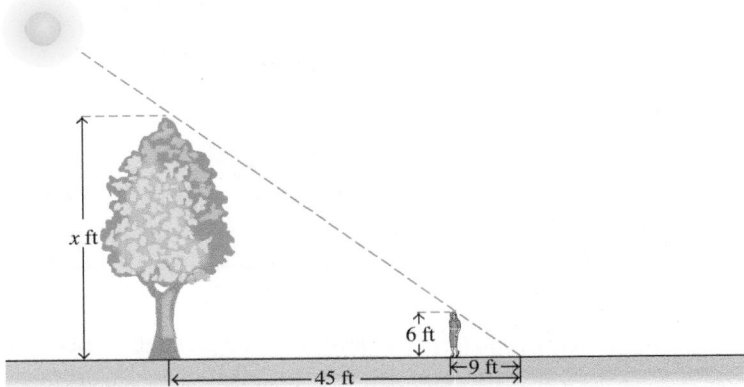

Figure 8.22

Solution We will let x represent the height of the tree. From Fig. 8.22, we can see that the triangle formed by the sun's rays, Monique, and her shadow is similar to the triangle formed by the sun's rays, the tree, and its shadow. To determine the height of the tree, we will set up and solve the following proportion:

$$\frac{\text{height of the tree}}{\text{height of Monique}} = \frac{\text{length of tree's shadow}}{\text{length of Monique's shadow}}$$

$$\frac{x}{6} = \frac{45}{9}$$

$$9x = 270$$

$$x = 30$$

Therefore, the tree is 30 ft tall. ∎

Congruent Figures

If the corresponding sides of two similar figures are the same length, the figures are called *congruent figures*. Corresponding angles of congruent figures have the same measure, and the corresponding sides are equal in length. Two congruent figures coincide when placed one upon the other.

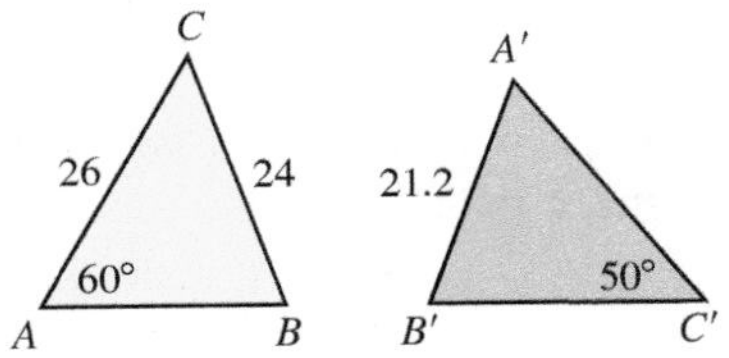

Figure 8.23

Example 4 *Congruent Triangles*

Triangles ABC and $A'B'C'$ in Fig. 8.23 are congruent. Determine

a) the length of side $\overline{A'C'}$.
b) the length of side $\overline{AB}$.

c) $m\angle C'A'B'$.
d) $m\angle ACB$.

e) $m\angle ABC$.

Solution Because $\triangle ABC$ is congruent to $\triangle A'B'C'$, we know that the corresponding side lengths are equal and corresponding angle measures are equal.

a) $A'C' = AC = 26$
b) $AB = A'B' = 21.2$

c) $m\angle C'A'B' = m\angle CAB = 60°$
d) $m\angle ACB = m\angle A'C'B' = 50°$

e) The sum of the angles of a triangle is 180°. Since $m\angle BAC = 60°$ and $m\angle ACB = 50°$, $m\angle ABC = 180° - 60° - 50° = 70°$. ∎

Earlier we learned that *quadrilaterals* are four-sided polygons, the sum of whose interior angle measures is 360°. Quadrilaterals may be classified according to their characteristics, as illustrated in the summary box below.

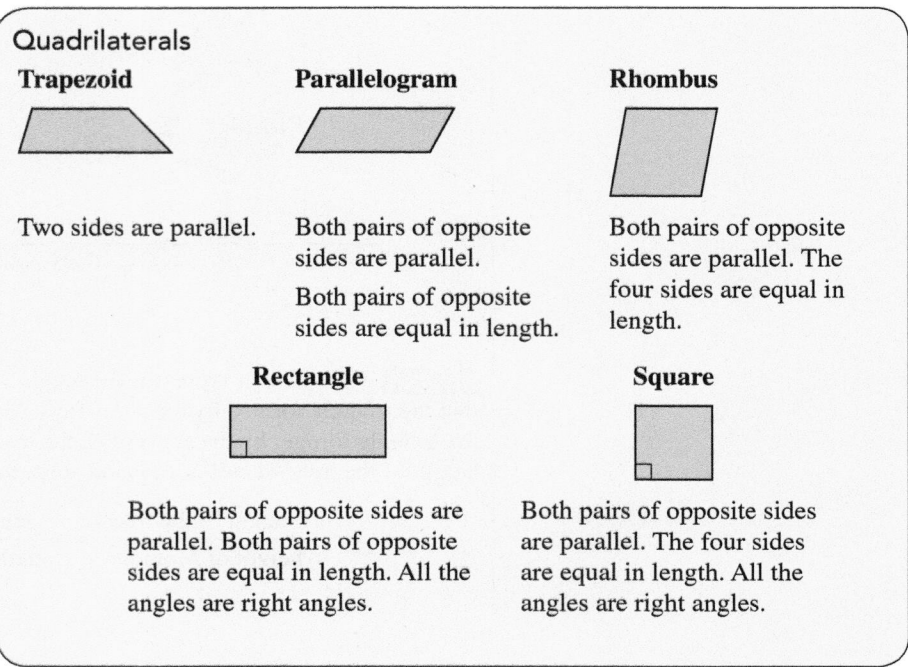

Quadrilaterals

Trapezoid

Two sides are parallel.

Parallelogram

Both pairs of opposite sides are parallel.

Both pairs of opposite sides are equal in length.

Rhombus

Both pairs of opposite sides are parallel. The four sides are equal in length.

Rectangle

Both pairs of opposite sides are parallel. Both pairs of opposite sides are equal in length. All the angles are right angles.

Square

Both pairs of opposite sides are parallel. The four sides are equal in length. All the angles are right angles.

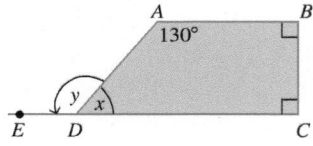

Figure 8.24

Example 5 *Angles of a Trapezoid*

Trapezoid *ABCD* is shown in Fig. 8.24.

a) Determine the measure of the interior angle, *x*.

b) Determine the measure of the exterior angle, *y*.

Solution

a) Because of the right angle symbols, we know that each of the two right angles in trapezoid *ABCD* has a measure of 90°. We also know that the sum of the interior angles in any quadrilateral is 360°. Therefore, we have

$$m \angle DAB + m \angle ABC + m \angle BCD + m \angle x = 360°$$
$$130° + 90° + 90° + m \angle x = 360°$$
$$310° + m \angle x = 360°$$
$$m \angle x = 50°$$

Thus, the measure of the interior angle, *x*, is 50°.

b) Angle *x* and angle *y* are supplementary angles. Therefore, $m \angle x + m \angle y = 180°$ and $m \angle y = 180° - m \angle x = 180° - 50° = 130°$. Thus, the measure of the exterior angle, *y*, is 130°.   ■

Note that in Example 5 part (b) we could also have determined the measure of angle *y* as follows. By the definition of a trapezoid, sides $\overline{AB}$ and $\overline{CD}$ must be parallel. Therefore, side $\overline{AD}$ may be considered a transversal and $\angle BAD$ and $\angle ADE$ are alternate interior angles. Recall from Section 8.1 that alternate interior angles are equal. Thus, $m \angle BAD = m \angle ADE$ and $m \angle y = 130°$.

SECTION 8.2 Exercises

Warm Up Exercises

In Exercises 1–6, fill in the blanks with an appropriate word, phrase, or symbol(s).

1. A closed figure in a plane determined by three or more straight line segments is called a(n) _____.

2. A polygon whose sides are all the same length and whose interior angles all have the same measure is called a(n) _____ polygon.

3. Two polygons are similar if their corresponding angles have the same measure and the lengths of their corresponding sides are in _____.

4. The sum of the measures of the interior angles of a triangle is _____.

5. If the corresponding sides of two similar figures are the same length, the figures are called _____ figures.

6. The sum of the measures of the interior angles of a quadrilateral is _____.

Practice the Skills

In Exercises 7–14, (a) name the polygon. (b) State whether or not the polygon is a regular polygon.

7.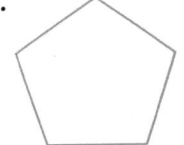

8.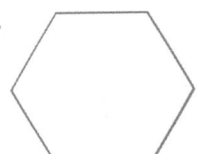

9.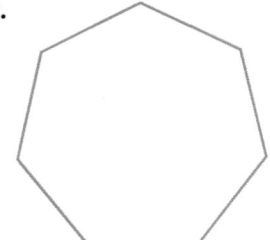

10.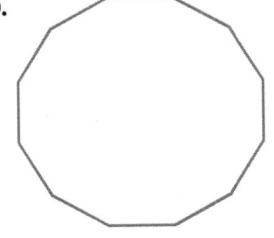

11.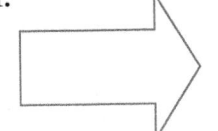

12.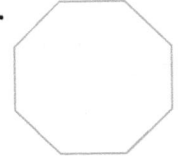

13.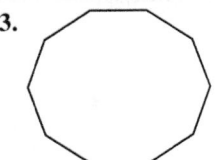

14.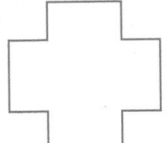

In Exercises 15–22, identify the triangle as (a) scalene, isosceles, or equilateral and as (b) acute, obtuse, or right. The parallel markings (the two small parallel lines) on two or more sides indicate that the marked sides are of equal length.

15.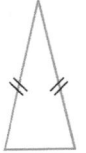

16.

17.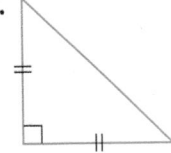

18.

19.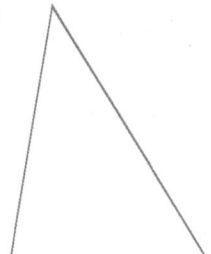

20.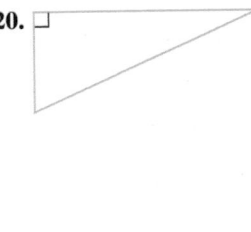

21.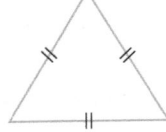

22.

In Exercises 23–28, identify the quadrilateral.

23.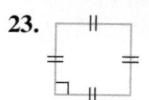

24.

25.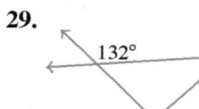

26.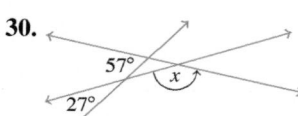

27.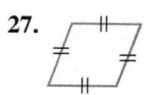

28.

In Exercises 29–32, determine the measure of ∡x.

29.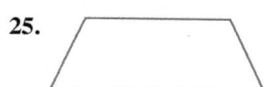

30.

31.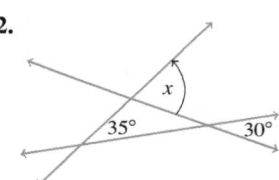

32.

In Exercises 33 and 34, lines l_1 and l_2 are parallel. Determine the measures of ∡1 through ∡12.

33.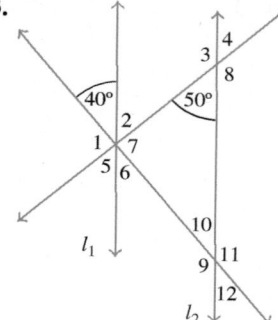

34.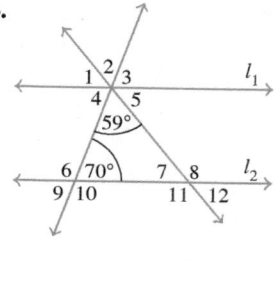

In Exercises 35–40, determine the sum of the measures of the interior angles of the indicated polygon.

35. Pentagon

36. Heptagon

37. Octagon

38. Nonagon

39. Dodecagon

40. Icosagon

In Exercises 41–46, (a) determine the measure of an interior angle of the named regular polygon. (b) If a side of the polygon is extended, determine the supplementary angle of an interior angle. See Example 1.

41. Triangle

42. Quadrilateral

43. Hexagon

44. Octagon

45. Nonagon

46. Dodecagon

In Exercises 47–52, the figures are similar. Determine the length of side x and side y.

47.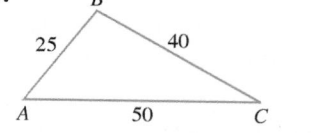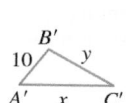

48.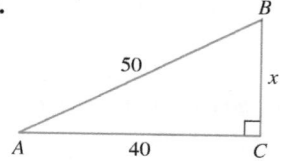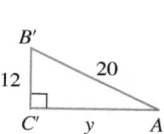

49.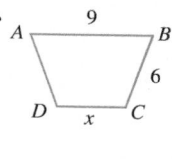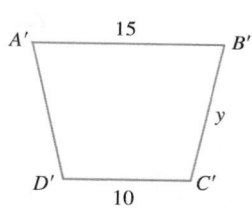

50.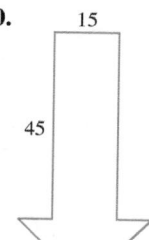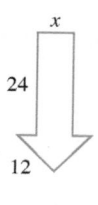

51.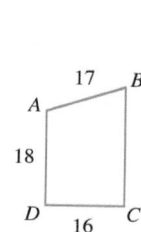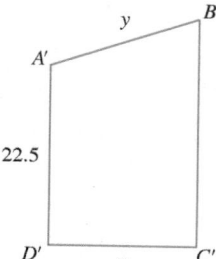

52.

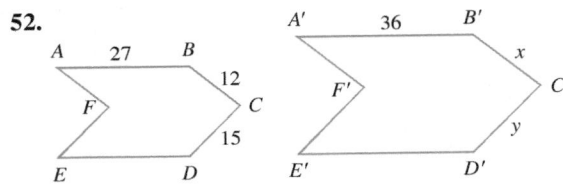

In Exercises 53–56, triangles ABC and DEC are similar figures. Determine the length of

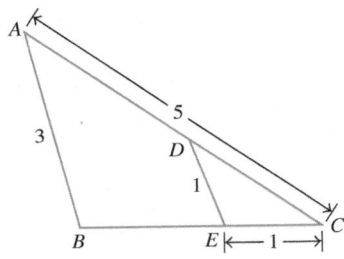

53. side $\overline{BC}$.

54. side $\overline{DC}$.

55. side $\overline{AD}$.

56. side $\overline{BE}$.

In Exercises 57–62, determine the length of the sides and the measures of the angles for the congruent triangles ABC and A'B'C'.

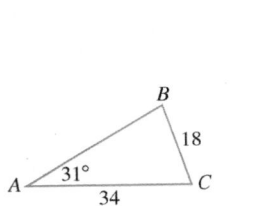
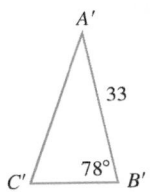

57. The length of side $\overline{AB}$

58. The length of side $\overline{A'C'}$

59. The length of side $\overline{B'C'}$

60. $m\angle ABC$

61. $m\angle ACB$

62. $m\angle A'C'B'$

In Exercises 63–68, determine the length of the sides and the measures of the angles for the congruent quadrilaterals ABCD and A'B'C'D'.

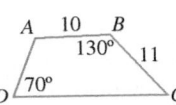
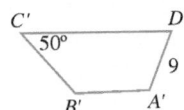

63. The length of side $\overline{AD}$

64. The length of side $\overline{B'C'}$

65. The length of side $\overline{A'B'}$

66. $m\angle BCD$

67. $m\angle DAB$

68. $m\angle A'D'C'$

Problem Solving

69. ***Distances in Texas*** A triangle can be formed by drawing line segments on a map of Texas connecting the cities of Austin, Houston, and San Antonio (see figure below). If the actual distance from Austin to Houston is approximately 160 miles, use the lengths of the line segments indicated in the figure along with similar triangles to approximate

a) the actual distance from Austin to San Antonio.

b) the actual distance from San Antonio to Houston.

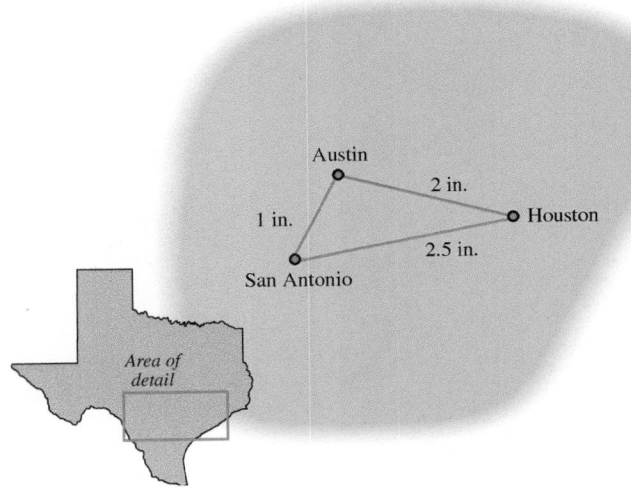

70. ***Distances in Missouri*** A triangle can be formed by drawing line segments on a map of Missouri connecting the cities of Kansas City, Springfield, and St. Louis (see figure below). If the actual distance from Kansas City to St. Louis is approximately 248 miles, use the lengths of the line segments indicated in the figure along with similar triangles to approximate

a) the actual distance from Kansas City to Springfield.

b) the actual distance from Springfield to St. Louis.

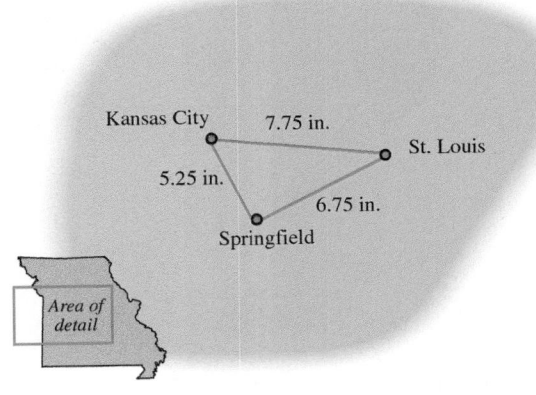

71. Distances in Minnesota A triangle can be formed by
drawing line segments on a map of Minnesota connecting
the cities of Austin, Rochester, and St. Paul (see figure
below). If the actual distance from Austin to Rochester is
approximately 44 miles, use the lengths of the line segments indicated in the figure along with similar triangles
to approximate (to the nearest tenth of a mile)

a) the actual distance from St. Paul to Austin.

b) the actual distance from St. Paul to Rochester.

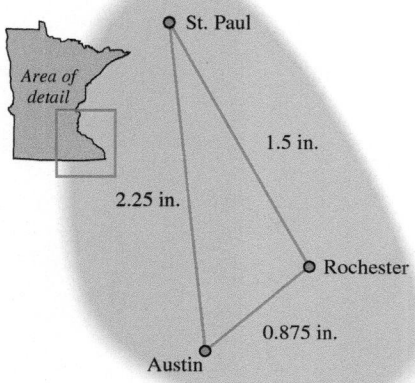

72. Distances in Illinois A triangle can be formed by drawing
line segments on a map of Illinois connecting the cities of
Rockford, Chicago, and Bloomington (see figure below).
If the actual distance from Chicago to Rockford is approximately 85.5 miles, use the lengths of the line segments
indicated in the figure along with similar triangles to approximate (to the nearest tenth of a mile)

a) the actual distance from Chicago to Bloomington.

b) the actual distance from Bloomington to Rockford.

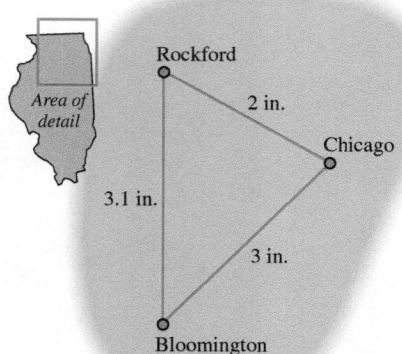

73. Height of a Silo Steve is buying a farm and needs to determine the height of a silo on the farm. Steve, who is 6 ft tall,
notices that when his shadow is 9 ft long, the shadow of the
silo is 105 ft long (see diagram above and to the right). How
tall is the silo? Note that the diagram is not to scale.

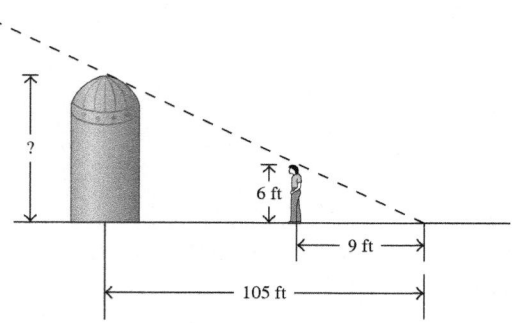

74. Angles on a Picnic Table The legs of a picnic table
form an isosceles triangle as indicated in the figure.
If $m \angle ABC = 80°$, determine $m \angle x$ and $m \angle y$ so that
the top of the table will be parallel to the ground.

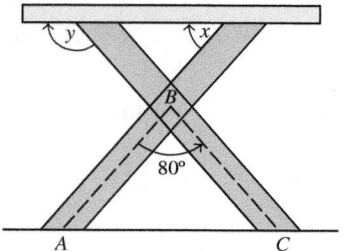

*In Exercises 75–78, determine the measure of the angle. In the
figure, $\angle ABC$ makes an angle of 125° with the floor and l_1
and l_2 are parallel.*

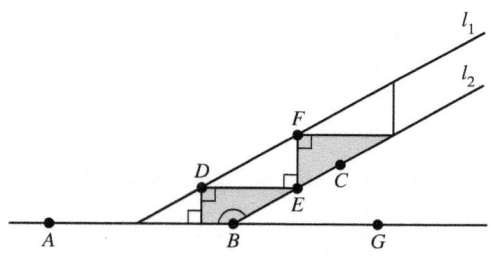

75. $m \angle GBC$

76. $m \angle EDF$

77. $m \angle DFE$

78. $m \angle DEC$

Challenge Problem/Group Activity

79. Height of a Wall You wish to measure the height of an
inside wall of a warehouse. No ladder tall enough to
measure the height is available. You place a mirror on
the floor. You then move away from the mirror until you
can see the reflection of the top of the wall in it, as shown
in the figure on page 467.

a) Explain why triangle *HFM* is similar to triangle *TBM*.
(*Hint*: In the reflection of light the angle of *incidence*,
$\angle HMF$, equals the angle of *reflection*, $\angle TMB$. Thus,
$\angle HMF = \angle TMB$.)

b) If your eyes are $5\frac{1}{2}$ ft above the floor and you are $2\frac{1}{2}$ ft from the mirror and the mirror is 20 ft from the wall, how high is the wall?

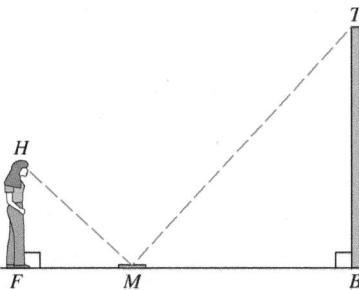

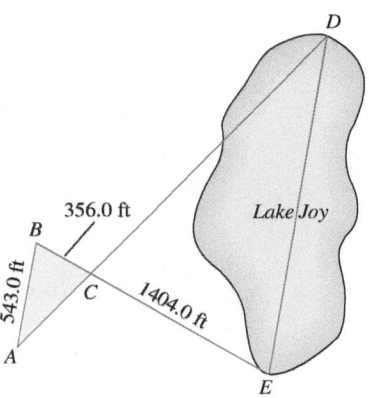

Recreational Mathematics

80. *Distance Across a Lake*

a) In the figure on the right, $m\angle CED = m\angle ABC$. Explain why triangles *ABC* and *DEC* must be similar.

b) Determine the distance across the lake, *DE*.

Research Activities

81. *Theodolite* Write a paper on the history and use of the theodolite, a surveying instrument.

82. *Photographic Process* Write a paper on the use of geometry in the photographic process. Include discussions on the use of similar figures.

SECTION 8.3   Perimeter and Area

Upon completion of this section, you will be able to:

- Understand and use formulas for the perimeter and area of polygons.
- Solve problems involving the Pythagorean theorem.
- Understand and use formulas for the circumference and area of circles.
- Solve problems involving square unit conversions.

If you were to walk around the outside edge of a volleyball court, how far would you walk? If you had to place floor tiles that each measured 1 foot by 1 foot on a basketball court, how many tiles would you need? In this section, we will study the geometric concepts of perimeter and area that are used to answer these and other questions.

Why This Is Important The concepts of area and perimeter are involved in many real-life applications of geometry. These include calculating the cost of flooring for your home, calculating the amount of fencing needed for your yard, and determining the best value when purchasing pizza!

Perimeter and Area

The *perimeter*, *P*, of a two-dimensional figure is the sum of the lengths of the sides of the figure. In Figs. 8.25 and 8.26 on page 468, the sums of the lengths of the red line segments are the perimeters. Perimeters are measured in the same units as the sides. For example, if the sides of a figure are measured in feet, the perimeter will be measured in feet.

The *area*, *A*, is the measure of the region within the boundaries of the figure. For example, in Fig. 8.25 the *area* of the pentagon refers to the measure of the blue region within the boundary of the pentagon. Area is measured in square units. For example, if the sides of a figure are measured in inches, the area of the figure will be measured in square inches (in.^{2}). See Table 7.7 on page 431 for common units of area in the U.S. customary and metric systems.

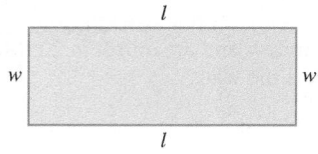

Figure 8.25   Figure 8.26

Consider the rectangle in Fig. 8.26. Two sides of the rectangle have length l, and two sides of the rectangle have width w. Thus, if we add the lengths of the four sides to get the perimeter, we determine $P = l + w + l + w = 2l + 2w$.

Perimeter of a Rectangle

$$P = 2l + 2w$$

Consider a rectangle of length 5 units and width 3 units (Fig. 8.27). Counting the number of 1-unit by 1-unit squares within the figure, we obtain the area of the rectangle, 15 square units. The area can also be obtained by multiplying the number of units of length by the number of units of width, or 5 units $\times$ 3 units = 15 square units. We can determine the area of a rectangle by the formula area = length $\times$ width.

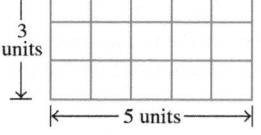

Figure 8.27

Area of a Rectangle

$$A = l \times w$$

Using the formula for the area of a rectangle, we can determine the formulas for the areas of other figures.

A square (Fig. 8.28) is a rectangle that contains four equal sides. Therefore, the length equals the width. If we call both the length and the width of the square s, then

$$A = l \times w, \quad \text{so} \quad A = s \times s = s^2$$

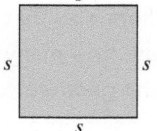

Figure 8.28

Area of a Square

$$A = s^2$$

A parallelogram with height h and base b is shown in Fig. 8.29(a).

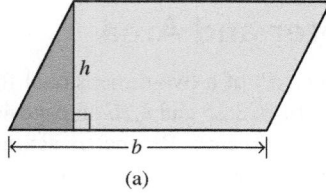   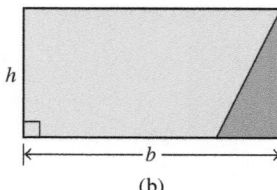

(a)   (b)

Figure 8.29

If we were to cut off the red portion of the parallelogram on the left, Fig. 8.29(a), and attach it to the right side of the figure, the resulting figure would be a rectangle, Fig. 8.29(b). Since the area of the rectangle is $b \times h$, the area of the parallelogram is also $b \times h$.

Area of a Parallelogram

$$A = b \times h$$

Consider the triangle with height, h, and base, b, shown in Fig. 8.30(a). Using this triangle and a second identical triangle, we can construct a parallelogram, Fig. 8.30(b). The area of the parallelogram is bh. The area of the triangle is one-half that of the parallelogram. Therefore, the area of the triangle is $\frac{1}{2} \times$ base $\times$ height.

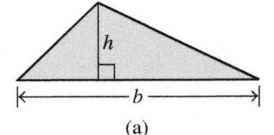

(a)

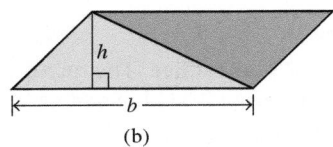

(b)

Figure 8.30

Area of a Triangle

$$A = \tfrac{1}{2} bh$$

Now consider the trapezoid shown in Fig. 8.31(a). We can partition the trapezoid into two triangles by drawing diagonal $\overline{DB}$, as in Fig. 8.31(b). One triangle has base $\overline{AB}$ (called b_2) with height $\overline{DE}$, and the other triangle has base $\overline{DC}$ (called b_1) with height $\overline{FB}$. Note that the line used to measure the height of the triangle need not be inside the triangle. Because heights $\overline{DE}$ and $\overline{FB}$ are equal, both triangles have the same height, h. The area of triangles DCB and ADB are $\frac{1}{2} b_1 h$ and $\frac{1}{2} b_2 h$, respectively. The area of the trapezoid is the sum of the areas of the triangles, $\frac{1}{2} b_1 h + \frac{1}{2} b_2 h$, which can be written $\frac{1}{2} h(b_1 + b_2)$.

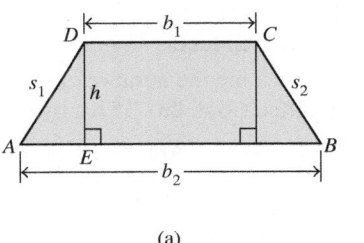

(a)

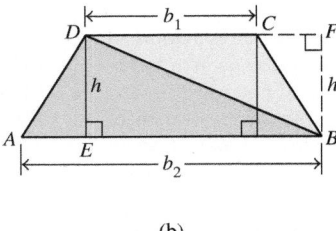

(b)

Figure 8.31

Area of a Trapezoid

$$A = \tfrac{1}{2} h(b_1 + b_2)$$

Following is a summary of the perimeters and areas of selected figures.

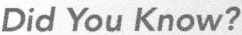

Perimeters and Areas

Triangle	Square	Rectangle
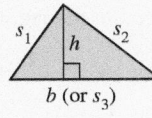	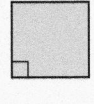	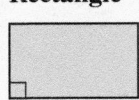
$P = s_1 + s_2 + s_3 \ (s_3 = b)$	$P = 4s$	$P = 2l + 2w$
$A = \frac{1}{2} bh$	$A = s^2$	$A = lw$

(Continued)

Pythagorean Theorem

Parallelogram

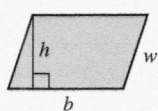

$$P = 2b + 2w$$

$$A = bh$$

Trapezoid

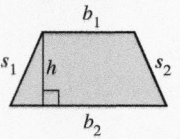

$$P = s_1 + s_2 + b_1 + b_2$$

$$A = \tfrac{1}{2} h(b_1 + b_2)$$

Example 1 *Office Flooring*

Malik wishes to install mahogany hardwood flooring in his home office. The rectangular office floor measures 18.5 feet by 12.5 feet. The flooring is sold in cartons that cost $220, and each carton contains 20 square feet of flooring. Only whole cartons of flooring may be purchased.

a) Determine the area of the office floor.

b) Determine the number of cartons of flooring Malik will need to purchase.

c) Determine the cost of the flooring purchased.

Solution

a) The area of the office floor is

$$A = l \cdot w = (18.5\,\text{ft})(12.5\,\text{ft}) = 231.25\,\text{ft}^2$$

The area is measured in square feet (ft^2) because both length and width are measured in feet.

b) To determine the number of cartons of flooring Malik needs, divide the area of the office floor, $231.25\,\text{ft}^2$, by the area of flooring in one carton, $20\,\text{ft}^2$.

$$\frac{\text{Area of office floor}}{\text{Area of flooring in one carton}} = \frac{231.25\,\text{ft}^2}{20\,\text{ft}^2} = 11.5625$$

Since only whole cartons are sold, Malik needs to purchase 12 cartons of flooring.

c) Malik needs 12 cartons of flooring, and each carton costs $220; therefore, the cost of the flooring is $12 \cdot \$220$, or $2640. ∎

Pythagorean Theorem

We introduced the Pythagorean theorem in Chapter 5. Because this theorem is an important tool for determining the perimeter and area of triangles, we restate it here.

Pythagorean Theorem
The sum of the squares of the lengths of the legs of a right triangle equals the square of the length of the hypotenuse.

$$\text{leg}^2 + \text{leg}^2 = \text{hypotenuse}^2$$

Symbolically, if a and b represent the lengths of the legs and c represents the length of the hypotenuse (the side opposite the right angle), then

$$a^2 + b^2 = c^2$$

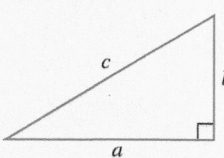

Example 2 *Crossing a Moat*

The moat surrounding a castle is 18 ft wide and the wall by the moat of the castle is 24 ft high (see Fig. 8.32). If an invading army wishes to use a ladder to cross the moat and reach the top of the wall, how long must the ladder be?

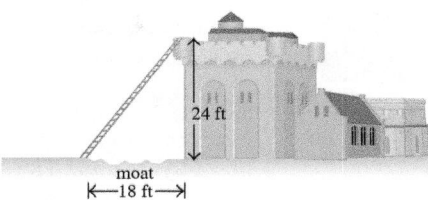

24 ft

moat
$\longleftarrow$18 ft$\longrightarrow$

Figure 8.32

Solution The moat, the castle wall, and the ladder form a right triangle. The moat and the castle wall form the legs of the triangle (sides a and b), and the ladder forms the hypotenuse (side c). By the Pythagorean theorem,

$$c^2 = a^2 + b^2$$
$$c^2 = (18)^2 + (24)^2$$
$$c^2 = 324 + 576$$
$$c^2 = 900$$
$$\sqrt{c^2} = \sqrt{900} \qquad \text{Take the square root of both sides of the equation.}$$
$$c = 30$$

Therefore, the ladder would need to be at least 30 ft long. ∎

Circumference and Area of Circles

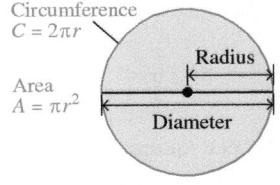

Circumference
$C = 2\pi r$

Radius

Area
$A = \pi r^2$

Diameter

Figure 8.33

A commonly used plane figure that is not a polygon is a *circle*. A *circle* is a set of points equidistant from a fixed point called the center. A *radius*, r, of a circle is a line segment from the center of the circle to any point on the circle (Fig. 8.33). A *diameter*, d, of a circle is a line segment through the center of a circle with both end points on the circle. Note that the diameter of the circle is twice its radius. The *circumference* is the length of the simple closed curve that forms the circle. The formulas for the area and circumference of a circle are given in the box below.

Circumference and Area of a Circle

r

$$C = 2\pi r$$
$$A = \pi r^2$$

The symbol pi, π, was introduced in Chapter 5. Recall that π is approximately 3.14. However, when solving problems, you should use the $\boxed{\pi}$ key on your calculator to get a more accurate answer.

Example 3 *Comparing Pizzas*

Jana wishes to order a large cheese pizza. She can choose among three pizza parlors in town: Antonio's, Steve's, and Dorsey's. Antonio's large cheese pizza is a round 16-in.-diameter pizza that sells for $15. Steve's large cheese pizza is a round 14-in.-diameter pizza that sells for $12. Dorsey's large cheese pizza is a square 12-in. by 12-in. pizza that sells for $10. All three pizzas have the same thickness. To get the most for her money, from which pizza parlor should Jana order her pizza?

Solution To determine the best value, we will calculate the cost per square inch of pizza for each of the three pizzas. To do so, we will divide the cost of each pizza by its area. The areas of the two round pizzas can be determined using the formula for the area of a circle, $A = \pi r^2$. Since the radius is half the diameter, we will use $r = 8$ and $r = 7$ for Antonio's and Steve's large pizzas, respectively, and we will use the $\boxed{\pi}$ key on our calculator. The area for the square pizza can be determined using the formula for the area of a square, $A = s^2$. We will use $s = 12$.

$$\text{Area of Antonio's pizza} = \pi r^2 = \pi(8)^2 = \pi(64) \approx 201.06 \text{ in.}^2$$

$$\text{Area of Steve's pizza} = \pi r^2 = \pi(7)^2 = \pi(49) \approx 153.94 \text{ in.}^2$$

$$\text{Area of Dorsey's pizza} = s^2 = (12)^2 = 144 \text{ in.}^2$$

Now, to determine the cost per square inch of pizza, we will divide the cost of the pizza by the area of the pizza.

$$\text{Cost per square inch of Antonio's pizza} \approx \frac{\$15}{201.06 \text{ in.}^2} \approx \$0.0746$$

Thus, Antonio's pizza costs about $0.0746, or about 7.5 cents, per square inch.

$$\text{Cost per square inch of Steve's pizza} \approx \frac{\$12}{153.94 \text{ in.}^2} \approx \$0.0780$$

Thus, Steve's pizza costs about $0.0780, or about 7.8 cents, per square inch.

$$\text{Cost per square inch of Dorsey's pizza} = \frac{\$10}{144 \text{ in.}^2} \approx \$0.0694$$

Thus, Dorsey's pizza costs about $0.0694, or about 6.9 cents, per square inch.

Since the cost per square inch of pizza is the lowest for Dorsey's pizza, Jana would get the most pizza for her money by ordering her pizza from Dorsey's. ∎

Example 4 *Determine the Shaded Area*

Determine the shaded area of the figure below. Use the $\boxed{\pi}$ key on your calculator and round your answer to the nearest hundredth.

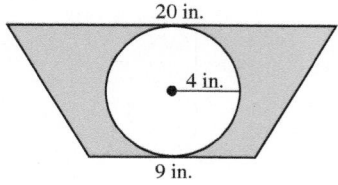

Solution To determine the area of the shaded region, we will subtract the area of the circle from the area of the trapezoid. Notice that the height of the trapezoid is equal to the diameter of the circle. Since the diameter of a circle is twice the radius, the height of the trapezoid is $2 \cdot 4$ in., or 8 in.

$$\text{Area of the trapezoid} = \frac{1}{2}h\,(b_1 + b_2)$$

$$= \frac{1}{2} \cdot 8(20 + 9) = 4(29) = 116\,\text{in.}^2$$

$$\text{Area of the circle} = \pi r^2$$

$$= \pi(4)^2 = \pi \cdot 16 \approx 50.27\,\text{in.}^2$$

$$\text{Area of the shaded region} = \text{Area of the trapezoid} - \text{Area of the circle}$$

$$\approx 116 - 50.27 \approx 65.73\,\text{in.}^2$$

Thus, the area of the shaded region is approximately $65.73\,\text{in.}^2$. ∎

Example 5 Applying Lawn Fertilizer

Altay plans to fertilize his lawn. The shapes and dimensions of his lot, house, driveway, pool, and rose garden are shown in Fig. 8.34. One bag of fertilizer costs \$29.95 and covers $5000\,\text{ft}^2$. Determine how many bags of fertilizer Altay needs and the total cost of the fertilizer. Assume only full bags may be purchased.

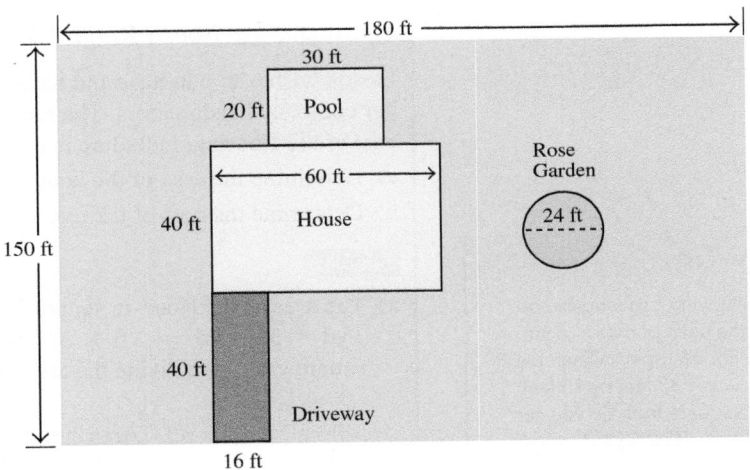

Figure 8.34

Solution The total area of the lot is $150 \cdot 180$, or $27{,}000\,\text{ft}^2$. To determine the area to be fertilized, subtract the area of house, driveway, pool, and rose garden from the total area.

$$\text{Area of house} = 60 \cdot 40 = 2400\,\text{ft}^2$$

$$\text{Area of driveway} = 40 \cdot 16 = 640\,\text{ft}^2$$

$$\text{Area of pool} = 20 \cdot 30 = 600\,\text{ft}^2$$

The diameter of the rose garden is 24 ft, so its radius is 12 ft.

$$\text{Area of rose garden} = \pi r^2 = \pi(12)^2 = \pi(144) \approx 452.39\,\text{ft}^2$$

The total area of the house, driveway, pool, and rose garden is approximately $2400 + 640 + 600 + 452.39$, or $4092.39\,\text{ft}^2$. The area to be fertilized is $27{,}000 - 4092.39\,\text{ft}^2$, or $22{,}907.61\,\text{ft}^2$. The number of bags of fertilizer is determined by dividing the total area to be fertilized by the number of square feet covered per bag.

The number of bags of fertilizer is $\dfrac{22{,}907.61}{5000}$, or about 4.58 bags. Therefore, Altay needs five bags. At \$29.95 per bag, the total cost is $5 \times \$29.95$, or \$149.75. ∎

Square Unit Conversions

Example 6 *Converting Between Square Feet and Square Inches*

a) Convert 1 ft² to square inches.   b) Convert 37 ft² to square inches.
c) Convert 432 in.² to square feet.

Solution

a) 1 ft = 12 in. Therefore, 1 ft² = 12 in. × 12 in. = 144 in.²
b) From part (a), we know that 1 ft² = 144 in.². Therefore,
   37 ft² = 37 × 144 in.² = 5328 in.².
c) In part (b), we converted from square feet to square inches by *multiplying* the number of square feet by 144. Now, to convert from square inches to square feet we will *divide* the number of square inches by 144. Therefore, 432 in.² = $\frac{432}{144}$ ft² = 3 ft². ∎

Example 7 *Board Room Flooring*

Ursula wishes to purchase and have marble flooring installed in the board room at her company headquarters. The rectangular board room measures 33 ft × 15 ft. The cost of the flooring, including installation, is \$245 per square yard.

a) Determine the area of the board room floor in square yards.
b) Determine the cost of the marble flooring for the board room.

Solution

a) The area of the floor, in square feet, is 33 ft × 15 ft = 495 ft². Since 1 yd = 3 ft, 1 yd² = 3 ft × 3 ft = 9 ft². To determine the area of the floor in square yards, we divide the area of the floor in square feet by 9 ft².

$$\text{Area in square yards} = \frac{495}{9} = 55$$

Therefore, the area of the board room floor is 55 yd².

b) The cost of 55 yd² of the marble flooring, including installation, is
   55 · \$245 = \$13,475. ∎

Timely Tip

When multiplying two lengths, be sure that the units of measure are the same. For example, in Example 7 you can multiply 33 feet by 15 feet to get 495 square feet. Or you can multiply 11 yards by 5 yards to get 55 square yards. However, you cannot get a valid answer if you multiply a length measured in feet by a length measured in yards.

SECTION 8.3 Exercises

Warm Up Exercises

In Exercises 1–4, fill in the blanks with an appropriate word, phrase, or symbol(s).

1. a) The sum of the lengths of the sides of a two-dimensional figure is called the _____ of the figure.

   b) The measure of the region within the boundaries of a two-dimensional figure is called the _____ of the figure.

2. In a right triangle, the side that is opposite the right angle is called the _____.

3. A set of points equidistant from a fixed point is called a(n) _____.

4. a) A line segment from the center of a circle to any point on the circle is called the _____ of the circle.

b) A line segment through the center of the circle with both endpoints on the circle is called the _____ of the circle.

c) The length of the simple closed curve that forms a circle is called the _____ of the circle.

Practice the Skills

In Exercises 5–8, determine the area of the triangle.

5.
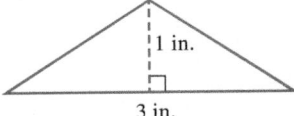
1 in.
3 in.

6.
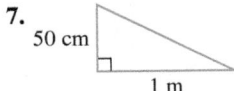
5 cm
7 cm

7.
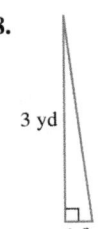
50 cm
1 m

8.
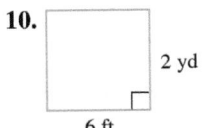
3 yd
1 ft

In Exercises 9–14, determine (a) the area and (b) the perimeter of the quadrilateral.

9.
4 ft
8 ft

10.
2 yd
6 ft

11.
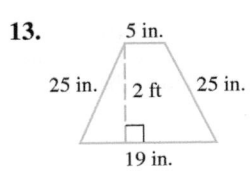
24 cm
0.25 m
25 cm

12.
6 in.
8 in.
13 in.

13.
5 in.
25 in.   2 ft   25 in.
19 in.

14.
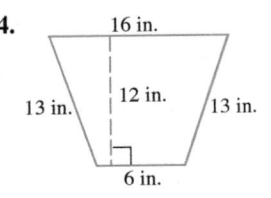
16 in.
13 in.   12 in.   13 in.
6 in.

In Exercises 15–18, determine (a) the area and (b) the circumference of the circle. Use the [π] key on your calculator and round your answer to the nearest hundredth.

15.
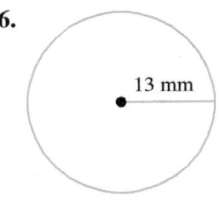
8 in.

16.
13 mm

17.
13 ft

18.
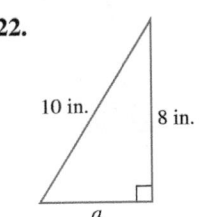
16 yd

In Exercises 19–22, (a) use the Pythagorean theorem to determine the length of the unknown side of the triangle, (b) determine the perimeter of the triangle, and (c) determine the area of the triangle.

19.
c 15 yd
8 yd

20.
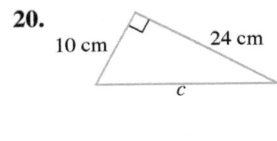
10 cm   24 cm
c

21.
5 km   13 km
b

22.
10 in.   8 in.
a

Problem Solving

In Exercises 23–32, determine the shaded area. When appropriate, use the [π] key on your calculator and round your answer to the nearest hundredth.

23.
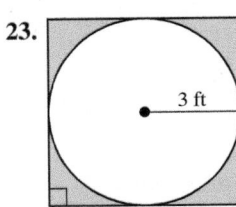
3 ft

24.
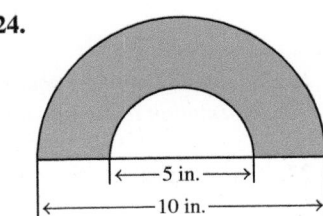
5 in.
10 in.

25.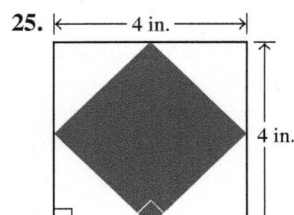

26.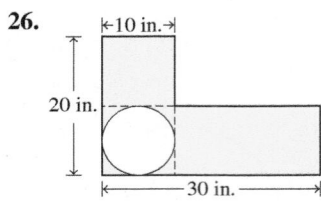

27.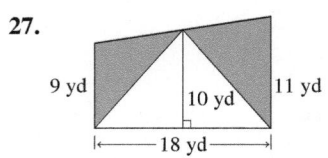

28.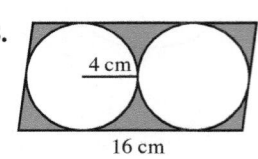

29.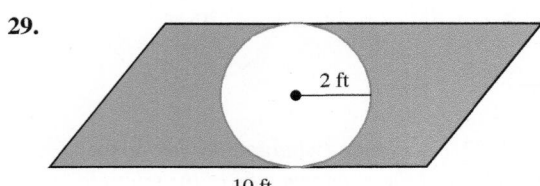

30.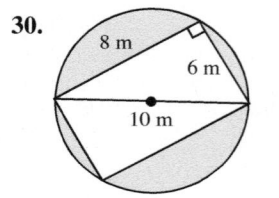

31.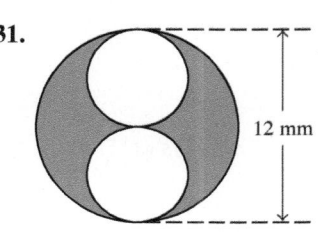

32.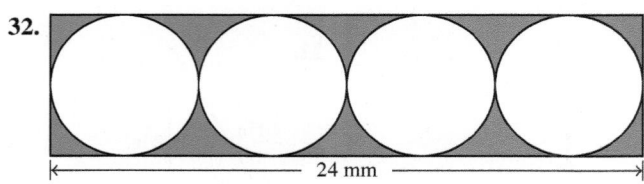

In Exercises 33–36, one square yard equals 9 ft². Use this information to convert the following.

33. 63 ft² to square yards

34. 112.5 ft² to square yards

35. 13.5 yd² to square feet

36. 48.1 yd² to square feet

In Exercises 37–40, one square meter equals 10,000 cm². Use this information to convert the following.

37. 7 m² to square centimeters

38. 0.75 m² to square centimeters

39. 4072 cm² to square meters

40. 8372 cm² to square meters

In Exercises 41–46, Nancy has just purchased a new house that is in need of new flooring. Use the measurements given on the floor plans of Nancy's house to obtain the answer.

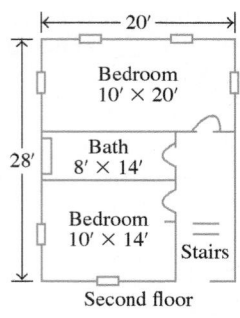

Second floor

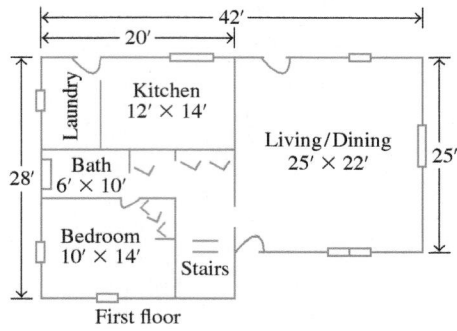

First floor

41. *Cost of Hardwood Flooring* The cost of walnut hardwood flooring is $11.99 per square foot if Nancy installs the flooring herself or $15.99 per square foot if she has the flooring installed by the flooring company. Determine the cost for hardwood flooring in the living/dining room if

a) Nancy installs it herself.

b) Nancy has it installed by the flooring company.

42. *Cost of Laminate Flooring* The cost of oak laminate flooring is $9.75 per square foot if Nancy installs the flooring herself or $12.75 per square foot if she has the flooring installed by the flooring company. Determine the cost for the flooring in the living/dining room if

a) Nancy installs it herself.

b) Nancy has it installed by the flooring company.

43. *Cost of Ceramic Tile* The cost of ceramic tile is $10.49 per square foot. This price includes the cost of installation. Determine the cost for Nancy to have ceramic tile installed in the kitchen and in both bathrooms.

44. *Cost of Linoleum* The cost of linoleum is $3.49 per square foot. This price includes the cost of installation. Determine the cost for Nancy to have linoleum installed in the kitchen and in both bathrooms.

45. *Cost of Berber Carpeting* The cost of Berber carpeting is $9.99 per square foot. This price includes the cost of installation. Determine the cost for Nancy to have this carpeting installed in all three bedrooms.

46. *Cost of Saxony Carpeting* The cost of Saxony carpeting is $7.49 per square foot. This price includes the cost of installation. Determine the cost for Nancy to have this carpeting installed in all three bedrooms.

47. *Cost of a Lawn Service* Clarence and Rose's home lot is illustrated here. Clarence and Rose wish to hire Picture Perfect Lawn Service to cut their lawn. How much will it cost Clarence and Rose to have their lawn cut if Picture Perfect charges $0.03 per square yard?

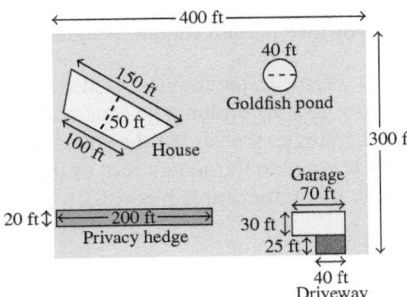

48. *Cost of a Lawn Service* Chloe and Zion's home lot is illustrated here. They wish to hire a lawn service to cut their lawn. M&M Lawn Service charges $0.03 per square yard of lawn. How much will it cost them to have their lawn cut?

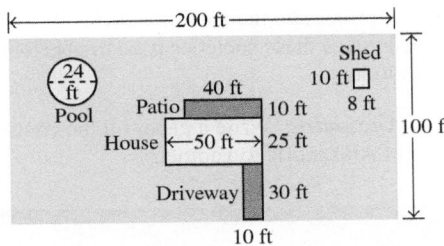

49. *Area of a Basketball Court* A National Basketball Association (NBA) basketball court is a rectangle that is 94 ft long and 50 ft wide.

a) If you were to walk around the outside edge of a basketball court, how far would you walk?

b) If you had to place floor tiles that each measured 1 foot by 1 foot on a basketball court, how many tiles would you need?

50. *Quartz Countertops* Larry wishes to have three quartz countertops installed in his new kitchen. The countertops are rectangular and have the following dimensions: $3\frac{1}{2}$ ft × 6 ft, $2\frac{1}{2}$ ft × 8 ft, and 3 ft × $11\frac{1}{2}$ ft. The cost of the countertops is $84 per square foot, which includes the cost of installation.

a) Determine the total area of the three countertops.

b) Determine the total cost to have all three countertops installed.

51. *Ladder on a Wall* Lorrie places a 29-ft ladder against the side of a building with the bottom of the ladder 20 ft away from the building (see figure). How high up on the wall does the ladder reach?

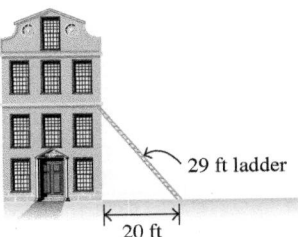

52. *Docking a Boat* Brian is bringing his boat into a dock that is 9 ft above the water level (see figure). If a 41-ft rope is attached to the dock on one side and to the boat on the other side, determine the horizontal distance from the dock to the boat.

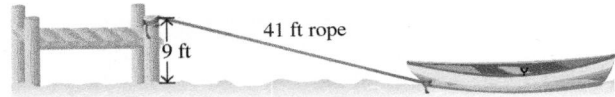

53. *The Green Monster* In Fenway Park, home of baseball's Boston Red Sox, the left field wall is known as the *Green Monster*. The distance from home plate down the third baseline to the bottom of the wall is 310 feet (see photo). In left field, at the end of the baseline, the Green Monster is perpendicular to the ground and is 37 feet tall. Determine the distance from home plate to the top of the Green Monster along the third baseline. Round your answer to the nearest foot.

54. *OLED Television* The screen of an OLED television is in the shape of a rectangle with a diagonal of length 43 in. If the height of the screen is 21 in., determine the width of the screen. Round your answer to the nearest hundredth.

Challenge Problems/Group Activities

55. *Heron's Formula* A second formula for determining the area of a triangle (called Heron's formula) is

$$A = \sqrt{s(s-a)(s-b)(s-c)}$$

where $s = \frac{1}{2}(a + b + c)$ and a, b, and c are the lengths of the sides of the triangle. Use Heron's formula to determine the area of right triangle ABC and check your answer using the formula $A = \frac{1}{2}ab$.

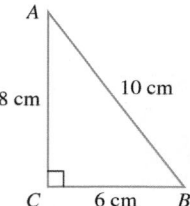

56. **Expansion of $(a + b)^2$** In the figure, one side of the largest square has length $a + b$. Therefore, the area of the largest square is $(a + b)^2$. Answer the following questions to determine a formula for the expansion of $(a + b)^2$.

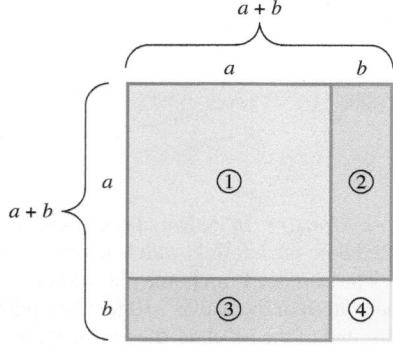

a) What is the area of the square marked ①?

b) What is the area of the rectangle marked ②?

c) What is the area of the rectangle marked ③?

d) What is the area of the square marked ④?

e) Add the four areas determined in parts (a) through (d) to write a formula for the expansion of $(a + b)^2$.

Recreational Mathematics

57. **Sports Areas** Research the official dimensions of the playing surface of your favorite sport and determine the area of the playing surface in both square feet and square yards.

58. **Scarecrow's Error** In the movie *The Wizard of Oz*, once the scarecrow gets his diploma he states the following: "In an isosceles triangle, the sum of the square roots of the two equal sides is equal to the square root of the third side." Discuss why this statement is incorrect.

Research Activities

59. **Garfield's Proof** Research the proof of the Pythagorean theorem provided by President James Garfield. Write a brief paper and make a poster of this proof and the associated diagrams.

60. **Babylonians and Egyptians** The early Babylonians and Egyptians did not know about π and had to devise techniques to approximate the area of a circle. Write a paper on the techniques these societies used to approximate the area of a circle.

61. **Heron of Alexandria** Write a paper on the contributions of Heron of Alexandria to geometry.

SECTION 8.4   Volume and Surface Area

Upon completion of this section, you will be able to:

- Understand and use formulas to determine the volume and surface area of rectangular solids, cylinders, cones, and spheres.

- Understand and use formulas to determine the volume of polyhedra, prisms, and pyramids.

- Solve problems involving cubic unit conversions.

Denise is painting her apartment and needs to estimate the amount of paint she will need. The paint can label states, "One gallon will cover about 400 sq. ft." This sentence refers to the two main geometric topics that we will cover in this section. *One gallon* refers to volume of the paint in the can, and *400 sq. ft.* refers to the surface area that the paint will cover.

Why This Is Important Volume and surface area are important geometric concepts that have many real-life applications. These include calculating surface areas to be painted, the amount of sand needed for a volleyball court, the size of a car engine, and the size of an air conditioner needed to cool a home. Understanding volume and surface area is a helpful problem-solving tool for these and many other applications.

When discussing a one-dimensional figure such as a line, we can determine its length. When discussing a two-dimensional figure such as a rectangle, we can determine its area and its perimeter. When discussing a three-dimensional figure such as a cube, we can determine its volume and its surface area. *Volume* is a measure of the capacity of a three-dimensional figure. *Surface area* is the sum of the areas of the surfaces of a three-dimensional figure. Volume refers to the amount of material that you can put *inside* a three-dimensional figure, and surface area refers to the total area that is on the *outside* surface of the figure.

Solid geometry is the study of three-dimensional solid figures, also called *space figures*. Volumes of three-dimensional figures are measured in cubic units such as cubic feet or cubic meters. Surface areas of three-dimensional figures are measured in square units such as square feet or square meters.

Rectangular Solids, Cylinders, Cones, and Spheres

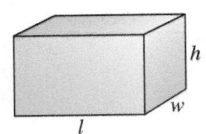

Rectangular Solid If the length of a *rectangular solid* is 5 units, the width is 2 units, and the height is 3 units, the total number of cubes that can be formed is 30 (Fig. 8.35). Thus, the volume is 30 cubic units. The volume of a rectangular solid can also be determined by multiplying its length times width times height; in this case, 5 units $\times$ 2 units $\times$ 3 units = 30 cubic units. In general, the volume of any rectangular solid is $V = l \times w \times h$.

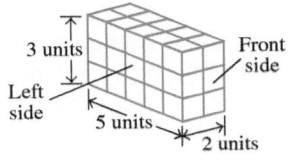

Figure 8.35

The surface area of the rectangular solid in Fig. 8.35 is the sum of the areas of the surfaces of the rectangular solid. Notice that each surface of the rectangular solid is a rectangle. The left and right sides of the rectangular solid each has an area of 5 units $\times$ 3 units, or 15 square units. The front and back sides of the rectangular solid each has an area of 2 units $\times$ 3 units, or 6 square units. The top and bottom sides of the rectangular solid each has an area of 5 units $\times$ 2 units, or 10 square units. Therefore, the surface area of the rectangular solid is $2(5 \times 3) + 2(2 \times 3) + 2(5 \times 2) = 2(15) + 2(6) + 2(10) = 30 + 12 + 20 = 62$ square units. In general, the surface area of any rectangular solid is $SA = 2lw + 2wh + 2lh$.

> **Volume and Surface Area of a Rectangular Solid**
>
> $$V = lwh \qquad SA = 2lw + 2wh + 2lh$$

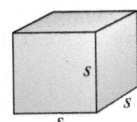

Cube A *cube* is a rectangular solid with the same length, width, and height (see the figure to the left). If we call the length of the side of the cube s and use the volume and surface area formulas for a rectangular solid, substituting s in for l, w, and h, we obtain $V = s \cdot s \cdot s = s^3$ and $SA = 2 \cdot s \cdot s + 2 \cdot s \cdot s + 2 \cdot s \cdot s = 2s^2 + 2s^2 + 2s^2 = 6s^2$.

> **Volume and Surface Area of a Cube**
>
> $$V = s^3 \qquad SA = 6s^2$$

Cylinder Now consider the right circular cylinder shown in Fig. 8.36(a) on page 480. *When we use the term cylinder in this book, we mean a right circular cylinder.* The volume of the cylinder is determined by multiplying the area of the circular base, πr^2, by the height, h, to get $V = \pi r^2 h$.

The surface area of the cylinder is the sum of the area of the top, the area of the bottom, and the area of the side of the cylinder. Both the top and bottom of the cylinder are circles with an area of πr^2. To determine the area of the side of the cylinder, examine Fig. 8.36(b) and (c).

Notice how when flattened, the side of the cylinder is a rectangle whose length is the circumference of the base of the cylinder and whose width is the height of the cylinder. Thus, the side of the cylinder has an area of $2\pi r \cdot h$, or $2\pi rh$. Therefore, the surface area of the cylinder is $\pi r^2 + \pi r^2 + 2\pi rh$, or $SA = 2\pi rh + 2\pi r^2$.

Volume of a Cylinder

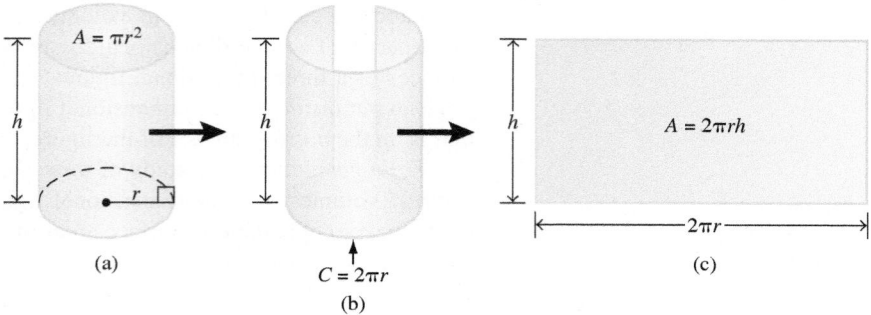

(a)

$C = 2\pi r$
(b)

$A = 2\pi rh$

$2\pi r$

(c)

Figure 8.36

> ### Volume and Surface Area of a Cylinder
> $$V = \pi r^2 h \qquad SA = 2\pi rh + 2\pi r^2$$

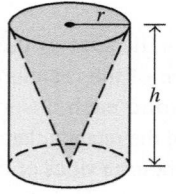

Figure 8.37

Now consider the right circular cone illustrated in Fig. 8.37. *When we use the term cone in this book, we mean a right circular cone.* The volume of a cone is less than the volume of a cylinder that has the same base and the same height. In fact, the volume of the cone is one-third the volume of the cylinder. The formula for the surface area of a cone is the sum of the area of the circular base of the cone, πr^2, and the area of the side of the cone, $\pi r \sqrt{r^2 + h^2}$, or $SA = \pi r^2 + \pi r \sqrt{r^2 + h^2}$. The derivation of the area of the side of the cone is beyond the scope of this book.

> ### Volume and Surface Area of a Cone
> $$V = \tfrac{1}{3}\pi r^2 h \qquad SA = \pi r^2 + \pi r \sqrt{r^2 + h^2}$$

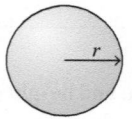

Sphere Baseballs, tennis balls, and so on have the shape of a *sphere* (see the figure to the left). The formulas for the volume and surface area of a sphere are as follows. The derivation of the volume and surface area of a sphere are beyond the scope of this book.

> ### Volume and Surface Area of a Sphere
> $$V = \tfrac{4}{3}\pi r^3 \qquad SA = 4\pi r^2$$

Following is a summary of the formulas for the volumes and surface areas of the three-dimensional figures we have discussed thus far.

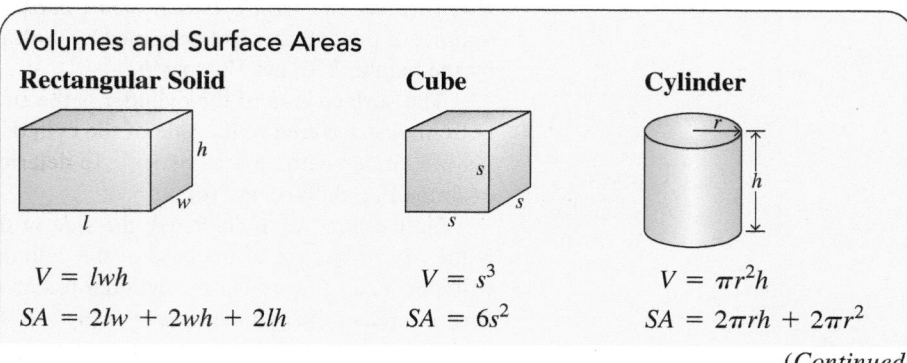

Volumes and Surface Areas

Rectangular Solid	Cube	Cylinder
$V = lwh$	$V = s^3$	$V = \pi r^2 h$
$SA = 2lw + 2wh + 2lh$	$SA = 6s^2$	$SA = 2\pi rh + 2\pi r^2$

(Continued)

Cone

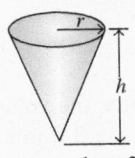

$$V = \tfrac{1}{3}\pi r^2 h$$
$$SA = \pi r^2 + \pi r \sqrt{r^2 + h^2}$$

Sphere

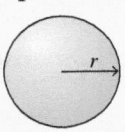

$$V = \tfrac{4}{3}\pi r^3$$
$$SA = 4\pi r^2$$

Example 1 *Volume and Surface Area*

Determine the volume and surface area of each of the following three-dimensional figures. When appropriate, use the $\boxed{\pi}$ key on your calculator and round your answer to the nearest hundredths.

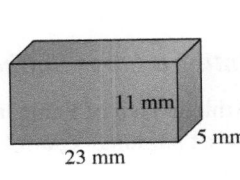

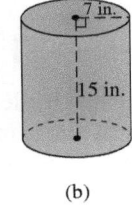

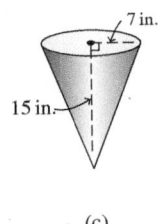

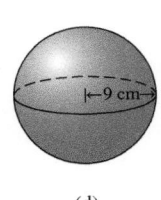

(a)   (b)   (c)   (d)

Solution

a) $V = lwh = 23 \cdot 5 \cdot 11 = 1265\,\text{mm}^3$
   $SA = 2lw + 2wh + 2lh = 2 \cdot 23 \cdot 5 + 2 \cdot 5 \cdot 11 + 2 \cdot 23 \cdot 11$
   $\qquad\qquad = 230 + 110 + 506 = 846\,\text{mm}^2$

b) $V = \pi r^2 h = \pi \cdot 7^2 \cdot 15 = 735\pi \approx 2309.07\,\text{in.}^3$
   $SA = 2\pi rh + 2\pi r^2 = 2 \cdot \pi \cdot 7 \cdot 15 + 2 \cdot \pi \cdot 7^2$
   $\qquad\qquad = 210\pi + 98\pi = 308\pi \approx 967.61\,\text{in.}^2$

c) $V = \dfrac{1}{3}\pi r^2 h = \dfrac{1}{3} \cdot \pi \cdot 7^2 \cdot 15 = \dfrac{1}{3} \cdot \pi \cdot 49 \cdot 15 = 245\pi \approx 769.69\ \text{in.}^3$
   $SA = \pi r^2 + \pi r \sqrt{r^2 + h^2} = \pi \cdot 7^2 + \pi \cdot 7 \cdot \sqrt{7^2 + 15^2}$
   $\qquad\qquad = 49 \cdot \pi + 7 \cdot \pi \cdot \sqrt{49 + 225}$
   $\qquad\qquad = 49\pi + 7\pi \sqrt{274} \approx 517.96\,\text{in.}^2$

d) $V = \dfrac{4}{3}\pi r^3 = \dfrac{4}{3} \cdot \pi \cdot 9^3 = \dfrac{4}{3} \cdot \pi \cdot 729 = 972\pi \approx 3053.63\,\text{cm}^3$
   $SA = 4\pi r^2 = 4 \cdot \pi \cdot 9^2 = 4 \cdot \pi \cdot 81 = 324\pi \approx 1017.88\,\text{cm}^2$ ∎

In Example 1(b), the volume of the cylinder is about $2309.07\,\text{in.}^3$. In Example 1(c), the volume of the cone with the same radius and height as the cylinder is about $769.69\,\text{in.}^3$. Notice that the volume of the cylinder is about three times the volume of the cone, which is what we expect.

Example 2 *Replacing a Sand Volleyball Court*

Maya needs to replace the sand in her rectangular sand volleyball court. The court is 60 ft long by 30 ft wide, and the sand has a uniform depth of 18 in. (see figure below). Sand sells for $27 per cubic yard.

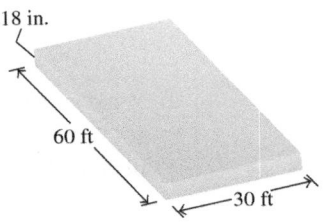

18 in.

60 ft

30 ft

a) How many cubic yards of sand does Maya need?

b) How much will the sand cost?

Solution

a) Since we are asked to determine the volume in cubic yards, we will convert each measurement to yards. There are 3 ft in a yard. Thus, 60 ft equals $\frac{60}{3}$, or 20 yd, and 30 ft equals $\frac{30}{3}$, or 10 yd. There are 36 in. in a yard, so 18 in. equals $\frac{18}{36}$, or $\frac{1}{2}$ yd. The amount of sand needed is determined using the formula for the volume of a rectangular solid, $V = l \cdot w \cdot h$. In this case, the height of the rectangular solid can be considered the depth of the sand.

$$V = l \cdot w \cdot h = 20 \cdot 10 \cdot \tfrac{1}{2} = 100 \text{ yd}^3$$

Note that since the measurements for length, width, and height are each in terms of yards, the answer is in terms of cubic yards.

b) One cubic yard of sand costs $27, so 100 yd^3 will cost 100 × $27, or $2700. ∎

Example 3 *Painting a Large Softball*

The Faber College Athletic Hall of Fame building features a statue of a large softball that is in need of painting. The softball has a diameter of 8.5 ft.

a) Determine the surface area of the softball.

b) If one quart of paint covers about 100 ft^2, how many quarts of paint will the college need to buy to paint the softball?

Solution

a) The softball has the shape of a sphere with a diameter of 8.5 feet. Since the radius is half of the diameter, $r = \tfrac{1}{2} \cdot 8.5 = 4.25$. Using the formula for the surface area of a sphere:

$$SA = 4\pi r^2 = 4\pi \cdot (4.25)^2 = 72.25\pi \approx 226.98 \text{ ft}^2$$

b) Since each quart of paint will cover about 100 ft^2, it will take about $\frac{226.98}{100} \approx 2.27$ quarts of paint. Since you cannot buy a portion of a quart of paint, the college will need to buy 3 quarts of paint to paint the softball. ∎

Example 4 *Silage Storage*

Cletus has three silos on his farm. The silos are each in the shape of a right circular cylinder (see Fig. 8.38). One silo has a 12-ft diameter and is 40 ft tall. The second silo has a 14-ft diameter and is 50 ft tall. The third silo has an 18-ft diameter and is 60 ft tall.

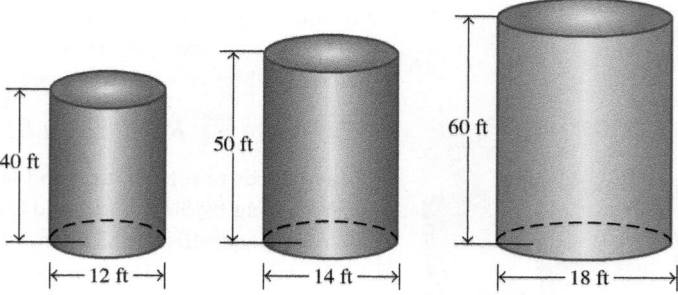

40 ft   50 ft   60 ft

|←— 12 ft —→| |←— 14 ft —→| |←——— 18 ft ———→|

Figure 8.38

a) What is the total capacity of all three silos in cubic feet?

b) If Cletus fills all three of his silos and then feeds his cattle 150 ft^3 of silage per day, in how many days will all three silos be empty?

Solution

a) The capacity of each silo can be determined using the formula for the volume of a right circular cylinder, $V = \pi r^2 h$. Since the radius is half the diameter, the radii for the three silos are 6 ft, 7 ft, and 9 ft, respectively. Now let's determine the volumes.

$$\text{Volume of the first silo} = \pi r^2 h = \pi \cdot 6^2 \cdot 40$$
$$= \pi \cdot 36 \cdot 40 = 1440\pi \approx 4523.89 \text{ ft}^3$$
$$\text{Volume of the second silo} = \pi r^2 h = \pi \cdot 7^2 \cdot 50$$
$$= \pi \cdot 49 \cdot 50 = 2450\pi \approx 7696.90 \text{ ft}^3$$
$$\text{Volume of the third silo} = \pi r^2 h = \pi \cdot 9^2 \cdot 60$$
$$= \pi \cdot 81 \cdot 60 = 4860\pi \approx 15{,}268.14 \text{ ft}^3$$

Therefore, the total capacity of all three silos is about

$$4523.89 + 7696.90 + 15{,}268.14 \approx 27{,}488.93 \text{ ft}^3$$

b) To determine how long it takes to empty all three silos, we will divide the total capacity by 150 ft³, the amount fed to Cletus's cattle per day.

$$\frac{27{,}488.93}{150} \approx 183.26$$

Thus, the silos will be empty in about 183 days. ∎

Polyhedra, Prisms, and Pyramids

A *polyhedron* (plural, *polyhedra*) is a closed surface formed by the union of polygonal regions. Figure 8.39 illustrates some polyhedra.

Polyhedra

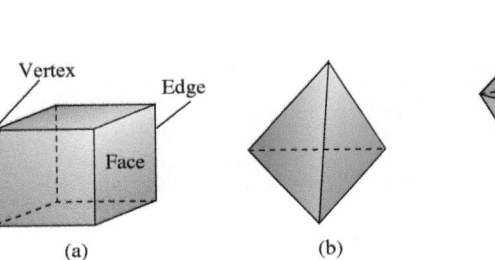

Vertex   Edge

Face

(a)   (b)   (c)   (d)

Figure 8.39

Each polygonal region is called a *face* of the polyhedron. The line segment formed by the intersection of two faces is called an *edge*. The point at which two or more edges intersect is called a *vertex*. In Fig. 8.39(a), there are 6 faces, 12 edges, and 8 vertices.

Note that:

$$\text{Number of vertices} - \text{number of edges} + \text{number of faces} = 2$$
$$8 \quad - \quad 12 \quad + \quad 6 \quad = 2$$

This formula, credited to Leonhard Euler, is true for any polyhedron.

> ### Euler's Polyhedron Formula
> $$\text{Number of vertices} - \text{number of edges} + \text{number of faces} = 2$$

We suggest that you verify that this formula holds for Fig. 8.39(b), (c), and (d).

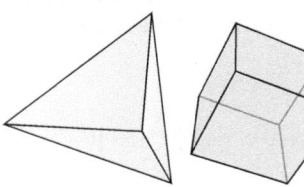
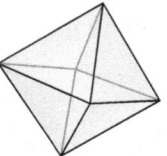 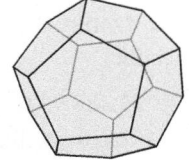
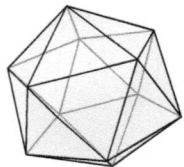
Example 5   *Using Euler's Polyhedron Formula*

A certain polyhedron has 20 vertices and 12 faces. Determine the number of edges on the polyhedron.

Solution Since we are seeking the number of edges, we will let x represent the number of edges on the polyhedron. Next, we will use Euler's polyhedron formula to set up an equation:

Number of vertices − number of edges + number of faces = 2

$$20 - x + 12 = 2$$
$$32 - x = 2$$
$$-x = -30$$
$$x = 30$$

Therefore, the polyhedron has 30 edges. ∎

A *platonic solid*, also known as a *regular polyhedron*, is a polyhedron whose faces are all regular polygons of the same size and shape. There are exactly five platonic solids. All five platonic solids are illustrated in the *Did You Know?* at left.

A *prism* is a special type of polyhedron whose bases are congruent polygons and whose sides are parallelograms. These parallelogram regions are called the *lateral faces* of the prism. If all the lateral faces are rectangles, the prism is said to be a *right prism*. The prisms illustrated in Fig. 8.40 are all right prisms. *When we use the word prism in this book, we are referring to a right prism.*

Prisms

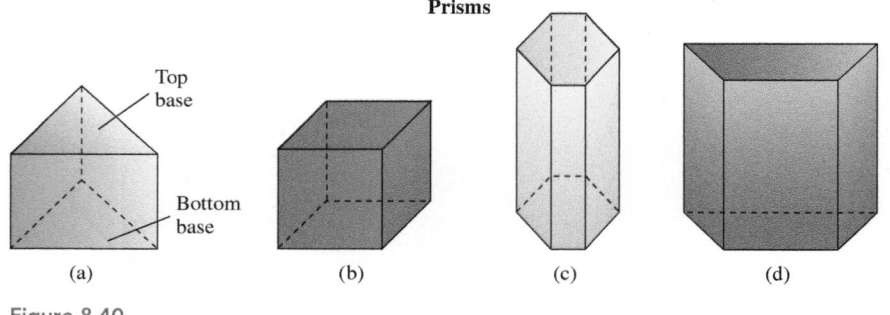

Top base

Bottom base

(a)   (b)   (c)   (d)

Figure 8.40

The volume of any prism can be determined by multiplying the area of the base, B, by the height, h, of the prism.

Volume of a Prism
$$V = Bh$$
where B is the area of a base and h is the height.

Example 6   *Volume of a Hexagonal Prism Fish Tank*

Frank's fish tank is in the shape of a hexagonal prism, as shown in Fig. 8.41 on page 485. Use the dimensions shown in the figure and the fact that 1 gal = 231 in.³ to

a) determine the volume of the fish tank in cubic inches.

b) determine the volume of the fish tank in gallons (round your answer to the nearest gallon).

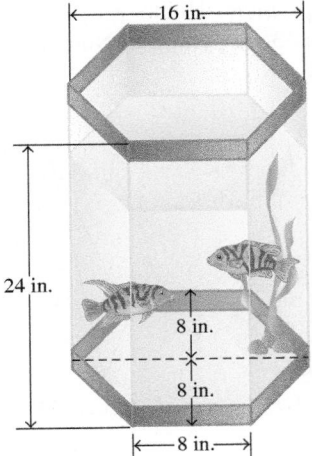

Figure 8.41

a) First we will need to calculate the area of the hexagonal base of the fish tank. Notice from Fig. 8.41 that by drawing a diagonal as indicated, the base can be divided into two identical trapezoids. To determine the area of the hexagonal base, we will calculate the area of one of these trapezoids and then multiply by 2.

$$\text{Area of one trapezoid} = \tfrac{1}{2}h(b_1 + b_2)$$
$$= \tfrac{1}{2}(8)(16 + 8) = 96 \text{ in.}^2$$
$$\text{Area of the hexagonal base} = 2(96) = 192 \text{ in.}^2$$

Now to determine the volume of the fish tank, we will use the formula for the volume of a prism, $V = Bh$. We already determined that the area of the base, B, is 192 in.^{2}.

$$V = B \cdot h = 192 \cdot 24 = 4608 \text{ in.}^3$$

In the above calculation, the area of the base, B, was measured in square inches and the height was measured in inches. The product of square inches and inches is cubic inches, or in.^{3}.

b) To determine the volume of the fish tank in gallons, since 1 gal = 231 in.^{3}, we will divide the volume of the fish tank in cubic inches by 231.

$$V = \frac{4608}{231} \approx 19.95 \text{ gal}$$

Thus, the volume of the fish tank is approximately 20 gal. ∎

Example 7   _Volumes Involving Prisms_

Determine the volume of the remaining solid after the cylinder, triangular prism, and square prism have been cut from the solid (Fig. 8.42).

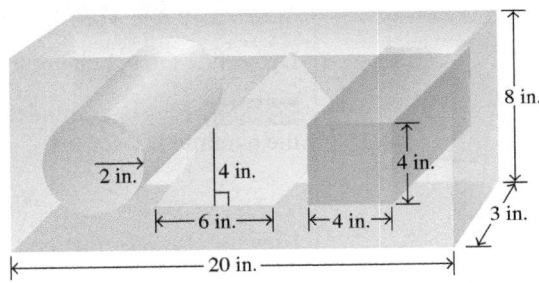

Figure 8.42

Solution To determine the volume of the remaining solid, first determine the volume of the rectangular solid. Then subtract the volume of the two prisms and the cylinder that were cut out.

$$\text{Volume of rectangular solid} = l \cdot w \cdot h$$
$$= 20 \cdot 3 \cdot 8 = 480 \text{ in.}^3$$
$$\text{Volume of circular cylinder} = \pi r^2 h$$
$$= \pi(2^2)(3)$$
$$= \pi(4)(3) = 12\pi \approx 37.70 \text{ in.}^3$$
$$\text{Volume of triangular prism} = \text{area of the base} \cdot \text{height}$$
$$= \tfrac{1}{2}(6)(4)(3) = 36 \text{ in.}^3$$
$$\text{Volume of square prism} = s^2 \cdot h$$
$$= 4^2 \cdot 3 = 48 \text{ in.}^3$$
$$\text{Volume of solid} \approx 480 - 37.70 - 36 - 48$$
$$\approx 358.30 \text{ in.}^3$$

Thus, the volume of the remaining solid is about 358.30 in.^{3}. ∎

Another special category of polyhedra is the *pyramid*. Unlike prisms, pyramids have only one base. The figures illustrated in Fig. 8.43 are pyramids. Note that all but one face of a pyramid intersect at a common vertex.

Pyramids

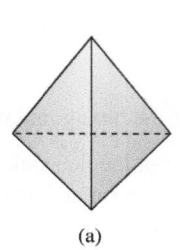   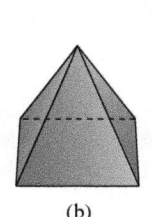   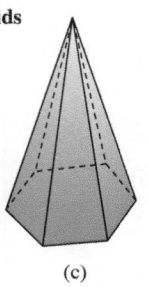   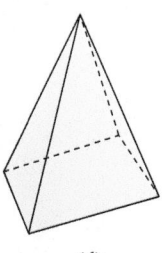

(a)   (b)   (c)   (d)

Figure 8.43

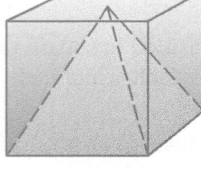

Figure 8.44

If a pyramid is drawn inside a prism, as shown in Fig. 8.44, the volume of the pyramid is less than that of the prism. In fact, the volume of the pyramid is one-third the volume of the prism.

> **Volume of a Pyramid**
> $$V = \tfrac{1}{3} Bh$$
> where B is the area of the base and h is the height.

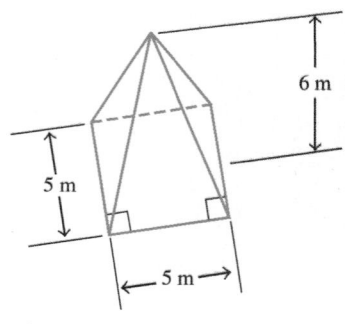

Figure 8.45

Example 8 *Volume of a Pyramid*

Determine the volume of the pyramid shown in Fig. 8.45.

Solution   First determine the area of the base, B, of the pyramid. Since the base of the pyramid is a square,

$$\text{Area of base} = s^2 = 5^2 = 25 \text{ m}^2$$

Now use this information to determine the volume of the pyramid.

$$V = \tfrac{1}{3} \cdot B \cdot h$$
$$= \tfrac{1}{3} \cdot 25 \cdot 6$$
$$= 50 \text{ m}^3$$

Thus, the volume of the pyramid is 50 m³.   ∎

Cubic Unit Conversions

In certain situations, converting volume from one cubic unit to a different cubic unit might be necessary. For example, when purchasing topsoil you might have to change the amount of topsoil from cubic feet to cubic yards prior to placing your order. Example 9 shows how that may be done.

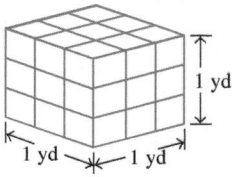

Figure 8.46

Example 9 *Cubic Yards and Cubic Feet*

a) Convert 1 yd³ to cubic feet. (See Fig. 8.46.)
b) Convert 5.7 yd³ to cubic feet.
c) Convert 843.75 ft³ to cubic yards.

Solution

a) We know that 1 yd = 3 ft. Thus, 1 yd^3 = 3 ft × 3 ft × 3 ft = 27 ft^3.

b) In part (a), we learned that 1 yd^3 = 27 ft^3. Thus, 5.7 yd^3 = 5.7 × 27 = 153.9 ft^3.

c) In part (b), we converted from cubic yards to cubic feet by *multiplying* the number of cubic yards by 27. Now, to convert from cubic feet to cubic yards we will *divide* the number of cubic feet by 27. Therefore, 843.75 ft^3 = $\frac{843.75}{27}$ = 31.25 yd^3. ∎

Example 10   *Filling in a Swimming Pool*

Tanzina recently purchased a home with a rectangular swimming pool. The pool is 30 ft long and 15 ft wide, and it has a uniform depth of 4.5 ft. Tanzina plans to fill the pool in with dirt to make a flower garden. How many cubic yards of dirt will Tanzina have to purchase to fill in the swimming pool?

Solution   To determine the amount of dirt, we will use the formula for the volume of a rectangular solid:

$$V = lwh$$
$$= (30)(15)(4.5)$$
$$= 2025 \text{ ft}^3$$

Now we must convert this volume from cubic feet to cubic yards. In Example 9, we learned that 1 yd^3 = 27 ft^3. Therefore, 2025 ft^3 = $\frac{2025}{27}$ = 75 yd^3. Thus, Tanzina needs to purchase 75 yd^3 of dirt to fill in her swimming pool. ∎

SECTION 8.4   *Exercises*

Warm Up Exercises

In Exercises 1–6, fill in the blanks with an appropriate word, phrase, or symbol(s).

1. A measure of the capacity of a three-dimensional figure is called the figure's _____.

2. The sum of the areas of the surfaces of a three-dimensional figure is called the figure's _____ area.

3. A regular polyhedron, whose faces are all regular polygons of the same size and shape, is also called a(n) _____ solid.

4. A polyhedron whose bases are congruent polygons and whose sides are parallelograms is called a(n) _____.

5. A prism whose sides are rectangles is called a(n) _____ prism.

6. Euler's polyhedron formula states that, for any polyhedron, the number of vertices minus the number of edges plus the number of faces equals _____.

Practice the Skills

In Exercises 7–14, determine (a) the volume and (b) the surface area of the three-dimensional figure. When appropriate, use the ⊡π key on your calculator and round your answer to the nearest hundredth.

7.
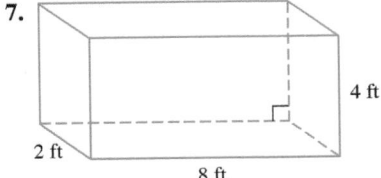
4 ft
2 ft
8 ft

8.
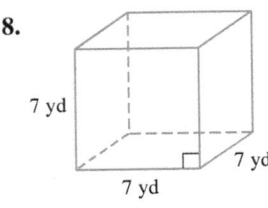
7 yd
7 yd
7 yd

9.
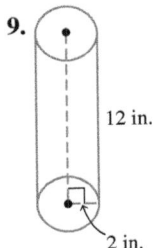
12 in.
2 in.

10.
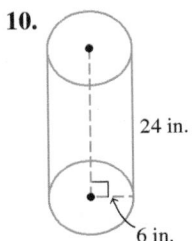
24 in.
6 in.

11.
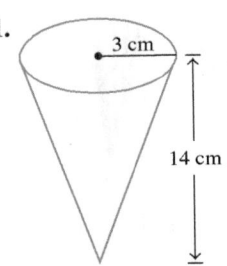
3 cm
14 cm

12.

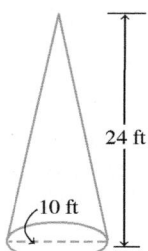

24 ft

10 ft

13.

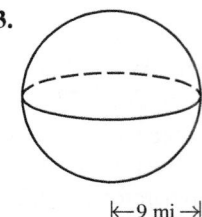

|←9 mi →|

14.

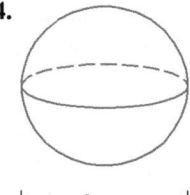

|←   9 cm   →|

In Exercises 15–18, determine the volume of the three-dimensional figure. When appropriate, round your answer to the nearest hundredth.

15.

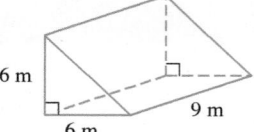

6 m

6 m

9 m

16.

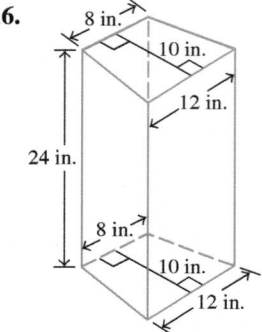

8 in.

10 in.

12 in.

24 in.

8 in.

10 in.

12 in.

17.

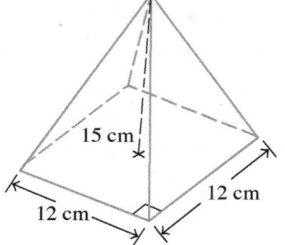

15 cm

12 cm

12 cm

18.

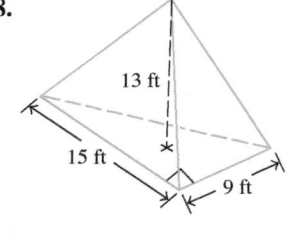

13 ft

15 ft

9 ft

Problem Solving

In Exercises 19–26, determine the volume of the shaded region. When appropriate, use the $\boxed{\pi}$ *key on your calculator and round your answer to the nearest hundredth.*

19.

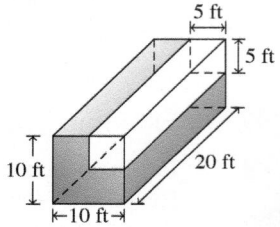

5 ft

5 ft

20 ft

10 ft

←10 ft→

20.

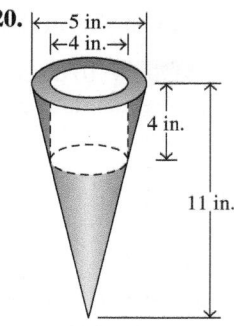

|←5 in.—→|

|←4 in.→|

4 in.

11 in.

21.

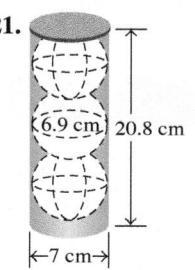

6.9 cm   20.8 cm

|←7 cm→|

22.

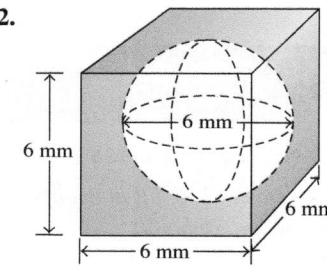

6 mm

←6 mm→

6 mm

6 mm

23.

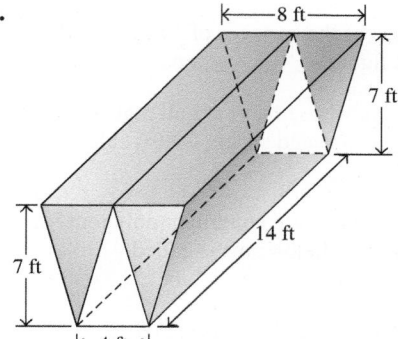

|←——8 ft——→|

7 ft

14 ft

7 ft

|←4 ft→|

24.

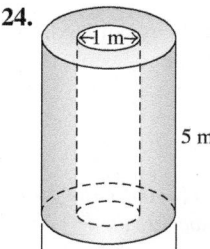

|←1 m→|

5 m

|←——3 m——→|

25.

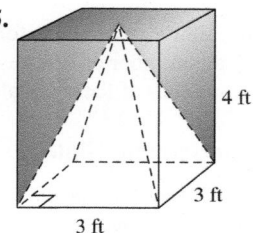

4 ft

3 ft

3 ft

26.

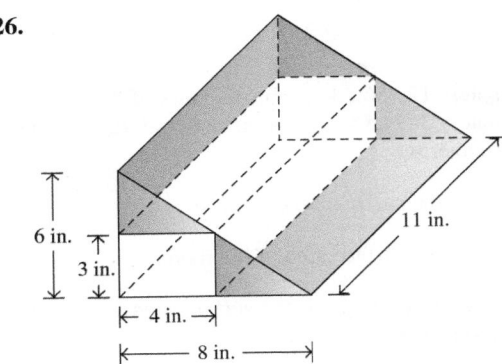

In Exercises 27–30, use the fact that 1 yd³ equals 27 ft³ to make the conversion.

27. 7 yd³ to cubic feet

28. 15.75 yd³ to cubic feet

29. 453.6 ft³ to cubic yards

30. 278.1 ft³ to cubic yards

In Exercises 31–34, use the fact that 1 m³ equals 1,000,000 cm³ to make the conversion.

31. 1.5 m³ to cubic centimeters

32. 13.7 m³ to cubic centimeters

33. 500,000 cm³ to cubic meters

34. 9,160,000 cm³ to cubic meters

In Exercises 35–46, solve the given problem. When appropriate, use the [π] key on your calculator and round your answer to the nearest hundredth.

35. *Playground Mulch* Davidson is building a backyard playground for his grandchildren and wishes to put down rubber mulch to provide safety from falls. Davidson wishes to put the mulch in a pit in the shape of a rectangular solid 20 ft long, 15 ft wide, and 9 in. deep.

a) Determine the volume, in cubic feet, of mulch Davidson will need.

b) If mulch costs $11 per cubic foot, what will the cost of the mulch be?

36. *Rose Garden Topsoil* Marisa wishes to plant a rose garden in her backyard. The rose garden will be in the shape of a 9 ft by 18 ft rectangle. Marisa wishes to add a 4-in. layer of organic topsoil on top of the rectangular area. The topsoil sells for $42 per cubic yard.

a) Determine how many cubic yards of topsoil Marisa will need.

b) Determine how much the topsoil will cost.

37. *A Fish Tank*

a) How many cubic centimeters of water will a fish tank in the shape of a rectangular solid hold if the tank is 80 cm long, 50 cm wide, and 30 cm high?

b) If 1 cm³ holds 1 mℓ of liquid, how many milliliters will the tank hold?

c) If 1 ℓ = 1000 mℓ, how many liters will the tank hold?

38. *Volume of a Freezer* The dimensions of the interior of an upright freezer are height 46 in., width 25 in., and depth 25 in. Determine the volume of the freezer

a) in cubic inches.   **b)** in cubic feet.

39. *Baseball Display Case* A baseball is displayed in a cube-shaped glass case. The diameter of the baseball and the length of each side of the cube are both 7.5 cm. What is the volume of the air outside the baseball that is inside the closed case?

40. *Softball Display Case* A softball is displayed in a cube-shaped glass case. See Exercise 39 for a similar description involving a baseball. The diameter of the softball and the length of each side of the cube are both 3.8 inches. What is the volume of the air outside the softball that is inside the closed cube?

41. *WNBA Basketball* A regulation WNBA basketball has a diameter of 9.15 inches.

a) Determine the volume of air inside the basketball

b) Determine the surface area of the basketball.

42. *NBA Basketball* A regulation NBA basketball has a diameter of 9.47 inches.

a) Determine the volume of air inside the basketball.

b) Determine the surface area of the basketball.

43. *Comparing Cake Pans* When baking a cake, you can choose between a circular pan with a 9-in. diameter and a 7 in. × 9 in. rectangular pan.

a) Determine the area of the base of each pan.

b) If both pans are 2 in. deep, determine the volume of each pan.

c) Which pan has the larger volume?

44. *Ice-Cream Comparison* The Louisburg Creamery packages its homemade ice cream in tubs and in boxes. The tubs are in the shape of a right cylinder with a radius of 3 in. and height of 5 in. The boxes are in the shape of a cube with each side measuring 5 in. Determine the volume of each container.

45. *Cake Icing* A bag used to apply icing to a cake is in the shape of a cone with a diameter of 3 in. and a height of 6 in. How much icing will this bag hold when full?

46. *Pool Toys* A Wacky Noodle Pool Toy, frequently referred to as a "noodle," is a cylindrical flotation device made from cell foam (see photo). One style of noodle is a cylinder that has a diameter of 2.5 in. and a length of 5.5 ft. Determine the volume of this style of noodle in

a) cubic inches.   **b)** cubic feet.

47. *Engine Capacity* The engine in a 1957 Chevrolet Corvette has eight cylinders. Each cylinder is a right cylinder with a bore (diameter) of 3.875 in. and a stroke (height) of 3 in. Determine the total displacement (volume) of this engine.

48. *Thunderbird Engine Capacity* The engine in a 1961 Ford Thunderbird has eight cylinders. Each cylinder is a right cylinder with a bore (diameter) of 4.05 in. and a stroke (height) of 3.78 in. Determine the total displacement (volume) of this engine.

In Exercises 49–54, determine the missing value indicated by the question mark. Use Euler's Polyhedron formula.

$$\left(\begin{array}{c}\text{Number of}\\ \text{vertices}\end{array}\right) - \left(\begin{array}{c}\text{number of}\\ \text{edges}\end{array}\right) + \left(\begin{array}{c}\text{number}\\ \text{of faces}\end{array}\right) = 2$$

	Number of Vertices	Number of Edges	Number of Faces
49.	8	12	?
50.	4	6	?
51.	?	12	8
52.	?	30	12
53.	12	?	20
54.	18	?	20

Challenge Problems/Group Activities

55. *Earth and Moon Comparisons* The diameter of Earth is approximately 12,756.3 km. The diameter of the moon is approximately 3474.8 km. Assume that both Earth and the moon are spheres.

a) Determine the surface area of Earth.

b) Determine the surface area of the moon.

c) How many times larger is the surface area of Earth than the surface area of the moon?

d) Determine the volume of Earth.

e) Determine the volume of the moon.

f) How many times larger is the volume of Earth than the volume of the moon?

56. *Flower Box* The flower box shown below is 4 ft long, and its ends are in the shape of a trapezoid. The upper and lower bases of the trapezoid measure 12 in. and 8 in., respectively, and the height is 9 in. Determine the volume of the flower box

a) in cubic inches.   **b)** in cubic feet.

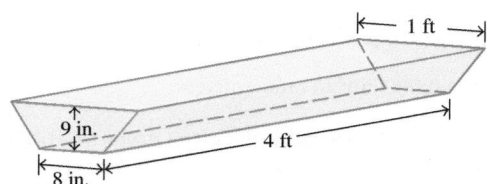

57. *Binomial Cubed*
a) What is the volume in terms of *a* and *b* of each numbered piece in the figure?

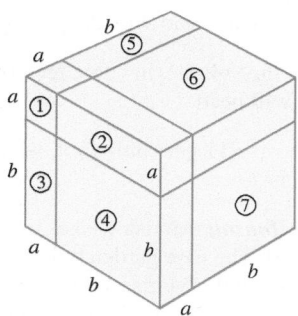

b) An eighth piece is not illustrated. What is its volume?
c) Explain how to demonstrate, using the cube shown that

$$(a + b)^3 = a^3 + 3a^2b + 3ab^2 + b^3$$

58. If the side of a cube is doubled, how is the volume of the cube affected?

59. If the radius of a sphere is doubled, how is the volume of the sphere affected?

Recreational Mathematics

60. *More Pool Toys* Wacky Noodle Pool Toys (see Exercise 46 on page 490) come in many different shapes and sizes.

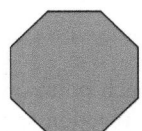 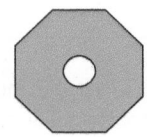

Base for part (a)   Base for part (b)

a) Determine the volume, in cubic inches, of a noodle that is in the shape of a 5.5-ft-long solid octagonal prism whose base has an area of 5 in.^{2}.

b) Determine the volume, in cubic inches, of a hollow noodle that has the same shape as the noodle described in part (a) except that a right circular cylinder of diameter 0.75 in. has been removed from the center.

Research Activities

61. *Air-Conditioner Selection* Calculate the volume of the room in which you sleep or study. Go to a store that sells room air conditioners and determine how many cubic feet can be cooled by the different models available. Describe the model that would be the proper size for your room. What is the initial cost? How much does that model cost to operate? If you moved to a room that had twice the amount of floor space and the same height, would the air conditioner you selected still be adequate? Explain.

62. *Pappus of Alexandria* Pappus of Alexandria (ca. A.D. 350) was the last of the well-known ancient Greek mathematicians. Write a paper on his life and his contributions to mathematics.

63. *Platonic Solids* Construct cardboard models of one or more of the platonic solids. Search the Internet for patterns to follow.

SECTION 8.5 — Transformational Geometry, Symmetry, and Tessellations

Upon completion of this section, you will be able to:

- Perform reflections of geometric figures.
- Perform translations of geometric figures.
- Perform rotations of geometric figures.
- Perform glide reflections of geometric figures.
- Identify symmetries of geometric figures.
- Identify tessellations of geometric figures.

Consider the capital letters of the alphabet. Now consider a kindergarten-age child who is practicing to write capital letters. Usually, children will write some of the letters "backwards." However, some letters cannot be made backwards. For example, the capital letter A would look the same backwards as it does forwards, but the capital letters B and C would look different backwards than they do forwards. The capital letters that appear the same forwards as they do backwards have a property called *symmetry* that we will define and study in this section. Symmetry is part of another branch of geometry known as *transformational geometry*.

Why This Is Important Transformational geometry is used to describe movement of geometric figures. For this reason, transformational geometry is used to model many problems in physics, chemistry, biology, and other scientific areas. Furthermore, transformational geometry plays an important role in the branch of mathematics called *group theory*, which we will study in Chapter 9. When taken together, group theory and transformational geometry can be used to understand any physical phenomenon that involves symmetry and patterning.

Thus far we have discussed *Euclidean geometry*. We will now introduce a second type of geometry called *transformational geometry*. In *transformational geometry*, we study various ways to move a geometric figure without altering the shape or size of the figure. When discussing transformational geometry, we often use the term *rigid motion*.

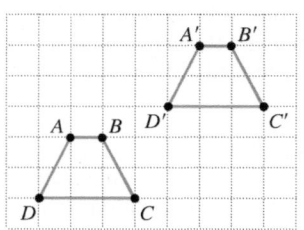

Figure 8.47

Definition: Rigid Motion or Transformation

The act of moving a geometric figure from some starting position to some ending position without altering its shape or size is called a **rigid motion** (or **transformation**).

Consider trapezoid *ABCD* in Fig. 8.47. If we move each point on this trapezoid 4 units to the right and 3 units up, the trapezoid is in the location specified by trapezoid *A′B′C′D′*. This figure illustrates one type of rigid motion. When studying rigid motions, we are concerned only about the starting and ending positions of the figure. When discussing rigid motions of two-dimensional figures, we note there are four basic types of rigid motions: reflections, translations, rotations, and glide reflections. We call these four types of rigid motions the *basic rigid motions in a plane*. After we discuss the four rigid motions, we will discuss symmetry of geometric figures and tessellations.

Reflections

In our everyday life, we are quite familiar with the concept of reflection. In transformational geometry, a reflection is an image of a geometric figure that appears on the opposite side of a designated line.

Definition: Reflection

A **reflection** is a rigid motion that moves a geometric figure to a new position such that the figure in the new position is a mirror image of the figure in the starting position. In two dimensions, the figure and its mirror image are equidistant from a line called the **reflection line** or the **axis of reflection**.

Figure 8.48 shows trapezoid *ABCD*, a reflection line *l*, and the reflected trapezoid *A′B′C′D′*. Notice that vertex *A* is 6 units to the *left* of reflection line *l* and that vertex *A′* is 6 units to the *right* of reflection line *l*. Next notice that vertex *B* is 2 units to the *left* of *l* and that vertex *B′* is 2 units to the *right* of *l*. A similar relationship holds true for vertices *C* and *C′* and for vertices *D* and *D′*. It is important to see that the trapezoid is not simply *moved* to the other side of the reflection line, but instead it is *reflected*. Notice in the trapezoid *ABCD* that the longer base $\overline{BC}$ is on the *right* side of the trapezoid, but in the reflected trapezoid *A′B′C′D′* the longer base $\overline{B′C′}$ is on the *left* side of the trapezoid. Finally, notice the colors of the sides of the two trapezoids. Side $\overline{AB}$

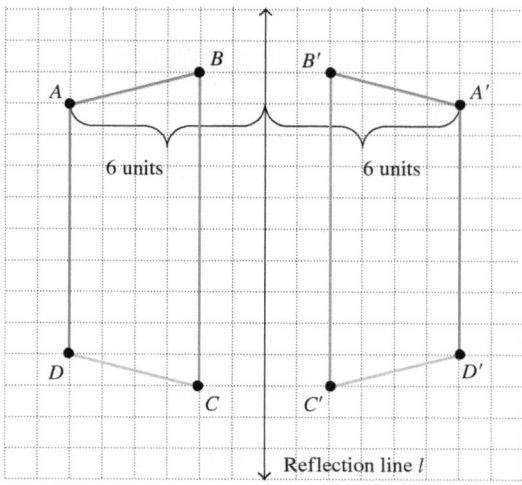

Figure 8.48

in trapezoid $ABCD$ and side $\overline{A'B'}$ in the reflected trapezoid are both blue. Side $\overline{BC}$ and side $\overline{B'C'}$ are both red, sides $\overline{CD}$ and $\overline{C'D'}$ are both gold, and sides $\overline{DA}$ and $\overline{D'A'}$ are both green. In this section, we will occasionally use such color coding to help you visualize the effect of a rigid transformation on a figure.

Example 1 *Reflection of a Triangle*

Construct the reflection of triangle ABC, shown in Fig. 8.49, about reflection line l.

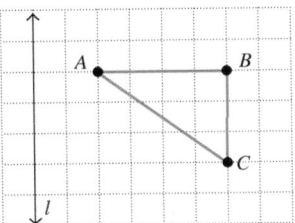

Figure 8.49

Solution The reflection of triangle ABC will be called $A'B'C'$. To determine the position of the reflection, we first examine vertex A in Fig. 8.49. Notice that vertex A is 2 units to the *right* of line l. Thus, in the reflected triangle $A'B'C'$, vertex A' must also be 2 units away from, but to the *left* of, reflection line l (see Fig 8.50). Next, notice that vertex B is 6 units to the *right* of line l. Thus, in the reflected triangle $A'B'C'$, vertex B' must also be 6 units away from, but to the *left* of, reflection line l. Next, notice that vertex C is 6 units to the *right* of line l. Thus, in the reflected triangle $A'B'C'$, vertex C' must also be 6 units away from, but to the *left* of, reflection line l. Figure 8.50 shows vertices A', B', and C'. Finally, we draw line segments between vertices A' and B', between B' and C', and between A' and C' to form the sides of the reflection triangle $A'B'C'$, as illustrated in Fig. 8.50.

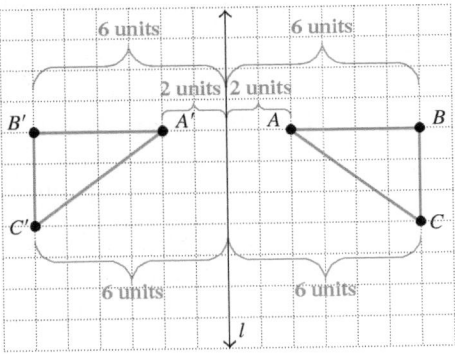

Figure 8.50

In Example 1, the reflection line did not intersect the figure being reflected. We will now study an example in which the reflection line goes directly through the figure to be reflected.

Example 2 *Reflection of a Hexagon*

Construct the reflection of hexagon $ABCDEF$, shown in Fig. 8.51 on page 494, about reflection line l.

Solution From Fig. 8.51 we see that vertex A in hexagon $ABCDEF$ is 2 units to the left of reflection line l. Thus, vertex A' in the reflected hexagon will be 2 units to the right of l (see Fig. 8.52). Notice that vertex A' of the reflected hexagon is in the same location as vertex B of hexagon $ABCDEF$ in Fig. 8.51.

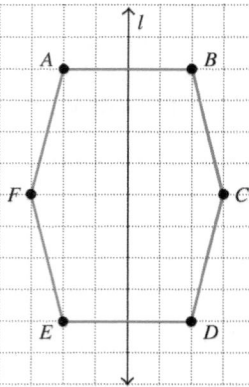

Figure 8.51

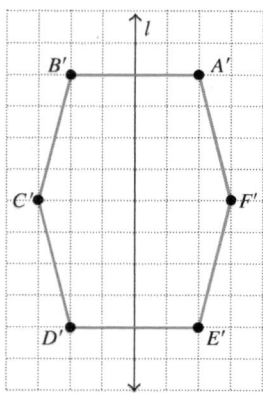

Figure 8.52

We next see that vertex *B* in hexagon *ABCDEF* is 2 units to the right of *l*. Thus, vertex *B'* in the reflected hexagon will be 2 units to the left of *l*. Notice that vertex *B'* of the reflected hexagon is in the same location as vertex *A* of hexagon *ABCDEF*. We continue this process to determine the locations of vertices *C'*, *D'*, *E'*, and *F'* of the reflected hexagon. Notice that each vertex of the reflected hexagon is in the same location as a vertex of hexagon *ABCDEF*. Finally, we draw the line segments to complete the reflected hexagon *A'B'C'D'E'F'* (see Fig. 8.52). For this example, we see that other than the vertex labels, the positions of the hexagon before and after the reflection are identical. ∎

In Example 2, the reflection line was in the center of the hexagon in the original position. As a result, the reflection line was also in the center of the reflected hexagon. In this particular case the reflected hexagon lies directly on top of the hexagon in its original position. We will revisit reflections such as the one in Example 2 again when we discuss *reflective symmetry* later in this section.

Now consider hexagon *ABCDEF* in Fig. 8.53 and its reflection about line *m*, hexagon *A'B'C'D'E'F'* in Fig. 8.54. Notice that the positions of the hexagon before and after the reflection, relative to line *m*, are not the same. If we line up reflection line *m* in Fig. 8.53 and Fig. 8.54, we see that hexagon *ABCDEF* and hexagon *A'B'C'D'E'F'* are in different positions.

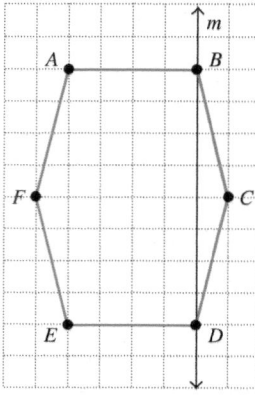

Figure 8.53

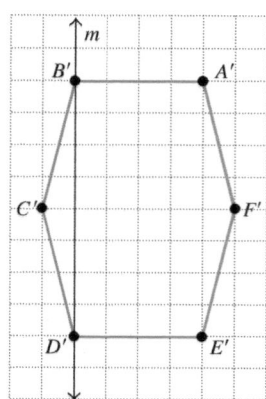

Figure 8.54

Translations

The next rigid motion we will discuss is the *translation*. In a translation, we simply move a figure along a straight line to a new position.

> **Definition: Translation or Glide**
> A **translation**, or **glide**, is a rigid motion that moves a geometric figure by sliding it along a straight line segment in the plane. The direction and length of the line segment completely determine the translation.

After conducting a translation, we say the figure was *translated* to a new position.

A concise way to indicate the direction and the distance that a figure is moved during a translation is with a *translation vector*. In mathematics, vectors are typically represented with boldface letters. For example, in Fig. 8.55 we see trapezoid *ABCD* and a translation vector, **v**, which is pointing to the right and upward. This translation vector indicates a translation of 9 units to the right and 4 units upward. Notice that in Fig. 8.55 the translated vector appears on the right side of the polygon. The placement of the translation vector does not matter.

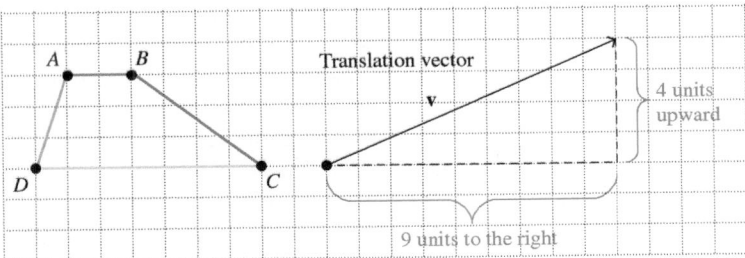

Figure 8.55

When trapezoid *ABCD* is translated using **v**, every point on trapezoid *ABCD* is moved 9 units to the right and 4 units upward. This movement is demonstrated for vertex *A* in Fig. 8.56(a). Figure 8.56(b) shows trapezoid *ABCD* and the translated trapezoid *A′B′C′D′*. Notice in Fig. 8.56(b) that every point on trapezoid *A′B′C′D′* is 9 units to the right and 4 units up from its corresponding point on trapezoid *ABCD*. Also notice that trapezoid *A′B′C′D′* is the same size and shape as trapezoid *ABCD*.

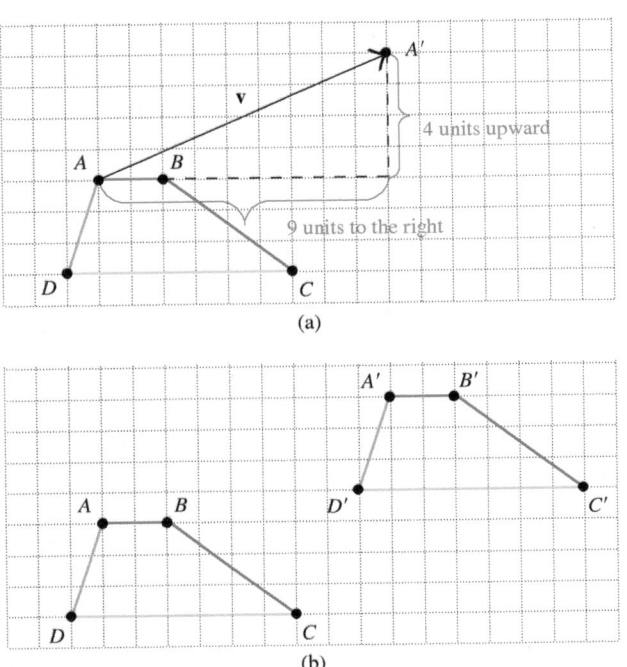

Figure 8.56

Example 3 *A Translated Square*

Given square $ABCD$ and translation vector **v**, shown in Fig. 8.57, construct the translated square $A'B'C'D'$.

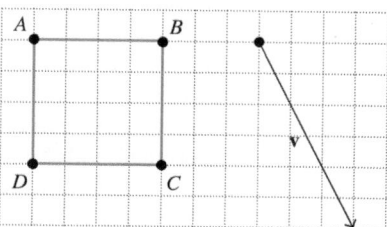

Figure 8.57

Solution The translated figure will be a square of the same size and shape as square $ABCD$. We notice that the translation vector, **v**, points 6 units downward and 3 units to the right. To determine the location of vertex A' of the translated square, start at vertex A of square $ABCD$ and move down 6 units and to the right 3 units. We label this vertex A' (see Fig. 8.58a). We determine vertices B', C', and D' in a similar manner by moving down 6 units and to the right 3 units from vertices B, C, and D, respectively. Figure 8.58(b) shows square $ABCD$ and the translated square $A'B'C'D'$. Notice in Fig. 8.58(b) that every point on square $A'B'C'D'$ is 6 units down and 3 units to the right of its corresponding point on square $ABCD$.

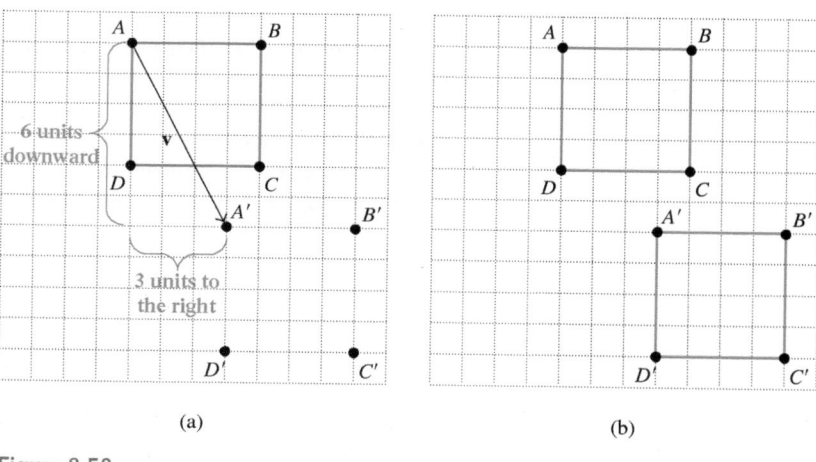

(a)   (b)

Figure 8.58   ■

Rotations

The next rigid motion we will discuss is *rotation*. To help visualize a rotation, examine Fig. 8.59, which shows right triangle ABC and point P, about which right triangle ABC is to be rotated.

Imagine that this page was removed from this book and attached to a bulletin board with a single pin through point P. Next imagine rotating the page 90° in the *counterclockwise* direction. The triangle would now appear as triangle $A'B'C'$ shown in Fig. 8.60 on page 497. Next, imagine rotating the original triangle 180° in a counterclockwise direction. The triangle would now appear as triangle $A''B''C''$ shown in Fig. 8.61.

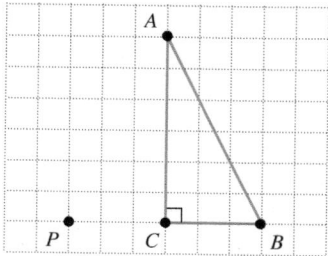

Figure 8.59

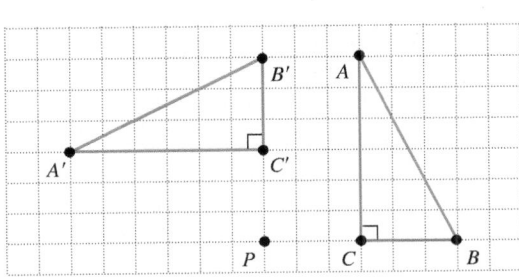

Figure 8.60

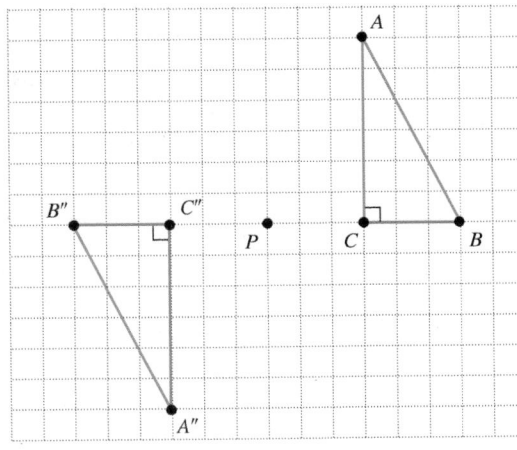

Figure 8.61

Now that we have an intuitive idea of how to determine a rotation, we give the definition of rotation.

> ### Definition: Rotation
> A **rotation** is a rigid motion performed by rotating a geometric figure in the plane about a specific point, called the **rotation point** or the **center of rotation**. The angle through which the object is rotated is called the **angle of rotation**.

We will measure angles of rotation using degrees. In mathematics, generally, *counterclockwise angles have positive degree measures and clockwise angles have negative degree measures.*

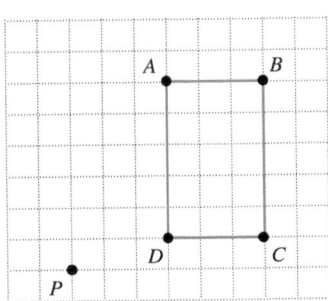

Figure 8.62

Example 4 *A Rotated Rectangle*

Given rectangle *ABCD* and rotation point *P*, shown in Fig. 8.62, construct rectangles that result from rotations through

a) 90°.   b) 180°.   c) 270°.

Solution

a) First, since 90 is a *positive* number, we will rotate the figure in a counterclockwise direction. We also note that the rotated rectangle will be the same size and shape as rectangle *ABCD*. To get an idea of what the rotated rectangle will look like, pick up this book and rotate it counterclockwise 90°. If you are viewing this page electronically, print out the page and then rotate the page 90°. Figure 8.63

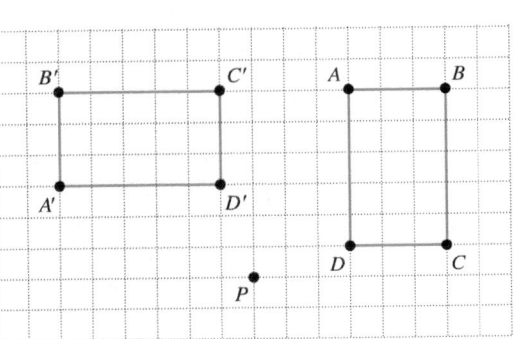

Figure 8.63

shows rectangle *ABCD* and rectangle *A'B'C'D'*, which is rectangle *ABCD* rotated 90°about point *P*. Notice how line segment *AB* in rectangle *ABCD* is horizontal, but in the rotated rectangle in Fig. 8.63 line segment *A'B'* is vertical. Also notice that in rectangle *ABCD* vertex *D* is 3 units to the *right* and 1 unit *above* rotation point *P*, but in the rotated rectangle, vertex *D'* is 3 units *above* and 1 unit to the *left* of rotation point *P*.

b) To gain some perspective on a 180° rotation, again pick up this book or a printed page, but this time rotate it 180° in the counterclockwise direction. The rotated rectangle *A"B"C"D"* is shown along with the rectangle *ABCD* in Fig. 8.64.

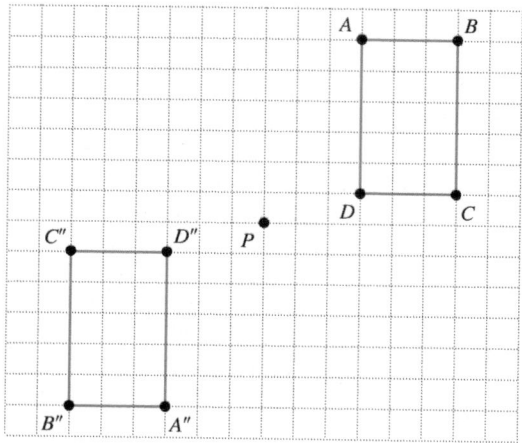

Figure 8.64

c) To gain some perspective on a 270° rotation, rotate this book or a printed page 270° in the counterclockwise direction. The rotated rectangle *A'''B'''C'''D'''* is shown along with rectangle *ABCD* in Fig. 8.65.   ∎

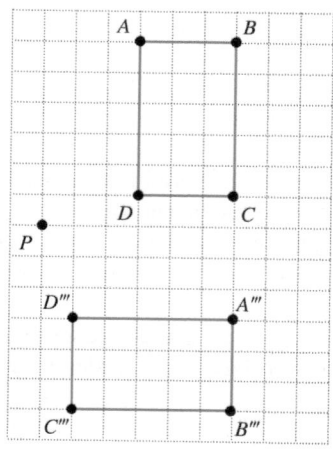

Figure 8.65

Thus far in our examples of rotations, the rotation point was outside the figure being rotated. We now will study an example in which the rotation point is inside the figure to be rotated.

Example 5 *A Rotation Point Inside a Parallelogram*

Given parallelogram *ABCD* and rotation point *P*, shown in Fig. 8.66, construct parallelograms that result from rotations through

a) 90°.   b) 180°.

a) We will rotate the parallelogram 90° in a counterclockwise direction. The resulting parallelogram will be the same size and shape as parallelogram *ABCD*. To visualize what the rotated parallelogram will look like, pick up this book or a printed page and rotate it counterclockwise 90°. Figure 8.67 shows the original parallelogram, *ABCD*, in pale blue, and the rotated parallelogram, *A'B'C'D'*, in darker blue. Notice how line segments $\overline{AB}$ and $\overline{CD}$ in parallelogram *ABCD* are *horizontal*, but in the rotated parallelogram *A'B'C'D'*, line segments $\overline{A'B'}$ and $\overline{C'D'}$ are *vertical*. Also notice in the original parallelogram *ABCD* that line segment $\overline{AB}$ is one unit *above* the rotation point *P*, but in the rotated parallelogram *A'B'C'D'*, line segment $\overline{A'B'}$ is 1 unit to the *left* of rotation point *P*.

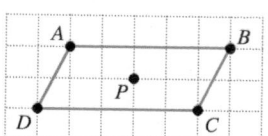

Figure 8.66

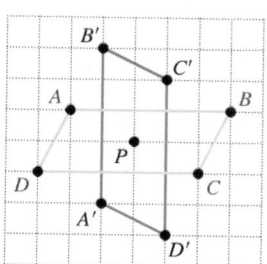

Figure 8.67

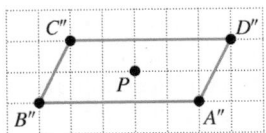

Figure 8.68

b) To visualize the parallelogram obtained through a 180° rotation, pick up this book or a printed page and rotate it 180° in the counterclockwise direction. Notice from Fig. 8.68 that vertex A'' of the rotated parallelogram is in the same position as vertex C of the original parallelogram. In fact, each of the vertices of the rotated parallelogram $A''B''C''D''$ is in the same position as a different vertex in the original parallelogram $ABCD$. From Fig. 8.68 we see that other than vertex labels, the position of rotated parallelogram $A''B''C''D''$ is the same as the position of the original parallelogram $ABCD$. ∎

The parallelogram used in Example 5 will be discussed again later when we address *rotational symmetry*. The three rigid motions we have discussed thus far are reflection, translation, and rotation. Now we discuss the fourth rigid motion, *glide reflection*.

Glide Reflections

> **Definition: Glide Reflection**
> A **glide reflection** is a rigid motion formed by performing a *translation* (or *glide*) followed by a *reflection*. In a glide reflection, the reflection line and the translation vector must be parallel to each other.

Consider triangle ABC (shown in blue), translation vector **v**, and reflection line l in Fig. 8.69. Note that the translation vector **v** and reflection line l are parallel to each other. The translation of triangle ABC using translation vector **v** is triangle $A'B'C'$ (shown in red). The reflection of triangle $A'B'C'$ about reflection line l is triangle $A''B''C''$ (shown in green). Thus, triangle $A''B''C''$ *is the glide reflection* of triangle ABC, using translation vector **v** and reflection line l.

Notice from Fig. 8.69 that had we performed the reflection first, followed by the translation, the triangle $A''B''C''$ would still end up in the same final position. In a glide reflection, the order in which the translation and the reflection are performed does not matter.

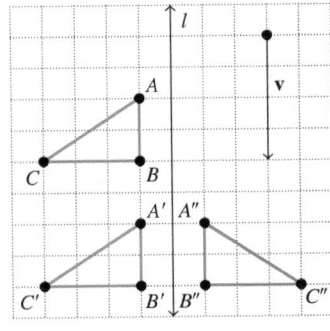

Figure 8.69

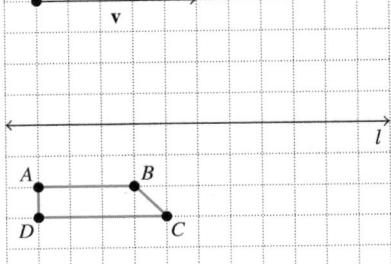

Figure 8.70

┌ Example **6** *A Glide Reflection of a Trapezoid*

Construct a glide reflection of trapezoid $ABCD$, given in Fig. 8.70, using translation vector **v** and reflection line l.

Symmetry in Nature

Symmetry can be found every-where in nature. One type of symmetry in nature is reflective symmetry, or *bilateral* symmetry. For example, if you draw a line down the center of a maple leaf, you will often find that one half has the same shape as the other half.

Rotational symmetry, or *radial symmetry*, is also found in nature. Starfish, sand dollars, and many flowers all display rotational sym-metry. For example, if you rotate a daisy 90°, 180°, or 270°, the rotated daisy will look identical to the origi-nal daisy. There are many other examples of symmetry in nature. In Exercise 56 on page 510 you are asked to find other examples of reflective symmetry and rotational symmetry.

Solution To construct the glide reflection of trapezoid *ABCD*, first translate the trapezoid 5 units to the right, as indicated by translation vector **v**. The translated trapezoid is labeled $A'B'C'D'$, shown in red in Fig. 8.71(a). Next, reflect trapezoid $A'B'C'D'$ about reflection line *l*. This reflected trapezoid, labeled $A''B''C''D''$, is shown in green in Fig. 8.71(b). The glide reflection of trapezoid *ABCD* is trapezoid $A''B''C''D''$.

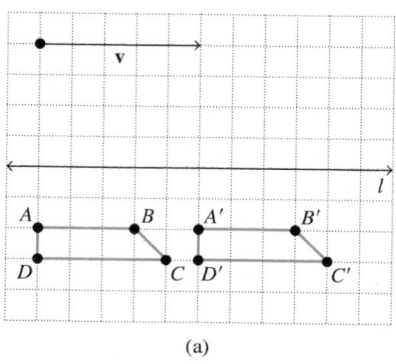

(a)

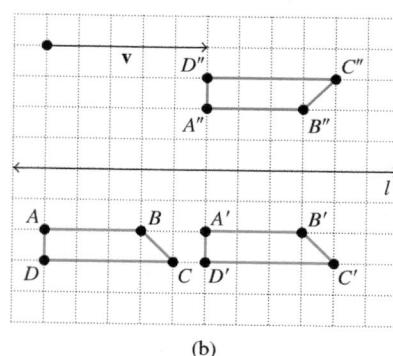

(b)

Figure 8.71

There's a Geometry App for That!

Many of the concepts that we discuss throughout this section and throughout this entire chapter can be further explored with the use of an app on your smartphone or tablet com-puter. Several of these apps allow users to construct both two- and three-dimensional ob-jects. Users are then allowed to explore these figures further, including calculating area and perimeter of polygons, area and circumference of circles, and volume and surface area of three-dimensional figures.

Several apps also allow users to explore the transformational geometry that we are studying in this section. Once a two-dimensional figure is constructed, the user is then able to perform reflections, translations, rotations, and glide reflections on the figure. As always, consult with your instructor before using these apps when completing work related to your course.

Symmetry

We are now ready to discuss symmetry. Our discussion of symmetry involves a rigid motion of an object.

Definition: Symmetry

A **symmetry** of a geometric figure is a rigid motion that moves the figure back onto itself. That is, the beginning position and ending position of the figure must be identical.

Suppose we start with a figure in a specific position and perform a rigid motion on this figure. If the position of the figure after the rigid motion is identical to the position of the figure before the rigid motion (if the beginning and ending positions of the figure coincide), then the rigid motion is a symmetry and we say that the figure has symmetry. For a two-dimensional figure, there are four types of symmetries: reflective symmetry, rotational symmetry, translational symmetry, and glide reflective symmetry. In this textbook, however, we will discuss only reflective symmetry and rotational symmetry.

Consider the polygon and reflection line l shown in Fig. 8.72(a). If we use the rigid motion of reflection and reflect the polygon $ABCDEFGH$ about line l, we get polygon $A'B'C'D'E'F'G'H'$. Notice that the ending position of the polygon is identical to the starting position, as shown in Fig. 8.72(b). Compare Fig. 8.72(a) with Fig. 8.72(b). Although the vertex labels are different, the reflected polygon is in the same position as the polygon in the original position. Thus, we say that the polygon has *reflective symmetry* about line l. We refer to line l as a *line of symmetry*. Notice the line of symmetry is also the reflection line.

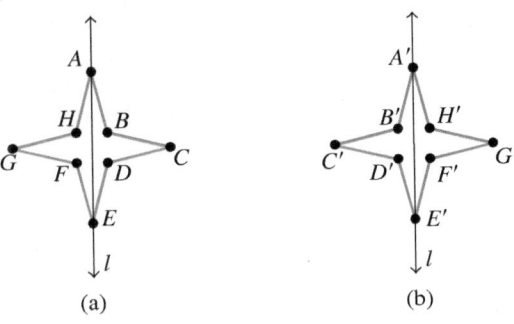

Figure 8.72

Recall Example 2 on pages 493 and 494, in which hexagon $ABCDEF$ was reflected about reflection line l. Examine the hexagon in the original position (Fig. 8.51) and the hexagon in the final position after being reflected about line l (Fig. 8.52). Other than the labels of the vertices, the beginning and ending positions of the hexagon are identical. Therefore, hexagon $ABCDEF$ has reflective symmetry about line l.

Example 7 *Reflective Symmetries of Polygons*

Determine whether the polygon shown in Fig. 8.73 has reflective symmetry about each of the following lines.

a) Line l b) Line m

Solution

a) Examine the reflection of the polygon about line l as seen in Fig. 8.74(a) on page 502. Notice that other than the vertex labels, the beginning and ending positions of the polygon are identical. Thus, the polygon has reflective symmetry about line l.

b) Examine the reflection of the polygon about line m as seen in Fig. 8.74(b). Notice that the position of the reflected polygon is different from the original position of the polygon. Thus, the polygon does not have reflective symmetry about line m.

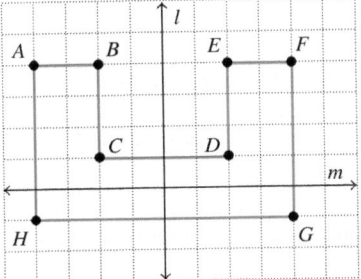

Figure 8.73

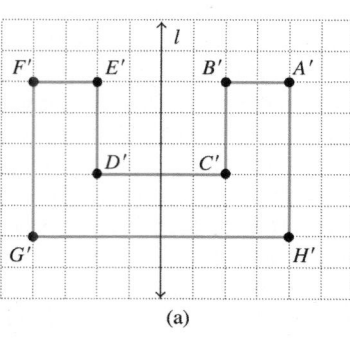
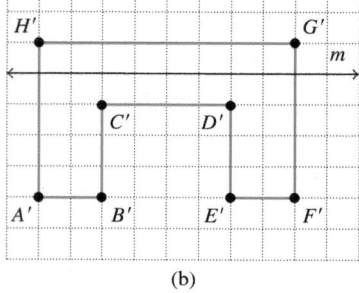

(a)   (b)

Figure 8.74

We will now discuss a second type of symmetry, rotational symmetry. Consider the polygon and rotation point P shown in Fig. 8.75(a). The rigid motion of rotation of polygon $ABCDEFGH$ through a 90° angle about point P gives polygon $A'B'C'D'E'F'G'H'$ shown in Fig. 8.75(b). Compare Fig. 8.75(a) with Fig. 8.75(b). Although the vertex labels are different, the position of the polygon before and after the rotation is identical. Thus, we say that the polygon has 90° *rotational symmetry* about point P. We refer to point P as the *point of symmetry*.

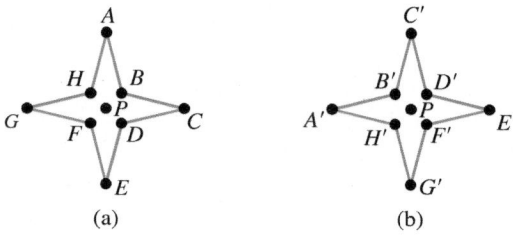

(a)   (b)

Figure 8.75

Recall Example 5 from page 498, in which parallelogram $ABCD$ was rotated 90° about point P in part (a) and 180° in part (b). First examine the parallelogram in the original position in Fig. 8.66 and the 90° rotated parallelogram in Fig. 8.67. Notice that the position of the parallelogram after the 90° rotation is different than the original position of the parallelogram. Therefore, parallelogram $ABCD$ in Fig. 8.66 does not have 90° rotational symmetry about point P. Next, examine the 180° rotated parallelogram in Fig. 8.68. Notice that other than the vertex labels, the positions of the two parallelograms $ABCD$ and $A'B'C'D'$ are identical with respect to rotation about point P. Therefore, parallelogram $ABCD$ in Fig. 8.66 has 180° rotational symmetry about point P.

Example 8 *Rotational Symmetries*

Determine whether the polygon shown in Fig. 8.76 has rotational symmetry about point P for rotations through each of the following angles.

a) 90°   b) 180°

Solution

a) To determine whether the polygon has 90° counterclockwise rotational symmetry about point P, we rotate the polygon 90° as shown in Fig. 8.77(a) on page 503. Compare Fig. 8.77(a) with Fig. 8.76. Notice that the position of the polygon after the rotation in Fig. 8.77(a) is different than the original position of the polygon (Fig. 8.76). Therefore, the polygon does not have 90° rotational symmetry about point P.

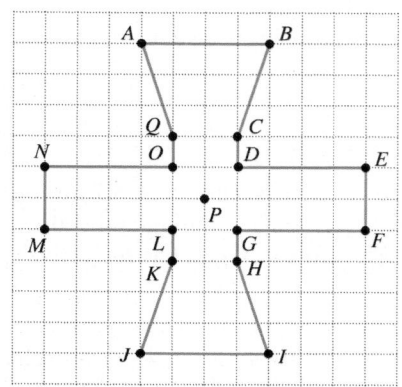

Figure 8.76

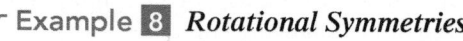

Exploring Reflective
and Rotational Symmetries

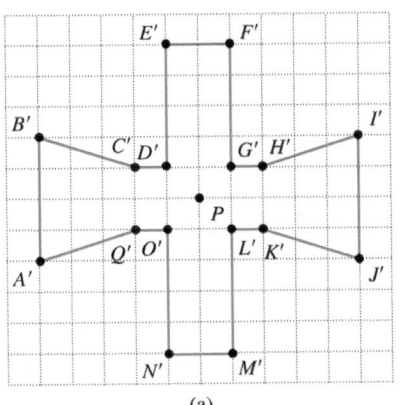
(a)

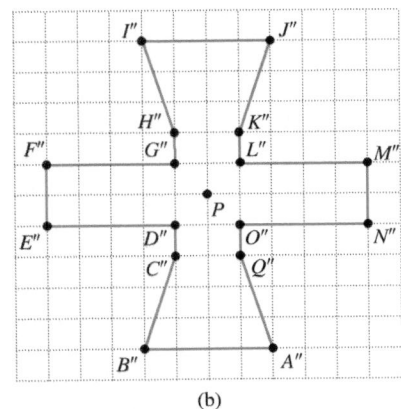
(b)

Figure 8.77

b) To determine whether the polygon has 180° counterclockwise rotational symmetry about the point P, we rotate the polygon 180° as shown in Fig. 8.77(b). Compare Fig. 8.77(b) with Fig. 8.76. Notice that other than vertex labels, the position of the polygon after the rotation in Fig. 8.77(b) is identical to the position of the polygon before the rotation (Fig. 8.76). Therefore, the polygon has 180° rotational symmetry about point P. ∎

Tessellations

A fascinating application of transformational geometry is the creation of *tessellations*.

> **Definition: Tessellation or Tiling**
> A **tessellation**, or **tiling**, is a pattern consisting of the repeated use of the same geometric figures to entirely cover a plane, leaving no gaps. The geometric figures used are called the **tessellating shapes** of the tessellation.

Figure 8.78 shows an example of a tessellation from ancient Egypt. Perhaps the most famous person to incorporate tessellations into his work is M. C. Escher (*see Profiles in Mathematics*).

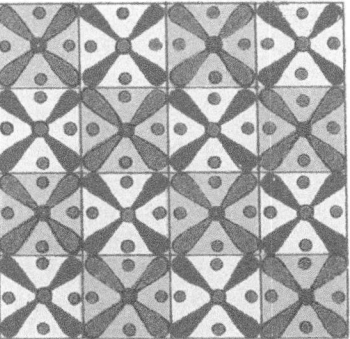

Figure 8.78

The simplest tessellations use one single regular polygon as the tessellating shape. Recall that a *regular polygon* is one whose sides are all the same length and whose interior angles all have the same measure. A tessellation that uses one single regular polygon as the tessellating shape is called a *regular tessellation*. It can be shown that only three regular tessellations exist: those that use an equilateral triangle, a square, or a regular hexagon as the tessellating shape. Figure 8.79 on page 504 shows each of these regular tessellations. Notice that each tessellation can be obtained from a single tessellating shape through the use of reflections, translations, or rotations.

Exploring Tessellations

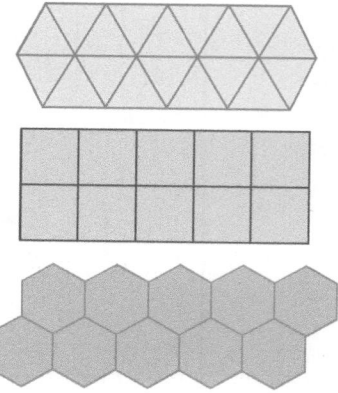

Figure 8.79

We will now learn how to create unique tessellations. We will do so by constructing a unique tessellating shape from a square. We could also construct other tessellating shapes, using an equilateral triangle or a regular hexagon. If you wish to follow along with our construction, you will need some lightweight cardboard, a ruler, cellophane tape, and a pair of scissors. We will start by measuring and cutting out a 2 in. by 2 in. square from the cardboard. We next cut the square into two parts by cutting it from top to bottom using any kind of cut. One example is shown in Fig. 8.80. We then rearrange the pieces and tape the two vertical edges together as shown in Fig. 8.81. Next we cut this new shape into two parts by cutting it from left to right using any kind of cut as shown in Fig. 8.82. We then rearrange the pieces and tape the two horizontal edges together as shown in Fig. 8.83. This completes our tessellating shape.

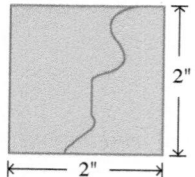

2"

2"

Figure 8.80

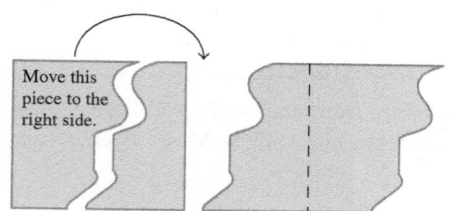

Move this piece to the right side.

Figure 8.81

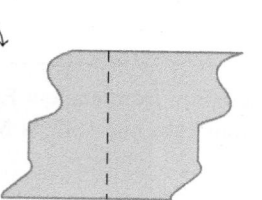

Figure 8.82

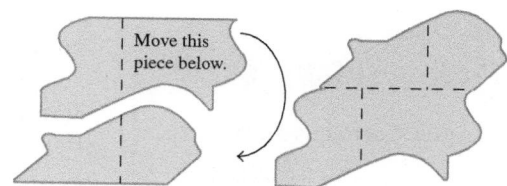

Move this piece below.

Figure 8.83

Figure 8.84

We now set the cardboard tessellating shape in the middle of a blank piece of paper (the tessellating shape can be rotated to any position as a starting point) and trace the outline of the shape onto the paper. Next move the tessellating shape so that it lines up with the figure already drawn and trace the outline again. Continue to do that until the page is completely covered. Once the page is covered with the tessellation, we can add some interesting colors or even some unique sketches to the tessellation. Figure 8.84 shows one tessellation created using the tessellation shape in Fig. 8.83. In Fig. 8.84, the tessellation shape was rotated about 45° counterclockwise.

An infinite number of different tessellations can be created using the method described by altering the cuts made. We could also create different tessellations using an equilateral triangle, a regular hexagon, or other types of polygons. There are also other, more complicated ways to create the tessellating shape. The Internet has many sites devoted to the creation of tessellations by hand. Many computer programs that generate tessellations are also available.

SECTION 8.5   *Exercises*

Warm Up Exercises

In Exercises 1–8, fill in the blanks with an appropriate word, phrase, or symbol(s).

1. A rigid motion that moves a geometric figure to a new position such that the new position is a mirror image of the figure in the starting position is called a(n) _____.

2. A rigid motion that moves a geometric figure by sliding it along a straight line segment in the plane is called a(n) _____.

3. A concise way to indicate the direction and the distance that a figure is moved during a translation is with a translation _____.

4. A rigid motion performed by rotating a geometric figure in the plane about a specific point is called a(n) _____.

5. The angle through which a geometric figure is rotated during a rotation is called the angle of _____.

6. A rigid motion formed by performing a translation followed by a reflection is called a(n) _____ reflection.

7. A rigid motion that moves the geometric figure back onto itself is called a(n) _____.

8. A pattern consisting of the repeated use of the same geometric figures to entirely cover a plane, leaving no gaps, is called a(n) _____.

Practice the Skills/Problem Solving

In Exercises 9–16, use the given figure and lines of reflection to construct the indicated reflections. Show the figure in the positions both before and after the reflection.

In Exercises 9 and 10, use the following figure. Construct

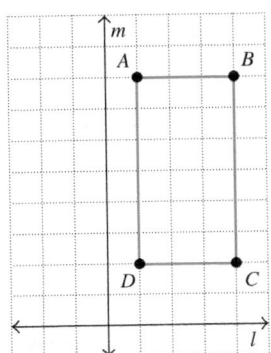

9. the reflection of rectangle *ABCD* about line *l*.

10. the reflection of rectangle *ABCD* about line *m*.

In Exercises 11 and 12, use the following figure. Construct

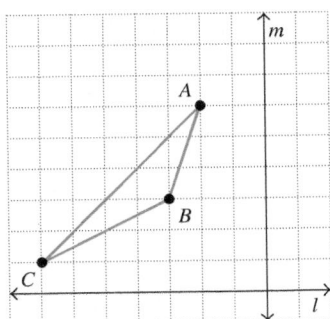

11. the reflection of triangle *ABC* about line *l*.

12. the reflection of triangle *ABC* about line *m*.

In Exercises 13 and 14, use the following figure. Construct

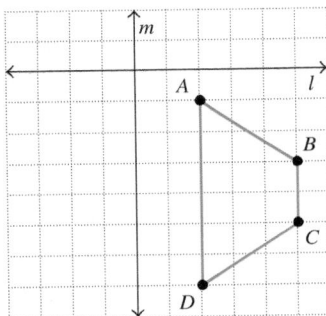

13. the reflection of trapezoid *ABCD* about line *m*.

14. the reflection of trapezoid *ABCD* about line *l*.

In Exercises 15 and 16, use the following figure. Construct

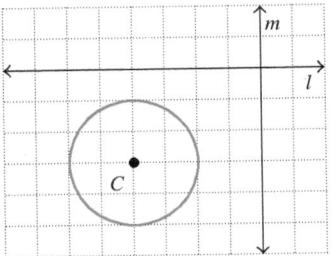

15. the reflection of circle *C* about line *m*.

16. the reflection of circle *C* about line *l*.

In Exercises 17–24, use the translation vectors **v** *and* **w,** *to construct the translations indicated in the exercises. Show the figure in the positions both before and after the translation.*

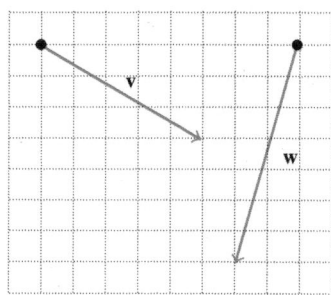

In Exercises 17 and 18, use the following figure. Construct

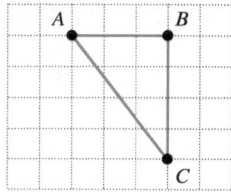

17. the translation of triangle *ABC* using translation vector **v.**

18. the translation of triangle *ABC* using translation vector **w.**

In Exercises 19 and 20, use the following figure. Construct

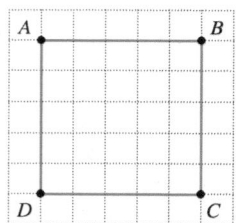

19. the translation of square *ABCD* using translation vector **w.**

20. the translation of square *ABCD* using translation vector **v.**

In Exercises 21 and 22, use the following figure. Construct

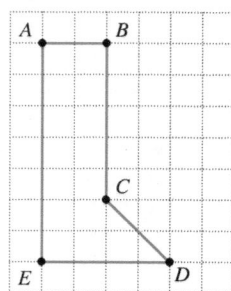

21. the translation of polygon *ABCDE* using translation vector **v.**

22. the translation of polygon *ABCDE* using translation vector **w.**

In Exercises 23 and 24, use the following figure. Construct

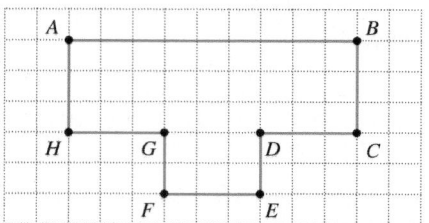

23. the translation of polygon *ABCDEFGH* using translation vector **w.**

24. the translation of polygon *ABCDEFGH* using translation vector **v.**

In Exercises 25–32, use the given figure and rotation point P to construct the indicated rotations. Show the figure in the positions both before and after the rotation.

In Exercises 25 and 26, use the following figure. Construct

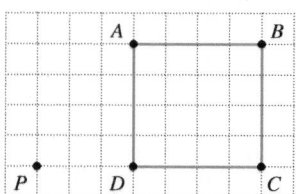

25. a 90° rotation of square *ABCD* about point *P.*

26. a 180° rotation of square *ABCD* about point *P.*

In Exercises 27 and 28, use the following figure. Construct

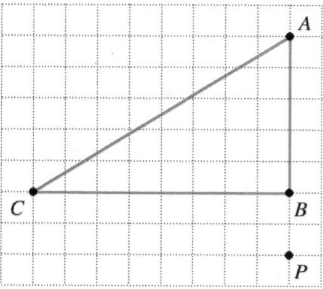

27. a 180° rotation of triangle *ABC* about point *P.*

28. a 270° rotation of triangle *ABC* about point *P.*

In Exercises 29 and 30, use the following figure. Construct

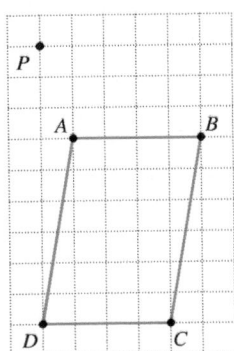

29. a 270° rotation of parallelogram *ABCD* about point *P*.

30. a 180° rotation of parallelogram *ABCD* about point *P*.

In Exercises 31 and 32, use the following figure. Construct

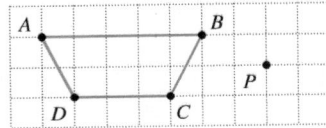

31. a 90° rotation of trapezoid *ABCD* about point *P*.

32. a 270° rotation of trapezoid *ABCD* about point *P*.

In Exercises 33– 40, use the given figure, translation vectors **v** *and* **w**, *and reflection lines l and m to construct the indicated glide reflections. Show the figure in the positions both before and after the glide reflection.*

In Exercises 33 and 34, use the following figure. Construct

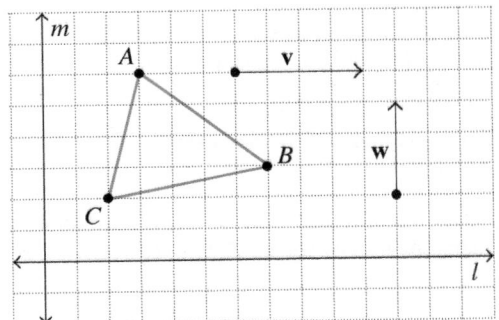

33. a glide reflection of triangle *ABC* using vector **v** and reflection line *l*.

34. a glide reflection of triangle *ABC* using vector **w** and reflection line *m*.

In Exercises 35 and 36, use the following figure. Construct

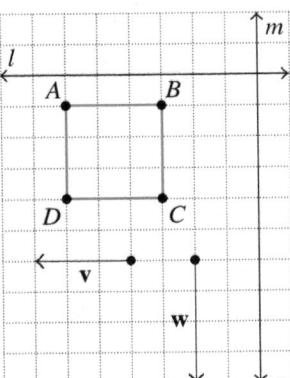

35. a glide reflection of square *ABCD* using vector **w** and reflection line *m*.

36. a glide reflection of square *ABCD* using vector **v** and reflection line *l*.

In Exercises 37 and 38, use the following figure. Construct

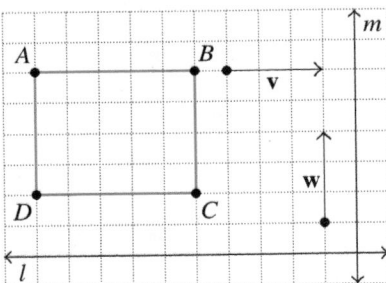

37. a glide reflection of rectangle *ABCD* using vector **v** and reflection line *l*.

38. a glide reflection of rectangle *ABCD* using vector **w** and reflection line *m*.

In Exercises 39 and 40, use the following figure. Construct

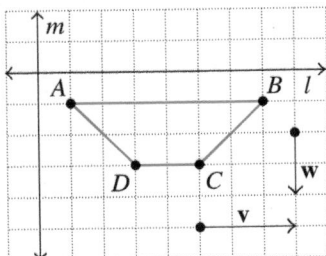

39. a glide reflection of trapezoid *ABCD* using vector **w** and reflection line *m*.

40. a glide reflection of trapezoid *ABCD* using vector **v** and reflection line *l*.

41. a) Reflect triangle *ABC*, shown below, about line *l*. Label the reflected triangle *A'B'C'*.

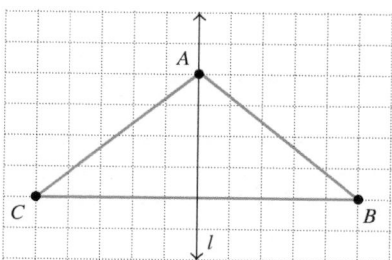

b) Other than vertex labels, is the position of triangle *A'B'C'* identical to the position of triangle *ABC*?

c) Does triangle *ABC* have reflective symmetry about line *l*?

42. a) Reflect quadrilateral *ABCD*, shown below, about line *l*. Label the reflected quadrilateral *A'B'C'D'*.

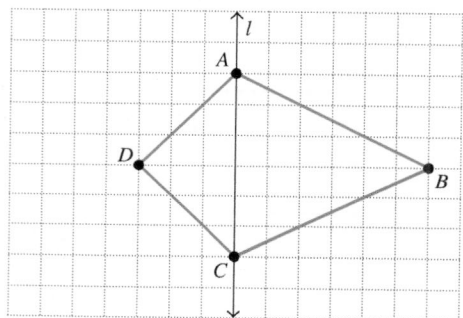

b) Other than vertex labels, is the position of quadrilateral *A'B'C'D'* identical to the position of quadrilateral *ABCD*?

c) Does quadrilateral *ABCD* have reflective symmetry about line *l*?

43. a) Reflect parallelogram *ABCD*, shown below, about line *l*. Label the reflected parallelogram *A'B'C'D'*.

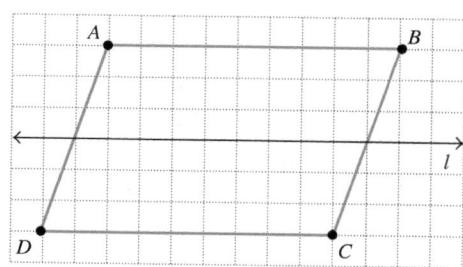

b) Other than vertex labels, is the position of parallelogram *A'B'C'D'* identical to the position of parallelogram *ABCD*?

c) Does parallelogram *ABCD* have reflective symmetry about line *l*?

44. a) Reflect pentagon *ABCDE*, shown below, about line *l*. Label the reflected pentagon *A'B'C'D'E'*.

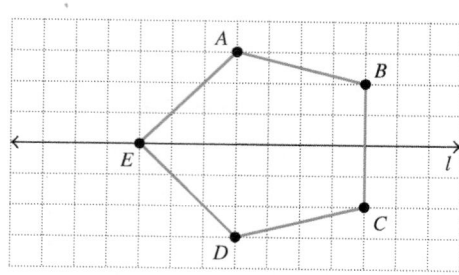

b) Other than vertex labels, is the position of pentagon *A'B'C'D'E'* identical to the position of pentagon *ABCDE*?

c) Does pentagon *ABCDE* have reflective symmetry about line *l*?

45. a) Rotate rectangle *ABCD*, shown below, 90° about point *P*. Label the rotated rectangle *A'B'C'D'*.

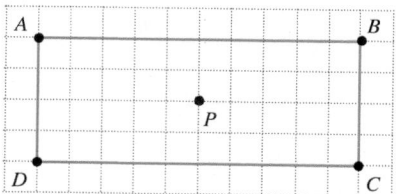

b) Other than vertex labels, is the position of rectangle *A'B'C'D'* identical to the position of rectangle *ABCD*?

c) Does rectangle *ABCD* have 90° rotational symmetry about point *P*?

d) Now rotate the rectangle in the original position, rectangle *ABCD*, 180° about point *P*. Label the rotated rectangle *A"B"C"D"*.

e) Other than vertex labels, is the position of rectangle *A"B"C"D"* identical to the position of rectangle *ABCD*?

f) Does rectangle *ABCD* have 180° rotational symmetry about point *P*?

46. a) Rotate square *ABCD*, shown below, 90° about point *P*. Label the rotated square *A'B'C'D'*.

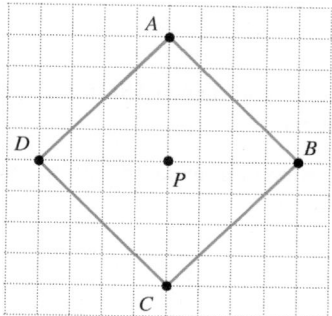

b) Other than vertex labels, is the position of square $A'B'C'D'$ identical to the position of square $ABCD$?

c) Does square $ABCD$ have 90° rotational symmetry about point P?

d) Now rotate the square in the original position, square $ABCD$, shown above, 180° about point P. Label the rotated square $A''B''C''D''$.

e) Other than vertex labels, is the position of square $A''B''C''D''$ identical to the position of square $ABCD$?

f) Does square $ABCD$ have 180° rotational symmetry about point P?

47. Consider the following figure.

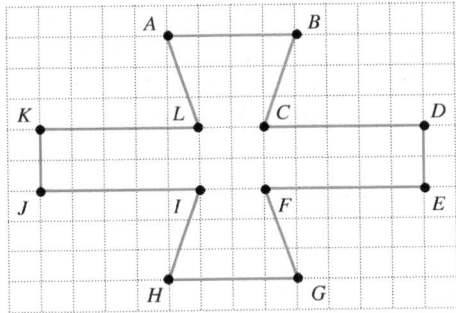

a) Insert a vertical line m through the figure so the figure has reflective symmetry about line m.

b) Insert a horizontal line l through the figure so the figure has reflective symmetry about line l.

c) Insert a point P within the figure so the figure has 180° rotational symmetry about point P.

d) Is it possible to insert a point P within the figure so the figure has 90° rotational symmetry about point P? Explain your answer.

48. Consider the following figure.

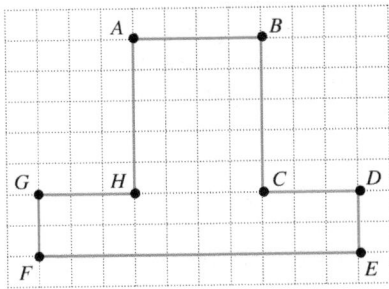

a) Insert a vertical line m through the figure so the figure has reflective symmetry about line m.

b) Is it possible to insert a horizontal line l through the figure so the figure has reflective symmetry about line l? Explain your answer.

c) Is it possible to insert a point P within the figure so the figure has 90° rotational symmetry about point P? Explain your answer.

d) Is it possible to insert a point P within the figure so the figure has 180° rotational symmetry about point P? Explain your answer.

Challenge Problems/Group Activities

49. *Tessellation with a Square* Create a unique tessellation from a square piece of cardboard by using the method described on page 504. Be creative, using color and sketches to complete your tessellation.

50. *Tessellation with a Hexagon* Using the method described on page 504, create a unique tessellation using a regular hexagon like the one shown below. Be creative, using color and sketches to complete your tessellation.

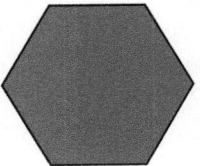

51. *Tessellation with an Octagon?*

Trace the regular octagon, shown below, onto a separate piece of paper.

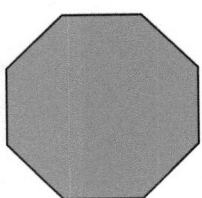

Try to create a regular tessellation by tracing this octagon repeatedly. Attempt to cover the entire piece of paper where no two octagons overlap each other. What conclusion can you draw about using a regular octagon as a tessellating shape?

52. *Tessellation with a Pentagon?* Repeat Exercise 51 using the regular pentagon below instead of a regular octagon.

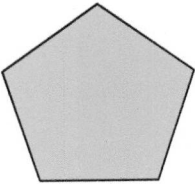

Recreational Mathematics

53. *Reflective Symmetry* Examine each capital letter in the alphabet and determine which letters have reflective symmetry about a horizontal line through the center of the letter.

54. *Reflective Symmetry* Examine each capital letter in the alphabet and determine which letters have reflective symmetry about a vertical line through the center of the letter.

55. *Rotational Symmetry* Examine each capital letter in the alphabet and determine which letters have 180° rotational symmetry about a point in the center of the letter.

Research Activities

56. *Bilateral and Radial Symmetry* In the study of biology, reflective symmetry is called *bilateral symmetry* and rotational symmetry is called *radial symmetry*. Write a report on the role symmetry plays in the study of biology.

57. *M. C. Escher* Write a paper on the mathematics displayed in the artwork of M. C. Escher. Include such topics as tessellations, optical illusions, perspective, and non-Euclidean geometry.

SECTION 8.6   Topology

Upon completion of this section, you will be able to:

- Learn about topological objects including Möbius strips, Klein bottles, maps, and Jordan curves.

- Understand the topological equivalence of two or more objects.

Examine the outline of the map of the continental United States shown on the left. Now suppose you were given four crayons, each of a different color. Could you color this map with the four crayons in a way so that no two bordering states have the same color? In this section, we will discuss this question and many other questions that are relevant to the branch of mathematics known as *topology*.

Why This Is Important When combined with other branches of mathematics, topology plays an important role in mathematical and scientific research. Advanced research in probability, statistics, measurement theory, game theory, and mathematical computing relies on the basic concepts of topology. Research in computer science, biology, genetics, and robotics depends on topology to connect many abstract concepts. Some of the products in our world that were developed with the aid of topology include Global Positioning Systems (GPS), Facebook, Twitter, and many of the other applications that we use with our smartphones.

Topology

The branch of mathematics called *topology* is the study of geometric properties and spatial relations unaffected by the continuous change of shape or size of figures. Topology is sometimes referred to as "rubber sheet geometry" because it deals with the bending and stretching of geometric figures. In this section, we will study several interesting objects and how topology is used to study these objects.

Möbius Strip

One of the first pioneers of topology was the German astronomer and mathematician August Ferdinand Möbius (1790–1866). A student of Gauss (see *Profile in Mathematics* on page 269), Möbius was the director of the University of Leipzig's observatory. He spent a great deal of time studying geometry and he played an essential part in the systematic development of projective geometry. He is best known for his studies of the properties of one-sided surfaces, including the one called the Möbius strip.

If you place a pencil on one surface of a sheet of paper and do not remove it from the sheet, you must cross the edge to get to the other surface. Thus, a sheet of paper has one edge and two surfaces. The sheet retains these properties even when crumpled

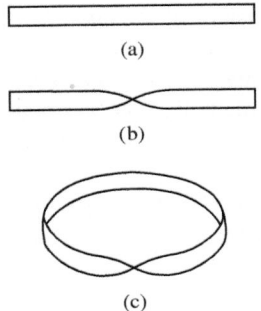

(a)

(b)

(c)

Figure 8.85

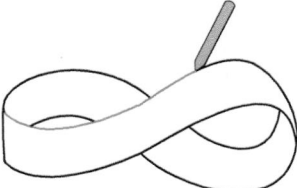

Figure 8.86

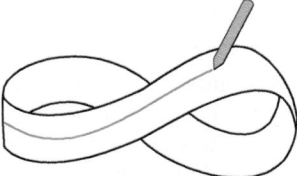

Figure 8.87

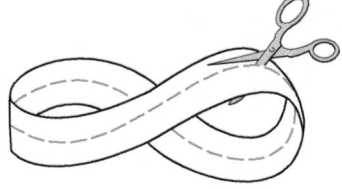

Figure 8.88

Figure 8.89

into a ball. The *Möbius strip*, also called a *Möbius band*, is a one-sided, one-edged surface. You can construct one, as shown in Fig. 8.85, by (a) taking a strip of paper, (b) giving one end a half twist, and (c) taping the ends together.

The Möbius strip has some very interesting properties. To better understand these properties, perform the following experiments.

Experiment 1 Make a Möbius strip using a strip of paper and tape as illustrated in Fig. 8.85. Place the point of a felt-tip pen on the edge of the strip (Fig. 8.86). Pull the strip slowly so that the pen marks the edge; do not remove the pen from the edge. Continue pulling the strip and observe what happens.

Experiment 2 Make a Möbius strip. Place the tip of a felt-tip pen on the surface of the strip (Fig. 8.87). Pull the strip slowly so that the pen marks the surface. Continue and observe what happens.

Experiment 3 Make a Möbius strip. Use scissors to make a small slit in the middle of the strip. Starting at the slit, cut along the strip, keeping the scissors in the middle of the strip (Fig. 8.88). Continue cutting and observe what happens.

Experiment 4 Make a Möbius strip. Make a small slit at a point about one-third of the width of the strip. Cut along the strip, keeping the scissors the same distance from the edge (Fig. 8.89). Continue cutting and observe what happens.

If you give a strip of paper several half twists, you get variations on the Möbius strip. To a topologist, the important distinction is between an odd number of half twists, which leads to a one-sided surface, and an even number of half twists, which leads to a two-sided surface. All strips with an odd number of half twists are topologically the same as a Möbius strip, and all strips with an even number of half twists are topologically the same as an ordinary cylinder without the top and bottom, which has no twists.

Klein Bottle

Another topological object is the punctured *Klein bottle*; see Fig. 8.90. This object, named after German mathematician Felix Klein (1849–1925), resembles a bottle but its surface has only one side.

A punctured Klein bottle can be made by stretching a hollow piece of glass tubing. The neck is then passed through a hole and joined to the base.

Look closely at the model of the Klein bottle shown in Fig. 8.90. The punctured Klein bottle has only one surface and no outside or inside because its surface has just one side. Figure 8.91 shows a Klein bottle blown in glass by Alan Bennett of Bedford, England.

Limericks from unknown writers:
"A mathematician confided
That a Möbius band is one-sided,
And you'll get quite a laugh
If you cut one in half
For it stays in one piece when divided."

"A mathematician named Klein
Thought the Möbius band was divine.
He said, 'If you glue
the edges of two
You'll get a weird bottle like mine.'"

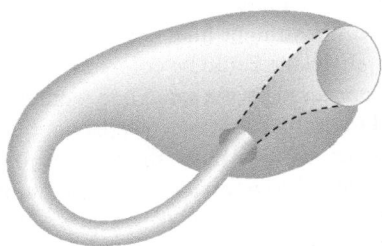

Figure 8.90

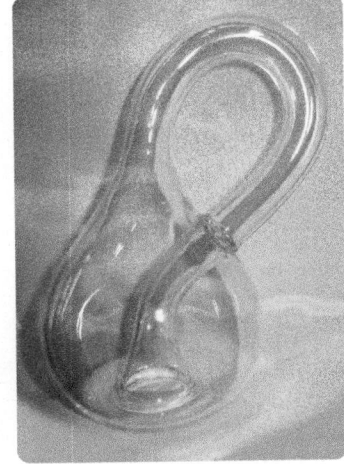

Figure 8.91 *Klein bottle,* a one-sided surface, blown in glass by Alan Bennett.

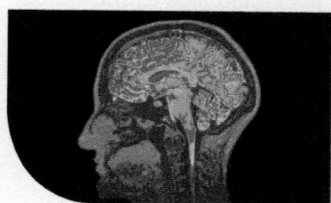
Imagine trying to paint a Klein bottle. You start on the "outside" of the large part and work your way down the narrowing neck. When you cross the self-intersection, you have to pretend temporarily that it is not there, so you continue to follow the neck, which is now inside the bulb. As the neck opens up, to rejoin the bulb, you find that you are now painting the inside of the bulb! What appear to be the inside and outside of a Klein bottle connect together seamlessly, since it is one-sided.

If a Klein bottle is cut along a curve, the results are two Möbius strips; see Fig. 8.92. Thus, a Klein bottle could also be made by gluing together two Möbius strips along the edges.

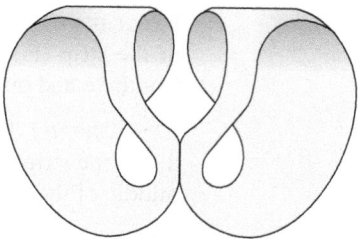

Figure 8.92 Two Möbius strips result from cutting a Klein bottle along a curve.

Maps

Mapmakers have known for a long time that regardless of the complexity of the map and whether it is drawn on a flat surface or a sphere, only four colors are needed to differentiate each country (or state) from its immediate neighbors. Thus, every map can be drawn by using only four colors, and no two countries with a common border will have the same color. Regions that meet at only one point (such as the states of Arizona, Colorado, Utah, and New Mexico) are not considered to have a common border. In Fig. 8.93(a), no two states with a common border are marked with the same color.

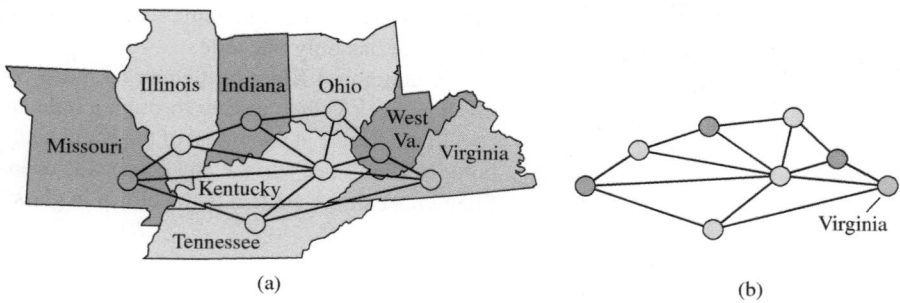

Figure 8.93

The four-color problem was first suggested by a student of Augustus De Morgan in 1852. In 1976, Kenneth Appel and Wolfgang Haken of the University of Illinois—using their ingenuity, logic, and 1200 hours of computer time—succeeded in proving that only four colors are needed to draw a map. They solved the four-color map problem by reducing any map to a series of points and connecting line segments. They replaced each country with a point. They connected two countries having a common border with a straight line; see Fig. 8.93(b). They then showed that the points of any graph in the plane could be colored by using only four colors in such a way that no two points connected by the same line were the same color. The four-color problem is now referred to as the *four-color theorem*.

Mathematicians have shown that, on different surfaces, more than four colors may be needed to draw a map. For example, a map drawn on a Möbius strip requires a maximum of six colors, as in Fig. 8.94(a) on page 513. A map drawn on a torus (the shape of a doughnut) requires a maximum of seven colors, as in Fig. 8.94(b).

In addition to the Möbius strip, many other interesting surfaces can be made using paper, scissors, and cellophane tape. Shown above is a surface that shares some of the same characteristics as a Möbius strip, but is not topologically equivalent to a Möbius strip. Notice that, like a Möbius strip, this surface has 1 side and 1 edge.

Construct the surface shown above using two strips of paper, scissors, and tape. Attempt to cut the surface "in half" by making a small slit along a dashed line in the middle of the paper surface. Then cut along the dashed line shown in the figure, keeping the scissors the same distance from the edge. What happens? Exercise 43 on page 516 describes another topological construction.

Many Internet websites are devoted to topological constructions. One such website demonstrates a Möbius strip made with a zipper. This process allows you to "cut" the Möbius strip and then put the two parts back together again.

Four Color Challenge

(a)

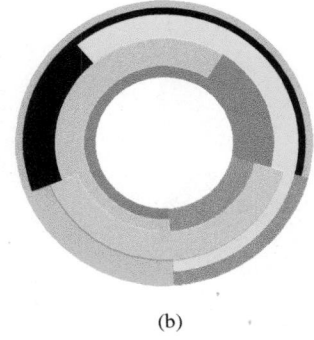

(b)

Figure 8.94

Jordan Curves

A *Jordan curve* is a topological object that can be thought of as a circle twisted out of shape; see Fig. 8.95 (a)–(d). Like a circle, a Jordan curve has an inside and an outside. To get from a point on the inside of the curve to a point on the outside of the curve, or vice versa, at least one edge of the curve must be crossed. Consider the Jordan curve in Fig. 8.95(d). In our next example, we will determine whether points A and B are inside or outside the Jordan curve.

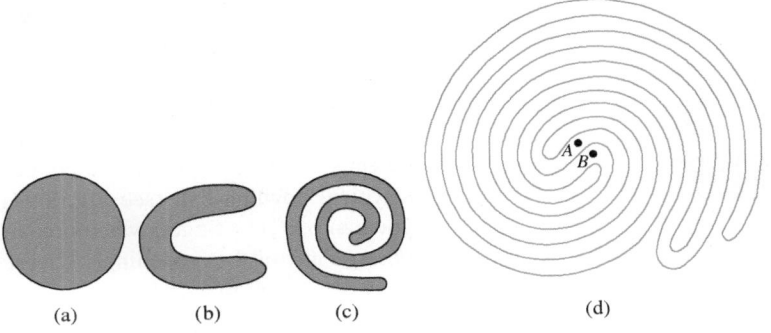

(a)　　(b)　　(c)　　(d)

Figure 8.95

Example 1 *A Jordan Curve*

Consider the Jordan curve in Fig. 8.95(d).

a) Is point A on the inside or the outside of the Jordan curve?
b) Is point B on the inside or the outside of the Jordan curve?

Solution

a) To determine whether a point is inside or outside a Jordan curve, draw a straight line from the point in question to a point that is clearly outside the Jordan curve. In the figure below, we draw two such lines, one from point A and one from point B.

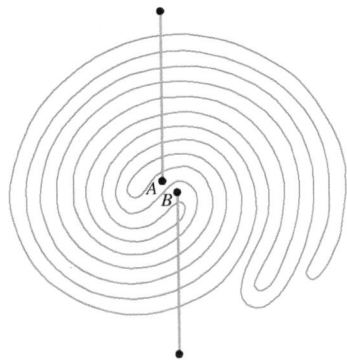

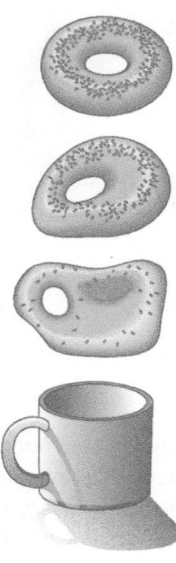

Figure 8.96

Bowling Ball

Figure 8.97

If the line crosses the Jordan curve an even number of times, the point in question is *outside* the Jordan curve. If the line crosses the Jordan curve an odd number of times, the point in question is *inside* the Jordan curve. In the figure, the line connecting point A to a point outside the curve crosses the Jordan curve 10 times. Since 10 is an even number, point A is outside the Jordan curve.

b) In the figure, the line connecting point B to a point outside the curve crosses the Jordan curve 7 times. Since 7 is an odd number, point B is inside the Jordan curve. ∎

Topological Equivalence

Someone once said that a topologist is a person who does not know the difference between a doughnut and a coffee cup. Two geometric figures are said to be *topologically equivalent* if one figure can be elastically twisted, stretched, bent, or shrunk into the other figure without puncturing or ripping the original figure. If a doughnut is made of elastic material, it can be stretched, twisted, bent, shrunk, and distorted until it resembles a coffee cup with a handle, as shown in Fig. 8.96. Thus, the doughnut and the coffee cup are topologically equivalent.

In topology, figures are classified according to their *genus*. The *genus* of an object is determined by the number of holes that go *through* the object. A cup with a handle and a doughnut each have one hole and are of genus 1 (and are therefore topologically equivalent). Notice that the cup's handle is considered a hole, whereas the opening at the rim of the cup is not considered a hole. For our purposes, we will consider an object's opening a hole if you could pour liquid *through* the opening. For example, a typical bowling ball (see Fig. 8.97) has three openings in the surface into which you can put your fingers when preparing to roll the ball, but liquid cannot be poured *through* any of these openings. Therefore, a bowling ball has genus 0 and is topologically equivalent to a marble.

Example 2 *Genus*

Name an object that has the given genus.
a) Genus 0   b) Genus 1   c) Genus 2   d) Genus 3 or more

Solution The answers are shown in columns below.

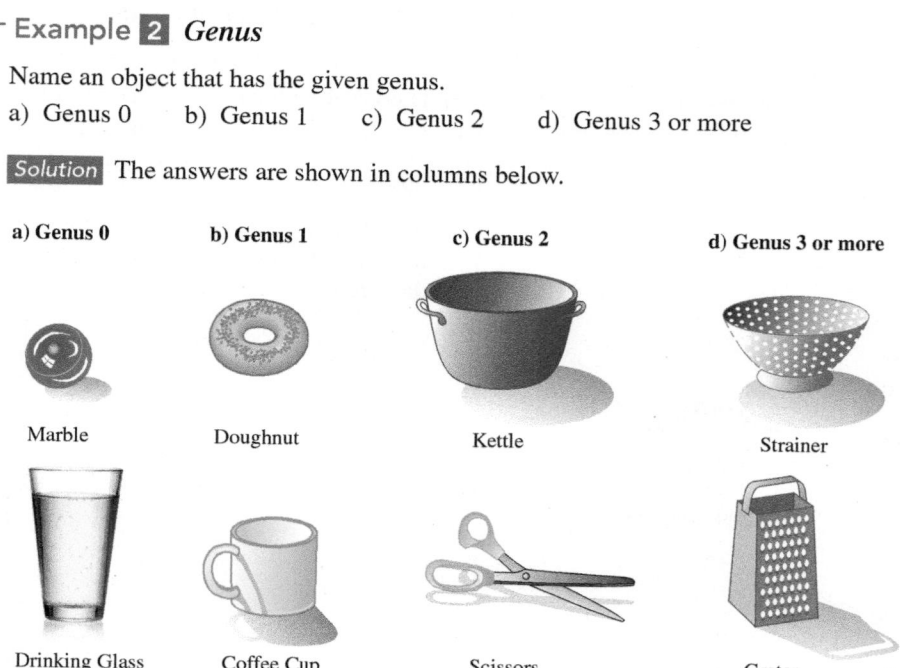

a) Genus 0	b) Genus 1	c) Genus 2	d) Genus 3 or more
Marble	Doughnut	Kettle	Strainer
Drinking Glass	Coffee Cup	Scissors	Grater

∎

SECTION 8.6 Exercises

Warm Up Exercises

In Exercises 1–6, fill in the blanks with an appropriate word, phrase, or symbol(s).

1. Because it deals with bending and stretching of geometric figures, topology is sometimes referred to as _____ sheet geometry.

2. A one-sided, one-edged surface is a(n) _____ strip.

3. A topological object that resembles a bottle but has only one side is a(n) _____ bottle.

4. If you color a map of the United States, the maximum number of colors needed so that no two states that share a common border have the same color is _____.

5. A topological object that can be thought of as a circle twisted out of shape is a(n) _____ curve.

6. The number of holes that go through an object determines the _____ of the object.

Practice the Skills

In Exercises 7–10, color the map by using a maximum of four colors so that no two regions with a common border have the same color.

7.

8.
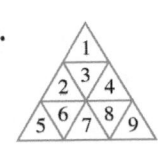

9.
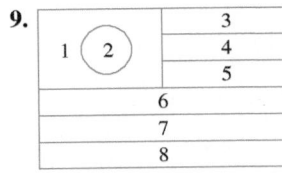

10.

Using the Four-Color Theorem

In Exercises 11–14, maps show certain areas of the United States, Canada, and Mexico. Shade in the states (or provinces) using a maximum of four colors so that no two states (or provinces) with a common border have the same color.

11.

12.

13.

14.

Jordan Curve

In Exercises 15–20, determine if the point is inside or outside the Jordan curve.

15. Point *A*

16. Point *B*

17. Point *C*

18. Point *D*

19. Point *E*

20. Point *F*

Genus

In Exercises 21–32, give the genus of the object. If the object has a genus larger than 5, write "larger than 5."

21.

22.

23.

24.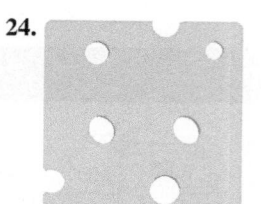

25.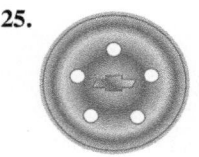

26.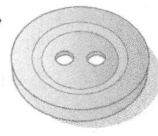

27.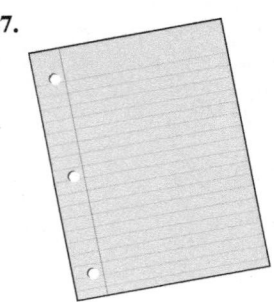

28.

29.

30.

31.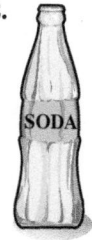

32.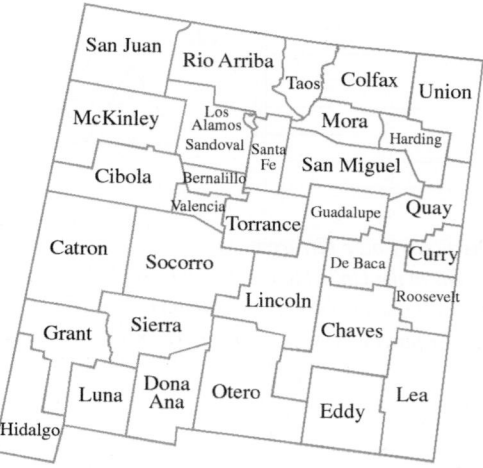

33. Name at least three objects not mentioned in this section that have

   a) genus 0.

   b) genus 1.

   c) genus 2.

   d) genus 3 or more.

34. Use the result of Experiment 1 on page 511 to determine the number of edges on a Möbius strip.

35. Use the result of Experiment 2 on page 511 to determine the number of surfaces on a Möbius strip.

36. How many separate strips are obtained in Experiment 3 on page 511?

37. How many separate strips are obtained in Experiment 4 on page 511?

38. Make a Möbius strip. Cut it one-third of the way from the edge, as in Experiment 4 on page 511. You should get two loops, one going through the other. Determine whether either (or both) of these loops is itself a Möbius strip.

39. a) Take a strip of paper, give it one full twist, and connect the ends. Is the result a Möbius strip with only one side? Explain.

   b) Determine the number of edges, as in Experiment 1.

   c) Determine the number of surfaces, as in Experiment 2.

   d) Cut the strip down the middle. What is the result?

40. Take a strip of paper, make one whole twist and another half twist, and then tape the ends together. Test by a method of your choice to determine whether this has the same properties as a Möbius strip.

Challenge Problems/Group Activities

41. Using clay (or glazing compound), make a doughnut. Without puncturing or tearing the doughnut, reshape it into a topologically equivalent figure, a cup with a handle.

42. Using at most four colors, color the map of the counties of New Mexico, shown below. Do not use the same color for any two counties that share a common border.

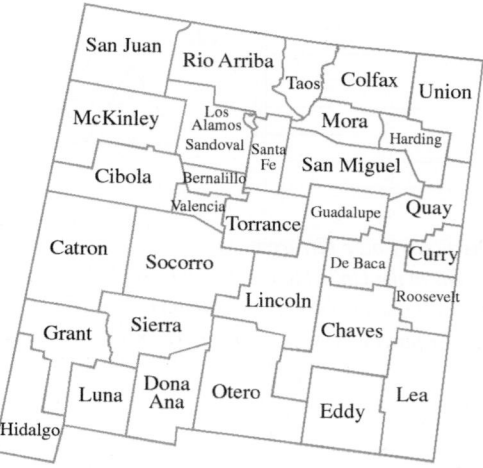

Recreational Math

43. *Topological Paper Constructions* Using paper, scissors, and tape, perform the construction described in the *Recreational Mathematics* box on page 513. Once you have completed the construction, cut along the dashed line as instructed. Set the result aside.

(Continued)

In this exercise we will construct another interesting surface. We begin by constructing a "cross" shape from two strips of paper, as shown below, using scissors and tape. Note the red dashed line and the green dashed line and the ends of the strips labeled A or B.

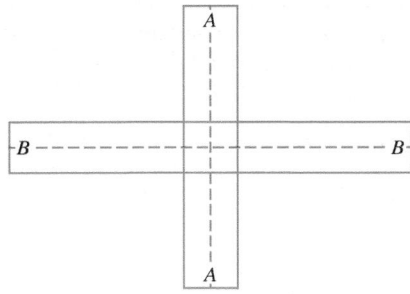

Next, using tape connect the two ends labeled A *without* twisting the ends. Then, connect the two ends labeled B by giving one end a half twist. The strip that connects the B ends should resemble a Möbius strip. Finally, cut the object first along the green dashed line and then along the red dashed line. Compare the result with that from the construction on page 513. What do you notice?

Research Activities

44. *Counties in Your State* Print out a map of the state you live in that shows the outline of all of the counties. Using at most four colors, color the map so that no two counties that share a common border are the same color.

45. *Paul Bunyan* The short story *Paul Bunyan versus the Conveyor Belt* (1947) by William Hazlett Upson focuses on a conveyor belt in the shape of a Möbius strip. The story can be found in several books that include mathematical essays. Read Upson's short story and write a 200-word description of what Paul Bunyan does to the conveyor belt. Confirm the outcome of the story by repeating Paul's actions with a paper Möbius strip.

SECTION 8.7 Non-Euclidean Geometry and Fractal Geometry

Upon completion of this section, you will be able to:

- Describe non-Euclidean geometry including elliptical and hyperbolic geometry.
- Describe fractal geometry.

Consider the following question: Given a line *l* and a point *P* not on the line *l*, how many lines can you draw through *P* that are parallel to *l*?

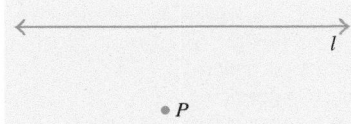

You may answer that only one line may be drawn through *P* parallel to *l*. This answer would be correct *provided* the setting of the problem is in a plane and not on the surface of a curved object. The study of this question led to the development of several new branches of geometry. It is now believed these branches of geometry, taken together, can be used to accurately represent space. In this section, we will study the geometry of surfaces other than the geometry of the plane.

> **Why This Is Important** While Euclidean geometry works well for most applications we encounter on Earth, scientists believe that non-Euclidean geometry is necessary to model the universe. Einstein's *General Theory of Relativity* is based on a theory that space is curved. The non-Euclidean geometry that we discuss in this section is also curved and provides a better scientific basis for studying both time and space. Additionally, non-Euclidean geometry is used in many technology applications including computer storage, medical imaging, and many of our smartphone apps.

Origins of Non-Euclidean Geometry

In Section 8.1, we stated that postulates or axioms are statements to be accepted as true. In his book *Elements*, Euclid's fifth postulate was, "If a straight line falling on two straight lines makes the interior angles on the same side less than two right angles,

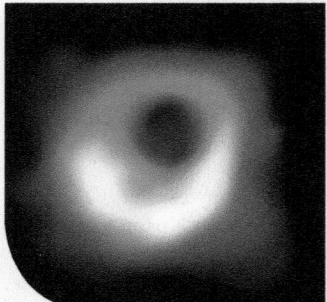
the two straight lines, if produced indefinitely, meet on that side on which the angles are less than the two right angles."

Euclid's fifth postulate may be better understood by observing Fig. 8.98. The sum of angles *A* and *B* is less than the sum of two right angles (180°). Therefore, the two lines will meet if extended.

In 1795, John Playfair (1748–1819), a Scottish physicist and mathematician, gave a logically equivalent interpretation of Euclid's fifth postulate. This version is often referred to as Playfair's postulate or the Euclidean parallel postulate.

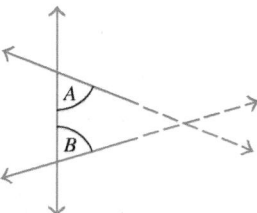

Figure 8.98

The Euclidean Parallel Postulate
Given a line and a point not on the line, one and only one line can be drawn through the given point parallel to the given line (see Fig. 8.99).

The Euclidean parallel postulate may be better understood by looking at Fig. 8.99.

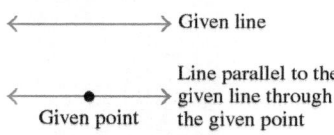

Figure 8.99

Many mathematicians after Euclid believed that this postulate was not as self-evident as the other nine postulates given by Euclid. Others believed that this postulate could be proved from the other nine postulates and therefore was not needed at all. Over the years, many mathematicians worked on the frustrating activity of trying to prove the parallel postulate used in Euclidean geometry. Eventually, Carl Friedrich Gauss (see *Profile in Mathematics* page 269) and Bernhard Riemann (1826–1866), and other mathematicians, realized that if the postulate is not applied, then two different types of geometry, collectively referred to as *non-Euclidean geometry*, can be developed. The two basic types of non-Euclidean geometries are *elliptical geometry* and *hyperbolic geometry*. The major differences among the three geometries lie in the fifth axiom, which we summarize here.

The Fifth Axiom of Geometry

Euclidean	**Elliptical**	**Hyperbolic**
Given a line and a point not on the line, one and only one line can be drawn parallel to the given line through the given point.	Given a line and a point not on the line, no line can be drawn through the given point parallel to the given line.	Given a line and a point not on the line, two or more lines can be drawn through the given point parallel to the given line.

To understand the fifth axiom of the two non-Euclidean geometries, remember that the term *line* is undefined. Thus, a line can be interpreted differently in different geometries. A model for Euclidean geometry is a plane, such as a blackboard (Fig. 8.100a). A model for elliptical geometry is a sphere (Fig. 8.100b). A model for hyperbolic geometry is a pseudosphere (Fig. 8.100c). A pseudosphere is similar to two trumpets placed bell to bell. Obviously, a line on a plane cannot be the same as a line on either of the other two figures.

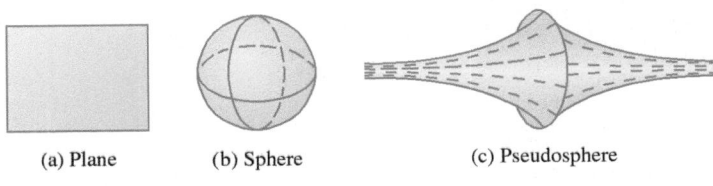

(a) Plane   (b) Sphere   (c) Pseudosphere

Figure 8.100

Elliptical Geometry

A circle on the surface of a sphere is called a great circle if it divides the sphere into two equal parts. If we were to cut through a sphere along a great circle, we would have two identical pieces. If we interpret a line to be a great circle, then the two red curves in Fig. 8.101(a) are lines. Figure 8.101(a) shows that the fifth axiom of elliptical geometry is true. Two great circles on a sphere must intersect; hence, there can be no parallel lines (Fig. 8.101a).

If we were to construct a triangle on a sphere, the sum of its angles would be greater than 180° (Fig. 8.101b). The theorem "The sum of the measures of the angles of a triangle is greater than 180°" has been proven by means of the axioms of elliptical geometry. The sum of the measures of the angles varies with the area of the triangle and gets closer to 180° as the area decreases.

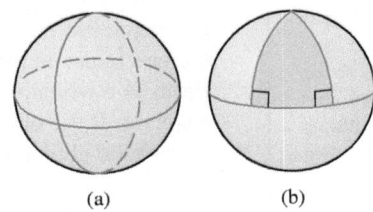

(a)   (b)

Figure 8.101

Hyperbolic Geometry

Lines in hyperbolic geometry* are represented by geodesics on the surface of a pseudosphere. A *geodesic* is the shortest and least-curved arc between two points on a curved surface. Figure 8.102 shows two different lines represented by geodesics, colored in red, on the surface of a pseudosphere. For simplicity of the diagrams, we show only one of the "bells" of the pseudosphere.

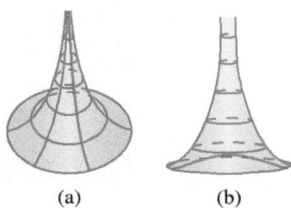

(a)   (b)

Figure 8.102

---

* A formal discussion of hyperbolic geometry is beyond the scope of this text.

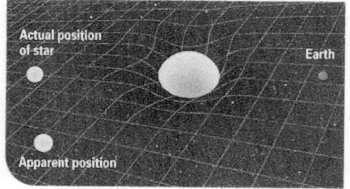
Figure 8.103(a) illustrates the fifth axiom of hyperbolic geometry. The diagram illustrates one way that, through the given point, two lines are drawn parallel to the given line. If we were to construct a triangle on a pseudosphere, the sum of the measures of the angles would be less than 180° (Fig. 8.103b). The theorem "The sum of the measures of the angles of a triangle is less than 180°" has been proven by means of the axioms of hyperbolic geometry.

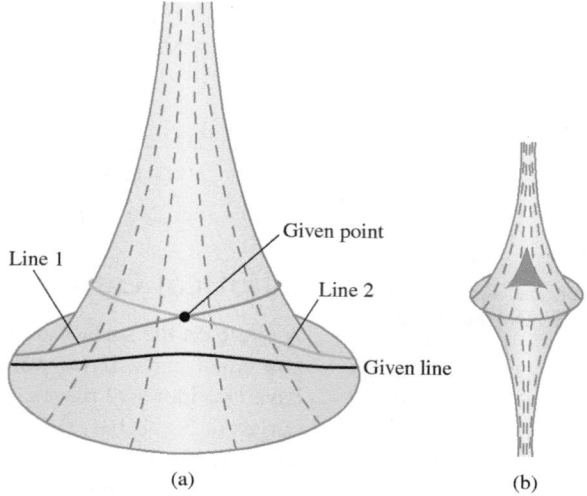

Figure 8.103

We have stated that the sum of the measures of the angles of a triangle is 180°, is greater than 180°, and is less than 180°. Which statement is correct? Each statement is correct *in its own geometry*. The many theorems based on the fifth postulate may differ in each geometry. It is important to realize that each theorem proved is true *in its own geometry*. No one system is the "best" system. Euclidean geometry may appear to be the one to use in the classroom, where the blackboard is flat. In discussions involving Earth as a whole, however, elliptical geometry may be the most useful, since Earth is a sphere. If the object under consideration has the shape of a saddle or pseudosphere, hyperbolic geometry may be the most useful.

Fractal Geometry

We are familiar with one-, two-, and three-dimensional figures. Many objects, however, are difficult to categorize as one-, two-, or three-dimensional. For example, how would you classify the irregular shapes we see in nature, such as a coastline, or the bark on a tree, or a mountain, or a path followed by lightning? For a long time mathematicians assumed that making realistic geometric models of natural shapes and figures was almost impossible, but the development of *fractal geometry* now makes it possible. Both images below were made by using fractal geometry.

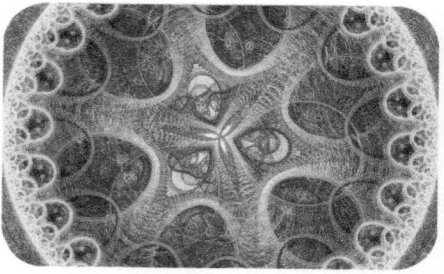

▲ *Fractal images*

The word *fractal* (from the Latin word *fractus*, "broken up, fragmented") was first used in the mid-1970s by mathematician Benoit Mandelbrot to describe shapes that had several common characteristics, including some form of "self-similarity," as will be seen shortly in the Koch snowflake. Fractals are developed by applying the same rule over and over again, with the end point of each simple step becoming the starting point for the next step, in a process called *recursion*.

Using the recursive process, we will develop a famous fractal called the *Koch snowflake* named after Niels Fabian Helge von Koch, a Swedish mathematician who first discovered its remarkable characteristics. The Koch snowflake illustrates a property of all fractals called *self-similarity*; that is, each smaller piece of the curve resembles the whole curve.

To develop the Koch snowflake:

1. Start with an equilateral triangle (Step 1, Fig. 8.104).
2. Whenever you see an edge ——— replace it with ⌒⌃⌒ (Steps 2–4).

What is the perimeter of the snowflake in Fig. 8.104, and what is its area? A portion of the boundary of the Koch snowflake known as the Koch curve, or the snowflake curve, is represented in Fig. 8.105.

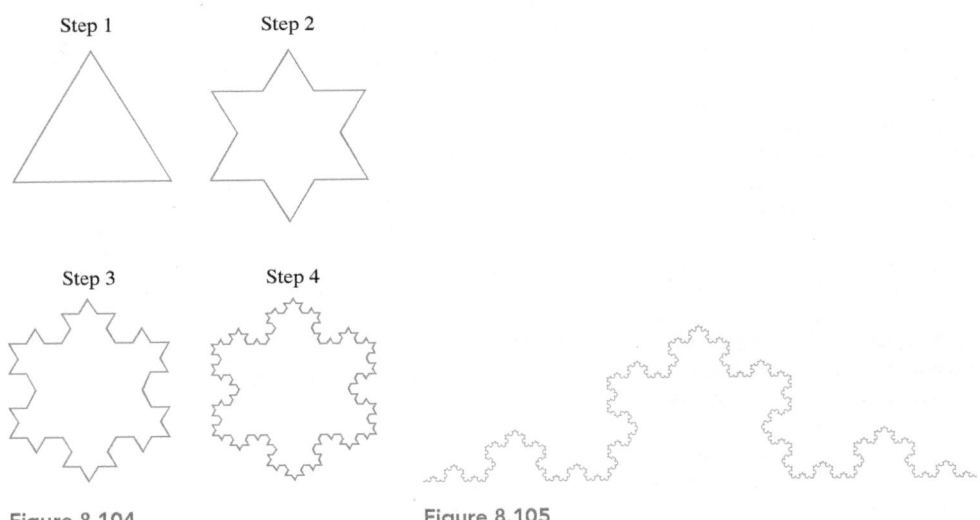

Step 1   Step 2

Step 3   Step 4

Figure 8.104   Figure 8.105

The Koch curve consists of infinitely many pieces of the form ⌒⌃⌒. It can be shown that after each step, the perimeter is $\frac{4}{3}$ times the perimeter of the previous step. Therefore, the Koch snowflake has an infinite perimeter. It can be shown that the area of the snowflake is 1.6 times the area of the starting equilateral triangle. Thus, the area of the snowflake is finite. The Koch snowflake has a finite area enclosed by an infinite boundary! This fact may seem difficult to accept, but it is true. However, the Koch snowflake, like other fractals, is not an everyday run-of-the-mill geometric shape.

Let us look at a few more fractals made using the recursive process. We will now construct what is known as a *fractal tree*. Start with a tree trunk (Fig. 8.106a on page 522). Draw two branches, each one a bit smaller than the trunk (Fig. 8.106b). Draw two branches from each of those branches, and continue; see Fig. 8.106(c) and (d). Ideally, we continue the process forever.

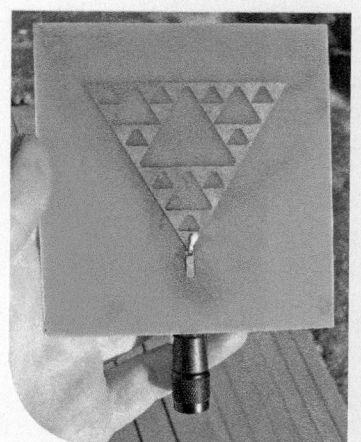
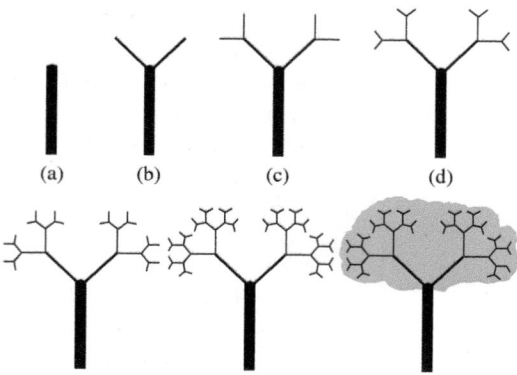

Figure 8.106 The fractal tree

If you take a little piece of any branch and zoom in on it, it will look exactly like the original tree. Fractals are *scale independent*, which means that you cannot really tell whether you are looking at something very big or something very small because the fractal looks the same whether you are close to it or far from it.

In Figs. 8.107 and 8.108, we develop two other fractals through the process of recursion. Figure 8.107 shows a fractal called the Sierpinski triangle, and Fig. 8.108 shows a fractal called the Sierpinski carpet. Both fractals are named after Polish mathematician Waclaw Sierpinski.

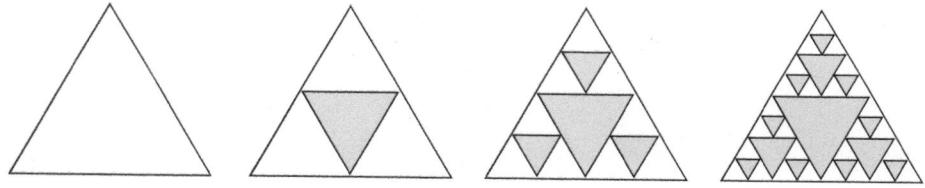

Figure 8.107 Sierpinski triangle

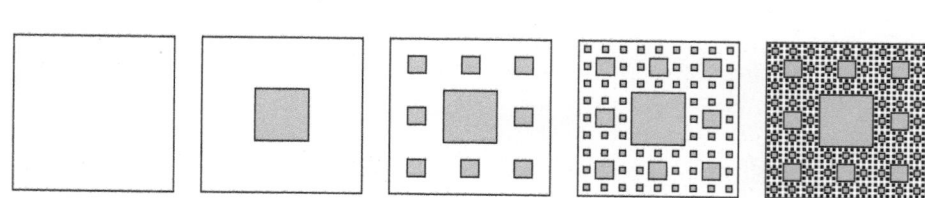

Figure 8.108 Sierpinski carpet

Fractals provide a way to study many natural forms, including coastlines, trees, mountains, rivers, weather patterns, galaxies, molecular structures, brain functions, and lung functions. Fractals help organize structures and processes that appear chaotic, such as the blood flow through veins, capillaries, and arteries. The study of such chaotic processes is called *chaos theory*. Fractals have the potential to revolutionize many fields of science, including medicine, engineering, meteorology, and astronomy. In short, fractals may alter how we study life itself within the vast universe!

SECTION 8.7 *Exercises*

Warm Up Exercises

In Exercises 1–8, fill in the blanks with an appropriate word, phrase, or symbol(s).

1. The fifth axiom of Euclidean geometry states that given a line and a point not on the line, one and only one line can be drawn through the given point _____ to the given line.

2. The fifth axiom of elliptical geometry states that given a line and a point not on the line, _____ line can be drawn through the given point parallel to the given line.

3. The fifth axiom of hyperbolic geometry states that given a line and a point not on the line, _____ or more lines can be drawn through the given point parallel to the given line.

4. A model for Euclidean geometry is a(n) _____.

5. A model for elliptical geometry is a(n) _____.

6. A model for hyperbolic geometry is a(n) _____.

7. The shortest and least-curved arc between two points on a curved surface is a(n) _____.

8. The study of chaotic processes is known as _____ theory.

Practice the Skills

In Exercises 9–12, we show a fractal-like figure made using a recursive process with the letter "M." Use this fractal-like figure as a guide in constructing fractal-like figures with the letter given. Show three steps, as is done here.

9. I   10. E   11. H   12. W

13. a) Develop a fractal by beginning with a square and replacing each side ——— with a ⌐_. Repeat this process twice.

   b) If you continue this process, will the fractal's perimeter be finite or infinite? Explain.

   c) Will the fractal's area be finite or infinite? Explain.

Problem Solving/Group Activity

14. In forming the Koch snowflake in Fig. 8.104 on page 521, the perimeter becomes greater at each step in the process. If each side of the original triangle is 1 unit, a general formula for the perimeter, L, of the snowflake at any step, n, may be determined by the formula

$$L = 3\left(\frac{4}{3}\right)^{n-1}$$

For example, at the first step when $n = 1$, the perimeter is 3 units, which can be verified by the formula as follows:

$$L = 3\left(\frac{4}{3}\right)^{1-1} = 3\left(\frac{4}{3}\right)^{0} = 3 \cdot 1 = 3$$

At the second step, when $n = 2$, we determine the perimeter as follows:

$$L = 3\left(\frac{4}{3}\right)^{2-1} = 3\left(\frac{4}{3}\right) = 4$$

Thus, at the second step the perimeter of the snowflake is 4 units.

a) Use the formula to complete the following table.*

Step	Perimeter
1	
2	
3	
4	
5	
6	

b) Use the results of your calculations to explain why the perimeter of the Koch snowflake is infinite.

c) Explain how the Koch snowflake can have an infinite perimeter, but a finite area.

Concept/Writing Exercises

15. What do we mean when we say that no one axiomatic system of geometry is "best"?

16. List the three types of curvature of space and the types of geometry that correspond to them.

17. List at least five natural forms that appear chaotic that we can study using fractals.

18. State the theorem concerning the sum of the measures of the angles of a triangle in

    a) Euclidean geometry.

    b) Hyperbolic geometry.

    c) Elliptical geometry.

Research Activities

19. *Poincaré Disk* To complete his masterpiece *Circle Limit III*, M. C. Escher studied a model of hyperbolic geometry called the *Poincaré disk*. Write a paper on the Poincaré disk and how it was used in Escher's art. Include representations of *infinity* and the concepts of *point* and *line* in hyperbolic geometry.

20. *M.C. Escher* To transfer his two-dimensional tiling known as *Symmetry Work 45* to a sphere, M. C. Escher used the spherical geometry of Bernhard Riemann. Write a paper on Escher's use of geometry to complete this masterpiece.

21. *Fantastic Fractals* Go to the website *Fantastic Fractals* at www.fantastic-fractals.com and study the information about fractals given there. Print color copies of the Mandlebrot set and the Julia set.

CHAPTER 8   *Summary*

Important Facts and Concepts

Examples and Discussion

Section 8.1

Point, line, plane, ray, half line, line segment, angle, and related terms and definitions are discussed throughout Section 8.1.

Discussion, pages 446–447, Examples 1 and 2, pages 447–449

Two angles are **complementary angles** if the sum of their measures is 90°. Two angles are **supplementary angles** if the sum of their measures is 180°.

Examples 3 and 4, pages 450–451

Vertical angles, alternate interior angles, alternate exterior angles, and **corresponding angles** and related terms are discussed in Section 8.1.

Discussion, pages 452–453, Example 7, page 453

Section 8.2

The sum of the measures of the interior angles of an n-sided polygon is $(n - 2)180°$.

Discussion, page 458, Example 1, page 458

Two polygons are **similar** if their corresponding angles have the same measure and the lengths of their corresponding sides are in proportion.

Discussion, page 459, Examples 2–3, pages 460–461

Section 8.3

Perimeter, P, and Area, A

TRIANGLE

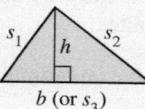

$b \text{ (or } s_3)$

$P = s_1 + s_2 + s_3 \quad (s_3 = b)$

$A = \frac{1}{2}bh$

SQUARE

$P = 4s$

$A = s^2$

RECTANGLE

$P = 2l + 2w$

$A = lw$

Discussion, pages 468–470, Example 1, page 470, Examples 3–5, pages 472–474

PARALLELOGRAM

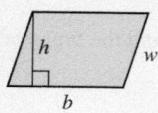

$P = 2b + 2w$

$A = bh$

TRAPEZOID

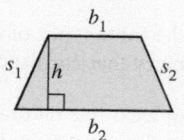

$P = s_1 + s_2 + b_1 + b_2$

$A = \frac{1}{2}h(b_1 + b_2)$

Pythagorean Theorem

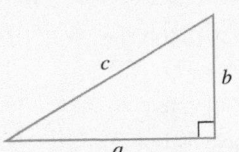

$$a^2 + b^2 = c^2$$

Example 2, page 471

Area, A, and Circumference, C, of a Circle

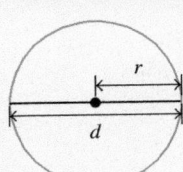

$$A = \pi r^2; C = 2\pi r \text{ or } C = \pi d$$

Discussion, page 471,
Examples 3–5, pages 472–474

Section 8.4

Volume, V, and Surface Area, SA

Discussion, pages 479–481; 483–484,

Examples 1–8, pages 481–486

CUBE

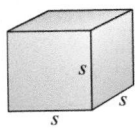

$V = s^3$

$SA = 6s^2$

RECTANGULAR SOLID

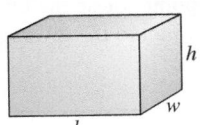

$V = lwh$

$SA = 2lw + 2wh + 2lh$

CYLINDER

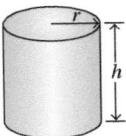

$V = \pi r^2 h$

$SA = 2\pi rh + 2\pi r^2$

CONE

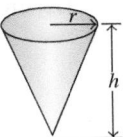

$V = \frac{1}{3}\pi r^2 h$

$SA = \pi r^2 + \pi r\sqrt{r^2 + h^2}$

SPHERE

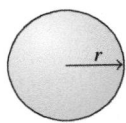

$V = \frac{4}{3}\pi r^3$

$SA = 4\pi r^2$

PYRAMID

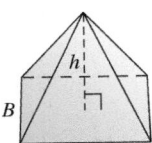

$V = \frac{1}{3}Bh$, where
B is the area of the base

PRISM

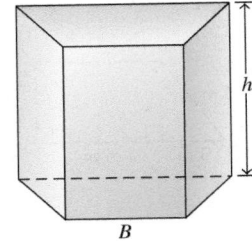

$V = Bh$, where B is the area of the base

Section 8.5

Rigid Motions

The **rigid motions: Reflections, translations, rotations,** and **glide reflections** are defined and discussed throughout Section 8.5.	Discussion, pages 491–500, Examples 1–6, pages 493–500
Reflective symmetry, rotational symmetry, and **tessellations** are described in Section 8.5.	Discussion, pages 500–503, Examples 7–8, pages 501–503

Section 8.6

Topology

Möbius strip, Klein bottle, maps, Jordan curves, and **topological equivalence** are described in Section 8.6.	Discussion, pages 510–514 Example 1-2, pages 513–514

Section 8.7

The Fifth Axiom

Fifth axiom in Euclidean geometry: Given a line and a point not on the line, one and only one line can be drawn through the given point parallel to the given line.	Discussion, pages 517–520
Fifth axiom in elliptical geometry: Given a line and a point not on the line, no line can be drawn through the given point parallel to the given line.	
Fifth axiom in hyperbolic geometry: Given a line and a point not on the line, two or more lines can be drawn through the given point parallel to the given line.	
Fractal Geometry	Discussion, pages 520–522

CHAPTER 8   *Review Exercises*

8.1

In Exercises 1–6, use the figure shown to determine the following. Many answers are possible.

1. $\overrightarrow{BF} \cup \overrightarrow{BC}$

2. $\measuredangle CBF \cap \measuredangle BCF$

3. $\overline{BF} \cup \overline{FC} \cup \overline{BC}$

4. $\overset{\circ}{BH} \cup \overset{\circ}{HB}$

5. $\overleftrightarrow{HI} \cap \overleftrightarrow{EG}$

6. $\overset{\circ}{CF} \cap \overset{\circ}{CG}$

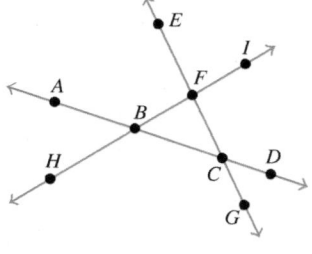

7. $m\measuredangle A = 27.6°$. Determine the measure of the complement of $\measuredangle A$.

8. $m\measuredangle B = 100.5°$. Determine the measure of the supplement of $\measuredangle B$.

8.2

In Exercises 9–12, use the similar triangles ABC and A′B′C shown to determine the following.

9. The length of $\overline{BC}$

10. The length of $\overline{A'B'}$

11. $m\measuredangle BAC$

12. $m\measuredangle ABC$

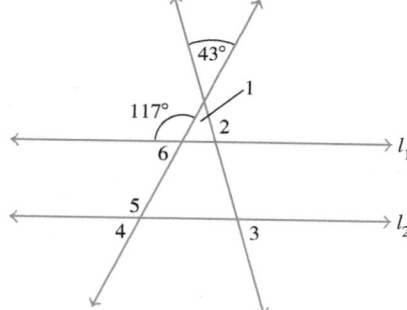

13. In the following figure, l_1 and l_2 are parallel lines. Determine $m\measuredangle 1$ through $m\measuredangle 6$.

14. Determine the sum of the measures of the interior angles of an octagon.

8.3

In Exercises 15–18, determine (a) the area and (b) the perimeter of the figure.

15.

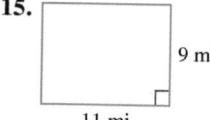

16.

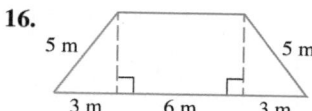

17.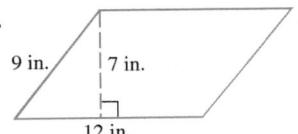

18.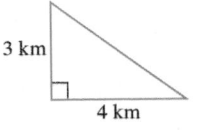

19. Determine (a) the area and (b) the circumference of the circle. Use the π key on a calculator and round your answer to the nearest hundredth.

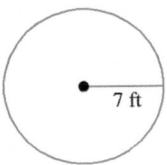

In Exercises 20 and 21, determine the shaded area. When appropriate, use the π key on your calculator and round your answer to the nearest hundredth.

20.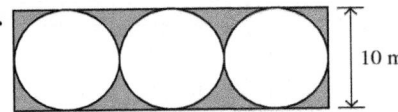

21.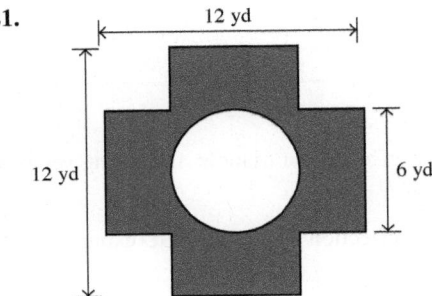

22. *Cost of Hardwood Flooring* Determine the total cost of covering a rectangular 14-ft by 16-ft kitchen floor with hardwood flooring. The cost of the flooring selected is $9.75 per square foot.

8.4

In Exercises 23–26, determine (a) the volume and (b) the surface area of the figure. When appropriate, use the π key on your calculator and round your answer to the nearest hundredth.

23.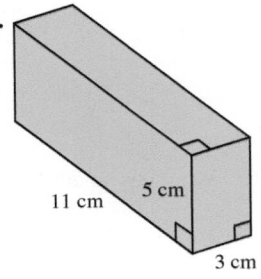

24.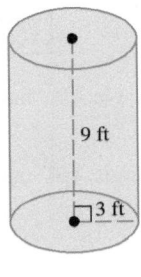

25.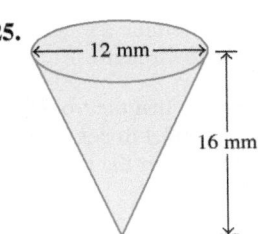

26.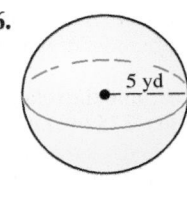

In Exercises 27 and 28, determine the volume of the figure.

27.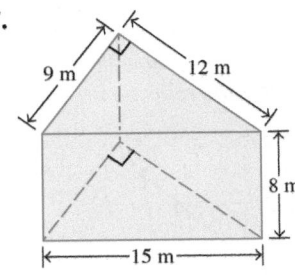

28.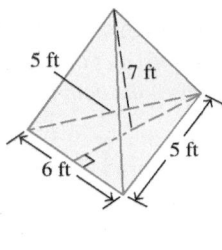

In Exercises 29 and 30, determine the volume of the shaded region. When appropriate, use the π key on your calculator and round your answer to the nearest hundredth.

29.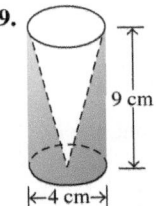

30.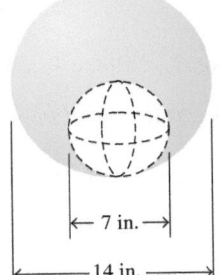

31. *Water Trough* Dale has a water trough whose ends are trapezoids and whose sides are rectangles, as illustrated. He is afraid that the base it is sitting on will not support the weight of the trough when it is filled with water. He knows that the base will support 4800 lb.

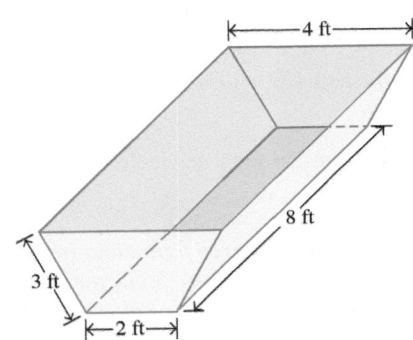

(Continued)

a) If the trough is filled with water, determine the number of cubic feet of water contained in the trough.

b) Determine the total weight, assuming that the trough weighs 375 lb and the water weighs 62.4 lb per cubic foot. Is the base strong enough to support the trough filled with water?

c) If 1 gal of water weighs 8.3 lb, how many gallons of water will the trough hold?

8.5

In Exercises 32 and 33, use the given triangle and reflection lines to construct the indicated reflections. Show the triangle in the positions both before and after the reflection.

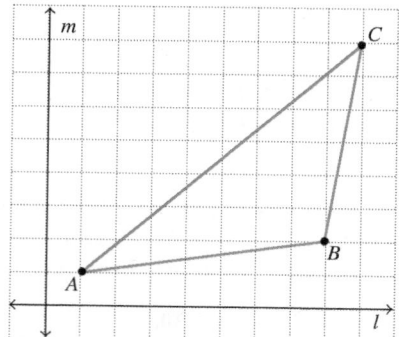

32. Construct the reflection of triangle *ABC* about line *l*.

33. Construct the reflection of triangle *ABC* about line *m*.

In Exercises 34 and 35, use translation vectors **v** *and* **w** *to construct the indicated translations. Show the rectangle in the positions both before and after the translation.*

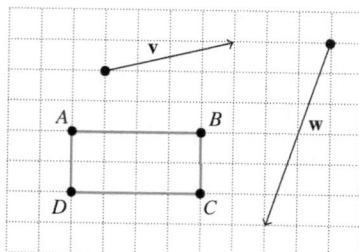

34. Construct the translation of rectangle *ABCD* using translation vector **v**.

35. Construct the translation of rectangle *ABCD* using translation vector **w**.

In Exercises 36–38, use the given figure and rotation point P to construct the indicated rotations. Show the trapezoid in the positions both before and after the rotation.

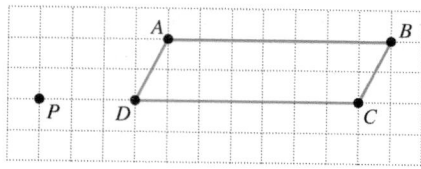

36. Construct a 90° rotation of parallelogram *ABCD* about point *P*.

37. Construct a 180° rotation of parallelogram *ABCD* about point *P*.

38. Construct a 270° rotation of parallelogram *ABCD* about point *P*.

In Exercises 39 and 40, use the given figure, translation vectors **v** *and* **w**, *and reflection lines l and m to construct the indicated glide reflections. Show the triangle in the positions both before and after the glide reflection.*

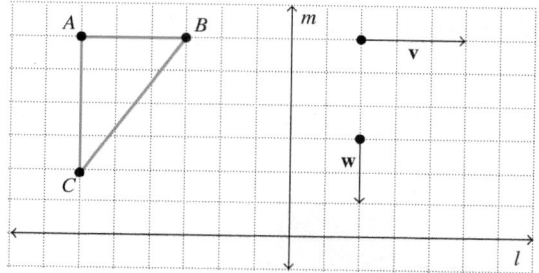

39. Construct a glide reflection of triangle *ABC* using vector **v** and reflection line *l*.

40. Construct a glide reflection of triangle *ABC* using vector **w** and reflection line *m*.

In Exercises 41 and 42, use the following figure to answer the following questions.

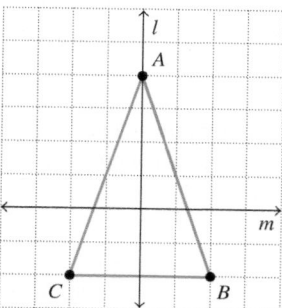

41. Does triangle *ABC* have reflective symmetry about line *l*? Explain.

42. Does triangle *ABC* have reflective symmetry about line *m*? Explain.

In Exercises 43 and 44, use the following figure to answer the following questions.

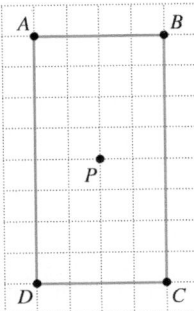

43. Does rectangle *ABCD* have 90° rotational symmetry about point *P*? Explain.

44. Does rectangle *ABCD* have 180° rotational symmetry about point *P*? Explain.

8.6

45. Give an example of an object that has

a) genus 0.   b) genus 1.

c) genus 2.   d) genus 3 or more.

46. Color the map by using a maximum of four colors so that no two regions with a common border have the same color.

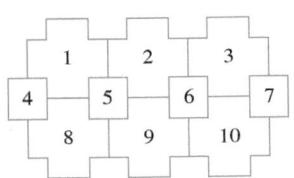

47. Determine whether point *A* is inside or outside the Jordan curve.

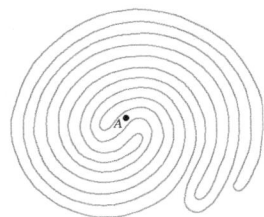

8.7

48. State the fifth axiom of Euclidean, elliptical, and hyperbolic geometry.

49. Develop a fractal by beginning with a square and replacing each side ⸺ with ⌐⌐. Repeat this process twice.

50. Construct a Koch snowflake by beginning with an equilateral triangle and replacing each side with a ⌐⌐. Repeat this process twice.

CHAPTER 8   *Test*

In Exercises 1–4, use the figure to describe the following. Many answers are possible.

1. $\overrightarrow{AB} \cup \overrightarrow{AE}$

2. $\overline{BC} \cup \overline{CD} \cup \overline{BD}$

3. $\sphericalangle EDF \cap \sphericalangle BDC$

4. $\overleftrightarrow{AC} \cup \overrightarrow{BA}$

5. $m\sphericalangle A = 41.8°$. Determine the measure of the complement of $\sphericalangle A$.

6. $m\sphericalangle B = 73.5°$. Determine the measure of the supplement of $\sphericalangle B$.

7. In the figure, determine the measure of $\sphericalangle x$.

8. Determine the sum of the measures of the interior angles of a hexagon.

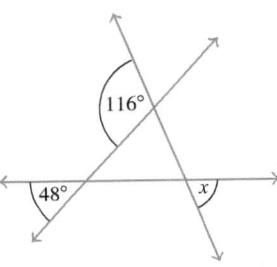

9. Triangles *ABC* and *A'B'C'* are similar figures. Determine the length of side $\overline{B'C'}$.

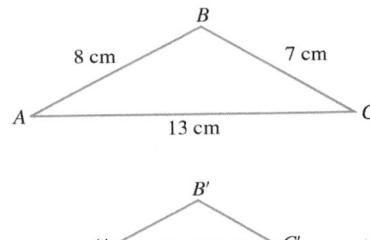

10. Right triangle *ABC* has one leg of length 12 in. and hypotenuse of length 13 in. (See the figure on the top of page 530)

a) Determine the length of the other leg.

b) Determine the perimeter of the triangle.

c) Determine the area of the triangle.

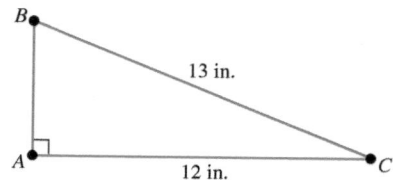

11. Determine (a) the volume and (b) the surface area of a sphere of diameter 14 cm.

12. Determine the volume of the shaded region. Use the $\boxed{\pi}$ key on your calculator and round your answer to the nearest hundredth.

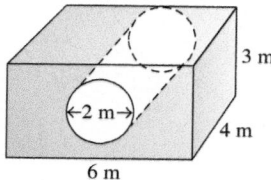

13. Determine the volume of the pyramid.

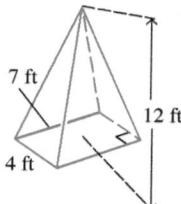

14. Construct a reflection of rectangle *ABCD*, shown below, about line *l*. Show the rectangle in the positions both before and after the reflection.

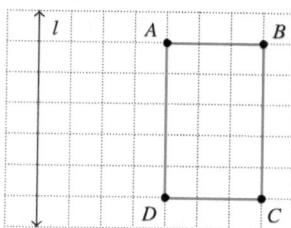

15. Construct a translation of quadrilateral *ABCD*, shown below, using translation vector **v**. Show the quadrilateral in the positions both before and after the translation.

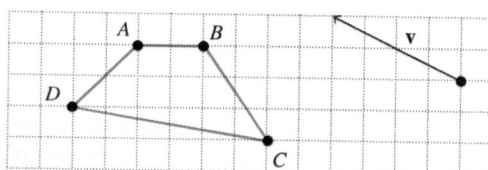

16. Construct a 180° rotation of triangle *ABC*, shown below, about rotation point *P*. Show the triangle in the positions both before and after the rotation.

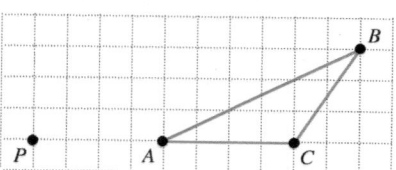

17. Construct a glide reflection of rectangle *ABCD*, shown below, using translation vector **v** and reflection line *l*. Show the rectangle in the positions both before and after the glide reflection.

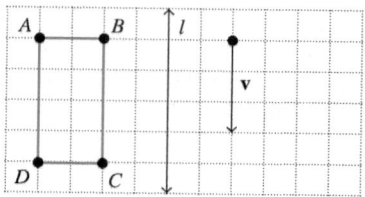

18. Use the figure below to answer the following questions.

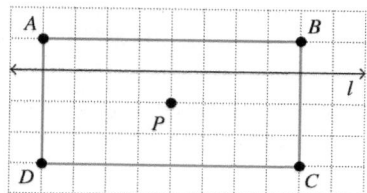

a) Does rectangle *ABCD* have reflective symmetry about line *l*? Explain.

b) Does rectangle *ABCD* have 180° rotational symmetry about point *P*? Explain.

19. What is a Möbius strip?

20. a) Name an object of genus 1.

   b) Name an object of genus 2.

◀ Mathematical systems have many applications in many scientific fields including chemistry and physics.

9

Mathematical Systems

What You Will Learn

- Mathematical systems including groups and their properties
- Finite mathematical systems
- Modular arithmetic
- Matrices

Why This Is Important

Mathematical systems can be used to solve a variety of problems that involve timekeeping, work schedules, cryptography (the study of secret codes), Internet security, and the safety of financial transactions. One branch of mathematics of particular importance to this study is group theory. Physicists and chemists rely on group theory to study the behavior of the tiniest particles that make up an atom in a scientific field known as *quantum mechanics*. Matrices are essential to a branch of mathematics called linear algebra. Matrices and linear algebra are used in virtually all computer programs and electronic devices we use in our lives daily. Group theory also plays an important role in robotics, computer graphics, weather forecasting, musical theory, and medical image analysis.

SECTION 9.1   Groups

Upon completion of this section, you will be able to:

- Understand and identify mathematical systems.
- Understand and use the commutative and associative properties.
- Understand and identify closure.
- Understand and identify identity elements and inverses.
- Show whether a mathematical system is a group or a commutative group.

How big is the universe? Is there an end to the universe? Does the universe continue to expand? Such questions are addressed by the scientific theory known as the *big bang theory*. In this section, we will introduce a mathematical structure known as a *group*. Many modern scientific theories, including the big bang theory, make use of groups and other mathematical topics discussed in this book.

Why This Is Important   The more we understand about our universe, the better prepared we are to answer questions regarding the history and future of planet Earth.

Mathematical Systems

We begin our discussion by introducing a mathematical system. As you will learn shortly, you already know and use many mathematical systems.

> **Definition: Mathematical System**
> A **mathematical system** consists of a set of elements and at least one binary operation.

Timely Tip

Throughout this section, we will work with many different sets of numbers, such as natural numbers, whole numbers, integers, rational numbers, and real numbers. A review of the numbers that make up each of these sets can be found on page 253 in Section 5.5.

In the above definition, we mention binary operation. A *binary operation* is an operation, or rule, that can be performed on two and only two elements of a set. The result is a single element. When we add *two* integers, the sum is *one* integer. When we multiply *two* integers, the product is *one* integer. Thus, addition and multiplication are both binary operations. Is determining the reciprocal of a number a binary operation? No, it is an operation on a single element of a set.

When you learned how to add integers, you were introduced to a mathematical system. The set of elements is the set of integers, and the binary operation is addition. When you learned how to multiply integers, you became familiar with a second mathematical system. The set of integers with the operation of subtraction and the set of integers with the operation of division are two other examples of mathematical systems, since subtraction and division are also binary operations.

Some mathematical systems are used in solving everyday problems, such as planning work schedules. Others are more abstract and are used primarily in research, chemistry, physical structure, matter, the nature of genes, and other scientific fields.

Commutative and Associative Properties

Once a mathematical system is defined, its structure may display certain properties. Consider the set of integers

$$I = \{\ldots, -3, -2, -1, 0, 1, 2, 3, \ldots\}$$

Recall that the ellipsis, the three dots at each end of the set, indicates that the set continues in the same manner.

The set of integers can be studied with the operations of addition, subtraction, multiplication, and division as separate mathematical systems. For example, when we study the set of integers under the operations of addition or multiplication, we see that the commutative and associative properties hold. The general forms of these properties are shown on the next page.

> **Commutative and Associative Properties**
> For any elements a, b, and c
>
> | | *Addition* | *Multiplication* |
> |---|---|---|
> | **Commutative property** | $a + b = b + a$ | $a \cdot b = b \cdot a$ |
> | **Associative property** | $(a + b) + c = a + (b + c)$ | $(a \cdot b) \cdot c = a \cdot (b \cdot c)$ |

The integers *are commutative* under the operations of *addition and multiplication.* For example,

Addition		*Multiplication*
$2 + 4 \overset{?}{=} 4 + 2$	and	$2 \cdot 4 \overset{?}{=} 4 \cdot 2$
$6 = 6$		$8 = 8$

The integers, however, *are not commutative* under the operations of *subtraction and division.* For example,

Subtraction		*Division*
$4 - 2 \overset{?}{=} 2 - 4$	and	$4 \div 2 \overset{?}{=} 2 \div 4$
$2 \neq -2$		$2 \neq \frac{1}{2}$

The integers *are associative* under the operations of *addition and multiplication.* For example,

Addition		*Multiplication*
$(1 + 2) + 3 \overset{?}{=} 1 + (2 + 3)$	and	$(1 \cdot 2) \cdot 3 \overset{?}{=} 1 \cdot (2 \cdot 3)$
$3 + 3 \overset{?}{=} 1 + 5$		$2 \cdot 3 \overset{?}{=} 1 \cdot 6$
$6 = 6$		$6 = 6$

The integers, however, *are not associative* under the operations of *subtraction and division.* See Exercises 17 and 18 on page 538.

To say that a set of elements is commutative under a given operation means that the commutative property holds for *any* elements a and b in the set. Similarly, to say that a set of elements is associative under a given operation means that the associative property holds for *any* elements a, b, and c in the set.

Consider the mathematical system consisting of the set of integers under the operation of addition. Because the set of integers is infinite, this mathematical system is an example of an *infinite mathematical system.* We will study certain properties of this mathematical system. The first property we will examine is closure.

Closure

The sum of any two integers is an integer. Therefore, the set of integers is said to be *closed*, or to satisfy the *closure property*, under the operation of addition.

> **Definition: Closure**
> If a binary operation is performed on any two elements of a set and the result is an element of the set, then that set is **closed** (or has **closure**) under the given binary operation.

Is the set of integers closed under the operation of multiplication? The answer is yes. When any two integers are multiplied, the product will be an integer.

Is the set of integers closed under the operation of subtraction? Again, the answer is yes. The difference of any two integers is an integer.

Is the set of integers closed under the operation of division? The answer is no because two integers may have a quotient that is not an integer. For example, if we select the integers 2 and 3, the quotient of 2 divided by 3 is $\frac{2}{3}$, which is not an integer. Thus, the integers are not closed under the operation of division.

We showed that the set of integers was not closed under the operation of division by determining two integers whose quotient was not an integer. A specific example illustrating that a specific property is not true is called a *counterexample*. Mathematicians and scientists often try to determine a counterexample to confirm that a specific property is not always true.

Identity Element

Is there an element in the set of integers that, when added to any given integer, results in a sum that is the given integer? The answer is yes. The sum of 0 and any integer is the given integer. For example, $1 + 0 = 0 + 1 = 1$, $-4 + 0 = 0 + (-4) = -4$, and so on. For this reason, we call 0 the *additive identity element* for the set of integers. Note that for any integer a, $a + 0 = 0 + a = a$.

> **Definition: Identity Element**
> An **identity element** is an element in a set such that when a binary operation is performed on it and any given element in the set, the result is the given element.

Is there an identity element for the set of integers under the operation of multiplication? The answer is yes; it is the number 1. Note that $2 \cdot 1 = 1 \cdot 2 = 2$, $3 \cdot 1 = 1 \cdot 3 = 3$, and so on. For any integer a, $a \cdot 1 = 1 \cdot a = a$. For this reason, 1 is called the *multiplicative identity element* for the set of integers.

Inverses

What integer, when added to 4, gives a sum of 0; that is, $4 + \underline{} = 0$? The shaded area is to be filled in with the integer -4: $4 + (-4) = 0$. We say that -4 is the additive inverse of 4 and that 4 is the additive inverse of -4. Note that the sum of the element and its additive inverse gives the additive identity element 0. What is the additive inverse of 12? Since $12 + (-12) = 0$, -12 is the additive inverse of 12.

Other examples of integers and their additive inverses are

Element	+	Additive Inverse	=	Identity Element
0	+	0	=	0
3	+	(-3)	=	0
-5	+	5	=	0

Note that for the operation of addition, every integer a has a unique inverse, $-a$, such that $a + (-a) = -a + a = 0$.

> **Definition: Inverse**
> When a binary operation is performed on two elements in a set and the result is the identity element for the binary operation, each element is said to be the **inverse** of the other.

Does every integer have an inverse under the operation of multiplication? For multiplication, the product of an integer and its inverse must yield the multiplicative identity element, 1. What is the multiplicative inverse of 2? That is, 2 times what number gives 1?

$$2 \cdot \boxed{} = 1 \qquad 2 \cdot \tfrac{1}{2} = 1$$

However, since $\tfrac{1}{2}$ is not an integer, 2 does not have a multiplicative inverse in the set of integers. Therefore, not every integer has a multiplicative inverse.

Group

Let's review what we have learned about the mathematical system consisting of the set of integers under the operation of addition.

1. The set of integers is *closed* under the operation of addition.
2. The set of integers has an *identity element* under the operation of addition.
3. Each element in the set of integers has an *inverse* under the operation of addition.
4. The *associative property* holds for the set of integers under the operation of addition.

The set of integers under the operation of addition is an example of a *group*. The properties of a group can be summarized as follows.

Properties of a Group

Any mathematical system that meets the following four requirements is called a **group**.
1. The set of elements is *closed* under the given operation.
2. An *identity element* exists for the set under the given operation.
3. Every element in the set has an *inverse* under the given operation.
4. The set of elements is *associative* under the given operation.

It is often very time consuming to show that the associative property holds for all cases. In many of the examples that follow, we will state that the associative property holds for the given set of elements under the given operation.

Commutative Group

The commutative property does not need to hold for a mathematical system to be a group. However, if a mathematical system meets the four requirements of a group and is also commutative under the given operation, the mathematical system is a *commutative* (or *abelian*) *group*. The abelian group is named after Niels Abel (see the *Profiles in Mathematics* in the margin).

Definition: Commutative Group

A group that satisfies the commutative property is called a **commutative group (or abelian group)**.

Because the commutative property holds for the set of integers under the operation of addition, the set of integers under the operation of addition is not only a group, but it is also a commutative group.

Rubik's Cube

Group theory may be viewed as the study of the algebra of symmetry and transformations (see Section 8.5 for a discussion of transformational geometry). One of the best, and perhaps most entertaining, examples of this view is *Rubik's cube*. In 1974, Erno Rubik, a Hungarian teacher of architecture and design, presented the world with his popular puzzle. Each face of the cube is divided into nine squares, and each row and column of each face can rotate. The result is approximately 43 quintillion different arrangements of the colors on the cube. Rubik's cube is the most popular and best-selling puzzle in human history. Although the popularity of the puzzle peaked during the 1980s, clubs and annual contests are still devoted to solving the puzzle. For more information, visit the website www.rubiks.com.

Properties of a Commutative Group

A mathematical system is a **commutative group** if all five of the following conditions hold.

1. The set of elements is *closed* under the given operation.
2. An *identity element* exists for the set under the given operation.
3. Every element in the set has an *inverse* under the given operation.
4. The set of elements is *associative* under the given operation.
5. The set of elements is *commutative* under the given operation.

To determine whether a mathematical system is a group under a given operation, *check, in the following order*, whether (1) the system is closed under the given operation, (2) there is an identity element in the set for the given operation, (3) every element in the set has an inverse under the given operation, and (4) the associative property holds under the given operation. If *any* of these four requirements is *not* met, stop and state that the mathematical system is not a group. If asked to determine whether the mathematical system is a commutative group, you also need to determine whether the commutative property holds for the given operation.

Example 1 *Whole Numbers Under Addition*

Determine whether the mathematical system consisting of the set of whole numbers under the operation of addition forms a group.

Solution Recall from Chapter 5 that the set of whole numbers is $\{0, 1, 2, 3, \ldots\}$. We will check the properties required, using the operation of addition, to determine if the mathematical system is a group.

1. *Closure*: The sum of any two whole numbers is a whole number. Therefore, the set of whole numbers is closed under the operation of addition.
2. *Identity element*: The additive identity for the set of whole numbers is 0. For example, $1 + 0 = 0 + 1 = 1$, and $2 + 0 = 0 + 2 = 2$, and so on. For any whole number a, $a + 0 = 0 + a = a$. Thus, the mathematical system contains an identity element.
3. *Inverse elements*: For the mathematical system to be a group, each element must have an additive inverse *in the set*. Remember that the additive inverse of a number is the *opposite* of the number. For example, the additive inverse of 1 is -1, the additive inverse of 2 is -2, the additive inverse of 3 is -3, and so on. Since the numbers $-1, -2, -3, \ldots$ are *not* in the set of whole numbers, not every number has an inverse in the set. Therefore, this mathematical system is not a group.

Because we have already shown that the set of whole numbers under the operation of addition is not a group, there is no need to check the associative property. ∎

Example 2 *Rational Numbers Under Multiplication*

Determine whether the set of rational numbers under the operation of multiplication forms a group.

Solution Recall from Chapter 5 that the rational numbers are the set of numbers of the form $\frac{p}{q}$, where p and q are integers and $q \neq 0$. The set of rational numbers includes all fractions and integers.

1. *Closure*: The product of any two rational numbers is a rational number. Therefore, the rational numbers are closed under the operation of multiplication.

2. *Identity element*: The multiplicative identity element for the set of rational numbers is 1. Note, for example, that $3 \cdot 1 = 1 \cdot 3 = 3$, and $\frac{3}{8} \cdot 1 = 1 \cdot \frac{3}{8} = \frac{3}{8}$. For any rational number a, $a \cdot 1 = 1 \cdot a = a$.

3. *Inverse elements*: For the mathematical system to be a group under the operation of multiplication, *each and every* rational number must have a multiplicative inverse in the set of rational numbers. Remember that for the operation of multiplication, the product of a number and its inverse must give the multiplicative identity element, 1. Let's check a few rational numbers:

Rational Number	·	Inverse	=	Identity Element
7	·	$\frac{1}{7}$	=	1
$\frac{2}{3}$	·	$\frac{3}{2}$	=	1
$-\frac{1}{5}$	·	-5	=	1

Looking at these examples you might deduce that each rational number does have an inverse. However, one rational number, 0, does not have an inverse.

$$0 \cdot ? = 1$$

Because there is no rational number that when multiplied by 0 gives 1, 0 does not have a multiplicative inverse. Since not *every* rational number has an inverse, this mathematical system is not a group.

There is no need at this point to check the associative property because we have already shown that the mathematical system of rational numbers under the operation of multiplication is not a group. ∎

Timely Tip

Identity Element and Inverse Elements

When checking to see if a mathematical system is a group, students often confuse the identity element and the inverse elements. Remember, there can be only one identity element in the set, but each element in the set must have an inverse. For example, consider the integers under addition. The additive identity element is the number 0. The additive inverse of every integer is its opposite. For example, the additive inverse of 1 is -1, the additive inverse of 2 is -2, and so on. In a group, there is only one identity but each element has its own inverse.

Example 3 *Real Numbers Under Addition*

Determine whether the set of real numbers under the operation of addition forms a commutative group.

Solution Recall from Chapter 5 that the real numbers can be thought of as the numbers that correspond to *each* point on the real number line. The set of real numbers includes all rational numbers and all irrational numbers.

1. *Closure*: The sum of any two real numbers is a real number. Therefore, the real numbers are closed under the operation of addition.

2. *Identity element*: The additive identity element for the set of real numbers is 0. For example, $\frac{2}{3} + 0 = 0 + \frac{2}{3} = \frac{2}{3}$ and $\sqrt{5} + 0 = 0 + \sqrt{5} = \sqrt{5}$. For any real number a, $a + 0 = 0 + a = a$.

3. *Inverse elements*: For the set of real numbers under the operation of addition to be a commutative group, each and every real number must have an additive inverse in the set of real numbers. For the operation of addition, the sum of a real number and its additive inverse, or *opposite*, must be the additive identity element, 0. From Chapter 5, we saw that every real number a has an opposite $-a$ such that $a + (-a) = -a + a = 0$. Recall also that $0 + 0 = 0$; therefore, 0 is its own additive inverse. Thus, all real numbers have an additive inverse.

4. *Associative property*: From Chapter 5, we know that the real numbers are associative under the operation of addition. That is, for any real numbers a, b, and c, $(a + b) + c = a + (b + c)$.

5. *Commutative property*: From Chapter 5, we know that the real numbers are commutative under the operation of addition. That is, for any real numbers a and b, $a + b = b + a$.

> Since all five properties (closure, identity element, inverse elements, associative property, and commutative property) hold, the set of real numbers under the operation of addition forms a commutative group. ∎

SECTION 9.1
Exercises

Warm Up Exercises

In Exercises 1–10, fill in the blanks with an appropriate word, phrase, or symbol(s).

1. A mathematical system consists of a set of elements and at least one _____ operation.

2. A binary operation is an operation or rule that can be performed on exactly two elements of a set, with the result being a(n) _____ element.

3. If a binary operation is performed on any two elements of a set and the result is an element of the set, then that set is _____ under the given binary operation.

4. A specific example illustrating that a property is not true is called a(n) _____.

5. Since the sum of 0 and any integer is the given integer, we say that 0 is the additive _____ element for the set of the integers under the operation of addition.

6. Since the product of 1 and any integer is the given integer, we say that 1 is the _____ identity element for the set of the integers under the operation of multiplication.

7. When a binary operation is performed on two elements in a set and the result is the identity element for the binary operation, each element is said to be the _____ of the other.

8. If a mathematical system possesses the following properties—closure, identity element, inverses, and the associative property—then the mathematical system is a(n) _____.

9. A group that also satisfies the commutative property is called a(n) _____ (or abelian) group.

10. Another name for a commutative group is a(n) _____ group.

Practice the Skills

11. Give the commutative property of addition and illustrate the property with an example.

12. Give the commutative property of multiplication and illustrate the property with an example.

13. Give the associative property of multiplication and illustrate the property with an example.

14. Give the associative property of addition and illustrate the property with an example.

15. Give an example to show that the commutative property does not hold for the set of integers under the operation of subtraction.

16. Give an example to show that the commutative property does not hold for the set of integers under the operation of division.

17. Give an example to show that the associative property does not hold for the set of integers under the operation of division.

18. Give an example to show that the associative property does not hold for the set of integers under the operation of subtraction.

19. Consider the set of integers under the operation of addition.

   a) Is the system closed? Explain.

   b) Is there an identity element? If so, what is it?

   c) Does each element in the set have an inverse? Explain.

   d) The associative property holds. Provide one example to demonstrate the associative property.

   e) The commutative property holds. Provide one example to demonstrate the commutative property.

   f) Is this mathematical system a commutative group? Explain.

20. Consider the set of real numbers under the operation of addition.

   a) Is the system closed? Explain.

   b) Is there an identity element? If so, what is it?

   c) Does each element in the set have an inverse? Explain.

d) The associative property holds. Provide one example to demonstrate the associative property.

e) The commutative property holds. Provide one example to demonstrate the commutative property.

f) Is this mathematical system a commutative group? Explain.

21. Consider the set of positive integers under the operation of addition.

    a) Is the system closed? Explain.

    b) Is there an Identity element? If so, what is it?

    c) Does each element in the set have an inverse? Explain.

    d) The associative property holds. Provide one example to demonstrate the associative property.

    e) The commutative property holds. Provide one example to demonstrate the commutative property.

    f) Is this mathematical system a commutative group? Explain.

22. Consider the set of whole numbers under the operation of addition.

    a) Is the system closed? Explain.

    b) Is there an identity element? If so, what is it?

    c) Does each element in the set have an inverse? Explain.

    d) The associative property holds. Provide one example to demonstrate the associative property.

    e) The commutative property holds. Provide one example to demonstrate the commutative property.

    f) Is this mathematical system a commutative group? Explain.

Problem Solving

In Exercises 23–32, explain your answer.

23. Is the set of rational numbers a group under the operation of addition?

24. Is the set of positive rational numbers a commutative group under the operation of multiplication?

25. Is the set of negative integers a group under the operation of multiplication?

26. Is the set of negative integers a group under the operation of division?

27. Is the set of positive real numbers a commutative group under the operation of multiplication?

28. Is the set of whole numbers a commutative group under the operation of multiplication?

29. Is the set of negative integers a commutative group under the operation of addition?

30. Is the set of integers a group under the operation of multiplication?

31. Is the set of rational numbers a commutative group under the operation of division?

32. Is the set of rational numbers a group under the operation of subtraction?

Challenge Problems/Group Activities

In Exercises 33 and 34, explain your answer.

33. Is the set of irrational numbers a group under the operation of addition?

34. Is the set of irrational numbers a group under the operation of multiplication?

35. Create a mathematical system with two binary operations. Select a set of elements and two binary operations so that one binary operation with the set of elements meets the requirements for a group and the other binary operation does not. Explain why the one binary operation with the set of elements is a group. For the other binary operation and the set of elements, provide counterexamples to show that it is not a group.

Research Activity

36. *Rings and Fields* There are other classifications of mathematical systems besides groups. For example, there are *rings* and *fields*. Determine the requirements that must be met for a mathematical system to be (a) a ring and (b) a field. (c) Is the set of real numbers, under the operations of addition and multiplication, a field?

SECTION 9.2   Finite Mathematical Systems

Upon completion of this section, you will be able to:

- Determine whether a finite mathematical system defined by clock arithmetic is a group.
- Determine whether a finite mathematical system without numbers is a group.

Suppose you decide to use your slow cooker to cook a meal for a dinner party. You would like to eat at 6:30 P.M., and it will take 10 hours for the meal to cook. What time should you begin to cook the meal in the slow cooker? Or suppose you need to take medicine every eight hours. You last took the medicine at 7:45 A.M. When should you take the medicine again? Questions such as these involve the use of mathematical systems that we will study in this section.

Why This Is Important Timekeeping is just one application of *finite mathematical systems*.

In the preceding section, we presented infinite mathematical systems. In this section, we present some finite mathematical systems. A *finite mathematical system* is one whose set contains a finite number of elements.

Clock Arithmetic

Figure 9.1

Let's develop a finite mathematical system called *clock 12 arithmetic*. The set of elements in this system will be the hours on a clock: $\{1, 2, 3, 4, 5, 6, 7, 8, 9, 10, 11, 12\}$. The binary operation that we will use is addition, which we define as movement of the hour hand in a clockwise direction. Assume that it is 4 o'clock. What time will it be in 9 hours? (See Fig. 9.1.) If we add 9 hours to 4 o'clock, the clock will read 1 o'clock. Thus, $4 + 9 = 1$ in clock arithmetic. Would $9 + 4$ be the same as $4 + 9$? Yes it will because $4 + 9 = 9 + 4 = 1$.

Table 9.1 below is the addition table for clock arithmetic. Its elements are based on the definition of addition as previously illustrated. For example, the sum of 4 and 9 is 1, so we put a 1 in the table where the row to the right of the 4 intersects the column below the 9 (see the pink shading). Likewise, the sum of 11 and 10 is 9, so we put a 9 in the table where the row to the right of the 11 intersects the column below the 10 (see the red outline).

Table 9.1   Clock 12 Arithmetic

+	1	2	3	4	5	6	7	8	9	10	11	12
1	2	3	4	5	6	7	8	9	10	11	12	1
2	3	4	5	6	7	8	9	10	11	12	1	2
3	4	5	6	7	8	9	10	11	12	1	2	3
4	5	6	7	8	9	10	11	12	1	2	3	4
5	6	7	8	9	10	11	12	1	2	3	4	5
6	7	8	9	10	11	12	1	2	3	4	5	6
7	8	9	10	11	12	1	2	3	4	5	6	7
8	9	10	11	12	1	2	3	4	5	6	7	8
9	10	11	12	1	2	3	4	5	6	7	8	9
10	11	12	1	2	3	4	5	6	7	8	9	10
11	12	1	2	3	4	5	6	7	8	9	10	11
12	1	2	3	4	5	6	7	8	9	10	11	12

The binary operation of this mathematical system, $+$, is defined by Table 9.1 on page 540. To determine the value of $a + b$, where a and b are any two numbers in the set, find a in the left-hand column and find b along the top row. Assume that there is a horizontal line through a and a vertical line through b; the point of intersection of these two lines is where you find the value of $a + b$. For example, $10 + 4 = 2$ has been circled in Table 9.1. Note that $4 + 10$ also equals 2, but this result will not necessarily hold for all examples in this chapter.

Example 1 *A Commutative Group?*

Determine whether the clock 12 arithmetic system under the operation of addition is a commutative group.

Solution Check the five requirements that must be satisfied for a commutative group.

1. *Closure*: Is the set of elements in clock 12 arithmetic closed under the operation of addition? Yes it is, since Table 9.1 contains only the elements in the set $\{1, 2, 3, 4, 5, 6, 7, 8, 9, 10, 11, 12\}$. If Table 9.1 had contained an element other than the numbers 1 through 12, the set would not have been closed under addition.

2. *Identity element*: Is there an identity element for clock 12 arithmetic? If the time is currently 4 o'clock, how many hours have to pass before it is 4 o'clock again? Twelve hours: $4 + 12 = 12 + 4 = 4$. In fact, given any hour, in 12 hours the clock will return to the starting point. Therefore, 12 is the additive identity element in clock 12 arithmetic.

   In examining Table 9.1, we see that the row of numbers next to the 12 in the far-left column is identical to the top row of numbers. We also see that the column of numbers under the 12 in the top row is identical to the column of numbers on the far left. The search for such a column and row is one technique for determining whether an identity element exists for a system defined by a table.

3. *Inverse elements*: Is there an inverse for the number 4 in clock 12 arithmetic for the operation of addition? Recall that the identity element in clock 12 arithmetic is 12. What number when added to 4 gives 12, that is, $4 + = 12$? Table 9.1 shows that $4 + 8 = 12$ and also that $8 + 4 = 12$. Thus, 8 is the additive inverse of 4, and 4 is the additive inverse of 8.

   To determine the additive inverse of 7, find 7 in the far-left column of Table 9.1. Look to the right of the 7 until you come to the identity element 12. Then determine the number at the top of this column. The number is 5. Since $7 + 5 = 5 + 7 = 12$, 5 is the inverse of 7 and 7 is the inverse of 5. The other inverses can be determined in the same way. Table 9.2 shows each element in clock 12 arithmetic and its inverse. Note that each element in the set has an *inverse*.

Table 9.2 Clock 12 Inverses

Element	+	Inverse	=	Identity Element
1	+	11	=	12
2	+	10	=	12
3	+	9	=	12
4	+	8	=	12
5	+	7	=	12
6	+	6	=	12
7	+	5	=	12
8	+	4	=	12
9	+	3	=	12
10	+	2	=	12
11	+	1	=	12
12	+	12	=	12

4. *Associative property*: Now consider the associative property. Does $(a + b) + c = a + (b + c)$ for all values a, b, and c of the set? Remember to always evaluate the values within the parentheses first. Let's select some values for a, b, and c.

Let $a = 2$, $b = 6$, and $c = 8$. Then

$$(2 + 6) + 8 \stackrel{?}{=} 2 + (6 + 8)$$
$$8 + 8 \stackrel{?}{=} 2 + 2$$
$$4 = 4 \quad \text{True}$$

Let $a = 5$, $b = 12$, and $c = 9$. Then

$$(5 + 12) + 9 \stackrel{?}{=} 5 + (12 + 9)$$
$$5 + 9 \stackrel{?}{=} 5 + 9$$
$$2 = 2 \quad \text{True}$$

Randomly selecting *any* elements a, b, and c of the set reveals $(a + b) + c = a + (b + c)$. Thus, the system of clock 12 arithmetic is associative under the operation of addition. Note that if there is just one set of values a, b, and c such that $(a + b) + c \neq a + (b + c)$, the system is not associative. Usually, you will not be asked to check every case to determine whether the associative property holds. *However, if there is a row or a column that does not contain every element in the system, then the associative property must be checked carefully.*

5. *Commutative property*: Does the commutative property hold under the given operation? Does $a + b = b + a$ for all elements a and b of the set? Let's randomly select some values for a and b to determine whether the commutative property appears to hold. Let $a = 5$ and $b = 8$; then Table 9.1 shows that

$$5 + 8 \stackrel{?}{=} 8 + 5$$
$$1 = 1 \quad \text{True}$$

Let $a = 9$ and $b = 6$; then

$$9 + 6 \stackrel{?}{=} 6 + 9$$
$$3 = 3 \quad \text{True}$$

The commutative property holds for these two specific cases. In fact, if we were to select *any* values for a and b, we would find that $a + b = b + a$. Thus, the commutative property of addition is true in clock 12 arithmetic. Note that if there is just one set of values a and b such that $a + b \neq b + a$, the system is not commutative.

This system satisfies the five properties required for a mathematical system to be a commutative group. Thus, clock 12 arithmetic under the operation of addition is a commutative, or abelian, group.   ∎

Table 9.3 Symmetry about the Main Diagonal

+	0	1	2	3	4
0	0	1	2	3	4
1	1	2	3	4	0
2	2	3	4	0	1
3	3	4	0	1	2
4	4	0	1	2	3

One method that can be used to determine whether a system defined by a table is commutative under the given operation is to determine whether the elements in the table are symmetric about the *main diagonal*. The main diagonal is the diagonal from the upper left-hand corner to the lower right-hand corner of the table. In Table 9.3, the main diagonal is shaded in pink.

If the elements are symmetric about the main diagonal, then the system is commutative. If the elements are not symmetric about the main diagonal, then the system is not commutative. If you examine the system in Table 9.3, you see that its elements are symmetric about the main diagonal because the same numbers appear in the same relative positions on opposite sides of the main diagonal. Therefore, this mathematical system is commutative.

It is possible to have groups that are not commutative. Such groups are called *noncommutative* or *nonabelian groups*. However, a *noncommutative group must have at least six distinct elements*. Nonabelian groups are illustrated in Exercises 67 through 69 on page 549.

Now we will look at another finite mathematical system.

Table 9.4   A Four-Element System

©	2	4	6	8
2	6	8	2	4
4	8	2	4	6
6	2	4	6	8
8	4	6	8	2

Example 2   *A Finite Mathematical System*

Consider the mathematical system defined by Table 9.4. Assume that the associative property holds for the given operation.

a) List the elements in the set of this mathematical system.

b) Identify the binary operation.

c) Determine whether this mathematical system is a commutative group.

Solution

a) The set of elements for this mathematical system consists of the elements found in both the top row and the far-left column of the table: $\{2, 4, 6, 8\}$.

b) The binary operation is ©.

c) We will determine whether the five requirements for a commutative group are satisfied.

1. *Closure*: All the elements in the table are in the original set of elements $\{2, 4, 6, 8\}$, so the mathematical system is closed.

2. *Identity element*: The identity is 6. Note that the row of elements to the right of the 6 is identical to the top row *and* the column of elements under the 6 is identical to the far-left column.

3. *Inverse elements*: When an element operates on its inverse element, the result is the identity element. For this example, the identity is 6. To determine the inverse of 2, determine which element from the set, $\{2, 4, 6, \text{or } 8\}$ replaces the shaded area in the following equation and gives a true statement:

$$2 \ © \ \blacksquare = 6$$

From Table 9.4, we see that $2 © 2 = 6$. Thus, 2 is its own inverse.

     To determine the inverse of 4, determine which element replaces the shaded area in the following equation and gives a true statement:

$$4 \ © \ \blacksquare = 6$$

From Table 9.4, we see that $4 © 8 = 8 © 4 = 6$. Therefore, 4 is the inverse of 8, and 8 is the inverse of 4. The elements and their inverses are shown in Table 9.5. Note that every element in the mathematical system has an inverse.

Table 9.5   Inverses Under ©

Element	©	Inverse	=	Identity Element
2	©	2	=	6
4	©	8	=	6
6	©	6	=	6
8	©	4	=	6

4. *Associative property*: It is given that the associative property holds for this binary operation, ©. One example of the associative property is

$$(2 © 8) © 4 \stackrel{?}{=} 2 © (8 © 4)$$
$$4 © 4 \stackrel{?}{=} 2 © 6$$
$$2 = 2 \qquad \text{True}$$

5. *Commutative property*: The elements in Table 9.4 are symmetric about the main diagonal, so the commutative property holds for the operation of ©. One example of the commutative property is

$$2 © 4 \stackrel{?}{=} 4 © 2$$
$$8 = 8 \qquad \text{True}$$

The five necessary properties hold. Thus, the mathematical system is a commutative group. ∎

Timely Tip

When determining whether a finite mathematical system is a group, remember the following tips.

Closure: The top row and the far-left column of the table will show the elements in the set. If any other elements appear in the body of the table, then the system is not closed.

Identity: Look for a row in the table that matches the top row and a column that matches the far-left column. The element that corresponds to this row and this column is the identity element.

Inverses: To determine the inverse of an element, *a*, first find *a* in the far-left column of the table. Next, scan across the row that corresponds to *a* until you find the identity element. The number at the top of this column is the inverse of *a*.

Associative Property: If there is a row or column of the table that does not contain every element in the system, you must check the associative property carefully.

Commutative Property: If the elements are symmetric about the main diagonal, then the system is commutative.

Mathematical Systems Without Numbers

Thus far, all the systems we have discussed have been based on sets of numbers. Example 3 illustrates a mathematical system that uses symbols rather than numbers.

Table 9.6 A System Without Numbers

◇	e	ʊ	𝓛
e	𝓛	e	ʊ
ʊ	e	ʊ	𝓛
𝓛	ʊ	𝓛	e

Example 3 *Investigating a System Without Numbers*

Use the mathematical system defined by Table 9.6 to determine

a) the set of elements.
b) the binary operation.
c) whether the system is closed.
d) the identity element.
e) the inverse of e.
f) $(e \diamond 𝓛) \diamond e$ and $e \diamond (𝓛 \diamond e)$.
g) $ʊ \diamond 𝓛$ and $𝓛 \diamond ʊ$

Solution

a) The set of elements of this mathematical system is $\{e, ʊ, 𝓛\}$.

b) The binary operation is $\diamond$.

c) Because the table does not contain any symbols other than e, ʊ, and 𝓛, the system is closed under the binary operation, $\diamond$.

d) The identity element is ʊ. Locate the ʊ in the far-left column of Table 9.6. Note that the row of elements to the right of the ʊ matches the top row of elements. Also, locate the ʊ in the top row of elements. Note that the column of elements below the ʊ matches the far-left column of elements. We see that

$$e \diamond ʊ = ʊ \diamond e = e, \quad 𝓛 \diamond ʊ = ʊ \diamond 𝓛 = 𝓛, \quad \text{and } ʊ \diamond ʊ = ʊ$$

e) To determine the inverse of e, locate e in the far left column of Table 9.6. Now scan to the right of e until you find the identity element ʊ. At the top of this column we see the element 𝓛. Since $e \diamond 𝓛 = 𝓛 \diamond e = ʊ$, 𝓛 is the inverse of e.

f) We first evaluate the expressions within the parentheses.

$$(e \diamond 𝓛) \diamond e = ʊ \diamond e = e \quad \text{and} \quad e \diamond (𝓛 \diamond e) = e \diamond ʊ = e$$

g) $ʊ \diamond 𝓛 = 𝓛$ and $𝓛 \diamond ʊ = 𝓛$ ∎

Table 9.7

♫	α	β	γ	δ
α	δ	α	β	γ
β	α	β	γ	δ
γ	β	γ	δ	α
δ	γ	δ	α	β

Table 9.8 Inverses Under ♫

Element	♫	Inverse	=	Identity Element
α	♫	γ	=	β
β	♫	β	=	β
γ	♫	α	=	β
δ	♫	δ	=	β

Example 4 *Is the System a Commutative Group?*

Determine whether the mathematical system defined by Table 9.7 is a commutative group under the operation of ♫. Assume that the associative property holds for the given operation.

Solution

1. *Closure*: The system is closed.

2. *Identity element*: The identity element is β.

3. *Inverse elements*: Each element has an inverse, as illustrated in Table 9.8.

4. *Associative property*: It is given that the associative property holds. An example illustrating the associative property is

$$(\gamma \, ♫ \, \delta) \, ♫ \, \alpha \stackrel{?}{=} \gamma \, ♫ \, (\delta \, ♫ \, \alpha)$$
$$\alpha \, ♫ \, \alpha \stackrel{?}{=} \gamma \, ♫ \, \gamma$$
$$\delta = \delta \quad \text{True}$$

5. *Commutative property*: By examining Table 9.7, we can see that the elements are symmetric about the main diagonal. Thus, the system is commutative under the given operation, ♫. One example of the commutative property is

$$\alpha \, ♫ \, \delta \stackrel{?}{=} \delta \, ♫ \, \alpha$$
$$\gamma = \gamma \quad \text{True}$$

All five properties are satisfied. Thus, the system is a commutative group. ∎

Table 9.9

☺	$	%	@
$	$	@	%
%	@	%	$
@	%	$	@

Example 5 *Another System to Study*

Determine whether the mathematical system defined by Table 9.9 is a commutative group under the operation of ☺.

Solution

1. The system is closed.
2. No row is identical to the top row, so there is no identity element. Therefore, this mathematical system is *not a group*. There is no need to go any further, but for practice, let's look at a few more items.
3. Since there is no identity element, there can be no inverses.
4. The associative property does not hold. The following counterexample illustrates the associative property does not hold for every case.

$$(\$\,☺\,\%)\,☺\,@ \overset{?}{=} \$\,☺\,(\%\,☺\,@)$$
$$@\,☺\,@ \overset{?}{=} \$\,☺\,\$$$
$$@ \neq \$$$

5. The table is symmetric about the main diagonal. Therefore, the commutative property does hold for the operation of ☺.

   Note that the associative property does not hold even though the commutative property does hold. This outcome can occur when there is no identity element and not every element has an inverse, as in this example. ∎

Example 6 *Is the System a Commutative Group?*

Determine whether the mathematical system defined by Table 9.10 is a commutative group under the operation of ▲.

Table 9.10

▲	A	B	C
A	A	B	C
B	B	B	A
C	C	A	C

Solution

1. The system is closed.
2. There is an identity element, *A*.
3. Each element has an inverse; *A* is its own inverse, and *B* and *C* are inverses of each other.
4. Every element in the set does not appear in every row and every column of the table, so we need to check the associative property carefully. There are many specific cases in which the associative property does hold. However, the following counterexample illustrates that the associative property does not hold for every case.

$$(B\,▲\,B)\,▲\,C \overset{?}{=} B\,▲\,(B\,▲\,C)$$
$$B\,▲\,C \overset{?}{=} B\,▲\,A$$
$$A \neq B$$

5. The commutative property holds because there is symmetry about the main diagonal.

   Since we have shown that the associative property does not hold under the operation of ▲, this system is not a group. Therefore, it cannot be a commutative group. ∎

Did You Know?

Creating Patterns by Design

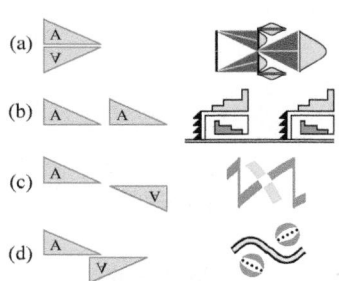

(a) (b) (c) (d)

Patterns from *Symmetries of Culture* by Dorothy K. Washburn and Donald W. Crowe (University of Washington Press, 1988)

▲ Patterns from Symmetries of Culture by Dorothy K. Washburn and Donald W. Crowe (University of Washington Press, 1988)

What makes group theory such a powerful tool is that it can be used to reveal the underlying structure of just about any physical phenomenon that involves symmetry and patterning, such as wallpaper or quilt patterns. Interest in the formal study of symmetry in design came out of the Industrial Revolution in the late nineteenth century. The new machines of the Industrial Revolution could vary any given pattern almost indefinitely. Designers needed a way to describe and manipulate patterns systematically. Shown here are the four geometric motions that generate all two-dimensional patterns: (a) reflection, (b) translation, (c) rotation, and (d) glide reflection. How these motions are applied, or not applied, is the basis of pattern analysis. The geometric motions, called *rigid motions*, were discussed in Section 8.5.

SECTION 9.2   *Exercises*

Warm Up Exercises

In Exercises 1–10, fill in the blanks with an appropriate word, phrase, or symbol(s).

1. In the mathematical system of clock 12 arithmetic, under the operation of addition, the set of elements is _____.

2. Since clock 12 arithmetic, under the operation of addition, contains only the elements $\{1, 2, 3, 4, 5, 6, 7, 8, 9, 10, 11, 12\}$, the mathematical system is _____.

3. In clock 12 arithmetic, since $a + 12 = 12 + a = a$, for any element a, we say that 12 is the additive _____ element.

4. In clock 12 arithmetic, since $1 + 11 = 11 + 1 = 12$, we say that 1 and 11 are additive _____.

5. In clock 12 arithmetic, for any elements a, b, and c, $(a + b) + c = a + (b + c)$; therefore, the _____ property holds under the operation of addition.

6. In clock 12 arithmetic, for any elements a and b, $a + b = b + a$; therefore, the _____ property holds under the operation of addition.

7. In clock 12 arithmetic under the operation of addition, since the system is closed, there is an identity element, each element has an inverse, and the associative and commutative properties hold, the mathematical system is a(n) _____ group.

8. In a finite mathematical system, if every element does not appear in every row and every column of the table, you need to check the _____ property carefully.

9. In a finite mathematical system defined by a table, if the elements are symmetric about the main diagonal, then the system is _____.

10. Groups that are not commutative are called noncommutative or _____ groups.

Practice the Skills

In Exercises 11–18, use Table 9.1 on page 540 to determine the sum in clock 12 arithmetic.

11. $7 + 6$

12. $9 + 8$

13. $8 + 7$

14. $11 + 7$

15. $4 + 12$

16. $12 + 12$

17. $3 + (8 + 9)$

18. $(8 + 7) + 6$

In Exercises 19–26, determine the difference in clock 12 arithmetic by starting at the first number and counting counterclockwise on the clock the number of units given by the second number.

19. $10 - 3$

20. $11 - 6$

21. $4 - 10$

22. $1 - 12$

23. $6 - 10$

24. $5 - 8$

25. $5 - 5$

26. $12 - 12$

27. Use the following figure to develop an addition table for clock 6 arithmetic. The figure will also be used in Exercises 28–34.

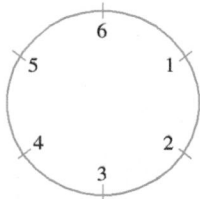

In Exercises 28–34, determine the sum or difference in clock 6 arithmetic.

28. $4 + 3$

29. $6 + 2$

30. $4 + 6$

31. $5 - 2$

32. $4 - 5$

33. $(3 - 5) - 6$

34. $2 + (1 - 3)$

35. Use the following figure to develop an addition table for clock 7 arithmetic. The figure will also be used in Exercises 36–43.

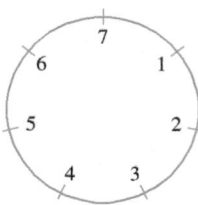

In Exercises 36–42, determine the sum or difference in clock 7 arithmetic.

36. $3 + 6$

37. $4 + 4$

38. $6 + 7$

39. $5 + 5$

40. $2 - 3$

41. $(4 - 5) - 6$

42. $3 - (2 - 6)$

43. Determine whether clock 7 arithmetic under the operation of addition is a commutative group. Explain.

44. Determine whether clock 6 arithmetic under the operation of addition is a commutative group. Explain.

In Exercises 45 and 46, determine if the system is closed. Explain how you determined your answer.

45.

⊗	x	y	z
x	x	y	z
y	y	z	z
z	z	t	y

46.

▽	a	b	c
a	a	b	c
b	b	c	a
c	c	a	b

In Exercises 47 and 48, determine if the system has an identity element. If so, state the identity element. Explain how you determined your answer.

47.

W	A	B	C
A	C	B	A
B	B	C	B
C	A	B	C

48.

⊖	□	⊙	△
□	△	□	⊙
⊙	⊙	△	□
△	□	⊙	△

In Exercises 49 and 50, the identity element is C. Determine the inverse, if it exists, of (a) A, (b) B, and (c) C.

49.

⫟	A	B	C
A	C	B	A
B	B	C	B
C	A	B	C

50.

⊗	A	B	C
A	A	A	A
B	A	C	B
C	A	B	C

In Exercises 51 and 52, determine if the system is commutative. Explain how you determined your answer.

51.

I	P	A	L
P	L	P	A
A	P	L	A
L	A	L	P

52.

⊟	A	⊗	W
A	⊗	W	A
⊗	W	A	⊗
W	A	⊗	⊗

53. Consider the mathematical system defined by the following table. Assume that the associative property holds for the given operation.

Q	0	2	4	6
0	0	2	4	6
2	2	4	6	0
4	4	6	0	2
6	6	0	2	4

a) What are the elements of the set in this mathematical system?

b) What is the binary operation?

c) Is the system closed? Explain.

d) Is there an identity element in the set? If so, what is it?

e) Does every element in the set have an inverse? If so, give each element and its corresponding inverse.

f) Give an example to illustrate the associative property.

g) Is the system commutative? Give an example to verify your answer.

h) Is the mathematical system a commutative group? Explain.

In Exercises 54–56, repeat parts (a)–(h) of Exercise 53 for the mathematical system defined by the given table. Assume that the associative property holds for the given operation.

54.

®	♣	♦	♥	♠
♣	♥	♠	♣	♦
♦	♠	♣	♦	♥
♥	♣	♦	♥	♠
♠	♦	♥	♠	♣

55.

$	4	5	L
4	5	L	4
5	L	4	5
L	4	5	L

56.

t	3	5	8	4
3	5	8	4	3
5	8	4	3	5
8	4	3	5	8
4	3	5	8	4

57. Consider the mathematical system defined by the following table.

☆	G	O	L	D
G	L	G	D	O
O	G	O	L	D
L	D	L	D	G
D	O	D	G	L

a) Is the system closed? Explain.

b) Is there an identity element in the set? If so, what is it?

c) For each element in the set, give the corresponding inverse element, if it exists.

d) Evaluate $(L \, ☆ \, L) \, ☆ \, D$ and $L \, ☆ \, (L \, ☆ \, D)$.

e) Does the associative property hold for the system?

f) Does the commutative property hold for the system?

g) Is the mathematical system a commutative group? Explain.

58. Consider the mathematical system defined by the following table.

✦	P	S	T	L
P	S	L	P	S
S	L	P	S	T
T	P	S	T	L
L	S	T	L	P

a) Is the system closed? Explain.

b) Is there an identity element in the set? If so, what is it?

c) For each element in the set, give the corresponding inverse element, if it exists.

d) Evaluate $(P ✦ L) ✦ L$ and $P ✦ (L ✦ L)$.

e) Does the associative property hold for the system?

f) Does the commutative property hold for the system?

g) Is the mathematical system a commutative group? Explain.

In Exercises 59–64, for the mathematical system given, determine which of the five properties of a commutative group do not hold.

59.

β	Δ	Φ	Γ
Δ	Δ	Φ	Γ
Φ	Φ	Γ	Δ
Γ	Γ	Δ	Π

60.

⊗	⊡	M	🔔
⊡	⊡	M	🔔
M	M	⊡	🔔
🔔	🔔	M	⊡

61.

⊡	⊙	C	?	T	P
⊙	T	P	⊙	C	C
C	P	⊙	C	⊙	T
?	⊙	C	?	T	P
T	C	⊙	T	P	?
P	C	T	P	?	⊙

62.

☺	a	b	π	0	Δ
a	Δ	a	b	π	0
b	a	b	π	0	Δ
π	b	π	0	Δ	a
0	π	0	Δ	a	b
Δ	0	Δ	π	b	a

63.

F	a	b	c	d	e
a	c	d	e	a	b
b	d	e	a	b	c
c	e	a	b	c	d
d	a	b	c	e	d
e	b	c	d	e	a

64.

▽	0	1	2	3	4	5
0	0	0	0	0	0	0
1	0	1	2	3	4	5
2	0	2	4	0	2	4
3	0	3	0	3	0	3
4	0	4	2	0	4	2
5	0	5	4	3	2	1

Problem Solving

65. a) Consider the set consisting of two elements $\{E, O\}$, where E stands for an even number and O stands for an odd number. For the operation of addition, complete the table.

+	E	O
E		
O		

b) Determine whether this mathematical system forms a commutative group under the operation of addition. Explain your answer.

66. a) Let E and O represent even numbers and odd numbers, respectively, as in Exercise 65. Complete the table for the operation of multiplication.

×	E	O
E		
O		

b) Determine whether this mathematical system forms a commutative group under the operation of multiplication. Explain your answer.

In Exercises 67 and 68 the tables shown below are examples of noncommutative, or nonabelian, groups. For each exercise, do the following.

a) *Show that the system under the given operation is a group. (It would be very time consuming to prove that the associative property holds, but give some examples to show that it appears to hold.)*

b) *Determine a counterexample to show that the commutative property does not hold.*

67.

?	1	2	3	4	5	6
1	5	3	4	2	6	1
2	4	6	5	1	3	2
3	2	1	6	5	4	3
4	3	5	1	6	2	4
5	6	4	2	3	1	5
6	1	2	3	4	5	6

68.

⊖	A	B	C	D	E	F
A	E	C	D	B	F	A
B	D	F	E	A	C	B
C	B	A	F	E	D	C
D	C	E	A	F	B	D
E	F	D	B	C	A	E
F	A	B	C	D	E	F

Challenge Problems/Group Activities

69. Book Arrangements—A Nonabelian Group Suppose that three books numbered 1, 2, and 3 are placed next to one another on a shelf. If we remove volume 3 and place it before volume 1, the new order of books is 3, 1, 2. Let's call this replacement R. We can write

$$R = \begin{pmatrix} 1 & 2 & 3 \\ 3 & 1 & 2 \end{pmatrix}$$

which indicates the books were switched in order from 1, 2, 3 to 3, 1, 2. Other possible replacements are S, T, U, V, and I, as indicated.

$$S = \begin{pmatrix} 1 & 2 & 3 \\ 2 & 1 & 3 \end{pmatrix} \quad T = \begin{pmatrix} 1 & 2 & 3 \\ 3 & 2 & 1 \end{pmatrix} \quad U = \begin{pmatrix} 1 & 2 & 3 \\ 1 & 3 & 2 \end{pmatrix}$$
$$V = \begin{pmatrix} 1 & 2 & 3 \\ 2 & 3 & 1 \end{pmatrix} \quad I = \begin{pmatrix} 1 & 2 & 3 \\ 1 & 2 & 3 \end{pmatrix}$$

Replacement set I indicates that the books were removed from the shelves and placed back in their original order. Consider the mathematical system with the set of elements $\{R, S, T, U, V, I\}$ with the operation of $*$.

To evaluate $R * S$, write

$$R * S = \begin{pmatrix} 1 & 2 & 3 \\ 3 & 1 & 2 \end{pmatrix} * \begin{pmatrix} 1 & 2 & 3 \\ 2 & 1 & 3 \end{pmatrix}$$

As shown in Fig. 9.2, R replaces 1 with 3 and S replaces 3 with 3 (no change), so $R * S$ replaces 1 with 3. R replaces 2 with 1 and S replaces 1 with 2, so $R * S$ replaces 2 with 2 (no change). R replaces 3 with 2 and S replaces 2 with 1, so $R * S$ replaces 3 with 1. $R * S$ replaces 1 with 3, 2 with 2, and 3 with 1.

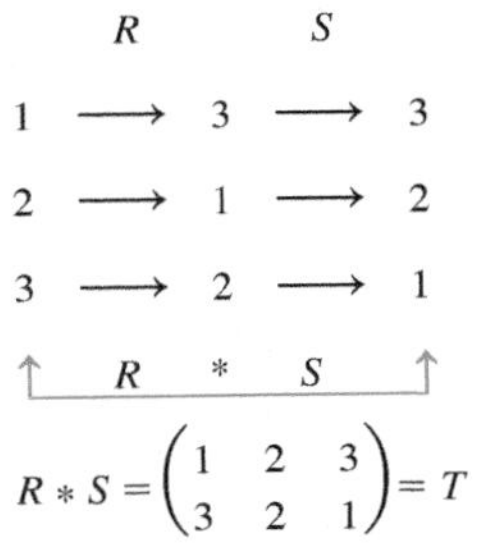

$$R * S = \begin{pmatrix} 1 & 2 & 3 \\ 3 & 2 & 1 \end{pmatrix} = T$$

Figure 9.2

Since this result is the same as replacement set T, we write $R * S = T$.

a) Complete the table for the operation using the procedure outlined.

*	R	S	T	U	V	I
R		T				
S						
T						
U						
V						
I						

b) Is this mathematical system a group? Explain.

c) Is this mathematical system a commutative group? Explain.

Recreational Mathematics

70. A Tiny Group Consider the mathematical system with the single element $\{1\}$ under the operation of multiplication.

a) Is this system closed?

b) Is there an identity element in the set?

c) Does 1 have an inverse?

d) Does the associative property hold?

e) Does the commutative property hold?

f) Is $\{1\}$ under the operation of multiplication a commutative group?

Research Activity

71. Twenty-Four-Hour Clock System Do research on the 24-hour clock system (see the *Did You Know?* box on page 541). Write a paper on its use throughout the world, including in the U.S. military and in medical facilities.

SECTION 9.3   Modular Arithmetic

Upon completion of this section, you will be able to:

- Solve problems involving modulo *m* systems.
- Determine whether a mathematical system defined by a modulo *m* system is a commutative group.

If your birthday is on a Monday this year, on what day of the week will it fall next year? What if next year is a leap year? When will your birthday next fall on a Saturday? These questions can be addressed with the type of arithmetic that we will study in this section.

Why This Is Important   *Modular arithmetic* is a fundamental part of many branches of mathematics. A wide variety of applications of modular arithmetic can be found in cryptography (the study of writing in secret codes), computer science, chemistry, music, and economics.

Modulo *m* Systems

Recall the mathematical system of clock arithmetic that we studied in Section 9.2. If we were to replace the number 12 with the number 0, the elements in the system could be rewritten as $\{0, 1, 2, 3, 4, 5, 6, 7, 8, 9, 10, 11\}$. This set together with the operation of addition forms a *modulo 12* or a *mod 12* arithmetic system. In this section we will study such mathematical systems.

A *modulo m system* consists of a set of *m* elements, $\{0, 1, 2, 3, \ldots, m - 1\}$, and a binary operation. To help explain how modulo *m* systems work, consider the following question: If today is Sunday, what day of the week will it be in 23 days? The answer, Tuesday, is arrived at by dividing 23 by 7. The quotient is 3 and the remainder is 2. Twenty-three days represent 3 weeks plus 2 days. Since we are interested only in the day of the week on which the twenty-third day will fall, the 3-week segment is unimportant to the answer. The remainder of 2 indicates the answer will be 2 days later than Sunday, which is Tuesday.

If we place the days of the week on a clock face as shown in Fig. 9.3, then, starting on Sunday, in 23 days the hand would have made three complete revolutions and end on Tuesday. If we replace the days of the week with numbers, then a modulo 7 arithmetic system will result. See Fig. 9.4: Sunday = 0, Monday = 1, Tuesday = 2, and so on. If we start at 0 and move the hand 23 places, we will end at 2.

Table 9.11   Modular 7 Addition

+	0	1	2	3	4	5	6
0	0	1	2	3	4	5	6
1	1	2	3	4	5	6	0
2	2	3	4	5	6	0	1
3	3	4	5	6	0	1	2
4	4	5	6	0	1	2	③
5	5	6	0	1	2	3	4
6	6	0	1	2	3	4	5

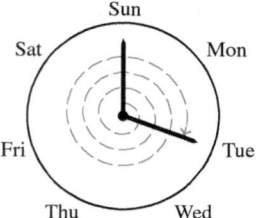

Figure 9.3

Figure 9.4

Figure 9.5

When working with a modulo *m* system it is often desirable to construct a table, similar to the ones used in Section 9.2. A modulo 7 addition table is shown in Table 9.11. Suppose we wish to calculate 4 + 6 in modulo 7. Consider the clock shown in Fig. 9.5. If we start at 4 and count clockwise 6 hours we end at 3. This number is circled in Table 9.11. The other numbers in Table 9.11 can be obtained in a similar manner.

A second method can be used to determine the sum of $4 + 6$ in modulo 7. We add 4 and 6 to get the number 10. Next, divide 10 by 7 and observe the remainder.

$$10 \div 7 = 1, \text{ remainder } 3$$

Therefore, the sum $4 + 6$ is 3 in a modulo 7 arithmetic system. We say that 10 is *congruent to* 3 in modulo 7. The concept of *congruence* is important in modular arithmetic.

> **Definition: Congruent**
> a is **congruent** to b modulo m, written, $a \equiv b \pmod{m}$, if a and b have the same remainder when divided by m.

We can show, for example, that $10 \equiv 3 \pmod 7$ by dividing both 10 and 3 by 7 and observing that we obtain the same remainder in each case.

$$10 \div 7 = 1, \quad \text{remainder } 3 \qquad \text{and} \qquad 3 \div 7 = 0, \quad \text{remainder } 3$$

Since the remainders are the same, 3 in each case, 10 is congruent to 3 modulo 7, and we write $10 \equiv 3 \pmod 7$.

Now consider $37 \equiv 5 \pmod 8$. If we divide both 37 and 5 by 8, each has the same remainder, 5.

In any modulo system, we can develop a set of *modulo classes* by placing all numbers with the same remainder in the appropriate modulo class. In a modulo 7 system, every number must have a remainder of either 0, 1, 2, 3, 4, 5, or 6. Thus, a modulo 7 system has seven modulo classes. The seven classes are presented in Table 9.12.

Every number is congruent to a number from 0 to 6 in modulo 7. For example, $24 \equiv 3 \pmod 7$ because 24 is in the same modulo class as 3.

The solution to a problem in modular arithmetic, if it exists, will always be a number from 0 through $m - 1$, where m is the *modulus* of the system. For example, in a modulo 7 system, because 7 is the modulus, the solution will be a number from 0 through 6.

Table 9.12 Modulo 7 Classes

0	1	2	3	4	5	6
0	1	2	3	4	5	6
7	8	9	10	11	12	13
14	15	16	17	18	19	20
21	22	23	24	25	26	27
28	29	30	31	32	33	34
⋮	⋮	⋮	⋮	⋮	⋮	⋮

Example 1 Congruence in Modulo 7

Determine which number, 0 through 6, the following numbers are equivalent to in modulo 7.

a) 44 b) 91 c) 165

Solution Although we could determine the answers by listing more entries in Table 9.12, this method becomes impractical with larger numbers. We will divide the given number by 7 and observe the remainder.

a) To determine the number 44 is equivalent to in mod 7, we divide 44 by 7 and observe the remainder.

$$44 \div 7 = 6, \text{ remainder } 2$$

Therefore, 44 is congruent to 2 in mod 7. We write $44 \equiv 2 \pmod 7$.

b) Divide 91 by 7 and observe the remainder.

$$91 \div 7 = 13, \text{ remainder } 0$$

Thus, $91 \equiv 0 \pmod 7$.

c) Divide 165 by 7 and observe the remainder.

$$165 \div 7 = 23 \text{ remainder } 4$$

Thus, $165 \equiv 4 \pmod 7$.

RECREATIONAL MATH

Cryptography

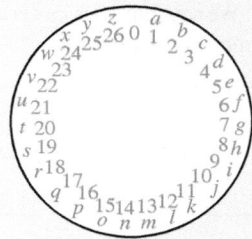

Cryptography is the science of communicating in secret codes. Modern cryptography involves the disciplines of mathematics, computer science, and engineering. Cryptography is used whenever we use an ATM, log on to a computer account, or purchase items on the Internet. Modular arithmetic has long been used in cryptography (See the *Did You Know?* box on page 554). One example of a coding circle is shown above.

To use it, the person sending and receiving the message must know the code key to decipher the code. The code key for the following message given by the 11 numbers below is the letter *j*. Can you decipher this code? (*Hint:* Subtract the code key from the code numbers.) The answer is upside down at the bottom of this box.

23 11 3 18 10
19 2 10 16 4 24

Exercise 79 on page 558 asks you to decipher another message.

Math is fun.

Example 2 *Congruence in Modulo 9*

Evaluate each of the following in modulo 9.

a) $7 + 8$ b) $8 - 5$ c) $7 \cdot 5$

Solution Since we are working in modulo 9, the answer in each part will be a number from 0 through 8.

a) The sum, $7 + 8$, is 15. Since 15 is not a number from 0 through 8, we must determine the number in modulo 9 to which 15 is equivalent. $15 \div 9 = 1$, remainder 6. Therefore, $7 + 8 \equiv 6 \pmod 9$.

b) The difference, $8 - 5$, is 3. Since 3 is a number from 0 through 8, we have $8 - 5 \equiv 3 \pmod 9$.

c) The product, $7 \cdot 5$, is 35. Since 35 is not a number from 0 through 8, we must determine the number in modulo 9 to which 35 is equivalent. Because $35 \div 9 = 3$, remainder 8, we write $7 \cdot 5 \equiv 8 \pmod 9$. ∎

Note in Table 9.12 on page 551 that every number in the same modulo class differs by a multiple of the modulus, in this case a multiple of 7. Adding (or subtracting) a multiple of the modulus to (or from) a given number does not change the modulo class or congruence of the given number. For example, $3, 3 + 1(7), 3 + 2(7), 3 + 3(7), \dots, 3 + n(7)$ are all in the same modulo class, namely, 3. We use this fact in the solution to Example 3.

Example 3 *Using Modulo Classes in Subtraction*

Determine the positive number replacement (less than the modulus) for the question mark that makes the statement true.

a) $3 - 5 \equiv ? \pmod 7$ b) $? - 4 \equiv 3 \pmod 5$ c) $5 - ? \equiv 7 \pmod 8$

Solution In each part, we wish to replace the question mark with a positive number less than the modulus. Therefore, in part (a) we wish to replace the ? with a positive number less than 7. In part (b), we wish to replace the ? with a positive number less than 5. In part (c), we wish to replace the ? with a positive number less than 8.

a) In mod 7, adding 7, or a multiple of 7, to a number results in a sum that is in the same modulo class. Thus, if we add 7, 14, 21, ... to 3, the result will be a number in the same modulo class. We want to replace 3 with an equivalent mod 7 number that is greater than 5. Adding 7 to 3 yields a sum of 10, which is greater than 5.

$$3 - 5 \equiv ? \pmod 7$$
$$(3 + 7) - 5 \equiv ? \pmod 7$$
$$10 - 5 \equiv ? \pmod 7$$
$$5 \equiv ? \pmod 7$$
$$5 \equiv 5 \pmod 7$$

Therefore, $? = 5$ and $3 - 5 \equiv 5 \pmod 7$.

b) We wish to replace the ? with a positive number less than 5. We know that $7 - 4 \equiv 3 \bmod 5$ because $3 \equiv 3 \bmod 5$. Therefore, we need to determine what positive number, less than 5, the number 7 is congruent to in mod 5. If we subtract the modulus, 5, from 7, we obtain 2. Thus, 2 and 7 are in the same modular class. Therefore, $? = 2$.

$$? - 4 \equiv 3 \pmod 5$$
$$7 - 4 \equiv 3 \pmod 5$$
$$2 - 4 \equiv 3 \pmod 5$$

Notice in the last equivalence that if we add 5 to 2, we get $7 - 4 \equiv 3 \pmod 5$, which is a true statement.

c) $5 - ? \equiv 7 \ (\text{mod } 8)$

In mod 8, adding 8, or a multiple of 8, to a number results in a sum that is in the same modulo class. Thus, we can add 8 to 5 so that the statement becomes

$$(8 + 5) - ? \equiv 7 \ (\text{mod } 8)$$
$$13 - ? \equiv 7 \ (\text{mod } 8)$$

We can see that $13 - 6 = 7$. Therefore, $? = 6$ and $5 - 6 \equiv 7 \ (\text{mod } 8)$. ∎

Example 4 *Using Modulo Classes in Multiplication*

Determine all positive number replacements, less than the modulus, for the question mark that make the statement true.

a) $2 \cdot ? \equiv 3 \ (\text{mod } 5)$ b) $3 \cdot ? \equiv 0 \ (\text{mod } 6)$ c) $3 \cdot ? \equiv 2 \ (\text{mod } 6)$

Solution

a) One method of determining the solution is to replace the question mark with the numbers 0, 1, 2, 3, and 4 and then determine the equivalent modulo class of the product. We use the numbers $0 - 4$ because we are working in modulo 5.

$$2 \cdot ? \equiv 3 \ (\text{mod } 5)$$
$$2 \cdot 0 \equiv 0 \ (\text{mod } 5)$$
$$2 \cdot 1 \equiv 2 \ (\text{mod } 5)$$
$$2 \cdot 2 \equiv 4 \ (\text{mod } 5)$$
$$2 \cdot 3 \equiv 1 \ (\text{mod } 5)$$
$$2 \cdot 4 \equiv 3 \ (\text{mod } 5)$$

Therefore, the question mark can be replaced with 4, since $2 \cdot 4 \equiv 3 \ (\text{mod } 5)$.

b) Since we are working in modulo 6, replace the question mark with the numbers $0 - 5$ and follow the procedure used in part (a).

$$3 \cdot ? \equiv 0 \ (\text{mod } 6)$$
$$3 \cdot 0 \equiv 0 \ (\text{mod } 6)$$
$$3 \cdot 1 \equiv 3 \ (\text{mod } 6)$$
$$3 \cdot 2 \equiv 0 \ (\text{mod } 6)$$
$$3 \cdot 3 \equiv 3 \ (\text{mod } 6)$$
$$3 \cdot 4 \equiv 0 \ (\text{mod } 6)$$
$$3 \cdot 5 \equiv 3 \ (\text{mod } 6)$$

Therefore, replacing the question mark with 0, 2, or 4 results in true statements. The answers are 0, 2, and 4.

c) $3 \cdot ? \equiv 2 \ (\text{mod } 6)$

Examining the products in part (b) shows there are no values that satisfy the statement. The answer is "no solution." ∎

Whenever a process is repetitive, modular arithmetic may be helpful in answering some questions about the process. Now let's look at an application of modular arithmetic.

Example 5 *Work Schedule*

Jackson is a nurse at a hospital. His working schedule is to work for 6 days, and then he gets 2 days off. If today is the third day that Jackson has been working, determine the following.

a) Will he be working 60 days from today?

b) Will he be working 82 days from today?

c) Was he working 124 days ago?

Did You Know?

The Enigma

A cipher, or secret code, actually has group properties: A well-defined operation turns plain text into a cipher, and an inverse operation allows it to be deciphered. During World War II, the Germans used an encrypting device based on a modulo 26 system (for the 26 letters of the alphabet) known as the Enigma. It has four rotors, each with the 26 letters. The rotors were wired to one another so that if an A were typed into the machine, the first rotor would contact a different letter on the second rotor, which contacted a different letter on the third rotor, and so on. The French secret service obtained the wiring instructions, and Polish and British cryptoanalysts were able to crack the ciphers using group theory. This breakthrough helped to keep the Allies abreast of the deployment of the German Navy. The 2014 movie, *the Imitation Game* is a true story about breaking the Enigma.

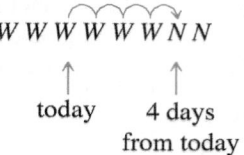

a) Since Jackson works for 6 days and then gets 2 days off, his working schedule may be considered a modular 8 system. That is, 8, 16, 24, … days from today will be just like today, the third day of the 8-day cycle.

If we divide 60 by 8, we obtain

$$\begin{array}{r} 7 \\ 8\overline{)60} \\ \underline{56} \\ 4 \end{array} \leftarrow \text{remainder}$$

Therefore, in 60 days Jackson will go through 7 complete cycles and will be 4 days further into the next cycle. If we let W represent a working day and N represent a nonworking day, then Jackson's cycle may be represented as follows:

$$W\ W\ W\ W\ W\ W\overset{\frown\frown\frown}{\ N\ N}$$

$$\underset{\text{today}}{\uparrow} \qquad \underset{\substack{\text{4 days} \\ \text{from today}}}{\uparrow}$$

Notice that 4 days from today will be his first nonworking day of this cycle. Therefore, Jackson will not be working 60 days from today.

b) We work this part in the same way we worked part (a). Divide 82 by 8 and determine the remainder.

$$\begin{array}{r} 10 \\ 8\overline{)82} \\ \underline{80} \\ 2 \end{array} \leftarrow \text{remainder}$$

Thus, in 82 days it will be 2 days later in the cycle than it is today. Because he is currently in day 3 of his cycle, Jackson will be in day 5 of his cycle and will be working.

c) This part is worked in the same way as parts (a) and (b), but once we determine the remainder we must move backward in the cycle.

$$\begin{array}{r} 15 \\ 8\overline{)124} \\ \underline{120} \\ 4 \end{array} \leftarrow \text{remainder}$$

Thus, 124 days ago is equivalent to 4 days earlier in the cycle. Marking day 3 of the cycle (indicated by the word *today*) and then moving 4 days backwards brings us to the first nonworking day. The two N's shown at the beginning of the letters below are actually the end days of the previous cycle.

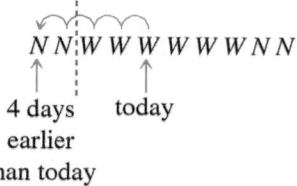

$$N\ N\ \vdots\ W\ W\ W\ W\ W\ W\ N\ N$$

$$\underset{\substack{\text{4 days} \\ \text{earlier} \\ \text{than today}}}{\uparrow} \qquad \underset{\text{today}}{\uparrow}$$

Therefore, 124 days ago Jackson was not working. ∎

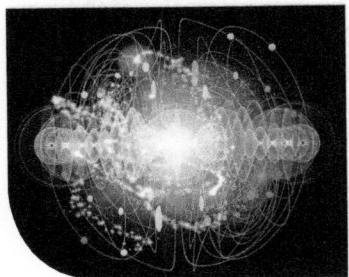
Modular Arithmetic and Groups

Modular arithmetic systems under the operation of addition are commutative groups, as illustrated in Example 6.

Example 6 *A Commutative Group*

Construct a mod 5 addition table and show that the mathematical system is a commutative group. Assume that the associative property holds for the given operation.

Solution The set of elements in modulo 5 arithmetic is $\{0, 1, 2, 3, 4\}$; the binary operation is $+$.

+	0	1	2	3	4
0	0	1	2	3	4
1	1	2	3	4	0
2	2	3	4	0	1
3	3	4	0	1	2
4	4	0	1	2	3

For this system to be a commutative group, it must satisfy the five properties of a commutative group.

1. *Closure*: Every entry in the table is a member of the set $\{0, 1, 2, 3, 4\}$, so the system is closed under addition.

2. *Identity element*: An easy way to determine whether there is an identity element is to look for a row in the table that is identical to the elements at the top of the table. Note that the row next to 0 is identical to the top of the table, which indicates that 0 *might be* the identity element. Now look at the column under the 0 at the top of the table. If this column is identical to the far-left column, then 0 is the identity element. Since the column under 0 is the same as the far-left column, 0 is the additive identity in modulo 5 arithmetic.

Element	+	Identity	=	Element
0	+	0	=	0
1	+	0	=	1
2	+	0	=	2
3	+	0	=	3
4	+	0	=	4

3. *Inverse elements*: Does every element have an inverse? Recall that an element plus its inverse must equal the identity element. In this example, the identity element is 0. Therefore, for each of the given elements 0, 1, 2, 3, and 4, we must find the element that when added to it results in a sum of zero. These elements will be the corresponding inverses.

Did You Know?

The Enormous Theorem

The largest theorem known to the mathematical world involves the classification of a special type of group. The official name of the theorem is the *classification of finite, simple groups*, but it is commonly referred to by mathematicians as the *enormous theorem*. The full proof consists of more than 15,000 pages in about 500 separate publications written by more than 100 mathematicians. In addition, many parts of the proof were completed by computer and some parts are so complex or massive that no particular mathematician can claim to have a complete understanding of the entire proof. Instead, mathematicians must rely on the reputed expertise of the individual contributors to the theorem, including computer programmers and chip designers.

Element	+	Inverse	=	Identity	
0	+	?	=	0	Since $0 + 0 \equiv 0$, 0 is its own inverse.
1	+	?	=	0	Since $1 + 4 \equiv 0$, 4 is the inverse of 1.
2	+	?	=	0	Since $2 + 3 \equiv 0$, 3 is the inverse of 2.
3	+	?	=	0	Since $3 + 2 \equiv 0$, 2 is the inverse of 3.
4	+	?	=	0	Since $4 + 1 \equiv 0$, 1 is the inverse of 4.

Note that each element has an inverse.

4. *Associative property*: It is given that the associative property holds. One example that illustrates the associative property is

$$(2 + 3) + 4 \stackrel{?}{=} 2 + (3 + 4)$$
$$0 + 4 \stackrel{?}{=} 2 + 2$$
$$4 = 4 \quad \text{True}$$

5. *Commutative property*: Is $a + b = b + a$ for *all* elements a and b of the given set? The table shows that the system is commutative because the elements are symmetric about the main diagonal. We will give one example to illustrate the commutative property.

$$4 + 2 \stackrel{?}{=} 2 + 4$$
$$1 = 1 \quad \text{True}$$

All five properties are satisfied. Thus, modulo 5 arithmetic under the operation of addition is a commutative group. ■

SECTION 9.3   Exercises

Warm Up Exercises

In Exercises 1–6, fill in the blanks with an appropriate word, phrase, or symbol(s).

1. The elements in a modulo m system consist of the m numbers, 0 through _____.

2. In a modulo m system, in addition to the m elements, there is also a(n) _____ operation.

3. In any modulo m system, we can develop a set of modulo classes by dividing the numbers by m and placing all numbers with the same _____ in the appropriate modulo class.

4. In a modulo m system, the number m is called the _____ of the system.

5. If a and b have the same remainder when divided by m, we say that a is _____ to b modulo m.

6. To determine the value that 45 is congruent to in modulo 7, we _____ 45 by 7 and determine the remainder.

Practice the Skills

In Exercises 7–14, assume that Sunday is represented as day 0, Monday is represented as day 1, and so on. If today is Thursday (day 4), determine the day of the week it will be in the specified number of days. Assume no leap years.

7. 17 days

8. 58

9. 105 days

10. 365 days

11. 3 years

12. 4 years, 34 days

13. 400 days

14. 5 years, 25 days

In Exercises 15–22, consider the 12 months to be a modulo 12 system with January being month 0. If it is currently October, determine the month it will be in the specified number of months.

15. 22 months

16. 49 months

17. 2 years, 10 months

18. 4 years, 8 months

19. 83 months

20. 7 years

21. 105 months

22. 5 years, 9 months

In Exercises 23–32, determine what number the sum, differ-ence, or product is congruent to in modulo 5.

23. $4 + 1$ **24.** $2 + 6$ **25.** $10 - 1$

26. $12 - 5$ **27.** $8 \cdot 9$ **28.** $8 \cdot 7$

29. $4 - 8$ **30.** $2 - 4$ **31.** $(15 \cdot 4) - 8$

32. $(4 - 9) \cdot 7$

In Exercises 33–42, determine the modulo class to which each number belongs for the indicated modulo system.

33. 15, mod 2   **34.** 27, mod 4   **35.** 77, mod 8

36. 43, mod 6   **37.** 41, mod 9   **38.** 75, mod 8

39. −1, mod 7   **40.** −7, mod 4   **41.** −27, mod 8

42. −39, mod 7

In Exercises 43–56, determine all positive number replace-ments (less than the modulus) for the question mark that make the statement true.

43. $? + 5 \equiv 3 \pmod 6$ **44.** $? + 3 \equiv 2 \pmod 5$

45. $6 + ? \equiv 2 \pmod 7$ **46.** $4 + ? \equiv 3 \pmod 6$

47. $4 - ? \equiv 5 \pmod 6$ **48.** $3 - ? \equiv 7 \pmod 9$

49. $5 \cdot ? \equiv 7 \pmod 9$ **50.** $4 \cdot ? \equiv 3 \pmod 7$

51. $2 \cdot ? \equiv 1 \pmod 6$ **52.** $3 \cdot ? \equiv 5 \pmod 6$

53. $4 \cdot ? \equiv 4 \pmod{10}$ **54.** $3 \cdot ? \equiv 3 \pmod{12}$

55. $? - 8 \equiv 9 \pmod{12}$ **56.** $? - 3 \equiv 4 \pmod 8$

Problem Solving

57. *Presidential Elections* In the United States, presidential elections have been held every four years starting in 1788. Each of these years is congruent to 0 in modulo 4.

   a) List the first five presidential election years that oc-curred after 1788.

   b) List the first five presidential election years that will occur after 2020.

   c) List the presidential election years that will occur be-tween 2550 and 2575.

58. *Governors' Elections* Wisconsin gubernatorial elections (elec-tions for governor) have been held every four years starting in 1970. Each of these years is congruent to 2 in modulo 4.

   a) List the first five Wisconsin gubernatorial election years held after 1970.

   b) List the first five Wisconsin gubernatorial election years that will occur after 2020.

   c) List the five Wisconsin gubernatorial election years that will occur between 2100 and 2120.

59. *Golf Pro Schedule* Jahlil is a golf pro who works at two different golf courses. His schedule is to work at Longboat Key golf course for 4 days, then work at The Meadows golf course for 3 days, and then rest for 2 days. He then repeats this cycle. If Jahlil is currently on his first day of rest, determine what he will be doing

   a) 30 days from today.

   b) 100 days from today.

   c) 366 days from today.

60. *Flight Schedules* A pilot is scheduled to fly for 5 consecu-tive days and rest for 3 consecutive days. If today is the second day of her rest shift, determine whether she will be flying or resting

   a) 60 days from today.   **b)** 90 days from today.

   c) 240 days from today.   **d)** Was she flying 7 days ago?

   e) Was she flying 20 days ago?

61. *The Weekend Off* A manager of a retail store has both Saturday and Sunday off every 7 weeks. It is week 2 of the 7 weeks.

   a) Determine the number of weeks before he will have both Saturday and Sunday off.

   b) Will he have both Saturday and Sunday off 25 weeks from this week?

   c) What is the first week, after 50 weeks from this week, that he will have both Saturday and Sunday off?

62. *Physical Therapy* A man has an Achilles' tendon injury and is receiving physical therapy. He must have physical therapy twice a day for 5 days, physical therapy once a day for 3 days, and then 2 days of rest; then the cycle begins again. If he is in his second day of his twice-a-day therapy cycle, determine what he will be doing

   a) 20 days from today.

   b) 46 days from today.

   c) 107 days from today.

   d) Will the man have a day off 78 days from today?

63. *Nursing Shifts* A nurse's work pattern at Community Hospital consists of working the 7 A.M. –3 P.M. shift for 3 weeks and then the 3 P.M. –11 P.M. shift for 2 weeks.

a) If it is the third week of the pattern, what shift will the nurse be working 6 weeks from now?

b) If it is the fourth week of the pattern, what shift will the nurse be working 7 weeks from now?

c) If it is the first week of the pattern, what shift will the nurse be working 11 weeks from now?

64. *Restaurant Rotation* A truck driver works both daytime and evening shifts. He works daytime for 5 consecutive days, then evenings for 3 consecutive days, then daytime for 4 consecutive days, then evenings for 2 consecutive days. Then the rotation starts again. If it is day 2 of the 5-day consecutive daytime shift, determine whether he will be working the daytime or evening shift

a) 20 days from today.

b) 52 days from today.

c) 365 days from today.

65. a) Construct a modulo 3 addition table.

b) Is the system closed? Explain.

c) Is there an identity element in the set? If so, what is it?

d) Does every element in the set have an inverse? If so, list the elements and their inverses.

e) The associative property holds for the system. Give an example.

f) Does the commutative property hold for the system? Give an example.

g) Is the system a commutative group?

h) Will every modulo system under the operation of addition be a commutative group? Explain.

66. Construct a modulo 8 addition table. Repeat parts (b)–(h) in Exercise 65.

67. a) Construct a modulo 4 multiplication table.

b) Is the system closed under the operation of multiplication?

c) Is there an identity element in the set? If so, what is it?

d) Does every element in the set have an inverse? Make a list showing the elements that have a multiplicative inverse and list the inverses.

e) The associative property holds for the system. Give an example.

f) Does the commutative property hold for the system? Give an example.

g) Is this mathematical system a commutative group? Explain.

68. Construct a modulo 7 multiplication table. Repeat parts (b)–(g) in Exercise 67.

Challenge Problems/Group Activities

We have not discussed division in modular arithmetic. With what number or numbers, if any, can you replace the question marks to make the statement true? The question mark must be a positive number less than the modulus. (Hint: Use the fact that $\frac{a}{b} \equiv c \ (mod \ m)$ means $a \equiv b \cdot c \ (mod \ m)$ and use trial and error to obtain your answer.)

69. $? \div 3 \equiv 4 \ (mod \ 5)$

70. $3 \div 4 \equiv ? \ (mod \ 7)$

71. $? \div ? \equiv 1 \ (mod \ 4)$

72. $1 \div 2 \equiv ? \ (mod \ 5)$

In Exercises 73–75, solve for x where k is any counting number.

73. $8k \equiv x \ (mod \ 8)$

74. $7k + 1 \equiv x \ (mod \ 7)$

75. $4k - 2 \equiv x \ (mod \ 4)$

76. Determine the smallest positive number divisible by 5 to which 2 is congruent in modulo 6.

77. *Birthday Question* During a certain year, Clarence's birthday is on Monday, April 18.

a) If next year is *not* a leap year, on what day of the week will Clarence's birthday fall next year?

b) If next year *is* a leap year, on what day of the week will Clarence's birthday fall next year?

78. *More Birthday Questions* During a certain leap year, Josephine's birthday is on Tuesday, May 18. What is the minimum number of years in which Josephine's birthday will fall on a

a) Friday?   **b)** Saturday?

Recreational Math

79. *Cryptography* Read the *Recreational Mathematics* box on page 552. Using the same coding circle and a code key of the letter *e*, decipher the following message.

16	19	20	1	17	10	9	12	10
5	14	24	5	21	20	1	10	23

80. **Rolling Wheel** The wheel shown is to be rolled. Before the wheel is rolled, it is resting on number 0. The wheel will be rolled at a uniform rate of one complete roll every 4 minutes. In exactly 1 year (not a leap year), what number will be at the bottom of the wheel?

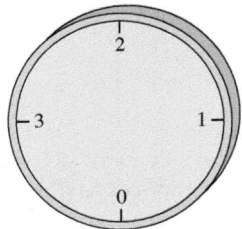

Research Activities

81. **The Enigma Cipher Machine** Do research on the Enigma ciphering machine (see the *Did You Know?* box on page 554). Write a paper on the role the machine played prior to and during World War II.

82. **Carl Friedrich Gauss** The concept and notation for modular systems were introduced by Carl Friedrich Gauss in 1801. Write a paper on Gauss's contribution to modular systems.

SECTION 9.4   Matrices

Upon completion of this section, you will be able to:

- Add matrices.
- Subtract matrices.
- Multiply matrices by a real number.
- Multiply two matrices.
- Show that matrices can be used to form a commutative group.

Five hundred students at the University of Delaware were asked if they were in favor of or opposed to an increase in their student fees to pay for building a new meeting room for student clubs. Each student was also asked to indicate whether he or she was a freshman, sophomore, junior, or senior. How can the responses from the survey be displayed? In this section, we will introduce a method used to display information such as responses from a survey.

Why This Is Important We often need an efficient way to display information such as responses from a survey, a company's current inventory, or when creating a budget.

So far in this chapter we have discussed various mathematical systems that are also groups. In this section we will introduce mathematical systems that involve *matrices*. We will also show that certain sets of matrices, along with the operation of addition, form commutative groups. We begin with some basic definitions.

Matrices

A *matrix* is a rectangular array of real numbers, symbols, or expressions. An array is a systematic arrangement of numbers or symbols in rows and columns. Matrices (the plural of matrix) may be used to display information and to solve systems of linear equations. The following matrix displays the responses from the survey of 500 students at the University of Delaware regarding an increase in their student fees.

	Freshmen	Sophomores	Juniors	Seniors
In favor	102	93	22	35
Opposed	82	94	23	49

The numbers, symbols, or expressions in the rows and columns of a matrix are called the *elements* of the matrix. The matrix given above contains eight elements. The *dimensions* of a matrix may be indicated with the notation $m \times n$, where m is the number of rows and n is the number of columns in the matrix. Because the given

MATHEMATICS TODAY

Matrices Are Everywhere

You are already familiar with matrices, although you may not be aware of it. A matrix is a good way to display numerical data, as illustrated on this trail sign in Yosemite National Park.

Why This Is Important Most charts and tables you see daily, including spreadsheets, are matrices.

matrix has 2 rows and 4 columns, it is a 2 by 4, written 2×4, matrix. In this text, from this point onward, we use brackets, $[\ \ ]$, to indicate a matrix. Consider the two matrices below. A matrix that contains the same number of rows and columns is called a *square matrix*. Following is an example of a 2×2 square matrix and a 3×3 square matrix.

$$\begin{bmatrix} 2 & 3 \\ 5 & 2 \end{bmatrix} \qquad \begin{bmatrix} 4 & 6 & -1 \\ 2 & 3 & 0 \\ 5 & 2 & 1 \end{bmatrix}$$

Two matrices are equal if and only if they have the same elements in the same relative positions. We use capital letters to represent matrices as illustrated in Example 1.

Example 1 *Equal Matrices*

Given $A = B$, determine x and y.

$$A = \begin{bmatrix} 2 & 5 \\ 7 & -1 \end{bmatrix}, \qquad B = \begin{bmatrix} x & 5 \\ 7 & y \end{bmatrix}$$

Solution Since the matrices are equal, the corresponding elements must be the same, so $x = 2$ and $y = -1$. ∎

Addition of Matrices

Two matrices can be added only if they have the same dimensions; that is, the same number of rows and the same number of columns. To obtain the sum of two matrices with the same dimensions, add the corresponding elements of the two matrices.

Example 2 *Adding Matrices*

Determine $A + B$ if

$$A = \begin{bmatrix} 4 & 6 \\ 3 & -3 \end{bmatrix} \quad \text{and} \quad B = \begin{bmatrix} 1 & 8 \\ 7 & 0 \end{bmatrix}$$

Solution $A + B = \begin{bmatrix} 4 & 6 \\ 3 & -3 \end{bmatrix} + \begin{bmatrix} 1 & 8 \\ 7 & 0 \end{bmatrix}$

$$= \begin{bmatrix} 4 + 1 & 6 + 8 \\ 3 + 7 & -3 + 0 \end{bmatrix} = \begin{bmatrix} 5 & 14 \\ 10 & -3 \end{bmatrix}$$ ∎

Matrix Addition

Example 3 MODELING —*Sales of Bicycles*

Towpath Bicycle Corporation owns and operates two stores, one in Virginia and one in North Carolina. The number of mountain bicycles (MB) and racing bicycles (RB) sold in each store during January through June and during July through December are indicated in the matrices that follow. We will call the matrices A and B.

	Virginia		North Carolina	
	MB	RB	MB	RB
Jan.–June	$\begin{bmatrix} 515 & 425 \\ 290 & 250 \end{bmatrix} = A$		$\begin{bmatrix} 520 & 350 \\ 180 & 271 \end{bmatrix} = B$	
July–Dec.				

Determine the total number of each type of bicycle sold by the corporation during each time period.

Solution To solve the problem, we add matrices A and B.

$$\begin{array}{c}\text{Jan.–June}\\\text{July–Dec.}\end{array}\begin{array}{cc}MB & RB\\\left[\begin{array}{cc}515+520 & 425+350\\290+180 & 250+271\end{array}\right]\end{array}=\begin{array}{cc}MB & RB\\\left[\begin{array}{cc}1035 & 775\\470 & 521\end{array}\right]\end{array}$$

We can see from the sum matrix that during the period from January through June, a total of 1035 mountain bicycles and 775 racing bicycles were sold. During the period from July through December, a total of 470 mountain bicycles and 521 racing bicycles were sold. ∎

Subtraction of Matrices

Only matrices with the same dimension may be subtracted. To do so, we subtract each entry in one matrix from the corresponding entry in the other matrix.

Example 4 *Subtracting Matrices*

Determine $A - B$ if

$$A = \begin{bmatrix}3 & 6\\5 & -1\end{bmatrix} \quad \text{and} \quad B = \begin{bmatrix}2 & -4\\8 & -3\end{bmatrix}$$

Solution

$$A - B = \begin{bmatrix}3 & 6\\5 & -1\end{bmatrix} - \begin{bmatrix}2 & -4\\8 & -3\end{bmatrix}$$
$$= \begin{bmatrix}3-2 & 6-(-4)\\5-8 & -1-(-3)\end{bmatrix} = \begin{bmatrix}1 & 10\\-3 & 2\end{bmatrix}$$ ∎

Multiplying a Matrix by a Real Number

A matrix may be multiplied by a real number by multiplying each entry in the matrix by the real number. When we multiply a matrix by a real number, we call that real number a *scalar*.

Example 5 *Multiplying a Matrix by a Scalar*

For matrices A and B, determine (a) $3A$ and (b) $3A - 2B$.

$$A = \begin{bmatrix}1 & 4\\-3 & 5\end{bmatrix}, \quad B = \begin{bmatrix}-1 & 3\\5 & 6\end{bmatrix}$$

Solution

a) The number 3 is a scalar, since we are multiplying matrix A by 3.

$$3A = 3\begin{bmatrix}1 & 4\\-3 & 5\end{bmatrix} = \begin{bmatrix}3(1) & 3(4)\\3(-3) & 3(5)\end{bmatrix} = \begin{bmatrix}3 & 12\\-9 & 15\end{bmatrix}$$

b) We determined $3A$ in part (a). Now we find $2B$.

$$2B = 2\begin{bmatrix}-1 & 3\\5 & 6\end{bmatrix} = \begin{bmatrix}2(-1) & 2(3)\\2(5) & 2(6)\end{bmatrix} = \begin{bmatrix}-2 & 6\\10 & 12\end{bmatrix}$$

$$3A - 2B = \begin{bmatrix}3 & 12\\-9 & 15\end{bmatrix} - \begin{bmatrix}-2 & 6\\10 & 12\end{bmatrix}$$
$$= \begin{bmatrix}3-(-2) & 12-6\\-9-10 & 15-12\end{bmatrix} = \begin{bmatrix}5 & 6\\-19 & 3\end{bmatrix}$$ ∎

Multiplication of Matrices

Multiplication of matrices is slightly more difficult than addition of matrices. Multiplication of matrices is possible only when the number of *columns* of the first matrix, A, is the same as the number of *rows* of the second matrix, B. We use the notation

$$A$$
$$3 \times 4$$

to indicate that matrix A has three rows and four columns. Suppose matrix A is a 3×4 matrix and matrix B is a 4×5 matrix. Then

$$A \qquad B$$
$$3 \times 4 \qquad 4 \times 5$$

Same

AB is a 3×5 matrix

This notation indicates that matrix A has four columns and matrix B has four rows. Therefore, we can multiply these two matrices. The product matrix, $A \times B$ or simply AB, will have the same number of rows as matrix A and the same number of columns as matrix B. Thus, the dimensions of the product matrix, AB, are 3×5.

Timely Tip

Matrices can be added or subtracted only if they have the same dimensions.

Matrices can be multiplied only if the number of *columns* in the first matrix is the same as the number of *rows* in the second matrix.

Example 6 Can These Matrices Be Multiplied?

Determine which of the following pairs of matrices can be multiplied to determine the product matrix AB.

a) $A = \begin{bmatrix} 3 & 2 \\ 5 & 7 \end{bmatrix}$, $\quad B = \begin{bmatrix} 0 & 6 \\ 4 & 1 \end{bmatrix}$

b) $A = \begin{bmatrix} 2 & 3 \\ 5 & 6 \end{bmatrix}$, $\quad B = \begin{bmatrix} 2 & 4 & -1 \\ 6 & 8 & 0 \end{bmatrix}$

c) $A = \begin{bmatrix} 2 & 1 & 4 \\ 3 & 2 & 8 \end{bmatrix}$, $\quad B = \begin{bmatrix} 2 & 1 & 3 \\ 1 & 0 & -2 \end{bmatrix}$

Solution

a)

$$A \qquad B$$
$$2 \times 2 \qquad 2 \times 2$$

Same

Because matrix A has two columns and matrix B has two rows, the two matrices can be multiplied. The product matrix, AB, is a 2×2 matrix.

b)

$$A \qquad B$$
$$2 \times 2 \qquad 2 \times 3$$

Same

Because matrix A has two columns and matrix B has two rows, the two matrices can be multiplied. The product matrix, AB, is a 2×3 matrix.

c)

$$A \qquad B$$
$$2 \times 3 \qquad 2 \times 3$$

Not same

Because matrix A has three columns and matrix B has two rows, the two matrices cannot be multiplied.

To explain matrix multiplication, let's use matrices A and B that follow.

$$A = \begin{bmatrix} 3 & 2 \\ 5 & 7 \end{bmatrix} \quad \text{and} \quad B = \begin{bmatrix} 0 & 6 \\ 4 & 1 \end{bmatrix}$$

Since matrix A contains two rows and matrix B contains two columns, the product matrix will contain two rows and two columns. To multiply two matrices, we use a row–column scheme of multiplying. The numbers in the *first row* of matrix A are multiplied by the numbers in the *first column* of matrix B. These products are then added to determine the entry in the product matrix.

$$A \times B = \begin{bmatrix} 3 & 2 \\ 5 & 7 \end{bmatrix} \begin{bmatrix} 0 & 6 \\ 4 & 1 \end{bmatrix}$$

First row   First column

$$\begin{bmatrix} 3 & 2 \\ 5 & 7 \end{bmatrix} \quad \begin{bmatrix} 0 & 6 \\ 4 & 1 \end{bmatrix}$$

$$(3 \times 0) + (2 \times 4) = 0 + 8$$
$$= 8$$

The 8 is placed in the first-row, first-column position of the product matrix, $A \times B$. The other numbers in the product matrix are obtained similarly, as illustrated in the matrix that follows.

First row   First column   First row   Second column

$$\begin{bmatrix} 3 & 2 \\ 5 & 7 \end{bmatrix} \quad \begin{bmatrix} 0 & 6 \\ 4 & 1 \end{bmatrix} \qquad \begin{bmatrix} 3 & 2 \\ 5 & 7 \end{bmatrix} \quad \begin{bmatrix} 0 & 6 \\ 4 & 1 \end{bmatrix}$$

$$(3 \times 0) + (2 \times 4) = 8 \qquad (3 \times 6) + (2 \times 1) = 20$$

$$A \times B = \begin{bmatrix} 8 & 20 \\ 28 & 37 \end{bmatrix} \longleftarrow \text{Product Matrix}$$

Second row   First column   Second row   Second column

$$\begin{bmatrix} 3 & 2 \\ 5 & 7 \end{bmatrix} \quad \begin{bmatrix} 0 & 6 \\ 4 & 1 \end{bmatrix} \qquad \begin{bmatrix} 3 & 2 \\ 5 & 7 \end{bmatrix} \quad \begin{bmatrix} 0 & 6 \\ 4 & 1 \end{bmatrix}$$

$$(5 \times 0) + (7 \times 4) = 28 \qquad (5 \times 6) + (7 \times 1) = 37$$

We can shorten the procedure as follows.

$$A \times B = \begin{bmatrix} 3 & 2 \\ 5 & 7 \end{bmatrix} \begin{bmatrix} 0 & 6 \\ 4 & 1 \end{bmatrix}$$

$$= \begin{bmatrix} 3(0) + 2(4) & 3(6) + 2(1) \\ 5(0) + 7(4) & 5(6) + 7(1) \end{bmatrix}$$

$$= \begin{bmatrix} 8 & 20 \\ 28 & 37 \end{bmatrix}$$

In general, if

$$A = \begin{bmatrix} a & b \\ c & d \end{bmatrix} \quad \text{and} \quad B = \begin{bmatrix} e & f \\ g & h \end{bmatrix}$$

then

$$A \times B = \begin{bmatrix} a & b \\ c & d \end{bmatrix} \begin{bmatrix} e & f \\ g & h \end{bmatrix}$$

$$= \begin{bmatrix} ae + bg & af + bh \\ ce + dg & cf + dh \end{bmatrix}$$

Let's do one more multiplication of matrices.

Matrix multiplication

Example 7 *Multiplying Matrices*

Determine $A \times B$ if

$$A = \begin{bmatrix} 3 & 1 \\ 4 & 2 \end{bmatrix} \quad \text{and} \quad B = \begin{bmatrix} 5 & -1 & 3 \\ 2 & 8 & 0 \end{bmatrix}$$

Solution Matrix A contains two columns, and matrix B contains two rows. Thus, the matrices can be multiplied. Since matrix A contains two rows and matrix B contains three columns, the product matrix will contain two rows and three columns.

$$A \times B = \begin{bmatrix} 3 & 1 \\ 4 & 2 \end{bmatrix}\begin{bmatrix} 5 & -1 & 3 \\ 2 & 8 & 0 \end{bmatrix}$$

$$= \begin{bmatrix} 3(5) + 1(2) & 3(-1) + 1(8) & 3(3) + 1(0) \\ 4(5) + 2(2) & 4(-1) + 2(8) & 4(3) + 2(0) \end{bmatrix}$$

$$= \begin{bmatrix} 17 & 5 & 9 \\ 24 & 12 & 12 \end{bmatrix}$$

It should be noted that multiplication of matrices *is not* commutative; that is, $A \times B \neq B \times A$, except in special instances.

Example 8 *A Manufacturing Application*

The Fancy Frock Company manufactures three types of women's outfits: a dress, a two-piece suit (skirt and jacket), and a three-piece suit (skirt, jacket, and a vest). On a particular day, the firm produces 20 dresses, 30 two-piece suits, and 50 three-piece suits. Each dress requires 4 units of material and 1 hour of work to produce, each two-piece suit requires 5 units of material and 2 hours of work to produce, and each three-piece suit requires 6 units of material and 3 hours to produce. Use matrix multiplication to determine the total number of units of material and the total number of hours needed for that day's production.

Solution Let matrix A represent the number of each type of women's outfits produced.

$$A = \begin{bmatrix} \overset{\text{Dress}}{20} & \overset{\text{Two-piece}}{30} & \overset{\text{Three-piece}}{50} \end{bmatrix}$$

The units of material and time requirements for each outfit are indicated in matrix B.

$$B = \begin{bmatrix} \overset{\text{Material}}{4} & \overset{\text{Hours}}{1} \\ 5 & 2 \\ 6 & 3 \end{bmatrix} \begin{matrix} \text{Dress} \\ \text{Two-piece} \\ \text{Three-piece} \end{matrix}$$

The product of A and B, or $A \times B$, will give the total number of units of material and the total number of hours of work needed for that day's production.

$$A \times B = \begin{bmatrix} 20 & 30 & 50 \end{bmatrix}\begin{bmatrix} 4 & 1 \\ 5 & 2 \\ 6 & 3 \end{bmatrix}$$

$$= \begin{bmatrix} 20(4) + 30(5) + 50(6) & 20(1) + 30(2) + 50(3) \end{bmatrix}$$

$$= \begin{bmatrix} 530 & 230 \end{bmatrix}$$

Thus, a total of 530 units of material and a total of 230 hours of work are needed that day.

Matrix Addition and Groups

Consider an infinite set consisting of all matrices that have the same given dimensions—for example, all 2 × 3 matrices. This set, along with the operation of matrix addition, is a commutative group. We will illustrate one such commutative group in Example 9.

Example 9 A Commutative Group

Consider the set of all 2 × 3 matrices under the operation of matrix addition. Show that this mathematical system is a commutative group. Assume that the associative property holds for matrix addition.

Solution For this system to be a commutative group, it must satisfy the five properties of a commutative group discussed on page 536.

1. *Closure*: The sum of any two 2 × 3 matrices is always another 2 × 3 matrix. Therefore, the system is closed under the operation of matrix addition.

2. *Identity*: Let $\begin{bmatrix} a & b & c \\ d & e & f \end{bmatrix}$ represent any 2 × 3 matrix. Next, consider the 2 × 3 matrix $\begin{bmatrix} 0 & 0 & 0 \\ 0 & 0 & 0 \end{bmatrix}$. Note that

$$\begin{bmatrix} a & b & c \\ d & e & f \end{bmatrix} + \begin{bmatrix} 0 & 0 & 0 \\ 0 & 0 & 0 \end{bmatrix} = \begin{bmatrix} a & b & c \\ d & e & f \end{bmatrix} \text{ and}$$

$$\begin{bmatrix} 0 & 0 & 0 \\ 0 & 0 & 0 \end{bmatrix} + \begin{bmatrix} a & b & c \\ d & e & f \end{bmatrix} = \begin{bmatrix} a & b & c \\ d & e & f \end{bmatrix}.$$

Therefore, $\begin{bmatrix} 0 & 0 & 0 \\ 0 & 0 & 0 \end{bmatrix}$ is the identity matrix under matrix addition, called the *additive identity matrix*.

3. *Inverses*: Again, let $\begin{bmatrix} a & b & c \\ d & e & f \end{bmatrix}$ represent any 2 × 3 matrix. Next, consider the matrix formed by taking the opposite of each element within this matrix, $\begin{bmatrix} -a & -b & -c \\ -d & -e & -f \end{bmatrix}$. Note that

$$\begin{bmatrix} a & b & c \\ d & e & f \end{bmatrix} + \begin{bmatrix} -a & -b & -c \\ -d & -e & -f \end{bmatrix} = \begin{bmatrix} 0 & 0 & 0 \\ 0 & 0 & 0 \end{bmatrix} \text{ and}$$

$$\begin{bmatrix} -a & -b & -c \\ -d & -e & -f \end{bmatrix} + \begin{bmatrix} a & b & c \\ d & e & f \end{bmatrix} = \begin{bmatrix} 0 & 0 & 0 \\ 0 & 0 & 0 \end{bmatrix}.$$

Therefore, every 2 × 3 matrix has an additive inverse.

4. *Associative Property*: It is given that the associative property holds. One example that illustrates the associative property follows.

$$\left(\begin{bmatrix} 1 & 3 & -1 \\ -2 & 5 & 0 \end{bmatrix} + \begin{bmatrix} 0 & -4 & 3 \\ 7 & -2 & 6 \end{bmatrix}\right) + \begin{bmatrix} 2 & 5 & 0 \\ -3 & 2 & 8 \end{bmatrix}$$

$$\stackrel{?}{=} \begin{bmatrix} 1 & 3 & -1 \\ -2 & 5 & 0 \end{bmatrix} + \left(\begin{bmatrix} 0 & -4 & 3 \\ 7 & -2 & 6 \end{bmatrix} + \begin{bmatrix} 2 & 5 & 0 \\ -3 & 2 & 8 \end{bmatrix}\right)$$

$$\begin{bmatrix} 1 & -1 & 2 \\ 5 & 3 & 6 \end{bmatrix} + \begin{bmatrix} 2 & 5 & 0 \\ -3 & 2 & 8 \end{bmatrix} \stackrel{?}{=} \begin{bmatrix} 1 & 3 & -1 \\ -2 & 5 & 0 \end{bmatrix} + \begin{bmatrix} 2 & 1 & 3 \\ 4 & 0 & 14 \end{bmatrix}$$

$$\begin{bmatrix} 3 & 4 & 2 \\ 2 & 5 & 14 \end{bmatrix} = \begin{bmatrix} 3 & 4 & 2 \\ 2 & 5 & 14 \end{bmatrix}$$

5. *Commutative Property:* Let $\begin{bmatrix} a & b & c \\ d & e & f \end{bmatrix}$ and $\begin{bmatrix} g & h & i \\ j & k & l \end{bmatrix}$ represent any two 2×3 matrices. Note that

$$\begin{bmatrix} a & b & c \\ d & e & f \end{bmatrix} + \begin{bmatrix} g & h & i \\ j & k & l \end{bmatrix} = \begin{bmatrix} a+g & b+h & c+i \\ d+j & e+k & f+l \end{bmatrix} \text{ and}$$

$$\begin{bmatrix} g & h & i \\ j & k & l \end{bmatrix} + \begin{bmatrix} a & b & c \\ d & e & f \end{bmatrix} = \begin{bmatrix} g+a & h+b & i+c \\ j+d & k+e & l+f \end{bmatrix}$$

Next, $a + g = g + a, b + h = h + b, c + i = i + c$, and so on. Therefore, $\begin{bmatrix} a+g & b+h & c+i \\ d+j & e+k & f+l \end{bmatrix} = \begin{bmatrix} g+a & h+b & i+c \\ j+d & k+e & l+f \end{bmatrix}$ and the commutative property holds.

All five properties of a commutative group are satisfied. Thus, the set of all 2×3 matrices under the operation of matrix addition is a commutative group. ∎

It can also be shown that certain sets of square matrices, along with the operation of matrix multiplication, form a noncommutative group.

SECTION 9.4   Exercises

Warm Up Exercises

In Exercises 1–8, fill in the blank with an appropriate word, phrase, or symbol(s).

1. A rectangular array of real numbers, symbols, or expressions is called a(n) _____.

2. A matrix that contains the same number of rows and columns is called a(n) _____ matrix.

3. The number of columns in a 3 × 2 matrix is _____.

4. The number of rows in a 4 × 3 matrix is _____.

5. To add or subtract two matrices, the matrices must have the same _____.

6. When multiplying a matrix by a real number, the real number is called a(n) _____.

7. Multiplication of matrices is possible only when the number of _____ of the first matrix is the same as the number of _____ of the second matrix.

8. The set of all 2 × 3 matrices under the operation of matrix addition forms a commutative _____.

Practice the Skills

In Exercises 9–12, determine A + B.

9. $A = \begin{bmatrix} 3 & 8 \\ 2 & -6 \end{bmatrix}, \quad B = \begin{bmatrix} -4 & 1 \\ 5 & -3 \end{bmatrix}$

10. $A = \begin{bmatrix} 6 & 2 & -1 \\ 4 & 5 & 8 \end{bmatrix}, \quad B = \begin{bmatrix} 3 & -1 & 2 \\ 2 & -3 & 9 \end{bmatrix}$

11. $A = \begin{bmatrix} 5 & 2 \\ -1 & 4 \\ 7 & 0 \end{bmatrix}, \quad B = \begin{bmatrix} -3 & 3 \\ -4 & 0 \\ 1 & 6 \end{bmatrix}$

12. $A = \begin{bmatrix} 2 & 6 & 3 \\ -1 & -6 & 4 \\ 3 & 0 & 5 \end{bmatrix}, \quad B = \begin{bmatrix} -1 & 3 & 1 \\ 7 & -2 & 1 \\ 2 & 3 & 8 \end{bmatrix}$

In Exercises 13–16, determine A − B.

13. $A = \begin{bmatrix} 4 & 5 \\ 1 & -3 \end{bmatrix}, \quad B = \begin{bmatrix} -2 & 1 \\ 6 & 7 \end{bmatrix}$

14. $A = \begin{bmatrix} 10 & 1 \\ 12 & 2 \\ -3 & -9 \end{bmatrix}, \quad B = \begin{bmatrix} -3 & 3 \\ 2 & 6 \\ -1 & 5 \end{bmatrix}$

15. $A = \begin{bmatrix} -5 & 1 \\ 8 & 6 \\ 1 & -5 \end{bmatrix}, \quad B = \begin{bmatrix} -6 & -8 \\ -10 & -11 \\ 3 & -7 \end{bmatrix}$

16. $A = \begin{bmatrix} 4 & 7 & 2 \\ 5 & 3 & 1 \\ -5 & -1 & 6 \end{bmatrix}, \quad B = \begin{bmatrix} -2 & -4 & 9 \\ 0 & -2 & 4 \\ 4 & 3 & 6 \end{bmatrix}$

In Exercises 17–20, let

$$A = \begin{bmatrix} 1 & 2 \\ 0 & 5 \end{bmatrix}, \quad B = \begin{bmatrix} 3 & 2 \\ 5 & 0 \end{bmatrix}, \quad \text{and} \quad C = \begin{bmatrix} -2 & 3 \\ 4 & 0 \end{bmatrix}.$$

Determine the following.

17. $2B$

18. $-3A$

19. $2B + 3C$

20. $3B - 2C$

In Exercises 21–26, determine $A \times B$.

21. $A = \begin{bmatrix} 2 & 7 \\ 1 & 3 \end{bmatrix}, \quad B = \begin{bmatrix} 1 & 0 \\ 3 & 8 \end{bmatrix}$

22. $A = \begin{bmatrix} 2 & -2 \\ 4 & 3 \end{bmatrix}, \quad B = \begin{bmatrix} 1 & 0 \\ -3 & 5 \end{bmatrix}$

23. $A = \begin{bmatrix} 2 & 3 & -1 \\ 0 & 4 & 6 \end{bmatrix}, \quad B = \begin{bmatrix} 2 \\ 4 \\ 1 \end{bmatrix}$

24. $A = \begin{bmatrix} 1 & 0 \\ 4 & -2 \\ -1 & 3 \end{bmatrix}, \quad B = \begin{bmatrix} 2 \\ -3 \end{bmatrix}$

25. $A = \begin{bmatrix} 2 & -5 & 0 \end{bmatrix}, \quad B = \begin{bmatrix} 4 & 1 \\ -1 & 0 \\ -2 & 6 \end{bmatrix}$

26. $A = \begin{bmatrix} 3 & 2 \end{bmatrix}, \quad B = \begin{bmatrix} -2 & 5 & 0 \\ 0 & -7 & 1 \end{bmatrix}$

In Exercises 27–34, determine $A + B$ and $A \times B$. If an operation cannot be performed, explain why.

27. $A = \begin{bmatrix} -2 & 5 \\ 0 & 4 \end{bmatrix}, \quad B = \begin{bmatrix} 1 & 0 \\ -6 & 3 \end{bmatrix}$

28. $A = \begin{bmatrix} -2 & 3 \\ 6 & 7 \end{bmatrix}, \quad B = \begin{bmatrix} 1 & 3 \\ 3 & -4 \end{bmatrix}$

29. $A = \begin{bmatrix} 1 & 3 & 0 \\ 2 & 4 & -1 \end{bmatrix}, \quad B = \begin{bmatrix} 7 & -2 & 3 \\ 2 & -1 & 1 \end{bmatrix}$

30. $A = \begin{bmatrix} 6 & 5 \\ 4 & 3 \\ 2 & 1 \end{bmatrix}, \quad B = \begin{bmatrix} 6 & 5 \\ 4 & 3 \\ 2 & 1 \end{bmatrix}$

31. $A = \begin{bmatrix} 2 & 5 & 1 \\ 8 & 3 & 6 \end{bmatrix}, \quad B = \begin{bmatrix} 3 & 2 \\ 4 & 6 \\ -2 & 0 \end{bmatrix}$

32. $A = \begin{bmatrix} 1 & 2 \\ 3 & 4 \end{bmatrix}, \quad B = \begin{bmatrix} -3 \\ 2 \end{bmatrix}$

33. $A = \begin{bmatrix} 3 & 4 \end{bmatrix}, \quad B = \begin{bmatrix} 5 \\ 9 \\ -8 \end{bmatrix}$

34. $A = \begin{bmatrix} 6 & 4 & -1 \\ 2 & 3 & 4 \end{bmatrix}, \quad B = \begin{bmatrix} 1 & 0 \\ 4 & -1 \end{bmatrix}$

In Exercises 35–38, show the commutative property of addition, $A + B = B + A$, holds for matrices A and B.

35. $A = \begin{bmatrix} 3 & 1 \\ -1 & 4 \end{bmatrix}, B = \begin{bmatrix} 4 & 1 \\ 7 & -3 \end{bmatrix}$

36. $A = \begin{bmatrix} 9 & 4 \\ 1 & 7 \end{bmatrix}, \quad B = \begin{bmatrix} 2 & 0 \\ -1 & 6 \end{bmatrix}$

37. $A = \begin{bmatrix} 0 & -1 \\ 3 & -4 \end{bmatrix}, \quad B = \begin{bmatrix} 8 & 1 \\ 3 & -4 \end{bmatrix}$

38. $A = \begin{bmatrix} 4 & -2 & 9 \\ 6 & -1 & 0 \\ -3 & 5 & 7 \end{bmatrix}, B = \begin{bmatrix} -11 & 2 & -7 \\ 2 & 0 & -5 \\ 1 & -4 & 10 \end{bmatrix}$

In Exercises 39–42, show that the associative property of addition, $(A + B) + C = A + (B + C)$, holds for matrices A, B, and C.

39. $A = \begin{bmatrix} 6 & 5 \\ -1 & 3 \end{bmatrix}, \quad B = \begin{bmatrix} 3 & 4 \\ -2 & 7 \end{bmatrix}, \quad C = \begin{bmatrix} -2 & 4 \\ 5 & 0 \end{bmatrix}$

40. $A = \begin{bmatrix} 7 & 4 \\ 9 & -36 \end{bmatrix}, \quad B = \begin{bmatrix} 5 & 6 \\ -1 & -4 \end{bmatrix}, \quad C = \begin{bmatrix} -7 & -5 \\ -1 & 3 \end{bmatrix}$

41. $A = \begin{bmatrix} 3 & 2 \\ 2 & 0 \\ -5 & 9 \end{bmatrix}, \quad B = \begin{bmatrix} 1 & 8 \\ 0 & 6 \\ 4 & -4 \end{bmatrix}, \quad C = \begin{bmatrix} 0 & 4 \\ 1 & 9 \\ 9 & -6 \end{bmatrix}$

42. $A = \begin{bmatrix} -5 & 2 & 1 \\ -11 & -3 & 0 \end{bmatrix}, \quad B = \begin{bmatrix} 4 & 1 & 8 \\ 6 & 4 & 0 \end{bmatrix},$

$C = \begin{bmatrix} 0 & -5 & 2 \\ 3 & 1 & -9 \end{bmatrix}$

In Exercises 43–46, determine whether $A \times B = B \times A$ holds for matrices A and B.

43. $A = \begin{bmatrix} -2 & 1 \\ -3 & 4 \end{bmatrix}, \quad B = \begin{bmatrix} -3 & 1 \\ 4 & 2 \end{bmatrix}$

44. $A = \begin{bmatrix} 4 & 2 \\ 1 & -3 \end{bmatrix}, \quad B = \begin{bmatrix} 2 & 4 \\ -3 & 1 \end{bmatrix}$

45. $A = \begin{bmatrix} 2 & 0 \\ -4 & 1 \end{bmatrix}, \quad B = \begin{bmatrix} 2 & 0 \\ 8 & 4 \end{bmatrix}$

46. $A = \begin{bmatrix} 2 & -1 \\ 3 & 3 \end{bmatrix}, \quad B = \begin{bmatrix} 3 & 1 \\ -3 & 2 \end{bmatrix}$

In Exercises 47–50, show that the associative property of multiplication, $(A \times B) \times C = A \times (B \times C)$, holds for matrices A, B, and C.

47. $A = \begin{bmatrix} 1 & 3 \\ 4 & 0 \end{bmatrix}$, $B = \begin{bmatrix} 4 & 2 \\ 3 & 1 \end{bmatrix}$, $C = \begin{bmatrix} 2 & 1 \\ 3 & 0 \end{bmatrix}$

48. $A = \begin{bmatrix} -2 & 3 \\ 0 & 4 \end{bmatrix}$, $B = \begin{bmatrix} 4 & 0 \\ 3 & 5 \end{bmatrix}$, $C = \begin{bmatrix} 3 & 4 \\ -2 & 5 \end{bmatrix}$

49. $A = \begin{bmatrix} 4 & 3 \\ -6 & 2 \end{bmatrix}$, $B = \begin{bmatrix} 1 & 2 \\ 0 & 1 \end{bmatrix}$, $C = \begin{bmatrix} 4 & 3 \\ 0 & -2 \end{bmatrix}$

50. $A = \begin{bmatrix} 3 & 4 \\ -1 & -2 \end{bmatrix}$, $B = \begin{bmatrix} 0 & 1 \\ 1 & 0 \end{bmatrix}$, $C = \begin{bmatrix} 2 & 0 \\ 3 & 0 \end{bmatrix}$

Problem Solving

51. MODELING—*High School Play* North Shore High School sold tickets for the musical *The Lion King.* Matrix *A* indicates the number of student, adult, and senior citizen tickets sold for the matinee and evening shows on day 1. Matrix *B* indicates the number of student, adult, and senior citizen tickets sold for the matinee and evening shows on day 2.

$$A = \begin{array}{c} \\ \\ \text{Matinee} \\ \text{Evening} \end{array} \quad \begin{array}{ccc} \text{Student} & \text{Adult} & \text{Senior citizen} \\ \begin{bmatrix} 85 & 150 & 50 \\ 95 & 162 & 41 \end{bmatrix} \end{array}$$

$$B = \begin{array}{ccc} \text{Student} & \text{Adult} & \text{Senior citizen} \\ \begin{bmatrix} 73 & 130 & 45 \\ 120 & 200 & 53 \end{bmatrix} \begin{array}{c} \text{Matinee} \\ \text{Evening} \end{array} \end{array}$$

Use matrix addition to determine the total number of student, adult, and senior citizen tickets sold by North Shore High School for both days for the matinee and evening shows.

52. MODELING—*T-shirt Inventory* Hollister sells two types of T-shirts—male and female. Matrix *A* indicates the stock on hand of each type and size of T-shirt at the beginning of a given day. Matrix *B* indicates the stock on hand of each type and size of T-shirt at the end of the same given day.

$$A = \begin{array}{ccc} \text{Male} & \text{Female} \\ \begin{bmatrix} 31 & 18 \\ 39 & 16 \\ 41 & 22 \\ 34 & 21 \end{bmatrix} \begin{array}{c} \text{Small} \\ \text{Medium} \\ \text{Large} \\ \text{Extra large} \end{array} \end{array} \quad B = \begin{array}{ccc} \text{Male} & \text{Female} \\ \begin{bmatrix} 14 & 9 \\ 18 & 9 \\ 19 & 15 \\ 15 & 9 \end{bmatrix} \begin{array}{c} \text{Small} \\ \text{Medium} \\ \text{Large} \\ \text{Extra large} \end{array} \end{array}$$

Use matrix subtraction to determine the number of T-shirts sold in the given day for each size of each type.

53. MODELING—*Cookie Company Costs* Mrs. Keller's Cookies bakes and sells four types of cookies: chocolate chip, sugar, molasses, and peanut butter. Matrix *A* indicates the number of units of various ingredients used in baking a dozen of each type of cookie.

$$A = \begin{array}{c} \\ \\ \\ \\ \\ \end{array} \begin{array}{cccc} \text{Sugar} & \text{Flour} & \text{Milk} & \text{Eggs} \\ \begin{bmatrix} 2 & 2 & \frac{1}{2} & 1 \\ 3 & 2 & 1 & 2 \\ 0 & 1 & 0 & 3 \\ \frac{1}{2} & 1 & 0 & 0 \end{bmatrix} \begin{array}{c} \text{Chocolate chip} \\ \text{Sugar} \\ \text{Molasses} \\ \text{Peanut butter} \end{array} \end{array}$$

The cost, in cents per cup or per egg, for each ingredient when purchased in small quantities and in large quantities is indicated in matrix *B* below.

$$B = \begin{array}{c} \\ \\ \\ \\ \end{array} \begin{array}{cc} \text{Large} & \text{Small} \\ \text{quantities} & \text{quantities} \\ \begin{bmatrix} 10 & 12 \\ 5 & 8 \\ 8 & 8 \\ 4 & 6 \end{bmatrix} \begin{array}{c} \text{Sugar} \\ \text{Flour} \\ \text{Milk} \\ \text{Eggs} \end{array} \end{array}$$

Use matrix multiplication to determine a matrix representing the comparative cost per item for small and large quantities purchased.

In Exercises 54 and 55, use the information given in Exercise 53. Suppose a typical day's order consists of 40 dozen chocolate chip cookies, 30 dozen sugar cookies, 12 dozen molasses cookies, and 20 dozen peanut butter cookies.

54. *Cookie Orders*
   a) Express these orders as a 1×4 matrix.

   b) Use matrix multiplication to determine the amount of each ingredient needed to fill the day's order.

55. *Cookie Costs* Use matrix multiplication to determine the cost, in dollars, under the two purchase options (small and large quantities) to fill the day's order.

56. MODELING—*Food Prices* To raise money for a local charity, the Spanish Club at Montclair High School sold hot dogs, soft drinks, and candy bars for 3 days in the student lounge. The sales for the 3 days are summarized in matrix *A*.

$$A = \begin{array}{c} \\ \\ \\ \end{array} \begin{array}{ccc} \text{Hot dogs} & \text{Soft drinks} & \text{Candy bars} \\ \begin{bmatrix} 52 & 50 & 75 \\ 48 & 43 & 60 \\ 62 & 57 & 81 \end{bmatrix} \begin{array}{c} \text{Day 1} \\ \text{Day 2} \\ \text{Day 3} \end{array} \end{array}$$

The cost and revenue (in dollars) for hot dogs, soft drinks, and candy are summarized in matrix *B* on the next page.

$$B = \begin{bmatrix} 0.30 & 0.75 \\ 0.25 & 0.50 \\ 0.15 & 0.45 \end{bmatrix} \begin{matrix} \text{Hot dogs} \\ \text{Soft drinks} \\ \text{Candy bars} \end{matrix}$$

Cost   Revenue

Multiply the two matrices to form a 3 × 2 matrix that shows the total cost and revenue for each item.

57. Consider the mathematical system consisting of the set of all 3 × 2 matrices under the operation of matrix addition.

a) Is the system closed?

b) Is there an additive identity matrix? If so, what is it?

c) Does each matrix in the set have an additive inverse matrix?

d) The associative property holds for matrix addition. Give an example.

e) Does the commutative property hold for matrix addition?

f) Does this mathematical system form a commutative group?

58. Consider the mathematical system consisting of the set of all 2 × 4 matrices under the operation of matrix addition.

a) Is the system closed?

b) Is there an additive identity matrix? If so, what is it?

c) Does each matrix in the set have an additive inverse matrix?

d) The associative property holds for matrix addition. Give an example.

e) Does the commutative property hold for matrix addition?

f) Does this mathematical system form a commutative group?

Concept/Writing Exercises

59. Create two matrices, A and B, with the same dimensions and show that $A + B = B + A$.

60. Create three matrices, A, B, and C, with the same dimensions and show that $(A + B) + C = A + (B + C)$.

61. Create two square matrices, A and B, with the same dimensions and determine whether $A \times B = B \times A$.

62. Create three 2 × 2 matrices, A, B, and C, and show that $(A \times B) \times C = A \times (B \times C)$.

Challenge Problems/Group Activities

63. For 2 × 2 matrices, the *multiplicative identity matrix* is $I = \begin{bmatrix} 1 & 0 \\ 0 & 1 \end{bmatrix}$. Show, for any 2 × 2 matrix, $A = \begin{bmatrix} a & b \\ c & d \end{bmatrix}$, that $A \times I = I \times A = A$.

64. Two 2 × 2 matrices, A and B, whose product is the multiplicative identity matrix (see Exercise 63) are said to be *multiplicative inverses*. That is, if $A \times B = B \times A = I$, then A and B are multiplicative inverses. In parts (a) and (b) below, show that matrices A and B are multiplicative inverses.

a) $A = \begin{bmatrix} 5 & -2 \\ -2 & 1 \end{bmatrix}$, $B = \begin{bmatrix} 1 & 2 \\ 2 & 5 \end{bmatrix}$

b) $A = \begin{bmatrix} 7 & 3 \\ 2 & 1 \end{bmatrix}$, $B = \begin{bmatrix} 1 & -3 \\ -2 & 7 \end{bmatrix}$

Research Activities

65. *Matrices and Systems of Equations* In Section 6.7 we solved systems of linear equations algebraically. Systems of linear equations may also be solved using matrix multiplication. Do research and show how the system of equations

$$3x + 2y = 5$$
$$4x - 3y = -16$$

can be solved using matrix multiplication.

66. *Matrix Display* Find an example from a magazine, a newspaper, or the Internet that displays information using a matrix. Write a paper explaining how to interpret the information provided by the matrix.

67. *Secret Messages* The study of encoding and decoding messages is called *cryptography*. (See the *Recreational Mathematics* on page 552.) Write a paper on how matrices and matrix multiplication are currently used in cryptography.

CHAPTER 9   Summary

Important Facts and Concepts	Examples and Discussion
Section 9.1	
A **mathematical system** consists of a set of elements and at least one binary operation.	Discussion, page 532
Commutative Property	
Addition: $a + b = b + a$	Discussion, pages 532–533
Multiplication: $a \cdot b = b \cdot a$	
Associative Property	
Addition: $(a + b) + c = a + (b + c)$	Discussion, pages 532–533
Multiplication: $(a \cdot b) \cdot c = a \cdot (b \cdot c)$	
Properties of a Group and a Commutative Group	
A mathematical system is a group if the first four following conditions hold, and a commutative group if all five conditions hold.	Discussion, pages 533–535, Examples 1–2, pages 536–537, Examples 1–6, pages 541–545, Example 6, pages 555–556
1. The set of elements is *closed* under the given operation.	Example 9, pages 565–566
2. An *identity element* exists for the set.	
3. Every element in the set has an *inverse*.	
4. The set of elements is *associative* under the given operation.	
5. The set of elements is *commutative* under the given operation.	
Section 9.2, 9.3	
If the elements of a finite mathematical system are symmetric about the main diagonal, then the system is *commutative*.	Discussion, page 542, Examples 1–6, pages 541–545, Example 6, pages 555–556
Section 9.3	
a is **congruent** to b modulo m, written $a \equiv b \ (mod \ m)$, if a and b have the same remainder when divided by m.	Discussion, pages 550-551, Examples 1–6, pages 551–556
Section 9.4	
A **matrix** is a rectangular array of real numbers, symbols, or expressions.	Discussion, pages 559–560
Matrix Operations	
Addition and Subtraction	Examples 2–4, pages 560–561
Multiplication by a Scalar	Example 5, page 561
Multiplication	Examples 6 and 7, pages 562–564
Matrices as a Commutative Group	
See Section 9.1 for the five properties of a commutative group.	Example 9, pages 565–566

CHAPTER 9   Review Exercises

9.1, 9.2

1. What is a binary operation?

2. List the parts of a mathematical system.

3. Is the set of whole numbers closed under the operation of division? Explain.

4. Is the set of real numbers closed under the operation of subtraction? Explain.

Determine the sum or difference in clock 12 arithmetic.

5. $9 + 8$

6. $10 + 8$

7. $8 - 10$

8. $7 - 4 + 6$

9. List the properties of a group and explain what each property means.

In Exercises 10–13, explain your answer.

10. Determine whether the set of integers under the operation of addition forms a group.

11. Determine whether the set of integers under the operation of multiplication forms a group.

12. Determine whether the set of natural numbers under the operation of addition forms a group.

13. Determine whether the set of rational numbers under the operation of multiplication forms a group.

In Exercises 14–16, for the given mathematical system, determine which of the five properties of a commutative group do not hold.

14.

∴	ρ	σ	ω
ρ	ρ	ω	σ
σ	ω	σ	ρ
ω	σ	ρ	ω

15.

?	4	#	L	P
4	P	4	#	L
#	4	#	L	P
L	#	L	P	4
P	L	P	4	L

16.

□	!	?	△	*p*
!	?	△	!	p
?	△	p	?	!
△	!	?	△	p
p	p	!	p	△

17. Consider the following mathematical system. Assume the associative property holds for the given operation.

⌂	☺	●	♀	♂
☺	☺	●	♀	♂
●	●	♀	♂	☺
♀	♀	♂	☺	●
♂	♂	☺	●	♀

a) What are the elements of the set in this mathematical system?

b) What is the binary operation?

c) Is the system closed? Explain.

d) Is there an identity element in the set?

e) Does every element in the set have an inverse? If so, give each element and its corresponding inverse.

f) Give an example to illustrate the associative property.

g) Is the system commutative? Give an example.

h) Is this mathematical system a commutative group? Explain.

9.3

In Exercises 18–23, determine the modulo class to which the number belongs for the indicated modulo system.

18. 19, mod 7

19. 24, mod 6

20. 47, mod 9

21. 34, mod 5

22. 71, mod 12

23. 54, mod 14

In Exercises 24–31, determine all positive number replacements (less than the modulus) for the question mark that make the statement true.

24. $2 + 5 \equiv ? \pmod 6$

25. $2 - ? \equiv 3 \pmod 7$

26. $7 \cdot ? \equiv 5 \pmod 8$

27. $? \cdot 4 \equiv 0 \pmod 8$

28. $10 \cdot 7 \equiv ? \pmod{11}$

29. $? \cdot 7 \equiv 3 \pmod{10}$

30. $? \cdot 3 = 5 \pmod 6$

31. $7 \cdot ? \equiv 2 \pmod 9$

32. Construct a modulo 6 addition table. Then determine whether the modulo 6 system forms a commutative group under the operation of addition.

33. Construct a modulo 4 multiplication table. Then determine whether the modulo 4 system forms a commutative group under the operation of multiplication.

34. *Police Officer Shifts* Julie, a police officer, has the following work pattern. She works 3 days, has 4 days off, then works 4 days, then has 3 days off, and then the pattern repeats. Assume that today is the first day of the work pattern.

a) Will Julie be working 30 days from today?

b) Julie's niece is getting married 45 days from today. Will Julie have the day off?

9.4

In Exercises 35–39, given $A = \begin{bmatrix} 2 & -1 \\ -3 & 0 \end{bmatrix}$ *and* $B = \begin{bmatrix} 3 & 5 \\ 1 & 2 \end{bmatrix}$, *determine the following.*

35. $A + B$

36. $A - B$

37. $3A - 2B$

38. $A \times B$

39. $B \times A$

CHAPTER 9   Test

1. What is a mathematical system?

2. List the requirements needed for a mathematical system to be a commutative group.

3. Is the set of integers a commutative group under the operation of addition? Explain your answer.

4. Is the set of natural numbers a commutative group under the operation of subtraction? Explain your answer.

5. Develop a clock 5 arithmetic addition table.

6. Is clock 5 arithmetic under the operation of addition a commutative group? Assume that the associative property holds. Explain your answer.

7. Consider the mathematical system

□	W	S	T	R
W	T	R	W	S
S	R	W	S	T
T	W	S	T	R
R	S	T	R	W

a) What is the binary operation?

b) Is this system closed? Explain.

c) Is there an identity element for this system under the given operation? Explain.

d) What is the inverse of the element R?

e) What is $(T \square R) \square W$?

In Exercises 8 and 9, determine whether the mathematical system is a commutative group. Explain your answer completely.

8.

*	a	b	c
a	a	b	c
b	b	b	a
c	c	a	d

9.

?	1	2	3
1	3	1	2
2	1	2	3
3	2	3	1

10. Determine whether the mathematical system is a commutative group. Assume that the associative property holds. Explain your answer.

○	@	$	&	%
@	%	@	$	&
$	@	$	&	%
&	$	&	%	@
%	&	%	@	$

In Exercises 11 and 12, determine the modulo class to which the number belongs for the indicated modulo system.

11. 59, mod 4

12. 96, mod 11

In Exercises 13–16, determine all positive number replacements for the question mark, less than the modulus, that make the statement true.

13. $? + 5 \equiv 3 \pmod 6$

14. $? - 3 \equiv 4 \pmod 5$

15. $3 - ? \equiv 7 \pmod 9$

16. $3 \cdot ? \equiv 2 \pmod 6$

17. a) Construct a modulo 5 multiplication table.

b) Is this mathematical system a commutative group? Explain your answer completely.

In Exercises 18–20, given $A = \begin{bmatrix} 2 & 1 \\ 3 & -6 \end{bmatrix}$ and $B = \begin{bmatrix} -2 & 2 \\ 5 & 3 \end{bmatrix}$, determine the following.

18. $A + B$

19. $2A - 3B$

20. $A \times B$

◀ Knowledge of consumer mathematics may help you reach your long-term goals.

10

Consumer Mathematics

What You Will Learn

- Percents, including percent increase, percent decrease, and percent markup and markdown
- Simple interest
- Compound interest
- Fixed and open-ended installment loans
- Mortgages
- Annuities, sinking funds, and retirement investments

Why This Is Important

Managing your money takes much thought and planning. All your daily needs must be met, but at the same time you should consider your long-term goals such as purchasing a house, saving for college expenses, and investing for retirement. If you need to borrow money, which type of loan would be the best for you? How much money should you invest in order to have a certain amount of money in the future? This chapter will provide information to make sound financial decisions to help you reach these goals and answer these questions.

SECTION 10.1   Percent

Upon completion of this section, you will be able to:

- Convert between a percent, a fraction, and a decimal number.
- Solve problems involving percent change.
- Solve problems involving percent markup and markdown.
- Solve other percent problems.

Your clock radio turns on in the morning and you hear a weather report that says there is a 75% chance of rain. Based on this information, you decide to carry your umbrella to your class. In this section, we will investigate the meaning of percent and learn how percents relate to mathematics and to daily life.

Why This Is Important Many decisions that we make every day involve applications of percent. An understanding of percent is essential in making informed decisions. For example, if we need to determine how much a meal will cost including tax and tip, we need an understanding of percent.

Percent, fractions, and decimal numbers all play an important role in consumer mathematics.

Percent, Fractions, and Decimal Numbers

A basic topic necessary for understanding the material in this chapter is percent. The word *percent* comes from the Latin *per centum*, meaning "per hundred." A *percent* is simply a ratio of some number to 100. Thus, $\frac{15}{100} = 15\%$, and $\frac{x}{100} = x\%$.

Percents are useful in making comparisons. Consider Ross, who took two chemistry tests. On the first test Ross answered 18 of the 20 questions correctly, and on the second test he answered 23 of 25 questions correctly. On which test did he have the higher score? One way to compare the results is to write ratios of the number of correct answers to the number of questions on the test and then convert the ratios to percents. We can determine the grades in percent for each test by (a) writing a ratio of the number of correct answers to the total number of questions, (b) rewriting these ratios with a denominator of 100, and (c) expressing the ratios as percents.

$$\text{Test 1} \qquad \text{(a)} \qquad \text{(b)} \qquad \text{(c)}$$
$$\frac{\text{Number of correct answers}}{\text{Number of questions on the test}} = \frac{18}{20} = \frac{18 \times 5}{20 \times 5} = \frac{90}{100} = 90\%$$

$$\text{Test 2} \qquad \text{(a)} \qquad \text{(b)} \qquad \text{(c)}$$
$$\frac{\text{Number of correct answers}}{\text{Number of questions on the test}} = \frac{23}{25} = \frac{23 \times 4}{25 \times 4} = \frac{92}{100} = 92\%$$

By changing the results of both tests to percents, we have a common standard for comparison. The results show that Ross scored 90% on the first test and 92% on the second test. Thus, he had a higher score on the second test.

Another procedure to change a fraction to a percent follows.

PROCEDURE   CHANGING FRACTIONS TO PERCENTS

1. Divide the numerator by the denominator to obtain a decimal number.
2. Multiply the decimal number by 100 (which has the effect of moving the decimal point two places to the right).
3. Add a percent sign.

Percent Versus Percentage

When should the word *percent* be used and when should the word *percentage* be used? Although writing style manuals are not always consistent on the use of these words, a general rule is given below.

- *Percent* (or the symbol %) accompanies a specific number. The word *percent* may also be used when a question is seeking a specific percent.
  Example 1—Eighty percent of the people liked the movie.
  Example 2—More than 80% (or 80 percent) of the boxes weighed more than 1 pound.
  Example 3—If a worker's salary increased from $35,000 to $38,000 what is the percent increase?
- *Percentage* is used without a specific number.
  Example 4—A large percentage of the population has had their flu shot.
  Example 5—The percentage of the population that has been exposed to the virus is between 60% and 70%.

Some style manuals state that the word *percent* can be used in place of the word *percentage* on tables and graphs, and other places where space is limited. Also, in the United States, *percent* is usually written as one word, but in Europe and elsewhere it is often written using two words, per cent. Note, too, that when writing a percent at the beginning of a sentence, as in Example 1 in this Mathematics Today box above, the words should be written out.

Why This Is Important This information may be helpful if you need to write reports for school or for your job.

This procedure has the effect of converting the fraction to a decimal number and then moving the decimal point in the decimal number two places to the right and adding a percent sign.

Note that Steps 2 and 3 together are the equivalent of multiplying by 100%. Since $100\% = 100/100 = 1$, we are not changing the *value* of the decimal number—we are simply changing the decimal number to a percent.

Example 1 *Converting Fractions to Percents*

Change each of the following fractions to a percent.

a) $\dfrac{67}{100}$ b) $\dfrac{11}{20}$ c) $\dfrac{15}{16}$

Solution For each fraction, follow the three steps in the procedure box on page 574.

a) 1. $67 \div 100 = 0.67$ 2. $0.67 \times 100 = 67$ 3. 67%

Thus, $\dfrac{67}{100} = 67\%$.

b) 1. $11 \div 20 = 0.55$ 2. $0.55 \times 100 = 55$ 3. 55%

Thus, $\dfrac{11}{20} = 55\%$.

c) 1. $15 \div 16 = 0.9375$ 2. $0.9375 \times 100 = 93.75$ 3. 93.75%

Thus, $\dfrac{15}{16} = 93.75\%$. ∎

In future examples, when changing a fraction to a percent, we will follow the three-step procedure given on page 574 without listing the individual steps. We will do this in Example 4.

To change a decimal number to a percent, use the following procedure.

PROCEDURE CHANGING DECIMAL NUMBERS TO PERCENTS

1. Multiply the decimal number by 100.
2. Add a percent sign.

The procedure for changing a decimal number to a percent is equivalent to moving the decimal point two places to the right and adding a percent sign.

Example 2 *Converting Decimal Numbers to Percents*

Change each of the following decimal numbers to a percent.

a) 0.14   b) 0.893   c) 0.7625

Solution

a) $0.14 = (0.14 \times 100)\% = 14\%$. Thus, $0.14 = 14\%$.
b) $0.893 = (0.893 \times 100)\% = 89.3\%$. Thus, $0.893 = 89.3\%$.
c) $0.7625 = (0.7625 \times 100)\% = 76.25\%$. Thus, $0.7625 = 76.25\%$. ∎

To change a number given as a percent to a decimal number, follow this procedure.

PROCEDURE CHANGING PERCENTS TO DECIMAL NUMBERS

1. Remove the percent sign.
2. Divide the number by 100.

The procedure for changing a number given as a percent to a decimal number is equivalent to moving the decimal point two places to the left and removing the percent sign. Another way to remember that is: Percent means per hundred (or to *divide by 100*).

┌─
Example **3** *Converting a Percent to a Decimal Number*

a) Change 35% to a decimal number.
b) Change 69.8% to a decimal number.
c) Change $\frac{1}{2}$ % to a decimal number.

Solution

a) $35\% = \dfrac{35}{100} = 0.35$. Thus, $35\% = 0.35$.

b) $69.8\% = \dfrac{69.8}{100} = 0.698$. Thus, $69.8\% = 0.698$.

c) $\dfrac{1}{2}\% = 0.5\% = \dfrac{0.5}{100} = 0.005$. Thus, $\dfrac{1}{2}\% = 0.005$.   ∎
└─

┌─
Example **4** *Apple Production*

According to the U.S. Department of Agriculture, in 2018 the United States produced about 256 million bushels of apples. Of these apples, 155 million bushels were produced in Washington, 31 million bushels were produced in New York, 12 million bushels were produced in Pennsylvania, and 58 million bushels were produced in other states. Determine the percent of apples produced in each of the following: Washington, New York, Pennsylvania, and other states. Round answers to the nearest tenth of a percent.

Solution   To determine each percent, divide the number of apples produced in each state by the total number of apples produced in the United States. Because all numbers are given in millions of pounds, we will calculate each percent using the number of millions of pounds.

Percent of apples produced in Washington $= \frac{155}{256} \approx 0.60547 \approx 60.5\%$

Percent of apples produced in New York $= \frac{31}{256} \approx 0.12109 \approx 12.1\%$

Percent of apples produced in Pennsylvania $= \frac{12}{256} \approx 0.04688 \approx 4.7\%$

Percent of apples produced in other states $= \frac{58}{256} \approx 0.22656 \approx 22.7\%$

The sum of the percents, $60.5\% + 12.1\% + 4.7\% + 22.7\%$, equals 100%.   ∎
└─

Percent Change

The percent increase or decrease, or percent change, over a period of time is determined by the formula:

┌───
Percent Change

$$\text{Percent change} = \frac{\left(\begin{array}{c}\text{amount in} \\ \text{latest period}\end{array}\right) - \left(\begin{array}{c}\text{amount in} \\ \text{previous period}\end{array}\right)}{\text{amount in previous period}} \times 100\%$$
└───

If the amount in the latest period is greater than the amount in the previous period, the answer will be positive and will indicate a percent increase. If the amount in the latest period is smaller than the amount in the previous period, the answer will be negative and will indicate a percent decrease.

Example 5 *Most Improved Baseball Team*

▲ *Khris Davis of the Oakland Athletics*

In 2018, the Major League Baseball team with the most improved record for winning games was the Oakland Athletics. In 2017, the Athletics won 75 games. In 2018, the Athletics won 97 games. Determine the percent increase in the number of games won by the Athletics from 2017 to 2018.

Solution The previous period is 2017, and the latest period is 2018.

$$\text{Percent change} = \frac{\left(\begin{array}{c}\text{amount in}\\\text{latest period}\end{array}\right) - \left(\begin{array}{c}\text{amount in}\\\text{previous period}\end{array}\right)}{\text{amount in previous period}} \times 100\%$$

$$= \frac{97 - 75}{75} \times 100\%$$

$$= \frac{22}{75} \times 100\%$$

$$\approx 0.2933 \times 100\%$$

$$\approx 29.3\%$$

Therefore, there was about a 29.3% increase in the number of games won by the Athletics from 2017 to 2018. ■

Example 6 *Walmart Stock Price*

On February 22, 2019, the price of one share of Walmart stock was $99.55. On March 28, 2019, the price of one share of Walmart stock had fallen to $97.13. Determine the percent change in the price of one share of Walmart stock during this period.

Solution The amount in the latest period is $97.13, and the amount in the previous period is $99.55.

$$\text{Percent change} = \frac{\left(\begin{array}{c}\text{amount in}\\\text{latest period}\end{array}\right) - \left(\begin{array}{c}\text{amount in}\\\text{previous period}\end{array}\right)}{\text{amount in previous period}} \times 100\%$$

$$= \frac{97.13 - 99.55}{99.55} \times 100\%$$

$$= \frac{-2.42}{99.55} \times 100\%$$

$$\approx -0.0243 \times 100\%$$

$$\approx -2.4\%$$

Thus, the price of one share of Walmart stock fell about 2.4% from February 22, 2019, to March 28, 2019 ■

Percent Markup and Markdown

A similar formula is used to calculate percent markup or markdown on cost. A positive answer indicates a markup, and a negative answer indicates a markdown.

Percent Markup or Markdown

$$\text{Percent markup (or markdown) on cost} = \frac{\text{selling price} - \text{dealer's cost}}{\text{dealer's cost}} \times 100\%$$

Example 7 *Determining Percent Markup*

Holdren Hardware stores pay $48.76 for glass fireplace screens. They regularly sell them for $79.88. At a sale they sell them for $69.99. Determine

a) the percent markup on the regular price.

b) the percent markup on the sale price.

Solution

a) We determine the percent markup on the regular price as follows.

$$\text{Percent markup} = \frac{\text{selling price} - \text{dealer's cost}}{\text{dealer's cost}} \times 100\%$$

$$= \frac{79.88 - 48.76}{48.76} \times 100\%$$

$$\approx 0.6382 \times 100\%$$

$$\approx 63.8\%$$

Thus, the percent markup on the regular price was about 63.8%.

b) We determine the percent markup on the sale price as follows.

$$\text{Percent markup} = \frac{69.99 - 48.76}{48.76} \times 100\%$$

$$\approx 0.4354 \times 100\%$$

$$\approx 43.5\%$$

Thus, the percent markup on the sale price was about 43.5%. ∎

Example 8 *Determining Percent Markdown*

To help reduce the inventory of its remaining 2020 calendars, Walgreens reduces the selling price to $4.00. If Walgreens cost per calendar was $5.00, determine the percent markdown on the calendars.

Solution We will use the same formula that we used for percent markup in Example 7.

$$\text{Percent markdown} = \frac{\text{selling price} - \text{dealer's cost}}{\text{dealer's cost}} \times 100\%$$

$$= \frac{4.00 - 5.00}{5.00} \times 100\%$$

$$= -0.20 \times 100\%$$

$$= -20\%$$

Thus, the percent markdown on the calendars was 20%. ∎

Other Percent Problems

In daily life, we may need to know how to solve any one of the following three types of problems involving percent:

1. What is a 15% tip on a restaurant bill of $45.71? The problem can be restated as

   15% of $45.71 is what number?

2. If Nancy sells a car for $19,000 and receives a commission of $475, what percent of the sale price is the commission? The problem can be restated as

   What percent of $19,000 is $475?

3. If the price of a laptop computer is reduced by 35% or $341.25, what was the original price of the computer? The problem can be restated as

   35% of what number is $341.25?

To answer these three questions, we will translate the words of the problem into an equation and then solve the equation. Some common words or phrases that we encounter are shown below, along with the symbols used when writing the equation.

Word or phrase	Symbol in the equation
what number	x
what percent	x
is, are, was, or *were*	$=$
of	$\cdot$

To solve each problem given, we change the percent into a decimal number and change the words of the sentence into symbols of an equation. We then solve the equation for x.

1. 15% of $45.71 is what number?
   0.15 · 45.71 = x

   $$0.15 \cdot 45.71 = x$$
   $$6.8565 = x$$

   Since 15% of $45.71 is $6.8565, the tip would be $6.86.

2. What percent of $19,000 is $475?
   x · 19,000 = 475

   $$x \cdot 19{,}000 = 475$$
   $$19{,}000x = 475$$
   $$x = \frac{475}{19{,}000} = 0.025 = 2.5\%$$

   Since 2.5% of $19,000 is $475, the commission is 2.5% of the sale price.

3. 35% of what number is $341.25?
   0.35 · x = 341.25

   $$0.35 \cdot x = 341.25$$
   $$0.35x = 341.25$$
   $$x = \frac{341.25}{0.35} = 975$$

   Since 35% of $975 is $341.25, the original price of the computer was $975.

Example 9   *Down Payment on a Condominium Home*

Melissa wishes to buy a condominium home for $189,000. To obtain the mortgage loan, she must pay 20% of the selling price as a down payment. Determine the amount of Melissa's down payment.

Solution We will let x be the amount of the down payment. This problem can be restated as "20% of the selling price is what number?" We will translate this sentence into an equation and then solve the equation.

$$\underbrace{20\%}\ \underbrace{\text{of}}\ \underbrace{\text{the selling price}}\ \underbrace{\text{is}}\ \underbrace{\text{what number?}}$$
$$0.20\ \cdot\qquad 189{,}000\qquad =\qquad x$$
$$0.20 \cdot 189{,}000 = x$$
$$37{,}800 = x$$

Melissa's down payment will be $37,800. ■

Example 10 Scholastic Chess Membership

The U.S. Chess Federation (USCF) has about 77,000 members. Of these members, about 28,000 are considered *scholastic* members, meaning they are currently attending some level of school from kindergarten through college. What percent of USCF members are scholastic members?

Solution Let $x =$ the percent of USCF members who are scholastic members. We will translate the words of the problem into an equation and then solve this equation.

$$\underbrace{\text{What percent}}\ \underbrace{\text{of}}\ \underbrace{\text{USCF members}}\ \underbrace{\text{are}}\ \underbrace{\text{scholastic members?}}$$
$$x\qquad\cdot\qquad 77{,}000\qquad =\qquad 28{,}000$$
$$77{,}000x = 28{,}000$$
$$x = \frac{28{,}000}{77{,}000} \approx 0.3636 \approx 36.4\%$$

Thus, about 36.4% of USCF members are scholastic members. ■

Example 11 Population of Italy

About 7,997,000, or about 13.5%, of Italy's population is younger than 15 years old. What is the approximate population of Italy?

Solution Let x be the population of Italy. This problem can be restated as "13.5% of what number is 7,997,000?" We will translate this sentence into an equation and then solve the equation.

$$\underbrace{13.5\%}\ \underbrace{\text{of}}\ \underbrace{\text{what number}}\ \underbrace{\text{is}}\ \underbrace{\text{7,997,000?}}$$
$$0.135\ \cdot\qquad x\qquad =\ 7{,}997{,}000$$
$$0.135x = 7{,}997{,}000$$
$$x = \frac{7{,}997{,}000}{0.135} = 59{,}237{,}037.04$$

Thus, the population of Italy is about 59,237,037 people. ■

SECTION 10.1
Exercises

Warm Up Exercises

In Exercises 1–6, fill in the blanks with an appropriate word, phrase, or symbol(s).

1. A percent is a ratio of some number to _____ .

2. To change a fraction to a percent, first divide the numerator by the denominator to obtain a decimal number.

Then multiply the decimal number by 100 and add a(n) _____ sign.

3. To change a decimal number to a percent, multiply the decimal number by _____ and add a percent sign.

4. To change a percent to a decimal number, divide the number by 100 and _____ the percent sign.

5. To compute the percent change, first determine the difference between the amount in the latest period and the amount in the previous period, then divide this difference by the amount in the _____ period.

6. To compute the percent markup or markdown on cost, first determine the difference between selling price and the dealer's cost, then divide this difference by the _____ cost.

Practice the Skills

In Exercises 7–14, change the number to a percent. Express your answer to the nearest tenth of a percent.

7. $\dfrac{2}{5}$ **8.** $\dfrac{3}{25}$ **9.** $\dfrac{7}{20}$

10. $\dfrac{7}{8}$ **11.** 0.007654   **12.** 0.5688

13. 3.78   **14.** 13.678

In Exercises 15–22, change the number given as a percent to a decimal number.

15. 9%   **16.** 43%   **17.** 7.24%

18. 0.75%   **19.** $\dfrac{1}{4}$%   **20.** $\dfrac{3}{8}$%

21. 135.9%   **22.** 298.7%

Problem Solving

For Exercises 23–26, where appropriate, round answers to the nearest tenth of a percent unless specified otherwise.

23. _Niacin_ The U.S. Department of Agriculture's recommended daily allowance (USRDA) of niacin for adults is 20 mg. One serving of Fruity Cheerios provides 5 mg of niacin. What percent of the USRDA of niacin does one serving of Fruity Cheerios provide?

24. _Potassium_ The U.S. Department of Agriculture's recommended daily allowance (USRDA) of potassium for adults is 3500 mg. One serving of Fruity Cheerios provides 95 mg of potassium. What percent of the USRDA of potassium does one serving of Fruity Cheerios provide?

25. _Reduced Fat Milk_ One serving of whole milk contains 8 g of fat. Reduced-fat milk contains 41.25% less fat per serving than whole milk. How many grams of fat does one serving of reduced-fat milk contain?

26. _Baked Lays_ One serving of Lays Classic Potato Chips contains 10 grams of fat. Lays Oven Baked Potato Chips contains 65% less fat per serving than Lays Classic Potato Chips. How many grams of fat does one serving of Lays Oven Baked Potato Chips contain?

**DVD Collection** In Exercises 27–30, use the circle graph to answer the questions. The Lafayette County Library classifies its DVDs into one of the categories shown below. The library owns 1743 DVDs. Round your answers to the nearest whole number.

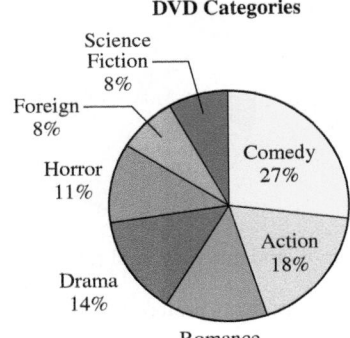

DVD Categories

Science Fiction 8%
Foreign 8%
Horror 11%
Comedy 27%
Action 18%
Romance 14%
Drama 14%

27. Determine the number of comedy DVDs.

28. Determine the number of romance DVDs.

29. Determine the number of horror DVDs.

30. Determine the number of foreign DVDs.

**Seawater Composition** In Exercises 31–34, use the table to answer the questions. The elements that compose seawater, along with the corresponding percents, are shown in the table below. Where appropriate, round your answers to the nearest hundredth.

Seawater Elements	
Chemical	**Percent**
Oxygen	85.84%
Hydrogen	10.82%
Chlorine	1.94%
Sodium	1.08%
Magnesium	0.13%
Other elements	0.19%

31. If a sample of seawater contains 25 mℓ, how many mℓ of oxygen are in the sample?

32. If a saltwater aquarium contains 55 gallons of seawater, how many gallons of hydrogen are in the aquarium?

33. Ms. Buller collects a 500-mℓ sample of seawater for her chemistry class. How many mℓ of chlorine are in the sample?

34. Mr. Kimball collects a 6000-mℓ sample of seawater for his marine biology class. How many mℓ of magnesium are in the sample?

Music Preferences *In Exercises 35–38, use the circle graph to answer the questions. Students at Bayshore High School were polled to determine the type of music they preferred. There were 1960 students who completed the poll. Their responses are represented in the circle graph below. Round your answers to the nearest tenth of a percent.*

Music Preferences

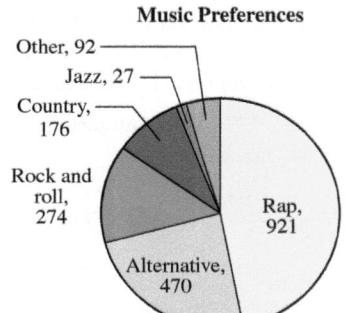

35. What percent of students preferred rap music?

36. What percent of students preferred alternative music?

37. What percent of students preferred rock and roll music?

38. What percent of students preferred country music?

In Exercises 39–66, where appropriate, round your answer to the nearest tenth of percent.

39. **Russian Population** The population of Russia has declined since the fall of the Soviet Union. In 1990, the population of Russia was about 148 million people. In 2018, the population of Russia was about 146.8 million people. Determine the percent decrease in Russia's population during this time period.

40. **U.S. Population** Unlike the population of Russia (see Exercise 39), the population of the United States has been increasing since 1990. In 1990, the population of the United States was about 249 million people. In 2018, the population of the United States was about 327.2 million people. Determine the percent increase in the U.S. population during this time period.

41. **Box Office Sales** The graph above and to the right, created by **StatCrunch**, shows the U.S. gross box office earnings for movies, in millions of dollars, for the years 2013–2018.

a) Determine the percent decrease in gross box office earnings from 2013 to 2014.

b) Determine the percent increase in gross box office earnings from 2014 to 2015.

c) Determine the percent increase in gross box office earnings from 2015 to 2016.

d) Determine the percent decrease in gross box office earnings from 2016 to 2017.

e) Determine the percent increase in gross box office earnings from 2017 to 2018.

U.S. Gross Box Office Earnings for Movies

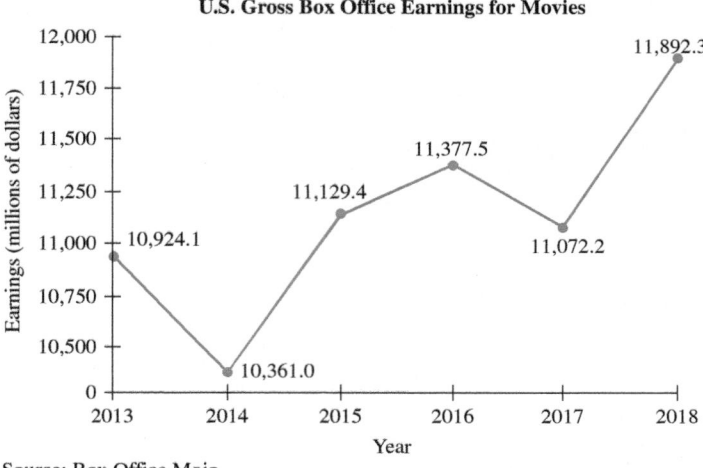

Source: Box Office Mojo

42. **Presidential Salary** The following graph shows the annual salary for the president of the United States for selected years from 1789 to 2001.

Annual U.S. Presidential Salary

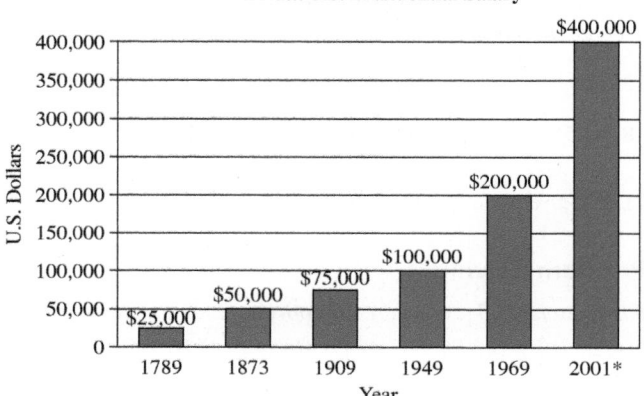

Source: *Congressional Quarterly*
*As of 2020, the annual presidential salary is still $400,000.

a) Determine the percent increase in annual salary from 1789 to 1873.

b) Determine the percent increase in annual salary from 1909 to 1949.

c) Determine the percent increase in annual salary from 1969 to 2001.

d) Determine the percent increase in annual salary from 1789 to 2001.

In Exercises 43–54, determine the answer to the question.

43. What is 15% of $75,000?

44. What is 7.5% of $200,000?

45. Eight percent of 32 is what number?

46. Thirty-two percent of 50 is what number?

47. What percent of 96 is 24?

48. What percent of 75 is 15?

49. Twelve is what percent of 150?

50. One hundred thirty is what percent of 208?

51. Five percent of what number is 15?

52. Ten percent of what number is 75?

53. Forty-two is 28% of what number?

54. Eighteen is 20% of what number?

55. *Tax and Tip* According to the Original Tipping Page, www.tipping.org, it is appropriate to tip waiters and waitresses 15%–20% of the pre-tax restaurant bill. The Leesebergs' dinner costs $63.50 before tax, and the tax rate is 6%.

   a) What is the tax on the Leesebergs' dinner?

   b) If the Leesebergs decide to tip 15% of the pre-tax bill, how much is the tip?

   c) What is the total cost of the dinner including the tax and tip?

56. *Tax and Tip* The Brants' dinner (see Exercise 55) cost $68.00 before tax, and the tax rate was 7.5%.

   a) What is the tax on the Brants' dinner?

   b) If the Brants decide to tip 20% of the pre-tax bill, how much is the tip?

   c) What is the total cost of the dinner, including the tax and the tip?

57. *Pretax Amount* The cost of a jacket at Macy's is $54.27, including 8% sales tax. What is the pretax cost of the jacket?

58. *Pretax Amount* The cost of a book at Barnes & Noble is $32.10, including 7% sales tax. What is the pretax cost of the book?

59. *Percentage of A's* In a mathematics class, eighteen students received an A on the third test, which is 150% of the students who received an A on the second test. How many students received an A on the second test?

60. *Employee Increase* The Fastlock Company hired 57 new employees, which increased its staff by 30%. What was the original number of employees?

61. *Salary Increase* Mr. Brown's present salary is $39,500. He is getting an increase of 7% in his salary next year. What will his new salary be?

62. *Salary Increase* Maya's present salary is $42,500. As part of a promotion her salary will increase by 4% next year. What will Maya's new salary be?

63. *Vacuum Cleaner Sales* A Kirby vacuum cleaner dealership sold 430 units in 2018 and 407 units in 2019.

Determine the percent increase or decrease in the number of units sold.

64. *Vacuum Cleaner Markup* If the Kirby dealership in Exercise 63 pays $320 for each unit and sells a unit for $699, what is the percent markup in the price of each unit?

65. *Television Sale* The regular price of a Phillips color TV is $539.62. During a sale, Hill TV is selling the TV for $439. Determine the percent decrease in the price of this TV.

66. *Weight Loss* On January 1, Juan weighed 235 pounds and decided to diet and exercise. On June 30, Juan weighed 210 pounds. Determine the percent decrease in Juan's weight from January 1 to June 30.

67. *Soccer Ball Markup* The cost of a soccer ball to Soccer Line Sports Goods is $35. They sell the same soccer ball for $49. Determine the percent markup in the price of the soccer ball.

68. *Restaurant Markup* The cost of a fish dinner to the owner of the Golden Wharf restaurant is $7.95. The fish dinner is sold for $11.95. Determine the percent markup.

Concept/Writing Exercises

69. *Reselling a Car* Quincy purchased a used car for $1000. He decided to sell the car for 10% above his purchase price. Quincy could not sell the car so he reduced his asking price by 10%. If he sells the car at the reduced price, will he have a profit or a loss or will he break even? Explain how you arrived at your answer.

70. *Furniture Sale* Kane's Furniture Store advertised a table at a 15% discount. The original selling price was $115, and the sale price was $100. Was the sale price consistent with the ad? Explain.

Challenge Problems/Group Activities

71. *Comparing Markdowns*

   a) A coat is marked down 10%, and the customer is given a second discount of 15%. Is that the same as a single discount of 25%? Explain.

   b) The regular price of a chair is $189.99. Determine the sale price of the chair if the regular price is reduced by 10% and this price is then reduced another 15%.

   c) Determine the sale price of the chair if the regular price of $189.99 is reduced by 25%.

   d) Examine the answers obtained in parts (b) and (c). Does your answer to part (a) appear to be correct? Explain.

72. Selling Ties The Tie Shoppe paid $5901.79 for a shipment of 500 ties and wants to make a profit of 40% of the cost on the entire shipment. The store is having two special sales. At the first sale it plans to sell 100 ties for $9.00 each, and at the second sale it plans to sell 150 ties for $12.50 each. What should be the selling price of the other 250 ties for the Tie Shoppe to make a 40% profit on the entire shipment?

73. In parts (a) through (c), determine which is the greater amount and by how much.

a) $100 increased by 25% or $200 decreased by 25%.

b) $100 increased by 50% or $200 decreased by 50%.

c) $100 increased by 100% or $200 decreased by 100%.

Research Activity

74. Circle Graphs Find two circle graphs in newspapers, in magazines, or on the Internet whose data are not given in percents. Redraw the graphs and label them with percents.

SECTION 10.2   Personal Loans and Simple Interest

Upon completion of this section, you will be able to:

- Use the simple interest formula to calculate ordinary interest.
- Use the United States rule to solve simple interest problems.

Consider the following dilemma. Your car breaks down and the mechanic tells you that the repairs will cost $450. You currently do not have the $450, but your family desperately needs to have a car. One option for you would be to borrow the money from a friend, a family member, a bank, or other lending institution. In this section, we will discuss personal loans and the cost of obtaining such loans.

Why This Is Important If you need to borrow money, you may need to consider different options for your loan. Understanding the cost of borrowing money will help you make informed decisions about your personal finances.

The money a bank or other lender is willing to lend you is called the amount of *credit* extended or the *principal of the loan*. The amount of credit and the interest rate that you may obtain depend on the assurance you can give the lender that you will be able to repay the loan. Your credit is determined by your reputation for repaying loans, by your earning power, and by what you can pledge as security to cover the loan. *Security* (or *collateral*) is anything of value pledged by the borrower that the lender may sell or keep if the borrower does not repay the loan. Acceptable security may be a business, a mortgage on a property, the title to an automobile, savings accounts, or stocks or bonds.

Bankers sometimes grant loans without security, but they require the signature of one or more other persons, called *cosigners*, who guarantee the loan will be repaid. For either of the two types of loans, the secured loan or the cosigner loan, the borrower (and cosigner, if there is one) must sign an agreement called a *personal note*. This document states the terms and conditions of the loan.

The most common ways for individuals to borrow money are through an installment loan or using a credit card. (Installment loans and credit cards are discussed in Section 10.4.)

The concept of simple interest is essential to the understanding of installment buying. *Interest* is the money the borrower pays for the use of the lender's money. One type of interest is called simple interest. *Simple interest* is based on the entire amount of the loan for the total period of the loan. The formula used to determine simple interest follows.

Simple Interest Formula

$$\text{Interest} = \text{principal} \times \text{rate} \times \text{time}$$
$$i = prt$$

In the simple interest formula, the *principal*, p, is the amount of money borrowed (or loaned), the *rate*, r, is the rate of interest expressed as a decimal number, and the

Paying off debt

time, t, is the number of days, months, or years for which the money will be lent. *Time is expressed in the same period as the rate.* For example, if the rate is 2% per month, the time must be expressed in months. Typically, rate means the annual rate unless otherwise stated. Principal and interest are expressed in dollars in the United States.

Ordinary Interest

The most common type of simple interest is called *ordinary interest*. When computing ordinary interest, we will use the *Banker's rule*. The *Banker's rule* states that a month has 30 days, a year has 12 months, and therefore, a year has 30 × 12 or 360 days. When computing ordinary interest using an annual simple interest rate, if the time given is 5 months then we will use $t = \frac{5}{12}$. If the time given is 73 days, then we will use $t = \frac{73}{360}$. In this textbook, simple interest will mean ordinary interest unless stated otherwise.

Example 1 *A New Deck*

Zawfar needs to borrow $8500 to replace the deck at his home. From his credit union, Zawfar obtains a 30-month loan with an annual simple interest rate of 4.9%.

a) Calculate the simple interest he is charged on the loan.

b) Determine the amount, principal plus interest, Zawfar will pay the credit union at the end of the 30 months to pay off his loan.

Solution

a) To determine the interest on the loan, we use the formula $i = prt$. We know that $p = \$8500$ and $r = 4.9\%$, which written as a decimal number is 0.049. Since the time is 30 months, the time in years is $t = \frac{30}{12} = 2.5$. We substitute the appropriate values in the simple interest formula

$$i = p \times r \times t$$
$$= \$8500 \times 0.049 \times 2.5$$
$$= \$1041.25$$

The simple interest on $8500 at 4.9% for 30 months is $1041.25.

b) The amount to be repaid is equal to the principal, $8500, plus the interest, $1041.25.

$$A = p + i$$
$$= \$8500 + \$1041.25$$
$$= \$9541.25$$

To pay off his loan, Zawfar will pay the credit union $9541.25 at the end of 30 months.   ∎

Example 2 *Determining the Annual Rate of Interest*

Philip agrees to lend $850 to his friend Joe to help Joe travel to Cincinnati to attend a family wedding. Nine months later, Joe repaid the original $850 plus $51 interest. What annual rate of interest did Joe pay to Philip?

Solution We need to solve for the interest rate, *r*. Since the time is 9 months, the time in years is $\frac{9}{12}$, or 0.75. Using the formula $i = prt$, we get

$$\$51 = \$850 \times r \times 0.75$$
$$51 = 637.5r$$
$$\frac{51}{637.5} = \frac{637.5r}{637.5} \qquad \text{Divide both sides of the equation by 637.5.}$$
$$0.08 = r$$

To change this decimal number to a percent, multiply by 100 and add a percent sign. Thus the annual rate of interest paid by Joe to Philip is 8%.   ∎

Example 3 *A Pawn Loan*

To obtain money to pay some medical bills, Kevin decides to pawn his bicycle. Kevin borrows $300 and after 30 days gets his bicycle back by paying the pawn broker $355. What annual rate of interest did Kevin pay?

Solution Kevin paid $355 − $300 = $55 in interest, and the length of the loan is for 30 days or $\frac{30}{360} = \frac{1}{12}$ of a year.

Using the formula $i = prt$, we get

$$\$55 = \$300 \times r \times \frac{1}{12}$$

$$55 = 25r$$

$$\frac{55}{25} = r$$

$$2.2 = r$$

The annual rate of interest as a decimal number is 2.2. To change this number to a percent, multiply the number by 100 and add a percent sign. Thus, the annual rate of interest paid is 220%.   ∎

In Examples 1, 2, and 3, we illustrated a simple interest loan, for which the interest and principal are paid on the due date of the note. There is another type of loan, the *discount note*, for which the interest is paid at the time the borrower receives the loan. The interest charged in advance is called the *bank discount*. A Federal Reserve Treasury bill is a bank discount note issued by the U.S. government. Example 4 illustrates a discount note.

Example 4 *True Interest Rate of a Discount Note*

Siegrid took out a $500 loan using a 10% discount note for a period of 3 months. Determine

a) the interest she must pay to the bank on the date she receives the loan.
b) the net amount of money she receives from the bank.
c) the actual rate of interest for the loan.

Solution

a) To determine the interest, use the simple interest formula.

$$i = prt$$

$$= \$500 \times 0.10 \times \frac{3}{12}$$

$$= \$12.50$$

Since the $12.50 is charged in advance, it is the bank discount for the note.

b) Since Siegrid must pay $12.50 interest when she first receives the loan, the net amount she receives is $500 − $12.50, or $487.50.

c) We calculate the actual rate of interest charged using the simple interest formula. In the formula for the principal, we use the amount Siegrid actually received from the bank, $487.50. For the interest, we use the interest calculated in part (a).

$$i = prt$$

$$\$12.50 = \$487.50 \times r \times \frac{3}{12}$$

$$12.50 = 121.875 \times r$$

$$\frac{12.50}{121.875} = r$$

$$0.1026 \approx r$$

Thus, the actual rate of interest is about 10.3% rather than the quoted 10%.   ∎

Example 5 *Loan Choices for Farm Equipment*

Rodney owns a farm. He needs to purchase new equipment for the farm but does not have the $1600 purchase price. The equipment dealership has two payment options. With option 1, Rodney can pay $800 as a down payment and then pay $850 in 6 months. With option 2, Rodney can pay $400 as a down payment and then pay $1380 in 9 months. Which option has a higher annual simple interest rate?

Solution Before we compute the annual simple interest rates, we first need to compute the principle and the interest.

Option 1: To determine the principal of the loan, subtract the down payment from the purchase price of the equipment. Therefore, $p = \$1600 - \$800 = \$800$. To determine the interest charged, subtract the purchase price of the equipment from the total amount paid. Therefore, $i = (\$800 + \$850) - \$1600 = \$50$. To determine the time, divide 6 by 12. Therefore $t = \frac{6}{12} = 0.5$. Then we use the simple interest formula.

$$i = p \times r \times t$$
$$50 = 800 \times r \times 0.5$$
$$50 = 400r$$
$$\frac{50}{400} = r$$
$$0.125 = r$$

Option 1 has a 12.5% annual simple interest rate.

Option 2: The principal is $p = \$1600 - \$400 = \$1200$, the interest is $i = (\$400 + \$1380) - \$1600 = \$180$, and the time is $t = \frac{9}{12} = 0.75$. Then

$$i = p \times r \times t$$
$$180 = 1200 \times r \times 0.75$$
$$180 = 900r$$
$$\frac{180}{900} = r$$
$$0.2 = r$$

Option 2 has a 20.0% annual simple interest rate. Therefore, option 2 charges a higher annual simple interest rate than option 1. ∎

The United States Rule

A loan has a date of maturity, at which time the principal and interest are due. It is possible to make payments on a loan before the date of maturity. A payment that is less than the full amount owed and made prior to the due date is known as a *partial payment*. A Supreme Court decision specified the method by which these payments are credited. The procedure is called the *United States rule*.

The United States rule states that if a partial payment is made on the loan, interest is computed on the principal from the first day of the loan until the date of the partial payment. The partial payment is used to pay the interest first; then the rest of the payment is used to reduce the principal. An individual can make as many partial payments as he or she wishes; the procedure is repeated for each payment. The balance due on the date of maturity is determined by computing interest due since the last partial payment and adding this interest to the unpaid principal.

Did You Know?

Selling Time

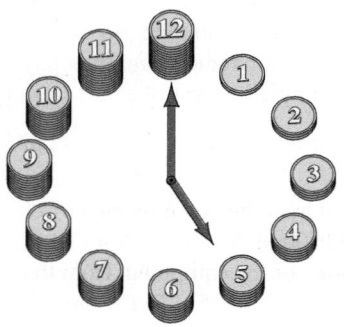

For most of human history, the practice of charging interest on money borrowed, or *usury*, was considered not only immoral, but a crime. This attitude arose in ancient times, in part because a peasant or farmer could literally be enslaved if payments could not be made. Charging interest over time was considered "selling time," something not ours to sell. What was especially objectionable was the idea of "pure interest," a fixed payment set at the time of the loan, even when there was no risk to the lender. Today, borrowing money has become a fact of life for individuals, businesses, and governments. In 2018, the U.S. government paid over $500 billion in interest alone on a debt of over $21.97 trillion.

Today the word *usury* means lending money at a rate of interest that is excessive or unlawfully high. A person who engages in such loans is often referred to as a "loan shark."

We will use the Banker's rule to calculate the simple interest when applying the United States rule. Recall that the Banker's rule considers a year to have 360 days and any fractional part of a year to be the exact number of days divided by 360.

To determine the exact number of days in a period, we can use Table 10.1 below. Example 6 on page 589 illustrates how to use the table.

Table 10.1

Days in Each Month											
Day of Month	**31 Jan.**	**28 Feb.**	**31 Mar.**	**30 Apr.**	**31 May**	**30 June**	**31 July**	**31 Aug.**	**30 Sept.**	**31 Oct.**	**30 Nov.**	**31 Dec.**
Day 1	1	32	60	91	121	152	182	213	244	274	305	335
Day 2	2	33	61	92	122	153	183	214	245	275	306	336
Day 3	3	34	62	93	123	154	184	215	246	276	307	337
Day 4	4	35	63	94	124	155	185	216	247	(277)	308	338
Day 5	5	36	64	95	125	156	186	217	248	278	309	339
Day 6	6	37	65	96	126	157	187	218	249	279	310	340
Day 7	7	38	66	(97)	127	158	188	219	250	280	311	341
Day 8	8	39	67	98	128	159	189	220	251	281	312	342
Day 9	9	40	68	99	129	160	190	221	252	282	313	343
Day 10	10	41	69	100	130	161	191	222	253	283	314	344
Day 11	11	42	70	101	131	162	192	223	254	284	315	345
Day 12	12	43	71	102	132	163	193	224	255	285	316	346
Day 13	13	44	72	103	133	164	194	225	256	286	317	347
Day 14	14	45	73	104	134	165	195	226	257	287	318	348
Day 15	15	46	74	105	135	166	196	227	258	288	319	349
Day 16	16	47	75	106	136	167	197	228	259	289	320	350
Day 17	17	48	76	107	137	168	198	229	260	290	321	351
Day 18	18	49	77	108	138	169	199	230	261	291	322	352
Day 19	19	50	78	109	139	170	200	231	262	292	323	353
Day 20	20	51	79	110	140	171	201	232	263	293	324	354
Day 21	21	52	80	111	141	172	202	233	264	294	325	355
Day 22	22	53	81	112	142	173	203	234	265	295	326	356
Day 23	23	54	82	113	143	174	204	235	266	296	327	357
Day 24	24	55	83	114	144	175	205	236	267	297	328	358
Day 25	25	56	84	115	145	176	206	237	268	298	329	359
Day 26	26	57	85	116	146	177	207	238	269	299	330	360
Day 27	27	58	86	117	147	178	208	239	270	300	331	361
Day 28	28	59	87	118	148	179	209	240	271	301	332	362
Day 29	29		88	119	149	180	210	241	272	302	333	363
Day 30	30		89	120	150	181	211	242	273	303	334	364
Day 31	31		90		151		212	243		304		365

Add 1 day for leap year if February 29 falls between the two dates under consideration.

Example 6 *Loan Due Date*

Determine (a) the due date of a loan made on April 7 for 180 days and (b) the number of days from March 15 to November 18.

Solution

a) We will use Table 10.1 to determine the due date of the loan. In the table, first locate Day 7 in the far-left column and then move directly to the right four columns, to the April column. We can see that April 7 corresponds to the number 97, circled in red, in the table. Thus, April 7 is the 97th day of the year. Since the loan is due 180 days after April 7, we add 180 to 97. Note that $97 + 180 = 277$. The due date is the 277th day of the year. From Table 10.1 we see that the number 277, circled in blue, corresponds to Day 4 of October. Thus, the due date of the loan is October 4.

b) To determine the number of days from March 15 to November 18, we first note from Table 10.1 that March 15 is the 74th day of the year and that November 18 is the 322nd day of the year. Next, we determine the difference between 322 and 74. Note that $322 - 74 = 248$. Thus, the number of days from March 15 to November 18 is 248 days. ∎

Example 7 *Using the Banker's Rule*

Using the Banker's rule, determine the simple interest that will be paid on a $2700 loan at an interest rate of 2% for the period of March 15 to November 18.

Solution The principal is $2700 and the rate, written as a decimal number, is 0.02. In Example 6(b), we determined that the number of days from March 15 to November 18 is 248 days. Therefore, the period of time in years is $\frac{248}{360}$. Substituting into the simple interest formula gives

$$i = p \times r \times t$$
$$= \$2700 \times 0.02 \times \frac{248}{360}$$
$$= \$37.20$$

Thus, the simple interest is $37.20. ∎

A borrower may wish to pay back part of a loan prior to the due date. The next example illustrates how a partial payment is credited under the United States rule. Making partial payments reduces the amount of interest paid and thus the cost of the loan.

Example 8 *Using the United States Rule*

Mrs. Panik is a mathematics teacher, and she plans to attend a national teachers conference. To pay for her airfare, on November 1, 2018, she takes out a 120-day loan for $400 at an interest rate of 12.5%. Mrs. Panik uses some birthday gift money to make a partial payment of $150 on January 5, 2019. She makes a second partial payment of $100 on February 2, 2019.

a) Determine the due date of the loan.
b) Determine the interest and the amount credited to the principal on January 5.
c) Determine the interest and the amount credited to the principal on February 2.
d) Determine the amount that Mrs. Panik must pay on the due date.

Solution

a) Using Table 10.1, we see that November 1 is the 305th day of the year. Next, we note that the sum of 305 and 120 is 425. Since this due date will extend into the next year, we subtract 365 from 425 to get 60. From Table 10.1, we see that the 60th day of the year is March 1. Therefore, the loan due date is March 1, 2019. Had 2019 been a leap year, the due date would have been February 29, 2019.

b) Using Table 10.1, January 5 is the 5th day of the year and November 1 is the 305th day of the year. The number of days from November 1 to January 5 can be computed as follows: $(365 - 305) + 5 = 65$. Then, using $i = prt$ and the Banker's rule, we get

$$i = \$400 \times 0.125 \times \frac{65}{360}$$
$$\approx \$9.03$$

The interest of $9.03 that is due January 5, 2019, is deducted from the payment of $150. The remaining payment of $150 − $9.03, or $140.97, is then credited to the principal. Therefore, the adjusted principal is now $400 − $140.97, or $259.03.

c) Since there was a second partial payment made, we use the Banker's rule to calculate the interest on the unpaid principal for the period from January 5 to February 2. According to Table 10.1, the number of days from January 5 to February 2 is $33 - 5$, or 28 days.

$$i = \$259.03 \times 0.125 \times \frac{28}{360}$$
$$\approx \$2.52$$

The interest of $2.52 that is due February 2, 2019, is deducted from the payment of $100. The remaining payment of $100 − $2.52, or $97.48, is then credited to the principal. Therefore, the new adjusted principal is now $259.03 − $97.48, or $161.55.

d) The due date of the loan is March 1. Using Table 10.1, we see that there are $60 - 33$, or 27, days from February 2 to March 1. The interest is computed on the remaining balance of $161.51 by using the simple interest formula.

$$i = \$161.55 \times 0.125 \times \frac{27}{360}$$
$$\approx \$1.51$$

Therefore, the balance due on the maturity date of the loan is the sum of the principal and the interest, $161.55 + $1.51, or $163.06. *Note:* The sum of the days in the three calculations, $65 + 28 + 27$, equals the total number of days in the loan, 120. ∎

Note that in Example 8(b), if the partial payment of $150 made on January 5 had been the only partial payment made, then the balance due on March 1, 2019, would be calculated by determining the interest on the balance, $259.03, for the remainder of the loan, 55 days, and adding this interest to the principal of $259.03. If this payment had been the only partial payment made, then the balance due on March 1 would be

$$\text{Balance due} = \text{principal} + \text{interest}$$
$$\approx 259.03 + \left(259.03 \times 0.125 \times \frac{55}{360}\right)$$
$$\approx 259.03 + 4.95$$
$$\approx 263.98$$

SECTION 10.2 Exercises

Warm Up Exercises

In Exercises 1–8, fill in the blanks with an appropriate word, phrase, or symbol(s).

1. The money a bank or other lending institution is willing to lend you is called the amount of credit extended or the _____ of the loan.

2. Anything of value pledged by the borrower that the lender may sell or keep if the borrower does not repay the loan is called the _____ or collateral.

3. The money the borrower pays for the use of the lender's money is called the _____.

4. When using the simple interest formula, the interest rate, r, is expressed as a(n) _____ number.

5. When using the simple interest formula, time, t, is expressed in the same period as the _____.

6. A payment that is less than the full amount owed and made prior to the due date is known as a(n) _____ payment.

7. If a partial payment is made on a loan, interest is computed on the principal from the first day of the loan until the date of the partial payment. This is known as the _____ rule.

8. The Banker's rule considers a year to have _____ days.

Practice the Skills

In Exercises 9–18, determine the simple interest. Unless noted otherwise, assume the rate is an annual rate. Assume 360 days in a year. Round answers to the nearest cent.

9. $p = \$375, r = 2.25\%, t = 4$ years

10. $p = \$620, r = 3.05\%, t = 1.5$ years

11. $p = \$1100, r = 8.75\%, t = 90$ days

12. $p = \$6742.75, r = 6.05\%, t = 90$ days

13. $p = \$587, r = 0.045\%$ per day, $t = 2$ months

14. $p = \$2500, r = 3\frac{7}{8}\%, t = 9$ months

15. $p = \$550.31, r = 8.9\%, t = 67$ days

16. $p = \$2756.78, r = 10.15\%, t = 103$ days

17. $p = \$1372.11, r = 1\frac{3}{8}\%$ per month, $t = 6$ months

18. $p = \$41,864, r = 0.0375\%$ per day, $t = 60$ days

In Exercises 19–24, use the simple interest formula to determine the missing value. If necessary, round all dollar figures to the nearest cent.

19. $p = \$2000, r = ?, t = 3$ years, $i = \$82.80$

20. $p = ?, r = 2.1\%, t = 135$ days $i = \$37.80$

21. $p = ?, r = 6\%, t = 5$ months, $i = \$175.00$

22. $p = \$800.00, r = 4\%, t = ?, i = \$64.00$

23. $p = \$957.62, r = 6.5\%, t = ?, i = \$124.49$

24. $p = \$1650.00, r = ?, t = 6.5$ years, $i = \$343.20$

Problem Solving

In Exercises 25–32, if necessary round all dollar figures to the nearest cent and round percents to the nearest hundredth of a percent.

25. **Small Business Administration Loan** Joshua uses the Small Business Administration (SBA) Loan Program to obtain a loan to help expand his health food store. Joshua borrows $45,000 for 2 years with a simple interest rate of 1.5%. Determine the amount of money Joshua must repay the SBA after 2 years.

26. **Business Loan** The city of Charlottesville is offering simple interest loans to start-up businesses at a rate of 3.5%. Mr. Cannata obtains one of these loans of $6000 for 3 years to help pay for start-up costs of his restaurant, Ortygia. Determine the amount of money Mr. Cannata must repay the city after 3 years.

27. **Bank Personal Note** Karen borrowed $3500 from a bank for 6 months. Her friend Ms. Harris was cosigner of Karen's personal note. The bank collected $7\frac{1}{2}\%$ simple interest on the date of maturity.

   a) How much did Karen pay for the use of the money?

   b) Determine the amount she repaid to the bank on the due date of the note.

28. Personal Note Janice owns a hair salon and needs $20,000 to renovate her salon. She borrows the money on a 30-month personal bank note. The simple interest rate charged is 5.75%.

a) How much interest does Janice pay?

b) Determine the amount Janice must repay the bank on the due date of the note.

29. Bank Discount Note Julie borrowed $3650 from her bank for 18 months using a 7.6% discount note.

a) How much interest did Julie pay the bank for the use of its money?

b) How much did she receive from the bank?

c) What was the actual rate of interest she paid?

30. Bank Discount Note Kwame borrowed $2500 for 5 months from a bank using a 2% discount note.

a) How much interest did Kwame pay the bank for the use of its money?

b) How much did he receive from the bank?

c) What was the actual rate of interest he paid?

31. Pawning a Saxophone Elisa needs money to repair her home air conditioner, so she pawns her saxophone. The pawnbroker loans Elisa $270. Seven days later, Elisa gets her saxophone back by paying the pawnbroker $280.50. What annual simple interest rate did the pawnbroker charge Elisa?

32. A Pawn Loan Jeffrey wants to take his mother out for dinner on her birthday, so he pawns his watch. The pawnbroker loans Jeffrey $75. Fourteen days later, Jeffrey gets his watch back by paying the pawnbroker $80.25. What annual simple interest rate did the pawnbroker charge Jeffrey?

In Exercises 33–38, determine the exact time from the first date to the second date. Use Table 10.1 on page 588. Assume the year is not a leap year unless otherwise indicated.

33. February 25 to May 5

34. July 5 to November 18

35. February 2 to October 31 (the loan is due in a leap year)

36. June 14 to January 24

37. August 24 to May 15

38. December 21 to April 28

In Exercises 39–42, determine the due date of the loan, using the exact time, if the loan is made on the given date for the given number of days.

39. May 31 for 150 days

40. March 13 for 120 days

41. November 25 for 120 days (the loan is due in a leap year)

42. July 5 for 210 days

In Exercises 43–52, a partial payment is made on the date(s) indicated. Use the United States rule to determine the balance due on the note at the date of maturity. The Effective Date is the date the note was written. Assume the year is not a leap year. Where appropriate, round interest calculations to the nearest cent.

	Prin-cipal	Rate	Effective Date	Partial Payment(s) Amount	Partial Payment(s) Date(s)	Maturity Date
43.	$3200	3%	June 1	$1200	July 1	Sep. 29
44.	$5400	4%	Jan. 20	$2500	Apr. 15	May 31
45.	$2400	5.5%	Feb. 1	$1000	May 1	Aug. 31
46.	$8500	11.5%	Sep. 1	$4250	Oct. 31	Dec. 31
47.	$9000	6%	July 15	$4000	Dec. 27	Feb. 1
48.	$1000	12.5%	Jan. 1	$300	Jan. 15	Feb. 15
49.	$1800	15%	Aug. 1	$500	Sept. 1	Nov. 1
				$500	Oct. 1	
50.	$5000	14%	Oct. 15	$800	Nov. 15	Jan. 1
				$800	Dec. 15	
51.	$11,600	6%	Mar. 1	$2000	Aug. 1	Dec. 1
				$4000	Nov. 15	
52.	$21,000	$4\frac{3}{8}$%	July 12	$8000	Oct. 10	Jan. 30
				$6000	Dec. 8	

53. *Company Loan* On March 1, the Zwick Balloon Company signed a $6500 note with simple interest of $10\frac{1}{2}$ % for 180 days. The company made payments of $1750 on May 1 and $2350 on July 1. How much will the company owe on the date of maturity?

54. *Flag Company Loan* On April 3, the Mears Flag Company borrowed $15,000 to pay for start-up costs for its new showroom. The loan had a simple interest rate of 8.25% and was for 270 days. The company was able to make partial payments of $6000 on May 28 and $3500 on October 12. How much will the company owe on the date of maturity of the loan?

Challenge Problems/Group Activities

55. *U.S. Treasury Bills* The U.S. government borrows money by selling Treasury bills. Treasury bills are discounted notes issued by the U.S. government. On November 14, 2018, Kris purchased a 364-day, $1000 U.S. Treasury bill at a 0.15% discount. On the date of maturity, Kris will receive $1000.

a) What is the date of maturity of the Treasury bill?

b) How much did Kris actually pay for the Treasury bill?

c) How much interest did the U.S. government pay Kris on the date of maturity?

d) What is the actual rate of interest of the Treasury bill? Round the answer to the nearest ten-thousandths of a percent.

▲ *U.S. Treasury building, Washington, D.C.*

56. *U.S. Treasury Bills* On December 3, 2018, Trinity purchased a 364-day, $6000 U.S. Treasury bill at a 0.12% discount. (See Exercise 55.)

a) What is the date of maturity of the Treasury bill?

b) How much did Trinity actually pay for the Treasury bill?

c) How much interest did the U.S. government pay Trinity on the date of maturity?

d) What is the actual rate of interest of the Treasury bill? Round the answer to the nearest thousandths of a percent.

57. *Tax Preparation Loan* Many tax preparation organizations will prepay customers' tax refunds if they pay a one-time finance charge. In essence, the customer is borrowing the money (the refund minus the finance charge) from the tax preparer, prepaying the interest (as in a discount note), and then repaying the loan with the tax refund. This procedure allows customers access to their tax refund money without having to wait. Joy had a tax refund of $743.21 due. She was able to get her tax refund immediately by paying a finance charge of $39.95. What annual simple interest rate, to the nearest hundredth of a percent, is Joy paying for this loan assuming

a) the tax refund check would be available in 5 days?

b) the tax refund check would be available in 10 days?

c) the tax refund check would be available in 20 days?

58. *Prime Interest Rate* Nick borrowed $600 for 3 months. The banker said that Nick must repay the loan at the rate of $200 per month plus interest. The bank was charging a rate of 2 percentage points above the prime interest rate. The *prime interest rate* is the rate charged to preferred customers of the bank. During the first month the prime interest rate was 4.75%, during the second month it was 5%, and during the third month it was 5.25%.

a) Determine the amount Nick paid the bank at the end of the first month, at the end of the second month, and at the end of the third month.

b) What was the total amount of interest Nick paid the bank?

59. *Banker's Rule* The Banker's rule considers a year to have 360 instead of 365 days. This exercise highlights the advantage the Banker's rule provides to banks and other lenders. Professor Acks obtains a bank loan to open a health food store. She borrows $25,000 for 180 days and pays a 3% simple interest rate on the loan.

a) Using the Banker's rule and $t = \frac{180}{360}$, calculate the simple interest Professor Acks pays on this loan.

b) Next, calculate the simple interest Professor Acks would pay on this loan if the simple interest were calculated using $t = \frac{180}{365}$ instead.

c) How much more simple interest does the bank receive by using a denominator of 360 rather than 365?

d) Determine the percent increase in simple interest the bank receives by using a denominator of 360 rather than 365 by dividing the answer to part (c) by the answer to part (b). Round your answer to the nearest hundredth of a percent.

Recreational Mathematics

60. *Columbus Investment* On August 3, 1492, Christopher Columbus set sail on a voyage that would eventually lead him to the Americas. If on this day Columbus had invested $1 in a 5% simple interest account, determine the amount of interest the account would have earned by the following dates. Use a scientific calculator and disregard leap years in your calculations.

a) December 11, 1620 (Pilgrims land on Plymouth Rock)

b) July 4, 1776 (Declaration of Independence)

c) December 7, 1941 (United States enters World War II)

d) Today's date

Research Activity

61. *Loan Sources* Consider the following places where a loan may be obtained: banks, savings and loans, credit unions, and pawnshops. Write a report that includes the following information:

a) Describe the ownership of each.

b) Historically, what need is each fulfilling?

c) What are the advantages and disadvantages of obtaining a loan from each of those places listed?

SECTION 10.3   Compound Interest

Upon completion of this section, you will be able to:

- Solve problems involving compound interest.
- Solve problems involving the present value of an investment.

Albert Einstein said of compound interest, "The most powerful force in the universe is compound interest." What is this marvelous concept that warrants such a strong statement from one of the most intelligent human beings of all time? Another very intelligent person, Benjamin Franklin, described compound interest quite well when he said, "Money makes money and the money that money makes makes more money." In this section, we will introduce compound interest.

Why This Is Important Investments that involve compound interest may play an important role in reaching some of your long-term financial goals. For example, if you want to make a down payment on a house in 5 years, understanding compound interest will help you determine how much money you need to invest now so that you reach this goal.

Compound Interest

An *investment* is the use of money or capital for income or profit. We can divide investments into two classes: fixed investments and variable investments. In a *fixed investment*, the amount invested as principal may be guaranteed and the interest is computed at a fixed rate. *Guaranteed* means that the exact amount invested will be paid back together with any accumulated interest. Examples of a fixed investment are savings accounts, money market deposit accounts, and certificates of deposit. Another fixed investment is a government savings bond. In a *variable investment*, neither the principal nor the interest is guaranteed. Examples of variable investments are stocks, mutual funds, and commercial bonds.

Simple interest, introduced earlier in the chapter, is calculated once for the period of a loan or investment using the formula $i = prt$. The interest paid on savings accounts at most banks is compound interest. A bank computes the interest periodically (for example, daily or quarterly) and adds this interest to the original principal. The interest for the following period is computed by using the new principal (original principal plus interest). In effect, the bank is computing interest on interest, which is called *compound interest*.

Definition: Compound Interest

Interest that is computed on the principal and any accumulated interest is called **compound interest**.

Example 1 *Computing Compound Interest*

Mrs. Bond works as a sales representative for Neutrogena. She recently received a $5000 bonus for reaching her annual sales quota. Mrs. Bond invests the $5000 in a 1-year certificate of deposit paying 3% interest compounded quarterly. Determine the amount, A, to which the $5000 will grow in 1 year.

Solution Compute the interest for the first quarter using the simple interest formula. Add this interest to the principal to determine the amount at the end of the first quarter. In our calculation, time is $\frac{1}{4}$ of a year, or $t = 0.25$.

$$i = prt$$
$$= \$5000 \times 0.03 \times 0.25 = \$37.50$$
$$A = \$5000 + \$37.50 = \$5037.50$$

Now repeat this process for the second quarter, this time using a principal of $5037.50.

$$i = \$5037.50 \times 0.03 \times 0.25 \approx \$37.78$$
$$A = \$5037.50 + \$37.78 = \$5075.28$$

For the third quarter, use a principal of $5075.28.

$$i = \$5075.28 \times 0.03 \times 0.25 \approx \$38.06$$
$$A = \$5075.28 + \$38.06 = \$5113.34$$

For the fourth quarter, use a principal of $5113.34.

$$i = \$5113.34 \times 0.03 \times 0.25 \approx \$38.35$$
$$A = \$5113.34 + \$38.35 = \$5151.69$$

Hence, the $5000 grows to a final value of $5151.69 over the 1-year period. ∎

This example shows the effect of earning interest on interest, or compounding interest. In 1 year, the amount of $5000 has grown to $5151.69, compared with $5150 that would have been obtained with a simple interest rate of 3%. Thus, in 1 year alone the gain was $1.69 more with compound interest than with simple interest.

A simpler and less time-consuming way to calculate compound interest is to use the compound interest formula and a calculator.

Did You Know?

Cost of Manhattan

In 1626, Peter Du Minuit, representing the Dutch West India Company, traded beads and blankets to the Native American inhabitants of Manhattan Island for the island. The trade was valued at 60 Dutch guilders, about $24. If at that time the $24 had been invested at 6% interest compounded annually, the investment would be worth about $224,244,683,838 in 2020. As a comparison, had the $24 been invested for the 394 years in a 6% simple interest account, the investment would be worth only $591.36. Had the $24 been invested in a continuously compounded interest account at 6%, the investment would be worth about $443,540,001,802. These results dramatically demonstrate the power of compounding interest over time.

Compound Interest Formula

$$A = p\left(1 + \frac{r}{n}\right)^{(n \cdot t)}$$

where A is the amount that accumulates in the account, p is the principal, r is the annual interest rate as a decimal number, n is the number of compounding periods per year, and t is the time in years.

We will now use this formula to show how an investment can grow using compound interest.

Example 2   *Using the Compound Interest Formula*

Kathy invested $3000 in a savings account with an interest rate of 1.8% compounded monthly. If Kathy makes no other deposits into this account, determine the amount in the savings account after 2 years.

Solution   We will use the formula for compound interest, $A = p\left(1 + \frac{r}{n}\right)^{(n \cdot t)}$. The principal, p, is the amount of money invested, so $p = \$3000$. The interest rate, r, is 1.8%, so $r = 0.018$. Because the interest is compounded monthly, there are 12 compounding periods per year and $n = 12$. Because the money is invested for 2 years, $t = 2$.

$$A = 3000\left(1 + \frac{0.018}{12}\right)^{(12 \cdot 2)}$$
$$= 3000(1 + 0.0015)^{24}$$
$$= 3000(1.0015)^{24}$$
$$\approx 3000(1.0366279)$$
$$\approx 3109.88$$

Thus, the amount in the account after 2 years would be about $3109.88.   ∎

TECHNOLOGY TIP

Scientific Calculator

There are various ways to determine compound interest using a scientific calculator. In Example 2, we had to evaluate

$$3000\left(1 + \frac{0.018}{12}\right)^{(12 \cdot 2)}$$

To evaluate this expression on a scientific calculator, enter the following keys.

3000 $\times$ (1 + 0.018 ÷ 12) y^x (12 $\times$ 2) =

After the ═ key is pressed, the answer 3109.883655 is displayed. This answer checks with the answer obtained in Example 2. Note that a scientific calculator will automatically apply the order of operations discussed in Section 6.1.

(Continued)

Compound interest

Microsoft Excel

The software program Excel can also be used to determine compound interest. Excel is a spreadsheet, so once you insert a formula, you can change the value of the variables and the new answer will be automatically displayed. If you are not familiar with Excel, you can get help at www.office.microsoft.com.

To use Excel to determine the compound interest, set up the following columns.

	A	B	C	D	E
1	p	r	n	t	A
2	3000	0.018	12	2	
3					

⏮ ◀ ▶ ⏭ \ Sheet1 ╱ Sheet2 ╱ Sheet3 ╱   ◄

Ready

Then, to set up the formula in cell E2, go to the formula bar on top and enter

$$= A2*(1 + B2/C2)\wedge(C2*D2)$$

Then press the Enter key. The answer, 3109.88, will appear in cell E2.

Suppose you now wanted to determine the compound interest for $p = 6000$, $r = 0.05$, $n = 24$, and $t = 6$. If you were to change the values in cells A2 through D2 to these values respectively, the new answer, 8096.626, would be displayed in cell E2.

Example 3   *Calculating Compound Interest*

Calculate the interest on $750 at 4% compounded semiannually for 8 years, using the compound interest formula.

Solution   Since the interest is compounded semiannually, there are two periods per year. Thus, $n = 2$, $r = 0.04$, and $t = 8$. Substituting into the formula, we can determine the accumulated amount in the account, A.

$$A = p\left(1 + \frac{r}{n}\right)^{(n \cdot t)}$$

$$= 750\left(1 + \frac{0.04}{2}\right)^{(2 \cdot 8)}$$

$$= 750(1.02)^{16}$$

$$\approx 750(1.3727857)$$

$$\approx 1029.59$$

Since the accumulated amount is $1029.59 and the original principal is $750, the interest is the difference, $1029.59 − $750, or $279.59. ∎

When making financial decisions, investors usually want to choose the investment that provides the highest interest rate. However, it can sometimes be difficult to compare interest rates. For example, if one investment states a compound interest rate and another investment states a simple interest rate, it may be difficult to choose the better investment. To help make such comparisons, we can compute the *effective annual yield*.

There's a Compound Interest App for That!
Many of the concepts that we discuss throughout this section and throughout this entire chapter can be further explored with the use of an app on your smartphone or tablet computer. Several of these apps allow you to calculate the accumulated amount in an account using the compound interest formula. Users enter the principal, annual interest rate, number of compounding periods, and the time in years, and the app will calculate the total amount accumulated.

As always, consult with your instructor before using these apps when completing work related to your course.

Suppose in Example 3, we had invested $1 at 4% compounded semiannually for 1 year.

$$A = 1\left(1 + \frac{0.04}{2}\right)^{(2 \cdot 1)} = 1.0404$$

To get the interest earned for 1 year, subtract the initial investment of $1:

$$i = 1.0404 - 1 = 0.0404$$

The amount of interest earned on $1 for 1 year, written as a percent, is 4.04%. For this investment we say that the *effective annual yield* is 4.04%. Most financial institutions refer to the effective annual yield as the *annual percentage yield (APY)* An investment with an interest rate of 4% compounded semiannually has an APY of 4.04%. This means an investment with an interest rate of 4% compounded semiannually provides the same amount of interest as an investment with a simple interest rate of 4.04%.

> **Definition: Effective Annual Yield**
> The **effective annual yield** or **annual percentage yield (APY)** is the simple interest rate that gives the same amount of interest as a compound rate over the same period of time.

Lakeview Credit Union

CD Rates

Type	Rate	APY*
12 mo	0.750%	0.753%
24 mo	1.240%	1.248%
36 mo	1.640%	1.653%
48 mo	1.880%	1.898%

* Annual Percentage Yield

Many banks compound interest daily. When computing the effective annual yield, they use 360 for the number of periods in a year. To determine the effective annual yield for any interest rate, calculate the amount using the compound interest formula where p is $1. Then subtract $1 from that amount. The difference, written as a percent, is the effective annual yield, as illustrated in Example 4 on page 599.

The sign in the margin shows interest rates and the corresponding annual percentage yields (APY) that were available for certificates of deposit (CDs) from Lakeview Credit Union on December 5, 2018. To determine the effective annual yield (or APY) for the 48-month CD, calculate the amount of interest earned on $1 for 1 year compounded daily.

$$A = 1\left(1 + \frac{0.0188}{360}\right)^{(360 \cdot 1)} \approx 1.0189773 \approx 1.01898$$

From this result, subtract 1 to obtain $1.01898 - 1 = 0.01898$, or 1.898%. Confirm that the other annual percentage yields shown on the sign are correct.

Example **4** *Annual Percentage Yield*

Determine the annual percentage yield or the effective annual yield for $1 invested for 1 year at the following interest rates. Round your answers to the nearest hundredth of a percent.

a) 2% compounded quarterly   b) 3% compounded monthly

Solution

a) With quarterly compounding, $n = 4$.

$$A = p\left(1 + \frac{r}{n}\right)^{(n \cdot t)}$$

$$= 1\left(1 + \frac{0.02}{4}\right)^{(4 \cdot 1)}$$

$$\approx 1.0202$$

$$i = A - 1$$

$$\approx 1.0202 - 1 \approx 0.0202$$

Thus, when the interest is 2% compounded quarterly, the annual percentage yield or the effective annual yield is about 2.02%.

b) With monthly compounding, $n = 12$.

$$A = p\left(1 + \frac{r}{n}\right)^{(n \cdot t)}$$

$$= 1\left(1 + \frac{0.03}{12}\right)^{(12 \cdot 1)}$$

$$\approx 1.0304$$

$$i = A - 1$$

$$\approx 1.0304 - 1 \approx 0.0304$$

Thus, when the interest is 3% compounded monthly, the annual percentage yield or effective annual yield is about 3.04%. ∎

There are numerous types of savings accounts. Many savings institutions compound interest daily. Some pay interest from the day of deposit to the day of withdrawal, and others pay interest from the first of the month on all deposits made before the tenth of the month. In each of these accounts in which interest is compounded daily, interest is entered into the depositor's account only once each month or quarter.

Present Value

You may wonder about what amount of money you must deposit in an account today to have a certain amount of money in the future. For example, how much must you deposit in an account today at a given rate of interest so that it will accumulate to $25,000 to pay your child's college costs in 4 years? The principal, p, that would have to be invested now is called the *present value*. If we solve the compound interest formula for p, we have the following formula for determining the present value.

Present Value Formula

$$p = \frac{A}{\left(1 + \frac{r}{n}\right)^{(n \cdot t)}}$$

where p is the present value, or the principal to invest now, A is the amount to be accumulated in the account, r is the annual interest rate as a decimal number, n is the number of compounding periods per year, and t is the time in years.

Example 5   *Savings for College*

Will would like his daughter to attend college in 6 years when she finishes high school. Will would like to invest enough money in a certificate of deposit (CD) now to pay for his daughter's college expenses. If Will estimates that he will need $30,000 in 6 years, how much should he invest now in a CD that has a rate of 2.5% compounded quarterly?

Solution To answer this question, we will use the present value formula with $A = \$30,000$, $r = 0.025$, $n = 4$, and $t = 6$.

$$p = \frac{A}{\left(1 + \dfrac{r}{n}\right)^{(n \cdot t)}}$$

$$= \frac{30{,}000}{\left(1 + \dfrac{0.025}{4}\right)^{(4 \cdot 6)}}$$

$$= \frac{30{,}000}{(1.00625)^{24}}$$

$$\approx \frac{30{,}000}{1.1612920}$$

$$\approx 25{,}833.30$$

Will needs to invest approximately $25,833.30 now at 2.5% compounded quarterly to have $30,000 in 6 years.   ∎

TECHNOLOGY TIP

Scientific Calculator

To determine the present value in Example 5, we had to evaluate the expression

$$\frac{30{,}000}{\left(1 + \dfrac{0.025}{4}\right)^{(4 \cdot 6)}}$$

This expression can be evaluated on a scientific calculator as follows.

30000 ÷ (1 + 0.025 ÷ 4) y^x (4 × 6) =

After the = key is pressed, the answer 25833.29562 is displayed. This answer agrees with the answer obtained in Example 5.

Excel

Read the Technology Tip on compound interest on pages 596–597. To determine the present value using Excel, enter A, r, n, t, and p in cells A1 through E1, respectively. In cells A2 through D2, enter the values of A, r, n, and t, respectively. (See the table in the Technology Tip on pages 596–597 for an example of the row and column setup.) When cell E2 is highlighted, in the formula box at the top, enter

= A2/(1 + B2/C2)^(C2*D2)

After the Enter key is pressed, the answer, 25833.30, is displayed in cell E2.

Excel also has many built-in financial functions, including Present Value (PV), Future Value (FV), and many others that are discussed in this section and chapter.

SECTION 10.3 *Exercises*

Warm Up Exercises

In Exercises 1–6, fill in the blanks with an appropriate word, phrase, or symbol(s).

1. An investment is the use of money or capital for income or _____.

2. In a fixed investment the interest is computed at a fixed rate and the amount invested as principal may be _____.

3. If neither the principal nor the interest is guaranteed, the investment is called a(n) _____ investment.

4. Interest that is computed on the principal and any accumulated interest is called _____ interest.

5. The effective annual yield or the annual percentage yield is the simple interest rate that gives the same amount of interest as a(n) _____ interest rate over the same period of time.

6. The principal that you would have to invest now to have a certain amount of money in the future is called the _____ value.

Practice the Skills

In Exercises 7–14, a) use the compound interest formula to compute the total amount accumulated and b) determine the interest earned. Round all answers to the nearest cent.

7. $850 for 4 years at 2.4% compounded semiannually.

8. $2500 for 3 years at 4% compounded semiannually.

9. $3000 for 6 years at 3% compounded quarterly.

10. $6500 for 5 years at 2.9% compounded quarterly.

11. $10,000 for 2 years at 4.5% compounded monthly

12. $1500 for 4 years at 1.2% compounded monthly

13. $8000 for 2 years at 4% compounded daily (use $n = 360$)

14. $5600 for 7 years at 2.7% compounded daily

In Exercises 15–18, use the present value formula to determine the amount to be invested now, or the present value needed.

15. The desired accumulated amount is $25,000 after 10 years invested in an account with 3.6% interest compounded monthly.

16. The desired accumulated amount is $90,000 after 8 years invested in an account with 4% interest compounded semi-annually.

17. The desired accumulated amount is $100,000 after 4 years invested in an account with 4% interest compounded quarterly.

18. The desired accumulated amount is $15,000 after 30 years invested in an account with 3% interest compounded monthly.

Problem Solving

In Exercises 19–40, if necessary, round all dollar figures to the nearest cent and round percents to the nearest hundredth of a percent.

19. *Inheritance* Emanuel receives an inheritance of $8500. He decides to invest this money in a 5-year certificate of deposit (CD) that pays 2.57% interest compounded semi-annually. How much money will Emanuel receive when he redeems the CD at the end of 5 years?

20. *Contest Winnings* Mary wins $2500 in a singing contest and invests the money in a 4-year CD that pays 3% interest compounded monthly. How much money will Mary receive when she redeems the CD at the end of the 4 years?

21. *New Equipment* To help pay for new costumes for a play, the Manlius Theater invests $1500 in a 30-month CD paying 2.8% interest compounded monthly. Determine the amount the theater will receive when it cashes in the CD after 30 months.

22. *Investing Prize Winnings* Marcella wins third prize in the Clearinghouse Sweepstakes and receives a check for $250,000. After spending $10,000 on a vacation, she decides to invest the rest in a money market account that pays 1.5% interest compounded monthly. How much money will be in the account after 10 years?

23. *House Down Payment* Karen invested $10,000 in a money market account with an interest rate of 1.75% compounded semiannually. Five years later, Karen withdrew the full amount to put toward the down payment on a new house. How much did Karen withdraw from the account?

24. *Investing a Signing Bonus* Joe just started a new job and has received a $5000 signing bonus. Joe decides to invest this money now so that he can buy a new motor scooter in 5 years. If Joe invests in a 5-year CD paying 3.35% interest compounded quarterly, how much money will he receive from his CD in 5 years?

25. *Personal Loan* Brent borrowed $3000 from his brother Dave. He agreed to repay the money at the end of 2 years, giving Dave the same amount of interest that he would have received if the money had been invested at 1.75% compounded quarterly. How much money did Brent repay his brother?

26. *Forgoing Interest* Rikki borrowed $2500 from her daughter, Lynette. She repaid the $2500 at the end of 2 years. If Lynette had left the money in a bank account that paid an interest rate of 1.5% compounded monthly, how much interest would she have accumulated?

27. *Investing Gifts and Scholarships* Cliff just graduated from high school and has received $800 in gifts of cash from friends and relatives. In addition, Cliff received three scholarships in the amounts of $150, $300, and $1000. If Cliff takes all his gift and scholarship money and invests it in a 24-month CD paying 2% interest compounded daily, how much money will he have when he cashes in the CD at the end of the 24 months?

28. *Little League* Braden River Little League held various fund raisers and received the following amounts of money: $450 from a car wash, $278 from a bake sale, and $327 from a used equipment sale. The league decides to invest the total of these amounts in a 3-year CD that pays 1.8% interest compounded daily. How much money will the league receive from the CD in 3 years?

29. *Determining Effective Annual Yield* Determine the effective annual yield for $1 invested for 1 year at 3.5% compounded semiannually.

30. *Determining Effective Annual Yield* Determine the effective annual yield for $1 invested for 1 year at 4.75% compounded monthly.

31. *Money Market Choice* Prospero Bank advertises a money market account that pays 1.9% interest compounded quarterly. Fuerza Bank advertises a money market account that pays 1.8% compounded daily.

   a) Determine the annual percentage yield for the Prospero Bank money market account.

   b) Determine the annual percentage yield for the Fuerza Bank money market account.

   c) Assuming all other factors are equal, which bank's money market account would be the better investment?

32. *Money Market Choice* Summit Credit Union advertises a money market account that pays 0.750% compounded daily. ESL Credit Union advertises a money market account that pays 0.749% compounded semiannually.

   a) Determine the annual percentage yield for the Summit Credit Union money market account.

   b) Determine the annual percentage yield for the ESL Credit Union money market account.

   c) Assuming all other factors are equal, which credit union's money market would be the better investment?

33. *Investment Choice* M&T Bank advertises a CD that pays 2.23% compounded quarterly. Key Bank advertises a CD that pays 2.25% compounded semiannually.

   a) Determine the annual percentage yield for the M&T Bank CD.

   b) Determine the annual percentage yield for the Key Bank CD.

   c) Assuming all other factors are equal, which bank's CD would be the better investment?

34. *Investment Choice* Confiance Bank advertises a 1-year CD that pays 1.75% compounded semiannually. Firme Bank advertises a 1-year CD that pays 1.74% compounded monthly.

   a) Determine the annual percentage yield for the Confiance Bank CD.

   b) Determine the annual percentage yield for the Firme Bank CD.

   c) Assuming all other factors are equal, which bank's CD would be the better investment?

35. *Verifying APY* Suppose you saw a sign at your local bank that said, "2.4% rate compounded monthly—2.6% Annual Percentage Yield (APY)." Is there anything wrong with the sign? Explain.

36. *Verifying APY* Suppose you saw an advertisement in the newspaper for a financial planner who was recommending a certificate of deposit that paid 4.5% interest compounded quarterly. In the fine print at the bottom of the advertisement, it stated that the APY on the CD was 4.58%. Was this advertisement accurate? Explain.

37. *Starting a New Business* Corinne's goal is to have $55,000 to start a new cake decorating business when she retires in 15 years. How much should she invest now in a

CD that pays 4% interest compounded quarterly to reach her goal?

38. Starting a New Business Mr. Fink's goal is to have $26,000 to start a new lawn mower repair business when he retires in 15 years. How much should he invest now in a CD that pays 2% interest compounded quarterly to reach his goal?

39. A New Car Adam wishes to have $25,000 available in 18 years to purchase a new car for his son as a gift for his high school graduation. To accomplish this goal, how much should Adam invest now in a CD that pays 3.5% interest compounded quarterly?

40. Kitchen Remodel Diane plans to remodel her kitchen in 5 years. How much should Diane invest in a money market account that pays 2.1% interest compounded quarterly in order to have $8000 in 5 years?

41. A New Water Tower The village of Kieler recently completed the construction of a new water tower. The entire cost of the water tower was $925,000, and the state paid $370,000 of the total cost through the awarding of a grant. In addition, the village can delay paying the balance of the cost for 30 years (without paying any interest during the 30 years). To finance the balance, the village board will at this time assess its 598 homeowners a one-time flat fee surcharge and then invest this money in a 30-year CD paying 7.5% interest compounded monthly.

a) What is the balance due on the water tower?

b) How much will the village of Kieler need to invest at this time in the CD to raise the balance due in 30 years?

c) What amount should each homeowner pay as a surcharge?

42. After seeing its neighboring village obtain a new water tower (see Exercise 41), the city board of East Dubuque begins planning to replace its water towers. The board estimates that it will need $1,750,000 to build the new water towers in 20 years. At this time, the city board plans to assess its 2753 homeowners with a one-time flat fee surcharge and then invest the money received in a CD paying 9% interest compounded daily (use $n = 360$) for 20 years.

a) How much money will the board need to raise at this time to meet the city's water tower needs at the end of the 20 years?

b) Before applying the surcharge, the city board receives a federal grant of $100,000 toward the water tower investment. Taking this grant into account, how much should the surcharge be on each homeowner?

Concept/Writing Exercises

43. Doubling the Rate Determine the total amount and the interest paid on $1000 with interest compounded semiannually for 2 years at

a) 2%.   **b)** 4%.   **c)** 8%.

d) Is there a predictable outcome in either the amount or the interest when the rate is doubled? Explain.

44. Doubling the Principal Compute the total amount and the interest paid at 12% compounded monthly for 2 years for the following principals.

a) $100   **b)** $200   **c)** $400

d) Is there a predictable outcome in the interest when the principal is doubled? Explain.

45. Interest Comparison You are given a choice of taking the simple interest on $100,000 invested for 4 years at a rate of 5% or the interest on $100,000 invested for 4 years at an interest rate of 5% compounded daily. Which would you select? Explain your answer and give the difference in the two investments.

Challenge Problems/Group Activities

46. Inflation The average cost for a loaf of bread in 2018 was $2.27. Assuming an annual inflation rate of 0.5% per year, what will be the cost for a loaf of bread in 2023?

47. Determining the Interest Rate For a total accumulated amount of $3586.58, a principal of $2000, and a time period of 5 years, use the compound interest formula to determine r if interest is compounded monthly.

48. Rule of 72 A simple formula can help you estimate the number of years required to double your money. It's called the *rule of 72*. You simply divide 72 by the interest rate (without the percent sign). For example, with an interest rate of 4%, your money would double in approximately $72 \div 4$, or 18 years. In (a)–(d), determine the approximate number of years it will take for $1000 to double at the given interest rate.

a) 3%   **b)** 6%   **c)** 8%   **d)** 12%

e) If $120 doubles in approximately 22 years, estimate the rate of interest.

49. *Determining the Interest Rate* Richard borrowed $2000 from Linda. The terms of the loan are as follows: The period of the loan is 3 years, and the rate of interest is 8% compounded semiannually. What rate of simple interest would be equivalent to the rate Linda charged Richard?

Research Activities

50. *Certificate of Deposit* Imagine you have $4000 to invest and you need this money to grow to $5000 by investing in a CD. Contact a local bank or credit union to obtain the

following CD information: the interest rate, the length of the term, and the number of times per year the CD is compounded. Determine how long it would take you to reach your goal with the institution selected. Write a report summarizing your findings.

51. *History of Interest* Write a paper on the history of simple interest and compound interest. Answer the questions: When was simple interest first charged on loans? When was compound interest first given on investments?

SECTION 10.4   Installment Buying

Upon completion of this section, you will be able to:

- Solve problems involving fixed installment loans.
- Solve problems involving open-ended installment loans.

Your friends are planning a trip for spring break, and you would like to join them. Although you don't have enough money to pay for the trip, you do have a credit card. You must make an important decision. Do you charge the trip to your credit card and worry about paying for it later, or do you decide you can't afford the trip and choose not to go? In this section, we will look at the real cost of using a credit card to make purchases. We will also study other common types of loans frequently used to purchase big-ticket items such as cars, appliances, home improvements, and vacations.

Why This Is Important An important part of establishing a financial plan is understanding the cost of borrowing money to make purchases. Installment loans and credit cards are two very common ways to borrow money for making such purchases. In order to make sound financial decisions, it is important to understand the costs involved with obtaining a loan and using a credit card.

In Section 10.2, we discussed personal notes and discounted notes. When borrowing money by either of these methods, the borrower normally repays the loan as a single payment at the end of the specified time period. There may be circumstances under which it is more convenient for the borrower to repay the loan on a weekly or monthly basis or to use some other convenient time period. One method of doing so is to borrow money on an *installment plan.*

There are two types of installment loans: fixed payment and open-end. A *fixed installment loan* is one on which you pay a fixed amount of money for a set number of payments. Examples of items purchased with fixed-payment installment loans are college tuition loans and loans for cars, boats, appliances, and furniture. These loans are generally repaid in 24, 36, 48, or 60 equal monthly payments. An *open-end installment loan* is a loan on which you can make variable payments each month. Credit cards, such as MasterCard, Visa, Discover, and some American Express cards are actually open-end installment loans, used to purchase items such as clothing, textbooks, and meals.

Credit Rating Services give any individual wishing to borrow money or purchase goods or services on an installment plan a *credit rating.* The lenders use this credit rating to determine if the borrower is likely to repay the loan. The lending institution determines whether the applicant is a good "credit risk" by examining the individual's income, assets, liabilities, and history of repaying debts.

The advantage of installment buying is that the buyer has the use of an article while paying for it. If the article is essential, installment buying may serve a real need. A disadvantage is that some people buy more on the installment plan than they can afford. Another disadvantage is the interest the borrower pays for the loan.

To provide the borrower with a way to compare interest rates from various lenders, Congress passed the Truth in Lending Act of 1968. The law requires lending institutions to tell the borrower two things: the *annual percentage rate* and the *finance charge*. The *annual percentage rate (APR)* is the true rate of interest charged for the loan. The APR provides consumers the ability to compare loans without having to do calculations to determine the true interest rate they are being charged. The *finance charge* is the total amount of money the borrower must pay for borrowing the money. The finance charge includes the interest plus any additional fees charged to obtain the loan. *For the rest of this section, we will include the interest only when we refer to the finance charge.* In this section we will also discuss the *total installment price*. The *total installment price* is the sum of all the monthly payments and the down payment, if any.

Fixed Installment Loans

We begin our discussion of installment loans by calculating the monthly payment for the same loan by using (a) a table in Example 1 and (b) a formula in Example 2.

┌ Example **1** *Window Blinds*

Kristin wishes to purchase new window blinds for her house at a cost of $1500. The home improvement store has an advertised finance option of no down payment and 6% APR for 24 months.

a) Determine the finance charge.

b) Determine Kristin's monthly payment.

Solution

a) Table 10.2 on page 606 gives the finance charge per $100 of the amount financed. The table shows that the finance charge per $100 for 24 months at 6% is $6.37 (circled in red). Because Kristin is financing $1500, the number of hundreds of dollars financed is $\frac{1500}{100} = 15$. To determine the total finance charge, multiply the finance charge per $100 by the number of hundreds of dollars financed.

$$\text{Total finance charge} = \$6.37 \times 15 = \$95.55$$

Therefore, Kristin will pay a total finance charge of $95.55.

b) To determine the monthly payments, first calculate the total installment price by adding the finance charge to the purchase price.

$$\text{Total installment price} = \$1500 + \$95.55 = \$1595.55$$

Next divide the total installment price by the number of payments.

$$\text{Monthly payment} = \frac{\$1595.55}{24} \approx \$66.48$$

Kristin will have 24 monthly payments of $66.48. ■

The installment payment formula on page 606 can be used to directly calculate the installment payment rather than using Table 10.2.

Table 10.2   Annual Percentage Rate Table for Monthly Payment Plans

Number of payments	Annual Percentage Rate												
	4.0%	4.5%	5.0%	5.5%	6.0%	6.5%	7.0%	7.5%	8.0%	8.5%	9.0%	9.5%	10.0%
	(Finance charge per $100 of amount financed)												
6	1.17	1.32	1.46	1.61	1.76	1.90	2.05	2.20	2.35	2.49	2.64	2.79	2.93
12	2.18	2.45	2.73	3.00	3.28	3.56	3.83	4.11	4.39	4.66	4.94	5.22	5.50
18	3.20	3.60	4.00	4.41	4.82	5.22	5.63	6.04	6.45	6.86	7.28	7.69	8.10
24	4.22	4.75	5.29	5.83	(6.37)	6.91	7.45	8.00	8.54	9.09	9.64	10.19	10.75
30	5.25	5.92	6.59	7.26	7.94	8.61	9.30	9.98	10.66	11.35	12.04	12.74	13.43
36	6.29	7.09	(7.90)	8.71	9.52	10.34	11.16	11.98	12.81	13.64	14.48	15.32	16.16
48	8.38	9.46	10.54	11.63	12.73	13.83	14.94	(16.06)	17.18	18.31	19.45	20.59	21.74
60	10.50	11.86	13.23	14.61	16.00	17.40	18.81	20.23	21.66	23.10	24.55	26.01	27.48

Installment Payment Formula

$$m = \frac{p\left(\dfrac{r}{n}\right)}{1 - \left(1 + \dfrac{r}{n}\right)^{(-n \cdot t)}}$$

In the formula, m is the installment payment, p is the amount financed, r is the annual percentage rate as a decimal number, n is the number of payments per year, and t is the time in years.

Example 2   Using the Installment Payment Formula

Reread Example 1 on page 605. Use the installment payment formula to determine Kristin's monthly payment.

Solution   In this example, Kristin is borrowing $1500, so $p = 1500$; the APR is 6%, so $r = 0.06$; she is making monthly payments, so $n = 12$; and time of the loan is 24 months, so the time in years is $t = \frac{24}{12} = 2$. We substitute these values into the installment payment formula below.

$$m = \frac{p\left(\dfrac{r}{n}\right)}{1 - \left(1 + \dfrac{r}{n}\right)^{(-n \cdot t)}}$$

$$= \frac{1500\left(\dfrac{0.06}{12}\right)}{1 - \left(1 + \dfrac{0.06}{12}\right)^{(-12 \cdot 2)}}$$

$$= \frac{1500(0.005)}{1 - (1.005)^{-24}}$$

$$\approx \frac{7.5}{1 - 0.88718567}$$

$$\approx \frac{7.5}{0.11281433}$$

$$\approx 66.48$$

Thus, Kristin's monthly payment would be $66.48. Note this matches the payment calculated in Example 1. ∎

Examples 1 and 2 show that either Table 10.2 or the installment payment formula can be used to calculate a monthly payment for an installment loan. The formula has the advantage that it can be used for any APR and any number of payments, not just those given in Table 10.2. One disadvantage of the formula is that if we are given certain information about a loan and we wish to determine the APR, the calculations may be extremely difficult. For this reason, we will use Table 10.2 in Example 3 to show how to determine the APR on an installment loan.

TECHNOLOGY TIP

Scientific Calculator
In Example 2 on page 606 we determined the monthly payment by evaluating the following expression:

$$\frac{1500\left(\dfrac{0.06}{12}\right)}{1 - \left(1 + \dfrac{0.06}{12}\right)^{(-12 \cdot 2)}}$$

There are various ways to evaluate this expression on a scientific calculator. One method is to press the following sequence of keys.

$$(\boxed{1500} (\boxed{0.06} \div \boxed{12})) \div (\boxed{1} - (\boxed{1} + \boxed{0.06} \div \boxed{12}) \boxed{y^x} (\boxed{(-)} \boxed{12} \times \boxed{2})) \boxed{=}$$

After the $\boxed{=}$ key is pressed, the answer 66.48091538 is displayed.

Microsoft Excel
Read the Technology Tip on page 597. To determine the monthly payment using Excel, place the letters P, r, n, t, and m in cells A1 through E1, respectively. In cells A2 through D2, place the values 1500, 0.06, 12, and 2, respectively. In cell E2, use the formula box to add the formula

$$= (A2*(B2/C2))/(1 - (1 + B2/C2)^{\wedge}(-C2*D2))$$

After the Enter key is pressed, the result is as follows.

	A	B	C	D	E
1	p	r	n	t	m
2	1500	0.06	12	2	66.48092
3					

◄ ◄ ► ► Sheet1 / Sheet2 / Sheet3 /

Ready

Example 3 *Determining the APR*

Bob is purchasing a motorcycle that sells for $26,500, including taxes and fees. Bob decides to make a $5000 down payment and finance the balance, $21,500, through his bank. The loan officer informs him that his monthly payment will be $644.40 for 36 months.

a) Determine the finance charge.

b) Determine the APR.

Solution

a) To determine the finance charge, we first determine the total installment price. The total installment price is the down payment plus the total monthly installment payments.

$$\text{Total installment price} = \$5000 + (36 \times \$644.40)$$
$$= \$5000 + \$23{,}198.40 = \$28{,}198.40$$

The finance charge is the total installment price minus the selling price.

$$\text{Finance charge} = \$28{,}198.40 - \$26{,}500 = \$1698.40$$

Therefore, Bob will pay a finance charge of $1698.40

b) To determine the annual percentage rate, use Table 10.2 on page 606. First divide the finance charge by the amount financed and multiply the quotient by 100. The result is the finance charge per $100 of the amount financed.

$$\frac{\text{Finance charge}}{\text{Amount financed}} \times 100 = \frac{1698.40}{21{,}500} \times 100 \approx 0.07899535 \times 100 \approx 7.90$$

Thus, Bob pays about $7.90 for each $100 being financed. To use Table 10.2, look for 36 in the far-left column under the heading Number of Payments. Next, move across the table to the right until you find the value that is closest to $7.90. In this case, $7.90 is in the table, circled in blue. At the top of this column is the value of 5.0%. Therefore, the annual percentage rate is approximately 5.0%. ∎

Much more complete APR tables similar to Table 10.2 are available at your local lending institution or on the Internet. Also available on the Internet and on smartphone apps are loan "calculators" that will calculate your payment for you if you provide the loan amount, the length of the loan, and the APR. One such loan calculator can be found at www.bankrate.com.

Example 4 *Financing a Restored Car*

Tino wants to purchase a classic 1966 Ford Mustang that sells for $9800. To purchase the car, he takes out a loan. He does not recall the APR of the loan but remembers that there are 48 payments of $237. If he did not make a down payment on the car, determine the APR.

Solution First determine the finance charge by subtracting the cash price of the car from the total amount paid.

$$\text{Finance charge} = (237 \times 48) - 9800$$
$$= 11{,}376 - 9800$$
$$= \$1576$$

Now divide the finance charge by the amount of the loan and multiply this quotient by 100.

$$\frac{1576}{9800} \times 100 \approx 16.08$$

Next find 48 payments in the left column of Table 10.2. Move to the right until you find the value that is closest to 16.08. The value closest to 16.08 is 16.06 (circled in green). At the top of the column is the APR of 7.5%. Thus, the APR for Tino's loan is approximately 7.5%. ∎

In Example 4, if Tino made all 48 payments, his finance charge would be $1576. If he decides to repay the loan after making 30 payments, must he pay the total finance charge? The answer is no. By paying off his loan early, Tino is not obligated to pay the entire finance charge. The amount of the reduction of the finance charge from paying off a loan early is called the *unearned interest*. We will learn about the most common way of calculating unearned interest, the *actuarial method*. Now we give the formula for determining the unearned interest using the actuarial method.

Actuarial Method for Unearned Interest

$$u = \frac{n \cdot P \cdot V}{100 + V}$$

where u is the unearned interest, n is the number of remaining monthly payments (excluding the current payment), P is the monthly payment, and V is the value from the APR table that corresponds to the annual percentage rate for the number of remaining payments (excluding the current payment).

Example 5 illustrates the actuarial method for calculating unearned interest when an installment loan is paid off early.

Example 5 *Using the Actuarial Method*

In Example 4, we determined the APR of Tino's loan to be 7.5%. Instead of making his 30th payment of his 48-payment loan, Tino wishes to pay his remaining balance and terminate the loan.

a) Use the actuarial method to determine how much interest Tino will save (the unearned interest, u) by repaying the loan early.

b) What is the total amount due to pay off the loan early on the day he makes his final payment?

Solution

a) Recall from Example 4 that Tino's monthly payments are $237.00. After 30 payments have been made, 18 payments remain. Thus, $n = 18$ and $P = \$237$. To determine V, use the APR table (Table 10.2). In the Number of Payments column, find the number of remaining payments, 18, and then look to the right until you reach the column headed by 7.5%, the APR. This row and column intersect at 6.04. Thus, $V = 6.04$. Now use the actuarial method formula to determine the unearned interest, u.

$$u = \frac{n \cdot P \cdot V}{100 + V}$$

$$= \frac{(18)(237)(6.04)}{100 + 6.04}$$

$$\approx 242.99$$

By paying off the loan early, Tino will save $242.99 in interest by the actuarial method.

b) Because the remaining payments total $18(\$237) = \$4266$, Tino's remaining balance, excluding his 30th monthly payment, is

$4266.00	Total of remaining payments (which includes interest)
− 242.99	Interest saved (unearned interest)
$4023.01	Balance due (excluding the 30th monthly payment)

A payment of $4023.01 plus the 30th monthly payment of $237 will terminate Tino's installment loan.

$4023.01	Balance due (excluding the 30th monthly payment)
+ 237.00	30th monthly payment
$4260.01	Total amount due

The total amount due is $4260.01.   ■

Open-End Installment Loan

A credit card is a popular way of making purchases or borrowing money. Use of a credit card is an example of an open-end installment loan. A typical credit card charge account with a bank or store may have the terms given in Table 10.3.

Table 10.3   Credit Card Terms

Type of Charge	Daily Periodic Rate*	Annual Percentage Rate*
Purchases	0.05173%	18.88%
Cash advances	0.06885%	25.13%

*These rates vary with different credit card accounts and localities.

Typically, credit card monthly statements contain the following information: balance at the beginning of the period, balance at the end of the period (or new balance), the transactions for the period, statement closing date (or billing date), payment due date, and the minimum payment due. In addition, the Credit Card Accountability Responsibility and Disclosure (CARD) Act of 2009 required credit card companies to provide additional information to their customers on their monthly statements (see *Mathematics Today* on page 609). For *purchases*, there is no finance or interest charge if there is no previous balance due and you pay the entire new balance by the payment due date. The period between when a purchase is made and when the credit card company begins charging interest is called the *grace period* and is usually 20 to 25 days. However, if you use a credit card to borrow money, called a *cash advance*, there generally is no grace period and a finance charge is applied from the date you borrowed the money until the date you repay the money. When you make purchases or obtain cash advances, the minimum monthly payment will generally be any new fees and interest plus at least 1% of the outstanding principal. This sum is then rounded up to the nearest whole dollar. Credit card companies often have $20 as the lowest possible minimum monthly payment. These guidelines will vary among the different credit card companies. If you currently have a credit card, you can determine how your minimum monthly payment is calculated by reading the back of your monthly statement or by reading the literature given to you when you obtained the credit card.

Example 6 *Calculating the Minimum Monthly Payments*

Ms. Johnson's credit card company determines her minimum monthly payment by adding any new interest to 1.2% of the outstanding principal. The credit card company charges an interest rate of 0.05271% per day for purchases. On April 15, Ms. Johnson used her credit card to purchase a new recliner and sofa for $3800. She made no other purchases in April.

a) Assuming Ms. Johnson owed no interest, determine her minimum payment due on May 1, her billing date.

b) On May 1, instead of making the minimum payment, Ms. Johnson makes a payment of $950. Assuming there are no additional charges or cash advances, determine her minimum payment due on June 1.

Solution

a) Since there is no new interest due, Ms. Johnson's minimum monthly payment due on May 1 is determined by taking 1.2% of the outstanding principal, or $0.012 \times \$3800 = \$45.60$. Rounding up to the nearest whole dollar, we determine that her minimum monthly payment is $46.

b) Ms. Johnson's minimum monthly payment due on June 1 will be the sum of the new interest plus 1.2% of her outstanding principal. To calculate her new interest charges, we will use the simple interest formula, $i = prt$. The principal is $\$3800 - \$950 = \$2850$; the rate r is 0.05271%, or 0.0005271 per day, and the time is 31 days since May has 31 days.

$$i = prt$$
$$= \$2850 \times 0.0005271 \times 31 \approx \$46.57$$

To determine 1.2% of the outstanding principal, multiply 0.012 by $2850 to get $34.20. Next we add the new interest, $46.57, to 1.2% of the outstanding principal, $34.20, to get $\$34.20 + \$46.57 = \$80.77$. Rounding up to the nearest whole dollar, we determine that Ms. Johnson's minimum monthly payment on June 1 is $81. ∎

In Example 6, Ms. Johnson made no additional charges during the month. When additional charges are made during the month, the finance charges on open-end installment loans or credit cards are generally calculated in one of two ways: the *previous balance method* or the *average daily balance method*. Example 7 illustrates the previous balance method, and Example 8 illustrates the average daily balance method.

With the *previous balance method*, the borrower is charged interest or a finance charge on the previous balance from the previous charge period.

Example 7 *Finance Charges Using the Previous Balance Method*

In October, Peter charged all the supplies for his Halloween party to his Visa card. On November 5, the billing date, Peter had a balance due of $275. From November 5 through December 4, he did some shopping and charged items totaling $320, and he also made a payment of $145.

a) Determine the finance charge due on December 5, Peter's next billing date, using the previous balance method. Assume that the interest rate charged is 1.3% per month.

b) Determine the new account balance on December 5.

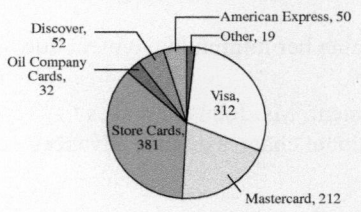
Solution

a) The finance charge is based on the $275 balance due on November 5. To determine the finance charge due on December 5, we used the simple interest formula with a time of 1 month.

$$i = \$275 \times 0.013 \times 1 \approx \$3.58$$

The finance charge due on December 5 is $3.58.

b) The balance due on December 5 is determined by adding the costs of the new purchases and the calculated interest to the balance due on November 5, and then subtracting the payment made from this sum.

$$\$275 + \$320 + \$3.58 - \$145 = \$453.58$$

The balance due on December 5 is $453.58. The finance charge on January 5 is based on the December 5 balance of $453.58. ∎

Many lending institutions use the *average daily balance method* of calculating the finance charge because they believe that it is fairer to the customer. With the average daily balance method, a balance is determined each day of the billing period for which there is a transaction in the account. The average daily balance method is illustrated in Example 8.

Example 8 Finance Charges Using the Average Daily Balance Method

The balance on Min's credit card account on July 1, the billing date, was $375.80. The following transactions occurred during the month of July.

July 5	Payment	$150.00
July 10	Charge: Toy store	74.35
July 18	Charge: Garage	123.50
July 28	Charge: Restaurant	42.50

a) Determine the average daily balance for the billing period.
b) Determine the finance charge to be paid on August 1, Min's next billing date. Assume that the interest rate is 1.3% per month.
c) Determine the balance due on August 1.

Solution

a) To determine the average daily balance, we do the following. (i) Determine the balance due for each transaction date.

July 1	$375.80				
July 5	$375.80	−	$150	=	$225.80
July 10	$225.80	+	$74.35	=	$300.15
July 18	$300.15	+	$123.50	=	$423.65
July 28	$423.65	+	$42.50	=	$466.15

(ii) Determine the number of days that the balance did not change between each transaction. Count the first day in the period but not the last day. Note that the time period from July 28 through August 1, the beginning of the next billing cycle, is 4 days. (iii) Multiply the balance due by the number of days the balance did not change. (iv) Determine the sum of the products.

Date	(i) Balance Due	(ii) Number of Days Balance Did Not Change	(iii) (Balance)(Days)
July 1	$375.80	4	($375.80)(4) = $1503.20
July 5	$225.80	5	($225.80)(5) = $1129.00
July 10	$300.15	8	($300.15)(8) = $2401.20
July 18	$423.65	10	($423.65)(10) = $4236.50
July 28	$466.15	4	($466.15)(4) = $1864.60
		31	(iv) Sum = $11,134.50

(v) Divide this sum by the number of days in the billing cycle (in the month). The number of days may be determined by adding the days in column (ii).

$$\frac{\$11,134.50}{31} = \$359.18$$

Thus, the average daily balance is $359.18.

b) The finance charge for the month is determined using the simple interest formula with the average daily balance as the principal.

$$i = \$359.18 \times 0.013 \times 1 = \$4.67$$

c) Since the finance charge for the month is $4.67, the balance owed on August 1 is $466.15 + $4.67, or $470.82. ∎

The calculations in Example 8 are tedious. These calculations, however, are made almost instantaneously with computers.

Example 9 illustrates how a credit card may be used to borrow money.

Example 9 *Credit Card Cash Advance*

Thomasina wishes to purchase an antique bookcase for $1500. Since the antique dealer accepts cash only, Thomasina uses her credit card to obtain a cash advance of $1500. She borrowed the money on July 10 and repaid it on July 31. If Thomasina is charged an interest rate of 0.06164% per day, how much did Thomasina have to repay the credit card company on July 31?

Solution The amount Thomasina pays is the original principal, $1500, plus any accrued interest. Interest on cash advances is generally calculated for the exact number of days of the loan, starting with the day the money is obtained. The time of this loan is 21 days. Using the simple interest formula, we get the following:

$$i = prt$$
$$= \$1500 \times 0.0006164 \times 21$$
$$\approx \$19.42$$

Therefore, on July 31, Thomasina must repay the credit card company $1500 + $19.42 = $1519.42. ∎

Anyone purchasing a car or other costly items should consider a number of different sources for a loan. Example 10 illustrates one method of making a comparison.

Example 10   *Comparing Loan Sources*

Franz purchased carpeting costing $2400 with his credit card. When the bill comes due on February 1, Franz realizes that he can pay $350 per month until the debt is paid off. His credit card charges a 1.5% interest rate per month.

a) Assuming Franz makes no other purchases with this credit card, how many payments are necessary to retire this debt?

b) What is the total interest Franz will pay?

c) How much money could Franz have saved by obtaining a fixed installment loan of $2400 with an annual percentage rate of 6% interest with six equal monthly payments?

Solution

a) Franz would make his first monthly payment of $350 on February 1, resulting in a new balance of $2400 − $350 = $2050. His next bill reflects the $2050 balance plus the monthly interest. He continues to make payments until the debt is retired. For each date indicated, the amount on the far right represents the amount due on that date.

February 1	$2400 − $350 = $2050
March 1	$2050 + 0.015($2050) = $2080.75; $2080.75 − $350 = $1730.75
April 1	$1730.75 + 0.015($1730.75) = $1756.71; $1756.71 − $350 = $1406.71
May 1	$1406.71 + 0.015($1406.71) = $1427.81; $1427.81 − $350 = $1077.81
June 1	$1077.81 + 0.015($1077.81) = $1093.98; $1093.98 − $350 = $743.98
July 1	$743.98 + 0.015($743.98) = $755.14; $755.14 − $350 = $405.14
August 1	$405.14 + 0.015($405.14) = $411.22; $411.22 − $350 = $61.22
September 1	$61.22 + 0.015($61.22) = $62.14

After eight payments — seven for $350 and one for $62.14 — Franz has paid off his credit card bill for his carpeting.

b) To calculate the total interest paid, we add up all his payments and then subtract the cost of the carpeting:

$$\text{Total of all payments} = 7(\$350) + \$62.14$$
$$= \$2512.14$$
$$\text{Total interest} = \$2512.14 - \$2400$$
$$= \$112.14$$

c) To determine the interest that Franz pays with the installment loan, we will use Table 10.2 on page 606. From Table 10.2, we see that a fixed installment loan with an APR of 6% for 6 months corresponds to a finance charge of $1.76 per $100 of the amount financed. So the interest, or finance charge, is

$$\text{Finance charge} = 1.76\left(\frac{2400}{100}\right)$$
$$= 1.76(24)$$
$$= \$42.24$$

Therefore, Franz would save $112.14 − $42.24 = $69.90 in interest by using an installment loan instead of a credit card. ∎

SECTION 10.4 Exercises

Warm Up Exercises

In Exercises 1–6, fill in the blanks with an appropriate word, phrase, or symbol(s).

1. A loan on which you pay a fixed amount of money for a set number of payments is called a fixed _____ loan.

2. A loan on which you make variable payments each month is called a(n) _____ installment loan.

3. The true rate of interest charged for a loan is called the _____ percentage rate.

4. The total amount of money the borrower must pay for borrowing the money is called the _____ charge.

5. The sum of all the monthly payments and the down payment, if any, is called the total _____ price.

6. The period between when a purchase is made and when the credit card company begins charging interest is called the _____ period.

Practice the Skills

In Exercises 7–10, determine the monthly payment for the installment loan. Unless told otherwise by your instructor, you may either use Table 10.2 or the installment payment formula (see Examples 1 and 2).

	Amount Financed (P)	Annual Percentage Rate (r)	Number Payments per Year (n)	Time in Years (t)
7.	$ 825	6%	12	2
8.	$ 2200	9%	12	3
9.	$15,000	6.5%	12	4
10.	$27,000	9.5%	12	5

Problem Solving

11. *Business Loan* Kara is opening up a new daycare business in her home. She needs $6500 to purchase toys and a backyard playset. Kara makes a 10% down payment and finances the balance with a 48-month fixed installment loan with an APR of 4.5%.

a) Determine Kara's finance charge.

b) Determine Kara's monthly payment.

12. *A New Air Conditioner* Juan paid $7000 for a new central air-conditioning unit for his house. He paid 20% as a down payment and financed the balance with a 36-month fixed installment loan with an APR of 5%.

a) Determine Juan's finance charge.

b) Determine Juan's monthly payment.

13. *Financing a New Roof* Jaime needs a new roof on her house. The cash cost is $7500. She decides to finance the project by paying 20% down, with the balance paid in 36 monthly payments of $189.40.

a) What finance charge will Jaime pay?

b) What is the APR to the nearest half percent?

14. *Financing a Computer* Ilga purchased a new computer on a monthly purchase plan. The computer sold for $1495. Ilga paid 5% down and $64 a month for 24 months.

a) What finance charge did Ilga pay?

b) What is the APR to the nearest half percent?

15. *Boat Loan* Tarik is purchasing a new boat that sells for $60,714, including taxes and fees. He receives $21,500 as a trade-in on his old boat and uses the money as a down payment. Tarik finances the balance and has monthly payments of $767.29 for 60 months.

a) What finance charge does Tarik pay?

b) What is the APR to the nearest half percent?

16. *Financing Furniture* Mr. and Mrs. Chan want to buy furniture that has a cash price of $3450. On the installment plan, they must pay 25% of the cash price as a down payment and make six monthly payments of $437.

a) What finance charge will the Chans pay?

b) What is the APR to the nearest half percent?

17. *Early Repayment of a Loan* Ray took out a 60-month fixed installment loan of $12,000 to open a new pet store. He paid no money down and began making monthly payments of $232. Ray's business does better than expected and instead of making his 24th payment, Ray wishes to repay his loan in full.

a) Determine the APR of the installment loan.

b) How much interest will Ray save by paying off his loan early?

c) What is the total amount due to pay off the loan?

18. *Early Repayment of a Loan* Jeslie Ann has a 48-month installment loan with a fixed monthly payment of $83.81. The amount she borrowed was $3500. Instead of making her 18th payment, Jeslie Ann is paying the remaining balance on the loan.

a) Determine the APR of the installment loan.

b) How much interest will Jeslie Ann save by paying off the loan early?

c) What is the total amount due to pay off the loan?

19. *Early Repayment of a Loan* Nina buys a new sport utility vehicle for $32,000. She trades in her old truck and receives $10,000, which she uses as a down payment. She finances the balance at 8% APR over 36 months. Before making her 24th payment, she decides to pay off the loan.

a) Use Table 10.2 to determine the total interest Nina would pay if all 36 payments were made.

b) What were Nina's monthly payments?

c) How much interest will Nina save by paying off the loan early?

d) What is the total amount due to pay off the loan?

20. *Early Repayment of a Loan* Jeff buys a new motorcycle for $12,000. He receives $4500 as a trade-in on his old motorcycle and uses the money as a down payment. Jeff finances the balance at 6.5% APR over 24 months. Before making his 12th payment, he decides to pay off the loan.

a) Use Table 10.2 to determine the total interest Jeff would pay if all 24 payments were made.

b) What were Jeff's monthly payments?

c) How much money will Jeff save by paying off the loan early?

d) What is the total amount due to pay off the loan?

21. *Credit Card Minimum Monthly Payment* Shiing's credit card company determines his minimum monthly payment by adding all new interest to 1% of the outstanding principal. The credit card company charges an interest rate of 0.04247% per day. On March 17, Shiing uses his credit card to purchase airline tickets for his family for $3000. He makes no other purchases during March.

a) Assuming Shiing had no new interest, determine Shiing's minimum payment due on April 1, his billing date.

b) On April 1, instead of making the minimum payment, Shiing makes a payment of $500. Assuming there are no additional charges or cash advances, determine Shiing's minimum payment due on May 1.

22. *Credit Card Minimum Monthly Payment* Evelyn's credit card company determines her minimum monthly payment by adding all new interest to 2% of the outstanding principal. The credit card company charges an interest rate of 0.04238% per day. On September 15, Evelyn uses her credit card to purchase a new picture window for her house for $3200. She makes no other purchases during September.

a) Assuming Evelyn had no new interest, determine Evelyn's minimum payment due on October 1, her billing date.

b) On October 1, instead of making the minimum payment, Evelyn makes a payment of $1200. Assuming there are no additional charges or cash advances, determine Evelyn's minimum payment due on November 1.

23. *Business Expenses* Kevin's credit card company determines his minimum monthly payment by adding all new interest to 1.5% of the outstanding principal. The credit card company charges an interest rate of 0.05163% per day. On November 12, Kevin used his credit card to pay for the following business expenses: van repairs ($677), equipment maintenance ($452), office supplies ($139), and dinner with clients ($141).

a) Assuming Kevin had no new interest, determine his minimum payment due on December 1, his billing date.

b) On December 1, instead of making his minimum payment, Kevin makes a payment of $300. Assuming there are no additional charges or cash advances, determine Kevin's minimum payment on January 1.

24. *Vacation Expenses* Harry's credit card company determines his minimum monthly payment by adding all new interest to 2.5% of the outstanding principal. The credit card company charges an interest rate of 0.03164% per day. On August 21, while on vacation in Costa Rica, Harry used his credit card to pay for the following expenses: airfare ($359), car rental ($273), hotel ($653), meals ($315), and surfboard rental ($225).

▲ *Costa Rica*

a) Assuming Harry had no new interest, determine Harry's minimum payment due on September 1, his billing date.

b) On September 1, instead of making his minimum payment, Harry makes a payment of $750. Assuming there are no additional charges or cash advances, determine Harry's minimum payment on October 1.

25. *Previous Balance Method* On the September 5 billing date, Tasia had a balance of $2302.65 on her credit card. From September 5 through October 5, Tasia charged an additional $106.72 and made a payment of $550.

a) Determine the finance charge on October 5, using the previous balance method. The interest rate is 1.3% per month.

b) Determine the new balance due on October 5.

26. *Previous Balance Method* On September 5, the billing date, Verna had a balance due of $567.20 on her credit card. The transactions during the following month were

September 8	Payment	$275.00
September 21	Charge: Airline ticket	330.00
September 27	Charge: Hotel bill	190.80
October 2	Charge: Clothing	84.75

a) Determine the finance charge on October 5, using the previous balance method. The interest rate is 1.1% per month.

b) Determine the new balance on October 5.

27. *Previous Balance Method* On February 3, the billing date, Carol Ann had a balance due of $124.78 on her credit card. Her bank charges an interest rate of 1.25% per month. She made the following transactions during the month.

February 8	Charge: Art supplies	$25.64
February 12	Payment	100.00
February 14	Charge: Flowers delivered	67.23
February 25	Charge: Music CD	13.90

a) Determine the finance charge on March 3, using the previous balance method.

b) Determine the new balance on March 3.

28. *Previous Balance Method* On April 15, the billing date, Gabrielle had a balance due of $57.88 on her credit card. She is redecorating her apartment and has the following transactions.

April 16	Charge: Paint	$64.75
April 20	Payment	45.00
May 3	Charge: Curtains	72.85
May 10	Charge: Chair	135.50

a) Determine the finance charge on May 15, using the previous balance method. Assume that the interest rate is 1.35% per month.

b) Determine the new balance on May 15.

29. *Average Daily Balance Method* The balance on the Razazadas' credit card on May 12, their billing date, was $378.50. For the period ending June 12, they had the following transactions.

May 13	Charge: Toys	$129.79
May 15	Payment	50.00
June 1	Charge: Clothing	135.85
June 8	Charge: Housewares	37.63

a) Determine the average daily balance for the billing period.

b) Determine the finance charge to be paid on June 12. Assume an interest rate of 1.3% per month.

c) Determine the balance due on June 12.

30. *Average Daily Balance Method* The Levys' credit card statement shows a balance due of $1578.25 on March 23, the billing date. For the period ending April 23, they had the following transactions.

March 26	Charge: Party supplies	$79.98
March 30	Charge: Restaurant meal	52.76
April 3	Payment	250.00
April 15	Charge: Clothing	190.52
April 22	Charge: Car repairs	190.85

a) Determine the average daily balance for the billing period.

b) Determine the finance charge to be paid on April 23. Assume an interest rate of 1.3% per month.

c) Determine the balance due on April 23.

31. *Average Daily Balance Method* Refer to Exercise 27. Instead of the previous balance method, suppose that Carol Ann's bank uses the average daily balance method.

a) Determine Carol Ann's average daily balance for the billing period from February 3 to March 3. Assume it is not a leap year.

b) Determine the finance charge to be paid on March 3.

c) Determine the balance due on March 3.

d) Compare these answers with those in Exercise 27.

32. *Average Daily Balance Method* Refer to Exercise 26. Instead of the previous balance method, suppose Verna's bank uses the average daily balance method.

a) Determine Verna's average daily balance for the billing period from September 5 to October 5.

b) Determine the finance charge to be paid on October 5.

c) Determine the balance due on October 5.

d) Compare these answers with those in Exercise 26.

33. *A Cash Advance* Dov borrowed $875 against his charge account on September 12 and repaid the loan on October 14 (32 days later). Assume that the interest rate is 0.04273% per day.

a) How much interest did Dov pay on the loan?

b) What amount did he pay the bank when he repaid the loan?

34. *A Cash Advance* Tanya uses her credit card to obtain a cash advance of $600 to pay for her eyeglasses. The interest rate charged for the loan is 0.05477% per day. Tanya repays the money plus the interest after 27 days.

a) Determine the interest charged for the cash advance.

b) When she repaid the loan, how much did she pay the credit card company?

35. *Comparing Loan Sources* Grisha needs to borrow $1000 for an automobile repair. State National Bank charges 5% simple interest on the amount borrowed for the duration of the loan and requires the loan to be repaid in six equal monthly payments. Consumer's Credit Union offers loans of $1000 to be repaid in 12 monthly payments of $86.30.

a) How much interest is charged by the State National Bank?

b) How much interest is charged by the Consumer's Credit Union?

c) What is the APR, to the nearest half percent, on the State National Bank loan?

d) What is the APR, to the nearest half percent, on the Consumer Credit Union loan?

36. *Comparing Loan Options* Sara wants to purchase a new television set. The purchase price is $890. If she purchases the set today and pays cash, she must take money out of her savings account. Another option is to charge the TV on her credit card, take the set home today, and pay next month. Next month she will have cash and can pay her credit card balance without paying any interest. The simple interest rate on her savings account is $5\frac{1}{4}$%. How much is she saving by using the credit card instead of taking the money out of her savings account?

Challenge Problems/Group Activities

37. *Determining Purchase Price* Ken bought a new car, but now he cannot remember the original purchase price. His payments are $379.50 per month for 36 months. He remembers that the salesperson said the simple interest rate for the period of the loan was 6%. He also recalls he was allowed $2500 on his old car. Determine the original purchase price.

38. *Comparing Loans* Suppose the Chans in Exercise 16 use a credit card rather than an installment plan to purchase furniture. Assume that they make the same down payment, have no finance charge the first month, make no additional purchases on their credit card, and pay $432 per month plus the finance charge starting with the second month. The interest rate is 1.3% per month.

a) How many months will it take them to repay the loan?

b) How much interest will they pay on the loan?

c) Which method of borrowing will cost the Chans the least amount of interest, the installment loan in Exercise 16 or using the credit card?

Research Activities

39. *Paying Off Your Credit Card Debt* Suppose you purchase a new wardrobe for $1000 with a credit card. You decide to pay for your new wardrobe by making the minimum monthly payment each month. You do not charge anything else to this credit card until you have the wardrobe paid off. Your credit card company determines your minimum monthly payment by adding all new interest to 1% of the outstanding principal. The credit card company charges an annual percentage rate of 18%. Go to an Internet credit card payment calculator web site such as www.bankrate.com to answer the following questions.

a) How long will it take you to pay off the entire credit card debt if you make the minimum monthly payments?

b) How much interest will you pay?

c) Adding the interest paid to the cash price, determine the total cost of your wardrobe.

40. *Car Lease* Write a brief report giving the advantages and disadvantages of leasing a car. Determine all the individual costs involved with leasing a car. Indicate why you would prefer to lease or purchase a car at the present time.

41. *Washer and Dryer* Suppose your family currently does not own a washer and dryer and you do not have the money needed to purchase the appliances. Do research to determine if it would be less expensive to borrow money on a 5-year installment loan and purchase the appliances or to continue to do laundry at the local coin-operated laundry until you have enough money to pay cash for the appliances. Be sure to include the utility costs of running your own appliances and the cost of gas used to drive to the coin-operated laundry.

42. *Rule of 78s* When an installment loan is repaid early, there is a second method, in addition to the actuarial method, used for calculating unearned interest called the *rule of 78s*. Although rarely used today, there are still some lenders using this method. Write a brief report describing the rule of 78s and comparing it to the actuarial method.

SECTION 10.5   Buying a House with a Mortgage

Upon completion of this section, you will be able to:

- Solve problems involving conventional mortgages.
- Solve problems involving adjustable-rate mortgages.

Part of the American dream is to own your own home. Owning your home means that you don't pay rent to someone else, but rather, in a way, you pay rent to yourself. Often you are also rewarded with several financial benefits. In this section, we will discuss the many advantages of owning your own home. We will also discuss the many financial obligations that must be met prior to obtaining a home loan as well as the obligations that must be met after you obtain the loan.

Why This Is Important Most people obtain a mortgage to purchase a new home. Understanding mortgages may help you be financially prepared when you are ready to purchase a house.

When purchasing a home, buyers usually seek a *mortgage* from a bank or other lending institution. Before approving a mortgage, which is a long-term loan, the bank will most likely require the buyer to have a specified minimum amount for the down payment. The *down payment* is the amount of cash the buyer must pay to the seller before the lending institution will grant the buyer a mortgage. If the buyer has the down payment and meets the other criteria for the mortgage, the lending institution prepares a written agreement called the mortgage, stating the terms of the loan. The loan specifies the repayment schedule, the duration of the loan, whether the loan can be assumed by another party, and the penalty if payments are late. The final step of purchasing a home is called the *closing*. At the closing, the home is transferred from the seller to the buyer and the buyer obtains the mortgage from the lender to finance the home purchase.

> **Definition: Homeowner's Mortgage**
> A **homeowner's mortgage** is a long-term loan in which the property is pledged as security for payment of the difference between the sale price and the down payment.

The two most popular types of mortgage loans available today are the *conventional loan* and the *adjustable-rate loan* (or *variable-rate loan*). The major difference between the two is that the interest rate for a conventional loan is fixed for the duration of the loan, whereas the interest rate for the variable-rate loan may change every period, as specified in the loan. We will first discuss the requirements that are the same for both types of loans.

The size of the down payment required by the lender depends on many factors, including the source of the loan, the current economic environment, the buyer's credit score (see the *Mathematics Today* box on page 621), and the location and age of the property. The amount of the down payment required can vary from 0% for some Veterans Affairs loans (see *Mathematics Today* box on page 626) to as high as 50% of the purchase price. In addition to a down payment, the lender may require the buyer to pay one or more *points*. *Points* are interest prepaid by the buyer and may be used to reduce the stated interest rate the lender charges. One point is equal to 1% of the loan amount. This reduction in the interest rate allows the lender to reduce the size of the monthly mortgage payment, which enables more people to purchase homes. However, because points are considered interest, the rate of interest that lenders state when you are applying for a mortgage is not the annual percentage rate (APR) for the loan. Determining the APR involves a number of steps, including adding the amount paid for points to the total interest paid. The APR can then be determined using an APR table, a calculator, or an Internet website. Often, when shopping for a mortgage, a buyer may have two choices: pay points and get a lower interest rate, or not pay points and get a higher interest rate. Websites such as www.hsh.com allow potential buyers to research whether paying points is to their advantage.

Conventional Loans

Example 1 illustrates purchasing a house with a conventional mortgage loan.

Example 1 *Down Payment and Points*

The Martins wish to purchase a house selling for $249,000. They plan to obtain a loan from their bank. The bank requires a 15% down payment, payable to the seller, and a payment of 2 points, payable to the bank, at the time of closing.
a) Determine the Martins' down payment.
b) Determine the amount of the Martins' mortgage.
c) Determine the cost of the 2 points paid by the Martins on their mortgage.

Solution
a) The down payment is 15% of $249,000, or

$$0.15 \times \$249,000 = \$37,350$$

b) The mortgage on the Martins' new home is the selling price minus the down payment.

$$\$249,000 - \$37,350 = \$211,650$$

c) Each point equals 1% of the mortgage amount, so 2 points equals 2% of the mortgage amount.

$$0.02 \times \$211,650 = \$4233$$

At the closing, the Martins will pay the down payment of $37,350 to the seller and the 2 points, or $4233, to the bank. ∎

Lenders typically use a formula to determine the maximum monthly payment that they believe the buyer can reasonably pay based on the buyer's monthly income and the buyer's other monthly debt payments. Other monthly debt payments considered are usually car payments, student loan payments, other fixed installment loan payments, and credit card payments. The buyer's monthly income before any deductions is called the *gross monthly income*. From the gross monthly income subtract any monthly debt payments with *more than 10 monthly payments remaining* to obtain the buyer's *adjusted monthly income*. Generally, a mortgage loan officer will multiply the adjusted monthly income by 28%. In general, this product is the *maximum monthly house payment* the lending institution believes that the purchaser can afford to pay. This payment must cover principal, interest, property taxes, and insurance. Taxes and insurance are not necessarily paid to the bank; they may be paid directly to the tax collector and the insurance company, respectively. Example 2 shows how a bank uses the formula to determine whether a prospective buyer qualifies for a mortgage.

Example 2 *Qualifying for a Mortgage*

Suppose the Martins' (see Example 1) gross monthly income is $7250 and they have 23 remaining monthly payments of $225 on their car loan, 17 remaining monthly payments of $175 on their daughter's orthodontic braces, and 11 remaining monthly payments of $45 on a loan used to purchase new furniture. The property taxes and homeowners' insurance on the house they wish to buy are $165 and $115 per month, respectively. Their bank will approve a loan that has a total monthly mortgage payment of principal, interest, property taxes, and homeowners' insurance that is less than or equal to 28% of their adjusted monthly income.

a) Determine 28% of the Martins' adjusted monthly income.

b) The Martins want a 30-year, $211,650 mortgage. If the interest rate is 4.0%, determine the total monthly house payment (including principal, interest, property taxes, and homeowners' insurance) for this mortgage.

c) Determine whether the Martins qualify for this mortgage.

Solution

a) To determine the Martins' adjusted monthly income, subtract the sum of their monthly payments, $225 + $175 + $45 = $445, from their gross monthly income, $7250.

$7250   Gross monthly income
-445 Monthly payments
$6805   Adjusted monthly income

Next, take 28% of the adjusted monthly income.

$$0.28 \times \$6805 = \$1905.40$$

Thus, 28% of the Martins' adjusted monthly income is $1905.40.

b) To determine the total monthly house payment, we first need to determine the monthly principal and interest payments. Then we add the monthly property taxes and homeowners' insurance. We will use a table to calculate the monthly principal and interest payments the Martins will need to pay. Table 10.4 on page 622 gives monthly principal and interest payments per $1000 of mortgage. With an interest rate of 4.0% and a 30-year mortgage, the Martins would have a monthly principal and interest payment of $4.77415 (circled in blue) per thousand dollars of mortgage.

To determine the Martins' monthly principal and interest payment on their mortgage, first divide the mortgage amount by $1000, which will give the number of thousands of dollars of the mortgage.

$$\frac{\$211,650}{\$1000} = 211.65$$

Then, to determine the monthly principal and interest payment, multiply the number of thousands of dollars of mortgage, 211.65, by the value found in Table 10.4, $4.77415.

$$211.65 \times \$4.77415 \approx \$1010.45$$

Thus, the monthly principal and interest payment is $1010.45. To this amount, we add the monthly property taxes, $165, and the monthly homeowners' insurance, $115.

$$\$1010.45 + \$165 + \$115 = \$1290.45$$

Therefore, the Martins' total monthly mortgage payment is $1290.45.

c) In part (a), we determined that 28% of the Martins' adjusted monthly income is $1905.40. In part (b), we determined that the Martins' total monthly house payment is $1290.45. Because their total monthly house payment is less than or equal to 28% of their adjusted monthly income, the Martins would most likely qualify for the mortgage.   ■

Table 10.4 Monthly Principal and Interest Payment per $1000 of Mortgage

Rate %	Number of Years				
	10	15	20	25	30
3.0	$9.65607	$6.90582	$5.54598	$4.74211	$4.21604
3.5	9.88859	7.14883	5.79960	5.00624	4.49045
4.0	10.12451	7.39688	6.05980	5.27837	4.77415
4.5	10.36384	7.64993	6.32649	5.55832	5.06685
5.0	10.60655	7.90794	6.59956	5.84590	5.36822
5.5	10.85263	8.17083	6.87887	6.14087	5.67789
6.0	11.10205	8.43857	7.16431	6.44301	5.99551
6.5	11.35480	8.71107	7.45573	6.75207	6.32068
7.0	11.61085	8.98828	7.75299	7.06779	6.65302
7.5	11.87018	9.27012	8.05593	7.38991	6.99215
8.0	12.13276	9.55652	8.36440	7.71816	7.33765
8.5	12.39857	9.84740	8.67823	8.05227	7.68913
9.0	12.66758	10.14267	8.99726	8.39196	8.04623
9.5	12.93976	10.44225	9.32131	8.73697	8.40854
10.0	13.21507	10.74605	9.65022	9.08701	8.77572
10.5	13.49350	11.05399	9.98380	9.44182	9.14739
11.0	13.77500	11.36597	10.32188	9.80113	9.52323

Animation

Mortgage Calculator

In Example 2 we used Table 10.4 to calculate the monthly principal and interest payment on a mortgage. We may also use a formula to calculate this payment. The formula is virtually the same one we used in Section 10.4 to calculate installment payments. We will now restate this formula and use it to calculate the monthly principal and interest payments on a mortgage.

Principal and Interest Payment Formula

$$m = \frac{p\left(\dfrac{r}{n}\right)}{1 - \left(1 + \dfrac{r}{n}\right)^{(-n \cdot t)}}$$

(Continued)

In the formula, m is the principal and interest payment, p is the amount of the mortgage, r is the interest rate as a decimal number, n is the number of mortgage payments per year, and t is the time in years.

In Example 3, we use the formula to calculate the monthly principal and interest payment for a mortgage.

Example 3 *Using the Principal and Interest Payment Formula*

Reread Example 2 on pages 621–622. Use the principal and interest payment formula to calculate the Martins' monthly principal and interest payment. Recall that the Martins are seeking a 30-year, $211,650 mortgage with an interest rate of 4%.

Solution In this example the mortgage amount is $211,650, so $p = 211{,}650$; the interest rate is 4%, so $r = 0.04$; the payments will be made monthly, so $n = 12$; and the length of the mortgage is 30 years, so $t = 30$. We substitute these values into the principal and interest formula below.

$$m = \frac{p\left(\dfrac{r}{n}\right)}{1 - \left(1 + \dfrac{r}{n}\right)^{(-n \cdot t)}}$$

$$= \frac{211{,}650\left(\dfrac{0.04}{12}\right)}{1 - \left(1 + \dfrac{0.04}{12}\right)^{(-12 \cdot 30)}}$$

$$\approx \frac{211{,}650(0.003333333)}{1 - (1.003333333)^{-360}}$$

$$\approx \frac{705.4999295}{1 - 0.301795912}$$

$$\approx \frac{705.4999295}{0.6982040988}$$

$$\approx 1010.45$$

Thus, the Martins' monthly principal and interest payment is $1010.45. Notice, this is the same payment that was determined in Example 2 part (b) using Table 10.4. To see how to determine the monthly principal and interest payment on a scientific calculator, or by using Excel, see the Technology Tip on page 607. ∎

Examples 2 and 3 demonstrated that either Table 10.4 or the principal and interest payment formula can be used to calculate the principal and interest payment. The formula's advantage is that it can be used for any interest rate or any length of mortgage—not just those shown in Table 10.4. Principal and interest payments can also be calculated on many websites, such as www.bankrate.com.

What is the effect on the total monthly mortgage payments when only the period of time of the mortgage has been changed? The total monthly mortgage payments, including principal, interest, property taxes, and homeowners' insurance for a 4.0% mortgage for the Martins in Example 2, would be about $2422.85 for 10 years, $1845.55 for 15 years, $1562.56 for 20 years, $1397.17 for 25 years, and $1290.45 for 30 years. (You should verify these numbers for yourself.) Increasing the length of time decreases the monthly payment but increases the total amount of interest paid because the borrower is paying for a longer period of time (see Exercise 27 on page 629). The longer the term of the mortgage, the more expensive the total cost of the house. Although the total cost of a house with the shorter-term mortgage is less expensive overall, it is not always possible for the buyer to obtain the shorter-term mortgage. Note that based on their current adjusted monthly income, the Martins most likely would not have qualified for a 10-year mortgage.

TECHNOLOGY TIP

There's a Mortgage Loan Calculator App for That!

Many of the concepts that we discuss throughout this section and throughout this entire chapter can be further explored with the use of an app on your smartphone or tablet computer. Several of these apps allow you to calculate the monthly mortgage payment based on the cost of your home, down payment, mortgage rate, and length of the mortgage. Many apps also allow you to include property taxes and homeowners' insurance to calculate your total monthly payment. Some mortgage loan calculator apps also calculate other types of payments including car loans and credit card payments.

As always, consult with your instructor before using these apps when completing work related to your course.

Example 4 *The Total Cost of a House*

The Martins of Examples 1, 2, and 3 purchased a house selling for $249,000. They made a 15% down payment of $37,350 and obtained a 30-year conventional mortgage for $211,650 at 4.0%. They also paid 2 points at closing. Their monthly principal and interest payment on their mortgage is $1010.45. Recall that points are considered prepaid interest.

a) Determine the total amount, including principal, interest, down payment, and points, the Martins will pay for their house over 30 years.

b) How much of the cost in part (a) is interest?

c) How much of the first mortgage payment is applied to the principal?

Solution

a) To determine the total amount the Martins will pay for their house, we first note that on a 30-year mortgage, there will be $30 \times 12 = 360$ monthly payments. We then perform the following computations.

	$1010.45	Monthly principal and interest payment
×	360	Number of monthly payments
	$363,762.00	Total principal and interest paid
+	37,350.00	Down payment, from Example 1, part (a)
+	4,233.00	2 points, from Example 1, part (c)
	$405,345.00	Total cost of the house

Note that the result might not be the *exact* cost of the house because the final mortgage payment might be slightly more or less than the rest of the monthly mortgage payments.

b) To determine the amount of the interest paid over 30 years, subtract the purchase price of the house from the total price of the house.

$405,345.00	Total cost of the house
−249,000.00	Purchase price
$156,345.00	Total interest (including the 2 points)

With this mortgage, the Martins will pay $156,345.00 in interest, including the 2 points.

"Being good in business is the most fascinating kind of art."

Andy Warhol

c) To determine the amount of the first payment that is applied to the principal, subtract the amount of interest on the first payment from the monthly principal and interest payment. We will use the simple interest formula, $i = prt$, to determine the interest on the first payment.

$$i = prt$$
$$= \$211{,}650 \times 0.04 \times \frac{1}{12}$$
$$\approx \$705.50$$

Now subtract the interest for the first month from the monthly principal and interest payment. The difference will be the amount paid on the principal for the first month.

$1010.45	Monthly principal and interest payment
−705.50	Interest paid for the first month
$304.95	Principal paid for the first month

Thus, the first monthly principal and interest payment consists of $705.50 in interest and $304.95 in principal. The $304.95 is applied to reduce the outstanding balance due on the loan. Thus, the balance due on the loan after the first monthly payment is made is $211,650.00 − $304.95, or $211,345.05. ∎

By repeatedly using the simple interest formula month to month on the unpaid balance, you could calculate the principal and the interest for all the payments, which is a tedious task. However, a list containing the payment number, payment on the interest, payment on the principal, and balance of the loan can be prepared using a computer. Such a list is called a loan *amortization schedule*. One way to obtain an amortization schedule is by using a computer spreadsheet program. Another way is to access an amortization "calculator" program on the Internet or on a smartphone app. A part of the amortization schedule for the Martins' loan in Example 3 is given in Table 10.5. This schedule was generated from an Internet site called *Monthly Mortgage Payment Calculator (www.hsh.com)*.

Table 10.5   Amortization Schedule

Annual % Rate: 4.0 Loan: $211,650 Periods: 360		Monthly Payment: $1010.45 Term: Years 30, Months 0	
Payment Number	Interest	Principal	Balance of Loan
1	$705.50	$304.95	$211,345.05
2	$704.48	$305.97	$211,039.08
3	$703.46	$306.99	$210,732.10
4	$702.44	$308.01	$210,424.09
11	$695.18	$315.27	$208,239.09
12	$694.13	$316.32	$207,922.77
119	$558.84	$451.61	$167,199.37
120	$557.33	$453.12	$166,746.25
239	$337.17	$673.28	$100,477.79
240	$334.93	$675.52	$99,802.27
359	$6.70	$1003.75	$1007.09
360	$3.36	$1007.09	$0.00

Adjustable-Rate Mortgages

A mortgage in which the interest rate is not fixed for the entire length of the loan is called an *adjustable-rate mortgage (ARM)* or a *variable-rate mortgage*. Generally, an ARM rate is fixed for an initial period of time, called the *initial rate period*. Thereafter, the rate may go up or down based on movements in the interest rate market. Most ARMs have an initial rate period of 5 or 7 years. The initial rate is usually lower than the rate for conventional mortgages, thus making the loan attractive to buyers. However, after the initial rate period, the rate may rise and cause the monthly mortgage payments to also rise. Typically, after the initial rate period, the ARM rate is adjusted once a year. In the United States, the interest rates of ARMs are often tied to one-year U.S. Treasury bills. The stated rate on the Treasury bill is called the *base rate* of the ARM. A lender then adds on a certain percentage called the *add-on rate* or *margin*. For example, if the current one-year Treasury bill rate is 0.22% and the add-on rate is 3.0%, then the ARM rate would be 3.22%. Should the one-year Treasury bill rate rise or fall, then so, too, would the ARM rate.

Example 5 *An Adjustable Rate Mortgage*

The Ghiselins purchased a condominium for $275,000 with a down payment of $115,000. They obtained a 15-year adjustable rate mortgage with the following terms. The interest rate is based on the one-year Treasury bill rate, which currently is 0.5%, and the add-on rate, which is 3.0%. The initial rate period is 5 years, and thereafter the interest rate is adjusted once a year and a new monthly mortgage payment is calculated.

a) Determine the Ghiselins' initial ARM rate.

b) Determine the Ghiselins' initial monthly payment for principal and interest.

c) If, after the 5-year initial rate period, the rate of the one-year Treasury bill rises to 1.5%, determine the Ghiselins' new ARM rate.

Solution

a) The ARM rate is the sum of the one-year Treasury bill rate, 0.5%, and the add-on rate, 3.0%. Thus, the Ghiselins' initial ARM rate is 0.5% + 3.0%, or 3.5%.

b) We will use Table 10.4 on page 622. The amount of the loan is $275,000 − $115,000, or $160,000. Divide the amount of the loan by $1000 to get $160,000/$1000, or 160. Now, looking at Table 10.4 with $r = 3.5\%$ for 15 years, we determine the value 7.14883.

$$160 \times \$7.14883 \approx \$1143.81$$

Thus, the initial monthly payment for principal and interest is $1143.81. Note that we could have also used the principal and interest payment formula to calculate the monthly payment for principal and interest.

c) The sum of the new one-year Treasury bill rate, 1.5%, and the add-on rate, 3.0%, is 4.5%. Thus, the Ghiselins' new ARM rate is 4.5%. ∎

To prevent rapid increases in interest rates of adjustable rate mortgages, some banks have a rate cap. A *rate cap* limits the maximum amount the interest rate may change. A *periodic rate cap* limits the amount the interest rate may increase in any one period. For example, your mortgage could provide that your rate can go up only 1% per year. An *aggregate rate cap* limits the interest rate increase and decrease over the entire life of the loan. If the initial interest rate is 6% and the aggregate rate cap is 2%, the interest rate could go no higher than 8% and no lower than 4% over the life of the mortgage.

Conventional mortgages and adjustable-rate mortgages are not the only methods of financing the purchase of a home. The *Mathematics Today* box, left, discusses some other types of mortgages that are available to consumers. More information about other types of mortgages can be found online at www.makinghomeaffordable.gov.

SECTION 10.5
Exercises

Warm Up Exercises

In Exercises 1–6, fill in the blanks with an appropriate word, phrase, or symbol(s).

1. A long-term loan in which the property is pledged as security for payment of the difference between the sale price and the down payment is known as a homeowner's _____.

2. The amount of cash the buyer must pay to the seller before the lending institution will grant the buyer a mortgage is known as the _____ payment.

3. Interest prepaid by the buyer, which may be used to reduce the stated interest rate the lender charges, are known as _____.

4. The final step of purchasing a home is called the _____.

5. The buyer's gross monthly income minus any fixed monthly payments with more than 10 months remaining is known as the buyer's _____ monthly income.

6. A list containing the payment number, payment on interest, payment on principal, and the balance of the loan is called a(n) _____ schedule.

Practice the Skills

In Exercises 7–10, determine the monthly principal and interest payment for the mortgage. Unless told otherwise by your instructor, you may use either Table 10.4 or the principal and interest payment formula (See Examples 2 and 3).

	Amount Financed (P)	Annual Percentage Rate (r)	Number of Payments per Year (n)	Time in Years (t)
7.	$95,000	5%	12	15
8.	$132,000	3%	12	30
9.	$236,000	5.5%	12	25
10.	$316,000	4.5%	12	20

Problem Solving

11. *Buying a House* Meghan is buying a house selling for $275,000. To obtain the mortgage, Meghan is required to make a 15% down payment. Megan obtains a 30-year mortgage with an interest rate of 4%.

  a) Determine the amount of the required down payment.

  b) Determine the amount of the mortgage.

  c) Determine the monthly payment for principal and interest.

12. *Buying a Condominium* Eraj is buying a condominium selling for $165,000. To obtain the mortgage, Eraj is required to make a 20% down payment. Eraj obtains a 25-year mortgage with an interest rate of 5%.

  a) Determine the amount of the required down payment.

  b) Determine the amount of the mortgage.

  c) Determine the monthly payment for principal and interest.

13. *Buying a Brownstone Townhouse* Phillip is purchasing a brownstone townhouse in Boston for $2,337,500. To obtain the mortgage, Phillip is required to make a 20% down payment. Phillip obtains a 20-year mortgage with an interest rate of 4.5%

  a) Determine the amount of the required down payment.

  b) Determine the amount of the mortgage.

  c) Determine the monthly payment for principal and interest.

14. *Buying a First House* The Firszts are purchasing their first home for $199,900. They are obtaining an FHA mortgage through their credit union and are required to make a 3% down payment. They obtain a 30-year mortgage with an interest rate of 4.5%.

  a) Determine the amount of the required down payment.

  b) Determine the amount of the mortgage.

  c) Determine the monthly payment for principal and interest.

15. *Paying Points* Martha is buying a house selling for $195,000. The bank is requiring a minimum down payment of 20%. To obtain a 20-year mortgage at 6% interest, she must pay 2 points at the time of closing.

  a) What is the required down payment?

  b) What is the amount of the mortgage?

  c) What is the cost of the 2 points?

16. *Down Payment and Points* The Nicols are buying a house selling for $245,000. They pay a down payment of $45,000 from the sale of their current house. To obtain a

15-year mortgage at a 4.5% interest rate, the Nicols must pay 1.5 points at the time of closing.

a) What is the amount of the mortgage?

b) What is the cost of the 1.5 points?

17. *Qualifying for a Mortgage* The Guenthers' gross monthly income is $3200. They have 25 remaining car payments of $335. The Guenthers are applying for a 15-year, $150,000 mortgage at 5% interest to buy a new house. The taxes and insurance on the house total $225 per month. Their credit union will approve a loan that has a total monthly house payment of principal, interest, property taxes, and home-owners' insurance that is less than or equal to 28% of their adjusted monthly income.

a) Determine 28% of the Guenthers' adjusted monthly income.

b) Determine the Guenthers' total monthly house payment, including principal, interest, taxes, and homeowners' insurance.

c) Do the Guenthers qualify for this mortgage?

18. *Qualifying for a Mortgage* The Zhengs' gross monthly income is $4100. They have 18 remaining boat payments of $505. The Zhengs are applying for a 20-year, $275,000 mortgage at a 9% interest rate to buy a new house. The taxes and insurance on the house total $425 per month. Their bank will approve a loan that has a total monthly house payment of principal, interest, property taxes, and homeowners' insurance that is less than or equal to 28% of their adjusted monthly income.

a) Determine 28% of the Zhengs' adjusted monthly income.

b) Determine the Zhengs' total monthly house payment, including principal, interest, taxes, and insurance.

c) Do the Zhengs qualify for this mortgage?

19. *A 30-Year Conventional Mortgage* Ingrid obtains a 15-year, $63,750 conventional mortgage at a 6.5% rate on a house selling for $75,000. Her monthly payment, including principal and interest, is $555.33. She pays 0 points.

a) Determine the total amount Ingrid will pay for her house.

b) How much of the cost will be interest?

c) How much of the first payment on the mortgage is applied to the principal?

20. *A 25-Year Conventional Mortgage* The Bells obtain a 25-year, $110,000 conventional mortgage at a 5.5% rate

on a house selling for $160,000. Their monthly mortgage payment, including principal and interest, is $675.50. They also pay 2 points at closing.

a) Determine the total amount the Bells will pay for their house.

b) How much of the cost will be interest (including the 2 points)?

c) How much of the first payment on the mortgage is applied to the principal?

21. *Evaluating a Loan Request* The Nejems found a house selling for $550,000. The taxes on the house are $5634 per year and the insurance is $2325 per year. The Nejems are requesting a conventional loan from a local bank. The bank requires a 20% down payment and 3 points at the closing. The Nejems are trying to qualify for a 30-year mortgage with an interest rate of 5.5%. Their gross monthly income is $15,375. They have more than 10 monthly payments remaining on a car loan, student loans, and a furniture loan. The total of these monthly payments is $995. Their bank will approve a loan that has a total monthly house payment of principal, interest, property taxes, and homeowners' insurance that is less than or equal to 28% of their adjusted monthly income.

a) Determine the required down payment.

b) Determine the cost of the 3 points.

c) Determine 28% of the Nejems' adjusted monthly income.

d) Determine the monthly payment for principal and interest.

e) Determine their total monthly house payment, including insurance and taxes.

f) Do the Nejems qualify for the loan?

g) Determine how much of the first mortgage payment is applied to the principal.

22. *Evaluating a Loan Request* Kathy wants to buy a condominium selling for $95,000. The taxes on the property are $1500 per year, and homeowners' insurance is $336 per year. Kathy's gross monthly income is $4000. She has 15 monthly payments of $135 remaining on her van. The bank is requiring 20% down and is charging a 9.5% interest rate with no points. Her bank will approve a loan that has a total monthly mortgage payment of principal, interest, property taxes, and homeowners' insurance that is less than or equal to 28% of her adjusted monthly income.

a) Determine the required down payment.

b) Determine 28% of her adjusted monthly income.

c) Determine the monthly payment of principal and interest for a 25-year loan.

d) Determine her total monthly payment, including homeowners' insurance and taxes.

e) Does Kathy qualify for the loan?

f) Determine how much of the first payment on the mortgage is applied to the principal.

g) Determine the total amount she pays for the condominium with a 25-year conventional loan. (Do not include taxes or homeowners' insurance.)

h) Determine the total interest paid for the 25-year loan.

23. *Comparing Loans* The Riveras are negotiating with two banks for a mortgage to buy a house selling for $105,000. The terms at bank A are a 10% down payment, an interest rate of 4%, a 30-year conventional mortgage, and 3 points to be paid at the time of closing. The terms at bank B are a 20% down payment, an interest rate of 5.5%, a 25-year conventional mortgage, and no points. Which loan should the Riveras select for the total cost of the house to be the least?

24. *Comparing Loans* Paul is negotiating with two credit unions for a mortgage to buy a condominium selling for $525,000. The terms at Grant County Teacher's Credit Union are a 20% down payment, an interest rate of 7.5%, a 15-year mortgage, and 1 point to be paid at the time of closing. The terms at Sinnipee Consumer's Credit Union are a 15% down payment, an interest rate of 8.5%, a 20-year mortgage, and no points. Which loan should Paul select for the total cost of the down payment, points, and total mortgage payments of the house to be the least?

25. *An Adjustable Rate Mortgage* The Bhatts purchased a new home for $235,000 with a down payment of $47,000. They obtained a 20-year adjustable rate mortgage with the following terms. The interest rate is based on the one-year Treasury bill rate, which is currently at 1.5%, and the add-on rate, which is 2.5%. The initial rate period is 5 years, and thereafter the interest rate is adjusted once a year and a new monthly mortgage payment is calculated.

a) Determine the Bhatts' initial ARM rate.

b) Determine the Bhatts' initial monthly payment for principal and interest.

c) If, after the 5-year initial rate period, the rate of the one-year Treasury bill rises to 3.0%, determine the Bhatts' new ARM rate.

26. *An Adjustable Rate Mortgage* The Pourans purchased a new home for $378,000 with a down payment of $127,000. They obtained a 10-year adjustable rate mortgage with the following terms. The interest rate is based on the one-year Treasury bill rate, which is currently at 2.5%, and the add-on rate, which is 3.5%. The initial rate period is 5 years, and thereafter the interest rate is adjusted once a year and a new monthly mortgage payment is calculated.

a) Determine the Pourans' initial ARM rate.

b) Determine the Pourans' initial monthly payment for principal and interest.

c) If, after the 5-year initial rate period, the rate of the one-year Treasury bill falls to 1.5%, determine the Pourans' new ARM rate.

Challenge Problems/Group Activities

27. *Changing Lengths of Mortgages* Rose and George are purchasing a house for $450,000. Their bank requires them to pay a down payment of 20%. The current mortgage rate is 10%, and they are required to pay 1 point at the time of closing. Determine the total amount Rose and George will pay for their house, including principal, interest, down payment, and points (do not include taxes and homeowners' insurance) if the length of their mortgage is

a) 10 years.

b) 20 years.

c) 30 years.

28. *Comparing Payment Frequency* Janet is purchasing a new house for $315,000. Her credit union requires her to make a 20% down payment, and the current mortgage rate is 6%. Janet is exploring different 10-year mortgage payment options. Use the principal and interest formula to determine Janet's principal and interest payment if she makes her payments

a) Monthly.

b) Bimonthly (use $n = 24$).

c) Weekly (use $n = 52$).

29. *Comparing Mortgages* The Hassads are applying for a $90,000 mortgage. They can choose between a 30-year conventional mortgage and a 30-year variable-rate mortgage. The interest rate on a 30-year conventional mortgage is 9.5%. The terms of the variable-rate mortgage are 6.5% interest rate the first year, an annual cap of 1%, and an aggregate cap of 6%. The interest rates and the mortgage payments are adjusted annually. Assume that the interest rates for the variable-rate mortgage increase by the maximum amount each year. Then the monthly mortgage payments for the variable-rate mortgage for years 1–6 are $568.86, $629.29, $692.02, $756.77, $823.27, and $891.26, respectively.

a) Knowing that they will be in the house for only 6 years, which mortgage, the conventional mortgage or the variable-rate mortgage, will be the least expensive for that period?

b) How much will they save by choosing the less expensive mortgage?

Research Activities

30. *Dream Home* Search in your area to find your "dream home" that is for sale and note the asking price. Next, search the websites of local lenders to find the best interest rates for both a 15-year fixed mortgage and a 30-year fixed mortgage. Assume that you make a 20% down payment; then calculate and compare the monthly payments of principal and interest for both the 15-year mortgage and the 30-year mortgage. Using an amortization calculator (see page 625), compare the total amount of interest paid on the 15-year mortgage and on the 30-year mortgage. Write a report summarizing your findings.

31. *Closing Costs* The closing of a house involves many additional costs for the both the buyer and the seller. Select a home that is for sale and note the asking price. Do research and use this asking price to calculate the closing costs in your area. A partial list of closing costs include: title search fee, title insurance fee, credit report fee, loan origination fee, attorney's fees, inspection fee, appraisal fee, survey fee, escrow account deposits, pest inspection fee, recording fee, and underwriting fees.

SECTION 10.6 Ordinary Annuities, Sinking Funds, and Retirement Investments

Upon completion of this section, you will be able to:

- Solve problems involving ordinary annuities.
- Solve problems involving sinking funds.
- Understand other annuities and retirement savings options.

In Sections 10.2 and 10.3, we studied simple and compound interest, respectively. In both sections, we answered questions involving the investment of one lump sum of money. Although such questions may be relevant to our everyday financial matters, more appropriate questions might involve investing smaller amounts of money on a regular basis over a longer period of time. For example, the Brabsons, who are planning for retirement, might ask the question: If we deposit $100 a month at a 5% interest rate compounded monthly, how much money will we accumulate after 30 years when we're ready to retire? Also consider the Weismans, who are saving for their child's college education. They might ask the question: If we know we will need $50,000 to send our child to college in 10 years, how much should we invest each month beginning now in an account paying a 6% interest rate compounded monthly? To answer these and similar questions, we will study two investments, the *annuity* and the *sinking fund*.

Why This Is Important It's never too early to start planning for retirement. In order to meet your retirement goals, you will want to make wise financial investments. Annuities and sinking funds are an important part of a long-term investment strategy.

___

We begin this section with a discussion of annuities.

> **Definition: Annuity**
> An **annuity** is an account into which, or out of which, a sequence of scheduled payments is made.

There are many different types of annuities. An annuity may be an investment account that you have with a bank, insurance company, or financial management firm. Annuities may contain investments in stocks, bonds, mutual funds, money market accounts, and other types of investments. Annuities are often used to save for long-term goals such as saving money for college or for retirement. An annuity can also be used to provide long-term regular payments to individuals. Lottery jackpots and professional athletes' salaries are often paid out over time from annuities. Retirees may invest some of their retirement savings into an annuity and then receive monthly payments that come from that annuity. We will focus primarily on two basic types of annuities that are used as investment accounts, *ordinary annuities* and *sinking funds*. Later in this section, we will discuss other types of annuities.

Ordinary Annuities

> **Definition: Ordinary Annuity**
> An annuity into which equal payments are made at regular intervals, with the interest compounded at the end of each interval and with a fixed interest rate for each compounding period, is called an **ordinary annuity** or a **fixed annuity**.

For example, the Brabsons, mentioned in the opening paragraph of this section, are asking a question that may involve an ordinary annuity. They plan to make $100 payments each month into an account that pays a 5% interest rate compounded monthly. With an ordinary annuity, the payment period and frequency of the compounding are the same, so the Brabsons make *monthly* payments *and* the interest is compounded *monthly*. They would like to know how much money would accumulate in this annuity after 30 years. The amount of money that is present in an ordinary annuity after t years is known as the *accumulated amount* or the *future value* of an annuity.

To determine the accumulated amount of an ordinary annuity, we could make many individual calculations using the compound interest formula discussed in Section 10.3. A better way is to use a computer to generate a spreadsheet. A portion of such a spreadsheet is shown in Table 10.6, below. This spreadsheet was generated using the data from the example involving the Brabsons. Note that the Brabsons would make monthly payments for 30 years for a total of 360 payments. Also note that had the Brabsons simply put $100 in their cookie jar each month, their total amount at the end of 30 years would have been $100 × 360, or $36,000. Instead, by investing

Table 10.6 Ordinary Annuity Growth

Accumulated Amount in an Ordinary Annuity with $100 Monthly Investments with Interest Rate 5% Compounded Monthly			
Payment Number	**Monthly Investment**	**Interest**	**Accumulated Amount**
1	$100.00	$ —	$ 100.00
2	$100.00	$ 0.42	$ 200.42
3	$100.00	$ 0.84	$ 301.25
4	$100.00	$ 1.26	$ 402.51
5	$100.00	$ 1.68	$ 504.18
6	$100.00	$ 2.10	$ 606.28
7	$100.00	$ 2.53	$ 708.81
8	$100.00	$ 2.95	$ 811.76
9	$100.00	$ 3.38	$ 915.15
10	$100.00	$ 3.81	$ 1018.96
⋮	⋮	⋮	⋮
351	$100.00	$328.58	$79,287.40
352	$100.00	$330.36	$79,717.76
353	$100.00	$332.16	$80,149.92
354	$100.00	$333.96	$80,583.88
355	$100.00	$335.77	$81,019.65
356	$100.00	$337.58	$81,457.23
357	$100.00	$339.41	$81,896.63
358	$100.00	$341.24	$82,337.87
359	$100.00	$343.07	$82,780.94
360	$100.00	$344.92	$83,225.86

in an interest-bearing annuity, they will have accumulated $83,225.86 after 30 years. This amount is circled in blue in Table 10.6.

Another way, and perhaps a more efficient way, to calculate the accumulated amount in an ordinary annuity is to use the ordinary annuity formula.

Ordinary Annuity Formula

The accumulated amount, A, of an ordinary annuity with payments of p dollars made n times per year, for t years, at interest rate r as a decimal number, compounded at the end of each payment period is given by the formula

$$A = \frac{p\left[\left(1 + \dfrac{r}{n}\right)^{(n \cdot t)} - 1\right]}{\left(\dfrac{r}{n}\right)}$$

Example 1 *Using the Ordinary Annuity Formula*

Bill and Megan are depositing $250 each quarter in an ordinary annuity that pays 4% interest compounded quarterly. Determine the accumulated amount in this annuity after 35 years.

Solution We will use the ordinary annuity formula. The payment, p, is $250, the interest rate, r, is 4% or 0.04, the account is compounded quarterly and the payments are made quarterly, so n is 4, and the number of years, t, is 35. We substitute these values into the formula to obtain the following.

$$A = \frac{p\left[\left(1 + \dfrac{r}{n}\right)^{(n \cdot t)} - 1\right]}{\left(\dfrac{r}{n}\right)}$$

$$= \frac{250\left[\left(1 + \dfrac{0.04}{4}\right)^{(4 \cdot 35)} - 1\right]}{\left(\dfrac{0.04}{4}\right)}$$
  Substitute the given values into the formula.

$$= \frac{250\left[1.01^{140} - 1\right]}{0.01}$$
  To evaluate 1.01^{140}, we use the $\boxed{y^x}$ or the $\boxed{\wedge}$ key on a scientific calculator.

$$\approx \frac{250\left[4.027099 - 1\right]}{0.01}$$
  We rounded 1.01^{140} to six decimal places.

$$\approx \frac{250(3.027099)}{0.01}$$

$$\approx \frac{756.77475}{0.01} \approx 75{,}677.475$$

Thus, there will be about $75,677.48 in Bill and Megan's annuity after 35 years. ∎

Example 1 illustrated the power of establishing and maintaining a consistent investment plan and how using an ordinary annuity can help us save money.

TECHNOLOGY TIP

Scientific Calculator

In Example 1, we determined the accumulated amount by evaluating the expression

$$\frac{250\left[\left(1 + \frac{0.04}{4}\right)^{(4\cdot35)} - 1\right]}{\left(\frac{0.04}{4}\right)}$$

There are various ways to evaluate this expression on a scientific calculator. One method is to press the following sequence of keys.

(250 × (((1 + 0.04 ÷ 4) y^x (4 × 35) − 1)) ÷ (0.04 ÷ 4)) =

After the = key is pressed, the answer 75677.48042 is displayed.

Microsoft Excel

Read the Technology Tip on pages 596–597. To determine the accumulated amount using Excel, place the letters *p*, *r*, *n*, *t*, and *A* in cells A1 through E1, respectively. In cells A2 through D2, place the values for *p*, *r*, *n*, and *t*, respectively. In cell E2, use the formula box to add the formula

= (A2*((1 + B2/C2)^(C2*D2) − 1))/(B2/C2)

After the Enter key is pressed, the result is as follows.

	A	B	C	D	E
1	*p*	*r*	*n*	*t*	*A*
2	250	0.04	4	35	75677.48

Sheet1 / Sheet2 / Sheet3 /
Ready

Next we will study a specific type of annuity called a *sinking fund*.

Sinking Funds

We now return to the second question asked at the beginning of this section. Recall that the Weismans have a goal of saving $50,000 in 10 years for their child's college education and want to know how much they should begin to invest each month in an account paying a 6% interest rate compounded monthly. The Weismans can help reach this goal by investing in a special kind of annuity called a *sinking fund*.

> **Definition: Sinking Fund**
> A **sinking fund** is a type of annuity in which the goal is to save a specific amount of money in a specific amount of time.

Many types of sinking funds are used for many different purposes by individuals, corporations, and governments. Historically, sinking funds were used by governments to set aside money to pay off bonds that were used to borrow large sums of money for major civic projects such as sewers, roads, or bridges. The term *sinking fund* referred to the *sinking* of the debt that was owed to repay the bonds. Our discussion of sinking funds will focus on answering questions similar to the question facing the Weismans.

Because they are interested in saving a specific amount of money, $50,000, in a specific amount of time, 10 years, their investment can be considered a sinking fund.

To determine the amount of money to be invested in a sinking fund, we can solve the ordinary annuity formula, given on page 632, for the payment, p. Solving this formula for p gives us the following formula.

Sinking Fund Payment Formula

$$p = \frac{A\left(\frac{r}{n}\right)}{\left(1 + \frac{r}{n}\right)^{(n \cdot t)} - 1}$$

In the formula, p is the payment needed to reach the accumulated amount, A. Payments are made n times per year, for t years, into a sinking fund with interest rate r as a decimal number, compounded n times per year.

Example 2 *Using the Sinking Fund Payment Formula*

The Burnettes would like to have $55,000 in 10 years, when they retire, to purchase a pontoon boat. The Burnettes decide to invest in a sinking fund that pays 3.3% interest compounded monthly. How much should the Burnettes invest each month in the sinking fund to accumulate $55,000 in 10 years?

Solution Since we are looking for how much the Burnettes should invest each month to obtain an accumulated amount, we will use the sinking fund payment formula. The accumulated amount, A, is $55,000; the interest rate, r, is 3.3%, or 0.033; the payments are made monthly and the interest is compounded monthly, so n is 12; and the number of years, t, is 10.

$$p = \frac{A\left(\frac{r}{n}\right)}{\left(1 + \frac{r}{n}\right)^{(n \cdot t)} - 1}$$

$$= \frac{55{,}000\left(\frac{0.033}{12}\right)}{\left(1 + \frac{0.033}{12}\right)^{(12 \cdot 10)} - 1}$$

$$\approx \frac{151.25}{0.39033827}$$

$$\approx 387.48442$$

We often round to the nearest cent, or in this case, to $387.48. However, because $387.48 is slightly *less* than the answer we obtained, we will round our answer *up* to $387.49 to ensure that the Burnettes reach their goal of $55,000. ∎

As we did in Example 2, *when calculating a sinking fund payment, we will round our final answers up to the nearest cent to ensure that we reach the desired accumulated amount in the sinking fund.*

In Example 2, we can check our answer by substituting $387.49 back into the ordinary annuity formula. Doing this calculation with a scientific calculator gives us an accumulated amount of about $55,000.79. If we substitute $387.48 for A into the ordinary annuity formula, the accumulated amount would be about $54,999.37, which is slightly short of the desired goal of $55,000.

Scientific Calculator

To determine the sinking fund payment in Example 2, we evaluated the expression

$$\frac{55{,}000\left(\dfrac{0.033}{12}\right)}{\left(1+\dfrac{0.033}{12}\right)^{(12\cdot10)}-1}$$

This expression can be evaluated on a scientific calculator, as follows:

$55000 \times (\ 0.033 \div 12\ ) \div (\ (\ 1 + 0.033 \div 12\ ) \ y^x \ (\ 12 \times 10\ ) - 1\ )$

After the $=$ key is pressed, the answer 387.48442 is displayed.

Microsoft Excel

Read the Technology Tip on page 633 and follow a similar procedure using the variables A, r, n, t, and p, respectively. To evaluate the sinking fund payment, in cell E2 enter the formula

$$= A2*(B2/C2)/((1 + B2/C2)\wedge(C2*D2) - 1)$$

After the Enter Key is pressed, the answer 387.4844 is displayed in cell E2.

Other Annuities and Retirement Savings Options

Thus far we have discussed ordinary annuities and sinking funds. Next, we will briefly discuss *variable annuities*, *immediate annuities*, and other types of retirement savings options.

A *variable annuity* is an annuity that is invested in stocks, bonds, mutual funds (see the *Did You Know?* boxes on pages 631, 632, and 633, respectively) or other investments that do not provide a guaranteed interest rate. The term *variable* is used because the value of the annuity will vary depending on the performance of the investment options chosen by the investor. Like an ordinary annuity, an investor usually invests in a variable annuity by making regular periodic payments. Unlike an ordinary annuity, however, a variable annuity does not have a fixed rate of interest. Most variable annuities are investments made to save for retirement.

An *immediate annuity* is an annuity that is established with a lump sum of money for the purpose of providing the investor with regular, usually monthly, payments for the rest of the investor's life. In exchange for giving the investment company a lump sum of money, the investor is guaranteed to receive a monthly income for the duration of the investor's life.

A *deferred income annuity*, or *deferred annuity*, provides you or your spouse with a guaranteed income for life. A deferred annuity commences only after a lapse of some specific time after the final purchase premium on the policy has been made. This type of annuity has two phases: the saving phase, which is when the account holder invests money into the account, and an income phase, which is when the plan is converted into an annuity and begins paying the account owner. Deferred annuities can be variable or fixed. Deferred annuities are often part of a retirement plan.

Annuities can be used to save for retirement, but depending on the investor's circumstances, other investment options for retirement savings may be more advantageous. *Individual retirement accounts*, also known as *IRAs*, are accounts in which individuals may invest up to a certain amount of money each year for the purpose of saving for retirement. IRAs have distinct tax advantages over non-IRA accounts. There are several different kinds of IRAs, but most can be classified as either a *traditional IRA* or a *Roth IRA*. A *traditional IRA* is an IRA into which *pretax* money is invested, meaning that any money invested into a traditional IRA is not subject to income taxes. However, when money is withdrawn from a traditional IRA, the money *is* subject to income taxes. A *Roth IRA* is an IRA into which *post-tax* money is invested, meaning that the investor has already paid income taxes on the money invested. When money is withdrawn from a Roth IRA after the investor reaches retirement age, the money *is not* subject to income taxes.

▲ *It is never too early to start saving for retirement.*

A *401k plan* is a retirement savings plan in which employees of private companies can make contributions of pre–income tax dollars that are then pooled with other employees' money. These funds are then invested in a variety of stocks, bonds, money markets, and mutual funds. Although 401k plans share many of the same tax advantages as IRAs, they also have several distinct advantages. First, employees are generally allowed to invest more money annually in a 401k plan than in an IRA. A second advantage is that some employers will match employee contributions up to a certain limit. A *Roth 401k plan* is a retirement savings plan that differs from a 401k plan in that an investor first pays income tax on the money used to invest in a Roth 401k plan. Like a Roth IRA, money withdrawn from a Roth 401k plan after the individual reaches retirement age is not subject to income tax.

A *403b plan* is a retirement plan similar to a 401k plan, but it is available only to employees of schools, civil governments, hospitals, charities, and other not-for-profit organizations. A 403b plan shares many of the same advantages as 401k plans. One major difference, though, is that a 403b plan is not necessarily established by the employer as is a 401k plan. Rather, a 403b plan is frequently established independently by the employee with a representative of an investment or insurance company.

As you can see from this brief discussion of annuities, IRAs, 401k plans, and 403b plans, many important details can affect your decisions as you begin to invest for retirement. For more information, and for other types of retirement accounts not mentioned here, consult a financial advisor or visit a library or Internet website. Regardless of the source of your information, it is very important to do your own research and ask many questions prior to investing.

SECTION 10.6
Exercises

Warm Up Exercises

In Exercises 1–8, fill in the blanks with an appropriate word, phrase, or symbol(s).

1. An account into which, or out of which, a sequence of scheduled payments is made, is called a(n) _____ .

2. The amount of money that is present in an ordinary annuity after t years is known as the accumulated amount or the _____ value of the annuity.

3. A type of annuity in which the goal is to save a specific amount of money in a specific amount of time is called a(n) _____ fund.

4. A variable annuity is an annuity that is invested in stocks, bonds, mutual funds, or other investments that do not provide a(n) _____ interest rate.

5. An annuity that is established with a lump sum for the purpose of providing the investor with regular payments for the rest of the investor's life is called a(n) _____ annuity.

6. Accounts in which individuals may invest up to a certain amount of money each year for the purpose of saving for retirement are called _____ retirement accounts.

7. A retirement savings plan in which employees of private companies can make contributions of pre–income tax dollars that are then pooled with other employees' money is called a(n) _____ plan.

8. A retirement plan similar to a 401k plan, but available only to employees of schools, governments, and other not-for-profit organizations, is called a(n) _____ plan.

Practice the Skills

In Exercises 9–12, use the ordinary annuity formula

$$A = \frac{p\left[\left(1 + \frac{r}{n}\right)^{(n \cdot t)} - 1\right]}{\left(\frac{r}{n}\right)}$$

to determine the accumulated amount in each annuity. Round all answers to the nearest cent.

9. $5200 invested annually for 30 years at a 3.5% interest rate compounded annually

10. $800 invested semiannually for 20 years at a 5% interest rate compounded semiannually

11. $400 invested quarterly for 35 years at a 8% interest rate compounded quarterly

12. $200 invested monthly for 40 years at a 6% interest rate compounded monthly

In Exercises 13–16, use the sinking fund formula

$$p = \frac{A\left(\frac{r}{n}\right)}{\left(1 + \frac{r}{n}\right)^{(n \cdot t)} - 1}$$

to determine the payment needed to reach the accumulated amount. To ensure that enough is invested each period, round each answer up to the next cent.

13. Semiannual payments with a 5.5% interest rate compounded semiannually for 25 years to accumulate $40,000

14. Weekly payment with a 6.5% interest rate compounded weekly for 30 years to accumulate $800,000

15. Monthly payments with a 6% interest rate compounded monthly for 35 years to accumulate $250,000

16. Quarterly payments with a 10% interest rate compounded quarterly for 40 years to accumulate $1,000,000

Problem Solving

In Exercises 17–20, round all answers to the nearest cent.

17. *Retirement Savings* To save for retirement Simon invests $150 each month in an ordinary annuity with a 3.3% interest rate compounded monthly. Determine the accumulated amount in Simon's annuity after 40 years.

18. *Starting a Business* To save money to start a new business, Sandra invests $200 each quarter in an ordinary annuity with a 5% interest rate compounded quarterly. Determine the accumulated amount in Sandra's annuity after 20 years.

19. *Saving for Graduate School* To save for graduate school, Ena invests $1500 semiannually in an ordinary annuity with a 6% interest rate compounded semiannually. Determine the accumulated amount in Ena's annuity after 10 years.

20. *Grandchild's College Expenses* To save for their grandchildren's college expenses, Ken and Laura invest $500 each month in an ordinary annuity with a 6% interest rate compounded monthly. Determine the accumulated amount in the their annuity after 18 years.

In Exercises 21–24, round all answers up to the next cent.

21. *Saving for a New Motorcycle* Nick would like to have $24,000 to buy a new motorcycle in 6 years. To accumulate $24,000 in 6 years, how much should Nick invest monthly in a sinking fund with a 7.5% interest rate compounded monthly?

22. *Campaign Fund-Raising* Ashley would like to run for Congress in 10 years and would like to have $500,000 to spend on campaign expenses. How much should she invest monthly in a sinking fund with a 6% interest rate compounded monthly to accumulate $500,000 in 10 years?

23. *Becoming a Millionaire* Becky would like to be a millionaire in 20 years. How much would she need to invest quarterly in a sinking fund paying an 8% interest rate compounded quarterly to accumulate $1,000,000 in 20 years?

24. *Repaying a Debt* The city of San Francisco wishes to repay a debt of $25,000,000 in 50 years. How much should the city invest semiannually in a sinking fund paying a 7.5% interest rate compounded semiannually to accumulate $25,000,000 in 50 years?

▲ *Golden Gate Bridge, San Francisco*

Challenge Problems/Group Activities

25. *Don't Wait to Invest* The purpose of this exercise is to demonstrate that it is much more beneficial to begin to invest for retirement early than it is to wait. We will compare the investments of two investors, Alberto and Zachary. Alberto begins to invest immediately after graduating from college, whereas Zachary waits 10 years to begin investing. For the purposes of this demonstration, we will assume that the interest rate remains the same over the entire period of time.

a) From age 25 to 35, Alberto invests $100 per month in an annuity that pays a 12% interest rate compounded monthly. Determine the accumulated amount in Alberto's annuity after 10 years.

b) At age 35, Alberto stops making monthly payments into his annuity, but he takes the money he accumulated in his annuity and invests it in a savings account with an interest rate of 12% compounded monthly. Using the compound interest formula $A = p\left(1 + \dfrac{r}{n}\right)^{(n \cdot t)}$, determine the amount Alberto has in his savings account 30 years later when he retires at age 65.

c) From age 35 to 65, Zachary invests $100 per month in an annuity that pays 12% interest compounded monthly. Determine the accumulated amount in Zachary's annuity after 30 years.

d) Determine the total amount Alberto invested during the 10 years from age 25 to 35.

e) Determine the total amount Zachary invested during the 30 years from age 35 to 65.

f) Which investor has more money from the investments described here upon retirement?

Research Activities

26. ***Investing in Stocks*** Write a paper on investing in stocks (see the *Did You Know?* on page 631). Include in your paper a description of the following terms: *New York Stock Exchange, National Association of Securities Dealers Automated Quotations (NASDAQ), dividends,* and *capital gains.*

27. ***Investing in Bonds*** Write a paper on investing in bonds (see the *Did You Know?* on page 632). Include in your paper a description of the following terms: *Treasury bonds, corporate bonds,* and *municipal bonds.*

28. ***Investing in Mutual Funds*** Write a paper on investing in mutual funds (see the *Did You Know?* on page 633). Include in your paper a description of the following terms: *net asset value (NAV), portfolio manager,* and *expense ratios.*

29. ***Financial Advisor Interview*** Annuities, IRAs, 401k plans, 403b plans, SEP accounts, Keogh accounts, and defined benefit plans are all types of investments that can be used to save for retirement. Interview a financial advisor to discuss each of these investment types. Write a paper in which you explain the differences among the different investment types and describe the advantages of each type of investment over the others.

CHAPTER 10 *Summary*

Important Facts and Concepts	Examples and Discussion
Section 10.1	
Percent Change	
Percent change $= \dfrac{\left(\begin{array}{c}\text{amount in}\\\text{latest period}\end{array}\right) - \left(\begin{array}{c}\text{amount in}\\\text{previous period}\end{array}\right)}{\text{amount in previous period}} \times 100$	Examples 5–6, pages 577
Percent Markup on Cost	
Percent markup $= \dfrac{\text{selling price} - \text{dealer's cost}}{\text{dealer's cost}} \times 100$	Examples 7–8, pages 578
Section 10.2	
Simple Interest Formula	
$$i = prt$$	Examples 1–5, pages 585–587
United States Rule	
If a partial payment is made on a loan, interest is computed on the principal from the first day of the loan until the date of the partial payment. The partial payment is used to pay the interest first; then the rest of the payment is used to reduce the principal.	Discussion, page 587, Example 8, pages 589–590
Banker's Rule	
When computing interest, a month is considered to have 30 days, a year is considered to have 360 days and any part of a year is the exact number of days.	Discussion, page 585, Examples 7–8, pages 589–590
Section 10.3	
Compound Interest Formula	
$$A = p\left(1 + \frac{r}{n}\right)^{(n \cdot t)}$$	Examples 2–4, pages 596–599
Present Value Formula	
$$p = \frac{A}{\left(1 + \dfrac{r}{n}\right)^{(n \cdot t)}}$$	Discussion page 599, Example 5, page 600

Section 10.4

Installment Payment Formula

$$m = \frac{p\left(\dfrac{r}{n}\right)}{1 - \left(1 + \dfrac{r}{n}\right)^{(-n \cdot t)}}$$

Example 2, pages 606–607

Actuarial Method

$$u = \frac{n \cdot P \cdot V}{100 + V}$$

Example 5, pages 609–610

Section 10.5

Mortgage Principal and Interest Payment Formula

$$m = \frac{p\left(\dfrac{r}{n}\right)}{1 - \left(1 + \dfrac{r}{n}\right)^{(-n \cdot t)}}$$

Example 3, pages 623

Section 10.6

Ordinary Annuity Formula

$$A = \frac{p\left[\left(1 + \dfrac{r}{n}\right)^{(n \cdot t)} - 1\right]}{\left(\dfrac{r}{n}\right)}$$

Discussion, pages 631–632, Example 1, page 632

Sinking Fund Payment Formula

$$p = \frac{A\left(\dfrac{r}{n}\right)}{\left(1 + \dfrac{r}{n}\right)^{(n \cdot t)} - 1}$$

Discussion, pages 633–634, Example 2, page 634

CHAPTER 10   Review Exercises

10.1

In Exercises 1–6, change the number to a percent. Round your answer to the nearest tenth of a percent.

1. $\dfrac{7}{20}$ **2.** $\dfrac{7}{12}$ **3.** $\dfrac{5}{8}$

4. 0.041   **5.** 0.0098   **6.** 3.141

In Exercises 7–12, change the percent to a decimal number.

7. 8%   **8.** 22.9%   **9.** 123%

10. $\dfrac{1}{4}$%   **11.** $\dfrac{5}{6}$%   **12.** 0.00045%

13. *Salary Increase* Charlotte had a salary of $41,500 during the first year of a new job and a salary of $42,745 during the second year. Determine the percent increase in Charlotte's salary from the first year to the second year.

14. *Calorie Decrease* Until he began his new wellness program, Mr. Davis consumed 3200 calories per day. He now consumes 2800 calories per day. Determine the percent decrease in the number of calories consumed per day by Mr. Davis.

In Exercises 15–17, solve for the unknown quantity.

15. What percent of 80 is 16?

16. Forty-four is 55% of what number?

17. What is 17% of 540?

18. **Tipping** At Mamason's, Anuj and Renuka's bill comes to $52.19, pretax. If they wish to leave a 15% tip on the pretax amount, how much should they tip the waiter?

19. **Increased Membership** If the number of people in your yoga club increased by 20%, or 8 people, what was the original number of people in the club?

20. **Jelly Beans** A jar contains 1050 jelly beans. Of these jelly beans, 126 are green. What percent of the jelly beans in the jar are green?

10.2

In Exercises 21–24, determine the missing quantity by using the simple interest formula.

21. $p = \$4400$, $r = 3.2\%$, $t = 4$ years, $i = ?$

22. $p = \$2700$, $r = ?$, $t = 100$ days, $i = \$37.50$

23. $p = ?$, $r = 5.5\%$, $t = 3$ years, $i = \$280.50$

24. $p = \$3600$, $r = 3\frac{1}{4}\%$, $t = ?$, $i = \$555.75$

25. **Roof Loan** To replace the roof on his house, Matt borrows $7500 from his brother. The loan was for 24 months and had a simple interest rate of 2.5%. Determine the amount Matt paid his brother on the date of maturity of the loan.

26. **Bank Loan** Nikos borrowed $6000 for 24 months from his bank using an $11\frac{1}{2}\%$ discount note.

   a) How much interest did Nikos pay the bank for the use of the money?

   b) How much did he receive from the bank?

   c) What was the actual rate of interest, to the nearest tenth of a percent?

27. **Bakery Loan** On April 1, The Red Velvet Bakery signed a $8400 bank note with a simple interest rate of 4.5% for 180 days. The company made a partial payment of $2900 on June 1. How much will the bakery owe on the date of maturity?

28. **Garage Loan** On May 15, The Underground Repair Service Garage signed a $6700 bank note with a simple interest rate of 6.1% for 210 days. The company made a partial payment of $3250 on August 31. How much will the garage owe on the date of maturity?

10.3

29. **Comparing Compounding Periods** Determine the amount and the interest when $5000 is invested for 5 years at a 6% interest rate

   a) compounded annually.

   b) compounded semiannually.

   c) compounded quarterly.

   d) compounded monthly.

   e) compounded daily (use $n = 360$).

30. **Certificate of Deposit** Seth invests $2500 in a 2-year certificate of deposit (CD) that pays 1.75% interest rate compounded monthly. What will be the value of the CD in 2 years?

31. **Effective Annual Yield** Determine the effective annual yield of an investment if the interest is compounded daily at an annual rate of 5.6%.

32. **Present Value** How much money should Maggie invest in a 5-year CD that pays a 3.1% interest rate compounded quarterly so the CD will be worth $5500 in 5 years?

10.4

33. **Car Loan** Denise purchased a new car selling for $26,000. Denise paid 30% as a down payment and financed the balance with a 60-month fixed installment loan with an APR of 5.0%.

   a) Determine the down payment.

   b) Determine the amount financed.

   c) Determine the finance charge.

   d) Determine the monthly payment.

34. **Office Equipment** Donna is starting a consulting business and purchased new office equipment and furniture selling for $13,220. Donna paid 20% as a down payment and financed the balance with a 36-month installment loan with an APR of 6%.

   a) Determine the down payment.

   b) Determine the amount financed.

   c) Determine the finance charge.

   d) Determine the monthly payment.

35. **Financing an Airplane** Deepankar bought a new Piper Sport airplane for $140,000. He made a 40% down payment and financed the balance with the dealer on a 60-month installment loan. The monthly payments were $1624.

   a) Determine the down payment.

   b) Determine the amount to be financed.

   c) Determine the total finance charge.

   d) Determine the APR.

36. **Repaying a Loan Early** Bill has a 48-month installment loan with a fixed monthly payment of $176.14. The amount borrowed was $7500. Instead of making his 24th payment, Bill is paying the remaining balance on the loan.

   a) Determine the APR of the installment loan.

b) How much interest will Bill save, computed by the actuarial method?

c) What is the total amount due to pay off the loan?

37. *Installment Loan* Dara's cost for a new wardrobe was $3420. She made a down payment of $860 and financed the balance on a 24-month fixed payment installment loan. The monthly payments are $111.73. Instead of making her 12th payment, Dara decides to pay the total remaining balance and terminate the loan.

a) Determine the APR of the installment loan.

b) How much interest will Dara save, computed by the actuarial method?

c) What is the total amount due to pay off the loan?

38. *Finance Charge Comparison* On June 1, the billing date, Tim had a balance due of $485.75 on his credit card. Tim's transactions during the month of June were

June 4	Payment	$375.00
June 8	Charge: Car repair	370.00
June 21	Charge: Airline ticket	175.80
June 28	Charge: Clothing	184.75

a) Determine the finance charge on July 1 by using the previous balance method. Assume that the interest rate is 1.3% per month.

b) Determine the new account balance on July 1 using the finance charge found in part (a).

c) Determine the average daily balance for the period.

d) Determine the finance charge on July 1 by using the average daily balance method. Assume that the interest rate is 1.3% per month.

e) Determine the new account balance on July 1 using the finance charge found in part (d).

39. *Finance Charge Comparison* On August 5, the billing date, Pat had a balance due of $185.72 on her credit card. Pat's transactions during the month of August were

August 8	Charge: Shoes	$85.75
August 10	Payment	75.00
August 15	Charge: Dry cleaning	72.85
August 21	Charge: Textbooks	275.00

a) Determine the finance charge on September 5 by using the previous balance method. Assume that the interest rate is 1.4% per month.

b) Determine the new account balance on September 5 using the finance charge found in part (a).

c) Determine the average daily balance for the period.

d) Determine the finance charge on September 5 by using the average daily balance method. Assume that the interest rate is 1.4% per month.

e) Determine the new account balance on September 5 using the finance charge found in part (d).

10.5

40. *Building a House* The Drummonds have decided to build a new house. The contractor quoted them a price of $135,700. The taxes on the house will be $3450 per year, and homeowners' insurance will be $350 per year. They have applied for a conventional loan from a local bank. The bank is requiring a 25% down payment, and the interest rate on the loan is 4.5%. The Drummonds' annual income is $64,000. They have more than 10 monthly payments remaining on each of the following: $218 on a car, $120 on new furniture, and $190 on a camper. Their bank will approve a loan that has a total monthly house payment of principal, interest, property taxes, and homeowners' insurance that is less than or equal to 28% of their adjusted monthly income. Determine

a) the required down payment.

b) 28% of their adjusted monthly income.

c) the monthly payment of principal and interest for a 30-year loan.

d) their total monthly payment, including insurance and taxes.

e) Do the Drummonds qualify for the mortgage?

41. *Thirty-Year Mortgage* James purchased a home selling for $89,900 with a 15% down payment. The period of the mortgage is 30 years, and the interest rate is 11.5% with no points. Determine the

a) amount of the down payment.

b) monthly payment of principal and interest.

c) amount of the first payment applied to the principal.

d) total cost of the house.

e) total interest paid.

42. *Adjustable-Rate Mortgage* The Cunninghams purchased a new home for $375,000 with a down payment of $140,000. They obtained a 15-year adjustable rate mortgage with the following terms. The interest rate is based on the one-year Treasury bill rate, which is currently at 1.0%, and the add-on rate, which is 3.5%. The initial rate period is 5 years, and thereafter the interest rate is adjusted once a year and a new monthly mortgage payment is calculated.

a) Determine the Cunninghams' initial ARM rate.

b) Determine the Cunninghams' initial monthly payment for principal and interest.

c) If, after the 5-year initial rate period, the rate of the one-year Treasury bill rises to 2.5%, determine the Cunninghams' new ARM rate.

10.6

43. *Vacation Savings* To save money to take an around- the-world vacation, Kim invests $250 monthly in an ordinary annuity with a 9% interest rate compounded monthly.

Determine the accumulated amount in Kim's annuity after 10 years.

44. *Saving for a New Boat* Raja would like to save $80,000 to buy a new Chris Craft Silver Bullet boat in 5 years. To accumulate $80,000 in 5 years, how much should Raja invest quarterly in a sinking fund with a 3.6% interest rate compounded quarterly?

CHAPTER 10   *Test*

1. Determine the missing quantity by using the simple Interest formula.

   a) $p = \$1600$, $r = 2.7\%$, $t = 27$ months, $i = ?$

   b) $p = \$4200$, $r = 3.2\%$, $t = ?$, $i = \$268.80$

In Exercises 2 and 3, Patty borrowed $1900 from a bank for 18 months to buy a new office computer. The simple interest rate charged is 3.15%.

2. How much interest did Patty pay for the use of the money?

3. What is the amount Patty repaid to the bank on the due date of the loan?

In Exercises 4 and 5, Yolanda received a $5400 loan with a 12.5% interest rate for 90 days on August 1. Yolanda made a payment of $3000 on September 15.

4. How much did she owe the bank on the date of maturity?

5. What total amount of interest did she pay on the loan?

6. Use the compound interest formula to compute the accumulated amount and the interest earned.

	Principal	Time	Rate	Compounded
a)	$7500	2 years	3%	Quarterly
b)	$2500	3 years	6.5%	Monthly

A New Bicycle In Exercises 7–9, Don purchases a new bicycle that sells for $2350 using a bank loan. To finance the loan the bank will require a down payment of 15% and monthly payments of $90.79 for 24 months.

7. How much money will Don borrow from the bank?

8. What finance charge will Don pay the bank?

9. What is the APR?

10. *Actuarial Method* Gino borrowed $7500. To repay the loan, he was scheduled to make 36 monthly installment

payments of $223.10. Instead of making his 24th payment, Gino decides to pay off the loan.

   a) Determine the APR of the installment loan.

   b) How much interest will Gino save (use the actuarial method)?

   c) What is the total amount due to pay off the loan?

11. *Previous Balance Method* Michael's credit card statement shows a balance due of $878.25 on March 23, the billing date. For the period ending on April 23, he had the following transactions.

March 26	Charge: Groceries	$95.89
March 30	Charge: Restaurant bill	68.76
April 3	Payment	450.00
April 15	Charge: Clothing	90.52
April 22	Charge: Eyeglasses	450.85

   a) Determine the finance charge on March 23 by using the previous balance method. Assume that the interest rate is 1.4% per month.

   b) Determine the new account balance on April 23 using the finance charge found in part (a).

   c) Determine the average daily balance for the period.

   d) Determine the finance charge on April 23 by using the average daily balance method. Assume that the interest rate is 1.4% per month.

   e) Determine the new account balance on April 23 using the finance charge found in part (d).

Building a House In Exercises 12–18, the Leungs decided to build a new house. The contractor quoted them a price of $215,000, including the lot. The taxes on the house would be $3200 per year, and homeowners' insurance would cost $450 per year. They have applied for a conventional loan from a bank. The bank is requiring a 15% down payment, and the interest rate is 5.5% with 2 points. The Leung's annual income is $122,740. They have more than 10 monthly payments remaining on each of the following: $220 for a *car, $175 for new furniture, and $210 on a college education loan. Their bank will approve a loan that has a total monthly house payment of principal, interest, property taxes, and homeowners' insurance that is less than or equal to 28% of their adjusted monthly income.*

12. What is the required down payment?

13. Determine the amount paid for points.

14. Determine 28% of their adjusted monthly income.

15. Determine the monthly payments of principal and interest for a 30-year loan.

16. Determine their total monthly payments, including homeowners' insurance and taxes.

17. Do the Leungs meet the requirements for the mortgage?

18. a) Determine the total cost of the house (excluding homeowners' insurance and taxes) after 30 years.

b) How much of the total cost is interest, including points?

19. *Saving for Musical Instruments* To raise money to buy new musical instruments, the rock band *White Noise Reverie* decides to invest $400 monthly in an ordinary annuity with a 3% interest rate compounded monthly. Determine the accumulated amount in the band's annuity after 5 years.

20. *Saving for a New Car* Elisa is a freshman in high school and would like to have $15,000 in 4 years to buy a car for college. How much should she invest monthly in a sinking fund with a 6% interest rate compounded monthly to accumulate $15,000 in 4 years?

◄ In addition to being used to assess gaming outcomes, probability is also vital to many areas of mathematics and science.

11

Probability

What You Will Learn

- Empirical and theoretical probabilities
- Odds
- Expected value
- Tree diagrams
- Probability problems involving the words *or* and *and*
- Conditional probability
- The fundamental counting principle and permutations
- Combinations
- Solving probability problems involving combinations
- Binomial probability

Why This Is Important

Probability is commonly associated with buying a lottery ticket, playing poker, spinning the roulette wheel, or some other form of gambling. Indeed, the branch of mathematics we call *probability* traces its beginnings to games of chance in the sixteenth and seventeenth centuries. While probability is still used to assess gaming outcomes, probability is also vital to many other branches of mathematics and to many fields of science. Probability theory is the foundation for the study of statistics, which we will discuss in Chapter 12, and probability plays a key role in the modern study of chemistry, physics, genetics, computer science, artificial intelligence, and quantum mechanics. Probability is also the basis for actuarial science, which is used throughout the insurance and financial industries, and probability is involved in minimizing risk to businesses when faced with difficult decisions. Anytime analysis involves making a difficult choice, probability can be used to help make the best choice. Throughout this chapter, we will introduce many basic aspects of probability that can help us make informed decisions in our daily lives.

SECTION 11.1   Empirical and Theoretical Probabilities

Upon completion of this section, you will be able to:

- Understand the history and nature of probability.
- Understand empirical probability and the law of large numbers.
- Understand theoretical probability.

How do researchers determine if a new potential drug is superior to an existing drug? How do automobile manufacturers determine the chance that a particular auto part will fail before the warranty expires? Which casino games provide the greatest chance of winning? These and similar questions can be answered once you have an understanding of probability.

Why This Is Important Understanding probability is essential to our daily lives. Probability is used to determine insurance premiums, the length of a warranty, and the expected gain or loss of business ventures. Probability is also important in medical research as well as many other areas.

In this section we introduce some of the basic concepts of probability including empirical and theoretical probability. We also discuss the law of large numbers and how probability was used in the foundation of genetics. We begin with the history of probability.

Profile in Mathematics

Jacob Bernoulli (1654–1705)

Swiss mathematician Jacob Bernoulli was considered a pioneer of probability theory. He was part of a famous family of mathematicians and scientists. In *Ars Conjectandi*, published posthumously in 1713, he proposed that an increased degree of accuracy can be obtained by increasing the number of trials of an experiment. This theorem, called Bernoulli's theorem (of probability), is also known as the law of large numbers. Bernoulli's theorem of fluid dynamics, used in aircraft wing design, was developed by Daniel Bernoulli (1700–1782). Daniel was one of Jacob's younger brother Johann's three sons.

The History and Nature of Probability

The study of probability originated from the study of games of chance. Mathematical problems relating to games of chance were studied by a number of mathematicians of the Renaissance. Italy's Girolamo Cardano (1501–1576) in his *Liber de Ludo Aleae* (book on the games of chance) presents one of the first systematic computations of probabilities. Although it is basically a gambler's manual, many consider it the first book ever written on probability. A short time later, two French mathematicians, Blaise Pascal (1623–1662) and Pierre de Fermat (1601–1665), worked together studying "the geometry of the die." In 1657, Dutch mathematician Christian Huygens (1629–1695) published *De Ratiociniis in Luno Aleae* (on ratiocination in dice games), which contained the first documented reference to the concept of mathematical expectation (see Section 11.3). Swiss mathematician Jacob Bernoulli (1654–1705), whom many consider the founder of probability theory, is said to have fused pure mathematics with the empirical methods used in statistical experiments. The works of Pierre-Simon de Laplace (1749–1827) dominated probability throughout the nineteenth century. Today, probability is involved in many areas of mathematics and science and is used in industrial and business applications.

Next, we introduce some of the definitions essential to our understanding of the nature of probability.

> **Definition: Experiment**
> An **experiment** is a controlled operation that yields a set of results.

Some of the experiments we will study in this section include flipping a coin, rolling a single die, and drawing a card from a standard 52-card deck of playing cards. Another example of an experiment is a medical researcher administering an experimental drug to patients to determine their reactions.

▲ *A die (one of a pair of dice) contains six surfaces, called faces. Each face contains a unique number of dots, from 1 to 6. The sum of the dots on opposite surfaces is 7.*

StatCrunch

Applets Experiment Flip Coin

> **Definition: Outcomes**
> The possible results of an experiment are called its **outcomes**.

The outcomes of flipping a coin are *heads* or *tails*. The outcomes of rolling a single die are the numbers 1, 2, 3, 4, 5, and 6. The outcomes of drawing a single card from a standard deck of playing cards are the 52 different cards. The possible outcomes from administering an experimental drug may be a favorable reaction, no reaction, or an adverse reaction.

> **Definition: Event**
> An **event** is a subcollection of the outcomes of an experiment.

When a coin is flipped, one example of an event would be getting a head. A different event would be getting a tail. When a single die is rolled, there are many different events such as rolling an even number, rolling a number less than 4, rolling a 5, and rolling a number greater than 2. When a card is drawn from a standard deck of playing cards, there are many different events such as drawing an ace, drawing a face card, drawing a club, or drawing a red card. When a researcher administers an experimental drug, one example of an event is the patient has a favorable reaction to the drug.

Once we understand the terms *experiment*, *outcome*, and *event*, we can discuss probability. The *probability* of an event is a numeric measure of the likelihood that the event will occur. In probability, we often symbolize an event with the letter E. Given an event E, we will symbolize the probability of E by $P(E)$, which is read "P of E."

Probability is classified as either *empirical* (experimental) or *theoretical* (mathematical). *Empirical probability* is the relative frequency of occurrence of an event and is determined by actual observations of an experiment. Determining the chance of something happening in the future by observing past results is called empirical probability. *Theoretical probability* is determined through a study of the possible *outcomes* that can occur for the given experiment.

Empirical Probability and the Law of Large Numbers

We will first briefly discuss empirical probability. The emphasis later in this section and in the remaining sections will be on theoretical probability. Following is the formula for computing empirical probability, or relative frequency.

> **Empirical Probability (Relative Frequency)**
> The **empirical probability** of an event, $P(E)$, can be determined by the following formula.
>
> $$P(E) = \frac{\text{number of times event } E \text{ has occurred}}{\text{total number of times the experiment has been performed}}$$

The probability of an event, whether empirical or theoretical, is always a number between 0 and 1, inclusive, and may be expressed as a decimal number or a fraction. An empirical probability of 0 indicates that the event has never occurred. An empirical probability of 1 indicates that the event has always occurred.

Did You Know?

Batting Averages

▲ *Mookie Betts*

If Mookie Betts of the Boston Red Sox gets three hits in his first three at bats of the season, he is batting a thousand (1.000). Over the course of the 162 games of the season (with three or four at bats per game), however, his batting average will fall closer to 0.346 (his 2018 major league leading batting average). In 2018, out of 520 at bats, Betts had 180 hits, an average above all other players' but much less than 1.000. His batting average is a relative frequency (or empirical probability) of hits to at bats. It is only the long-term average that we take seriously because it is based on the law of large numbers.

Example 1 *Selecting a Beverage*

At a Little League concession stand, the last 40 beverage purchases were as follows: 16 bottles of water, 14 bottles of sports drink, and 10 bottles of soda. Using this information, determine the empirical probability that the next beverage purchased is a bottle of

a) water.   b) sports drink.   c) soda.

Solution

a) Let E be the event that a bottle of water is purchased. Then

$$P(E) = \frac{\text{number of times } E \text{ has occured}}{\text{total number of times the experiment has been performed}}$$

$$= \frac{16}{40} = \frac{2}{5}$$

b) Let E be the event that a bottle of sports drink is purchased. Then

$$P(E) = \frac{14}{40} = \frac{7}{20}$$

c) Let E be the event that a bottle of soda is purchased. Then

$$P(E) = \frac{10}{40} = \frac{1}{4}$$

In Example 1, we wrote our answers as fractions in lowest terms. In our next example, we will write our answers using decimal numbers. Throughout this chapter, we will be working with both fractions and decimal numbers. Section 5.3 provides a review of fractions and decimal numbers.

Example 2 *Blood Pressure Reduction*

A pharmaceutical company is testing a drug that is supposed to help reduce high blood pressure. The drug is given to 500 individuals with the following outcomes.

Blood pressure reduced	Blood pressure unchanged	Blood pressure increased
379	62	59

If this drug is given to an individual, determine the empirical probability that the person's blood pressure is (a) reduced, (b) unchanged, (c) increased. Write your answer as a decimal number.

Solution

a) Let E be the event that the blood pressure is reduced.

$$P(E) = \frac{379}{500} = 0.758$$

b) Let E be the event that the blood pressure is unchanged.

$$P(E) = \frac{62}{500} = 0.124$$

c) Let E be the event that the blood pressure is increased.

$$P(E) = \frac{59}{500} = 0.118$$

In general, empirical probability is used when probabilities cannot be theoretically calculated. Empirical probabilities are used when there is enough data from past events to reasonably predict similar future events. For example, life insurance companies use data gathered over many years and empirical probability to determine the likelihood that an individual in a certain profession, with certain risk factors, lives to be age 75. Empirical probability is related to theoretical probability, which we will discuss later in this section, through the law of large numbers, which we will define shortly.

It is reasonable to accept that if a "fair coin" is tossed many, many times, it will be heads approximately half of the time. Intuitively, we can guess that the probability that a fair coin will be heads is $\frac{1}{2}$. Does that mean that if a coin is tossed twice, it will be heads exactly once? If a fair coin is tossed 10 times, will there necessarily be five heads? The answer is clearly no. What, then, does it mean when we state that the probability that a fair coin will be heads is $\frac{1}{2}$? To answer this question, let's examine Table 11.1, which shows what may occur when a fair coin is tossed a given number of times.

"The laws of probability, so true in general, so fallacious in particular."
Edward Gibbon, 1796

StatCrunch

Applets Simulation Coin Flipping

Table 11.1

Number of Tosses	Expected Number of Heads	Actual Number of Heads Observed	Relative Frequency of Heads
10	5	4	$\frac{4}{10} = 0.4$
100	50	45	$\frac{45}{100} = 0.45$
1000	500	546	$\frac{546}{1000} = 0.546$
10,000	5000	4852	$\frac{4852}{10,000} = 0.4852$
100,000	50,000	49,770	$\frac{49,770}{100,000} = 0.49770$

The far right column of Table 11.1, the relative frequency of heads, is a ratio of the number of heads observed to the total number of tosses of the coin. The relative frequency is the empirical probability, as defined earlier. Note that as the number of tosses increases, the relative frequency of heads gets closer and closer to $\frac{1}{2}$, or 0.5, which is what we expect.

The nature of probability is summarized by the law of large numbers.

> **Definition: Law of Large Numbers**
> The **law of large numbers** states that probability statements apply in practice to a large number of trials, not to a single trial. It is the relative frequency over the long run that is accurately predictable, not individual events or precise totals.

What does it mean to say that the probability of rolling a 2 with a single die is $\frac{1}{6}$? It means that over the long run, on the average, one of every six rolls will result in a 2.

Next, we will study how empirical probability and the law of large numbers were used in the founding of the study of genetics.

Empirical Probability in Genetics

Using empirical probability, Gregor Mendel (1822–1884) developed the laws of heredity by crossbreeding different types of "pure" pea plants and observing the relative frequencies of the resulting offspring. These laws became the foundation for the study of genetics. For example, when he crossbred a pure yellow pea plant and a pure green pea plant, the resulting offspring (the first generation) were always yellow; see Fig. 11.1(a). When he crossbred a pure round-seeded pea plant and a pure wrinkled-seeded pea plant, the resulting offspring (the first generation) were always round; see Fig. 11.1(b).

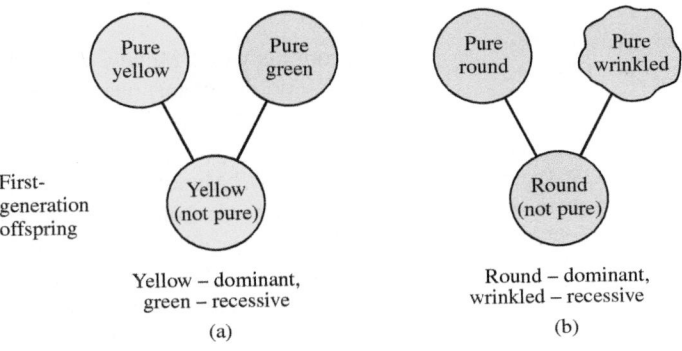

Figure 11.1

Mendel called traits such as yellow color and round seeds *dominant* because they overcame or "dominated" the other trait. He labeled the green color and the wrinkled traits *recessive*.

Mendel then crossbred the offspring of the first generation. The resulting second-generation offspring had both the dominant and the recessive traits of their grandparents; see Fig. 11.2(a) and 11.2(b). What's more, these traits always appeared in approximately a 3 to 1 ratio of dominant to recessive.

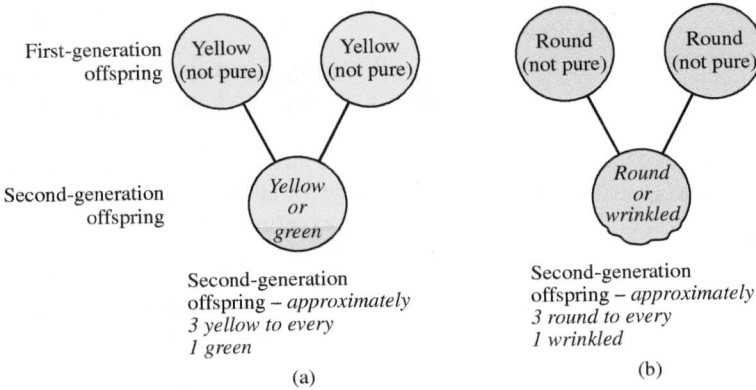

Figure 11.2

Table 11.2 on page 650 lists some of the actual results of Mendel's experiments with pea plants. Note that the ratio of dominant trait to recessive trait in the second-generation offspring is about 3 to 1 for each experiment. The empirical probability of the dominant trait has also been calculated. How would you determine the empirical probability of the recessive trait?

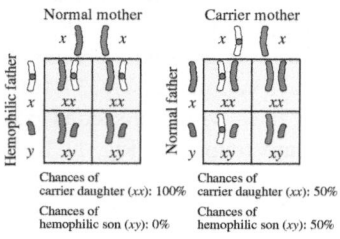

Normal mother · Carrier mother

Chances of carrier daughter (xx): 100% · Chances of carrier daughter (xx): 50%
Chances of hemophilic son (xy): 0% · Chances of hemophilic son (xy): 50%

The effect of genetic inheritance is dramatically demonstrated by the occurrence of hemophilia in some of the royal families of Europe. Great Britain's Queen Victoria (1819–1901) was the initial carrier. The disease was subsequently introduced into the royal lines of Prussia, Russia, and Spain through the marriages of her children.

Hemophilia, a recessive sex-linked disease, keeps blood from clotting. As a result, even a minor bruise or cut can be dangerous. Females have a second gene that enables the blood to clot, which blocks the effects of the recessive carrier gene; males do not. So even though both males and females may be carriers, the disease afflicts only males.

In humans, genes are located on 23 pairs of *chromosomes*. Each parent contributes one member of each pair to a child. The gene that affects blood clotting is carried on the x chromosome. Females have two x chromosomes; males have one x chromosome and one y chromosome. Thus, on the diagram above, xx represents a female and xy represents a male. A female with the defective gene, symbolized by ⬚, will be a carrier, whereas a male with the defective gene will have hemophilia.

Table 11.2 Second-Generation Offspring

Dominant Trait	Number with Dominant Trait	Recessive Trait	Number with Recessive Trait	Ratio of Dominant to Recessive	P (Dominant Trait)
Yellow seeds	6022	Green seeds	2001	3.01 to 1	$\frac{6022}{8023} \approx 0.75$
Round seeds	5474	Wrinkled seeds	1850	2.96 to 1	$\frac{5474}{7324} \approx 0.75$

From his work, Mendel concluded that the sex cells (now called gametes) of the pure yellow (dominant) pea plant carried some factor that caused the offspring to be yellow and that the gametes of the green variety had a variant factor that "induced the development of green plants." In 1909, Danish geneticist W. Johannsen called these factors "genes." Mendel's work led to the understanding that each pea plant contains two genes for color, one that comes from the mother and the other from the father. If the two genes are alike—for instance, both for yellow plants or both for green plants—the plant will be that color. If the genes for color are different, the plant will grow the color of the dominant gene. Thus, if one parent contributes a gene for the plant to be yellow (dominant) and the other parent contributes a gene for the plant to be green (recessive), the plant will be yellow.

Theoretical Probability

Recall that the possible results of an experiment are called outcomes. When you roll a die, the possible outcomes are 1, 2, 3, 4, 5, and 6. If the die is a fair die it is *equally likely* that you will roll any one of the six possible numbers.

> **Definition: Equally Likely Outcomes**
> If each outcome of an experiment has the same chance of occurring as any other outcome, we say that the outcomes are **equally likely outcomes**.

When we flip a fair coin, two outcomes are possible: heads or tails. Since the chance of flipping a head is the same as flipping a tail, the results of this experiment are equally likely outcomes. When we draw a single card from a standard deck of cards, there are 52 different cards that can be drawn. Since the chance of drawing any one card is the same for all 52 cards, the results of this experiment are equally likely outcomes.

If an experiment has *equally likely outcomes*, the theoretical probability of event E, symbolized by P(E), may be determined with the following formula.

> **Theoretical Probability**
> The **theoretical probability** of an event, P(E), can be determined using the following formula.
> $$P(E) = \frac{\text{number of outcomes favorable to } E}{\text{total number of possible outcomes}}$$

In the remainder of this chapter, the word probability will refer to theoretical probability.

Example 3 illustrates how to use the theoretical probability formula.

Example 3 *Determining Theoretical Probabilities*

A fair die is rolled. Determine the probability of rolling

a) a 2.

b) an even number.

c) a number greater than 3.

d) a 7.

e) a number less than 7.

StatCrunch

Applets Experiment Roll Die

Solution

a) There are six possible equally likely outcomes: 1, 2, 3, 4, 5, or 6. The event of rolling a 2 can occur in only one outcome.

$$P(2) = \frac{\text{number of outcomes that will result in a 2}}{\text{total number of possible outcomes}} = \frac{1}{6}$$

b) The event of rolling an even number can occur in three outcomes: 2, 4, or 6.

$$P(\text{even number}) = \frac{\text{number of outcomes that result in an even number}}{\text{total number of possible outcomes}}$$

$$= \frac{3}{6} = \frac{1}{2}$$

c) The event of rolling a number greater than 3 can occur in three outcomes: 4, 5, or 6.

$$P(\text{number greater than 3}) = \frac{3}{6} = \frac{1}{2}$$

d) The event of rolling a 7 can occur in zero outcomes. Thus, the event cannot occur and the probability is 0.

$$P(7) = \frac{0}{6} = 0$$

e) The event of rolling a number less than 7 can occur in 6 outcomes: 1, 2, 3, 4, 5, or 6. Thus, the event must occur and the probability is 1.

$$P(\text{number less than 7}) = \frac{6}{6} = 1$$

Four important facts about probability follow.

Important Probability Facts

1. The probability of an event that cannot occur is 0.
2. The probability of an event that must occur is 1.
3. Every probability is a number between 0 and 1 inclusive; that is, $0 \leq P(E) \leq 1$.
4. The sum of the probabilities of all possible outcomes of an experiment is 1.

Example 4 *Choosing One Bird from a List*

The names of 9 birds and their food preferences are listed in Table 11.3. Each of the 9 birds' names is listed on a slip of paper, and the 9 slips are placed in a bag. One slip is to be randomly selected from the bag. Determine the probability that the slip contains the name of

a) a finch (any type listed).

b) a bird that has a high attractiveness to cracked corn.

c) a bird that has a low attractiveness to peanut kernels, *and* a low attractiveness to cracked corn, *and* a high attractiveness to black-striped sunflower seeds.

d) a bird that has a medium attractiveness to either peanut kernels *or* cracked corn (or both).

Table 11.3 Birds and Their Food Preferences

Bird	Peanut Kernels	Cracked Corn	Black-Striped Sunflower Seeds
American goldfinch	L	L	H
Blue jay	H	M	H
Chickadee	M	L	H
Common grackle	M	H	H
Evening grosbeak	L	L	H
House finch	M	L	H
House sparrow	L	M	M
Mourning dove	L	M	M
Northern cardinal	L	L	H

Source: *How to Attract Birds* (Ortho Books)

Note: H = high attractiveness; M = medium attractiveness; L = low attractiveness.

Solution

a) Two of the 9 birds listed are finches (American goldfinch and house finch).

$$P(\text{finch}) = \frac{2}{9}$$

b) One of the 9 birds listed has a high attractiveness to cracked corn (common grackle).

$$P(\text{high attractiveness to cracked corn}) = \frac{1}{9}$$

c) Reading across the rows reveals that 3 birds have a low attractiveness to peanut kernels, a low attractiveness to cracked corn, and a high attractiveness to black-striped sunflower seeds (American goldfinch, evening grosbeak, and northern cardinal).

$$P\left(\begin{array}{l}\text{low attractiveness to peanuts, and low to} \\ \text{corn, and high to black-striped sunflower sweeds}\end{array}\right) = \frac{3}{9} = \frac{1}{3}$$

d) Six birds have a medium attractiveness to either peanut kernels or to cracked corn (or both). They are the blue jay, chickadee, common grackle, house finch, house sparrow, and mourning dove.

$$P(\text{medium attractiveness to peanut kernels or cracked corn}) = \frac{6}{9} = \frac{2}{3}$$ ∎

In any experiment, an event must either occur or not occur. *The sum of the probability that an event will occur and the probability that it will not occur is 1.* Thus, for any event *A* we conclude that

---

The Sum of the Probabilities Equals 1

$$P(A) + P(\text{not } A) = 1$$

or

$$P(\text{not } A) = 1 - P(A)$$

For example, if the probability that event *A* will occur is $\frac{5}{12}$, the probability that event *A* will not occur is $1 - \frac{5}{12}$, or $\frac{7}{12}$. Similarly, if the probability that event *A* will not occur is 0.3, the probability that event *A* will occur is $1 - 0.3 = 0.7$, or $\frac{7}{10}$. We make use of this concept in Example 5.

StatCrunch

Applets Experiment Draw Cards

Example 5 *Drawing One Card from a Deck*

A standard deck of 52 playing cards is shown in Fig. 11.3. The deck consists of four suits: hearts, clubs, diamonds, and spades. Each suit has 13 cards, including numbered cards ace (1) through 10 and three face cards, the jack, the queen, and the king. Hearts and diamonds are red cards; clubs and spades are black cards. There are 12 face cards, consisting of 4 jacks, 4 queens, and 4 kings. One card is to be randomly drawn from the deck of cards. Determine the probability that the card drawn is

a) an 8.

b) not an 8.

c) a club.

d) a jack *or* queen *or* king (a face card).

e) a heart *and* a spade.

f) a card greater than 5 *and* less than 9.

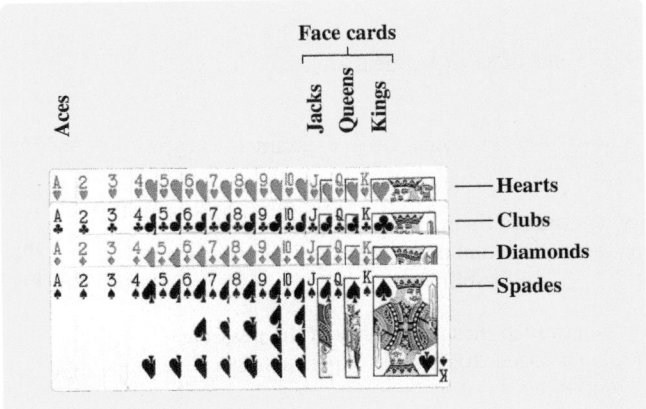

Figure 11.3

Solution

a) There are four 8's in a deck of 52 cards.

$$P(8) = \frac{4}{52} = \frac{1}{13}$$

b) $P(\text{not an } 8) = 1 - P(8) = 1 - \frac{1}{13} = \frac{12}{13}$

This probability could also have been determined by noting that there are 48 cards that are not 8's in a deck of 52 cards.

$$P(\text{not an } 8) = \frac{48}{52} = \frac{12}{13}$$

c) There are 13 clubs in the deck.

$$P(\text{club}) = \frac{13}{52} = \frac{1}{4}$$

d) There are 4 jacks, 4 queens, and 4 kings, or a total of 12 face cards.

$$P(\text{jack } or \text{ queen } or \text{ king}) = \frac{12}{52} = \frac{3}{13}$$

e) The word *and* means that *both* events must occur. Since it is not possible to select one card that is both a heart and a spade the probability is 0.

$$P(\text{heart and spade}) = \frac{0}{52} = 0$$

f) The cards that are both greater than 5 and less than 9 are 6's, 7's, and 8's. There are four 6's, four 7's, and four 8's, or a total of 12 cards.

$$P(\text{greater than } 5 \text{ } and \text{ less than } 9) = \frac{12}{52} = \frac{3}{13}$$

SECTION 11.1
Exercises

Warm Up Exercises

In Exercises 1–10, fill in the blank with an appropriate word, phrase, or symbol(s).

1. A controlled operation that yields a set of results is called a(n) _____.

2. The possible results of an experiment are called its _____.

3. A subcollection of the outcomes of an experiment is called a(n) _____.

4. Probability determined by the relative frequency of occurrence of an event, or actual observations of an experiment is called _____ probability.

5. Probability determined through a study of the possible outcomes that can occur for a given experiment is called _____ probability.

6. If each outcome of an experiment has the same chance of occurring as any other outcome, the outcomes are _____ likely outcomes.

7. a) The probability of an event that cannot occur is _____.

b) The probability of an event that must occur is _____.

c) Every probability must be a number between _____ and _____ inclusive.

8. The sum of the probabilities of all possible outcomes of an experiment is _____.

9. For any event A, $P(A) + P(\text{not } A) = $ _____.

10. If the probability that an event occurs is 0.2, the probability that the event does not occur is _____.

Practice the Skills/Problem Solving

In Exercises 11–14, write your answer as a fraction in lowest terms.

11. *Boat Rentals* The last 30 boat rentals at Green Lakes Boat Rentals were 14 sunfish, 10 kayaks, and 6 rowboats. Use this information to determine the empirical probability that the next boat rental is a

a) sunfish.   b) kayak.   c) rowboat.

12. ***Music Purchases*** At the Jackpot Records store in Portland, Oregon, 60 people entering the store were randomly selected and were asked to choose their favorite type of music. Of the 60, 12 chose rock, 16 chose country, 8 chose classical, and 24 chose something other than rock, country, or classical. Determine the empirical probability that the next person entering the store favors

   a) rock music.   b) country music.

   c) something other than rock, country, or classical music.

13. ***Veterinarian*** In a given week, a veterinarian treated the following animals.

Animal	Number Treated
Dog	45
Cat	40
Bird	15
Rabbit	5

   Determine the empirical probability that the next animal she treats is

   a) a dog.   b) a cat.   c) a rabbit.

14. ***Bicycle Repair*** In a given week, Paul's Bike Shop repaired bicycles with the following brand names.

Bicycle Brand	Number Repaired
Giant	36
Trek	33
Specialized	18
Cannondale	15

   Determine the empirical probability that the next bicycle repaired is a

   a) Giant.   b) Trek.   c) Cannondale.

For Exercises 15–18, write your answer as a decimal number rounded to four decimal places. In each exercise, assume the trend continues.

15. ***Top-Selling Video Games*** The following table shows the number of video games sold worldwide for the five highest selling video games, all formats combined, in 2018.

Title	Units Sold
Red Redemption 2	19,711,000
Call of Duty: Black Ops IIII	14,168,000
FIFA 19	12,119,000
Super Smash Bros.	8,951,000
Spider-Man	8,758,000
Total	63,707,000

Source: VGCartz.com

If one of these five video games is purchased, determine the empirical probability the video game purchased is

   a) *Red Redemption 2.*

   b) *FIFA 19.*

   c) *Spider-Man.*

16. ***Top Movies*** The following table shows the top five attended movies, at movie theaters, in the United States during 2018.

Movie	2018 Movie Attendance
Black Panther	70,006,000
Avengers: Infinity War	67,881,500
Incredibles 2	60,858,200
Jurassic World: Fallen Kingdom	41,772,000
Aquaman	33,506,200
Total	274,023,900

Source: BoxOfficeMojo.com

If a person attends one of these five movies, determine the empirical probability the movie is

   a) *Black Panther.*   b) *Incredibles 2.*

   c) *Aquaman.*

17. ***U.S. Auto Sales*** The following graph was created using StatCrunch with data from AutomobileMag.com. The graph shows U.S. automobile sales, in automobiles sold, by manufacturer, for the year 2018. The total sales for the year 2018 were 17,300,000 automobiles.

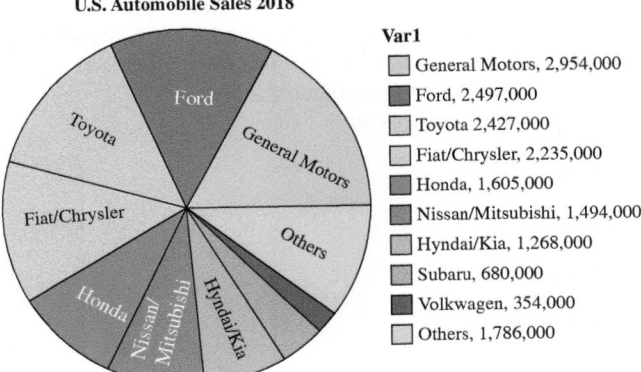

U.S. Automobile Sales 2018

Var1
- General Motors, 2,954,000
- Ford, 2,497,000
- Toyota 2,427,000
- Fiat/Chrysler, 2,235,000
- Honda, 1,605,000
- Nissan/Mitsubishi, 1,494,000
- Hyndai/Kia, 1,268,000
- Subaru, 680,000
- Volkwagen, 354,000
- Others, 1,786,000

Source: AutomobileMag.com

If an automobile is sold in the United States, determine the empirical probability that the automobile sold is manufactured by

a) General Motors.

b) a manufacturer other than General Motors.

c) Honda.

d) a manufacturer other than Honda.

18. *National Park Visits* The following graph was created using **StatCrunch** with data from the website NPS.gov.

The graph shows the number of visits to the top 10 most visited national parks for the year 2018. The total number of visits to these 10 parks was 47,933,000 visits.

National Park Visits 2018

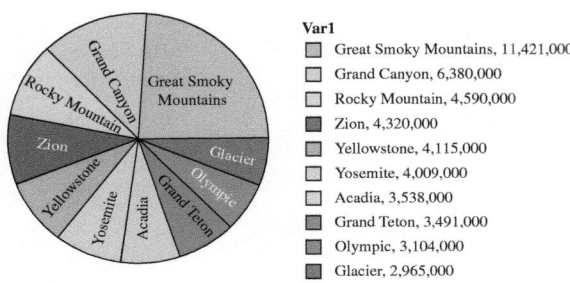

Var1
- Great Smoky Mountains, 11,421,000
- Grand Canyon, 6,380,000
- Rocky Mountain, 4,590,000
- Zion, 4,320,000
- Yellowstone, 4,115,000
- Yosemite, 4,009,000
- Acadia, 3,538,000
- Grand Teton, 3,491,000
- Olympic, 3,104,000
- Glacier, 2,965,000

Source: NPS.gov

If a park on this list is visited, determine the empirical probability that the park visited is

a) Great Smoky Mountains National Park.

b) a park other than Great Smoky Mountains National Park.

c) Yellowstone National Park.

d) a park other than Yellowstone National Park.

19. *Medical Research* A medical researcher administers an experimental medical treatment to 200 patients. The patients in the study are categorized by blood types *A, B, AB,* and *O*. The researcher observed that the treatment had a favorable outcome for 24 of the 80 patients with blood type *A*, 11 of the 22 patients with blood type *B*, 8 of the 8 patients with blood type *AB*, and none of the 90 patients with blood type *O*. Determine the empirical probability of a favorable outcome for those patients with

a) blood type *A*.

b) blood type *B*.

c) blood type *AB*.

d) blood type *O*.

20. *Alternative Veterinarian Medicine* Eva Marie, a veterinarian researcher, is working on a new non-antibiotic medicine to treat bacterial infections in animals. While conducting a study, Eva Marie observed that the medicine worked on 14 of 20 cats, 18 of 30 dogs, 0 of 24 horses, and 35 of 35 cows. Determine the empirical probability that the medicine will work for

a) cats. **b)** dogs.

c) horses. **d)** cows.

21. *Cell Biology Experiment* An experimental serum was injected into 500 guinea pigs. Initially, 150 of the guinea pigs had circular cells, 250 had elliptical cells, and 100 had irregularly shaped cells. After the serum was injected, none of the guinea pigs with circular cells were affected, 50 with elliptical cells were affected, and all those with irregular cells were affected. Determine the empirical probability that a guinea pig with (a) circular cells, (b) elliptical cells, and (c) irregular cells will be affected by injection of the serum.

22. *Mendel's Experiment* In one of his experiments (see pages 649–650), Mendel crossbred not pure purple flower pea plants. These purple pea plants had two traits for flowers, purple (dominant) and white (recessive). The result of this crossbreeding was 705 second-generation plants with purple flowers and 224 second-generation plants with white flowers. Determine the empirical probability of a second-generation plant having

a) white flowers. **b)** purple flowers.

In Exercises 23–70, write the theoretical probability answer as a fraction in lowest terms.

23. *Multiple-Choice Question* A multiple-choice test question has five possible choices. If you randomly select one of the choices, what is the probability that you select

a) the correct choice? **b)** an incorrect choice?

If you can eliminate two of the five choices and randomly select one of the remaining choices, what is the probability that you select

c) the correct choice? **d)** an incorrect choice?

24. *Ball Choice* A bag has 3 red, 2 green, and 4 orange balls. If a ball is randomly selected from the bag, what is the theoretical probability that it is

a) green? **b)** not green?

c) orange? **d)** not orange?

Drawing a Card In Exercises 25–34, one card is drawn from a standard 52-card deck. Determine the probability that the card selected is

25. a club.

26. a queen.

27. not a club.

28. not a queen.

29. a face card.

30. not a face card.

31. a heart and a jack.

32. a queen and a club.

33. a red card and a king.

34. a jack and a black card.

Spin the Spinner In Exercises 35–38, assume that the spinner cannot land on a line. Determine the probability that the spinner lands on (a) red, (b) green, (c) yellow, (d) not yellow.

35.

36.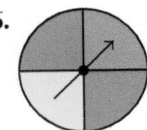

37.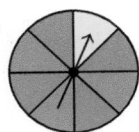

38.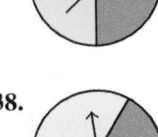

Klondike Bars In Exercises 39–44, a cooler contains 100 Klondike ice cream bars. There are 40 Vanilla bars, 20 Mint Chocolate Chip bars, 30 Krunch bars, and 10 Rocky Road bars. If one bar is randomly selected determine the probability that the bar selected is

39. a Mint Chocolate Chip bar.

40. not a Mint Chocolate Chip bar.

41. a Rocky Road bar.

42. not a Rocky Road bar.

43. a Vanilla, Mint Chocolate Chip, or Krunch bar.

44. a Vanilla, Krunch, or Rocky Road bar.

Wheel of Fortune In Exercises 45–50, use the small replica of the Wheel of Fortune.

If the wheel is spun, determine the probability that the indicated sector will stop under the pointer.

45. $2500

46. A number greater than $500

47. Lose a Turn or Bankrupt

48. Not Bankrupt

49. A number less than $3000

50. Surprise, Lose a Turn, or Bankrupt

TENNESSEE In Exercises 51–56, each individual letter of the word TENNESSEE is placed on a piece of paper and all 9 pieces of paper are placed in a hat. If one letter is randomly selected from the hat, determine the probability that

51. the letter *S* is selected.

52. the letter *S* is not selected.

53. a consonant is selected.

54. the letter *T* or *E* is selected.

55. the letter *W* is selected.

56. the letter *V* is not selected.

Soda In Exercises 57–62, a cooler contains 66 cans of soda. There are 24 cans of Coca-Cola, 12 cans of Diet Coke, 9 cans of Dr Pepper, 15 cans of Mountain Dew, and 6 cans of Diet Pepsi. If one can of soda is randomly selected, determine the probability that the can is

57. Coca-Cola

58. Diet Pepsi.

59. not Coca-Cola.

60. not Diet Pepsi.

61. a diet soda.

62. Dr Pepper or Mountain Dew.

Salsa In Exercises 63–70, refer to the table, on page 658, which contains information about 36 jars of salsa available at the Meals for Millions Food Pantry.

Brand	Mild	Medium	Hot	Total
Frontera	6	3	1	10
Herdez	5	4	2	11
La Victoria	7	8	0	15
Total	18	15	3	36

If a jar is randomly selected, determine the probability that the salsa is

63. Frontera.

64. a medium salsa.

65. not Frontera.

66. not a medium salsa.

67. a mild or medium salsa.

68. Herdez or La Victoria.

69. La Victoria hot salsa.

70. Herdez hot salsa.

Concept/Writing Exercises

71. The theoretical probability of a coin flip being heads is $\frac{1}{2}$. Does this probability mean that if a coin is flipped two times, one flip will be heads? If not, what does it mean?

72. The theoretical probability of rolling a 4 on a fair die is $\frac{1}{6}$. Does this probability mean that if a die is rolled six times one 4 will appear? If not, what does it mean?

73. a) Explain how you would determine the empirical probability of rolling a 5 on a die.

b) What do you believe is the empirical probability of rolling a 5?

c) Determine the empirical probability of rolling a 5 by rolling a die 40 times.

74. To determine premiums, life insurance companies must compute the probable date of death. On the basis of a great deal of research, Mr. Duncan, age 36, is expected to live another 43.21 years. Does this determination mean that Mr. Duncan will live until he is 79.21 years old? If not, what does it mean?

Challenge Problems/Group Activities

Bean Bag Toss *In Exercises 75–82, a bean bag is randomly thrown onto the square table top shown below and does not touch a line.*

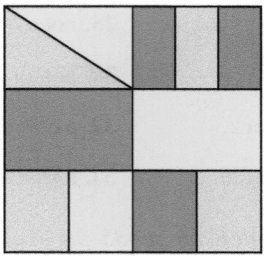

Determine the probability that the bean bag lands on

75. a red area.

76. a green area.

77. an area that is not red.

78. an area that is not green.

79. a yellow area.

80. a red or green area.

81. a yellow or green area.

82. a red or yellow area.

83. ***Flip a Coin*** Flip a coin 50 times and record the results. Determine the empirical probability of flipping

a) a head.

b) a tail.

c) Does the probability of flipping a head appear to be the same as flipping a tail?

84. ***Pair of Dice*** Roll a pair of dice 60 times and record the sums. Determine the empirical probability of rolling a sum of

a) 2.

b) 7.

c) Does the probability of rolling a sum of 2 appear to be the same as the probability of rolling a sum of 7?

85. ***Roll a Die*** Roll a die 50 times and record the results. Determine the empirical probability of rolling

a) a 1.

b) a 4.

c) Does the probability of rolling a 1 appear to be the same as the probability of rolling a 4? Explain.

86. ***Two Coins*** Flip two coins 50 times and record the number of times exactly one head was obtained. Determine the empirical probability of flipping exactly one head.

Recreational Mathematics

87. ***Cola Preference*** Can people distinguish Coke from Pepsi? Which do they prefer?

a) Design an experiment to determine the empirical probability that a randomly selected person can select Coke when given samples of both Coke and Pepsi.

b) Perform the experiment in part (a) and determine the empirical probability that a person can select Coke.

c) Determine the empirical probability that a person randomly selected will prefer Coke over Pepsi.

88. *Dice* On a die, the sum of the dots on the opposite faces is seven. Two six-sided dice are placed together on top of one another, on a table, as shown in the figure below. The top and bottom faces of the bottom die and the bottom face of the top die cannot be seen. If you walk around the table, what is the sum of all the dots on all the visible faces of the dice?

Research Activities

89. *Insurance* Write a paper on how insurance companies use empirical probabilities in determining insurance premiums. An insurance agent may be able to direct you to a source of information.

90. *Bernoulli Family* On page 645, we briefly discuss Jacob Bernoulli. The Bernoulli family produced several prominent mathematicians, including Jacob I, Johann I, and Daniel. Write a paper on the Bernoulli family, indicating some of the accomplishments of each of the three Bernoullis named and their relationship to one another. Indicate which Bernoulli the Bernoulli numbers are named after, which Bernoulli the Bernoulli theorem in statistics is named after, and which Bernoulli the Bernoulli theorem of fluid dynamics is named after.

SECTION 11.2 Odds

Upon completion of this section, you will be able to:

- Understand odds against an event.
- Understand odds in favor of an event.
- Understand how to obtain probabilities from odds and odds from probabilities.

The odds against winning the Mega Millions lottery are about 300 million to 1. The odds against being audited by the IRS this year are about 174 to 1. The odds against the Chicago Cubs winning the World Series this year may be 12 to 1. We see the word *odds* daily in newspapers and magazines and often use it ourselves. Yet there is widespread misunderstanding of its meaning. In this section, we will explain the meaning of odds. We will also discuss how to determine odds against an event and how to determine odds in favor of an event.

Why This Is Important In addition to applications involving sports and gambling, odds play an important role in many areas of society. Companies use odds when making difficult business decisions. Odds play an important role throughout the financial services industry, especially in all types of insurance. Odds are used when studying the spread of disease and the effectiveness of medical treatments. Odds are also used by social researchers when analyzing the results of surveys. A basic understanding of odds is important for many areas of employment.

Odds Against an Event

The odds given at gambling venues are usually given as *odds against* winning unless they are otherwise specified. The *odds against* an event is a ratio of the probability that the event will fail to occur (failure) to the probability that the event will occur (success).

> ### Odds Against an Event
> The following formula may be used to determine the odds against an event.
> $$\text{Odds against event} = \frac{P(\text{event fails to occur})}{P(\text{event occurs})} = \frac{P(\text{failure})}{P(\text{success})}$$

Example 1 *Rolling a 4*

Determine the odds against rolling a 4 on one roll of a fair die.

Solution Before we can determine the odds, we must first determine the probability of rolling a 4 (success) and the probability of not rolling a 4 (failure). When a die is rolled there are six possible outcomes: 1, 2, 3, 4, 5, and 6.

$$P(\text{rolling a 4}) = \frac{1}{6} \qquad P(\text{failure to roll a 4}) = \frac{5}{6}$$

Now that we know the probabilities of success and failure, we can determine the odds against rolling a 4.

$$\text{Odds against rolling a 4} = \frac{P(\text{failure to roll a 4})}{P(\text{rolling a 4})}$$

$$= \frac{\frac{5}{6}}{\frac{1}{6}} = \frac{5}{\cancel{6}} \cdot \frac{\cancel{6}}{1} = \frac{5}{1}$$

The ratio $\frac{5}{1}$ is commonly written as $5:1$ and is read "5 to 1." Thus, the odds against rolling a 4 are 5 to 1. ∎

Timely Tip
The denominators of the probabilities in an odds problem will always divide out, as was shown in Example 1.

In Example 1, when a die is rolled there are 6 possible equally likely outcomes: 1, 2, 3, 4, 5, and 6. If all the equally likely outcomes of an experiment are known, the odds against an event can be determined without calculating probabilities by using the following formula.

> ### Odds Against an Event
> In an experiment with equally likely outcomes, the odds against an event E can also be determined using the following formula.
> $$\text{Odds against event } E = \frac{\text{number of outcomes unfavorable to } E}{\text{number of outcomes favorable to } E}$$

In Example 1, using probabilities we determined the odds against rolling a 4 when a die is rolled. We can also determine the odds against rolling a 4 using the above definition. There are 5 outcomes that are unfavorable to rolling a 4 (1, 2, 3, 5, and 6). There is 1 outcome favorable to rolling a 4 (the 4 itself). Thus,

$$\text{Odds against rolling a 4} = \frac{\text{number of outcomes unfavorable to rolling a 4}}{\text{number of outcomes favorable to rolling a 4}} = \frac{5}{1}$$

Therefore, the odds against rolling a 4 are 5 to 1. If all of the equally likely outcomes of an experiment are known, this alternate procedure for determining the odds against an event is often easier than using probabilities.

In Example 1, we considered the possible outcomes of the die: 1, 2, 3, 4, 5, 6. Over the long run, one of every six rolls will result in a 4, and five of every six rolls will result in a number other than 4. Therefore, if a person is gambling, for each dollar bet in favor of the rolling of a 4, $5 should be bet against the rolling of a 4 if the person is to break even. The person betting in favor of the rolling of a 4 will either lose $1 (if a number other than a 4 is rolled) or win $5 (if a 4 is rolled). The person betting against the rolling of a 4 will either win $1 (if a number other than a 4 is rolled) or lose $5 (if a 4 is rolled). If this game is played for a long enough period, each player theoretically will break even.

Example 2 involves a circle graph that involves percents; see Fig. 11.4, below. Before we discuss Example 2, let us briefly discuss percents. Recall that probabilities are numbers between 0 and 1, inclusive. We can change a percent between 0% and 100% to a probability by writing the percent as a fraction or a decimal number as discussed in Section 10.1. In Fig. 11.4, we see that the green sector of the circle, *Lee*, corresponds to 13%. To change 13% to a probability, we can write $\frac{13}{100}$, or 0.13. Note that both the fraction and the decimal number are numbers between 0 and 1, inclusive.

Example 2 *Blue Jeans*

In a recent survey, students at Etheridge University were asked to name the brand of blue jeans they preferred. The circle graph in Fig. 11.4, which was created using **StatCrunch**, shows the results of the survey. If a student from the survey is randomly selected, determine the odds against the individual preferring

Favorite Brand of Blue Jeans

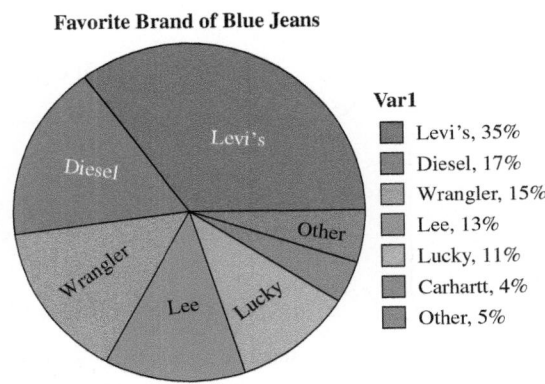

Var1
- Levi's, 35%
- Diesel, 17%
- Wrangler, 15%
- Lee, 13%
- Lucky, 11%
- Carhartt, 4%
- Other, 5%

Figure 11.4

a) Levi's.

b) Wrangler

Solution

a) From the graph, we see that 35%, or $\frac{35}{100} = \frac{7}{20}$, of the students surveyed prefer Levi's blue jeans. Therefore, the probability of a student preferring Levi's blue jeans is $\frac{7}{20}$. The probability of a student not preferring Levi's blue jeans is $1 - \frac{7}{20} = \frac{13}{20}$.

$$\text{Odds against a student preferring Levi's} = \frac{P(\text{student not preferring Levi's})}{P(\text{student preferring Levi's})} = \frac{\frac{13}{20}}{\frac{7}{20}}$$

$$= \frac{13}{20} \cdot \frac{20}{7} = \frac{13}{7}, \text{ or } 13:7$$

Thus, the odds against a student preferring Levi's blue jeans are 13 : 7.

b) From the graph, we see that 15%, or $\frac{15}{100} = \frac{3}{20}$, of the students surveyed prefer Wrangler blue jeans. Therefore, the probability of a student preferring Wrangler blue jeans is $\frac{3}{20}$. The probability of a student not preferring Wrangler blue jeans is $1 - \frac{3}{20} = \frac{17}{20}$.

$$\text{Odds against a student preferring Wrangler} = \frac{P(\text{student not preferring Wrangler})}{P(\text{student preferring Wrangler})} = \frac{\frac{17}{20}}{\frac{3}{20}}$$

$$= \frac{17}{20} \cdot \frac{20}{3} = \frac{17}{3}, \text{ or } 17:3$$

Thus, the odds against a student preferring Wrangler blue jeans are $17:3$. ∎

Odds in Favor of an Event

Although odds are generally given against an event, at times they may be given in favor of an event. The *odds in favor of* an event are expressed as a ratio of the probability that the event will occur to the probability that the event will fail to occur.

> **Odds in Favor of an Event**
> The following formula may be used to determine the odds in favor of an event.
> $$\text{Odds in favor of event} = \frac{P(\text{event occurs})}{P(\text{event fails to occur})} = \frac{P(\text{success})}{P(\text{failure})}$$

If the odds *against* an event are $a:b$, the odds *in favor of* the event are $b:a$.

Example 3 *Selecting a Card*

One card is drawn from a standard 52-card deck. Determine
a) the odds in favor of selecting a queen.
b) the odds against selecting a queen.

Solution

a) There are 52 cards in a standard deck of cards, 4 of which are queens. Therefore, the probability of selecting a queen is $\frac{4}{52}$, or $\frac{1}{13}$. The probability the card selected is not a queen is $1 - \frac{1}{13} = \frac{12}{13}$.

$$\text{Odds in favor of selecting a queen} = \frac{P(\text{selecting a queen})}{P(\text{not selecting a queen})} = \frac{\frac{1}{13}}{\frac{12}{13}}$$

$$= \frac{1}{13} \cdot \frac{13}{12} = \frac{1}{12}, \text{ or } 1:12$$

Thus, the odds in favor of selecting a queen are $1:12$.
b) The odds against selecting a queen are $12:1$. ∎

The odds in favor of an event can also be determined without calculating probabilities by using the following formula.

Odds in Favor of an Event

In an experiment with equally likely outcomes, the odds in favor of an event E can also be determined using the following formula.

$$\text{Odds in favor of event } E = \frac{\text{number of outcomes favorable to } E}{\text{number of outcomes unfavorable to } E}$$

In Example 3(a), the number of outcomes favorable to selecting a queen is 4 (the four queens), and the number of outcomes unfavorable to selecting a queen is 48 (all the other cards in the deck). Thus,

$$\text{Odds in favor of selecting a queen} = \frac{\text{number of outcomes favorable to selecting a queen}}{\text{number of outcomes unfavorable to selecting a queen}} = \frac{4}{48} = \frac{1}{12}$$

Notice both procedures give us the same odds in favor of selecting a queen, 1 to 12.

Determining Probabilities from Odds

When odds are given, either in favor of or against a particular event, it is possible to determine the probability that the event occurs and the probability that the event does not occur. The denominators of the probabilities are determined by adding the numbers in the odds statement. The numerators of the probabilities are the numbers given in the odds statements.

Example 4   Determining Probabilities from Odds

The odds against Ting being admitted to the college of her choice are $9:2$. Determine the probability that (a) Ting is admitted and (b) Ting is not admitted.

Solution

a) We have been given odds against and have been asked to determine probabilities.

$$\text{Odds against being admitted} = \frac{P(\text{fails to be admitted})}{P(\text{is admitted})}$$

Since the odds statement is $9:2$, the denominators of both the probability of success and the probability of failure must be $9 + 2$, or 11. To get the odds ratio of $9:2$, the probabilities must be $\frac{9}{11}$ and $\frac{2}{11}$. Since odds against is a ratio of failure to success, the $\frac{9}{11}$ and $\frac{2}{11}$ represent the probabilities of failure and success, respectively. Thus, the probability that Ting is admitted (success) is $\frac{2}{11}$.

b) The probability that Ting is not admitted (failure) is $\frac{9}{11}$.   ∎

Odds and probability statements are sometimes stated incorrectly. For example, consider the statement, "The odds of being selected to represent the district are 1 in 5." Odds are given using the word *to*, not *in*. Thus, there is a mistake in this statement. The correct statement might be, "The odds of being selected to represent the district are 1 to 5" or "The probability of being selected to represent the district is 1 in 5." Without additional information, it is not possible to tell which statement is the correct interpretation.

SECTION 11.2

Exercises

Warm Up Exercises

In Exercises 1–8, fill in the blank with an appropriate word, phrase, or symbol(s).

1. The ratio of the probability that an event will fail to occur to the probability that the event will occur is called the odds _____ an event.

2. The ratio of the probability that the event will occur to the probability that the event will fail to occur is called the odds _____ an event.

3. If the odds in favor of Haikal winning a horse race are $1:4$, then the odds against Haikal winning the race are _____.

4. If the odds against winning at Splendor are $5:1$, then the odds in favor of winning at Splendor are _____.

5. If the probability that an event will occur is $\frac{2}{3}$ then the probability that the event will not occur is $\frac{1}{3}$, and the odds in favor of the event occurring are _____.

6. If the probability that an event will fail to occur is $\frac{1}{5}$ then the probability that the event will occur is $\frac{4}{5}$, and the odds against the event occurring are _____.

7. If the odds against an event are $1:3$, the probability that the event will fail to occur is _____.

8. If the odds in favor of an event are $2:5$, then the probability that the event occurs is _____.

Practice the Skills/Problem Solving

9. *Raffle Drawing* Tito purchased one raffle ticket as part of a Meals on Wheels fundraiser. The grand prize–winning ticket will be randomly drawn from 100 tickets sold. Determine

  a) the probability that Tito wins the grand prize.

  b) the probability that Tito does not win the grand prize.

  c) the odds against Tito winning the grand prize.

  d) the odds in favor of Tito winning the grand prize.

10. *Smartphone Chargers* A store has a bin full of smartphone chargers of various colors. There are 8 red, 10 white, 6 blue, and 12 pink chargers in the bin. If Kat randomly selects one of the chargers, determine

  a) the probability that Kat selects a pink charger.

  b) the probability that Kat does not select a pink charger.

  c) the odds against Kat selecting a pink charger.

  d) the odds in favor of Kat selecting a pink charger.

11. *Liberal Arts Students* In a class of 30 students, 18 are liberal arts majors. If a student is randomly selected, determine

  a) the probability the student is majoring in liberal arts.

  b) the probability the student is not majoring in liberal arts.

  c) the odds against the student majoring in liberal arts.

  d) the odds in favor of the student majoring in liberal arts.

12. *Making a Donation* In her wallet, Tasha has 12 bills. Six are $1 bills, two are $5 bills, three are $10 bills, and one is a $20 bill. She passes a volunteer seeking donations for a charity and decides to randomly select one bill from her wallet and donate it. Determine

  a) the probability that she selects a $5 bill.

  b) the probability that she does not select a $5 bill.

  c) the odds in favor of her selecting a $5 bill.

  d) the odds against her selecting a $5 bill.

Toss a Die In Exercises 13–16, a fair die is tossed. Determine the odds against rolling

13. a 2.

14. an even number.

15. a number less than 5.

16. a number greater than 4.

Deck of Cards In Exercises 17–20, one card is drawn from a standard 52-card deck. Determine the odds against and the odds in favor of selecting

17. a jack.

18. a diamond.

19. a face card.

20. a card greater than 5 (ace is low).

Spin the Spinner *In Exercises 21–24, assume that the spinner cannot land on a line. Determine the odds against the spinner landing on the color red.*

21.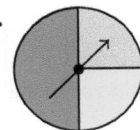

22.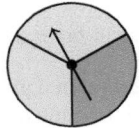

23.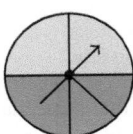

24.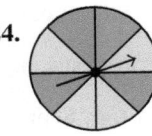

Billiard Balls *In Exercises 25–30, use the rack of 15 billiard balls shown.*

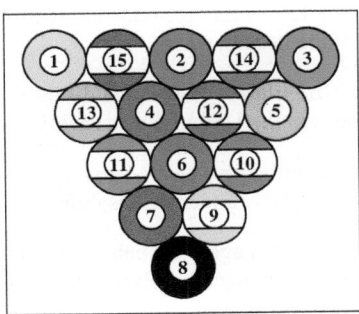

25. If one ball is randomly selected, determine the odds against it containing a stripe. (Balls numbered 9 through 15 contain stripes.)

26. If one ball is randomly selected, determine the odds in favor of it being a ball other than the 8 ball.

27. If one ball is randomly selected, determine the odds in favor of it being an even-numbered ball.

28. If one ball is randomly selected, determine the odds against it containing any red coloring (solid or striped).

29. If one ball is randomly selected, determine the odds against it containing a number greater than or equal to 9.

30. If one ball is randomly selected, determine the odds in favor of it containing two digits.

Playing Bingo *In Exercises 31–36, refer to the following information. When playing bingo, 75 balls are placed in a bin and balls are randomly selected. Each ball is marked with a letter and number as indicated in the following chart.*

B	I	N	G	O
1–15	16–30	31–45	46–60	61–75

For example, there are balls marked B1, B2, up to B15; I16, I17, up to I30; and so on (see photo). Assuming one bingo ball is randomly selected, determine

31. the probability that it contains the letter *B*.

32. the probability that it does not contain the letter *B*.

33. the odds in favor of it containing the letter *B*.

34. the odds against it containing the letter *B*.

35. the odds against it being *G*50.

36. the odds in favor of it being *G*50.

Blood Types in the United States *In Exercises 37–42, the following circle graph, which was created using* **StatCrunch** *with data from the website www.healthline.com, shows the percentages of Americans with the various blood types.*

Blood Types of Americans

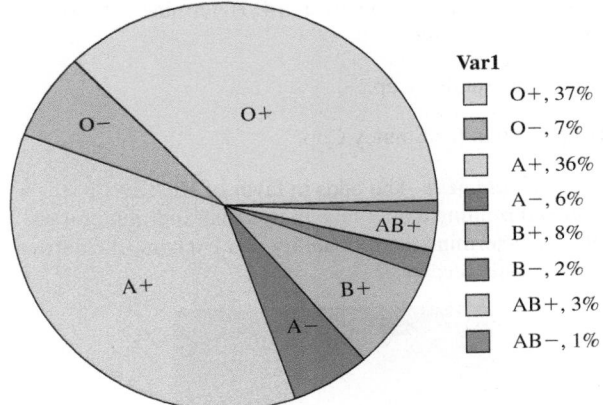

Source: www.healthline.com

If one American is randomly selected, determine

37. the probability that the person has blood type A+.

38. the probability that the person has blood type B+.

39. the odds against the person having blood type A+.

40. the odds in favor of the person having blood type B+.

41. the odds in favor of a person having blood type A−, B−, AB−, or O−.

42. The odds against a person having blood type A+, B+, AB+, or O+.

43. **Rock Concert** The odds against a rock concert selling out are 2 : 9. Determine the probability the concert

    a) sells out.   b) does not sell out.

44. Getting Hired The odds against Ishaq getting hired for a job are 3 : 8. Determine the probability

a) Ishaq gets hired.   b) Ishaq does not get hired.

45. Winning an Election The odds in favor of Wendy winning an election are 7 : 4. Determine the probability that

a) Wendy wins.   b) Wendy does not win.

46. Selling a Motorcycle The odds in favor of Sam selling his motorcycle are 3 : 10. Determine the probability that Sam will

a) sell his motorcycle.   b) not sell his motorcycle.

47. Super Bowl Odds The odds against the Cincinnati Bengals winning the Super Bowl are 100 : 1. Determine the probability that the Bengals

a) win the Super Bowl.

b) do not win the Super Bowl.

48. Stanley Cup Odds The odds against the St. Louis Blues winning the Stanley Cup are 17 : 4. Determine the probability that the Blues

a) win the Stanley Cup.

b) do not win the Stanley Cup.

49. Four-Minute Mile The odds in favor of Hicham El Guerrouj running a mile race in less than four minutes are 197 : 2. Determine the probability that Hicham El Guerrouj runs the mile race in

▲ Hicham El Guerrouj

a) less than four minutes.

b) four minutes or greater.

50. Bowling a Strike The odds in favor of Aleta bowling a strike in her next frame are 23 : 2. Determine the probability that in her next frame Aleta

a) bowls a strike.   b) does not bowl a strike.

51. Diabetes In 2019, approximately 9% of Americans had diabetes. If an American is randomly selected, determine the

a) odds against the person having diabetes.

b) odds in favor of the person having diabetes.

52. Hurricane In a given year, there is a 27% chance that a major hurricane hits Florida. Determine

a) odds in favor of a major hurricane hitting Florida.

b) odds against a major hurricane hitting Florida.

53. Fixing a Car Suppose that the probability that a mechanic fixes a car correctly is 0.9. Determine the odds against the mechanic fixing a car correctly.

54. Bookcase Assembly Suppose that the probability that all the parts needed to assemble a bookcase are included in the carton is $\frac{7}{8}$. Determine the odds in favor of the carton including all the needed parts.

55. Congenital Disabilities Congenital disabilities affect 1 in 33 babies born in the United States each year.

a) What is the probability that a baby born in the United States will have a congenital disability?

b) What are the odds against a baby born in the United States having a congenital disability?

56. IRS Audit One in 175 individual tax returns will be audited by the IRS. Determine

a) the probability that an individual tax return will be audited.

b) the odds against an individual tax return being audited.

Challenge Problems/Group Activities

57. Horse Racing Racetracks quote the approximate odds against each horse winning on a large board called a *tote board*. The odds quoted on a tote board for a race with four horses are as follows.

Horse Number	Odds
2	15 : 1
3	1 : 1
4	3 : 1
5	13 : 3

Determine the probability of each horse winning the race.

58. Roulette Observe the roulette wheel illustrated on page 677. If the wheel is spun, determine

a) the probability that the ball lands on red.

b) the odds against the ball landing on red.

c) the probability that the ball lands on 0 or 00.

d) the odds in favor of the ball landing on 0 or 00.

Recreational Mathematics

59. *Multiple Births* Multiple births make up about 3.5% of births per year in the United States. The following artwork illustrates the number and type of multiple births in the U. S. in 2017.

Multiple Births in the United States in 2017

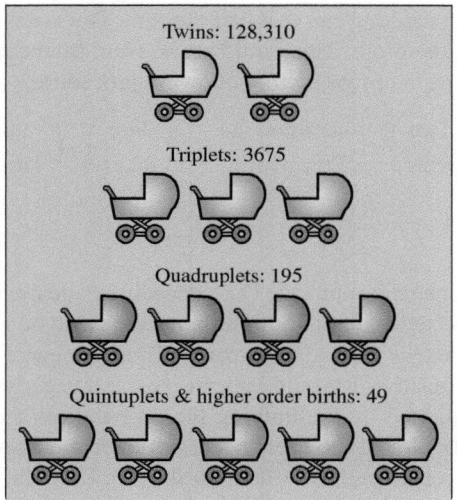

Twins: 128,310

Triplets: 3675

Quadruplets: 195

Quintuplets & higher order births: 49

Source: National Center for Health Statistics

Determine an estimate for the odds against a birth being a multiple birth in 2017.

Research Activities

60. *State Lottery* Choose a state lottery and do research and write a paper indicating

a) the probability of winning the grand prize.

b) the odds against winning the grand prize.

c) Explain, using real objects such as pennies or table tennis balls, what these odds actually mean.

61. *Casino Advantages* There are many types of games of chance to choose from at casinos. The casino has the advantage in each game, but the advantages differ according to the game. Do research and write a paper indicating

a) the games available at a typical casino.

b) the games for which the casino has the smallest advantage of winning.

c) the games for which the casino has the greatest advantage of winning.

SECTION 11.3   Expected Value (Expectation)

Upon completion of this section, you will be able to:

- Determine expected value.
- Determine fair price.

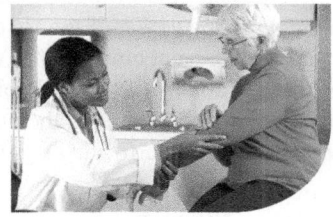

When a company determines the price to charge for a life insurance policy, it takes many factors regarding the insured person into account. These factors may include the insured person's age, gender, health, family medical history, and lifestyle factors such as whether the person uses tobacco or drinks alcohol. Using large amounts of data, insurance companies can assign probabilities to each of these factors. These probabilities measure how likely the insured person is to die before a certain age. Such calculations use *expected value* or *expectation*. Many businesses and governments use expected value when making important decisions.

Why This Is Important Expectation is used throughout the financial services industry and especially the insurance industry to set rates for products and services. Expectation is used by manufacturers when determining the length of the warranty on their products. Expectation is also used by many businesses whenever difficult decisions must be made. Expectation provides a tool that can be used to analyze all possible outcomes of a decision.

Expected Value

Expected value, also called *expectation*, is often used to determine the expected results of an experiment or business venture *over the long term*. People use expectation to make important decisions in many different areas. For example, expectation is used in business to predict future profits of a new product. The insurance industry uses expectation to determine premiums so insurance companies make a profit. Expectation is also used to predict the expected gain or loss in games of chance.

Consider the following situation. Tim tells Barbara that he will give her $1 if she can roll an even number on a single die. If she fails to roll an even number, she must give Tim $1. Who would win money in the long run if this game were played many times? In this situation, we would expect in the long run that half the time Tim would win $1 and half the time he would lose $1; therefore, Tim would break even. Mathematically, we could determine Tim's expected gain or loss by the following procedure:

$$\text{Tim's expected gain or loss} = P\!\left(\begin{array}{c}\text{Tim}\\\text{wins}\end{array}\right) \cdot \left(\begin{array}{c}\text{amount}\\\text{Tim wins}\end{array}\right) + P\!\left(\begin{array}{c}\text{Tim}\\\text{loses}\end{array}\right) \cdot \left(\begin{array}{c}\text{amount}\\\text{Tim loses}\end{array}\right)$$

$$= \frac{1}{2}(\$1) + \frac{1}{2}(-\$1) = \$0$$

Note that the loss is written as a negative number. This procedure indicates that Tim has an expected gain or loss (or expected value) of $0. The expected value of zero indicates that he would indeed break even, as we had anticipated. Thus, the game is a *fair game*. If Tim's expected value were positive, it would indicate a gain; if negative, a loss.

The expected value, E, is calculated by multiplying the probability of an event occurring by the *net* amount gained or lost if the event occurs. If there are a number of different events and amounts to be considered, we use the following formula.

Expected Value

$$E = P_1 \cdot A_1 + P_2 \cdot A_2 + P_3 \cdot A_3 + \ldots + P_n \cdot A_n$$

The symbol P_1 represents the probability that the first event will occur, and A_1 represents the net amount won or lost if the first event occurs. P_2 is the probability of the second event, and A_2 is the net amount won or lost if the second event occurs, and so on. The sum of these products of the probabilities and their respective amounts is the expected value. The expected value is the average (or mean) result that would be obtained if the experiment were performed a great many times.

Example 1 *A New Flight Route*

Spirit Airlines is considering adding a route from Pittsburgh to Nashville. Before making a decision, the company needs to consider many factors, including potential profit or loss. After considerable research, Spirit estimates that if it adds the route, there is a 60% chance of making an annual $800,000 profit, a 10% chance of breaking even, and a 30% chance of losing $1,000,000. How much can Spirit expect to make annually on this new route?

Solution Three amounts are to be considered: a gain of $800,000, breaking even at $0, and a loss of $1,000,000. The probability of gaining $800,000 is 0.6, the probability of breaking even is 0.1, and the probability of losing $1,000,000 is 0.3.

$$\begin{aligned}\text{Spirit's expectation} &= P_1A_1 + P_2A_2 + P_3A_3\\ &= (0.6)(\$800,000) + (0.1)(\$0) + (0.3)(-\$1,000,000)\\ &= \$180,000\end{aligned}$$

Spirit Airlines has an expectation, or expected average annual gain, of $180,000 for adding this particular route. Thus, if the company opened routes like this one, with these particular probabilities and amounts, in the long run it would have an average annual gain of $180,000 per route. However, you must remember that there is a 30% chance that Spirit Airlines will lose $1,000,000 on this particular route or any particular route with these probabilities and amounts. ∎

Example 2 *Test-Taking Strategy*

Maria is taking a multiple-choice exam in which there are five possible answers for each question. The instructions indicate that she will be awarded 2 points for each correct response, that she will lose $\frac{1}{2}$ point for each incorrect response, and that no points will be added or subtracted for answers left blank.

a) If Maria does not know the correct answer to a question, is it to her advantage or disadvantage to guess at an answer?

b) If she can eliminate one of the possible choices, is it to her advantage or disadvantage to guess at an answer?

Solution

a) Let's determine the expected value if Maria guesses at an answer. Only one of five possible answers is correct.

$$P(\text{guesses correctly}) = \frac{1}{5} \qquad P(\text{guesses incorrectly}) = \frac{4}{5}$$

$$\text{Maria's expectation} = \overbrace{P_1 \cdot A_1}^{\text{Guesses correctly}} + \overbrace{P_2 \cdot A_2}^{\text{Guesses incorrectly}}$$

$$= \frac{1}{5}(2) + \frac{4}{5}\left(-\frac{1}{2}\right)$$

$$= \frac{2}{5} - \frac{2}{5} = 0$$

Thus, Maria's expectation is zero when she guesses. Therefore, over the long run she will neither gain nor lose points by guessing.

b) If Maria can eliminate one possible choice, one of four answers will be correct.

$$P(\text{guesses correctly}) = \frac{1}{4} \qquad P(\text{guesses incorrectly}) = \frac{3}{4}$$

$$\text{Maria's expectation} = \overbrace{P_1 \cdot A_1}^{\text{Guesses correctly}} + \overbrace{P_2 \cdot A_2}^{\text{Guesses incorrectly}}$$

$$= \frac{1}{4}(2) + \frac{3}{4}\left(-\frac{1}{2}\right)$$

$$= \frac{2}{4} - \frac{3}{8} = \frac{4}{8} - \frac{3}{8} = \frac{1}{8}$$

Since the expectation is a positive $\frac{1}{8}$, Maria will, on average, gain $\frac{1}{8}$ point each time she guesses when she can eliminate one possible choice. ∎

Example 3 *Expected Attendance*

For an outdoor concert, event organizers estimate that 20,000 people will attend if it is not raining and 12,000 people will attend if it is raining. On the day of the concert, meteorologists predict a 30% chance of rain. Determine the expected number of people who will attend this concert.

Solution The amounts in this example are the number of people who will attend the concert. Since there is a 30% chance of rain on the day of the concert, the

Did You Know?

Expected Value and Decisions

Expected value can be used to help evaluate the consequences of decisions. These decisions can range from routine decisions such as deciding where to park your car to life-changing decisions such as deciding whether to go back to college. For each decision we make, we must consider the probabilities and the gains or losses associated with each possible outcome. For example, if you decide to park your car illegally, the probability of getting caught may be low, but the loss associated with this decision, a parking ticket, may be high. If you decide to go back to college, there is a probability that you might not succeed. Associated with this probability is the loss of time and money for books and tuition. However, there is also a probability that you will succeed and graduate. Associated with this probability is the ability to attain a higher-paying and more rewarding career! Expected value can be used to help make wise decisions that help us in the long run.

chance it will not rain is 100% − 30% = 70%. When written as probabilities, 30% and 70% are 0.3 and 0.7, respectively.

$$E = P(\text{does not rain}) \cdot (\text{number of people}) \ + \ P(\text{rain}) \cdot (\text{number of people})$$
$$= \ 0.7(20{,}000) \ + \ 0.3(12{,}000)$$
$$= \ 14{,}000 \ + \ 3600$$
$$= \ 17{,}600$$

Thus, the average, or expected, number of people who will attend the concert is 17,600. ∎

In the expected value formula, the amounts refer to *net amounts*, which are the actual amounts gained or lost. Our next three examples illustrate how net amounts are used in applications.

Example 4 *Real Estate Listing*

Jayden is a real estate broker who has a contract to sell a house. Such a contract is called a *listing*. Jayden estimates that this listing will cost her $2500 for various expenses including advertising. Jayden must pay these expenses whether she sells the house or not. If Jayden sells the house within 6 months, she will receive a commission of $15,000. If she does not sell the house within 6 months, the listing will expire, and she will not receive a commission, but she still must pay the $2500 for various expenses. Jayden estimates the probability that she sells the house within 6 months is 75%. Determine Jayden's expectation for this listing.

Solution There is a 75% probability that Jayden sells the house within 6 months; thus, $P_1 = 0.75$. If she sells the house within 6 months, she will gain the commission of $15,000 but will have $2500 in expenses. Thus, $A_1 = \$15{,}000 - \$2500$, or $A_1 = \$12{,}500$. There is a 25% probability that Jayden does not sell the house within 6 months; thus, $P_2 = 0.25$. If she does not sell the house within 6 months, she will not receive a commission, but she still must pay the $2500 in expenses. Thus, $A_2 = -\$2500$. Substituting $P_1, A_1, P_2,$ and A_2 into the expected value formula gives us the following.

$$\begin{aligned}\text{Jayden's expectation} &= P_1A_1 \ + \ P_2A_2 \\ &= (0.75)(\$12{,}500) \ + \ (0.25)(-\$2500) \\ &= \$8750\end{aligned}$$

Thus, Jayden has an expectation of $8750 for this listing. This means if Jayden has more listings like this one, with these probabilities and amounts, in the long run Jayden would average a net gain of $8750 per house. However, it is important to remember that there is a 25% chance that Jayden will lose $2500 on this listing or on similar listings with these probabilities and amounts. ∎

Example 5 *Winning a Door Prize*

When Josh attends a charity event, he is given a free ticket for the $50 door prize. A total of 100 tickets will be given out. Determine his expectation of winning the door prize.

Solution The probability of winning the door prize is $\frac{1}{100}$, since Josh has 1 of 100 tickets. If he wins, his net or actual winnings will be $50, since he did not pay for the ticket. The probability that Josh loses is $\frac{99}{100}$. If Josh loses, the amount he loses is $0 because he did not pay for the ticket.

$$\text{Expectation} = P(\text{Josh wins}) \cdot (\text{amount won}) + P(\text{Josh loses}) \cdot (\text{amount lost})$$
$$= \frac{1}{100}(50) + \frac{99}{100}(0) = \frac{50}{100} = 0.50$$

Thus, Josh's expectation is $0.50, or 50 cents. ∎

Now we will consider a problem similar to Example 5, but this time we will assume that Josh must purchase the ticket for the door prize.

Example 6 *Winning a Door Prize*

When Josh attends a charity event, he is given the opportunity to purchase a ticket for the $50 door prize. The cost of the ticket is $2, and 100 tickets will be sold. Determine Josh's expectation if he purchases one ticket.

Solution As in Example 5, Josh's probability of winning is $\frac{1}{100}$. However, if he does win, his actual or net winnings will be $48. The $48 is obtained by subtracting the cost of the ticket, $2, from the amount of the door prize, $50. There is also a probability of $\frac{99}{100}$ that Josh will not win the door prize. If he does not win the door prize, he has lost the $2 that he paid for the ticket. Therefore, we must consider two amounts when we determine Josh's expectation, winning $48 and losing $2.

$$\text{Expectation} = P(\text{Josh wins}) \cdot (\text{amount won}) + P(\text{Josh loses}) \cdot (\text{amount lost})$$

$$= \frac{1}{100}(48) + \frac{99}{100}(-2)$$

$$= \frac{48}{100} - \frac{198}{100} = -\frac{150}{100} = -1.50$$

Josh's expectation is $-1.50 when he purchases one ticket. ∎

In Example 6, we determined that Josh's expectation was $-$1.50 when he purchased one ticket. If he purchased two tickets, his expectation would be $2(-\$1.50)$, or $-\$3.00$. We could also compute Josh's expectation if he purchased two tickets as follows:

$$E = \frac{2}{100}(46) + \frac{98}{100}(-4) = -3.00$$

This answer, $-\$3.00$, checks with the answer obtained by multiplying the expectation for a single ticket by 2.

Let's look at one more example in which a person must pay for a chance to win a prize. In the following example, there will be more than two amounts to consider.

Example 7 *Raffle Tickets*

One thousand raffle tickets are sold for $1 each. One grand prize of $500 and two consolation prizes of $100 will be awarded. The tickets are placed in a bin. The winning tickets will be selected from the bin. Assuming that the probability that any given ticket selected for the grand prize is $\frac{1}{1000}$ and the probability that any given ticket selected for a consolation prize is $\frac{2}{1000}$, determine

a) Irene's expectation if she purchases one ticket.

b) Irene's expectation if she purchases five tickets.

Solution

a) Three amounts are to be considered: the net gain in winning the grand prize, the net gain in winning one of the consolation prizes, and the loss of the cost of the ticket. If Irene wins the grand prize, her net gain is $499 ($500 minus $1 spent for the ticket). If Irene wins one of the consolation prizes, her net gain is $99 ($100 minus $1). The probability that Irene wins the grand prize is $\frac{1}{1000}$ and the probability that she wins a consolation prize is $\frac{2}{1000}$. The probability that she does not win a prize is $1 - \frac{1}{1000} - \frac{2}{1000} = \frac{997}{1000}$.

$$E = P_1 \cdot A_1 + P_2 \cdot A_2 + P_3 \cdot A_3$$

$$= \frac{1}{1000}(499) + \frac{2}{1000}(99) + \frac{997}{1000}(-1)$$

$$= \frac{499}{1000} + \frac{198}{1000} - \frac{997}{1000} = -\frac{300}{1000} = -0.30$$

Timely Tip

In any expectation problem, the sum of the probabilities of all the events should always be 1. Note in Example 7 that the sum of the probabilities is

$$\frac{1}{1000} + \frac{2}{1000} + \frac{997}{1000} = \frac{1000}{1000}$$
$$= 1.$$

Thus, Irene's expectation is −$0.30 per ticket purchased.

b) On average, Irene loses 30 cents on each ticket purchased. On five tickets, her expectation is $(-0.30)(5)$, or −$1.50. ∎

Fair Price

In Example 6, we determined that Josh's expectation was −$1.50. Now we will determine the price that should have been charged for a ticket so that his expectation would be $0. If Josh's expectation were to be $0, he could be expected to break even over the long run. Suppose that Josh paid 50 cents, or $0.50, for the ticket. His expectation, if paying $0.50 for the ticket, would be calculated as follows.

$$\text{Expectation} = P\,(\text{Josh wins}) \cdot (\text{amount won}) + P\,(\text{Josh loses}) \cdot (\text{amount lost})$$
$$= \frac{1}{100}(49.50) + \frac{99}{100}(-0.50)$$
$$= \frac{49.50}{100} - \frac{49.50}{100} = 0$$

Thus, if Josh paid 50 cents per ticket, his expectation would be $0. The 50 cents, in this case, is called the fair price of the ticket. The *fair price* is the amount to be paid that will result in an expected value of $0. The fair price may be determined by adding the *cost to play* to the *expected value*.

> **Fair Price**
> To determine the fair price, use the following formula.
>
> $$\textbf{Fair price} = \text{expected value} + \text{cost to play}$$

In Example 6, the cost to play was $2 and the expected value was determined to be −$1.50. The fair price for a ticket in Example 5 may be determined as follows.

$$\text{Fair price} = \text{expected value} + \text{cost to play}$$
$$= -1.50 + 2.00 = 0.50$$

We obtained a fair price of $0.50. If the tickets were sold for the fair price of $0.50 each, Josh's expectation would be $0, as shown above.

In our next example, we will determine the fair price of the raffle tickets sold in Example 7.

Example 8 *Fair Price of a Raffle Ticket*

In Example 7, the cost of a raffle ticket was $1, and we determined that the expected value was −$0.30.

a) Determine the fair price for this raffle ticket.

b) Verify that if the raffle tickets were sold for the fair price determined in part (a), the expectation would be $0.00.

Solution

a) We use the formula

$$\text{Fair price} = \text{expected value} + \text{cost to play}$$
$$= -0.30 + 1.00$$
$$= 0.70$$

Thus, the fair price for the raffle ticket is $0.70, or 70 cents.

b) When calculating the expected value, the probabilities will be the same as in Example 7, only the net amounts change. If the raffle tickets were sold for $0.70, then A_1, the net amount gained by winning the grand prize would be $500 − $0.70, or $499.30. A_2, the net amount gained by winning one of the consolation prizes would be $100 − $0.70, or $99.30. A_3, the amount lost by not winning a prize is the price of the ticket, or −$0.70. Thus, we have

$$\text{Expected value} = P_1A_1 + P_2A_2 + P_3A_3$$

$$= \frac{1}{1000}(499.30) + \frac{2}{1000}(99.30) + \frac{997}{1000}(-0.70)$$

$$= \frac{499.3}{1000} + \frac{198.6}{1000} - \frac{697.9}{1000}$$

$$= \frac{499.3 + 198.6 - 697.9}{1000} = \frac{0}{1000} = 0$$

Thus, if the raffle tickets were sold for $0.70, the expectation would be $0.00. ∎

Example 9 *Expectation and Fair Price*

At a game of chance, the expected value is determined to be −$1.50, and the cost to play the game is $4.00. Determine the fair price to play the game.

Solution We use the formula

$$\text{Fair price} = \text{expected value} + \text{cost to play}$$

$$= -1.50 + 4.00$$

$$= 2.50$$

Thus, the fair price to play this game would be $2.50. If the cost to play was $2.50 instead of $4.00, the expected value would be $0. ∎

SECTION 11.3 *Exercises*

Warm Up Exercises

In Exercises 1–4, fill in the blank with an appropriate word, phrase, or symbol(s).

1. The expected gain or loss of an experiment over the long run is called the _____ value.

2. In an experiment, if there is a loss in the long run, the expected value is _____.

3. In an experiment, if there is a gain in the long run, the expected value is _____.

4. In an experiment, if there is neither a gain nor a loss in the long run, the expected value is _____.

Practice the Skills/Problem Solving

5. *Selling Hot Dogs* An outdoor hot dog vendor sells an average of 60 hot dogs per day in dry weather and an average of 20 per day in rainy weather. If the weather in this area is rainy 20% of the time, determine the expected number of hot dogs sold per day.

6. *Tax Law Changes* Jonalynn, an investment counselor, is advising her client on a particular investment. She estimates that if the tax law does not change, the client will make $15,000 but if the tax law changes, the client will lose $4000. Determine the client's expected value if there is a 20% chance the tax law will change.

7. *Investment Club* The Triple L investment club is considering purchasing a certain stock. After considerable research, the club members determine that there is a 70% chance of making $15,000, a 10% chance of breaking even, and a 20% chance of losing $8500. Determine the expectation of this purchase.

8. *Antiques Businesss* Zelda is opening an antiques shop. In the first year, she estimates that there is a 25% chance that she will make $20,000, a 40% chance that she will break even, and a 35% chance that she will lose $30,000. Determine Zelda's expected value for the first year of her business.

9. *Corn Stalk Growth* In July in Benton Township, corn stalks grow 2.5 in. per day on sunny days and 1.5 in. per day on cloudy days. If in Benton Township in July, 70% of the days are sunny and 30% are cloudy,

a) determine the expected amount of corn stalk growth on a typical day in July in Benton Township.

b) determine the expected amount of corn stalk growth in July in Benton Township.

10. **Seattle Greenery** In July in Seattle, the grass grows $\frac{1}{2}$ in. a day on a sunny day and $\frac{1}{4}$ in. a day on a cloudy day. If in Seattle in July, 75% of the days are sunny and 25% are cloudy

a) determine the expected amount of grass growth on a typical day in July in Seattle.

b) determine the expected total grass growth in the month of July in Seattle.

11. **Fortune Cookies** At the Pan Pacific restaurant, fortune cookies contain a slip of paper indicating a dollar amount that will be subtracted from your restaurant bill. You are to randomly select one out of ten fortune cookies. Of the ten cookies, seven contain "$1 off," two contain "$2 off," and one contains "$5 off." If your restaurant bill was $45, determine

a) the expected dollar amount to be deducted from your bill.

b) the expected dollar amount you will pay for your meal.

12. **Clothing Sale** At a special clothing sale at the Crescent Oaks Country Club, after the cashier rings up your purchase, you select a slip of paper from a box. The slip of paper indicates the dollar amount, either $5 or $10, that is deducted from your purchase price. The probability of selecting a slip indicating $5 is $\frac{7}{10}$, and the probability of selecting a slip indicating $10 is $\frac{3}{10}$. If your original purchase before you select the slip of paper is $200, determine

a) the expected dollar amount to be deducted from your purchase.

b) the expected dollar amount you will pay for your purchase.

13. **Black Chips and White Chips** A bag contains 7 black chips and 3 white chips. Dexter and Thanh play the following game. Dexter randomly selects one chip from the bag. If the chip is black, Dexter gives Thanh $5. If the chip is white, Thanh gives Dexter $10.

a) Determine Dexter's expectation.

b) Determine Thanh's expectation.

14. **Blue Chips and Red Chips** A bag contains 4 blue chips and 6 red chips. Chi and Dolly play the following game. Chi selects one chip from the bag. If Chi selects a blue chip, Dolly gives Chi $6. If Chi selects a red chip, Chi gives Dolly $5.

a) Determine Chi's expectation.

b) Determine Dolly's expectation.

15. **Roll a Die** Alyssa and Gabriel play the following game. Alyssa rolls a die. If she rolls a 1, 2, or 3, Gabriel gives Alyssa $3. If Alyssa rolls a 4 or 5, Gabriel gives Alyssa $2. However, if Alyssa rolls a 6, she gives Gabriel $14.

a) Determine Alyssa's expectation.

b) Determine Gabriel's expectation.

16. **Pick a Card** Mike and Dave play the following game. Mike draws one card from a standard 52-card deck. If he draws an ace, Dave gives him $5. If not, he gives Dave $2.

a) Determine Mike's expectation.

b) Determine Dave's expectation.

17. **Multiple-Choice Test** A multiple-choice exam has five possible answers for each question. For each correct answer, you are awarded 5 points. For each incorrect answer, 1 point is subtracted from your score. For answers left blank, no points are added or subtracted.

a) If you do not know the correct answer to a particular question, is it to your advantage to guess? Explain.

b) If you do not know the correct answer but can eliminate one possible choice, is it to your advantage to guess? Explain.

18. **Multiple-Choice Test** A multiple-choice exam has four possible answers for each question. For each correct answer, you are awarded 5 points. For each incorrect answer, 2 points are subtracted from your score. For answers left blank, no points are added or subtracted.

a) If you do not know the correct answer to a particular question, is it to your advantage to guess? Explain.

b) If you do not know the correct answer but can eliminate one possible choice, is it to your advantage to guess? Explain.

Spinners In Exercises 19 and 20, assume that a person spins the pointer and is awarded the amount indicated by the pointer. Determine the person's expectation.

19. **20.**

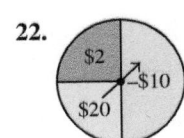

Spinners In Exercises 21 and 22, assume that a person spins the pointer and is awarded the amount indicated if the pointer points to a positive number but must pay the amount indicated if the pointer points to a negative number. Determine the person's expectation if the person plays the game.

21. **22.**

Selecting an Envelope In Exercises 23–26, a person randomly selects one of the six envelopes shown below. Each envelope contains a check that the person gets to keep. Determine the person's expectation if the checks in the envelopes are as indicated in the exercises.

23. Three envelopes contain a $500 check, and three envelopes contain a $1000 check.

24. Four envelopes contain a $1000 check, and two envelopes contain a $5000 check.

25. The six envelopes contain checks for $100, $200, $300, $400, $500, and $1000, respectively.

26. Two envelopes contain a check for $600, two envelopes contain a check for $2000, and two envelopes contain a check for $5000.

Selecting an Envelope In Exercises 27–30, a person randomly selects one of the four envelopes shown above and to the right. Each envelope contains a check that the person gets to keep. However, before the person can select an envelope, he or she

must pay $15 to play. Determine the person's expectation if the checks in the envelopes are as indicated in the exercises.

27. Two envelopes contain a $10 check, and two envelopes contain a $25 check.

28. Three envelopes contain a $20 check, and one envelope contains a $0 check.

29. The checks in the envelopes are for $0, $2, $5, and $20, respectively.

30. The checks in the envelopes are for $2, $2, $100, and $400, respectively.

31. *Job Openings* There is a 20% chance that next year the total number of employees in the Menards stores in Milwaukee will increase by 300. There is a 70% chance that the increase will be 400 employees, and a 10% chance the total number of employees will decrease by 500. Determine the expected gain or loss in the number of employees for next year.

32. *Life Insurance* According to Bristol Mutual Life Insurance's mortality table, the probability that a 20-year-old woman will survive 1 year is 0.994 and the probability that she will die within 1 year is 0.006. If a 20-year-old woman buys a $10,000 1-year policy for $100, what is Bristol Mutual's expected gain or loss?

33. *Employee Hiring* The academic vice president at Irving Valley College has requested that new academic programs be added to the college curriculum. If the college's Board of Trustees approves the new programs, the college will hire 75 new employees. If the new programs are not approved, the college will hire only 20 new employees. If the probability that the new programs will be approved is 0.65, what is the expected number of new employees to be hired by Irving Valley College? Round your answer to the nearest whole number of employees.

34. *Completing a Project* A mechanical contractor is preparing for a construction project. He determines that if he completes the project on schedule, his net profit will be $450,000. If he completes the project between 0 and 3 months late, his net profit decreases to $120,000. If he completes the project more than 3 months late, his net loss is $275,000. The probability that he completes the project on schedule is 0.6, the probability that he completes the project between 0 and 3 months late is 0.3, and the probability that he completes the project more than 3 months late is 0.1. Determine his expected gain or loss for this project.

35. New Store Hungry Howie's is opening a new restaurant. The company estimates that there is a 85% chance the restaurant will have a profit of $200,000, a 5% chance the restaurant will break even, and a 10% chance the restaurant will lose $30,000. Determine the expected gain or loss for this restaurant.

36. China Cabinet The owner of an antique store estimates that there is a 40% chance she will make $2000 when she sells an antique china cabinet, a 50% chance she will make $750 when she sells the cabinet, and a 10% chance she will break even when she sells the cabinet. Determine the expected amount she will make when she sells the cabinet.

37. Road Service On a clear day in Boston, the Automobile Association of American (AAA) makes an average of 110 service calls for motorist assistance, on a rainy day it makes an average of 160 service calls, and on a snowy day it makes an average of 210 service calls. If the weather in Boston is clear 200 days of the year, rainy 100 days of the year, and snowy 65 days of the year, determine the expected number of service calls made by the AAA in a given day.

38. Lawsuit Don is considering bringing a lawsuit against the Dummote Company. His lawyer estimates that there is a 70% chance Don will make $60,000, a 10% chance Don will break even, and a 20% chance they will lose the case and Don will need to pay $30,000 in legal fees. Estimate Don's expected gain or loss if he proceeds with the lawsuit.

39. Real Estate Listing See Example 4. Darya is a real estate broker with a house listing. Darya estimates that this listing will cost her $2400 for various expenses. If Darya sells the house within 6 months, she will receive a commission of $20,000. If she does not sell the house within 6 months, the listing will expire, and she will not receive a commission, but she still must pay the $2400 for various expenses. Darya estimates the probability that she sells the house within 6 months is 80%. Determine Darya's expectation for this listing.

40. Real Estate Listing See Example 4. Shelby is a real estate broker with a house listing. Shelby estimates that this listing will cost him $3000 for various expenses. If Shelby sells the house within 6 months, he will receive a commission of $25,000. If he does not sell the house within 6 months, the listing will expire, and he will not receive a commission, but he still must pay the $3000 for various expenses. Shelby estimates the probability that he sells the house within 6 months is 85%. Determine Shelby's expectation for this listing.

Dart Board In Exercises 41 and 42, assume that you are blindfolded and throw a dart at the dart board shown above and to the right. Assume that your dart sticks in the dart board, and not on a line.

a) *Determine the probabilities that the dart lands on $1, $10, $20, and $100, respectively.*

b) *If you win the amount of money indicated by the section of the board where the dart lands, determine your expectation when you throw the dart.*

41.

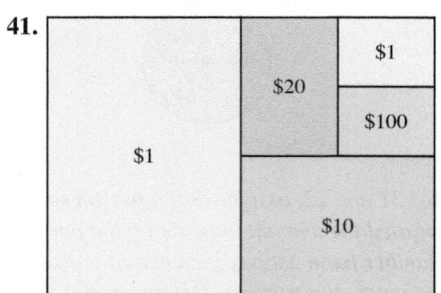

42.

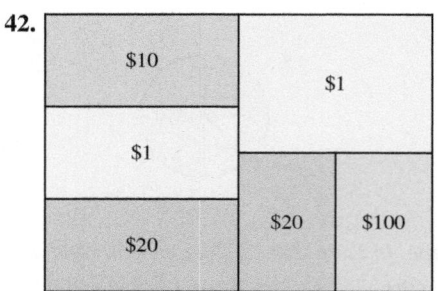

43. Fair Price The expected value of a carnival game is −$4.50, and the cost to play the game is $10.00. Determine the fair price to play the game.

44. Fair Price The expected value when you purchase a lottery ticket is −$2.00, and the cost of the ticket is $5.00. Determine the fair price of the lottery ticket.

45. Raffle Tickets Five hundred raffle tickets are sold for $3 each. One prize of $500 is to be awarded. Raul purchases one ticket.

a) Determine his expected value.

b) Determine the fair price of a ticket.

46. Raffle Tickets One thousand raffle tickets are sold for $2 each. One prize of $600 is to be awarded. Rena purchases one ticket.

a) Determine her expected value.

b) Determine the fair price of a ticket.

47. Raffle Tickets Two thousand raffle tickets are sold for $3 each. Three prizes will be awarded: one for $1000 and two for $500. Assume that the probability that any given ticket is selected for the $1000 prize is $\frac{1}{2000}$ and the probability that any given ticket is selected for the $500 prize is $\frac{2}{2000}$. Jeremy purchases one of these tickets.

a) Determine his expected value.

b) Determine the fair price of a ticket.

48. *Raffle Tickets* Ten thousand raffle tickets are sold for $5 each. Four prizes will be awarded: one for $5000, one for $2500, and two for $1000. Assume that the probability that any given ticket is selected for the $5000 prize is $\frac{1}{10,000}$, the probability that any given ticket is selected for the $2500 prize is $\frac{1}{10,000}$, and the probability that any given ticket is selected for a $1000 prize is $\frac{2}{10,000}$. Sidhardt purchases one of these tickets.

a) Determine his expected value.

b) Determine the fair price of a ticket.

Spinners *In Exercises 49–52, assume that a person spins the pointer and is awarded the amount indicated by the pointer. If it costs $2 to play the game, determine*

a) *the expectation of a person who plays the game.*

b) *the fair price to play the game.*

49.
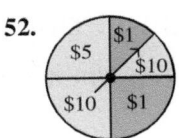

50.

51.

52.

Challenge Problems/Group Activities

53. *Term Life Insurance* An insurance company will pay the face value of a term life insurance policy if the insured person dies during the term of the policy. For how much should an insurance company sell a 10-year term policy with a face value of $40,000 to a 30-year-old man for the company to make a profit? The probability of a 30-year-old man living to age 40 is 0.97. Explain your answer. Remember that the customer pays for the insurance before the policy becomes effective.

54. *Lottery Ticket* Is it possible to determine your expectation when you purchase a lottery ticket? Explain.

Roulette *In Exercises 55 and 56, use the roulette wheel illustrated above and to the right. A roulette wheel typically contains slots with numbers 1–36 and slots marked 0 and 00. A ball is spun on the wheel and comes to rest in one of the 38 slots. Eighteen numbers are colored red, and 18 numbers are*

colored black. The 0 and 00 are colored green. If you bet on one particular number and the ball lands on that number, the casino pays off odds of 35 to 1. If you bet on a red number or black number and win, the casino pays 1 to 1 (even money).

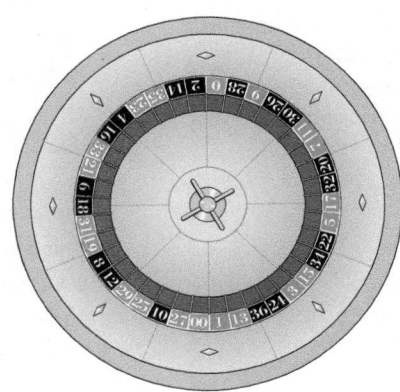

55. Determine the expected value of betting $1 on a particular number.

56. Determine the expected value of betting $1 on red.

Recreational Mathematics

57. *Wheel of Fortune* The following is a miniature version of the Wheel of Fortune. When Dave spins the wheel, he is awarded the amount on the wheel indicated by the pointer. If the wheel points to Bankrupt, he loses the total amount he has accumulated and also loses his turn. Assume that the wheel stops on a position and that each position is equally likely to occur.

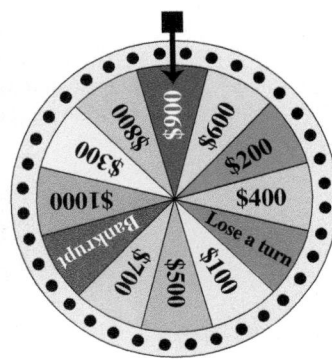

a) Determine Dave's expectation when he spins the wheel at the start of the game (he has no money to lose if he lands on Bankrupt).

b) If Dave presently has a balance of $1800, determine his expectation when he spins the wheel.

SECTION 11.4   Tree Diagrams

Upon completion of this section, you will be able to:

- Understand the fundamental counting principle.
- Construct tree diagrams to determine probabilities.

At the coffee station at work, you have several options when making a cup of coffee. You can choose to add no cream, regular cream, or hazelnut-flavored cream. You can also choose to add no sweetener, sugar, or aspartame. How many different ways can you prepare yourself a cup of coffee? If you were to randomly prepare a cup of coffee, what is the probability that you prepared a cup of coffee with regular cream and sugar? In this section, we will answer these questions using an important formula called the *fundamental counting principle*. We will also use a helpful diagram called a *tree diagram* to help us analyze the possible outcomes when multiple experiments are involved in problems.

Why This Is Important Being able to determine the number of outcomes when multiple experiments are being performed is essential to our understanding of probability concepts. Throughout this section we will explore many real-world applications that involve both the fundamental counting principle and tree diagrams. Such applications may appear in a wide variety of occupations.

The Fundamental Counting Principle

We now introduce the fundamental counting principle. We will also use the fundamental counting principle in Section 11.7.

> **Fundamental Counting Principle**
> If a first experiment has M distinct outcomes and a second experiment has N distinct outcomes, then the two experiments in that specific order have $M \cdot N$ distinct outcomes.

If we wanted to determine the number of possible outcomes when a coin is tossed and a die is rolled, we could reason that the coin has two possible outcomes, heads and tails. The die has six possible outcomes: 1, 2, 3, 4, 5, and 6. Thus, the two experiments together have $2 \cdot 6$, or 12, possible outcomes.

⌐ Example **1** *Drawing Two Cards*

If two cards are drawn from a standard 52-card deck, determine the number of possible outcomes if the cards are drawn

a) with replacement.

b) without replacement.

Solution

a) *With replacement* means that after the first card is drawn, it is replaced in the deck. The first card drawn can be any one of the 52 cards in the deck. Since the first card is then replaced in the deck, the second card can also be any one of the 52 cards in the deck. Using the fundamental counting principle, we have the following.

$$\text{Number of ways to draw two cards } \textit{with} \text{ replacement} = 52 \cdot 52 = 2704$$

Thus, there are 2704 ways to draw two cards with replacement from a standard deck of cards.

b) *Without replacement* means that after the first card is drawn, it is not replaced in the deck. The first card drawn can be any one of the 52 cards in the deck. Since the first card is not replaced in the deck, the second card can be any one of the 51 cards remaining in the deck. Using the fundamental counting principle, we have the following.

$$\text{Number of ways to draw two cards } \textit{without} \text{ replacement} = 52 \cdot 51 = 2652$$

Thus, there are 2652 ways to draw two cards without replacement from a standard deck of cards.   ∎

Tree Diagrams

A list of all the possible outcomes of an experiment is called a *sample space*. Each individual outcome in the sample space is called a *sample point*. *Tree diagrams* are helpful in determining sample spaces.

A tree diagram illustrating all the possible outcomes when a coin is tossed and a die is rolled (see Fig. 11.5) has two initial branches, one for each possible outcome of the coin. Each of these branches will have six branches emerging from them, one for each possible outcome of the die. That will give a total of 12 branches, the same number of possible outcomes determined by using the fundamental counting principle. We can obtain the sample space by listing all the possible combinations of branches. Note that this sample space consists of 12 sample points.

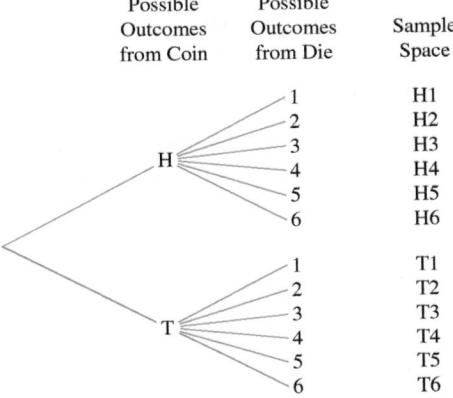

Figure 11.5

Example 2 *Selecting Ticket Winners*

A radio station has two tickets to give away to a Monster Jam truck rally. It held a contest and narrowed the possible recipients down to four people: Christine (C), Mike (M), Larry (L), and Phyllis (P). The names of two of these four people will be randomly selected from a hat, and the two people selected will be awarded the tickets.
a) Determine the number of sample points in the sample space.
b) Construct a tree diagram and list the sample space.
c) Determine the probability that Christine is selected.
d) Determine the probability that Christine is selected and then Mike is selected.

Solution

a) The first selection may be any one of the four people; see Fig. 11.6. Once the first person is selected, only three people remain for the second selection. Thus, there are 4 · 3, or 12, sample points in the sample space.

b)

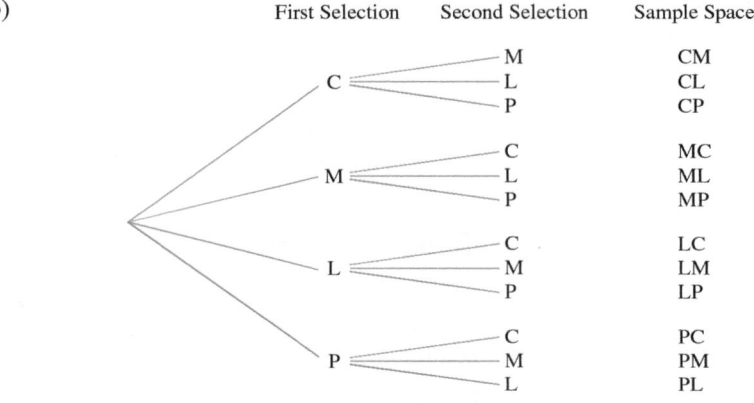

Figure 11.6

c) If we know the sample space, we can compute probabilities using the formula

$$P(E) = \frac{\text{number of outcomes favorable to } E}{\text{total number of outcomes}}$$

The total number of outcomes will be the number of points in the sample space. From Fig 11. 6, we determine that there are 12 possible outcomes when two people are selected. Six of the outcomes have Christine. They are CM, CL, CP, MC, LC, PC.

$$P(\text{Christine is selected}) = \frac{6}{12} = \frac{1}{2}$$

d) One possible outcome meets the criteria when Christine is selected and then Mike is selected: CM.

$$P(\text{Christine is selected and then Mike is selected}) = \frac{1}{12} \qquad \blacksquare$$

The fundamental counting principle can be extended to any number of experiments, as illustrated in Example 3.

Example 3 *Lunch Choices*

At Tops Diner, each lunch special consists of a sandwich, a salad, and a beverage. The sandwich choices are roast beef (R), ham (H), or turkey (T). The salad choices are macaroni (M) or potato (P). The beverage choices are coffee (C) or soda (S).

a) Determine the number of different lunch specials offered by this restaurant.

b) Construct a tree diagram and list the sample space.

c) If a customer randomly selects one of the lunch specials, determine the probability that a roast beef sandwich and soda are selected.

d) If a customer randomly selects one of the lunch specials, determine the probability that neither macaroni salad nor coffee is selected.

Solution

a) There are 3 choices for a sandwich, 2 choices for a salad, and 2 choices for a beverage. Using the fundamental counting principle, we can determine that there are $3 \cdot 2 \cdot 2$, or 12, different lunch specials.

b) The tree diagram illustrating the 12 lunch specials is given in Fig. 11.7.

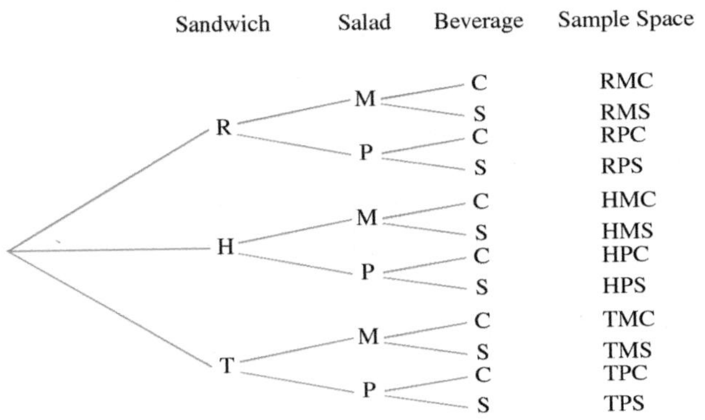

Sandwich	Salad	Beverage	Sample Space
	M	C	RMC
		S	RMS
R	P	C	RPC
		S	RPS
	M	C	HMC
		S	HMS
H	P	C	HPC
		S	HPS
	M	C	TMC
		S	TMS
T	P	C	TPC
		S	TPS

Figure 11.7

c) Of the 12 lunch specials, 2 contain both a roast beef sandwich and soda (RMS, RPS).

$$P(\text{roast beef sandwich and soda are selected}) = \frac{2}{12} = \frac{1}{6}$$

d) Of the 12 lunch specials, 3 contain neither macaroni salad nor coffee (RPS, HPS, TPS).

$$P(\text{neither macaroni salad nor coffee are selected}) = \frac{3}{12} = \frac{1}{4}$$ ∎

Example 4 uses the phrase *with replacement*. As in example 1, this phrase tells us that once an item is selected, it can be selected again, making it possible to select the same item twice.

Example 4 *Selecting Balls with Replacement*

Two balls are to be selected *with replacement* from a bag that contains one red, one blue, one green, and one orange ball (see Fig. 11.8).

a) Determine the number of sample points in the sample space.

b) Construct a tree diagram and list the sample space.

c) Determine the probability that the red ball is selected twice.

d) Determine the probability that neither ball selected is orange.

e) Determine the probability that the orange ball is selected at least once.

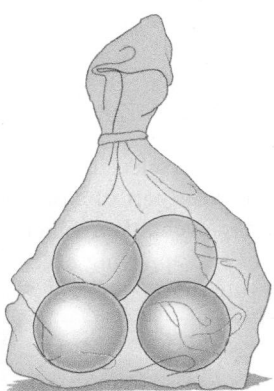

Figure 11.8

Solution

a) The first selection can be any of the four balls. Since the first ball is replaced, all four balls remain for the second selection. Thus, there are 4 · 4, or 16, sample points in the sample space.

b) The first ball selected can be red, blue, green, or orange. Since this experiment is done with replacement, the same colored ball can be selected twice. The tree diagram and sample space are shown in Fig. 11.9. The sample space contains 16 points. That result checks with the answer obtained in part (a) using the fundamental counting principle.

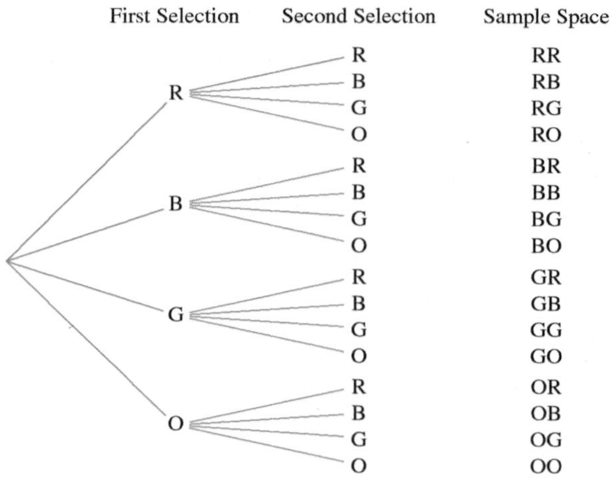

First Selection	Second Selection	Sample Space
R	R	RR
	B	RB
	G	RG
	O	RO
B	R	BR
	B	BB
	G	BG
	O	BO
G	R	GR
	B	GB
	G	GG
	O	GO
O	R	OR
	B	OB
	G	OG
	O	OO

Figure 11.9

c) Of the 16 points in the sample space, one has the red ball chosen twice, RR.

$$P(\text{two red balls}) = \frac{1}{16}$$

d) Of the 16 points in the sample space, nine do not have an orange ball. They are RR, RB, RG, BR, BB, BG, GR, GB, and GG.

$$P(\text{neither ball selected is orange}) = \frac{9}{16}$$

e) Selecting the orange ball at least once means the orange ball is selected one or more times. There are 7 points in the sample space which list the orange ball at least once (RO, BO, GO, OR, OB, OG, OO).

$$P(\text{at least one orange ball is selected}) = \frac{7}{16}$$ ∎

In Example 4, if you add the probability of no orange ball being selected (part d) with the probability of the orange ball being selected at least once (part e), you get $\frac{9}{16} + \frac{7}{16} = \frac{16}{16}$, or 1. In any probability problem, if E is a specific event, then either E happens at least one time or it does not happen at all. Thus, $P(E$ happening at least once$) + P(E$ does not happen$) = 1$, which leads to the following rule.

Probability of an Event Happening at Least Once

The probability of an event happening at least once can be determined by the following formula.

$$P(\text{event happening at least once}) = 1 - P(\text{event does not happen})$$

For example, if you are planting flowers, suppose that the probability of not getting any red flowers from the seeds that are planted is $\frac{2}{7}$. Then the probability of getting at least one red flower from the seeds that are planted is $1 - \frac{2}{7} = \frac{5}{7}$. We will use this rule in later sections.

In all the tree diagrams in this section, the outcomes were always equally likely; that is, each outcome had the same probability of occurrence. Consider a rock that has 4 faces such that each face has a different surface area and the rock is not uniform in density (see Fig. 11.10). When the rock is dropped, the probability that the rock lands on face 1 will not be the same as the probability that the rock lands on face 2. In fact, the probabilities that the rock lands on face 1, face 2, face 3, and face 4 may all be different. Therefore, the outcomes of the rock landing on face 1, face 2, face 3, and face 4 are not equally likely outcomes. Because the outcomes are not equally likely and we are not given additional information, we cannot determine the theoretical probability of the rock landing on each individual face. However, we can still determine the sample space indicating the faces that the rock may land on when the rock is dropped twice. The tree diagram and sample space are shown in Fig. 11.11.

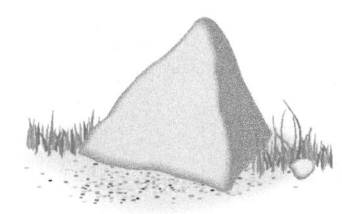

Figure 11.10

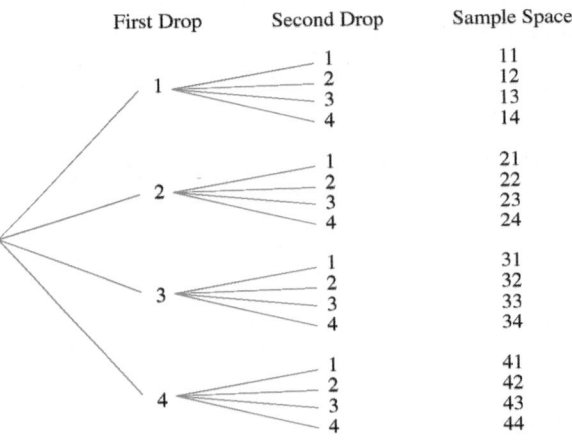

Figure 11.11

Since the outcomes are not equally likely, the probability of each of the 16 sample points in the sample space occurring cannot be determined. If the outcomes were equally likely, then each of the 16 points in the sample space would have a probability of $\frac{1}{16}$. See Exercises 27 and 28 on pages 687–688, which deal with outcomes that are not equally likely.

SECTION 11.4 *Exercises*

Warm Up Exercises

In Exercises 1–4, fill in the blank with an appropriate word, phrase, or symbol(s).

1. A list of all possible outcomes of an experiment is called a(n) _____ space.

2. Each individual outcome in a sample space is called a sample _____.

3. If a first experiment has 3 distinct outcomes and a second experiment has 8 distinct outcomes, then the two experiments in that specific order have _____ distinct outcomes.

4. A helpful method to determine a sample space is to construct a(n) _____ diagram.

Practice the Skills

5. *Selecting Letters* If two lower case letters are randomly selected from the English alphabet, use the fundamental counting principle to determine the number of possible outcomes if the letters are selected

    a) with replacement.   b) without replacement.

6. *Selecting Dates* If two dates are randomly selected from the 365 days of the year, use the fundamental counting

principle to determine the number of possible outcomes if the dates are selected

a) with replacement.

b) without replacement.

7. *Euchre Deck* Euchre is a card game played with a deck that contains 24 cards. If two cards are drawn from a euchre deck, determine the number of possible outcomes if the cards are drawn

a) with replacement.

b) without replacement.

8. *Pinochle Deck* Pinochle is a card game played with a deck that contains 48 cards. If two cards are drawn from a pinochle deck, determine the number of possible outcomes if the cards are drawn

a) with replacement.

b) without replacement.

Problem Solving

In Exercises 9–26, use the fundamental counting principle to determine the answer to part (a). Assume that each event is equally likely to occur.

9. *Coin Toss* Two coins are tossed.

a) Determine the number of sample points in the sample space.

b) Construct a tree diagram and list the sample space.

Determine the probability that

c) two tails are tossed.

d) exactly one tail is tossed.

e) no tails are tossed.

f) at least one tail is tossed.

10. *Boys and Girls* A couple plans to have two children.

a) Determine the number of sample points in the sample space of the possible arrangements of boys and girls.

b) Construct a tree diagram and list the sample space.

Assuming that boys and girls are equally likely, determine the probability that the couple has

c) two girls.

d) a girl and then a boy.

e) no girls

f) at least one girl.

11. *Cards* A box contains three cards. On one card there is a sun, on another card there is a question mark, and on the third card there is an apple. Use S for sun, Q for question mark, and A for apple.

Two cards are to be randomly selected with replacement.

a) Determine the number of sample points in the sample space.

b) Construct a tree diagram and list the sample space.

Determine the probability that

c) two cards containing apples are selected.

d) No cards containing an apple are selected.

e) at least one card that contains an apple is selected.

12. *Cards* Repeat Exercise 11 but assume that the cards are drawn without replacement.

13. *Marble Selection* A hat contains four marbles: 1 yellow, 1 red, 1 blue, and 1 green. Two marbles are to be randomly selected without replacement from the hat.

a) Determine the number of sample points in the sample space.

b) Construct a tree diagram and list the sample space.

Determine the probability of selecting

c) exactly 1 red marble.

d) at least 1 marble that is not red.

e) no green marbles.

14. *Three Coins* Three coins are tossed.

a) Determine the number of sample points in the sample space.

b) Construct a tree diagram and list the sample space.

Determine the probability that

c) no heads are tossed.

d) at least one head is tossed.

e) exactly one head is tossed.

f) three heads are tossed.

15. *Baby Name Choices* The Sannons are choosing a name for their baby girl. They have narrowed down their options, and they will choose one first name and one middle name from the list as shown in the table below.

First Name	Middle Name
Sasha	Aisha
Jade	Wynona
Cali	Opal

a) Determine the number of sample points in the sample space.

b) Construct a tree diagram and list the sample space.

Determine the probability that they select

c) Jade.

d) Jade Opal.

e) a name other than Jade.

16. *Coffee Choices* At the coffee station at work, you have several options when making a cup of coffee. You can choose to add no cream (N), regular cream (R), or hazelnut-flavored cream (H). You can also choose to add no sweetener (O), sugar (S), or aspartame (A).

a) Determine the number of points in the sample space.

b) Construct a tree diagram and list the sample space.

If you were to randomly prepare one cup of coffee, determine the probability that

c) you prepared a cup of coffee with regular cream and sugar?

d) you prepared a cup of coffee with no cream and no sweetener?

e) you prepared a cup of coffee with sugar?

17. *Rolling Dice* A single fair die is rolled twice.

a) Determine the number of sample points in the sample space.

b) Construct a tree diagram and list the sample space.

Determine the probability that

c) a double (a 1, 1 or 2, 2, etc.) is rolled.

d) a sum of 8 is rolled.

e) a sum of 2 is rolled.

f) Are you as likely to roll a sum of 2 as you are of rolling a sum of 8?

18. *Bonsai Trees* Selby Gardens sells bonsai trees. Customers can choose a juniper, maple, fir, or boxwood bonsai tree. Customers can also choose a circular, oval, or square base in which to plant the tree. The Chens are going to randomly select one of the bonsai trees and one of the bases.

a) Determine the number of sample points in the sample space.

b) Construct a tree diagram and list the sample space.

Determine the probability that they select

c) the juniper tree.

d) the maple tree and the square base.

e) a tree other than the boxwood.

19. *Shopping* From the list below, Steve will purchase one appetizer, one beverage, and one dessert at random to bring to a party.

Appetizer	Beverage	Dessert
Vegetable tray	Soda	Pie
Guacamole	Lemonade	Brownies
		Cake

a) Determine the number of sample points in the sample space.

b) Construct a tree diagram and list the sample space.

Determine the probability that Steve selects

c) guacamole for the appetizer.

d) lemonade and brownies.

e) a dessert other than pie.

20. *Gift Cards* Three different people are to be randomly selected and each will be given one gift card. There is one card from Home Depot, one from Best Buy, and one from Red Lobster. The first person selected gets to choose one of the cards. The second person selected gets to choose between the two remaining cards. The third person selected gets the third card.

a) If each card is equally likely to be selected, determine the number of sample points in the sample space.

b) Construct a tree diagram and list the sample space.

Determine the probability that

c) the Best Buy card is selected first.

d) the Home Depot card is selected first and the Red Lobster card is selected last.

e) The cards are selected in this order: Best Buy, Red Lobster, Home Depot.

21. *Courses* You decide to take a science course, an English course, and a mathematics course during the next term. The available courses that you can take are listed below.

Science	English	Mathematics
Biology	Composition	Algebra
Geology	Literature	Statistics
Physics		
Oceanography		

If the courses are randomly selected,

a) determine the number of sample points in the sample space.

b) construct a tree diagram and list the sample space.

Determine the probability that

c) physics is selected.

d) geology and literature are selected.

e) oceanography is not selected.

22. *A New Home Theater System* For your new home theater system you are going to purchase a Blu-ray player, a receiver, and a speaker system at random from among the following brands.

Blu-ray Player	Receiver	Speaker System
LG	Yamaha	Bose
Toshiba	Onkyo	JBL
Sony	Pioneer	
Insignia		

a) Determine the number of sample points in the sample space.

b) Construct a tree diagram and list the sample space.

Determine the probability of selecting

c) a Sony Blu-ray player.

d) Bose speakers.

e) a Sony Blu-ray player and Bose speakers.

23. *Landscaping* The Franks just moved into a new home and need to do some landscaping. They are going to purchase one tree, one shrub, and one lilac bush at random from the list below.

Trees	Shrubs	Lilac Bushes
Dogwood	Forsythia	Common
Maple	Holly	Sensation
Birch	Juniper	Primrose

a) Determine the number of sample points in the sample space.

b) Construct a tree diagram and list the sample space.

Determine the probability that the Franks select

c) a maple tree.

d) a dogwood tree and a holly shrub.

e) a lilac bush other than Sensation.

24. *Literature Choices* You decide to take a literature course. A requirement for the course is that you must read one classic book, one nonfiction book, and one science fiction book from the list below.

Classic	Nonfiction	Science Fiction
Ficciones (F)	*John Adams (J)*	*Dune (D)*
Go Tell It on the Mountain (G)	*Angela's Ashes (A)*	*Ender's Game (E)*
Tom Sawyer (T)		
Moby-Dick (M)		

If the books are randomly selected,

a) determine the number of sample points in the sample space.

b) construct a tree diagram and list the sample space.

Determine the probability that

c) *John Adams* is selected.

d) either *Ficciones* or *Tom Sawyer* is selected.

e) *Moby-Dick* is not selected.

25. *Clothing Options* Brisson is getting dressed for work and is considering his options for a shirt, a tie, and a pair of slacks to wear. He has shirts available in white, pink, and yellow. He has ties available in red, gray, and blue. He has pairs of slacks available in navy and khaki.

a) How many different options for a shirt, a tie, and a pair of slacks does Brisson have?

b) Construct a tree diagram and list the sample space.

Assuming all options are equally likely, what is the probability that Brisson chooses

c) a white shirt, a red tie, and a navy pair of slacks?

d) a gray tie?

e) a shirt other than a white shirt?

26. *Mendel Revisited* A pea plant must have exactly one of each of the following pairs of traits: short (*s*) or tall (*t*); round (*r*) or wrinkled (*w*) seeds; yellow (*y*) or green (*g*) peas; and white (*wh*) or purple (*p*) flowers (for example, short, wrinkled, green pea with white flowers).

a) How many different classifications of pea plants are possible?

b) Use a tree diagram to list all the classifications possible.

If each characteristic is equally likely

c) determine the probability that the pea plant will have round peas.

d) determine the probability that the pea plant will be short, have wrinkled seeds, have yellow seeds, and have purple flowers.

Challenge Problems/Group Activities

27. *Three Chips* Suppose that a bag contains one white chip and two red chips. Two chips are going to be randomly selected from the bag *with replacement*.

a) What is the probability of selecting a white chip from the bag on the first selection?

b) What are the odds against selecting a white chip from the bag on the first selection?

c) What is the probability of selecting a red chip from the bag on the first selection?

d) Are the outcomes of selecting a white chip and selecting a red chip on the first selection equally likely? Explain.

e) The sample space when two chips are selected from the bag with replacement is ww, wr, rw, rr. Do you believe that the probability of selecting ww is greater than, equal to, or less than the probability of selecting rr? Explain.

28. *Thumbtacks* A thumbtack is dropped on a concrete floor. Assume that the thumbtack can land only point up (u) or point down (d), as shown in the figure below.

If two thumbtacks are dropped, one after the other, the following tree diagram can be used to show the possible outcomes.

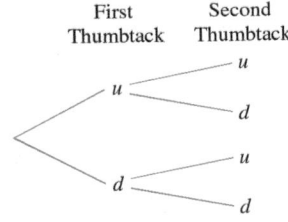

a) Do you believe that the outcomes of the thumbtack landing point up and the thumbtack landing point down are equally likely? Explain.

b) List the sample points in the sample space of this experiment.

c) Do you believe that the probability that both thumbtacks land point up (*uu*) is the same as the probability that both thumbtacks land point down (*dd*)? Explain.

d) Can you compute the theoretical probability of a thumbtack landing point up and the theoretical probability of a thumbtack landing point down? Explain.

e) Obtain a box of thumbtacks and drop the thumbtacks out of the box with care. Determine the empirical probability of a thumbtack landing point up when dropped and the empirical probability of a thumbtack landing point down when dropped.

Recreational Mathematics

29. *Ties* All my ties are red except two. All my ties are blue except two. All my ties are brown except two. How many ties do I have?

30. *Rock Faces* An experiment consists of 3 parts: flipping a coin, tossing a rock, and rolling a die. If the sample space consists of 60 sample points, determine the number of faces on the rock.

JUMBLE *For Exercises 31 and 32, refer to the Recreational Math box on page 680. Determine*

a) *the number of possible arrangements of the letters given.*

b) *the word when the letters are placed in the proper order.*

31. ROTES

32. STEABK

SECTION 11.5   *OR* and *AND* Problems

Upon completion of this section, you will be able to:

- Understand and solve probability problems that involve the word *or*.
- Understand and solve probability problems that involve the word *and*.

The words *or* and *and* are part of our everyday vocabulary. These words also play an important role in the study of probability. For example, consider the following two questions related to drawing a single card from a standard 52-card deck. What is the probability of drawing a king *or* a heart? What is the probability of drawing a king *and* a heart? In this section, we will discuss these and similar probability problems that involve the words *or* and *and*.

Why This Is Important Probability involving the words *or* and *and* is used in a wide variety of science, business, computer science, and engineering applications. A basic understanding of the probability we present in this section is an important skill that can be applied to a wide variety of occupations.

In Section 11.4, we showed how to work probability problems by constructing sample spaces. Often it is inconvenient or too time consuming to solve a problem by first constructing a sample space. For example, if an experiment consists of drawing two cards with replacement from a standard 52-card deck, there would be 52 · 52, or 2704, points in the sample space. Trying to list all these sample points could take hours. In this section, we learn how to solve *compound probability* problems that contain the words *or* or *and* without constructing a sample space.

Or Problems

The *or probability problem* requires obtaining a "successful" outcome for *at least one* of the given events. For example, suppose that we roll one die and we are interested in determining the probability of rolling an even number *or* a number greater than 4. For this situation, rolling either a 2, 4, or 6 (an even number) or a 5 or 6 (a number greater than 4) would be considered successful. Note that the number 6 satisfies both criteria. Since 4 of the 6 numbers meet the criteria (the 2, 4, 5, and 6), the probability of rolling an even number *or* a number greater than 4 is $\frac{4}{6}$ or $\frac{2}{3}$.

A formula for determining the probability of event A or event B, symbolized $P(A \text{ or } B)$, follows.

> **Probability of A or B**
> To determine the probability of A or B, use the following formula.
> $$P(A \text{ or } B) = P(A) + P(B) - P(A \text{ and } B)$$

Since we add (and subtract) probabilities to determine $P(A \text{ or } B)$, this formula is sometimes referred to as the *addition formula*. We explain the use of the *or* formula in Example 1.

Example 1 *Using the Addition Formula*

Each of the numbers 1, 2, 3, 4, 5, 6, 7, 8, 9, and 10 is written on a separate piece of paper. The 10 pieces of paper are then placed in a hat, and one piece is randomly selected. Determine the probability that the piece of paper selected contains an even number or a number greater than 6.

Solution We are asked to determine the probability that the number selected is *even* or is *greater than 6*. Let's use set A to represent the statement "the number is even" and set B to represent the statement "the number is greater than 6." Figure 11.12 is a Venn diagram, as introduced in Chapter 2, with sets A (even) and B (greater than 6).

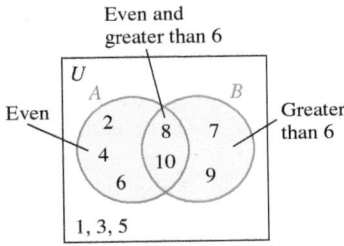

Figure 11.12

There are a total of 10 numbers, of which five are even (2, 4, 6, 8, and 10). Thus, the probability of selecting an even number is $\frac{5}{10}$. Four numbers are greater than 6: the 7, 8, 9, and 10. Thus, the probability of selecting a number greater than 6 is $\frac{4}{10}$. Two numbers are both even and greater than 6: the 8 and 10. Thus, the probability of selecting a number that is both even and greater than 6 is $\frac{2}{10}$.

If we substitute the appropriate statements for A and B in the formula, we obtain

$$P(A \text{ or } B) = P(A) + P(B) - P(A \text{ and } B)$$

$$P\left(\begin{array}{c}\text{even or}\\\text{greater than 6}\end{array}\right) = P(\text{even}) + P\left(\begin{array}{c}\text{greater}\\\text{than 6}\end{array}\right) - P\left(\begin{array}{c}\text{even and}\\\text{greater than 6}\end{array}\right)$$

$$= \frac{5}{10} + \frac{4}{10} - \frac{2}{10}$$

$$= \frac{7}{10}$$

Thus, the probability of selecting an even number or a number greater than 6 is $\frac{7}{10}$. The seven numbers that are even or greater than 6 are 2, 4, 6, 7, 8, 9, and 10. ∎

Example 1 illustrates that when determining the probability of A or B, we add the probabilities of events A and B and then subtract the probability of both events occurring simultaneously.

Example 2 *Using the Addition Formula*

Consider the same sample space, the numbers 1 through 10, as in Example 1. If one piece of paper is selected, determine the probability that it contains a number less than 5 or a number greater than 8.

Solution Let A represent the statement "the number is less than 5" and B represent the statement "the number is greater than 8." A Venn diagram illustrating these statements is shown in Figure 11.13.

$$P(\text{number is less than 5}) = \frac{4}{10}$$

$$P(\text{number is greater than 8}) = \frac{2}{10}$$

Since there are no numbers that are *both* less than 5 and greater than 8, P (number is less than 5 and greater than 8) $= 0$. Therefore,

$$P\begin{pmatrix}\text{number is}\\\text{less than 5}\\\text{or greater}\\\text{than 8}\end{pmatrix} = P\begin{pmatrix}\text{number is}\\\text{less than 5}\end{pmatrix} + P\begin{pmatrix}\text{number is}\\\text{greater than 8}\end{pmatrix} - P\begin{pmatrix}\text{number is}\\\text{less than 5}\\\text{and greater}\\\text{than 8}\end{pmatrix}$$

$$= \frac{4}{10} + \frac{2}{10} - 0 = \frac{6}{10} = \frac{3}{5}$$

Thus, the probability of selecting a number less than 5 or greater than 8 is $\frac{3}{5}$. The six numbers that are less than 5 or greater than 8 are 1, 2, 3, 4, 9, and 10. ∎

In Example 2, it is impossible to select a number that is both less than 5 *and* greater than 8 when only one number is to be selected. Events such as these are said to be *mutually exclusive*.

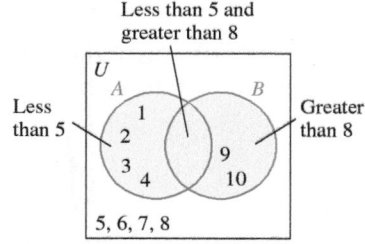

Figure 11.13

> **Definition: Mutually Exclusive**
> Two events A and B are **mutually exclusive** if it is impossible for both events to occur simultaneously.

If events A and B are mutually exclusive, then $P(A \text{ and } B) = 0$ and the addition formula simplifies to $P(A \text{ or } B) = P(A) + P(B)$.

Example 3 *Probability of A or B*

One card is drawn from a standard 52-card deck. Determine whether the following pairs of events are mutually exclusive and determine P $(A$ or $B)$.

a) A = a jack, B = a queen
b) A = an ace, B = a heart
c) A = a red card, B = a black card
d) A = a face card, B = a red card

Solution

a) There are four jacks and four queens in a standard deck of 52 cards. It is impossible to select both a jack and a queen when only one card is selected. Therefore, these events are mutually exclusive.

$$P(\text{jack or queen}) = P(\text{jack}) + P(\text{queen}) = \frac{4}{52} + \frac{4}{52} = \frac{8}{52} = \frac{2}{13}$$

b) There are 4 aces and 13 hearts in a standard 52-card deck. One card, the ace of hearts, is both an ace and a heart. Therefore, these events are not mutually exclusive.

$$P(\text{ace}) = \frac{4}{52} \qquad P(\text{heart}) = \frac{13}{52} \qquad P(\text{ace and heart}) = \frac{1}{52}$$

$$P(\text{ace or heart}) = P(\text{ace}) + P(\text{heart}) - P(\text{ace and heart})$$

$$= \frac{4}{52} + \frac{13}{52} - \frac{1}{52}$$

$$= \frac{16}{52} = \frac{4}{13}$$

The ace of hearts is
both an ace and a heart.

c) There are 26 red cards and 26 black cards in a standard 52-card deck. It is impossible to select one card that is both a red card and a black card. Therefore, the events are mutually exclusive.

$$P(\text{red or black}) = P(\text{red}) + P(\text{black})$$

$$= \frac{26}{52} + \frac{26}{52} = \frac{52}{52} = 1$$

Since $P(\text{red or black}) = 1$, a red card or a black card must be selected.

d) There are 12 face cards in a standard 52-card deck. Six of the 12 face cards are red (the jacks, queens, and kings of hearts and diamonds). Thus, selecting a face card and a red card are not mutually exclusive.

$$P\left(\begin{array}{c}\text{face card}\\ \text{or red card}\end{array}\right) = P\left(\begin{array}{c}\text{face}\\ \text{card}\end{array}\right) + P\left(\begin{array}{c}\text{red}\\ \text{card}\end{array}\right) - P\left(\begin{array}{c}\text{face card}\\ \text{and red card}\end{array}\right)$$

$$= \frac{12}{52} + \frac{26}{52} - \frac{6}{52}$$

$$= \frac{32}{52} = \frac{8}{13}$$

And Problems

A second type of probability problem is the *and probability problem*, which requires obtaining a favorable outcome in *each* of the given events. For example, suppose that *two* cards are to be drawn from a standard 52-card deck and we are interested in the probability of selecting two aces (one ace *and* then a second ace). Only if *both* cards selected are aces would this experiment be considered successful. A formula for determining the probability of events A and B, symbolized $P(A \text{ and } B)$, follows.

Probability of A and B
To determine the probability of A and B, use the following formula.

$$P(A \text{ and } B) = P(A) \cdot P(B)$$

Since we multiply to determine $P(A \text{ and } B)$, this formula is sometimes referred to as the *multiplication formula*. **When using the multiplication formula, we always assume that event A has occurred when calculating $P(B)$** because we are determining the probability of obtaining a favorable outcome in both of the given events.*

Unless we specify otherwise, $P(A \text{ and } B)$ indicates that we are determining the probability that event A occurs *and then* event B occurs (in that order). Consider a

---

*$P(B)$, assuming that event A has occurred, may be denoted $P(B \mid A)$, which is read "the probability of B, given A." We will discuss this type of probability (conditional probability) further in Section 11.6.

bag that contains three chips: 1 red chip (r), 1 blue chip (b), and 1 green chip (g). Suppose that two chips are selected from the bag with replacement. The tree diagram and sample space for the experiment are shown in Fig. 11.14. There are nine possible outcomes for the two selections, as indicated in the sample space.

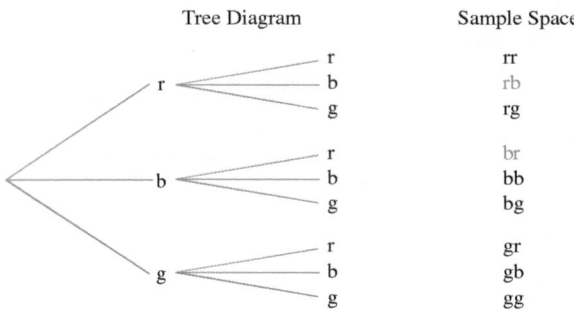

Figure 11.14

Note that the probability of selecting a red chip followed by a blue chip (rb), indicated by P(red and blue), is $\frac{1}{9}$. The probability of selecting a red chip and a blue chip, in any order (rb or br), is $\frac{2}{9}$. In this section, when we ask for $P(A \text{ and } B)$, it means the probability of event A occurring *and then* event B occurring, in that order.

Example 4 *An Experiment with Replacement*

Two cards are to be drawn *with replacement* from a standard 52-card deck. Determine the probability that two spades will be selected.

Solution Since the deck of 52 cards contains thirteen spades, the probability of selecting a spade on the first draw is $\frac{13}{52}$. The card selected is then returned to the deck. Therefore, the probability of selecting a spade on the second draw remains $\frac{13}{52}$.

If we let A represent the selection of the first spade and B represent the selection of the second spade, the formula may be written as follows.

$$P(A \text{ and } B) = P(A) \cdot P(B)$$

$$P(2 \text{ spades}) = P(\text{spade 1 } and \text{ spade 2}) = P(\text{spade 1}) \cdot P(\text{spade 2})$$

$$= \frac{13}{52} \cdot \frac{13}{52}$$

$$= \frac{1}{4} \cdot \frac{1}{4} = \frac{1}{16}$$ ∎

Example 5 *An Experiment without Replacement*

Two cards are to be drawn *without replacement* from a standard 52-card deck. Determine the probability that two spades will be selected.

Solution This example is similar to Example 4. However, this time we are doing the experiment without replacing the first card selected to the deck before selecting the second card.

The probability of selecting a spade on the first draw is $\frac{13}{52}$. When calculating the probability of selecting the second spade, we must assume that the first spade has been selected. Once this first spade has been selected, only 51 cards, including 12 spades, remain in the deck. The probability of selecting a spade on the

StatCrunch

Applets Experiment Birthdays

second draw becomes $\frac{12}{51}$. The probability of selecting two spades without replacement is

$$P(2 \text{ spades}) = P(\text{spade 1}) \cdot P(\text{spade 2})$$
$$= \frac{13}{52} \cdot \frac{12}{51}$$
$$= \frac{1}{4} \cdot \frac{4}{17} = \frac{1}{17}$$

Now we introduce *independent events*.

> ### Definition: Independent Events
> Event A and event B are **independent events** if the occurrence of either event in no way affects the probability of occurrence of the other event.

Rolling dice and tossing coins are examples of independent events. In Example 4, the events are independent, since the first card was returned to the deck. The probability of selecting a spade on the second draw was not affected by the first selection. The events in Example 5 are not independent, since the probability of the selection of the second spade was affected by removing the first spade selected from the deck. Such events are called *dependent events*. *Experiments done with replacement will result in independent events, and those done without replacement will result in dependent events.*

Example 6 *Independent or Dependent Events?*

At a fundraiser for the Sarasota Pet Shelter, a raffle is held in which 100 tickets are sold. Three tickets are randomly drawn from a bin without replacement, and the winners each receive a prize. Are the events of selecting the three winners independent or dependent events?

Solution The tickets are selected *without* replacement. Thus, once a ticket is drawn from the bin, the ticket cannot be drawn again a second or third time. Therefore, the events are dependent, since each time a ticket is drawn, the probability changes for the subsequent drawing. In the first drawing, the probability that a specific ticket is selected is $\frac{1}{100}$. If that specific ticket is not drawn, then in the second drawing, the probability that it is drawn is now $\frac{1}{99}$. If that specific ticket is still not drawn, then in the third drawing, the probability that it is drawn is now $\frac{1}{98}$. In general, in any experiment in which two or more items are selected *without* replacement, the events will be dependent.

The multiplication formula may be extended to more than two events, as illustrated in Example 7.

Example 7 *Medical Research*

A medical research study on a new medicine for rheumatoid arthritis is being conducted with 25 patients. After the study was concluded, it was determined that 19 patients reacted favorably to the medicine, 2 reacted unfavorably, and 4 were unaffected. If 3 of the patients are randomly selected, determine the probability of each of the following.

a) All three reacted favorably.

b) The first patient reacted favorably, the second patient reacted unfavorably, and the third patient was unaffected.

c) No patient reacted favorably.

d) At least one patient reacted favorably.

Solution Each time a patient is selected, the number of patients remaining decreases by 1.

a) The probability that the first patient reacted favorably is $\frac{19}{25}$. If the first patient reacted favorably, of the 24 remaining patients only 18 patients are left who reacted favorably. The probability of selecting a second patient who reacted favorably is $\frac{18}{24}$. If the second patient reacted favorably, only 17 patients are left who reacted favorably. The probability of selecting a third patient who reacted favorably is $\frac{17}{23}$.

$$P\begin{pmatrix}\text{three patients}\\\text{reacted}\\\text{favorably}\end{pmatrix}=P\begin{pmatrix}\text{first patient}\\\text{reacted}\\\text{favorably}\end{pmatrix}\cdot P\begin{pmatrix}\text{second patient}\\\text{reacted}\\\text{favorably}\end{pmatrix}\cdot P\begin{pmatrix}\text{third patient}\\\text{reacted}\\\text{favorably}\end{pmatrix}$$

$$=\frac{19}{25}\cdot\frac{18}{24}\cdot\frac{17}{23}=\frac{969}{2300}$$

b) The probability that the first patient reacted favorably is $\frac{19}{25}$. Once a patient is selected, there are only 24 patients remaining. Two of the remaining 24 patients reacted unfavorably. Thus, the probability that the second patient reacted unfavorably is $\frac{2}{24}$. After the second patient is selected, there are 23 remaining patients, of which 4 were unaffected. The probability that the third patient was unaffected is therefore $\frac{4}{23}$.

$$P\begin{pmatrix}\text{first patient reacted favorably, the}\\\text{second patient reacted unfavorably, and}\\\text{the third patient was unaffected}\end{pmatrix}$$

$$=P\begin{pmatrix}\text{first patient}\\\text{reacted}\\\text{favorably}\end{pmatrix}\cdot P\begin{pmatrix}\text{second patient}\\\text{reacted}\\\text{unfavorably}\end{pmatrix}\cdot P\begin{pmatrix}\text{third patient}\\\text{was}\\\text{unaffected}\end{pmatrix}$$

$$=\frac{19}{25}\cdot\frac{2}{24}\cdot\frac{4}{23}=\frac{19}{1725}$$

c) If none of the patients reacted favorably, the patients either reacted unfavorably or were unaffected. Six patients did not react favorably (2 reacted unfavorably and 4 were unaffected). The probability that the first patient selected did not react favorably is $\frac{6}{25}$. After the first patient is selected, 5 of the remaining 24 patients did not react favorably. After the second patient is selected, 4 of the remaining 23 patients did not react favorably.

$$P\begin{pmatrix}\text{none}\\\text{reacted}\\\text{favorably}\end{pmatrix}=P\begin{pmatrix}\text{first patient}\\\text{did not react}\\\text{favorably}\end{pmatrix}\cdot P\begin{pmatrix}\text{second patient}\\\text{did not react}\\\text{favorably}\end{pmatrix}\cdot P\begin{pmatrix}\text{third patient}\\\text{did not react}\\\text{favorably}\end{pmatrix}$$

$$=\frac{6}{25}\cdot\frac{5}{24}\cdot\frac{4}{23}=\frac{1}{115}$$

d) In Section 11.4, we learned that

$$P(\text{event happening at least once}) = 1 - P(\text{event does not happen})$$

Timely Tip

Which formula to use
It is sometimes difficult to determine when to use the *or* formula and when to use the *and* formula. The following information may be helpful in deciding which formula to use.

Or formula
Or problems will almost always contain the word *or* in the statement of the problem. For example, determine the probability of selecting a heart *or* a 6. *Or* problems generally involve only *one* selection. For example, "one card is selected" or "one die is rolled."

And formula
And problems often do *not* use the word *and* in the statement of the problem. For example, "determine the probability that both cards selected are red" or "determine the probability that none of those selected is a banana" are both *and*-type problems. *And* problems will generally involve *more* than one selection. For example, the problem may read "two cards are selected" or "three coins are flipped."

In part (c), we determined that the probability of selecting three patients none of whom reacted favorably was $\frac{1}{115}$. Therefore, the probability that at least one of the patients selected reacted favorably can be determined as follows.

$$P\left(\begin{array}{c}\text{at least one of the three}\\\text{patients reacted favorably}\end{array}\right) = 1 - P\left(\begin{array}{c}\text{none of the three}\\\text{patients reacted favorably}\end{array}\right)$$

$$= 1 - \frac{1}{115} = \frac{115}{115} - \frac{1}{115} = \frac{114}{115} \qquad \blacksquare$$

SECTION 11.5   *Exercises*

Warm Up Exercises

In Exercises 1–10, fill in the blank with an appropriate word, phrase, or symbol(s).

1. Probability problems that contain the words *and* or *or* are considered _____ probability problems.

2. Probability problems that require obtaining a successful outcome for at least one of the given events are _____ probability problems.

3. Probability problems that require obtaining a favorable outcome in each of the given events are _____ probability problems.

4. If it is impossible for two events A and B to occur simultaneously, then the events are considered to be _____ exclusive.

5. For two events A and B, if the occurrence of either event in no way affects the probability of the occurrence of the other event, then the two events are considered to be _____ events.

6. For two events A and B, if the occurrence of either event has an effect on the probability of the occurrence of the other event, then the two events are considered to be _____ events.

7. Experiments done with replacement will result in _____ events.

8. Experiments done without replacement will result in _____ events.

9. The formula for determining the probability of event A or event B is $P(A \text{ or } B) =$ _____.

10. The formula for determining the probability of event A and event B is $P(A \text{ and } B) =$ _____.

Practice the Skills

Number Drawing *In Exercises 11–12, each of the numbers 1, 2, 3, 4, 5, 6, 7, 8, 9, and 10 are written on a separate piece of paper and placed in a hat. One piece of paper is randomly*

selected from the hat. Use the Venn diagram shown to help you determine the probability that the number selected is

11. odd or greater than 5.

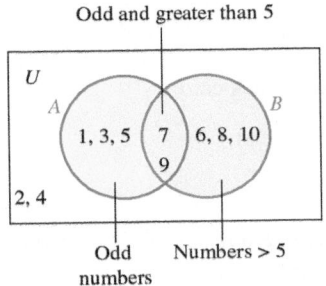

Odd and greater than 5

12. even or less than or equal to 4.

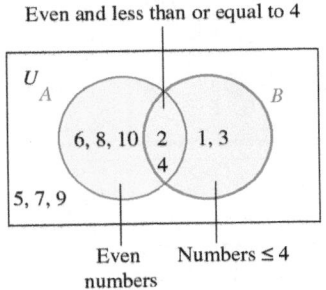

Even and less than or equal to 4

In Exercises 13–18, determine the indicated probability.

13. If $P(A) = 0.3, P(B) = 0.5$, and $P(A \text{ and } B) = 0.2$, determine $P(A \text{ or } B)$.

14. If $P(A) = 0.7, P(B) = 0.8$, and $P(A \text{ and } B) = 0.6$, determine $P(A \text{ or } B)$.

15. If $P(A \text{ or } B) = 0.9, P(A) = 0.8$, and $P(B) = 0.2$, determine $P(A \text{ and } B)$.

16. If $P(A \text{ or } B) = 0.65, P(A) = 0.35$, and $P(B) = 0.45$, determine $P(A \text{ and } B)$.

17. If $P(A \text{ or } B) = 0.7, P(A) = 0.6$, and $P(A \text{ and } B) = 0.3$, determine $P(B)$.

18. If $P(A \text{ or } B) = 0.6, P(B) = 0.3$, and $P(A \text{ and } B) = 0.1$, determine $P(A)$.

Roll a Die *In Exercises 19–22, a single die is rolled one time. Determine the probability of rolling*

19. a 5 or 6.

20. a number greater than 4 or less than 2.

21. an odd number or a number greater than 4.

22. an even number or a number greater than 1.

Select One Card *In Exercises 23–28, one card is drawn from a standard 52-card deck. Determine the probability of selecting*

23. a 2 or a 3.

24. a face card or an ace.

25. a 4 or a diamond.

26. a 5 or a red card.

27. a face card or a black card.

28. a face card or a heart.

Problem Solving

Select Two Cards *In Exercises 29–36, a board game uses the deck of 20 cards shown.*

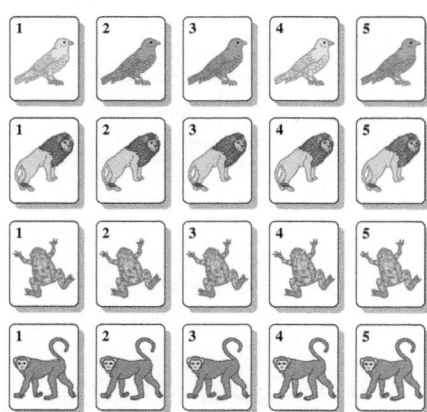

Two cards are randomly selected from this deck. Determine the probability of the following

    a) with replacement.

    b) without replacement.

29. They both show birds.

30. They both show the number 3.

31. The first shows a bird, and the second shows a monkey.

32. The first shows a 2, and the second shows a 4.

33. The first shows a yellow bird, and the second shows a lion.

34. The first shows a red bird and the second shows a frog.

35. Neither shows an even number.

36. At least one of the cards shows an even number

Select One Card *In Exercises 37–40, use the deck of cards given in Exercises 29–36. If one card is selected, determine the probability that the card shows*

37. a lion or an even number.

38. a yellow bird or a number less than 3.

39. a lion or a 3.

40. a red bird or an even number.

Two Spins *If the pointer in Fig. 11.15 is spun twice, determine the probability that the pointer lands on*

41. green and then red.

42. red on both spins.

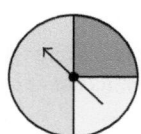

Figure 11.15

Two Spins *If the pointer in Fig. 11.16 is spun twice, determine the probability that the pointer lands on*

43. red on both spins.

44. a color other than green on both spins.

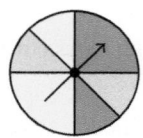

Figure 11.16

Two Spins *In Exercises 45–48, assume that the pointer in Fig. 11.15 is spun and then the pointer in Fig. 11.16 is spun. Determine the probability of the pointers landing on*

45. red on both spins.

46. yellow on the first spin and red on the second spin.

47. a color other than red on both spins.

48. yellow on the first spin and a color other than yellow on the second spin.

Selecting an Envelope *In Exercises 49–58, consider the colored envelopes shown below.*

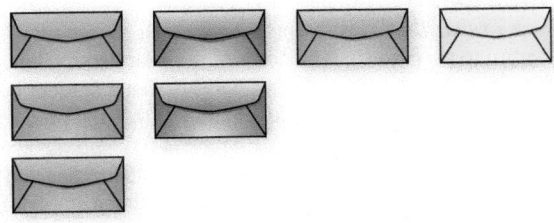

If one of the envelopes is randomly selected, determine the probability that

49. a yellow or a blue envelope is selected.

50. an envelope other than a red envelope is selected.

If two envelopes are randomly selected, with replacement, determine the probability that

51. neither are red envelopes.

52. both are red envelopes.

53. at least one envelope is a red envelope.

54. the first is a blue envelope and the second is a yellow envelope.

If three envelopes are randomly selected, without replacement, determine the probability that

55. they are all red envelopes.

56. none is a red envelope.

57. at least one envelope is a red envelope.

58. the first is a red envelope, the second is a blue envelope, and the third is a blue envelope.

Boys and Girls *In Exercises 59–62, a couple has three children. Assuming independence and that the probability of a boy is $\frac{1}{2}$, determine the probability that*

59. all three children are girls.

60. all three children are boys.

61. at least one child is a girl.

62. the youngest child is a boy and the two older children are girls.

63. a) *Five Children* The Martinos plan to have five children. Determine the probability that all their children will be boys. (Assume that $P(\text{boy}) = \frac{1}{2}$ and assume independence.)

   b) If their first four children are boys and they are expecting another child, what is the probability that the fifth child will be a boy?

64. a) *The Probability of a Girl* The Bronsons plan to have eight children. Determine the probability that all their children will be girls. (Assume that $P(\text{girl}) = \frac{1}{2}$ and assume independence.)

   b) If their first seven children are girls and they are expecting another child, what is the probability that the eighth child will be a girl?

Baseball Bucket *In Exercises, 65–68, Scott has a bucket containing baseballs. He has 18 Diamond baseballs, 14 Rawlings baseballs, and 4 Wilson baseballs. Scott will randomly select two baseballs from the bucket. Determine the probability of selecting each of the following*

   a) *with replacement.*

   b) *without replacement.*

65. a Diamond baseball and then a Wilson baseball

66. a Wilson baseball and then a Rawlings baseball

67. no Diamond baseballs

68. at least one Diamond baseball

Health Insurance *In Exercises 69–72, a sample of 75 people yielded the following information about their health insurance.*

Number of People	Type of Insurance
33	Managed care plan
27	Traditional insurance
15	No insurance

Two people who provided information for the table were randomly selected, without replacement. Determine the probability that

69. neither had traditional insurance.

70. they both had a managed care plan.

71. at least one had traditional insurance.

72. the first had traditional insurance and the second had a managed care plan.

Hiring An Attorney In Exercises 73–76, a sample of 30 people who recently hired an attorney yielded the following information about their attorneys.

Number of People	Would You Recommend Your Attorney to a Friend
23	Yes
3	No
4	Not sure

Three people who provided information for the table were randomly selected. Determine the probability that

73. they would all recommend their attorneys.

74. the first would not recommend the attorney, but the second and third would recommend their attorneys.

75. the first two would not recommend their attorneys, and the third is not sure if he or she would recommend the attorney.

76. the first would recommend the attorney, but the second and third would not recommend their attorneys.

Medical Research In Exercises 77–80, a medical research study on a new medicine for multiple sclerosis is being conducted with 24 patients. After the study was concluded, it was determined that 16 patients reacted favorably to the medicine, 5 reacted unfavorably, and 3 were unaffected. If three of the patients are randomly selected, determine the probability of each of the following.

77. All the patients reacted favorably.

78. All the patients reacted unfavorably.

79. None of the patients reacted favorably.

80. At least one of the patients reacted favorably.

Two Wheels In Exercises 81–84, suppose that you spin the double wheel shown below.

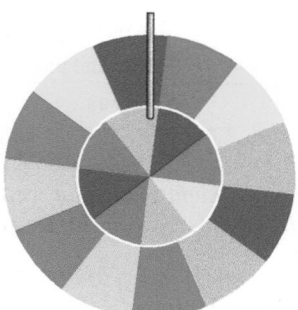

Assuming that the wheels are independent and each sector is equally likely, determine the probability that you get

81. blue on both wheels.

82. red on the outer wheel and blue on the inner wheel.

83. a color other than red on both wheels.

84. red on at least one wheel.

85. *Exam Preparation* Professor Nguyen is in charge of a program to prepare students for a high school equivalency exam. Records show that the probability that a student in the program needs help in mathematics is 0.7, the probability that a student needs help in English is 0.6, and the probability that a student needs help in both mathematics and English is 0.55. Determine the probability that a student in the program needs help in mathematics or English.

86. *Car Repair* The manager at Arango Automotive has determined that the probability that a car brought into the shop requires an oil change is 0.6, the probability that a car brought into the shop requires brake repair is 0.2, and the probability that a car requires both an oil change and brake repair is 0.1. For a car brought into the shop, determine the probability that the car will require an oil change or brake repair.

Hitting a Target In Exercises 87–92, the probability that a heat-seeking torpedo will hit its target is 0.4. If the first torpedo hits its target, the probability that the second torpedo will hit the target increases to 0.9 because of the extra heat generated by the first explosion.

87. Are the events of firing two heat-seeking torpedoes at a target independent events?

88. If one heat-seeking torpedo is fired at a target, determine the probability that it hits the target.

If two heat-seeking torpedoes are fired at a target, determine the probability that

89. neither hits the target.

90. the first hits the target and the second misses the target.

91. both hit the target.

92. the first misses the target and the second hits the target.

Polygenic Disabilities In Exercises 93–98, certain congenital disabilities are polygenic in nature. Typically, the chance that a child will be born with a polygenic disability is small. However, once a child is born with a polygenic disability, the probability that future children from the same parents will be born with the same disability increases. Assume that the probability of a child being born with polygenic disability A is 0.001. If a couple has a child with this disability, then the probability that the same couple has another child with the same disability becomes 0.04.

93. Are the events of children being born, from the same parents, having the polygenic disability independent events?

94. A couple plans to have one child. Determine the probability that the child will be born with the disability.

If a couple plans to have two children, determine the probability that

95. both children will be born with the disability.

96. the first will be born with the disability and the second will not.

97. the first will not be born with the disability and the second will.

98. neither will be born with the disability.

Challenge Problems/Group Activities

99. *Picking Chips*   A bag contains five red chips, three blue chips, and two yellow chips. Two chips are selected from the bag without replacement. If each chip is equally likely to be selected, determine the probability that two chips of the same color are selected.

100. *Peso Coins*   Miguel has ten coins from Mexico: three 1-peso coins, one 2-peso coin, two 5-peso coins, one 10-peso coin, and three 20-peso coins. He selects two coins without replacement. Assuming that each coin is equally likely to be selected, determine the probability that Miguel selects at least one 1-peso coin.

101. *A Fair Game?*   Two playing cards are dealt to you from a standard 52-card deck. If either card is a diamond or if both are diamonds, you win; otherwise, you lose. Determine whether this game favors you, is fair, or favors the dealer. Explain your answer.

102. *Face Card Probability*   You have three cards: an ace, a king, and a queen. A friend shuffles the cards, randomly selects two of them, and discards the third. You ask your friend to show you a face card, and she turns over the king. What is the probability that she also has the queen?

Recreational Mathematics

*A **Different Die**   For Exercises 103–106, consider a six-sided die that has 1 dot on one side, 2 dots on two sides, and 3 dots on three sides.*

If the die is rolled twice, determine the probability of rolling

103. two 2's.

104. two 3's.

If the die is rolled only once, determine the probability of rolling

105. an even number or a number less than 3.

106. an odd number or a number greater than 1.

107. *The Birthday Problem*   Read the Did You Know? on page 693. Use this information to calculate the probability that at least two people in your class have the same birthday. Poll your classmates to determine whether two or more of them have the same birthday.

Research Activity

108. *Girolamo Cardano*   Girolamo Cardano (1501–1576) wrote *Liber de Ludo Aleae*, which is considered to be the first book on probability. Cardano had a number of different vocations. Do research and write a paper on the life and accomplishments of Girolamo Cardano.

SECTION 11.6   Conditional Probability

Upon completion of this section, you will be able to:

■ Solve conditional probability problems.

The probability of an event often depends on other conditions related to the event. For example, the probability that it rains on a given day depends on many conditions such as humidity level, atmospheric pressure, wind speed, and temperature. The probability that a new electric car model sells well may depend on the price of the car, the overall state of the economy, the advertising budget for the car, and the sales of competing cars. Probability problems such as these, where the outcomes of related events affect the outcome of a given event, are called *conditional probability* problems.

Why This Is Important   In addition to applications in weather forecasting and business modeling, conditional probability has many applications in science, engineering, and finance.

In Section 11.5, we learned that when two events are *dependent*, the occurrence of the first event, *A*, affected the probability of the second event, *B*, occurring. When we calculated $P(A \text{ and } B)$, when in determining the probability of event *B*, we assumed that event *A* occurred. That is, we calculated the probability of event *B*, given event *A*. The probability of event *B*, given event *A*, is called a *conditional probability*. The definition of conditional probability follows.

Definition: Conditional Probability

In general, the probability of event E_2 occurring, given that an event E_1 has happened (or will happen; the time relationship does not matter), is called a **conditional probability** and is written $P(E_2 | E_1)$.

The symbol $P(E_2 | E_1)$, read "the probability of E_2, given E_1," represents the probability of E_2 occurring, assuming that E_1 has already occurred (or will occur).

Example 1 Conditional Probability

One card is drawn from a standard 52-card deck. Determine the probability that the card drawn is a

a) king.

b) king given that the card is a face card.

c) diamond.

d) diamond given that the card is a red card.

Solution

a) There are 4 kings in a standard 52-card deck. Therefore,

$$P(\text{king}) = \frac{4}{52} = \frac{1}{13}.$$

b) We are asked to determine the probability that the card is a king given that the card is a face card. Since we know the card is a face card, we no longer consider all 52 cards as possible outcomes. Instead we only consider the 12 face cards as possible outcomes. Of these 12 face cards, 4 are kings. Therefore,

$$P(\text{king} \mid \text{face card}) = \frac{4}{12} = \frac{1}{3}.$$

c) There are 13 diamonds in a standard 52-card deck. Therefore,

$$P(\text{diamond}) = \frac{13}{52} = \frac{1}{4}.$$

d) We are asked to determine the probability that the card is a diamond given that the card is a red card. Since we know the card is a red card, we no longer consider all 52 cards as possible outcomes. Instead, we only consider the 26 red cards as possible outcomes. Of these red cards, 13 are diamonds. Therefore,

$$P(\text{diamond} \mid \text{red card}) = \frac{13}{26} = \frac{1}{2}.$$

Chance of Showers

A t one time or another, you probably have been caught in a downpour on a day the weather forecaster had predicted sunny skies. Short-term (24-hour) weather forecasts are correct nearly 85% of the time, a level of accuracy achieved through the use of conditional probability. Computer models are used to analyze data taken on the ground and in the air and make predictions of atmospheric pressures at some future time, say 10 minutes ahead. Based on these predicted conditions, another forecast is then computed. This process is repeated until a weather map has been generated for the next 12, 24, 36, and 48 hours. Since each new prediction relies on the previous prediction being correct, the margin of error increases as the forecast extends further into the future.

Why This Is Important Weather forecasting is one of many real-life applications of conditional probability.

Example 2 *Girls in a Family*

A family has two children. Assuming that boys and girls are equally likely, determine the probability that the family has

a) two girls.

b) two girls given that at least one of the children is a girl.

c) two girls given that the older child is a girl.

Solution

a) To determine the probability that the family has two girls, we can determine the sample space of a family with two children. Then, from the sample space we can determine the probability that both children are girls. The sample space of two children can be determined by a tree diagram (see Fig. 11.17).

1st Child	2nd Child	Sample Space
B	B	BB
B	G	BG
G	B	GB
G	G	GG

Figure 11.17

There are four possible equally likely outcomes: BB, BG, GB, and GG. Only one of the outcomes has two girls, GG. Thus,

$$P(2 \text{ girls}) = \frac{1}{4}$$

b) We are given that at least one of the children is a girl. Therefore, for this example the sample space is BG, GB, GG. There are three possibilities, of which only one has two girls, GG. Thus,

$$P(\text{both girls} \mid \text{at least one is a girl}) = \frac{1}{3}$$

c) If the older child is a girl, the sample space reduces to GB, GG. Thus,

$$P(\text{both girls} \mid \text{older child is a girl}) = \frac{1}{2}$$

A number of formulas can be used to determine conditional probabilities. We will use the following formula.

> **Conditional Probability**
> For any two events, E_1 and E_2, the conditional probability, $P(E_2 \mid E_1)$, is determined as follows.
>
> $$P(E_2 \mid E_1) = \frac{n(E_1 \text{ and } E_2)}{n(E_1)}$$

In the formula, $n(E_1 \text{ and } E_2)$ represents the number of sample points common to both event 1 and event 2, and $n(E_1)$ is the number of sample points in event E_1, the given event. Since the intersection of E_1 and E_2, symbolized $E_1 \cap E_2$, represents the sample points common to both E_1 and E_2, the formula can also be expressed as

$$P(E_2 \mid E_1) = \frac{n(E_1 \cap E_2)}{n(E_1)}$$

The Monty Hall Problem

▲ *Monty Hall and Wayne Brady, hosts of Let's Make a Deal*

The Monty Hall problem is a prob-ability problem based on the tele-vision game show *Let's Make a Deal* and is named after the original host, Monty Hall. The problem became fa-mous in 1990 in Marilyn vos Savant's *Ask Marilyn* column in *Parade* maga-zine. The problem can be stated as follows. You are a contestant on *Let's Make a Deal,'* and there are three doors to choose from: *A, B,* and *C.* Behind one door is a new car, and behind two doors are goats. You choose door *A.* Monty then reveals that behind door *B* is a goat. He then asks if you'd like to stay with your original choice, door *A,* or change your choice to door *C.* What should you do, stay with your original choice or change your choice?

The calculations of the probabil-ities involved in this problem involve a formula called *Bayes' theorem.* Although the calculation of these probabilities is beyond our level of discussion, there are several web-sites and smartphone apps that can simulate the Monty Hall problem. It can be shown, using Bayes' theorem, that the probability of winning the new car if you stay with your original choice is $\frac{1}{3}$. However, the probability of winning the new car if you change your choice is $\frac{2}{3}$! While these prob-abilities may be surprising, long term repeated trials of the problem using computer simulations confirm these results. We will discuss the Monty Hall problem and Bayes' theorem again in Exercise 84 on page 707.

Figure 11.18 is helpful in explaining conditional probability.

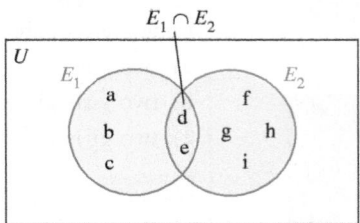

Figure 11.18

Here, the number of elements in E_1 is five, the number of elements in E_2 is six, and the number of elements in both E_1 and E_2, or $E_1 \cap E_2$, is two.

$$P(E_2|E_1) = \frac{n(E_1 \text{ and } E_2)}{n(E_1)} = \frac{2}{5}$$

Thus, for this situation, the probability of selecting an element from E_2, given that the element is in E_1, is $\frac{2}{5}$.

Example 3 *Using the Conditional Probability Formula*

Two hundred and fifty patients who had knee, hip, or heart surgery were asked whether they were satisfied or dissatisfied regarding the results of their surgery. The responses are given in the table below.

Surgery	Satisfied	Dissatisfied	Total
Knee	75	20	95
Hip	90	15	105
Heart	45	5	50
Total	210	40	250

If one person from the 250 patients surveyed is randomly selected, determine the probability that the person

a) was satisfied with the results of the surgery.

b) was satisfied with the results of the surgery, given that the person had knee surgery.

c) was dissatisfied with the results of the surgery, given that the person had hip surgery.

d) had heart surgery, given that the person was dissatisfied with the results of the surgery.

Solution

a) The total number of patients is 250, of which 210 were satisfied with the results of the surgery.

$$P(\text{satisfied with the results of the surgery}) = \frac{210}{250} = \frac{21}{25}$$

b) We are given that the person had knee surgery. Thus, we have a conditional probability problem. Let E_1 be the given information "the person had knee sur-gery." Let E_2 be "the person was satisfied with the results of the surgery." We are being asked to determine $P(E_2|E_1)$. The number of people who had knee surgery, $n(E_1)$, is 95. The number of people who had knee surgery and were satisfied with the results of the surgery, $n(E_1 \text{ and } E_2)$, is 75. Thus,

$$P(E_2 | E_1) = \frac{n(E_1 \text{ and } E_2)}{n(E_1)} = \frac{75}{95} = \frac{15}{19}$$

c) We are given that the person had hip surgery. Thus, we have a conditional probability problem. Let E_1 be the given information "the person had hip surgery." Let E_2 be "the person was dissatisfied with the results of the surgery." We are asked to determine $P(E_2 | E_1)$. The number of people who had hip surgery, $n(E_1)$, is 105. The number of people who had hip surgery and were dissatisfied with the results of the surgery, $n(E_1 \text{ and } E_2)$, is 15. Thus,

$$P(E_2 | E_1) = \frac{n(E_1 \text{ and } E_2)}{n(E_1)} = \frac{15}{105} = \frac{1}{7}$$

d) We are given that the person was dissatisfied with the results of the surgery. Thus, we have a conditional probability problem. Let E_1 be the given information "the person was dissatisfied with the results of the surgery." Let E_2 be "the person had heart surgery." We are asked to determine $P(E_2 | E_1)$. The number of people who were dissatisfied with the results of the surgery, $n(E_1)$, is 40. The number of people who were dissatisfied with the results of the surgery and had heart surgery, $n(E_1 \text{ and } E_2)$, is 5. Thus,

$$P(E_2 | E_1) = \frac{n(E_1 \text{ and } E_2)}{n(E_1)} = \frac{5}{40} = \frac{1}{8}$$ ∎

In many of the examples, we used the words *given that*. Other words may be used instead. For example, in Example 3(b), the question could have been worded "was satisfied with their surgery *if* the person had knee surgery."

SECTION 11.6 *Exercises*

Warm Up Exercises

In Exercises 1–4, fill in the blank with an appropriate word, phrase, or symbol(s).

1. The probability of event E_2 occurring, given that event E_1 has happened, is called a(n) _____ probability.

2. The notation for the probability of E_2, given E_1, is

_____.

3. If $n(E_1 \text{ and } E_2) = 3$ and $n(E_1) = 7$, then $P(E_2 | E_1) =$ _____.

4. If $n(E_1 \text{ and } E_2) = 4$ and $n(E_1) = 16$, then $P(E_2 | E_1) =$ _____.

Practice the Skills

Drawing a Card In Exercises 5–8, one card is drawn from a standard 52-card deck. Determine the probability that the card is a

5. a) a club.

b) a club, given that the card is a black card.

6. a) a queen.

b) a queen, given that the card is a face card.

7. a) a jack or a king.

b) a jack or a king, given that the card is a face card.

8. a) a 5 or a 7.

b) a 5 or a 7, given that the card is not a face card.

Tennis Balls In Exercises 9–14, consider the tennis balls in the figure below.

Assume that one tennis ball is randomly selected. Determine the probability that the ball selected shows

9. a) a 5.

b) a 5 given that the ball is orange.

10. a) a 6.

    b) a 6, given that the ball is yellow.

11. a) an even number.

    b) an even number, given that the ball is orange.

12. a) an odd number.

    b) an odd number, given that the ball is orange.

13. a) a red number.

    b) a red number, given that the ball is yellow.

14. a) a black number.

    b) a black number, given that the ball is yellow.

Select a Number *In Exercises 15–20, consider the following figures.*

Assume that one figure is randomly selected and each figure is equally likely to be selected. Determine the probability of selecting

15. a circle, given that an odd number is selected.

16. a triangle, given that a number greater than or equal to 5 is selected.

17. an odd number, given that a circle is selected.

18. a number greater than or equal to 5, given that a triangle is selected.

19. a circle or square, given that a number less than 4 is selected.

20. a triangle, given that an odd number is selected.

Spin the Wheel *In Exercises 21–28, consider the following wheel.*

If the wheel is spun and each section is equally likely to stop under the pointer, determine the probability that the pointer lands on

21. an even number, given that the color is purple.

22. an odd number, given that the color is red.

23. purple, given that the number is even.

24. red, given that the number is odd.

25. a number greater than 4, given that the color is purple.

26. an even number, given that the color is red or purple.

27. gold, given that the number is greater than 5.

28. gold, given that the number is greater than 10.

Money from a Hat *In Exercises 29–32, assume that a hat contains four bills: a $1 bill, a $5 bill, a $10 bill, and a $20 bill. Each bill is equally likely to be selected. Two bills are to be randomly selected with replacement. Construct a sample space as was done in Example 2 and determine the probability that*

29. both bills are $1 bills.

30. both bills are $1 bills if the first selected is a $1 bill.

31. both bills are $5 bills if at least one of the bills is a $5 bill.

32. both bills have a value greater than a $5 bill if the second bill is a $10 bill.

Two Dice *In Exercises 33–38, two fair dice are rolled one after the other. Construct a sample space and determine the probability that the sum of the dots on the dice total*

33. 3.

34. 3 if the first die is a 1.

35. 3 if the first die is a 3.

36. an even number if the second die is a 2.

37. a number greater than 7 if the second die is a 5.

38. a 7 or 11 if the first die is a 5.

Problem Solving

Taste Test *In Exercises 39–42, use the following results of a coffee taste test given at a local grocery store.*

	Prefers Manhattan's Finest Coffee	Prefers John's Famous Coffee	Total
Men	60	40	100
Women	50	75	125
Total	110	115	225

If one of these individuals is randomly selected, determine the probability the individual selected

39. a) prefers John's Famous Coffee.

   b) prefers John's Famous Coffee, given that a woman is selected.

40. a) prefers Manhattan's Finest Coffee.

   b) prefers Manhattan's Finest Coffee, given that a man is selected.

41. a) is a woman.

   b) is a woman, given that the person prefers John's Famous Coffee.

42. a) is a man.

   b) is a man, given that the person prefers Manhattan's Finest Coffee.

E-Z Pass *In Exercises 43–46, use the following table, which shows the number of cars and trucks that used the Pennsylvania Turnpike on a particular day. The number of cars and trucks that used, and did not use, E-Z Pass on that same day was also recorded.*

E-Z Pass	Cars	Trucks	Total
Used	527	316	843
Did not use	935	683	1618
Total	1462	999	2461

If one of these vehicles is randomly selected, determine the probability (as a decimal number rounded to four decimal places) that the vehicle

43. a) is a car.

   b) is a car, given that the vehicle used E-Z Pass.

44. a) is a truck.

   b) is a truck, given that the vehicle did not use E-Z Pass.

45. a) used E-Z Pass.

   b) used E-Z Pass, given that the vehicle is a car.

46. a) did not use E-Z Pass.

   b) did not use E-Z Pass, given that the vehicle is a truck.

Sales Effectiveness *In Exercises 47–52, use the following information. Sales representatives at a car dealership were split into two groups. One group used an aggressive approach to sell a customer a new automobile. The other group used a passive approach. The following table summarizes the records for 650 customers.*

Approach	Sale	No Sale	Total
Aggressive	100	250	350
Passive	220	80	300
Total	320	330	650

If one of these customers is randomly selected, determine the probability

47. that the aggressive approach was used.

48. of a sale.

49. of no sale, given that the passive approach was used.

50. of a sale, given that the aggressive approach was used.

51. of a sale, given that the passive approach was used.

52. of no sale, given that the aggressive approach was used.

Car Sales *In Exercises 53–58, use the following table, which shows the number of Chevrolet Malibus and Chevrolet Impalas sold at Kieler Garage during August 2019. The number of coupes and sedans is also shown.*

Style	Malibu	Impala	Total
Coupe	34	21	55
Sedan	72	59	131
Total	106	80	186

If one of these vehicles is randomly selected, determine the probability that the vehicle is

53. a Malibu, given that the vehicle selected is a coupe.

54. a coupe, given that the vehicle selected is a Malibu.

55. a sedan, given that the vehicle selected is an Impala.

56. an Impala, given that the vehicle selected is a sedan.

57. a Malibu, given that the vehicle selected is a sedan.

58. a coupe, given that the vehicle selected is an Impala.

Quality Control *In Exercises 59–64, Orpa, a quality control inspector, is checking a sample of lightbulbs for defects. The following table summarizes her findings.*

Wattage	Good	Defective	Total
20	80	15	95
50	100	5	105
100	120	10	130
Total	300	30	330

If one of these lightbulbs is randomly selected, determine the probability that the lightbulb is

59. good.

60. good, given that it is 50 watts.

61. defective, given that it is 20 watts.

62. good, given that it is 100 watts.

63. good, given that it is 50 or 100 watts.

64. defective, given that it is not 50 watts.

Smartphone Ownership *In Exercises 65–70, the faculty of Macomb Community College were surveyed to determine the brand of smartphone they owned. The results are shown in the table below.*

	Apple	Samsung	LG	Other	Total
Men	127	74	16	25	242
Women	281	68	26	20	395
Total	408	142	42	45	637

If a faculty member is randomly selected, determine the probability (as a decimal number rounded to four decimal places) that the faculty member

65. owns an Apple smartphone, given the faculty member is a man.

66. is a man, given the faculty member owns an Apple smartphone.

67. is a woman, given the faculty member owns an LG smartphone.

68. owns an LG smartphone, given the faculty member is a woman.

69. owns a Samsung smartphone, given the faculty member is a woman.

70. is a man, given the faculty member owns a Samsung smartphone.

Challenge Problems/Group Activities

Mutual Fund Holdings *In Exercises 71–74, use the following information. Mutual funds often hold many stocks. Each stock may be classified as a value stock, a growth stock, or a blend of the two. The stock may also be categorized by how large the company is. It may be classified as a large company stock, medium company stock, or small company stock. A selected mutual fund contains 200 stocks as illustrated in the following chart.*

	Value	Blend	Growth	
	28	23	42	Large
	19	15	18	Medium
	26	12	17	Small

Equity Investment Style

If one stock is randomly selected from the mutual fund, determine the probability that it is

71. a large company stock.

72. a value stock.

73. a blend, given that it is a medium company stock.

74. a large company stock, given that it is a blend stock.

75. *Venn Diagram* Consider the Venn diagram below. The numbers in the regions of the circle indicate the number of items that belong to that region. For example, 60 items are in set *A* but not in set *B*.

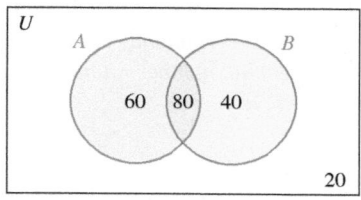

Determine **a)** $n(A)$ **b)** $n(B)$ **c)** $P(A)$ **d)** $P(B)$

Use the formula on page 701 to determine

e) $P(A\,|\,B)$.   **f)** $P(B\,|\,A)$.

g) Explain why $P(A\,|\,B) \neq P(A) \cdot P(B)$.

76. A formula we gave for conditional probability is

$$P(E_2\,|\,E_1) = \frac{n\,(E_1 \text{ and } E_2)}{n\,(E_1)}$$

This formula may be derived from the formula

$$P(E_2\,|\,E_1) = \frac{P(E_1 \text{ and } E_2)}{P(E_1)}$$

Can you explain why? [*Hint:* Consider what happens to the denominators of $P(E_1 \text{ and } E_2)$ and $P(E_1)$, when they are expressed as fractions and the fractions are divided out.]

77. Given that $P(A) = 0.3$, $P(B) = 0.5$, and $P(A \text{ and } B) = 0.15$, use the formula

$$P(E_2\,|\,E_1) = \frac{P(E_1 \text{ and } E_2)}{P(E_1)}$$

to determine

a) $P(A\,|\,B)$.   **b)** $P(B\,|\,A)$.

c) Are A and B independent? Explain.

Recreational Mathematics

In Exercises 78–83, suppose that each circle is equally likely to be selected. One circle is randomly selected.

Determine the probability indicated.

78. $P(\text{green circle} \,|\, + \text{ obtained})$

79. $P(+\,|\, \text{ orange circle obtained})$

80. $P(\text{yellow circle} \,|\, - \text{ obtained})$

81. $P(\text{green} + \,|\, + \text{ obtained})$

82. $P(\text{green or yellow circle} \,|\, \text{green} + \text{ obtained})$

83. $P(\text{yellow circle with green} + \,|\, + \text{ obtained})$

Research Activity

84. *Bayes' Theorem* Read the Recreational Mathematics box on page 702 regarding the Monty Hall problem. Do research and write a paper on Bayes' theorem. Include a discussion of how Bayes' theorem can be used to solve the Monty Hall problem.

SECTION 11.7

The Fundamental Counting Principle and Permutations

Upon completion of this section, you will be able to:

- Solve problems using the fundamental counting principle.
- Solve problems involving permutations.
- Solve problems involving permutations of duplicate items.

Sumiko works at an animal shelter that has seven dogs available for adoption. Sumiko is preparing a poster that will feature a photograph of each of the seven dogs. How many different arrangements of the seven photographs are possible? If one of these arrangements of photographs is randomly selected, what is the probability that the photographs are in alphabetical order by the dogs' names? In this section, we will answer questions like these and discuss many other applications involving the number of ordered arrangements of a set of objects.

Why This Is Important The number of ordered arrangements of a set of objects is important to many applications including computer passwords, zip codes, telephone numbers, and Social Security numbers. Probability related to such arrangements is important to many branches of mathematics, science, engineering, and business.

The Fundamental Counting Principle

In Section 11.4, we introduced the fundamental counting principle, which is repeated here for your convenience.

> Definition: **Fundamental Counting Principle**
> If a first experiment has M distinct outcomes and a second experiment has N distinct outcomes, then the two experiments in that specific order have $M \cdot N$ distinct outcomes.

The fundamental counting principle is illustrated in Examples 1, 2, and 3.

Example 1   *License Plate Numbers*

A license plate "number" consists of two uppercase letters followed by four digits. Determine how many different license plate numbers are possible if

a) repetition of letters and digits is permitted.

b) repetition of letters and digits is not permitted.

c) the first letter must be a vowel (*a, e, i, o, u*) and the first digit cannot be a 0, and repetition of letters and digits is not permitted.

Solution There are 26 letters and 10 digits (0–9). We have six positions to fill, as indicated.

$$\underline{L}\,\underline{L}\,\underline{D}\,\underline{D}\,\underline{D}\,\underline{D}$$

a) Since repetition is permitted, there are 26 possible choices for both the first and second positions. There are 10 possible choices for the third, fourth, fifth, and sixth positions.

$$\frac{26}{L}\,\frac{26}{L}\,\frac{10}{D}\,\frac{10}{D}\,\frac{10}{D}\,\frac{10}{D}$$

Since $26 \cdot 26 \cdot 10 \cdot 10 \cdot 10 \cdot 10 = 6{,}760{,}000$, there are 6,760,000 different possible arrangements.

b) There are 26 possibilities for the first position. Since repetition of letters is not permitted, there are only 25 possibilities for the second position. The same reasoning is used when determining the number of digits for positions 3 through 6.

$$\frac{26}{L}\,\frac{25}{L}\,\frac{10}{D}\,\frac{9}{D}\,\frac{8}{D}\,\frac{7}{D}$$

Since $26 \cdot 25 \cdot 10 \cdot 9 \cdot 8 \cdot 7 = 3{,}276{,}000$, there are 3,276,000 different possible arrangements.

c) Since the first letter must be an *a, e, i, o,* or *u*, there are five possible choices for the first position. The second position can be filled by any of the letters except for the vowel selected for the first position. Therefore, there are 25 possibilities for the second position.

Since the first digit cannot be a 0, there are nine possibilities for the third position. The fourth position can be filled by any digit except the one selected for the third position. Thus, there are nine possibilities for the fourth position. Since the fifth position cannot be filled by any of the two digits previously used, there are eight possibilities for the fifth position. The last position can be filled by any of the seven remaining digits.

$$\underline{\underset{L}{5}}\ \underline{\underset{L}{25}}\ \underline{\underset{D}{9}}\ \underline{\underset{D}{9}}\ \underline{\underset{D}{8}}\ \underline{\underset{D}{7}}$$

Since $5 \cdot 25 \cdot 9 \cdot 9 \cdot 8 \cdot 7 = 567{,}000$, there are 567,000 different arrangements that meet the conditions specified. ∎

Example 2 *Fundamental Counting Principle: T-Shirt Colors*

At Old Navy, a supply of solid-colored T-shirts has just been received. The T-shirts come in the following colors: green, blue, white, yellow, and red. Preston, the floor manager, decides to display one of each color T-shirt in a row on a shelf.

a) In how many different ways can he display the five different color T-shirts on a shelf?

b) If he wants to place the blue T-shirt in the middle, in how many different ways can he arrange the T-shirts?

c) If Preston wants the white T-shirt to be the first T-shirt and the blue T-shirt to be the last T-shirt, in how many different ways can he arrange the T-shirts?

Solution

a) There are five positions to fill, using the five colors. In the first position, on the left, he can use any one of the five colors. In the second position, he can use any of the four remaining colors. In the third position, he can use any of the three remaining colors, and so on. The number of distinct possible arrangements is

$$\underline{5} \cdot \underline{4} \cdot \underline{3} \cdot \underline{2} \cdot \underline{1} = 120$$

b) We begin by satisfying the specified requirements stated. In this case, the blue T-shirt must be placed in the middle. Therefore, there is only one possibility for the middle position.

$$\underline{\ }\ \underline{\ }\ \underline{1}\ \underline{\ }\ \underline{\ }$$

For the first position, there are now four possibilities. For the second position, there will be three possibilities. For the fourth position, there will be two possibilities. Finally, in the last position, there is only one possibility.

$$\underline{4} \cdot \underline{3} \cdot \underline{1} \cdot \underline{2} \cdot \underline{1} = 24$$

Thus, under the condition stated, there are 24 different possible arrangements.

c) For the first T-shirt, there is only one possibility, the white T-shirt. For the last T-shirt, there is only one possibility, the blue T-shirt.

$$\underline{1}\ \underline{\ }\ \underline{\ }\ \underline{\ }\ \underline{1}$$

The second position can be filled by any of the three remaining T-shirts. The third position can be filled by any of the two remaining T-shirts. There is only one T-shirt left for the fourth position. Thus, the number of possible arrangements is

$$\underline{1} \cdot \underline{3} \cdot \underline{2} \cdot \underline{1} \cdot \underline{1} = 6$$

There are only six possible arrangements that satisfy the given conditions. ∎

Example 3 *Garage Door Codes*

Many garage doors come equipped with a keypad that allows residents to open the garage door by entering a four-digit code. Repetition of digits in the code is allowed.

a) How many different four-digit codes are possible?

b) If one of these codes is randomly selected, what is the probability that the code has no repeated digits?

Solution

a) Since repetition of digits is allowed, there are 10 choices for each of the four digits. Therefore, using the fundamental counting principle, there are $10 \cdot 10 \cdot 10 \cdot 10 = 10{,}000$ possible four-digit codes.

b) Of these 10,000 possible codes, we need to determine how many codes have no repeated digits. There are 10 choices for the first digit, and since the digits do not repeat, there are 9 choices for the second digit, 8 choices for the third digit, and 7 choices for the fourth digit. Using the fundamental counting principle, there are $10 \cdot 9 \cdot 8 \cdot 7 = 5040$ codes with no repeated digits. We calculate the probability as follows.

$$P(\text{no repeated digits}) = \frac{\text{number of codes with no repeated digits}}{\text{number of possible codes}}$$

$$= \frac{5040}{10{,}000} = \frac{63}{125} = 0.504$$

Thus, the probability that a randomly selected four-digit garage code has no repeated digits is 0.504. ∎

Permutations

Now we introduce the definition of a permutation.

> Definition: **Permutation**
> A **permutation** is any *ordered arrangement* of a given set of objects.

"Superman, Batman, Wonder Woman" and "Wonder Woman, Superman, Batman" represent two different ordered arrangements or two different permutations of the same three characters. In Example 2(a), there are 120 different ordered arrangements, or permutations, of the five colored T-shirts. In Example 2(b), there are 24 different ordered arrangements, or permutations possible, if the blue T-shirt must be displayed in the middle.

When determining the number of permutations possible, we assume that repetition of an item is not permitted. To help you understand and visualize permutations, we illustrate the various permutations possible when a triangle, rectangle, and circle are to be placed in a line; see Fig. 11.19.

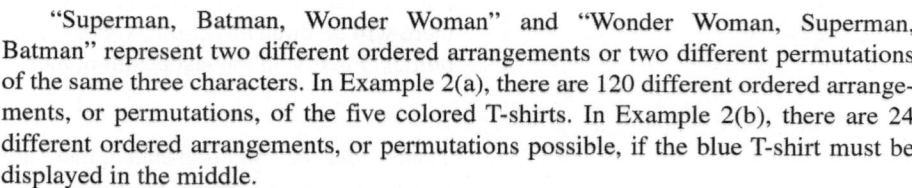

Figure 11.19

For this set of three shapes, six different arrangements, or six permutations, are possible. We can obtain the number of permutations by using the fundamental counting principle. For the first position, there are three choices. There are then two choices for the second position, and only one choice is left for the third position.

$$\text{Number of permutations} = 3 \cdot 2 \cdot 1 = 6$$

The product $3 \cdot 2 \cdot 1$ is referred to as 3 factorial, and is written 3!. Thus,

$$3! = 3 \cdot 2 \cdot 1 = 6$$

> **Number of Permutations**
> The number of permutations of n distinct items is n factorial, symbolized $n!$, where
> $$n! = n(n-1)(n-2) \cdots (3)(2)(1)$$

It is important to note that 0! is defined to be 1. Many calculators have the ability to determine factorials. Often to determine factorials you need to press the 2nd or INV key. If necessary, do an internet search using your model of calculator to determine how to calculate factorials on your calculator.

Example 4 *Dogs on a Poster*

Sumiko works at an animal shelter that has seven dogs available for adoption. Sumiko is preparing a poster that will feature a photograph of each of the seven dogs.

a) How many different arrangements of the seven photographs are possible?

b) If one of these arrangements of photographs is randomly selected, what is the probability that the photographs are in alphabetical order by the dogs' names? Assume the dogs all have different names.

Solution

a) Since there are seven different photographs, the number of permutations is 7!

$$7! = 7 \cdot 6 \cdot 5 \cdot 4 \cdot 3 \cdot 2 \cdot 1 = 5040$$

The seven photographs can be arranged in 5040 different ways.

b) Since the dogs all have different names, there is only one of 5040 arrangements of photographs in which the dogs' names are in alphabetical order. Therefore,

$$P(\text{alphabetical order}) = \frac{1}{5040}.$$

Example 5 illustrates how to use the fundamental counting principle to determine the number of permutations possible when only a part of the total number of items is to be selected and arranged.

Some of the many permutations of 3 of the 5 letters

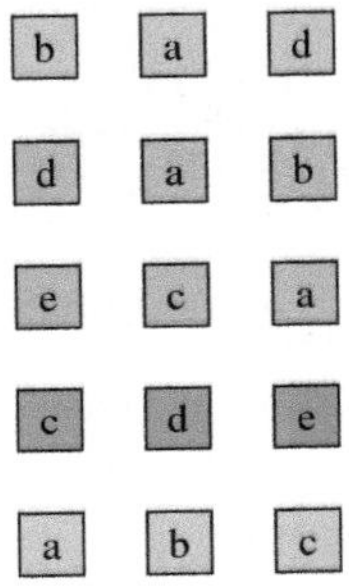

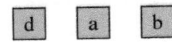

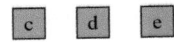

Figure 11.20

Example 5 *Permutations of Three of Five Letters*

Consider the five letters a, b, c, d, e. In how many distinct ways can three letters be selected and arranged if repetition is not allowed?

Solution We are asked to select and arrange only three of the five possible letters. Figure 11.20 shows some possibilities. Using the fundamental counting principle, we determine that there are five possible letters for the first choice, four possible letters for the second choice, and three possible letters for the third choice:

$$5 \cdot 4 \cdot 3 = 60$$

Thus, there are 60 different possible ordered arrangements, or permutations. On the left, we show 5 of the 60 possible permutations.

In Example 5, we determined the number of different ways in which we could select and arrange three of the five items. We can indicate that result by using the notation $_5P_3$. The notation $_5P_3$ is read "the number of permutations of five items taken three at a time." The notation $_nP_r$ is read "the number of permutations of n items taken r at a time."

We use the fundamental counting principle below to evaluate $_8P_4$, $_9P_3$, and $_{10}P_5$. Note the relationship between the number preceding the P, the number following the P, and the last number in the product.

$$_8P_4 = 8 \cdot 7 \cdot 6 \cdot 5 \quad \text{— One more than } 8 - 4$$
$$_9P_3 = 9 \cdot 8 \cdot 7 \quad \text{— One more than } 9 - 3$$
$$_{10}P_5 = 10 \cdot 9 \cdot 8 \cdot 7 \cdot 6 \quad \text{— One more than } 10 - 5$$

To evaluate $_nP_r$, we begin with n and form a product of r consecutive decreasing factors. For example, to evaluate $_{10}P_5$, we start with 10 and form a product of five consecutive decreasing factors (see the preceding illustration).

In general, the number of permutations of n items taken r at a time, $_nP_r$, may be determined by the formula

$$_nP_r = n(n-1)(n-2) \cdots \overbrace{(n-r+1)}^{\text{One more than } n - r}$$

Therefore, when evaluating $_{20}P_{15}$, we would determine the product of consecutive decreasing integers from 20 to $(20 - 15 + 1)$ or 6, which is written as $20 \cdot 19 \cdot 18 \cdot 17 \cdots 6$.

Now let's develop an alternative formula that we can use to determine the number of permutations possible when r objects are selected from n objects:

$$_nP_r = n(n-1)(n-2) \cdots (n-r+1)$$

Now multiply the expression on the right side of the equals sign by $\dfrac{(n-r)!}{(n-r)!}$, which is equivalent to multiplying the expression by 1.

$$_nP_r = n(n-1)(n-2) \cdots (n-r+1) \times \frac{(n-r)!}{(n-r)!}$$

For example,

$$_{10}P_5 = 10 \cdot 9 \cdot \cdots \cdot 6 \times \frac{5!}{5!}$$

or

$$_{10}P_5 = \frac{10 \cdot 9 \cdot \cdots \cdot 6 \times 5!}{5!}$$

Since $(n-r)!$ means $(n-r)(n-r-1) \cdots (3)(2)(1)$, the expression for $_nP_r$ can be rewritten as

$$_nP_r = \frac{n(n-1)(n-2) \cdots (n-r+1)\overbrace{(n-r)(n-r-1) \cdots (3)(2)(1)}^{(n-r)!}}{(n-r)!}$$

Since the numerator of this expression is $n!$, we can write

$$_nP_r = \frac{n!}{(n-r)!}$$

For example,

$$_{10}P_5 = \frac{10!}{(10-5)!}$$

Now we give the permutation formula.

Stock Ticker Symbols

Stock ticker symbols on the New York Stock Exchange (NYSE) typically consist of one, two, or three letters. For example, T represents AT & T, GE represents General Electric, and IBM represents International Business Machines. Stocks on the National Association of Security Dealers Automated Quotation System (NASDAQ) typically contain four letters. For example, AAPL represents Apple and GOOG represents Alphabet, the parent company of Google. Are there more possible NYSE ticker symbols or more NASDAQ ticker symbols? Why?

Answer: For the NYSE there are $26 + (26 \times 26) + (26 \times 26 \times 26) = 18{,}278$ different possible ticker symbols. For the NASDAQ there are $26 \times 26 \times 26 \times 26 = 456{,}976$ different possible ticker symbols.

> **Permutation Formula**
> The number of permutations possible when r objects are selected from n objects is determined by the **permutation formula**
> $$_nP_r = \frac{n!}{(n-r)!}$$

In Example 5, we determined that when selecting three of five letters, there were 60 permutations. We can obtain the same result using the permutation formula:

$$_5P_3 = \frac{5!}{(5-3)!} = \frac{5!}{2!} = \frac{5 \cdot 4 \cdot 3 \cdot 2 \cdot 1}{2 \cdot 1} = 60$$

Example 6 *Fishing Club Officers*

The fishing club at Central New Mexico Community College has nine members. To determine club officers, nine slips of paper each containing the name of a member are placed into a hat. The club advisor randomly selects a first name from the hat to become the president, a second name to become secretary, and a third name to become treasurer. How many different arrangements of club officers are possible?

Solution There are nine people, $n = 9$, of which three are to be selected; thus, $r = 3$.

$$_9P_3 = \frac{9!}{(9-3)!} = \frac{9!}{6!} = \frac{9 \cdot 8 \cdot 7 \cdot 6 \cdot 5 \cdot 4 \cdot 3 \cdot 2 \cdot 1}{6 \cdot 5 \cdot 4 \cdot 3 \cdot 2 \cdot 1} = 504$$

Thus, with nine people there can be 504 different arrangements for president, secretary, and treasurer.

In Example 6, the fraction

$$\frac{9 \cdot 8 \cdot 7 \cdot 6 \cdot 5 \cdot 4 \cdot 3 \cdot 2 \cdot 1}{6 \cdot 5 \cdot 4 \cdot 3 \cdot 2 \cdot 1} = 504$$

can be also expressed as

$$\frac{9 \cdot 8 \cdot 7 \cdot 6!}{6!} = 504$$

The solution to Example 6, like other permutation problems, can also be obtained using the fundamental counting principle.

Example 7 *Bicycle Club*

The Rainbow bicycle club has 10 different routes that members wish to travel exactly once, but they have only 6 specific dates for their trips. In how many ways can the different routes be assigned to the dates scheduled for their trips?

Solution There are 10 possible routes but only 6 specific dates scheduled for the trips. Since traveling route A on day 1 and traveling route B on day 2 is different than traveling route B on day 1 and traveling route A on day 2, we have a permutation problem. There are 10 possible routes; thus, $n = 10$. There are 6 routes that are going to be selected and assigned to different days; thus, $r = 6$. Now we calculate the number of different permutations of selecting and arranging the dates for 6 out of 10 possible routes.

$$_{10}P_6 = \frac{10!}{(10-6)!} = \frac{10!}{4!} = \frac{10 \cdot 9 \cdot 8 \cdot 7 \cdot 6 \cdot 5 \cdot 4!}{4!} = 151{,}200$$

There are 151,200 different ways that 6 routes can be selected and scheduled from the 10 possible routes.

Example 7 could also be worked using the fundamental counting principle because we are discussing an *ordered arrangement* (a permutation) that is done *without replacement*. For the first date scheduled, there are 10 possible outcomes. For the second date selected, there are 9 possible outcomes. By continuing this process we would determine that the number of possible outcomes for the 6 different trips is $10 \cdot 9 \cdot 8 \cdot 7 \cdot 6 \cdot 5 = 151,200$.

We have worked permutation problems (selecting and arranging, without replacement, *r* items out of *n distinct* items) by using the fundamental counting principle and using the permutation formula. When you are given a permutation problem, unless specified by your instructor, you may use either technique to determine its solution.

TECHNOLOGY TIP

Evaluating Permutations

Most scientific and all graphing calculators can evaluate permutations. While the commands on these calculators may vary depending on the model, look for a key or function that has the notation nPr. If you cannot locate this key or function, conduct an Internet search with your calculator model and the word *permutation*. For example, if you search for "Casio FX-82MS permutation" you will be directed to several websites including online videos that can assist you. Also, there are many apps for smartphones and tablets that can be used to calculate permutations.

To evaluate $_{10}P_6$ on a TI-84 Plus calculator, enter the number 10. Then press the $\boxed{\text{MATH}}$ key. Use the right arrow key to scroll to PRB, then scroll to nPr. Press the $\boxed{\text{ENTER}}$ key. Next, press 6. Press the $\boxed{\text{ENTER}}$ key again. The calculator will display 151200, which agrees with our answer in Example 7. The screenshots below show the calculation depending on the version of the TI-84 Plus.

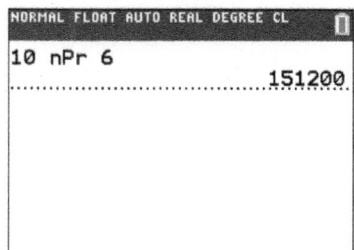
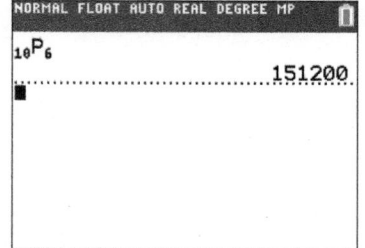

Permutations of Duplicate Items

So far, all the examples we have discussed in this section have involved arrangements with distinct items. Now we will consider permutation problems in which some of the items to be arranged are duplicates. For example, the name BOB contains three letters, of which the two Bs are duplicates. How many permutations of the letters in the name BOB are possible? If the two Bs were distinguishable (one red and the other blue), there would be six permutations.

BOB   BBO   OBB

BOB   BBO   OBB

However, if the Bs are not distinguishable (replacing all colored Bs with black print), we see that there are only three permutations.

BOB   BBO   OBB

The number of permutations of the letters in BOB can be computed as

$$\frac{3!}{2!} = \frac{3 \cdot 2 \cdot 1}{2 \cdot 1} = 3$$

where 3! represents the number of permutations of three letters, assuming that none are duplicates, and 2! represents the number of ways the two items that are duplicates can be arranged (BB or BB). In general, we have the following rule.

Permutations of Duplicate Items
The number of distinct permutations of n objects where n_1 of the objects are identical, n_2 of the objects are identical, ..., n_r of the objects are identical is determined by using

$$\frac{n!}{n_1! n_2! \cdots n_r!}$$

Example 8 Duplicate Letters

In how many different ways can the letters of the word "CINCINNATI" be arranged?

Solution Of the 10 letters, three are I's, three are N's, and two are C's. The number of possible arrangements is

$$\frac{10!}{3!3!2!} = \frac{10 \cdot 9 \cdot \overset{4}{\cancel{8}} \cdot 7 \cdot \cancel{6} \cdot 5 \cdot 4 \cdot 3 \cdot 2 \cdot 1}{\cancel{3 \cdot 2 \cdot 1} \cdot \cancel{3} \cdot \cancel{2} \cdot 1 \cdot \cancel{2} \cdot 1} = 10 \cdot 9 \cdot 4 \cdot 7 \cdot 5 \cdot 4 = 50{,}400$$

There are 50,400 different possible arrangements of the letters in the word "CINCINNATI." ∎

SECTION 11.7
Exercises

Warm Up Exercises

In Exercises 1–8, fill in the blank with an appropriate word, phrase, or symbol(s).

1. Any ordered arrangement of a given set of objects is called a(n) _____.

2. To determine the number of distinct outcomes when two or more experiments are performed, the fundamental _____ principle can be used.

3. The symbol for n factorial is _____.

4. The formula for the number of permutations of n distinct items is $n! =$ _____.

5. The formula for the number of permutations when r objects are selected from n objects is $_nP_r =$ _____.

6. The number of permutations of n objects, where n_1 of the items are identical, n_2 of the items are identical, ..., n_r are identical, is determined by _____.

7. The notation used to express the number of permutations of five items taken three at a time is _____.

8. The notation used to express the number of permutations of seven items taken two at a time is _____.

Practice the Skills

In Exercises 9–20, evaluate the expression.

9. $6!$ 10. $8!$ 11. $0!$ 12. $1!$

13. $_3P_2$ 14. $_4P_3$ 15. $_8P_0$ 16. $_6P_0$

17. $_4P_4$ 18. $_3P_3$ 19. $_8P_4$ 20. $_9P_5$

Problem Solving

21. *ATM Codes* To use an automated teller machine, you generally must enter a four-digit code, using the digits 0–9. How many four-digit codes are possible if repetition of digits is permitted?

22. *Worldwide ATM Codes* In many countries outside the United States, the ATM code is a six-digit code instead of a four-digit code (see Exercise 21). How many six-digit codes are possible if repetition of digits is permitted?

23. *Outfits* Juan has three ties, four shirts, and two pairs of pants. How many different outfits can he wear if he chooses one tie, one shirt, and one pair of pants for each outfit?

24. *Choosing Classes* Kai plans to enroll in four classes: sociology, chemistry, economics, and humanities. There are seven sociology classes, four chemistry classes, three economics classes, and four humanities classes that fit his schedule. How many different ways can he select his four classes?

25. *Computer Systems* At a computer store, Imelda is considering 5 different computers, 4 different monitors, 7 different printers, and 2 different soundbars. Assuming that each of the components is compatible with one another and that one of each is to be selected, determine the number of different computer systems possible.

26. *Selecting Furniture* The Johnsons just moved into their new home and are selecting furniture for the family room. They are considering 5 different sofas, 2 different chairs, and 6 different tables. They plan to select one item from each category. Determine the number of different ways they can select the furniture.

27. *Smartphone Passcode* A passcode on a smartphone consists of 6 digits, and repetition of digits is allowed.

  a) Determine the number of possible six-digit passcodes.

  b) If a person finds a smartphone and randomly enters 6 digits, what is the probability that the correct passcode is entered?

28. *Car Door Locks* Some doors on cars can be opened by pressing a correct sequence of buttons. A display of the five buttons by the door handle of a car follows.*

---

* On most cars, although each key lists two numbers, the key acts as a single number. Therefore, if your code is 1, 6, 8, 5, 3, the code 2, 5, 7, 6, 4 will also open the lock.

The correct sequence of five buttons must be pressed to unlock the door.

  a) How many different sequences of five buttons are possible (repetition is permitted)?

  b) If a sequence of five buttons is randomly pressed, determine the probability that the sequence unlocks the door.

29. *Social Security Numbers* A social security number consists of nine digits. How many different social security numbers are possible if repetition of digits is permitted?

30. *License Plate Numbers* In Indiana, a standard car license plate number consists of 3 digits followed by 3 uppercase letters. How many different license plate numbers are possible if

  a) repetition of characters is allowed?

  b) repetition of characters is not allowed?

31. *Geometric Shapes* Consider the five figures shown.

In how many different ways can the figures be arranged

  a) from left to right?

  b) from top to bottom if placed one under the other?

  c) from left to right if the triangle is to be placed on the far right?

  d) from left to right if the circle is to be placed on the far left and the triangle is to be placed on the far right?

32. *Arranging Pictures* The six pictures shown are to be placed side by side along a wall.

In how many ways can they be arranged from left to right if

  a) they can be arranged in any order?

  b) the bird must be on the far left?

  c) the bird must be on the far left and the giraffe must be next to the bird?

  d) a four-legged animal must be on the far right?

33. Team Photo The eight members of the Model United Nations team from Pima Community College are lining up for a photo.

   a) How many different ways can the eight team members line up?

   b) What is the probability that the members randomly line up in alphabetical order?

34. Family Portrait The seven members of the Snipes family are lining up for a family portrait.

   a) How many different ways can the seven family members line up?

   b) What is the probability that the family members randomly line up in order from youngest to oldest?

35. Car Collection Kalin's sports car collection consists of six different cars manufactured by Bentley, Bugatti, Ferrari, Lamborghini, McLaren, and Mercedes, respectively. Kalin wishes to line up the cars on her driveway to show her friends.

   a) How many different ways can Kalin line up the cars on her driveway?

   b) What is the probability that Kalin randomly lines up the cars with the Lamborghini in the first position and the Mercedes in the last position?

36. Colors of the Spectrum Kayode is building an art display on the campus of Metropolitan Community College. The display will feature arrangements of flags in the colors of the spectrum as shown below.

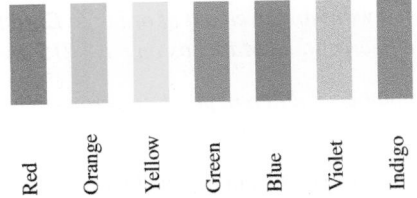

   a) How many different arrangements of the flags are possible?

   b) If one of these arrangements is randomly selected, what is the probability that the green flag is in the first position and the blue flag is in the second position?

37. License Plates A license plate is to consist of three digits followed by two uppercase letters. Determine the number of different license plates possible if

   a) repetition of numbers and letters is permitted.

   b) repetition of numbers and letters is not permitted.

   c) the first and second digits must be odd, and repetition of numbers and letters is not permitted.

   d) the first digit cannot be zero, and repetition of numbers and letters is not permitted.

38. Possible Phone Numbers A telephone number consists of seven digits with the restriction that the first digit cannot be 0 or 1. Repetition of digits is permitted.

   a) How many distinct telephone numbers are possible?

   b) How many distinct telephone numbers are possible with three-digit area codes preceding the seven-digit number, where the first digit of the area code is not 0 or 1?

   c) With the increasing use of cell phones, our society is beginning to run out of usable phone numbers. Various phone companies are developing phone numbers that use 8 digits instead of 7. How many distinct phone numbers can be made using a three digit area code followed by an eight digit phone number, where the first digit of the area code and the phone number cannot be 0 or 1?

39. DJ A disc jockey (DJ) has 9 songs to play. Five are slow songs, and 4 are fast songs. Each song is to be played only once. In how many ways can the DJ play the 9 songs if

   a) the songs can be played in any order?

   b) the first song must be a slow song and the last song must be a slow song?

   c) the first two songs must be fast songs?

40. Meal Delivery Service Fajar has six different meals from a meal delivery service. Three meals include chicken, two meals include beef, and one meal is vegetarian. Fajar will eat one of these meals per day for the next six days. How many different arrangements of meals are possible if

   a) the meals can be eaten in any order?

   b) the first meal must include chicken and the last meal must include beef?

   c) the first two meals must include beef and the last meal must be vegetarian?

41. Club Officers If a club consists of 10 members, how many different arrangements of president, vice president, and secretary are possible?

42. Winning the Trifecta The trifecta at most racetracks consists of selecting the first-, second-, and third-place finishers in a particular race in their proper order. If there are seven entries in the trifecta race, how many tickets must you purchase to guarantee a win?

43. Flag Messages Five different colored flags will be placed on a pole, one beneath another. The arrangement of the colors indicates the message. How many messages are possible if five flags are to be selected from nine different colored flags?

44. Jay Leno's Garage Jay Leno owns 117 motorcycles of which he must select 4 to line up in his garage for his television show, *Jay Leno's Garage*. How many different arrangements of 4 motorcycles are possible?

▲ Jay Leno

45. Skiing Event A skiing event has 16 participants for the slalom. The 5 participants with the fastest speeds will be listed, in the order of their speed, on the leader board. How many different ways are there for the five names to be listed?

46. Swimming Event A swimming event has fifteen participants. The 4 swimmers with the fastest speeds will be listed, in order, on the leader board. How many different ways are there for the names to be listed?

47. Determine the number of permutations of the letters of the word *EDUCATION*.

48. Determine the number of permutations of the letters in the word *IMPORTANCE*.

49. In how many ways can the letters in the word *STATISTICS* be arranged?

50. In how many ways can the letters in the word *MANAGEMENT* be arranged?

51. In how many ways can the letters in the word *KISSIMMEE* be arranged?

52. In how many ways can the letters in the word *CHATTAHOOCHEE* be arranged?

53. In how many ways can the digits in the number 4,568,865 be arranged?

54. In how many ways can the digits in the number 2,142,332 be arranged?

Challenge Problems/Group Activities

55. Car Keys Door keys for a certain automobile are made from a blank key on which five cuts are made. Each cut may be one of five different depths.

a) How many different keys can be made?

b) If 400,000 of these automobiles are made such that each of the keys determined in part (a) opens the same number of cars, determine how many cars can be opened by a specific key.

c) If one of these cars is randomly selected, what is the probability that the key randomly selected will unlock the door?

56. Voting On a ballot, each committee member is asked to rank three of eight candidates for recommendation for promotion, giving first, second, and third choices (no ties). What is the minimum number of ballots that must be cast to guarantee that at least two ballots are the same?

57. Scrabble Nancy, who is playing Scrabble with Dale, has seven different letters. She decides to test each five-letter permutation before her next move. If each permutation takes 5 sec, how long will it take Nancy to check all the permutations?

58. Scrabble In Exercise 57, assume, of Nancy's seven letters, that three are identical and two are identical. How long will it take Nancy to try all different permutations of her seven letters?

59. Does $_nP_r = {_nP_{(n-r)}}$ for all whole numbers, where $n \geq r$?

Recreational Mathematics

60. Stations There are eight bus stations from town A to town B. How many different single tickets must be printed so that a passenger may purchase a ticket from any station to any other station?

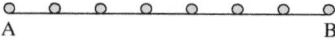

A   B

61. Bus Loop How many tickets with different points of origin and destination can be sold on a bus line that travels a loop with 25 stops?

JUMBLE Many newspapers contain JUMBLE puzzles, where the letters of a word are given out of order. In Exercises 62 and 63, read the Recreational Math box on page 715 and determine

a) *the number of different possible arrangements of the letters given.*

b) *the word when the letters are placed in the proper order.*

62. HEICOC

63. DUTENST

Research Activity

64. ISBN When a book is published, it is assigned a 10-digit code number called the International Standard Book Number (ISBN). Do research and write a report on how this coding system works.

Combinations

Upon completion of this section, you will be able to:

- Solve problems involving combinations.

Isaac is the faculty advisor for the Brain Bowl team at Northeast Texas Community College. For an upcoming tournament, Isaac selects a team of four players from six eligible students. How many different teams of four players are possible? After the tournament, Isaac takes the team out for pizza. They order an extra-large pizza and must choose three toppings from a list of 11 toppings. How many different three-topping pizzas are possible? In this section, we will answer these and similar questions that involve forming an arrangement of objects (or people) without regard to the order of the objects. Such arrangements are called *combinations*.

Why This Is Important Combinations are used in many branches of mathematics including probability, statistics, analysis, operations research, and abstract algebra. Mathematical applications that involve combinations play a role in data storage and data mining. Combinations are also used in chemistry, physics, and biology. A basic understanding of combinations may be helpful to you in future mathematics and science classes.

In Section 11.7, we considered problems in which selections of objects were made where the order of the selections was important. These problems involved *permutations*. In this section, we will consider problems in which selections of objects are made where the order of the selections is not important. Two such problems were presented in the section opener. These problems involve *combinations*.

For example, *a, b, c* and *b, c, a* are two different permutations of the same letters because the ordering of the three letters is different. The letters *a, b, c* and *b, c, a* represent the same combination of letters because the *same letters* are used in each set. However, the letters *a, b, c* and *a, b, d* represent two different combinations of letters because the letters contained in each set are different.

> **Definition: Combination**
> A **combination** is a distinct group (or set) of objects without regard to their arrangement.

Example 1 *Permutation or Combination*

The Earth Club at Joliet Junior College has members Jaspar, Carla, Opal, David, and Malcolm. Determine whether the description given below regarding a selection involving the Earth Club members describes a permutation or a combination.

a) The club will select a president and a treasurer.

b) Of the five club members, two will be attending a Sierra Club meeting together.

Solution

a) Since the president's position is different from the treasurer's position, the order in which the selection is made is important. For example, Jaspar as president and Carla as treasurer is different from Carla as president and Jaspar as treasurer. Since the order of the selection is important, this is a permutation.

b) The order in which two club members are selected to attend a meeting does not matter. For example, selecting Opal and then David is the same as selecting David and then Opal. Since the order of the selection is not important, this is a combination. ∎

In Section 11.7, you learned that $_nP_r$ represents the number of permutations when r items are selected from n distinct items. *Similarly, $_nC_r$ represents the number of combinations when r items are selected from n distinct items.*

Consider the set of elements $\{a, b, c, d, e\}$. The number of permutations of two letters from the set is represented as $_5P_2$, and the number of combinations of two letters from the set is represented as $_5C_2$. Twenty permutations of two letters and 10 combinations of two letters are possible from these five letters. Thus, $_5P_2 = 20$ and $_5C_2 = 10$, as shown.

Permutations	Combinations
$\left.\begin{array}{l} ab,\ ba,\ ac,\ ca,\ ad,\ da,\ ae,\ ea,\ bc,\ cb, \\ bd,\ db,\ be,\ eb,\ cd,\ dc,\ ce,\ ec,\ de,\ ed \end{array}\right\}20$	$\left.\begin{array}{l} ab,\ ac,\ ad,\ ae,\ bc, \\ bd,\ be,\ cd,\ ce,\ de \end{array}\right\}10$

When discussing both combination and permutation problems, we always assume that the experiment is performed without replacement. That is why duplicate letters such as *aa* or *bb* are not included in the preceding example.

Note that from one combination of two letters, two permutations can be formed. For example, the combination *ab* gives the permutations *ab* and *ba*, or twice as many permutations as combinations. Thus, for this example we may write

$$_5P_2 = 2 \cdot (_5C_2)$$

Since $2 = 2!$, we may write

$$_5P_2 = 2!(_5C_2)$$

If we repeated this same process for comparing the number of permutations in $_nP_r$ with the number of combinations in $_nC_r$, we would determine that

$$_nP_r = r!(_nC_r)$$

Dividing both sides of the equation by $r!$ gives

$$_nC_r = \frac{_nP_r}{r!}$$

Since $_nP_r = \dfrac{n!}{(n-r)!}$, the combination formula may be expressed as

$$_nC_r = \frac{n!/(n-r)!}{r!} = \frac{n!}{(n-r)!r!}$$

Combination Formula

The number of combinations possible when r objects are selected from n objects is determined by the **combination formula**

$$_nC_r = \frac{n!}{(n-r)!r!}$$

Timely Tip

When evaluating combinations, the following are true for any positive integer, n.

$$_nC_0 = 1$$
$$_nC_1 = n$$
$$_nC_{n-1} = n$$
$$_nC_n = 1$$

Also, remember that
$$0! = 1 \text{ and } 1! = 1.$$

Example 2 *Book Selection*

In your American History class, you are assigned to read any 3 books from a list of 8 books. In how many ways can you select 3 of the 8 books to read?

Solution This problem is a combination problem because the order in which the three books are selected and read does not matter.

$$_8C_3 = \frac{8!}{(8-3)!3!} = \frac{8!}{5!3!} = \frac{8 \cdot 7 \cdot 6 \cdot 5 \cdot 4 \cdot 3 \cdot 2 \cdot 1}{5 \cdot 4 \cdot 3 \cdot 2 \cdot 1 \cdot 3 \cdot 2 \cdot 1} = 56$$

There are 56 different ways that you can select 3 of the 8 books to read. ∎

TECHNOLOGY TIP

Evaluating Combinations

Most scientific and all graphing calculators can evaluate combinations. While the commands on these calculators may vary depending on the model, look for a key or function that has the notation *nCr*. If you cannot locate this key or function, conduct an Internet search with your calculator model and the word *combination*. For example, if you search for "Casio FX-82MS combination" you will be directed to several websites including online videos that can assist you. Also, there are many apps for smartphones and tablets that can be used to calculate combinations.

To evaluate $_8C_3$ on a TI-84 Plus calculator, enter the number 8. Then press the MATH key. Use the right arrow key to scroll to PRB, and then scroll to nCr. Press the ENTER key. Next, press 3. Press the ENTER key again. The calculator will display 56, which agrees with our answer in Example 2. The screenshots below show the calculation depending on the version of the TI-84 Plus.

```
NORMAL FLOAT AUTO REAL DEGREE CL
8 nCr 3
                              56
```

```
NORMAL FLOAT AUTO REAL DEGREE MP
8C3
                              56
```

Example 3 *Brain Bowl Decisions*

Isaac is the faculty advisor for the Brain Bowl Team at Northeast Texas Community College. For an upcoming tournament, Isaac selects a team of 4 players from 6 eligible students.

a) How many different teams of four players are possible?

b) After the tournament, Isaac takes the team out for pizza. They order an extra-large pizza and must choose 3 toppings from a list of 11 toppings. How many different three-topping pizzas are possible?

Solution

a) Since the order in which the team members are selected does not matter, this is a combination problem. Isaac must choose four players from 6 eligible students.

$$_6C_4 = \frac{6!}{(6-4)!4!} = \frac{6!}{2!4!} = \frac{6 \cdot 5 \cdot 4 \cdot 3 \cdot 2 \cdot 1}{2 \cdot 1 \cdot 4 \cdot 3 \cdot 2 \cdot 1} = \frac{30}{2} = 15$$

Thus, there are 15 different teams of four players possible.

b) Since the order in which the pizza toppings are selected does not matter, this is a combination problem. There are 11 toppings, and they must choose 3.

$$_{11}C_3 = \frac{11!}{(11-3)!3!} = \frac{11!}{8!3!} = \frac{11 \cdot 10 \cdot 9 \cdot 8 \cdot 7 \cdot 6 \cdot 5 \cdot 4 \cdot 3 \cdot 2 \cdot 1}{8 \cdot 7 \cdot 6 \cdot 5 \cdot 4 \cdot 3 \cdot 2 \cdot 1 \cdot 3 \cdot 2 \cdot 1}$$

$$= \frac{990}{6} = 165$$

Thus, there are 165 different 3-topping pizzas possible.   ∎

Example 4 *Dinner Combinations*

At the Tokyo Bangkok restaurant, dinner consists of selecting 3 items from column A, selecting 4 items from column B, and selecting 3 items from column C from the menu. If columns A, B, and C have 5, 7, and 6 items, respectively, how many different dinner combinations are possible?

Solution For column A, 3 of 5 items must be selected, which can be represented as $_5C_3$. For column B, 4 of 7 items must be selected, which can be represented as $_7C_4$. For column C, 3 of 6 items must be selected, or $_6C_3$.

$$_5C_3 = 10 \quad _7C_4 = 35 \quad \text{and} \quad _6C_3 = 20$$

Using the fundamental counting principle, we can determine the total number of dinner combinations by multiplying the number of choices from columns A, B, and C:

$$\text{Total number of dinner choices} = _5C_3 \cdot _7C_4 \cdot _6C_3$$
$$= 10 \cdot 35 \cdot 20 = 7000$$

Therefore, 7000 different combinations are possible under these conditions.   ∎

We have presented various counting methods, including the fundamental counting principle, permutations, and combinations. You often need to decide which method to use to solve a problem. Table 11.4 may help you in selecting the procedure to use.

Table 11.4 Summary of Counting Methods

Fundamental Counting Principle: If a first experiment has M distinct outcomes and a second experiment has N distinct outcomes, then the two experiments in that specific order have $M \cdot N$ distinct outcomes.	Determining the Number of Ways of Selecting r Items from n Items (Repetition Is Not Permitted)	
	Permutations	**Combinations**
The fundamental counting principle may be used with or without repetition of items. It is used when determining the number of distinct outcomes when two or more experiments are performed. It is also used when there are specific placement requirements, such as the first digit must be a 0 or 1.	Permutations are used when order is important.	Combinations are used when order is not important.
	For example, a, b, c and b, c, a are two different permutations of the same three letters.	For example, a, b, c and b, c, a are the same combination of three letters. But $a, b, c,$ and a, b, d are two different combinations of three letters.
	$$_nP_r = \frac{n!}{(n-r)!}$$	$$_nC_r = \frac{n!}{(n-r)!r!}$$
	Problems solved with the permutation formula may also be solved by using the fundamental counting principle.	

SECTION 11.8 *Exercises*

Warm Up Exercises

In Exercises 1–6, fill in the blank with an appropriate word, phrase, or symbol(s).

1. A distinct group of objects without regard to their arrangement is called a(n) _____.

2. The symbol for the number of combinations when r items are selected from n distinct items is _____.

3. If we want to select r items from n items, and the order of the arrangement is important, then _____ are used.

4. If we want to select r items from n items, and the order of the arrangement is not important, then _____ are used.

Practice the Skills

In Exercises 5–6, determine whether the description given describes a permutation or a combination.

5. **Softball Team** Jocelyn is the manager of the Central Carolina Community College softball team. The team has 16 players.

   a) Jocelyn selects 9 players for the batting order of today's game.

   b) Jocelyn selects 5 players to volunteer at the school's open house.

6. **Smartphone Photos** Sundar has 20,291 photos on his smartphone.

   a) Sundar selects and deletes 500 photos to have enough storage to download a new app.

   b) Sundar selects 25 photos to place in order for a Power Point presentation.

In Exercises 7–20, evaluate the expression.

7. $_5C_3$

8. $_8C_2$

9. a) $_6C_4$ b) $_6P_4$ 10. a) $_7C_2$ b) $_7P_2$

11. a) $_8C_0$ b) $_8P_0$ 12. a) $_{12}C_8$ b) $_{12}P_8$

13. a) $_{10}C_3$ b) $_{10}P_3$ 14. a) $_5C_5$ b) $_5P_5$

15. $\dfrac{_3C_2}{_{13}C_2}$ 16. $\dfrac{_5C_3}{_9C_3}$ 17. $\dfrac{_8C_5}{_{14}C_5}$

18. $\dfrac{_4C_2}{_8C_2}$ 19. $\dfrac{_4C_3}{_{10}C_3}$ 20. $\dfrac{_6C_4}{_{10}C_4}$

Problem Solving

21. **Hiring** There are 10 qualified applicants for 3 security positions at a hospital. In how many ways can the positions be filled?

22. **Banana Split** An ice-cream parlor has 20 different flavors. Cynthia orders a banana split and has to select 3 different flavors. How many different selections are possible?

23. **Test Essays** Tina must select and answer in any order four of six essay questions on a test. In how many ways can she do so?

24. **Software Packages** During a special promotion at Dell, a customer purchasing a computer and a printer is given the choice of 2 free software packages. If there are 9 different software packages from which to select, how many different ways can the 2 packages be selected?

25. **Scholarships** A scholarship committee has received 8 applications for a $500 scholarship. The committee has decided to select 3 of the 8 candidates for further consideration. In how many ways can the committee do so?

26. **Quinella Bet** A quinella bet consists of selecting the first- and second-place winners, in any order, in a particular event. For example, suppose you select a 2–5 quinella. If 2 wins and 5 finishes second, or if 5 wins and 2 finishes second, you win. Mr. Smith goes to a jai alai match. In the match, 8 jai alai teams compete. How many quinella tickets must Mr. Smith purchase to guarantee a win?

▲ Jai Alai

27. **Pennsylvania Lottery** Match 6 Lotto is a Pennsylvania lottery game in which players select 6 different numbers, in any order, from 1 to 49, inclusively. How many different sets of 6 numbers can be chosen?

28. **Texas Lottery** Lotto Texas is a Texas lottery game in which players select 6 different numbers, in any order, from 1 to 54, inclusively. How many different sets of 6 numbers can be chosen?

29. **Shirts** Matthew has 12 shirts in his closet and must select 4 to take with him on vacation. In how many ways can Matthew select the shirts?

30. **Shoe Donation** Dirk owns 45 pairs of shoes and would like to donate 10 pairs to Goodwill. How many different ways can Dirk select the pairs of shoes?

31. **Full House** If 5 cards are dealt from a standard 52-card deck, how many different ways can 3 kings and 2 queens be dealt?

32. **Four of a Kind** If 5 cards are dealt from a standard 52-card deck, how many different ways can 4 aces and 1 other card that is not an ace be dealt?

33. **Choosing Volunteers** Kipsie is the athletic director at Owens Community College. She must select 5 tennis players and 5 basketball players to volunteer at the school's community showcase. She can choose from 9 tennis players and 15 basketball players. How many different ways can she select the volunteers?

34. **Test Question** On an English test, Tito must write an essay for three of the five questions in Part 1 and four of the six questions in Part 2 in any order. How many different combinations of questions can he answer?

35. **Strollers and Car Seats** A Target store has 9 different strollers and 8 different car seats in stock. The store's manager wishes to place 3 strollers and 2 car seats on sale. In how many ways can this be done?

36. **Newsletter** An animal shelter has 15 cats and 12 dogs available for adoption. The shelter wants to select 6 cats and 4 dogs to feature in the adoption newsletter. In how many ways can this be done?

37. **Real Estate Flier** Bobby is selling his house and is preparing a flier to give to potential buyers. On the flier, Bobby wishes to put 2 photos of the outside of the house and 4 photos of the inside of the house. He has 7 photos of the outside and 11 photos of the inside to choose from. How many different ways can the flier be prepared?

38. **Forming a Committee** The Philadelphia City Council is forming a committee to explore ways to improve public transporation in the city. The committee will consist of 4 representatives from the city council and 3 representatives from a citizens advisory board. If there are 7 city council members and 5 citizens advisory board members from which to choose, how many different ways can the committee be formed?

39. **Selecting Soda** Josef is sent to the store to get 5 different bottles of regular soda and 3 different bottles of diet soda. If there are 10 different types of regular sodas and 7 different types of diet sodas to choose from, how many different choices does Josef have?

40. **Chili Recipe** Christoph is making chili. He wishes to include 3 types of beans, and his choices are kidney, pinto, red, black, garbanzo, fava, and navy. He wishes to include 4 types of peppers, and his choices are red bell, green bell, orange bell, yellow bell, jalapeno, habanero, poblano, and datil. Based on these two choices, how many different varieties of chili can Christoph make?

41. **Selecting Mutual Funds** Hamid recently graduated from college and now has a job that provides a retirement investment plan. Hamid wants to diversify his investments, so he wants to invest in four stock mutual funds and two bond mutual funds. If he has a choice of eight stock mutual funds and five bond mutual funds, how many different selections of mutual funds does he have?

42. **Door Prize** As part of a door prize, Mary won three tickets to a baseball game and three tickets to a theater performance. She decided to give all the tickets to friends. For the baseball game she is considering six different friends, and for the theater she is considering eight different friends. In how many ways can she distribute the tickets?

43. **New Products** Nabisco is testing 5 chocolate chip cookies, 4 crackers, and 6 reduced-fat cookies. If it plans to market 2 of the chocolate chip cookies, 3 of the crackers, and 2 of the reduced-fat cookies, how many different combinations are possible?

44. **Catering Service** A catering service is making up trays of hors d'oeuvres. The hors d'oeuvres are categorized as inexpensive, average, and expensive. If the client must select three of the eight inexpensive, five of the nine average, and two of the four expensive hors d'oeuvres, how many different choices are possible?

Challenge Problems/Group Activities

45. **Test Answers** Consider a 10-question test in which each question can be answered either correctly or incorrectly.

   a) How many different ways are there to answer the questions so that eight are correct and two are incorrect?

   b) How many different ways are there to answer the questions so that at least eight are correct?

46. a) **A Dinner Toast** Four people at dinner make a toast. If each person is to tap glasses with each other person one at a time, how many taps will take place?

   b) Repeat part (a) with five people.

   c) How many taps will there be if there are n people at the dinner table?

47. *Pascal's Triangle* The notation $_nC_r$ may be written $\binom{n}{r}$.

   a) Use this notation to evaluate each of the combinations in the following array. Form a triangle of the results, similar to the one given, by placing the answer to each combination in the same relative position in the triangle.

$$\binom{0}{0}$$

$$\binom{1}{0} \qquad \binom{1}{1}$$

$$\binom{2}{0} \qquad \binom{2}{1} \qquad \binom{2}{2}$$

$$\binom{3}{0} \qquad \binom{3}{1} \qquad \binom{3}{2} \qquad \binom{3}{3}$$

$$\binom{4}{0} \qquad \binom{4}{1} \qquad \binom{4}{2} \qquad \binom{4}{3} \qquad \binom{4}{4}$$

   b) Using the number pattern in part (a), determine the next row of numbers of the triangle (known as *Pascal's triangle*).

48. *Lottery Combinations* Determine the number of combinations possible in a state lottery where you must select

   a) 6 of 46 numbers.

   b) 6 of 47 numbers.

   c) 6 of 48 numbers.

   d) 6 of 49 numbers.

   e) Does the number of combinations increase by the same amount going from part (a) to part (b) as from part (b) to part (c)?

49. a) *Table Seating Arrangements* How many distinct ways can four people be seated in a row?

   b) How many distinct ways can four people be seated at a circular table?

50. Show that $_nC_r = \, _nC_{(n-r)}$.

51. *Forming a Committee* A group of 15 people wants to form a committee consisting of a chair, vice chair, and three additional members. How many different committees can be formed?

52. *Card Hands* In popular card games, there is such a variety of possible combinations of cards that a player rarely gets the same hand twice. Use the following information about card games to determine the number of different hands.

   a) *Poker*: 5 cards are dealt from a standard 52-card deck.

   b) *Gin Rummy*: 10 cards are dealt from a standard 52-card deck.

   c) *Bridge, Hearts,* or *Spades*: 13 cards are dealt from a standard 52-card deck.

   d) *Euchre*: 5 cards are dealt from a 24-card deck.

   e) *Pinochle*: 12 cards are dealt from a 48-card deck.

Recreational Mathematics

53. a) *Combination Lock* To open a combination lock, you must know the lock's three-number sequence in its proper order. Repetition of numbers is permitted. Why is this lock more like a permutation lock than a combination lock? Why is it not a true permutation problem?

   b) Assuming that a combination lock has 40 numbers, determine how many different three-number arrangements are possible if repetition of numbers is allowed.

   c) Answer the question in part (b) if repetition is not allowed.

Research Activity

54. *Combinatorics* The area of mathematics called combinatorics is the science of counting. Do research and write a paper on combinatorics and its many applications.

SECTION 11.9   Solving Probability Problems by Using Combinations

Upon completion of this section, you will be able to:

■ Solve probability problems using combinations.

Fedor is playing poker and has been dealt a hand of 5 cards. What is the probability that Fedor has been dealt a royal flush? What is the probability that Fedor has been dealt a pair aces and a pair of 8's? In this section, we will use combinations to answer questions such as these.

Why This Is Important Probability problems that involve combinations are part of a branch of mathematics called *combinatorics*. Combinatorics has many applications in computer science, cyber security, engineering, and other branches of mathematics. Combinatorics is also used in the social sciences, genetics, and many business applications.

In Section 11.5, we used the *and* probability formula to solve probability problems. Here we use the *and* formula to determine the probability of drawing 2 face cards, without replacement, from a standard 52-card deck.

$$P(2\text{ face cards}) = P(\text{first face card}) \cdot P(\text{second face card})$$

$$= \frac{12}{52} \cdot \frac{11}{51} = \frac{3}{13} \cdot \frac{11}{51} = \frac{33}{663} = \frac{11}{221}$$

The order in which the 2 cards are drawn does not matter, thus; this, problem can also be considered a combination problem. To determine the probability of drawing 2 face cards, we divide the number of different successful outcomes (drawing any 2 of the 12 face cards) by the number of total possible outcomes (drawing any 2 of the 52 cards in the deck).

The number of ways in which two face cards can be selected from the 12 face cards in a deck is $_{12}C_2$, or

$$_{12}C_2 = \frac{12!}{(12-2)!2!} = \frac{\overset{6}{\cancel{12}} \cdot 11 \cdot \cancel{10!}}{\cancel{10!} \cdot \cancel{2} \cdot 1} = 66$$

The number of ways in which two cards can be selected from a standard 52-card deck is $_{52}C_2$, or

$$_{52}C_2 = \frac{52!}{(52-2)!2!} = \frac{\overset{26}{\cancel{52}} \cdot 51 \cdot \cancel{50!}}{\cancel{50!} \cdot \cancel{2} \cdot 1} = 1326$$

Thus,

$$P(\text{selecting 2 face cards}) = \frac{_{12}C_2}{_{52}C_2} = \frac{66}{1326} = \frac{11}{221}$$

Note that the same answer is obtained with either method. To give you more exposure to counting techniques, we will work the problems in this section using combinations.

Example 1 *Math Club Committee*

The Math Club at Clackamas Community College has 5 freshmen and 7 sophomores. Joy, the club advisor, randomly selects 4 club members to attend the college's interclub council. What is the probability that Joy selects 4 sophomores?

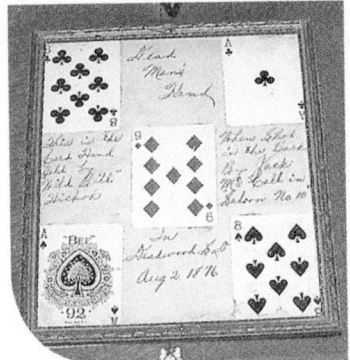
Solution The order in which the 4 members are selected is not important. Therefore, we will work this problem using combinations.

$$P(\text{selecting 4 sophomores}) = \frac{\text{number of ways to select any 4 sophomores}}{\text{number of ways to select any 4 club members}}$$

Since there are 7 sophomores, the number of ways to select any 4 sophomores is $_7C_4 = 35$. Since there are a total of 12 club members, the number of ways to select any 4 club members is $_{12}C_4 = 495$.

$$P(\text{selecting 4 sophomores}) = \frac{_7C_4}{_{12}C_4} = \frac{35}{495} = \frac{7}{99}$$

The probability that Joy randomly selects 4 sophomores is $\frac{7}{99}$. ∎

Example 2 A Heart Flush

A flush in the game of poker is five cards of the same suit (5 hearts, 5 diamonds, 5 clubs, or 5 spades). If you are dealt a five-card hand, determine the probability that you will be dealt a heart flush.

Solution The order in which the five hearts are dealt is not important. Therefore, we may work this problem using combinations.

$$P(\text{heart flush}) = \frac{\text{number of possible 5-card heart flushes}}{\text{total number of possible 5-card hands}}$$

Since there are 13 hearts in a deck of cards, the number of possible five-card heart flush hands is $_{13}C_5 = 1287$. The total number of possible five-card hands in a deck of 52 cards is $_{52}C_5 = 2,598,960$.

$$P(\text{heart flush}) = \frac{_{13}C_5}{_{52}C_5} = \frac{1287}{2,598,960} = \frac{33}{66,640}$$

The probability of being dealt a heart flush is $\frac{33}{66,640}$, or ≈ 0.000495. ∎

Example 3 DVD Selection

Jennifer has 4 *Scooby Doo* DVDs, 5 *Magic School Bus* DVDs, and 3 *PAW Patrol* DVDs. If Jennifer randomly selects 4 DVDs for her children to take on a road trip, determine the probability that

a) no *Scooby Doo* DVDs are selected.

b) at least 1 *Scooby Doo* DVD is selected.

c) 2 *Scooby Doo* DVDs and 2 *Magic School Bus* DVDs are selected.

Solution

a) If no *Scooby Doo* DVDs are to be selected, then only *Magic School Bus* and *PAW Patrol* DVDs must be selected. Eight DVDs are either *Magic School Bus* or *PAW Patrol*. Thus, the number of ways that 4 *Magic School Bus* or *PAW Patrol* DVDs may be selected from the possible 8 *Magic School Bus* or *PAW Patrol* DVDs is $_8C_4$. The total number of possible selections is $_{12}C_4$.

$$P(\text{no } Scooby\ Doo \text{ DVDs}) = \frac{_8C_4}{_{12}C_4} = \frac{70}{495} = \frac{14}{99}$$

b) When 4 DVDs are selected, the choice must contain either no *Scooby Doo* DVDs or at least 1 *Scooby Doo* DVD. Since one of these outcomes must occur, the sum of the probabilities must be 1, or

$$P(\text{no } Scooby\ Doo \text{ DVDs}) + P(\text{at least 1 } Scooby\ Doo \text{ DVD}) = 1$$

▲ *Scooby Doo and Friends*

Note that the probability of selecting no *Scooby Doo* DVDs, $\frac{14}{99}$, was determined in part (a).

Therefore,

$$P(\text{at least 1 } Scooby\ Doo \text{ DVD}) = 1 - P(\text{no } Scooby\ Doo \text{ DVDs})$$

$$= 1 - \frac{14}{99} = \frac{99}{99} - \frac{14}{99} = \frac{85}{99}$$

c) The number of ways of selecting 2 *Scooby Doo* DVDs out of 4 *Scooby Doo* DVDs is $_4C_2$, which equals 6. The number of ways of selecting 2 *Magic School Bus* DVDs out of 5 *Magic School Bus* DVDs is $_5C_2$, which equals 10. The total number of possible selections when 4 DVDs are selected from the 12 choices is $_{12}C_4$, which equals 495. Since both the 2 *Scooby Doo* and 2 *Magic School Bus* DVDs must be selected, the probability is calculated as follows.

$$P(2\ Scooby\ Doo \text{ and } 2\ Magic\ School\ Bus) = \frac{_4C_2 \cdot {}_5C_2}{_{12}C_4} = \frac{6 \cdot 10}{495} = \frac{60}{495} = \frac{4}{33} \blacksquare$$

Example 4 *Rare Coins*

Conner's rare coin collection is made up of 8 silver dollars, 7 quarters, and 5 dimes. Conner plans to sell 8 of his 20 coins to finance part of his college education. If he selects the coins at random, what is the probability that 3 silver dollars, 2 quarters, and 3 dimes are selected?

Solution The number of ways that Conner can select 3 out of 8 silver dollars is $_8C_3$. The number of ways he can select 2 out of 7 quarters is $_7C_2$. The number of ways he can select 3 out of 5 dimes is $_5C_3$. He will select 8 coins from a total of 20 coins. The number of ways he can do so is $_{20}C_8$. The probability that Conner selects 3 silver dollars, 2 quarters, and 3 dimes is calculated as follows.

$$P(3 \text{ silver dollars, 2 quarters, and 3 dimes}) = \frac{_8C_3 \cdot {}_7C_2 \cdot {}_5C_3}{_{20}C_8}$$

$$= \frac{56 \cdot 21 \cdot 10}{125{,}970} = \frac{11{,}760}{125{,}970} = \frac{392}{4199} \blacksquare$$

SECTION 11.9   *Exercises*

Practice the Skills

In Exercises 1–8, set up the problem as if it were to be solved, but do not solve. Assume that each problem is to be done without replacement.

1. **Pens** A bin full of promotional pens for the University of Florida contains 12 blue pens and 8 orange pens. If Lindsay randomly selects 4 pens from the bin, determine the probability that all 4 pens are blue.

2. **Stock Car Race** The Crawford County Speedway held a race with 18 Ford cars and 12 Chevrolet cars. If 8 of these cars are randomly selected, determine the probability that all 8 are Chevrolet cars.

▲ *See Exercise 2*

3. **Kings** Determine the probability of being dealt 3 kings from a standard 52-card deck when 3 cards are dealt.

4. **Letters** Three letters are to be randomly selected from 26 letters. Determine the probability that 3 vowels out of the 5 vowels (a, e, i, o, u) are selected.

5. *Kindergarten* Mrs. Lopez's kindergarten class has 15 students, of whom 8 are from Manatee County. If 5 students are randomly selected, determine the probability that all 5 are from Manatee County.

6. *Movie Attendance* Of 80 people attending a movie, 38 have a college degree. If 5 people at the movie are randomly selected, determine the probability that each of the 5 has a college degree.

7. *College Orientation* Of 120 students attending a college orientation, 97 had completed the mathematics placement test. If 5 students at the orientation are randomly selected, determine the probability that all 5 had *not* completed the mathematics placement test.

8. *October Birthdays* A class of 16 people contains 4 people whose birthdays are in October. If 3 people from the class are randomly selected, determine the probability that *none* of those selected has an October birthday.

Problem Solving

In Exercises 9–18, the problems are to be done without replacement. Use combinations to determine probabilities.

9. *Selecting Soda* A refrigerator contains 3 cans of diet soda and 10 cans of regular soda. If two cans of soda are randomly selected, determine the probability that both cans selected are diet soda.

10. *Green and Red Balls* A bag contains four red balls and five green balls. You randomly select three balls. Determine the probability of selecting three green balls.

11. *Loose Change* Ashton's car console has 6 nickels and 8 quarters. Ashton randomly selects 5 coins to pay a toll. Assume all coins are equally likely to be selected. What is the probability that Ashton selects 5 quarters?

12. *Bills* Duc's wallet contains 8 bills of the following denominations: four $5 bills, two $10 bills, one $20 bill, and one $50 bill. If Duc randomly selects two bills, determine the probability that he selects two $5 bills.

13. *Lottery Selection* A lottery involves randomly selecting 3 digits from 0–9, inclusively, without replacement. What is the probability that all 3 digits selected are greater than 5.

14. *Sock Drawer* Mateo's sock drawer has 12 black socks and 8 white socks all mixed together. If Mateo randomly selects 4 socks from the drawer, what is the probability that all 4 socks are white?

15. *Candy* Cassandra's candy dish has 8 peppermint candies and 5 spearmint candies. If Cassandra randomly selects 3 candies, what is the probability that she selects 1 peppermint candy and 2 spearmint candies?

16. *Motorcycle Club* The Matadors motorcycle club has 15 members. Eleven members ride Harley-Davidson motorcycles, and 4 members ride Victory motorcycles. If 5 members of the club are randomly selected, what is the probability that 3 ride Harley-Davidson motorcycles and 2 ride Victory motorcycles?

Game Show *In Exercises 17–20, a game show has 5 doors concealing prizes. There are new cars behind 2 doors and consolation prizes behind 3 doors. Ginny, the contestant, selects 2 doors. Determine the probability that Ginny selects*

17. no cars.   **19.** at least one car.

18. both cars.   **20.** exactly one car.

Football Roster *In Exercises 21–24, the Green Bay Packers roster consists of 26 players on offense, 24 players on defense, and 3 players on special teams. If 10 players are randomly selected, determine the probability (as decimal number rounded to four decimal places) that of the players selected,*

21. 5 play offense and 5 play defense.

22. 3 play offense and 7 play defense.

23. 4 play offense, 4 play defense, and 2 play special teams.

24. 8 play offense, 1 plays defense, and 1 plays special teams.

New Breakfast Cereal *In Exercises 25–28, General Mills is testing 12 new cereals for possible production. They are testing 3 oat cereals, 4 wheat cereals, and 5 rice cereals. If each of the 12 cereals has the same chance of being produced, and 4 new cereals will be produced, determine the probability that the 4 new cereals that will be produced are as follows.*

25. 2 are oat cereals and 2 are rice cereals.

26. 1 is a wheat cereal and 3 are rice cereals.

27. 2 are oat cereals, 1 is a wheat cereal, and 1 is a rice cereal.

28. at least one is an oat cereal.

29. Baseball Roster The Auburn University baseball roster consists of 12 pitchers, 2 catchers, 6 infielders, and 5 outfielders. If the team manager randomly selects 9 players, what is the probability that he selects 1 pitcher, 1 catcher, 4 infielders, and 3 outfielders?

30. Restaurant Staff The staff at The Captain's Lair consists of 5 cooks, 11 servers, and 4 bartenders. If the restaurant manager randomly selects 10 staff members, what is the probability that she selects 2 cooks, 7 servers, and 1 bartender?

31. A Royal Flush A royal flush consists of an ace, king, queen, jack, and 10 all in the same suit. If 7 cards are dealt from a standard 52-card deck, determine the probability of getting a

a) royal flush in spades.

b) royal flush in any suit.

32. Poker Probability A full house in poker consists of three of one kind and two of another kind in a five-card hand. For example, if a hand contains three kings and two 5's, it is a full house. If 5 cards are dealt from a standard 52-card deck, determine the probability of getting three kings and two 5's.

33. Poker Hand Ronnie is playing poker and is dealt his hand of 5 cards from a standard 52-card deck. What is the probability that Ronnie is dealt 1 club, 1 diamond, 1 heart, and 2 spades?

34. Poker Hand Bianca is playing poker and is dealt her hand of 5 cards from a standard 52-card deck. What is the probability that Bianca is dealt 2 face cards, 2 aces, and 1 card that is neither a face card nor an ace?

35. Joker's Included When a standard deck of cards is purchased, the deck usually includes 2 jokers. This means if the jokers are included, the deck contains 54 cards. Teresa is playing a game in which she is dealt 8 cards from a deck that includes the jokers. What is the probability that Teresa's hand includes at least one joker?

36. Euchre Hand A deck of cards for playing euchre contains 24 cards—9, 10, jack, queen, king, and ace of each suit. Sergio is playing euchre and is dealt a 5-card hand. What is the probability that Sergio's hand includes at least one ace?

37. "Dead Man's Hand" A pair of aces and a pair of 8's is often known as a "dead man's hand." (See the *Did You Know?* on page 727.)

a) Determine the probability of being dealt a dead man's hand (any two aces, any two 8's, and one other card that is not an ace or an 8) when 5 cards are dealt, without replacement, from a standard 52-card deck.

b) The actual cards "Wild Bill" Hickok was holding when he was shot were the aces of spades and clubs, the 8's of spades and clubs, and the 9 of diamonds. If you are dealt five cards, determine the probability of being dealt this exact hand.

38. Poker Straights A straight in poker is 5 cards *of any suit* in sequence. A jack is considered an 11, a queen is considered a 12, a king is considered a 13, and an ace is considered as either a 1 or a 14. The following are examples of straights: ace, 2, 3, 4, 5; 6, 7, 8, 9, 10; 8, 9, 10, jack, queen; 10, jack, queen, king, ace. Remember the suit does not matter when forming a straight. Molly is playing poker and is dealt 5 cards from a standard 52-card deck.

a) What is the probability that Molly is dealt a straight consisting of an ace, a 2, a 3, a 4, and a 5?

b) What is the probability that Molly is dealt any straight? *Hint*: A straight may start with an ace, 2, 3, 4, 5, 6, 7, 8, 9, or 10. Therefore, multiply the answer from part (a) by 10.

Challenge Problems/Group Activities

39. Selecting Officers A club consists of 15 people including Ali, Kendra, Ted, Alice, Marie, Dan, Linda, and Frank. From the 15 members, a president, vice president, and treasurer will be randomly selected. An advisory committee of 5 other individuals will also be randomly selected.

a) Determine the probability that Ali is selected president, Kendra is selected vice president, Ted is selected treasurer, and the other 5 individuals named above form the advisory committee.

b) Determine the probability that 3 of the 8 individuals named are selected for the three officers' positions and the other 5 are selected for the advisory board.

40. Alternate Seating If three men and three women are to be randomly assigned to six seats in a row at a theater, determine the probability that they will alternate by gender.

Recreational Mathematics

41. *Hair* When the Isle of Flume took its most recent census, the population was 100,002 people. Everyone on the island has fewer than 100,001 hairs on his or her head. Determine the probability that at least two people have exactly the same number of hairs on their head.

Research Activity

42. *Poker Probabilities* Throughout this section we have discussed the probabilities of several poker hands. Do research and write a paper on how to calculate the probabilities for all poker hands. Include in your report one pair, two pairs, three of a kind, straight, flush, full house, four of a kind, straight flush, and royal flush.

SECTION 11.10 Binomial Probability Formula

Upon completion of this section, you will be able to:

■ Solve problems using the binomial probability formula.

Suppose that you are a server at a restaurant and have learned from past experience that 80% of your customers leave a tip. If you wait on 6 customers, what is the probability that all 6 customers will leave a tip? If you wait on 8 customers, what is the probability that at least 5 of them will leave a tip? In this section, we will learn how to use the binomial probability formula to answer these and similar questions.

Why This Is Important Many applications from business and industry involve the use of binomial probability. Insurance companies use binomial probability when setting insurance rates. Pharmacological companies use binomial probability to assess the effectiveness of their drugs. Binomial probability also plays a vital role in other branches of mathematics—especially statistics—and in computer science, biology, chemistry, and physics.

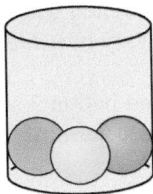

Figure 11.21

StatCrunch

Applets Simulation Urn Sampling

Suppose that a basket contains three identical balls, except for their color. One is red, one is blue, and one is yellow (Fig. 11.21). Suppose further that we are going to randomly select three balls *with replacement* from the basket. We can determine specific probabilities by examining the tree diagram shown in Fig. 11.22. Note that 27 different selections are possible, as indicated in the sample space.

Sample Space

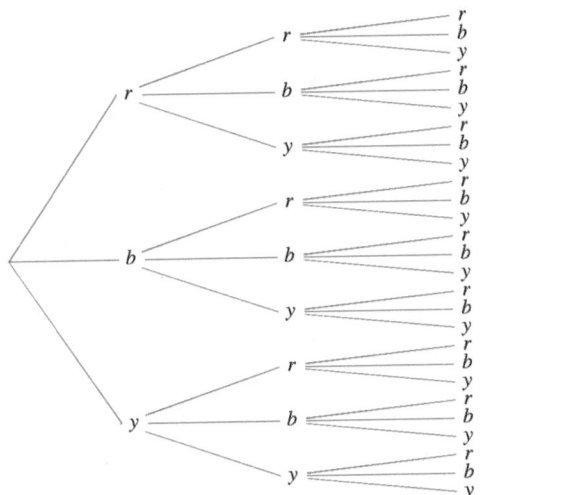

Figure 11.22

Our three selections may yield 0, 1, 2, or 3 red balls. We can determine the probability of selecting exactly 0, 1, 2, or 3 red balls by using the sample space. To determine the probability of selecting 0 red balls, we count those outcomes that do not contain a red ball. There are 8 of them (*bbb, bby, byb, byy, ybb, yby, yyb, yyy*). Thus, the probability of obtaining exactly 0 red balls is 8/27. We determine the probability of selecting exactly 1 red ball by counting the sample points that contain exactly 1 red ball. There are 12 of them. Thus, the probability is $\frac{12}{27}$, or $\frac{4}{9}$.

We can determine the probability of selecting exactly 2 red balls and exactly 3 red balls in a similar manner. The probabilities of selecting exactly 0, 1, 2, and 3 red balls are illustrated in Table 11.5.

Table 11.5   A Probability Distribution for Three Balls Selected with Replacement

Number of Red Balls Selected, (x)	Probability of Selecting the Number of Red Balls, $P(x)$
0	$\dfrac{8}{27}$
1	$\dfrac{12}{27}$
2	$\dfrac{6}{27}$
3	$\dfrac{1}{27}$
	Sum $= \dfrac{27}{27} = 1$

Note that the sum of the probabilities in Table 11.5 is 1. This table is an example of a *probability distribution*, which shows the probabilities associated with each specific outcome of an experiment. *In a probability distribution, every possible outcome must be listed and the sum of the probabilities must be 1.*

Let us specifically consider the probability of selecting 1 red ball in 3 selections. We see from Table 11.5 that this probability is $\frac{12}{27}$, or $\frac{4}{9}$. Can we determine this probability without developing a tree diagram? The answer is yes.

Suppose that we consider selecting a red ball success, S, and selecting a non–red ball failure, F. Furthermore, suppose that we let p represent the probability of success and q the probability of failure on any trial. Then $p = \frac{1}{3}$ and $q = \frac{2}{3}$. We can obtain 1 success in 3 selections in the following ways:

$$\text{SFF} \qquad \text{FSF} \qquad \text{FFS}$$

We can compute the probabilities of each of these outcomes using the multiplication formula because each of the selections is independent since the selections are done with replacement.

$$P(\text{SFF}) = P(S) \cdot P(F) \cdot P(F) = p \cdot q \cdot q = pq^2 = \frac{1}{3}\left(\frac{2}{3}\right)^2 = \frac{4}{27}$$

$$P(\text{FSF}) = P(F) \cdot P(S) \cdot P(F) = q \cdot p \cdot q = pq^2 = \frac{1}{3}\left(\frac{2}{3}\right)^2 = \frac{4}{27}$$

$$P(\text{FFS}) = P(F) \cdot P(F) \cdot P(S) = q \cdot q \cdot p = pq^2 = \frac{1}{3}\left(\frac{2}{3}\right)^2 = \frac{4}{27}$$

$$\text{Sum} = \frac{12}{27} = \frac{4}{9}$$

We obtained an answer of $\frac{4}{9}$, the same answer that was obtained using the tree diagram. Note that each of the 3 sets of outcomes has 1 success and 2 failures. Rather than listing all the possibilities containing 1 success and 2 failures, we can use the combination formula to determine the number of possible combinations of 1 success in 3 trials. To do so, evaluate $_3C_1$.

$$_3C_1 = \frac{3!}{(3-1)!1!} = \frac{3 \cdot 2 \cdot 1}{2 \cdot 1 \cdot 1} = 3$$

Number of trials   Number of successes

Thus, we see that there are 3 ways the 1 success could occur in 3 trials. To compute the probability of 1 success in 3 trials, we can multiply the probability of success in any one trial, $p \cdot q^2$, by the number of ways the 1 success can be arranged among the 3 trials, $_3C_1$. Thus, the probability of selecting 1 red ball, $P(1)$, in 3 trials may be determined as follows.

$$P(1) = (_3C_1)p^1q^2 = 3\left(\frac{1}{3}\right)^1\left(\frac{2}{3}\right)^2 = \frac{12}{27} = \frac{4}{9}$$

The binomial probability formula, which we introduce shortly, explains how to obtain expressions like $P(1) = (_3C_1)p^1q^2$ and is very useful in determining certain types of probabilities.

To use the binomial probability formula, the following three conditions must hold.

To Use the Binomial Probability Formula

1. There are n repeated independent trials.
2. Each trial has two possible outcomes, *success* and *failure.*
3. For each trial, the probability of success (and failure) remains the same.

Before going further, let's discuss why we can use the binomial probability formula to determine the probability of selecting a specific number of red balls when three balls are selected with replacement. First, since each trial is performed *with replacement*, the three trials are independent of each other. Second, we may consider selecting a red ball as success and selecting any ball of another color as failure. Third, for each selection, the probability of success (selecting a red ball) is $\frac{1}{3}$ and the probability of failure (selecting a ball of another color) is $\frac{2}{3}$. Now let's discuss the binomial probability formula.

Binomial Probability Formula

The probability of obtaining exactly x successes, $P(x)$, in n independent trials is given by the binomial probability formula

$$P(x) = (_nC_x)p^xq^{n-x}$$

where p is the probability of success on a single trial and q is the probability of failure on a single trial.

In the formula, p will be a number between 0 and 1, inclusive, and $q = 1 - p$. Therefore, if $p = 0.2$, then $q = 1 - 0.2 = 0.8$. If $p = \frac{3}{5}$, then $q = 1 - \frac{3}{5} = \frac{2}{5}$. Note that $p + q = 1$ and the values of p and q remain the same for each independent trial. The combination $_nC_x$ is called the *binomial coefficient.*

In Example 1, we use the binomial probability formula to solve the same problem we recently solved by using a tree diagram.

Example 1 *Selecting Colored Balls with Replacement*

A basket contains 3 balls: 1 red, 1 blue, and 1 yellow. Three balls are going to be randomly selected with replacement from the basket. Determine the probability that

a) no red balls are selected.

b) exactly 1 red ball is selected.

c) exactly 2 red balls are selected.

d) exactly 3 red balls are selected.

Solution

a) We will consider selecting a red ball a success and selecting a ball of any other color a failure. Since only 1 of the 3 balls is red, the probability of success on any single trial, p, is $\frac{1}{3}$. The probability of failure on any single trial, q, is $1 - \frac{1}{3} = \frac{2}{3}$. We are determining the probability of selecting 0 red balls, or 0 successes. Since x represents the number of successes, we let $x = 0$. There are 3 independent selections (or trials), so $n = 3$. In our calculations, we will need to evaluate $\left(\frac{1}{3}\right)^0$. Note that any nonzero number raised to a power of 0 is 1. Thus, $\left(\frac{1}{3}\right)^0 = 1$. We determine the probability of 0 successes, or $P(0)$, as follows.

$$P(x) = (_nC_x)p^x q^{n-x}$$

$$P(0) = (_3C_0)\left(\frac{1}{3}\right)^0 \left(\frac{2}{3}\right)^{3-0}$$

$$= (1)(1)\left(\frac{2}{3}\right)^3$$

$$= \left(\frac{2}{3}\right)^3 = \frac{8}{27}$$

b) We are determining the probability of obtaining exactly 1 red ball or exactly 1 success in 3 independent selections. Thus, $x = 1$ and $n = 3$. We determine the probability of exactly 1 success, or $P(1)$, as follows.

$$P(x) = (_nC_x)p^x q^{n-x}$$

$$P(1) = (_3C_1)\left(\frac{1}{3}\right)^1 \left(\frac{2}{3}\right)^{3-1}$$

$$= 3\left(\frac{1}{3}\right)\left(\frac{2}{3}\right)^2$$

$$= 3\left(\frac{1}{3}\right)\left(\frac{4}{9}\right) = \frac{4}{9}$$

c) We are determining the probability of selecting exactly 2 red balls in 3 independent trials. Thus, $x = 2$ and $n = 3$. We determine $P(2)$ as follows.

$$P(x) = (_nC_x)p^x q^{n-x}$$

$$P(2) = (_3C_2)\left(\frac{1}{3}\right)^2 \left(\frac{2}{3}\right)^{3-2}$$

$$= 3\left(\frac{1}{3}\right)^2 \left(\frac{2}{3}\right)^1$$

$$= 3\left(\frac{1}{9}\right)\left(\frac{2}{3}\right) = \frac{2}{9}$$

d) We are determining the probability of selecting exactly 3 red balls in 3 independent trials. Thus, $x = 3$ and $n = 3$. We determine $P(3)$ as follows.

$$P(x) = (_nC_x)p^x q^{n-x}$$

$$P(3) = (_3C_3)\left(\frac{1}{3}\right)^3 \left(\frac{2}{3}\right)^{3-3}$$

$$= 1\left(\frac{1}{3}\right)^3 \left(\frac{2}{3}\right)^0$$

$$= 1\left(\frac{1}{27}\right)(1) = \frac{1}{27} \qquad \blacksquare$$

All the probabilities obtained in Example 1 agree with the answers obtained by using the tree diagram. Whenever you obtain a value for $P(x)$, you should obtain a value between 0 and 1, inclusive. If you obtain a value greater than 1, you have made a mistake.

Example 2 *Rolling a Die*

A single die is rolled 5 times.

a) Write the binomial probability formula used to determine the probability that x out of the 5 rolls will result in the number 4.

b) Write the binomial probability formula used to determine the probability that 2 out of the 5 rolls will result in the number 4. Evaluate the formula using a scientific or graphing calculator. Round your answer to five decimal places.

StatCrunch

Applets Simulation Dice Rolling

Solution

a) The general binomial probability formula is

$$P(x) = (_nC_x)p^x q^{n-x}$$

Here, the number of trials of the experiment, n, is 5. A success is considered rolling a 4 on a single trial; therefore, the probability of success, p, is $\frac{1}{6}$, and the probability of failure, q, is $\frac{5}{6}$. Substituting these values into the general binomial probability formula, we have

$$P(x) = (_5C_x)\left(\frac{1}{6}\right)^x \left(\frac{5}{6}\right)^{5-x}$$

b) We are asked to write the binomial probability formula used to determine the probability that 2 out of the 5 rolls will result in the number 4. We use the formula from part a) with $x = 2$.

$$P(2) = (_5C_2)\left(\frac{1}{6}\right)^2 \left(\frac{5}{6}\right)^{5-2}$$

$$= 10\left(\frac{1}{6}\right)^2 \left(\frac{5}{6}\right)^3$$

$$= 10\left(\frac{1}{36}\right)\left(\frac{125}{216}\right) \approx 0.16075$$

Thus, the probability that 2 out of the 5 rolls will result in the number 4 is about 0.16075. $\qquad \blacksquare$

TECHNOLOGY TIP

Evaluating the Binomial Probability Formula

A scientific calculator can be used to evaluate binomial probabilities. In Example 2, the expression $_5C_2$ can first be evaluated using the procedure described in the Technology Tip on page 721. Note that $_5C_2 = 10$. Then to evaluate the expression $10(\frac{1}{6})^2(\frac{5}{6})^3$, the following key sequence can be entered.

10 (1 ÷ 6) ^ 2 (5 ÷ 6) ^ 3 ENTER

To evaluate the binomial probability formula using a TI-84 Plus calculator, we can also use the function *binompdf*, which stands for *binomial probability distribution function*. To use this function, we need to enter the following binompdf(n, p, x), where $n, p,$ and x are from the binomial probability formula. For Example 2, we will enter binompdf(5, 1/6, 2). To do this, we press 2nd VARS, then use the down arrow key to scroll to binompdf(, and then press ENTER. Then type in the following 5 , 1 ÷ 6 , 2) ENTER. The answer displayed rounds to 0.16075, which matches that obtained in Example 2. The screenshot below shows the calculation on the TI-84 Plus.

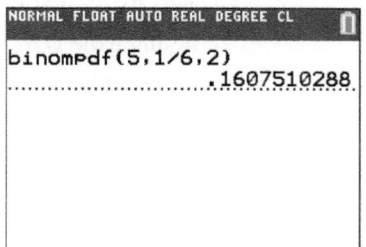

```
NORMAL FLOAT AUTO REAL DEGREE CL
binompdf(5,1/6,2)
                    .1607510288
```

┌ **Example 3 *Spring Break***

According to a survey conducted by OnCampus Research, 20% of college students reported they will work during spring break. Determine the probability that

a) exactly 3 of 5 college students randomly selected will work during spring break.
b) exactly 4 of 4 college students randomly selected will work during spring break.

Solution

a) We want to determine the probability that exactly 3 of 5 college students randomly selected will work during spring break. Therefore a college student working during spring break is considered a success. Thus, $x = 3$ and $n = 5$. The probability of success, p, is 20%, or 0.2. The probability of failure, q, is $1 - 0.2$, or 0.8. Substituting these values into the binomial formula yields

$$P(x) = (_nC_x)p^x q^{n-x}$$
$$P(3) = (_5C_3)(0.2)^3(0.8)^{5-3}$$
$$= 10(0.2)^3(0.8)^2$$
$$= 10(0.008)(0.64)$$
$$= 0.0512$$

Thus, the probability that exactly 3 of 5 randomly selected college students will work during spring break is 0.0512.

b) We want to determine the probability that 4 of 4 college students randomly selected will work during spring break. Thus, $x = 4$ and $n = 4$. We wish to determine $P(4)$.

$$P(x) = (_nC_x)p^xq^{n-x}$$
$$P(4) = (_4C_4)(0.2)^4(0.8)^{4-4}$$
$$= 1(0.2)^4(0.8)^0$$
$$= 1(0.0016)(1)$$
$$= 0.0016$$

Thus, the probability that exactly 4 of 4 randomly selected college students will work during spring break is 0.0016.   ■

Example 4 *Planting Trees*

The probability that a tree planted by Forever Green Landscaping will survive is 0.8. Determine the probability that

a) none of four trees planted will survive.

b) at least one of four trees planted will survive.

Solution

a) Success is a tree that survives. Thus, $p = 0.8$ and $q = 1 - p = 1 - 0.8 = 0.2$. We want to determine the probability of 0 successes in 4 trials. Thus, $x = 0$ and $n = 4$. We determine the probability of 0 successes, or $P(0)$, as follows.

$$P(x) = (_nC_x)p^xq^{n-x}$$
$$P(0) = (_4C_0)(0.8)^0(0.2)^{4-0}$$
$$= 1(1)(0.2)^4$$
$$= 1(1)(0.0016)$$
$$= 0.0016$$

Thus, the probability that none of the four trees planted will survive is 0.0016.

b) The probability that at least one tree of the four trees planted will survive can be determined by subtracting from 1 the probability that none of the four trees planted survives. We worked problems of this type in earlier sections of the chapter, including Sections 11.5 and 11.9.

In part (a), we determined the probability that none of the four trees planted survives is 0.0016. Thus,

$$P(\text{at least one tree planted will survive}) = 1 - P\binom{\text{none of the four trees}}{\text{planted will survive}}$$
$$= 1 - 0.0016$$
$$= 0.9984$$ ■

SECTION 11.10 *Exercises*

Warm Up Exercises

In Exercises 1–4, fill in the blank with an appropriate word, phrase, or symbol(s).

1. In a binomial experiment, there are n repeated, independent _____.

2. Each trial in a binomial experiment has two possible outcomes, success and _____.

3. In the binomial probability formula, p represents the probability of _____.

4. In the binomial probability formula, q represents the probability of _____.

Practice the Skills

In Exercises 5–10, assume that each of the n trials is independent and that p is the probability of success on a given trial. Use the binomial probability formula to determine $P(x)$. When necessary, round your answers to five decimal places.

5. $n = 6, x = 3, p = 0.2$

6. $n = 3, x = 2, p = 0.6$

7. $n = 5, x = 2, p = \frac{1}{3}$

8. $n = 6, x = 3, p = \frac{3}{7}$

9. $n = 6, x = 0, p = 0.5$

10. $n = 5, x = 0, p = 0.6$

Problem Solving

11. **A Dozen Eggs** An egg distributor determines that the probability that any individual egg has a crack is 0.14.

    a) Write the binomial probability formula used to determine the probability that exactly x eggs of n eggs are cracked.

    b) Write the binomial probability formula used to determine the probability that exactly 2 eggs in a one-dozen egg carton are cracked.

    c) Evaluate the formula written in part (b) using a scientific or graphing calculator. Round your answer to five decimal places.

12. **Social Media** According to a survey conducted by Gerber, 37% of babies have a social media account before their first birthday.

    a) Write the binomial formula used to determine the probability that exactly x out of n randomly selected babies have a social media account before their first birthday.

    b) Write the binomial formula used to determine the probability that exactly 5 out of 25 randomly selected babies have a social media account before their first birthday.

    c) Evaluate the formula written in part (b) using a scientific or graphing calculator. Round your answer to five decimal places.

In Exercises 13–26, use the binomial probability formula to answer the question. When necessary round answers to five decimal places.

13. **Flipping a Coin** A fair coin is flipped 10 times. Determine the probability that the coin is heads exactly

    a) one time.

    b) five times.

    c) nine times.

14. **Flipping a Coin** A fair coin is flipped 5 times. Determine the probability that the coin is heads

    a) zero times.

    b) at least one time.

15. **Balls in a Bin** A bin contains 5 balls colored red, yellow, blue, orange, and purple, respectively. Anika randomly selects 4 balls with replacement. Determine the probability that she selects

    a) the red ball zero times.

    b) the red ball at least one time.

16. **Balls in a Bin** A bin contains 5 balls colored red, yellow, blue, orange, and purple, respectively. Tristan randomly selects 7 balls with replacement. Determine the probability that he selects the yellow ball exactly

    a) one time.

    b) three times.

    c) five times.

17. **Distracted Driving** According to a survey conducted by Progressive Insurance, 53% of adults stated they are distracted by their smartphone while driving. If 10 adult drivers are randomly selected, determine the probability that exactly 6 of them are distracted by their smartphone while driving.

18. *Haircut Tip* Marcus is a barber and has learned that 75% of his customers tip him when they are paying for a haircut. One day, Marcus gives haircuts to 12 customers. What is the probability that exactly 9 of his customers tip him when they are paying for their haircut?

19. *Auto Insurance* The Truman Insurance company reported that 60% of all automobile damage claims were made by people under the age of 25. If eight automobile damage claims were randomly selected, determine the probability that exactly five of them were made by someone under the age of 25.

20. *Bank Loans* Records from a specific bank show that 70% of car loan applications are approved. If eight car loan applications from this bank are randomly selected, determine the probability that exactly five of the applications are approved.

21. *Dolphin Drug Care* When treated with the antibiotic resonocyllin, 92% of all dolphins are cured of brucellosis. If six dolphins with brucellosis are treated with resonocyllin, determine the probability that exactly four are cured.

22. *Medical Research* When treated with a new drug to treat crohn's disease, 80% of all patients saw an improvement in their symptoms. If 25 patients are treated with the new drug, determine the probability that 20 will see an improvement in their symptoms.

23. *Cricket Bats* D'Yard Corporation makes cricket bats. The quality control manager determines that 99.0% of the bats made at D'Yard meet the International Cricket Council (ICC) safety standards. Montgomery College wishes to purchase 100 bats. What is the probability that all 100 meet the ICC safety standards?

24. *Quality Improvement* See Exercise 23. After undergoing a quality improvement program, D'Yard improves the bat

quality, so that 99.9% of the bats made meet ICC safety standards. What is the probability that of the next 100 bats produced, all 100 meet the ICC safety standards?

25. *Online Purchases* According to www.retaildive.com, of all clothing purchases made, 27% are made online. If 20 purchases of clothing are made, determine the probability that

a) 10 of the purchases are made online.

b) none of the purchases are made online.

c) at least 1 purchase is made online.

26. *Guessing* Dell is taking a 15-question multiple-choice test. Dell forgot to study, and he must guess at each of the questions. Each question has three options to choose from. Determine the probability that Dell answers

a) 9 questions correctly.

b) 0 questions correctly.

c) at least 1 question correctly.

Challenge Problems/Group Activities

In Exercises 27–28, when necessary, round answers to 5 decimal places.

27. *Selecting 6 Cards* Six cards are randomly selected from a standard 52-card deck with replacement. Determine the probability that

a) exactly three face cards are obtained.

b) exactly two spades are obtained.

28. *Office Visit* The probability that a person visiting Dr. Suarez's office is more than 60 years old is 0.7. Determine the probability that

a) exactly three of the next five people visiting the office are more than 60 years old.

b) at least three of the next five people visiting the office are more than 60 years old.

Recreational Mathematics

29. *Aruba* The island of Aruba is well known for its beaches and predictable warm, sunny weather. In fact, Aruba's weather is so predictable that the daily newspapers don't even bother to print a forecast. Strangely enough, however, on New Year's Eve, as the islanders were counting down the last 10 sec of 2019, it began to rain. What is the probability, from 0 to 1, that 72 hr later the sun will be shining?

CHAPTER 11   *Summary*

Important Facts and Concepts	Examples and Discussion
Section 11.1	
Empirical Probability	
$P(E) = \dfrac{\text{number of times event } E \text{ has occurred}}{\left(\begin{array}{c}\text{total number of times the} \\ \text{experiment has been performed}\end{array}\right)}$	Examples 1–2, page 647
The Law of Large Numbers	
Probability statements apply in practice to a large number of trials, not to a single trial. It is the relative frequency over the long run that is accurately predictable, not individual events or precise totals.	Discussion, pages 646–648
Theoretical Probability	
$P(E) = \dfrac{\text{number of outcomes favorable to } E}{\text{total number of possible outcomes}}$	Examples 3–5, pages 651–654
Probability Facts	
The probability of an event that cannot occur is 0. The probability of an event that must occur is 1. Every probability must be a number between 0 and 1 inclusively; that is,	Discussion, page 651
$$0 \le P(E) \le 1$$	
The sum of the probabilities of all possible outcomes of an event is 1.	Example 5, pages 653–654
$$P(A) + P(\text{not } A) = 1$$	
Section 11.2	
Odds Against an Event	
$\text{Odds against} = \dfrac{P(\text{event fails to occur})}{P(\text{event occurs})} = \dfrac{P(\text{failure})}{P(\text{success})}$	Examples 1–3, pages 660–662
$\text{Odds against event } E = \dfrac{\text{number of outcomes unfavorable to } E}{\text{number of outcomes favorable to } E}$	
Odds in Favor of an Event	
$\text{Odds in favor} = \dfrac{P(\text{event occurs})}{P(\text{event fails to occur})} = \dfrac{P(\text{success})}{P(\text{failure})}$	Example 3, page 662
$\text{Odds in favor of event } E = \dfrac{\text{number of outcomes favorable to } E}{\text{number of outcomes unfavorable to } E}$	
Determining Probabilities from Odds	
The denominators of the probabilities are determined by adding the numbers in the odds statement. The numerators of the probabilities are the numbers given in the odds statements.	Example 4, page 663
Section 11.3	
Expected Value	
$E = P_1 A_1 + P_2 A_2 + P_3 A_3 + \cdots + P_n A_n$	Examples 1–7, pages 668–672
Fair Price	
Fair price = expected value + cost to play	Examples 8–9, pages 672–673

Section 11.4

Fundamental Counting Principle

If a first experiment has M distinct outcomes and a second experiment has N distinct outcomes, then the two experiments in that specific order have $M \cdot N$ distinct outcomes.

Examples 1–3, pages 678–681

Tree Diagrams

A list of all possible outcomes of an experiment is called a sample space. Tree diagrams are helpful in determining sample spaces.

Examples 2–4, pages 680–682

Section 11.5

OR and AND problems

$$P(A \text{ or } B) = P(A) + P(B) - P(A \text{ and } B)$$

$$P(A \text{ and } B) = P(A) \cdot P(B)$$

Examples 1–3, pages 689–691

Examples 4–7, pages 692–695

Section 11.6

Conditional Probability

$$P(E_2 \mid E_1) = \frac{n(E_1 \text{ and } E_2)}{n(E_1)}$$

Examples 1–3, pages 700–703

Section 11.7

The number of permutations of n items is $n!$.

$$n! = n(n-1)(n-2) \cdots (3)(2)(1)$$

Example 4, page 711

Permutation Formula

$${}_nP_r = \frac{n!}{(n-r)!}$$

Examples 5–7, pages 711–713

The number of different permutations of n objects where $n_1, n_2, \ldots, n_r$ of the objects are identical is

$$\frac{n!}{n_1! n_2! \cdots n_r!}$$

Example 8, page 715

Section 11.8, 11.9

Combination Formula

$${}_nC_r = \frac{n!}{(n-r)! \, r!}$$

Examples 2–4, pages 721–722

Examples 1–4, pages 726 – 728

Section 11.10

Binomial Probability Formula

$$P(x) = ({}_nC_x)p^x q^{n-x}$$

Examples 1–4, pages 734–737

CHAPTER 11   *Review Exercises*

11.1–11.10

1. In your own words, explain the law of large numbers.

2. Explain how empirical probability can be used to determine whether a die is "loaded" (not a fair die).

3. *Phones* Of 45 people who purchase a phone at an electronics store, 40 purchased a smartphone. Determine the empirical probability that the next person who purchases a phone from that store purchases a smartphone.

4. *Cards* Draw one card from a standard 52-card deck 40 times with replacement and compute the empirical probability of selecting a heart.

5. *Television News* In a small town, 250 people were asked whether they watched ABC, CBS, NBC, Fox, or MSNBC news. The results are indicated below.

Network	Number of People
ABC	80
CBS	35
NBC	60
Fox	35
MSNBC	40

Determine the empirical probability that the next person randomly selected from the town watches ABC news.

Digits In Exercises 6–9, each of the digits 0, 1, 2, 3, 4, 5, 6, 7, 8, 9 is written on a piece of paper and all the pieces of paper are placed in a hat. One number is randomly selected. Determine the probability that the number selected is

6. even.

7. odd or greater than 3.

8. greater than 2 or less than 6.

9. even and greater than 4.

Cheese Preference In Exercises 10–13, a taste test is given to 60 customers at a supermarket. The customers are asked to taste 4 types of cheese and to list their favorite. The results are summarized below.

Type	Number of People
Cheddar	21
Monterey Jack	17
Gouda	13
Mozzarella	9

If one person who participated in the taste test is randomly selected, determine the probability that the person's favorite cheese was

10. cheddar.

11. Gouda.

12. either cheddar or Monterey Jack.

13. a cheese other than mozzarella.

14. *High Blood Pressure* According to the U.S. Centers for Disease Control and Prevention, 31% of adults in the United States have high blood pressure. If an adult was randomly selected, determine the odds

a) against the adult having high blood pressure.

b) in favor of the adult having high blood pressure.

15. *Vegetable Mix-up* Nicholas, a mischievous little boy, has removed labels on the eight cans of vegetables in the cabinet. Nicholas's father knows that there are three cans of corn, three cans of beans, and two cans of carrots. If the father randomly selects and opens one can, determine the odds against his selecting a can of corn.

16. *Horseracing* The odds against Fedora winning the Triple Crown in horse racing are 82 : 3. Determine the probability that Fedora wins the Triple Crown.

17. *Pizza Restaurant Success* The probability that a new pizza restaurant will succeed at a given location is 0.75. Determine the odds in favor of the pizza restaurant succeeding.

18. *Raffle Tickets* One thousand raffle tickets are sold at $2 each. Three prizes of $200 and two prizes of $100 will be awarded. Assume that the probability that any given ticket is selected for a $200 prize is $\frac{3}{1000}$ and the probability that any given ticket is selected for a $100 prize is $\frac{2}{1000}$.

a) Determine the expectation of a person who purchases a ticket.

b) Determine the expectation of a person who purchases three tickets.

19. *Expectation of a Card* If Cameron selects a face card from a standard 52-card deck, Lindsey will give him $9. If Cameron does not select a face card, he must give Lindsey $3.

a) Determine Cameron's expectation.

b) Determine Lindsey's expectation.

c) If Cameron plays this game 100 times, how much can he expect to lose or gain?

20. *Expected Attendance* If the day is sunny, 1000 people will attend the baseball game. If the day is cloudy, only 500 people will attend. If it rains, only 100 people will attend. The local meteorologist states that the probability of a sunny day is 0.4, of a cloudy day is 0.5, and of a rainy day is 0.1. Determine the number of people that are expected to attend.

21. *Fair Price* At a game of chance, the expected value is determined to be −$2.50, and the cost to play the game is $6.50. Determine the fair price to play the game.

22. *Fair Price* The expected value when you purchase a lottery ticket is −$1.50, and the cost of the lottery ticket is $5.00. Determine the fair price of the lottery ticket.

23. *Club Officers* Tina, Jake, Gina, and Carla form a club. They plan to select a president and a vice president.

  a) Construct a tree diagram showing all the possible outcomes.

  b) List the sample space.

  c) Determine the probability that Gina is selected president and Jake is selected vice president.

24. *A Coin and a Number* A coin is flipped and then a number from 1 through 4 is randomly selected from a bag.

  a) Construct a tree diagram showing all the possible outcomes.

  b) List the sample space.

  c) Determine the probability that a head is flipped and an odd number is selected.

  d) Determine the probability that a head is flipped or an odd number is selected.

Spinning Two Wheels *In Exercises 25–30, the outer and inner wheels are spun.*

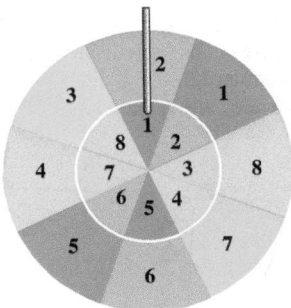

Assuming that the wheels are independent and the sectors are equally likely, determine the probability of obtaining

25. even numbers on both wheels.

26. numbers greater than 5 on both wheels.

27. an odd number on the outer wheel and a number less than 6 on the inner wheel.

28. an even number or a number less than 6 on the outer wheel.

29. an even number or a color other than green on the inner wheel.

30. gold on the outer wheel and a color other than gold on the inner wheel.

Soft Drink Selection *In Exercises 31–34, assume that Sonya is going to purchase three bottles of soft drinks at a store. The store has 12 different bottles of soft drinks, including 5 different bottles of cola, 4 different bottles of root beer, and 3 different bottles of ginger ale. If she randomly selects 3 of these bottles, determine the probability she selects*

31. 3 bottles of cola.

32. no bottles of root beer.

33. at least one bottle of root beer.

34. a bottle of cola, a bottle of root beer, and a bottle of ginger ale in that order.

Spinner Probabilities *In Exercises 35–38, assume that the spinner cannot land on a line.*

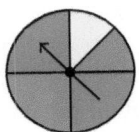

If spun once, determine

35. the probability that the spinner does not land on green.

36. the odds in favor of and the odds against the spinner landing on green.

37. A person wins $10 if the spinner lands on green, wins $5 if the spinner lands on red, and loses $20 if the spinner lands on yellow. Determine the expectation of a person who plays this game.

38. If the spinner is spun three times, determine the probability that at least one spin lands on red.

Restaurant Service *In Exercises 39–42, use the results of a survey regarding the service at the Lake View Diner, which is summarized as follows.*

Meal	Service Rated Good	Service Rated Poor	Total
Breakfast	95	15	110
Lunch	75	15	90
Total	170	30	200

If one person who completed the survey is randomly selected, determine the probability that the person indicated that the

39. service was rated poor.

40. service was rated good, given that the meal was lunch.

41. service was rated poor, given that the meal was breakfast.

42. meal was breakfast, given that the service was rated poor.

Neuroscience *In Exercises 43–46, assume that in a neuroscience course the students perform an experiment. Tests are given to determine if people are right brained, are left brained, or have no predominance. It is also recorded whether they are right handed or left handed. The following chart shows the results obtained.*

	Right Brained	Left Brained	No Predominance	Total
Right handed	40	130	60	230
Left handed	120	30	20	170
Total	160	160	80	400

If one person who completed the survey is randomly selected, determine the probability the person selected is

43. right handed.

44. left brained, given that the person is left handed.

45. right handed, given that the person has no predominance.

46. right brained, given that the person is left handed.

47. ***Game Show*** Four contestants are on a game show. There are four different-colored rubber balls in a box, and each contestant gets to pick one ball from the box. Inside each ball is a slip of paper indicating the amount the contestant has won. The amounts are $10,000, $5000, $2000, and $1000.

  a) In how many different ways can the contestants select the balls?

  b) What is the expectation of a contestant?

48. ***Spelling Bee*** Five finalists remain in a high school spelling bee. Two will receive $50 each, two will receive $100 each, and one will receive $500. How many different arrangements of prizes are possible?

49. ***Cereal Selection*** Mrs. Williams takes her 3 children shopping. Each of her children gets to select a different type of cereal that only that child will eat. At the store, there are only 10 boxes of cereal left and each is a different type. In how many ways can the 3 children select the cereal?

50. ***Astronaut Selection*** Three of nine astronauts must be selected for a mission. One will be the captain, one will be the navigator, and one will perform scientific experiments. Assuming each of the nine astronauts can perform any of the tasks, in how many ways can a three-person crew be selected so that each person has a different assignment?

51. ***Vaccine*** Dr. Goldberg has three doses of vaccine for influenza type A. Six patients in the office require the vaccine. In how many different ways could Dr. Goldberg dispense the vaccine?

52. ***Dogsled*** Ten huskies are to be selected to pull a dogsled. How many different arrangements of the 10 huskies on a dogsled are possible?

53. ***Mega Millions*** To play the lottery game Mega Millions, you select 5 numbers, in any order, from 1 through 70 and 1 Mega Ball number from 1 through 25. If you match all 6 numbers, you win the jackpot!

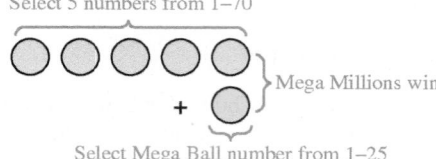

Select 5 numbers from 1–70

} Mega Millions win

+

Select Mega Ball number from 1–25

  a) What is the probability that you match the first 5 numbers?

  b) What is the probability that you match the first 5 numbers and the Mega Ball?

54. ***Parent-Teacher Committee*** A committee of 7 is to be formed from 10 parents and 12 teachers. If the committee is to consist of 2 parents and 5 teachers, how many different committees are possible?

55. ***Selecting Test Subjects*** In a psychology research laboratory, one room contains eight men and another room contains five women. Three men and two women are to be randomly selected to be given a psychological test. How many different selections of these people are possible?

56. ***Choosing Two Aces*** Two cards are dealt without replacement, from a standard 52-card deck. Determine the probability that two aces are selected (use combinations).

Color Chips *In Exercises 57–60, a bag contains five red chips, three white chips, and two blue chips. Three chips are to be randomly selected, without replacement. Determine the probability that*

57. all are red.

58. the first two are red and the third is blue.

59. the first is red, the second is white, and the third is blue.

60. at least one is red.

Magazines *In Exercises 61–64, on a table in a doctor's office are five Motor Trend magazines, six Parenting magazines, and three Sports Illustrated magazines. If Ramona randomly selects three magazines, determine the probability that*

61. three *Motor Trend* magazines are selected.

62. two *Parenting* magazines and one *Sports Illustrated* magazine are selected.

63. no *Parenting* magazine is selected.

64. at least one *Parenting* magazine is selected.

65. *New Homes* In the community of Spring Hill, 60% of the homes purchased cost more than $160,000.

a) Write the binomial probability formula to determine the probability that exactly x of the next n homes purchased in Spring Hill cost more than $160,000.

b) Write the binomial probability formula to determine the probability that exactly 50 of the next 100 homes purchased in Spring Hill cost more than $160,000.

c) Evaluate the formula written in part b) using a scientific or graphing calculator. Round your answer to five decimal places.

66. *Long-Stemmed Roses* At Floyd's Flower Shop, $\frac{1}{5}$ of people ordering flowers select long-stemmed roses. Determine the probability that exactly 3 of the next 5 customers ordering flowers select long-stemmed roses.

67. *Taking a Math Course* During any semester at City College, 60% of the students are taking a mathematics course. Determine the probability that of five students randomly selected,

a) none are taking a mathematics course this semester.

b) at least one is taking a mathematics course this semester.

CHAPTER 11 *Test*

1. *Doughnuts* Of the last 25 people who purchased a doughnut at Dunkin' Donuts, 11 purchased a glazed doughnut. Determine the empirical probability that the next person who purchases a doughnut at Dunkin' Donuts purchases a glazed doughnut.

One Sheet of Paper *In Exercises 2–4, each of the numbers 1–9 is written on a sheet of paper and the nine sheets of paper are placed in a hat. If one sheet of paper is randomly selected from the hat, determine the probability that the number selected is*

2. greater than 5.

3. odd.

4. even or less than 4.

Two Sheets of Paper *In Exercises 5–7, if two of the same nine sheets of paper mentioned above are selected, without replacement, from the hat, determine the probability that*

5. both numbers are less than 3.

6. the first number is odd and the second number is even.

7. neither of the numbers is greater than 6.

8. One card is randomly selected from a standard 52-card deck. Determine the probability that the card selected is a red card or a face card.

One Chip and One Die *In Exercises 9–13, one colored chip—red, blue, or green—is randomly selected and a fair die is rolled.*

9. Use the fundamental counting principle to determine the number of sample points in the sample space.

10. Construct a tree diagram illustrating all the possible outcomes and list the sample space.

In Exercises 11–13, by observing the sample space of the chips and die mentioned above, determine the probability of obtaining

11. a green chip and the number 2.

12. a red chip or the number 1.

13. a chip other than a red chip or an even number.

14. *Passwords* A personal password for an Internet brokerage account is to consist of an uppercase letter, followed by two digits, followed by two uppercase letters. Determine the number of personal codes possible if the first digit cannot be zero and repetition is permitted.

15. *Puppies* A litter of collie puppies consists of four males and five females. If one of the puppies is randomly selected, determine the odds

a) against the puppy being male.

b) in favor of the puppy being female.

16. *Pick a Card* You get to randomly select one card from a standard 52-card deck. If you pick a club, you win $8. If you pick a heart, you win $4. If you pick any other suit, you lose $6. Determine your expectation for this game.

17. *Cars and SUVs* The number of cars and the number of SUVs going through the toll gates of the George Washington Bridge and the Golden Gate Bridge is recorded. The results are shown below.

Bridge	Cars	SUVs	Total
George Washington	120	106	226
Golden Gate	94	136	230
Total	214	242	456

If one of these vehicles going through the toll gate of one of these two bridges is randomly selected, determine the probability that

a) it is an SUV.

b) it is going through the toll gate of the George Washington Bridge.

c) it is an SUV, given that it is going through the toll gate of the George Washington Bridge.

d) it is going through the toll gate of the Golden Gate Bridge, given that it is a car.

18. *Awarding Prizes* Three of six people are to be selected and given small prizes. One will be given a book, one will be given a calculator, and one will be given an iTunes card. In how many different ways can these prizes be awarded?

19. *Quality Control* A bin contains a total of 20 batteries, of which 6 are defective. If you randomly select 2 batteries, without replacement, determine the probability that

a) both of batteries are defective.

b) at least one battery is not defective.

20. *University Admission* The probability that a person is accepted for admission to a specific university is 0.6. Determine the probability that exactly two of the next five people who apply to that university get accepted.

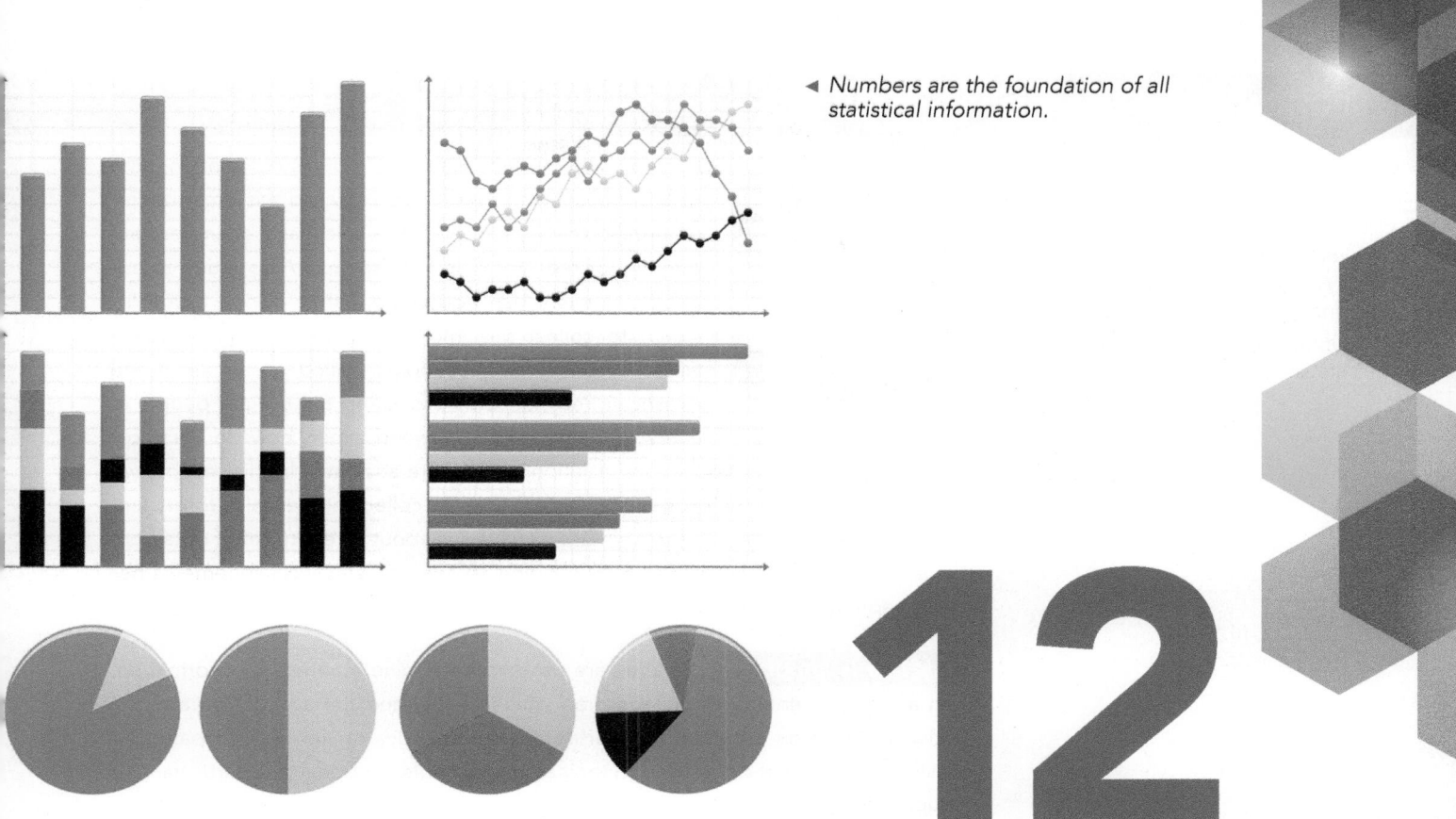

◀ *Numbers are the foundation of all statistical information.*

12

Statistics

What You Will Learn

- Sampling techniques and misuses of statistics
- Frequency distributions and statistical graphs
- Measures of central tendency and position
- Measures of dispersion
- The normal curve
- Linear correlation and regression

Why This Is Important

We live in an increasingly data-driven society. Regardless of what profession you choose, it is likely that you will need an understanding of statistics to either read a report containing statistics or understand a graphical display of data. Businesses use statistics in a variety of ways including where to locate an office, marketing a product, and determining how many units of an item to produce in order to maximize profit. Economists use statistics when determining the inflation rate and determining per capita income. Statistics is used in the medical field when developing new treatments for diseases. Statistics is also used to help meteorologists predict weather patterns. Since many important decisions are made using data, it is crucial to have an understanding of how to collect, interpret, and analyze data. This chapter will help you see the valuable uses of statistics and also recognize when statistical information is being manipulated and misused.

SECTION 12.1 Sampling Techniques and Misuses of Statistics

Upon completion of this section, you will be able to:

- Understand sampling techniques, including: random, systematic, cluster, stratified, and convenience sampling.
- Understand the misuses of statistics.

According to the American Pet Products Association, 68% of U.S. households own a pet. How did this organization determine this percentage? It would have been very expensive for the American Pet Products Association to ask people in every U.S. household if they owned a pet. Instead, to collect this information, the polling company asked a subset of households, called a *sample*. If the American Pet Products Association did not ask people in every household if they owned a pet, how do we know that the American Pet Products Association's results are accurate? In this section, we will discuss different techniques statisticians use to collect numerical information, called data, and how they make accurate conclusions about an entire set of data from the sample collected. We will also discuss how to examine statistical statements before accepting them as fact.

Why This Is Important Samples are used to determine a variety of information, such as the percentage of college graduates in a city and the age of viewers of a particular television program. If advertisers know the ages of viewers of a particular television program, they can design their advertisements to appeal to a particular age group.

Sampling Techniques

The study of statistics was originally used by governments to manage large amounts of numerical information. The use of statistics has grown significantly and today is applied in all walks of life, including estimating the unemployment rate and the cost of living, comparing achievements of individuals, and providing the results of different polls. Statistics is used in scores of other professions; in fact, it is difficult to find any profession that does not depend on some aspect of statistics.

Before we discuss different techniques used to collect numerical information, we will first introduce a few important definitions. *Statistics* is the art and science of gathering, analyzing, and making inferences (predictions) from numerical information obtained in an experiment. The numerical information so obtained is referred to as *data*. Statistics is divided into two main branches: descriptive and inferential. *Descriptive statistics* is concerned with the collection, organization, and analysis of data. *Inferential statistics* is concerned with making generalizations or predictions from the data collected.

Probability and statistics are closely related. Someone in the field of probability is interested in computing the chance of occurrence of a particular event when all the possible outcomes are known. A statistician's interest lies in drawing conclusions about possible outcomes through observations of only a few particular events.

If a probability expert and a statistician find identical boxes, the probability expert might open the box, observe the contents, replace the cover, and proceed to compute the probability of randomly selecting a specific object from the box. The statistician might select a few items from the box without looking at the contents and make a prediction as to the total contents of the box.

The entire contents of the box constitute the *population*. A population consists of all items or people of interest. The statistician often uses a subset of the population, called a *sample*, to make predictions concerning the population. It is important to

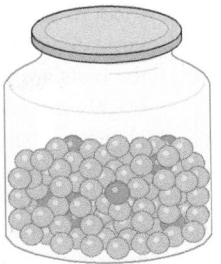

Figure 12.1

understand the difference between a population and a sample. A population includes *all* items of interest. A sample includes *some* of the items in the population.

When a statistician draws a conclusion from a sample, there is always the possibility that the conclusion is incorrect. For example, suppose that a jar contains 90 blue marbles and 10 red marbles, as shown in Fig. 12.1. If the statistician selects a random sample of five marbles from the jar and they are all blue, he or she may wrongly conclude that the jar contains all blue marbles. If the statistician takes a larger sample, say, 15 marbles, he or she is likely to select some red marbles. At that point, the statistician may make a prediction about the contents of the jar based on the sample selected. Of course, the most accurate result would occur if every object in the jar, the entire population, were observed. However, in most statistical experiments, observing the entire population is not practical.

Statisticians use samples instead of the entire population for two reasons: (a) It is often impossible to obtain data on an entire population, and (b) sampling is less expensive because collecting the data takes less time and effort. For example, suppose that you wanted to determine the number of each species of all the fish in a lake. To do so would be almost impossible without using a sample. If you did try to obtain this information from the entire population, the cost would be astronomical. Or suppose that you wanted to test soup cans for spoilage. If every can produced by the company was opened and tested, the company wouldn't have any product left to sell. Instead of testing the entire population of soup cans, a sample is selected. The results obtained from the sample of soup cans selected are used to make conclusions about the entire population of soup cans.

Consider the task of determining the political strength of a certain candidate running in a national election. It is not possible for pollsters to ask each of the approximately 245.5 million eligible voters his or her preference of a candidate. Thus, pollsters must select and use a sample of the population to obtain their information. How large a sample do you think they use to make predictions about an upcoming national election? You might be surprised to learn that pollsters use only about 1600 registered voters in their national sample. How can a pollster using such a small percentage of the population make an accurate prediction?

The answer is that when pollsters select a sample, they use sophisticated statistical techniques to obtain an unbiased sample. An *unbiased sample* is one that is a small replica of the entire population with regard to income, education, gender, race, religion, political affiliation, age, and so on. The procedures statisticians use to obtain unbiased samples are quite complex. The following sampling techniques will give you a brief idea of how statisticians obtain unbiased samples.

Random Sampling

If a sample is drawn in such a way that each time an item is selected each item in the population has an equal chance of being drawn, the sample is said to be a *random sample*. When using a random sample, one combination of a specified number of items has the same probability of being selected as any other combination. When all the items in the population are similar with regard to the specific characteristic we are interested in, a random sample can be expected to produce satisfactory results. For example, consider a large container holding 300 tennis balls that are identical except for color. One-third of the balls are yellow, one-third are white, and one-third are green. If the balls can be thoroughly mixed between each draw of a tennis ball so that each ball has an equally likely chance of being selected, randomness is not difficult to achieve. However, if the objects or items are not all the same size, shape, or texture, it might be impossible to obtain a random sample by reaching into a container and selecting an object.

The best procedure for selecting a random sample is to use a random number generator. A random number generator is a device, usually a calculator or computer program, that produces a list of random numbers. To select a random sample, first assign a number to each element in the population. Numbers are usually assigned in order. Then select the number of random numbers needed, which is determined by the sample size. Each numbered element from the population that corresponds to a selected random number becomes part of the sample.

Systematic Sampling

A *systematic sample* is obtained by selecting a random starting point and then selecting every *n*th item in a population. For example, in a factory producing bicycles using an assembly line, a systematic sample could be obtained by randomly selecting a bicycle and then selecting every 20th bicycle thereafter coming off of the assembly line.

It is important that the list from which a systematic sample is chosen includes the entire population being studied. See the *Did You Know?* called "Don't Count Your Votes Until They're Cast" on page 751. Another problem that must be avoided when this method of sampling is used is the constantly recurring characteristic. For example, on an assembly line, every 10th item could be the work of robot X. If only every 10th item is checked for defects, the work of other robots doing the same job may not be checked and may be defective.

Cluster Sampling

A *cluster sample* is sometimes referred to as an *area sample* because it is frequently applied on a geographical basis. Essentially, the sampling consists of a random selection of groups of units. To select a cluster sample, we divide a geographic area into sections. Then we randomly select the sections or clusters. Either each member of the selected cluster is included in the sample or a random sample of the members of each cluster is used. For example, geographically we might randomly select city blocks to use as a sample unit. Then either every member of each selected city block would be used or a random sample from each selected city block would be used. Another example is to select *x* boxes of screws from a whole order, count the number of defective screws in the *x* boxes selected, and use this number to determine the expected number of defective screws in the whole order.

Stratified Sampling

When a population is divided into parts, called strata, for the purpose of drawing a sample, the procedure is known as *stratified sampling*. Stratified sampling involves dividing the population by characteristics called *stratifying factors*, such as gender, race, religion, or income. When a population has varied characteristics, it is desirable to separate the population into classes with similar characteristics and then take a random sample from each stratum (or class). For example, we could separate the population of undergraduate college students into strata called freshmen, sophomores, juniors, and seniors.

The use of stratified sampling requires some knowledge of the population. For example, to obtain a cross section of voters in a city, we must know where various groups are located and the approximate number of voters in each location.

Convenience Sampling

A *convenience sample* uses data that are easily or readily obtained. Occasionally, data that are conveniently obtained may be all that is available. In some cases, some information is better than no information at all. Nevertheless, convenience sampling can be extremely biased. For example, suppose that a town wants to raise taxes to build a new elementary school. The local newspaper wants to obtain the opinion of some of the residents and sends a reporter to a senior citizens center. The first 10 people who exit the building are asked if they are in favor of raising taxes to build a new school. This sample could be biased against raising taxes for the new school. Most senior citizens would not have school-age children and may not be interested in paying increased taxes to build a new school. Although a convenience sample may be very easy to select, one must be very cautious when using the results obtained by this method.

Example 1 *Identifying Sampling Techniques*

Identify the sampling technique used to obtain a sample in the following. Explain your answer.

a) Every 20th smartwatch coming off an assembly line is checked for defects.

b) A $50 gift certificate is given away at the National Nurses Convention. Tickets are placed in a bin, and the tickets are mixed up. Then the winning ticket is selected by a blindfolded person.

c) Children in a large city are classified based on the neighborhood school they attend. A random sample of five schools is selected. All the children from each selected school are included in the sample.

d) The first 50 people entering a zoo are asked if they support an increase in taxes to support a zoo expansion.

e) All registered vehicles in the state of California are classified according to type: subcompact, compact, mid-size, full-size, SUV, and truck. A random sample of vehicles from each category is selected.

Solution

a) Systematic sampling. The sample is obtained by drawing every nth item. In this example, every 20th item on an assembly line is selected.

b) Random sampling. Every ticket has an equal chance of being selected.

c) Cluster sampling. A random sample of geographic areas is selected.

d) Convenience sampling. The sample is selected by picking data that are easily obtained.

e) Stratified sampling. The vehicles are divided into strata based on their classification. Then random samples are selected from each strata.   ∎

Misuses of Statistics

Statistics, when used properly, is a valuable tool to society. However, many individuals, businesses, and advertising firms misuse statistics to their own advantage. You should examine statistical statements very carefully before accepting them as fact. You should ask yourself two questions: Was the sample used to gather the statistical data unbiased and of sufficient size? Is the statistical statement ambiguous; that is, can it be interpreted in more than one way?

Let's examine two advertisements. "Four out of five dentists recommend sugarless gum for their patients who chew gum." In this advertisement, we do not know the sample size and the number of times the experiment was performed to obtain the desired results. The advertisement does not mention that possibly only 1 out of 100 dentists recommended gum at all.

MATHEMATICS TODAY

Creative Displays

How Employers Make Workers Happy

Stress reduction	21%
Massage therapy	8%
Nap during workday	1%

Visual graphics are often used to "dress up" what might otherwise be considered boring statistics. Although visually appealing, such creative displays of numerical data can be misleading. The graph above shows the percentage of employers that offer "perks" such as stress reduction, massage therapy, or a nap during the workday to make workers happy. This graph is misleading because the lengths of the bars are not proportional to one another as they should be to accurately reflect the percent of employers offering each of the named perks. For example, the bar for massage therapy should be eight times as long as the bar for nap during workday instead of being approximately four times as long, as the graph shows.

Why This Is Important In order to correctly interpret data from a graph, it is important to be aware that a graph may be misleading.

In an advertisement for golf balls, a *Driver* golf ball is hit and another brand golf ball is hit in the same manner. We are told that the *Driver* golf ball travels farther and we are supposed to conclude that the *Driver* golf ball is the better golf ball. The advertisement does not mention the number of times the experiment was previously performed or the results of the earlier experiments. Possible sources of bias include (1) wind speed and direction, (2) that no two swings are identical, and (3) that the ball may land on a rough or smooth surface.

Vague or ambiguous words also lead to statistical misuses or misinterpretations. The word *average* is one such culprit. There are many different averages used in the study of statistics. In Section 12.3 we discuss four of these averages. Each is calculated differently, and each may have a different value for the same sample. During contract negotiations, it is not uncommon for an employer to state publicly that the average salary of its employees is $45,000, whereas the employees' union states that the average salary is $40,000. Who is lying? Actually, both sides may be telling the truth. Each side will use the average that best suits its needs to present its case. Advertisers also use the average that most enhances their products. Consumers often misinterpret this average as the one with which they are most familiar.

Another vague word is *largest*. For example, Hardings claims that it is the largest department store in the United States. Does that mean largest profit, largest sales, largest building, largest staff, largest acreage, or largest number of outlets?

Still another deceptive technique used in advertising is to state a claim from which the public may draw irrelevant conclusions. For example, a disinfectant manufacturer claims that its product killed 40,760 germs in a laboratory in 5 seconds. "To prevent colds, use disinfectant A." It may well be that the germs killed in the laboratory were not related to any type of cold germ. In another example, company C claims that its paper towels are heavier than its competition's towels. Therefore, they will hold more water. Is weight a measure of absorbency? A rock is heavier than a sponge, yet a sponge is more absorbent.

An insurance advertisement claims that in Duluth, Minnesota, 212 people switched to insurance company Z. One may conclude that this company is offering something special to attract these people. What may have been omitted from the advertisement is that 415 people in Duluth, Minnesota, dropped insurance company Z during the same period.

A foreign car manufacturer claims that 9 of every 10 of a popular-model car it sold in the United States during the previous 10 years were still on the road. From this statement, the public is to conclude that this foreign car is well manufactured and would last for many years. The commercial neglects to state that this model has been selling in the United States for only a few years. The manufacturer could just as well have stated that 9 of every 10 of these cars sold in the United States in the previous 100 years were still on the road.

Charts and graphs can also be misleading or deceptive. In Fig. 12.2, the two graphs show the performance of two stocks over a 6-month period. Based on the graphs, which stock would you purchase? Actually, the two graphs present identical

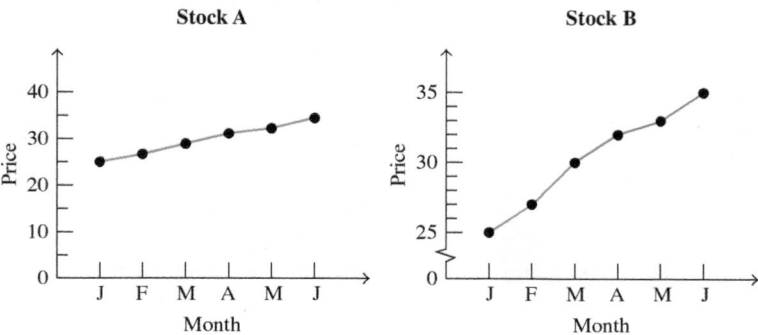

Figure 12.2

information; the only difference is that the vertical scale of the graph for stock B has been exaggerated.

The two graphs in Fig. 12.3 show the same change. However, the graph in part (a) appears to show a greater increase than the graph in part (b), again because of a different scale.

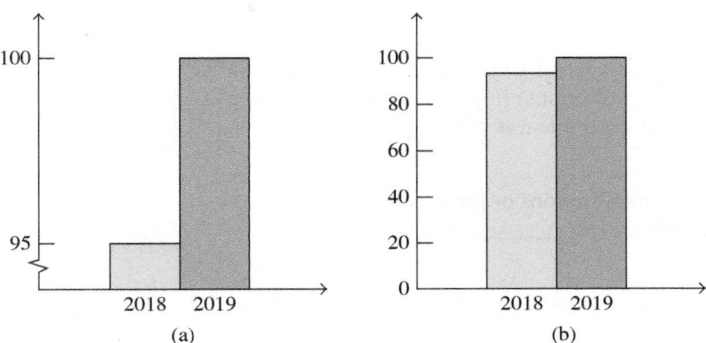

Figure 12.3

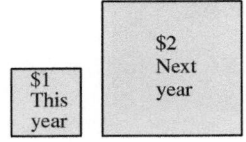

Figure 12.4

Consider a claim that if you invest $1, by next year you will have $2. This type of claim is sometimes misrepresented, as in Fig. 12.4. Actually, your investment has only doubled, but the area of the square on the right is four times that of the square on the left. By expressing the amounts as cubes (Fig. 12.5), you increase the volume eightfold.

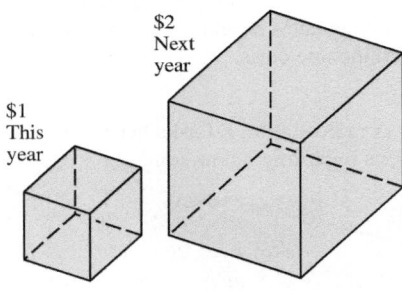

Figure 12.5

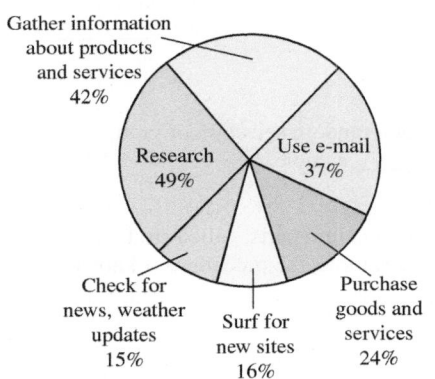

Figure 12.6

The graph in Fig. 12.6 is an example of a circle graph. We will discuss circle graphs further in Section 12.2. In a circle graph, the total circle represents 100%. Therefore, the sum of the parts should add up to 100%. This graph is misleading, since the sum of its parts is 183%. A graph other than a circle graph should have been used to display the top six reasons Americans say they use the Internet.

Despite the examples presented in this section, you should not be left with the impression that statistics is used solely for the purpose of misleading or cheating the consumer. As stated earlier, there are many important and necessary uses of statistics. Most statistical reports are accurate and useful. You should realize, however, the importance of being an aware consumer.

SECTION 12.1   *Exercises*

Warm Up Exercises

In Exercises 1–10, fill in the blank with an appropriate word, phrase, or symbol(s).

1. The art and science of gathering, analyzing, and making inferences (predictions) from numerical information obtained in an experiment is called _____.

2. Making generalizations or predictions from the data collected is called _____ statistics.

3. The collection, organization, and analysis of data is called _____ statistics.

4. All items or people of interest in an experiment are collectively called a(n) _____.

5. A subset of a population used by statisticians to make predictions about a population is called a(n) _____.

6. When a sample is obtained by drawing every *n*th item, the sample is called a(n) _____ sample.

7. If a sample is drawn in such a way that each time an item is selected, each item in the population has an equal chance of being drawn, the sample is called a(n) _____ sample.

8. A sample that consists of a random selection of groups or units is called a(n) _____ sample.

9. When a population is divided into parts, called strata, for the purpose of drawing a sample, the procedure is known as _____ sampling.

10. A sample that uses data that are easily or readily obtained is called a(n) _____ sample.

Practice the Skills/Problem Solving

Sampling Techniques In Exercises 11–20, identify the sampling technique used to obtain a sample.

11. *TV Viewers* Viewers of Hulu are classified according to age. A random sample of viewers from each age group is selected.

12. *Assembly Line* Every 10th refrigerator coming off an assembly line is checked for defects.

13. *Selecting Counties* A state is divided into counties. A random sample of 12 counties is selected. A random sample from each of the 12 selected counties is selected.

14. *Door Prize* A door prize is given away at a home improvement seminar. Tickets are placed in a bin, and the tickets are mixed up. Then a ticket is selected by a blindfolded person.

15. *Buying Tickets* Every 20th person in line to buy tickets to a concert is asked his or her age.

16. *Town Services* The Oneida town supervisor wants to determine residents' opinions regarding town services. She divides the residents into three income classes: working-class households, middle-class households, and upper-class households. She then takes a random sample of households from each income class.

17. *Listeners Opinions* A radio station DJ asks her listeners to call in their opinions regarding a new song.

18. *Customers Opinions* To determine customer opinion of its carry-on luggage policy, American Airlines randomly selects 50 flights and surveys all passengers of those flights.

19. *Satisfaction Survey* The manager at Planet Fitness wants to administer a satisfaction survey to its current members. Using its membership roster, the manager randomly selects 40 members and sends them an email with a link to an online survey.

20. *Voting* The Student Senate at the University of North Carolina is electing a new president. The first 25 people leaving the library are asked for whom they will vote.

Misinterpretations of Statistics In Exercises 21–34, discuss the statement and tell what possible misuses or misinterpretations may exist.

21. *Cold Remedy* In a study of patients with cold symptoms, each patient was found to have improved symptoms after taking honey. Therefore, honey cures the common cold.

22. *Driving* Most accidents occur on Saturday night. That means people do not drive carefully on Saturday night.

23. *Mathematics Scores* At Central High School, half the students are below average in mathematics. Therefore, the school should receive more federal aid to raise student scores.

24. *Snacks* Healthy Snacks cookies are fat free. So eat as many as you like and you will not gain weight.

25. *Toothpaste* Seventy-five percent of all dentists recommend Shiny Smile toothpaste. Therefore, this toothpaste is better than all other types of toothpaste.

26. *Department Store* Morgan's is the largest department store in New York. So shop at Morgan's and save money.

27. *Automobile Accidents* Eighty percent of all automobile accidents occur within 10 miles of the driver's home. Therefore, it is safer to take long trips.

28. *Asthma* Arizona has the highest death rate for asthma in the United States. Therefore, it is unsafe to go to Arizona if you have asthma.

▲ Sedona, Arizona

29. *Recommending a Professor* Thirty students said that they would recommend Professor Fogal to a friend. Twenty students said that they would recommend Professor Bond to a friend. Therefore, Professor Fogal is a better teacher than Professor Bond.

30. *Mattress* Sleep Number mattresses cost more than Therapedic mattresses. Therefore, you will sleep better if you buy a Sleep Number mattress than if you buy a Therapedic mattress.

31. *Lawn Tractors* John Deere lawn tractors cost more than Toro lawn tractors. Therefore, John Deere lawn tractors will last longer than Toro lawn tractors.

32. *Automobile Accidents* More men than women are involved in automobile accidents. Therefore, women are better drivers.

33. *Depth of a Pond* The average depth of the pond is only 3 ft, so it is safe to go wading.

34. *Popular Musicians* Billy Joel has appeared in concert at Madison Square Garden more than Elton John. Therefore, Billy Joel must be more popular than Elton John.

35. *Price of a Movie Ticket* The following table shows the average price of a movie ticket in the United States for the years 2014–2018.

Year	Average Ticket Price
2014	$8.17
2015	$8.43
2016	$8.65
2017	$8.97
2018	$9.11

Source: Box Office Mojo

Draw a line graph that makes the increase in the average price of a movie ticket for the years shown appear to be

a) small.   **b)** large.

36. *Nike Revenue* The following table shows the worldwide revenue for Nike, in billions of dollars, for the years 2015–2018.

Year	Revenue (billions of dollars)
2015	30.60
2016	32.38
2017	34.35
2018	36.40

Source: Nike

Draw a line graph that makes the increase in revenue for Nike appear to be

a) small.   **b)** large.

First Marriage *In Exercises 37 and 38, use the following table.*

Median Age at First Marriage

Male		Female	
Year	Age	Year	Age
2008	27.6	2008	25.9
2010	28.2	2010	26.1
2012	28.6	2012	26.6
2014	29.3	2014	27.0
2016	29.5	2016	27.4
2018	29.8	2018	27.8

Source: U.S. Census Bureau

37. a) Draw a bar graph that appears to show a small increase in the median age at first marriage for males.

b) Draw a bar graph that appears to show a large increase in the median age at first marriage for males.

38. a) Draw a bar graph that appears to show a small increase in the median age at first marriage for females.

b) Draw a bar graph that appears to show a large increase in the median age at first marriage for females.

39. *Households with a Pet* The following graph shows the percent of U.S. household that had a pet for the years 2015 and 2017.

Percent of U.S. Households That Had a Pet

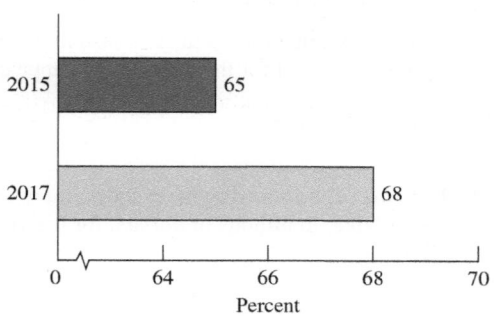

Source: American Pet Products Association

a) Draw a bar graph that shows the entire scale from 0 to 100 percent.

b) Does the new graph give a different impression? Explain.

Concept/Writing Exercises

40. *Advertising* Find five advertisements or commercials that may be statistically misleading. Explain why each may be misleading.

41. *Driving* The following circle graph shows the percentage of commuters who are frustrated by particular driving situations. Is the graph misleading? Explain.

Driving Situations Frustrating to Commuters

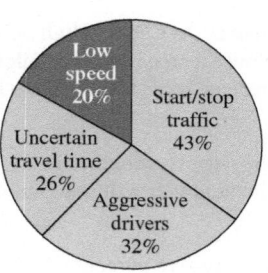

Challenge Problems/Group Activities

42. *Random Sampling*

a) Select and indicate a topic and population of interest to which a random sampling technique can be applied to obtain data.

b) Explain how you or your group would obtain a random sample for your population of interest.

c) Actually obtain the sample by the procedure stated in part (b).

43. *Data from Questionnaire* Some subscribers of *Consumer Reports* respond to an annual questionnaire regarding their satisfaction with new appliances, cars, and other items. The information obtained from these questionnaires is then used as a sample from which frequency of repairs and other ratings are made by the magazine. Are the data obtained from these returned questionnaires representative of the entire population, or are they biased?

44. *U.S. Population* Consider the graph on page 757, which shows the U.S. population in 2000 and the projected U.S. population in 2050.

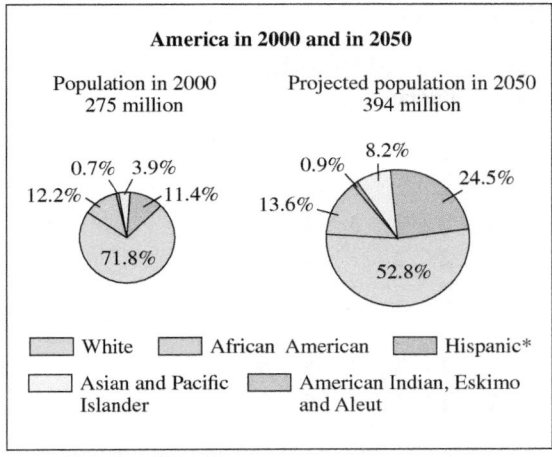

America in 2000 and in 2050

Population in 2000
275 million

0.7% 3.9%
12.2%   11.4%

71.8%

Projected population in 2050
394 million

8.2%
0.9%   24.5%
13.6%

52.8%

☐ White   ☐ African American   ☐ Hispanic*

☐ Asian and Pacific   ☐ American Indian, Eskimo
   Islander   and Aleut

Source: U.S. Census Bureau

a) Compute the projected percent increase in population from 2000 to 2050 by using the formula given on page 576.

b) Measure the radius and then compute the area of the circle representing 2000. Use $A = \pi r^2$.

c) Repeat part (b) for the circle representing 2050.

d) Compute the percent increase in the size of the area of the circle from 2000 to 2050.

e) Are the circle graphs misleading?

Research Activities

45. *Sampling Techniques* Select one type of sampling technique and write a report on how statisticians obtain that type of sample. Also indicate when that type of sampling technique may be preferred. Describe two examples of when the sampling technique may be used.

46. *Manipulating Statistics* Read the book *How to Lie with Statistics* by Darrell Huff and write a book report on it. Select three illustrations from the book that show how people manipulate statistics.

SECTION 12.2 # Frequency Distributions and Statistical Graphs

Upon completion of this section, you will be able to:

- Construct frequency distributions.
- Construct histograms and frequency polygons.
- Construct stem-and-leaf displays.
- Understand circle graphs.

You are creating a budget and decide to record the amount of money you spend on food weekly. After keeping a list of your weekly food expenses for many months, you realize you have a large set of data and want to condense your data into a more manageable form. In this section, we will learn a method to organize and summarize data. We will also introduce four types of graphs that can be used to display information in a meaningful way.

Why This Is Important Organizing and summarizing data and using graphs can make large sets of information more useful and help to determine important characteristics of the data.

Frequency Distributions

It is not uncommon for statisticians and others to have to analyze thousands of pieces of data. A *piece of data* is a single response to an experiment. The word *data* is plural. This means that the word *data* can represent either a single piece of data or a group of data. When the amount of data is large, it is usually advantageous to construct a frequency distribution. A *frequency distribution* is a listing of the observed values and the corresponding frequency of occurrence of each value. Example 1 shows how we construct a frequency distribution.

"Statistical thinking will one day be as necessary for efficient citizenship as the ability to read and write."

H. G. Wells

Example 1 *Frequency Distribution*

The number of children per family is recorded for 64 families surveyed. Construct a frequency distribution of the following data:

0	1	1	2	2	3	4	5
0	1	1	2	2	3	4	5
0	1	1	2	2	3	4	6
0	1	2	2	2	3	4	6
0	1	2	2	2	3	4	7
0	1	2	2	3	3	4	8
0	1	2	2	3	3	5	8
0	1	2	2	3	3	5	9

Solution Listing the number of children (observed values) and the number of corresponding families (frequency) gives the following frequency distribution.

Number of Children (Observed Values)	Number of Families (Frequency)
0	8
1	11
2	18
3	11
4	6
5	4
6	2
7	1
8	2
9	$\dfrac{1}{64}$

Eight families had no children, 11 families had one child, 18 families had two children, and so on. Note that the sum of the frequencies is equal to the original number of pieces of data, 64. ∎

Often data are grouped in classes to provide information about the distribution that would be difficult to observe if the data were ungrouped. Graphs called *histograms* and *frequency polygons* can be made of grouped data, as will be explained later in this section. These graphs also provide a great deal of useful information.

When data are grouped in classes, certain rules should be followed.

RECREATIONAL MATH

Can You Count the F's?

Statistical errors often result from careless observations. To see how such errors can occur, consider the statement below. How many F's do you count in the statement?

FINISHED FILES ARE THE RESULT OF YEARS OF SCIENTIFIC STUDY COMBINED WITH THE EXPERIENCE OF YEARS.

Answer: There are 6 F's.

PROCEDURE   RULES FOR DATA GROUPED BY CLASSES

1. The classes should be the same "width."
2. The classes should not overlap.
3. Each piece of data should belong to only one class.

In addition, it is often suggested that a frequency distribution should be constructed with 5 to 12 classes. If there are too few or too many classes, the distribution may become difficult to interpret. For example, if you use fewer than 5 classes, you risk losing too much information. If you use more than 12 classes, you may gain more detail but you risk losing clarity. Let the spread of the data be a guide in deciding the number of classes to use.

To understand these rules, let's consider a set of observed integral values that go from a low of 0 to a high of 26. Let's assume that the first class is arbitrarily selected to go from 0 through 4. Thus, any of the data with values of 0, 1, 2, 3, 4 would belong in this class. We say that the *class width* is 5, since there are five integral values that belong to the class. This first class ended with 4, so the second class must start with 5. If this class is to have a width of 5, at what value must it end? The answer is 9 (5, 6, 7, 8, 9). The second class is 5–9. Continuing in the same manner, we obtain the following set of classes.

Classes

Lower class limits $\left\{ \begin{array}{l} 0\text{--}4 \\ 5\text{--}9 \\ 10\text{--}14 \\ 15\text{--}19 \\ 20\text{--}24 \\ 25\text{--}29 \end{array} \right\}$ Upper class limits

We need not go beyond the 25–29 class because the largest value we are considering is 26. The classes meet our three criteria: They have the same width, there is no overlap among the classes, and each of the values from a low of 0 to a high of 26 belongs to one and only one class.

The choice of the first class, 0–4, was arbitrary. If we wanted to have more classes or fewer classes, we would make the class widths smaller or larger, respectively.

The numbers 0, 5, 10, 15, 20, 25 are called the *lower class limits*, and the numbers 4, 9, 14, 19, 24, 29 are called the *upper class limits*. Each class has a width of 5. Note that the class width, 5, can be obtained by subtracting the first lower class limit from the second lower class limit: $5 - 0 = 5$. The difference between any two consecutive lower class or upper class limits is also 5.

Example 2 *A Frequency Distribution of Novels*

Table 12.1 in the margin of this page and the next page shows the number of copies sold for the 30 top-selling novels of all time. The number of copies sold is rounded to the nearest million. Construct a frequency distribution of the data, letting the first class be 50–94 million.

Solution The 30 pieces of data are given in *descending* order from highest to lowest. We are given that the first class is to be 50–94. The second class must therefore start at 95. This is the lower class limit of the second class. To determine the class width, we subtract 50, the lower class limit of the first class, from 95, the lower class limit of the second class, to obtain a class width of $95 - 50 = 45$. The upper class limit of the second class is determined by adding the class width, 45, to the upper class limit of the first class, 94. Therefore, the upper class limit of the second class is $94 + 45 = 139$. Thus,

$$50\text{--}94 \qquad \text{first class}$$

$$95\text{--}139 \qquad \text{second class}$$

The other classes are determined using a similar technique. They are 140–184, 185–229, 230–274, 275–319, 320–364, 365–409, 410–454, 455–499, 500–544.

Table 12.1

Novel	Copies Sold (millions)
Don Quixote	500
A Tale of Two Cities	200
The Lord of the Rings	150
The Little Prince	140
Harry Potter and the Sorcerer's Stone	120
And Then There Were None	100
The Hobbit	100
Dream of the Red Chamber	100
Alice's Adventures in Wonderland	100
The Lion, the Witch, and the Wardrobe	85

(Continued)

Table 12.1 (*Continued*)

Novel	Copies Sold (millions)
She: A History of Adventure	83
The Adventures of Pinocchio	80
Vardi Wala	80
The Da Vinci Code	80
Harry Potter and the Chamber of Secrets	77
Harry Potter and the Prisoner of Azkaban	65
Harry Potter and the Goblet of Fire	65
Harry Potter and the Order of the Phoenix	65
Harry Potter and the Half-Blood Prince	65
Harry Potter and the Deathly Hallows	65
The Alchemist	65
The Catcher in the Rye	65
Think and Grow Rich	60
The Bridges of Madison County	60
The Adventures of Sherlock Holmes	60
20,000 Leagues Under the Sea	60
One Hundred Years of Solitude	50
Lolita	50
Heidi's Years of Living and Travel	50
Anne of Green Gables	50

Source: *The New York Times*

Since the highest value in the data is 500, there is no need to go any further. Note that each two consecutive lower class limits differ by 45, as do each two consecutive upper class limits. There are 21 pieces of data in the 50–94 class, 5 in the 95–139 class, 2 in the 140–184 class, 1 in the 185–229 class, 0 in the 230–274 class, 0 in the 275–319 class, 0 in the 320–364 class, 0 in the 365–409 class, 0 in the 410–454 class, 0 in the 455–499 class, and 1 in the 500–544 class. The complete frequency distribution of the 11 classes is given below. The number of novels totals 30, so we have included each piece of data.

Copies Sold (millions)	Number of Novels
50–94	21
95–139	5
140–184	2
185–229	1
230–274	0
275–319	0
320–364	0
365–409	0
410–454	0
455–499	0
500–544	1
	30

The *modal class* of a frequency distribution is the class with the greatest frequency. In Example 2, the modal class is 50–94. A frequency distribution may have more than one modal class. The *midpoint of a class*, also called the *class mark*, is determined by adding the lower and upper class limits and dividing the sum by 2. The midpoint of the first class, and also the class mark of the first class, in Example 2 is

$$\frac{50 + 94}{2} = \frac{144}{2} = 72$$

The midpoint of the second class in Example 2 is

$$\frac{95 + 139}{2} = \frac{234}{2} = 117$$

Note that the difference between successive class marks, $117 - 72$, is the class width 45. The class mark of the second class can therefore be obtained by adding the class width, 45, to the class mark of the first class, 72. The sum is $72 + 45 = 117$. Note that 117 checks with the class mark obtained by adding the lower class limit and the upper class limit of the second class and dividing the sum by 2.

Example 3 *A Frequency Distribution of Family Income*

The following set of data represents the family income (in thousands of dollars, rounded to the nearest hundred) of 15 randomly selected families.

46.5	31.8	45.8	44.7	40.9
65.2	52.4	44.6	53.7	48.8
35.5	40.3	39.8	56.3	50.7

Construct a frequency distribution with a first class of 31.5–37.6.

Solution First rearrange the data from lowest to highest so that the data will be easier to categorize.

31.8	40.3	44.7	48.8	53.7
35.5	40.9	45.8	50.7	56.3
39.8	44.6	46.5	52.4	65.2

The first class goes from 31.5 to 37.6. Since the data are in tenths, the class limits will also be given in tenths. The first class ends with 37.6; therefore, the second class must start with 37.7. The class width of the first class is 37.7 − 31.5, or 6.2. The upper class limit of the second class must therefore be 37.6 + 6.2, or 43.8. The frequency distribution is given below.

Income ($1000)	Number of Families
31.5–37.6	2
37.7–43.8	3
43.9–50.0	5
50.1–56.2	3
56.3–62.4	1
62.5–68.6	1
	15

Note in Example 3 that the class width is 6.2, the modal class is 43.9–50.0, and the class mark of the first class is $\dfrac{31.5 + 37.6}{2}$, or 34.55.

We have discussed how to organize and summarize data. Now we will introduce graphs that can be used to display information. We will consider four types of graphs: the histogram, the frequency polygon, the stem-and-leaf graph, and the circle graph.

Histograms and Frequency Polygons

Histograms and frequency polygons are statistical graphs used to illustrate frequency distributions. A *histogram* is a graph with observed values on its horizontal scale and frequencies on its vertical scale. A bar is constructed above each observed value (or class when classes are used), indicating the frequency of that value (or class). The horizontal scale need not start at zero, and the calibrations on the horizontal and vertical scales do not have to be the same. The vertical scale must start at zero. It may be necessary to break the vertical scale, as was done in displaying Stock B in Fig. 12.2 on page 752, to accommodate large frequencies on the vertical scale. Because histograms and other bar graphs are easy to interpret visually, they are used a great deal in newspapers and magazines.

Example 4 *Construct a Histogram*

The frequency distribution developed in Example 1, on page 758, is repeated on the top of page 762. Construct a histogram of this frequency distribution.

Number of Children (Observed Values)	Number of Families (Frequency)
0	8
1	11
2	18
3	11
4	6
5	4
6	2
7	1
8	2
9	1

Solution The vertical scale must extend at least to the number 18, since that is the greatest recorded frequency. The horizontal scale must include the numbers 0–9, the number of children observed. Eight families have no children. We indicate that by constructing a bar above the number 0, centered at 0, on the horizontal scale extended up to 8 on the vertical scale (Fig. 12.7). Eleven families have one child, so we construct a bar extending to 11 above the number 1, centered at 1, on the horizontal scale. We continue this procedure for each observed value. Both the horizontal and vertical scales should be labeled, the bars should be the same width and centered at the observed value, and the histogram should have a title. In a histogram, the bars should always touch.

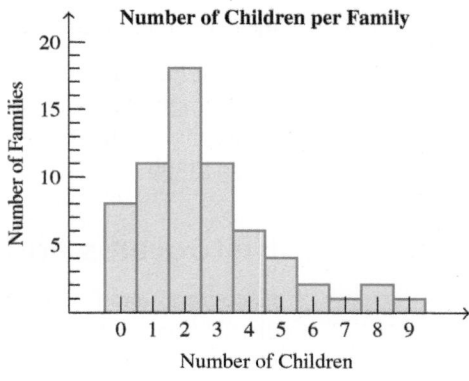

Figure 12.7

Frequency polygons are line graphs with scales the same as those of the histogram; that is, the horizontal scale indicates observed values and the vertical scale indicates frequency. To construct a frequency polygon, place a dot at the corresponding frequency above each of the observed values. Then connect the dots with straight-line segments. When constructing frequency polygons, always put in two additional class marks, one at the lower end and one at the upper end on the horizontal scale (values for these added class marks are not needed on the frequency polygon). Since the frequency at these added class marks is 0, the end points of the frequency polygon will always be on the horizontal scale.

Timely Tip

When constructing a histogram or frequency polygon, be sure to label both scales of the graph.

Example 5 *Construct a Frequency Polygon*

Construct a frequency polygon of the frequency distribution in Example 1 on page 758.

Table 12.2

Distance (miles)	Number of Workers
1–7	15
8–14	24
15–21	13
22–28	8
29–35	5
36–42	5

Solution Since eight families have no children, place a mark above the 0 at 8 on the vertical scale, as shown in Fig. 12.8. Because there are 11 families with one child, place a mark above the 1 on the horizontal scale at the 11 on the vertical scale, and so on. Connect the dots with straight-line segments and bring the end points of the graph down to the horizontal scale, as shown.

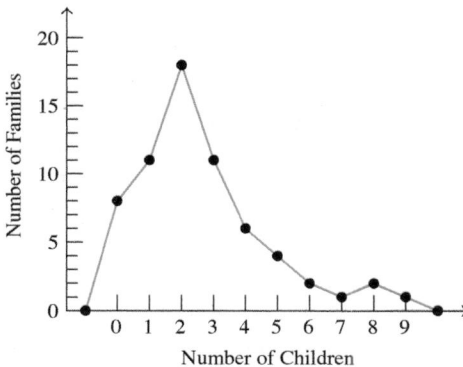

Figure 12.8

Example 6 *Commuting Distances*

The frequency distribution of the one-way commuting distances for 70 workers is listed in Table 12.2. Construct a histogram and then construct a frequency polygon.

Solution The histogram can be constructed with either class limits or class marks (class midpoints) on the horizontal scale. Frequency polygons are constructed with class marks on the horizontal scale. Since we will construct a frequency polygon on the histogram, we will use class marks. Recall that class marks are determined by adding the lower class limit and upper class limit and dividing the sum by 2. For the first class, the class mark is $\frac{1 + 7}{2}$, or 4. Since the class widths are seven units, the class marks will also differ by seven units (see Fig. 12.9).

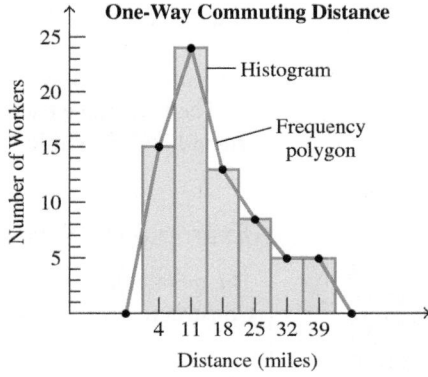

Figure 12.9

Example 7 *Carry-on Luggage Weights*

The histogram in Fig. 12.10 on page 764 shows the weights of selected pieces of carry-on luggage at an airport. Construct the frequency distribution from the histogram in Fig. 12.10.

Table 12.3

Weight (pounds)	Number of Pieces of Luggage
1–5	8
6–10	10
11–15	7
16–20	5
21–25	6
26–30	3
31–35	1
36–40	2

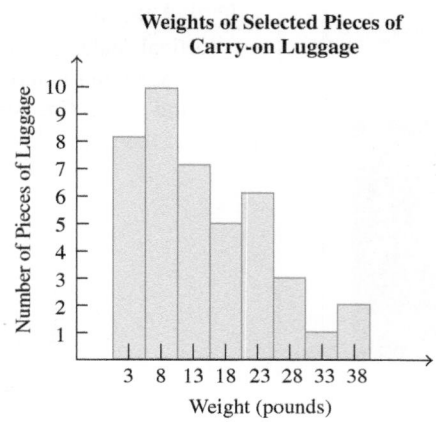

Figure 12.10

> **Solution** There are five units between class midpoints, so each class width must also be five units. Since three is the midpoint of the first class, there must be two units below and two units above it. The first class must be 1–5. The second class must therefore be 6–10. The frequency distribution is given in Table 12.3. ∎

Stem-and-Leaf Displays

Frequency distributions and histograms provide very useful tools to organize and summarize data. However, if the data are grouped, we cannot identify specific data values in a frequency distribution and in a histogram. For example, in Example 7, we know that there are eight pieces of luggage in the class of 1 to 5 pounds, but we don't know the specific weights of those eight pieces of luggage.

A *stem-and-leaf display* is a tool that organizes and groups the data while allowing us to see the actual values that make up the data. To construct a stem-and-leaf display, each value is represented with two different groups of digits. The left group of digits is called the *stem*. The remaining group of digits on the right is called the *leaf*. There is no rule for the number of digits to be included in the stem. Usually the units digit is the leaf and the remaining digits are the stem. For example, the number 53 would be broken up into 5 and 3. The 5 would be the stem and the 3 would be the leaf. The number 417 would be broken up into 41 and 7. The 41 would be the stem and the 7 would be the leaf. The number 6, which can be represented as 06, would be broken up into 0 and 6. The stem would be the 0 and the leaf would be the 6. With a stem-and-leaf display, the stems are listed, in ascending order, to the left of a vertical line. Then we place each leaf to the right of its corresponding stem, to the right of the vertical line.* Example 8 illustrates this procedure.

▲ *Zion National Park*

Example 8 *Stem-and-Leaf Display*

The table below indicates the ages of a sample of 20 visitors to Zion National Park. Construct a stem-and-leaf display.

29	31	39	43	56
60	62	59	58	32
47	27	50	28	71
72	44	45	44	68

---

* In stem-and-leaf displays, the leaves are sometimes listed from lowest digit to greatest digit, but that is not necessary.

Solution By quickly glancing at the data, we can see the ages consist of two-digit numbers. Let's use the first digit, the tens digit, as our stem and the second digit, the units digit, as the leaf. For example, for an age of 62, the stem is 6 and the leaf is 2. Our values are numbers in the 20s, 30s, 40s, 50s, 60s, and 70s. Therefore, the stems will be 2, 3, 4, 5, 6, 7 as shown below.

```
2|
3|
4|
5|
6|
7|
```

Next we place each leaf on its stem. We will do so by placing the second digit of each value next to its stem, to the right of the vertical line. Our first value is 29. The 2 is the stem and the 9 is the leaf. Therefore, we place a 9 next to the stem of 2 and to the right of the vertical line.

$$2|9$$

The next value is 31. We will place a leaf of 1 next to the stem of 3.

```
2|9
3|1
```

The next value is 39. Therefore, we will place a leaf of 9 after the leaf of 1 that is next to the stem of 3.

```
2|9
3|1  9
```

We continue this process until we have listed all the leaves on the display. The diagram below shows the stem-and-leaf display for the ages of the visitors. In our display, we will also include a legend to indicate the values represented by the stems and leaves. For example, 5|6 represents 56.

5|6 represents 56

```
2|9 7 8
3|1 9 2
4|3 7 4 5 4
5|6 9 8 0
6|0 2 8
7|1 2
```
Stems _____   _____ Leaves

Every piece of the original data can be seen in a stem-and-leaf display. From the above diagram, we can see that five of the visitors' ages were in the 40s. Only two visitors were older than 70. Note that the stem-and-leaf display gives the same visual impression as a sideways histogram.

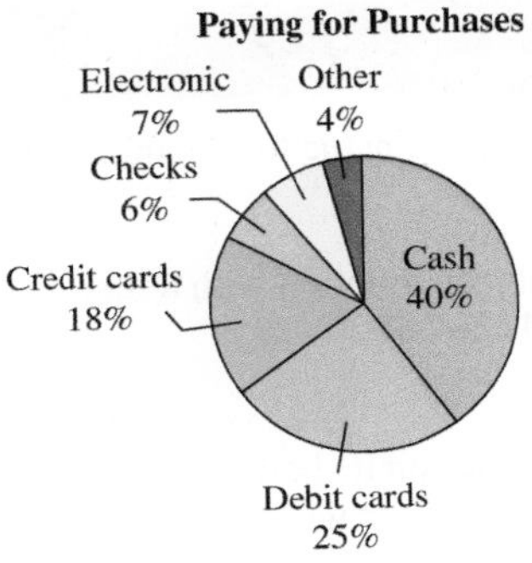

Circle Graphs

Circle graphs (also known as pie charts) are often used to compare parts of one or more components of the whole to the whole. A random sample of 500 consumers were asked how they pay for their purchases. The circle graph in Fig. 12.11 shows their responses. Since the total circle represents 100%, the sum of the sectors should be 100%, and it is.

In the next example, we will discuss a circle graph.

Paying for Purchases
Electronic 7%, Other 4%, Checks 6%, Credit cards 18%, Cash 40%, Debit cards 25%

Figure 12.11

⌐ Example **9** *Circus Performances*

Eight hundred people who attended a circus were asked to indicate their favorite performance. The circle graph in Fig. 12.12 shows the percentage of respondents that answered tigers, elephants, acrobats, jugglers, and other. Determine the number of respondents for each category.

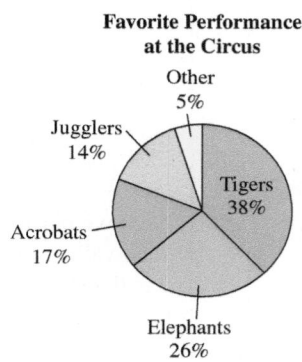

**Favorite Performance
at the Circus**

Figure 12.12

Solution To determine the number of respondents in a category, we multiply the percentage for each category, written as a decimal number, by the total number of people, 800. Table 12.4 indicates the results.

Table 12.4

Performance	Percent of Respondents	Percent Written as a Decimal Number	Number of People
Tigers	38%	0.38	$0.38 \times 800 = 304$
Elephants	26%	0.26	$0.26 \times 800 = 208$
Acrobats	17%	0.17	$0.17 \times 800 = 136$
Jugglers	14%	0.14	$0.14 \times 800 = 112$
Other	5%	0.05	$0.05 \times 800 = \underline{40}$
Total			800

As we can see from the table, 304 people indicated that the tigers were their favorite performance, 208 indicated elephants, 136 people indicated the acrobats, 112 people indicated the jugglers, and 40 people indicated some other performance. ∎

SECTION 12.2
Exercises

Warm Up Exercises

In Exercises 1–8, fill in the blank with an appropriate word, phrase, or symbol(s).

1. A listing of observed values and the corresponding frequency of occurrence of each value is called a(n) _____ distribution.

2. In a frequency distribution, the class with the greatest frequency is called the _____ class.

3. In a frequency distribution, another name for the midpoint of a class is the class _____.

4. The class width of a frequency distribution with a first class of 10–19 and a second class of 20–29 is _____.

5. A bar graph with observed values on its horizontal scale and frequencies on its vertical scale is called a(n) _____.

6. A line graph with observed values on its horizontal scale and frequencies on its vertical scale is called a frequency _____.

7. In a stem-and-leaf display

   a) the left group of digits is called the _____.

   b) the right group of digits is called the _____.

8. Graphs that are often used to compare parts of one or more components of the whole to the whole are called pie charts or _____ graphs.

Practice the Skills/Problem Solving

In Exercises 9 and 10, use the frequency distribution to determine

   a) *the total number of observations.*

   b) *the width of each class.*

   c) *the midpoint of the second class.*

   d) *the modal class (or classes).*

   e) *the class limits of the next class if an additional class were to be added.*

9.

Class	Frequency
9 – 15	4
16 – 22	7
23 – 29	1
30 – 36	0
37 – 43	3
44 – 50	5

10.

Class	Frequency
40 – 49	7
50 – 59	5
60 – 69	3
70 – 79	2
80 – 89	7
90 – 99	1

11. Visits to the Library Glendale Community College is planning to expand its library. Forty students were asked how many times they visited the library during the previous semester. Their responses are given below. Construct a frequency distribution, letting each class have a width of 1 (as in Example 1 on page 758).

0	1	1	3	4	5	7	8
0	1	2	3	5	5	7	8
0	1	2	3	5	5	7	9
1	1	2	3	5	6	8	10
1	1	3	4	5	6	8	10

12. Hot Dog Sales A hot dog vendor is interested in the number of hot dogs he sells each day at his hot dog cart. The number of hot dogs sold is indicated above and to the right

for 32 consecutive days. Construct a frequency distribution, letting each class have a width of 1.

15	16	19	20	21	22	24	27
15	18	19	20	21	22	25	27
15	18	19	20	21	23	25	28
16	18	19	21	21	23	26	29

Note that there were no days in which the vendor sold 17 hot dogs. However, it is customary to include a missing value as an observed value and assign to it a frequency of 0.

Top-selling Novels *In Exercises 13–16, use the data given in Table 12.1 on pages 759 and 760 to construct a frequency distribution with a first class (in millions) of*

13. 0–99

14. 0–74

15. 50–99

16. 50–149

City Population *In Exercises 17–20, the data represent the approximate 2018 population, in millions, of the 20 most populous cities in the world.*

38.0	21.0	17.6	13.3
25.7	20.4	16.6	13.1
23.7	20.2	15.2	12.9
21.1	18.8	14.9	12.9
21.0	18.6	14.2	12.5

Source: *United Nations*

Use these data to construct a frequency distribution with a first class of

17. 12.5–17.4

18. 12.5–15.4

19. 12.0–17.9

20. 12.0–15.9

▲ *Shanghai–the third most populated city in the world*

Cost of Living *In Exercises 21–24, the data in the table on page 768 indicate the cost of living in 2018, by state. The cost of living is a measure of the average price paid for housing, utilities, groceries, healthcare, transportation, and miscellaneous expenses. The national average cost of living is 100. The data can be used to compare a state to the national average and to other states.*

State	Cost of Living	State	Cost of Living
AK	130.6	MT	104.0
AL	89.5	NC	94.0
AR	88.4	ND	98.7
AZ	97.7	NE	93.3
CA	138.7	NH	109.3
CO	105.5	NJ	122.5
CT	128.8	NM	92.8
DE	106.0	NV	108.3
FL	98.9	NY	135.7
GA	91.2	OH	92.8
HI	190.1	OK	88.1
IA	91.8	OR	131.2
ID	94.2	PA	98.6
IL	95.7	RI	122.5
IN	90.1	SC	98.3
KS	89.7	SD	98.5
KY	91.8	TN	89.5
LA	93.6	TX	91.3
MA	133.8	UT	98.2
MD	131.3	VA	102.0
ME	117.2	VT	118.7
Ml	89.3	WA	109.5
MN	101.5	WI	95.8
MO	88.8	WV	94.7
MS	85.7	WY	90.5

Source: Bureau of Labor Statistics

Use these data to construct a frequency distribution with a first class of

21. 80.0−94.9

22. 80.0−99.9

23. 85.7−100.6

24. 85.7−105.6

25. *Jogging Distances* Twenty members of a health club who jog were asked how many miles they jog per week. The responses are as follows. Construct a stem-and-leaf display. For single digit data, use a stem of 0.

12	15	4	7	12	25	21
33	18	6	8	27	40	22
19	13	23	34	17	16	

▲ *See Exercise 25.*

26. *College Credits* Eighteen students in a geology class were asked how many college credits they had earned. The responses are as follows. Construct a stem-and-leaf display.

10	15	24	36	48	45
42	53	60	17	24	30
33	45	48	62	54	60

In Exercises 27–30, part (c), construct the frequency polygon on the same axes as the histogram. See Fig 12.9 on page 763.

27. *Movies* The following table shows the ticket sales of the 15 highest grossing movies of all time in the United States, adjusted for inflation, in millions of dollars, as of May 5, 2019.

Movie	Ticket Sales (millions of dollars)
Gone with the Wind	1824
Star Wars	1635
The Sound of Music	1284
E.T. the Extra-Terrestrial	1278
Titanic	1221
The Ten Commandments	1180
Jaws	1154
Doctor Zhivago	1118
The Exorcist	996
Snow White and the Seven Dwarfs	982
Star Wars: The Force Awakens	974
101 Dalmatians	900
Star Wars: The Empire Strikes Back	884
Ben-Hur	883
Avatar	877

Source: Box Office Mojo

a) Construct a frequency distribution with a first class of 877–996.

b) Construct a histogram.

c) Construct a frequency polygon.

28. Starting Salaries Starting salaries, in thousands of dollars, for social workers with a bachelor of science degree and no experience are shown for a random sample of 25 different social workers.

32	33	34	36	38
33	33	34	36	38
33	33	35	37	38
33	34	35	37	39
33	34	35	37	39

a) Construct a frequency distribution. Let each class have a width of one.

b) Construct a histogram.

c) Construct a frequency polygon.

d) Construct a stem-and-leaf display.

29. Visiting an Art Museum The ages of a random sample of 40 people visiting an art museum are

20	26	31	34	39	45	50	62
20	29	31	35	40	47	51	63
23	30	32	35	40	49	51	66
23	30	33	37	40	49	54	69
26	30	34	38	42	49	57	72

a) Construct a frequency distribution with a first class of 20–30.

b) Construct a histogram.

c) Construct a frequency polygon.

d) Construct a stem-and-leaf display.

30. Volunteers The ages of 44 volunteers at a food pantry are as follows.

57	57	49	52	50	42	54	55	64
61	61	64	56	47	51	51	56	46
57	54	50	46	55	56	60	61	54
57	68	48	54	55	55	62	52	47
58	51	65	49	54	51	43	69	

a) Construct a frequency distribution with a first class of 42–47.

b) Construct a histogram.

c) Construct a frequency polygon.

In Exercises 31 and 32, use class marks on the horizontal axis.

31. Histogram and Frequency Polygon Use the frequency table from Exercise 9 to create a

a) histogram

b) frequency polygon

32. Histogram and Frequency Polygon Use the frequency table from Exercise 10 to create a

a) histogram

b) frequency polygon

33. Number of Televisions per Home Use the histogram below to answer the following questions.

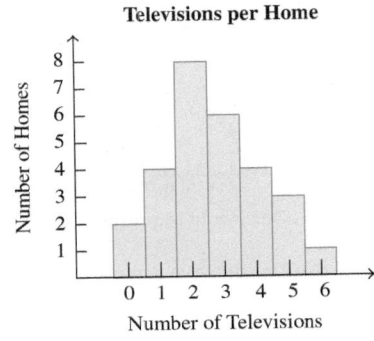

a) How many homes were included in the survey?

b) In how many homes were four televisions observed?

c) What is the modal class?

d) How many televisions were observed?

e) Construct a frequency distribution from this histogram.

34. Car Insurance Use the histogram below to answer the questions on page 770.

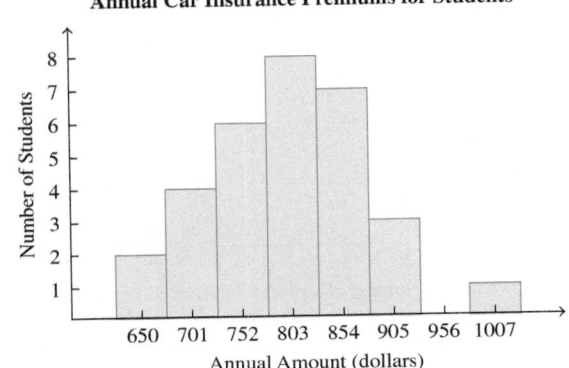

a) How many students were surveyed?

b) What are the lower and upper class limits of the first and second classes?

c) How many students have an annual car insurance premium in the class with a class mark of $752?

d) What is the class mark of the modal class?

e) Construct a frequency distribution from this histogram. Use a first class of 625–675.

35. *E-mail Messages* Use the frequency polygon below to answer the following questions.

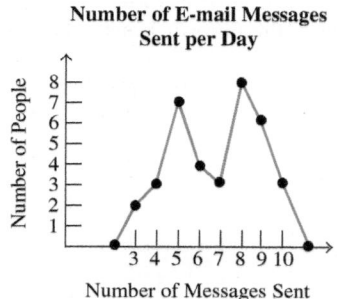

Number of E-mail Messages Sent per Day

a) How many people sent five e-mail messages?

b) How many people sent six or fewer e-mail messages?

c) How many people were included in the survey?

d) Construct a frequency distribution from the frequency polygon.

e) Construct a histogram from the frequency distribution in part (d).

36. *San Diego Zoo* Use the frequency polygon below to answer the following questions.

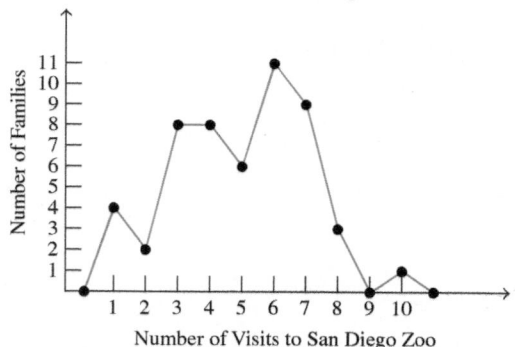

Number of Visits Selected Families Have Made to the San Diego Zoo

a) How many families visited the San Diego Zoo four times?

b) How many families visited the San Diego Zoo at least six times?

c) How many families were surveyed?

d) Construct a frequency distribution from the frequency polygon.

e) Construct a histogram from the frequency distribution in part (d).

37. *College Costs* The 2017–2018 cost for Texas residents to attend the University of Texas at Austin was $25,440. The circle graph below shows the percentages of that cost for tuition, books/supplies, room/board, and other expenses. Determine the cost, in dollars, for each category.

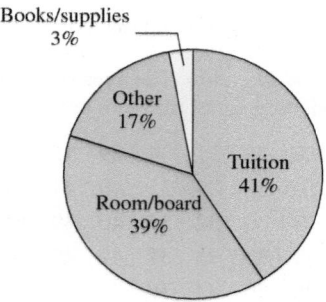

Cost to Attend the University of Texas at Austin for 2017–2018

Source: University of Texas at Austin

38. *Automobile Accessories* A sample of 600 people were asked which one automobile accessory they would most prefer to have on a family road trip. The following circle graph shows the percentage of respondents that answered Bluetooth capability, DVD player, extra cup holders, roof rack, and other. Determine the number of respondents for each category.

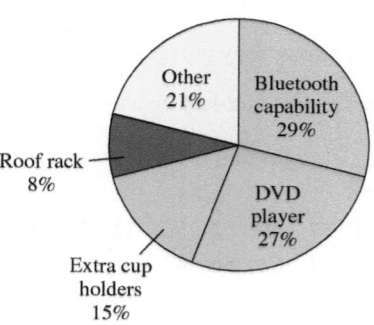

Automobile Accessories for a Family Road Trip

Challenge Problems/Group Activities

39. a) *Birthdays* What do you believe a histogram of the months in which the students in your mathematics class were born (January is month 1 and December is month 12) would look like? Explain.

b) By asking, determine the month in which the students in your class were born (include yourself).

c) Construct a frequency distribution containing 12 classes.

d) Construct a histogram from the frequency distribution in part (c).

e) Construct a frequency polygon of the frequency distribution in part (c).

40. *Social Security Numbers* Repeat Exercises 39 for the last digit of the students' Social Security numbers. Include classes for the digits 0–9.

Recreational Mathematics

41. a) *Count the F's* Count the number of F's in the sentence at the bottom of the Recreational Mathematics box on page 758.

b) Can you explain why so many people count the number of F's incorrectly?

42. *Vitamins* In what month do people take the least number of daily vitamins?

Research Activity

43. *Social Security* Over the years many changes have been made in the U.S. Social Security System.

a) Do research and determine the number of people receiving Social Security benefits for the years 1945, 1950, 1955, 1960, . . ., 2015, 2020. Then construct a frequency distribution and histogram of the data.

b) Determine the maximum amount that self-employed individuals had to pay into Social Security (the FICA tax) for the years 1945, 1950, 1955, 1960, . . ., 2015, 2020. Then construct a frequency distribution and a histogram of the data.

SECTION 12.3   Measures of Central Tendency and Position

Upon completion of this section, you will be able to:

- Calculate measures of central tendency of a set of data, including the mean, median, mode, and midrange.
- Calculate the measures of position of a set of data including percentiles and quartiles.

Most people have an intuitive idea of what is meant by an "average." The term is used daily in many familiar ways. "This car averages 19 miles per gallon," "The average test grade was 82," and "The average height of a blooming sunflower is 5.5 ft" are three examples. In this section, we will introduce four different averages and discuss the circumstances in which each average is used.

Why This Is Important An average is one of the most common ways to represent a set of data. Often, a set of data will have a different value for each different type of average. Understanding the appropriate average to use in each circumstance will enable you to make accurate conclusions about the data.

Measures of Central Tendency

An *average* is a number that is representative of a group of data. There are at least four different averages: the mean, the median, the mode, and the midrange. Each is calculated differently and may yield different results for the same set of data. Each will result in a number near the center of the data; for this reason, averages are commonly referred to as *measures of central tendency*.

The *arithmetic mean*, or simply the *mean*, is symbolized either by $\bar{x}$ (read "x bar") or by the Greek letter mu, μ. The symbol $\bar{x}$ is used when the mean of a *sample* of the population is calculated. The symbol μ is used when the mean of the *entire population* is calculated. Unless otherwise indicated, we will assume that the data featured in this book represent samples; therefore, we will use $\bar{x}$ for the mean.

The Greek letter sigma, Σ, is used to indicate "summation." The notation Σx, read "the sum of x," is used to indicate the sum of all the data. For example, consider the following data: 4, 6, 1, 0, 5. Then $\Sigma x = 4 + 6 + 1 + 0 + 5 = 16$.

Profile in Mathematics

Florence Nightingale (1820–1910)

While Florence Nightingale is known as the founder of modern nursing, she was also a passionate statistician and a pioneer in statistical graphs. During the Crimean War, Nightingale systematized medical records and collected data for analysis. Partially due to her analysis of data, the death rate at her hospital dropped from 42% to 2%. After returning to Britain, she asked Queen Victoria to appoint a Royal Commission on the Health of the Army so that the data she had collected, and other medical data, could be analyzed for medical reform. To support the case for reform, she created a clever graphical representation of the data as shown above.

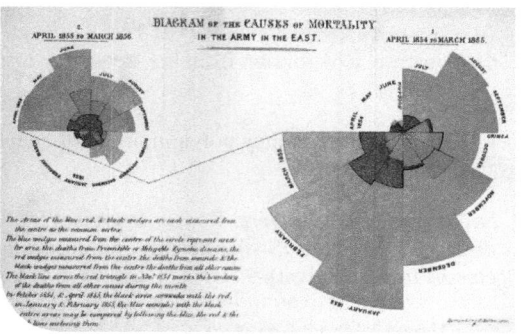

Nightingale continued to use statistics to reform medical record-keeping in Britain and to reform sanitary living conditions in rural India. In 1858, she became the first woman elected as a fellow of the Royal Statistical Society. She was also elected an honorary member of the American Statistical Association in 1874.

Now we can discuss the procedure for determining the mean of a set of data.

> **Definition: Mean**
> The **mean**, $\bar{x}$, is the sum of the data divided by the number of pieces of data. The formula for calculating the mean is
> $$\bar{x} = \frac{\Sigma x}{n}$$
> where Σx represents the sum of all the data and n represents the number of pieces of data.

The most common use of the word *average* is the mean.

Example 1 Determine the Mean

Determine the mean age of a group of firefighters if the ages of the individuals are 30, 21, 51, 37, and 51.

Solution

$$\bar{x} = \frac{\Sigma x}{n} = \frac{30 + 21 + 51 + 37 + 51}{5} = \frac{190}{5} = 38$$

Therefore, the mean, $\bar{x}$, is 38 years.

The mean represents "the balancing point" of a set of data. For example, if a seesaw were pivoted at the mean and uniform weights were placed at points corresponding to the ages in Example 1, the seesaw would balance. Figure 12.13 shows the five ages given in Example 1 and the calculated mean.

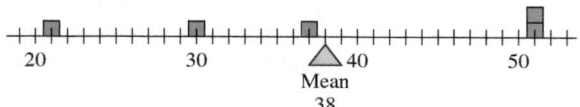

Figure 12.13

A second average is the *median*. To determine the median of a set of data, *rank the data* from smallest to largest, or largest to smallest, and determine the value in the middle of the set of *ranked data*. This value will be the median.

> **Definition: Median**
> The **median** is the value in the middle of a set of *ranked data*.

Timely Tip

Data must be ranked before determining the median. A common error made when determining the median is neglecting to arrange the data in ascending (increasing) or in descending (decreasing) order.

Example 2 *Determine the Median*

Determine the median of the firefighters' ages in Example 1 on page 772.

Solution Ranking the data from smallest to largest gives 21, 30, 37, 51, and 51. Since 37 is the value in the middle of this set of ranked data (two pieces of data above it and two pieces below it), 37 years is the median.

When there are an even number of pieces of data, the median is halfway between the two middle pieces. In this case, to determine the median, add the two middle pieces and divide this sum by 2.

Example 3 *Determine the Median of an Even Number of Data*

Determine the median of the following sets of data.
a) 14, 19, 21, 22, 16, 21, 16, 17
b) 11, 12, 12, 12, 13, 14

Solution

a) Ranking the data gives 14, 16, 16, 17, 19, 21, 21, 22. There are eight pieces of data. Therefore, the median will lie halfway between the two middle pieces, the 17 and the 19. The median is $\frac{17 + 19}{2}$, or $\frac{36}{2}$, or 18.

b) There are six pieces of data, and they are already ranked. Therefore, the median lies halfway between the two middle pieces. Both middle pieces are 12's. The median is $\frac{12 + 12}{2}$, or $\frac{24}{2}$, or 12.

A third average is the *mode*.

MATHEMATICS TODAY

Buying the American Dream

▲ *San Jose, California*

One of the biggest dreams for most people is to own their own home. Yet depending on where you live, the American dream may be hard to achieve. Among major metropolitan housing markets in 2018, San Jose, California, had the highest median home price, $1,100,000, and Youngstown, Ohio, had the lowest median home price, $77,000.

▲ *Youngstown, Ohio*
Source: National Association of Realtors

> **Definition: Mode**
> The **mode** is the piece of data that occurs most frequently.

Example 4 *Determine the Mode*

Determine the mode of the firefighters' ages in Example 1 on page 772.

Solution The ages are 30, 21, 51, 37, and 51. The age 51 is the mode because it occurs twice and the other values occur only once.

If each piece of data occurs only once, the set of data has no mode. For example, the set of data 1, 2, 3, 4, 5 has no mode. If two values in a set of data occur more often than all the other data, we consider both these values as modes and say that the data are *bimodal**, which means there are two modes for the set of data. For example, the set of data 1, 1, 2, 3, 3, 5 has two modes, 1 and 3.

---

* Some textbooks say that sets of data such as 1, 1, 2, 3, 3, 5 have no mode.

The last average we will discuss is the midrange. The *midrange* is the value half-way between the lowest (L) and highest (H) values in a set of data. It is determined by adding the lowest and highest values and dividing the sum by 2. A formula for determining the midrange follows.

> **Midrange**
> The **midrange** of a set of data can be calculated using the following formula.
> $$\text{Midrange} = \frac{\text{lowest value} + \text{highest value}}{2}$$

Example 5 *Determine the Midrange*

Determine the midrange of the firefighters' ages given in Example 1 on page 772.

Solution The ages of the firefighters are 30, 21, 51, 37, and 51. The lowest age is 21, and the highest age is 51.

$$\text{Midrange} = \frac{\text{lowest} + \text{highest}}{2} = \frac{21 + 51}{2} = \frac{72}{2} = 36 \text{ years} \quad \blacksquare$$

The "average" of the ages 30, 21, 51, 37, 51 can be considered any one of the following values: 38 (mean), 37 (median), 51 (mode), or 36 (midrange). Which average do you feel is most representative of the ages? We will discuss this question later in this section.

Example 6 *Measures of Central Tendency*

The salaries of eight social workers rounded to the nearest thousand dollars are 40, 25, 28, 35, 42, 60, 60, and 73. For this set of data, determine the (a) mean, (b) median, (c) mode, and (d) midrange. Then (e) list the measures of central tendency from lowest to highest.

Solution

a) $\bar{x} = \dfrac{\Sigma x}{n} = \dfrac{40 + 25 + 28 + 35 + 42 + 60 + 60 + 73}{8} = \dfrac{363}{8} = 45.375$

The mean is 45.375

b) Ranking the data from the smallest to largest gives

$$25, 28, 35, 40, 42, 60, 60, 73$$

Since there are an even number of pieces of data, the median is halfway between 40 and 42. The median $= \dfrac{40 + 42}{2} = \dfrac{82}{2} = 41$.

c) The mode is the piece of data that occurs most frequently. The mode is 60.

d) The midrange $= \dfrac{L + H}{2} = \dfrac{25 + 73}{2} = \dfrac{98}{2} = 49$.

e) The averages from lowest to highest are the median, mean, midrange, and mode. Their values are 41, 45.375, 49, and 60, respectively. $\quad \blacksquare$

At this point, you should be able to calculate the four measures of central tendency: mean, median, mode, and midrange. Now let's examine the circumstances in which each should be used.

The mean is used when each piece of data is to be considered and "weighed" equally. It is the most commonly used average. It is the only average that can be affected by *any* change in the set of data; for this reason, it is the most sensitive of all the measures of central tendency (see Exercise 23).

Data Sets:
Mean, Median, Mode

Occasionally, one or more pieces of data may be much greater or much smaller than the rest of the data. When this situation occurs, these "extreme" values have the effect of increasing or decreasing the mean significantly so that the mean will not be representative of the set of data. Under these circumstances, the median should be used instead of the mean. The median is often used in describing average family incomes because a relatively small number of families have extremely large incomes. These few incomes would inflate the mean income, making it nonrepresentative of the millions of families in the population.

Consider a set of exam scores from a mathematics class: 0, 16, 19, 65, 65, 65, 68, 69, 70, 72, 73, 73, 75, 78, 80, 85, 88, 92. Which average would best represent these grades? The mean is 64.06. The median is 71. Since only 3 of the 18 scores fall below the mean, the mean would not be considered a good representative score. The median of 71 probably would be the better average to use.

The mode is the piece of data, if any, that occurs most frequently. Builders planning houses are interested in the most common family size. Retailers ordering shirts are interested in the most common shirt size. An individual purchasing a thermometer might choose one, from those on display, whose temperature reading is the most common reading among those on display. These examples illustrate how the mode may be used.

The midrange is sometimes used as the average when the item being studied is constantly fluctuating. Average daily temperature, used to compare temperatures in different areas, is calculated by adding the lowest and highest temperatures for the day and dividing the sum by 2. The midrange is actually the mean of the high value and the low value of a set of data. Occasionally, the midrange is used to estimate the mean, since it is much easier to calculate.

Sometimes an average itself is of little value, and care must be taken in interpreting its meaning. For example, Jim is told that the average depth of Willow Pond is only 3 feet. He is not a good swimmer but decides that it is safe to go out a short distance in this shallow pond. After he is rescued, he exclaims, "I thought this pond was only 3 feet deep." Jim didn't realize that an average does not indicate extreme values or the spread of the values. The spread of data is discussed in Section 12.4.

TECHNOLOGY TIP

There's an App for That!

Many of the concepts that we discuss throughout this section and throughout this entire chapter can be further explored with the use of an app on your smartphone or tablet computer. Several of these apps allow you to calculate the mean, median, and mode. Although these can be helpful, be sure to read the directions carefully. Some apps will not support decimal numbers.

As always, consult with your instructor before using these apps when completing work related to your course.

Measures of Position

Measures of position are used to describe the position of a piece of data in relation to the entire set of data. If you took the SAT before applying to college, your score was described as a measure of position rather than a measure of central tendency. *Measures of position* are often used to make comparisons, such as comparing the scores of individuals from different populations, and are generally used when the amount of data is large.

Two measures of position are *percentiles* and *quartiles*. There are 99 percentiles dividing a set of data into 100 equal parts; see Fig. 12.14 on page 776. For example, suppose that you scored 520 on the mathematics portion of the SAT, and the score of

520 was reported to be in the 78th percentile of high school students. This wording *does not* mean that 78% of your answers were correct; it *does* mean that you outperformed about 78% of all those taking the exam. In general, a score in the nth percentile means that you outperformed about n% of the population who took the test and that about $(100 - n)$% of the people taking the test performed better than you did.

Percentiles

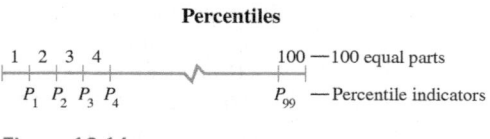

Figure 12.14

Example 7 *Heights of 5-Year-Old Children*

Mrs. Judge takes her 5-year-old son to the pediatrician for a checkup. The pediatrician tells Mrs. Judge her son's height is at the 95th percentile for 5-year-old boys. Explain what this means.

Solution If a height is at the 95th percentile, it means that about 95% of the heights are below that height. Therefore, Mrs. Judge's son is taller than about 95% of all 5-year-old boys. Also, about 5% of all 5-year-old boys are taller than Mrs. Judge's son. ∎

Quartiles are another measure of position. Quartiles divide data into four equal parts: The first quartile is the value that is higher than about $\frac{1}{4}$, or 25%, of the population. It is the same as the 25th percentile. The second quartile is the value that is higher than about $\frac{1}{2}$ the population and is the same as the 50th percentile, or the median. The third quartile is the value that is higher than about $\frac{3}{4}$ of the population and is the same as the 75th percentile; see Fig. 12.15.

Quartiles

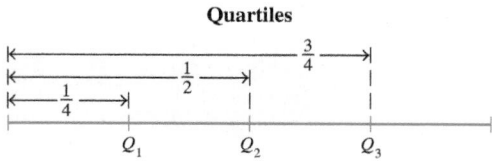

Figure 12.15

Note that a quartile is a single number, not a range of values. For example, we say a piece of data lies above or below the first quartile, but not in the first quartile.

PROCEDURE DETERMINING THE QUARTILES OF A SET OF DATA

1. List the data from smallest to largest.

2. Determine the median, or the 2nd quartile, of the set of data. If there is an odd number of pieces of data, the median is the middle value. If there is an even number of pieces of data, the median will be halfway between the two middle pieces of data.

3. The first quartile, Q_1, is the median of the lower half of the data; that is, Q_1 is the median of the data less than Q_2.

4. The third quartile, Q_3, is the median of the upper half of the data; that is, Q_3 is the median of the data greater than Q_2.

Example 8 *Determining Quartiles*

Foot Locker is concerned about the high turnover of its sales staff. A survey was done to determine how long (in months) the sales staff had been in their current positions. The responses of 27 sales staff follow. Determine Q_1, Q_2, and Q_3.

25	3	7	15	31	36	17	21	2
11	42	16	23	19	21	9	20	5
8	12	27	14	39	24	18	6	10

Solution First we list the data from smallest to largest.

2	3	5	6	7	8	9	10	11
12	14	15	16	17	18	19	20	21
21	23	24	25	27	31	36	39	42

Next we determine the median. Since there are 27 pieces of data, an odd number, the median will be the middle value. The middle value is 17, with 13 pieces of data less than 17 and 13 pieces of data greater than 17. Therefore, the median, Q_2, is 17, shown in red.

To determine Q_1, the median of the lower half of the data, we need to determine the median of the 13 pieces of data that are less than Q_2. The middle value of the lower half of the data is 9. There are 6 pieces of data less than 9 and 6 pieces of data greater than 9. Therefore, Q_1 is 9, shown in blue.

To determine Q_3, the median of the upper half of the data, we need to determine the median of the 13 pieces of data that are greater than 17, or Q_2. The middle value of the upper half of the data is 24. There are 6 pieces of data less than 24 and 6 pieces of data greater than 24. Therefore, Q_3 is 24, shown in blue. ∎

TECHNOLOGY TIP

Several computer software programs and calculators can be used to determine the mean of a set of data. These programs and calculators can also provide other types of statistical information that we will discuss in this chapter. In this *Technology Tip* we will provide the instructions for using a software program called Microsoft Excel as well as information on how to use Texas Instruments TI-84 Plus graphing calculator and StatCrunch. In the example, we will use the data from Example 1 on page 772, which represent the ages of firefighters.

Excel

In our discussion, we will use the symbol > to indicate the next item to be selected from the list or menu of items.

Begin by entering the five pieces of data in column A. Press the Enter key after each piece of data is entered. Next select

Insert > Function ... > Statistical > AVERAGE

Then click the OK box at the bottom. The program will then generate a gray box, where you need to enter the data. In the area to the right of **Number1**, you need to enter the data for which you want to determine the mean. Since you have already entered the data in column A, rows 1 to 5, if A1:A5 is not already listed, you can enter A1:A5 in the area to the right of **Number1**. Then, at the bottom of the gray box, *Formula Results* = 36 is displayed. The 36 is the mean of the set of data. If you then press the OK box, the mean will be displayed in the cell where the cursor was located. If you do not want the mean displayed in a cell, press CANCEL. To determine the median or mode you follow a similar procedure, except that instead of selecting AVERAGE you would select MEDIAN or MODE, respectively.

TI-84 Plus Graphing Calculators

To enter the data, press STAT. Then highlight **1: Edit** and press the ENTER key. If any data are currently listed in column **L1**, move the cursor to **L1** and press CLEAR and then the ⎡ENTER⎤ key. This step will eliminate all data from column **L1**. Now enter the data in column **L1**. After you enter each piece of data, press the ⎡ENTER⎤ key. After all the data have been entered, press ⎡STAT⎤. Then highlight **CALC**. At this point, **1: 1-Var Stats** should be highlighted. Press the ⎡ENTER⎤ key twice. You will now see $\bar{x} = 36$. Thus, the mean is 36. Several other descriptive statistics that we will discuss shortly are also shown. If you scroll down, you will eventually see the values of Q_1, the median, and Q_3.

StatCrunch (available in MyLab Math and separately)

1. To compute the mean and median, go to the StatCrunch website and then select Open StatCrunch in the top menu.
2. Enter your data into the spreadsheet under var 1.
3. Click **Applets** in the top menu. Select Mean/SD vs Median/IQR from the dropdown menu.
4. Click on the radio button **Data Table**. From the select column dropdown menu, select var 1. Make sure mean and median are selected under Statistics Shown.
5. Click **Compute**. You will see that the mean is 38 and the median is 37.

SECTION 12.3 Exercises

Warm Up Exercises

In Exercises 1–10, fill in the blank with an appropriate word, phrase, or symbol(s).

1. A number that is representative of a set of data is called a(n) _____.

2. Averages are referred to as measures of central _____.

3. The average that is determined by summing the data and then dividing the sum by the number of pieces of data is called the _____.

4. The value in the middle of a set of ranked data is called the _____.

5. The piece of data that occurs most frequently in a set of data is called the _____.

6. The value halfway between the lowest and highest values in a set of data is called the _____.

7. The measures of position that divide a set of data into four equal parts are called _____.

8. The measures of position that divide a set of data into 100 equal parts are called _____.

9. a) The symbol for the sample mean is _____.

   b) The symbol for the population mean is _____.

10. a) Another name for the 25th percentile is the _____ quartile.

    b) Another name for the 50th percentile is the _____ quartile.

    c) Another name for the 75th percentile is the _____ quartile.

Many of the exercises in this section are designed so that they can be done by using an appropriate form of technology (graphing calculator, spreadsheet, or StatCrunch).

Practice the Skills

In Exercises 11–20, determine the mean, median, mode, and midrange of the set of data. Where appropriate, round your answer to the nearest tenth.

11. 11, 12, 12, 14, 16, 16, 16, 27, 29

12. 5, 7, 8, 8, 8, 10, 10

13. 61, 67, 79, 40, 85, 37, 81

14. 1, 1, 1, 1, 4, 4, 4, 4, 6, 8, 10, 12, 15, 21

15. 1, 3, 5, 7, 9, 11, 13, 15

16. 2, 2, 2, 2, 5, 5, 5, 5, 8, 10, 15, 17, 21, 25

17. 1, 7, 11, 27, 36, 14, 12, 9, 1

18. 16, 14, 20, 22, 21, 29, 385, 50, 47, 18

19. 6, 8, 12, 13, 11, 13, 15, 17

20. 5, 15, 5, 15, 5, 15

Problem Solving

For Exercises 21–46, where appropriate, round your answer to the nearest tenth.

21. **Basketball Scores** In the last 10 games of the 2019 season, the Syracuse University women's basketball team scored the following points: 76, 67, 92, 66, 70, 64, 94, 68, 90, 77. Determine the mean, median, mode, and midrange.

▲ Miranda Drummond, Syracuse University
(See Exercise 21 on page 778.)

DS 22. ***Daily Tips*** The amount of money Stephanie collected in tips in each of seven days is $51, $70, $54, $28, $105, $64, $83. Determine the mean, median, mode, and midrange.

23. ***Change in the Data*** The mean is the "most sensitive" average because it is affected by any change in the data.

a) Determine the mean, median, mode, and midrange for 1, 2, 3, 5, 5, 7, 11.

b) Change the 7 to 10 in part (a). Determine the mean, median, mode, and midrange.

c) Which averages were affected by changing the 7 to 10?

d) Which averages will be affected by changing the 11 to 10 in part (a)?

24. ***Change in the Data***

a) Determine the mean, median, mode, and midrange for 16, 26, 27, 27, 29, 31, 33, 39.

b) Change the 33 in part (a) to 37. Determine the mean, median, mode, and midrange.

c) Which average(s) were affected by changing the 33 to 37?

d) Which average(s) will be affected by changing the 39 to 38 in part (a)?

DS 25. ***Employee Salaries*** The salaries of 10 employees of a small company follow.

$34,000	$70,000
31,000	30,000
37,000	33,000
32,000	87,000
32,000	35,000

Determine the

a) mean.

b) median.

c) mode.

d) midrange.

e) If the employees wanted to demonstrate the need for a raise, which average would they use to show they are being underpaid: the mean or the median? Explain.

f) If the management did not want to give the employees a raise, which average would they use: the mean or the median? Explain.

DS 26. ***House Prices*** The prices of 9 houses sold in a neighborhood follow.

$175,000	$185,000
$210,000	$190,000
$199,000	$178,000
$200,000	$188,000
$500,000	

Determine the

a) mean.

b) median.

c) mode.

d) midrange.

e) Which measure of central tendency, the mean or the median, best represents the typical price of the houses sold?

DS 27. ***National Parks*** The 10 national parks with the greatest number of visitors, in millions, in 2018 are listed below.

Park	Visitors (millions)
Great Smoky Mountains	11.4
Grand Canyon	6.4
Rocky Mountain	4.6
Zion	4.3
Yellowstone	4.1
Yosemite	4.0
Acadia	3.5
Grand Teton	3.5
Olympic	3.1
Glacier	3.0

Source: nps.gov

Determine, to the nearest tenth, the

a) mean.

b) median.

c) mode.

d) midrange.

▲ Great Smoky Mountains National Park

DS **28.** *Employment Websites* The top 10 employment websites with the highest number of monthly visitors, in millions, in January 2019, are listed below.

Website	Number of Visitors (millions)
Indeed	55.0
Monster	35.0
GlassDoor	30.0
CareerBuilder	20.4
SimplyHired	12.0
Aol Jobs	10.0
JobDiagnosis	9.5
Beyond	4.8
ZipRecruiter	4.5
USAJobs	4.3

Source: ebizmba.com

Determine, to the nearest tenth, the

a) mean.   **c)** median.

b) mode.   **d)** midrange.

29. *Lunch Expenses* The amounts of money Tino spent on lunch for seven days are as follows:

$6.99 $7.50 $8.25 $10.15 $12.75 $9.30 $7.25

Determine, to the nearest cent, the

a) mean.   **b)** median.

c) mode.   **d)** midrange.

30. *Living Expenses* Bob's monthly living expenses for 1 year are as follows:

$1200	$1050	$1570	$1600
2000	1050	1550	1450
1800	1100	1310	1430

Determine, to the nearest cent, the

a) mean.   **c)** mode.

b) median.   **d)** midrange.

31. *Exam Average* Malcolm's mean average on five exams is 81. Determine the sum of his scores.

32. *Exam Average* Jeremy's mean average on six exams is 92. Determine the sum of his scores.

33. *Water Park* For the 2019 season, 35,000 people visited the Blue Lagoon Water Park. The park was open 120 days for water activities. The highest number of visitors on a single day was 500. The lowest number of visitors on a single

day was 50. Determine whether it is possible to determine the following with the given information.

a) the mean number of visitors per day

b) the median number of visitors per day

c) the mode number of visitors per day

d) the midrange number of visitors per day

34. *Ski Resort* For the 2019 season, 9500 people skied at Bristol Mountain Ski Resort. The resort was open 100 days for skiing. The highest number of skiers on a single day was 175. The lowest number of skiers on a single day was 65. Determine whether it is possible to determine the following with the given information.

a) the mean number of skiers per day

b) the median number of skiers per day

c) the mode number of skiers per day

d) the midrange number of skiers per day

35. *A Grade of B* A mean average of 80 or greater for five exams is needed for a final grade of B in a course. Jorge's first four exam grades are 73, 69, 85, and 80. What grade does Jorge need on the fifth exam to get a B in the course?

36. *Grade of A* A mean average of 90 or greater for 6 exams is needed for a final grade of A in a course. Sandy's first five exam grades are 92, 90, 87, 93, and 96. What grade does Sandy need on the sixth exam to get an A in the course?

37. *Grading Methods* A mean average of 60 or greater on seven exams is needed to pass a course. On her first six exams, Sheryl received grades of 52, 72, 80, 65, 57, and 69.

a) What grade must she receive on her last exam to pass the course?

b) An average of 70 is needed to get a C in the course. Is it possible for Sheryl to get a C? If so, what grade must she receive on the seventh exam?

c) If her lowest grade of the exams already taken is to be dropped, what grade must she receive on her last exam to pass the course?

d) If her lowest grade of the exams already taken is to be dropped, what grade must she receive on her last exam to get a C in the course?

38. *Grading Methods* A mean average of 70 or greater on six exams is needed to get a C in a course. On his first 5 exams, Quincy received grades of 74, 82, 76, 70, and 54.

a) What grade must he receive on his last exam to get a C?

b) A mean average of 80 is needed to get a B in the course. Is it possible for Quincy to get a B? If so, what grade must he receive on the sixth exam?

c) If the lowest grade of the exams already taken is to be dropped, what grade must he receive on his last exam to get a C in the course?

d) If the lowest grade of the exams already taken is to be dropped, would it be possible for Quincy to get a B? If so, what grade would he need to receive on his last exam?

39. Grocery Expenses The Taylors have recorded their weekly grocery expenses for the past 12 weeks and determined that the mean weekly expense was $85.20. Later Mrs. Taylor discovered that 1 week's expense of $74 was incorrectly recorded as $47. What is the correct mean?

40. Entertainment Expenses The Comstocks have recorded their monthly entertainment expenses for the past 6 months and determined that the mean monthly expense was $250. Later, Mr. Comstock discovered that one month's expense of $175 was incorrectly recorded as $157. What is the correct mean?

41. Quartiles The prices of the 19 top-rated all-season tires, as reported in the November 2018, issue of *Consumer Reports,* for a specific tire size, are as follows:

$87 $115 $97 $79 $81 $91 $92 $88 $92 $80
$108 $113 $100 $95 $85 $91 $76 $99 $90

Determine

a) $Q_2.$ **b)** $Q_1.$ **c)** $Q_3.$

42. Quartiles The prices of the 20 top-rated 60-inch televisions, as reported in the December 2018, issue of *Consumer Reports,* are as follows:

$2800 $3000 $3800 $3500 $2200 $3800 $2700
$2200 $2600 $1700 $1800 $2200 $3000 $2200
$1580 $1700 $1800 $1600 $1200 $1250

Determine

a) $Q_2.$ **b)** $Q_1.$ **c)** $Q_3.$

43. College Admissions Jonathan took an admission test for the University of California and scored in the 85th percentile. The following year, Jonathan's sister Kendra took a similar admission test for the University of California and scored in the 90th percentile.

a) Is it possible to determine which of the two answered the higher percent of questions correctly on their respective exams?

b) Is it possible to determine which of the two was in a better relative position with regard to their respective populations? Explain.

44. Physical Education Class Samantha took a flexibility test in her physical education class and scored at the 75th percentile. In a different class, Jose took the same flexibility test and scored at the 60th percentile.

a) Is it possible to determine which of the two had the higher flexibility score on their respective exams?

b) Is it possible to determine which of the two was in a better relative position with regard to their respective classes? Explain.

45. Employee Salaries The following statistics represent weekly salaries at the Midtown Construction Company:

Mean	$620	First quartile	$580
Median	$610	Third quartile	$645
Mode	$600	83rd percentile	$685

a) What is the most common salary?

b) What salary did half the employees' salaries surpass?

c) About what percent of employees' salaries surpassed $645?

d) About what percent of employees' salaries were less than $580?

e) About what percent of employees' salaries surpassed $685?

f) If the company has 100 employees, what is the total weekly salary of all employees?

46. Employee Salaries The following statistics represent annual salaries of physical therapists at Lattimore Physical Therapy:

Mean	$88,000	First quartile	$72,000
Median	$86,520	Third quartile	$93,000
Mode	$80,000	90th percentile	$99,000

a) What is the most common salary?

b) What salary did half of the physical therapists' salaries surpass?

c) About what percent of physical therapists' salaries surpassed $93,000?

d) About what percent of physical therapists' salaries were less than $72,000?

e) About what percent of physical therapists' salaries surpassed $99,000?

f) If Lattimore Physical Therapy employs 25 physical therapists, what is the total annual salary of all employees?

Concept/Writing Exercises

47. Life Expectancy In 2016, the National Center for Health Statistics indicated an "average life expectancy" of 78.6 years for the total U.S. population. The average life expectancy for men was 76.1 years, and for women it was 81.1 years. Which "average" do you think the National Center for Health Statistics is using?

48. A Grade of B To get a grade of B, a student must have a mean average of 80 or greater. Jim has a mean average of 79 for 10 quizzes. He approaches his teacher and asks for a B, reasoning that he missed a B by only one point. What is wrong with Jim's reasoning?

49. Creating a Data Set Construct a set of five pieces of data in which the mode has a lower value than the median and the median has a lower value than the mean.

50. Creating a Data Set Construct a set of six pieces of data with a mean, median, and midrange of 75 and where no two pieces of data are the same.

51. Creating a Data Set Construct a set of six pieces of data with a mean of 84 and where no two pieces of data are the same.

52. Creating a Data Set Construct a set of six pieces of data such that if only one piece of data is changed, the mean, median, and mode will all change.

53. Central Tendencies Which of the measures of central tendency *must* be an actual piece of data in the distribution?

54. Changing One Piece of Data Is it possible to construct a set of six different pieces of data such that by changing only one piece of data you cause the mean, median, mode, and midrange to change? Explain.

55. Percentiles For any set of data, what must be done to the data before percentiles can be determined?

56. Percentiles Josie scored in the 73rd percentile on the verbal part of her College Board test. What does that mean?

57. Percentiles When a national sample of heights of kindergarten children was taken, Kevin was told that he was in the 35th percentile. Explain what that means.

58. Quartiles A union leader is told that, when all workers' salaries are considered, the first quartile is $27,750. Explain what that means.

59. The 50th Percentile Give the names of two other statistics that have the same value as the 50th percentile.

Challenge Problems/Group Activities

60. The Mean of the Means Consider the following five sets of values.

   i) 5 6 7 7 8 9 14

   ii) 3 6 8 9

   iii) 1 1 1 2 5

   iv) 6 8 9 12 15

   v) 50 51 55 60 80 100

  a) Compute the mean of each of the five sets of data.

  b) Compute the mean of the five means in part (a).

c) Determine the mean of the 27 pieces of data.

d) Compare your answer in part (b) to your answer in part (c). Are the values the same? Does your answer make sense?

61. Ruth Versus Mantle The tables below compare the batting performances for selected years for two well-known former baseball players, Babe Ruth and Mickey Mantle.

▲ Babe Ruth
Boston Red Sox 1914–1919
New York Yankees 1920–1934

Year	At Bats	Hits	AVG
1925	359	104	
1930	518	186	
1933	459	138	
1916	136	37	
1922	406	128	
Total	1878	593	

▲ Mickey Mantle
New York Yankees 1951–1968

Year	At Bats	Hits	Pct.
1954	543	163	
1957	474	173	
1958	519	158	
1960	527	145	
1962	377	121	
Total	2440	760	

a) For each player, compute the batting average percent (pct.) for each year by dividing the number of hits by the number of at bats. Round to the nearest thousandth. Place the answers in the pct. column.

b) Going across each of the five horizontal lines (for example Ruth, 1925, vs. Mantle, 1954), compare the percents (pct.) and determine which is greater in each case.

c) For each player, compute the mean batting average percent for the 5 given years by dividing the total hits by the total at bats. Which is greater, Ruth's or Mantle's?

d) Based on your answer in part (b), does your answer in part (c) make sense? Explain.

e) Determine the mean percent for each player by adding the five pcts. and dividing by 5. Which is greater, Ruth's or Mantle's?

f) Why do the answers obtained in parts (c) and (e) differ? Explain.

g) Who would you say has the better batting average percent for the 5 years selected? Explain.

62. *Employee Salaries* The following table gives the distribution of annual salaries for employees at Kulzer's Home Improvement.

Annual Salary	Number Receiving Salary
$100,000	1
85,000	2
24,000	6
21,000	4
18,000	5
17,000	7

Using the information provided in the table, determine the

a) mean annual salary.

b) median annual salary.

c) mode annual salary.

d) midrange annual salary.

e) Which do you believe is the best measure of central tendency for this set of data? Explain your answer.

Weighted Average *In Exercises 63 and 64 we use weighted averages. Sometimes when we wish to determine an average, we may wish to assign more importance, or weight, to some of the pieces of data. To calculate a **weighted average**, we use the formula:*

$$\text{weighted average} = \frac{\Sigma xw}{\Sigma w}, \text{ where } w \text{ is the weight of the}$$

piece of data, x; Σxw is the sum of the products of each piece of data multiplied by its weight; and Σw is the sum of the weights. For example, suppose that students in a class need to submit a report that counts for 20% of their grade, they need to take a midterm exam that counts for 30% of their grade, and they need to take a final exam that counts for 50% of their grade. Suppose that a student got a 72 on the report, an 85 on the midterm exam, and a 93 on the final exam. To determine this student's weighted average, first determine Σxw: $\Sigma xw = 72(0.20) + 85(0.30) + 93(0.50) = 86.4$. Next determine Σw, the sum of the weights:

$\Sigma w = 0.20 + 0.30 + 0.50 = 1.00$. *Now determine the weighted average as follows.*

$$\text{Weighted average} = \frac{\Sigma xw}{\Sigma w} = \frac{86.4}{1.00} = 86.4$$

Thus, the weighted average is 86.4. Note that Σw does not always have to be 1.00. In Exercises 63 and 64, use the weighted average formula.

63. *Course Average* Suppose that your final grade for a course is determined by a midterm exam and a final exam. The midterm exam is worth 40% of your grade, and the final exam is worth 60%. If your midterm exam grade is 84 and your final exam grade is 94, calculate your final weighted average.

64. *Grade Point Average* In a four-point grade system, an A corresponds to 4.0 points, a B corresponds to 3.0 points, a C corresponds to 2.0 points, and a D corresponds to 1.0 points. No points are awarded for an F. Last semester, Tanya received a B in a four-credit hour course, an A in a three-credit hour course, a C in a three-credit hour course, and an A in another three-credit hour course. A grade point average (GPA) is calculated as a weighted average using the credit hours as weights and the number of points corresponding to the grade as pieces of data. Calculate Tanya's GPA for the previous semester. (Round your answer to the nearest hundredth.)

Recreational Mathematics

65. *Average of Your Exams*

a) Calculate the mean, median, mode, and midrange of your exam grades in your mathematics course.

b) Which measure of central tendency best represents your average grade?

c) Which measure of central tendency would you rather use as your average grade?

66. *Purchases* Matthew purchased some items at Staples each day for five days. The mode of the number of items Matthew purchased is higher than the median of the number of items he purchased. The median of the number of items Matthew purchased is higher than the mean of the number of items he purchased. He purchased at least two items but no more than seven items each day.

a) How many items did Matthew purchase each day? (*Note:* There is more than one correct answer.)

b) Determine the mean, median, and mode for your answer to part (a).

Research Activity

67. *Stanines and Deciles* Two other measures of location that we did not mention in this section are *stanines* and *deciles*. Use statistics books, books on educational testing and measurements, and Internet websites to write a report on what stanines and deciles are and when percentiles, quartiles, stanines, and deciles are used.

SECTION 12.4   Measures of Dispersion

Upon completion of this section, you will be able to:

- Calculate the range of a set of data.
- Calculate the standard deviation of a set of data.

Measures of central tendency by themselves do not always give sufficient information to analyze a situation and make decisions. For example, suppose Apple is considering two companies to produce batteries for its iPads. Testing shows that Company A batteries have a mean life of 10 hours. Company B batteries have a mean life of 9.5 hours. If both manufacturers' batteries cost the same, which one should be purchased? The average battery life may not be the most important factor. If half of Company A batteries last only 5 hours, while half last 15 hours, there is a large variability in the life of the batteries. If all of Company B batteries last between 9.0 and 10.0 hours, the batteries are more consistent and reliable. This example illustrates the importance of knowing something about the *spread*, or *variability*, of the data. In this section we will discuss two measures of variability or dispersion.

Why This Is Important   Knowing the measures of central tendency of a set of data is important, but knowing the measures of dispersion is just as important. As we will see in this section, measures of dispersion help us better understand the data and help us draw accurate conclusions about the data.

Measures of dispersion are used to measure the variability of the data, including the *spread of the data*, and how the data vary about the mean. The range and standard deviation[*] are the measures of dispersion that will be discussed in this section.

Range

The *range* is the difference between the highest and lowest values; it indicates the total spread of the data.

> **Range**
> The range of a set of data can be calculated using the following formula.
>
> $$\textbf{Range} = \text{highest value} - \text{lowest value}$$

Example 1   *Determine the Range*

The amounts of calories in a slice of cheese pizza from nine different brands are given below. Determine the range of these data.

$$320, 295, 290, 250, 258, 234, 244, 320, 285$$

Solution   Range = highest value − lowest value = 320 − 234 = 86. The range of the amount of calories is 86. ∎

---

[*]*Variance*, another measure of dispersion, is the square of the standard deviation.

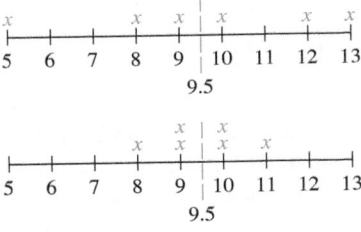

Figure 12.16

Standard Deviation

The second measure of dispersion we discuss in this section, the *standard deviation*, measures how much the data *differ from the mean*. It is symbolized either by the letter *s* or by the Greek lowercase letter sigma, σ. The *s* is used when the standard deviation of a *sample* is calculated. The σ is used when the standard deviation of the entire *population* is calculated. *Since we are assuming that all data presented in this section are for samples, we use s to represent the standard deviation* (note, however, that on the height and weight charts on page 790, σ is used).

The larger the variability of the data about the mean, the larger the standard deviation is. Consider the following two sets of data.

$$5, 8, 9, 10, 12, 13 \qquad 8, 9, 9, 10, 10, 11$$

Both have a mean of 9.5. Which set of values on the whole do you believe differs less from the mean of 9.5? Figure 12.16, which shows a visual picture of the two sets of data, may make the answer more apparent. The scores in the second set of data are closer to the mean and therefore have a smaller standard deviation. You will soon be able to verify such relationships yourself.

Sometimes only a very small standard deviation is desirable or acceptable. Consider a cereal box that is to contain 8 oz of cereal. If the amount of cereal put into the boxes varies too much—sometimes underfilling, sometimes overfilling—the manufacturer will soon be in trouble with consumer groups and government agencies.

At other times, a larger spread of data is desirable or expected. For example, intelligence quotients (IQs) are expected to exhibit a considerable spread about the mean because everyone is different. The following procedure explains how we determine the standard deviation of a set of data.

PROCEDURE DETERMINING THE STANDARD DEVIATION OF A SET OF DATA

1. Determine the mean of the set of data.
2. Make a chart having three columns:

   Data   Data − Mean $(\text{Data} - \text{Mean})^2$

3. List the data vertically under the column marked Data.
4. Subtract the mean from each piece of data and place the difference in the Data − Mean column.
5. Square the values obtained in the Data − Mean column and record these values in the $(\text{Data} - \text{Mean})^2$ column.
6. Determine the sum of the values in the $(\text{Data} - \text{Mean})^2$ column.
7. Divide the sum obtained in Step 6 by $n - 1$, where n is the number of pieces of data.*
8. Determine the square root of the number obtained in Step 7. This number is the standard deviation of the set of data.

Example 2 illustrates the procedure to follow to determine the standard deviation of a set of data.

Example 2 *Determine the Standard Deviation*

A veterinarian in an animal hospital recorded the following life spans of selected Labrador retrievers (to the nearest year):

$$7, 9, 11, 15, 18, 12$$

Determine the standard deviation of the life spans.

---

*To determine the standard deviation of a sample, divide the sum of $(\text{Data} - \text{Mean})^2$ column by $n - 1$. To determine the standard deviation of a population, divide the sum by n. In this book, we assume that the set of data represents a sample and divide by $n - 1$. The quotient obtained in Step 7 represents a measure of dispersion called the *variance*.

Solution First determine the mean:

$$\bar{x} = \frac{\Sigma x}{n} = \frac{7 + 9 + 11 + 15 + 18 + 12}{6} = \frac{72}{6} = 12$$

Next construct a table with three columns, as illustrated in Table 12.5, and list the data in the first column (it is often helpful to list the data in ascending or descending order). Complete the second column by subtracting the mean, 12 in this case, from each piece of data in the first column.

The sum of the values in the Data − Mean column should always be zero; if not, you have made an error. (If a rounded value of $\bar{x}$ is used, the sum of the values in the Data − Mean column will not always be exactly zero; however, the sum will be very close to zero.)

Table 12.5

Data	Data − Mean	(Data − Mean)^2
7	7 − 12 = −5	
9	9 − 12 = −3	
11	11 − 12 = −1	
12	12 − 12 = 0	
15	15 − 12 = 3	
18	18 − 12 = 6	
	0	

Next square the values in the second column and place the squares in the third column (Table 12.6).

Table 12.6

Data	Data − Mean	(Data − Mean)^2
7	−5	$(-5)^2 = (-5)(-5) = 25$
9	−3	$(-3)^2 = (-3)(-3) = 9$
11	−1	$(-1)^2 = (-1)(-1) = 1$
12	0	$(0)^2 = (0)(0) = 0$
15	3	$(3)^2 = (3)(3) = 9$
18	6	$(6)^2 = (6)(6) = 36$
	0	80

Add the squares in the third column. In this case, the sum is 80. Divide this sum by one less than the number of pieces of data $(n - 1)$. In this case, the number of pieces of data is 6. Therefore, we divide by 5 and get

$$\frac{80}{5} = 16^*$$

Finally, take the square root of this number. Since $\sqrt{16} = 4$, the standard deviation, symbolized s, is 4. ∎

Now we will develop a formula for determining the standard deviation of a set of data. If we call the individual data x and the mean $\bar{x}$, we could write the three column heads Data, Data − Mean, and (Data − Mean)^2 in Table 12.5 as

$$x \qquad x - \bar{x} \qquad (x - \bar{x})^2$$

---

*16 is the variance, symbolized s^2, of this set of data.

Let's follow the procedure we used to obtain the standard deviation in Example 2. We determined the sum of the $(\text{Data} - \text{Mean})^2$ column, which is the same as the sum of the $(x - \bar{x})^2$ column. We can represent the sum of the $(x - \bar{x})^2$ column by using the summation notation, $\Sigma(x - \bar{x})^2$. Thus, in Table 12.6, $\Sigma(x - \bar{x})^2 = 80$. We then divided this number by 1 less than the number of pieces of data, $n - 1$. Therefore, we have

$$\frac{\Sigma(x - \bar{x})^2}{n - 1}$$

Finally, we took the square root of this value to obtain the standard deviation.

> **Standard Deviation**
> The standard deviation, s, of a set of data can be calculated using the following formula.
>
> $$s = \sqrt{\frac{\Sigma(x - \bar{x})^2}{n - 1}}$$

Example 3 *Determine the Standard Deviation of Stock Prices*

The following are the prices of nine stocks on the New York Stock Exchange. Determine the standard deviation of the prices.

$$\$17, \$28, \$32, \$36, \$50, \$52, \$66, \$74, \$104$$

Solution The mean, $\bar{x}$, is

$$\bar{x} = \frac{\Sigma x}{n} = \frac{17 + 28 + 32 + 36 + 50 + 52 + 66 + 74 + 104}{9} = \frac{459}{9} = 51$$

The mean is $51.

Data Sets:
Mean and Standard Deviation

Table 12.7

x	$x - \bar{x}$	$(x - \bar{x})^2$
17	−34	1156
28	−23	529
32	−19	361
36	−15	225
50	−1	1
52	1	1
66	15	225
74	23	529
104	53	2809
	0	5836

Table 12.7 shows us that $\Sigma(x - \bar{x})^2 = 5836$. Since there are nine pieces of data, $n - 1 = 9 - 1$, or 8.

$$s = \sqrt{\frac{\Sigma(x - \bar{x})^2}{n - 1}} = \sqrt{\frac{5836}{8}} = \sqrt{729.5} \approx 27.01$$

The standard deviation, to the nearest hundredth, is $27.01.

Standard deviation will be used in Section 12.5 to determine the percent of data between any two values in a normal curve. Standard deviations are also often used in determining norms for a population (see Exercise 29).

TECHNOLOGY TIP

In this *Technology Tip*, we will explain how to determine the standard deviation using Excel the Texas Instrument TI-84 Plus graphing calculator, as well as StatCrunch. In our illustration, we will use the data from Example 3 on page 787, which represent the prices of nine stocks on the New York Stock Exchange.

Excel
The instructions used to determine the standard deviation are very similar to those used to determine the mean in the *Technology Tip* on page 777 in Section 12.3. Please read that material now. Then enter the nine pieces of data in columns A1–A9 and press the ENTER key. Now select the following:

Insert > Function... > Statistical > STDEV

Then click the OK box. The program will then generate a gray box where you need to enter the data. In the area to the right of **Number1** you need to enter the data for which you want to determine the standard deviation. Since you have already entered the data in column A, rows 1 to 9, if A1:A9 is not already listed, you can enter A1:A9 in the area to the right of **Number1**. At the bottom of the gray box, *Formula Results* = 27.00925767, which is the standard deviation, is displayed. If you click OK, Excel will place the standard deviation in cell A10.

TI-84 Plus Graphing Calculators
To determine the standard deviation on Texas Instruments graphing calculators, follow the instructions for determining the mean in the *Technology Tip* on page 777 in Section 12.3. As explained there, press STAT > EDIT > ENTER. Remove existing data by highlighting **L1** and then pressing CLEAR > ENTER. Then enter the nine pieces of data, pressing the ENTER key after each entry. Then press STAT > CALC > ENTER > ENTER. The fourth statistic down is $S_x = 27.00925767$. This value is the standard deviation.

StatCrunch (available in Mylab Math and separately)
To compute this standard deviation, follow the steps in the *Technology Tip* on page 778 except make sure that standard deviation is selected in Step 4. You will see the standard deviation is 27.009.

SECTION 12.4 *Exercises*

Warm Up Exercises

In Exercises 1–6, fill in the blank with an appropriate word, phrase, or symbol(s).

1. Measures of dispersion are used to indicate the spread or _____ of the data.

2. The difference between the highest and lowest values in a set of data is called the _____.

3. The measure of dispersion that measures how much the data differ from the mean is called the _____.

4. The symbol σ is used to indicate the standard deviation of a(n) _____.

5. The symbol s is used to indicate the standard deviation of a(n) _____.

6. The standard deviation of a set of data in which all the data values are the same is _____.

Many of the exercises in this section are designed so that they can be done by hand or by using an appropriate form of technology (graphing calculator, spreadsheet, or StatCrunch).

Practice the Skills

In Exercises 7–14, determine the range and standard deviation of the set of data. When appropriate, round standard deviations to the nearest hundredth.

7. 7, 5, 2, 8, 13

8. 12, 12, 16, 18, 10, 10

9. 150, 151, 152, 153, 154, 155, 156

10. 12, 16, 17, 21, 9, 18, 20, 21, 15, 11

11. 4, 8, 9, 11, 13, 15

12. 9, 9, 9, 9, 9, 9, 9

13. 7, 9, 7, 9, 9, 10, 12

14. 60, 58, 62, 67, 48, 51, 72, 70

Problem Solving

In Exercises 15–20, where appropriate, round standard deviations to the nearest hundredth.

DS 15. *Wireless Headphones* Determine the range and standard deviation of the following prices of selected wireless headphones: $90, $30, $50, $40, $85, $60, $90, $75.

DS 16. *Sugar in Cereal* Determine the range and standard deviation of the following number of grams of sugar in selected cereals: 13, 6, 15, 2, 14, 6, 20, 0, 14.

DS 17. *Camping Tents* Determine the range and standard deviation of the following prices of selected camping tents: $109, $60, $80, $60, $210, $250, $60, $100, $115.

DS 18. *Prescription Prices* The amount of money seven people spent on prescription medication in a year are as follows: $600, $100, $850, $350, $250, $140, $300. Determine the range and standard deviation of the amounts.

19. *Count Your Money* Six people were asked to determine the amount of money they were carrying, to the nearest dollar. The results were

$$\$32, \$60, \$14, \$25, \$5, \$68$$

a) Determine the range and standard deviation of the amounts.

b) Add $10 to each of the six amounts. How do you expect the range and standard deviation of the new set of data to change?

c) Determine the range and standard deviation of the new set of data. Do the results agree with your answer to part (b)? If not, explain why.

20. a) *Five Numbers* Pick any five numbers. Compute the mean and the standard deviation of this set of data.

b) Add 20 to each of the numbers in your original set of data and compute the mean and the standard deviation of this new set of data.

c) Subtract 5 from each number in your original set of data and compute the mean and standard deviation of this new set of data.

d) What conclusions can you draw about changes in the mean and the standard deviation when the same number is added to or subtracted from each piece of data in a set of data?

e) How will the mean and standard deviation of the numbers 8, 9, 10, 11, 12, 13, 14 differ from the mean and standard deviation of the numbers 648, 649, 650, 651, 652, 653, 654? Determine the mean and standard deviation of both sets of numbers.

21. a) *Multiplying Each Number* Pick any five numbers. Compute the mean and standard deviation of this set of data.

b) Multiply each number in your set of data by 3 and compute the mean and the standard deviation of this new set of data.

c) Multiply each number in your original set of data by 9 and compute the mean and the standard deviation of this new set of data.

d) What conclusions can you draw about changes in the mean and the standard deviation when each value in a set of data is multiplied by the same number?

e) The mean and standard deviation of the set of data 1, 3, 4, 4, 5, 7 are 4 and 2, respectively. Use the conclusion drawn in part (d) to determine the mean and standard deviation of the set of data

$$5, 15, 20, 20, 25, 35$$

22. *Waiting in Line* Consider the illustrations below and on page 790 of two bank-customer waiting systems.

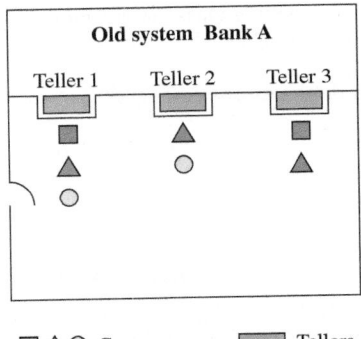

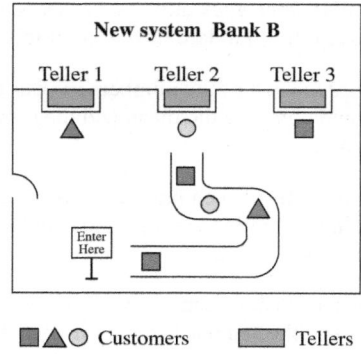

New system Bank B

Teller 1   Teller 2   Teller 3

Enter
Here

■▲◯ Customers   Tellers

a) How would you expect the mean waiting time in Bank A to compare with the mean waiting time in Bank B?

b) How would you expect the standard deviation of waiting times in Bank A to compare with the standard deviation of waiting times in Bank B?

Concept/Writing Exercises

23. Can you think of any situations in which a large standard deviation may be desirable?

24. Can you think of any situations in which a small standard deviation may be desirable?

25. Without actually doing the calculations, decide which, if either, of the following two sets of data will have the greater standard deviation. Explain why.

    15, 18, 19, 20, 22, 26   21, 22, 22, 23, 23, 24

26. Without actually doing the calculations, decide which, if either, of the following two sets of data will have the greater standard deviation. Explain why.

    2, 4, 6, 8, 10   102, 104, 106, 108, 110

27. By studying the standard deviation formula, explain why the standard deviation of a set of data will always be greater than or equal to 0.

28. Patricia teaches two statistics classes, one in the morning and the other in the evening. On the midterm exam, the morning class had a mean of 75.2 and a standard deviation of 5.7. The evening class had a mean of 75.2 and a standard deviation of 12.5.

    a) How do the means compare?

    b) If we compare the set of scores from the first class with those in the second class, how will the distributions of the two sets of scores compare?

Challenge Problems/Group Activities

29. *Height and Weight Distribution* The chart shown above and to the right uses the symbol σ to represent the standard deviation. Note that 2σ represents the value that is two standard deviations above the mean; -2σ represents

the value that is two standard deviations below the mean. The unshaded areas, from two standard deviations below the mean to two standard deviations above the mean, are considered the normal range. For example, the average (mean) 8-year-old boy has a height of about 50 inches, but any heights between approximately 45 inches and 55 inches are considered normal for 8-year-old boys. Refer to the chart below to answer the following questions.

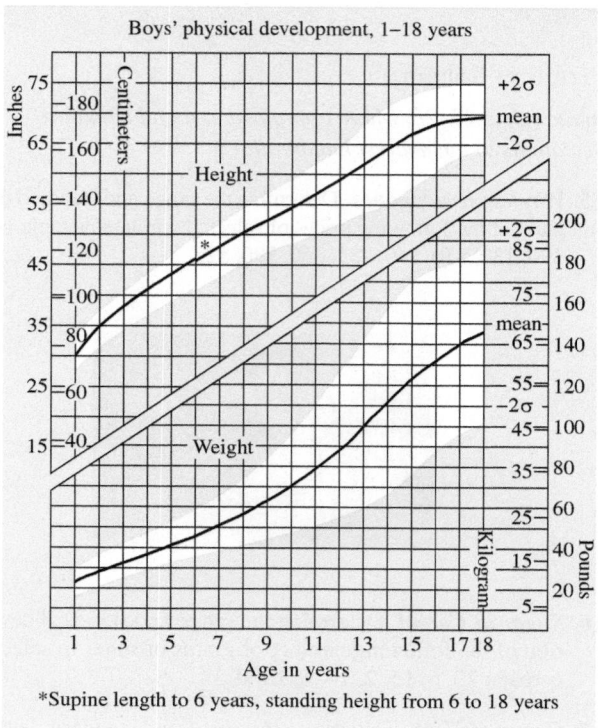

Boys' physical development, 1–18 years

*Supine length to 6 years, standing height from 6 to 18 years

a) What happens to the standard deviation for weights of boys as the age of boys increases? What is the significance of this fact?

b) Determine the mean weight and normal range for boys at age 13.

c) Determine the mean height and normal range for boys at age 13.

d) At age 17, what is the mean weight, in pounds, of boys?

e) What is the approximate standard deviation of boys' weights at age 17?

f) Assuming that this chart was constructed so that approximately 95% of all boys are always in the normal range, determine what percentage of boys are not in the normal range.

30. *Athletes' Salaries* The tables on page 791 list the 10 highest-paid athletes in Major League Baseball and in the National Football League.

Major League Baseball (2018 Season)

Player	Salary (millions of dollars)
1. Stephen Strasburg	38.3
2. Max Scherzer	37.4
3. Zack Greinke	34.5
4. Mike Trout	34.1
5. David Price	31.0
6. Clayton Kershaw	31.0
7. Miguel Cabrera	30.0
8. Yeonis Cespedes	29.0
9. Justin Verlander	28.0
10. Albert Pujols	28.0

Source: Major League Baseball

National Football League (2019–2020 Season)

Player	Salary (millions of dollars)
1. Russell Wilson	35.0
2. Ben Roethlisberger	34.0
3. Aaron Rodgers	33.5
4. Matt Ryan	30.0
5. Kirk Cousins	28.0
6. Jimmy Garoppolo	27.5
7. Matthew Stafford	27.0
8. Derek Carr	25.0
9. Drew Brees	25.0
10. Andrew Luck	24.6

Source: National Football League

a) Without doing any calculations, which do you believe is greater, the mean salary of the 10 baseball players or the mean salary of the 10 football players?

b) Without doing any calculations, which do you believe is greater, the standard deviation of the salary of the 10 baseball players or the standard deviation of the salary of the 10 football players?

c) Compute the mean salary of the 10 baseball players and the mean salary of the 10 football players and determine whether your answer in part (a) was correct.

d) Compute the standard deviation of the salary of the 10 baseball players and the standard deviation of the salary of the 10 football players and determine whether your answer in part (b) is correct. Round each mean to the nearest tenth to determine the standard deviation.

31. *Oil Change* Jiffy Lube has franchises in two different parts of a city. The number of oil changes made daily, for 25 days, is given below.

East Store

33	59	27	30	42
19	42	25	22	32
43	27	57	37	52
40	67	38	44	43
15	31	49	41	35

West Store

38	46	38	38	30
38	38	37	39	31
39	36	40	37	47
30	34	42	45	29
31	46	28	45	48

a) Construct a frequency distribution for each store with a first class of 15–20.

b) Draw a histogram indicating the number of oil changes at each store.

c) Using the histograms, determine which store appears to have a greater mean, or do the means appear about the same? Explain.

d) Using the histogram, determine which store appears to have the greater standard deviation. Explain.

e) Calculate the mean for each store and determine whether your answer in part (c) was correct.

f) Calculate the standard deviation for each store and determine whether your answer in part (d) was correct.

Recreational Mathematics

32. *Measures of Dispersion* Calculate the range and standard deviation of your exam grades in this mathematics course. Round the mean to the nearest tenth to calculate the standard deviation.

33. *Measures of Central Tendency* Construct a set of five pieces of data with a mean, median, mode, and midrange of 6 and a standard deviation of 0.

Research Activity

34. *Salaries* Using the Internet, determine the current 10 highest-paid CEOs in each of the following: Tech companies, Healthcare companies, and Retailers. Determine the mean and the standard deviation of each group.

SECTION 12.5 The Normal Curve

Upon completion of this section, you will be able to:

- Understand properties of a normal distribution.
- Calculate a z-score and use it to determine the area under a normal curve.
- Calculate the percentage of data between any two values in a normal distribution.

Suppose your mathematics teacher states that exam scores for the previous exam followed a bell-shaped distribution and that your score was 1.5 standard deviations above the mean. How does your exam grade compare with the exam grades of your classmates? What percentage of students in your class had an exam grade below your exam grade? In this section, we will discuss sets of data whose histograms approximate bell-shaped distributions and learn how to determine the percentage of data that fall below a particular piece of data in the set of data.

Why This Is Important There are many real-life applications, such as IQ scores, heights and weights of males, heights and weights of females, and wearout mileage of automobile tires, whose histograms approximate a bell-shaped distribution. Understanding the properties of bell-shaped distributions will help us determine the percentage of data that fall in certain intervals of the distributions of these many real-life applications.

When examining data using a histogram, we can refer to the overall appearance of the histogram as the *shape* of the distribution of the data. Certain shapes of distributions of data are more common than others. In this section, we will illustrate and discuss a few of the more common ones. In each case, the vertical scale is the frequency and the horizontal scale is the observed values.

In a *rectangular distribution* (Fig. 12.17), all the observed values occur with about the same frequency. If a die is rolled many times, we would expect the numbers 1–6 to occur with about the same frequency. The distribution representing the outcomes of the die is rectangular.

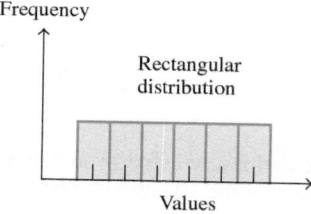

Figure 12.17

In *J-shaped distributions*, the frequency is either constantly increasing (Fig. 12.18(a)) or constantly decreasing (Fig. 12.18(b)). The number of hours studied

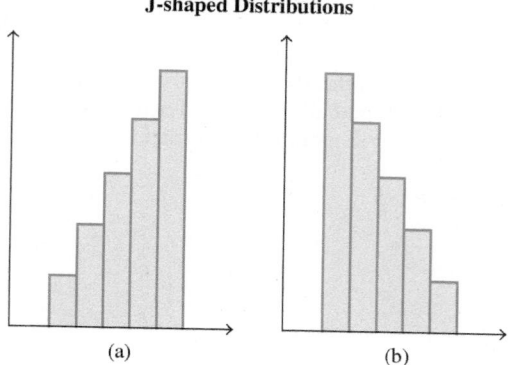

Figure 12.18

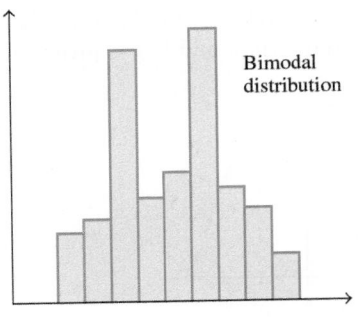

Bimodal
distribution

Figure 12.19

per week by students may have a distribution like that in Fig. 12.18(b). The bars might represent (from left to right) 0–5 hours, 6–10 hours, 11–15 hours, and so on.

A *bimodal distribution* (Fig. 12.19) is one in which two nonadjacent values occur more frequently than any other values in a set of data. For example, if an equal number of men and women were weighed, the distribution of their weights would probably be bimodal, with one mode for the women's weights and the second for the men's weights. For a distribution to be considered bimodal, both modes need not have the same frequency but they must both have a frequency greater than the frequency of each of the other values in the distribution.

The life expectancy of lightbulbs has a bimodal distribution: a small peak very near 0 hours of life, resulting from the bulbs that burned out very quickly because of a manufacturing defect, and a much higher peak representing the nondefective bulbs. A bimodal frequency distribution generally means that you are dealing with two distinct populations, in this case, defective and nondefective lightbulbs.

Another distribution, called a *skewed distribution*, has more of a "tail" on one side than the other. A skewed distribution with a tail on the right (Fig. 12.20(a)) is said to be skewed to the right. If the tail is on the left (Fig. 12.20(b)), the distribution is referred to as skewed to the left.

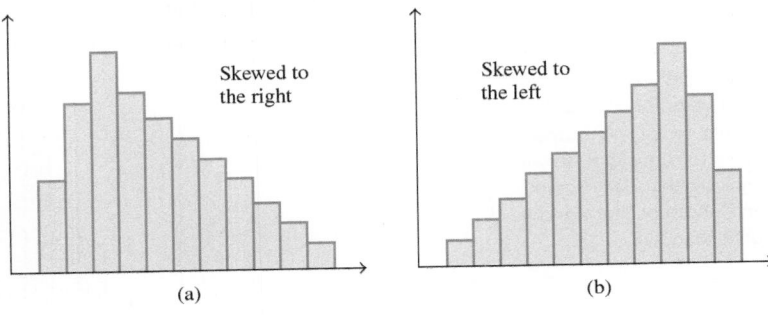

Skewed to
the right

Skewed to
the left

(a)   (b)

Figure 12.20

The number of children per family might be a distribution skewed to the right. Some families have no children, more families may have one child, the greatest percentage may have two children, fewer may have three children, still fewer may have four children, and so on.

Since few families have high incomes, distributions of family incomes might also be skewed to the right.

Smoothing the histograms of the skewed distributions shown in Fig. 12.20 to form curves gives the curves illustrated in Fig. 12.21.

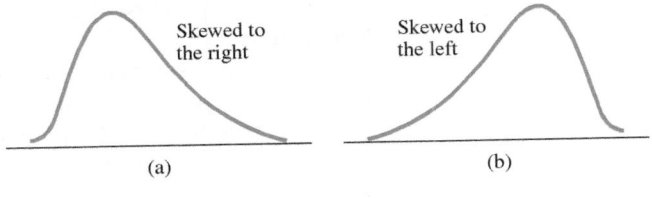

Skewed to
the right

Skewed to
the left

(a)   (b)

Figure 12.21

In Fig. 12.21(a), the greatest frequency appears on the left side of the curve and the frequency decreases from left to right. Since the mode is the value with the greatest frequency, the mode would appear on the left side of the curve.

Every value in the set of data is considered in determining the mean. The values on the far right side of the curve in Fig. 12.21(a) would tend to increase the value of the mean. Thus, the value of the mean would be farther to the right than the mode.

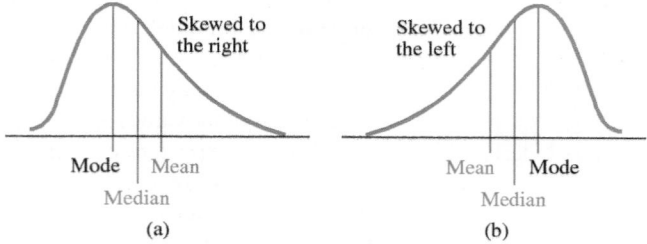
The median would be between the mode and the mean. The relationship between the mean, median, and mode for curves that are skewed to the right and left is given in Fig. 12.22.

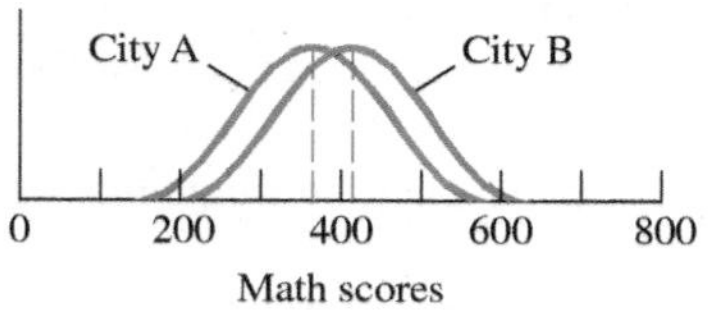

Skewed to the right

Skewed to the left

Mode   Mean

Mean   Mode

Median

Median

(a)

(b)

Figure 12.22

Normal Distributions

Each of the distributions previously mentioned is useful in describing sets of data. However, the most important distribution is the *normal* or *Gaussian distribution*, named for German mathematician Carl Friedrich Gauss. (See the Profile in Mathematics on page 269). The histogram of a normal distribution is illustrated in Fig. 12.23.

Normal distribution

Figure 12.23

The normal distribution is important because many sets of data are normally distributed or closely resemble a normal distribution. Such distributions include intelligence quotients, heights and weights of males, heights and weights of females, lengths of full-grown boa constrictors, weights of watermelons, wearout mileage of automobile brakes, and life spans of refrigerators, to name just a few.

The normal distribution is symmetric about the mean. If you were to fold the histogram of a normal distribution down the middle, the left side would fit the right side exactly. **In a normal distribution, the mean, median, and mode all have the same value.**

When the histogram of a normal distribution is smoothed to form a curve, the curve is bell-shaped. The bell may be high and narrow or short and wide. Each of the three curves in Fig. 12.24 represents a normal curve. Curve 12.24(a) has the smallest standard deviation (spread from the mean); curve 12.24(c) has the largest.

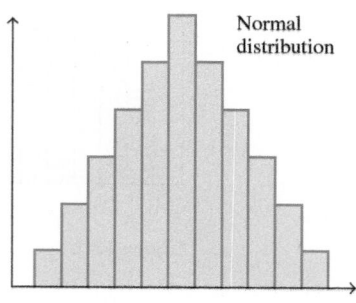

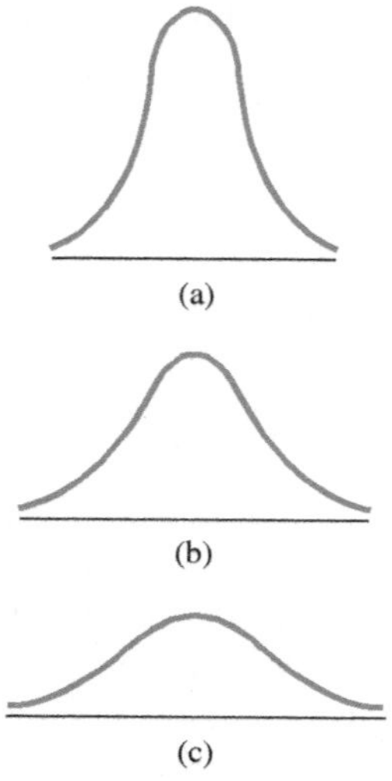

(a)

(b)

(c)

Figure 12.24

A Normal Distribution
- The graph of a normal distribution is called a normal curve.
- The normal curve is bell-shaped and symmetric about the mean.
- The mean, median, and mode of a normal distribution all have the same value and all occur at the center of the distribution.

Since the curve representing the normal distribution is symmetric, 50% of the data always fall above (to the right of) the mean and 50% of the data fall below (to the left of) the mean. In addition, every normal distribution has approximately 68% of the data between the value that is one standard deviation below the mean, and the value that is one standard deviation above the mean; see Fig. 12.25. Approximately 95% of the data fall between the value that is two standard deviations below the mean and the value that is two standard deviations above the mean. Approximately 99.7% of the data fall between the value that is three standard deviations below the mean and the value that is three standard deviations above the mean. These three percentages, 68%, 95%, and 99.7%, are used in what is referred to as the *empirical rule*.

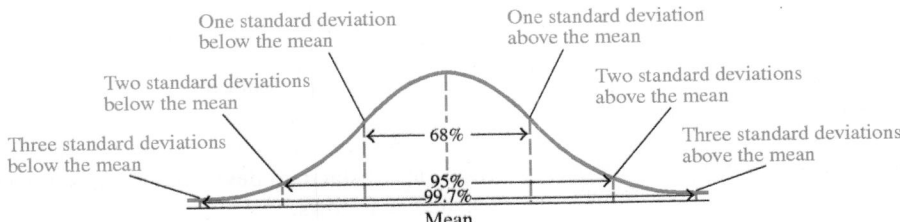

Figure 12.25

Thus, if a normal distribution has a mean of 100 and a standard deviation of 10, then approximately 68% of all the data fall between $100 - 10$ and $100 + 10$, or between 90 and 110. Approximately 95% of all the data fall between $100 - 20$ and $100 + 20$, or between 80 and 120, and approximately 99.7% of all the data fall between $100 - 30$, and $100 + 30$, or between 70 and 130.

The empirical rule is summarized as follows.

> ## Empirical Rule
> In any normal distribution,
> - Approximately 68% of all the data lie within one standard deviation of the mean (in both directions).
> - Approximately 95% of all the data lie within two standard deviations of the mean (in both directions).
> - Approximately 99.7% of all the data lie within three standard deviations of the mean (in both directions).

Example 1

Heights for males that are the same age are normally distributed. In a random sample of 200 males of the same age, determine the approximate number of males in the sample who have a height

a) within one standard deviation of the mean.

b) within two standard deviations of the mean.

Solution

a) By the empirical rule, about 68% of males have a height within one standard deviation of the mean. Since there are 200 males in the sample, the number of males expected to have a height within one standard deviation of the mean is

$$68\% \times 200 = 0.68 \times 200 = 136$$

Therefore, about 136 of the 200 males are expected to have a height within one standard deviation of the mean.

b) By the empirical rule, about 95% of males have a height within two standard deviations of the mean. Since there are 200 males in the sample, the number of males expected to have a height within two standard deviations of the mean is

$$95\% \times 200 = 0.95 \times 200 = 190$$

Therefore, about 190 of the 200 males are expected to have a height within two standard deviations of the mean. ∎

z-Scores

Now we turn our attention to z-scores. We use *z-scores* (or *standard scores*) to determine how far, in terms of standard deviations, a given data value is from the mean of the distribution. For example, a data value that has a z-score of 1.5 indicates the data value is 1.5 standard deviations above the mean. The z-score or standard score is calculated as follows.

> ### z-Scores or Standard Scores
> The formula for determining **z-scores** or standard scores is
> $$z = \frac{\text{value of the piece of data } - \text{ mean}}{\text{standard deviation}}$$

In this book, the notation z_x represents the z-score, or standard score, of the data value x. For example, z_{110} represents the z-score, or standard score, of the data value 110. If a normal distribution has a mean of 86 with a standard deviation of 12, a data value of 110 has a z-score or standard score of

$$z_{110} = \frac{110 - 86}{12} = \frac{24}{12} = 2$$

Therefore, a data value of 110 in this distribution has a z-score of 2 and is two standard deviations above the mean.

Data values below the mean will always have negative z-scores; data values above the mean will always have positive z-scores. The mean will always have a z-score of 0.

Example 2 Determining z-Scores

A normal distribution has a mean of 80 and a standard deviation of 14. Determine z-scores for the following data values.
a) 94   b) 115   c) 80   d) 59

Solution

a)
$$z = \frac{\text{value } - \text{ mean}}{\text{standard deviation}}$$

$$z_{94} = \frac{94 - 80}{14} = \frac{14}{14} = 1$$

A data value of 94 has a z-score of 1. Therefore, a data value of 94 is one standard deviation above the mean.

Exploring z-scores

b)
$$z_{115} = \frac{115 - 80}{14} = \frac{35}{14} = 2.5$$

A data value of 115 has a z-score of 2.5, and is 2.5 standard deviations above the mean.

c)
$$z_{80} = \frac{80 - 80}{14} = \frac{0}{14} = 0$$

A data value of 80 has a z-score of 0. The mean always has a z-score of 0.

d)
$$z_{59} = \frac{59 - 80}{14} = \frac{-21}{14} = -1.5$$

A data value of 59 has a z-score of -1.5 and is 1.5 standard deviations below the mean. ∎

Determine the Percent of Data Between Any Two Data Values in a Normal Distribution

If we are given any normal distribution with a known mean and standard deviation, it is possible through the use of Table 12.8 on pages 798 and 799 (the z-table) to determine the percent of data between any two given values. The total area under any normal curve is 1.00. Table 12.8 will be used to determine the cumulative area under the normal curve that lies to the *left of a specified z-score*. We will use Table 12.8(a) when we wish to determine area to the left of a *negative z-score*, and we will use Table 12.8(b) when we wish to determine area to the left of a *positive z-score*.

Example 3 illustrates the procedure to follow when using Table 12.8 to determine the area under the normal curve. When you are determining the area under the normal curve, it is often helpful to draw a picture and shade the area to be determined.

Example 3 *Determining the Area Under the Normal Curve*

Determine the area under the normal curve

a) to the left of $z = -1.00$.

b) to the left of $z = 1.19$.

c) to the right of $z = 1.19$.

d) between $z = -1.62$ and $z = 2.57$.

Solution

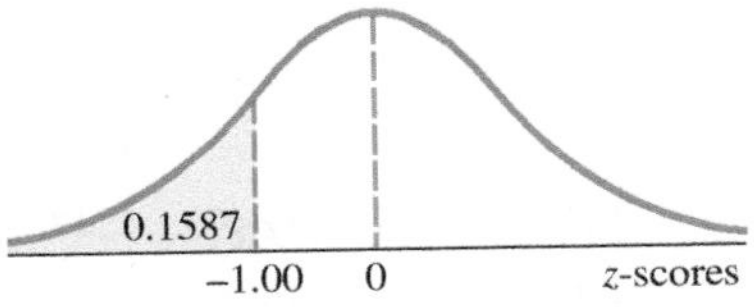

a) To determine the area under the normal curve to the left of $z = -1.00$, as illustrated in Fig. 12.26, we use Table 12.8(a), since we are looking for an area to the left of a negative z-score. In the upper-left corner of the table, we see the letter z. The column under z gives the units and the tenths value for z. To locate the hundredths value of z, we use the column headings to the right of z. In this case, the hundredths value of $z = -1.00$ is 0, so we use the first column labeled .00. To determine the area to the left of $z = -1.00$, we use the row labeled -1.0 and move to the column labeled .00. The table entry, which is .1587, is circled in blue. Therefore, the total area to the left of $z = -1.00$ is 0.1587.

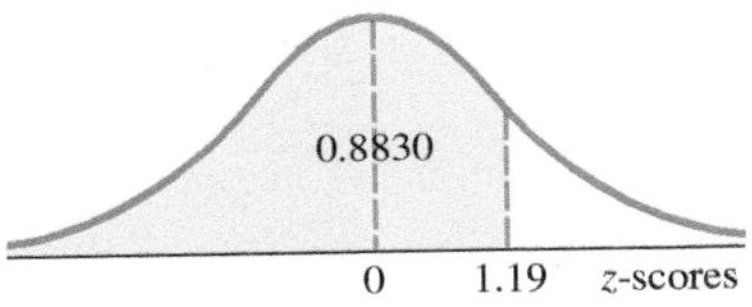

b) To determine the area under the normal curve to the left of $z = 1.19$ (Fig. 12.27), we use Table 12.8(b), since we are looking for an area to the left of a positive z-score. We first look for 1.1 in the column under z. Since the hundredths value of

Figure 12.26

Figure 12.27

(Example 3 continued on page 800)

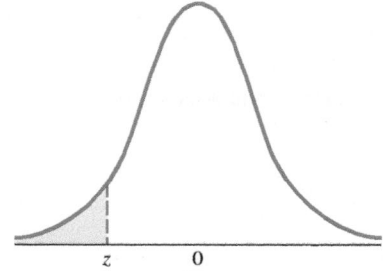

▲ *Table entry for z is the area to the left of z.*

Table 12.8 (a)   Areas of a Standard Normal Distribution

Table of Areas to the Left of z When z Is Negative

z	.00	.01	.02	.03	.04	.05	.06	.07	.08	.09
−3.4	.0003	.0003	.0003	.0003	.0003	.0003	.0003	.0003	.0003	.0002
−3.3	.0005	.0005	.0005	.0004	.0004	.0004	.0004	.0004	.0004	.0003
−3.2	.0007	.0007	.0006	.0006	.0006	.0006	.0006	.0005	.0005	.0005
−3.1	.0010	.0009	.0009	.0009	.0008	.0008	.0008	.0008	.0007	.0007
−3.0	.0013	.0013	.0013	.0012	.0012	.0011	.0011	.0011	.0010	.0010
−2.9	.0019	.0018	.0018	.0017	.0016	.0016	.0015	.0015	.0014	.0014
−2.8	.0026	.0025	.0024	.0023	.0023	.0022	.0021	.0021	.0020	.0019
−2.7	.0035	.0034	.0033	.0032	.0031	.0030	.0029	.0028	.0027	.0026
−2.6	.0047	.0045	.0044	.0043	.0041	.0040	.0039	.0038	.0037	.0036
−2.5	.0062	.0060	.0059	.0057	.0055	.0054	.0052	.0051	.0049	.0048
−2.4	.0082	.0080	.0078	.0075	.0073	.0071	.0069	.0068	.0066	.0064
−2.3	.0107	.0104	.0102	.0099	.0096	.0094	.0091	.0089	.0087	.0084
−2.2	.0139	.0136	.0132	.0129	.0125	.0122	.0119	.0116	.0113	.0110
−2.1	.0179	.0174	.0170	.0166	.0162	.0158	.0154	.0150	.0146	.0143
−2.0	.0228	.0222	.0217	.0212	.0207	.0202	.0197	.0192	.0188	.0183
−1.9	.0287	.0281	.0274	.0268	.0262	.0256	.0250	.0244	.0239	.0233
−1.8	.0359	.0351	.0344	.0336	.0329	.0322	.0314	.0307	.0301	.0294
−1.7	.0446	.0436	.0427	.0418	.0409	.0401	.0392	.0384	.0375	.0367
−1.6	.0548	.0537	.0526	.0516	.0505	.0495	.0485	.0475	.0465	.0455
−1.5	.0668	.0655	.0643	.0630	.0618	.0606	.0594	.0582	.0571	.0559
−1.4	.0808	.0793	.0778	.0764	.0749	.0735	.0721	.0708	.0694	.0681
−1.3	.0968	.0951	.0934	.0918	.0901	.0885	.0869	.0853	.0838	.0823
−1.2	.1151	.1131	.1112	.1093	.1075	.1056	.1038	.1020	.1003	.0985
−1.1	.1357	.1335	.1314	.1292	.1271	.1251	.1230	.1210	.1190	(.1170)
−1.0	(.1587)	.1562	.1539	.1515	.1492	.1469	.1446	.1423	.1401	.1379
−0.9	.1841	.1814	.1788	.1762	.1736	.1711	.1685	.1660	.1635	.1611
−0.8	.2119	.2090	.2061	.2033	.2005	.1977	.1949	.1922	.1894	.1867
−0.7	.2420	.2389	.2358	.2327	.2296	.2266	.2236	.2206	.2177	.2148
−0.6	.2743	.2709	.2676	.2643	.2611	.2578	.2546	.2514	.2483	.2451
−0.5	.3085	.3050	.3015	.2981	.2947	.2912	.2877	.2843	.2810	.2776
−0.4	.3446	.3409	.3372	.3336	.3300	.3264	.3228	.3192	.3156	.3121
−0.3	.3821	.3783	.3745	.3707	.3669	.3632	.3594	.3557	.3520	.3483
−0.2	.4207	.4168	.4129	.4090	.4052	.4013	.3974	.3936	.3897	.3859
−0.1	.4602	.4562	.4522	.4483	.4443	.4404	.4364	.4325	.4286	.4247
−0.0	.5000	.4960	.4920	.4880	.4840	.4801	.4761	.4721	.4681	.4641

For z-scores less than −3.49, use 0.000 to approximate the area.

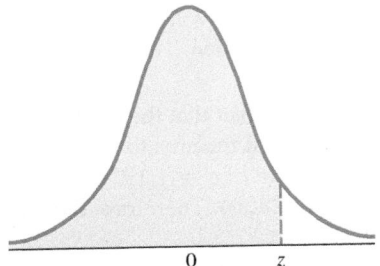

▲ Table entry for z is the area to the left of z.

Table 12.8 (b) Areas of a Standard Normal Distribution (*continued*)

Table of Areas to the Left of z When z Is Positive

z	.00	.01	.02	.03	.04	.05	.06	.07	.08	.09
0.0	.5000	.5040	.5080	.5120	.5160	.5199	.5239	.5279	.5319	.5359
0.1	.5398	.5438	.5478	.5517	.5557	.5596	.5636	.5675	.5714	.5753
0.2	.5793	.5832	.5871	.5910	.5948	.5987	.6026	.6064	.6103	.6141
0.3	.6179	.6217	.6255	.6293	.6331	.6368	.6406	.6443	.6480	.6517
0.4	.6554	.6591	.6628	.6664	.6700	.6736	.6772	.6808	.6844	.6879
0.5	.6915	.6950	.6985	.7019	.7054	.7088	.7123	.7157	.7190	.7224
0.6	.7257	.7291	.7324	.7357	.7389	.7422	.7454	.7486	.7517	.7549
0.7	.7580	.7611	.7642	.7673	.7704	.7734	.7764	.7794	.7823	.7852
0.8	.7881	.7910	.7939	.7967	.7995	.8023	.8051	.8078	.8106	.8133
0.9	.8159	.8186	.8212	.8238	.8264	.8289	.8315	.8340	.8365	.8389
1.0	.8413	.8438	.8461	.8485	.8508	.8531	.8554	.8577	.8599	.8621
1.1	.8643	.8665	.8686	.8708	.8729	.8749	.8770	.8790	.8810	(.8830)
1.2	.8849	.8869	.8888	.8907	.8925	.8944	.8962	.8980	.8997	.9015
1.3	.9032	.9049	.9066	.9082	.9099	.9115	.9131	.9147	.9162	.9177
1.4	.9192	.9207	.9222	.9236	.9251	.9265	.9279	.9292	.9306	.9319
1.5	.9332	.9345	.9357	.9370	.9382	.9394	.9406	.9418	.9429	.9441
1.6	.9452	.9463	.9474	.9484	.9495	.9505	.9515	.9525	.9535	.9545
1.7	.9554	.9564	.9573	.9582	.9591	.9599	.9608	.9616	.9625	.9633
1.8	.9641	.9649	.9656	.9664	.9671	.9678	.9686	.9693	.9699	.9706
1.9	.9713	.9719	.9726	.9732	.9738	.9744	.9750	.9756	.9761	.9767
2.0	.9772	.9778	.9783	.9788	.9793	.9798	.9803	.9808	.9812	.9817
2.1	.9821	.9826	.9830	.9834	.9838	.9842	.9846	.9850	.9854	.9857
2.2	.9861	.9864	.9868	.9871	.9875	.9878	.9881	.9884	.9887	.9890
2.3	.9893	.9896	.9898	.9901	.9904	.9906	.9909	.9911	.9913	.9916
2.4	.9918	.9920	.9922	.9925	.9927	.9929	.9931	.9932	.9934	.9936
2.5	.9938	.9940	.9941	.9943	.9945	.9946	.9948	.9949	.9951	.9952
2.6	.9953	.9955	.9956	.9957	.9959	.9960	.9961	.9962	.9963	.9964
2.7	.9965	.9966	.9967	.9968	.9969	.9970	.9971	.9972	.9973	.9974
2.8	.9974	.9975	.9976	.9977	.9977	.9978	.9979	.9979	.9980	.9981
2.9	.9981	.9982	.9982	.9983	.9984	.9984	.9985	.9985	.9986	.9986
3.0	.9987	.9987	.9987	.9988	.9988	.9989	.9989	.9989	.9990	.9990
3.1	.9990	.9991	.9991	.9991	.9992	.9992	.9992	.9992	.9993	.9993
3.2	.9993	.9993	.9994	.9994	.9994	.9994	.9994	.9995	.9995	.9995
3.3	.9995	.9995	.9995	.9996	.9996	.9996	.9996	.9996	.9996	.9997
3.4	.9997	.9997	.9997	.9997	.9997	.9997	.9997	.9997	.9997	.9998

For z-scores greater than 3.49, use 1.000 to approximate the area.

$z = 1.19$ is 9, we move to the column labeled .09. Using the row labeled 1.1 and the column labeled 0.09, the table entry is .8830, circled in green. Therefore, the total area to the left of $z = 1.19$ is 0.8830.

c) To determine the area to the right of $z = 1.19$, we use the fact that the total area under the normal curve is 1. In part (b), we determined that the area to the left of $z = 1.19$ was 0.8830. To determine the area to the right of $z = 1.19$, we can subtract the area to the left of $z = 1.19$ from 1 (Fig. 12.28(a)). Therefore, the area to the right of $z = 1.19$ is $1 - 0.8830$, or 0.1170.

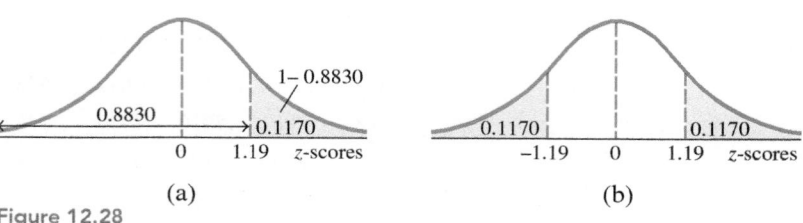

Figure 12.28

(a)   (b)

Another way to determine the area to the right of $z = 1.19$ is to use the fact that the normal curve is symmetric about the mean. Therefore, the area to the left of $z = -1.19$ is equal to the area to the right of $z = 1.19$ (Fig. 12.28(b)). Using Table 12.8(a), we see that the area to the left of $z = -1.19$ is .1170. This value is circled in red in the table. Therefore, the area to the right of $z = 1.19$ is also .1170. This answer agrees with our answer obtained by subtracting the area to the left of $z = 1.19$ from 1.

d) To determine the area between two z-scores, we subtract the smaller area from the larger area (Fig. 12.29). Using Table 12.8(b), we see that the area to the left of $z = 2.57$ is .9949 (Fig. 12.29(a)). Using Table 12.8(a), we see that the area to the left of $z = -1.62$ is .0526 (Fig. 12.29(b)). Thus, the area between $z = -1.62$ and $z = 2.57$ is $0.9949 - 0.0526$, or 0.9423 (Fig. 12.29(c)).

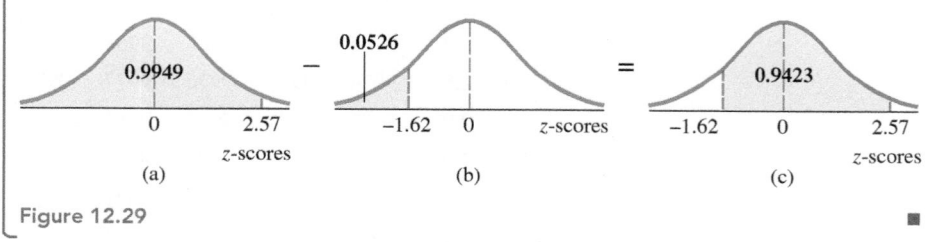

(a)   (b)   (c)

Figure 12.29   ∎

To change the area under the normal curve to a percent, multiply the area by 100%. In Example 3(a), we determined the area to the left of $z = -1.00$ to be 0.1587. To change this area to a percent, multiply 0.1587 by 100%.

$$0.1587 = 0.1587 \times 100\% = 15.87\%$$

Therefore 15.87% of the normal curve is less than a score that is one standard deviation below the mean.

Below, we summarize the procedure to determine the percent of data for any interval under the normal curve.

PROCEDURE DETERMINING THE PERCENT OF DATA BETWEEN ANY TWO VALUES IN A NORMAL DISTRIBUTION

1. Draw a diagram of the normal curve, indicating the area or percent to be determined.

2. Use the formula $z = \dfrac{\text{value of the piece of data} \ - \ \text{mean}}{\text{standard deviation}}$ to convert the given values to z-scores. Indicate these z-scores on the diagram.

3. Look up the areas that correspond to the specified z-scores in Table 12.8.

   a) When determining the area to the left of a negative z-score, use Table 12.8(a).

   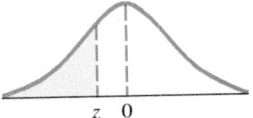

   b) When determining the area to the left of a positive z-score, use Table 12.8(b).

   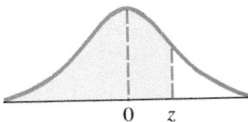

   c) When determining the area to the right of a z-score, subtract the percent of data to the left of the specified z-score from 100%.

   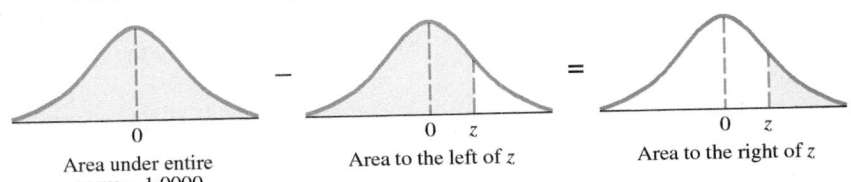

   Area under entire curve = 1.0000   Area to the left of z Area to the right of z

   Or, use the symmetry of a normal distribution.

   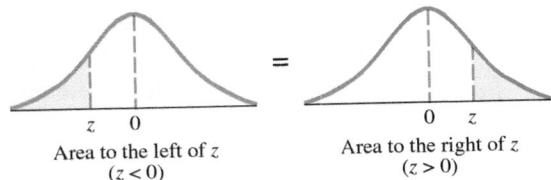

   Area to the left of z ($z < 0$)   Area to the right of z ($z > 0$)

   d) When determining the area between two z-scores, subtract the smaller area from the larger area.
      In the figure below, we let z_1 represent the smaller z-score and z_2 represent the larger z-score.

   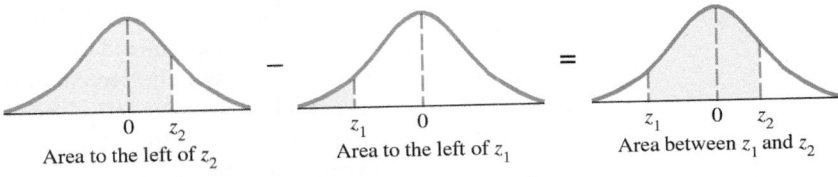

   Area to the left of z_2 Area to the left of z_1 Area between z_1 and z_2

4. Change the areas you determined in Step 3 to percents as explained on page 800.

Example **4** *IQ Scores*

Intelligence quotients (IQ scores) are normally distributed with a mean of 100 and a standard deviation of 15. Determine the percent of individuals with IQ scores

a) below 115.

b) below 130.

c) below 70.

d) between 70 and 115.

e) between 115 and 130.

f) above 122.5.

Solution

a) We want to determine the area under the normal curve below the value of 115, as illustrated in Fig. 12.30(a). Converting 115 to a z-score yields a z-score of 1.00.

$$z_{115} = \frac{115 - 100}{15} = \frac{15}{15} = 1.00$$

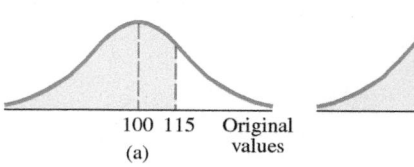 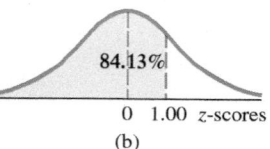

100 115   Original
            values
(a)

84.13%

0 1.00 z-scores
(b)

Figure 12.30

The percent of individuals with IQ scores below 115 is the same as the percent of data below a z-score of 1.00 (Fig. 12.30(b)). Since our z-score is positive, we use Table 12.8(b). From Table 12.8(b), we determine that the area to the left of a z-score of 1.00 is .8413. Therefore, 84.13% of all the IQ scores are below a z-score of 1.00. Thus, 84.13% of individuals have IQ scores below 115.

b) Begin by determining the z-score for 130.

$$z_{130} = \frac{130 - 100}{15} = \frac{30}{15} = 2.00$$

The percent of data below a z-score of 130 is the same as the percent of data below a z-score of 2.00 (Fig. 12.31). Using Table 12.8(b), we determine that the area to the left of a z-score of 2.00 is .9772. Therefore, 97.72% of the IQ scores are below a z-score of 2.00. Thus, 97.72% of all individuals have IQ scores below 130.

c) Begin by determining the z-score for 70.

$$z_{70} = \frac{70 - 100}{15} = \frac{-30}{15} = -2.00$$

The percent of data below a score of 70 is the same as the percent of data below a z-score of -2.00 (Fig. 12.32). Since the z-score is negative, we use Table 12.8(a). Using the table, we determine that the area to the left of $z = -2.00$ is .0228. Therefore, 2.28% of the data are below a z-score of -2.00. Thus, 2.28% of all individuals have IQ scores below 70.

d) In part (a), we determined that $z_{115} = 1.00$, and in part (c), we determined that $z_{70} = -2.00$. The percent of data below a z-score of 1.00 is 84.13% (Fig. 12.33(a) on page 803). The percent of data below a z-score of -2.00 is 2.28% (Fig. 12.33(b)). Since we want to determine the percent of data between two z-scores, we subtract the smaller percent from the larger percent: $84.13\% - 2.28\% = 81.85\%$ (Fig. 12.33(c)). Thus, 81.85% of all individuals have IQ scores between 70 and 115.

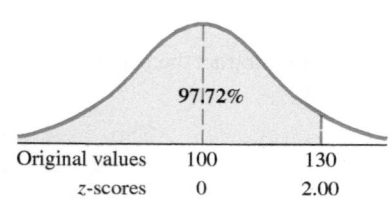

97.72%

Original values   100   130
   z-scores   0   2.00

Figure 12.31

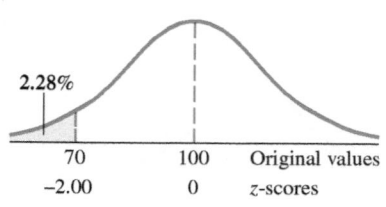

2.28%

70   100   Original values
-2.00   0 z-scores

Figure 12.32

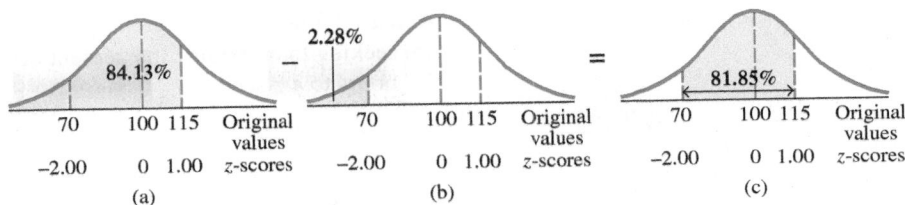

Figure 12.33

e) In part (a), we determined that $z_{115} = 1.00$, and in part (b), we determined $z_{130} = 2.00$. The percent of data below a z-score of 1.00 is 84.13%. The percent of data below a z-score of 2.00 is 97.72%. Since we want to determine the percent of data between two z-scores, we subtract the smaller percent from the larger percent: $97.72\% - 84.13\% = 13.59\%$ (Fig. 12.34). Thus, 13.59% of all individuals have IQ scores between 115 and 130.

f) Begin by determining a z-score for 122.5.

$$z_{122.5} = \frac{122.5 - 100}{15} = \frac{22.5}{15} = 1.50$$

The percent of IQ scores above 122.5 is the same as the percent of data above $z = 1.50$ (Fig. 12.35). To determine the percent of data above $z = 1.50$, we can determine the percent of data below $z = 1.50$ and subtract this percent from 100%. In Table 12.8(b), we see that the area to the left of $z = 1.50$ is .9332. Therefore, 93.32% of the IQ scores are below $z = 1.50$. The percent of IQ scores above $z = 1.50$ are $100\% - 93.32\%$, or 6.68%. Thus, 6.68% of all IQ scores are greater than 122.5.

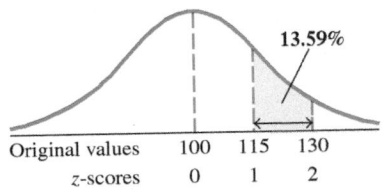

Figure 12.34

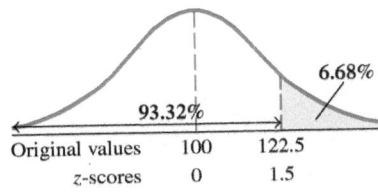

Figure 12.35

Example 5 Horseback Rides

Assume that the length of time for a horseback ride on the trail at Triple R Ranch is normally distributed with a mean of 3.2 hours and a standard deviation of 0.4 hour.

a) What percent of horseback rides last at least 3.2 hours?

b) What percent of horseback rides last less than 2.8 hours?

c) What percent of horseback rides last at least 3.7 hours?

d) What percent of horseback rides last between 2.8 hours and 4.0 hours?

e) In a random sample of 500 horseback rides at Triple R Ranch, how many last at least 3.7 hours?

Solution

a) In a normal distribution, half the data are always above the mean. Since 3.2 hours is the mean, half, or 50%, of the horseback rides last at least 3.2 hours.

b) Convert 2.8 hours to a z-score.

$$z_{2.8} = \frac{2.8 - 3.2}{0.4} = -1.00$$

Use Table 12.8(a) to determine the area of the normal curve that lies below a z-score of -1.00. The area to the left of $z = -1.00$ is 0.1587. Therefore, the percent of horseback rides that last less than 2.8 hours is 15.87% (Fig. 12.36).

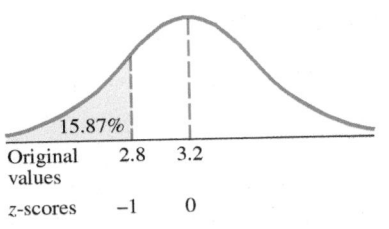

Figure 12.36

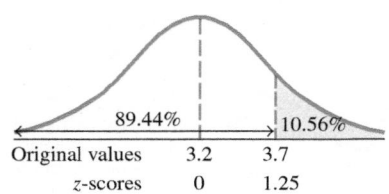

89.44%   10.56%
| Original values | 3.2 | 3.7 |
| z-scores | 0 | 1.25 |

Figure 12.37

Timely Tip

Following is a summary of some important items presented in this section.

- The normal curve is symmetric about the mean.
- A negative z-score indicates that the corresponding value in the original distribution is less than the mean.
- A positive z-score indicates that the corresponding value in the original distribution is greater than the mean.
- A z-score of 0 indicates that the corresponding value in the original distribution is the mean.
- Table 12.8 provides the area to the left of a specified z-score.
- When using Table 12.8 to determine the area to the left of a specified z-score, locate the units value and tenths value of your specified z-score under the column labeled z. Then move to the column containing the hundredths value of your specified z-score to obtain the area.

c) At least 3.7 hours means greater than or equal to 3.7 hours. Therefore, we are seeking to determine the percent of data to the right of 3.7 hours. Convert 3.7 hours to a z-score.

$$z_{3.7} = \frac{3.7 - 3.2}{0.4} = 1.25$$

From Table 12.8(b), we determine that the area to the left of $z = 1.25$ is .8944. Therefore, 89.44% of the data are below $z = 1.25$. The percent of data above $z = 1.25$ (or to the right of $z = 1.25$) is $100\% - 89.44\%$, or 10.56% (Fig. 12.37). Thus, 10.56% of horseback rides last at least 3.7 hours.

d) Convert 4.0 to a z-score.

$$z_{4.0} = \frac{4.0 - 3.2}{0.4} = 2.00$$

From Table 12.8(b), we determine that the area to the left of $z = 2.00$ is .9772 (Fig. 12.38(a)). Therefore the percent of data below a z-score of 2.00 is 97.72%. From part (b), we determined that $z_{2.8} = -1.00$ and that the percent of data below a z-score of -1.00 is 15.87% (Fig. 12.38(b)). To determine the percent of data between a z-score of -1.00 and a z-score of 2.00, we subtract the smaller percent from the larger percent. Thus, the percent of horseback rides that last between 2.8 hours and 4.0 hours is $97.72\% - 15.87\%$, or 81.85% (Fig. 12.38(c)).

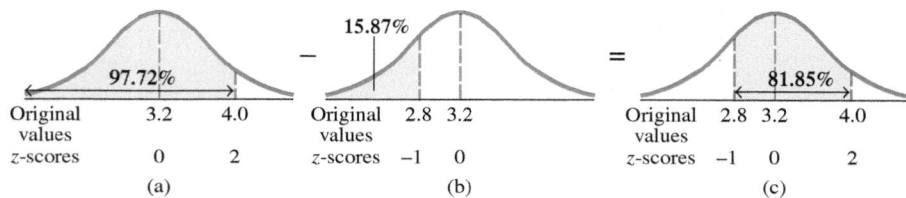

Original values	3.2	4.0
z-scores	0	2
(a)		

97.72%

15.87%

Original values	2.8	3.2
z-scores	-1	0
(b)		

81.85%

Original values	2.8	3.2	4.0
z-scores	-1	0	2
(c)			

Figure 12.38

e) In part (c), we determined that 10.56% of all horseback rides last at least 3.7 hours. We now multiply 0.1056 times the number in the random sample, 500, to determine the number of horseback rides that last at least 3.7 hours. There are $0.1056 \times 500 = 52.8$, or approximately 53, horseback rides that last at least 3.7 hours. ∎

SECTION 12.5   Exercises

Warm Up Exercises

In Exercises 1–10, fill in the blank with an appropriate word, phrase, or symbol(s).

1. A distribution in which all the data values have about the same frequency is called a(n) _____ distribution.

2. A distribution in which the frequency is either constantly increasing or constantly decreasing is called a(n) _____ distribution.

3. a) A distribution that has a "tail" on its right is skewed to the _____.

b) A distribution that has a "tail" on its left is skewed to the _____.

4. A distribution in which two nonadjacent data values occur more frequently than any other values in a set of data is called a(n) _____ distribution.

5. A normal distribution is a(n) _____ shaped distribution.

6. A measure of how far, in terms of standard deviations, a given data value is from the mean is called a z-score or a(n) _____ score.

7. The mean of a set of data will always have a z-score of _____.

8. a) A data value that has a negative z-score is _____ the mean.

b) A data value that has a positive z-score is _____ the mean.

9. According to the empirical rule, in a normal distribution,

a) approximately _____ of the data lie within plus or minus 1 standard deviation of the mean.

b) approximately _____ of the data lie within plus or minus 2 standard deviations of the mean, and

c) approximately _____ of the data lie within plus or minus 3 standard deviations of the mean.

10. In a normal distribution, the mean, median, and mode all have the same _____.

Practice the Skills

In Exercises 11–14, give an example of the type of distribution.

11. Rectangular

12. Skewed

13. J-shaped

14. Bimodal

For the distributions in Exercises 15–22, state whether you think the distribution would be normal, J-shaped, bimodal, rectangular, skewed left, or skewed right.

15. The wingspan of a full-grown red-tailed hawk

16. The numbers resulting from tossing a die many times

17. The number of people per household in the United States

18. The heights of a sample of high school seniors, where there are an equal number of males and females

19. Peak hours at a restaurant that only serves lunch and dinner

20. The wearout mileage of automobile tires

21. A spinner can land on equally sized sectors of a circle labeled 1–4, respectively. The number resulting from spinning the spinner many times

22. The number of children per family

In Exercises 23–34, use Table 12.8 on pages 798 and 799 to determine the specified area.

23. Above the mean

24. Below the mean

25. Between one standard deviation below the mean and two standard deviations above the mean

26. Between 1.30 and 1.70 standard deviations above the mean

27. To the left of $z = 1.62$

28. To the left of $z = 2.37$

29. To the left of $z = -2.19$

30. To the right of $z = -1.72$

31. To the right of $z = -2.25$

32. To the right of $z = 0.62$

33. Between $z = -1.64$ and $z = -1.32$

34. Between $z = 1.23$ and $z = 1.81$

In Exercises 35–44, use Table 12.8 on pages 798 and 799 to determine the percent of data specified.

35. Less than $z = 0.83$

36. Less than $z = -0.45$

37. Greater than $z = -1.90$

38. Greater than $z = 2.66$

39. Between $z = -1.84$ and $z = 1.84$

40. Between $z = -2.29$ and $z = -1.33$

41. Between $z = 1.96$ and $z = 2.14$

42. Between $z = -1.53$ and $z = -1.82$

43. Between $z = 0.72$ and $z = 2.14$

44. Between $z = -2.71$ and $z = 3.21$

Problem Solving

Heights of Girls *In Exercises 45 and 46, on page 806, assume that the heights of 7-year-old girls are normally distributed. The heights of 8 randomly selected 7-year-old girls are given in z-scores in the table on page 806.*

| Luisa | 0.9 | Jenny | 0.0 | Heather | −1.3 | Shenice | 0.0 |
| Sarah | 1.7 | Sadaf | −0.2 | Eleanor | 0.8 | Kim-Liu | −1.2 |

45. a) Which of these girls are taller than the mean?

    b) Which of these girls are at the mean?

    c) Which of these girls are shorter than the mean?

46. a) Which girl is the tallest?

    b) Which girl is the shortest?

In Exercises 47–54, assume the average amount of caffeine consumed daily by adults is normally distributed with a mean of 250 mg and a standard deviation of 48 mg.

47. Determine the percent of adults who consume more than 250 mg of caffeine daily.

48. Determine the percent of adults who consume more than 310 mg of caffeine daily.

49. Determine the percent of adults who consume less than 250 mg of caffeine daily.

50. Determine the percent of adults who consume less than 190 mg of caffeine daily.

51. Determine the percent of adults who consume between 202 mg of caffeine and 346 mg of caffeine daily.

52. Determine the percent of adults who consume between 154 mg of caffeine and 298 mg of caffeine daily.

53. In a random sample of 500 adults, how many consume at least 310 mg of caffeine daily?

54. In a random sample of 500 adults, how many consume between 202 mg of caffeine and 346 mg of caffeine daily?

Placement Exam In Exercises 55–60, assume that the mathematics scores on a placement test are normally distributed with a mean of 500 and a standard deviation of 100.

55. What percent of students who took the test have a mathematics score below 550?

56. What percent of students who took the test have a mathematics score above 680?

57. What percent of students who took the test have a mathematics score between 550 and 650?

58. What percent of students who took the test have a mathematics score between 400 and 575?

59. What percent of students who took the test have a mathematics score below 300?

60. What percent of students who took the test have a mathematics score above 480?

Vending Machine In Exercises 61–64, a vending machine is designed to dispense a mean of 7.6 oz of coffee into an 8-oz cup. If the standard deviation of the amount of coffee dispensed is 0.4 oz and the amount is normally distributed, determine the percent of times the machine will

61. dispense more than 7.0 oz.

62. dispense less than 7.0 oz.

63. dispense less than 7.7 oz.

64. result in the cup overflowing (therefore dispense more than 8 oz).

Automobile Speed In Exercises 65–70, assume that the speed of automobiles on an expressway during rush hour is normally distributed with a mean of 62 mph and a standard deviation of 5 mph.

65. What percent of cars are traveling faster than 62 mph?

66. What percent of cars are traveling between 58 mph and 66 mph?

67. What percent of cars are traveling slower than 56 mph?

68. What percent of cars are traveling faster than 70 mph?

69. If 200 cars are selected at random, how many will be traveling slower than 56 mph?

70. If 200 cars are selected at random, how many will be traveling faster than 70 mph?

In Exercises 71–74, assume that females have pulse rates that are normally distributed with a mean of 74.0 beats per minute and a standard deviation of 12.5 beats per minute.

71. Determine the percent of females with a pulse rate between 70 beats per minute and 78 beats per minute.

72. Determine the percent of females with a pulse rate greater than 88 beats per minute.

73. In a random sample of 500 females, how many will have a pulse rate less than 70 beats per minute?

74. In a random sample of 500 females, how many will have a pulse rate greater than 88 beats per minute?

Cost of Day Care In Exercises 75–80, assume the annual day care cost per child is normally distributed with a mean of $12,000 and a standard deviation of $2000.

75. What percent of day care costs are more than $13,200 per child annually?

76. What percent of day care costs are between $10,000 and $13,000 per child annually?

77. What percent of day care costs are more than $17,000 per child annually?

78. What percent of day care costs are less than $17,000 per child annually?

79. In a random sample of 120 families that use day care, how many pay more than $11,000 annually per child?

80. In a random sample of 120 families that use day care, how many pay between $10,000 and $13,000 annually per child?

81. *Weight Loss* A weight-loss clinic guarantees that its new customers will lose at least 5 lb by the end of their first month of participation or their money will be refunded. If the weight loss of customers at the end of their first month is normally distributed, with a mean of 6.7 lb and a standard deviation of 0.81 lb, determine the percent of customers who will be able to claim a refund.

82. *Battery Warranty* The warranty on a car battery is 36 months. If the breakdown times of this battery are normally distributed with a mean of 46 months and a standard

deviation of 8 months, determine the percent of batteries that can be expected to require repair or replacement under warranty.

83. *Coffee Machine* A vending machine that dispenses coffee does not appear to be working correctly. The machine rarely gives the proper amount of coffee. Some of the time the cup is underfilled, and some of the time the cup overflows. Does this variation indicate that the mean number of ounces dispensed has to be adjusted, or does it indicate that the standard deviation of the amount of coffee dispensed by the machine is too large?

84. *Grading on a Normal Curve* Mr. Sanderson marks his class on a normal curve. Those with z-scores with $z \geq 1.80$ will receive an A, those with $1.10 \leq z < 1.80$ will receive a B, those with $-1.20 \leq z < 1.10$ will receive a C, those with $-1.90 \leq z < -1.20$ will receive a D, those with $z < -1.90$ will receive an F. Determine the percent of grades that will be an A, B, C, D, and F.

85. Consider the following normal curve, representing a normal distribution, with points A, B, and C. One of these points corresponds to the mean, one point corresponds to the mean plus one standard deviation, and one point corresponds to the mean minus two standard deviations.

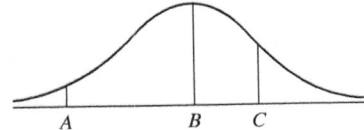

a) Which point corresponds to the mean?

b) Which point corresponds to the mean plus one standard deviation?

c) Which point corresponds to the mean minus two standard deviations?

86. Consider the following two normal curves.

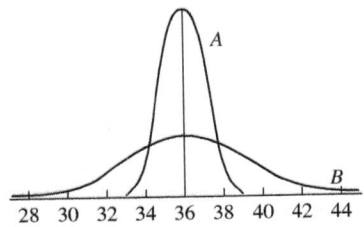

a) Do these distributions have the same mean? If so, what is the mean?

b) One of these curves corresponds to a normal distribution with a standard deviation of 1. The other curve corresponds to a normal distribution with a standard deviation of 3. Which curve, A or B, has a standard deviation of 3?

Concept/Writing Exercises

87. In a distribution that is skewed to the right, which has the greatest value: the mean, median, or mode? Which has the smallest value? Explain.

88. In a distribution skewed to the left, which has the greatest value: the mean, median, or mode? Which has the smallest value? Explain.

89. List three populations other than those given in the text that may be normally distributed.

90. List three populations other than those given in the text that may not be normally distributed.

Challenge Problems/Group Activities

91. *Salesperson Promotion* The owner at Kim's Home Interiors is reviewing the sales records of two managers who are up for promotion, Katie and Stella, who work in different stores. At Katie's store, the mean sales have been $23,200 per month, with a standard deviation of $2170. At Stella's store, the mean sales have been $25,600 per month, with a standard deviation of $2300. Last month, Katie's store sales were $28,408 and Stella's store sales were $29,510. At both stores, the distribution of monthly sales is normal.

a) Convert last month's sales for Katie's store and for Stella's store to z-scores.

b) If one of the two were to be promoted based solely on the sales last month, who should be promoted? Explain.

92. *Chebyshev's Theorem* How can you determine whether a distribution is approximately normal? A statistical theorem called *Chebyshev's theorem* states that the *minimum percent* of data between plus and minus K standard deviations from the mean $(K > 1)$ in *any distribution* can be determined by the formula

$$\text{Minimum percent} = 1 - \frac{1}{K^2}$$

Thus, for example, between ± 2 standard deviations from the mean there will always be a minimum of 75% of data. This minimum percent applies to any distribution. For $K = 2$,

$$\text{Minimum percent} = 1 - \frac{1}{2^2}$$

$$= 1 - \frac{1}{4} = \frac{3}{4}, \quad \text{or} \quad 75\%$$

Likewise, between ± 3 standard deviations from the mean there will always be a minimum of 89% of the data. For $K = 3$,

$$\text{Minimum percent} = 1 - \frac{1}{3^2}$$

$$= 1 - \frac{1}{9} = \frac{8}{9}, \quad \text{or} \quad 89\%$$

The following table lists the minimum percent of data in *any distribution* and the actual percent of data in *the normal distribution* between ± 1.1, ± 1.5, ± 2.0, and ± 2.5 standard deviations from the mean. The minimum percents of data in any distribution were calculated by using Chebyshev's theorem. The actual percents of data for the normal distribution were calculated by using the area given in the standard normal, or z, table.

	$K = 1.1$	$K = 1.5$	$K = 2$	$K = 2.5$
Minimum (for any distribution)	17.4%	55.6%	75%	84%
Normal distribution	72.9%	86.6%	95.4%	98.8%
Given distribution				

The percentages for the third row of the chart will be calculated in part (d).

Consider the following 30 pieces of data obtained from a quiz.

1, 1, 1, 1, 2, 2, 2, 2, 3, 3, 4, 4, 4, 5, 6,

6, 6, 7, 7, 7, 7, 8, 8, 8, 8, 9, 9, 9, 10, 10

a) Determine the mean of the set of scores.

b) Determine the standard deviation of the set of scores.

c) Determine the values that correspond to 1.1, 1.5, 2.0, and 2.5 standard deviations above the mean. Then determine the values that correspond to 1.1, 1.5, 2.0, and 2.5 standard deviations below the mean.

d) By observing the 30 pieces of data, determine the actual percent of quiz scores between

± 1.1 standard deviations from the mean.

± 1.5 standard deviations from the mean.

± 2 standard deviations from the mean.

± 2.5 standard deviations from the mean.

e) Place the percents determined in part (d) in the third row of the chart.

f) Compare the percents in the third row of the chart with the minimum percents in the first row and the normal percents in the second row, and then make a judgment as to whether this set of 30 scores is approximately normally distributed.

93. Determine a value of z such that $z \leq 0$ and 11.9% of the standard normal curve lies to the left of the z-value.

Recreational Mathematics

94. *Z-Scores* Ask your instructor for the class mean and class standard deviation for one of the exams taken by your class. For that exam, calculate the z-score for your exam grade. How many standard deviations is your exam grade away from the mean?

95. *Standard Deviation* If the mean score on a math quiz is 12.0 and 77% of the students in your class scored between 9.6 and 14.4, determine the standard deviation of the quiz scores.

Research Activity

96. *Collect A Set of Data* In this project, you actually become the statistician.

a) Select a project of interest to you in which data must be collected.

b) Write a proposal and submit it to your instructor for approval. In the proposal, discuss the aims of your project and how you plan to gather the data to make your sample unbiased.

c) After your proposal has been approved, gather 50 pieces of data by the method you proposed.

d) Rank the data from smallest to largest.

e) Compute the mean, median, mode, and midrange of the data.

f) Determine the range and standard deviation of the data. You may round the mean to the nearest tenth when computing the standard deviation.

g) Construct a frequency distribution, histogram, frequency polygon, and stem-and-leaf display of your data. Select your first class so that there will be between 5 and 12 classes. Be sure to label your histogram and frequency polygon.

h) Does your distribution appear to be normal? Explain your answer. Does it appear to be another type of distribution discussed? Explain.

i) Determine whether your distribution is approximately normal by using the technique discussed in Exercise 92.

SECTION 12.6   Linear Correlation and Regression

Upon completion of this section, you will be able to:

- Understand linear correlation.
- Calculate a linear correlation coefficient.
- Understand linear regression.
- Calculate the equation of the line of best fit and use the equation to make estimations or predictions.

Do you believe that there is a relationship between the time a person studies for an exam and the exam grade the person receives? Is there a relationship between the age of a car and the value of the car? Can we predict the value of a car based on the age of the car? In this section, we will learn how to determine whether there is a relationship between two quantities and, if so, how strong that relationship is. We will also learn how to determine the equation of the line that best describes the relationship between two quantities.

Why This Is Important   There are many real-life applications in which a relationship exists between two variables, such as the amount of time spent studying for an exam and the exam score. Once a linear relationship between the amount of time spent studying for an exam score and the exam score, we can use that relationship to predict an exam score based on the amount of time spent studying.

In this section, we discuss two important statistical topics: correlation and regression. *Correlation* is used to determine whether there is a relationship between two quantities and, if so, how strong the relationship is. *Regression* is used to determine the equation that relates the two quantities. Although there are other types of correlation and regression, in this section we discuss only linear correlation and linear regression. We begin by discussing linear correlation.

Linear Correlation

The *linear correlation coefficient*, r, is a unitless measure that describes the strength of the linear relationship between two variables. A positive value of r, or a positive correlation, means that as one variable increases, the other variable also increases. A negative value of r, or a negative correlation, means that as one variable increases, the other variable decreases. The correlation coefficient, r, will always be a value between -1 and 1 inclusive. A value of 1 indicates the strongest possible positive correlation, a value of -1 indicates the strongest possible negative correlation, and a value of 0 indicates no linear correlation (Fig. 12.39).

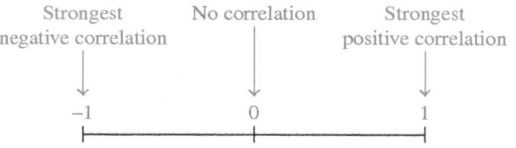

Figure 12.39

A visual aid used with correlation is the *scatter diagram*, a plot of data points. To help understand how to construct a scatter diagram, consider the following data from Egan Electronics. During a 6-day period, Egan Electronics kept daily records of the number of assembly line workers absent and the number of defective parts produced. The information is provided in the following chart.

Day	1	2	3	4	5	6
Number of workers absent	3	5	0	1	2	6
Number of defective parts	15	22	7	12	20	30

For each of the 6 days, two pieces of data are provided: number of workers absent and number of defective parts. Data, such as these, that involve two variables are called *bivariate data*. Often when we have a set of bivariate data, we can control one of the quantities. We generally denote the quantity that can be controlled, the *independent variable*, as x. The other variable, the *dependent variable*, is denoted as y. In this problem, we will assume that the number of defective parts produced is affected by the number of workers absent. Therefore, we will call the number of workers absent x and the number of defective parts produced y. When we plot bivariate data, the independent variable is marked on the horizontal axis and the dependent variable is marked on the vertical axis. Therefore, for this problem, number of workers absent is marked on the horizontal axis and number of defective parts is marked on the vertical axis. If we plot the six pieces of bivariate data in the Cartesian coordinate system, we get a scatter diagram, as shown in Fig. 12.40 on page 811.

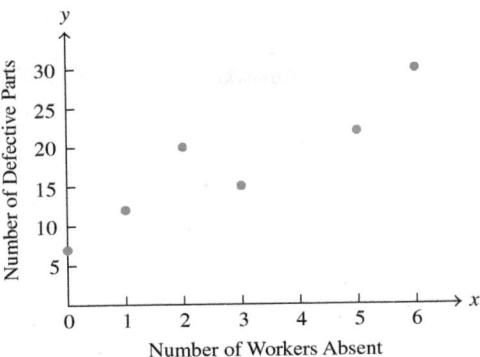

Figure 12.40

The figure shows that, generally, the more workers that are absent, the more defective parts are produced.

In Fig. 12.41, we show some scatter diagrams and indicate the corresponding strength of correlation between the quantities on the horizontal and vertical axes.

Earlier, we mentioned that r will always be a value between -1 and 1 inclusive. A value of $r = 1$ is obtained only when every point of the bivariate data on a scatter diagram lies in a straight line and the line is increasing from left to right (Fig. 12.41(a)). In other words, the line has a positive slope, as discussed in Section 6.6.

A value of $r = -1$ will be obtained only when every point of the bivariate data on a scatter diagram lies in a straight line and the line is decreasing from left to right. (Fig. 12.41(e)). In other words, the line has a negative slope.

The value of r is a measure of how far a set of points varies from a straight line. The greater the spread, the weaker the correlation and the closer the value of r is to 0. Figure 12.41 shows that the more the points diverge from a straight line, the weaker the correlation becomes.

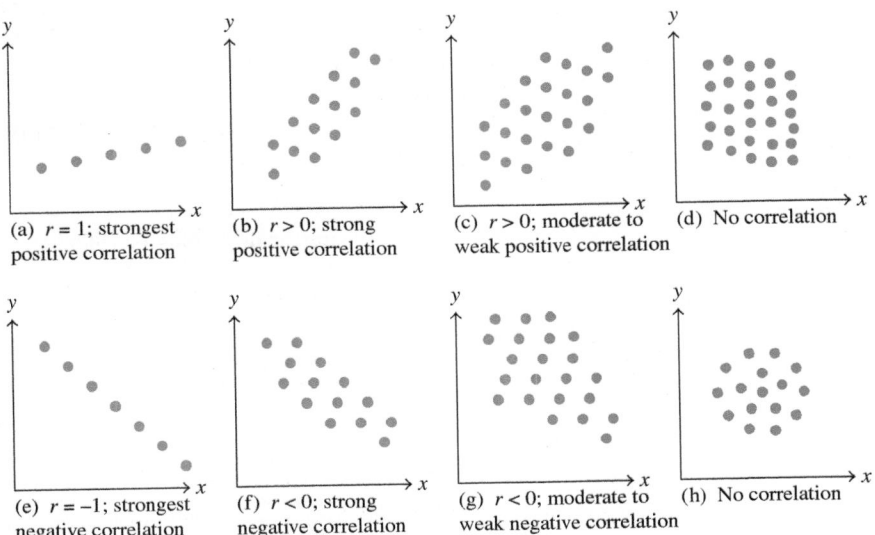

Figure 12.41

The following formula is used to calculate r, the linear correlation coefficient.

Linear Correlation Coefficient
The formula to calculate the **linear correlation coefficient, r,** is as follows.

$$r = \frac{n(\Sigma xy) - (\Sigma x)(\Sigma y)}{\sqrt{n(\Sigma x^2) - (\Sigma x)^2} \sqrt{n(\Sigma y^2) - (\Sigma y)^2}}$$

In this book, the *linear correlation coefficient* is sometimes just referred to as the *correlation coefficient*. To determine the *correlation coefficient, r,* and the equation of the line of best fit (to be discussed shortly), a statistical calculator may be used. In the Technology Tip box at the end of this section, we indicate the procedure to follow to use the computer software spreadsheet program Microsoft Excel, the TI-84 Plus calculator, and StatCrunch to determine the correlation coefficient.

In Example 1, we show how to determine r for a set of bivariate data without the use of a statistical calculator. We will use the same set of bivariate data given on page 810 that was used to make the scatter diagram in Fig. 12.40.

Example 1 *Number of Absences Versus Number of Defective Parts*

Egan Electronics provided the following daily records about the number of assembly line workers absent and the number of defective parts produced for 6 days. Determine the correlation coefficient between the number of workers absent and the number of defective parts produced.

Day	1	2	3	4	5	6
Number of workers absent	3	5	0	1	2	6
Number of defective parts	15	22	7	12	20	30

Solution We plotted this set of data on the scatter diagram in Fig. 12.40 on page 811. We will call the number of workers absent x. We will call the number of defective parts produced y. In the table below, we list the values of x and y and calculate the necessary sums: Σx, Σy, Σxy, Σx^2, Σy^2. We determine the values in the column labeled x^2 by squaring the x's (multiplying the x's by themselves). We determine the values in the column labeled y^2 by squaring the y's. We determine the values in the column labeled xy by multiplying each x-value by its corresponding y-value.

Number of Workers Absent	Number of Defective Parts			
x	y	x^2	y^2	xy
3	15	9	225	45
5	22	25	484	110
0	7	0	49	0
1	12	1	144	12
2	20	4	400	40
6	30	36	900	180
17	106	75	2202	387

Thus, $\Sigma x = 17$, $\Sigma y = 106$, $\Sigma x^2 = 75$, $\Sigma y^2 = 2202$, and $\Sigma xy = 387$. In the formula for r, we use both $(\Sigma x)^2$ and Σx^2. Note that $(\Sigma x)^2 = (17)^2 = 289$ and that $\Sigma x^2 = 75$. Similarly, $(\Sigma y)^2 = (106)^2 = 11{,}236$ and $\Sigma y^2 = 2202$.

The n in the formula represents the number of pieces of bivariate data. Here $n = 6$. Now let's determine r.

$$r = \frac{n(\Sigma xy) - (\Sigma x)(\Sigma y)}{\sqrt{n(\Sigma x^2) - (\Sigma x)^2}\ \sqrt{n(\Sigma y^2) - (\Sigma y)^2}}$$

$$= \frac{6(387) - (17)(106)}{\sqrt{6(75) - (17)^2}\ \sqrt{6(2202) - (106)^2}}$$

$$= \frac{2322 - 1802}{\sqrt{6(75) - 289}\ \sqrt{6(2202) - 11{,}236}}$$

$$= \frac{520}{\sqrt{450 - 289}\ \sqrt{13{,}212 - 11{,}236}}$$

$$= \frac{520}{\sqrt{161}\ \sqrt{1976}} \approx 0.922$$

Since the maximum possible value for r is 1.00, a correlation coefficient of 0.922 is a strong, positive correlation. This result implies that, generally, the more assembly line workers absent, the more defective parts produced. ∎

In Example 1, had we determined r to be a value greater than 1 or less than -1, it would have indicated that we had made an error. Also, from the scatter diagram, we should realize that r should be a positive value and not negative.

In Example 1, there appears to be a cause–effect relationship. That is, the more assembly line workers who are absent, the more defective parts are produced. *However, a correlation does not necessarily indicate a cause–effect relationship.* For example, there is a positive correlation between police officers' salaries and the cost of medical insurance over the past 10 years (both have increased), but that does not mean that the increase in police officers' salaries caused the increase in the cost of medical insurance.

Suppose in Example 1 that r had been 0.53. Would this value have indicated a correlation? What is the minimum value of r needed to assume that a correlation exists between the variables? To answer this question, we introduce the term *level of significance*. The *level of significance*, denoted α (alpha), is used to identify the cutoff between results attributed to chance and results attributed to an actual relationship between the two variables. Table 12.9 on page 814 gives *critical values* (or cutoff values) that are sometimes used for determining whether two variables are related. The table indicates two different levels of significance: $\alpha = 0.05$ and $\alpha = 0.01$. A level of significance of 5%, written $\alpha = 0.05$, means that there is a 5% chance that, when you say the variables are related, they actually are *not* related. Similarly, a level of significance of 1%, or $\alpha = 0.01$, means that there is a 1% chance that, when you say the variables are related, they actually are *not* related. More complete critical value tables are available in statistics books.

To explain the use of the table, we use *absolute value*, symbolized $|\ \ |$. The absolute value of a nonzero number is the positive value of the number, and the absolute value of 0 is 0. Therefore,

$$|3| = 3, \qquad |-3| = 3, \qquad |5| = 5, \qquad |-5| = 5, \qquad \text{and} \qquad |0| = 0$$

Table 12.9   Critical Values* for the Correlation Coefficient, r

n	$\alpha = 0.05$	$\alpha = 0.01$
4	0.950	0.990
5	0.878	0.959
6	(0.811)	0.917
7	0.754	0.875
8	0.707	0.834
9	0.666	0.798
10	0.632	0.765
11	0.602	0.735
12	0.576	0.708
13	0.553	0.684
14	0.532	0.661
15	0.514	0.641
16	0.497	0.623
17	0.482	0.606
18	0.468	0.590
19	0.456	0.575
20	0.444	0.561
22	0.423	0.537
27	0.381	0.487
32	0.349	0.449
37	0.325	0.418
42	0.304	0.393
47	0.288	0.372
52	0.273	0.354
62	0.250	0.325
72	0.232	0.302
82	0.217	0.283
92	0.205	0.267
102	0.195	0.254

The derivation of this table is beyond the scope of this text. It shows the critical values of the Pearson correlation coefficient.

If the absolute value of r, written $|r|$, is *greater than* the value given in the table under the specified α and appropriate sample size n, we assume that a correlation does exist between the variables. If $|r|$ is less than the table value, we assume that no correlation exists.

Returning to Example 1, if we want to determine whether there is a correlation at a 5% level of significance, we locate the critical value (or cutoff value) that corresponds to $n = 6$ (there are 6 pieces of bivariate data) and $\alpha = 0.05$. The value to the right of $n = 6$ and under the $\alpha = 0.05$ column is the critical value 0.811, circled in red in Table 12.9. From the formula, we had obtained $r = 0.922$. Since $|0.922| > 0.811$, or $0.922 > 0.811$, we assume that a correlation between the variables exists.

Note in Table 12.9 that the larger the sample size, the smaller is the value of r needed for a significant correlation.

Example 2   *Amount of Drug Remaining in the Bloodstream*

To test the length of time that an infection-fighting drug stays in a person's bloodstream, a doctor gives 300 milligrams of the drug to 10 patients, labeled 1–10 in the table below. Once each hour, for 10 hours, one of the 10 patients is selected at random and that person's blood is tested to determine the amount of the drug remaining in the bloodstream. The results are as follows.

Patient	1	2	3	4	5	6	7	8	9	10
Time (hr)	1	2	3	4	5	6	7	8	9	10
Drug remaining (mg)	250	230	200	210	140	120	210	100	90	85

Determine at a level of significance of 5% whether a correlation exists between the time elapsed and the amount of drug remaining.

Solution Let time be represented by x and the amount of drug remaining by y. We first draw a scatter diagram (Fig. 12.42).

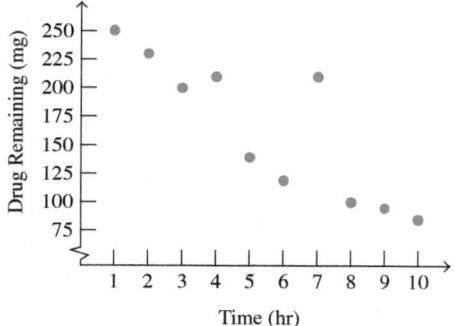

Figure 12.42

The scatter diagram suggests that, if a correlation exists, it will be negative. We now construct a table of values and calculate r.

---

*This table of values may be used only under certain conditions. If you take a statistics course, you will learn more about which critical values to use to determine whether a linear correlation exists.

x	y	x^2	y^2	xy
1	250	1	62,500	250
2	230	4	52,900	460
3	200	9	40,000	600
4	210	16	44,100	840
5	140	25	19,600	700
6	120	36	14,400	720
7	210	49	44,100	1470
8	100	64	10,000	800
9	90	81	8100	810
10	85	100	7225	850
55	1635	385	302,925	7500

$$r = \frac{n(\Sigma xy) - (\Sigma x)(\Sigma y)}{\sqrt{n(\Sigma x^2) - (\Sigma x)^2}\ \sqrt{n(\Sigma y^2) - (\Sigma y)^2}}$$

$$= \frac{10(7500) - (55)(1635)}{\sqrt{10(385) - (55)^2}\ \sqrt{10(302,925) - (1635)^2}}$$

$$= \frac{-14,925}{\sqrt{825}\ \sqrt{356,025}} \approx \frac{-14,925}{17,138.28} \approx -0.871$$

From Table 12.9, for $n = 10$ and $\alpha = 0.05$, we get 0.632. Since $|-0.871| = 0.871$ and $0.871 > 0.632$, a correlation exists. The correlation is negative, which indicates that the longer the time period, the smaller is the amount of drug remaining. ∎

Linear Regression

Let's now turn to regression. *Linear regression* is the process of determining the linear relationship between two variables. Our goal is to describe this relationship using a linear equation. Under certain conditions, if we are given a value for one variable we can use this linear equation to predict the corresponding value of the second variable.

Using a set of bivariate data, we will determine the equation of *the line of best fit*. The line of best fit is also called *the regression line*, or *the least squares line*. The *line of best fit* is the line such that the sum of the squares of the vertical distances from the line to the data points (on the scatter diagram) is a minimum, as shown in Fig. 12.43. In Fig. 12.43, the line of best fit minimizes the sum of the squares of d_1 through d_8. To determine the equation of the line of best fit, $y = mx + b$,* we must determine m and then b. Recall from Section 6.6 that the slope–intercept form of a straight line is $y = mx + b$, where m is the slope and $(0, b)$ is the y-intercept. The formulas for determining m and b are as follows.

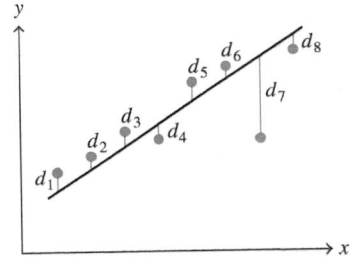

Figure 12.43

---

*Some statistics books use $y = ax + b$, $y = b_0 + b_1 x$, or something similar, for the equation of the line of best fit. In any case, the letter next to the variable x represents the slope of the line of best fit, and the other letter represents the y-coordinate of the y-intercept of the graph.

The Line of Best Fit
The equation of the line of best fit is

$$y = mx + b,$$

where $m = \dfrac{n(\Sigma xy) - (\Sigma x)(\Sigma y)}{n(\Sigma x^2) - (\Sigma x)^2}$ and $b = \dfrac{\Sigma y - m(\Sigma x)}{n}$

Note that the numerator of the fraction used to determine m is identical to the numerator used to determine r. Therefore, if you have previously determined r, you do not need to repeat this calculation. Also, the denominator of the fraction used to determine m is identical to the radicand of the first square root in the denominator of the fraction used to determine r.

Example 3 The Line of Best Fit

a) Use the data in Example 1 on pages 812–813 to determine the equation of the line of best fit that relates the number of workers absent on an assembly line and the number of defective parts produced.
b) Graph the equation of the line of best fit on a scatter diagram that illustrates the set of bivariate points.

Solution

a) In Example 1, we determined $n(\Sigma xy) - (\Sigma x)(\Sigma y) = 520$ and $n(\Sigma x^2) - (\Sigma x)^2 = 161$. Thus,

$$m = \frac{n(\Sigma xy) - (\Sigma x)(\Sigma y)}{n(\Sigma x^2) - (\Sigma x)^2} = \frac{520}{161} \approx 3.23$$

Now we determine b. In Example 1, we determined $n = 6$, $\Sigma x = 17$, and $\Sigma y = 106$.

$$b = \frac{\Sigma y - m(\Sigma x)}{n}$$

$$\approx \frac{106 - 3.23(17)}{6} \approx \frac{51.09}{6} \approx 8.52$$

Therefore, the equation of the line of best fit, with values rounded to the nearest hundredth, is

$$y = mx + b$$

$$y = 3.23x + 8.52$$

where x represents the number of workers absent and y represents the number of defective parts produced.

b) To graph $y = 3.23x + 8.52$, we need to plot at least two points. We will plot three points and then draw the graph.

$$y = 3.23x + 8.52$$

$x = 2$ $y = 3.23(2) + 8.52 = 14.98$

$x = 4$ $y = 3.23(4) + 8.52 = 21.44$

$x = 6$ $y = 3.23(6) + 8.52 = 27.90$

x	y
2	14.98
4	21.44
6	27.90

These three calculations indicate that if 2 assembly line workers are absent on the assembly line, the predicted number of defective parts produced is about 15. If 4 assembly line workers are absent, the predicted number of defective parts produced is about 21, and if 6 assembly line workers are absent, the predicted number of defective parts produced is about 28. Plot the three points (the three black points in Fig. 12.44) and then draw a straight line through the three points. The scatter diagram and graph of the equation of the line of best fit are plotted in Fig. 12.44.

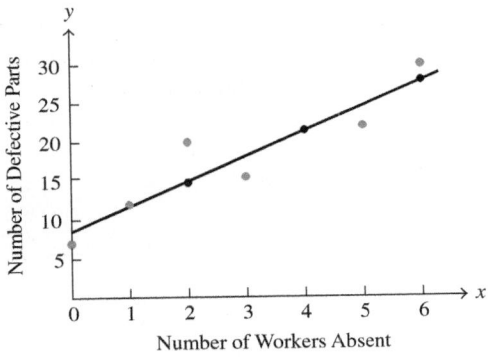

Figure 12.44

In Example 3, the line of best fit intersects the y-axis at 8.52, the value we determined for the y-intercept, b, in part (a).

Correlation and Regression

Example 4 *Line of Best Fit for Example 2*

a) Determine the equation of the line of best fit between the time elapsed and the amount of drug remaining in a person's bloodstream in Example 2 on pages 814–815.

b) If the average person is given 300 mg of the drug, how much will remain in the person's bloodstream after 5 hr?

Solution

a) From the scatter diagram on page 814, we see that the slope of the line of best fit, m, will be negative. In Example 2, we determined that $n(\Sigma xy) - (\Sigma x)(\Sigma y) = -14{,}925$ and that $n(\Sigma x^2) - (\Sigma x)^2 = 825$. Thus,

$$m = \frac{n(\Sigma xy) - (\Sigma x)(\Sigma y)}{n(\Sigma x^2) - (\Sigma x)^2}$$

$$= \frac{-14{,}925}{825}$$

$$\approx -18.09$$

From Example 2, $n = 10$, $\Sigma x = 55$, and $\Sigma y = 1635$.

$$b = \frac{\Sigma y - m\Sigma x}{n}$$

$$\approx \frac{1635 - (-18.09)(55)}{10}$$

$$\approx 263.00$$

Thus, the equation of the line of best fit, with values rounded to the nearest hundredth, is

$$y = mx + b$$
$$y = -18.09x + 263.00$$

where x is the elapsed time and y is the amount of drug remaining.

b) We evaluate $y = -18.09x + 263.00$ at $x = 5$.

$$y = -18.09x + 263.00$$
$$y = -18.09(5) + 263.00 = 172.55$$

Thus, after 5 hr, about 173 mg of the drug remains in the average person's bloodstream.

TECHNOLOGY TIP

We can use Microsoft EXCEL, the Texas Instruments TI-84 Plus graphing calculator, and StatCrunch to determine the correlation coefficient, r, and the equation of the line of best fit. Below we use the data from Example 1 on pages 812–813, where x represents the number of workers absent and y represents the number of defective parts for 6 selected days.

Excel

Begin by reading the *Technology Tip* on pages 777. The instructions for determining the correlation coefficient and equation of the line of best fit are similar to the instructions given on page 777. In column A, rows 1 through 6, enter the 6 values representing the number of workers absent. In column B, rows 1 through 6, enter the 6 values that represent the number of defective parts. In general, the items listed on the horizontal axis in the scatter diagram (the independent variable if there is one) are placed in column A and the items listed on the vertical axis are placed in column B. Then select the following:

Insert > Function...> Statistical

Then to determine the value of the correlation coefficient, select **CORREL** and click OK. Excel will then generate a gray box and ask you for information about your data. In the box to the right of **Array1**, type in A1:A6, and in the box to the right of **Array2**, type in B1:B6. Then at the bottom of the box you will see *Formula results* = *.921927852*, which is the correlation coefficient. If you then wish to determine the equation of the line of best fit, press CANCEL on the bottom of the gray box. Then select

Insert > Function...> Statistical > SLOPE > OK

Excel will then generate a gray box. In the box to the right of **Known_y's**, type in B1:B6. In the box to the right of **Known_x's**, type in A1:A6. At that point, at the bottom of the box you will see *Formula results* = *3.229813665*, which is the slope of the equation of the line of best fit. To determine the y-intercept, press CANCEL to remove the gray box. Then select

Insert > Function...> Statistical...> INTERCEPT > OK

Excel will then generate a gray box. In the box to the right of **Known_y's**, type in B1:B6. In the box to the right of **Known_x's**, type in A1:A6. At that point, at the bottom of the box you will see *Formula results* = *8.51552795*, which is the y-coordinate of the y-intercept of the equation of the line of best fit.

TI-84 Plus Graphing Calculators

First, press 2nd CATALOG. Use the down arrow key until the indicator is pointing at DiagnosticOn. Press ENTER twice. This will enable the calculator to display the correlation coefficient and the line of best fit at the same time. Next, press STAT. Then select **Edit**. If you have any data in **L1** or **L2**, delete the data by going to **L1** and pressing CLEAR and then ENTER. Follow a similar procedure to clear column **L2**. Now enter the 6 pieces of data representing the number of workers absent in column **L1** and enter the 6 pieces of data representing the number of defective parts in **L2**. Then press STAT and highlight **CALC**. At this point, **1: 1 − Var Stats** should be highlighted. Scroll down to highlight **4: LinReg (ax + b)**. Then press ENTER twice. You will get $r = .921927852$, which represents the correlation coefficient. On the screen, you will also see values for a and b. Note that a represents the slope and b represents the y-coordinate of the y-intercept of the equation of the line of best fit. Thus, the slope is 3.229813665 and the y-coordinate of the y-intercept is 8.51552795.

StatCrunch (available in MyLab Math and separately)

To compute this line of best fit, start by entering the data (see Step 1 in the *Technology Tip* on page 778).

1. Enter the values for x under var 1 and the values for y under var 2.

2. Click **Stat** in the top menu. Select **Regression** then **Simple Linear** from the dropdown menu.

3. You will see the equation of the line of best fit written as var 2 = 8.515528 + 3.2298137 var 1. You will also see r (correlation coefficient) = 0.92192785. If you click on the arrow at the bottom of the results window, you will see the scatterplot with the graph of the line of best fit.

There's an App for That!

Many of the concepts that we discuss throughout this section and throughout this entire chapter can be further explored with the use of an app on your smartphone or tablet computer. Several of these apps allow you to calculate the linear correlation coefficient and the equation of the line of best fit and allow you to estimate the predicted value of y from the line of best fit. As always, consult with your instructor before using these apps when completing work related to your course.

SECTION 12.6   *Exercises*

Warm Up Exercises

In Exercises 1–6, fill in the blank with an appropriate word, phrase, or symbol(s).

1. A unitless measure that describes the strength of the linear relationship between two variables is called the linear correlation _____.

2. The process of determining the linear relationship between two variables is called linear _____.

3. a) The value of r, the linear correlation coefficient, that represents the strongest positive correlation between two variables is _____.

   b) The value of r, the linear correlation coefficient, that represents the strongest negative correlation between two variables is _____.

   c) The value of r, the linear correlation coefficient, that represents no correlation between two variables is _____.

4. If one quantity increases as the other quantity decreases, the two variables are said to have a(n) _____ correlation.

5. If one quantity increases as the other quantity increases, the two variables are said to have a(n) _____ correlation.

6. A plot of data points is called a(n) _____ diagram.

In Exercises 7–10, indicate if you believe that a correlation exists between the quantities on the horizontal and vertical axes. If so, indicate if you believe that the correlation is a strong positive correlation, a strong negative correlation, a weak positive correlation, a weak negative correlation, or no correlation. Explain your answer.

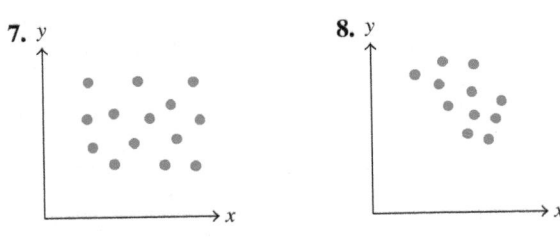

7.

8.

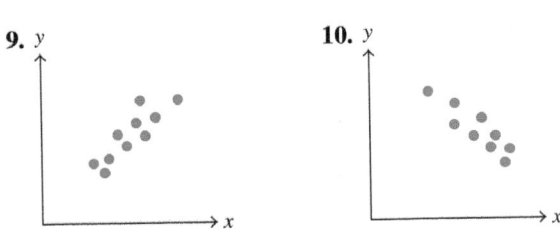

9.

10.

Many of the exercises in this section are designed so that they can be done by hand or by using an appropriate form of technology (graphing calculator, spreadsheet, or StatCrunch).

Practice the Skills

In Exercises 11–18, assume that a sample of bivariate data yields the correlation coefficient, r, indicated. Use Table 12.9 on page 814 for the specified sample size and level of significance to determine whether a linear correlation exists.

11. $r = 0.92$ when $n = 10$ at $\alpha = 0.01$

12. $r = 0.43$ when $n = 17$ at $\alpha = 0.01$

13. $r = -0.59$ when $n = 11$ at $\alpha = 0.05$

14. $r = -0.89$ when $n = 6$ at $\alpha = 0.05$

15. $r = -0.32$ when $n = 102$ at $\alpha = 0.01$

16. $r = -0.51$ when $n = 13$ at $\alpha = 0.05$

17. $r = 0.75$ when $n = 6$ at $\alpha = 0.01$

18. $r = 0.96$ when $n = 5$ at $\alpha = 0.01$

In Exercises 19–26, (a) draw a scatter diagram; (b) determine the value of r, rounded to the nearest thousandth; (c) determine whether a correlation exists at $\alpha = 0.05$; and (d) determine whether a correlation exists at $\alpha = 0.01$.

19.

x	y
4	8
5	10
6	12
7	12
10	16

20.

x	y
8	14
10	12
13	10
16	11
19	8

21.

x	y
23	30
35	38
31	27
43	21
49	40

22.

x	y
90	3
80	4
60	6
60	5
40	5
20	7

23.

x	y
5.3	10.3
4.7	9.6
8.4	12.5
12.7	16.2
4.9	9.8

24.

x	y
12	15
16	19
13	45
24	30
100	60
50	28

25.

x	y
100	2
80	3
60	5
60	6
40	6
20	8

26.

x	y
90	90
70	70
65	65
60	60
50	50
40	40
15	15

In Exercises 27–34, determine the equation of the line of best fit from the data indicated in the exercise. Round both m and b to the nearest hundredth.

27. Exercises 19

28. Exercises 20

29. Exercises 21

30. Exercises 22

31. Exercises 23

32. Exercises 24

33. Exercises 25

34. Exercises 26

Problem Solving

In Exercises 35–44, round both m and b the nearest hundredth.

35. **Commuting to a Pet Shelter** The following data show the commuting distances and commuting times of six volunteers working at a pet shelter.

Distance (miles)	2	15	16	9	21	5
Time (minutes)	5	25	30	20	35	10

a) Determine the correlation coefficient between the commuting distance and the commuting time.

b) Determine whether a correlation exists at $\alpha = 0.05$.

c) Determine the equation of the line of best fit for the commuting distance and the commuting time.

DS **36. Sales** The following table shows the average (mean) number of clients per month and the sales volume, in thousands of dollars, for 8 salespeople.

Average number of clients	12	14	16	20	23	46	50	48
Sales (thousands $)			16	25	30	30	75	82	100	90

a) Determine the correlation coefficient between the average number of clients and the sales.

b) Determine whether a correlation exists at $\alpha = 0.05$.

c) Determine the equation of the line of best fit for the average number of clients and the sales.

d) Use the equation in part (c) to estimate the sales for a person with an average number of clients per month of 25.

DS **37. Selling Cups of Hot Chocolate** The high temperature for seven January days in Chicago along with the number of cups of hot chocolate sold for those days at the C and C Coffee Shop are shown in the table below.

High temperature (°F)	42	38	20	30	28	35	17
Cups of hot chocolate sold	35	45	60	52	55	47	65

a) Determine the correlation coefficient between the high temperature and the number of cups of hot chocolate sold.

b) Determine whether a correlation exists at $\alpha = 0.01$.

c) Determine the equation of the line of best fit for the high temperature and number of cups of hot chocolate sold.

d) Use the equation in part (c) to estimate the number of cups of hot chocolate sold if the daily high temperature is 33 °F.

DS **38. Fuel Efficiency of Cars** The table above and to the right shows the weights, in hundreds of pounds, for six selected cars. Also shown is the corresponding fuel efficiency, in miles per gallon (mpg), for the car in city driving.

Weight (hundreds of pounds)	27	31	35	32	30	30
Fuel efficiency (mpg)		23	22	20	21	24	22

a) Determine the correlation coefficient between the weight of a car and the fuel efficiency.

b) Determine whether a correlation exists at $\alpha = 0.01$.

c) Determine the equation of the line of best fit for the weight of a car and the fuel efficiency of a car.

d) Use the equation in part (c) to estimate the fuel efficiency of a car that weighs 33 hundred pounds.

DS **39. Tips** The following table shows the amount of a dinner bill and the amount of money left as a tip for 6 customers at a restaurant.

Bill ($)	35.20	22.93	16.45	45.82	54.16	39.70
Tip ($)	5.50	4.50	2.00	9.00	10.00	5.75

a) Determine the correlation coefficient between the amount of a dinner bill and the amount of money left as a tip.

b) Determine whether a correlation exists at $\alpha = 0.05$.

c) Determine the equation of the line of best fit for the amount of a dinner bill and the amount of money left as a tip.

d) Use the equation in part (c) to estimate the amount of money left as a tip by a customer with a dinner bill of $25.00.

DS **40. Blood Pressure** The following table shows the systolic blood pressure and diastolic blood pressure readings for six adults.

Systolic	110	153	120	143	100	112
Diastolic	70	110	80	98	70	75

a) Determine the correlation coefficient between the systolic blood pressure and the diastolic blood pressure readings.

b) Determine whether a correlation exists at $\alpha = 0.05$.

c) Determine the equation of the line of best fit for the systolic blood pressure and the diastolic blood pressure readings.

d) Use the equation in part (c) to estimate the diastolic blood pressure for an adult with a systolic blood pressure reading of 115.

DS **41. Sports Drinks** The following table shows the energy provided (in kilocalories) and the amount of carbohydrates (in
grams) in a 100-mℓ bottle of a sports drink for six different brands.

Carbohydrates (grams)	6.5	7	6	2	6.4	6
Energy (kilocalories)	27	30	25	10	28	24

a) Determine the correlation coefficient between amount
of carbohydrates and the energy provided.

b) Determine whether a correlation exists at $\alpha = 0.05$.

c) Determine the equation of the line of best fit for the
amount of carbohydrates and the energy provided.

d) Use the equation in part (c) to estimate the amount of
energy provided in a sports drink with 5 grams of
carbohydrates.

DS **42. Crimes** The following chart shows the number of police
officers patrolling a section of a city and the number of
crimes reported for 8 successive days.

Police officers	20	12	18	15	22	10	20	12
Crimes reported	8	10	12	9	6	15	7	18

a) Determine the correlation coefficient between the
number of police officers and the number of crimes
reported.

b) Determine whether a correlation exists at $\alpha = 0.05$.

c) Determine the equation of the line of best fit for the
number of police officers and the number of crimes
reported.

d) Use the equation in part (c) to estimate the average
number of crimes reported when 14 police officers are
patrolling that section of the city.

43. Hitting the Brakes **a)** Examine the art at the top of the
column to the right. Do you believe that there is a positive correlation, a negative correlation, or no correlation
between the speed of a car and the stopping distance when
the brakes are applied?

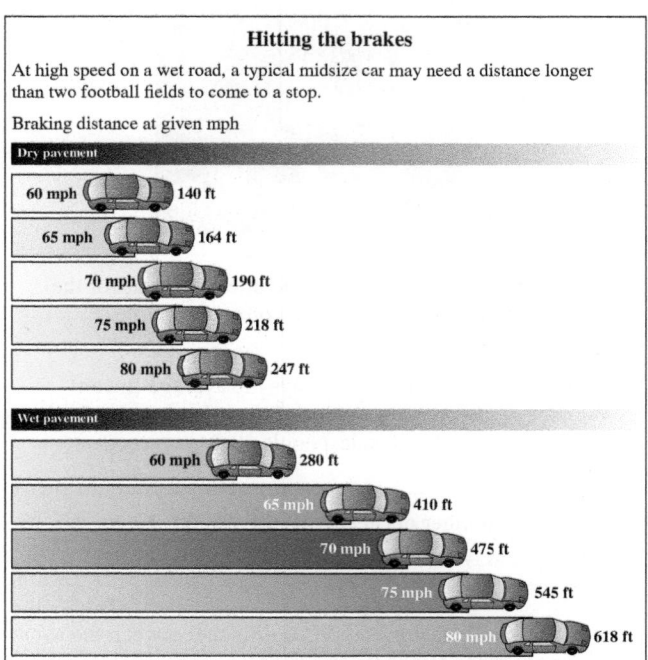

Hitting the brakes

At high speed on a wet road, a typical midsize car may need a distance longer
than two football fields to come to a stop.

Braking distance at given mph

Dry pavement

60 mph 140 ft
65 mph 164 ft
70 mph 190 ft
75 mph 218 ft
80 mph 247 ft

Wet pavement

60 mph 280 ft
65 mph 410 ft
70 mph 475 ft
75 mph 545 ft
80 mph 618 ft

Source: *Car and Driver*, American Automobile Association

b) Do you believe that there is a stronger correlation between the speed of a car and the stopping distance on
wet or dry roads?

c) Use the figure to construct two scatter diagrams, one
for dry pavement and the other for wet pavement. Place
the speed of the car on the horizontal axis.

d) Compute the correlation coefficient for the speed of the
car and the stopping distance for dry pavement.

e) Repeat part (d) for wet pavement.

f) Were your answers to parts (a) and (b) correct? Explain.

g) Determine the equation of the line of best fit for dry
pavement.

h) Repeat part (g) for wet pavement.

i) Use the equations in parts (g) and (h) to estimate the
stopping distance of a car going 77 mph on both dry
and wet pavements.

DS **44. Chlorine in a Swimming Pool** A gallon of chlorine is put
into a swimming pool. Each hour later for the following
6 hours the percent of chlorine that remains in the pool is
measured. The following information is obtained.

Time (hours)	1	2	3	4	5	6
Chlorine remaining (percent)	80.0	76.2	68.7	50.1	30.2	20.8

a) Determine the correlation coefficient for the time and the percent of chlorine remaining.

b) Determine whether a correlation exists at $\alpha = 0.01$.

c) Determine the equation of the line of best fit for the time and the amount of chlorine remaining.

d) Use the equation in part (c) to estimate the average amount of chlorine remaining after 4.5 hr.

Challenge Problems/Group Activities

45. *Interchanging Variables* a) Assume that a set of bivariate data yields a specific correlation coefficient. If the x- and y-values are interchanged and the correlation coefficient is recalculated, will the correlation coefficient change?

b) Make up a table of five pieces of bivariate data and determine r using the data. Then switch the values of the x's and y's and recompute the correlation coefficient. Has the value of r changed?

46. *Height vs. Length* a) Do you believe that a correlation exists between a person's height and the length of a person's arm?

b) Select 10 people from your class and measure (in inches) their height and the length of one of their arms.

c) Plot the 10 ordered pairs on a scatter diagram.

d) Calculate the correlation coefficient, r.

e) Determine the equation of the line of best fit.

f) Estimate the length of the arm of a person who is 58 in. tall.

47. *Calculating a Correlation Coefficient* a) Select a category of bivariate data that you think has a strong positive correlation. Designate the independent variable and the dependent variable. Explain why you believe that the bivariate data have a strong positive correlation.

b) Collect at least 10 pieces of bivariate data that can be used to determine the correlation coefficient. Explain how you chose these data.

c) Plot a scatter diagram.

d) Calculate the correlation coefficient.

e) Does there appear to be a strong positive correlation? Explain your answer.

f) Calculate the equation of the line of best fit.

g) Explain how the equation in part (f) may be used.

DS 48. *Consumer Price Index* The following table shows the consumer price index (CPI) for the years 2013–2018.

Year	2013	2014	2015	2016	2017	2018
CPI	233.0	236.7	237.0	240.0	245.1	251.1

a) Determine the linear correlation coefficient, r, between the year and the CPI.

b) If 2013 is subtracted from each year, the table obtained becomes:

Year	0	1	2	3	4	5
CPI	233.0	236.7	237.0	240.0	245.1	251.1

If r is calculated from these values, how will it compare with the r determined in part (a)? Explain.

c) Calculate r from the values in part (b) and compare the results with the value of r determined in part (a). Are they the same? If not, explain why.

49. a) There are equivalent formulas that can be used to determine the correlation coefficient and the equation of the line of best fit. A formula used in some statistics books to determine the correlation coefficient is

$$r = \frac{SS(xy)}{\sqrt{SS(x)SS(y)}}$$

where

$$SS(x) = \Sigma x^2 - \frac{(\Sigma x)^2}{n}$$

$$SS(y) = \Sigma y^2 - \frac{(\Sigma y)^2}{n}$$

$$SS(xy) = \Sigma xy - \frac{(\Sigma x)(\Sigma y)}{n}$$

Use this formula to determine the correlation coefficient of the set of bivariate data given in Example 1 on pages 812–813.

b) Compare your answer with the answer obtained in Example 1.

Research Activities

50. *Bivariate Data*

a) Obtain a set of bivariate data from a newspaper, magazine, or the Internet.

b) Plot the information on a scatter diagram.

c) Indicate whether you believe that the data show a positive correlation, a negative correlation, or no correlation. Explain your answer.

d) Calculate r and determine whether your answer to part (c) was correct.

e) Determine the equation of the line of best fit for the bivariate data.

51. *Scatter Diagram* Find a scatter diagram in a newspaper, magazine, or the Internet and write a paper on what the diagram indicates. Indicate whether you believe that the bivariate data show a positive correlation, a negative correlation, or no correlation and explain why.

CHAPTER 12   *Summary*

Important Facts and Concepts

Examples and Discussion

Important Facts and Concepts	Examples and Discussion
Section 12.1	
Sampling Techniques	
Random sampling	
Systematic sampling	Example 1, page 751
Cluster sampling	
Stratified sampling	
Convenience sampling	
Misuses of statistics	Discussion, pages 751–753
Section 12.2	
Frequency distribution	Examples 1–3, pages 758–761
Rules for Data Grouped by Classes	
1. The classes should be the same width.	Discussion, page 758; Examples 2–3, pages 759–761
2. The classes should not overlap.	
3. Each piece of data should belong to only one class.	
In addition, there should be between 5 and 12 classes, inclusive.	
Statistical Graphs	
Histogram	Examples 4, 6, 7, pages 761–764
Frequency polygon	Examples 5–6, pages 762–763
Stem-and-leaf display	Example 8, pages 764–765
Circle graph	Example 9, page 766

Section 12.3

Measures of Central Tendency

The **mean** is the sum of the data divided by the number of pieces of data: $\bar{x} = \dfrac{\Sigma x}{n}$.

Example 1, page 772

The **median** is the value in the middle of a set of ranked data.

Examples 2–3, 6, pages 773–774

The **mode** is the piece of data that occurs most frequently (if there is one).

Examples 4, 6, pages 773–774

The **midrange** is the value halfway between the lowest and highest values:

$$\text{midrange} = \frac{\text{lowest value} + \text{highest value}}{2}.$$

Examples 5, 6, page 774

Percentiles and **quartiles** are measures of position.

Examples 7–8, pages 776–777

Section 12.4

Measures of Dispersion

The **range** is the difference between the highest value and lowest value in a set of data.

Discussion, page 784;
Example 1, page 784

The **standard deviation**, s, is a measure of the spread of a set of data about the mean: $s = \sqrt{\dfrac{\Sigma(x - \bar{x})^2}{n - 1}}$.

Examples 2–3, pages 785–787

Section 12.5

z-SCORES

$$z = \frac{\text{value of the piece of data} - \text{mean}}{\text{standard deviation}}$$

Discussion, page 796;
Examples 2–5, pages 796–804

Section 12.6

Linear Correlation And Regression

Linear correlation coefficient, r, is

$$r = \frac{n(\Sigma xy) - (\Sigma x)(\Sigma y)}{\sqrt{n(\Sigma x^2) - (\Sigma x)^2}\sqrt{n(\Sigma y^2) - (\Sigma y)^2}}$$

r is always a value between -1 and 1, inclusive.

Discussion, pages 810–812;
Examples 1–2, pages 812–815

Equation of the Line of the Best Fit

$y = mx + b$, where

$$m = \frac{n(\Sigma xy) - (\Sigma x)(\Sigma y)}{n(\Sigma x^2) - (\Sigma x)^2}$$

$$b = \frac{\Sigma y - m(\Sigma x)}{n}$$

Discussion, pages 815–816;
Examples 3–4, pages 816–818

CHAPTER 12   *Review Exercises*

12.1

1. a) What is a population?

b) What is a sample?

2. a) What is a random sample?

b) What is a systematic sample?

c) What is a cluster sample?

d) What is a stratified sample?

e) What is a convenience sample?

In Exercises 3 and 4, tell what possible misuses or misinterpretations may exist in the statements.

3. The Stay Healthy Candy Bar indicates on its label that it has no cholesterol. Therefore, it is safe to eat as many of these candy bars as you want.

4. More copies of *Time* magazine are sold than are copies of *Kiplinger's* magazine. Therefore, *Time* is a more profitable magazine than *Kiplinger's*.

5. **Amazon Prime Subscriptions in the United States** In 2017, there were 90 million Amazon Prime subscribers in the United States. In 2018, there were 101 million Amazon Prime subscribers in the United States. Draw a bar graph that appears to show a

   a) small increase in the number of Amazon Prime subscribers from 2017 to 2018.

   b) large increase in the number of Amazon Prime subscribers from 2017 to 2018.

12.2

6. Consider the following set of data.

35	37	38	41	43
36	37	38	41	43
36	37	39	41	43
36	37	39	41	44
37	37	39	42	45

   a) Construct a frequency distribution letting each class have a width of 1.

   b) Construct a histogram of the frequency distribution.

   c) Construct a frequency polygon of the frequency distribution on the same set of axes as the histogram.

7. **Average Monthly High Temperature** Consider the following average monthly high temperature in July for 40 selected U.S. cities.

71	79	58	73	80	75	84	77
82	72	80	70	75	66	73	72
80	66	74	68	81	84	75	67
91	76	82	79	63	69	68	79
71	76	80	83	73	87	82	71

a) Construct a frequency distribution. Let the first class be 58–62.

b) Construct a histogram of the frequency distribution.

c) Construct a frequency polygon of the frequency distribution on the same set of axes as the histogram.

d) Construct a stem-and-leaf display.

12.3, 12.4

In Exercises 8–13, use the following test scores: 65, 72, 77, 81, 82, 91. Determine the

8. mean.

9. median.

10. mode.

11. midrange.

12. range.

13. standard deviation.

In Exercises 14–19, for the set of data: 9, 10, 17, 19, 24, 12, 17, 28, 12, 22, 20, 26, determine the

14. mean.

15. median.

16. mode.

17. midrange.

18. range.

19. standard deviation.

12.5

Police Response Time *In Exercises 20–24, assume that police response time to emergency calls is normally distributed with a mean of 9 minutes and a standard deviation of 2 minutes. Determine the percent of emergency calls with a police response time*

20. between 7 and 11 minutes.

21. between 5 and 13 minutes.

22. less than 12.2 minutes.

23. more than 12.2 minutes.

24. more than 7.8 minutes.

Pizza Delivery *In Exercises 25–28, assume that the amount of time to prepare and deliver a pizza from Pepe's Pizza is normally distributed with a mean of 20 minutes and standard deviation of 5 minutes. Determine the percent of pizzas that were prepared and delivered*

25. between 20 and 25 minutes.

26. in less than 18 minutes.

27. between 22 and 28 minutes.

28. If Pepe's Pizza advertises that the pizza is free if it takes more than 30 min to deliver, what percent of the pizza will be free?

12.6

29. *Hiking* The following table shows the number of hiking permits issued for a specific trail at Yellowstone National Park for selected years and the corresponding number of bears sighted by the hikers on that trail.

Hiking permits	765	926	1145	842	1485	1702
Bears	119	127	150	119	153	156

a) Construct a scatter diagram with hiking permits on the horizontal axis.

b) Use the scatter diagram in part (a) to determine whether you believe that a correlation exists between the number of hiking permits issued and the number of bears sighted by hikers. If so, is it a positive or negative correlation?

c) Calculate the correlation coefficient between the number of hiking permits issued and the number of bears sighted by hikers.

d) Determine whether a correlation exists at $\alpha = 0.05$.

e) Determine the equation of the line of best fit between the number of hiking permits issued and the number of bears sighted by hikers. Round both m and b to the nearest hundredth.

f) Assuming that this trend continues, use the equation of the line of best fit to estimate the number of bears sighted if 1500 hiking permits were issued.

30. *Daily Sales* For six weeks, Ace Hardware recorded the price of a particular item and the corresponding sales of that item as shown in the table below.

Price ($)	0.75	1.00	1.25	1.50	1.75	2.00
Number sold	200	160	140	120	110	95

a) Construct a scatter diagram with price on the horizontal axis.

b) Use the scatter diagram in part (a) to determine whether you believe that a correlation exists between the price of the item and number sold. If so, is it a positive or a negative correlation?

c) Determine the correlation coefficient between the weekly price and the number of items sold.

d) Determine whether a correlation exists at $\alpha = 0.05$.

e) Determine the equation of the line of best fit for the price and the number of items sold.

f) Use the equation in part (e) to estimate the number of items sold if the price is $1.60.

12.3–12.5

Men's Weight In Exercises 31–38, use the data below which was obtained from a study of the weights of adult men.

Mean	192 lb	First quartile	178 lb
Median	185 lb	Third quartile	232 lb
Mode	180 lb	86th percentile	239 lb
Standard deviation	23 lb		

31. What is the most common weight?

32. What weight did half of those surveyed exceed?

33. About what percent of those surveyed weighed more than 232 lb?

34. About what percent of those surveyed weighed less than 178 lb?

35. About what percent of those surveyed weighed more than 239 lb?

36. If 100 men were surveyed, what is the total weight of all the men?

37. What weight represents two standard deviations above the mean?

38. What weight represents 1.8 standard deviations below the mean?

12.2–12.5

For Exercises 39–50, use the following data, which show the number of children per family for 20 families.

0	3	1	2	1	4	0	0	7	6
2	2	2	2	3	3	0	1	0	1

In Exercises 39–44, determine the following.

39. Mean

40. Mode

41. Median

42. Midrange

43. Range

44. Standard deviation

45. Construct a frequency distribution.

46. Construct a histogram of the frequency distribution.

47. Construct a frequency polygon of the frequency distribution.

48. Does this distribution appear to be normal? Explain.

49. Do you think the number of children per family in the United States is a normal distribution? Explain.

50. Do you think this set of data is representative of the U.S. population? Explain.

CHAPTER 12   *Test*

In Exercises 1–6, for the set of data 22, 38, 38, 40, 47, determine the

1. mean.

2. median.

3. mode.

4. midrange.

5. range.

6. standard deviation.

In Exercises 7–9, use the set of data

26	28	35	46	49	56
26	30	36	46	49	58
26	32	40	47	50	58
26	32	44	47	52	62
27	35	46	47	54	66

to construct the following.

7. a frequency distribution; let the first class be 25–30

8. a histogram of the frequency distribution

9. a frequency polygon of the frequency distribution on the same set of axes as the histogram

Statistics on Salaries *In Exercises 10–15, use the following data on weekly salaries at Donovan's Construction Company.*

Mean	$820	First quartile	$770
Median	$790	Third quartile	$825
Mode	$815	79th percentile	$832
Standard deviation	$40		

10. What is the most common salary?

11. What salary did half the employees exceed?

12. About what percent of employees' salaries exceeded $770?

13. About what percent of employees' salaries was less than $832?

14. If the company has 100 employees, what is the total weekly salary of all employees?

15. What salary represents one standard deviation above the mean?

Anthropology *In Exercises 16–19, assume that anthropologists have determined that the akidolestes, a small primitive mammal believed to have lived with the dinosaurs, had a head circumference that was normally distributed with a mean of 42 cm and a standard deviation of 5 cm.*

16. What percent of head circumferences were between 36 and 53 cm?

17. What percent of head circumferences were greater than 35.75 cm?

18. What percent of head circumferences were greater than 48.25 cm?

19. What percent of head circumferences were less than 50 cm?

20. ***Time Spent Studying*** Six students provided the length of time they studied for a sociology exam and the grade they received.

Time studied (minutes)	20	40	50	60	80	100
Grade received (percent)	40	45	70	76	92	95

a) Construct a scatter diagram placing the time studied on the horizontal axis.

b) Use the scatter diagram in part (a) to determine whether you believe a correlation exists between the time studied and the grade received.

c) Determine the correlation coefficient between the time studied and the grade received.

d) Determine whether a correlation exists at $\alpha = 0.01$.

e) Determine the equation of the line of best fit, $y = mx + b$, between the time studied and the grade received. Round both m and b to the nearest hundredth.

f) Use the equation in part (e) to estimate the grade received for a student who studied for 75 minutes.

◄ *Graph theory can be used to determine efficient routes to make deliveries.*

13

Graph Theory

What You Will Learn

- Graphs, paths, and circuits including the Königsberg bridge problem
- Euler paths and Euler circuits
- Hamilton paths and Hamilton circuits including traveling salesman problems, the brute force method, and the nearest neighbor method
- Trees, spanning trees, and minimum-cost spanning trees

Graph theory is an important branch of mathematics that is used to represent and determine solutions to a variety of problems that affect businesses, governments, and other organizations. Anytime a problem involves making connections between various locations, graph theory can be used to help determine an efficient solution to the problem.

For example, graph theory can be used to determine the most efficient route for a FedEx driver to make multiple deliveries in the same city. Graph theory can be used by ride-sharing services, such as Uber and Lyft, to place its drivers in the optimal locations to maximize revenues. Graph theory can also be used by individuals when determining the quickest or least-expensive route when running errands daily.

Upon completion of this section, you will be able to:

- Represent problems using graphs.
- Understand paths, circuits, and bridges.

Sapreet is a Girl Scout leader who needs to deliver cookies to eight different homes in her town. Sapreet realizes there are many possible routes to accomplish this task. However, she would like to determine the shortest route to deliver the cookies to all eight homes and return to her home. In this section we will introduce a type of diagram called a *graph* that can be used to help analyze and solve such problems.

Why This Is Important   Graphs can be used to help solve everyday problems such as the one described above. Graphs also play an important role in solving problems from many branches of mathematics, science, business, and engineering. A basic understanding of graphs is a powerful problem-solving tool.

The study of graph theory can be traced back to the eighteenth century when people of the East Prussian town of Königsberg (now Kaliningrad, Russia) sought the solution to a popular problem. Königsberg was situated on both banks and two islands of the Prigel River. From Fig. 13.1, we see that the sections of town were connected with a series of seven bridges. The townspeople wondered if one could walk through town and cross all seven bridges without crossing any of the bridges twice. This question was presented to Swiss mathematician Leonhard Euler (pronounced "oiler," 1707–1783; see *Profile in Mathematics* on page 150). To study this problem, Euler reduced the problem to one that could be represented with a series of dots and lines. This problem came to be known as the Königsberg bridge problem.

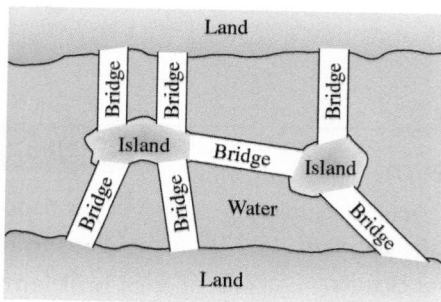

Figure 13.1

We will revisit the Königsberg bridge problem several times in this section and in Section 13.2. Our main focuses in this section are to introduce definitions of graph theory and to explain how these definitions can help represent problems from the physical world. In the next section, we will introduce some basic graph theory used to solve these problems. It should be noted that the *graphs* we discuss in this chapter are different from the graphs of equations and functions that we discussed in Chapter 6.

Graphs

A *graph* is a finite set of points called *vertices* (singular form is *vertex*) connected by line segments (not necessarily straight) called *edges* or *arcs*. An edge that connects a vertex to itself is called a *loop*. Vertices, edges, and a loop are displayed in the graph in Fig. 13.2. We generally refer to vertices with capitalized letters. The edge between two

Figure 13.2

A   *B*   — Loop

Not a vertex

— Edge

C   *D*   — Vertex

vertices will be referred to using the two vertices. For example, in Fig. 13.2, the edge that connects vertex *A* to vertex *B* is referred to as edge *AB* or as edge *BA*. The loop in Fig. 13.2 is referred to as edge *BB* or loop *BB*. Not every place where two edges cross is a vertex. A dot must be present to represent a vertex. For instance, in Fig. 13.2, even though edges *AD* and *BC* cross, since no dot is indicated, the place where these edges cross is not a vertex.

With these basic definitions, we can begin using graphs to represent physical settings.

RECREATIONAL MATH

Online Social Networks

Online social networks such as Facebook, Twitter, LinkedIn, Pinterest, Instagram, and Snapchat are Internet sites that share information among a network of people who are linked together. For example, if two Facebook users wish to be linked, they become *friends*. If a Facebook user posts any comments, pictures, or videos, the information is available to all of that user's friends. Social networks can be depicted with graphs where the people are represented with vertices and the friendship links are represented with edges. Such a graph for Facebook can be represented only with the help of a very large computer. As of June 2019, Facebook had over 2.38 billion users! Exercise 44 on page 840 asks you to draw a graph representing the friendships of six people on Facebook.

Example 1 *Representing the Königsberg Bridge Problem*

Using the definitions of vertex and edge, represent the Königsberg bridge problem with a graph.

Solution Note that each bridge in Fig. 13.1 on page 830 connects two pieces of land in a manner similar to an edge connecting two vertices. To begin representing the Königsberg bridge problem as a graph, we will label each piece of land with a capital letter, as shown in Fig. 13.3.

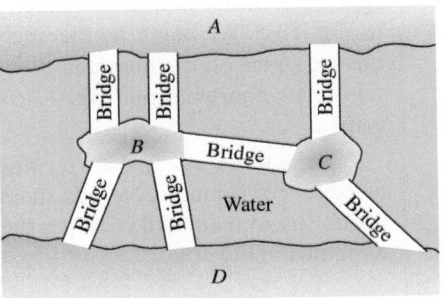

Figure 13.3

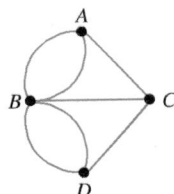

Figure 13.4

Next, we draw edges to represent the bridges. Notice that there are two bridges connecting land *A* to land *B* in Fig. 13.3. Therefore, we need two edges to connect vertex *A* to vertex *B* in our graph (see Fig. 13.4). Notice that there is one bridge connecting land *A* to land *C* in Fig. 13.3. Therefore, we need one edge to connect vertex *A* to vertex *C* in our graph. We continue to let edges represent bridges, and the resulting graph is given in Fig. 13.4.

Although the picture and the graph do not look very similar, the graph represents the key aspects of the problem: The land is represented with vertices, and the bridges are represented with edges. ∎

Graphs are powerful tools in problem solving because they allow us to focus on the key aspects of problems without getting distracted with unnecessary details. Our next three examples show other ways that graphs can be used to represent settings from the physical world.

Example 2 *Vacationing in the Southwest*

The Jugo family is vacationing in the southwestern portion of the United States. They plan to visit the states shown in Fig. 13.5 on page 832. Construct a graph to show the states that share a common border. States whose borders touch only at a single corner point will not be considered to share a common border.

▲ *The Grand Canyon*

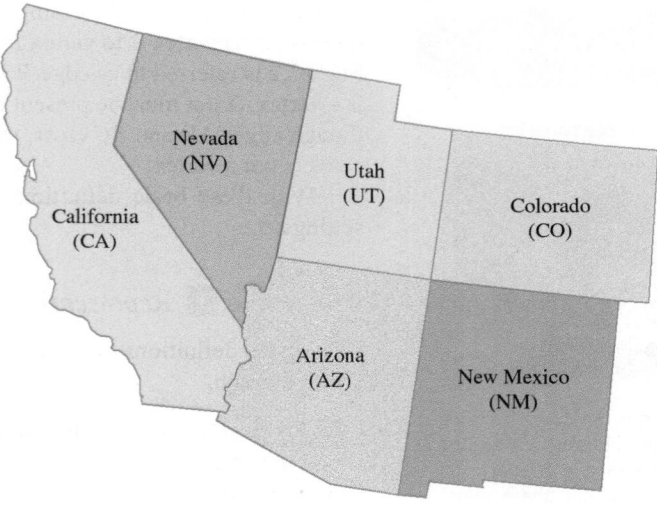

Figure 13.5

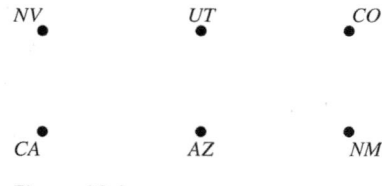

Figure 13.6

Solution In our graph, each vertex will represent one of the six states shown in Fig. 13.5. We begin by placing six vertices in the same relative positions as the six states on the map and then labeling each vertex with the corresponding two-letter abbreviation (Fig. 13.6). The exact placement of the vertices is not critical.

Next, if two states share a common border, connect the respective vertices with an edge. For example, Nevada shares a common border with Utah, Arizona, and California so there will be edges that connect vertex *NV* to vertices *UT*, *AZ*, and *CA*, as shown in Fig. 13.7.

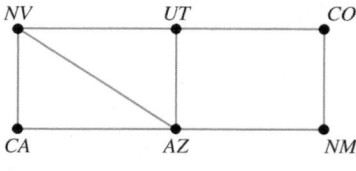

Figure 13.7

Next, note that Utah shares a common border with Nevada, Arizona, and Colorado. Therefore, the graph in Fig. 13.7 will have edges that connect *UT* to *NV*, *AZ*, and *CO*, respectively. Note that Utah and New Mexico have borders that touch only at a corner point and thus are not considered to share a common border. Therefore, in the graph shown in Fig. 13.7, no edge connects *UT* to *NM*. Likewise, Arizona and Colorado are not considered to share a common border; thus, no edge connects *AZ* to *CO*. To finish the graph, we continue adding edges until, for every two states in Fig. 13.5 that share a common border, there is an edge between their respective vertices in Fig. 13.7.   ■

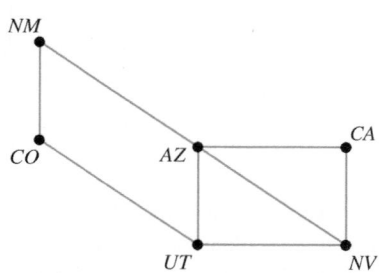

Figure 13.8

In Example 2, the graph shown in Fig. 13.7 is only one possible arrangement of vertices and edges that can show which states share a common border. Many other graphs that display the same relationship are equally valid. Figure 13.8 shows one such graph. Although the vertices in the graph in Fig. 13.8 do not resemble the respective locations of the states they represent, the relationship between the vertices is the same as in Fig. 13.7.

Example 3 *Representing a Floor Plan*

Figure 13.9 shows the floor plan of the Phenomenal Phitness gym. Use a graph to represent the floor plan.

Outside

Rae's Training Room	Locker Room A	Sierra's Training Room
Health/Weight Room		
Keegan's Training Room	Locker Room B	Chet's Training Room

Figure 13.9

Solution To create a graph, we will use vertices to represent each of the seven rooms and a separate vertex to represent the outside of the gym. To label the vertices, we will use the first letter of the trainers' names along with H for the health/weight room, A for locker room A, B for locker room B, and O for the outside of the gym. Next, connect the vertices with edges. To determine placement of the edges within the graph, visualize walking through various doorways in the building. For example, since a person can walk from Rae's training room directly into either the health/weight room or into locker room A, there is an edge between R and H, and an edge between R and A. The resulting graph is shown in Fig. 13.10. Notice that since the health/weight room has two separate doorways that lead to the outside of the building, there are two separate edges that connect vertices H and O.

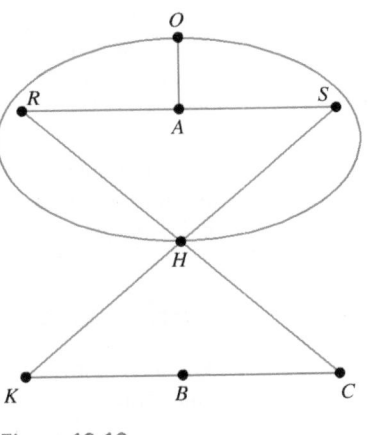

Figure 13.10

In Example 3, we placed the vertex O, for outside, near the top of the graph. If vertex O had been placed elsewhere, the graph would look different, but the graph would still represent the floor plan shown in Fig. 13.9.

Example 4 *Representing a Neighborhood*

Figure 13.11 on page 834 shows a sketch of the Havenwoods subdivision of homes. Use a graph to represent the streets of this neighborhood.

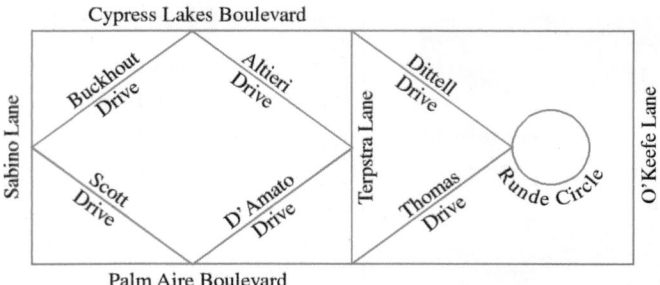

Figure 13.11

Solution In this problem, the vertices will represent the street intersections and the edges will represent the street blocks. We will use the letters $A, B, C, \ldots, K$ to name the vertices. The resulting graph is shown in Fig. 13.12.

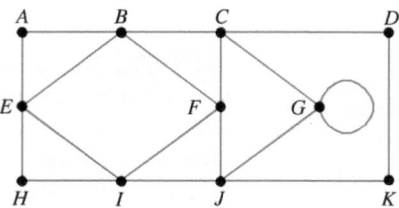

Figure 13.12

Now that we have some experience using graphs to represent real-life settings, we will introduce some additional vocabulary used in graph theory. We will begin with a method of classifying vertices. The *degree* of a vertex is the number of edges that connect to that vertex. For instance, in Fig. 13.13, vertex A has two edges connected to it. Therefore, the degree of vertex A is two. We also see from Fig. 13.13 that vertex B has three edges connected to it, as does vertex C. Therefore, vertices B and C each have degree three. Vertex D has a loop connected to it. A question arises: Should loop DD count as one edge or two edges when determining the degree of vertex D? We will agree to count each end of a loop when determining the degree of a vertex. Thus, vertex D from Fig. 13.13 will have degree four. A vertex with an even number of edges connected to it is an *even vertex*, and a vertex with an odd number of edges connected to it is an *odd vertex*. In Fig. 13.13, vertices A and D are even and vertices B and C are odd.

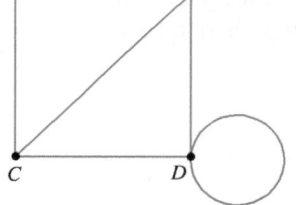

Figure 13.13

Paths, Circuits, and Bridges

We next introduce some definitions of graph theory that can be used to describe "movement."

Definition: **Path**
A **path** is a sequence of adjacent vertices and the edges connecting them.

By adjacent vertices, we mean two vertices that are connected by a common edge. For example, in Fig. 13.13, vertices A and B are adjacent vertices, since there is a common edge between vertex A and vertex B. Vertices A and D are not adjacent vertices, since there is not a common edge between vertex A and vertex D. An example of a path is given in Fig. 13.14 on page 835. It is important to recognize the difference between a graph and a path. In Fig. 13.14, the *graph* is the set of black vertices and the blue edges connecting them. The *path*, shown in red, can be thought of as movement from vertex C to vertex D to vertex A to vertex B. For convenience, and to help in explanations, paths will be referred to with the sequence of vertices separated by commas. For example, the path in Fig. 13.14 will be referred to as path C, D, A, B. Note that path C, D, A, B consists of three separate paths: C, D; D, A; and A, B.

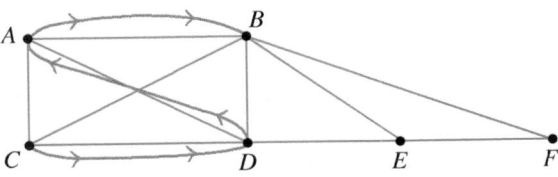

Figure 13.14

A path does not need to include every edge and every vertex of a graph. In addition, a path could include the same vertices and the same edges several times. For example, in Fig. 13.15, we see a graph with four vertices. The path $A, B, C, D, A, B, C, D, A, B, C, D, A, B, C$ starts at vertex A, "circles" the graph three times, and then goes through vertex B to vertex C.

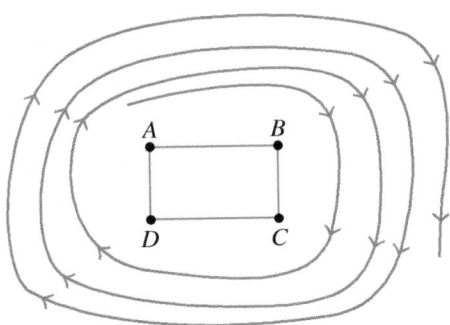

Figure 13.15

We now will discuss a special kind of path called a *circuit*.

Definition: Circuit
A **circuit** is a path that begins and ends at the same vertex.

Examine Fig. 13.16. The path given by A, C, B, D, A forms a circuit. The path given by B, D, E, B forms another circuit.

Note that every circuit is, by definition, a path; however, not every path is a circuit. A circuit needs to begin and end at the same vertex. For example, in Fig. 13.16, the path given by A, C, B, D is not a circuit, since it does not begin and end at the same vertex. In Sections 13.2 and 13.3, we will discuss paths and circuits in more depth.

Our next definitions classify graphs themselves. A graph is *connected* if, for any two vertices in the graph, there is a path that connects them. All the graphs we have studied thus far have been connected because a path existed between each pair of vertices in each graph. If a graph is not connected, it is *disconnected*. Figure 13.17 on page 836 shows three examples of disconnected graphs. In Fig. 13.17(c), although edge AD and edge BC cross, the graph is not connected because there is no vertex indicated where the edges cross.

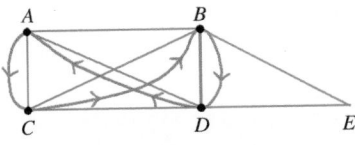

Figure 13.16

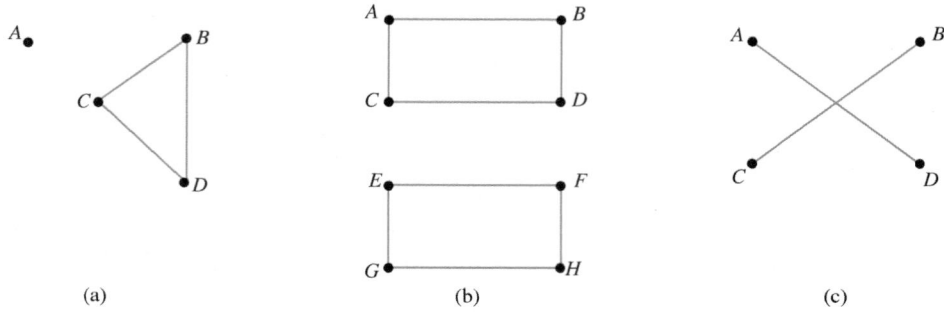

Figure 13.17

We next discuss an important type of edge called a *bridge*.

> **Definition: Bridge**
> A **bridge** is an edge that, if removed from a connected graph, would create a disconnected graph.

Figure 13.18 shows graphs with bridges indicated. Compare these graphs with those in Fig. 13.17. The graphs in Fig. 13.18 are the same graphs as those in Fig. 13.17 with a bridge added. Note in Fig. 13.18(c) that, in addition to edge *AB* that we added, edges *AD* and *BC* are also bridges. If edge *AD* were removed, vertex *D* would be isolated from the rest of the graph and we would have a disconnected graph. If edge *BC* were removed, vertex *C* would be isolated from the rest of the graph. Note that our graph theory definition of *bridge* is different from bridges as they are used in the Königsberg bridge problem. In the Königsberg bridge problem, none of the edges that represent real bridges is a bridge in the graph theory sense.

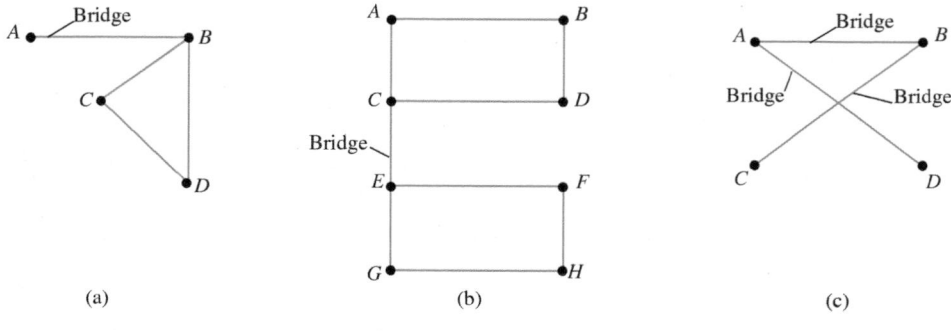

Figure 13.18

SECTION 13.1
Exercises

Warm Up Exercises

In Exercises 1–8, fill in the blanks with an appropriate word, phrase, or symbol(s).

1. A finite set of points connected by line segments is called a(n) _____.

2. A point in a graph is called a(n) _____.

3. A line segment in a graph is called a(n) _____.

4. An edge that connects a vertex to itself is called a(n) _____.

5. A sequence of adjacent vertices and the edges connecting them is called a(n) _____.

6. A path that begins and ends with the same vertex is called a(n) _____.

7. The number of edges that connect to a vertex is called the _____ of the vertex.

8. A bridge is an edge that, if removed from a connected graph, would create a(n) _____ graph.

Practice the Skills

In Exercises 9–14, create a graph with the given properties. There are many possible answers for each problem.

9. Two even and two odd vertices

10. Four odd vertices

11. Seven vertices and one bridge

12. Eight vertices and one bridge

13. Three even vertices and one loop

14. Four odd vertices and one loop

In Exercises 15–20, use the graph below to answer the following questions.

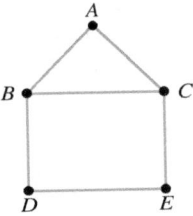

15. a) Is A, B, C, D, E a path? Explain.

 b) Is A, B, C, E, D a path? Explain.

 c) Is A, B, C, E, D a circuit? Explain.

 d) Is A, B, D, E, C, A a circuit? Explain.

16. a) Is A, C, B, E, D a path? Explain.

 b) Is D, E, C, B, A a path? Explain.

 c) Is D, E, C, B, A a circuit Explain.

 d) Is D, E, C, A, B, D a circuit? Explain.

17. a) On the graph, is it possible to determine a path that begins with vertex B, contains all the edges exactly once, and ends with vertex C? If so, determine one such path.

 b) On the graph, is it possible to determine a path that begins with vertex A, contains all the edges exactly once, and ends with vertex B? If so, determine one such path.

 c) On the graph, it possible to determine a circuit that begins and ends with vertex A and includes all the edges exactly one time? If so, determine one such circuit.

18. a) On the graph, is it possible to determine a path that begins with vertex C, contains all the edges exactly once, and ends with vertex B? If so, determine one such path.

 b) On the graph, is it possible to determine a path that begins with vertex D, contains all the edges exactly once, and ends with vertex E? If so, determine one such path.

 c) On the graph, is it possible to determine a circuit that begins and ends with vertex E and includes all the edges exactly one time? If so, determine one such circuit.

19. On the graph, is it possible to determine a circuit that begins and ends with vertex B and contains all the other vertices exactly once? If so, determine one such circuit.

20. On the graph, is it possible to determine a circuit that begins and ends with vertex C and contains all the other vertices exactly once? If so, determine one such circuit.

Problem Solving

Modified Königsberg Bridge Problems *In Exercises 21 and 22, suppose that the people of Königsberg decide to add several bridges to their city. Two such possibilities are shown. Create graphs that would represent the Königsberg bridge problem with these new bridges.*

21.

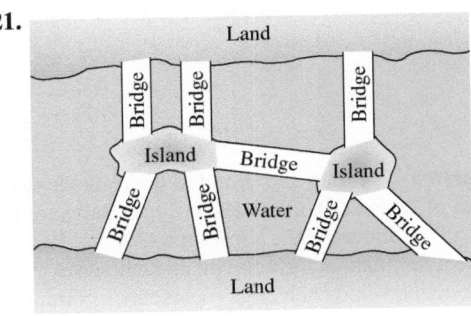

22.

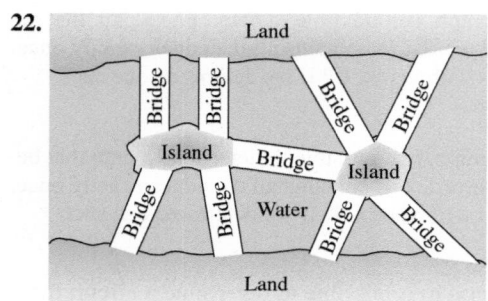

Vacations *In Exercises 23 and 24, the maps of states that the Jugo family (see Example 2 on page 831) is planning to visit on future vacations are shown. Represent each map as a graph where each vertex represents a state and each edge represents a common border between the states. Use the state abbreviations to label the vertices of the graph.*

23. Northwestern states

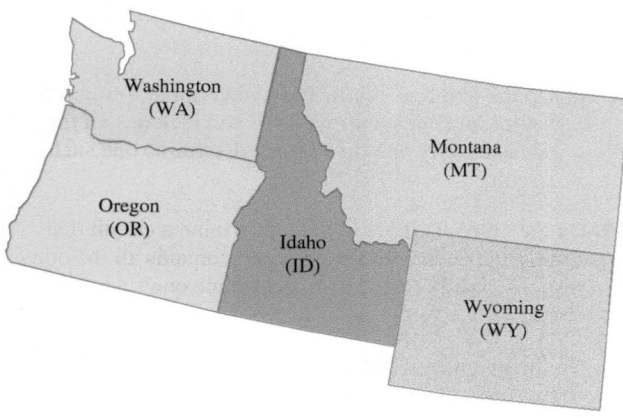

24. Southeastern states

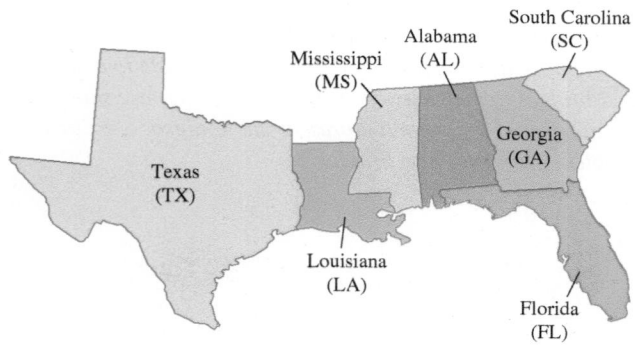

25. *Central America* The map above and to the right shows the countries of Belize (*B*), Costa Rica (*C*), El Salvador (*E*), Guatemala (*G*), Honduras (*H*), Nicaragua (*N*), and Panama (*P*). Represent the map as a graph where each vertex represents a country and each edge represents a common border between the countries. Use the letters indicated in parentheses to label the vertices.

26. *Northern Africa* The map below shows the countries of Algeria (*A*), Egypt (*E*), Libya (*L*), Morocco (*M*), Sudan (*S*), Tunisia (*T*), and Western Sahara (*W*). Represent the map as a graph where each vertex represents a country and each edge represents a common border between the countries. Use the letters indicated in parentheses to label the vertices.

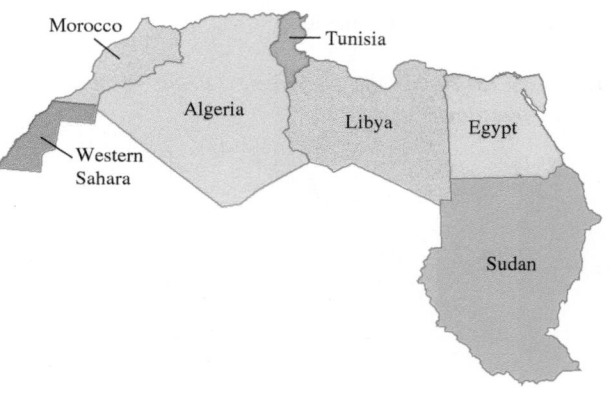

For Exercises 27–30, the first floors of the floor plans offered by Amani Builders are shown. Use a graph to represent each floor plan. Use the letter O, near the top or the bottom of the graph, to represent the outside of the building.

27. *Amore model*

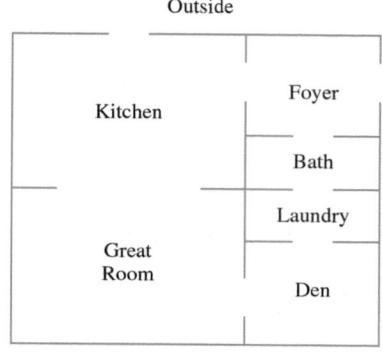

28. Esperanza model

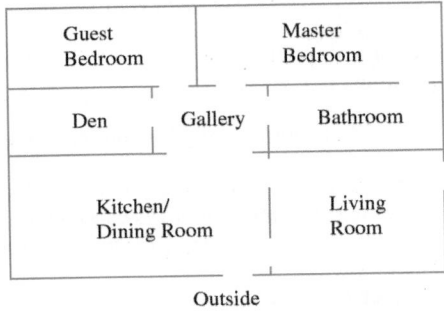

29. Caridad model

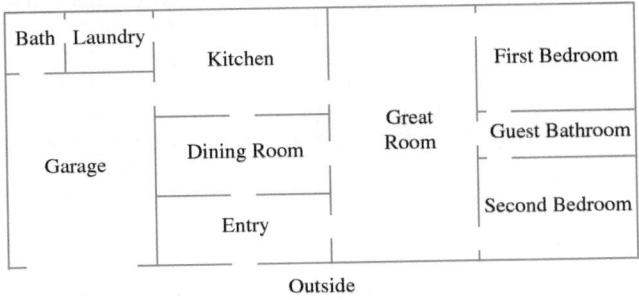

30. Algeria model

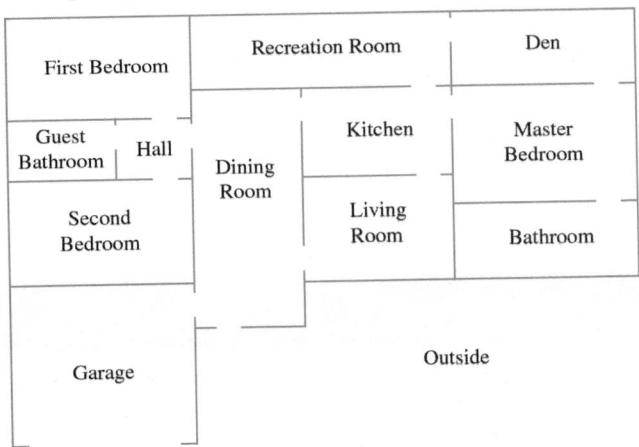

31. Representing a Neighborhood The map of the Marlow Heights subdivision is shown below. Use a graph to represent the streets in this subdivision. Use letters to label the vertices.

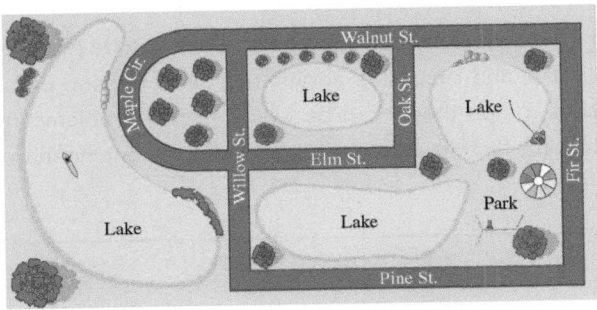

32. Representing a Neighborhood The map of the Temple Hills subdivision is shown below. Use a graph to represent the streets in this subdivision. Use letters to label the vertices.

In Exercises 33–36, determine whether the graph shown is connected or disconnected.

33.
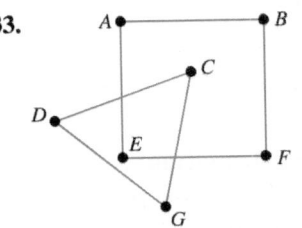

34.
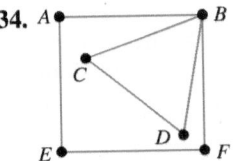

35.
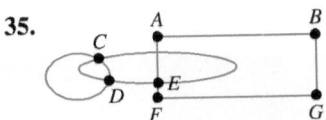

36.
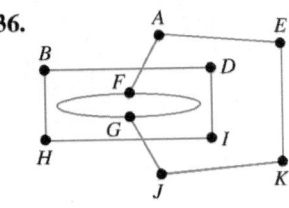

In Exercises 37–40, a connected graph is shown. For each graph, identify any (a) bridges and (b) loops.

37.
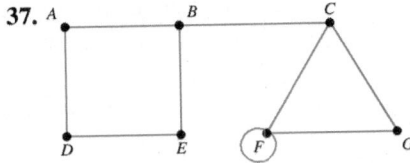

38.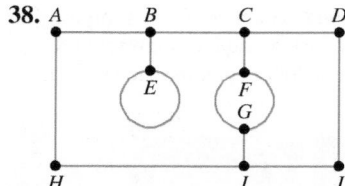

39.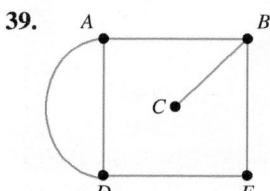

40.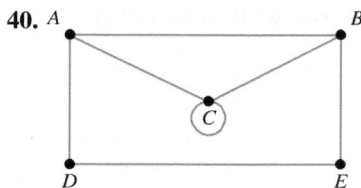

Challenge Problems/Group Activities

41. *Class Poll* Poll your entire class to determine which students knew each other prior to taking this course. Draw a graph that shows this relationship.

42. *Odd Vertices* Attempt to draw a graph that has an odd number of odd vertices. What conclusion can you draw from this exercise?

43. *Degrees* Draw four different graphs and then for each graph:

 a) determine the degree of each vertex.

 b) determine the sum of the degrees from all vertices from this same graph.

 c) determine the number of edges on this same graph.

 d) What conclusion relating the sum of the degrees of vertices to the number of edges can you draw from this exercise? Explain why this conclusion is true.

Recreational Mathematics

44. *Facebook Friends* Read the *Recreational Mathematics* box on page 831. Belinda, Dennis, Jeff, Kristie, Lucy, and Mike are all on Facebook. The following pairs of people are friends: Belinda and Dennis, Belinda and Kristie, Belinda and Lucy, Dennis and Kristie, Dennis and Lucy, Dennis and Mike, Jeff and Lucy, Kristie and Lucy, and Kristie and Mike. Represent the friendships with a graph.

45. *Home Floor Plan* Use a graph to represent

 a) the floor plan of your home.

 b) the streets in your neighborhood within $\frac{1}{2}$ mile from your home in all directions.

Research Activity

46. *Continent Graph* Choose a continent other than North America or Australia and create a graph showing the common boundaries between countries.

SECTION 13.2   Euler Paths and Euler Circuits

Upon completion of this section, you will be able to:

■ Solve problems involving Euler paths and Euler circuits.

A snowplow driver must plow every road in the township. To minimize the time and cost involved, the driver would like to establish a route that would allow the plow to travel over each road exactly one time. In this section, we will study paths and circuits that can be used to determine the best route for the snowplow driver.

Why This Is Important There are many real-life problems that can be represented and solved using graphs, paths, and circuits. In addition to transportation problems like the one described above, graph theory can be used to solve problems involving city planning, computer architecture, interior design, industrial engineering, and many other fields.

Section 13.1 provided us with some of the basic definitions of graph theory. Although we were able to represent physical settings with graphs, we have not yet discussed solving problems related to these settings. In this section, we will provide more details of graph theory, which allows us to reach solutions to the Königsberg bridge problem and other real-world problems. Before we do so, we need to give two more definitions.

> **Definition: Euler Path**
> An **Euler path** is a path that passes through each edge of a graph exactly one time.

If a graph has an Euler path, we say the graph is *traversable*. If a graph has an Euler path, or is traversable, it is possible to trace the entire graph without removing the pencil from the paper and without tracing an edge more than once. For our next definition, recall that a circuit is a path that begins and ends at the same vertex.

> **Definition: Euler Circuit**
> An **Euler circuit** is a circuit that passes through each edge of a graph exactly one time.

The difference between an Euler path and an Euler circuit is that an Euler circuit must start and end at the same vertex. An alternate definition for an Euler circuit is that it is an Euler path that begins and ends at the same vertex. To become familiar with Euler paths and Euler circuits, examine the graphs in Fig. 13.19(a) and (b). As we discuss each graph, trace each path or circuit with a pencil or with your finger, or draw the path or circuit on a separate sheet of paper.

In Fig. 13.19(a), the path $D, E, B, C, A, B, D, C, E$ is an Euler path, since each edge was traced only one time. However, since the path begins at vertex D and ends at a different vertex, E, the path is not a circuit and therefore is not an Euler circuit. In Fig. 13.19(b), the path $D, E, B, C, A, B, D, C, E, F, D$ is an Euler path, since each edge was traced only one time. Furthermore, since the path begins and ends with the same vertex, D, it is also an Euler circuit. Note that every Euler circuit is an Euler path, but not every Euler path is an Euler circuit.

Our next task is to determine whether a given graph has an Euler path, an Euler circuit, neither, or both. In the next example, we will use trial and error to determine Euler paths and Euler circuits. Our discussion will help us introduce *Euler's theorem*, which will be used to solve the Königsberg bridge problem and other problems involving graph theory.

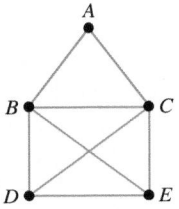

An Euler path $D, E, B, C, A, B, D, C, E$

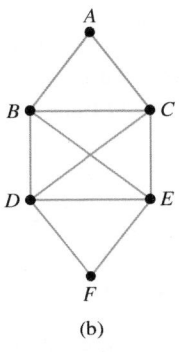

(b)

An Euler circuit $D, E, B, C, A, B, D, C, E, F, D$

Figure 13.19

Example 1 *Euler Paths and Circuits*

For the graphs shown in Fig. 13.20, determine if an Euler path, an Euler circuit, neither, or both exist.

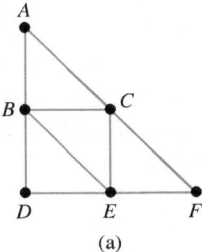
(a)

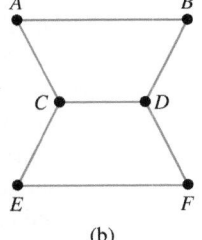
(b)

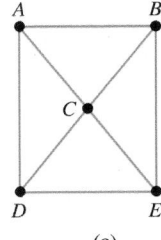
(c)

Figure 13.20

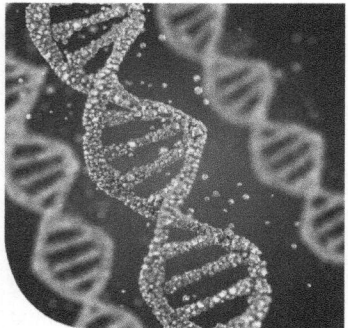
Graph theory is part of a larger branch of mathematics known as *topology*. Although topology involves concepts from other older branches of mathematics, such as geometry, algebra, and analysis, topology was not recognized as a separate branch of mathematics until the 1920s and 1930s when the writings of Solomon Lefschetz (see Profile in Mathematics on page 867) established many of the fundamental aspects of the field of study. Topology is frequently referred to as rubbersheet geometry. This descriptive name refers to topology's focus on geometric properties that remain consistent even when an object is bent, twisted, or stretched (see Section 8.6). One of the more interesting areas within topology is knot theory. Once a very obscure area of mathematics, knot theory is becoming increasingly important because of its applications to science. Among other things, knot theory is used to study DNA molecules, to discover new drugs, and to study how infectious diseases spread.

Solution

a) The graph in Fig. 13.20(a) has many Euler circuits, each of which is also an Euler path. One Euler circuit is *A, B, D, E, B, C, E, F, C, A*. Trace this circuit now. Notice that you traced each edge exactly one time and that you started and finished with vertex *A*. Notice that this graph has no odd vertices (and therefore has all even vertices). This graph has both Euler paths and Euler circuits.

b) Using trial and error, we can see that the graph in Fig. 13.20(b) has an Euler path but it does not have an Euler circuit. One Euler path is *C, A, B, D, C, E, F, D*. Another Euler path is *D, B, A, C, D, F, E, C*. There are actually several Euler paths for this graph, but each one must begin at either vertex *C* or vertex *D*. If the Euler path begins at vertex *C*, it must end at vertex *D*, and vice versa (you should try to determine a few more to confirm this observation). Notice that there are exactly two odd vertices, vertex *C* and vertex *D*.

c) Our attempts to trace the graph in Fig. 13.20(c) lead to either tracing at least one edge twice or to omitting at least one of the edges. We must therefore conclude that this graph has neither an Euler path nor an Euler circuit. Notice that this graph has more than two odd vertices. ■

We are now ready to introduce Euler's theorem, which is used to determine if a graph contains Euler paths and Euler circuits. As you will see, the number of *odd* vertices of a graph determines whether the graph has Euler paths and Euler circuits.

Euler's Theorem

For a connected graph, the following statements are true.

1. A graph with *no odd vertices* (all even vertices) has at least one Euler path, which is also an Euler circuit. An Euler circuit can be started at any vertex, and it will end at the same vertex.

2. A graph with *exactly two odd vertices* has at least one Euler path but no Euler circuits. Each Euler path must begin at one of the two odd vertices, and it will end at the other odd vertex.

3. A graph with *more than two odd vertices* has neither an Euler path nor an Euler circuit.

The proof of this theorem is beyond the scope of this book. In Example 2, we will reexamine the graphs used in Example 1. However, this time we will use Euler's theorem to determine whether the graph has Euler paths and Euler circuits.

Example 2 *Using Euler's Theorem*

Use Euler's theorem to determine whether an Euler path or an Euler circuit exists in Fig. 13.20(a), (b), and (c) on page 841.

Solution

a) The graph in Fig. 13.20(a) has no odd vertices (all the vertices are even). According to item 1 in Euler's theorem, at least one Euler circuit exists. An Euler circuit can be determined starting at any vertex. The Euler circuit will end at the vertex from which it started. Recall that each Euler circuit is also an Euler path.

b) We see that the graph in Fig. 13.20(b) has four even vertices (*A, B, E,* and *F*) and two odd vertices (*C* and *D*). Based on item 2 in Euler's theorem, we conclude that since there are exactly two odd vertices, at least one Euler path exists but no Euler circuit exists. Each Euler path must begin at one of the odd vertices and end at the other odd vertex.

c) The graph in Fig. 13.20(c) has four odd vertices (*A, B, D,* and *E*) and one even vertex (*C*). According to item 3 in Euler's theorem, since this graph has more than two odd vertices, the graph has neither an Euler path nor an Euler circuit. ■

The results determined in Example 2 using Euler's theorem are the same as the results determined in Example 1 using trial and error. Although both methods led to the same result, Euler's theorem is superior to trial and error because it gives us a tool to examine more complicated graphs.

Next, we will use Euler's theorem to solve questions regarding the Königsberg bridge problem as well as problems relating to the graphs we developed in Section 13.1.

Example 3 *Solving the Königsberg Bridge Problem*

In Section 13.1, Example 1 on page 831, we discussed the Königsberg bridge problem and drew a graph to represent the situation. The graph is repeated in Fig. 13.21. Could a walk be taken through Königsberg during which each bridge is crossed exactly one time?

Solution The vertices of the graph given in Fig. 13.21 represent the land areas, and the edges of the graph represent the bridges in the Königsberg bridge problem. The original problem posed by the townspeople of Königsberg, in graph theory terms, was whether or not an Euler path exists for the graph in Fig. 13.21. From the figure, we see that the graph has four odd vertices (*A*, *B*, *C*, and *D*). Thus, according to item 3 of Euler's theorem, no Euler path exists. Therefore, no walk could be taken through Königsberg in which each bridge was crossed exactly one time. ■

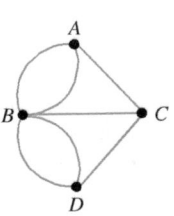

Figure 13.21

▲ *The Great Salt Lake*

Example 4 *Vacationing in the Southwest*

In Section 13.1, Example 2, on page 831, we discussed the Jugo family, who are vacationing in the Southwestern portion of the United States. We drew a graph showing the states they plan to visit that share a common border. The graph is repeated in Fig. 13.22.

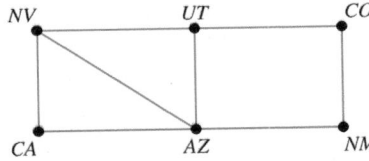

Figure 13.22

a) Is it possible for the Jugo Family to travel among these six states and cross each common state border (or use each edge) exactly one time?

b) If yes, determine a path that the Jugo Family could take to travel between these states and cross each common state border exactly one time.

Solution

a) Since the Jugo Family must use each edge (or common border) exactly one time, they are seeking an Euler path for the graph in Fig. 13.22. Notice that this graph has exactly two odd vertices, *NV* and *UT*. According to item 2 of Euler's theorem, at least one Euler path, but no Euler circuit, exists. Therefore, it *is* possible for them to travel among these states and cross each common state border exactly one time. However, since there is no Euler circuit, they must start their travel and end their travel in different states.

b) Many Euler paths exist, but each one must either start with *NV* and end with *UT* or start with *UT* and end with *NV*. One such path is *NV, UT, CO, NM, AZ, CA, NV, AZ, UT*. There are many other Euler paths on this graph. You should try to determine at least two now. ■

Euler's theorem can be used to determine whether a graph has an Euler path or an Euler circuit. If an Euler path or Euler circuit exists, how can we determine it? We can always attempt to determine Euler paths and Euler circuits using trial and error. This strategy might work well for simpler graphs. However, we would like to attempt this task with a more systematic approach. Our final topic in this section, Fleury's algorithm, will give us such an approach. An *algorithm* is a procedure for accomplishing some task.

Euler Circuits

PROCEDURE FLEURY'S ALGORITHM

To determine an Euler path or an Euler circuit:

1. Use Euler's theorem to determine whether an Euler path or an Euler circuit exists. If one exists, proceed with Steps 2–5.

2. If the graph has no odd vertices (therefore has at least one Euler circuit, which is also an Euler path), choose any vertex as the starting point. If the graph has exactly two odd vertices (therefore has only an Euler path), choose one of the two odd vertices as the starting point.

3. Begin to trace edges as you move through the graph. Number the edges as you trace them. Since you can't trace any edges twice in Euler paths and Euler circuits, once an edge is traced consider it "invisible."

4. When faced with a choice of edges to trace, if possible, choose an edge that is not a bridge (i.e., don't create a disconnected graph with your choice of edges).

5. Continue until each edge of the entire graph has been traced once.

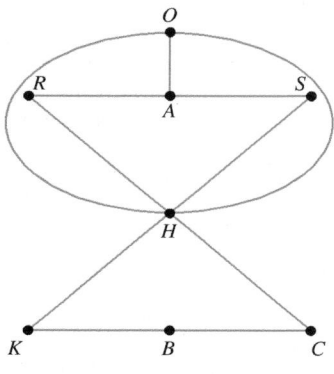

Figure 13.23

⌐ Example **5** *Locking a Fitness Center*

In Section 13.1, Example 3 on page 833, we introduced the Phenomenal Phitness gym and drew the graph that represented the floor plan. The graph is repeated in Fig. 13.23. Joe runs a gym-cleaning service and is responsible for cleaning the gym and locking all the doors after the gym closes each night. Joe wishes to determine a way to move through the gym using each doorway exactly once.

a) Is it possible for Joe to move through the gym using each doorway of the gym exactly one time? In other words, does an Euler path exist in the graph in Fig. 13.23?

b) If so, determine one such Euler path.

Solution

a) To determine if there is an Euler path, we will use Euler's theorem. The graph in Fig. 13.23 contains exactly two odd vertices, O and A. By item 2 of Euler's theorem, the graph has at least one Euler path (but no Euler circuit). By Euler's theorem, the Euler path would have to begin at one of the two odd vertices, either O or A, and end at the other odd vertex. Therefore, Joe could move through the gym and use each doorway of the gym exactly one time by starting at the outside (vertex O) or at locker room A (vertex A).

b) In part (a), we determined that an Euler path exists. Now we use Fleury's algorithm to determine one such path. We can start at either odd vertex, O or A. Let us choose to start at vertex O (see Fig. 13.23). As a matter of convention, once we trace an edge, we will redraw it using a dashed, instead of solid, edge. The dashes will indicate that the edge has already been traced and so it cannot be traced again. At vertex O, there are three choices of edges: the left-side edge OH, the right-side edge OH, or edge OA. None of these edges is a bridge. Thus, if we were to remove any one of these edges, we would not create a disconnected graph. Since none of these edges is a bridge, we can choose to trace any one of these edges. Let us choose the left-side edge OH as our first edge and number it 1 (see Fig. 13.24).

Now closely examine our position at vertex H. We now must make a choice among the following edges: HR, HS, HK, HC, and the right-side edge HO. None of these edges is a bridge. Thus, if we were to remove any of these edges, we would not create a disconnected graph. Since none of these edges is a bridge, we can choose to trace any one of these edges. Let us choose edge HK and label it edge 2 as shown in Fig. 13.24.

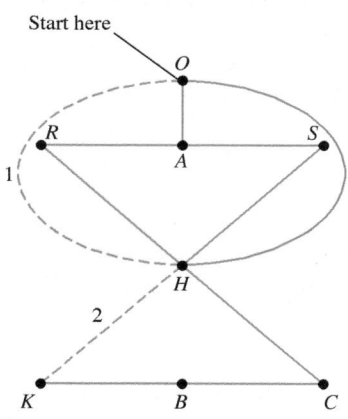

Figure 13.24

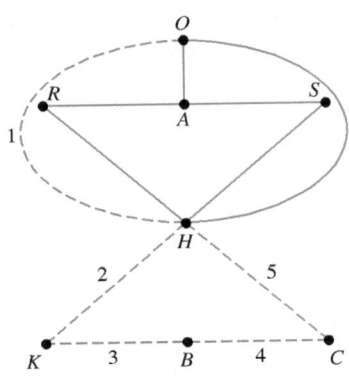

Figure 13.25

Now we are at vertex *K*. At vertex *K*, we have no choice but to select edge *KB*, followed by edge *BC*, followed by edge *CH*. We will choose these edges and label them edges 3, 4, and 5, respectively, as in Fig. 13.25. We are now back at vertex *H*.

Examine our position at vertex *H* in Fig. 13.25. There are three choices of edges: *HR*, *HS*, and right-side edge *HO*. None of these edges is a bridge, so we may choose to trace any one of these edges. Let us choose edge *HR* and label it 6. That will put us at vertex *R*, where our only choice is to trace edge *RA* and label it 7. We are now at vertex *A*. At vertex *A*, there is a choice between edges *AO* and *AS*. Because neither edge is a bridge, we may choose either one of these edges. Let us choose edge *AS* and label it 8. Our choices up to this point are shown in Fig. 13.26.

We are now at vertex *S*, where our only choice is to trace edge *SH* followed by edge *HO*, followed by edge *OA*. These edges are labeled 9, 10, and 11, respectively, and the completed Euler path is given in Fig. 13.27.

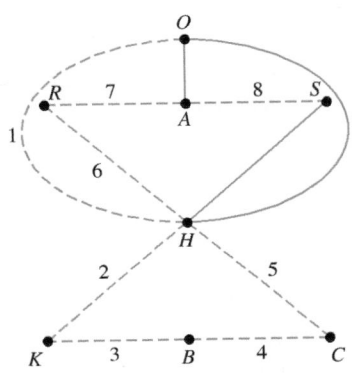

Figure 13.26

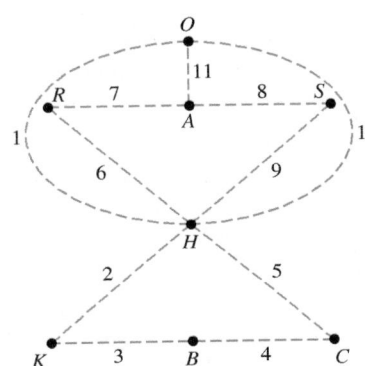

Figure 13.27

Notice that we started at one odd vertex, *O*, and ended at the other odd vertex, *A*. Therefore, if Joe followed the path *O, H, K, B, C, H, R, A, S, H, O, A*, he would travel through each doorway exactly one time. Many other Euler paths also exist for this problem. You will be asked to determine another one in Exercise 46 on page 852.

Example 6 *Crime Stoppers Problem*

In Section 13.1, Example 4, on page 833, we introduced the Havenwoods subdivision of homes and drew a graph to represent it. The graph is repeated in Fig. 13.28 below. The Havenwoods Neighborhood Association is planning to organize a crime stopper group in which residents take turns walking through the neighborhood with cell phones to report any suspicious activity to the police.

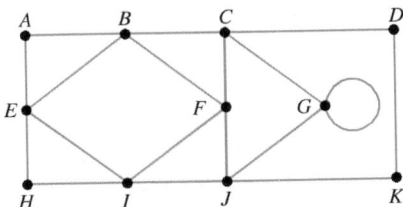

Figure 13.28

a) Can the residents of Havenwoods start at one intersection (or vertex) and walk each street block (or edge) in the neighborhood exactly once and return to the intersection where they started?

b) If yes, determine a circuit that could be followed to accomplish their walk.

> Solution

a) Since the residents wish to start at one intersection (or vertex) and return to the same intersection (or vertex), we need to determine if this graph has an Euler circuit. From Fig. 13.28 on page 845, we see that there are no odd vertices. Therefore, by item 1 of Euler's theorem, we determine that there is at least one Euler circuit.

b) In part (a), we determined that an Euler circuit exists. Now we use Fleury's algorithm to determine one such circuit. Euler's theorem states that we can start at any vertex to determine an Euler circuit. Let us choose to begin at vertex A. We face a choice of tracing edge AB or edge AE. Since neither edge AB nor edge AE is a bridge, we can choose either edge. Let us choose edge AB. Our first choice is indicated in Fig. 13.29.

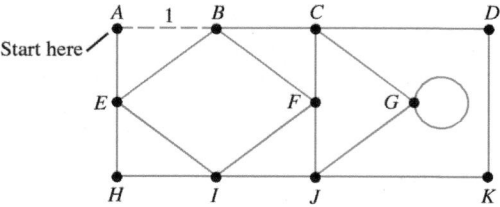

Figure 13.29

We will continue to trace from vertex to vertex around the outside of the graph, as indicated in Fig. 13.30. Notice that no edge chosen is a bridge.

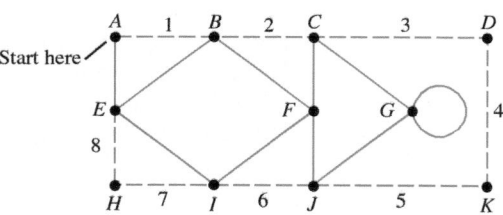

Figure 13.30

We are now at vertex E. Our choices are edge EA, edge EB, or edge EI. Edge EA is a bridge because removing edge EA would leave vertex A disconnected from the rest of the untraced graph. Neither edge EB nor edge EI is a bridge. Therefore, we need to choose either edge EB or edge EI. Let us choose edge EB. Now we are at vertex B and must choose edge BF. Our choices so far are shown in Fig. 13.31.

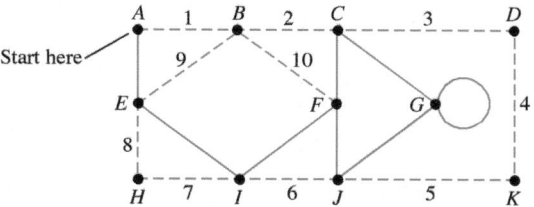

Figure 13.31

We are now at vertex *F* with choices of edge *FC*, edge *FJ*, or edge *FI*. Notice that edge *FI* is a bridge (it connects vertices *A*, *E*, and *I* to the rest of the untraced graph). We can therefore choose either edge *FC* or edge *FJ*. Let us choose edge *FC*, which will put us at vertex *C* (see Fig. 13.32). At vertex *C*, our only choice is edge *CG*.

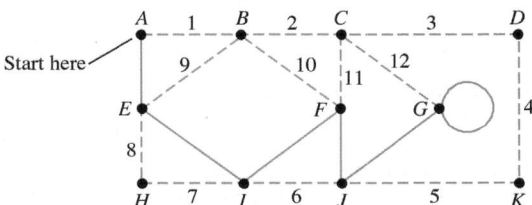

Figure 13.32

We are now at vertex *G* and have to choose between edge *GJ* and edge *GG* (which is a loop): see Fig. 13.32. Since *GJ* is a bridge, we choose the edge *GG*. We then are back at vertex *G*. Here and throughout the rest of our circuit we only have one choice at each of the remaining vertices. We trace each of these remaining edges to get the result given in Fig. 13.33.

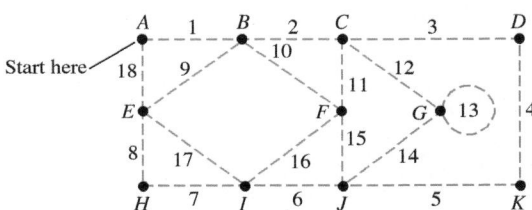

Figure 13.33

The circuit given in Fig. 13.33 gives one possible Euler circuit that would provide the residents of Havenwoods with one way to cover each street block exactly one time starting and ending at vertex *A*. Fleury's algorithm could be used to determine many alternative circuits. You will be asked to determine several such alternative circuits in Exercises 47 through 50 on page 852. ■

In our first two sections on graph theory, we introduced key concepts and theory that allowed us to represent physical settings as graphs and to solve some problems relating to these settings. We solved problems dealing with the bridges of Königsberg, a vacation in the southwestern United States, the doorways of a gym, and the street blocks involved in a neighborhood crime stopper program. Although these problems are very different, they are all related because in each case a graph is used to represent a physical setting.

There are many other examples of how Euler paths and Euler circuits can be applied to real-life problems, including problems that deal with the most efficient route for a snowplow or a street sweeper. Another example of a path problem is found in the classic arcade games Pac-Man and Ms. Pac-Man (Fig. 13.34). In both these games, the ideal path (disregarding the ghosts) is an Euler path.

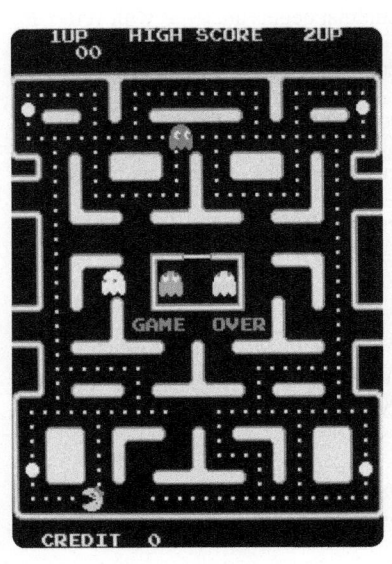

Figure 13.34 The game Ms. Pac-Man

SECTION 13.2   Exercises

Warm Up Exercises

In Exercises 1–6, fill in the blanks with an appropriate word, phrase, or symbol(s).

1. A path that passes through each edge of a graph exactly one time is called a(n) _____ path.

2. A circuit that passes through each edge of a graph exactly one time is called a(n) _____ circuit.

3. A connected graph has at least one Euler path that is also an Euler circuit, if the graph has _____ odd vertices.

4. A connected graph has at least one Euler path, but no Euler circuits, if the graph has exactly _____ odd vertices.

5. A connected graph has neither an Euler path nor an Euler circuit, if the graph has more than two _____ vertices.

6. If a connected graph has exactly two odd vertices, *A* and *B*, then each Euler path must begin at vertex *A* and end at vertex _____, or begin at vertex *B* and end at vertex *A*.

Practice the Skills

For Exercises 7–10, use the following graph.

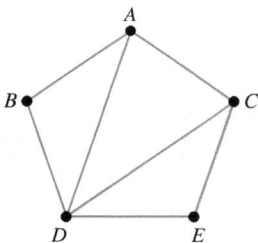

7. a) Is it possible to determine an Euler path that begins with vertex *A*? If so, determine one such Euler path.

b) Is it possible to determine an Euler path that begins with vertex *B*? If so, determine one such Euler path.

8. a) Is it possible to determine an Euler path that begins with vertex *C*? If so, determine one such Euler path.

b) Is It possible to determine an Euler path that begins with vertex *D*? If so, determine one such Euler path.

9. Is it possible to determine an Euler circuit that begins with vertex *E*? If so, determine one such Euler circuit.

10. Is it possible to determine an Euler circuit for this graph? If so, determine one such Euler circuit.

For Exercises 11–14, use the following graph.

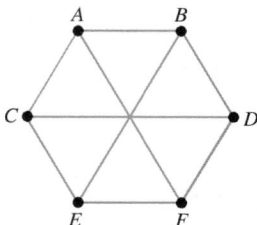

11. Is it possible to determine an Euler path that begins with vertex *A*? If so, determine one such Euler path.

12. Is it possible to determine an Euler path that begins with vertex *B*? If so, determine one such Euler circuit.

13. Is it possible to determine an Euler circuit that begins with vertex *C*? If so, determine one such Euler circuit.

14. Is it possible to determine an Euler path that begins with vertex *D*? If so, determine one such Euler path.

For Exercises 15–20, use the following graph.

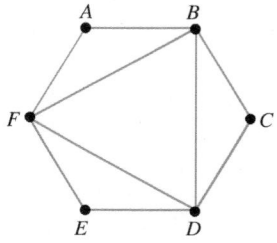

15. Determine an Euler circuit that begins and ends with vertex *A*.

16. Determine an Euler circuit that begins and ends with vertex *B*.

17. Determine an Euler circuit that begins and ends with vertex *C*.

18. Determine an Euler circuit that begins and ends with vertex *D*.

19. Determine an Euler circuit that begins and ends with vertex *E*.

20. Determine an Euler circuit that begins and ends with vertex *F*.

Problem Solving

Revisiting the Königsberg Bridge Problem *In Exercises 21 and 22, suppose that the people of Königsberg decide to add several bridges to their city. Two such possibilities are shown on page 849.*

a) *Would the townspeople be able to walk across all the bridges without crossing the same bridge twice?*

b) *If so, where should they begin and where would they end?*

21.

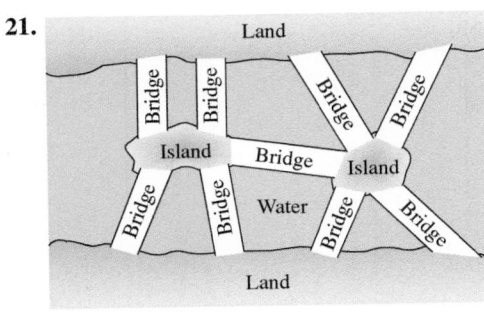

22.

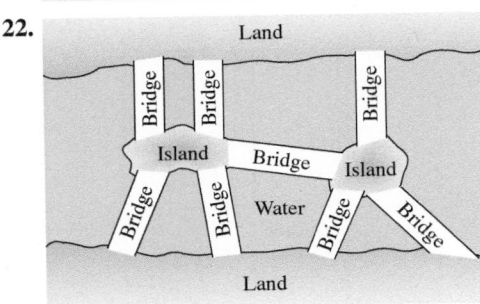

a) *Represent the map as a graph. Use the state abbreviations to label the vertices of the graph.*

b) *Determine (state yes or no) whether the graph in part (a) has an Euler path or an Euler circuit. If yes, give one such Euler path or circuit.*

Areas of the World *In Exercises 25–28, use each map shown.*

a) *Represent the map as a graph. Use the letter indicated in parentheses to label vertices of the graph.*

b) *Determine (state yes or no) whether the graph in part (a) has an Euler path. If yes, give one such Euler path.*

c) *Determine (state yes or no) whether the graph in part (a) has an Euler circuit. If yes, give one such Euler circuit.*

Vacations *In Exercises 23 and 24, the maps of states that the Jugo family (see Example 4 on page 843) is planning to visit on future vacations are shown. Use these graphs to complete the exercises.*

23. Northwestern states

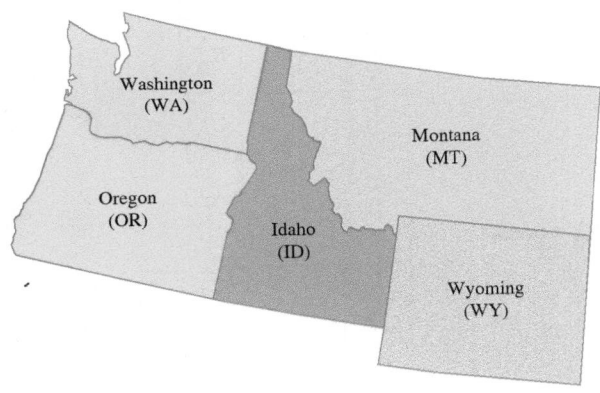

a) *Represent the map as a graph. Use the state abbreviations to label the vertices of the graph.*

b) *Determine (state yes or no) whether the graph in part (a) has an Euler circuit. If yes, give one such Euler circuit.*

24. Southeastern states

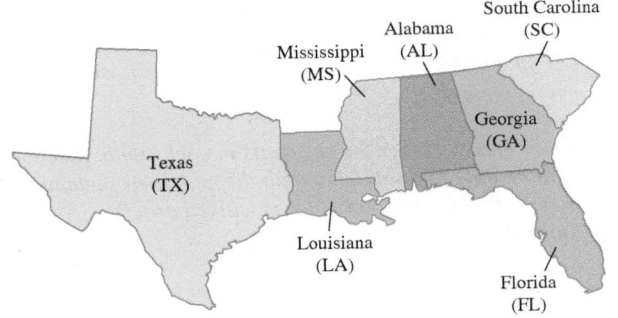

25. ***Southwest Asia*** Afghanistan (*A*), Iran (*N*), Iraq (*Q*), Jordan (*J*), Pakistan (*P*), Syria (*S*), and Turkey (*T*).

26. ***Central Europe*** Austria (*A*), Czech Republic (*C*), Germany (*G*), Italy (*I*), Poland (*P*), and Switzerland (*S*)

27. *Southeast Asia* Myanmar (*M*), Cambodia (*C*), Laos (*L*), Thailand (*T*), and Vietnam (*V*)

28. *South America* Argentina (*A*), Bolivia (*B*), Brazil (*Z*), Chile (*C*), Paraguay (*P*), and Uruguay (*U*)

29. Fit 4U

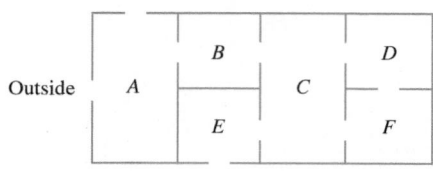

30. Pop Fitness

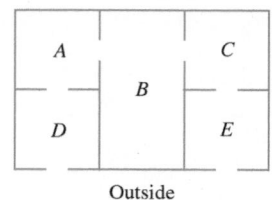

31. Silver's Gym

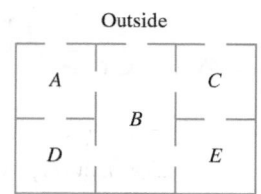

32. NY Fitness

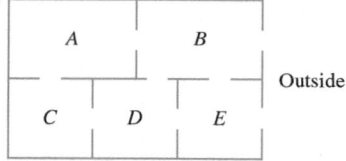

Locking Doors *Recall Joe from Example 5 on page 844 who runs a gym-cleaning service. Joe is also responsible for locking the doors of four other gyms. The floor plans for these gyms are shown in Exercises 29–32.*

a) *Represent each floor plan as a graph. Use the letters shown to label the vertices of the graph and use the letter O to label the vertex that represents the outside of the gym. Place vertex O near where the word Outside is placed in the diagram.*

b) *Determine (state yes or no) whether it is possible for Joe to move through the gym using each doorway of the gym exactly one time. In other words, does an Euler path exist in the graph?*

c) *If your answer to part (b) is yes, determine one such Euler path.*

Crime Stopper Routes *The Havenwoods crime stopper organization (see Example 6 on page 845) was so successful that the residents shared their strategies with friends living in the subdivisions of Marlow Heights and Temple Hills. In Exercises 33 and 34, the respective maps of these communities are given on page 851.*

a) *Determine whether the residents in each subdivision will be able to establish a path through their communities so that each street block is walked exactly one time.*

b) *If yes, where would the residents need to start their walk?*

33. Marlow Heights

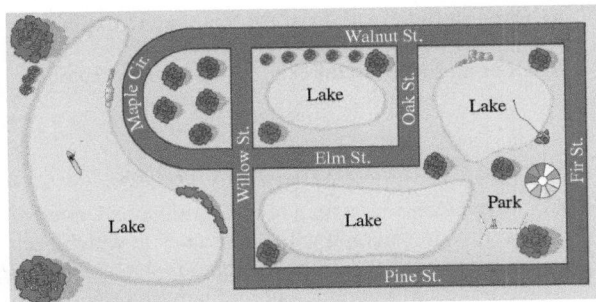

34. Temple Hills

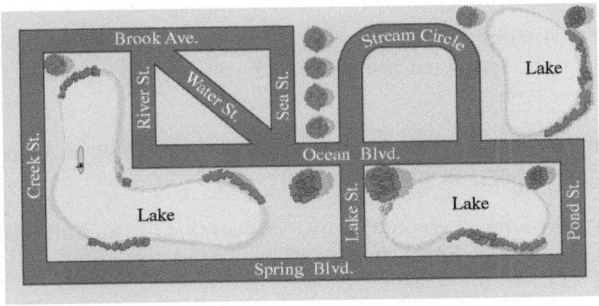

In Exercises 35–38, use Fleury's algorithm to determine an Euler path. There are many answers possible.

35.

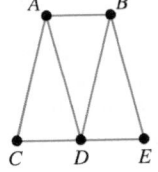

36.

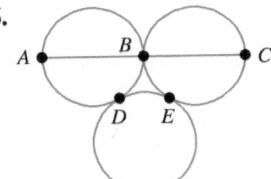

37.

38.

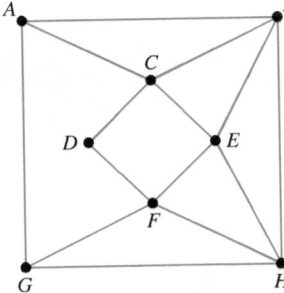

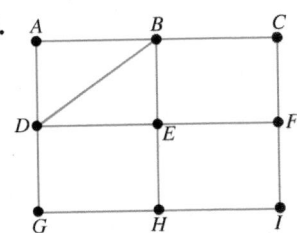

In Exercises 39–44, use Fleury's algorithm to determine an Euler circuit. There are many answers possible.

39.

40.

41.

42.

43.

44.

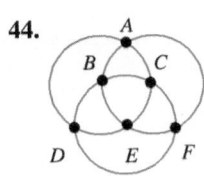

45. Determine an Euler path through the Southwestern states (see Example 4 on page 843) that begins with the state of Utah.

46. Determine an Euler path for the Phenomenal Phitness Gym (see Example 5 on page 844) that begins in locker room (vertex) *A*.

In Exercises 47–50, determine an Euler circuit for the Havenwoods crime stopper group (see Example 6 on page 845) that begins with

47. vertex *B* followed by vertex *A*.

48. vertex *B* followed by vertex *E*.

49. vertex *J* followed by vertex *G*.

50. vertex *J* followed by vertex *F*.

Concept/Writing Exercises

51. Imagine a very large connected graph that has 400 even vertices and no odd vertices.

 a) Does an Euler path exist for this graph? Explain.

 b) Does an Euler circuit exist for this graph? Explain.

52. Imagine a very large connected graph that has two odd vertices and 398 even vertices.

 a) Does an Euler path exist for this graph? Explain.

 b) Does an Euler circuit exist for this graph? Explain.

53. Imagine a very large connected graph that has 400 odd vertices and no even vertices.

 a) Does an Euler path exist for this graph? Explain.

 b) Does an Euler circuit exist for this graph? Explain.

54. Imagine a very large connected graph that has 200 odd vertices and 200 even vertices.

 a) Does an Euler path exist for this graph? Explain.

 b) Does an Euler circuit exist for this graph? Explain.

Challenge Problems/Group Activities

55. *U.S. Map* Consider a map of the contiguous United States. Imagine a graph with 48 vertices in which each vertex represents one of the contiguous states. Each edge would represent a common border between states.

 a) Would this graph have an Euler path?

 b) Explain why or why not.

56. Attempt to draw a graph with an Euler circuit that has a bridge. What conclusion can you develop from this exercise?

57. a) Draw a graph with one vertex that has both an Euler path and an Euler circuit.

 b) Draw a graph with two vertices that has an Euler path but no Euler circuit.

 c) Draw a graph with two vertices that has both an Euler path and an Euler circuit.

Research Activity

58. *Knot Theory* Write a paper on the history and development of the branch of mathematics known as *knot theory*. Include a discussion on its applications to science and technology. See the *Did You Know?* box on page 842.

SECTION 13.3   Hamilton Paths and Hamilton Circuits

Upon completion of this section, you will be able to:

- Understand Hamilton paths, Hamilton circuits, and complete graphs.
- Solve traveling salesman problems.

Ralph delivers mail for the United States Postal Service. He wishes to determine the shortest route that begins at the post office, goes to each mailbox in his delivery area exactly one time, and ends back at the post office. In this section, we will study a special type of problem like the one facing Ralph.

Why This Is Important As we saw in Section 13.2, there are many real-life problems that can be represented and solved using graphs, paths, and circuits. Anytime a problem involves visiting multiple locations, graph theory can be used to represent and solve the problem. The methods we study in this section can also be used in biology, physics, chemistry, and engineering and in many business applications.

In this section, we continue our study of graph theory by studying Hamilton paths and Hamilton circuits. These paths and circuits are named for Irish mathematician and astronomer William Rowan Hamilton (1805–1865); see the *Profile in Mathematics* on page 857. Before formally defining Hamilton paths and Hamilton circuits, we will introduce an example that will be examined at several points throughout this section.

Example 1 A Transportation Problem

Priya is the chief operating officer for Hard Rock Cafe International. Priya lives in Orlando, Florida, and needs to visit Hard Rock Cafe restaurants in Atlanta, Georgia; Memphis, Tennessee; and New Orleans, Louisiana. She would like to determine the least expensive way to visit each of these cities one time and return to her home in Orlando. To help analyze this problem, Priya searched online and obtained the least expensive one-way fares offered between each of the four cities (see Table 13.1).

Table 13.1

	Orlando	Atlanta	Memphis	New Orleans
Orlando (*O*)	*	$103	$161	$139
Atlanta (*A*)	$103	*	$229	$110
Memphis (*M*)	$161	$229	*	$171
New Orleans (*N*)	$139	$110	$171	*

Use this table to create a graph. Let the vertices represent the cities, then connect each pair of cities with an edge. List the airfare between each two cities on the respective edges.

Solution First we note that the cost of a one-way flight between two given cities in the table is the same regardless from which city Priya starts. For example, the one-way fare from New Orleans to Atlanta is $110 and the one-way fare from Atlanta to New Orleans is also $110. This information will allow us to list the flight cost along an edge with a single number. Next we draw a graph with four vertices that represent the four cities and six edges that represent the flights between each city. We also include the price of one-way flights along the appropriate edge. Two such graphs are shown in Fig. 13.35. Many other graphs could also display this information. ∎

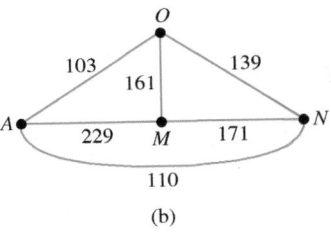

(a)

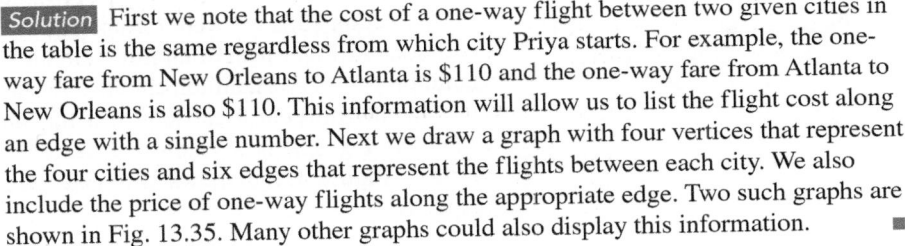

(b)

Figure 13.35

Problems like the one in Example 1 are called *traveling salesman problems*. There are usually variations to the situation, but these problems generally involve seeking the least expensive or shortest way to travel among several locations. To help analyze traveling salesman problems, the costs or distances to travel between locations are indicated along each edge of a graph. Such a graph, as in Fig. 13.35(a) or (b), is called a *weighted graph*. We will use weighted graphs to solve traveling salesman problems, but first we need to introduce Hamilton paths and Hamilton circuits.

Hamilton Paths, Hamilton Circuits, and Complete Graphs

Now we introduce another important definition of graph theory.

> **Definition: Hamilton Path**
> A **Hamilton path** is a path that contains each *vertex* of a graph exactly one time.

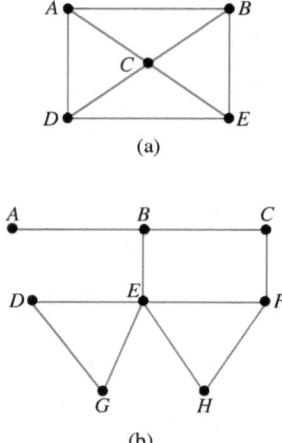

(a)

(b)

Figure 13.36

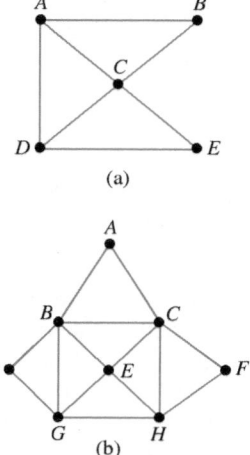

(a)

(b)

Figure 13.37

The graph in Fig. 13.36(a) has Hamilton path *A, B, C, E, D*. The graph in Fig. 13.36(a) also has Hamilton path *C, B, A, D, E*. The graph in Fig. 13.36(b) has a Hamilton path *A, B, C, F, H, E, G, D*. The graph in Fig. 13.36(b) also has Hamilton path *G, D, E, H, F, C, B, A*. Both of the graphs in Fig. 13.36 have other Hamilton paths. Can you determine some of them?

Notice that, unlike Euler paths, not every edge needs to be traced in a Hamilton path. To help distinguish between Euler paths and Hamilton paths, remember that *Euler paths are concerned with visiting all the edges, whereas Hamilton paths are concerned with visiting all the vertices.*

Now we introduce Hamilton circuits.

Definition: Hamilton Circuit

A **Hamilton circuit** is a path that begins and ends at the same vertex and passes through all other vertices of a graph exactly one time.

An alternate definition for a Hamilton circuit is that a Hamilton circuit is a Hamilton path that starts and ends at the same vertex. For example, the graph in Fig. 13.37(a) has Hamilton circuit *A, B, C, E, D, A*. The graph in Fig. 13.37(a) also has Hamilton circuit *B, A, D, E, C, B*. The graph in Fig. 13.37(b) has Hamilton circuit *A, B, D, G, E, H, F, C, A*. The graph in Fig. 13.37(b) also has Hamilton circuit *E, G, D, B, A, C, F, H, E*. Note that in an ***Euler*** *circuit*, the path followed must include every *edge* and must begin and end at the same vertex. In a ***Hamilton*** *circuit*, the path followed must include every *vertex* and must begin and end at the same vertex. Unlike the Euler circuit, however, a Hamilton circuit does not have to include every edge.

In Section 13.2, we were fortunate that Euler's theorem could tell us under what conditions an Euler path or an Euler circuit would exist. We are not as fortunate with Hamilton paths and Hamilton circuits; no such general theorem exists.

We now shift our focus to graphs that are guaranteed to have Hamilton circuits. A *complete graph* is a graph that has exactly one edge between each pair of its vertices. Figure 13.38 shows complete graphs with three, four, and five vertices, respectively.

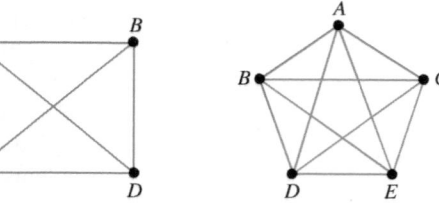

Figure 13.38

Complete graphs are important because *every complete graph has a Hamilton circuit*, and traveling salesman problems can be represented by complete graphs. Since each pair of vertices on a complete graph is connected by an edge, we can create a Hamilton circuit by simply starting at any vertex and moving from vertex to vertex until we have passed through each vertex exactly one time. Then to complete the circuit we simply move back to the vertex from which we started.

Example 2 *Determining Hamilton Circuits*

Determine a Hamilton circuit for the complete graph shown in Fig. 13.39.

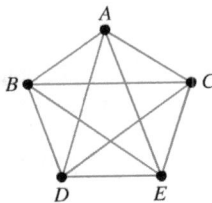

Figure 13.39

Solution We can determine many Hamilton circuits in this complete graph. For example, one is *A, B, C, D, E, A*. Another is *B, A, D, C, E, B*. Another is *E, B, C, D, A, E*. To build a Hamilton circuit from this complete graph, we can list all five vertices in any order and then return to the first vertex. ∎

A question that naturally arises from Example 2 is, "How many different Hamilton circuits are there in a complete graph?" Before we can give a formula that answers this question, we need to discuss *factorials*, which were presented in Section 11.7. We will review that material here. The symbol 5! is read "five factorial" and means to multiply 5 by each natural number less than 5. So $5! = 5 \cdot 4 \cdot 3 \cdot 2 \cdot 1 = 120$. Also $3! = 3 \cdot 2 \cdot 1 = 6$ and $7! = 7 \cdot 6 \cdot 5 \cdot 4 \cdot 3 \cdot 2 \cdot 1 = 5040$. Note that 0! is defined to be 1.

> ### Number of Unique Hamilton Circuits in a Complete Graph
> The number of unique Hamilton circuits in a complete graph with n vertices is $(n - 1)!$, where
> $$(n - 1)! = (n - 1)(n - 2)(n - 3) \cdots (3)(2)(1)$$

Example 3 *Number of Hamilton Circuits*

How many unique Hamilton circuits are in a complete graph with the following number of vertices?

a) Five   b) Eight   c) Ten   d) Twelve

Solution

a) Using the formula, for the number of unique Hamilton circuits, a complete graph with five vertices has $(5 - 1)! = 4! = 4 \cdot 3 \cdot 2 \cdot 1 = 24$ unique Hamilton circuits.

b) A complete graph with eight vertices has
   $(8 - 1)! = 7! = 7 \cdot 6 \cdot 5 \cdot 4 \cdot 3 \cdot 2 \cdot 1 = 5040$ unique Hamilton circuits.

c) A complete graph with ten vertices has
   $(10 - 1)! = 9! = 9 \cdot 8 \cdot 7 \cdot 6 \cdot 5 \cdot 4 \cdot 3 \cdot 2 \cdot 1 = 362{,}880$
   unique Hamilton circuits.

d) A complete graph with twelve vertices has
   $(12 - 1)! = 11! = 11 \cdot 10 \cdot 9 \cdot 8 \cdot 7 \cdot 6 \cdot 5 \cdot 4 \cdot 3 \cdot 2 \cdot 1 = 39{,}916{,}800$
   unique Hamilton circuits. ∎

Example 4 *American Idol Travel*

Lionel Richie, Katy Perry, Luke Bryan, and Ryan Seacrest, the cast of the television show *American Idol*, are in Hollywood (*H*). They need to travel to the following cities to judge contestants' auditions: San Antonio (*SA*), East Rutherford (*ER*), Birmingham (*B*), Memphis (*M*), and Seattle (*S*). In how many different

▲ *Lionel Richie, Katy Perry, and Luke Bryan*

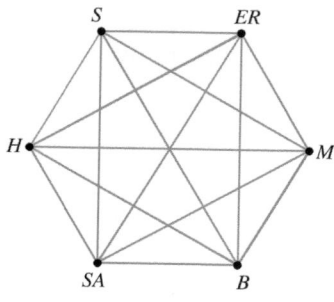

Figure 13.40

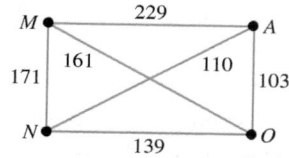

Figure 13.41

ways can the cast, traveling together, visit each of these cities and return to Hollywood?

Solution We can represent this problem with the complete graph in Fig. 13.40. In the graph, the six vertices represent the six locations and the edges represent the one-way flights between these locations.

To determine the number of possible routes, we need to determine the number of Hamilton circuits within this graph. We know that there are $(6 - 1)! = 5! = 5 \cdot 4 \cdot 3 \cdot 2 \cdot 1 = 120$ different Hamilton circuits within this graph. So the cast has 120 different ways to start in Hollywood, visit each of these five cities, and return to Hollywood. ∎

Now that we have been introduced to Hamilton circuits, we will apply this knowledge to solving traveling salesman problems.

Traveling Salesman Problems

In Examples 1 and 4, we saw how complete graphs can represent cities and the process of traveling between these cities. Our goal in a traveling salesman problem is to determine the least expensive or shortest way to visit each city once and return to the starting location. In terms of graph theory, our goal is to determine the Hamilton circuit with the lowest associated cost or distance. The Hamilton circuit with the lowest associated cost (or shortest distance, etc.) is called the *optimal solution*. In this section, we will discuss two methods, the *brute force method* and the *nearest neighbor method*. In both methods, we will use the term *complete, weighted graph*. A *complete, weighted graph* is a complete graph with the weights (or numbers) listed on the edges, as illustrated in Fig. 13.41. We now introduce the brute force method.

PROCEDURE   THE BRUTE FORCE METHOD OF SOLVING TRAVELING SALESMAN PROBLEMS

To determine the optimal solution:

1. Represent the problem with a complete, weighted graph.
2. List all possible Hamilton circuits for this graph.
3. Determine the cost (or distance) associated with each of these Hamilton circuits.
4. The Hamilton circuit with the lowest cost (or shortest distance) is the optimal solution.

Example 5 *Using the Brute Force Method*

In Example 1 on page 853, we introduced Priya, the chief operating officer for Hard Rock Cafe International. We now want to use the brute force method to determine the optimal solution for Priya to start at her home in Orlando (*O*), visit restaurants in Atlanta (*A*), Memphis (*M*), and New Orleans (*N*), and then return to her home in Orlando.

Solution We illustrated two complete, weighted graphs for the example in Fig. 13.35 on page 853. The graph in Fig. 13.35(a) is repeated in Fig. 13.41. The numbers shown on the graph represent the one-way fares, in dollars, between the two cities.

We know that since there are 4 cities that must be visited, there are $(4 - 1)! = 3! = 3 \cdot 2 \cdot 1 = 6$ possible unique Hamilton circuits we need to examine for cost. These Hamilton circuits are listed in the first column of Table 13.2.

Table 13.2

Hamilton Circuit	First Leg/Cost	Second Leg/Cost	Third Leg/Cost	Fourth Leg/Cost	Total Cost
O, A, M, N, O	O to A $\$103$	A to M $\$229$	M to N $\$171$	N to O $\$139$	$\$642$
O, A, N, M, O	O to A $\$103$	A to N $\$110$	N to M $\$171$	M to O $\$161$	$\$545$
O, M, A, N, O	O to M $\$161$	M to A $\$229$	A to N $\$110$	N to O $\$139$	$\$639$
O, M, N, A, O	O to M $\$161$	M to N $\$171$	N to A $\$110$	A to O $\$103$	$\$545$
O, N, A, M, O	O to N $\$139$	N to A $\$110$	A to M $\$229$	M to O $\$161$	$\$639$
O, N, M, A, O	O to N $\$139$	N to M $\$171$	M to A $\$229$	A to O $\$103$	$\$642$

The Hamilton circuits are listed in the column on the far left. In the next four columns, we place the cost associated with each leg of the given circuit. In the far-right column, we place the total cost of travel using the given circuit. From this last column, we see that two circuits have the lowest cost, $545 (circled). Priya has two choices for an optimal solution. She can either fly from Orlando to Atlanta to New Orleans to Memphis then back to Orlando, or she can fly from Orlando to Memphis to New Orleans to Atlanta then back to Orlando. These two Hamilton circuits are shown on the maps in Fig. 13.42. Notice that the second circuit involves visiting the cities in the exact reverse order of the first circuit. Either circuit provides Priya with the least expensive way to visit each city and return to her home in Orlando.

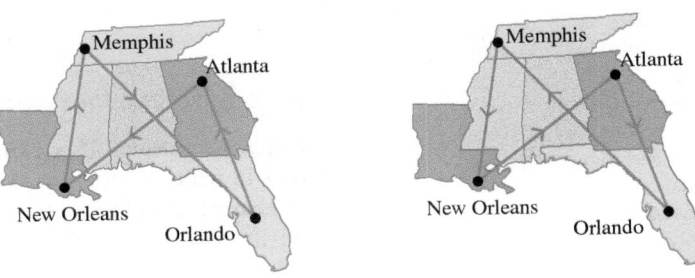

Figure 13.42

Although we used the brute force method in Example 5, it becomes impractical as the number of vertices increases. In fact, with more complex problems, the brute force method is impractical on even the world's fastest supercomputers. For example, suppose the president of the United States wished to visit all of the state capitals and return to Washington, DC. The president would have $(51 - 1)! = 50!$ choices to accomplish this trip. Furthermore, each choice requires adding the costs or distances of traveling the 51 legs of each trip. As of June 2019, it would take the world's fastest supercomputer about 1.23×10^{41} or 123,000,000,000,000,000,000,000,000,000,000,000,000,000,000 (123 duodecillion) *years* to perform the brute force method to determine the optimal solution! Computer scientists and mathematicians have developed much more efficient algorithms for approximating the optimal solution to traveling salesmen problems.

Did You Know?

Six Degrees of Kevin Bacon

▲ Halle Berry

The Six Degrees of Kevin Bacon is a trivia game started by Albright College students Craig Fass, Brian Turtle, and Mike Ginelli. The object is to link any actor to the actor Kevin Bacon through associations from different films. A Bacon number is the number of movies it takes to link an actor to Kevin Bacon. For example, Mark Wahlberg has a Bacon number of 1, since he was in *Patriots Day* (2016) with Kevin Bacon. Halle Berry has a Bacon number of 2, since she was in *New Year's Eve* (2011) with Sarah Jessica Parker and Sarah Jessica Parker was in *Footloose* (1984) with Kevin Bacon.

What does this game have to do with graph theory? A graph with actors as vertices and common movies as edges could be used to represent these associations. Such a graph would be very large and difficult to draw. However, computerized versions of the game can manage the associations quite nicely. The website www.oracleofbacon.org makes use of a 500-MB database that contains the casts of more than 500,000 movies to compute the Bacon number of more than 390,000 actors, ranging in Bacon number 0 (Bacon himself) to 10 (three individuals). Other online databases provide links among musicians (Six Degrees of Vince Gill), baseball players (The Oracle of Baseball), and mathematicians (Erdös Numbers).

Now we introduce a method for determining an *approximate solution* to a traveling salesman problem. Approximate solutions can be used in cases where determining the optimal solution is not reasonable. One method for determining an approximate solution is the *nearest neighbor method*. In this method, the salesperson begins at a given location. If the salesperson wishes to minimize the distance traveled, the salesperson first visits the city (or location) closest to his or her starting location. Then the salesperson visits the next city (or location) closest to his or her present location that has not already been visited. This process continues until all the cities (or locations) are visited. Thus, the salesperson moves to the *nearest neighbor*. Sometimes the salesperson wishes to minimize the cost of travel, such as airfare. In this case, from the starting location the salesperson first visits the city (or location) in which the cost of travel is a minimum. Then from the present location, the salesperson visits the city (or location) where the cost of travel is a minimum. This process continues until all the cities (or locations) are visited. Approximate solutions will always be Hamilton circuits.

> **PROCEDURE** NEAREST NEIGHBOR METHOD OF DETERMINING AN APPROXIMATE SOLUTION TO A TRAVELING SALESMAN PROBLEM
>
> To approximate the optimal solution:
> 1. Represent the problem with a complete, weighted graph.
> 2. Identify the starting vertex.
> 3. Of all the edges attached to the starting vertex, choose the edge that has the smallest weight. This edge is generally either the shortest distance or the lowest cost. Travel along this edge to the second vertex.
> 4. At the second vertex, choose the edge that has the smallest weight that does not lead to a vertex already visited. Travel along this edge to the third vertex.
> 5. Continue this process, each time moving along the edge with the smallest weight until all vertices are visited.
> 6. Travel back to the original vertex.

Example 6 *Using the Nearest Neighbor Method*

In Example 4 on page 855, we discussed the *American Idol* cast's plan to visit five cities in which auditions would take place and then return to Hollywood. Use the nearest neighbor method to determine an approximate solution for the cast's visits. The one-way per person flight prices between cities are given in Table 13.3.

Table 13.3

	Birmingham	East Rutherford	Hollywood	Memphis	San Antonio	Seattle
Birmingham (B)	*	$258	$114	$324	$274	$155
East Rutherford (ER)	$258	*	$134	$104	$355	$154
Hollywood (H)	$114	$134	*	$144	$129	$108
Memphis (M)	$324	$104	$144	*	$634	$299
San Antonio (SA)	$274	$355	$129	$634	*	$119
Seattle (S)	$155	$154	$108	$299	$119	*

Nearest Neighbor Method

Solution We will use the nearest neighbor method to determine the least expensive per person cost for the cast to complete the trip. We begin by modifying the graph from Example 4 by putting the one-way per person prices into place, making it a weighted graph (see Fig. 13.43). The cast starts in Hollywood and chooses the least expensive flight, which is to Seattle ($108). In Seattle, the cast chooses the least expensive flight to a location they have not already visited, San Antonio ($119). Next, in San Antonio, they choose the least expensive flight to a location they have not already visited, Birmingham ($274). In Birmingham, they choose the least expensive flight to a location they have not already visited, East Rutherford ($258). Once in East Rutherford, the cast has no choice but to fly to the only location not yet visited, Memphis ($104). Since they began in Hollywood, are now in Memphis, and have visited all five audition cities, the cast now returns to Hollywood ($144) to complete their trip. The nearest neighbor method would produce a Hamilton circuit of *H, S, SA, B, ER, M, H*. The cost per person is $108 + $119 + $274 + $258 + $104 + $144 = $1007. The total cost for all four cast members is $1007 × 4 = $4028.

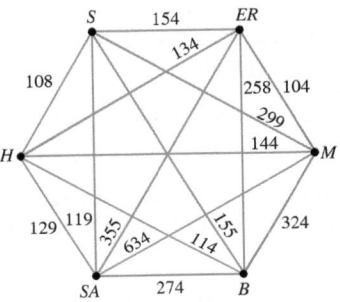

Figure 13.43

For the sake of comparison, let us look at four other randomly chosen Hamilton circuits from Example 6 and the per person costs associated with them (see Table 13.4).

Table 13.4

Randomly Chosen Hamilton Circuit	Cost Calculation	Per Person Cost
H, SA, ER, B, M, S, H	$129 + $355 + $258 + $324 + $299 + $108	$1473
H, M, S, B, ER, SA, H	$144 + $299 + $155 + $258 + $355 + $129	$1340
H, ER, S, M, SA, B, H	$134 + $154 + $299 + $634 + $274 + $114	$1609
H, ER, M, B, S, SA, H	$134 + $104 + $324 + $155 + $119 + $129	$ 965

From Table 13.4, we see that the Hamilton circuit *H, ER, M, B, S, SA, H* results in a per person cost of $965, which is less than the $1007 we obtained in Example 6. Thus, we see that the nearest neighbor method does not always produce the optimal solution. Without using the brute force method, we cannot determine if the Hamilton circuit *H, ER, M, B, S, SA, H* is the optimal solution. The nearest neighbor method produces an *approximation* for the optimal solution. Note also that the nearest neighbor method did produce a Hamilton circuit that was less expensive than three of the four randomly chosen Hamilton circuits.

SECTION 13.3   *Exercises*

Warm Up Exercises

In Exercises 1–8, fill in the blanks with an appropriate word, phrase, or symbol(s).

1. Problems that generally involve seeking the least expensive or shortest way to travel among several locations are called traveling _____ problems.

2. A graph that has numbers along the edges to indicate cost or distance is called a(n) _____ graph.

3. A path that begins and ends at the same vertex and passes through all other vertices of a graph exactly one time is called a(n) _____ circuit.

4. A graph that has exactly one edge between each pair of vertices is called a(n) _____ graph.

5. Every complete graph has a Hamilton circuit but not necessarily a(n) _____ circuit.

6. In a traveling salesman problem, the Hamilton circuit with the lowest associated cost or shortest distance is called the _____ solution.

7. The method discussed in this section that will determine the optimal solution to a traveling salesman problem is the brute _____ method.

8. The method discussed in this section that will determine an approximate solution to a traveling salesman problem is the nearest _____ method.

Practice the Skills

In Exercises 9–14, determine two different Hamilton paths in each of the following graphs.

9.
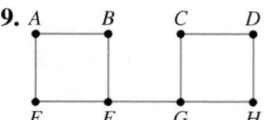

10.
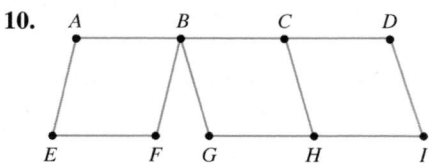

11.
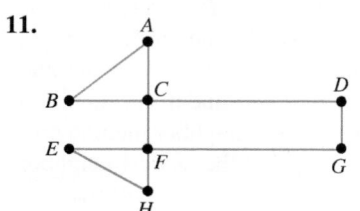

12.
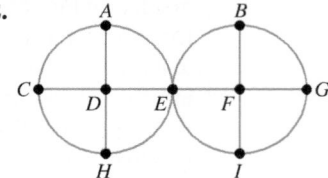

13.
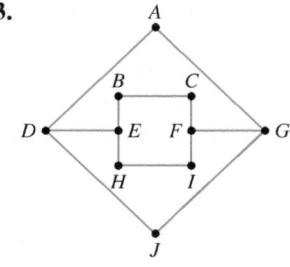

14.
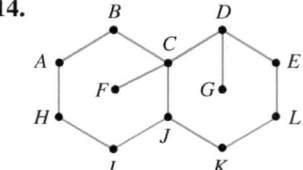

In Exercises 15–18, determine two different Hamilton circuits in each of the following graphs.

15.
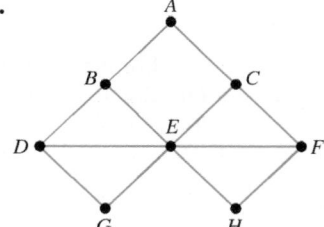

16.
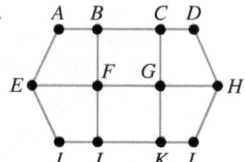

17.
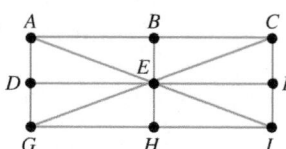

18.
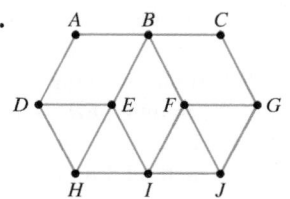

19. Draw a complete graph with four vertices.

20. Draw a complete graph with five vertices.

Problem Solving

21. *College Visits* Tramon is a high school student who lives in Bakersfield, CA, and he wishes to visit colleges in the following cities: St. Cloud, MN; Valparaiso, IN; Lindonville, VT; Asheville, NC; and Norman, OK. In how many different ways can Tramon visit each of these cities and return home to Bakersfield?

22. *Culver's Restaurants* Craig, the president of Culver's Restaurants, is at his home in Prairie du Sac, WI, and wishes to investigate locations for a new restaurant in the following cities: Robinson, IL; Erie, PA; Memphis, TN; Little Rock, AR; Tulsa, OK; Topeka, KS; and Dickinson, ND. In how many ways can Craig visit each of these cities and return to his home in Prairie du Sac?

23. *Florida State League* The Florida State League is a minor league baseball organization with teams in Bradenton, Charlotte County, Clearwater, Daytona, Dunedin, Fort Myers, Jupiter, Lakeland, Osceola County, Palm Beach, Port Saint Lucie, and Tampa. James, the league commissioner, would like to visit each team and then return to his home in Sarasota. In how many ways can James visit each team and return to his home?

24. *A Milk Truck Route* Dale is a milk truck driver for Swiss Valley Farms Cooperative in eastern Iowa. Dale has to start at the processing plant and pick up milk on 10 different farms. In how many ways can Dale visit each farm and return to the processing plant?

25. *Running Errands on Campus* Song needs to run errands on the campus of Pima Community College. She is in her office and needs to go to the duplicating center, student center, and library. She estimates the walking distances as follows: From her office to the duplicating center is 150 feet, from her office to the student center is 100 feet, from her office to the library is 400 feet, from the duplicating center to the student center is 125 feet, from the duplicating center to the library is 450 feet, and from the student center to the library is 250 feet.

a) Represent this traveling salesman problem with a complete, weighted graph showing the distances on the appropriate edges.

b) Use the brute force method to determine the shortest route for Song to accomplish her errands and then return back to her office.

c) What is the minimum distance Song can walk?

26. *Job Interviews* Christina is searching for a new job. She lives in Shreveport, Louisiana, and has interviews in Barrow, Alaska; Tucson, Arizona; and Rochester, New York. The costs of the one-way flights between these four cities are as follows: Shreveport to Barrow costs $855, Shreveport to Tucson costs $803, Shreveport to Rochester costs $113, Barrow to Tucson costs $393, Barrow to Rochester costs $337, and Tucson to Rochester costs $841.

a) Represent this traveling salesman problem with a complete, weighted graph showing the prices of the flights on the appropriate edges.

b) Use the brute force method to determine the least expensive route for Christina to travel to each city and return home to Shreveport.

c) What is the minimum cost she can pay?

27. *A Family Vacation* The Ackermans live in Tunkhannock, Pennsylvania (*T*), and are planning their summer vacation. They wish to visit the following national parks: Acadia (*A*), Cuyahoga Valley (*C*), and Great Smoky Mountains (*S*). The approximate distances among these locations are as follows: *T* to *A* is 582 miles, *T* to *C* is 342 miles, *T* to *S* is 663 miles, *A* to *C* is 917 miles, *A* to *S* is 1188 miles, and *C* to *S* is 521 miles.

a) Represent this traveling salesman problem with a complete, weighted graph showing the distances along the edges.

b) Use the brute force method to determine the shortest route for the Ackermans to visit each national park and return to their home.

c) What is the minimum distance the Ackermans can travel?

28. *Retirement Road Trip* To celebrate his retirement, Baron (*B*) wishes to ride his motorcycle to visit his friends Woody (*W*), Marcellus (*M*), and Simeon (*S*). The approximate distances between the homes of these friends are as follows: *B* to *W* is 940 miles, *B* to *M* is 507 miles, *B* to *S* is 374 miles, *W* to *M* is 799 miles, *W* to *S* is 575 miles, and *M* to *S* is 332 miles.

a) Represent this traveling salesman problem with a complete weighted graph showing the distances along the edges.

b) Use the brute force method to determine the shortest route for Baron to visit his friends and return home.

c) What is the minimum distance that Baron can travel?

29. *Package Delivery* Laurice works for FedEx and is in her office (*O*). She has packages to deliver to the following families: Moller (*M*), Ransford (*R*), Seifert (*S*), and Warren (*W*). The approximate distances among these locations are as follows: *O* to *M* is 4.8 miles, *O* to *R* is 5.6 miles, *O* to *S* is 2.3 miles, *O* to *W* is 1.7 miles, *M* to *R* is 3.1 miles, *M* to *S* is 2.4 miles, *M* to *W* is 4.3 miles, *R* to *S* is 7.2 miles, *R* to *W* is 6.4 miles, and *S* to *W* is 0.9 miles.

a) Represent this traveling salesman problem with a complete, weighted graph showing the distances along the edges.

b) Use the nearest neighbor method to approximate the optimal route for Laurice to deliver the packages and return to her office. Give the distance of the route determined.

c) Randomly select another route for Laurice to deliver the packages and return to her office. Compute the distance of this route. Compare this distance with the distance determined in part (b).

30. *Basketball Teams* Jasmine lives in Elko, Nevada (*E*), and is planning a trip to see the following basketball teams play on their home courts: Utah Jazz (*J*), Sacramento Kings (*K*), Portland Trailblazers (*T*), and Golden State Warriors (*W*). The approximate distances among these locations are as follows: *E* to *J* is 230 miles, *E* to *K* is 420 miles, *E* to *T* is 625 miles, *E* to *W* is 501 miles, *J* to *K* is 650 miles, *J* to *T* is 766 miles, *J* to *W* is 730 miles, *K* to *T* is 577 miles, *K* to *W* is 84 miles, and *T* to *W* is 631 miles.

a) Represent this traveling salesman problem with a complete, weighted graph showing the distances along the edges.

b) Use the nearest neighbor method to approximate the optimal route for Jasmine to visit each location and return to her home. Give the distance of the route determined.

c) Randomly select another route for Jasmine to visit the locations and return to her home and compute the distance of this route. Compare this distance with the distance determined in part (b).

31. *O'Reilly Auto Parts* David, the chief executive officer of O'Reilly Auto Parts, is at his office in Springfield, Missouri, and wishes to visit his company's distribution centers in the following locations: Bismarck, ND; Carson City, NV; Helena, MT; and Knoxville, TN. He wishes to visit each of these locations and return to his office in Springfield. The one-way flight prices are given in the table above and to the right.

	Bismarck	Carson City	Helena	Knoxville	Springfield
Bismarck	*	$431	$483	$476	$378
Carson City	$431	*	$144	$159	$505
Helena	$483	$144	*	$542	$492
Knoxville	$476	$159	$542	*	$459
Springfield	$378	$505	$492	$459	*

a) Represent this traveling salesman problem with a complete, weighted graph showing the prices of flights on the appropriate edges.

b) Use the nearest neighbor method to approximate the optimal route for David to travel to each city and return to Springfield. Give the cost of the route determined.

c) Randomly select another route for David to travel from Springfield to each of the other cities and return to Springfield and then compute the cost of this route. Compare this cost with the cost determined in part (b).

32. *Wayfair* Yilan lives in Boston, Massachusetts, and works for the e-commerce company Wayfair. Yilan wishes to visit company warehouses in the following locations: Madison, WI; Princeton, NJ; Salem, OR; and Walla Walla, WA. The one-way flight prices are given in the following table.

▲ Boston

	Boston	Madison	Princeton	Salem	Walla Walla
Boston	*	$131	$ 256	$298	$ 576
Madison	$131	*	$ 154	$356	$ 970
Princeton	$256	$154	*	$353	$1164
Salem	$298	$356	$ 353	*	$ 179
Walla Walla	$576	$970	$1164	$179	*

a) Represent this traveling salesman problem with a complete, weighted graph showing the prices of flights on the appropriate edges.

b) Use the nearest neighbor method to approximate the optimal route for Yilan to travel to each city and return to Boston. Give the cost of the route determined.

c) Randomly select another route for Yilan to travel from Boston to each of the other cities and return to Boston and then compute the cost of this route. Compare this cost with the cost determined in part (b).

Challenge Problems/Group Activities

33. *Visiting Five Cities* Come up with a list of five cities you would like to visit. Use the Internet to search for airline prices between these five cities. Make sure to include the city with the airport nearest your home from which you would start and end your trip.

    a) Draw a complete, weighted graph that represents these cities and the costs associated with flying between each pair of cities.

    b) Use the brute force method to determine the optimal solution to visiting each city and returning home.

    c) Use the nearest neighbor method to approximate the optimal solution.

    d) How much money does the optimal solution, obtained in part (b), save you over the approximation obtained in part (c)?

34. *Number of Circuits* The following tasks are intended to help you understand the formula for determining the number of unique Hamilton circuits in a complete graph.

    a) Draw a complete graph with three vertices labeled A, B, and C. Assume that you are starting at vertex A and wish to move to another vertex. How many choices do you have for moving to the second vertex? Once you choose the second vertex, how many choices do you have for moving to a third vertex? Multiply the number of choices you had from vertex A by the number of choices you had from the second vertex. Compare the number you obtained with the number of Hamilton circuits determined by using $(n - 1)!$.

    b) Draw a complete graph with four vertices labeled A, B, C, and D. Assume that you are starting at vertex A and wish to move to a second vertex. How many choices do you have for moving to this second vertex? Once you choose the second vertex, how many choices do you have for moving to the third vertex? Once you choose the third vertex, how many choices do you have for the fourth vertex? Multiply the number of choices you had from each vertex together. Compare the number you obtained with the number of Hamilton circuits determined by using $(n - 1)!$.

    c) Repeat this process for complete graphs with five and six vertices.

    d) Explain why $(n - 1)!$ gives the number of Hamilton circuits in a complete graph with n vertices.

Recreational Mathematics

35. *The Icosian Game* In 1857, William Rowan Hamilton invented the game called the Icosian Game. The game consisted of a round board with 20 holes and 20 numbered pegs. On the surface of the board was a pattern similar to the graphs we have studied in this chapter (see photo and artwork below).

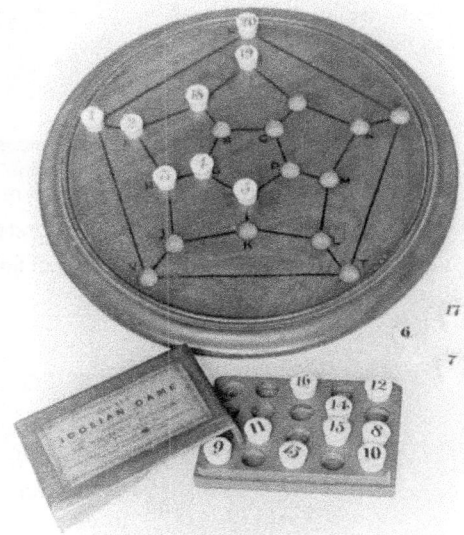

In essence, the object of the game was to use the pegs to form a Hamilton circuit on the graph. The graph below is taken from the Icosian Game. Determine a Hamilton circuit on this graph.

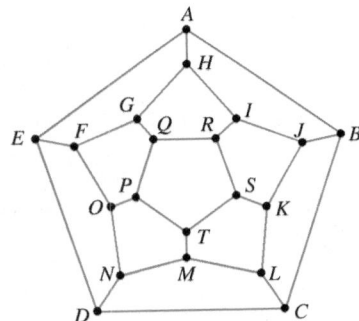

Research Activity

36. *Traveling Salesman Problems* Write a paper on traveling salesman problems. Include a brief history of the problems. Also include advances made toward reducing the computational time of determining the optimal solution as well as advances made toward determining better approximate solutions.

SECTION 13.4   Trees

Upon completion of this section, you will be able to:

- Use trees to represent real-life problems.
- Solve problems involving spanning trees.

Josefina wishes to install an irrigation system to water the five flower gardens in her backyard. How can Josefina determine the irrigation system that reaches each of her five gardens and has the lowest cost? In this section, we will introduce a type of graph, called a *tree*, that can help solve the problem facing Josefina.

Why This Is Important   As we will see shortly, trees can be used to represent a wide variety of real-life problems. Trees are used to represent relationships, such as family trees or corporate structures. Trees also are used to determine the cost-effective solutions in a variety of fields including engineering, computer design, architecture, and urban planning.

Introduction

In this chapter, we have introduced graphs as a means to represent problems from everyday life (Section 13.1). We used graph theory to solve a variety of problems by determining Euler paths and Euler circuits (Section 13.2). We also used graph theory to determine the optimal and approximate solutions to traveling salesman problems using Hamilton circuits (Section 13.3). We now turn our attention to another type of graph, called a *tree*, which is also frequently used to represent problems from everyday life. Before we define a tree, let's look at Example 1, in which we create a tree showing the supervisory relationships at a service company.

Example 1   *Business Structure*

Construct a graph to represent the supervisory relationships at Handi Henri's Repair Service. Henrietta is the president and she has three shift supervisors: Tommy, Patrick, and Susan. Tommy supervises two service specialists: Morris and Winton. Patrick supervises two service specialists: Alicia and Nigel. Susan supervises one service specialist: Rodney.

Solution   We will construct this graph with three layers, one for each level of employee. The vertices represent people and the edges represent supervisor-supervisee relationships. Start with Henrietta and make edges to each of her three shift supervisors, Tommy, Patrick, and Susan. Then make edges from Tommy to his two supervisees, from Patrick to his two supervisees, and from Susan to her supervisee.

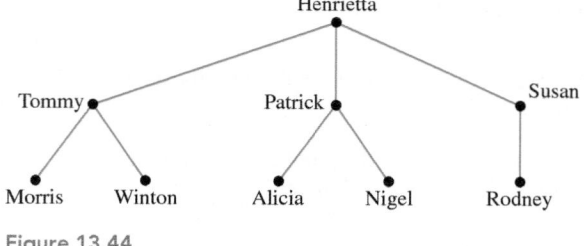

Figure 13.44

The graph in Figure 13.44 is an example of a *tree*.

> **Definition: Tree**
> A **tree** is a connected graph in which each edge is a bridge.

Recall from Section 13.1 that a bridge is an edge that if removed from a connected graph would create a disconnected graph. Thus, if you remove *any edge* in a tree, it creates a disconnected graph. Since each edge would create a disconnected graph if removed, a tree cannot have any Euler circuits or Hamilton circuits. Figure 13.45(a) gives four examples of graphs that are trees, and Fig. 13.45(b) gives four examples of graphs that are not trees.

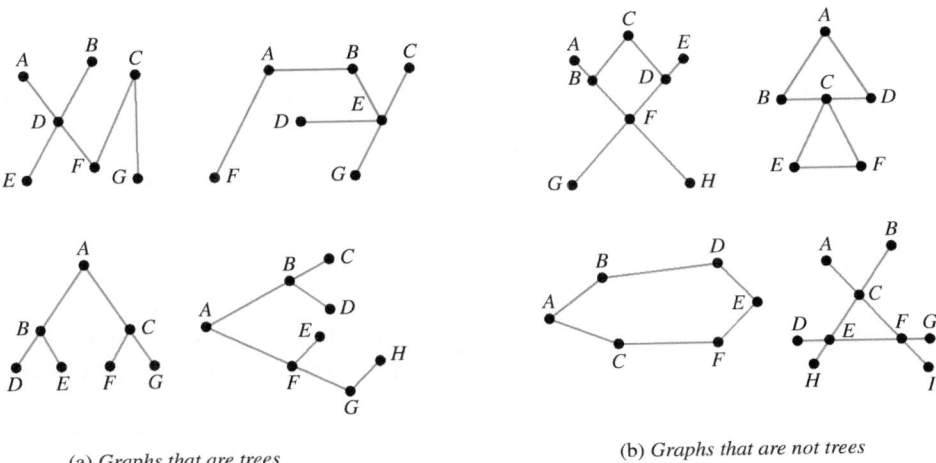

(a) *Graphs that are trees* (b) *Graphs that are not trees*

Figure 13.45

Spanning Trees

Some applications, like the business structure in Example 1, can be represented with a graph that is a tree. In other applications, the problem may initially be represented with a graph that is not a tree. In this section, to solve certain problems, we will need to remove edges from a graph that is not a tree to form a tree known as a *spanning tree*. We now define spanning tree.

> **Definition: Spanning Tree**
> A **spanning tree** is a tree that is created from another graph by removing edges while still maintaining a path to each vertex.

Since a spanning tree is a tree, and since a tree cannot have any circuits, a spanning tree cannot have circuits.

Example 2 *Determining Spanning Trees*

Determine two different spanning trees for each graph shown in Fig. 13.46.

Solution Each of the spanning trees we create will need to have a path connecting all vertices, but cannot have any circuits. To create a spanning tree from a graph, we remove edges one at a time while leaving all the vertices in place. When we remove an edge, we must make sure the edge is not a bridge. Removing a bridge would create a disconnected graph. We need to reduce our original graph to one that is still connected. Keeping these guidelines in mind, we continue removing edges until the remaining graph is a tree. Two spanning trees formed from each graph in Fig. 13.46(a) and (b) are given in Fig. 13.47(a) and (b) on page 866, respectively. Other spanning trees are possible in each case.

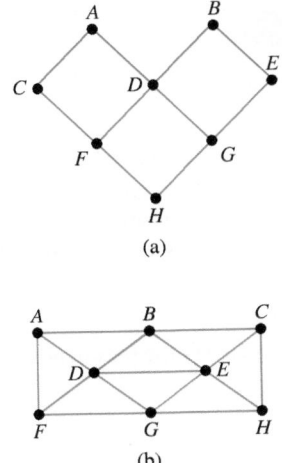

Figure 13.46

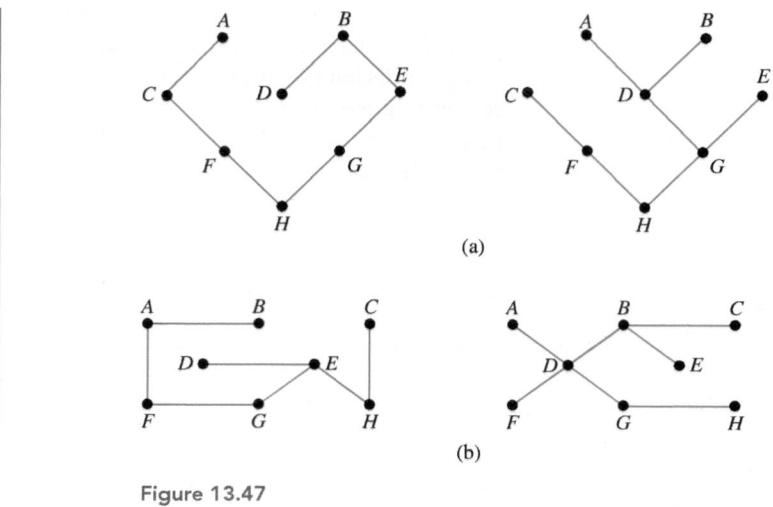

(a)

(b)

Figure 13.47

Notice in all four spanning trees in Example 2 that each edge is a bridge and that no circuits are present. Our next example provides an application of spanning trees.

Example 3 *A Spanning Tree Problem*

Kutztown University is considering adding awnings above its sidewalks to help shelter students from the snow and rain while they walk between some of the buildings on campus. A diagram of the buildings and the connecting sidewalks where the awnings are to be added is given in Fig. 13.48.

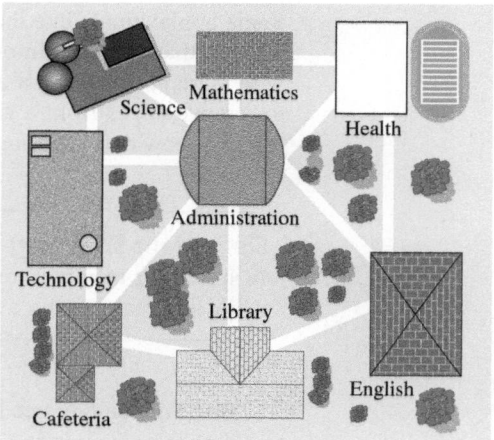

Figure 13.48

Originally, the president of the university wished to have awnings placed over all the sidewalks shown in Fig. 13.48, but that was determined to be too costly. Instead, the president has proposed to place just enough awnings over a select number of sidewalks so that, by moving from building to building, students would still be able to reach any location shown without being exposed to the elements.

a) Represent all the buildings and sidewalks shown with a graph.

b) Create three different spanning trees from this graph that would satisfy the president's proposal.

Solution

a) Using letters to represent the building names, vertices to represent the buildings, and edges to represent the sidewalks between buildings, we generate the graph in Fig. 13.49 on page 867.

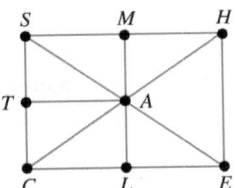

Figure 13.49

b) To create a spanning tree we remove nonbridge edges until a tree is created. Three possible spanning trees are given in Fig. 13.50; however, many others are possible.

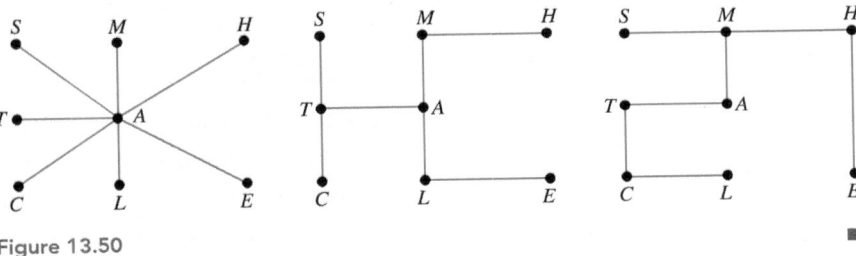

Figure 13.50

Example 3 shows us how spanning trees can be used to represent a real-life problem. However, the president of Kutztown University still has a problem: Which sidewalks in Figure 13.49 should be chosen to cover with awnings (or which spanning tree should be selected)? Problems like the one faced by Kutztown University usually have additional considerations that need to be included in the decision-making process. The most common consideration is the cost of the project. To help analyze these problems, weighted graphs—graphs with costs or distances associated with each edge—and minimum-cost spanning trees are used.

> **Definition: Minimum-Cost Spanning Tree**
> A **minimum-cost spanning tree** is the least expensive spanning tree of all spanning trees under consideration.

Example 4 will explain how we determine a minimum-cost spanning tree for a weighted graph with three edges.

Example 4 *Determining a Minimum-Cost Spanning Tree*

Examine the weighted graph in Fig. 13.51. This graph shows the costs, in dollars, associated with each edge. Determine the minimum-cost spanning tree for this graph.

Solution There are three spanning trees associated with this graph; they are shown in Fig. 13.52.

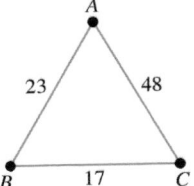

Figure 13.51

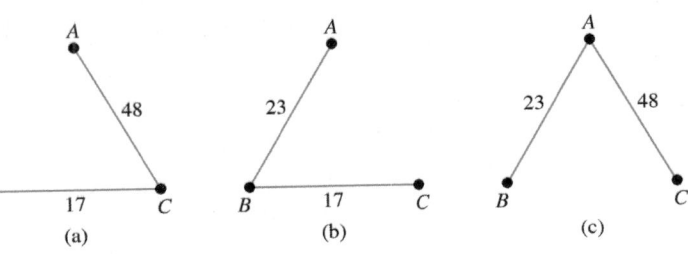

Figure 13.52

The spanning tree in Fig. 13.52(a) has a cost of $48 + $17 = $65. The spanning tree in Fig. 13.52(b) has a cost of $23 + $17 = $40. The spanning tree in Fig. 13.52(c) has a cost of $23 + $48 = $71. Therefore, the minimum-cost spanning tree is shown in Fig. 13.52(b). It has a cost of $40. ∎

The process used in Example 4 is similar to the brute force method used to determine the optimal solution to traveling salesman problems in Section 13.3. Although this process is easy to carry out for a small graph, such a process would be impossible for graphs with a large number of vertices. Fortunately, we have an algorithm, called *Kruskal's algorithm*, to determine the minimum-cost spanning tree.

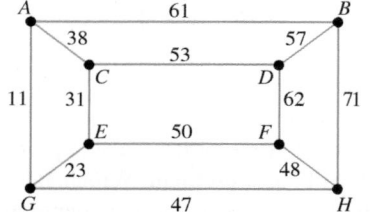

Figure 13.53

> **PROCEDURE** KRUSKAL'S ALGORITHM
>
> To construct the minimum-cost spanning tree from a weighted graph:
> 1. Select the lowest-cost edge on the graph.
> 2. Select the next lowest-cost edge that does not form a circuit with the first edge.
> 3. Select the next lowest-cost edge that does not form a circuit with previously selected edges.
> 4. Continue selecting the lowest-cost edges that do not form circuits with the previously selected edges.
> 5. When a spanning tree is complete, you have the minimum-cost spanning tree.

Example 5 *Using Kruskal's Algorithm*

Use Kruskal's algorithm to determine the minimum-cost spanning tree for the weighted graph shown in Fig. 13.53. The numbers along the edges in Fig. 13.53 represent dollars.

Solution We begin by selecting the lowest-cost edge of the graph, edge *AG*, which is $11 (see Fig. 13.54). Next we select the next lowest-cost edge that does not form a circuit, edge *GE*, which is $23. Once again, looking for the lowest-cost edge that does not form a circuit, we select edge *EC*, which is $31. The result of our first three selections is shown in Fig. 13.54.

Looking at the original weighted graph in Fig. 13.53, we see that the next lowest-cost edge is edge *AC*, which is $38. However, selecting edge *AC* would create a circuit among vertices *A*, *C*, *E*, and *G*, so we must not select edge *AC*. Instead, we select edge *GH*, which is $47, since it is the next lowest-cost edge and its selection does not lead to a circuit; see Fig. 13.55. Next, we select edge *HF*, which is $48.

The next lowest-cost edge not yet selected is edge *EF*, which is $50. However, this edge would create a circuit among vertices *E*, *F*, *H*, and *G*. Instead, we select edge *CD*, which is $53, followed by selecting edge *DB*, which is $57. The result of our selections thus far is shown in Fig. 13.56.

Notice that the graph in Fig. 13.56 is a spanning tree. According to Kruskal's algorithm, the tree in Fig. 13.56 is the minimum-cost spanning tree for the original weighted graph. ∎

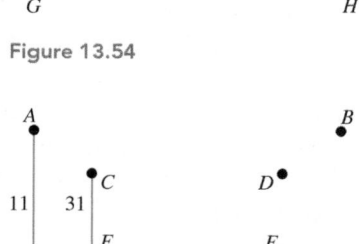

Figure 13.54

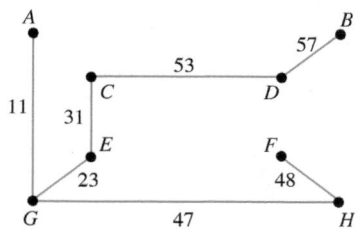

Figure 13.55

Figure 13.56

Note that in Example 5, all the edges selected were connected to edges we had already selected. However, that will not always be the case. When we select the edges in Example 6, some will be disconnected at the time we select them.

We now will apply our knowledge of minimum-cost spanning trees. Note that in examples and exercises, the graphs need not be drawn to scale.

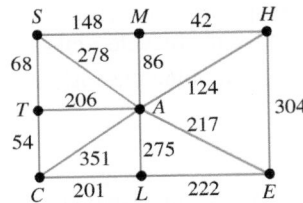

Figure 13.57

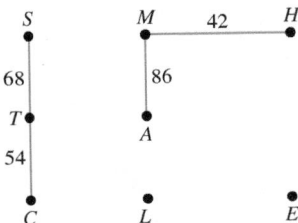

Figure 13.58

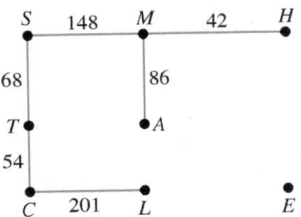

Figure 13.59

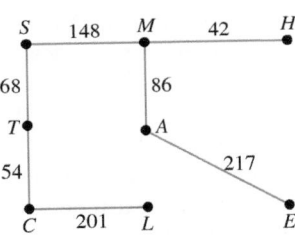

Figure 13.60

Kruskal's Algorithm

Example 6 An Application Using Kruskal's Algorithm

Recall from Example 3 on page 866 that Kutztown University is considering the construction of awnings over a select number of sidewalks on campus. The distances, in feet, between buildings are shown in the weighted graph in Fig. 13.57.

a) Determine the shortest series of sidewalks to cover so that students would be able to move between any buildings shown without being exposed to the elements.
b) The cost of the awnings is $51 per foot regardless of where they are placed on campus. What will it cost to cover the sidewalks determined in part (a)?

Solution

a) Kutztown University is seeking a minimum-cost spanning tree for the weighted graph shown in Fig. 13.57. To determine the minimum-cost spanning tree, we use Kruskal's algorithm. First, we select the edge *MH*, since it has the shortest distance and therefore has the lowest cost. We next select the edge *TC*, since it has the second shortest distance and it doesn't form a circuit with our previously selected edge (see Fig. 13.58). Our third and fourth choices will be edge *ST* and edge *MA*, respectively, since these edges have the next shortest distances and they do not form a circuit with previously selected edges. The result of our first four selections is shown in Fig. 13.58.

The next lowest-cost edge is edge *AH*. However, selection of edge *AH* would form a circuit with vertices *A, H,* and *M*. Instead, we choose edge *SM*, the next lowest-cost edge, since it does not form a circuit with the previously selected edges (see Fig. 13.59). Our next lowest-cost edge will be edge *CL*. Note that edge *CL* does not form a circuit with the previously selected edges. The result of all our selections so far is shown in Fig. 13.59.

We note that the next lowest-cost edge is edge *TA*, but selecting this edge would create a circuit with vertices *S, M, A,* and *T*. So we select edge *AE*. The result of our selections so far is shown in Fig. 13.60.

From Fig. 13.60, we can see that we have formed a spanning tree. According to Kruskal's algorithm, this graph is the minimum-cost spanning tree. Therefore, Fig. 13.60 shows which sidewalks should be covered with the awnings. Covering these sidewalks will provide protection to students at the lowest cost to Kutztown University.

b) From Fig. 13.60, we see that there are 148 + 42 + 68 + 86 + 54 + 217 + 201 = 816 feet of sidewalks that need to be covered. At $51 per foot, the cost to cover these sidewalks is $51 × 816 = $41,616. ■

We can also use Kruskal's algorithm to build a minimum-cost spanning tree from information that is not given in graph form. We demonstrate this in our next example.

Example 7 Colorado Bicycle Paths

The Colorado Parks and Recreation Association wishes to build bicycle trails that connect the cities of Aurora, Boulder, Denver, Englewood, and Golden. The approximate distances, in miles, between these cities are given in Table 13.5 below.

Table 13.5

	Aurora	Boulder	Denver	Englewood	Golden
Aurora	*	42	9	15	30
Boulder	42	*	34	41	26
Denver	9	34	*	8	16
Englewood	15	41	8	*	21
Golden	30	26	16	21	*

Did You Know?

Data Networks

In our current "information age," data networks support so many daily conveniences that we probably don't even realize a network is involved. Each time we use a smartphone, fly in an airplane, send an e-mail, or use the Internet, we are using some form of modern data network. Each of these networks can be represented as a graph with many vertices and edges. It might surprise you to know that data networks predate the twentieth century. The most well-known example of a primitive data network is the electromagnetic telegraph used in the nineteenth century in the United States and around the world prior to the invention of the telephone. An even older example of a data network is the optical telegraph (with nearly 1000 stations) used in the eighteenth century all across Europe. Perhaps the oldest, yet maybe the slowest, form of data network is one still in use today: the writing and sending of letters!

a) Use Kruskal's algorithm to determine the minimum-cost spanning tree that would link each city and create the shortest total distance for the bicycle trails.

b) If the cost of building the bicycle trails is $5500 per mile, determine the cost of building the bicycle trails determined in part (a).

Solution

a) To begin constructing the minimum-cost spanning tree, we draw five labeled vertices that represent the five cities as shown in Fig. 13.61. Note that the exact location of the vertices is unimportant and the distance between the vertices does not need to be drawn to scale.

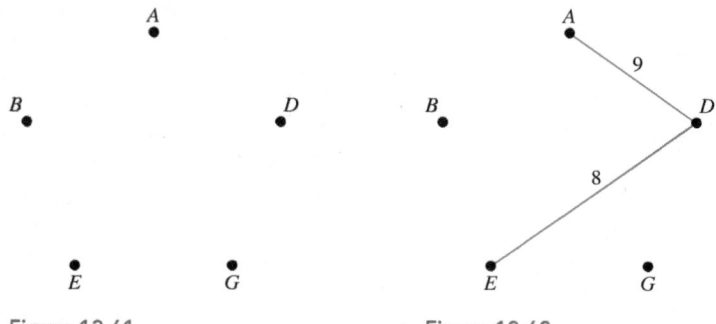

Figure 13.61   Figure 13.62

Next, study the distances given in Table 13.5 on page 869. The shortest distance listed in the table is between Denver and Englewood, 8 miles. The next shortest distance is between Aurora and Denver, 9 miles. Thus we include edge *DE* and edge *AD*, respectively, as shown in Fig. 13.62.

The next shortest distance is between Aurora and Englewood, 15 miles. However, inclusion of edge *AE* would create a circuit among vertices *A*, *D*, and *E*; therefore, we do not include edge *AE*. Instead, we note that the next shortest distance is between Denver and Golden, 16 miles, and that including edge *DG* does not create any circuits. Thus, we include edge *DG* as shown in Figure 13.63.

The next shortest distance is between Englewood and Golden, 21 miles. However, inclusion of edge *EG* would create a circuit among vertices *D*, *E*, and *G*; therefore, we do not include edge *EG*. Instead we note that the next shortest distance is between Boulder and Golden, 26 miles, and that including edge *BG* does not create any circuits. Thus, we include edge *BG* as shown in Fig. 13.64.

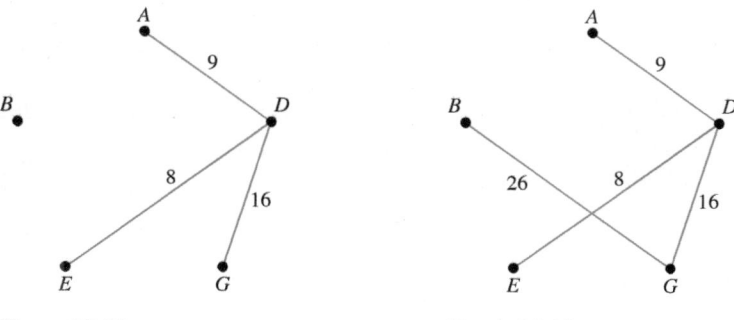

Figure 13.63   Figure 13.64

We now have a spanning tree. According to Kruskal's algorithm, Fig. 13.64 shows the minimum-cost spanning tree. Fig. 13.64 represents the bicycle trails that join the given cities with the shortest total distance.

b) From Fig. 13.64, we can see that there are 8 + 9 + 16 + 26, or 59 miles of bicycle trails needed. At a cost of $5500 per mile, the cost to create these trails is $5500 × 59, or $324,500. ∎

SECTION 13.4 Exercises

Warm Up Exercises

In Exercises 1–6, fill in the blanks with an appropriate word, phrase, or symbol(s).

1. A connected graph in which each edge is a bridge is called a(n) _____.

2. If you remove any edge in a tree, it creates a(n) _____ graph.

3. A tree does not have any Euler or Hamilton _____.

4. A tree that is created from another graph by removing edges while still maintaining a path to each vertex is called a(n) _____ tree.

5. The least expensive spanning tree of all spanning trees under consideration is called the _____ spanning tree.

6. To determine the minimum-cost spanning tree for a weighted graph, we can use _____ algorithm.

Practice the Skills

7. *Family Tree* Use a tree to show the parent–child relationships in the following family. Angela has two sons, Eli and Connor. Eli has four children: Jason, Max, Gabriella, and Tony. Connor has three children: Austin, Caleb, and Sandi.

8. *Family Tree* Use a tree to show the parent–child relationships in the following family. Cengiz has two children: Kimber and Robby. Kimber has three children: Pierre, William, and Samuel. Robby has three children: Burnette, Wasilea, and Arumugam.

9. *Corporate Structure* Use a tree to show the following supervisory relationships at DTX Industries. Kroger is president and has three vice presidents: Dorfman, Blutarsky, and Hoover. Dorfman has three managers: Stratton, Day, and Stork. Blutarsky has three managers: Schoenstein, De Pasto, and Liebowitz. Hoover has two managers: Jennings and Wormer.

10. *Employment Structure* Use a tree to show the following supervisory relationships at Halford College. President Downing supervises three division directors: Hill, Tipton, and Davis. Hill supervises two department chairs: Travis and Falkner. Tipton supervises three department chairs: Atkins, Ellis, and Campbell. Davis supervises four department chairs: Hinch, Moore, Binks, and Holland.

In Exercises 11–16, determine two different spanning trees for the given graph. There are many possible answers.

11.
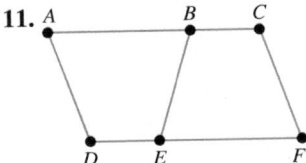

12.
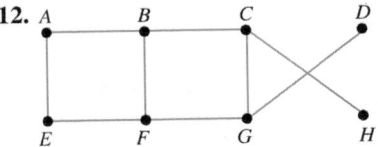

13.
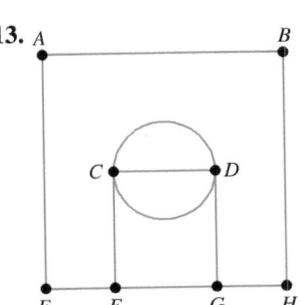

14.
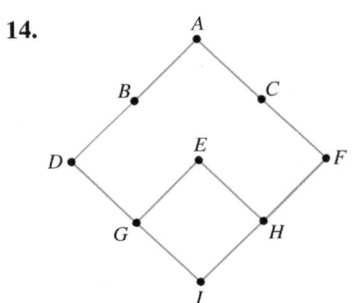

15.
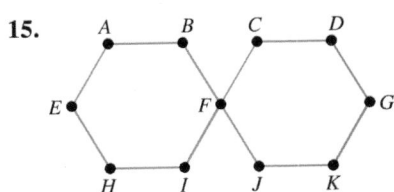

16.
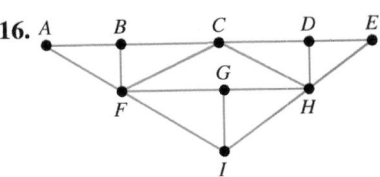

In Exercises 17–24, determine the minimum-cost spanning trees for the given graph.

17.
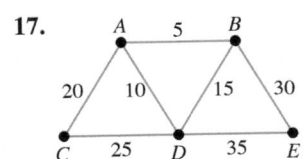

18.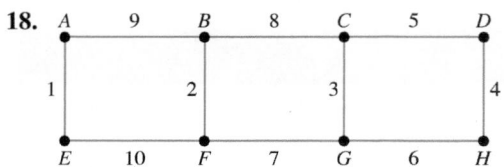

19.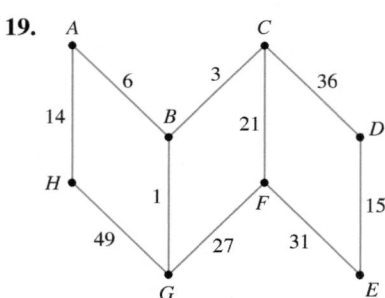

20.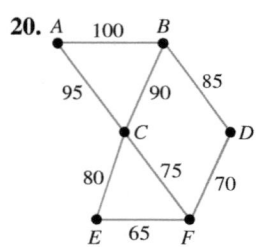

21.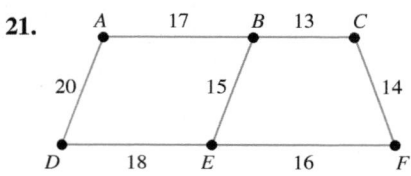

22.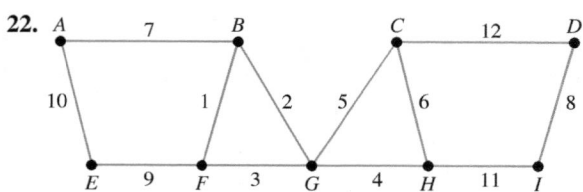

23.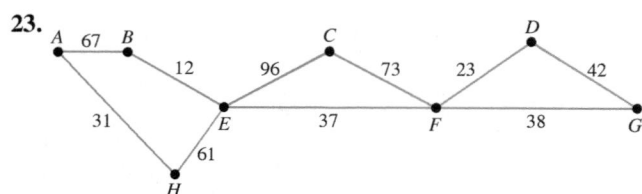

24.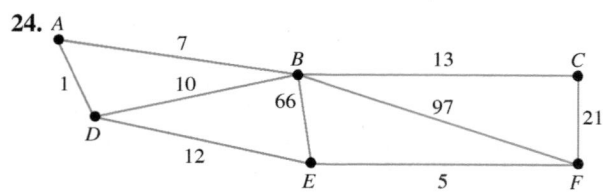

Problem Solving

25. *Irrigation System* Josefina wishes to install an irrigation system to water her five flower gardens. The distances in feet between her gardens are shown in the figure below.

a) Represent this information with a weighted graph.

b) Use Kruskal's algorithm to determine the minimum-cost spanning tree.

c) If the cost of installing irrigation pipe is $59 per foot, determine the minimum cost of installing the irrigation system from part (b).

26. *Art Sculpture* Joe is creating a modern art sculpture that needs to have electricity available at five different sites. The figure below shows the items included in the sculpture. The distances shown are in inches.

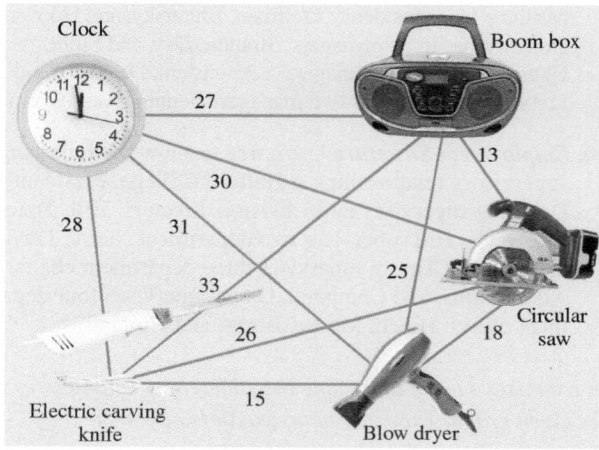

a) Represent this sculpture with a weighted graph.

b) Use Kruskal's algorithm to determine the minimum-cost spanning tree that would provide electricity to each site on the sculpture.

c) If the cost of installing wiring on the sculpture is $2.75 per inch, determine the minimum cost for wiring this sculpture.

27. ***Commuter Train System*** Several communities in eastern Pennsylvania wish to establish a commuter rail train system between the cities shown in the map below (distances are in miles).

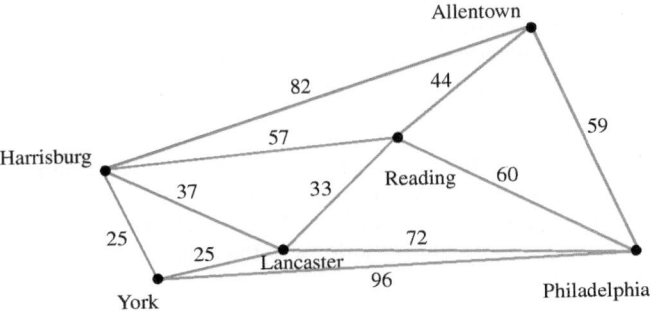

a) Use Kruskal's algorithm to determine the minimum-cost spanning tree that would link the cities using the shortest distance.

b) If it costs $1,300,000 per mile of railroad track, how much does the commuter rail system determined in part (a) cost?

28. ***Linking Campuses*** Five of the campuses in the University of Texas system would like to establish a high-speed telephone and data network between the campuses. The campuses are located in the following cities: Brownsville, Dallas, El Paso, San Antonio, and Tyler. The map below gives the approximate distances in miles between these five campuses.

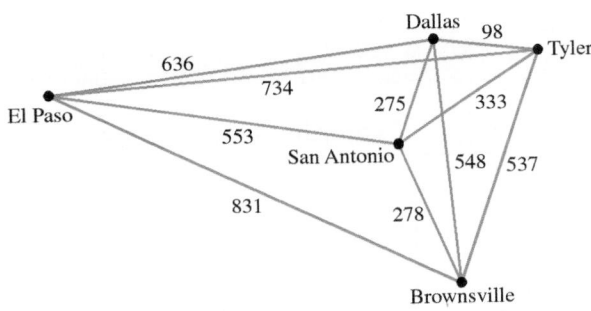

a) Use Kruskal's algorithm to determine the minimum-cost spanning tree that would link the five campuses to create the telephone and data network with the shortest distance.

b) If it costs $2500 per mile to install the network, how much will it cost to produce the network determined in part (a)?

29. ***Horse Trails*** The Darlington County, South Carolina, tourism office wishes to build a horse trail that connects the towns of Darlington, Hartsville, Lamar, and Society Hill.

The distances, in miles, between these cities are given in the table below.

	Darlington	Hartsville	Lamar	Society Hill
Darlington	*	12	14	15
Hartsville	12	*	13	16
Lamar	14	13	*	26
Society Hill	15	16	26	*

a) Use Kruskal's algorithm to determine the minimum-cost spanning tree that would link each city to create the least expensive horse trail.

b) If the cost of building such a trail is $11,500 per mile, what would be the cost of building the trail determined in part (a)?

30. ***Bicycle Trails*** The Ohio Recreation Council is interested in creating bicycle trails connecting the following cities: Akron, Cleveland, Dayton, Toledo, and Youngstown. The distances, in miles, between these cities are given in the table below.

	Akron	Cleveland	Dayton	Toledo	Youngstown
Akron	*	37	226	135	57
Cleveland	37	*	240	123	69
Dayton	226	240	*	163	265
Toledo	135	123	163	*	184
Youngstown	57	69	265	184	*

a) Use Kruskal's algorithm to determine the minimum-cost spanning tree that would link each city and create the shortest total distance for the bicycle trails.

b) If the cost of building bicycle trails is $13,000 per mile, what would be the cost of building the trails determined in part (a)?

31. ***Recreation Trail*** The Missouri Park and Recreation Association would like to build a recreation trail for biking and hiking that connects the cities of Jefferson City, Kansas City, Springfield, St. Joseph, and St. Louis. The distances in miles between these cities are given in the table on the top of page 874.

	Jefferson City	Kansas City	Springfield	St. Joseph	St. Louis
Jefferson City	*	158	138	215	134
Kansas City	158	*	191	56	249
Springfield	138	191	*	227	218
St. Joseph	215	56	227	*	307
St. Louis	134	249	218	307	*

a) Use Kruskal's algorithm to determine the minimum-cost spanning tree that would link each city to create the least expensive recreation trail.

b) If the cost of building such a trail is $9500 per mile, what would be the cost of building the trail determined in part (a)?

32. **Light-Rail Transit System** The state of Illinois is applying for a grant to connect several metropolitan areas of the state with a light-rail transport system. The cities involved are Urbana, Chicago, Peoria, Rockford, and Springfield. The distances in miles between these cities are given in the table below.

	Urbana	Chicago	Peoria	Rockford	Springfield
Urbana	*	135	89	181	86
Chicago	135	*	170	85	202
Peoria	89	170	*	129	74
Rockford	181	85	129	*	193
Springfield	86	202	74	193	*

a) Use Kruskal's algorithm to determine the minimum-cost spanning tree that would link each city using the shortest distance.

b) What would be the total distance of the light-rail transportation system determined in part (a)?

Challenge Problems/Group Activities

33. **Computer Network** Create a minimum-cost spanning tree that would serve as the least expensive way to create a computer network among the five largest cities in your state. Assume that the cost per mile is the same regardless of the path taken.

34. **College Structure** Create a tree that shows the administrative structure at your college or university. Start with the highest-ranking officer (that is, president, chancellor, provost) and work down to the department leader level.

35. **Sidewalk Covers** Create a minimum-cost spanning tree that would serve as the least expensive way to provide sheltered sidewalks between major buildings on your college campus (see Example 6 on page 869). Assume that the cost per foot is the same regardless of the path taken.

Research Activity

36. **Joseph Kruskal** Write a research paper on the life and work of Joseph Kruskal, who developed Kruskal's algorithm.

CHAPTER 13 *Summary*

Important Facts and Concepts

Examples and Discussion

Section 13.1	
Paths and Circuits	
A **path** is a sequence of adjacent vertices and the edges connecting them.	Discussion, pages 834–835
A **circuit** is a path that begins and ends at the same vertex.	

Section 13.2	
Paths and Circuits	
An **Euler path** is a path that passes through each *edge* of a graph exactly one time.	Discussion, page 841; Example 1, page 841
An **Euler circuit** is a circuit that passes through each *edge* of a graph exactly one time.	
Euler's Theorem	
For a connected graph, the following statements are true:	Examples 2–6, pages 842–847
1. A graph with *no odd vertices* (all even vertices) has at least one Euler path, which is also an Euler circuit. An Euler circuit can be started at any vertex, and it will end at the same vertex.	

2. A graph with *exactly two odd vertices* has at least one Euler path but no Euler circuits. Each Euler path must begin at one of the two odd vertices and end at the other odd vertex.

3. A graph with *more than two odd vertices* has no Euler paths or Euler circuits.

Fleury's Algorithm

To determine an Euler path or an Euler circuit:

Examples 5–6, pages 844–847

1. Use Euler's theorem to determine whether an Euler path or an Euler circuit exists. If one exists, proceed with Steps 2–5.

2. If the graph has no odd vertices (therefore has at least one Euler circuit that is also an Euler path), choose any vertex as the starting point. If the graph has exactly two odd vertices (therefore has only an Euler path), choose one of the two odd vertices as the starting point.

3. Begin to trace edges as you move through the graph. Number the edges as you trace them. Since you can't trace any edges twice in Euler paths and Euler circuits, once an edge is traced consider it "invisible."

4. When faced with a choice of edges to trace, if possible, choose an edge that is not a bridge (i.e., don't create a disconnected graph with your choice of edges).

5. Continue until each edge of the entire graph has been traced once.

Section 13.3

Paths and Circuits

A **Hamilton path** is a path that contains each *vertex* of a graph exactly once.

A **Hamilton circuit** is a path that begins and ends at the same vertex and passes through all *vertices* exactly one time.

Discussion, pages 853–854;
Example 2, page 855

Number of Unique Hamilton Circuits in a Complete Graph

The number of unique Hamilton circuits in a complete graph with n vertices is $(n - 1)!$, where

Examples 3–5, pages 855–857

$$(n - 1)! = (n - 1)(n - 2)(n - 3) \cdots (3)(2)(1)$$

The Brute Force Method of Solving Traveling Salesman Problems

To determine the optimal solution:

Discussion, page 856;
Example 5, pages 856–857

1. Represent the problem with a complete, weighted graph.

2. List all possible Hamilton circuits in this graph.

3. Determine the cost associated with each of these Hamilton circuits.

4. The Hamilton circuit with the lowest cost is the optimal solution.

The Nearest Neighbor Method for Determining an Approximate Solution to a Traveling Salesman Problem

To approximate the optimal solution:

Discussion, page 858;
Example 6, pages 858–859

1. Represent the problem with a complete, weighted graph.

2. Identify the starting vertex.

3. Of all the edges attached to the starting vertex, choose the edge that has the smallest weight. This edge is the shortest distance or the lowest cost. Travel along this edge to the second vertex.

4. At the second vertex, choose the edge that has the smallest weight that does not lead to a vertex already visited. Travel along this edge to the third vertex.

5. Continue this process, each time moving along the edge with the smallest weight until all vertices are visited.

6. Travel back to the original vertex.

Section 13.4

Kruskal's Algorithm

To construct the minimum-cost spanning tree from a weighted graph:

1. Select the lowest-cost edge on the graph.

2. Select the next lowest-cost edge that does not form a circuit with the first edge.

3. Select the next lowest-cost edge that does not form a circuit with previously selected edges.

4. Continue selecting the lowest-cost edges that do not form circuits with the previously selected edges.

5. When a spanning tree is complete, you have the minimum-cost spanning tree.

Examples 5–7, pages 868–870

CHAPTER 13   Review Exercises

13.1

In Exercises 1 and 2, create a graph with the given properties.

1. Create a graph with four even vertices, two odd vertices, a bridge, and a loop.

2. Create a graph with four even vertices and two loops.

In Exercises 3 and 4, use the following graph.

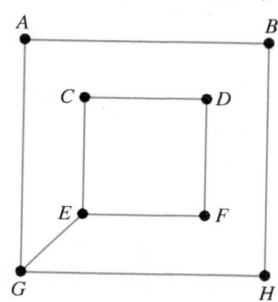

3. Determine a path that passes through each edge exactly one time.

4. Is it possible to have a path that begins at vertex *A*, includes all edges exactly once, and ends at vertex *B*?

5. *Western States* Represent the map as a graph where each vertex represents a state and each edge represents a common border between the states.

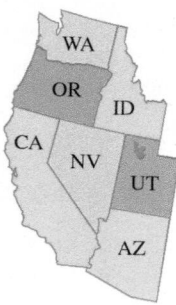

6. *School Floor Plan* The drawing below shows the floor plan of Raintree Montessori School. Construct a graph to represent the school using the letters shown to label the vertices. Use the letter *O* to label the vertex that represents the outside of the school. Place vertex *O* near the bottom of the graph.

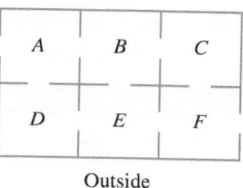

Outside

In Exercises 7 and 8, determine whether the graph shown is connected or disconnected.

7.

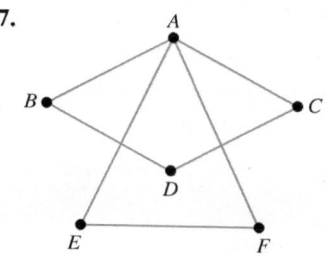

8.

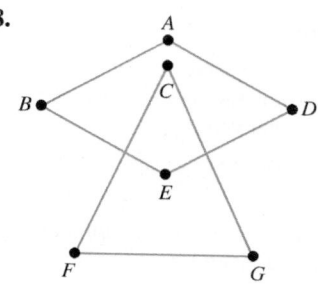

9. Identify any bridges in the graph below.

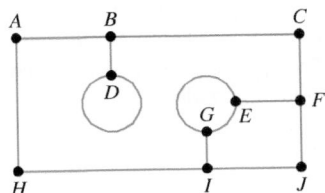

13.2

Use the following graph for Exercises 10 and 11.

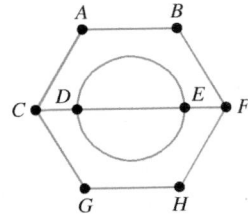

10. Determine an Euler path that begins with vertex C.

11. Determine an Euler path that begins with vertex F.

Use the following graph for Exercises 12 and 13.

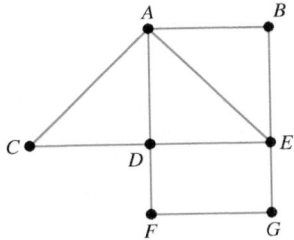

12. Determine an Euler circuit that begins with vertex A.

13. Determine an Euler circuit that begins with vertex E.

14. Consider the following map.

a) Represent the map as a graph.

b) Determine (state yes or no) whether the graph has an Euler path. If yes, give one such Euler path.

c) Determine (state yes or no) whether the graph has an Euler circuit. If yes, give one such Euler circuit.

15. a) The drawing below shows the floor plan of a single-story house. Construct a graph that represents the floor plan.

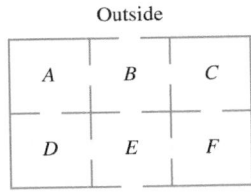

b) Is it possible for a person to walk through each doorway in the house without using any of the doorways twice?

c) If so, where can the person start and where will the person finish? Explain.

16. a) Can a police officer walk each street shown in the figure below without walking any street more than once?

b) If yes, where would the police officer have to start the walk?

17. Use Fleury's algorithm to determine an Euler path in the following graph.

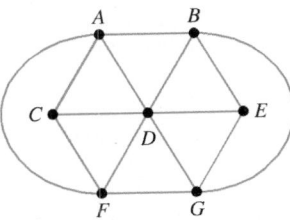

18. Use Fleury's algorithm to determine an Euler circuit in the following graph.

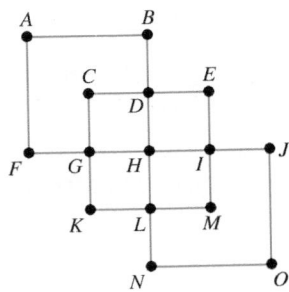

13.3

19. Determine two different Hamilton paths in the following graph.

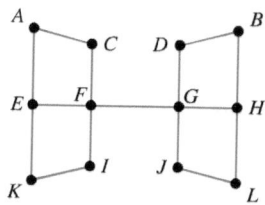

20. Determine two different Hamilton circuits in the following graph.

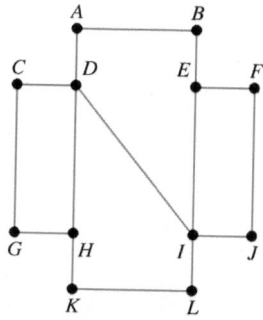

21. Draw a complete graph with five vertices.

22. Baseball The American League Eastern Division has the following teams: Baltimore Orioles, Boston Red Sox, New York Yankees, Tampa Bay Rays, and Toronto Blue Jays. How many different possible ways could a Boston Red Sox fan visit each team in the American League Eastern Division one time and return home to Boston?

23. Job Interviews Lance is searching for a new job. He lives in Portland, Oregon, and has interviews in Des Moines, Iowa; Charleston, West Virginia; and Montgomery, Alabama. The costs of the one-way flights between these four cities are as follows: Portland to Des Moines is $428, Portland to Charleston is $787, Portland to Montgomery is $902, Des Moines to Charleston is $449, Des Moines to Montgomery is $458, and Charleston to Montgomery is $415.

a) Represent this traveling salesman problem with a complete, weighted graph showing the prices of the flights on the appropriate edges.

b) Use the brute force method to determine the least expensive route for Lance to travel to each city and return home to Portland.

c) What is the total cost of airfare for Lance's trip in part (b)?

24. Visiting Sales Offices Jennifer is the sales manager for AT&T for the state of Missouri. There are major sales offices in Columbia, Kansas City, St. Joseph, St. Louis, and Springfield. The distances in miles between these cities are given in the following table.

	Columbia	Kansas City	St. Joseph	St. Louis	Springfield
Columbia	*	130	177	127	168
Kansas City	130	*	54	256	192
St. Joseph	177	54	*	304	224
St. Louis	127	256	304	*	210
Springfield	168	192	224	210	*

a) Represent this traveling salesman problem with a complete, weighted graph showing the prices of the flights on the appropriate edges.

b) Use the nearest neighbor method to approximate the optimal route for Jennifer to begin in St. Joseph, visit each city once, and return to St. Joseph.

c) Suppose that Jennifer starts in Springfield. Use the nearest neighbor method to approximate the optimal route for Jennifer to visit each city once and return to Springfield.

13.4

25. Supervisory Relationships Use a tree to show the supervisory relationships in the Hulka Consulting Corporation. Hulka is president. Winger and Ziskey are vice presidents who work for Hulka. Markowicz, Oxburger, and Soyer are district managers who work for Winger. Elmo, Hector, and Jenesky are district managers who work for Ziskey.

26. Determine two different spanning trees for the following graph.

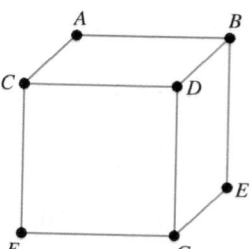

27. Determine the minimum-cost spanning tree for the following graph.

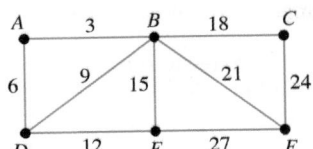

28. *Irrigation System* Renata wishes to install an irrigation system in her backyard to water all six of her flower beds. The flower beds include juniper (J), bougainvilleas (B), geraniums (G), orchids (O), azaleas (A), and petunias (P). The graph to the right shows the location of each flower bed and the distance between them in feet.

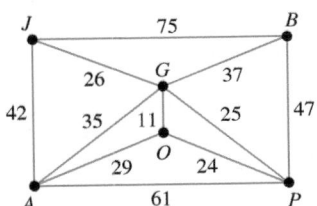

a) Use Kruskal's algorithm to determine the minimum-cost spanning tree

b) The cost for installing the irrigation system is $56 per foot. Determine the cost of installing the irrigation system.

CHAPTER 13   *Test*

1. Create a graph with seven vertices, a bridge, and a loop.

2. *Part of Africa* Represent the map below as a graph where each vertex represents a country and each edge represents a common border between the countries.

3. *Photography Studio* The drawing below shows the floor plan of Carol Ann's Photography Studio. Construct a graph to represent the studio using the letters shown to label the vertices. Use the letter O to label the vertex that represents the outside of the studio. Place vertex O near the top of the graph.

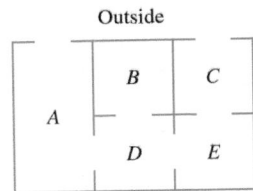

4. Draw a disconnected graph.

5. In the following graph, determine an Euler path.

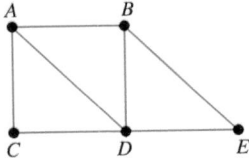

6. In the following graph, determine an Euler circuit.

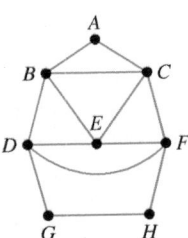

7. Is it possible for a person to walk through each doorway in the house, whose floor plan is below, without using any of the doorways twice? If so, indicate in which room the person may start and where he or she will end.

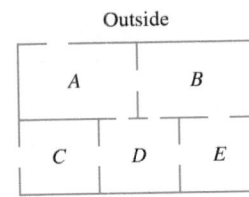

8. Use Fleury's algorithm to determine an Euler circuit in the following graph.

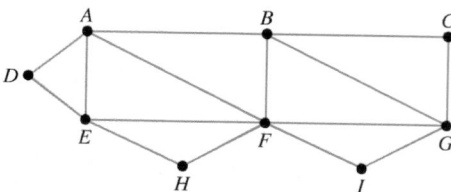

9. In the following graph, determine a Hamilton path.

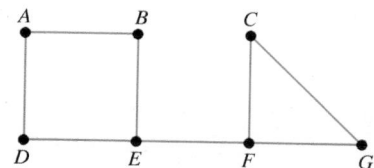

10. In the following graph, determine a Hamilton circuit.

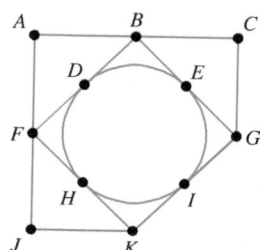

11. Draw a complete graph with five vertices.

12. *Senate Campaign* Tamiko is campaigning in Arizona for the U.S. Senate. Tamiko lives in Yuma and she wishes to travel to the following cities: Flagstaff, Phoenix, Tombstone, Tucson, and Winslow. How many different ways can Tamiko visit each city and return to her home in Yuma?

Business Travel In Exercises 13–15, use the following information.

Doug is a purchasing agent for Champs Sports *(C)*. For his job, he must travel to meet with representatives from the following sports companies: Nike *(N)*, Reebok *(R)*, and Under Armour *(U)*. The prices of one-way flights between the corresponding cities for these companies are as follows: Champs to Nike is $253, Champs to Reebok is $114, Champs to Under Armour is $122, Nike to Reebok is $199, Nike to Under Armour is $183, and Reebok to Under Armour is $112.

13. Represent this traveling salesman problem with a complete weighted graph showing the prices of flights on the appropriate edges.

14. Use the brute force method to determine the least expensive route for Doug to visit each company and return to Champs Sports. What is the cost of this route?

15. Use the nearest neighbor method to approximate the optimal route for Doug to visit each company and return to Champs Sports. What is the cost of this route?

16. *Family Tree* Use a tree to show the parent–child relationships in the following family. Beth has three children: Kevin, Demarius, and Teri. Kevin has one child, Chris. Demarius has two children: Michelle and April. Teri has one child, Lucy.

17. Determine a spanning tree for the graph shown below.

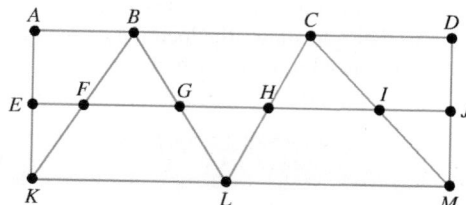

18. Determine the minimum-cost spanning tree for the following graph.

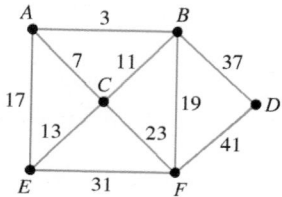

Irrigation System In Exercises 19 and 20, use the following information.

Daniel is planning a new irrigation system for his yard. His current system has valves already in place as shown in the figure below. The numbers shown are in feet.

19. Determine the minimum-cost spanning tree that reaches each valve.

20. If the new irrigation materials cost $1.25 per foot, what would be the cost of installing the system determined in part (a)?

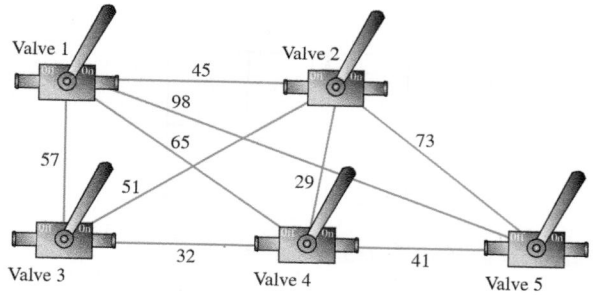

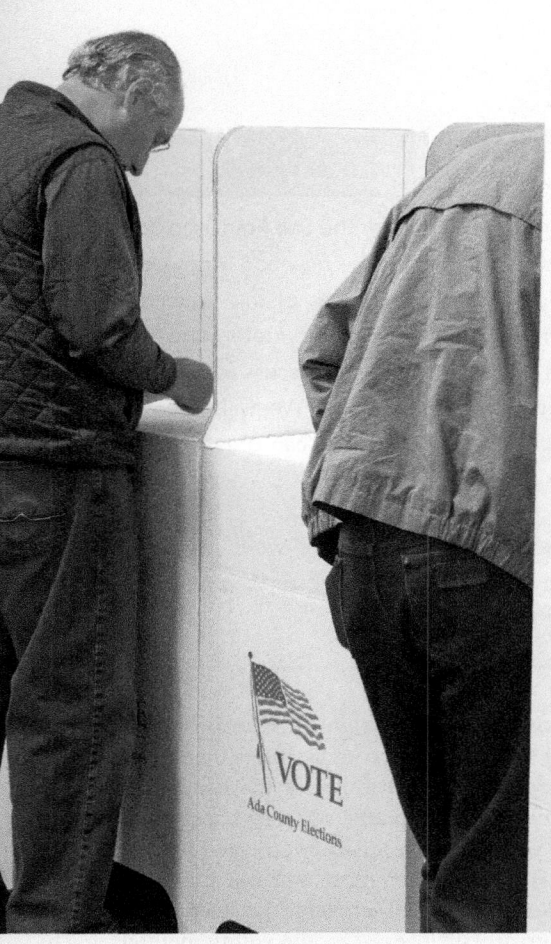

◀ Registering to vote and then voting in elections are part of our civic duties as citizens.

14

Voting and Apportionment

What You Will Learn

- Voting methods
- Flaws of voting methods
- Apportionment methods
- Flaws of apportionment methods

Why This Is Important

We are frequently told that it is our civic duty to vote in an election. Can one person's vote make a difference? How are voting decisions made? Could a candidate for student government president receive a majority of votes and still lose the election? Depending on the voting method used, determining the winner of an election can sometimes lead to different results. In this chapter, we will discuss different voting methods and some flaws of these methods. We will also discuss apportionment problems, such as how to determine the number of representatives each state has in the U.S. House of Representatives, and some flaws that can occur with apportionment methods.

Both voting methods and apportionment methods can be used in a variety of applications involving businesses, governments, schools, and other organizations.

SECTION 14.1   Voting Methods

Upon completion of this section, you will be able to:

- Use the plurality method to determine the winner of an election.
- Use the Borda count method to determine the winner of an election.
- Use the plurality with elimination method to determine the winner of an election.
- Use the pairwise comparison method to determine the winner of an election.
- Understand tie-breaking procedures for an election.

The planning committee for the San Francisco Zoo is voting on an animal for the new zoo logo. How should it determine the winner of this election? Is there more than one way to determine the winner? In this section, we will discuss four of the most popular voting methods. We will discover that the voting method chosen can greatly affect the results of an election.

Why This Is Important   Voting methods are used in a variety of real-world problems from voting for our elected representatives to voting for a zoo logo to voting for the president of a student government.

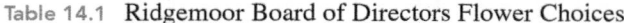

In a constitutional democracy, one of the most fundamental rights and responsibilities of its citizens is the right to vote. Voting gives us a voice in decisions that affect our lives. Conducting a fair election to select representatives lies at the heart of a democratic form of government. In general, elections for public officials seem to be a simple proposition: You vote for the person you wish to represent you, and the candidate who obtains the most votes is elected. However, when more than one choice is involved in the election or when voters are asked to rank their choices, the selection is not as simple. The outcome of the election is greatly influenced by the voting method used, as will be discussed in this section. Consider the following example.

The Ridgemoor Homeowners Association board of directors needs to choose one type of flower to plant at the main entrance to the housing development. The choices available are begonia, impatiens, marigold, or periwinkle. Each of the five directors is asked to rank the flowers according to preference. The results of this election are given in Table 14.1.

Table 14.1   Ridgemoor Board of Directors Flower Choices

Choice	Charlie	Denise	Edwardo	Gerry	Joe
First	Impatiens	Impatiens	Impatiens	Marigold	Marigold
Second	Marigold	Marigold	Marigold	Begonia	Begonia
Third	Periwinkle	Periwinkle	Periwinkle	Periwinkle	Periwinkle
Fourth	Begonia	Begonia	Begonia	Impatiens	Impatiens

If we were to run a simple election in which we only considered each director's first choice, impatiens would be chosen. It also can be said that impatiens received a *majority* of first-place votes. A majority simply refers to receiving more than 50% of the votes. However, looking at the table, it appears that marigolds might be a better choice since all the directors have marigolds listed either as their first or second choice and impatiens are listed last by two directors.

In this first section, we will describe four of the most popular voting methods and explain how they are implemented. In the next section, we will discuss the weaknesses and flaws of each method. We will discuss the following methods:

1. Plurality method
2. Borda count method
3. Plurality with elimination method
4. Pairwise comparison method

Before we discuss any of these voting methods, though, we will illustrate how the results of an election might be indicated in a table called a *preference table*. A *preference table* summarizes the results of an election and shows how often each particular outcome was selected. In Example 1, we explain how to construct a preference table.

Example 1 *Balloting by Committee Members*

The planning committee for the San Francisco Zoo is voting on an animal for the new zoo logo. The choices are a tiger (T), a kangaroo (K), and a giraffe (G). There are 15 planning committee members (labeled CM1, CM2, CM3, . . ., CM15). Their ballots are shown in Fig. 14.1.

CM1	CM2	CM3	CM4	CM5	CM6	CM7	CM8
1ST T	1ST T	1ST T	1ST G	1ST K	1ST K	1ST K	1ST T
2ND K	2ND G	2ND K	2ND T	2ND G	2ND T	2ND T	2ND G
3RD G	3RD K	3RD G	3RD K	3RD T	3RD G	3RD G	3RD K

CM9	CM10	CM11	CM12	CM13	CM14	CM15	
1ST K	1ST G	1ST T	1ST G	1ST G	1ST G	1ST G	
2ND T	2ND T	2ND G	2ND T	2ND T	2ND T	2ND K	
3RD G	3RD K	3RD K	3RD K	3RD K	3RD K	3RD T	

Figure 14.1

Construct a preference table to illustrate the results of the election.

Solution Notice that committee member 1 (CM1) listed T, K, G as his or her choice. Also, notice that committee member 3 (CM3) also listed T, K, G, in that order. None of the other committee members listed this order of selection. Therefore, there are two committee members who listed T, K, G in that order. This information is indicated in the column in boldface in Table 14.2. Next look at CM2. He or she listed T, G, K. Two other committee members, CM8 and CM11, listed T, G, K in that order. Therefore, three committee members listed the order T, G, K. This information is indicated in the column to the right of the boldface column in Table 14.2. Continue this process to obtain the preference table.

Table 14.2 Zoo Logo Preference Table

Number of Votes	2	3	5	1	3	1
First	T	T	G	K	K	G
Second	K	G	T	G	T	K
Third	G	K	K	T	G	T

Notice that the total number of votes indicated in the preference table is 15, which checks with the individual number of votes given. ∎

When an election has many voters, a preference table is an efficient way to report the results.

The initial purpose of Example 1 was to introduce the preference table. We will revisit the zoo logo election several times in this section. Example 2 introduces another election that will be used frequently in this section.

Example **2**   *Voting for the Veterans Club President*

Four students are running for president of the Veterans Club: Antoine (A), Betty (B), Carlos (C), and Don (D). The club members were asked to rank all candidates. The resulting preference table for this election is given in Table 14.3.

Table 14.3   Veterans Club Preference Table

Number of Votes	19	15	11	7	2
First	B	C	D	A	C
Second	A	A	C	D	D
Third	C	D	A	C	A
Fourth	D	B	B	B	B

a) How many students voted in the election?
b) How many students selected the candidates in this order: C, A, D, B?
c) How many students selected C as their first choice?

Solution

a) To determine the number of students who voted, we add the numbers in the row labeled Number of Votes.

$$19 + 15 + 11 + 7 + 2 = 54$$

Therefore, 54 students voted in the election.

b) Fifteen students voted in the given order (the second column of letters).

c) To determine the number who voted for C as their first choice, read across the row that says First. When you see a C in the row, record the number above it. Then determine the sum of the numbers. Thus,

$$15 + 2 = 17 \text{ students selected C as their first choice.} \qquad \blacksquare$$

Now we will begin to examine the results of the elections in Examples 1 and 2. Each election will be analyzed using each of the four different voting methods to be discussed in this section: plurality method, Borda count method, plurality with elimination method, and pairwise comparison method.

Plurality Method

When electing a candidate using the *plurality method*, each voter votes for only one candidate. The candidate receiving the most votes is declared the winner. With the plurality method, a preference table does not have to be used, since the candidates are not ranked. In this section, we will often use a preference table, since the data may often be given in that form from a previous example. The plurality method is the most commonly used method, and it is the easiest method to use when there are more than two candidates.

> **PROCEDURE**   **PLURALITY METHOD**
>
> 1. Each voter votes for one candidate.
> 2. The candidate receiving the most votes is declared the winner.

If a candidate has a majority of votes, he or she will win an election using the plurality method. However, it is possible for a candidate to win an election using the plurality method without the candidate having a majority of votes, as we will see in the next example.

Example 3 *Selecting a Zoo Logo Using the Plurality Method*

Consider the zoo logo election given in Example 1. Which animal is selected using the plurality method?

Solution Using the plurality method, the animal with the most votes is selected for the logo. In Example 1, the planning committee members ranked the animals. We will assume that each planning committee member would vote for the animal he or she listed first in Table 14.2 on page 883. From Table 14.2, we see that the tiger received $2 + 3 = 5$ votes, the kangaroo received $1 + 3 = 4$ votes, and the giraffe received $5 + 1 = 6$ votes. These results indicate that by using the plurality method, the giraffe is chosen as the animal for the zoo logo. Note that the giraffe received only $\frac{6}{15}$, or 40%, of the first-place votes, which is less than a majority. ∎

Example 4 *Electing the Veterans Club President by the Plurality Method*

Consider the Veterans Club election given in Example 2. Who is elected president using the plurality method?

Solution We will assume that each member would vote for the person he or she listed in first place in Table 14.3 on page 884. From Table 14.3, we can see that Antoine received 7 votes, Betty received 19 votes, Carlos received $15 + 2 = 17$ votes, and Don received 11 votes. Therefore, using the plurality method, Betty received the most votes and is elected president of the Veterans Club. Note that Betty received only $\frac{19}{54}$, or about 35%, of the first-place votes, which is less than a majority. ∎

As simple as the plurality method is, this voting method ignores voters' second and subsequent choices. The rest of the voting methods we will discuss take all voters' choices into consideration.

Borda Count Method

The *Borda count* method, developed by Jean-Charles de Borda, requires that voters rank candidates in order from most favorable to least favorable. The Borda count method then quantifies this ranking to determine a winner.

> **PROCEDURE** BORDA COUNT METHOD
>
> 1. Voters rank the candidates from the most favorable to the least favorable.
> 2. Each last-place vote is awarded one point, each next-to-last-place vote is awarded two points, each third-from-last-place vote is awarded three points, and so forth.
> 3. The candidate receiving the most points is the winner of the election.

The primary advantage of the Borda count method is that voters are able to provide more information than in the plurality method. Many well-known ranking systems use variations of the Borda count method. The Heisman trophy, given annually to the best college football player, is awarded by having the voters (sportswriters and former Heisman trophy winners) rank their first, second, and third choices out of possibly thousands of college football players. A player receives 3 points for each first-place vote, 2 points for each second-place vote, and 1 point for each third-place vote.

Example 5 *Selecting a Zoo Logo Using the Borda Count Method*

Let's reexamine the San Francisco Zoo election for a new logo as discussed in Example 1 on page 883. For convenience, the preference table is reproduced in Table 14.4. Recall that T represents tiger, K represents kangaroo, and G represents giraffe. Which animal is selected using the Borda count method?

Table 14.4 San Francisco Zoo Logo Preference Table

Number of Votes	2	3	5	1	3	1
First	T	T	G	K	K	G
Second	K	G	T	G	T	K
Third	G	K	K	T	G	T

Solution In this problem, there are three candidates, so a first-place vote is worth 3 points, a second-place vote is worth 2 points, and a third-place vote is worth 1 point. To award points, we examine each candidate starting with the tiger. Using Table 14.4, we see that the

Tiger receives

$2 + 3 = \mathbf{5}$ first-place votes producing $\mathbf{5} \times 3 = 15$ points,

$5 + 3 = \mathbf{8}$ second-place votes producing $\mathbf{8} \times 2 = 16$ points, and

$1 + 1 = \mathbf{2}$ third-place votes producing $\mathbf{2} \times 1 = 2$ points

Therefore, the tiger receives $15 + 16 + 2 = 33$ points.

Giraffe receives

$5 + 1 = \mathbf{6}$ first-place votes producing $\mathbf{6} \times 3 = 18$ points,

$3 + 1 = \mathbf{4}$ second-place votes producing $\mathbf{4} \times 2 = 8$ points, and

$2 + 3 = \mathbf{5}$ third-place votes producing $\mathbf{5} \times 1 = 5$ points

Therefore, the giraffe receives $18 + 8 + 5 = 31$ points.

Kangaroo receives

$1 + 3 = \mathbf{4}$ first-place votes producing $\mathbf{4} \times 3 = 12$ points,

$2 + 1 = \mathbf{3}$ second-place votes producing $\mathbf{3} \times 2 = 6$ points, and

$3 + 5 = \mathbf{8}$ third-place votes producing $\mathbf{8} \times 1 = 8$ points

Therefore, the kangaroo receives $12 + 6 + 8 = 26$ points.

The tiger receives 33 points, the giraffe receives 31 points, and the kangaroo receives 26 points. By using the Borda count method, the tiger is chosen for the zoo logo.   ∎

We will now reexamine the election for president of the Veterans Club by using the Borda count method.

Borda Count Method

Example 6 *Electing the Veterans Club President Using the Borda Count Method*

Use the Borda count method to determine the winner of the election for president of the Veterans Club discussed in Example 2 on page 884. Recall that the candidates are Antoine (A), Betty (B), Carlos (C), and Don (D). For convenience, the preference table is reproduced below in Table 14.5.

Table 14.5 Veterans Club Preference Table

Number of Votes	19	15	11	7	2
First	B	C	D	A	C
Second	A	A	C	D	D
Third	C	D	A	C	A
Fourth	D	B	B	B	B

Solution In this problem, there are four candidates, so a first-place vote is worth 4 points, a second-place vote is worth 3 points, a third-place vote is worth 2 points, and a fourth-place vote is worth 1 point.

Using Table 14.5, we see that:

Antoine receives

   7 first-place votes producing **7** $\times$ 4 = 28 points,

   19 + 15 = **34** second-place votes producing **34** $\times$ 3 = 102 points, and

   11 + 2 = **13** third-place votes producing **13** $\times$ 2 = 26 points

Antoine has no fourth-place votes. Therefore, Antoine receives 28 + 102 + 26 = 156 points.

Betty receives

   19 first-place votes producing **19** $\times$ 4 = 76 points, and

   15 + 11 + 7 + 2 = **35** fourth-place votes producing **35** $\times$ 1 = 35 points

Betty has no second- or third-place votes. Therefore, Betty receives 76 + 35 = 111 points.

Carlos receives

   15 + 2 = **17** first-place votes producing **17** $\times$ 4 = 68 points,

   11 second-place votes producing **11** $\times$ 3 = 33 points, and

   19 + 7 = **26** third-place votes producing **26** $\times$ 2 = 52 points

Carlos has no fourth-place votes. Therefore, Carlos receives 68 + 33 + 52 = 153 points.

Don receives

   11 first-place votes producing **11** $\times$ 4 = 44 points,

   7 + 2 = **9** second-place votes producing **9** $\times$ 3 = 27 points,

   15 third-place votes producing **15** $\times$ 2 = 30 points, and

   19 fourth-place votes producing **19** $\times$ 1 = 19 points

Therefore, Don receives 44 + 27 + 30 + 19 = 120 points.

Antoine, with 156 points, receives the most points using the Borda count method and is declared the winner.

Even though Antoine received the fewest number of first-place votes, he is declared the winner under the Borda count method.

Notice that with both the zoo logo election (Examples 3 and 5) and the Veterans Club president election (Examples 4 and 6), the plurality method and the Borda count method produced different winners. This result indicates the importance of choosing a voting method *prior to the election*. This will become even more evident when we examine the remaining voting methods. Next we will examine the plurality with elimination method, which also takes voters' second and subsequent preferences into account.

Plurality with Elimination

The next method we will discuss, *plurality with elimination*, may involve a series of elections. When using the plurality with elimination method each voter votes for only one candidate. If a candidate initially receives a majority of votes, that candidate is declared the winner. If no candidate receives a majority of votes, the candidate with the fewest number of votes is eliminated and a second election is held. This process continues until a candidate receives a majority of votes.

PROCEDURE   PLURALITY WITH ELIMINATION METHOD

1. Each voter votes for one candidate.

2. If a candidate receives a majority of votes, that candidate is declared the winner.

3. If no candidate receives a majority, eliminate the candidate with the fewest votes and hold another election. (If there is a tie for the fewest votes, eliminate all candidates tied for the fewest votes.)

4. Repeat this process until a candidate receives a majority.

One of the immediate disadvantages of the plurality with elimination method is the possible need for two or more elections, depending on the number of candidates. One way to limit the number of elections held is to begin with a *runoff election* in which voters choose among a field of candidates. In the second election, the voters only select among the top two candidates from the runoff election. Another way to avoid multiple elections entirely is to have voters rank candidates from most to least favorable just as we did with the Borda count method. We assume that in the first round, the voters would have selected their first choice. We must also assume that after a candidate is eliminated, the order of preference is not changed. For instance, let's say that a voter's original ranking order was E, G, B, D, F. Suppose that B is eliminated. This voter's ranking order now becomes E, G, D, F. *In this book, when working the examples and exercises, we will assume that the order preference remains the same in the second election as it was in the first election.* The next two examples will demonstrate the elimination process when used with the plurality with elimination method.

┌ Example **7** *Selecting a Zoo Logo Using the Plurality with Elimination Method*

Once again, we will consider the zoo logo election at the San Francisco Zoo. Recall that T represents tiger, K represents kangaroo, and G represents giraffe. Which animal is selected using the plurality with elimination method?

Solution Reexamine the preference table shown in Table 14.6.

Table 14.6 Zoo Logo Preference Table

Number of Votes	2	3	5	1	3	1
First	T	T	G	K	K	G
Second	K	G	T	G	T	K
Third	G	K	K	T	G	T

We first count the number of first-place votes for each animal:

Tiger: $2 + 3 = 5$ Giraffe: $5 + 1 = 6$ Kangaroo: $1 + 3 = 4$

To receive a majority, a candidate must have more than half of the votes. Since there are 15 people voting, 8 or more votes are needed to receive a majority. In this election, no candidate received a majority. Since the kangaroo received the fewest number of first-place votes, it is eliminated in the first round. Next, we assume that all preference orders remain the same. For example, the three committee members who had their preference order of K, T, and G will now have a preference order of T and G. Table 14.7 shows the new preference table after the kangaroo is eliminated.

Table 14.7 Zoo Logo Preference Table After the Kangaroo Is Eliminated

Number of Votes	2	3	5	1	3	1
First	T	T	G	G	T	G
Second	G	G	T	T	G	T

We see now that the tiger has $2 + 3 + 3 = 8$ first-place votes and the giraffe has $5 + 1 + 1 = 7$ first-place votes. The tiger has eight of 15 first-place votes, a majority. Using the plurality with elimination method, the tiger is chosen for the zoo logo. ∎

Example 8 *Electing the Veterans Club President Using the Plurality with Elimination Method*

Use the plurality with elimination method to determine the winner of the election for president of the Veterans Club from Example 2. The preference table is shown again in Table 14.8. Recall that A represents Antoine, B represents Betty, C represents Carlos, and D represents Don.

Table 14.8 Veterans Club Preference Table

Number of Votes	19	15	11	7	2
First	B	C	D	A	C
Second	A	A	C	D	D
Third	C	D	A	C	A
Fourth	D	B	B	B	B

Solution First we count the number of first-place votes for each candidate:

Antoine: 7   Betty: 19   Carlos: 17   Don: 11

Since 54 votes were cast, a candidate must have at least 28 first-place votes to receive a majority. In this election, no candidate received a majority. Antoine has the fewest number of first-place votes, so he is eliminated. From column one of Table 14.8, we see that 19 voters ranked their preference as B, A, C, D. With A eliminated, we assume that those same 19 voters would now rank their preference as B, C, D. From column two in Table 14.8, 15 voters ranked their preference as C, A, D, B. With A eliminated, we assume that those same 15 voters would now rank their preference as C, D, B. We eliminate A from the other columns of the table in a similar manner.

The new preference table with A eliminated is given in Table 14.9.

Table 14.9   Veterans Club Preference Table After Antoine Is Eliminated

Number of Votes	19	15	11	7	2
First	B	C	D	D	C
Second	C	D	C	C	D
Third	D	B	B	B	B

The number of first-place votes is now

Betty: 19   Carlos: 17   Don: 18

Still, no candidate received a majority. Carlos now has the fewest number of first-place votes, so he is eliminated. With Carlos eliminated, the new preference table is given in Table 14.10.

Table 14.10   Veterans Club Preference Table After Carlos Is Eliminated

Number of Votes	19	15	11	7	2
First	B	D	D	D	D
Second	D	B	B	B	B

The number of first-place votes is now

Betty: 19   Don: 35

Don has a majority of first-place votes and is declared the winner using the plurality with elimination method.   ∎

The Veterans Club elections in the previous examples are quite interesting. When the plurality method was used, Betty was declared the winner. When the Borda count method was used, Antoine was declared the winner. When the plurality with elimination method was used, Don was declared the winner. This dilemma, once again, underscores the importance of publicly stating the voting method to be used *before* the election takes place.

The selection of the city to host the Olympic Games occurs by using a type of plurality with elimination. The election process for the Academy Awards involves several stages. One stage involves a variation of the plurality with elimination method in which the motion picture nominee with the fewest number of first-place votes is eliminated.

Now we will discuss our next voting method, the pairwise comparison method.

Pairwise Comparison Method

To use the pairwise comparison method, each voter first ranks all the candidates. Then each candidate is compared with each of the other candidates using the rankings. It is treated like a round-robin tournament. For instance, if the candidates for an election were Bert, Ernie, and Louis, the following comparisons would be made: Bert versus Ernie, Bert versus Louis, and Ernie versus Louis. The winner of each individual comparison is awarded one point. If there is a tie, each candidate receives $\frac{1}{2}$ point. The candidate who obtains the most points in the one-to-one comparisons is declared the winner.

PROCEDURE PAIRWISE COMPARISON METHOD

1. Voters rank the candidates.

2. A series of comparisons in which each candidate is compared with each of the other candidates follows.

3. If candidate A is preferred to candidate B, A receives 1 point. If candidate B is preferred to candidate A, B receives 1 point. If the candidates tie, each receives $\frac{1}{2}$ point.

4. After making all comparisons among the candidates, the candidate receiving the most points is declared the winner.

Example 9 *Selecting a Zoo Logo Using the Pairwise Comparison Method*

We will again examine the San Francisco Zoo logo election. Which animal is selected if the pairwise comparison method is used?

Solution The preference table is shown again in Table 14.11. Recall that T stands for tiger, K stands for kangaroo, and G stands for giraffe.

Table 14.11 Zoo Logo Preference Table

Number of Votes	2	3	5	1	3	1
First	T	T	G	K	K	G
Second	K	G	T	G	T	K
Third	G	K	K	T	G	T

To determine the winner, it is necessary to make comparisons between the tiger and the kangaroo, between the tiger and the giraffe, and between the kangaroo and the giraffe.

We will begin by comparing the tiger and the kangaroo. We see from Table 14.11 that the 2 voters in the first column, the 3 voters in the second column, and the 5 voters in the third column all prefer the tiger to the kangaroo. We also see that the 1 voter in the fourth column, the 3 voters in the fifth column, and the 1 voter in the sixth column all prefer the kangaroo to the tiger.

The pairwise comparison of the **tiger versus the kangaroo** is

Tiger: $2 + 3 + 5 = 10$ votes   Kangaroo: $1 + 3 + 1 = 5$ votes

The tiger wins the pairwise comparison between the tiger and the kangaroo and is awarded 1 point.

The pairwise comparison of the **tiger versus the giraffe** is

Tiger: $2 + 3 + 3 = 8$ votes   Giraffe: $5 + 1 + 1 = 7$ votes

Timely Tip

When we compared the tiger and the kangaroo in Example 9, we determined the sum of the tiger's votes to be 10. Then we determined the sum of the kangaroo's votes to be 5. Notice that $10 + 5 = 15$, the number of votes in the preference table. If the tiger had 10 votes, the kangaroo must have $15 - 10$ or 5 votes. When we use the pairwise comparison method, if you determine the number of votes of one member in the comparison, the number of votes of the second member can be determined by subtracting the first number of votes determined from the total number of votes.

The tiger wins the pairwise comparison between the tiger and the giraffe and is awarded another point.

The pairwise comparison of the **kangaroo versus the giraffe** is

Kangaroo: $2 + 1 + 3 = 6$ votes   Giraffe: $3 + 5 + 1 = 9$ votes

The giraffe wins the pairwise comparison between the kangaroo and the giraffe and is awarded 1 point. Since the tiger received 2 points, the giraffe received 1 point, and the kangaroo received 0 points, using the pairwise comparison method the tiger is chosen for the zoo logo.   ∎

Let's look at how we arrived at the comparisons. Each candidate was compared one-to-one with all the other candidates. In Example 9, we started with the tiger. Since the tiger had to be compared with each of the other candidates, we compared the tiger with the kangaroo and then we compared the tiger with the giraffe. Next, we chose the kangaroo. Since the kangaroo had already been compared with the tiger, we only needed to compare the kangaroo with the giraffe. The last candidate was the giraffe. Since the giraffe was already compared with each of the other candidates, we have made all the necessary comparisons. This process is used regardless of the number of candidates.

In Example 9, there were 3 candidates and we made 3 comparisons. The formula to determine the number of comparisons that are needed when n candidates are being considered follows.

> ## Number of Comparisons
> The number of comparisons, c, needed when using the pairwise comparison method when there are n candidates is
> $$c = \frac{n(n-1)}{2}$$

For example, when there are 5 candidates, $n = 5$ and the number of comparisons is

$$c = \frac{n(n-1)}{2} = \frac{5(4)}{2} = \frac{20}{2} = 10$$

Example 10 *Electing the Veterans Club President Using the Pairwise Comparison Method*

Use the pairwise comparison method to determine the winner of the election for president of the Veterans Club that was originally discussed in Example 2 on page 884.

Solution The preference table is shown again in Table 14.12. Recall that A represents Antoine, B represents Betty, C represents Carlos, and D represents Don.

Table 14.12 Veterans Club Preference Table

Number of Votes	19	15	11	7	2
First	B	C	D	A	C
Second	A	A	C	D	D
Third	C	D	A	C	A
Fourth	D	B	B	B	B

Since we have 4 candidates, $n = 4$, the number of comparisons needed is

$$c = \frac{n(n-1)}{2} = \frac{4(3)}{2} = 6$$

▲ *Rutherford B. Hayes lost the popular vote, but won the electoral college and the election.*

Since the United States was founded in 1776, Americans and their elected representatives have had many disputes regarding election procedures. As you read through this chapter, you will see various voting methods that were used to try to make the voting procedures as fair as possible. You will also learn that there is no ideal voting method, that is, one that has no flaws. Although there may be flaws in the voting procedures, the American election procedure is still a model to democracies all around the world. One current debate is whether we should continue using our present electoral system or change to use the popular vote to decide presidential elections. As of this writing, five presidents were elected even though they lost the popular vote: John Quincy Adams (1824), Rutherford B. Hayes (1876), Benjamin Harrison (1888), George W. Bush (2000), and Donald Trump (2016).

The 6 comparisons we will make are A versus B, A versus C, A versus D, B versus C, B versus D, and C versus D.

1. The pairwise comparison of **Antoine versus Betty** is

    Antoine: $15 + 11 + 7 + 2 = 35$ votes   Betty: 19 votes

    Antoine wins this comparison and is awarded 1 point.

2. The pairwise comparison of **Antoine versus Carlos** is

    Antoine: $19 + 7 = 26$ votes   Carlos: $15 + 11 + 2 = 28$ votes

    Carlos wins this comparison and is awarded 1 point.

3. The pairwise comparison of **Antoine versus Don** is

    Antoine: $19 + 15 + 7 = 41$ votes   Don: $11 + 2 = 13$ votes

    Antoine wins this comparison and is awarded a second point.

4. The pairwise comparison of **Betty versus Carlos** is

    Betty: 19 votes   Carlos: $15 + 11 + 7 + 2 = 35$ votes

    Carlos wins this comparison and is awarded a second point.

5. The pairwise comparison of **Betty versus Don** is

    Betty: 19 votes   Don: $15 + 11 + 7 + 2 = 35$ votes

    Don wins this comparison and is awarded 1 point.

6. The pairwise comparison of **Carlos versus Don** is

    Carlos: $19 + 15 + 2 = 36$ votes   Don: $11 + 7 = 18$ votes

Carlos wins this comparison and is awarded a third point.

We have made all possible one-to-one comparisons. Antoine received 2 points, Betty received 0 points, Carlos received 3 points, and Don received 1 point. Since Carlos received 3 points, the most points from the pairwise comparison method, Carlos wins the election. ∎

As we have illustrated, using four different voting methods resulted in four different winners for the president of the Veterans Club. Using the plurality method, Betty was the winner. Using the Borda count method, Antoine was the winner. Using the plurality with elimination method, Don was the winner. Using the pairwise comparison method, Carlos was the winner. Each candidate could claim that he or she should be declared the winner. Thus, the voting method we choose to use could greatly affect the election results and therefore should be chosen prior to an election.

Tie Breaking

Just as the four methods we have examined can produce four different winners, each method could also produce a tie between two or more candidates. In each of the previous examples, the voting method used led to one winner, but that is not always the case. With each method described—plurality method, Borda count method, plurality with elimination method, and pairwise comparison method—the possibility of the method producing a tie is very real. A tie could be achieved simply by having only two candidates each with the same number of supporters.

Breaking a tie can be achieved either by making an arbitrary choice, such as flipping a coin, or by bringing in an additional voter. For example, under Robert's Rules of Order, the president of any group votes only when there is a tie or to create a tie. There are other ways of breaking a tie that are less arbitrary than flipping a coin. If a tie results from using the Borda count method, it could be broken by choosing the candidate with the most first-place votes. If a tie results from the pairwise comparison method, it could be broken by choosing the winner of the one-to-one comparison between the candidates involved in the tie. Different tie-breaking methods could produce different winners. Therefore, to remain fair, the tie-breaking method should be decided upon in advance, as is done in the next example.

▲ John Krasinski

Example 11 *Choosing a Commencement Speaker*

The University of Georgia Student Government Association (SGA) is trying to decide whom to invite as the keynote speaker for commencement exercises. The choices are Michael Jordan (J), John Krasinski (K), and Ellen DeGeneres (D). To make this difficult decision, the SGA decides to hold an election among graduating seniors. It also decides that the plurality method will be used to determine the winner and that the Borda count method will decide the winner in the event that the plurality method leads to a tie. The results of the election are represented in Table 14.13. Who will be asked to speak at the commencement ceremony?

Table 14.13   University of Georgia Commencement Speaker Election

Number of Votes	400	250	150
First	D	J	J
Second	K	K	D
Third	J	D	K

Solution Michael Jordan and Ellen DeGeneres each receive 400 first-place votes; therefore, the plurality method leads to a tie. Thus, the Borda count method is used to break the tie between Jordan and DeGeneres. From Table 14.13, we award the following points:

$$\text{Ellen DeGeneres: } 400(3) + 150(2) + 250(1) = 1750$$

$$\text{Michael Jordan: } 400(3) + 0(2) + 400(1) = 1600$$

Since the Borda count method breaks the tie, Ellen DeGeneres is asked to present the commencement address. ∎

Table 14.14 summarizes the voting methods we discussed in this section.

Table 14.14   Summary of Voting Methods

Voting Method	Description
Plurality method	Each voter votes for one candidate. The candidate receiving the most votes is declared the winner.
Borda count method	Voters rank candidates from the most favorable to the least favorable. Each last-place vote is awarded 1 point, each next-to-last-place vote is awarded 2 points, each third-from-last-place vote is awarded 3 points, and so forth. The candidate receiving the most points is the winner.
Plurality with elimination method	Each voter votes for one candidate. If a candidate receives a majority of first-place votes, that candidate is declared the winner. If no candidate receives a majority of first-place votes, eliminate the candidate with the fewest number of first-place votes and hold another election. (If there is a tie for the fewest number of first-place votes, eliminate all candidates tied for the fewest number of first-place votes.) Repeat this process until a candidate receives a majority of first-place votes.
Pairwise comparison method	Voters rank the candidates. A series of comparisons in which each candidate is compared with each of the other candidates follows. If candidate A is preferred to candidate B, A receives 1 point. If candidate B is preferred to candidate A, B receives 1 point. If the candidates tie, each receives $\frac{1}{2}$ point. The candidate receiving the most points is declared the winner.

SECTION 14.1 Exercises

Warm Up Exercises

In Exercises 1–8, fill in the blank with an appropriate word, phrase, or symbol(s).

1. A candidate who receives more than 50% of the votes in an election is said to have a(n) _____ of votes.

2. A table that summarizes the results of an election is called a(n) _____ table.

3. When using the pairwise comparison method, the number of comparisons, c, needed when there are n candidates is determined by the formula $c =$ _____.

4. When there are four candidates, the number of comparisons needed with the pairwise comparison method is _____.

5. The voting method in which each voter votes for one candidate and the candidate receiving the most votes is declared the winner is called the _____ method.

6. In the Borda _____ voting method, the candidates are ranked from most favored to least favored.

7. The voting method in which each candidate is compared with each of the other candidates is called the pairwise _____ method.

8. The voting method that may involve a series of elections until a candidate receives a majority of the votes is called the _____ with elimination method.

Practice the Skills

9. *Plurality Method* Three candidates are running for mayor of Austin. They receive the following votes:

   Li   102,503   Sanchez   76,431
   Johnson   91,863

   a) Determine the winner using the plurality method.

   b) Did this candidate receive a majority of votes?

10. *Plurality Method* Five candidates are running for president of the Student Senate. They receive the following votes:

    Michael 2192   Gerry   2562   Karen   1671
    Felicia 2863   Stephen 1959

    a) Determine the winner using the plurality method.

    b) Did this candidate receive a majority of votes?

11. *Preference Table for Ice Cream* Nine survey respondents are asked to rank three brands of ice cream: Ben & Jerry's (B), Haagen-Dazs (H), and Dairy Queen (D). The nine survey respondents turn in the following ballots showing their preferences in order:

First	B	D	B	H	H	H	B	D	H
Second	D	B	D	B	B	D	D	H	D
Third	H	H	H	D	D	B	H	B	B

Make a preference table for these ballots.

12. *Preference Table for Yogurt* Eight voters are asked to rank three brands of yogurt: Dannon (D), Breyers (B), and Columbo (C). The eight voters turn in the following ballots showing their preferences in order:

First	D	C	C	B	C	B	C	D
Second	B	D	D	D	B	D	D	B
Third	C	B	B	C	D	C	B	C

Make a preference table for these ballots.

Problem Solving

Logo Choice In Exercises 13–18, members of a music club are choosing a new club logo. The choices are a guitar (G), a keyboard (K), and a musical note (N). Each member ranks the three choices from first to third. The preference table below shows the results of the ballots.

Number of Votes	10	5	4	2
First	G	K	K	N
Second	K	G	N	K
Third	N	N	G	G

13. How many members voted?

14. Is there a majority winner?

15. Determine the winner using the plurality method.

16. Determine the winner using the Borda count method.

17. Determine the winner using the plurality with elimination method.

18. Determine the winner using the pairwise comparison method.

Choosing a Vacation Destination *In Exercises 19–22, the Sanchez family is trying to decide whether to go to Chicago (C), Nashville (N), or Ocean City (O) on their summer vacation. The two parents, five children, and two grandparents rank their three choices according to the preference table below.*

Number of Votes	3	2	1	2	1
First	C	O	C	N	O
Second	N	C	O	O	N
Third	O	N	N	C	C

19. Determine the winner using the Borda count method.

20. Determine the winner using the plurality method.

21. Determine the winner using the pairwise comparison method.

22. Determine the winner using the plurality with elimination method.

▲ Nashville, TN

NFL Expansion *In Exercises 23–26, the National Football League (NFL) is considering four cities for expansion: San Antonio (S), Portland (P), Honolulu (H), and Toronto (T). The 32 current owners rank the four candidates according to the following preference table*

Number of Votes	8	6	3	4	3	3	2	1	2
First	P	S	S	H	P	H	P	T	H
Second	S	H	P	S	S	S	H	H	P
Third	H	T	H	T	T	P	S	P	S
Fourth	T	P	T	P	H	T	T	S	T

23. Determine the winner using the plurality method.

24. Determine the winner using the Borda count method.

25. Determine the winner using the plurality with elimination method.

26. Determine the winner using the pairwise comparison method.

▲ Portland, Oregon

Board of Trustees Election *In Exercises 27–30, the 12 members of the executive committee of the Student Senate must vote for a student representative for the college board of trustees from among three candidates: Brownstein (B), Samuels (S), and Marquez (M). The preference table follows.*

Number of Votes	5	1	4	2
First	B	S	M	M
Second	S	M	B	S
Third	M	B	S	B

27. Determine the winner using the Borda count method.

28. Determine the winner using the plurality method.

29. Determine the winner using the pairwise comparison method.

30. Determine the winner using the plurality with elimination method.

Post Office Sites *In Exercises 31–34, the 11 members of the Henrietta Town Board must decide where to build a new post office. Their three choices are Lehigh Road (L), Erie Road (E), and Ontario Road (O). The preference table follows.*

Number of Votes	5	2	4
First	L	E	O
Second	E	O	E
Third	O	L	L

31. Determine the winner using the plurality method.

32. Determine the winner using the Borda count method.

33. Determine the winner using the plurality with elimination method.

34. Determine the winner using the pairwise comparison method.

35. *Choosing a Contractor* The IBM board of directors is selecting a contractor to build a corporate headquarters expansion. They are considering proposals from four contractors: Bishara (B), Davis (D), Chang (C), and Stewart (S). The 15 members rank the four choices according to the preference table below.

Number of Votes	6	4	3	2
First	B	B	C	D
Second	C	D	B	B
Third	S	C	S	C
Fourth	D	S	D	S

a) Determine which contractor is selected using the plurality method.

b) Determine which contractor is selected using the Borda count method.

c) Determine which contractor is selected using the plurality with elimination method.

d) Determine which contractor is selected using the pairwise comparison method.

36. *Prime-Time Programming* The programmers at the Retro-Vision cable television station are deciding which classic TV show should lead their prime-time programming. Their choices are *The Jeffersons* (J), *Seinfeld* (S), *Good Times* (G), and *Cheers* (C). The 17 programmers rank their choices according to the following preference table.

Number of Votes	8	4	3	2
First	G	J	S	C
Second	C	C	C	S
Third	S	S	J	J
Fourth	J	G	G	G

▲ Kramer and Jerry from Seinfeld

a) Determine which TV show is selected using the plurality method.

b) Determine which TV show is selected using the Borda count method.

c) Determine which TV show is selected using the plurality with elimination method.

d) Determine which TV show is selected using the pairwise comparison method.

37. *Flowers in a Garden* The flowers in a garden at a resort need to be replaced. The choices for the flowers are geraniums (G), impatiens (I), petunias (P), and zinnias (Z). The head gardener holds an election in which all employees get to vote. The results of the election are given in the preference table below.

Number of Votes	8	11	3	4	3
First	G	P	P	Z	I
Second	P	Z	I	I	G
Third	I	I	Z	G	P
Fourth	Z	G	G	P	Z

a) Determine which flowers are selected using the Borda count method.

b) Determine which flowers are selected using the plurality method.

c) Determine which flowers are selected using the plurality with elimination method.

d) Determine which flowers are selected using the pairwise comparison method.

38. *Choosing a Computer* The Wizards Computer Club is ordering five new computers. To obtain a discount price, the club has decided that all five should be the same brand. The choices they are considering are Apple (A), Hewlett-Packard (H), Dell (D), and Lenovo (L). The 142 members' choices are reflected in the preference table below.

Number of Votes	43	30	29	26	14
First	L	A	H	D	D
Second	A	D	D	H	H
Third	H	H	A	L	A
Fourth	D	L	L	A	L

a) Determine which brand of computer is selected using the pairwise comparison method.

b) Determine which brand of computer is selected using the plurality with elimination method.

c) Determine which brand of computer is selected using the Borda count method.

d) Determine which brand of computer is selected using the plurality method.

e) What does this example demonstrate about the need for choosing a voting method prior to the election?

Concept/Writing Exercises

39. Describe a benefit of the Borda count method over the plurality method.

40. Describe one way other than flipping a coin to settle a tied election.

Challenge Problems/Group Activities

41. Another way to determine the winner if the plurality with elimination method is used is to eliminate the candidate with the most *last*-place votes at each step. Using the preference table given for Exercises 27–30 on page 896, determine the winner if the plurality with elimination method is used and the candidate with the most *last*-place votes is eliminated at each step.

42. Determine the winner using the plurality with elimination method, with the preference table for Exercises 31–34, if the candidate with the most *last*-place votes is eliminated at each step.

43. *A Lost Preference Table* A ski club is having an election for president. There are three candidates and 15 voters. Each voter casts a ballot listing his or her preferences. Just before the winner is to be announced, the previous president announces that the preference table has been lost. The only information he can remember from the preference table is that there were only *two* columns in the table.

a) Explain why there must be a majority winner.

b) Explain why there must always be a majority winner, regardless of the number of candidates, when all voters vote, when there are an odd number of voters and only two columns in the preference table.

44. *Cat Competition* Friskies Cat Food is looking for a new cat for the cover of their 2021 calendar. Four cats are finalists: A, B, C, and D. They are ranked first, second, third, and fourth in each of the following four categories: looks, eye color, size, and attitude. The preference table is shown below.

	Categories			
Cat	**Looks**	**Eyes**	**Size**	**Attitude**
A	4	3	2	3
B	2	4	1	2
C	1	1	4	4
D	3	2	3	1

a) Which cat wins if a Borda count method is used that awards 4 points, 3 points, 2 points, and 1 point for first, second, third, and fourth place, respectively, in each category?

b) Which cat wins if a modified Borda count method is used where 5, 3, 1, and 0 points are awarded for first, second, third, and fourth place, respectively, in each category?

45. *Ranking the Swim Teams* The results of a swim meet between the Comets (C), Rams (R), Wildcats (W), and Tigers (T) are shown in the preference table below. Notice, for example, that in the sprint competition, the Comets (C) took both first and fourth place.

	Sprint	Distance	Relay	Medley
First	C	R	R	T
Second	W	T	W	R
Third	T	W	T	W
Fourth	C	W	W	C

a) Assume that the finishing positions are scored 4, 3, 2, 1 for first, second, third, and fourth place, respectively. How do the teams rank from highest to lowest if the Borda count method is used?

b) Assume that the finishing positions are scored 5, 3, 1, 0 points for first, second, third, and fourth place, respectively. How do the teams rank from highest to lowest if we modify the Borda count method to use these values?

46. *Using the Borda Count Method* Suppose that 20 voters rank three candidates.

a) What is the total number of points awarded to the candidates using the Borda count method?

b) Suppose that, using the Borda count method, candidate A receives 55 points and candidate C receives 25 points. How many points does candidate B receive?

c) Given the information from part (b), is it possible for candidate B to win an election if the Borda count method is used to determine the winner? Explain.

47. Suppose that 15 voters rank four candidates.

    a) What is the total number of points awarded to candidates using the Borda count method?

    b) Suppose that using the Borda count method, candidate A receives 35 points, candidate B receives 40 points, and candidate C receives 25 points. How many points does candidate D receive?

    c) Is it possible for candidate D to win? Explain.

48. *Using Approval Voting* In many corporations and professional societies, the members of the board of directors are elected by a method called *approval voting*. With approval voting, a voter does not cast a ranked ballot. Instead, the voter votes for as many candidates as he or she would support. If a voter does not approve a candidate, he or she simply does not cast a vote for that candidate. The candidate(s) with the most votes wins.

    The board of directors of CSC Computers is using approval voting to fill two vacancies on their executive committee. Twenty-one ballots are cast as follows for candidates A, B, C, and D.

	1	2	3	4	5	6	7	8	9	10	11	12	13	14	15	16	17	18	19	20	21
A	X	X		X				X		X	X			X	X					X	X
B			X	X	X			X		X						X	X				
C			X		X							X	X					X			
D			X	X	X					X	X			X	X	X					X

Which two candidates will win seats on the executive committee?

Recreational Mathematics

49. Construct a preference table showing 12 votes for 3 candidates, A, B, and C, where candidates A and B both have 5 first-place votes. Many answers are possible.

Research Activity

50. *Award Methods* Research and write a report on how voting is conducted on one of the following events:

    a) Academy Awards   **b)** Grammy Awards

    c) Heisman Trophy Award   **d)** Nobel Prizes

    e) Pulitzer Prize

Include in your report how nominees are picked, who votes, how the voting takes place, and who tabulates the results.

▲ *Mahershala Ali with his Academy Award*

SECTION 14.2   Flaws of the Voting Methods

Upon completion of this section, you will be able to:

- Determine if the results of an election violate the majority criterion.
- Determine if the results of an election violate the head-to-head criterion.
- Determine if the results of an election violate the monotonicity criterion.
- Determine if the results of an election violate the irrelevant alternatives criterion.

You are hosting a party and ask your guests to vote on whether they would like you to serve pizza, hot dogs, or hamburgers at the party. Each guest is asked to complete a ballot by ranking the three choices from most favorable to least favorable. You decide to use the Borda count method to determine the winner and announce that pizza wins. After reviewing the ballots, one of your friends quickly points out that hot dogs should win because a majority of voters ranked hot dogs as their first choice. This scenario indicates that there is a flaw with the Borda count method. In this section, we will discuss flaws with each of the voting methods introduced in Section 14.1.

Why This Is Important Understanding flaws of different voting methods can help determine which method to use for an election.

In Section 14.1, we discussed four voting methods: the plurality method, the Borda count method, the plurality with elimination method, and the pairwise comparison method. We discovered that the voting method chosen can be as important to the outcome as the preferences in the voting. Sometimes all four methods will result in the same winner. At other times, the four methods may produce four different winners. Although each of these methods seems to provide a reasonable means for determining a winner, we soon shall see that there are certain flaws with each method.

In this section, we will examine four criteria, known as the *fairness criteria*, that mathematicians and political scientists have agreed that a voting method should meet to be considered fair. The fairness criteria include the *majority criterion*, the *head-to-head criterion*, the *monotonicity criterion*, and the *irrelevant alternatives criterion*. The first fairness criterion we will discuss is the majority criterion.

Majority Criterion

If a single candidate is the first choice of a majority (more than 50%) of voters, most voters would agree that that candidate should be declared the winner. If that does not happen with a voting method, that voting method violates the majority criterion.

> ### Majority Criterion
> If a candidate receives a majority (more than 50%) of first-place votes, that candidate should be declared the winner.

Table 14.15   Property Maintenance Company Preference Table

Number of Votes	7	4	2
First	G	L	L
Second	L	J	G
Third	J	G	J

Example 1 *Park Maintenance Company Selection*

The members of the Fayetteville Town Board are holding an election to select a company to maintain the property at the town park. The choices are Green World Landscaping (G), Lawn and Garden Haven (L), and Jubilant Gardens (J). The 13 board members rank the three choices and then use the Borda count method to make their selection. The preference table is shown in Table 14.15.

a) Which company is chosen using the Borda count method?

b) Does the winner from part (a) have a majority of first-place votes?

Solution

a) Using the Borda count method, Lawn and Garden Haven is chosen. Verify this outcome yourself.

b) Lawn and Garden Haven has 6 out of 13 first-place votes, which is less than 50%. So the winner of the election, Lawn and Garden Haven does not have a majority of first-place votes. In fact, a majority of town board members, 7 out of 13, chose Green World Landscaping.   ■

In Example 1, although a majority of town board members preferred one choice, Green World Landscaping, the Borda count method yielded a different choice, Lawn and Garden Haven. This example demonstrates that *the Borda count method has the potential to violate the majority criterion.*

Table 14.16

Number of Votes	17	9	7
First	A	B	C
Second	B	C	B
Third	C	A	A

Example 2 *Applying the Majority Criterion*

Preference Table 14.16 shows the outcomes of 33 votes.

Decide which candidate would be chosen with each of the following voting methods. Also discuss whether or not the method violates the majority criterion.

a) The plurality method

b) The Borda count method

c) The plurality with elimination method

d) The pairwise comparison method

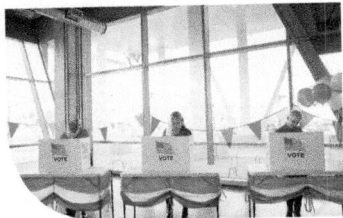
Solution

a) The plurality method awards the election to A. Notice that A also has a majority, 17 out of 33, of first-place votes. The plurality method does not violate the majority criterion. In general, a candidate who holds a majority of first-place votes also holds a plurality of first-place votes. Therefore, *the plurality method never violates the majority criterion.* However, that does not mean that the plurality method always produces a winner that has a majority of the votes. It does mean that if a candidate has a majority of first-place votes, that candidate will also have a plurality of votes and will be selected with the plurality method.

b) The Borda count method awards the election to candidate B. Since candidate A holds a majority of first-place votes, this method violates the majority criterion. This example is the second example in which the Borda count method violates the majority criterion. The Borda count method does not always violate the majority criterion; however, from our examples, we can conclude that *the Borda count method has the potential for violating the majority criterion.*

c) The plurality with elimination method awards the election to candidate A. Therefore, the majority criterion is not violated. In general, a candidate who holds a majority of first-place votes is awarded the election without having to hold a second election, or without having to consider eliminating a candidate and then realigning voters' choices. Therefore, *the plurality with elimination method never violates the majority criterion.*

d) The pairwise comparison method awards the election to candidate A. In general, if a candidate holds a majority of first-place votes, this candidate always wins every pairwise comparison, thereby granting this candidate the victory. Thus, *the pairwise comparison method never violates the majority criterion.* ∎

Of the four methods we discussed, only the Borda count method can violate the majority criterion. Therefore, the Borda count method has a flaw that none of the other voting methods has. An advantage of the Borda count method is that it takes into account voters' preferences by having all candidates ranked. However, a candidate with a majority of first-place votes can lose an election with the Borda count method. Next we will introduce a second fairness criterion and use it to uncover flaws in other voting methods.

Head-to-Head Criterion

Suppose that it is determined that one candidate is preferred over each of the other candidates using head-to-head comparisons. It would seem reasonable that this candidate should be chosen as the winner. That, however, is not always the case, and the head-to-head criterion will be used to uncover the next voting method flaw.

> ### Head-to-Head Criterion
> If a candidate is favored when compared head-to-head with every other candidate, that candidate should be declared the winner.

The head-to-head criterion is also known as the *Condorcet criterion*. This criterion is named after the Marquis de Condorcet; see the *Profile in Mathematics* on page 902.

Profile in Mathematics

Marquis de Condorcet

The Marquis de Condorcet (1743–1794) was one of the most influential mathematicians, economists, political scientists, and sociologists during the American and French revolutions. Condorcet believed that principles of fair government could be discovered mathematically. By analyzing voting methods, he discovered, to his disappointment, that occasionally there can be no clear and fair way to declare the winner of an election. The pairwise comparison method is also referred to as the Condorcet method. One of Condorcet's discoveries was that in an election with three candidates A, B, and C, it was possible for the voters to prefer A to B and B to C but also C to A.

Example 3 Applying the Head-to-Head Criterion

Four candidates are running for mayor of Springwater: Alvarez (A), Buchannon (B), Czechanski (C), and Davis (D). The election involves voters ranking the candidates with the results shown in Table 14.17.

Table 14.17 Springwater Mayor Preference Table

Number of Votes	129	90	87	78	42
First	D	A	B	C	C
Second	A	C	C	B	B
Third	B	B	A	D	A
Fourth	C	D	D	A	D

a) Is there one candidate who is favored over all others using a head-to-head comparison?

b) Who wins this election if the plurality method is used? Does that result violate the head-to-head criterion?

c) Who wins this election if the Borda count method is used? Does that result violate the head-to-head criterion?

d) Who wins this election if the plurality with elimination method is used? Does that result violate the head-to-head criterion?

e) Who wins this election if the pairwise comparison method is used? Does that result violate the head-to-head criterion?

Solution

a) Using Table 14.17, we can determine that Alvarez is favored, using head-to-head comparison, over Buchannon. Alvarez is favored over Buchannon in columns 1 and 2, giving Alvarez 129 + 90 or 219 votes. Buchannon is favored over Alvarez in columns 3, 4, and 5, giving Buchannon 87 + 78 + 42 or 207 votes. Alvarez is favored over Buchannon by a margin of 219 votes to 207 votes. Using this procedure, we can determine that Alvarez is also favored over Czechanski by a margin of 219 votes to 207 votes and Alvarez is also favored over Davis by a margin of 219 votes to 207 votes (you should verify these numbers). Therefore, Alvarez is favored over all the candidates, and according to the head-to-head criterion, Alvarez should be declared the winner.

b) Using the plurality method, Davis is elected by virtue of the 129 first-place votes. Therefore, this result violates the head-to-head criterion. This example demonstrates that *the plurality method has the potential to violate the head-to-head criterion.*

c) Using the Borda count method, Buchannon receives 1146 points, Czechanski receives 1140 points, Alvarez receives 1083 points, and Davis receives 891 points (you should verify these tallies). Therefore, Buchannon is elected using the Borda count method. Therefore, this result violates the head-to-head criterion. This example also demonstrates that *the Borda count method has the potential to violate the head-to-head criterion.*

d) Using plurality with elimination, we see that Buchannon is eliminated in the first round, Alvarez is eliminated in the second round, and in the third round Czechanski defeats Davis by a count of 297 to 129 (you should verify these results). Therefore, Czechanski is elected using the plurality with elimination method. Therefore, this result violates the head-to-head criterion. This example demonstrates that *the plurality with elimination method has the potential to violate the head-to-head criterion.*

e) Using the pairwise comparison method, we see that Alvarez is favored over Buchannon, Czechanski, and Davis, thereby giving Alvarez 3 points. Buchannon is favored using head-to-head comparison over Czechanski and Davis, thereby giving Buchannon 2 points. Czechanski is favored over Davis, thereby giving Czechanski 1 point. Davis is not favored over any of the other candidates, thereby giving Davis 0 points. (You should verify these results.) Therefore, Alvarez is elected using the pairwise comparison method. It should be clear that if a certain candidate were favored to all other candidates, then this candidate would be elected using the pairwise comparison method. Therefore, *the pairwise comparison method never violates the head-to-head criterion.* ■

By looking at Example 3, we can conclude that the *only voting method that does not have the potential to violate the head-to-head criterion is the pairwise comparison method.*

At this point in our discussion, we should note the differences between the *head-to-head fairness criterion* and the *pairwise comparison voting method.* The head-to-head criterion is one of the fairness criteria used to determine the fairness of a voting method and therefore is not used to determine the winner of an election. The pairwise comparison method is a voting method and is used to determine the winner of an election. When applying the head-to-head criterion, we are trying to determine if there is one candidate who is favored when compared head-to-head with every other candidate. If there is such a candidate, this candidate, according to the head-to-head criterion, should win an election, regardless of which voting method is used to determine the winner. As we have seen in Example 3, it is possible for a candidate to be favored when compared head-to-head with every other candidate yet lose an election. When a candidate is favored when compared with all other candidates and does not win the election, as in Example 3, we say that the voting method used violated the head-to-head criterion.

Example 3 highlighted how the head-to-head criterion uncovered flaws in the plurality method, Borda count method, and the plurality with elimination method. Now, we will introduce a third criterion for fair elections, the monotonicity criterion, and use it to uncover an additional flaw in the plurality with elimination method.

Monotonicity Criterion

The third criterion for fair elections is only relevant when an election is repeated. A repeated election refers to the same number of voters choosing among the same candidates. One example of a repeated election is a *straw vote* followed by another vote. Frequently, preceding an election, voters discuss the candidates' strengths and weaknesses. Prior to holding an official election, voters may take a straw vote to gain a preliminary measure of the candidates' strength. After discussions, some voters may decide to change their preferences. If a candidate gains votes at the expense of the other candidates, this candidate's chances of winning should increase. If that candidate was already leading, this candidate should still win with additional votes in his or her favor. The repeated election leads us to our next criterion, the monotonicity criterion. The monotonicity criterion uncovers a strange flaw in the plurality with elimination voting method. As we soon shall see, it is possible for the winner of the first election to gain additional support before the second election and then lose the second election!

> **Monotonicity Criterion**
> A candidate who wins a first election and then gains additional support without losing any of the original support should also win a second election.

Our next example demonstrates this unusual occurrence.

Example **4** *Choosing a New Service*

The marketing committee at Family Groceries is considering three new services to offer: delivery (D), self-serve checkout (S), and catering (C). To obtain an initial measure of the intentions of the committee, a straw vote is taken and the results are given in Table 14.18.

Table 14.18   Marketing Committee Preference Table:
First Election (straw vote)

Number of Votes	8	9	5	11
First	C	D	C	S
Second	D	S	S	C
Third	S	C	D	D

After several hours of debate and discussion, a second vote is taken. In the second election, the five people who previously voted C, S, D now vote S, C, D. This change has the effect of eliminating the C, S, D column of Table 14.18 and increasing the number voting for S, C, D from 11 to 16. The results of the second election are given in Table 14.19.

Table 14.19   Marketing Committee Preference Table:
Second Election

Number of Votes	8	9	16
First	C	D	S
Second	D	S	C
Third	S	C	D

a) Using the plurality with elimination method, which service wins the first election?

b) Using the plurality with elimination method, which service wins the second election?

c) Does this result violate the monotonicity criterion?

Solution

a) Refer to Table 14.18. Using the plurality with elimination method, the first election (the straw vote) results in delivery being eliminated and in self-serve checkout gaining a majority over catering by a vote of 20 to 13. Thus, self serve checkout would be chosen as the new service to offer.

b) Refer to Table 14.19 above. Using the plurality with elimination method, the second election results in catering being eliminated and in delivery gaining a majority over self-serve checkout by a vote of 17 to 16. Thus, delivery would be chosen as the new service to offer.

c) The five voters who changed their ballots did so in a way to add support to self-serve checkout. However, self serve checkout's victory was not repeated in the second election. Therefore, this result violates the monotonicity criterion. This occurrence shows that the *plurality with elimination method has the potential to violate the monotonicity criterion.*   ■

In Example 4, we showed that the plurality with elimination method has the potential to violate the monotonicity criterion. *Note that the Borda count method and the pairwise comparison method also have the potential to violate the monotonicity criterion.*

We have thus far shown how the majority criterion, the head-to-head criterion, and the monotonicity criterion have all uncovered flaws in our voting methods. Our next criterion, the irrelevant alternatives criterion, demonstrates a potential flaw with the pairwise comparison method.

Irrelevant Alternatives Criterion

The last criterion we will discuss, the irrelevant alternatives criterion, will address the result of removing a candidate from an election who has no chance of winning. Suppose that four candidates, A, B, C, and D, are involved in an election. The voting takes place and it is determined that candidate B is the winner. However, before the results of the election are announced, it is discovered that candidate C had dropped out of the election. We might conclude that candidate C's action should not have any effect on the outcome of the election. We shall soon see that such action can negatively affect the winner. We now have our fourth fairness criterion.

> **Irrelevant Alternatives Criterion**
> If a candidate is the winner of an election and in a second election one or more of the other candidates are removed, the previous winner should still be the winner.

Example 5 *Selecting a Kayaking Club President*

The members of the Adirondack Kayaking Club are holding an election to select a club president. The choices are Anderson (A), Bacari (B), Mendolson (M), Sanchez (S), and Tompkins (T). The election results are shown in Table 14.20.

Table 14.20 Adirondack Kayaking Club President Preference Table

Number of Votes	24	16	16	16	8	4	4
First	B	B	T	A	S	M	M
Second	S	S	S	M	T	B	T
Third	M	T	A	T	M	S	S
Fourth	T	A	M	B	B	T	B
Fifth	A	M	B	S	A	A	A

Prior to the announcement of the election results, it is discovered that Mendolson had dropped out of the election. The election officials then decide to eliminate Mendolson's name from the election results.

a) Using the pairwise comparison method, which candidate wins this election if Mendolson is included?

b) Using the pairwise comparison method, which candidate wins the election if Mendolson is eliminated from the preference table?

c) Does this result violate the irrelevant alternatives criterion?

Solution

a) Using the pairwise comparison method, since we have five candidates, $n = 5$ and the number of pairwise comparisons needed is

$$c = \frac{n(n-1)}{2} = \frac{5(4)}{2} = 10$$

We will do the comparisons in the following order: Anderson versus Bacari, Anderson versus Mendolson, Anderson versus Sanchez, Anderson versus Tompkins, Bacari versus Mendolson, Bacari versus Sanchez, Bacari versus Tompkins, Mendolson versus Sanchez, Mendolson versus Tompkins, and Sanchez versus Tompkins. These are the 10 possible comparisons. The following results can be verified from Table 14.20.

1. Bacari defeats Anderson (56 to 32), so Bacari gets 1 point.
2. Anderson defeats Mendolson (48 to 40), so Anderson gets 1 point.
3. Sanchez defeats Anderson (72 to 16), so Sanchez gets 1 point.
4. Tompkins defeats Anderson (72 to 16), so Tompkins gets 1 point.
5. Mendolson defeats Bacari (48 to 40), so Mendolson gets 1 point.
6. Bacari defeats Sanchez (60 to 28), so Bacari gets 1 point.
7. Bacari ties Tompkins (44 to 44), so Bacari gets $\frac{1}{2}$ point and Tompkins gets $\frac{1}{2}$ point.
8. Sanchez defeats Mendolson (64 to 24), so Sanchez gets 1 point.
9. Mendolson defeats Tompkins (48 to 40), so Mendolson gets 1 point.
10. Sanchez defeats Tompkins (52 to 36), so Sanchez gets 1 point.

We have made all possible one-to-one comparisons. Sanchez has 3 points, Bacari has $2\frac{1}{2}$ points, Mendolson has 2 points, Tompkins has $1\frac{1}{2}$ points, and Anderson has 1 point. Therefore, Sanchez is the winner when Mendolson is included.

b) Once Mendolson is eliminated from consideration, there is a new election and a new preference table. In the new election we will assume that the voters voted in the same order as in the first election. The new preference table is obtained by eliminating Mendolson from Table 14.20 and keeping the remaining candidates in the same order. See Table 14.21. Since we now have four candidates, $n = 4$ and the number of comparisons needed is

$$c = \frac{n(n-1)}{2} = \frac{4(3)}{2} = 6$$

Table 14.21 Adirondack Kayaking Club President Preference Table, after Mendolson drops out

Number of Votes	24	16	16	16	8	4	4
First	B	B	T	A	S	B	T
Second	S	S	S	T	T	S	S
Third	T	T	A	B	B	T	B
Fourth	A	A	B	S	A	A	A

The pairwise comparison method, using Table 14.21, produces the following 6 results:

1. Bacari defeats Anderson (56 to 32), so Bacari gets 1 point.
2. Sanchez defeats Anderson (72 to 16), so Sanchez gets 1 point.
3. Tompkins defeats Anderson (72 to 16), so Tompkins gets 1 point.
4. Bacari defeats Sanchez (60 to 28), so Bacari gets 1 point.

5. Bacari ties Tompkins (44 to 44), so Bacari gets $\frac{1}{2}$ point and Tompkins gets $\frac{1}{2}$ point.

6. Sanchez defeats Tompkins (52 to 36), so Sanchez gets 1 point.

We have made all 6 possible one-to-one comparisons. Now Bacari has $2\frac{1}{2}$ points, Sanchez has 2 points, Tompkins has $1\frac{1}{2}$ points, and Anderson has 0 points. Thus, Bacari is selected.

c) The first election produced Sanchez as the winner. Then, even though Mendolson was not the winning candidate, Mendolson's removal from the election produced Bacari as the winner. Therefore, this result violates the irrelevant alternatives criterion. This example demonstrates that the *pairwise comparison method has the potential for violating the irrelevant alternatives criterion.*   ■

In Example 5, we showed that the pairwise comparison method has the potential to violate the irrelevant alternatives criterion. *Note that each voting method we have discussed—the plurality method, the Borda count method, the plurality with elimination method, and the pairwise comparison method—has the potential to violate the irrelevant alternatives criterion.*

Table 14.22

Number of Votes	14	6	4
First	A	B	C
Second	B	C	B
Third	C	A	A

Example 6 *Satisfying the Four Fairness Criteria*

Suppose that the plurality method is used to determine the winner of an election. The results of the election are given in Table 14.22. Does this method satisfy the four fairness criteria, that is, the majority criterion, the head-to-head criterion, the monotonicity criterion, and the irrelevant alternatives criterion?

Solution By the plurality method, A is the winner with 14 votes. Since a total of 14 votes is also a majority of the 24 votes cast (58%), the majority criterion is satisfied. Examining pairwise comparison matchups, we see that A is favored to B (14 to 10) and that A is also favored to C (14 to 10). Therefore, the head-to-head criterion is satisfied. Since A holds a majority of first-place votes, the monotonicity criterion is satisfied when a second election is held in which A picks up additional votes; A still wins by plurality. Last, if B or C drops out, A still wins by plurality and the irrelevant alternatives criterion is satisfied. Therefore, this election satisfies all four fairness criteria.   ■

We saw in Example 6 that in a particular election, it is possible that all of the fairness criteria are satisfied. However, in a different election, with a different preference table, it is quite possible for the same voting method to violate one or more of the fairness criteria.

Table 14.23 summarizes the four fairness criteria, and Table 14.24 on page 908 summarizes which voting methods always satisfy the criteria.

Table 14.23   Summary of Fairness Criteria

Majority criterion	If a candidate receives a majority of first-place votes, that candidate should be declared the winner.
Head-to-head criterion	If a candidate is favored when compared individually with every other candidate, that candidate should be declared the winner.
Monotonicity criterion	A candidate who wins a first election and then gains additional support without losing any of the original support should also win a second election.
Irrelevant alternatives criterion	If a candidate is the winner of an election and in a second election one or more of the other candidates are removed, the previous winner should still be the winner.

Profile in Mathematics

Kenneth Arrow

Kenneth Arrow (1921–), an economist, was interested in the theory of corporations and how corporations make decisions when their leaders have different opinions. His particular interest in stockholders led to his discovery that voting methods can be flawed. While working at the RAND Corporation, he studied how mathematical concepts, game theory in particular, apply to decision making in the military and in diplomatic affairs. As a result, he developed a set of properties for any fair and reasonable voting method. After experimenting for days, he realized that there was no voting method that would satisfy all reasonable conditions. This conclusion is called the *Arrow impossibility theorem*, which he proved in 1951. In 1972, Arrow received the Nobel Prize in Economics for his exceptional work in the theory of general economic equilibrium.

Table 14.24   Summary of the Voting Methods and Whether They Satisfy the Fairness Criteria

	Plurality Method	Borda Count Method	Plurality with Elimination Method	Pairwise Comparison Method
Majority criterion	Always satisfies	May not satisfy	Always satisfies	Always satisfies
Head-to-head criterion	May not satisfy	May not satisfy	May not satisfy	Always satisfies
Monotonicity criterion	Always satisfies	May not satisfy	May not satisfy	May not satisfy
Irrelevant alternatives criterion	May not satisfy	May not satisfy	May not satisfy	May not satisfy

As we have shown, each of the voting methods introduced in Section 14.1 has the potential to violate at least one of the fairness criteria. Is there another voting method that does satisfy all four criteria? Mathematicians and political scientists struggled with this question for two centuries. Finally, in 1951, an economist, Kenneth Arrow, proved mathematically that there is no voting method that can satisfy all four fairness criteria. This result is called *Arrow's impossibility theorem*. See the *Profile in Mathematics* on page 907.

> **Arrow's Impossibility Theorem**
> It is mathematically impossible for any democratic voting method to simultaneously satisfy each of the following fairness criteria:
> The majority criterion
> The head-to-head criterion
> The monotonicity criterion
> The irrelevant alternatives criterion

SECTION 14.2   *Exercises*

Warm Up Exercises

In Exercises 1–8, fill in the blank with an appropriate word, phrase, or symbol(s).

1. If a candidate receives a majority of first-place votes in an election, that candidate should be declared the winner. This criterion is called the _____ criterion.

2. A candidate who wins a first election, then gains additional support without losing any of the original support, should also win a second election. This criterion is called the _____ criterion.

3. If a candidate is favored when compared head-to-head with every other candidate in an election, that candidate should be declared the winner. This criterion is called the _____ criterion.

4. If a candidate is the winner of an election and in a second election one or more of the other candidates is removed, the previous winner should still be the winner. This criterion is called the _____ criterion.

5. A voting method that may not satisfy any of the fairness criteria is the _____ method.

6. A voting method that always satisfies the majority criterion and head-to-head criterion, but may not satisfy any other criterion, is the _____ method.

7. A voting method that always satisfies the majority criterion but may not satisfy any other criterion is the _____ method.

8. A voting method that always satisfies the majority criterion and the monotonicity criterion, but may not satisfy any other criterion, is the _____ method.

Practice the Skills/Problem Solving

9. ***Annual Meeting*** Members of the board of directors of the American Nursing Association are voting to select a city to host the association's annual meeting. The board is considering the following cities: Portland (P), Orlando (O), and New Orleans (N). The preference table for the 19 members of the board of directors is shown below. Show that the Borda count method violates the majority criterion.

Number of Votes	10	5	4
First	O	N	N
Second	N	P	O
Third	P	O	P

▲ New Orleans, LA

10. ***Hiring a New Director*** Texas Tech University is hiring a new director of counseling. The hiring committee ranks the four candidates, Alvarez (A), Brown (B), Singleton (S), and Yu (Y), according to the preference table below. Suppose that the Borda count method is used to determine the winner. Is the majority criterion satisfied? Explain.

Number of Votes	5	4	1	1
First	A	B	B	A
Second	B	S	Y	B
Third	S	Y	S	Y
Fourth	Y	A	A	S

11. ***Restructuring a Company*** The board of directors at The Limited is considering three different administrative restructuring plans, A, B, and C. The 11 members of the board of directors rank the plans according to the preference table above and to the right.

a) Which restructuring plan is preferred to all the others in a head-to-head comparison?

b) Suppose that the plurality method is used to determine the winner. Is the head-to-head criterion satisfied? Explain.

Number of Votes	2	4	2	3
First	B	A	C	C
Second	A	B	A	B
Third	C	C	B	A

12. ***Party Theme*** The children in Ms. Cohn's seventh-grade class are voting on a theme for their class party. Their choices are beach (B), the world (W), superheroes (S), or rock and roll (R). The results are shown in the preference table below.

a) Which theme is preferred to all the other themes in a head-to-head comparison?

b) Suppose that the plurality method is used to determine the winner. Is the head-to-head criterion satisfied? Explain.

Number of Votes	12	6	4	3
First	B	S	R	W
Second	W	B	S	R
Third	S	W	B	B
Fourth	R	R	W	S

13. ***Residence Hall Improvements*** The administration at Erie Community College is considering three possible areas of improvement to the school's residence halls: parking (P), security (S), and lounge areas (L). The nine members of the executive committee of the Student Senate rank the choices according to the preference table shown below. Suppose that the Borda count method is used to determine the winner. Is the head-to-head criterion satisfied? Explain.

Number of Votes	4	1	1	1	2
First	P	L	S	S	L
Second	L	P	L	P	S
Third	S	S	P	L	P

14. *Design of a Technology Building* The planning committee of Tulsa Community College is considering three proposals, A, B, and C, for the design of a new technology building for the college. The nine members of the planning committee rank the proposals according to the preference table shown below. Suppose that the Borda count method is used to determine the winner. Is the head-to-head criterion satisfied? Explain.

Number of Votes	5	2	2
First	C	A	B
Second	A	B	A
Third	B	C	C

15. *A Taste Test* Twenty-seven people are surveyed in the Mall of America and asked to taste and rank four different brands of Spaghetti Sauce. The preference table below shows the rankings of the 27 voters. Suppose that the plurality with elimination method is used to determine the winner. Is the head-to-head criterion satisfied? Explain.

Number of Votes	11	3	5	8
First	A	B	D	C
Second	B	C	B	B
Third	D	A	C	A
Fourth	C	D	A	D

16. *Preference for Grape Jelly* Twenty-one people are asked to taste test and rank three different brands of grape jelly. The preference table below shows the rankings of the 21 voters. Suppose that the plurality with elimination method is used to determine the winner. Is the head-to-head criterion satisfied? Explain.

Number of Votes	8	7	6
First	B	C	A
Second	A	A	C
Third	C	B	B

17. *Plurality: Irrelevant Alternatives Criterion* Suppose that the plurality method is used on the following preference table. If B drops out, is the irrelevant alternatives criterion satisfied? Explain.

Number of Votes	10	5	6
First	A	C	B
Second	C	B	C
Third	B	A	A

18. *Plurality: Irrelevant Alternatives Criterion* Suppose that the plurality method is used on the following preference table. If C drops out, is the irrelevant alternatives criterion satisfied? Explain.

Number of Votes	7	5	4
First	B	C	A
Second	C	A	C
Third	A	B	B

19. *Borda Count: Irrelevant Alternatives Criterion* Suppose that the Borda count method is used on the following preference table. If C drops out, is the irrelevant alternatives criterion satisfied? Explain.

Number of Votes	10	9	8
First	B	C	A
Second	A	B	C
Third	C	A	B

20. *Borda Count: Irrelevant Alternatives Criterion* Suppose that the Borda count method is used on the following preference table. If B drops out, is the irrelevant alternatives criterion satisfied? Explain.

Number of Votes	7	6	5
First	A	C	B
Second	B	A	C
Third	C	B	A

21. *Plurality with Elimination: Monotonicity Criterion*
Suppose that the plurality with elimination method is used on the following preference table. If the four voters who voted A, C, B, in that order, change their vote to C, A, B, is the monotonicity criterion satisfied? Explain.

Number of Votes	7	8	4	10
First	A	B	A	C
Second	B	C	C	A
Third	C	A	B	B

22. *Plurality with Elimination: Monotonicity Criterion*
Suppose that the plurality with elimination method is used on the following preference table. If the three voters who voted C, B, A, in that order, change their vote to B, C, A, is the monotonicity criterion satisfied? Explain.

Number of Votes	5	6	3	7
First	C	A	C	B
Second	A	B	B	C
Third	B	C	A	A

23. *Pairwise Comparison Method: Monotonicity Criterion*
Suppose that the pairwise comparison method is used on the following preference table. If the five voters who voted C, D, A, B, in that order, change their votes to D, B, C, A, is the monotonicity criterion satisfied? Explain.

Number of Votes	8	7	6	5
First	D	B	B	C
Second	C	A	A	D
Third	A	C	D	A
Fourth	B	D	C	B

24. *Borda Count: Monotonicity Criterion* Suppose that the Borda count method is used on the following preference table. If the five voters who voted D, A, C, B, in that order, change their votes to A, B, D, C and the three voters who voted C, A, D, B, in that order, change their votes to A, B, C, D, is the monotonicity criterion satisfied? Explain.

Number of Votes	8	7	5	3
First	A	B	D	C
Second	B	D	A	A
Third	C	C	C	D
Fourth	D	A	B	B

25. *Pairwise Comparison: Irrelevant Alternatives Criterion*
Suppose that the pairwise comparison method is used on the following preference table. If A, C, and E drop out, is the irrelevant alternatives criterion satisfied? Explain.

Number of Votes	1	1	1	1	1
First	B	B	A	C	E
Second	A	C	D	D	D
Third	C	E	B	B	B
Fourth	E	A	C	E	A
Fifth	D	D	E	A	C

26. *Pairwise Comparison: Irrelevant Alternatives Criterion*
Suppose that the pairwise comparison method is used on the following preference table. If A, B, and E drop out, is the irrelevant alternatives criterion satisfied? Explain.

Number of Votes	2	1	1	1	1	1
First	C	C	D	A	E	B
Second	E	B	C	D	A	E
Third	A	A	B	C	D	D
Fourth	B	E	A	B	C	C
Fifth	D	D	E	E	B	A

27. *Borda Count: Majority Criterion* Suppose that the Borda count method is used on the following preference table. Is the majority criterion satisfied?

Number of Votes	4	2	1
First	A	B	C
Second	B	C	B
Third	C	A	A

28. *Borda Count: Majority Criterion* Suppose that the Borda count method is used on the following preference table. Is the majority criterion satisfied?

Number of Votes	5	1	4	1
First	B	B	C	A
Second	C	A	A	C
Third	A	C	B	B

29. Spring Trip The Quilting Club of Charlotte is voting on which city to visit for its spring trip. The choices are Charleston (C), Savannah (S), Philadelphia (P), and Boston (B). The preference table for the 23 members' choices is given below.

▲ *Charleston, SC*

Number of Votes	12	8	3
First	S	P	B
Second	C	C	S
Third	P	B	P
Fourth	B	S	C

a) Which city, if any, holds a majority of first-place votes?

b) Which city is selected using the plurality method?

c) Which city is selected using the Borda count method?

d) Which city is selected using the plurality with elimination method?

e) Which city is selected using the pairwise comparison method?

f) For this preference table, which voting method(s), if any, violate the majority criterion?

30. Investment Choices The Motley Crew Investing Club wishes to purchase stock in a pharmaceutical company. The choices are American Pharmacy (A), Burrows-Welcome (B), Chicago Labs (C), Dow Chemical (D), and Emerging Medicines (E). The choice rankings are shown in the following preference table.

Number of Votes	12	8	8	8	4	2	2
First	B	B	E	A	D	C	C
Second	D	D	D	C	E	B	E
Third	C	E	A	E	C	D	D
Fourth	E	A	C	B	B	E	B
Fifth	A	C	B	D	A	A	A

Before calculating the results of this election, the club learns that Chicago Labs has filed for bankruptcy.

a) Using the pairwise comparison method, which company is chosen if Chicago Labs is included?

b) Using the pairwise comparison method, which company is chosen if Chicago Labs is eliminated from the preference table?

c) Does this second result violate the irrelevant alternatives criterion?

31. Selecting a Spokesperson The Campbell Soup Company is selecting a new spokesperson for its advertisements. The choices are Adam Levine (L), Bradley Cooper (C), and Steve Harvey (H). At the annual meeting of its managers, a straw vote is taken to gauge the managers' initial reaction to the candidates. The results of this straw vote are given in the preference table below.

Number of Votes	28	24	20	10
First	H	C	L	L
Second	L	H	C	H
Third	C	L	H	C

▲ *Adam Levine*

After discussion about the candidates, eight people who had supported Adam Levine over Steve Harvey now switch their preferences to Steve Harvey over Adam Levine. This switch leads to the following preference table.

Number of Votes	36	24	20	2
First	H	C	L	L
Second	L	H	C	H
Third	C	L	H	C

a) Using the plurality with elimination method, which person is selected from the first (or straw) vote?

b) Using the plurality with elimination method, which person is selected from the second vote?

c) Does this second result violate the monotonicity criterion?

32. *Electing a Communications Director* The City of Phoenix is electing a communications director. The choices for the position are Allen (A), Brady (B), Conrad (C), Davis (D), and Evers (E). The election is conducted, and the preference table is shown below.

Number of Votes	60	40	40	40	20	10	10
First	B	B	E	A	D	C	C
Second	D	D	D	C	E	B	E
Third	C	E	A	E	C	D	D
Fourth	E	A	C	B	B	E	B
Fifth	A	C	B	D	A	A	A

Before calculating the results of this election, it is learned that Conrad has dropped out of the race for personal reasons.

a) Using the pairwise comparison method, which candidate is chosen if Conrad is included?

b) Using the pairwise comparison method, which candidate is chosen if Conrad is eliminated from the preference table?

c) Does this second result violate the irrelevant alternatives criterion?

Challenge Problems/Group Activities

33. Explain why the plurality method always satisfies the monotonicity criterion.

34. Construct a preference table with four candidates and four rankings of the candidates, similar to the preference table in Exercise 10 on page 909, such that the plurality method violates the head-to-head criterion. Have the total number of votes be 25.

35. Construct a preference table with three candidates and three rankings of the candidates, similar to the preference table in Exercise 9 on page 909, such that the Borda count method violates the majority criterion. Have the total number of votes be 19.

36. Construct a preference table with three candidates and four rankings of the candidates, similar to the preference table in Exercise 11 on page 909, such that the plurality with elimination method violates the monotonicity criterion. Have the total number of votes be 21.

37. Construct a preference table with four candidates and three rankings of the candidates, similar to the preference table in Exercise 29 on page 912, such that the pairwise comparison method violates the irrelevant alternatives criterion. Have the total number of votes be 6.

Not voting according to a voter's true preference, called insincere voting, is frequently practiced to affect the outcome of an election. Insincere voting is used in Exercise 38.

38. *Voting Strategy* Consider the preference table below. Assume that a majority is needed to win the election. Since no candidate has a majority but C has the most first-place votes, a runoff election is held between A and B. The winner of the runoff election will run against C.

Number of Votes	5	4	2
First	C	A	B
Second	A	C	A
Third	B	B	C

a) Using the plurality method, which candidate will win the runoff election, A or B?

b) Will the candidate determined in part (a) win the election against candidate C?

c) Suppose that the voters who support candidate C, the candidate with the most first-place votes, learn prior to the vote that C will not win if B is eliminated and C runs against A. How can the voters who support C vote insincerely to enable C to win?

Research Activities

39. *Choosing a Leader* Choose a country other than the United States and write a research paper on how the president or prime minister is chosen. Describe in detail the voting method used. Include the strengths and weaknesses of the method used.

40. *History of Voting Methods* Write a research paper on the history of voting methods. Include how the flaws in the voting methods discussed in this section were uncovered.

SECTION 14.3 Apportionment Methods

Upon completion of this section, you will be able to:

- Solve apportionment problems using Hamilton's method.
- Solve apportionment problems using Jefferson's method.
- Solve apportionment problems using Webster's method.
- Solve apportionment problems using Adams' method.

Suppose a college receives a grant to purchase 50 tablets to be distributed to four different libraries on campus. Since the libraries do not all serve the same number of students, the director of libraries decides to apportion the tablets based on the number of students served at each library. In this section, we will discuss some apportionment methods the college can use to distribute the tablets to the four libraries.

Why This Is Important Apportionment methods can be used by businesses, schools, and government agencies whenever a set of items must be allocated among various groups. Being informed about these methods is an important skill that can be applied to many occupations.

When the delegates for the original 13 states met in 1787 to draft a constitution, their most important discussion concerned the representation of the states in the legislature. Some states, particularly the small states, preferred that each state have the same number of representatives. The large states wanted a proportional representation based on population. The delegates resolved this issue by creating a Senate, in which each state received two Senators, and a House of Representatives, in which each state received a number of representatives based on its population. Article 1, Section 2, of the Constitution of the United States includes the statement "Representatives ... shall be apportioned among the several states ... according to their respective numbers." What the constitution did not say is how the apportionments are determined. In this section, we will discuss four different apportionment methods.

The goal of *apportionment* is to determine a method to allocate the total number of items to be apportioned in a fair manner. One of the most important examples of apportionment is in determining the representation of governing bodies. However, apportionment problems occur in many other places, from determining how many police officers should be assigned to each precinct in a city to determining how many nurses should be assigned to each shift at a hospital.

In this section, we will discuss four different apportionment methods:

1. Hamilton's method
2. Jefferson's method
3. Webster's method
4. Adams' method

Apportionment Problems

We begin by considering a problem facing First Physicians Organization, a healthcare provider that operates five medical clinics. The organization has recently hired 60 doctors. Since the clinics do not all serve the same number of patients, the organization decides to apportion the 60 doctors based on the number of patients who visit each clinic in a given week, as shown in Table 14.25.

Table 14.25 First Physicians Organization

Clinic	A	B	C	D	E	Total
Patients	246	201	196	211	226	1080

If the organization wants to distribute the doctors based on the number of patients served at each clinic, it would take each clinic's number of patients served and divide it by the total number of patients. For example, clinic A would be entitled to $\frac{246}{1080}$, or $22\frac{7}{9}\%$, of the doctors. Similarly, clinic B would be entitled to $\frac{201}{1080}$, or $18\frac{11}{18}\%$, of the doctors and so on. Dividing the number of patients at each clinic by the total number of patients leaves a fractional part of a doctor. Since a clinic cannot have a fraction of a doctor, the organization has a problem. How can it apportion the doctors so that each clinic receives its fair allotment?

Before we discuss how to apply apportionment methods, we will introduce some important terms used in this section, *standard divisor* and *standard quota*. The *standard divisor* is determined by dividing the total population under consideration by the number of items to be allocated. The total population in the previous illustration is the 1080 patients given in Table 14.25.

Standard Divisor

To obtain the standard divisor when determining apportionment, use the following formula.

$$\text{Standard divisor} = \frac{\text{total population}}{\text{number of items to be allocated}}$$

For our doctor problem, the standard divisor is determined as follows:

$$\text{Standard divisor} = \frac{\text{total number of patients}}{\text{number of doctors to be allocated}} = \frac{1080}{60} = 18$$

The *standard quota* for a particular group is determined by dividing the population of the group by the standard divisor.

Standard Quota

To obtain the standard quota when determining apportionment, use the following formula.

$$\text{Standard quota} = \frac{\text{population for the particular group}}{\text{standard divisor}}$$

For our doctor problem, the standard quota for clinic A is determined as follows:

$$\text{Standard quota for clinic A} = \frac{\text{number of patients at clinic A}}{\text{standard divisor}} = \frac{246}{18} \approx 13.67$$

We will round all standard divisors and standard quotas to the nearest hundredth.

Example 1 *Determining Standard Quotas*

Determine the standard quotas for clinics B, C, D, and E of the First Physicians Organization and complete Table 14.26. Use 18 as the standard divisor, as was determined above.

Table 14.26 First Physicians Organization

Clinic	A	B	C	D	E	Total
Patients	246	201	196	211	226	1080
Standard quota	13.67					

Did You Know?

Mathematical Contributions from Political Leaders

▲ *Alexander Hamilton (1757–1804)*

As a result of his distinctive service in the Revolutionary War, Hamilton became George Washington's aide-de-camp and personal secretary in 1777. In 1789, Washington appointed Hamilton as the first secretary of the treasury. Hamilton and Thomas Jefferson were strong political adversaries. In 1804, Hamilton was killed in a duel with Aaron Burr, who was serving as Jefferson's vice president.

Solution The standard quota for clinic B is $\frac{201}{18} = 11.17$, rounded to the nearest hundredth. The other standard quotas are determined in a similar manner. Table 14.27 shows the completed table.

Table 14.27   First Physicians Organization

Clinic	A	B	C	D	E	Total
Patients	246	201	196	211	226	1080
Standard quota	13.67	11.17	10.89	11.72	12.56	60.01

The standard quota represents the exact number of doctors each clinic would receive if we were able to divide a doctor into fractional parts. Notice that the sum of the standard quotas is slightly above 60, the total number of doctors due to rounding. ∎

Now we introduce two more important definitions, the *lower quota* and the *upper quota*. The *lower quota* is the standard quota rounded down to the nearest integer. The *upper quota* is the standard quota rounded up to the nearest integer. For example, clinic A has a lower quota of 13 and an upper quota of 14, clinic B has a lower quota of 11 and an upper quota of 12, and so on.

How should we round the standard quotas so that each clinic receives its fair share? If we were to use conventional rounding and round each clinic's standard quota to the nearest integer, clinic A would receive 14 doctors and clinic B would receive 11 doctors. Clinics C, D, and E would receive 11, 12, and 13 doctors, respectively. The total number of doctors needed would be 61. Since there are only 60 doctors to distribute, conventional rounding of standard quotas can be problematic. We need a more sophisticated apportionment method to avoid the problems that can occur with rounding standard quotas. We will now discuss the first of our four apportionment methods, Hamilton's method.

Hamilton's Method

Mathematically, Hamilton's method is the easiest to apply.

PROCEDURE HAMILTON'S METHOD

To use Hamilton's method for apportionment, do the following.

1. Calculate the standard divisor for the set of data.
2. Calculate each group's* standard quota.
3. Round each standard quota **down** to the nearest integer (the lower quota). Initially, each group receives its lower quota.
4. Distribute any leftover items to the groups with the largest fractional parts until all items are distributed.

---

* We use the word *group* in Steps 2, 3, and 4 as a general term. However, *group* could refer to a state, a school, or even an individual.

Example 2 *Using Hamilton's Method for Apportioning Doctors*

Use Hamilton's method to distribute the 60 doctors to the First Physicians Organization clinics discussed in Example 1 on page 915.

Solution Table 14.28 includes the standard quotas (row 2) calculated from Example 1, along with the lower quotas (row 3) and the apportionment using Hamilton's method (row 4).

Table 14.28 First Physicians Organization

Clinic	A	B	C	D	E	Total
Patients	246	201	196	211	226	1080
Standard quota	13.67	11.17	10.89	11.72	12.56	60.01
Lower quota	13	11	10	11	12	57
Hamilton's apportionment	14	11	11	12	12	60

The sum of the lower quotas is 57, leaving three additional doctors to be distributed. Since clinics C, D, and A, in this order, have the three highest fractional parts (0.89, 0.72, and 0.67, respectively) in the standard quota, each receives one of the additional doctors using Hamilton's method. Note that with Hamilton's method, each clinic receives either its lower quota or its upper quota of doctors. Also note that the total number of doctors apportioned by Hamilton's method is 60, which is what we expected. ∎

Let's look at another example using Hamilton's method.

Example 3 *Using Hamilton's Method for Apportioning Legislative Seats*

The Republic of Hydrangea needs to apportion 250 seats in the legislature. Suppose that the population of Hydrangea is 8,800,000 and that there are five states in Hydrangea, A, B, C, D, and E. The 250 seats are to be divided among the five states according to their respective populations, given in Table 14.29.

Table 14.29 Republic of Hydrangea Population

State	A	B	C	D	E	Total
Population	1,003,200	1,228,600	4,990,700	813,000	764,500	8,800,000

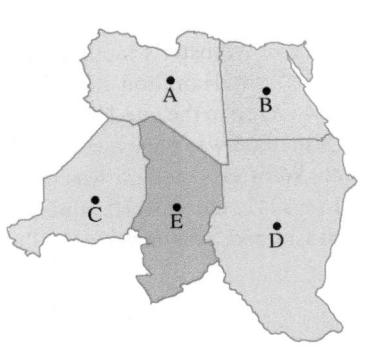

▲ *Republic of Hydrangea population 8,800,000*

Use Hamilton's method to apportion the seats.

Solution The standard divisor is determined by dividing the total population by the number of seats in the legislature.

$$\text{Standard divisor} = \frac{8,800,000}{250} = 35,200$$

The standard quotas are determined by dividing the population of each state by the standard divisor. Recall that we will round the standard quotas to the nearest hundredth. The results are shown in the second row in Table 14.30 on page 918.

Table 14.30 Republic of Hydrangea Population

State	A	B	C	D	E	Total
Population	1,003,200	1,228,600	4,990,700	813,000	764,500	8,800,000
Standard quota	28.50	34.90	141.78	23.10	21.72	
Lower quota	28	34	141	23	21	247
Hamilton's apportionment	28	35	142	23	22	250

The sum of the lower quotas is 247, leaving three additional seats to be distributed. Since states B, C, and E, in this order, have the three highest fractional parts in the standard quota (0.90, 0.78, and 0.72, respectively), they each receive the additional seats using Hamilton's method. ∎

As we can see from both examples, an advantage of Hamilton's method is that every apportionment is either the lower quota or the upper quota. If an apportionment method assigns either the lower or upper quota to every group under consideration, it is said to satisfy the *quota rule*.

> **The Quota Rule**
> An apportionment for every group under consideration should always be either the upper quota or the lower quota.

We have just discussed Alexander Hamilton's method for apportionment. In this section, we will also discuss Thomas Jefferson's method for apportionment, Daniel Webster's method for apportionment, and John Quincy Adams' method for apportionment. Here is a brief history of the use of these different methods to apportion members of the U.S. House of Representatives. Although Hamilton's method for apportionment was the first method approved by Congress after the 1790 census, it was not used until 1852. Shortly after Congress approved Hamilton's method, President George Washington vetoed it. Congress then adopted a method developed by Thomas Jefferson. As there were apportionment problems with Jefferson's method in 1822 and 1832, Webster's method replaced Jefferson's method in 1842. Webster's method had similar flaws as Jefferson's method and was replaced in 1852 by Hamilton's method, the first apportionment method proposed. When it was discovered in the late 1800s that there were also flaws with Hamilton's method, Webster's method was once again used to apportion the number of representatives for each state. Webster's method was used from 1900 until it was replaced by the current method in 1941, the Hill–Huntington method (see Exercise 51). Adams' method, although considered, was never actually used by Congress to apportion members.

Jefferson's Method

To determine the standard quota with Hamilton's method, the population of each group was divided by the standard divisor. Then each lower quota was used. Any leftover items were distributed to the groups with the highest fractional part of the standard quota. As we saw with Examples 2 and 3, each group received either its lower quota or its upper quota. A weakness that can occur with Hamilton's method is that the groups that received their upper quota gained an advantage over the groups that received their lower quota. Is there a way to modify the quotas to overcome this weakness in Hamilton's method? Jefferson's method uses a divisor other than the standard divisor, called the *modified divisor*, to obtain quotas. A *modified divisor* is a divisor that approximates the standard divisor. We will explain how to determine modified divisors shortly. *A modified divisor is used in Jefferson's method, Webster's method, and Adams' method, each of which will be discussed shortly.*

In each case, a modified divisor is used to obtain *modified quotas*. *Modified quotas* are obtained by dividing a group's population by a modified divisor. The differences in Jefferson's method, Webster's method, and Adams' method are in the modified divisor used to obtain the modified quotas and in the rounding process used to apportion the items. *Hamilton's method is the only method that uses the standard quota rather than a modified quota in determining apportionments.*

Now we will discuss Jefferson's method.

PROCEDURE JEFFERSON'S METHOD

To use Jefferson's method for apportionment, do the following.

1. Determine a modified divisor, d, such that when each group's modified quota is *rounded **down** to the nearest integer*, the total of the integers is the exact number of items to be apportioned. We will refer to the modified quotas that are rounded down as **modified lower quotas.**

2. Apportion to each group its modified lower quota.

With Jefferson's method, we use a modified divisor that is *less than* the standard divisor. This results in the modified quotas being slightly greater than the standard quotas. When using Hamilton's method, the standard quotas are *rounded down* and then leftover items are distributed. With Jefferson's method, the modified quotas are also *rounded down*. The modified divisor used with Jefferson's method must result in all items being distributed, with no items left over, when the modified quotas are rounded down. We will explain Jefferson's method in Example 4.

Example 4 *Using Jefferson's Method for Apportioning Legislative Seats*

Consider the Republic of Hydrangea from Example 3 on page 917. Use Jefferson's method to apportion the 250 legislature seats among the five states. Use the population from Table 14.31.

Table 14.31 Republic of Hydrangea Population

State	A	B	C	D	E	Total
Population	1,003,200	1,228,600	4,990,700	813,000	764,500	8,800,000

Solution In Example 3, the standard divisor was calculated to be 35,200. With Jefferson's method, the modified quota for each group needs to be slightly greater than the standard quota. To accomplish this we use a *modified divisor, which is slightly less than the standard divisor.* When you divide a quantity by a smaller number, the quotient becomes greater. There is no formula to determine this new modified divisor, and usually more than one divisor will work. We will use trial and error until we determine a divisor that apportions exactly 250 legislative seats using modified lower quotas.

Suppose that we try 35,000 as our modified divisor, d. To determine each modified quota, we will divide each state's population by 35,000. For example, the modified quota for state A is

$$\frac{1,003,200}{35,000} \approx 28.66$$

We will round modified quotas to the nearest hundredth as we did with standard quotas. Since the modified quota for state A is 28.66, the modified *lower* quota for state A is 28. This result is indicated in column 1, row 4, of Table 14.32.

Table 14.32 shows the results for all the states. In the table, we give the standard quotas as references, but they are not used in obtaining the apportionment using Jefferson's method.

Table 14.32 Republic of Hydrangea Population Using a Modified Divisor, $d = 35,000$

State	A	B	C	D	E	Total
Population	1,003,200	1,228,600	4,990,700	813,000	764,500	8,800,000
Standard quota	28.50	34.90	141.78	23.10	21.72	
Modified quota	28.66	35.10	142.59	23.23	21.84	
Modified lower quota	28	35	142	23	21	249

The sum of the modified lower quotas, 249, is less than the number of available seats, 250. If the sum of the modified lower quotas is too low (as it is in this case), use a lower modified divisor. If the sum of the modified lower quotas is too high, use a higher modified divisor. Since our sum is too low, we will try a lower modified divisor. Let's see what happens if we use 34,900 as our modified divisor. This modified divisor is a little lower than the modified divisor of 35,000 previously used.

Using 34,900 as the modified divisor, the modified quota for state A is

$$\frac{1,003,200}{34,900} \approx 28.74$$

Table 14.33 shows the modified quotas for all the states.

Table 14.33 Republic of Hydrangea Population Using a Modified Divisor, $d = 34,900$

State	A	B	C	D	E	Total
Population	1,003,200	1,228,600	4,990,700	813,000	764,500	8,800,000
Standard quota	28.50	34.90	141.78	23.10	21.72	
Modified quota	28.74	35.20	143.00	23.30	21.91	
Modified lower quota	28	35	143	23	21	250

The sum of the modified lower quotas is now 250, our desired sum. Each state is awarded the number of legislative seats listed in Table 14.33 under the category of modified lower quota. ∎

Whenever we use Jefferson's method, we first round the modified quota to the nearest hundredth and then determine the modified lower quota. In Example 4, we used a modified divisor of 34,900 to obtain the desired sum of 250, which is the number of legislative seats to be apportioned. There are an infinite number of modified divisors (in a small range) that could be used to obtain the modified lower quotas that we obtained. In fact, any modified divisor from about 34,758 to about 34,901 would result in the modified lower quotas we obtained in Table 14.33. If you selected a modified divisor lower than about 34,758, you would obtain modified lower quotas for which the sum of the seats is too high. For example, if you selected a modified divisor of 34,750, you would obtain modified lower quotas of 28, 35, 143, 23, and 22. The sum of these seats is 251, which is above the number of seats to be allocated. If you selected a modified divisor greater than about 34,901, you would obtain modified lower quotas for which the sum of the seats is too low. For example, if you selected a modified divisor of 34,925, you would obtain modified lower quotas of 28, 35, 142, 23, and 21. The sum of these seats is 249, which is below the number of seats to be allocated.

With Jefferson's method, we will always use a modified divisor that is less than the standard divisor used with Hamilton's method.

Recall from the discussion of Hamilton's method that if an apportionment method assigns either the lower or upper quota, it is said to satisfy the quota rule. Hamilton's method satisfies the quota rule, since it uses either the lower quota or upper quota of the standard quota. Notice that the standard quota for state C from Example 3 is 141.78. According to the quota rule, state C should receive either 141 or 142 seats. Using Jefferson's method, state C received 143 seats (see Table 14.33 on page 920). Therefore, Jefferson's method violates the quota rule.

The first case in the House of Representatives in which Jefferson's method led to a violation of the quota rule occurred in 1832. New York had a standard quota of 38.59 but was awarded 40 seats using Jefferson's method. Daniel Webster argued that this result was unconstitutional and suggested a compromise between Hamilton's method and Jefferson's method, leading to Webster's method. Let's discuss how Webster proposed to apportion the seats in the House of Representatives.

Webster's Method

Hamilton's method rounded *standard quotas down* to the nearest integer. Jefferson's method rounded *modified quotas down* to the nearest integer. With Webster's method, we round *modified quotas to* the nearest integer such that the rounding gives the exact sum to be apportioned.

> **PROCEDURE   WEBSTER'S METHOD**
>
> To use Webster's method for apportionment, do the following.
>
> 1. Determine a modified divisor, d, such that when each group's modified quota is *rounded to the nearest* integer, the total of the integers is the exact number of items to be apportioned. We will refer to modified quotas that are rounded to the nearest integer as **modified rounded quotas**.
>
> 2. Apportion to each group its modified rounded quota.

With Webster's method, we use a modified divisor that may be *less than, equal to,* or *greater than* the standard divisor and we use *modified rounded quotas*. With Webster's method, as with Jefferson's method, the modified divisor used must result in all the items being distributed, with no items left over. We will use Webster's method in Example 5.

Example 5   *Using Webster's Method for Apportioning Legislative Seats*

Once again, let's consider the Republic of Hydrangea and apportion the 250 seats among the five states using Webster's method. Use the population from Table 14.34.

Table 14.34   Republic of Hydrangea Population

State	A	B	C	D	E	Total
Population	1,003,200	1,228,600	4,990,700	813,000	764,500	8,800,000

Solution Our first task is to determine the modified divisor, d. We could choose a number either less than, greater than, or equal to the standard divisor. As with Jefferson's method, there is no formula to determine this new divisor and usually there is more than one divisor that works. Again, we must use trial and error until we determine a divisor that apportions the 250 legislative seats.

Let's use the results from Example 3 on pages 917–918 as a guide. If we round the standard quotas from Table 14.30 on page 918 to the nearest integer, the sum is $29 + 35 + 142 + 23 + 22$, or 251. This number is too high, since we want a sum of 250. If the sum is too high, as it is in this case, we need to use a larger divisor. If the sum is too low, we need to use a smaller divisor. In Example 3, we used the standard divisor of 35,200. Let's try using 35,250 as our modified divisor, d. Using 35,250 as the modified divisor, the modified quota for state A is

$$\frac{1,003,200}{35,250} \approx 28.46$$

Table 14.35 shows the results using $d = 35,250$.

Table 14.35   Republic of Hydrangea Population Using a Modified Divisor, $d = 35,250$

State	A	B	C	D	E	Total
Population	1,003,200	1,228,600	4,990,700	813,000	764,500	8,800,000
Standard quota	28.50	34.90	141.78	23.10	21.72	
Modified quota	28.46	34.85	141.58	23.06	21.69	
Modified rounded quota	28	35	142	23	22	250

Now round each modified quota to the nearest integer to get the modified rounded quotas, as shown in Table 14.35. Our sum of the modified rounded quotas is 250, as desired.

Therefore, using Webster's method, each state is awarded the number of legislative seats listed in Table 14.35 under the category of modified rounded quota. ■

Whenever we use Webster's method, we first round the modified quota to the nearest hundredth and then determine the modified rounded quota. In Example 5, we used a modified divisor of 35,250 to obtain the desired sum of 250. There are an infinite number of modified divisors (in a small range) that could be used to obtain the modified rounded quotas. In fact, any modified divisor from about 35,207 to about 35,271 would result in the modified rounded quotas we obtained in Table 14.35. If you selected a modified divisor lower than about 35,207, you would obtain modified rounded quotas for which the sum of the seats is too high. If you selected a modified divisor greater than about 35,271, you would obtain modified rounded quotas for which the sum of the seats is too low.

Occasionally, the results of Webster's method agree with the results from Hamilton's method. With Webster's method, we can use a modified divisor that is less than, greater than, or equal to the standard divisor used with Hamilton's method.

Webster's method is similar to Jefferson's method, since they both make use of a modified divisor. Webster's method, however, is a little more difficult in practice to apply than Jefferson's method, since the modified divisor could be less than, equal to, or greater than the standard divisor. Webster's method may seem the most reasonable because the modified quotas are rounded in the conventional way. Webster's method, however, does have a flaw in that it can violate the quota rule. Even though Example 5 did not uncover this flaw, there are examples in which Webster's method violates the quota rule. This violation happens more in theory than in practice. As a result, many experts consider Webster's method the best overall apportionment method.

Let's discuss one more apportionment method that was proposed (but never used by the House of Representatives) by John Quincy Adams around the same time that Webster proposed his method.

Adams' Method

When Jefferson's method was discovered to be problematic because it violated the quota rule, John Quincy Adams proposed a method that is exactly the opposite of Jefferson's method. Instead of using modified lower quotas as Jefferson proposed, Adams suggested using modified upper quotas.

> **PROCEDURE ADAMS' METHOD**
>
> To use Adams' method for apportionment, do the following.
> 1. Determine a modified divisor, d, such that when each group's modified quota is *rounded **up** to the nearest integer* the total of the integers is the exact number of items to be apportioned. We will refer to the modified quotas that are rounded up as **modified upper quotas.**
> 2. Apportion to each group its modified upper quota.

With Adams' method, we use a modified divisor that is *greater than* the standard divisor and we use quotas that are *rounded up*, or *modified upper quotas*. As with Jefferson's and Webster's methods, the modified divisor used must result in all items being distributed, with no items left over.

Let's see what happens when we apply Adams' method to the Republic of Hydrangea.

Example 6 *Using Adams' Method for Apportioning Legislative Seats*

Once again, consider the Republic of Hydrangea. Apportion the 250 seats among the five states using Adams' method. Use the population from Table 14.36.

Table 14.36 Republic of Hydrangea Population

State	A	B	C	D	E	Total
Population	1,003,200	1,228,600	4,990,700	813,000	764,500	8,800,000

Solution With Adams' method, the modified quota for each group needs to be slightly smaller than the standard quota. To accomplish that, we use a modified divisor that is slightly greater than the standard divisor. When you divide a quantity by a larger number, the quotient becomes smaller. Since the standard divisor is 35,200, let's try letting $d = 35,400$. Using 35,400 as the modified divisor, the modified quota for state A is

$$\frac{1,003,200}{35,400} \approx 28.34$$

Table 14.37 on page 924 shows the results.

Table 14.37 Republic of Hydrangea Population Using a Modified Divisor, $d = 35,400$

State	A	B	C	D	E	Total
Population	1,003,200	1,228,600	4,990,700	813,000	764,500	8,800,000
Standard quota	28.50	34.90	141.78	23.10	21.72	
Modified quota	28.34	34.71	140.98	22.97	21.60	
Modified upper quota	29	35	141	23	22	250

Using $d = 35,400$ and rounding up to the modified quotas, we have a sum of 250 seats, as desired. Therefore, each state will be awarded the number of legislative seats listed in Table 14.37 under the category of modified upper quota. ◼

Whenever we use Adams' method, we first round the modified quota to the nearest hundredth and then determine the modified upper quota. In Example 6, we used a modified divisor of 35,400 to obtain the desired sum of 250 seats. There are an infinite number of modified divisors (in a small range) that could be used to obtain the modified upper quotas that we obtained. In fact, any modified divisor from about 35,394 to about 35,646 would result in the modified upper quotas we obtained in Table 14.37. If you selected a modified divisor lower than about 35,394, you would obtain modified upper quotas for which the sum of the seats is too high. If you selected a modified divisor greater than about 35,646, you would obtain modified upper quotas for which the sum of the seats is too low.

With Adams' method, we will always use a modified divisor that is greater than the standard divisor used with Hamilton's method.

In some cases, the range of numbers that can be used as a modified divisor is very narrow and you may need to use a decimal number. For example, 34 may be too small for a modified divisor and 35 may be too large for a modified divisor. In that case, you will need to use a decimal number between 34 and 35 as the modified divisor.

Looking at the results from Example 6, we may want to conclude that Adams' method is the best, since the quota rule was not violated in this example. Unfortunately, that is not always the case. We have seen that Jefferson's method can result in a state receiving more than its upper quota of seats. Adams' method can lead to just the opposite: a state receiving fewer than its lower quota of seats. Although the Jefferson method favors large states, such as his state of Virginia, Adams' method favors small states, such as those in his area of New England. In addition, Hamilton's method favors large states and Webster's method favors small states.

Example 7 *Comparing Apportionment Methods*

The Tiger Republic is a small country with a population of 10,400,000 that consists of four states A, B, C, and D. There are 260 seats in the legislature that need to be apportioned among the four states. The population of each state is shown in Table 14.38.

Table 14.38

State	A	B	C	D	Total
Population	3,350,000	1,850,000	2,365,000	2,835,000	10,400,000

a) Determine each state's apportionment using Hamilton's method.
b) Determine each state's apportionment using Jefferson's method.
c) Determine each state's apportionment using Adams' method.
d) Determine each state's apportionment using Webster's method.

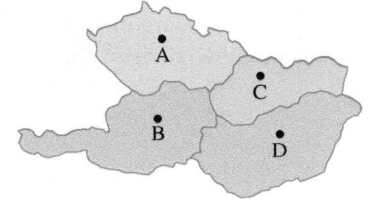
▲ *Tiger Republic population 10,400,000*

Solution

a) First we must determine the standard divisor. The standard divisor is determined by dividing the total population by the number of seats to be apportioned. The standard divisor is

$$\frac{10,400,000}{260} = 40,000$$

Next we determine the standard quotas by dividing each state's population by the standard divisor, 40,000. Recall that we round standard quotas to the nearest hundredth. With Hamilton's method, we round each standard quota down to obtain the lower quota. Then we distribute any leftover seats to the states with the largest fractional parts until all seats are distributed. Table 14.39 shows the apportionment using Hamilton's method.

Table 14.39 Hamilton's Method, Standard Divisor of 40,000

State	A	B	C	D	Total
Population	3,350,000	1,850,000	2,365,000	2,835,000	10,400,000
Standard quota	83.75	46.25	59.13	70.88	
Lower quota	83	46	59	70	258
Hamilton's apportionment	84	46	59	71	260

Using the lower quota, only 258 of the 260 seats were apportioned. Since states D and A, in that order, had the highest fractional parts of the standard quota (0.88 and 0.75, respectively), each of these states receives one additional seat.

b) With Jefferson's method, we use a modified divisor that is *less than* the standard divisor and use modified *lower quotas*. Table 14.40 shows the results using a modified divisor of 39,700.

Table 14.40 Jefferson's Method, Modified Divisor, $d = 39,700$

State	A	B	C	D	Total
Population	3,350,000	1,850,000	2,365,000	2,835,000	10,400,000
Modified quota	84.38	46.60	59.57	71.41	
Modified lower quota	84	46	59	71	260

Using $d = 39,700$, we have a sum of 260 seats, as desired. Therefore, each state receives the number of seats listed under the modified lower quotas in Table 14.40.

c) With Adams' method, we use a modified divisor that is *greater than* the standard divisor and use modified *upper quotas*. Table 14.41 shows the results using a modified divisor of 40,300.

Table 14.41 Adams' Method, Modified Divisor, $d = 40,300$

State	A	B	C	D	Total
Population	3,350,000	1,850,000	2,365,000	2,835,000	10,400,000
Modified quota	83.13	45.91	58.68	70.35	
Modified upper quota	84	46	59	71	260

Using $d = 40{,}300$, we have a sum of 260 seats, as desired. Therefore, each state receives the number of seats listed under the modified upper quotas in Table 14.41.

d) With Webster's method, we need to determine a modified divisor such that when each state's modified quota is rounded to the nearest integer, the total of the integers is 260. The modified divisor may be *less than, equal to,* or *greater than* the standard divisor. If we round the standard quotas from part (a) to the nearest integer, the sum of the rounded quotas is 260. Therefore, we will use the standard divisor as our modified divisor. Table 14.42 shows the results when we round the modified quotas to the nearest integer.

Table 14.42 Webster's Method, Modified Divisor, $d = 40{,}000$

State	A	B	C	D	Total
Population	3,350,000	1,850,000	2,365,000	2,835,000	10,400,000
Standard quota	83.75	46.25	59.13	70.88	
Modified quota	83.75	46.25	59.13	70.88	
Modified rounded quota	84	46	59	71	260

Each state receives the number of seats listed under the modified rounded quotas in Table 14.42. Note that in this example, all four methods led to the same apportionment. However, that is not always the case. ■

Of the four methods we have discussed in this section, *Hamilton's method uses standard quotas. Jefferson's method, Webster's method, and Adams' method all make use of a modified quota and can all lead to violations of the quota rule.* As we will see in the next section, Hamilton's method can also be problematic by producing paradoxes. Table 14.43 summarizes the four apportionment methods.

Table 14.43 Summary of Apportionment Methods

Method	Divisor	Apportionment
Hamilton's	Standard divisor $= \dfrac{\text{total population}}{\text{number of items to be allocated}}$	Round each standard quota down. Distribute any leftover items to the groups with the largest fractional parts until all items are distributed. **Favors large states.**
Jefferson's	The modified divisor is *less than* the standard divisor.	Each group's modified quota is *rounded down* to the nearest integer. Apportion to each group its modified lower quota. **Favors large states.**
Webster's	The modified divisor is *less than, greater than,* or *equal to,* the standard divisor.	Each group's modified quota is *rounded to the nearest* integer. Apportion to each group its modified rounded quota. **Favors small states.**
Adams'	The modified divisor is *greater than* the standard divisor.	Each group's modified quota is *rounded up* to the nearest integer. Apportion to each group its modified upper quota. **Favors small states.**

SECTION 14.3 *Exercises*

Warm Up Exercises

In Exercises 1–10, fill in the blank with an appropriate word, phrase, or symbol(s).

1. The total population under consideration divided by the number of items to be allocated is called the standard _____.

2. When each group's population is divided by the standard divisor, a standard _____ is obtained.

3. A standard quota rounded up to the nearest integer is called a(n) _____ quota.

4. A standard quota rounded down to the nearest integer is called a(n) _____ quota.

5. The rule stating that an apportionment should always be either the upper quota or the lower quota is called the _____ rule.

6. Jefferson's method, Webster's method, and Adams' method require using a(n) _____ quota.

7. The apportionment method that requires rounding the *standard quota down* to the lower quota is called _____ method.

8. a) The apportionment method that uses a modified divisor that is less than the standard divisor is _____ method.

   b) The apportionment method that uses a modified divisor that is greater than the standard divisor is _____ method.

   c) The apportionment method that uses a modified divisor that could be less than, greater than, or equal to the standard divisor is _____ method.

9. a) The apportionment method that uses a *modified quota* that is always rounded to the nearest integer is _____ method.

   b) The apportionment method that uses a *modified quota* that is always rounded up to the nearest integer is _____ method.

   c) The apportionment method that uses a *modified quota* that is always rounded down to the nearest integer is _____ method.

10. Jefferson's method, Webster's method, and Adams' method all make use of a modified quota and can all lead to violations of the _____ rule.

Practice the Skills/Problem Solving

In Exercises 11–49, when appropriate round quotas to the nearest hundredth.

Legislative Seats In Exercises 11–18, suppose that Turtlestan is a small country with a population of 7,500,000 that consists of four states, A, B, C, and D. There are 150 seats in the legislature that need to be apportioned among the four states. The population of each state is shown in the table below.

State	A	B	C	D	Total
Population	1,220,000	2,730,000	857,000	2,693,000	7,500,000

11. a) Determine the standard divisor.

   b) Determine each state's standard quota.

12. Determine each state's apportionment using Hamilton's method.

13. a) Determine each state's modified quota using the divisor 49,300.

   b) Determine each state's apportionment using Jefferson's method.

14. a) Determine each state's modified quota using the divisor 49,250.

   b) Determine each state's apportionment using Jefferson's method.

15. a) Determine each state's modified quota using the divisor 50,700.

   b) Determine each state's apportionment using Adams' method.

16. a) Determine each state's modified quota using the divisor 50,600.

   b) Determine each state's apportionment using Adams' method.

17. Determine each state's apportionment using Webster's method using the standard divisor.

18. a) Determine each state's modified quota using the divisor 49,900.

   b) Determine each state's apportionment using Webster's method.

Surf Boards In Exercises 19–26, a hotel chain in Hawaii needs to apportion 25 new surfboards for its visitors to rent among three hotels based on the numbers of rooms in each hotel as shown in the table below.

Hotel	A	B	C	Total
Number of rooms	306	214	155	675

19. a) Determine the standard divisor.

  b) Determine each hotel's standard quota.

20. Determine each hotel's apportionment using Hamilton's method.

21. a) Determine each hotel's modified quota using the divisor 25.8.

  b) Determine each hotel's apportionment using Jefferson's method.

22. a) Determine each hotel's modified quota using the divisor 25.7.

  b) Determine each hotel's apportionment using Jefferson's method.

23. a) Determine each hotel's modified quota using the divisor 29.

  b) Determine each hotel's apportionment using Adams' method.

24. a) Determine each hotel's modified quota using the divisor 28.

  b) Determine each hotel's apportionment using Adams' method.

25. Determine each hotel's apportionment using Webster's method using the standard divisor.

26. a) Determine each hotel's modified quota using the divisor 28.1.

  b) Determine each hotel's apportionment using Webster's method.

New Toboggans *In Exercises 27–34, Mountain View Resorts in Vermont plans to apportion 50 new toboggans among the four resorts based on the number of rooms in each resort as shown below.*

Resort	A	B	C	D	Total
Number of rooms	86	102	130	232	550

27. a) Determine the standard divisor.

  b) Determine each resort's standard quota.

  c) Determine each resort's apportionment using Hamilton's method.

28. Determine each resort's apportionment using Jefferson's method. (*Hint:* Some divisor between 10 and 11 will work.)

29. Determine each resort's apportionment using Adams' method. (*Hint:* Some divisor between 11 and 12 will work.)

30. Determine each resort's apportionment using Webster's method. (*Hint:* Some divisor between 10.5 and 11.5 will work.)

Scooter Rental Stations *In Exercises 31–34, a university is made up of five colleges: Arts and Sciences, Engineering, Business, Education, and Visual and Performing Arts. There are 250 scooter rental stations to be apportioned among the five colleges based on their enrollment shown in the table below. The total enrollment is 13,000.*

School	Arts and Science	Business	Engineering	Education	Visual and Performing Arts
Enrollment	1746	7095	2131	937	1091

31. a) Determine the standard divisor.

  b) Determine each college's standard quota.

  c) Determine each college's apportionment using Hamilton's method.

32. Determine each college's apportionment using Adams' method. (*Hint:* Some divisors between 52 and 53 will work.)

33. Determine each college's apportionment using Jefferson's method. (*Hint:* Some divisors between 51 and 52 will work.)

34. Determine each college's apportionment using Webster's method.

New Boats *In Exercises 35–38, a boat manufacturer has 120 new boats of a new model to be apportioned to four dealerships. The manufacturer decides to apportion the boats based on the number of boats each dealership sold in the previous year. The number of boats sold by each dealership is shown in the table below.*

Dealership	A	B	C	D	Total
Sales	3840	2886	2392	1682	10,800

35. a) Determine the standard divisor.

   b) Determine each dealership's standard quota.

   c) Determine each dealership's apportionment using Hamilton's method.

36. Determine each dealership's apportionment using Jefferson's method. (*Hint:* Some divisors between 88.5 and 89 will work.)

37. Determine each dealership's apportionment using Webster's method. (*Hint:* Some divisors between 90 and 91 will work.)

38. Determine each dealership's apportionment using Adams' method. (*Hint:* Some divisors between 91 and 92 will work.)

New Segways *In Exercises 39–42, the Police Department in Scranton, PA has 100 new segways to be apportioned among six precincts. The department decides to apportion the segways based on the population per precinct, as shown in the table below.*

Precinct	A	B	C	D	E	F	Total
Population	9070	15,275	12,810	5720	25,250	6875	75,000

39. a) Determine the standard divisor.

   b) Determine each precinct's standard quota.

   c) Determine each precinct's apportionment using Hamilton's method.

40. Determine each precinct's apportionment using Jefferson's method.

41. Determine each precinct's apportionment using Adams' method.

42. Determine each precinct's apportionment using Webster's method.

Nursing Shifts *In Exercises 43–46, a hospital has 200 nurses to be apportioned among four shifts: shifts A, B, C, and D. The hospital decides to apportion the nurses based on the average number of room calls reported during each shift. Room calls are shown in the table at the top of the right column.*

Shift	A	B	C	D	Total
Room calls	751	980	503	166	2400

43. a) Determine the standard divisor.

   b) Determine each shift's standard quota.

   c) Determine each shift's apportionment using Hamilton's method.

44. Determine each shift's apportionment using Adams' method. (*Note:* Divisors do not have to be whole numbers.)

45. Determine each shift's apportionment using Jefferson's method. (*Note:* Divisors do not have to be whole numbers.)

46. Determine each shift's apportionment using Webster's method. (*Hint:* Some divisors between 12 and 12.05 will work.)

Challenge Problems/Group Activities

47. The First Census In 1790, the first United States census was taken. The following table shows the population of the 15 states at that time. One hundred five seats in the United States House of Representatives were to be apportioned among the 15 states.

State	Population
Connecticut	236,841
Delaware	55,540
Georgia	70,835
Kentucky	68,705
Maryland	278,514
Massachusetts	475,327
New Hampshire	141,822
New Jersey	179,570
New York	331,589
North Carolina	353,523
Pennsylvania	432,879
Rhode Island	68,446
South Carolina	206,236
Vermont	85,533
Virginia	630,560
Total	3,615,920

a) Determine the apportionment that would have resulted if Hamilton's method had been used as the method originally approved by Congress.

b) Determine the apportionment that was used with Jefferson's method.

c) Compare the apportionments from parts (a) and (b). Which state(s) benefited from Jefferson's method? Which state(s) were at a disadvantage from Jefferson's method?

48. *Legislative Seats* A country with a population of 10,000,000 has 250 legislative seats to be apportioned among four states, where each state has a different population. Determine a population for each state in which Hamilton's method, Jefferson's method, Webster's method, and Adams' method all lead to the same apportionment of the 250 legislative seats. Many answers are possible.

49. *Police Officers* A police department has 210 new officers to apportion among six precincts. The department plans to apportion the officers based on the number of crimes committed during the previous year in each precinct. Suppose that the number of crimes committed in each precinct is different and that the total number of crimes committed in all six precincts was 2940. Determine the number of crimes committed in each precinct such that Hamilton's method, Jefferson's method, Webster's method, and

Adams' method all lead to the same apportionment of the 210 new officers. Many answers are possible.

Research Activities

50. *House of Representatives* Do research and write a report on the apportionment method used in the House of Representatives in 1872. Include in your report a description of the method used.

51. *Huntington–Hill Method* Do research and write a report on the Huntington–Hill method, the current method used to apportion the representatives in the House of Representatives. Include historical background and describe how the method works.

SECTION 14.4   Flaws of the Apportionment Methods

Upon completion of this section, you will be able to:

- Determine if a given apportionment demonstrates the Alabama paradox.
- Determine if a given apportionment demonstrates the population paradox.
- Determine if a given apportionment demonstrates the new-states paradox.

Recall from Section 14.3 the situation in which a college receives a grant to purchase 50 tablets that will be placed in four different libraries on campus. Suppose that before the tablets are purchased the price of the tablets decreases enough so that 51 tablets can now be purchased and apportioned to the four libraries. Because of a flaw in the apportionment method to be used, it is possible that one of the libraries would receive fewer tablets with 51 tablets being apportioned than with 50 tablets being apportioned. In this section, we will discuss this flaw and other flaws with apportionment methods.

Why This Is Important With each new census taken, the number of representatives in the United States Congress allocated to various states may change. How these representatives are assigned can make an important difference in the laws and bills that are passed.

Just as we have discovered that voting methods can have flaws, apportionment methods can also have flaws. In this section, as we did in Section 14.2 with voting methods, we will consider several reasonable properties that an apportionment method should have. Then we will see examples in which these properties are violated. Problems with apportionment can occur in a variety of ways. Population changes, changes in the number of items to be apportioned, and the addition of one or more groups can lead to problems with apportionment.

Recall from Section 14.3 that the quota rule states that an apportionment for every group under consideration should always be either the upper quota or the lower quota. Hamilton's method, which satisfies the quota rule, would appear to be a reasonable and fair apportionment method. As we will discuss in this section, though, Hamilton's method can result in some serious flaws. In this section, we will discuss three flaws of Hamilton's method: the *Alabama paradox*, the *population paradox*, and the *new-states paradox. These flaws apply only to Hamilton's method and not to Jefferson's method, Webster's method, or Adams' method.*

The Alabama Paradox

The first, and perhaps most serious, flaw of Hamilton's method occurs when an increase in the total number of items to be apportioned results in a loss of an item for one of the groups. This flaw first occurred in the apportionment of the House of Representatives in 1880 when a discussion occurred on whether to have 299 or 300 members in the House. Using Hamilton's method with 299 members, Alabama would receive eight seats. But if the total number of representatives were increased to 300, Alabama would receive only seven seats. As a result, this situation became known as the *Alabama paradox.*

> **Alabama Paradox**
> The **Alabama paradox** occurs when an increase in the total number of items to be apportioned results in a loss of an item for a group.

Example 1 illustrates the Alabama paradox.

Example 1 *Demonstrating the Alabama Paradox*

Consider Stanhope, a small country with a population of 18,000 people and three states A, B, and C (Table 14.44). There are 150 seats in the legislature that must be apportioned among the three states, according to their population. Show that the Alabama paradox occurs if the number of seats is increased to 151.

When appropriate, round standard divisors and standard quotas to the nearest hundredth.

Table 14.44

State	A	B	C	Total
Population	888	8076	9036	18,000

Solution With 150 seats in the legislature, the standard divisor is

$$\frac{18,000}{150} = 120$$

The standard quotas for each state and the apportionment for each state are shown in Table 14.45.

Table 14.45

State	A	B	C	Total
Population	888	8076	9036	18,000
Standard quota	7.40	67.30	75.30	
Lower quota	7	67	75	149
Hamilton's apportionment	8	67	75	150

With 151 seats in the legislature, the standard divisor is

$$\frac{18,000}{151} \approx 119.21$$

The standard quotas for each state and the apportionment for each state with 151 total seats are shown in Table 14.46.

Table 14.46

State	A	B	C	Total
Population	888	8076	9036	18,000
Standard quota	7.45	67.75	75.80	
Lower quota	7	67	75	149
Hamilton's apportionment	7	68	76	151

When the number of seats increased from 150 to 151, state A's apportionment actually decreased, from 8 to 7. This example illustrates the Alabama paradox. ∎

When the total number of items increases, each group's standard quota increases, but not by the same amount. Therefore, it is possible that the order of assignment of the items can change. As a result, some groups can lose items they already had. When the total number of items increases, usually the larger groups benefit at the expense of the smaller groups.

The next paradox we will discuss is the population paradox.

Population Paradox

Another paradox that can occur with Hamilton's method may occur when the population of one or more states changes. It was discovered in the early 1900s that under Hamilton's method, one state could lose a seat to another state even though its population is growing at a faster rate. At the time, Virginia lost a seat in the House of Representatives while Maine gained a seat, although Virginia's population was growing at a much faster rate than Maine's population. Thus, this paradox became known as the *population paradox*.

> **Population Paradox**
> The **population paradox** occurs when group A loses items to group B even though group A's population increased at a faster rate than group B's.

Example 2 illustrates the population paradox.

Example 2 *Demonstrating the Population Paradox*

Consider Alexandria, a small country with a population of 100,000 and three states A, B, and C. There are 100 seats in the legislature that must be apportioned among the three states. Using Hamilton's method, the apportionment is shown in Table 14.47.

Table 14.47

State	A	B	C	Total
Population	23,527	5548	70,925	100,000
Standard quota	23.53	5.55	70.93	
Lower quota	23	5	70	98
Hamilton's apportionment	23	6	71	100

Suppose that the population increases according to Table 14.48 and that the 100 seats are reapportioned. Show that the population paradox occurs when Hamilton's method is used.

Table 14.48

State	A	B	C	Total
Population	23,926	5648	71,110	100,684

Solution To calculate the percent increase, we use the procedure discussed in Section 10.1. State A has an increase of 399 people. Therefore, state A has a percent increase of

$$\frac{399}{23,527} \approx 0.01696 \approx 1.696\%$$

State B has an increase of 100 people. Therefore, state B has a percent increase of

$$\frac{100}{5548} \approx 0.01802 \approx 1.802\%$$

State C has an increase of 185 people. Therefore, state C has a percent increase of

$$\frac{185}{70,925} \approx 0.00261 \approx 0.261\%$$

All three states had an increase in their population, but state B increased at a faster rate than state A and state C.

The standard divisor using the new population is

$$\frac{100,684}{100} = 1006.84$$

Table 14.49 shows the reapportionment, using Hamilton's method with the standard divisor 1006.84.

Table 14.49

State	A	B	C	Total
Population	23,926	5648	71,110	100,684
Standard quota	23.76	5.61	70.63	
Lower quota	23	5	70	98
Hamilton's apportionment	24	5	71	100

State B has lost a seat to state A even though state B's population grew at a faster rate than state A's. As a result, we have an example of the population paradox. ∎

The next and final paradox we will discuss is the new-states paradox.

Timely Tip

In Example 2, the population paradox occurred because state B lost a seat to state A even though state B's population grew at a faster rate than state A's population. When checking for the population paradox, it is possible for a state to lose a seat to another state without the population paradox occurring. *Remember that for the population paradox to occur, the state that loses the seat must be growing at a faster rate than the state that gains the seat.*

The New-States Paradox

The new-states paradox was discovered in 1907 when Oklahoma was added as a state. When a new state is added, new seats must be added to the legislature. How do we determine the number of new seats to add? A reasonable answer would be to add the number of seats the new state would be entitled to based on its population. When reapportioning the House of Representatives with the additional five seats Oklahoma was entitled to, Maine's apportionment increased from three to four seats and New York's apportionment decreased from 38 to 37. By adding a new state and additional seats, New York was required to give a seat to Maine. This paradox became known as the *new-states paradox*.

> ### New-States Paradox
> The **new-states paradox** occurs when the addition of a new group, and additional items to be apportioned, reduces the previous apportionment of another group.

Example 3 *Demonstrating the New-States Paradox*

The Iowa Public Library System has received a grant to purchase 100 laptop computers to be distributed between two libraries A and B. The 100 laptops will be apportioned based on the population served by each library. The apportionment using Hamilton's method is shown in Table 14.50. The standard divisor is

$$\frac{10,000}{100} = 100$$

Table 14.50

Library	A	B	Total
Population	2145	7855	10,000
Standard quota	21.45	78.55	
Lower quota	21	78	99
Hamilton's apportionment	21	79	100

Suppose that an anonymous donor decides to donate money to purchase six more laptops provided that a third library, C, that serves a population of 625, is included in the apportionment. Show that the new-states paradox occurs when the laptops are reapportioned.

Solution The total population is now 10,000 + 625 or 10,625 and the total number of laptops to be apportioned is now 100 + 6 or 106. Therefore, when the third library is added, the new standard divisor is

$$\frac{10,625}{106} \approx 100.24$$

The new standard quotas and Hamilton's apportionment are shown in Table 14.51.

Table 14.51   Using a New Standard Divisor of 100.24

Library	A	B	C	Total
Population	2145	7855	625	10,625
Standard quota	21.40	78.36	6.24	
Lower quota	21	78	6	105
Hamilton's apportionment	22	78	6	106

Before library C was added, library B would receive 79 laptops. By adding a new library and increasing the total number of laptops to be apportioned, library B ended up losing a laptop to library A. Thus, we have a case of the new-states paradox.   ∎

When a new group is added to the apportionment, we must determine how many additional items should be added to the total to be apportioned. In Example 3, we were told that six additional laptops were to be added to the total apportioned when a new library was included in the apportionment. If we are given the new group's population but are not given the number of additional items to be added to the total to be apportioned, we calculate the number of items to be apportioned to the new group as follows. First determine the new group's standard quota. The number of additional items would be the new group's standard quota *rounded down* to the nearest integer. For example, in Example 3, the new library, library C, served a population of 625. The standard quota for library C would be

$$\frac{625}{100} = 6.25$$

Rounding 6.25 down to the nearest integer gives us six laptops to be added to the total to be apportioned, or 106 laptops.

As we have discovered in Section 14.3, Hamilton's method appears to be a fair and reasonable apportionment method, since it satisfies the quota rule. In this section, however, we discovered that Hamilton's method can produce paradoxes. Jefferson's, Adams', and Webster's methods can all violate the quota rule but do not produce paradoxes. Hamilton's and Jefferson's apportionment methods can favor large states, whereas Adams' and Webster's apportionment methods can favor small states. Is there a perfect apportionment method that satisfies the quota rule, does not produce any paradoxes, and favors neither large nor small states? In 1980, mathematicians Michel Balinski and H. Payton Young proved that there is no apportionment method that satisfies the quota rule while also avoiding all known paradoxes. Their theorem is called Balinski and Young's impossibility theorem.

> **Balinski and Young's Impossibility Theorem**
> There is no perfect apportionment method that satisfies the quota rule and avoids all known paradoxes.

Table 14.52 summarizes the four apportionment methods we have discussed in this chapter and indicates which methods may violate the quota rule and which methods may produce the paradoxes we have discussed.

Table 14.52   Comparison of Apportionment Methods

	Apportionment Method			
	Hamilton	Jefferson	Adams	Webster
May violate the quota rule (apportionment should always be either upper or lower quota)	No	Yes	Yes	Yes
May produce the Alabama paradox (an increase in the total number of items results in a loss of an item for a group)	Yes	No	No	No
May produce the population paradox (group A loses an item to group B although group A's population grew faster than group B's population)	Yes	No	No	No
May produce the new-states paradox (the addition of a new group reduces the apportionment of another group)	Yes	No	No	No
Apportionment method favors	Large states	Large states	Small states	Small states

Just as there is no perfect voting method, there is also no perfect apportionment method.

SECTION 14.4 Exercises

Warm Up Exercises

In Exercises 1–6, fill in the blank with an appropriate word, phrase, or symbol(s).

1. When group A loses an item or items to group B even though group A's population grew at a faster rate than group B's, the _____ paradox occurs.

2. When the addition of a new group and additional items to be apportioned reduces the prior apportionment of another group, the _____ paradox occurs.

3. When an increase in the total number of items to be apportioned results in a loss of an item for a group, the _____ paradox occurs.

4. Hamilton's and Jefferson's apportionment methods, favor _____ states.

5. Adams' and Webster's apportionment methods favor _____ states.

6. The apportionment method that satisfies the quota rule but may produce a paradox is called _____ method.

Practice the Skills/Problem Solving

In Exercises 7–18, when appropriate, round quotas and divisors to the nearest hundredth.

7. Doctors Consider the apportionment of 60 doctors for First Physicians Organization given in Example 2 on page 917 of Section 14.3. The apportionment using Hamilton's method is shown in the table below.

Clinic	A	B	C	D	E	Total
Patients	246	201	196	211	226	1080
Standard quota	13.67	11.17	10.89	11.72	12.56	60.01
Lower quota	13	11	10	11	12	57
Hamilton's apportionment	14	11	11	12	12	60

Does the Alabama paradox occur using Hamilton's method if the number of doctors is increased from 60 to 61? Explain your answer.

8. Color Printers A large company with offices in four cities must distribute 148 new color printers to the four offices. The printers will be apportioned based on the number of employees in each office as shown in the table at the top of the right column.

Office	A	B	C	D	Total
Employees	757	295	636	976	2664

a) Apportion the printers using Hamilton's method.

b) Does the Alabama paradox occur using Hamilton's method if the number of new printers is increased from 148 to 149? Explain your answer.

9. Legislative Seats A country with three states has 30 seats in the legislature. The population of each state is shown in the table below.

State	A	B	C	Total
Population	161	250	489	900

a) Apportion the seats using Hamilton's method.

b) Does the Alabama paradox occur using Hamilton's method if the number of seats is increased from 30 to 31? Explain your answer.

10. Legislative Seats A country with three states has 200 seats in the legislature. The population of each state is shown in the table below.

State	A	B	C	Total
Population	247,100	481,900	271,000	1,000,000

a) Apportion the seats using Hamilton's method.

b) Does the Alabama paradox occur using Hamilton's method if the number of seats is increased from 200 to 201? Explain your answer.

In Exercises 11–14, assume that the number of items to be apportioned does not change.

11. Promotions Spectrum has 30,000 employees in three cities as shown in the table below. It wishes to give promotions to 200 employees.

Cities	A	B	C	Total
Employees	9130	6030	14,840	30,000

a) Apportion the promotions using Hamilton's method.

b) Suppose that in 10 years the cities have the following number of employees and the company wishes to again give promotions to 200 employees. Does the population paradox occur using Hamilton's method?

Cities	A	B	C	Total
Employees	9150	6030	14,945	30,125

12. *Forklifts* Anabru Manufacturing has 100 forklifts to apportion among three factories. The forklifts are to be apportioned based on the number of employees at each factory as shown in the table below

Factory	A	B	C	Total
Employees	3822	7818	28,360	40,000

a) Apportion the forklifts using Hamilton's method.

b) Suppose that 1 year later the factories have the following number of employees. If the 100 forklifts are reapportioned to the factories, does the population paradox occur using Hamilton's method?

Factory	A	B	C	Total
Employees	3861	7896	28,360	40,117

13. *3-D Printers* A college with five divisions has funds for 54 3-D printers. The student population for each division is shown in the table below.

Division	A	B	C	D	E	Total
Population	733	1538	933	1133	1063	5400

a) Apportion the printers using Hamilton's method.

b) Suppose that 1 year later the divisions have the following populations. If the college wishes to apportion 54 printers, does the population paradox occur using Hamilton's method?

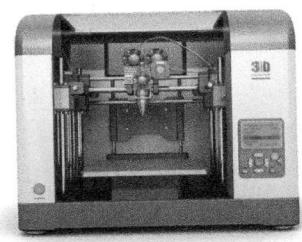

Division	A	B	C	D	E	Total
Population	733	1539	933	1133	1116	5454

14. *Legislative Seats* A country with three states has 250 seats in the legislature. The population of each state is shown in the table at the top of the right column.

State	A	B	C	Total
Population	459	10,551	18,990	30,000

a) Apportion the seats using Hamilton's method.

b) Suppose that in 10 years the states have the following populations. If the country reapportions 250 seats in the legislature, does the population paradox occur using Hamilton's method?

State	A	B	C	Total
Population	464	10,551	19,100	30,115

15. *Additional Employees* Cynergy Telecommunications has employees in Europe, labeled A, and in the United States, labeled B. The number of employees in each group is shown in the table below. There are 48 managers to be apportioned between the two groups.

Cynergy Telecommunications	A	B	Total
Employees	844	3956	4800

a) Apportion the managers using Hamilton's method.

b) Suppose that additional employees in Asia, labeled C, with the number of employees shown in the table below, are added with seven new managers. Does the new-states paradox occur using Hamilton's method?

Cynergy Telecommunications	A	B	C	Total
Employees	844	3956	724	5524

16. *Adding a Park* The town of Manlius purchased 25 new picnic tables to be apportioned between two parks. The picnic tables are to be apportioned based on the annual number of visitors to each park as shown below.

Park	A	B	Total
Visitors	3750	6250	10,000

a) Apportion the picnic tables using Hamilton's method.

b) Suppose that the town decides to purchase five additional picnic tables and include a third park with an annual number of visitors as shown in the table below. The town will now apportion 30 picnic tables among the three parks. Does the new-states paradox occur using Hamilton's method?

Park	A	B	C	Total
Visitors	3750	6250	2100	12,100

17. *Adding a State* A country with two states has 33 seats in the legislature. The population of each state is shown in the table below.

State	A	B	Total
Population	744	2556	3300

a) Apportion the states using Hamilton's method.

b) Suppose that a third state with the population shown in the table below is added, with seven additional seats. Does the new-states paradox occur using Hamilton's method?

State	A	B	C	Total
Population	744	2556	710	4010

18. *Adding a State* A country with three states has 60 seats in the legislature. The population of each state is shown in the table below.

State	A	B	C	Total
Population	62,700	230,700	606,600	900,000

a) Apportion the seats using Hamilton's method.

b) Suppose that a fourth state with the population shown in the table below is added, with five additional seats. Does the new-states paradox occur using Hamilton's method?

State	A	B	C	D	Total
Population	62,700	230,700	606,600	78,000	978,000

Research Activity

19. *An Apportionment method* Write a paper on which apportionment method you think is the best. Include reasons to support your choice, including the advantages and disadvantages of the method you have selected.

CHAPTER 14   *Summary*

Important Facts and Concepts

Examples and Discussion

Important Facts and Concepts	Examples and Discussion
Section 14.1	
Voting Methods	
Plurality	Discussion page 884, Examples 3–4, page 885
Borda count method	Discussion pages 885–886, Examples 5–6, pages 886–887
Plurality with elimination method	Discussion page 888, Examples 7–8, pages 888–890
Pairwise comparison method	Discussion pages 891–892, Examples 9–10, pages 891–893
Section 14.2	
Flaws of Voting Methods	
Majority criterion	Discussion page 900, Examples 1–2, 6, pages 900–901, 907
Head-to-head criterion	Discussion page 901, Examples 3, 6, pages 902–903, 907
Monotonicity criterion	Discussion page 903, Examples 4, 6, pages 904, 907
Irrelevant alternatives criterion	Discussion page 905, Examples 5, 6, pages 905–907
Section 14.3	
Standard quota	Example 1, pages 915–916
Standard divisor	Examples 2–3, pages 917–918
Apportionment Methods	
Hamilton's method	Discussion page 916, Examples 2–3, 7, pages 917–918, 924–926
Jefferson's method	Discussion pages 918–919, Examples 4, 7, pages 919–920, 924–926
Webster's method	Discussion page 921, Examples 5, 7, pages 921–922, 924–926
Adams' method	Discussion page 923 Examples 6–7, pages 923–926
Section 14.4	
Flaws of Apportionment Methods	
Alabama paradox	Discussion page 931 Example 1, pages 931–932
Population paradox	Discussion page 932 Example 2, pages 932–933
New-states paradox	Discussion pages 933–934 Example 3, page 934

CHAPTER 14   *Review Exercises*

14.1

1. *Electing the Club President* The Sailing Club of Lakeport is holding an election to choose the club president. The 42 votes were cast as follows: Comstock, 20 votes; Pendergast, 15 votes; and Hanzalik, 7 votes.

   a) Using the plurality method, which candidate is elected president?

   b) Did this candidate receive a majority of votes?

2. *Electing a Chairperson* Suffolk County Community College is holding an election to appoint a chairperson of the board of trustees. The 413 faculty members vote as follows: Michelle, 231 votes; Jeffrey, 155 votes; and Donald, 27 votes.

   a) Using the plurality method, who is elected?

   b) Did this candidate receive a majority of votes?

3. *Ranking Candidates* Ten voters are asked to rank four candidates. The 10 voters turn in the following ballots showing their preferences in order:

   B A D B A C B C D C
   A C C A C B A B A B
   C D A C D A C A B A
   D B B D B D D D C D

   Make a preference table for these ballots.

4. *Ranking Candidates* Seven voters are asked to rank three candidates. The seven voters turn in the following ballots showing their preferences in order:

   C C A C A B B
   A A C B C A A
   B B B A B C C

   Make a preference table for these ballots.

In Exercises 5–10, the members of the Student Council at Louisiana State University are planning to go out to dinner following an upcoming meeting. The restaurant choices are Chipotle Mexican Grill (C), Jimmy John's (J), Domino's Pizza (D), and Burger King (B). The members rank their choices according to the following preference table.

Number of Votes	5	3	1	2
First	C	D	D	B
Second	D	B	B	C
Third	B	J	C	J
Fourth	J	C	J	D

5. How many members voted?

6. Using the plurality method, which restaurant is chosen?

7. Using the Borda count method, which restaurant is chosen?

8. Using the plurality with elimination method, which restaurant is chosen?

9. Using the pairwise comparison method, which restaurant is chosen?

10. Which restaurant is chosen if the plurality with elimination method is used and the restaurant with the most *last*-place votes is eliminated at each step?

Sports Preferences In Exercises 11–16, the employees at Delphi Engineering must decide whether to play baseball (B), soccer (S), or volleyball (V) at their year-end picnic. The preference table follows.

Number of Votes	38	30	25	7	10
First	S	V	B	B	V
Second	B	S	V	S	B
Third	V	B	S	V	S

11. How many employees voted?

12. Determine the winner using the plurality method.

13. Determine the winner using the Borda count method.

14. Determine the winner using the plurality with elimination method.

15. Determine the winner using the pairwise comparison method.

16. Determine the winner if the plurality with elimination method is used and the candidate with the most *last*-place votes is eliminated at each step.

17. *Choosing a License Plate Style* Park Forest retirement community in Tennessee is purchasing a van to be used by the residents. The 372 residents are unable to agree on a style of license plate for the van and decide to hold an election in which the residents rank their choices from among the following: American Music (A), standard issue Tennessee license plate (T), Wildlife (W), and State Parks (S). The results of this election are given in the preference table below.

Number of Votes	161	134	65	12
First	A	A	T	W
Second	S	W	A	S
Third	T	S	W	T
Fourth	W	T	S	A

   a) Does any plate receive a majority of first-place votes? If so, which plate received a majority?

b) Using the plurality method, which plate is selected?

c) Using the Borda count method, which plate is selected?

d) Using the plurality with elimination method, which plate is selected?

e) Using the pairwise comparison method, which plate is selected?

18. *Accountants Convention* The National Association of Accountants held an election among its delegates to decide on the 2022 conference site. The 200 delegates ranked their choices among Chicago (C), Seattle (S), Dallas (D), and Las Vegas (L). The preference table giving the results of the election is shown below.

Number of Votes	60	55	45	30	10
First	S	L	D	C	S
Second	L	D	C	L	D
Third	C	C	L	D	L
Fourth	D	S	S	S	C

a) Does any city receive a majority of first-place votes? If so, which city received a majority?

b) Using the plurality method, which city is selected?

c) Using the Borda count method, which city is selected?

d) Using the plurality with elimination method, which city is selected?

e) Using the pairwise comparison method, which city is selected?

19. *Selecting a DVD Set* Park Street Library is planning to invest in a set of children's DVDs for the library, but the staff members differ regarding which set to buy. To settle the debate, the 16 staff members decide to rank the following choices: *Chronicles of Narnia* (C), *Family Classic* (F), *Anne of Green Gables* (A), and *Where on Earth Is Carmen Sandiego?* (W). Aware that sentiments seem to be evenly divided, the staff members agree to use the plurality with elimination method, with the Borda count method used in case of a tie. The results of this election are given in the following table.

Number of Votes	6	4	3	1	1	1
First	W	C	C	F	F	W
Second	A	A	W	W	C	F
Third	C	W	F	A	W	C
Fourth	F	F	A	C	A	A

a) Which DVD set is selected if the plurality with elimination method is used?

b) According to the rules described above, which DVD set is chosen?

c) If the staff members had agreed to break a tie using the pairwise comparison method, which DVD set is chosen?

14.2

Hiring a New Paralegal In Exercises 20 and 21, a law firm is hiring one new paralegal from among four candidates, A, B, C, and D. The search committee decides to use the Borda count method to determine the winner. The preference table follows.

Number of Votes	10	7	2
First	A	C	B
Second	B	B	D
Third	C	D	C
Fourth	D	A	A

20. Is the majority criterion satisfied?

21. Is the head-to-head criterion satisfied?

22. *Plurality with Elimination* Consider the following preference table.

Number of Votes	12	16	8	14
First	B	C	B	A
Second	A	B	C	C
Third	C	A	A	B

a) Who wins the election if the plurality with elimination method is used?

b) Assume that in a second election the eight voters who voted for B, C, A, in that order, all change their preference to C, B, A, in that order. If the plurality with elimination method is used, is the monotonicity criterion satisfied?

c) Using the preference table from part (a), assume that B drops out. Does the plurality with elimination method satisfy the irrelevant alternatives criterion?

23. *A Taste Test* In a taste test, 114 people are asked to taste and rank four different brands of spaghetti sauce. The choices are Ragu (R), Prego (P), Newman's Own (N), and Barilla (B). The preference table is shown on the next page.

Number of Votes	34	24	23	21	12
First	P	R	N	B	B
Second	R	B	B	N	N
Third	N	N	R	P	R
Fourth	B	P	P	R	P

a) Which brand is favored over all others in a head-to-head comparison?

b) Which brand wins if the plurality method is used?

c) Which brand wins if the Borda count method is used?

d) Which brand wins if the plurality with elimination method is used?

e) Which brand wins if the pairwise comparison method is used?

f) Which voting method(s) in parts (b) through (e) violate the head-to-head criterion?

24. **Selecting a Band** The Southwestern High School Class of 1986 is organizing its 35-year class reunion and is trying to pick out a band to play at the reunion. The choices are REO Speedwagon (R), Boston (B), Journey (J), and Fleetwood Mac (F). The class members have ranked their choices as indicated in the following preference table.

Number of Votes	34	25	15	9	4
First	B	F	R	J	J
Second	F	J	J	R	R
Third	R	R	F	B	F
Fourth	J	B	B	F	B

a) Is one band favored over all others in a head-to-head comparison?

b) Which band is chosen if the plurality method is used?

c) Which band is chosen if the Borda count method is used?

d) Which band is chosen if the plurality with elimination method is used?

e) Which band is chosen if the pairwise comparison method is used?

f) Which voting methods in parts (a) through (e) violate the head-to-head criterion?

25. **Violating the Majority Criterion** Which voting method(s) violate the majority criterion in the following election data?

Number of Votes	36	20	8	6
First	A	B	C	D
Second	B	C	B	B
Third	D	A	D	A
Fourth	C	D	A	C

26. **Violating the Monotonicity Criterion** Using the following tables, determine which voting method(s) violate the monotonicity criterion.

First Election

Number of Votes	28	24	20	10
First	A	B	C	C
Second	C	A	B	A
Third	B	C	A	B

Second Election

Number of Votes	36	24	20	2
First	A	B	C	C
Second	C	A	B	A
Third	B	C	A	B

27. **Violating the Irrelevant Alternatives Criterion** Using the following table, determine which voting method(s) violates the irrelevant alternatives criterion. Assume that candidate D drops out prior to the conclusion of the election.

Number of Votes	24	16	16	16	8	4	5
First	B	B	E	C	A	D	D
Second	A	A	A	D	E	B	E
Third	D	E	C	E	D	A	A
Fourth	E	C	D	B	B	E	B
Fifth	C	D	B	A	C	C	C

14.3, 14.4

Postal Service Apportionment In Exercises 28–32, a post office in the city of Riverside has three regions in which to distribute the mail and 10 new mail trucks. The trucks are to be apportioned based on the number of houses in each region, as shown in the table below.

Region	A	B	C	Total
Number of houses	2592	1428	1980	6000

28. Determine each region's apportionment using Hamilton's method.

29. Determine each region's apportionment using Jefferson's method.

30. Determine each region's apportionment using Adams' method.

31. Determine each region's apportionment using Webster's method.

32. Suppose that the post office purchases one additional truck. Using Hamilton's method, does the Alabama paradox occur if the number of trucks is increased from 10 to 11?

Apportioning Biology Sections *In Exercises 33–37, Miami Dade College plans to offer 23 sections of three different biology courses: Anatomy and Physiology (A), Introduction to Biology (B), and Biochemistry (C). The sections will be apportioned based on preregistration as shown in the table below.*

Course	A	B	C	Total
Number of students	311	219	160	690

33. Determine the apportionment for each course using Hamilton's method.

34. Determine the apportionment for each course using Jefferson's method.

35. Determine the apportionment for each course using Adams' method.

36. Determine the apportionment for each course using Webster's method.

37. Suppose that the table above and to the right shows the final registration for each course. If the 23 sections are reapportioned using Hamilton's method, does the population paradox occur?

Course	A	B	C	Total
Number of students	317	219	162	698

Apportioning Seats *In Exercises 38–42, a country has two states, A and B, and 50 seats in the legislature. The population of each state is shown in the table below.*

State	A	B	Total
Population	4420	45,580	50,000

38. Determine each state's apportionment using Hamilton's method.

39. Determine each state's apportionment using Jefferson's method.

40. Determine each state's apportionment using Adams' method.

41. Determine each state's apportionment using Webster's method.

42. Suppose that a third state, C, with the population shown in the table below, is added along with five new seats. Does the new-states paradox occur using Hamilton's method?

State	A	B	C	Total
Population	4420	45,580	5400	55,400

CHAPTER 14   *Test*

Lunch Choices *In Exercises 1–6, the employees of an advertising firm are planning to have lunch delivered to a meeting. Their choices are deli sandwiches (D), pizza (P), and burgers (B). The preference table follows.*

Number of Votes	4	3	3	2
First	D	P	B	P
Second	P	B	D	D
Third	B	D	P	B

1. How many members voted?

2. Does any choice have a majority of votes?

3. Determine the winner using the plurality method.

4. Determine the winner using the Borda count method.

5. Determine the winner using the plurality with elimination method.

6. Determine the winner using the pairwise comparison method.

Favorite Animal *In Exercises 7–11, the children of Happy Faces Preschool are voting on their favorite classroom animal. The choices are hamster (H), iguana (I), lemming (L), and salamander (S). The results of the election are given in the following preference table.*

Number of Votes	43	30	29	26	14
First	S	L	I	H	H
Second	L	H	H	I	I
Third	I	I	L	S	L
Fourth	H	S	S	L	S

7. How many children voted?

8. Which animal wins this election if the plurality method is used?

9. Which animal wins this election if the Borda count method is used?

10. Which animal wins this election if the plurality with elimination method is used?

11. Which animal wins this election if the pairwise comparison method is used?

12. **Violating the Head-to-Head Criterion** Which voting method(s)—plurality, Borda count, plurality with elimination, or pairwise comparison—violate the head-to-head criterion using the following election data?

Number of Votes	86	60	58	52	28
First	W	Y	Z	X	X
Second	Y	X	X	Z	Z
Third	Z	Z	Y	W	Y
Fourth	X	W	W	Y	W

13. **Voting for a Logo Design** The park rangers in Yosemite National Park are holding a contest to choose a new logo design. The four logo choices for their stationery are El Capitan (E), a sequoia tree (S), a mule deer (M), and a waterfall (W). The 35 rangers rank their choices according to the following preference table.

Number of Votes	18	10	4	3
First	E	M	W	S
Second	M	W	M	M
Third	S	E	S	E
Fourth	W	S	E	W

Using the data provided, does the Borda count method violate the majority criterion?

Apportioning Legislative Seats In Exercises 14–20, a country has three states and 30 seats in the legislature. The population of each state is shown above and to the right.

State	A	B	C	Total
Population	6933	9533	16,534	33,000

14. Determine each state's apportionment using Hamilton's method.

15. Determine each state's apportionment using Jefferson's method.

16. Determine each state's apportionment using Adams' method.

17. Determine each state's apportionment using Webster's method.

18. If the number of seats in the legislature increases to 31, does the Alabama paradox occur using Hamilton's method?

19. Suppose that in 10 years the states have the following population and 30 seats are apportioned. If the seats are reapportioned, does the population paradox occur using Hamilton's method?

State	A	B	C	Total
Population	7072	9724	17,030	33,826

20. Suppose that a fourth state with the population shown in the table below is added, with five additional seats for a total of 35 seats. Does the new-states paradox occur using Hamilton's method?

State	A	B	C	D	Total
Population	6933	9533	16,534	5100	38,100

Answers

Chapter 1

RECREATIONAL MATH, PAGE 24

The farmer crosses with the goat first. Then he can take the herbs across the river. But he must return the goat temporarily so that the goat doesn't eat the herbs. He then takes the wolf across the river. Then he returns for the goat. He will make seven crossings—four forward and three back.

The answer to the St. Ives riddle is 1.

SECTION 1.1, PAGE 6

1. Natural   3. Counterexample   5. Inductive

7. Deductive   9. $5 \times 3 = 15$

11. 1 5 10 10 5 1   13. ⟨○⟩   15. ⬠

17. 9, 11, 13   19. 5, −5, 5   21. $\frac{1}{5}, \frac{1}{6}, \frac{1}{7}$ 23. 21, 28, 36

25. 34, 55, 89   27. Y

29. a) 36, 49, 64   b) square 6, 7, 8, 9, and 10
   c) No, 72 is between 8^2 and 9^2, so it is not a square number.

31. Blue: 1, 5, 7, 10, 12; Purple: 2, 4, 6, 9, 11; Yellow: 3, 8

33. a) $3700
   b) We are using specific cases to make a prediction.

35.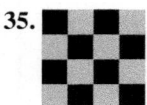

37. a) You should obtain the original number.
   b) You should obtain the original number.
   c) The result is the original number.
   d) $n, 3n, 3n + 6, \frac{3n + 6}{3} = n + 2, n + 2 - 2 = n$

39. a) 5
   b) You should obtain the number 5.
   c) The result is always the number 5.
   d) $n, n + 1, \frac{n + (n + 1) + 9}{2} = \frac{2n + 10}{2} = n + 5,$
   $n + 5 - n = 5$

41. $3 \times 5 = 15$, which is not divisible by 2.

43. $(3 + 2)/2 = 5/2$, which is not an even number.

45. $1 - 2 = -1$, which is not a counting number.

47. a) The sum of the measures of the interior angles should be 180°.
   b) Yes, the sum of the measures of the interior angles should be 180°.
   c) The sum of the measures of the interior angles of a triangle is 180°.

49. Inductive reasoning; a general conclusion is obtained from observation of specific cases.

51. 129, the numbers in positions are determined as follows:
   a b
   c $a + b + c$

53. (c)

SECTION 1.2, PAGE 15

1. Estimation   3. 460   5. 50,000   7. 18

9. 17   11. 1800   13. $30   15. $600   17. $53

19. 400 lb   21. $3.90   23. $1300   25. $120   27. $30

29. 90 mi   31. a) 175   b) 70   c) 140

33. a) 5 million   b) 98 million   c) 65 million   d) 280 million

35. a) 85%   b) 15%   c) 59,500,000 acres
   d) No, since we are not given the area of each state

37. 20   39. 120 bananas   41. 150°   43. 10%

45. 9 square units   47. 160 feet   49. Answers will vary.

51. Answers will vary.

53. Answers will vary.   55. Answers will vary.

57. a) Answers will vary.   b) Answers will vary.
   c) Answers will vary.

58. $1.40

59. There are 118 ridges around the edge.

60. a) Answers will vary.   b) 11.6 days

SECTION 1.3, PAGE 29

1. 51 mi   3. 38.4 ft   5. $493.81

7. a) $124.80   b) $358.80

9. 43 bus rides

11. $90   13. $46,470.60   15. a) 10,000   b) 1 in 10,000

17. ≈18.7 mpg   19. $82.08   21. Snow

23. a) $8760   b) $18,400

25. a) 4106.25 gal   b) $45.99

27. a) ≈35.61 gal   b) $106.83   c) ≈4,985,400,000 gal

29. $1039.36   31. $420

33. a) Water/milk: 3 cups; salt: $\frac{3}{8}$ tsp; Cream of Wheat: 9 tbsp
      (or $\frac{9}{16}$ cup)
   b) Water/milk: $2\frac{7}{8}$ cups; salt: $\frac{3}{8}$ tsp; Cream of Wheat: $\frac{5}{8}$ cup
      (or 10 tbsp)
   c) Water/milk: $2\frac{3}{4}$ cups; salt: $\frac{3}{8}$ tsp; Cream of Wheat: $\frac{9}{16}$ cup
      (or 9 tbsp)
   d) Differences exist in water/milk because the amount for 4 servings is not twice that for 2 servings. Differences also exist in Cream of Wheat because $\frac{1}{2}$ cup is not twice 3 tbsp.

35. a) 1 box of 20 and 1 box of 12   b) $420

37. 144 square inches   39. The area is 4 times as large.

41. The volume is 8 times as large.   43. 10¢   45. 3

47. a) refresh   b) workout

49.
```
    (4)
  (3) (2)
 (5)(1)(6)
```

51.
8	6	16
18	10	2
4	14	12

53. The sum of the four corners is 4 times the number in the center.

55. Multiply the center number by 9.   57. 6 ways

59.
```
      7
   3  1  4
   5  8  6
      2
```
Other answers are possible, but 1 and 8 must appear in the center.

61.
1	2	3	4	5
2	3	4	5	1
3	4	5	1	2
4	5	1	2	3
5	1	2	3	4

Other answers are possible.

63. Mark plays the drums.

65. 714 square units

67. Thomas would have opened the box labeled grapes and cherries. Because all the boxes are labeled incorrectly, whichever fruit he pulls from the grapes and cherries box will be the only fruit in that box. If he pulls a grape from the box, the box must be labeled grapes; if he pulls a cherry from the box, the box must be labeled cherries. That leaves two boxes whose original labels were incorrect. Because all labels must be changed, there will be only one way for Thomas to assign the two remaining labels.

REVIEW EXERCISES, PAGE 35

1. 23, 28, 33   **2.** 16, 13, 10   **3.** 64, −128, 256

4. 25, 32, 40   **5.** 10, 4, −3   **6.** $\frac{3}{8}, \frac{3}{16}, \frac{3}{32}$

7. ⊘ ⊟ ⊘   **8.** △ ○ ☐   **9.** (c)

10. a) The final number is twice the original number.
   b) The final number is twice the original number.
   c) The final number is twice the original number.
   d) $n, 10n, 10n + 5, \dfrac{10n + 5}{5} = 2n + 1,$
     $2n + 1 - 1 = 2n$

11. This process will always result in an answer of 3.

12. $1^2 + 2^2 = 5$ **13.** 800,000,000   **14.** 2000   **15.** 400

16. Answers will vary.   **17.** $200   **18.** $1600   **19.** 3 mph

20. $14.00   **21.** 2 mi   **22.** 40°F   **23.** 10°F

24. 13 square units   **25.** Length 22 ft; height 8 ft

26. $30   **27.** $2.97   **28.** Silvan is less expensive by $20.   **29.** $4

30. a) 288 lb   **b)** 12,500 ft^2 **31.** $661   **32.** 7.8 mg

33. $1078   **34.** 6 hr 45 min   **35.** July 26, 11:00 A.M.

36. a) 104 km/hr   **b)** 56.25 mi/hr   **37.** 201

38.

21	7	8	18
10	16	15	13
14	12	11	17
9	19	20	6

39.

23	25	15
13	21	29
27	17	19

40. 59 min 59 sec   **41.** 6   **42.**

$ 25	Room
$ 3	Friends
$ 2	Clerk
$ 30	

43. 52

44. Yes; 3 quarters and 4 dimes, or 1 half dollar, 1 quarter and 4 dimes, or 1 quarter and 9 dimes. Other answers are possible.

45. 216 cm^3

46. Place six coins in each pan with one coin off to the side. If it balances, the heavier coin is the one on the side. If the pan does not balance, take the six coins on the heavier side and split them into two groups of three. Select the three heavier coins and weigh two coins. If the pan balances, it is the third coin. If the pan does not balance, you can identify the heavier coin.

47. 125,250   **48.** 16 blue   **49.** 90

50. The fifth figure will be an octagon with sides of equal length. Inside the octagon will be a seven-sided figure with each side of equal length. The figure will have one antenna.

51. 61

52. Some possible answers are shown. Others are possible.

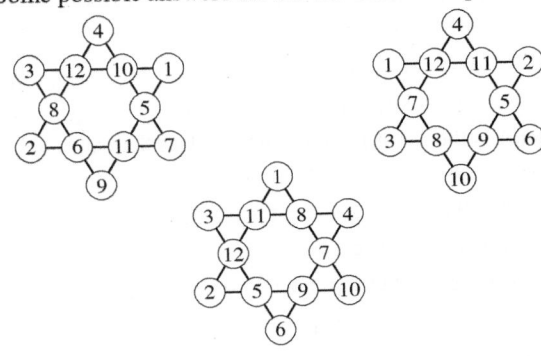

53. a) 2   **b)** 6   **c)** 24   **d)** 120
   e) $n(n-1)(n-2)\ldots 1$, (or n!), where n = the number of people in the line

CHAPTER TEST, PAGE 38

1. 26, 32, 38   **2.** $\frac{1}{5}, \frac{1}{6}, \frac{1}{7}$

3. a) The result is the original number plus 1.
   b) The result is the original number plus 1.
   c) The result will always be the original number plus 1.
   d) $n, 5n, 5n + 10, \frac{5n + 10}{5} = n + 2,$
     $n + 2 - 1 = n + 1$

4. 50,000   **5.** 2,000,000   **6.** 9 square units

7. a) ≈ 23.03 **b)** He is in the at-risk range.

8. a) 325,000 visitors   **b)** 100,000 visitors

9. 77 miles   **10.** 32 cans   **11.** $7\frac{1}{2}$ min

12. 7.5 miles   **13.** $49.00

14.

40	15	20
5	25	45
30	35	10

15. Less time if she had driven at 45 mph for the entire trip

16. $\frac{1}{2}$ tablespoon   **17.** 48 square meters   **18.** 243 jelly beans

19. a) $11.97   **b)** $11.81   **c)** Save 16 cents by using the 25% off coupon.

20. 24

Chapter 2

SECTION 2.1, PAGE 47

1. Set   **3.** Description, roster form, set-builder notation

5. Infinite   **7.** Equal   **9.** Empty or null

11. Not well defined   **13.** Well defined   **15.** Well defined

17. Infinite   **19.** Infinite   **21.** Infinite   **23.** {Hawaii}

25. $\{11, 12, 13, 14, \ldots, 177\}$ **27.** $B = \{2, 4, 6, 8, \ldots\}$

29. { } or ∅   **31.** $E = \{14, 15, 16, 17, \ldots, 84\}$

33. {Google Play Games, Pokémon Go}

35. {Solitaire by MobilityWare, Candy Crush Soda Saga, ApplsLoading}

37. {2015, 2016, 2017, 2018}   **39.** { } or ∅

41. $B = \{x \mid x \in N \text{ and } 6 < x < 15\}$ or
$B = \{x \mid x \in N \text{ and } 7 \le x \le 14\}$
43. $C = \{x \mid x \in N \text{ and } x \text{ is a multiple of } 3\}$
45. $E = \{x \mid x \in N \text{ and } x \text{ is odd}\}$
47. $C = \{x \mid x \text{ is February}\}$
49. Set A is the set of natural numbers less than or equal to 7.
51. Set V is the set of vowels in the English alphabet.
53. Set T is the set of species of trees.
55. Set S is the set of seasons.
57. {Facebook, Instagram, Facebook Messenger}
59. {Twitter, Pinterest, Snapchat}
61. {2016, 2017, 2018, 2019}
63. {2013, 2014, 2015}
65. False; $\{e\}$ is a set, and not an element of the set.
67. False; h is not an element of the set.
69. False; 3 is an element of the set.
71. True **73.** 4 **75.** 0 **77.** Both **79.** Neither **81.** Equivalent
83. a) Set A is the set of natural numbers greater than 2. Set B is the set of all numbers greater than 2.
b) Set A contains only natural numbers. Set B contains other types of numbers, including fractions and decimal numbers.
c) $A = \{3, 4, 5, 6, \ldots\}$
d) No; because there are an infinite number of elements between any two elements in set B, we cannot write set B in roster form.
85. Cardinal **87.** Ordinal **89.** Answers will vary.
91. Answers will vary.

SECTION 2.2, PAGE 54

1. Subset **3.** 2^n **5.** True
7. False; apple is not in the second set. **9.** True
11. False; no set is a proper subset of itself. **13.** True
15. False; {cookie} is a set, not an element.
17. True **19.** True
21. False; the set $\{\varnothing\}$ contains the element $\varnothing$.
23. False; the set $\{0\}$ contains the element 0.
25. False; 0 is a number and $\{\ \}$ is a set.
27. $B \subseteq A$, $B \subset A$ **29.** $B \subseteq A$, $B \subset A$ **31.** none
33. $A = B$, $A \subseteq B$, $B \subseteq A$
35. $\{\ \}$ **37.** $\{\ \}$, {cow}, {horse}, {cow, horse}
39. a) $\{\ \}$, $\{a\}$, $\{b\}$, $\{c\}$, $\{d\}$, $\{a, b\}$, $\{a, c\}$, $\{a, d\}$, $\{b, c\}$, $\{b, d\}$, $\{c, d\}$, $\{a, b, c\}$, $\{a, b, d\}$, $\{a, c, d\}$, $\{b, c, d\}$, $\{a, b, c, d\}$
b) $\{a, b, c, d\}$
41. False **43.** True **45.** True **47.** True **49.** True
51. True **53.** 2^4, or 16 **55.** 2^7, or 128 **57.** $E = F$
59. a) Yes **b)** No **c)** Yes **61.** 1 **62.** Yes **63.** Yes **64.** No

SECTION 2.3, PAGE 63

1. Complement **3.** Intersection **5.** Cartesian **7.** $m \times n$

9.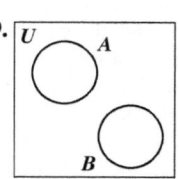

11.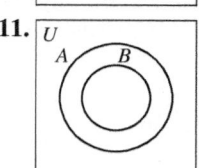

13.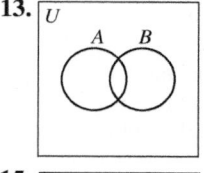

15.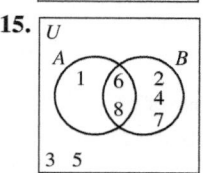

17.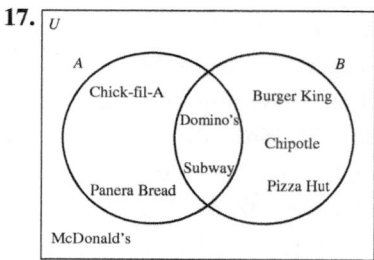

19.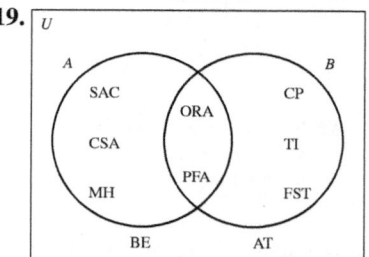

21. The set of retail stores in the United States that do not sell children's clothing
23. The set of cities in the United States that do not have a professional sports team
25. The set of cities in the United States that have a professional sports team or a symphony
27. The set of cities in the United States that have a professional sports team and do not have a symphony
29. The set of furniture stores that sell mattresses or leather furniture
31. The set of furniture stores that do not sell outdoor furniture and sell leather furniture

33. The set of furniture stores that sell mattresses or outdoor furniture or leather furniture

35. $\{a, b, e, g, h\}$ **37.** $\{b, e\}$

39. $\{a, b, c, e, f, g, h\}$ **41.** $\{d, i, j\}$

43. $\{a, g, h\}$ **45.** $\{1, 2, 7, 8, 9\}$

47. $\{1, 2, 3, 4, 5, 6, 7, 8, 9, 10, 11\}$

49. $\{2, 3, 4, 5, 6, 9, 10, 11\}$

51. $\{3, 5, 6\}$ **53.** $\{4, 10, 11\}$ **55.** $\{1, 2, 3, 4, 5, 6, 7\}$

57. $\{8\}$ **59.** $\{ \}$ **61.** $\{8\}$ **63.** $\{2, 3, 5, 6, 8\}$

65. $\{a, b, c, d, f, g, i, j\}$ **67.** $\{a, b, e, h, i, j, k\}$

69. $\{a, b, c, d, f, g, i, j\}$ **71.** $\{a, b, c, d, e, f, g, h, i, j, k\}$, or U

73. $\{c, d, e, g, h, k\}$

75. $\{(1, a), (1, b), (2, a), (2, b), (3, a), (3, b)\}$

77. No. The ordered pairs are not the same. For example, $(1, a) \neq (a, 1)$.

79. 6 **81.** $\{ \}$ **83.** $\{7, 9\}$ **85.** $\{1, 3, 5, 7, 9\}$, or A

87. $\{2,4\}$ **89.** $\{1,2,3,4,5\}$, or C

91. A set and its complement will always be disjoint. For example, if $U = \{1, 2, 3\}$ and $A = \{1, 2\}$, then $A' = \{3\}$, and $A \cap A' = \{ \}$.

93. 49

95. a) $8 = 4 + 6 - 2$
  b) Answers will vary.
  c) Answers will vary.

97. $\{1, 2, 3, 4, \ldots\}$, or A

99. $\{2, 4, 6, 8, \ldots\}$, or C

101. $\{2, 4, 6, 8, \ldots\}$, or C

103. $\{0, 1, 2, 3, 4, \ldots\}$, or U

105. $\{ \}$ **107.** A **109.** U **111.** $B \subseteq A$

113. A and B are disjoint sets.

SECTION 2.4, PAGE 72

1. 8 **3. a)** $A' \cap B'$ **b)** $A' \cup B'$ **5.** 8

7.
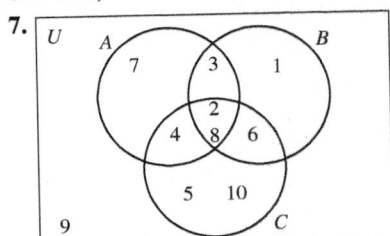

9.
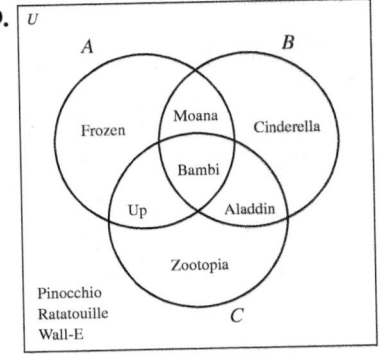

11.
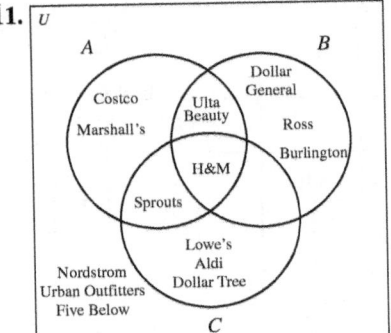

13.
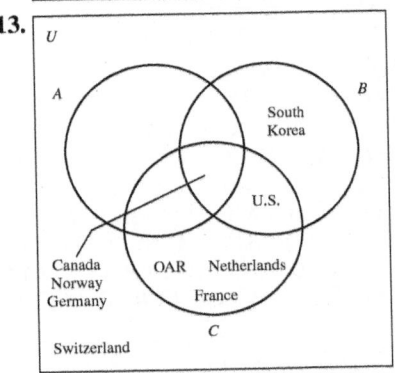

15. IV **17.** V **19.** VIII **21.** VI **23.** III **25.** III

27. V **29.** II **31.** VII **33.** I **35.** VIII **37.** VI

39. $\{1, 3, 4, 5, 9, 10\}$ **41.** $\{4, 5, 6, 8, 9, 11\}$

43. $\{3, 4, 5\}$ **45.** $\{1, 2, 3, 7, 9, 10, 11, 12, 13, 14\}$

47. $\{2, 7, 12, 13, 14\}$ **49.** $\{2, 3, 4, 5, 6, 7, 8, 11, 12, 13, 14\}$

51. Yes **53.** No **55.** No **57.** No **59.** Yes **61.** Yes **63.** Yes

65. $(A \cup B)'$ **67.** $(A \cup B) \cap C'$

69. a) Both equal $\{1, 2, 4, 5, 7, 8\}$
  b) Answers will vary.

71.
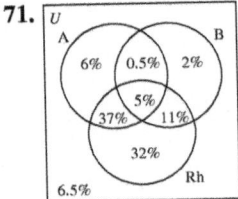

73. a)
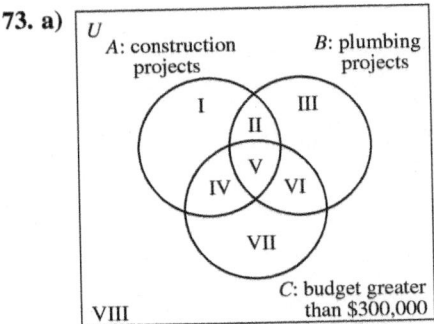

b) V; $A \cap B \cap C$ **c)** VI; $A' \cap B \cap C$
d) I; $A \cap B' \cap C'$

75. a)

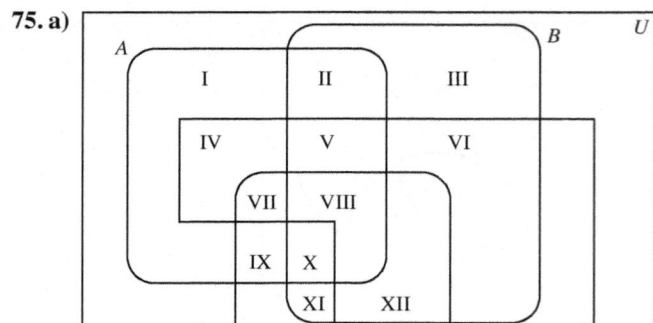

b)

Region	Set	Region	Set
I	$A \cap B' \cap C' \cap D'$	IX	$A \cap B' \cap C \cap D'$
II	$A \cap B \cap C' \cap D'$	X	$A \cap B \cap C \cap D'$
III	$A' \cap B \cap C' \cap D'$	XI	$A' \cap B \cap C \cap D'$
IV	$A \cap B' \cap C' \cap D$	XII	$A' \cap B \cap C \cap D$
V	$A \cap B \cap C' \cap D$	XIII	$A' \cap B' \cap C \cap D'$
VI	$A' \cap B \cap C' \cap D$	XIV	$A' \cap B' \cap C \cap D$
VII	$A \cap B' \cap C \cap D$	XV	$A' \cap B' \cap C' \cap D$
VIII	$A \cap B \cap C \cap D$	XVI	$A' \cap B' \cap C' \cap D'$

SECTION 2.5, PAGE 79

1.

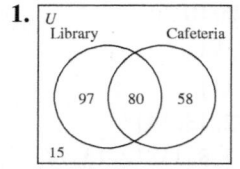

a) 97 **b)** 58 **c)** 15

3.

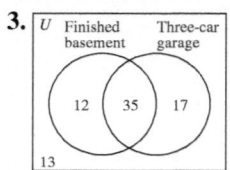

a) 12 **b)** 17 **c)** 64

5.

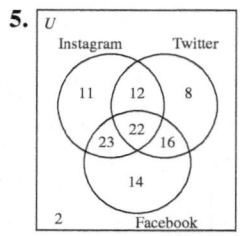

a) 11 **b)** 12 **c)** 92 **d)** 31 **e)** 51

7.

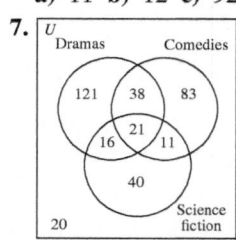

a) 20 **b)** 121 **c)** 244 **d)** 65 **e)** 290

9.

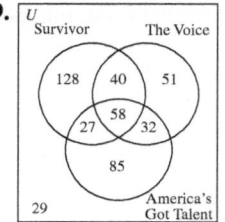

a) 450 **b)** 40 **c)** 85 **d)** 99 **e)** 421

11.

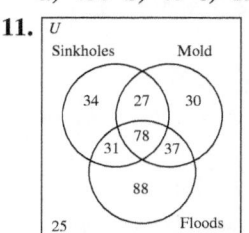

a) 67 **b)** 237 **c)** 37 **d)** 25

13. In a Venn diagram, regions II, IV, and V contain a total of 37 cars driven by women. This total is greater than the 35 cars driven by the women, as given in the exercise.

15.

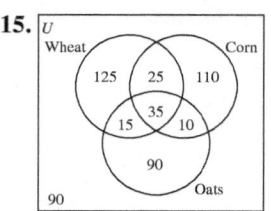

a) 410 **b)** 35 **c)** 90 **d)** 50

17. a) 10 **b)** 10 **c)** 6

SECTION 2.6, PAGE 86

1. Infinite

3. $\{5, 6, 7, 8, 9, \ldots, n + 4, \ldots\}$
$\downarrow \downarrow \downarrow \downarrow \downarrow \qquad \downarrow$
$\{6, 7, 8, 9, 10, \ldots, n + 5, \ldots\}$

5. $\{6, 8, 10, 12, 14, \ldots, 2n + 4, \ldots\}$
$\downarrow \downarrow \downarrow \downarrow \downarrow \qquad \downarrow$
$\{8, 10, 12, 14, 16, \ldots, 2n + 6, \ldots\}$

7. $\{5, 7, 9, 11, 13, \ldots, 2n + 3, \ldots\}$
$\downarrow \downarrow \downarrow \downarrow \downarrow \qquad \downarrow$
$\{7, 9, 11, 13, 15, \ldots, 2n + 5, \ldots\}$

9. $\left\{\dfrac{1}{2}, \dfrac{1}{4}, \dfrac{1}{6}, \dfrac{1}{8}, \dfrac{1}{10}, \ldots, \dfrac{1}{2n}, \ldots\right\}$
$\downarrow \downarrow \downarrow \downarrow \downarrow \qquad \downarrow$
$\left\{\dfrac{1}{4}, \dfrac{1}{6}, \dfrac{1}{8}, \dfrac{1}{10}, \dfrac{1}{12}, \ldots, \dfrac{1}{2n + 2}, \ldots\right\}$

11. $\left\{\dfrac{4}{11}, \dfrac{5}{11}, \dfrac{6}{11}, \dfrac{7}{11}, \dfrac{8}{11}, \ldots, \dfrac{n + 3}{11}, \ldots\right\}$
$\downarrow \downarrow \downarrow \downarrow \downarrow \qquad \downarrow$
$\left\{\dfrac{5}{11}, \dfrac{6}{11}, \dfrac{7}{11}, \dfrac{8}{11}, \dfrac{9}{11}, \ldots, \dfrac{n + 4}{11}, \ldots\right\}$

13. $\{1, 2, 3, 4, 5, \ldots, n, \ldots\}$
$\{3, 6, 9, 12, 15, \ldots, 3n, \ldots\}$

15. $\{1, 2, 3, 4, 5, \ldots, n, \ldots\}$
$\{4, 6, 8, 10, 12, \ldots, 2n + 2, \ldots\}$

17. $\{1, 2, 3, 4, 5, \ldots, n, \ldots\}$
$\{2, 5, 8, 11, 14, \ldots, 3n - 1, \ldots\}$

19. $\{1, 2, 3, 4, 5, \ldots, n, \ldots\}$
$\left\{\dfrac{1}{3}, \dfrac{1}{6}, \dfrac{1}{9}, \dfrac{1}{12}, \dfrac{1}{15}, \ldots, \dfrac{1}{3n}, \ldots\right\}$

21. $\{1, 2, 3, 4, 5, \ldots, n, \ldots\}$
$\left\{\dfrac{1}{3}, \dfrac{1}{4}, \dfrac{1}{5}, \dfrac{1}{6}, \dfrac{1}{7}, \ldots, \dfrac{1}{n + 2}, \ldots\right\}$

23. $\{1, 2, 3, 4, 5, \ldots, n, \ldots\}$
$\{1, 4, 9, 16, 25, \ldots, n^2, \ldots\}$

25. $\{1, 2, 3, 4, 5, \ldots, n, \ldots\}$
$\{3, 9, 27, 81, 243, \ldots, 3^n, \ldots\}$

27. = **28.** = **29.** = **30.** = **31.** =

32. a) Answers will vary. **b)** No

REVIEW EXERCISES, PAGE 88

1. True

2. False; the word *best* makes the statement not well-defined.

3. True **4.** False; no set is a proper subset of itself.

5. False; the elements 6, 12, 18, 24, . . . are members of both sets.

6. True

7. False; the sets do not contain exactly the same elements.

8. True **9.** True **10.** True **11.** True **12.** True **13.** True

14. True **15.** $A = \{7, 9, 11, 13, 15\}$

16. {Colorado, Nebraska, Missouri, Oklahoma}

17. $C = \{1, 2, 3, 4, \ldots, 174\}$

18. $D = \{9, 10, 11, 12, \ldots, 80\}$

19. $A = \{x \mid x \in N \text{ and } 50 < x < 150\}$ or
$A = \{x \mid x \in N \text{ and } 49 \leq x \leq 151\}$

20. $B = \{x \mid x \in N \text{ and } x > 42\}$

21. $C = \{x \mid x \in N \text{ and } x < 7\}$

22. $D = \{x \mid x \in N \text{ and } 27 \leq x \leq 51\}$

23. Set A is the set of capital letters in the English alphabet from E through M, inclusive.

24. Set B is the set of U.S. coins with a value of less than a dollar.

25. Set C is the set of the first three lowercase letters in the English alphabet.

26. Set D is the set of numbers greater than or equal to 3 and less than 9.

27. $\{3, 7\}$ **28.** $\{1, 2, 3, 4, 5, 6, 7, 8\}$ **29.** $\{9, 10\}$

30. $\{1, 2, 4, 6, 7, 8, 10\}$ **31.** $\{1, 5\}$ **32.** $\{1, 7\}$

33. $\{(1, 1), (1, 7), (1, 10), (3, 1), (3, 7), (3, 10), (5, 1), (5, 7), (5, 10), (7, 1), (7, 7), (7, 10)\}$

34. $\{(3, 1), (3, 3), (3, 5), (3, 7), (7, 1), (7, 3), (7, 5), (7, 7), (9, 1), (9, 3), (9, 5), (9, 7), (10, 1), (10, 3), (10, 5), (10, 7)\}$

35. 16 **36.** 15

37.

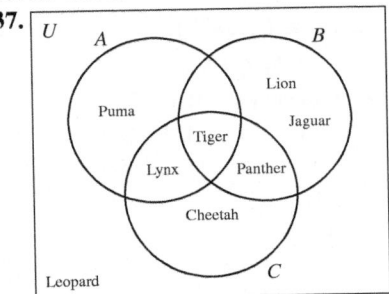

38. $\{b, c, d, e, f, h, k, l\}$ **39.** $\{e, k\}$

40. $\{a, b, c, d, e, f, g, h, k, l\}$ **41.** $\{f\}$ **42.** $\{c, e, f\}$

43. $\{d, f, l\}$ **44.** True **45.** True **46.** II **47.** III **48.** I

49. IV **50.** IV **51.** II **52.** II **53.** \$450

54.

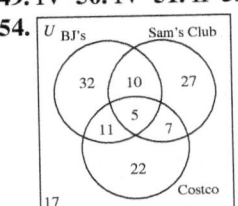

a) 131 **b)** 32 **c)** 10 **d)** 65

55.

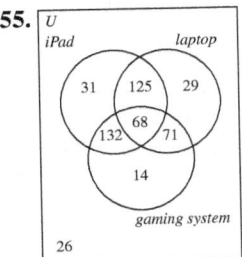

a) 31 **b)** 74 **c)** 71 **d)** 185 **e)** 328

56. $\{2, 4, 6, 8, 10, \ldots, 2n, \ldots\}$
$\{4, 6, 8, 10, 12, \ldots, 2n + 2, \ldots\}$

57. $\{3, 5, 7, 9, 11, \ldots, 2n + 1, \ldots\}$
$\{5, 7, 9, 11, 13, \ldots, 2n + 3, \ldots\}$

58. $\{1, 2, 3, 4, 5, \ldots, n, \ldots\}$
$\{5, 8, 11, 14, 17, \ldots, 3n + 2, \ldots\}$

59. $\{1, 2, 3, 4, 5, \ldots, \quad n, \ldots\}$

$\quad \downarrow \downarrow \downarrow \; \downarrow \; \downarrow \qquad\quad \downarrow$

$\quad \{4, 9, 14, 19, 24, \ldots, 5n - 1, \ldots\}$

CHAPTER TEST, PAGE 90

1. True

2. False; the sets do not contain exactly the same elements.

3. True

4. False; the second set does not contain the element 7.

5. False; the set has 2^4, or 16, subsets. **6.** True

7. False; for any set A, $A \cup A' = U$, not { }.

8. True **9.** $A = \{1, 2, 3, 4, 5, 6, 7, 8, 9, 10, 11\}$

10. Set A is the set of natural numbers less than 12.

11. $\{6, 8\}$ **12.** $\{2, 4, 6, 8, 12\}$ **13.** $\{6, 8\}$ **14.** 2 **15.** $\{2, 4\}$

16. $\{(2, 2), (2, 10), (2, 14), (4, 2), (4, 10), (4, 14), (6, 2), (6, 10),$
$(6, 14), (8, 2), (8, 10), (8, 14)\}$

17. 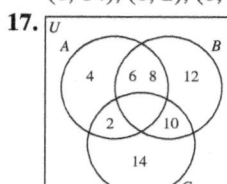   **18.** Equal

19.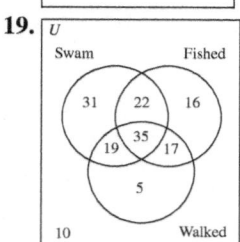

a) 58 b) 10 c) 145 d) 22 e) 69 f) 16

20. $\{7, 8, 9, 10, 11, \ldots, n + 6, \ldots\}$

$\quad \downarrow \downarrow \; \downarrow \; \downarrow \; \downarrow \qquad\quad \downarrow$

$\quad \{8, 9, 10, 11, 22, \ldots, n + 7, \ldots\}$

Chapter 3

Answer to Recreational Math on page 95

7	1	9	4	8	2	5	6	3
5	4	6	7	3	9	8	1	2
2	3	8	6	5	1	4	9	7
8	2	4	3	9	5	1	7	6
6	7	5	1	2	4	3	8	9
1	9	3	8	6	7	2	4	5
9	6	1	5	4	3	7	2	8
3	8	7	2	1	6	9	5	4
4	5	2	9	7	8	6	3	1

SECTION 3.1, PAGE 99

1. Statement

3. Compound

---

5. a) Not **b)** And **c)** Or

7. Simple statement

9. Compound; biconditional, $\leftrightarrow$

11. Compound; conjunction, $\wedge$

13. Compound; conditional, $\rightarrow$

15. Compound; negation, $\sim$

17. Some Eco Sun scooters are not made by Amigo.

19. All turtles have claws.

21. Some bicycles have three wheels.

23. No pedestrians are in the crosswalk.

25. Some mountain climbers are teachers.

27. $\sim p$ **29.** $\sim q \vee \sim p$

31. $\sim p \rightarrow \sim q$ **33.** $p \wedge \sim q$

35. $\sim q \leftrightarrow p$ **37.** $\sim (p \vee q)$

39. Brie does not have a MacBook.

41. Joe has an iPad and Brie has a MacBook.

43. If Joe does not have an iPad, then Brie has a MacBook.

45. Joe does not have an iPad or Brie does not have a MacBook.

47. It is false that Joe has an iPad and Brie has a MacBook.

49. $(p \wedge \sim q) \wedge r$ **51.** $(p \wedge q) \vee r$

53. $(r \wedge q) \rightarrow p$ **55.** $(r \leftrightarrow q) \wedge p$

57. The water is 70° or the sun is shining, and we do not go swimming.

59. The water is not 70°, and the sun is shining or we go swimming.

61. If we do not go swimming, then the sun is shining and the water is 70°.

63. If the sun is shining then we go swimming, and the water is 70°.

65. The sun is shining if and only if the water is 70°, and we go swimming.

67. Not permissible, you cannot have both soup and salad. The *or* used on menus is the *exclusive or*.

69. Not permissible, you cannot have both potatoes and pasta. The *or* used on menus is the *exclusive or*.

71. a) $b \wedge \sim m$ **b)** Conjunction

73. a) $\sim (w \rightarrow \sim g)$ **b)** Negation

75. a) $(f \vee v) \rightarrow h$ **b)** Conditional

77. a) $c \leftrightarrow (\sim f \vee p)$ **b)** Biconditional

79. a) $(c \leftrightarrow w) \vee s$ **b)** Disjunction

81. a) Answers will vary. **b)** Answers will vary.

82.

6	4	8	7	5	3	2	1	9
3	9	2	8	6	1	7	5	4
7	5	1	9	4	2	8	6	3
9	1	6	3	2	5	4	8	7
5	7	3	4	8	6	9	2	1
8	2	4	1	9	7	5	3	6
4	3	5	6	7	8	1	9	2
2	6	9	5	1	4	3	7	8
1	8	7	2	3	9	6	4	5

SECTION 3.2, PAGE 113

1. Opposite   **3.** False

5. F   **7.** T   **9.** F   **11.** T
F   F   T   F
          T   F   T
          T   F   T

13. T   **15.** T   **17.** F
T   F   T
T   T   F
T   T   T
T   T   T
F   F   F
T   T   T
T   F   T

19. $p \wedge q$ **21.** $p \vee \sim q$
T   T
F   T
F   F
F   T

23. $\sim(\sim p \wedge q)$ **25.** $(\sim p \wedge q) \vee r$
T   T
T   F
F   T
T   F
           T
           T
           T
           F

27. a) True **b)** True **29. a)** False **b)** False
31. a) True **b)** False **33. a)** False **b)** True
35. a) True **b)** True **37.** True **39.** True
41. True **43.** True **45.** True **47.** False
49. True **51.** False

53. $p \wedge \sim q$ **55.** $p \vee \sim q$
F   T
T   T
F   F
F   T
True in case 2   True in cases 1, 2, and 4

57. $(r \vee q) \wedge p$ **59.** $q \vee (p \wedge \sim r)$
T   T
T   T
T   F
F   T
F   T
F   T
F   F
F   F
True in cases 1, 2, and 3   True in cases 1, 2, 4, 5, and 6

61. a) Mr. Duncan and Mrs. Tuttle qualify.
b) Mrs. Rusinek does not qualify, since their combined income is less than $46,000.

63. a) Xavier qualifies; the other four do not.
b) Gina is returning on April 3. Kara is returning on a Monday. Christos is not staying over on a Saturday. Chang is returning on a Monday.

65. T   **67.** Yes
T
T
T
F
T
F
T

SECTION 3.3, PAGE 123

1. False   **3.** True   **5.** Self-contradiction

7. T   **9.** T   **11.** T   **13.** T   **15.** F
T   F   F   T   T
T   F   F   F   T
F   T   F   T   F

17. T   **19.** F   **21.** T   **23.** T
T   F   T   T
T   T   T   T
T   F   T   F
T   T   F   T
F   T   T   T
F   F   T   F
F   T   T   T

25. $p \to (q \wedge r)$

T
F
F
F
T
T
T
T

27. $(p \leftrightarrow q) \vee r$

T
T
T
F
T
F
T
T

29. $(\sim p \to q) \vee r$

T
T
T
T
T
T
T
F

31. Tautology   **33.** Self-contradiction   **35.** Neither

37. Not an implication   **39.** Implication   **41.** Implication

43. True   **45.** False   **47.** True   **49.** True   **51.** True

53. False   **55.** True   **57.** False   **59.** True   **61.** False

63. False   **65.** True   **67.** False

69. No, the statement only states what will occur if your sister gets straight A's. If your sister does not get straight A's, your parents may still get her a computer.

71. F F T T T F F F

73. a) TFFFTFFFTFFFTTTT
  b) TFFTTTTTTTTTTTT

74. Allen was born in January. Booker was born in February. Chris was born in March. Dennis was born in April.

75.

Tiger	Boots	Sam	Sue
Blue	Yellow	Pink	Green
Nine Lives	Whiskas	Friskies	Meow Mix

76. Katie was born last. Katie and Mary are saying the same thing.

SECTION 3.4, PAGE 135

Answer to Recreational Math on page 128

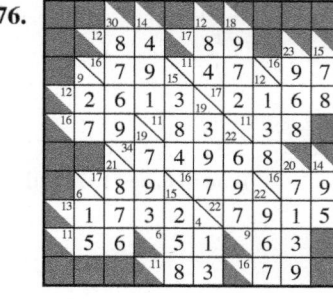

1. Equivalent   **3.** $\sim p \vee q$ **5.** $q \to p$ **7.** $\sim q \to \sim p$

9. Equivalent   **11.** Equivalent   **13.** Not equivalent

15. Equivalent   **17.** Not equivalent   **19.** Equivalent

21. Not equivalent   **23.** Equivalent

25. Jay-Z does not sing opera or Beyoncé does not sing country.

27. It is false that the dog was a bulldog or the dog was a boxer.

29. If I am late for class, then it is false that I will get bonus points and I will not fail the test.

31. I do not see a movie or I buy popcorn.

33. If Chase is hiding, then the pitcher is broken.

35. If Opal does not exercise daily, then she is not healthy.

37. We go to Chicago and we do not go to Navy Pier.

39. It is false that if I am cold then the heater is working.

41. Amazon has a sale and we will not buy $100 worth of books.

43. *Converse:* If you are on the president's list, then you got straight A's.

Inverse: If you did not get straight A's, then you are not on the president's list.

Contrapositive: If you are not on the president's list, then you did not get straight A's.

45. *Converse:* If I buy silver jewelry, then I go to Mexico.

Inverse: If I do not go to Mexico, then I do not buy silver jewelry.

Contrapositive: If I do not buy silver jewelry, then I do not go to Mexico.

47. *Converse:* If I do not stay on my diet, then the menu includes calzones.

Inverse: If the menu does not include calzones, then I stay on my diet.

Contrapositive: If I stay on my diet, then the menu does not include calzones.

49. If a natural number is not divisible by 7, then the natural number is not divisible by 14. True.

51. If a natural number is not divisible by 6, then the natural number is not divisible by 3. False.

53. If two lines are not parallel, then the two lines intersect in at least one point. True.

55. b) and **c)** are equivalent.   **57. a)** and **c)** are equivalent.

59. b) and **c)** are equivalent.   **61. b)** and **c)** are equivalent.

63. None are equivalent.   **65.** None are equivalent.

67. a) and **c)** are equivalent.

69. True. If $p \to q$ is false, it must be of the form T $\to$ F. Therefore, the converse must be of the form F $\to$ T, which is true.

71. False. A conditional statement and its contrapositive always have the same truth values.

73. Answers will vary.

75. a) 0.75   **b)** 0.80
  c) 0.20   **d)** 0.25
  e) 0.95   **f)** 0.95

76.

SECTION 3.5, PAGE 146

1. Valid   **3.** Fallacy   **5.** Valid   **7.** Syllogism

9. Inverse   **11.** Syllogism   **13.** Invalid   **15.** Valid

17. Valid   **19.** Invalid   **21.** Valid   **23.** Valid

25. Valid **27.** Invalid **29.** Invalid **31.** Valid

33. a) $p \to q$ **b)** Invalid   **35. a)** $p \to q$ **b)** Valid

$$\frac{\sim p}{\therefore \sim q}$$

$$\frac{p}{\therefore q}$$

37. a) $p \to q$ **b)** Valid   **39. a)** $p \to q$ **b)** Invalid

$$\frac{\sim q}{\therefore \sim p}$$

$$\frac{q}{\therefore p}$$

41. a) $p \vee q$ **b)** Valid   **43. a)** $p \to q$ **b)** Valid

$$\frac{\sim p}{\therefore q}$$

$$\frac{q \to r}{\therefore p \to r}$$

45. a) $p \wedge q$ **b)** Valid   **47. a)** $p \vee q$ **b)** Invalid

$$\frac{q \to r}{\therefore r \to p}$$

$$\frac{p \to r}{q \leftrightarrow r}$$

49. a) $p \to q$ **b)** Valid   **51. a)** $p \to q$ **b)** Invalid

$$\frac{\sim q}{\therefore \sim p}$$

$$\frac{\sim p}{\therefore \sim q}$$

53. a) $\sim p \to \sim q$ **b)** Valid   **55. a)** $p \wedge \sim q$ **b)** Invalid

$$\frac{\sim p}{\therefore \sim q}$$

$$\frac{q \to p}{\therefore q}$$

57. a) $p \to q$ **b)** Invalid

$$\frac{q \to \sim r}{\therefore p \to r}$$

59. Therefore, the radio is made by RCA.

61. Therefore, I am stressed out.

63. Therefore, you did not close the deal.

65. Yes, if the conclusion necessarily follows from the premises, the argument is valid, even if the conclusion is false.

67. Yes, if the conclusion does not necessarily follow from the premises, the argument is invalid, even if the premises are true.

69. Valid **71. a)** $p \to q$ **b)** No **c)** This argument is the fallacy of the inverse.

$$\frac{\sim p}{\therefore \sim q}$$

SECTION 3.6, PAGE 153

1. Euler **3.** Invalid **5.** No **7.** Valid **9.** Invalid

11. Valid **13.** Invalid **15.** Invalid **17.** Invalid

19. Valid **21.** Invalid **23.** Invalid **25.** Invalid **27.** Valid

29. Yes, if the conclusion necessarily follows from the premises, the argument is valid.

SECTION 3.7, PAGE 160

1. Series **3.** Closed

5. a) $p \vee q$
 b) The lightbulb will be on in all cases except when p is open and q is open.

7. a) $(p \vee q) \wedge \sim q$
 b) The lightbulb will be on only when p is closed and q is open.

9. a) $(p \wedge q) \wedge \big[(p \wedge \sim q) \vee r\big]$
 b) The lightbulb will be on only when p, q, and r are all closed.

11. a) $p \vee q \vee (r \wedge \sim p)$
 b) The lightbulb will be on in all cases except when p, q, and r are all open.

13.

15.

17.

19.

21. $p \vee \sim q$; $\sim p \wedge q$ not equivalent

23. $\big[(p \wedge q) \vee r\big] \wedge p$; $(q \vee r) \wedge p$; equivalent

25. $(p \vee \sim p) \wedge q \wedge r$; $p \wedge q \wedge r$; not equivalent

27. It is a series circuit; therefore, both switches must be closed for current to flow and the lightbulb to go on. When the p switch is closed, the $\overline{p}$ switch is open and no current will flow through the circuit. When the $\overline{p}$ switch is closed, the p switch is open and no current will flow through the circuit.

29. 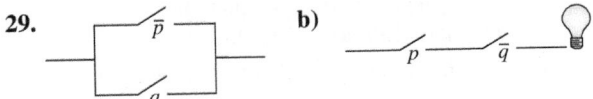 **b)**

REVIEW EXERCISES, PAGE 163

1. Some diamonds are not made of carbon.

2. Some pets are allowed in this park.

3. No women are presidents.

4. All pine trees are green.

5. The coffee is hot or the coffee is strong.

6. The coffee is not hot and the coffee is strong.

7. If the coffee is hot, then the coffee is strong and the coffee is not Maxwell House.

8. The coffee is Maxwell House if and only if the coffee is not strong.

9. The coffee is not Maxwell House, if and only if the coffee is strong and the coffee is not hot.

10. The coffee is Maxwell House or the coffee is not hot, and the coffee is not strong.

11. $q \wedge \sim r$ **12.** $r \rightarrow \sim p$ **13.** $(r \rightarrow q) \vee \sim p$
14. $(q \leftrightarrow p) \wedge \sim r$ **15.** $(r \wedge q) \vee \sim p$ **16.** $\sim (r \wedge q)$

17. F	**18.** T	**19.** T	**20.** T	**21.** F	**22.** F
F	F	T	F	T	T
T	F	T	T	F	T
F	F	T	T	F	T
		T	F	T	T
		F	F	T	T
		F	F	T	T
		T	F	T	T

23. True **24.** False **25.** False **26.** True
27. True **28.** True **29.** False **30.** False **31.** Not equivalent
32. Equivalent **33.** Equivalent **34.** Not equivalent
35. It is false that if a grasshopper is an insect then a spider is an insect.
36. If Lynn Swann did not play for the Steelers, then Jack Tatum played for the Raiders.
37. Altec Lansing does not produce only speakers and Harman Kardon does not produce only stereo receivers.
38. It is false that we did go to the beach or we did find sharks' teeth.
39. The temperature is above 32° or we will go ice fishing at O'Leary's Lake.
40. a) If Maya sits on the bench, then she plays basketball.

    b) If Maya does not play basketball, then she does not sit on the bench.

    c) If Maya does not sit on the bench, then she does not play basketball.

41. a) If we will learn the table's value, then we take the table to *Antiques Roadshow*.

    b) If we do not take the table to *Antiques Roadshow*, then we will not learn the table's value.

    c) If we will not learn the table's value, then we do not take the table to *Antiques Roadshow*.

42. a) If you do not sell more doughnuts, then you do not advertise.

    b) If you advertise, then you sell more doughnuts.

    c) If you sell more doughnuts, then you advertise.

43. a) If you fail the course, then you do not study for the math test.

    b) If you study for the math test, then you do not fail the course.

    c) If you do not fail the course, then you study for the math test.

44. a) If I let you attend the prom, then you will get straight A's on your report card.

    b) If you do not get straight A's on your report card, then I will not let you attend the prom.

    c) If I will not let you attend the prom, then you did not get straight A's on your report card.

45. a), b), and **c)** are equivalent. **46.** None are equivalent.

47. a) and **c)** are equivalent. **48.** None are equivalent.
49. Invalid **50.** Valid **51.** Valid **52.** Invalid
53. Invalid **54.** Valid **55.** Invalid **56.** Invalid
57. a) $p \wedge [(q \wedge r) \vee \sim p]$

    b) The lightbulb will be on only when p, q, and r are all closed.

58.

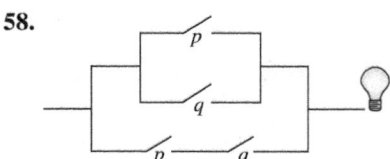

59. Equivalent

CHAPTER TEST, PAGE 165

1. $(\sim p \vee q) \wedge \sim r$
2. $(r \rightarrow q) \vee \sim p$
3. $\sim (r \leftrightarrow \sim q)$
4. Phobos is not a moon of Mars and Rosalind is a moon of Uranus, if and only if Callisto is not a moon of Jupiter.
5. If Phobos is a moon of Mars or Callisto is not a moon of Jupiter, then Rosalind is a moon of Uranus.

6. F	**7.** T
T	T
F	T
F	T
F	F
F	T
F	T
F	F

8. True **9.** True **10.** True **11.** True **12.** Equivalent
13. a) and **b)** are equivalent. **14. a)** and **b)** are equivalent.
15. $s \rightarrow f$ **16.** Invalid
    $f \rightarrow p$
    $\therefore s \rightarrow p$
    Valid
17. Some coffee beans do not contain caffeine.
18. Nick did not play football or Max did not play baseball.
19. *Converse:* If today is Saturday, then the garbage truck comes.

    Inverse: If the garbage truck does not come, then today is not Saturday.

    Contrapositive: If today is not Saturday, then the garbage truck does not come.

20.

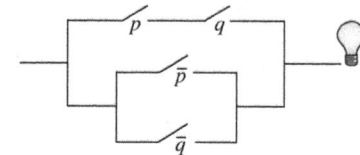

Chapter 4

1. How many   **3.** Hindu–Arabic   **5.** Subtract

7. Multiplicative   **9.** 225   **11.** 12,341   **13.** 334,214

15. ∩∩‖   **17.** 𓏢𓏢∩∩∩∩∩‖‖‖‖‖

19. ⌒𓏲𓏲𓏲𓏲𓏲𓏲𓏢𓏢𓏢99999999∩∩∩‖‖‖‖‖

21. 36   **23.** 194   **25.** 2642   **27.** 2946   **29.** 4499   **31.** 12,666

33. LXII   **35.** CCCXLIX   **37.** MCMXIV

39. $\overline{\text{IV}}$DCCXCIII   **41.** $\overline{\text{IX}}$CMXCIX   **43.** $\overline{\text{XLVI}}$CCLXXXI

45. 74   **47.** 4081   **49.** 8550   **51.** 4003

53. 五十三   **55.** 三百七十八   **57.** 四千二百六十   **59.** 七千零五十六

61. 45   **63.** 878   **65.** 2883   **67.** φμγ   **69.** ψπδ   **71.** ˊεε

73. 1021, MXXI, 一千零二十一, ˊακα

75. 527, 99999∩∩‖‖‖‖‖‖‖, DXXVII, φκζ

77. A number is a quantity, and it answers the question "How many?" A numeral is a symbol used to represent the number.

79. $\overline{\text{CMXCIX}}$CMXCIX

81. Turn the book upside down.   **82.** MM

83. 1888, MDCCCLXXXVIII

1. 10   **3.** Hundreds   **5.** Units   **7. a)** 1   **b)** 10   **c)** Subtraction

9. a) 0   **b)** 1   **c)** 5

11. $(2 \times 10) + (5 \times 1)$

13. $(7 \times 100) + (1 \times 10) + (2 \times 1)$

15. $(4 \times 1000) + (3 \times 100) + (8 \times 10) + (7 \times 1)$

17. $(1 \times 10{,}000) + (6 \times 1000) + (4 \times 100) + (0 \times 10) + (2 \times 1)$

19. $(3 \times 100{,}000) + (4 \times 10{,}000) + (6 \times 1000) + (8 \times 100) + (6 \times 10) + (1 \times 1)$

21. 14   **23.** 784   **25.** 4868   **27.** ⪡⪡‖‖‖

29. ⪡Ⲧ‖‖‖   ⪡⪡⪡⪡⪡Ⲧ   **31.** Ⲧ   ‖‖‖‖‖

33. 57   **35.** 2707   **37.** 4000

39. (Mayan numeral)   **41.** (Mayan numeral)   **43.** (Mayan numeral)   **45.** 1944, ⪡⪡⪡‖‖ ⪡⪡‖‖‖‖

47. $\left(\triangle \times \square^2 \right) + \left(\square \times \square \right) + \left(\diamond \times 1 \right)$

49. The Mayan system has a different base and the numerals are written vertically.

1. Base   **3.** 5   **5.** 5   **7.** 152   **9.** 161   **11.** 441

13. 1395   **15.** 53   **17.** 4012   **19.** 1549   **21.** 50,809

23. 111_2 **25.** 221_3 **27.** 2332_4 **29.** 402_5 **31.** 2112_8

33. 5264_{12} **35.** 24EF_{16} **37.** 2202212_3 **39.** 31041_5

41. 5615_7 **43.** 1205_{12} **45.** 13   **47.** 78

49. (base 5 figure)   **51.** (base 5 figure)

53. 7   **55.** 36

57. (base 4 figure)   **59.** (base 4 figure)   **61. a)** 10213_5 **b)** 1373_8

63. a) 2   **b)** 8   **c)** 16   **d)** 32   **e)** b **65.** $b = 6$

67. a) $1_3, 2_3, 10_3, 11_3, 12_3, 20_3, 21_3, 22_3, 100_3, 101_3, 102_3, 110_3, 111_3, 112_3, 120_3, 121_3, 122_3, 200_3, 201_3, 202_3$
    b) 1000_3

69. a) Answers will vary.   **b)** Answers will vary.

70. $5125

71. a) 876   **b)** (base 4 figure)

1. 12_5 **3.** 5   **5.** 22_5 **7.** 111_3 **9.** 221_5

11. 1450_7 **13.** 11001_2 **15.** BA03_{12} **17.** B9BF_{16}

19. 11_3 **21.** 54_9 **23.** 110_2 **25.** 3616_7 **27.** 1341_5

29. 81BB_{12} **31.** 121_3 **33.** 4103_8 **35.** 1017_8

37. 77676_{12} **39.** 10010_2 **41.** 6031_7 **43.** 123_4

45. $146_7 \text{ R}2_7$ **47.** $52_8 \text{ R}2_8$ **49.** $103_4 \text{ R}1_4$ **51.** $41_5 \text{ R}1_5$

53. $45_7 \text{ R}2_7$ **55.** (base 5 figure)   **57.** (base 5 figure)   **59.** (base 4 figure)

61. (base 4 figure)   **63.** (base 4 figure)   **65.** (base 4 figure)   **67.** 3EAC_{16}

69. a) 21252_8 **b)** 306 and 29   **c)** 8874   **d)** 8874   **e)** Yes

71. ○ = 0, ● = 1, ● = 2, ● = 3

1. a) Divided   **b)** Doubled   **3.** Three; Two   **5.** 209

7. 1015   **9.** 8260   **11.** 8649   **13.** 687   **15.** 2550   **17.** 900

19. 204,728   **21.** 104   **23.** 405   **25.** 625   **27.** 60,678

29. a) 253×46 **b)** 11,638   **31. a)** 4×382 **b)** 1528

33. 99∩∩∩∩∩∩∩‖‖‖‖‖   **35.** 1211_3

1. 232   **2.** 42,053   **3.** 200,421   **4.** 1,214,330

51. a) (Mayan numeral)   **b)** ‖‖‖‖ ⪡⪡⪡⪡Ⲧ‖‖‖ ⪡⪡⪡⪡‖‖‖‖‖‖ ⪡⪡⪡⪡Ⲧ‖

(Mayan numerals column)

53. ‖‖‖ ‖   **55.** (Mayan numeral)

5. 𓏲𓏲𓏲𓎖𓎖𓏺𓏺𓏺𓏺   **6.** 𓏲𓎖𓎖   **7.** ⌒𓏲𓏲𓏲𓏲𓎖𓎖𓎖𓈖𓈖𓏺

8. 𓏭𓏲𓏲𓏲𓏲𓈖𓈖𓈖𓏺𓏺   **9.** 14   **10.** 142   **11.** 2437

12. 5759   **13.** XXIX   **14.** DXLIII   **15.** MCMLXIV

16. V̄ICDXCI   **17.** 32   **18.** 45   **19.** 267   **20.** 3429

21. 二十三   **22.** 五十四   **23.** 四百九十二   **24.** 二千六百五十二   **25.** 81   **26.** 548

27. 605   **28.** 3334   **29.** κβ   **30.** ψογ   **31.** ωξζ   **32.** ͵δσιθ

33. 191   **34.** 189   **35.** 3673   **36.** 7388   **37.** (Babylonian symbols)

38. (Babylonian symbols)   **39.** (Babylonian symbols)   **40.** (Babylonian symbols)

41. 48   **42.** 1367   **43.** 2521   **44.** 15,923   **45.** (Maya symbols)   **46.** (Maya symbols)

47. (Maya symbols)   **48.** (Maya symbols)   **49.** 39   **50.** 7   **51.** 28   **52.** 1244

53. 1552 **54.** 186   **55.** 111001111_2 **56.** 13033_4 **57.** 3323_5
58. 717_8 **59.** 327_{12} **60.** $1CF_{16}$ **61.** 1103_4 **62.** 101111_2
63. 166_{12} **64.** $70F_{16}$ **65.** 12102_5 **66.** 12423_8 **67.** 122_4
68. 100_2 **69.** $9A6_{12}$ **70.** 3324_5 **71.** 450_8 **72.** $CC1_{16}$
73. 2133_6 **74.** 20223_4 **75.** 1314_5 **76.** 13632_8 **77.** 5656_{12}
78. $1D76_{16}$ **79.** $21_3 R1_3$ **80.** 130_4 **81.** $23_5 R1_5$ **82.** 433_6
83. $411_6 R1_6$ **84.** $664_8 R2_8$ **85.** 2875   **86.** 2875   **87.** 2875

CHAPTER TEST, PAGE 206

1. 2479   **2.** 1873   **3.** 8090   **4.** 969   **5.** 122,142   **6.** 2745

7. 𓏲𓏲𓎖𓈖𓈖𓏺𓏺𓏺   **8.** ͵βυοϛ   **9.** (Maya symbols)   **10.** (Babylonian symbols)   (Babylonian symbols)

11. MMDCCXLIX   **12.** 11   **13.** 1694   **14.** 100100_2
15. $B7A_{16}$ **16.** 11000_2 **17.** 2003_6 **18.** 220_5 **19.** 980
20. 8428

Chapter 5

SECTION 5.1, PAGE 215

1. Theory   **3.** Zero   **5.** Composite

7. Divisor   **9.** Conjecture

11. 2, 3, 5, 7, 11, 13, 17, 19, 23, 29, 31, 37, 41, 43, 47, 53, 59, 61, 67, 71, 73, 79, 83, 89, and 97.

13. False   **15.** False. 35 is a multiple of 7.

17. True   **19.** True   **21.** False

23. True   **25.** True   **27.** 3 and 5

29. 2, 3, 4, 6, 8, and 9.

31. 2, 3, 4, 5, 6, 8, and 10.

33. 60 (other answers are possible)   **35.** $2^2 \cdot 3$

37. $2^2 \cdot 3^2 \cdot 5$ **39.** $2^2 \cdot 83$ **41.** $3^3 \cdot 19$

43. $2^3 \cdot 167$ **45. a)** 3   **b)** 42

47. a) 5   **b)** 140   **49. a)** 20   **b)** 1800   **51. a)** 4   **b)** 5088

53. a) 8   **b)** 384   **55.** 28 days

57. $2 \times 60, 3 \times 40, 4 \times 30, 5 \times 24, 6 \times 20, 8 \times 15,$
$10 \times 12, 12 \times 10, 15 \times 8, 20 \times 6, 24 \times 5, 30 \times 4,$
$40 \times 3, 60 \times 2$

59. 35 cars   **61.** 30 trees   **63.** 30 days

65. a) Yes **b)** No **c)** No **d)** Yes **67.** 17, 19, and 29, 31

69. 5, 17, and 257 are all prime.

71. A number is divisible by 14 if both 2 and 7 divide the number.

73. 5   **75.** 35   **77.** 30   **79.** Yes   **81.** No

83. a) 12   **b)** 1, 2, 3, 4, 5, 6, 10, 12, 15, 20, 30, 60

85. For any three consecutive natural numbers, one of the numbers is divisible by 2 and another number is divisible by 3. Therefore, the product of the three numbers would be divisible by 6.

87. Yes

89. $8 = 2 + 3 + 3, 9 = 3 + 3 + 3, 10 = 2 + 3 + 5,$
$11 = 2 + 2 + 7, 12 = 2 + 5 + 5, 13 = 3 + 3 + 7,$
$14 = 2 + 5 + 7, 15 = 3 + 5 + 7, 16 = 2 + 7 + 7,$
$17 = 5 + 5 + 7, 18 = 2 + 5 + 11, 19 = 3 + 5 + 11,$
$20 = 2 + 7 + 11$

91. a) 5, 7, 11, 13, 17, 19, 23, and 29

b) Every prime number greater than 3 differs by 1 from a multiple of 6.

c) This conjecture should appear to be correct.

SECTION 5.2, PAGE 226

Answer to *Recreational Mathematics* box on page 223

$$0 = 4 + 4 - 4 - 4 \qquad\qquad 1 = \frac{4 + 4}{4 + 4}$$

$$2 = \frac{4 \cdot 4}{4 + 4} \qquad\qquad 3 = \frac{4 + 4 + 4}{4}$$

$$4 = \frac{4 - 4}{4} + 4 \qquad\qquad 5 = \frac{4 \cdot 4 + 4}{4}$$

$$6 = \frac{4 + 4}{4} + 4 \qquad\qquad 7 = 4 + 4 - \frac{4}{4}$$

$$8 = 4 + 4 + 4 - 4 \qquad\qquad 9 = 4 + 4 + \frac{4}{4}$$

Other answers are possible.

1. Whole   **3. a)** Positive   **b)** Negative

5. a) <   **b)** <   **c)** <   **d)** >   **7.** −9, −6, −3, 0, 3, 6

9. −6, −5, −4, −3, −2, −1   **11.** 3   **13.** −3   **15.** −5

17. 2   **19.** −2   **21.** −8   **23.** −2   **25.** −6   **27.** −10

29. 16   **31.** 96   **33.** −60   **35.** 2   **37.** −1   **39.** −7   **41.** −15

43. a) 9   **b)** 8   **45. a)** 25   **b)** −25   **47. a)** −16   **b)** 16

49. a) −64   **b)** −64   **51.** −1   **53.** 10   **55.** 12   **57.** −75   **59.** 1

61. 1100 feet under water   **63.** 65,233 ft

65. a) 11 hours   **b)** 2 hours   **c)** 14 hours   **d)** 15 hours   **67.** True

69. False; the difference of two negative integers may be a positive integer, a negative integer, or zero.

71. True   **73.** True

75. False; the sum of a positive integer and a negative integer may be a positive integer, a negative integer, or zero.

77. Division by zero is undefined because $\frac{a}{0} = x$ always leads to a false statement.   **79.** -1

81. $0 + 1 - 2 + 3 + 4 - 5 + 6 - 7 - 8 + 9 = 1$

82. a) $4\left(4 - \frac{4}{4}\right) = 12, 4 \cdot 4 - \frac{4}{4} = 15, \frac{4 \cdot 4 \cdot 4}{4} = 16,$

$4 \cdot 4 + \frac{4}{4} = 17, 4\left(4 + \frac{4}{4}\right) = 20$ **b)** $\frac{44 - 4}{4} = 10$

SECTION 5.3, PAGE 239

1. Integers   **3.** Denominator   **5.** Improper

7. Repeating   **9.** Ten-thousandths

11. $\frac{3}{4}$ **13.** $\frac{4}{7}$ **15.** $\frac{19}{25}$ **17.** $\frac{17}{6}$ **19.** $-\frac{23}{4}$ **21.** $-\frac{79}{16}$ **23.** $\frac{3}{2}$

25. $\frac{15}{8}$ **27.** $1\frac{5}{8}$ **29.** $-9\frac{1}{5}$ **31.** $-58\frac{8}{15}$ **33.** 0.4 **35.** $0.\overline{3}$

37. 0.375 **39.** $2.1\overline{6}$ **41.** $\frac{6}{10} = \frac{3}{5}$ **43.** $\frac{175}{1000} = \frac{7}{40}$

45. $\frac{295}{1000} = \frac{59}{200}$ **47.** $\frac{131}{10,000}$ **49.** $\frac{1}{9}$ **51.** $\frac{1}{1} = 1$ **53.** $\frac{15}{11}$

55. $\frac{37}{18}$ **57.** $\frac{4}{15}$ **59.** $\frac{2}{5}$ **61.** $\frac{49}{64}$ **63.** $\frac{36}{35}$ **65.** $\frac{20}{21}$ **67.** $\frac{13}{15}$

69. $\frac{9}{22}$ **71.** $\frac{25}{42}$ **73.** $\frac{17}{144}$ **75.** $-\frac{109}{600}$ **77.** $\frac{11}{14}$

79. $-\frac{1}{24}$ **81.** $\frac{13}{30}$ **83.** $-\frac{1}{40}$ **85.** $\frac{7}{32}$ **87.** $\frac{53}{77}$ **89.** $\frac{11}{10}$

91. $2\frac{3}{5}$ in.   **93.** $120\frac{3}{4}$ in.   **95.** $9\frac{15}{16}$ in.   **97.** $\frac{1}{10}$ **99.** $58\frac{7}{8}$ in.

101. a) $1\frac{49}{60}, 2\frac{48}{60}, 9\frac{6}{60}, 6\frac{3}{60}, 2\frac{9}{60}, \frac{22}{60}$

b) $22\frac{17}{60}$ hours; 22 hours 17 minutes

103. $26\frac{5}{32}$ in.   **105. a)** $29\frac{3}{16}$ in.   **b)** 33 in.   **c)** $32\frac{3}{4}$ in.

107. 0.215 **109.** -2.1755 **111.** $\frac{1}{2}$ **113.** $\frac{11}{200}$

115. a) $\frac{1}{1}$ or 1   **b)** $0.\overline{9}$ **c)** $\frac{1}{3} + \frac{2}{3} = 1, 0.\overline{3} + 0.\overline{6} = 0.\overline{9}$

d) $0.\overline{9} = 1$

SECTION 5.4, PAGE 250

1. Irrational   **3.** Radicand   **5.** Itself   **7.** Rationalized

9. Rational   **11.** Irrational   **13.** Irrational   **15.** Rational

17. Rational   **19.** 0   **21.** 5   **23.** -6 **25.** -10

27. Rational, integer   **29.** Irrational

31. Rational   **33.** Rational

35. Irrational   **37.** $2\sqrt{5}$ **39.** $2\sqrt{10}$ **41.** $3\sqrt{7}$ **43.** $2\sqrt{21}$

45. $7\sqrt{6}$ **47.** $\sqrt{2}$ **49.** $-13\sqrt{3}$ **51.** $4\sqrt{3}$ **53.** 6

55. $2\sqrt{15}$ **57.** 2   **59.** 3

61. $\frac{\sqrt{5}}{5}$ **63.** $\frac{\sqrt{21}}{7}$ **65.** $\frac{2\sqrt{15}}{3}$ **67.** $\frac{\sqrt{15}}{3}$

69. Between 6 and 6.5; $\sqrt{37} \approx 6.08$.

71. Between 9.5 and 10; $\sqrt{97} \approx 9.85$.

73. Between 13 and 13.5; $\sqrt{170} \approx 13.04$.

75. 36.0 inches

77. a) 2.5 sec   **b)** 5 sec   **c)** 7.5 sec   **d)** 10 sec

79. False. $\sqrt{c}$ may be a rational number or an irrational number for a composite number c. (For example, $\sqrt{25}$ is a rational number; $\sqrt{8}$ is an irrational number.)   **81.** True

83. False. The product of a rational number and an irrational number may be a rational number or an irrational number.

85. $\pi + (-\pi) = 0$ **87.** $\sqrt{2} \cdot \sqrt{3} = \sqrt{6}$

89. $\sqrt{2} \neq 1.414$ since $\sqrt{2}$ is irrational and 1.414 is rational.

91. No. π is irrational; it cannot equal $\frac{22}{7}$ or 3.14, both of which are rational.

93. $\sqrt{4 \cdot 9} = \sqrt{4} \cdot \sqrt{9}, \sqrt{36} = 2 \cdot 3, 6 = 6$

95. a) Rational. $\sqrt{0.04} = 0.2$, which is a rational number.

b) Irrational $\sqrt{0.7} = 0.8366600265 \ldots$. Since the decimal number is not a terminating or a repeating decimal number, this number is an irrational number

97. a) $(44 \div \sqrt{4}) \div \sqrt{4} = 11$ **b)** $(44 \div 4) + \sqrt{4} = 13$ **c)** $4 + 4 + 4 + \sqrt{4} = 14$ **d)** $\sqrt{4}(4 + 4) + \sqrt{4} = 18$; Other answers are possible.

SECTION 5.5, PAGE 256

Answer to *Recreational Mathematics* box on page 254

1. Real   **3.** Closed   **5.** Commutative   **7.** Associative

9. Yes   **11.** No   **13.** Yes   **15.** Yes   **17.** Yes   **19.** Yes

21. No   **23.** No   **25.** Yes   **27.** No

29. Commutative property of addition. The only difference between the expressions on both sides of the equal sign is the order of 5 and x.

31. No. $3 \div 4 \neq 4 \div 3$

33. $(-3)(-4) = (-4)(-3) = 12$

35. $[(-2) + (-3)] + (-4) = (-2) + [(-3) + (-4)] = -9$

37. No. $(16 \div 8) \div 2 \neq 16 \div (8 \div 2)$.

39. No. $(81 \div 9) \div 3 \neq 81 \div (9 \div 3)$.

41. Distributive property

43. Associative property of multiplication

45. Associative property of addition

47. Commutative property of multiplication

49. Distributive property

51. Commutative property of addition

53. $3b + 21$ **55.** $6c - 12$ **57.** $-8x + 2$ **59.** $2x - 1$

61. 2   **63.** $5\sqrt{2} + 5\sqrt{3}$ **65.** Yes   **67.** No   **69.** No   **71.** Yes

73. Yes   **75.** Answers will vary.

77. No. $0 \div a = 0$ (when $a \neq 0$), but $a \div 0$ is undefined.

78.

^{2}2	$^{2-}$3	$^{2\div}$1	$^{24\times}$4
$^{3-}$4	1	2	3
1	$^{7+}$4	3	2
$^{1-}$3	2	$^{5+}$4	1

79. a) No   **b)** No   **c)** Answers will vary.

SECTION 5.6, PAGE 265

1. x^5 **3.** 1   **5.** x^6 **7. a)** 32   **b)** -32

9. a) 25   **b)** 25   **11. a)** 1   **b)** -1 **13. a)** 1 **b)** 6

15. a) $\dfrac{1}{27}$ **b)** $\dfrac{1}{49}$ **17. a)** $-\dfrac{1}{81}$ **b)** $\dfrac{1}{81}$

19. a) 64   **b)** 729   **21. a)** 4   **b)** $\dfrac{1}{16}$ **23.** 5.03×10^5

25. 4.2×10^{-4} **27.** 5.6×10^{-1} **29.** 1.9×10^4

31. 1.86×10^{-4} **33.** 2300   **35.** 0.00168   **37.** 0.0000862

39. 2.01   **41.** 820,000   **43.** 0.0153   **45.** 250   **47.** 0.0021

49. 7.2×10^8 **51.** 4.5×10^{-7} **53.** 7.0×10^1

55. 2.0×10^{-7} **57.** 3.6×10^{-3}; 1.7; 9.8×10^2; 1.03×10^4

59. 8.3×10^{-5}; 0.00079; 4.1×10^3; 40,000

61. a) $55,932.20   **b)** $12,086.38

63. 0.043   **65.** $62,987.80   **67.** 11.95 hours

69. $8.64 \times 10^9 \text{ ft}^3$ **71.** 2.9×10^8 cells   **73.** 333,333 times

75. 1000   **77.** 2832 people

79. a) $1 \times 10^6, 1 \times 10^9, 1 \times 10^{12}$

   b) 1000 days (about 2.74 years)

   c) 1,000,000 days (about 2739.73 years)

   d) 1,000,000,000 days (about 2,739,726.03 years)

   e) 1000 times greater

81. a) 1024 bacteria   **b)** 1448 bacteria

SECTION 5.7, PAGE 273

1. Sequence   **3.** Arithmetic   **5.** Geometric   **7.** 3, 7, 11, 15, 19

9. 25, 20, 15, 10, 5   **11.** 5, 3, 1, -1, -3 **13.** 61   **15.** 125

17. $-\dfrac{91}{5}$ **19.** $a_n = n$ **21.** $a_n = 2n - 1$ **23.** $a_n = 3n + 2$

25. $s_{50} = 1275$ **27.** $s_{50} = 2500$ **29.** $s_8 = -52$

31. 2, 10, 50, 250, 1250 **33.** 5, 10, 20, 40, 80 **35.** $-3, 3, -3, 3, -3$

37. 160   **39.** 1   **41.** -3645 **43.** $a_n = 3^n$

45. $a_n = 2 \cdot \left(\dfrac{1}{2}\right)^{n-1}$ **47.** $a_n = -16 \cdot \left(\dfrac{1}{2}\right)^{n-1}$ **49.** 189

51. -4095 **53.** 14,348,906   **55.** 5050   **57.** 10,100

59. 78 times   **61. a)** 63 in.   **b)** 954 in.

63. 15,992 students   **65.** $\approx$ $45,667   **67.** 1.5 billion people

69. 161.4375   **71.** 267   **73.** 191.3568 ft

74. a) $320, $310   **b)** $10,240; $10,230

   c) It is risky because it is likely that you won't have enough money to continue doubling your previous bet if you lose several times in a row. Also, there may be a maximum amount you can bet in a game of chance.

SECTION 5.8, PAGE 280

1. Fibonacci   **3.** Divine   **5.** Ratio   **7.** Yes; 55, 89   **9.** No

11. Yes; 3, 5   **13.** Yes; 105, 170

15. 1, 1, 2, 3, 5, 8, 13, 21, 34, 55, 89, 144, 233, 377, 610

17. Answers will vary.   **19.** Answers will vary.

21. a) Answers will vary.   **b)** Answers will vary.

23. The sums of the numbers along the diagonals (parallel to the one shown) are Fibonacci numbers.

25. a) 1.618   **b)** 0.618   **c)** 1

27. Answers will vary.

29. The decimal expansion shows several terms of the Fibonacci sequence.

31. The ratio of the second to the first, and the fourth to the third measures estimates the golden ratio.

33. Answers will vary.

35. a) Answers will vary.   **b)** Answers will vary.

   c) Answers will vary.

REVIEW EXERCISES, PAGE 283

1. 2, 3, 4, 5, 6, 9, 10   **2.** 2, 3, 4, 6, 9   **3.** $2^3 \cdot 3 \cdot 5 \cdot 7$

4. $2^2 \cdot 3 \cdot 11^2$ **5.** 12; 72   **6.** 9; 756   **7.** 180 minutes

8. -4 **9.** 3   **10.** -6 **11.** -4 **12.** -9 **13.** 3   **14.** -24

15. 24   **16.** 5   **17.** -2 **18.** 23   **19.** 12   **20.** 35   **21.** -2738

22. 0.44   **23.** 3.25   **24.** $0.\overline{857142}$ **25.** $0.58\overline{3}$

26. $\dfrac{42}{10} = \dfrac{21}{5}$ **27.** $\dfrac{2}{3}$ **28.** $\dfrac{51}{99} = \dfrac{17}{33}$ **29.** $\dfrac{83}{1000}$ **30.** $\dfrac{18}{7}$

31. $-\dfrac{13}{4}$ **32.** $3\dfrac{4}{5}$ **33.** $-27\dfrac{1}{5}$ **34.** $-\dfrac{2}{15}$ **35.** $\dfrac{17}{12}$

36. $\dfrac{1}{4}$ **37.** $\dfrac{35}{54}$ **38.** $\dfrac{17}{33}$ **39.** $\dfrac{9}{20}$ **40.** $\dfrac{13}{15}$ **41.** $\dfrac{13}{34}$

42. $2\dfrac{7}{32}$ tsp   **43.** $2\sqrt{15}$ **44.** $-3\sqrt{2}$ **45.** $8\sqrt{2}$

46. $-20\sqrt{3}$ **47.** $5\sqrt{7}$ **48.** $3\sqrt{2}$ **49.** $4\sqrt{3}$ **50.** 10

51. $\dfrac{4\sqrt{3}}{3}$ **52.** $\dfrac{\sqrt{35}}{5}$ **53.** $6 + 3\sqrt{7}$ **54.** $4\sqrt{3} + 3\sqrt{2}$

55. Commutative property of addition

56. Commutative property of multiplication

57. Associative property of addition

58. Distributive property

59. Associative property of multiplication

60. No   **61.** Yes   **62.** No   **63.** Yes   **64.** No   **65.** 243

66. 36   **67.** 1

68. $\dfrac{1}{125}$ **69.** 4096   **70.** -1 **71.** $-\dfrac{1}{49}$ **72.** $\dfrac{3}{8}$

73. 3.62×10^7 **74.** 1.58×10^{-5} **75.** 280,000

76. 0.000139 **77.** 6.0×10^{-5} **78.** 3.0×10^0

79. 1,100,000,000,000 **80.** 120 **81.** 5 **82.** 388 times

83. $\approx$ \$5555.56 **84.** Arithmetic; 27, 33 **85.** Geometric; 8, 16

86. 10 **87.** 25 **88.** 48 **89.** $-19{,}683$ **90.** 3825 **91.** 57.5

92. 45 **93.** -21 **94.** Arithmetic; $a_n = 3n - 2$

95. Geometric; $a_n = 2(-1)^{n-1}$

96. 1, 1, 2, 3, 5, 8, 13, 21, 34, 55, 89, 144, 233, 377, 610

97. No **98.** Yes; $-8, -13$

CHAPTER TEST, PAGE 286

1. 2, 5, 10 **2.** $3^2 \cdot 5 \cdot 7$ **3.** 8 **4.** -175 **5.** $\dfrac{35}{11}$

6. 0.52 **7.** $\dfrac{129}{20}$ **8.** $-\dfrac{11}{60}$ **9.** $5\sqrt{7}$ **10.** $\dfrac{\sqrt{14}}{7}$

11. Yes; the product of any two integers is an integer.

12. Distributive property of multiplication over addition

13. 64 **14.** 1024 **15.** $\dfrac{1}{81}$ **16.** 8.0×10^{11}

17. $a_n = -4n + 2$ **18.** -187 **19.** $a_n = 3(2)^{n-1}$

20. 1, 1, 2, 3, 5, 8, 13, 21, 34, 55

Chapter 6

SECTION 6.1, PAGE 296

1. Variable **3.** Expression **5.** Terms **7.** Coefficient

9. Identity **11.** $13x$ **13.** $-4x - 8$ **15.** $3x + 11y$

17. $2x - 18y - 5$ **19.** $-x^2 + 8x - 5$ **21.** $8t$

23. $13.3x - 8.3$ **25.** $\frac{9}{10}x - 8$

27. $1.4x - 2.8$ **29.** $\frac{5}{6}x + 5$ **31.** -3 **33.** 13

35. 64 **37.** -25 **39.** 6 **41.** -12 **43.** 5 **45.** Yes **47.** No

49. Yes **51.** 2 **53.** -45 **55.** -9 **57.** -6 **59.** 3

61. -3 **63.** 2 **65.** $\frac{2}{3}$ **67.** $\frac{4}{27}$ **69.** 4 **71.** 29

73. No solution **75.** All real numbers **77.** \$6.32

79. a) 32°F **b)** 212°F **c)** 95°F **d)** 23°F

81. $(-1)^n = 1$ for any even number n, since an even number of factors of (-1) when multiplied will always be 1.

83. $1^n = 1$ for all natural numbers because 1 multiplied by itself any number of times will always be 1.

85. a) 310 ft **b)** 265 ft below sea level

SECTION 6.2, PAGE 301

1. Subscripts **3.** 40 **5.** 46 **7.** 8 **9.** 62.83 **11.** 10

13. 2 **15.** -32 **17.** 11.99 **19.** 44.6 **21.** 6 **23.** 4 **25.** 14

27. 7 **29.** 1462.79

31. $y = -2x + 3$ **33.** $y = \dfrac{1}{6}x - 2$ **35.** $y = \dfrac{2}{3}x + 2$

37. $r = \dfrac{d}{t}$ **39.** $a = p - b - c$ **41.** $r = \dfrac{C}{2\pi}$

43. $b = \dfrac{2A}{h}$ **45.** $x = \dfrac{y - a}{b}$ **47.** $w = \dfrac{P - 2l}{2}$

49. $C = \dfrac{5}{9}(F - 32)$ **51.** $r = \dfrac{A - P}{Pt}$

53. 6.5 hr **55.** 8%

57. a) 27.12 **b)** lose 8.27 1b

SECTION 6.3, PAGE 309

1. Ratio **3.** $x + 3$ **5.** $11x$ **7.** $6 - 4y$

9. $8w + 9$ **11.** $4x + 6$

13. $\dfrac{z}{13} - 4$ **15.** $\dfrac{12 - s}{4}$ **17.** $3(x + 7)$

19. $x + 8 = 23; 15$

21. $9x = 54; 6$ **23.** $4x - 10 = 42; 13$

25. $4x + 12 = 32; 5$ **27.** $x + 6 = 2x - 3; 9$

29. $x + 10 = 2(x + 3); 4$ **31.** $150 + 0.50x = 255; 210$ miles

33. $x + 0.05x = 42;$ \$40 per half hour

35. $400 + 0.06x = 790;$ \$6500

37. $x + 3x = 20{,}000;$ \$5000 in mutual funds and \$15,000 in stocks

39. $2w + 2(w + 2) = 52;$ width: 12 ft, length: 14 ft

41. $3w + 2(2w) = 140;$ width: 20 ft, length: 40 ft

43. a) \$25.40 **b)** 4890.97 gallons

45. 29,520,000 households

47. a) 1600 lb **b)** 40 bags **49.** 3.75 mℓ **51.** 0.875 cc

53. a) 4.5 months **b)** \$795 **54.** -40°

SECTION 6.4, PAGE 318

1. Direct **3.** Joint **5.** Direct **7.** Inverse **9.** Direct

11. Inverse **13.** Direct **15.** Inverse **17.** Direct **19.** Direct

21. Answers will vary.

23. a) $y = kx$ **b)** 108 **25. a)** $m = \dfrac{k}{n^2}$ **b)** 0.20

27. a) $A = \dfrac{kB}{C}$ **b)** 2.5 **29. a)** $F = kDE$ **b)** 210

31. a) $H = kL$ **b)** 3

33. a) $q = \dfrac{k}{w^2}$ **b)** 25 **35. a)** $J = k\sqrt{b}$ **b)** 3

37. a) $Z = kWY$ **b)** 100 **39. a)** $F = \dfrac{kM_1 M_2}{d^2}$ **b)** 20

41. a) $p = kn$ **b)** \$1012.50

43. a) $t = \dfrac{k}{T}$ **b)** 1.875 minutes

45. a) $d = kt^2$ **b)** 144 ft **47. a)** $R = \dfrac{kL}{A}$ **b)** 25 ohms

49. a) $W = kI^2R$ **b)** 9 watts

51. a) Directly **b)** 0.5

53. a) y will double. **b)** y will be half as large. **55.** $\dfrac{1}{9}$

SECTION 6.5, PAGE 325

1. Inequality **3.** Compound **5.** All **7.** Closed

9.

11.

13.

15.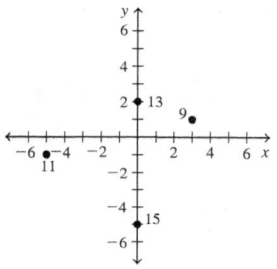

(number line with solid point, labeled −9)

17.

(number line, labeled 0, 4)

19.

(number line, labeled −1 0 1 2 3 4 5 6 7 8 9)

21. No solution

(number line, labeled 0)

23. All real numbers

(number line, labeled −5 −4 −3 −2 −1 0 1 2 3 4 5)

25.

(number line, labeled −1, 3)

27.

(number line, labeled 6, 10)

29.

(number line, labeled −5 −4 −3 −2 −1 0 1 2 3 4 5)

31.

(number line, labeled −1 0 1 2 3 4 5 6)

33.

(number line, labeled −11 −10 −9 −8 −7 −6 −5 −4)

35.

(number line, labeled −11 −10 −9 −8 −7 −6)

37.

(number line, labeled −2 −1 0 1 2 3 4 5)

39.

(number line, labeled −6 −5 −4 −3 −2 −1 0 1)

41.

(number line, labeled −5 −4 −3 −2 −1 0 1 2)

43.

(number line, labeled −1 0 1 2 3 4 5 6)

45.

(number line, labeled −3 −2 −1 0 1 2 3 4)

47. a) 2015, 2016   **b)** 2010, 2011, 2012
  c) 2010, 2011, 2012   **d)** 2014, 2015, 2016   **49.** 11 months

51. 350 hours   **53.** 6 hours   **55.** $0.5 < t < 1.5$

57. 96% or higher   **59.** $6.875 \le x \le 11$

61. The student's answer is $x \le -12$, whereas the correct
  answer is $x \ge -12$. Yes, -12 is in both solution sets.

SECTION 6.6, PAGE 337

1. Graph   **3.** x

5. Plotting points, using intercepts, and using the slope and
  the y-intercept

7. m

For odd Exercises 9–15, see the following figure.

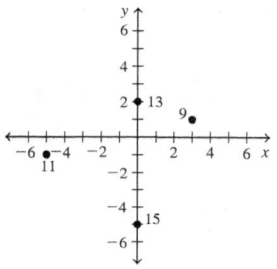

For odd Exercises 17–23, see the following figure.

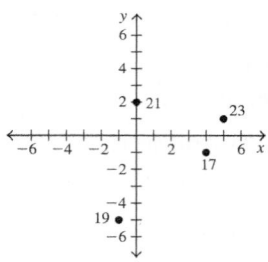

25. $(0, 2)$ **27.** $(-2, 0)$ **29.** $(-3, -4)$

31. $(2, -2)$ **33.** $(2, 2)$ **35.** $(0, 2), (3, 0)$

37. $(8, 7), \left(\frac{10}{3}, 0\right)$ **39.** $(4, 3)$ **41.** $(8, 0), (0, 3)$

43.   **45.**

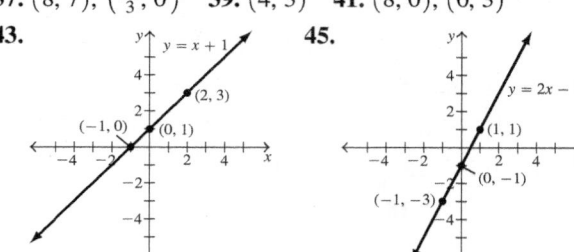

47.   **49.**

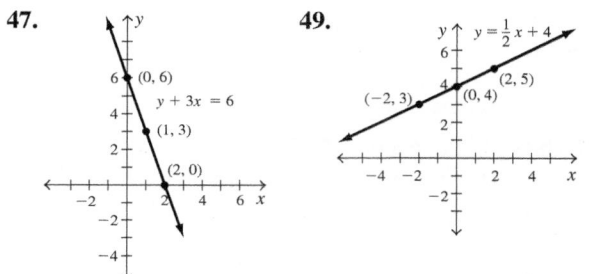

51.   **53.**

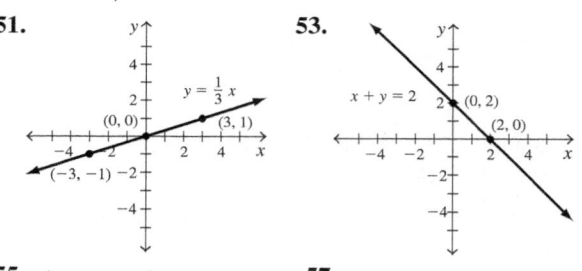

55.   **57.**

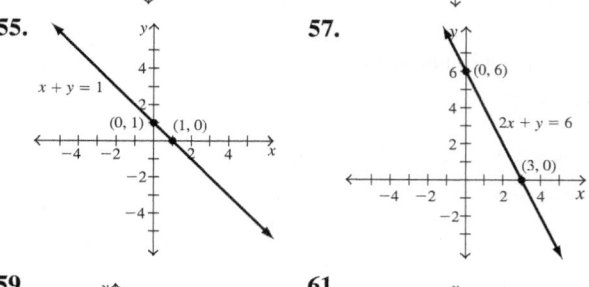

59.   **61.**

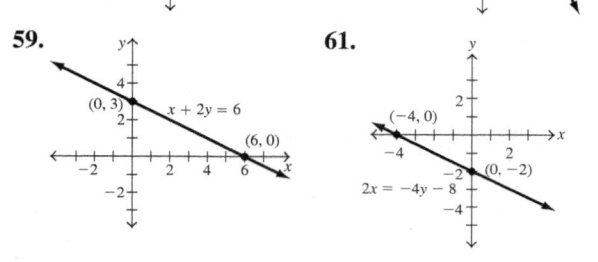

63. 2 **65.** $\frac{7}{3}$ **67.** 0 **69.** Undefined

71.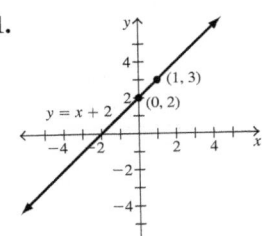

73.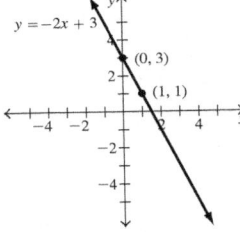

75.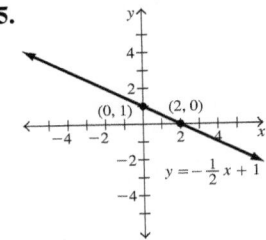

77.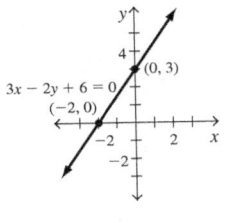

79. $y = -\dfrac{3}{4}x + 3$ **81.** $y = 3x + 2$

83.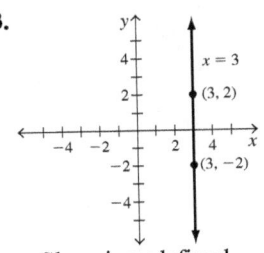

Slope is undefined

85.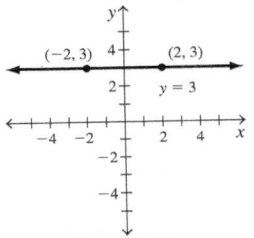

Slope is 0.

87. $y = -2$ **89.** $x = -3$

91. a) $D(4, 2)$ **b)** $A = 10$ square units

93. $(7, 2)$ or $(-1, 2)$ **95.** -3 **97.** 3

99. a)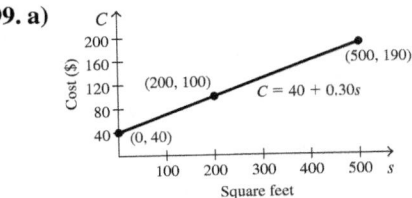

b) $130 **c)** 100 square feet

101. a)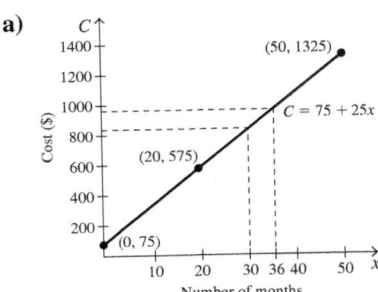

b) $825 **c)** 36 months

103. a) 2 **b)** $y = 2x + 9$ **c)** 15 defects **d)** 4 workers

105. a) 2.1 **b)** $y = 2.1x + 20$ **c)** 32.6 **d)** 2012

107. a) Solve the equations for y to put them in slope-intercept form. Then compare the slopes and y-intercepts. If the slopes are equal but the y-intercepts are different, then the lines are parallel.
   b) The lines are parallel.

SECTION 6.7, PAGE 350

1. System **3.** Inconsistent **5.** Dependent
7. One **9.** Infinite **11.** $(2, 6)$
13. $(0, 3)$ **15.** $(2, 1)$ **17.** $(-2, 3)$
19. No solution; inconsistent system
21. An infinite number of solutions; dependent system
23. $(1, 10)$ **25.** $(1, -1)$
27. No solution; inconsistent system
29. $(3, 1)$ **31.** $(3, 5)$
33. An infinite number of solutions; dependent system
35. $(3, 1)$ **37.** $(-4, 5)$ **39.** $(-2, 2)$ **41.** $(2, -3)$
43. An infinite number of solutions; dependent system
45. No solution; inconsistent system
47. $\left(\dfrac{28}{19}, \dfrac{34}{19}\right)$
49. a) Cost for U-Haul: $C = 0.79x + 30$
   Cost for Discount Rentals: $C = 0.85x + 24$

b) 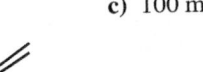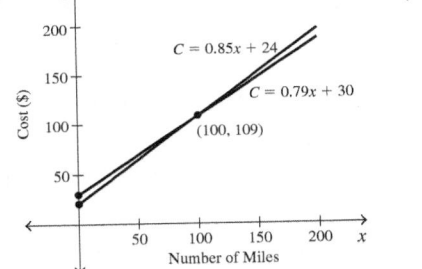 **c)** 100 miles

51. a) $C = 15x + 400$
   $R = 25x$

b) 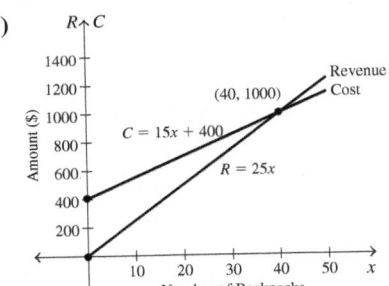 **c)** 40 backpacks

d) $P = 10x - 400$ **e)** Loss of $100 **f)** 140 backpacks

53. a) $C = 15x + 4050, R = 40x$

b)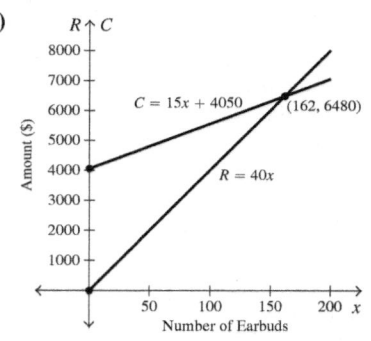

c) 162 units

d) $P = 25x - 4050$ **e)** Loss of \$175   **f)** 185 units

55. $x + y = 50$

    $73x + 85y = 4070$

15 shares of Under Armour, 35 shares of Hershey Company

57. $x + y = 15$

    $0.40x + 0.10y = 0.25(15)$

    7.5 l of 40%, 7.5 l of 10%

59. a) $y = 249 + 0.15x,$
    $y = 300 + 0.12x,$

    \$1700   **b)** Harbor Sales

61. $y = 2.65x + 468.75$
    $y = 3.10x + 412.50$
    a) 125 square feet   **b)** Home Depot

63. 7 hours

65. a) One unique solution; the lines intersect at one and only one point.

    b) No solution; the lines do not intersect.
    c) Infinitely many solutions; the lines coincide.

67. $\left(\frac{1}{2}, \frac{1}{3}\right)$

69. b) Answers will vary.   **71.** Answers will vary.

73. Answers will vary.

SECTION 6.8, PAGE 362

1. Half-plane   **3.** Solid   **5.** Ordered   **7.** Feasible   **9.** Objective

11. a) No   **b)** Yes   **c)** Yes   **d)** No

13.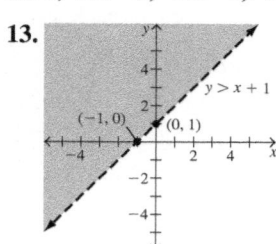

15.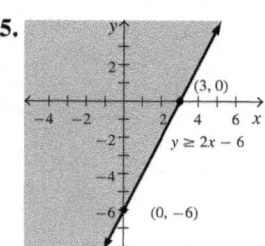

17.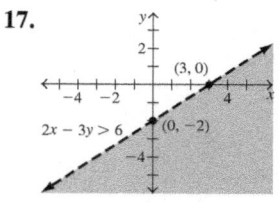

19.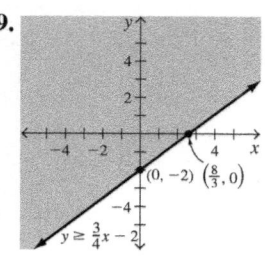

21.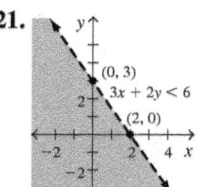

23.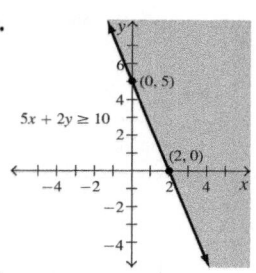

25.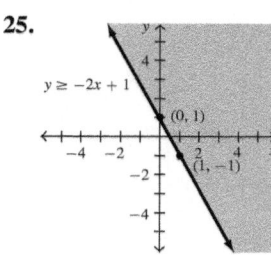

27.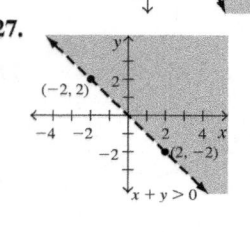

29.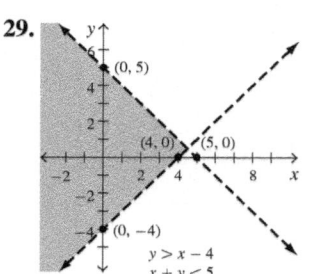

31.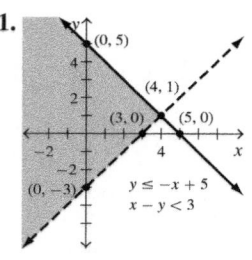

33.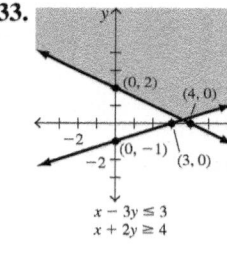

35.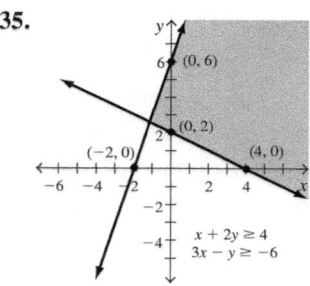

37.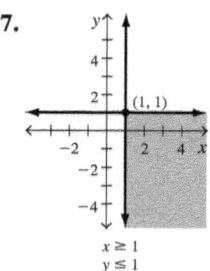

39. Maximum is 30 at $(5, 0)$, minimum is 0 at $(0, 0)$.

41. Maximum is 190 at $(50, 30)$, minimum is 70 at $(20, 10)$.

43. a)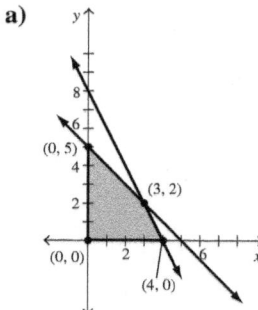

    b) Maximum is 18 at $(3, 2)$; minimum is 0 at $(0, 0)$.

45. a)

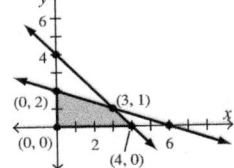

b) Maximum is 28 at $(4, 0)$; minimum is 0 at $(0, 0)$.

47. a)

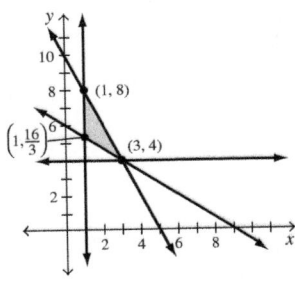

b) Maximum is 15.4 at $(1, 8)$; minimum is 11 at $\left(1, \frac{16}{3}\right)$.

49. a) $x + y < 500, x \geq 150, y \geq 150$

b)

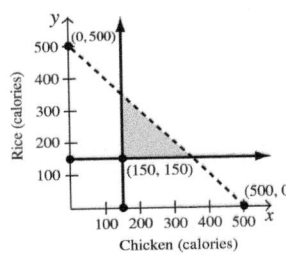

c) One example is $(220, 220)$; approximately 3.7 oz of chicken, 8.8 oz of rice.

51. a) $x + y \leq 20, x \geq 3, x \leq 6, y \geq 2$

b) $P = 25x + 20y$

c)

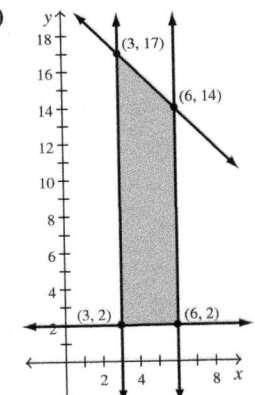

d) $(3, 17), (6, 14), (6, 2), (3, 2)$
e) Six skateboards and 14 pairs of in-line skates
f) $430

53. a) $3x + 4y \geq 60, 10x + 5y \geq 100, x \geq 0, y \geq 0$
b) $C = 28x + 33y$

c)

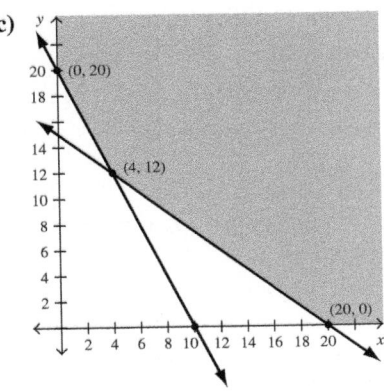

d) $(0, 20), (4, 12), (20, 0)$
e) 4 hours for Machine I and 12 hours for Machine II
f) $508

55. $x + \frac{1}{2}y \leq 200$
$\frac{1}{2}y \leq 150$
$x \geq 0$
$y \geq 0$

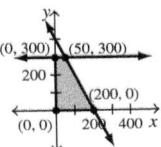

$P = 0.40x + 0.30y$

Maximum profits occurs at $(50, 300)$. Thus, the manufacturer should make 50 lb of the all-beef hot dogs and 300 lb of the regular hot dogs for a profit of $110.

57. a) No   **b)** One example: $x + y > 4$
$\qquad\qquad\qquad\qquad\qquad x + y < 1$

59. (a), (b), and (d)

SECTION 6.9, PAGE 374

1. Binomial   **3.** FOIL   **5.** Quadratic
7. $x^2 - 8x - 33$ **9.** $6x^2 + 7x - 3$ **11.** $48x^2 - 2x - 63$
13. $(x + 2)(x + 3)$ **15.** $(x - 3)(x + 2)$
17. $(x + 5)(x - 2)$ **19.** $(x + 1)(x - 3)$
21. $(x - 6)(x - 3)$ **23.** $(x + 7)(x - 4)$
25. $(2x + 1)(x + 3)$ **27.** $(3x - 2)(x + 1)$
29. $(3x - 2)(x - 5)$ **31.** $(4x + 3)(x + 2)$
33. $(4x - 3)(x - 2)$ **35.** $(4x - 3)(2x + 1)$
37. $6, \frac{4}{5}$ **39.** $-\frac{4}{3}, \frac{1}{2}$ **41.** $-5, -3$ **43.** $7, 1$ **45.** $5, -3$
47. $3, 1$ **49.** $-\frac{3}{2}, -\frac{5}{3}$ **51.** $\frac{2}{3}, -4$ **53.** $\frac{1}{3}, -2$ **55.** $-\frac{2}{3}, -1$
57. $-\frac{5}{4}, \frac{3}{5}$ **59.** $-3, 1$ **61.** $6, -3$ **63.** $2 \pm \sqrt{2}$
65. No real solution   **67.** $\frac{-9 \pm \sqrt{21}}{6}$ **69.** $\frac{4 \pm \sqrt{13}}{3}$
71. $\frac{5 \pm \sqrt{73}}{8}$ **73.** $-1, -\frac{5}{2}$ **75.** $\frac{9 \pm \sqrt{21}}{6}$

77. No real solution   **79.** 88.72 min
81. Width = 12 m, length = 22 m   **83.** 4 seconds
85. a) The zero-factor property cannot be used.   **b)** $\frac{11 \pm \sqrt{33}}{2}$
87. $x^2 - 2x - 3 = 0$, other answers are possible

SECTION 6.10, PAGE 388

 1. Relation   **3.** Domain   **5.** Upward   **7.** $x = \frac{-b}{2a}$
 9. Yes   **11.** No   **13.** Yes   **15.** No
17. Function, domain: $x = -2, -1, 1, 2, 3$;
    range: $y = -1, 1, 2, 3$
19. Function, domain: $\mathbb{R}$; range: $\mathbb{R}$ **21.** Not a function
23. Function, domain: $\mathbb{R}$; range: $y \geq -4$
25. Not a function   **27.** Function, domain: $\mathbb{R}$; range: $\mathbb{R}$
29. Function, domain: $0 \leq x \leq 10$; range: $-1 \leq y \leq 3$
31. 8   **33.** 6   **35.** 5   **37.** −23   **39.** 45

41.
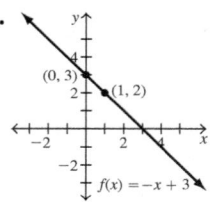

43.
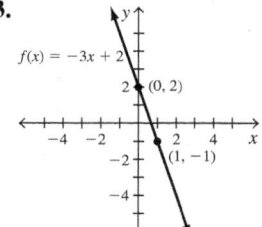

45.
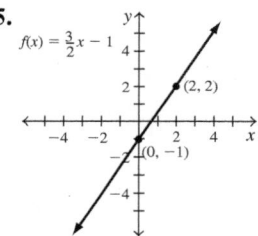

47. a) Upward   **b)** $x = 0$ **c)** $(0, -1)$ **d)** $(0, -1)$
    e) $(1, 0), (-1, 0)$ **f)**
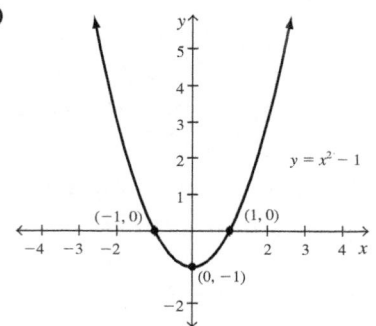
    g) Domain: $\mathbb{R}$; range: $y \geq -1$

49. a) Downward   **b)** $x = 0$ **c)** $(0, 4)$ **d)** $(0, 4)$
    e) $(-2, 0), (2, 0)$
    f)   **g)** Domain: $\mathbb{R}$; range: $y \leq 4$
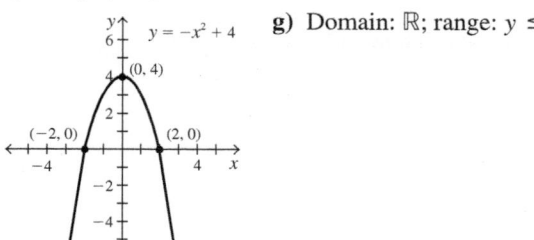

51. a) Downward   **b)** $x = 0$ **c)** $(0, -8)$ **d)** $(0, -8)$
    e) No x-intercepts
    f)
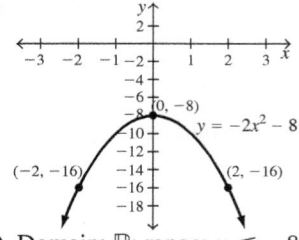
    g) Domain: $\mathbb{R}$; range: $y \leq -8$
53. a) Upward   **b)** $x = -2$ **c)** $(-2, -8)$ **d)** $(0, -4)$
    e) $(-4.83, 0), (0.83, 0)$
    f)
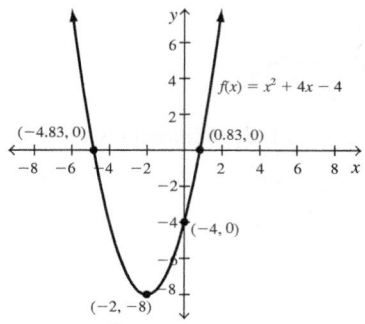
    g) Domain: $\mathbb{R}$; range: $y \geq -8$
55. a) Upward   **b)** $x = -\frac{5}{2}$ **c)** $(-2.5, -0.25)$ **d)** $(0, 6)$
    e) $(-3, 0), (-2, 0)$
    f)
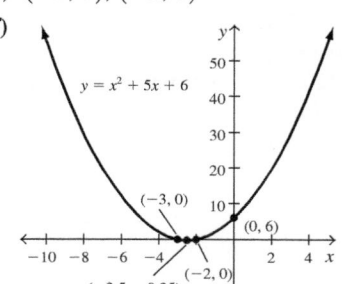
    g) Domain: $\mathbb{R}$; range: $y \geq -0.25$
57. a) Downward   **b)** $x = 2$ **c)** $(2, -2)$ **d)** $(0, -6)$
    e) No x-intercepts
    f)

    g) Domain: $\mathbb{R}$; range: $y \leq -2$
59. a) Downward   **b)** $x = \frac{3}{4}$ **c)** $\left(\frac{3}{4}, -\frac{7}{8}\right)$ **d)** $(0, -2)$
    e) No x-intercepts

f)

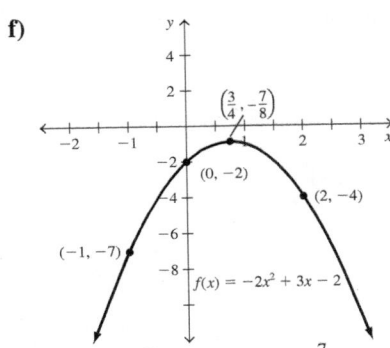

g) Domain: $\mathbb{R}$; range: $y \leq -\dfrac{7}{8}$

61. Domain: $\mathbb{R}$; range: $y > 0$

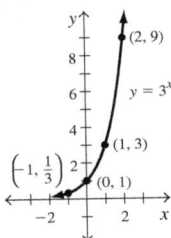

63. Domain: $\mathbb{R}$; range: $y > 0$

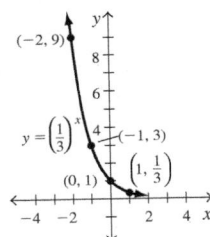

65. Domain: $\mathbb{R}$; range: $y > 1$

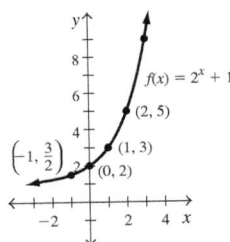

67. Domain: $\mathbb{R}$; range: $y > 1$

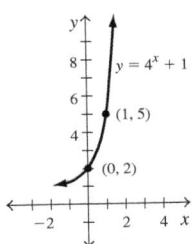

69. Domain: $\mathbb{R}$; range: $y > 0$

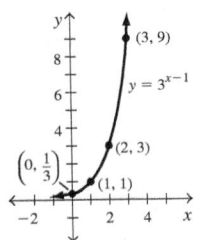

71. \$55,000   **73. a)** 74°F   **b)** 2 P.M.   **c)** 4 hours or 2 P.M; 75°F

75. 486,000 bacteria

77. a) 5200 people   **b)** $\approx$ 14,852 people

79. $\approx$ 1.46 billion people   **81.** $\$1.17 \times 10^{15}$

83. $\approx$ 6.52 grams

85. a) 23.2 cm   **b)** 55.2 cm   **c)** 69.5 cm

87. a) 170 beats per minute   **b)** $\approx$ 162 beats per minute
   c) $\approx$ 145 beats per minute   **d)** 136 beats per minute
   e) 120 years of age

REVIEW EXERCISES, PAGE 395

1. 39   **2.** −9   **3.** 15   **4.** 7   **5.** $4x - 3$ **6.** $12x - 6$ **7.** −5

8. −31   **9.** −13   **10.** 3   **11.** 200   **12.** 210   **13.** 101.5

14. 25   **15.** $y = 2x - 4$

16. $y = \dfrac{-2x + 17}{9}$ or $y = -\dfrac{2}{9}x + \dfrac{17}{9}$

17. $w = \dfrac{A}{l}$ **18.** $l = \dfrac{L - 2wh}{2h}$ or $l = \dfrac{L}{2h} - w$

19. $5 + 3x$ **20.** $\dfrac{9}{q} - 15$

21. $3 + 7x = 17; x = 2$ **22.** $3x + 8 = x - 6; x = -7$

23. $5(x - 4) = 45; x = 13$

24. $10x + 14 = 8(x + 12); x = 41$

25. $x + 2x = 15,000$; bonds: \$5000, mutual funds: \$10,000

26. $x + (x + 15,000) = 75,000$; \$30,000 for B and \$45,000 for A

27. $\dfrac{1}{2}$ cup   **28.** 250 min, or 4 hr 10 min   **29.** 72   **30.** 4

31. 20   **32.** $\approx$ 426.7   **33.** \$3150   **34.** \$119.88

35.

36.

37.

38.

39. – 42.

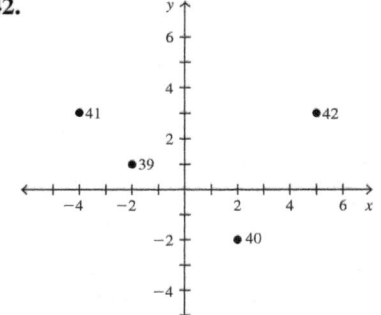

43.

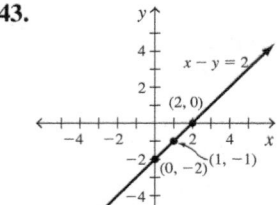

44.

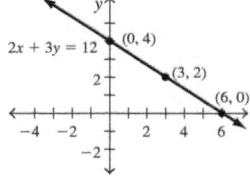

45.

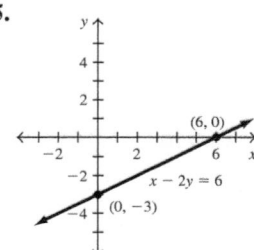

46.

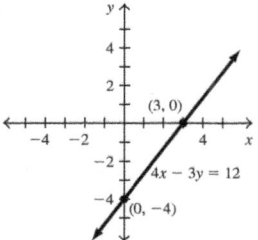

47. $\frac{4}{3}$ **48.** $-\frac{3}{2}$ **49.** $\frac{7}{3}$ **50.** Undefined

51.

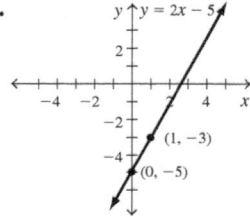

52.

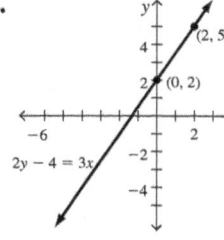

53. $y = 2x + 4$ **54.** $y = -x + 1$

55. a)

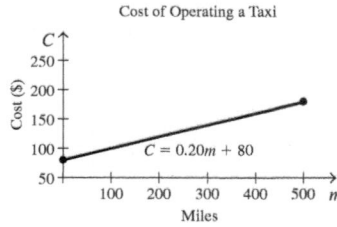

b) $110   **c)** 120 miles

56.

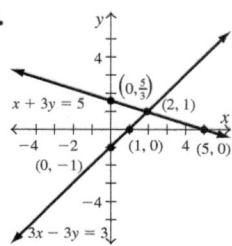

57.

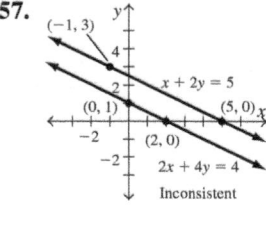

58. $(3, 1)$ **59.** $(-1, -5)$ **60.** $(-2, -8)$

61. No solution; inconsistent   **62.** $(4, -2)$

63. An infinite number of solutions; dependent

64. $(2, -1)$ **65.** $(2, 0)$ **66.** $250,000 at 3%, $150,000 at 6%

67. Mix $83\frac{1}{3}\ell$ of 80% acid solution with $16\frac{2}{3}\ell$ of 50% acid solution.

68. a) 32.5 months   **b)** Model 6070B

69. a) 3 hr   **b)** All-Day parking lot

70.

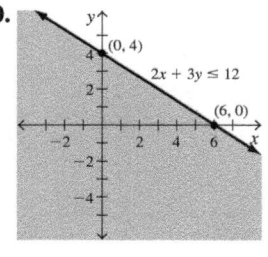

71.

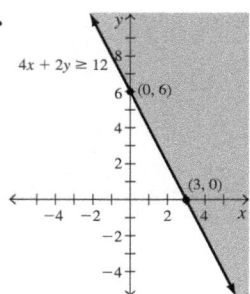

72.

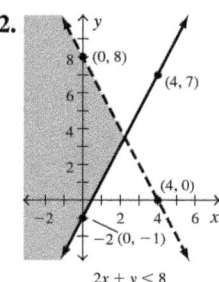

$2x + y < 8$
$y \geq 2x - 1$

73.

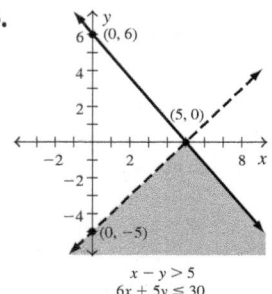

$x - y > 5$
$6x + 5y \leq 30$

74. a)

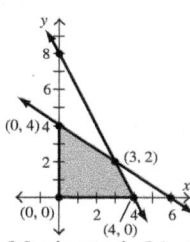

b) Maximum is 21 at $(3, 2)$; minimum is 0 at $(0, 0)$.

75. $(x + 3)(x + 3)$ **76.** $(x + 5)(x - 3)$

77. $(x - 6)(x - 4)$ **78.** $(3x - 1)(2x + 3)$ **79.** $-4, -5$

80. $-5, 2$ **81.** $\frac{2}{3}, 5$ **82.** $-2, -\frac{1}{3}$ **83.** $-2, 8$ **84.** $-1, \frac{3}{2}$

85. No real solution   **86.** $\frac{3 \pm \sqrt{17}}{2}$

87. Function, domain: $x = -2, -1, 2, 3$; range: $y = -1, 0, 2$

88. Not a function   **89.** Not a function

90. Function, domain: $\mathbb{R}$; range: $\mathbb{R}$ **91.** 19   **92.** 11

93. 66   **94.** -2

95. a) Downward   **b)** $x = -2$ **c)** $(-2, 25)$ **d)** $(0, 21)$

e) $(-7, 0), (3, 0)$ **f)**

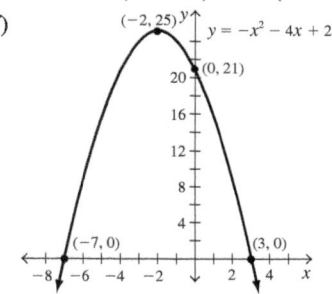

g) Domain: $\mathbb{R}$; range: $y \leq 25$

96. a) Upward   **b)** $x = -2$ **c)** $(-2, -2)$
d) $(0, 6)$ **e)** $(-3, 0)$, $(-1, 0)$

f)

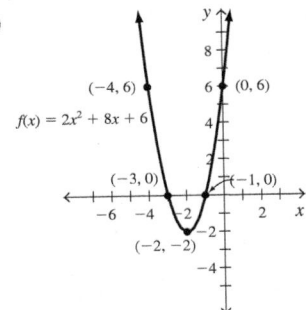

g) Domain: $\mathbb{R}$; range: $y \geq -2$

97. Domain: $\mathbb{R}$; range: $y > 0$

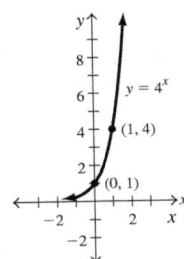

98. Domain: $\mathbb{R}$; range: $y > 0$

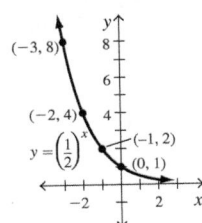

99. 22.8 mpg   **100.** $\approx 68.7\%$

CHAPTER TEST, PAGE 398

1. -1 **2.** 10   **3.** 18   **4.** $350 + 0.06x = 710$; $6000

5. 77   **6.** $y = \dfrac{-3x + 11}{5}$ or $y = -\dfrac{3}{5}x + \dfrac{11}{5}$ **7.** 6.75 ft

8. 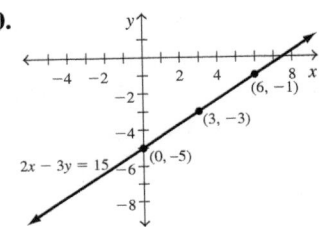   **9.** 2

10.

11.

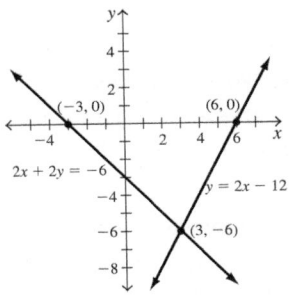

12. $(2, -3)$ **13.** $(-1, 3)$

14. Daily fee: $35; mileage charge: $0.18

15.

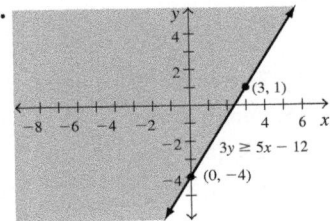

16. a)

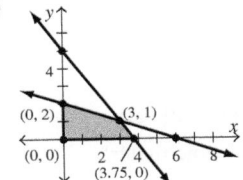

b) Maximum is 22.5 at $(3.75, 0)$; minimum is 0 at $(0, 0)$.

17. $-1, -6$

18. $\frac{4}{3}, -2$ **19.** 15

20. a) Upward   **b)** $x = 1$ **c)** $(1, 3)$
d) $(0, 4)$ **e)** No x-intercepts

f)

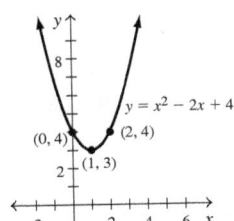

g) Domain: $\mathbb{R}$; range: $y \geq 3$

Chapter 7

SECTION 7.1, PAGE 400

1. Metric   **3. a)** Meter   **b)** Kilogram   **c)** Liter   **d)** Celsius
5. a) Deka   **b)** Deci   **7. a)** Kilo   **b)** Milli
9. a) 0°C   **b)** 100°C   **c)** 37°C   **11. b)**   **13. c)**   **15. f)**
17. a) 0.001   **b)** 100   **c)** 1000   **d)** 0.01   **e)** 10   **f)** 0.1

19. cg; 0.01 g   **21.** dg; 0.1 g   **23.** kg; 1000 g   **25.** 603 000 m/h

27. 9200   **29.** 0.000 057   **31.** 1.862   **33.** 6 500 000 mm

35. 2.4 km   **37.** 4.0302 daℓ **39.** 580 cm   **41.** 0.5 m

43. 1 270 000 g   **45.** 56 300 m/h   **47.** 620 cm, 4.4 dam, 0.52 km

49. 1.4 kg, 1600 g, 16 300 dg   **51.** 105 000 mm, 2.6 km, 52.6 hm

53. 1 hm in 10 min, 1 hm $>$ 1 dam

55. The pump that removes 1 daℓ per minute, 1 daℓ $>$ 1 dℓ

57. a) 526 m   **b)** 52 600 cm   **c)** 5.26 hm   **59. a)** 3500 mg
   b) 3.5 g   **61. a)** 2160 mℓ **b)** 2.16 ℓ **c)** $1.13 per liter

63. a) 108 m   **b)** 0.108 km   **c)** 108 000 mm   **65.** 1000

67. $1 \times 10^{24} = 1\ 000\ 000\ 000\ 000\ 000\ 000\ 000\ 000$

69. 9 dam   **71.** 6 mg   **73.** 2 daℓ

SECTION 7.2, PAGE 409

1. Length   **3.** Area   **5.** Length   **7.** Volume   **9.** Volume

11. Centimeters   **13.** Meters   **15.** Centimeters

17. Kilometers   **19.** (c)   **21.** (c)   **23.** (b)

25. Answers will vary.   **27.** Answers will vary.

29. Answers will vary.   **31.** Square meters

33. Hectares or square kilometers

35. Square centimeters or square millimeters

37. (a)   **39.** (b)   **41.** (a)   **43.** Answers will vary.

45. Answers will vary.   **47.** Answers will vary.

49. Milliliters   **51.** Kiloliters   **53.** Cubic meters

55. Cubic meters   **57.** (c)   **59.** (b)   **61.** (b)   **63.** (b)

65. a) Answers will vary.   **b)** 10 580 cm^3

67. a) Answers will vary.   **b)** 0.20 m^3

69. A cubic decimeter   **71.** A cubic centimeter

73. Longer side $\approx$ 4 cm, shorter side $\approx$ 2.2 cm,
   area $\approx$ 8.8 cm^2 **75.** $\approx$ 804.25 m^2

77. a) 3.2 km^2 **b)** 320 ha   **79. a)** 1378 m^2 **b)** 0.1378 ha

81. 1504 cm^2 **83. a)** 56 000 cm^3 **b)** 56 000 mℓ **c)** 56 ℓ

85. 100 times larger   **87.** 1000 times larger   **89.** 100

91. 0.0001   **93.** 1 000 000   **95.** 724   **97.** 189

99. 60 kℓ **101.** 5300

103. a) 4,014,489,600 sq in.   **b)** Answers will vary.

104. a) Answers will vary. The average use is about 400 ℓ/day.
   b) Answers will vary. The average use is about 75 ℓ/day.
   c) Answers will vary. The average use is about 10 ℓ/day.

SECTION 7.3, PAGE 420

1. Mass   **3.** Kilogram   **5.** Celsius   **7.** 0   **9.** Kilograms

11. Grams or Kilograms   **13.** Grams   **15.** Kilograms

17. Kilograms or metric tonnes   **19.** (c)   **21.** (a)   **23.** (b)

25. Answers will vary.   **27.** Answers will vary.   **29.** (c)   **31.** (b)

33. (c)   **35.** (b)   **37.** 77°F   **39.** -31.7°C   **41.** -17.8°C

43. 98.6°F   **45.** -12.2°C   **47.** 32°F   **49.** -40°C   **51.** 71.6°F

53. 67.46°F   **55.** 64.04°F to 74.3°F   **57.** 752 g

59. a) 70 euros   **b)** 35 euros   **c)** 300 g

61. a) 1600 m^3 **b)** 1600 kℓ **c)** 1600 t

63. 0.0035   **65.** 0.001 46

67. a) Yes; mass is a measure of the amount of matter in an
   object.   **b)** No; weight is a measure of gravitational force.

69. a) 1200 g   **b)** 1200 cm^3

71. 1500 g   **72. a)** -79.8°F   **b)** 36.5°F

SECTION 7.4, PAGE 429

1. Dimensional   **3.** $\dfrac{60\ \text{sec}}{1\ \text{min}}, \dfrac{1\ \text{min}}{60\ \text{sec}}$ **5.** $\dfrac{100\ \text{cm}}{1\ \text{m}}, \dfrac{1\ \text{m}}{100\ \text{cm}}$

7. a) $\dfrac{1\ \text{lb}}{0.45\ \text{kg}}$ **b)** $\dfrac{0.45\ \text{kg}}{1\ \text{lb}}$ **9. a)** $\dfrac{3.8\ \ell}{1\ \text{gal}}$ **b)** $\dfrac{1\ \text{gal}}{3.8\ \ell}$

11. 8.1 kg   **13.** 15.18 oz   **15.** 388.89 lb   **17.** 62.4 km

19. 1687.5 acres   **21.** 53.62 pints   **23.** 1.46 mi^2 **25.** 54 kg

27. 460 yd   **29.** 1170.49 ft   **31.** 50 mph   **33.** 120 m^2

35. 0.21 oz   **37.** 36 678.4 ha   **39. a)** 305 000 m   **b)** 305 km

41. a) 102 mℓ **b)** 3.4 oz   **43. a)** 3975 mi   **b)** 6360 km
   c) 7.63 mi   **d)** 15 mi   **45. a)** 7.4 ha   **b)** 74 000 m^2

47. 81 mg   **49.** 6840 mg, or 6.84 g   **51. a)** 472 mg   **b)** 3776 mg

53. a) Usain Bolt   **b)** The roadrunner

55. a) 6.08 mi   **b)** 443.13 mph   **c)** 113.13 mph   **d)** -49°F

57. a) 1.08 euros   **b)** $1.23

59. a) $\dfrac{745\ \text{kph}}{402\ \text{knots}} \approx \dfrac{1.85\ \text{kph}}{1\ \text{knot}}$ **b)** $\dfrac{402\ \text{knots}}{463\ \text{mph}} \approx \dfrac{0.87\ \text{knot}}{1\ \text{mph}}$
   c) $\dfrac{463\ \text{mph}}{745\ \text{kph}} \approx \dfrac{0.62\ \text{mph}}{1\ \text{kph}}$

61. 1.0 cc, or b)   **63. a)** 6200 cc   **b)** 378.28 in.^{3}

REVIEW EXERCISES, PAGE 441

1. $\dfrac{1}{100}$ of basic unit   **2.** 1000 $\times$ basic unit

3. 100 $\times$ basic unit   **4.** $\dfrac{1}{1000}$ of basic unit

5. 10 $\times$ basic unit   **6.** $\dfrac{1}{10}$ of basic unit   **7.** 0.08 g

8. 5200 dm   **9.** 0.57 hm   **10.** 1 kg   **11.** 4620 ℓ

12. 19 260 dg   **13.** 3000 mℓ, 14 630 cℓ, 2.67 kℓ

14. 0.047 km, 47 000 cm, 4700 m   **15.** Degrees Celsius

16. Centimeters   **17.** Square meters   **18.** Milliliters or liters

19. Kilograms or tonnes   **20.** Kilometers

21. a) Answers will vary.   **b)** Answers will vary.

22. a) Answers will vary.   **b)** Answers will vary.

23. (b)   **24.** (a)   **25.** (c)   **26.** (a)   **27.** (a)   **28.** (b)   **29.** 1.8 t

30. 9 200 000 g   **31.** 75.2°F   **32.** 20°C   **33.** ≈ -21.1°C

34. 102.2°F   **35.** $l = 4$ cm, $w = 1.6$ cm, $A = 6.4$ cm^2

36. $r = 1.5$ cm, $A \approx 7.07$ cm^2

37. a) ≈ 3.14 m^3 **b)** ≈ 3140 kg

38. a) 899.79 cm^2 **b)** 0.089 979 m^2

39. a) 180 000 cm^3 **b)** 0.18 m^3 **c)** 180 000 mℓ **d)** 0.18 kℓ

40. 10,000 times larger   **41.** 91.11   **42.** 7.09   **43.** 33.3

44. 111.11   **45.** 83.2   **46.** 28.42   **47.** 76   **48.** 78.95

49. 14.77   **50.** 3.8   **51. a)** ≈ 53.94 in.   **b)** ≈ 98.22 1b
52. 385.6 km   **53.** 32.4 m² **54. a)** 190 kℓ   **b)** 190 000 kg
55. a) 43.75 mi/hr   **b)** 70 000 m/hr
56. a) 252 ℓ   **b)** 252 kg   **57.** $1.58 per pound

CHAPTER TEST, PAGE 443

1. 2.4 kℓ   **2.** 0.004 62 hm   **3.** 100 times greater
4. 2.4 km   **5.** (b)   **6.** (a)   **7.** (c)   **8.** (b)   **9.** (c)
10. 10,000 times greater   **11.** 1,000,000 times greater
12. 39.6 kg   **13.** 555.05 ft   **14.** ≈ 20,508.1 mi²
15. ≈ 4.44°C   **16.** 23°F   **17. a)** 300 000 g   **b)** ≈ 666.67 lb
18. a) 3200 m³   **b)** 3 200 000 ℓ (or 3200 kℓ)
   c) 3 200 000 kg   **19.** $245
20. a) 6.12 euros per gallon   **b)** $7.03 per gallon

Chapter 8

SECTION 8.1, PAGE 453

1. Parallel   **3.** Angle   **5.** Supplementary   **7.** Right   **9.** Acute
11. Line segment, $\overline{AB}$ **13.** Ray, $\overrightarrow{AB}$ **15.** Ray, $\overrightarrow{BA}$
17. Line, $\overleftrightarrow{AB}$ **19.** $\overline{AD}$ **21.** $\overline{BD}$ **23.** $\overrightarrow{BC}$ **25.** $\overrightarrow{IC}$ **27.** $\triangle BCF$
29. $\{F\}$ **31.** $\angle CBG$ or $\angle GBC$ **33.** $\overrightarrow{BC}$ **35.** $\{B\}$ **37.** $\overline{BF}$
39. $\angle ABE$ or $\angle EBA$ **41.** $\angle CBG$ or $\angle GBC$ **43.** Acute
45. Obtuse   **47.** None of these angles   **49.** 77°   **51.** $57\frac{1}{4}°$
53. 25.3°   **55.** 151°   **57.** 159.5°   **59.** $136\frac{2}{7}°$ **61.** (c)   **63.** (b)
65. (a)   **67.** $m\angle 1 = 15°, m\angle 2 = 75°$ **69.** 141° and 39°
71. Angles 3, 4, and 7 each measure 130°; angles 1, 2, 5, and 6 each measure 50°.
73. Angles 1, 4, and 7 each measure 20°; angles 2, 3, 5, and 6 each measure 160°.
75. $m\angle 1 = 64°, m\angle 2 = 26°$ **77.** $m\angle 1 = 33°, m\angle 2 = 57°$
79. $m\angle 1 = 134°, m\angle 2 = 46°$
81. $m\angle 1 = 29°, m\angle 2 = 151°$
83. $\overleftrightarrow{EF}$ and $\overleftrightarrow{DG}$
85. Plane ABG and plane JCD
87. Plane $ABG \cap$ plane $ABC \cap$ plane $BCD = \{B\}$
89. $\overrightarrow{BC} \cap$ plane $ABG = \{B\}$
91. Always true. If any two lines are parallel to a third line, then they must be parallel to each other.
93. Sometimes true. Vertical angles are only complementary when each is equal to 45°.
95. Sometimes true. Alternate interior angles are only complementary when each is equal to 45°.
97. a) An infinite number   **b)** An infinite number
99. An infinite number
101. a) – d) Answers will vary.
103. No. Line l and line n may be parallel or skew.
105. a)

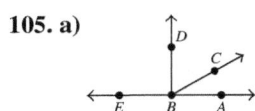

   Other answers are possible.
   b) 30° **c)** 60° **d)** 90°

SECTION 8.2, PAGE 463

1. Polygon   **3.** Proportion   **5.** Congruent
7. a) Pentagon   **b)** Regular
9. a) Heptagon   **b)** Regular
11. a) Heptagon   **b)** Not regular
13. a) Decagon   **b)** Regular
15. a) Isosceles   **b)** Acute   **17. a)** Isosceles   **b)** Right
19. a) Scalene   **b)** Acute   **21. a)** Equilateral   **b)** Acute
23. Square   **25.** Trapezoid   **27.** Rhombus
29. 95°   **31.** 58°
33. $m\angle 1 = 90°, m\angle 2 = 50°, m\angle 3 = 130°, m\angle 4 = 50°,$
   $m\angle 5 = 50°, m\angle 6 = 40°, m\angle 7 = 90°, m\angle 8 = 130°,$
   $m\angle 9 = 140°, m\angle 10 = 40°, m\angle 11 = 140°, m\angle 12 = 40°$
35. 540°   **37.** 1080°   **39.** 1800°   **41. a)** 60°   **b)** 120°
43. a) 120°   **b)** 60°   **45. a)** 140°   **b)** 40°   **47.** $x = 20, y = 16$
49. $x = 6, y = 10$ **51.** $x = 20, y = 21.25$ **53.** 3   **55.** $3\frac{1}{3}$
57. 33   **59.** 18   **61.** 71°   **63.** 9   **65.** 10   **67.** 110°
69. a) 80 miles   **b)** 200 miles   **71. a)** 113.1 mi   **b)** 75.4 mi
73. 70 ft   **75.** 55°   **77.** 35°
79. a) $m\angle HMF = m\angle TMB, m\angle HFM = m\angle TBM,$
   $m\angle MHF = m\angle MTB$
   b) 44 ft
80. a) $m\angle CED = m\angle ABC, m\angle ACB = m\angle DCE,$
   $m\angle BAC = m\angle CDE$
   b) ≈ 2141.49 ft

SECTION 8.3, PAGE 474

1. a) Perimeter   **b)** Area   **3.** Circle   **5.** 1.5 in.²
7. 2500 cm²   **9. a)** 32 ft²   **b)** 24 ft
11. a) 600 cm²   **b)** 100 cm
13. a) 288 in.²   **b)** 74 in.
15. a) 201.06 in.²   **b)** 50.27 in.
17. a) 132.73 ft²   **b)** 40.84 ft
19. a) 17 yd   **b)** 40 yd   **c)** 60 yd²
21. a) 12 km   **b)** 30 km   **c)** 30 km²
23. 7.73 ft²   **25.** 8 in.²   **27.** 90 yd²   **29.** 27.43 ft²
31. 56.55 mm²   **33.** 7 yd²   **35.** 121.5 ft²   **37.** 70,000 cm²
39. 0.4072 m²   **41. a)** $6594.50   **b)** $8794.50
43. $3566.60   **45.** $4795.20   **47.** $351.31
49. a) 288 ft   **b)** 4700 tiles   **51.** 21 ft   **53.** 312 ft   **55.** 24 cm²
57. Answers will vary.   **58.** Answers will vary.

SECTION 8.4, PAGE 487

1. Volume   **3.** Platonic   **5.** Right   **7. a)** 64 ft³   **b)** 112 ft²
9. a) 150.80 in.³   **b)** 175.93 in.²
11. a) 131.95 cm³   **b)** 163.22 cm²
13. a) 3053.63 mi³   **b)** 1017.88 mi²   **15.** 162 m³
17. 720 cm³   **19.** 1500 ft³   **21.** 284.46 cm³
23. 392 ft³   **25.** 24 ft³   **27.** 189 ft³   **29.** 16.8 yd³
31. 1,500,000 cm³   **33.** 0.5 m³
35. a) 225 ft³   **b)** $2475

37. a) 120,000 cm³ **b)** 120,000 mℓ **c)** 120 ℓ

39. 200.98 cm³ **41. a)** 401.11 in.³ **b)** 263.02 in.²

43. a) Circular pan base: 63.62 in.²; rectangular
pan base: 63 in.²
b) Circular pan volume: 127.24 in.³; rectangular
pan volume: 126 in.³
c) Circular pan

45. 14.14 in.³ **47.** 283.04 in.³ **49.** 6 faces

51. 6 vertices **53.** 30 edges

55. a) ≈5.11 × 10⁸ km² **b)** ≈3.79 × 10⁷ km²
c) ≈13 times larger **d)** ≈1.09 × 10¹² km³
e) ≈2.20 × 10¹⁰ km³ **f)** ≈50 times larger

57. a) $V_1 = a^3$; $V_2 = a^2b$; $V_3 = a^2b$; $V_4 = ab^2$;
$V_5 = a^2b$; $V_6 = ab^2$; $V_7 = b^3$
b) ab^2
c) Answers will vary.

59. The new volume will be 8 times the original volume.

60. a) 330 in.³ **b)** ≈330.84 in.³

SECTION 8.5, PAGE 505

1. Reflection **3.** Vector **5.** Rotation **7.** Symmetry

This figure contains the answers for Exercises 9 and 10.

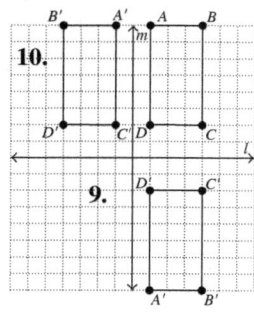

This figure contains the answers for Exercises 11 and 12.

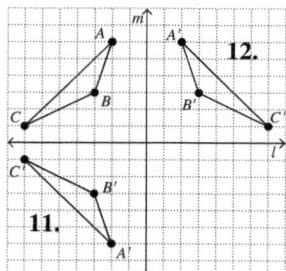

This figure contains the answers for Exercises 13 and 14.

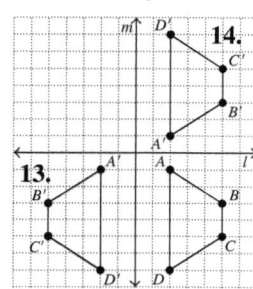

This figure contains the answers for Exercises 15 and 16.

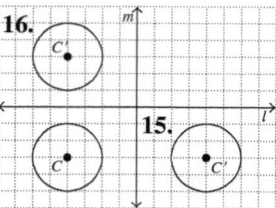

This figure contains the answers for Exercises 17 and 18.

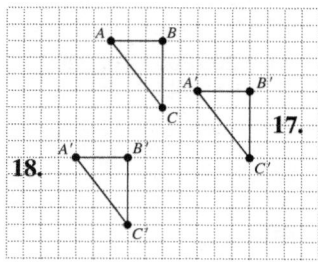

This figure contains the answers for Exercises 19 and 20.

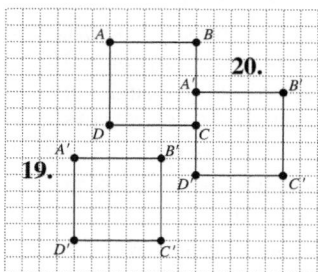

This figure contains the answers for Exercises 21 and 22.

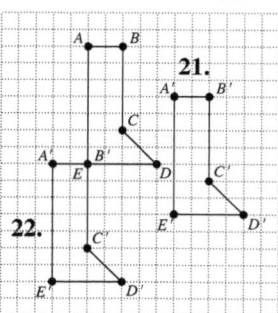

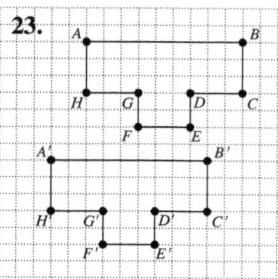

This figure contains the answers for Exercises 25 and 26.

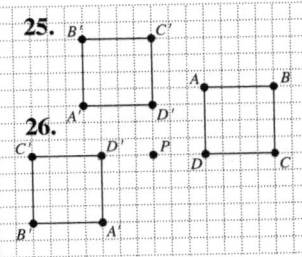

This figure contains the answers for Exercises 27 and 28.

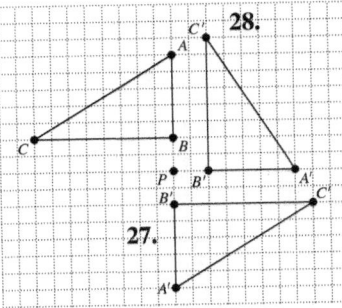

This figure contains the answers for Exercises 29 and 30.

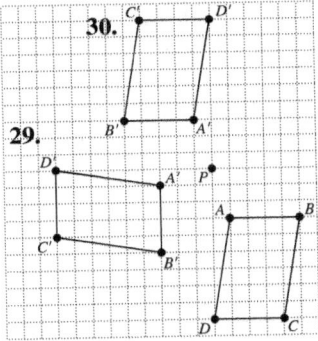

This figure contains the answers for Exercises 31 and 32.

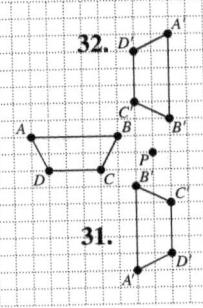

This figure contains the answers for Exercises 33 and 34.

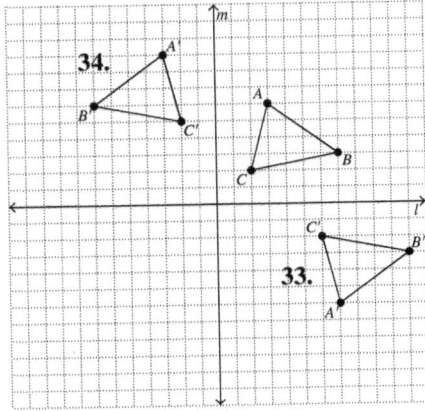

This figure contains the answers for Exercises 35 and 36.

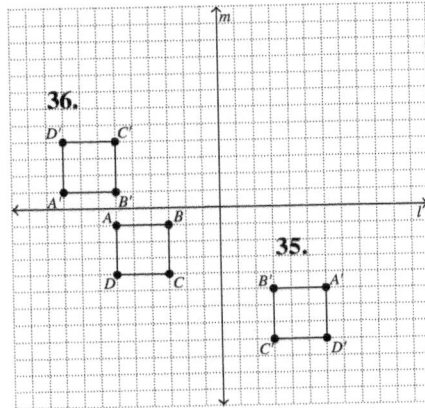

This figure contains the answers for Exercises 37 and 38.

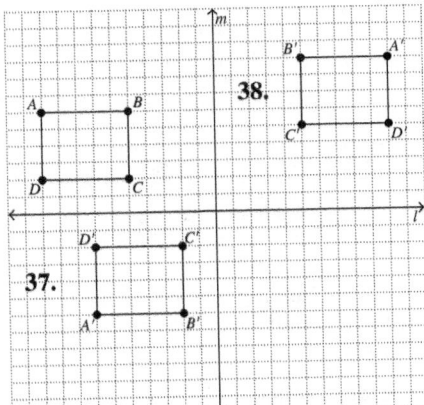

This figure contains the answers for Exercises 39 and 40.

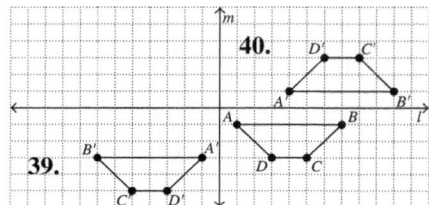

41. a) 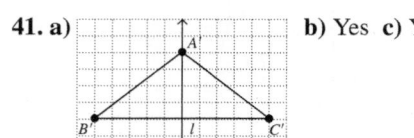 **b)** Yes **c)** Yes

43. a) 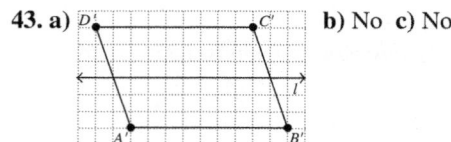 **b)** No **c)** No

45. a) 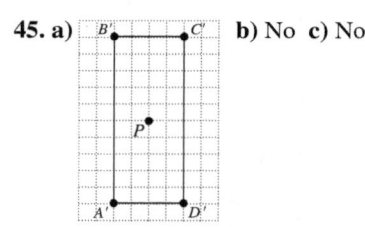 **b)** No **c)** No

d) 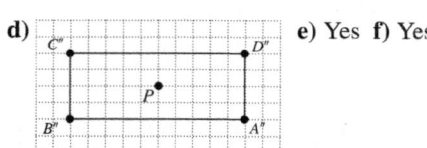 **e)** Yes **f)** Yes

47. a) a)–c)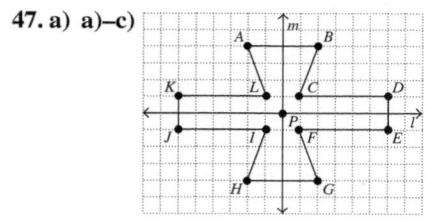

d) No. Any 90° rotation will result in the figure being in a different position than the starting position.

49. Answers will vary.

51. A regular octagon cannot be used as a tessellating shape.

53. Although answers will vary depending on the font, the following capital letters have reflective symmetry about a horizontal line drawn through the center of the letter: B, C, D, E, H, I, K, O, X.

54. Although answers will vary depending on the font, the following capital letters have reflective symmetry about a vertical line drawn through the center of the letter: A, H, I, M, O, T, U, V, W, X, Y.

55. Although answers will vary depending on the font, the following capital letters have 180° rotational symmetry about a point in the center of the letter: H, I, N, O, S, X, Z.

SECTION 8.6, PAGE 515

1. Rubber **3.** Klein **5.** Jordan

7. Answers will vary. **9.** Answers will vary.

11. Answers will vary. **13.** Answers will vary.

15. Outside **17.** Inside **19.** Inside **21.** 0 **23.** 1

25. 5 **27.** 3 **29.** Larger than 5 **31.** 4

33. a) Answers will vary. **b)** Answers will vary.
 c) Answers will vary. **d)** Answers will vary.

35. One **37.** Two

39. a) No, it has an inside and an outside. **b)** Two
 c) Two **d)** Two strips, one inside the other

41. Answers will vary. **43.** Answers will vary.

SECTION 8.7, PAGE 523

1. Parallel **3.** Two **5.** Sphere **7.** Geodesic

9.

11.

13. a)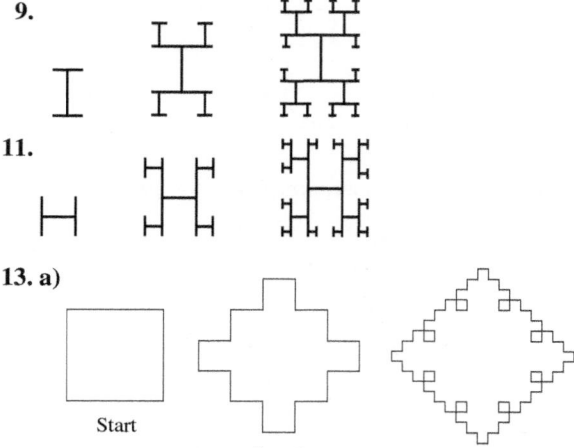

 b) Infinite **c)** Finite

15. Each type of geometry can be used in its own frame of reference.

17. Coastlines, trees, mountains, galaxies, rivers, weather patterns, brains, lungs, blood supply

REVIEW EXERCISES, PAGE 526

1. $\angle CBF$ or $\angle FBC$ **2.** $\overline{BC}$ **3.** $\triangle BFC$ **4.** $\overleftrightarrow{BH}$ **5.** $\{F\}$

6. $\{\ \}$ **7.** 62.4° **8.** 79.5° **9.** 10.2 in.

10. 2 in. **11.** 58° **12.** 92°

13. $m\angle 1 = 43°$, $m\angle 2 = 106°$, $m\angle 3 = 74°$, $m\angle 4 = 63°$, $m\angle 5 = 117°$, $m\angle 6 = 63°$

14. 1080° **15. a)** 99 mi² **b)** 40 mi

16. a) 36 m² **b)** 28 m **17. a)** 84 in.² **b)** 42 in.

18. a) 6 km² **b)** 12 km **19. a)** 153.94 ft² **b)** 43.98 ft

20. 64.38 m² **21.** 79.73 yd² **22.** $2184

23. a) 165 cm³ **b)** 206 cm²

24. a) 254.47 ft³ **b)** 226.19 ft²

25. a) 603.19 mm³ **b)** 435.20 mm²

26. a) 523.60 yd³ **b)** 314.16 yd² **27.** 432 m³ **28.** 28 ft³

29. 75.40 cm³ **30.** 1257.16 in.³

31. a) ≈ 67.88 ft³ **b)** ≈ 4610.71 lb; yes **c)** ≈ 510.33 gal

This figure contains the answers for Exercises 32 and 33.

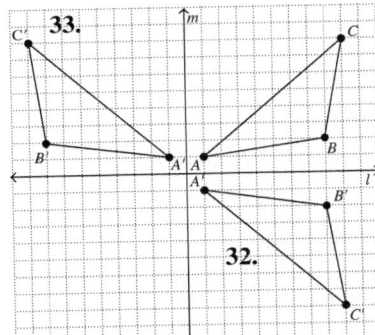

This figure contains the answers for Exercises 34 and 35.

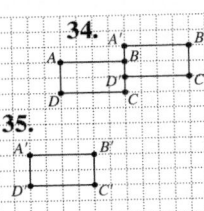

This figure contains the answers for Exercises 36–38.

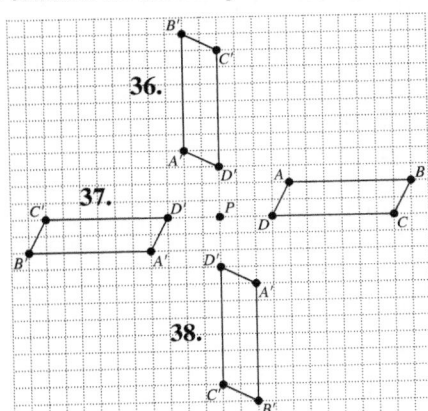

This figure contains the answers for Exercises 39 and 40.

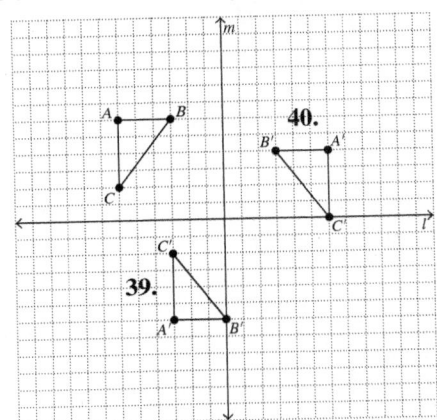

41. Yes **42.** No **43.** No **44.** Yes

45. a) Answers will vary. **b)** Answers will vary.
    c) Answers will vary. **d)** Answers will vary.

46. Answers will vary. **47.** Outside

48. Euclidean: Given a line and a point not on the line, one and only one line can be drawn parallel to the given line through the given point. Elliptical: Given a line and a point not on the line, no line can be drawn through the given point parallel to the given line. Hyperbolic: Given a line and a point not on the line, two or more lines can be drawn through the given point parallel to the given line.

49.

50.

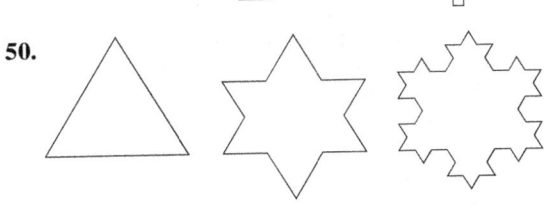

CHAPTER TEST, PAGE 529

1. $\angle BAE$ or $\angle EAB$ **2.** $\triangle BCD$ **3.** {D} **4.** $\overleftrightarrow{AC}$ **5.** 48.2°

6. 106.5° **7.** 68° **8.** 720° **9.** ≈ 2.69 cm

10. a) 5 in. **b)** 30 in. **c)** 30 in.²

11. a) ≈ 1436.76 cm³ **b)** ≈ 615.75 cm²

12. 59.43 m³ **13.** 112 ft³

14.

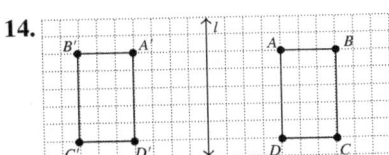

15.

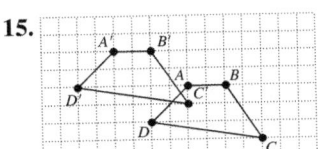

16.

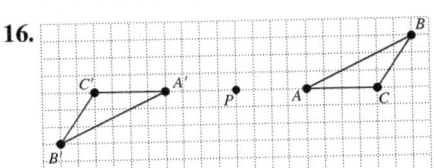

17.

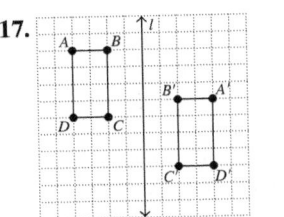

18. a) No **b)** Yes

19. A surface with one side and one edge

20. a) Answers will vary. **b)** Answers will vary.

Chapter 9

SECTION 9.1, PAGE 538

1. Binary   **3.** Closed   **5.** Identity   **7.** Inverse   **9.** Commutative

11. $a + b = b + a$ for any elements a and b; $3 + 4 = 4 + 3$

13. $(a \cdot b) \cdot c = a \cdot (b \cdot c)$ for any elements a, b, and c; $(2 \cdot 3) \cdot 4 = 2 \cdot (3 \cdot 4)$

15. $7 - 3 \stackrel{?}{=} 3 - 7$
$\quad\quad 4 \neq -4$

17. $(8 \div 4) \div 2 \stackrel{?}{=} 8 \div (4 \div 2)$
$\quad\quad\quad\quad 1 \neq 4$

19. a) Yes; the sum of any two integers is an integer.   **b)** Yes; 0.
c) Yes.   **d)** $(1 + 2) + 3 = (1 + 2) + 3$
e) $2 + 4 = 4 + 2$
f) Yes; it satisfies the five properties needed.

21. a) Yes; the sum of any two positive integers is a positive integer.   **b)** No.   **c)** Since there is no identity element, each element does not have an inverse.
d) $(2 + 3) + 4 = 2 + (3 + 4)$ **e)** $1 + 2 = 2 + 1$
f) No; there is no identity element.

23. Yes; it satisfies the four properties needed.

25. No; the system is not closed.

27. Yes; it satisfies the five properties needed.

29. No; there is no identity element.

31. No; the system is not closed. For example, $\frac{1}{0}$ is undefined.

33. No; the system is not closed. For example, $\sqrt{2} + (-\sqrt{2}) = 0$, which is rational. There is also no identity element.

35. Answers will vary.

SECTION 9.2, PAGE 546

1. $\{1, 2, 3, 4, 5, 6, 7, 8, 9, 10, 11, 12\}$ **3.** Identity

5. Associative   **7.** Commutative   **9.** Commutative

11. 1   **13.** 3   **15.** 4   **17.** 8   **19.** 7   **21.** 6   **23.** 8   **25.** 12

27.

+	1	2	3	4	5	6
1	2	3	4	5	6	1
2	3	4	5	6	1	2
3	4	5	6	1	2	3
4	5	6	1	2	3	4
5	6	1	2	3	4	5
6	1	2	3	4	5	6

29. 2   **31.** 3   **33.** 4

35.

+	1	2	3	4	5	6	7
1	2	3	4	5	6	7	1
2	3	4	5	6	7	1	2
3	4	5	6	7	1	2	3
4	5	6	7	1	2	3	4
5	6	7	1	2	3	4	5
6	7	1	2	3	4	5	6
7	1	2	3	4	5	6	7

37. 1   **39.** 3   **41.** 7

43. Yes; it satisfies the five required properties.

45. No

47. Yes; C is the identity element, since the row next to C is identical to the top row, and the column under C is identical to the far-left column.

49. a) The inverse of A is A, since $A \, \text{⊥⊤} \, A = C$.
b) The inverse of B is B, since $B \, \text{⊥⊤} \, B = C$.
c) The inverse of C is C, since $C \, \text{⊥⊤} \, C = C$.

51. No; the elements are not symmetric about the main diagonal.

53. a) $\{0, 2, 4, 6\}$ **b)** Q **c)** Yes   **d)** Yes; 0
e) Yes; 0–0, 2–6, 4–4, 6–2   **f)** $(2 \, Q \, 4) \, Q \, 6 = 2 \, Q \, (4 \, Q \, 6)$
g) Yes; $2 \, Q \, 6 = 6 \, Q \, 2$ **h)** Yes

55. a) $\{4, 5, L\}$ **b)** $\$$ **c)** Yes   **d)** Yes, L
e) Yes, 4–5, 5–4, L–L   **f)** $(4 \, \$ \, 5) \, \$ \, 5 = 4 \, \$ \, (5 \, \$ \, 5)$
g) Yes, $L \, \$ \, 4 = 4 \, \$ \, L$ **h)** Yes

57. a) Yes   **b)** Yes; O **c)** G–D, O–O, L does not have an inverse, D–G   **d)** $(L \, \text{☆} \, L) \, \text{☆} \, D = L; L \, \text{☆} \, (L \, \text{☆} \, D) = D$
e) No   **f)** Yes   **g)** No; not every element has an inverse and the associative property does not hold.

59. Not closed, not associative

61. No inverse for $\odot$ or for C, not associative

63. No identity element, no inverses, not associative, not commutative

65. a)

+	E	O
E	E	O
O	O	E

b) Yes, it is a commutative group; it satisfies the five properties.

67. a) It is closed; identity element is 6; inverses: 1–5, 2–2, 3–3, 4–4, 5–1, 6–6; is associative; for example,
$(2 \, ? \, 5) \, ? \, 3 \stackrel{?}{=} 2 \, ? \, (5 \, ? \, 3)$
$\quad\quad 3 \, ? \, 3 \stackrel{?}{=} 2 \, ? \, 2$
$\quad\quad\quad 6 = 6$
b) $3 \, ? \, 1 \stackrel{?}{=} 1 \, ? \, 3$
$\quad\quad 2 \neq 4$

69. a)

*	R	S	T	U	V	I
R	V	T	U	S	I	R
S	U	I	V	R	T	S
T	S	R	I	V	U	T
U	T	V	R	I	S	U
V	I	U	S	T	R	V
I	R	S	T	U	V	I

b) Yes. It is closed; identity element is I; inverses R–V, S–S, T–T, U–U, V–R, I–I; and the associative property will hold.
c) No, it is not commutative. For example, $R * S \neq S * R$.

70. a) Yes   **b)** Yes, 1   **c)** Yes, 1   **d)** Yes   **e)** Yes   **f)** Yes

SECTION 9.3, PAGE 556

1. $m - 1$ **3.** Remainder   **5.** Congruent   **7.** Sunday
9. Thursday   **11.** Sunday   **13.** Friday   **15.** August
17. August   **19.** September   **21.** July   **23.** 0   **25.** 4   **27.** 2
29. 1   **31.** 2   **33.** 1   **35.** 5   **37.** 5   **39.** 6   **41.** 5   **43.** 4
45. 3   **47.** 5   **49.** 5   **51.** { }   **53.** 1 and 6   **55.** 5
57. a) 1792, 1796, 1800, 1804, 1808
    b) 2024, 2028, 2032, 2036, 2040
    c) 2552, 2556, 2560, 2564, 2568, 2572
59. a) Working at Longboat Key
    b) Resting
    c) Working at The Meadows
61. a) 5   **b)** No   **c)** 54 weeks from this week
63. a) 3 P.M.–11 P.M.   **b)** 7 A.M.–3 P.M.   **c)** 7 A.M.–3 P.M.
65. a)

+	0	1	2
0	0	1	2
1	1	2	0
2	2	0	1

    b) Yes   **c)** Yes, 0   **d)** Yes; 0–0, 1–2, 2–1
    e) $(1 + 2) + 2 = 1 + (2 + 2)$
    f) Yes; $2 + 1 = 1 + 2$ **g)** Yes   **h)** Yes
67. a)

×	0	1	2	3
0	0	0	0	0
1	0	1	2	3
2	0	2	0	2
3	0	3	2	1

    b) Yes   **c)** Yes, 1
    d) No; no inverse for 0 or for 2, inverse of 1 is 1, inverse of 3 is 3
    e) $(1 \times 2) \times 3 = 1 \times (2 \times 3)$
    f) Yes; $2 \times 3 = 3 \times 2$
    g) No; 0 and 2 do not have an inverse.
69. 2   **71.** 1, 2, 3   **73.** 0   **75.** 2
77. a) Tuesday   **b)** Wednesday
79. Knowledge is power   **80.** 0

SECTION 9.4, PAGE 566

1. Matrix   **3.** 2   **5.** Dimensions   **7.** Columns, rows

9. $\begin{bmatrix} -1 & 9 \\ 7 & -9 \end{bmatrix}$ **11.** $\begin{bmatrix} 2 & 5 \\ -5 & 4 \\ 8 & 6 \end{bmatrix}$ **13.** $\begin{bmatrix} 6 & 4 \\ -5 & -10 \end{bmatrix}$

15. $\begin{bmatrix} 1 & 9 \\ 18 & 17 \\ -2 & 2 \end{bmatrix}$ **17.** $\begin{bmatrix} 6 & 4 \\ 10 & 0 \end{bmatrix}$ **19.** $\begin{bmatrix} 0 & 13 \\ 22 & 0 \end{bmatrix}$

21. $\begin{bmatrix} 23 & 56 \\ 10 & 24 \end{bmatrix}$ **23.** $\begin{bmatrix} 15 \\ 22 \end{bmatrix}$ **25.** $\begin{bmatrix} 13 & 2 \end{bmatrix}$

27. $A + B = \begin{bmatrix} -1 & 5 \\ -6 & 7 \end{bmatrix}$; $A \times B = \begin{bmatrix} -32 & 15 \\ -24 & 12 \end{bmatrix}$

29. $A + B = \begin{bmatrix} 8 & 1 & 3 \\ 4 & 3 & 0 \end{bmatrix}$; cannot be multiplied

31. Cannot be added; $A \times B = \begin{bmatrix} 24 & 34 \\ 24 & 34 \end{bmatrix}$

33. Cannot be added; cannot be multiplied

35. $A + B = B + A = \begin{bmatrix} 7 & 2 \\ 6 & 1 \end{bmatrix}$

37. $A + B = B + A = \begin{bmatrix} 8 & 0 \\ 6 & -8 \end{bmatrix}$

39. $(A + B) + C = A + (B + C) = \begin{bmatrix} 7 & 13 \\ 2 & 10 \end{bmatrix}$

41. $(A + B) + C = A + (B + C) = \begin{bmatrix} 4 & 14 \\ 3 & 15 \\ 8 & -1 \end{bmatrix}$

43. No

45. Yes

47. $(A \times B) \times C = A \times (B \times C) = \begin{bmatrix} 41 & 13 \\ 56 & 16 \end{bmatrix}$

49. $(A \times B) \times C = A \times (B \times C) = \begin{bmatrix} 16 & -10 \\ -24 & 2 \end{bmatrix}$

51.

	Student	Adult	Senior citizen	
	158	280	95	Matinee
	215	362	94	Evening

53.

	Large	Small	
	38	50	Sugar
	56	72	Flour
	17	26	Milk
	10	14	Eggs

55. $\begin{bmatrix} 36.04 & 47.52 \end{bmatrix}$

57. a) Yes   **b)** Yes, $\begin{bmatrix} 0 & 0 \\ 0 & 0 \\ 0 & 0 \end{bmatrix}$ **c)** Yes

d) Answers will vary.   **e)** Yes   **f)** Yes

59. Answers will vary.

61. Answers will vary.

63. $\begin{bmatrix} a & b \\ c & d \end{bmatrix} \begin{bmatrix} 1 & 0 \\ 0 & 1 \end{bmatrix} = \begin{bmatrix} 1 & 0 \\ 0 & 1 \end{bmatrix} \begin{bmatrix} a & b \\ c & d \end{bmatrix} = \begin{bmatrix} a & b \\ c & d \end{bmatrix}$

REVIEW EXERCISES, PAGE 570

1. A binary operation is an operation that can be performed on two and only two elements of a set. The result is a single element.

2. A mathematical system consists of a set of elements and at least one binary operation.

3. No; for example $2 \div 3 = \frac{2}{3}$ and $\frac{2}{3}$ is not a whole number.

4. Yes; the difference of any two real numbers is a real number.

5. 5   **6.** 6   **7.** 10   **8.** 9

9. Closure, identity element, inverses, and associative property

10. Yes

11. No; no inverse for any integer except 1 and −1

12. No; no identity element

13. No; no inverse for 0

14. No identity element; no inverses

15. Not every element has an inverse; not associative.
For example, $(P ? P) ? 4 \neq P ? (P ? 4)$.

16. Not associative. For example, $(! \square p) \square ? \neq ! \square (p \square ?)$.

17. a) {☺, ●, ♀, ♂}   **b)** △   **c)** Yes   **d)** Yes; ☺
e) Yes; ☺−☺, ●−♂, ♀−♀, ♂−●   **f)** (●△♀)△♂=●△(♀△♂)
g) Yes; ●△♂=♂△●   **h)** Yes

18. 5   **19.** 0   **20.** 2   **21.** 4   **22.** 11   **23.** 12   **24.** 1   **25.** 6

26. 3   **27.** 0, 2, 4, 6   **28.** 4   **29.** 9   **30.** No solution   **31.** 8

32.

+	0	1	2	3	4	5
0	0	1	2	3	4	5
1	1	2	3	4	5	0
2	2	3	4	5	0	1
3	3	4	5	0	1	2
4	4	5	0	1	2	3
5	5	0	1	2	3	4

Yes; it is a commutative group.

33.

×	0	1	2	3
0	0	0	0	0
1	0	1	2	3
2	0	2	0	2
3	0	3	2	1

No; no inverse for 0 or 2

34. a) Yes   **b)** Yes

35. $\begin{bmatrix} 5 & 4 \\ -2 & 2 \end{bmatrix}$ **36.** $\begin{bmatrix} -1 & -6 \\ -4 & -2 \end{bmatrix}$ **37.** $\begin{bmatrix} 0 & -13 \\ -11 & -4 \end{bmatrix}$

38. $\begin{bmatrix} 5 & 8 \\ -9 & -15 \end{bmatrix}$ **39.** $\begin{bmatrix} -9 & -3 \\ -4 & -1 \end{bmatrix}$

CHAPTER TEST, PAGE 572

1. A set of elements and a binary operation.

2. Closure, identity element, inverses, associative property, commutative property

3. Yes

4. No; not closed, not associative, not commutative,

5.

+	1	2	3	4	5
1	2	3	4	5	1
2	3	4	5	1	2
3	4	5	1	2	3
4	5	1	2	3	4
5	1	2	3	4	5

6. Yes; it is a commutative group.

7. a) □   **b)** Yes   **c)** Yes, T **d)** S **e)** S

8. No, not closed, not associative.

9. Yes; it is a commutative group.

10. Yes; it is a commutative group.   **11.** 3   **12.** 8

13. 4   **14.** 2   **15.** 5   **16.** { }

17. a)

×	0	1	2	3	4
0	0	0	0	0	0
1	0	1	2	3	4
2	0	2	4	1	3
3	0	3	1	4	2
4	0	4	3	2	1

b) No; no inverse for 0

18. $\begin{bmatrix} 0 & 3 \\ 8 & -3 \end{bmatrix}$ **19.** $\begin{bmatrix} 10 & -4 \\ -9 & -21 \end{bmatrix}$ **20.** $\begin{bmatrix} 1 & 7 \\ -36 & -12 \end{bmatrix}$

Chapter 10

SECTION 10.1, PAGE 580

1. 100   **3.** 100   **5.** Previous   **7.** 40%   **9.** 35%   **11.** 0.8%

13. 378.0%   **15.** 0.09   **17.** 0.0724   **19.** 0.0025   **21.** 1.359

23. 25%   **25.** 4.7 grams   **27.** 471 comedy DVDs

29. 192 horror DVDs   **31.** 21.46 mℓ of oxygen

33. 9.7 mℓ of chlorine   **35.** 47.0%   **37.** 14.0%

39. 0.8% decrease

41. a) 5.2% decrease
b) 7.4% increase
c) 2.2% increase
d) 2.7% decrease
e) 7.4% increase

43. $11,250   **45.** 2.56   **47.** 25%   **49.** 8%   **51.** 300   **53.** 150

55. a) $3.81   **b)** $9.53   **c)** $76.84

57. $50.25   **59.** 12 students   **61.** $42,265   **63.** 5.3% decrease

65. 18.6% decrease   **67.** 40%   **69.** He will have a loss of $10.

71. a) No, the 25% discount is greater.
b) $145.34   **c)** $142.49   **d)** Yes

73. a) $200 decreased by 25% is greater by $25.
b) $100 increased by 50% is greater by $50.
c) $100 increased by 100% is greater by $200.

SECTION 10.2, PAGE 591

1. Principal   **3.** Interest   **5.** Rate   **7.** United States
9. $33.75   **11.** $24.06   **13.** $15.85   **15.** $9.12   **17.** $113.20
19. 1.38%   **21.** $7000   **23.** 2 years   **25.** $46,350

27. a) $131.25   **b)** $3631.25

29. a) $416.10   **b)** $3233.90   **c)** 8.58%

31. 200%   **33.** 69 days   **35.** 272 days   **37.** 264 days

39. October 28   **41.** March 24   **43.** $2023.06   **45.** $1459.33

47. $5278.99   **49.** $850.64   **51.** $6086.82   **53.** $2646.24

55. a) November 13, 2019   **b)** $998.48
c) $1.52   **d)** 0.1506%

57. a) 409.01%   **b)** 204.50%   **c)** 102.25%

59. a) $375   **b)** $369.86   **c)** $5.14   **d)** 1.39%

60. a) $6.42   **b)** $14.20   **c)** $22.47   **d)** Answers will vary.

SECTION 10.3, PAGE 601

1. Profit   **3.** Variable   **5.** Compound

7. a) $935.11   **b)** $85.11

9. a) $3589.24   **b)** $589.24

11. a) $10,939.90   **b)** $939.90

13. a) $8666.26   **b)** $666.26

15. $17,451.31   **17.** $85,282.13   **19.** $9657.62   **21.** $1608.63

23. $10,910.27   **25.** $3106.62   **27.** $2341.82   **29.** 3.53%

31. a) 1.91%   **b)** 1.82%   **c)** Prospero Bank

33. a) 2.25%   **b)** 2.26%   **c)** Key Bank

35. Yes, the APY should be 2.43%.   **37.** $30,274.73

39. $13,351.33

41. a) $555,000   **b)** $58,907.61   **c)** $98.51

43. a) $1040.60, $40.60   **b)** $1082.43, $82.43
   c) $1169.86, $169.86   **d)** No

45. The simple interest is $20,000. The compound interest is
   $22,138.58. The compound interest is greater by $2138.58.

47. 11.74%   **49.** 8.84%

SECTION 10.4, PAGE 615

1. Installment   **3.** Annual   **5.** Installment   **7.** $36.56

9. $355.72   **11. a)** $553.41   **b)** $133.40

13. a) $818.40   **b)** 8.5%

15. a) $6823.40   **b)** 6.5%

17. a) 6.0%   **b)** $726.00   **c)** $7858.00

19. a) $2818.20   **b)** $689.39   **c)** $347.90   **d)** $8614.17

21. a) $30.00   **b)** $54.73, which rounds up to $55

23. a) $21.14, which rounds up to $22
   b) $34.39, which rounds up to $35

25. a) $29.93   **b)** $1889.30

27. a) $1.56   **b)** $133.11

29. a) $512.00   **b)** $6.66   **c)** $638.43

31. a) $121.78   **b)** $1.52   **c)** $133.07
   d) The interest charged using the average daily balance
      method is $0.04 less than the interest charged using the
      previous balance method.

33. a) $11.96   **b)** $886.96

35. a) $25   **b)** $35.60   **c)** 8.5%   **d)** 6.5%   **37.** $14,077.97

SECTION 10.5, PAGE 627

1. Mortgage   **3.** Points   **5.** Adjusted

7. $751.25   **9.** $1449.25

11. a) $41,250   **b)** $233,750   **c)** $1115.96

13. a) $467,500   **b)** $1,870,000   **c)** $11,830.54

15. a) $39,000   **b)** $156,000   **c)** $3120

17. a) $802.20   **b)** $1411.19   **c)** No

19. a) $111,209.40   **b)** $36,209.40   **c)** $210.02

21. a) $110,000   **b)** $13,200   **c)** $4026.40   **d)** $2498.27
   e) $3161.52   **f)** Yes   **g)** $481.60

23. Bank B   **25. a)** 4.0%   **b)** $1139.24   **c)** 5.5%

27. a) $664,491.60   **b)** $927,379.20   **c)** $1,230,933.60

29. a) The variable-rate mortgage   **b)** $2149.80

SECTION 10.6, PAGE 636

1. Annuity   **3.** Sinking   **5.** Immediate   **7.** 401k

9. $268,437.42   **11.** $299,929.32   **13.** $381.64

15. $175.48   **17.** $149,271.58   **19.** $40,305.56

21. $264.97   **23.** $5160.71

25. a) $23,003.87   **b)** $826,980.88   **c)** $349,496.41
   d) $12,000   **e)** $36,000   **f)** Alberto

REVIEW EXERCISES, PAGE 639

1. 35%   **2.** 58.3%   **3.** 62.5%   **4.** 4.1%
5. 0.98% ≈ 1.0%   **6.** 314.1%   **7.** 0.08   **8.** 0.229
9. 1.23   **10.** 0.0025   **11.** 0.008$\overline{3}$ **12.** 0.0000045
13. 3.0%   **14.** 12.5%   **15.** 20.0%   **16.** 80   **17.** 91.8
18. $7.83   **19.** 40 people   **20.** 12%   **21.** $563.20
22. 5.0%   **23.** $1700   **24.** 4.75 years   **25.** $7875

26. a) $1380   **b)** $4620   **c)** 14.9%

27. $5646.82   **28.** $3634.36

29. a) $6691.13; $1691.13   **b)** $6719.58; $1719.58
   c) $6734.28; $1734.28   **d)** $6744.25; $1744.25
   e) $6749.13; $1749.13

30. $2588.98   **31.** 5.76%   **32.** $4713.10

33. a) $7800   **b)** $18,200   **c)** $2407.86   **d)** $343.46

34. a) $2644   **b)** $10,576   **c)** $1006.84   **d)** $321.74

35. a) $56,000   **b)** $84,000   **c)** $13,440   **d)** 6.0%

36. a) 6.0%   **b)** $253.16   **c)** $4150.34

37. a) 4.5%   **b)** $32.06   **c)** $1420.43

38. a) $6.31   **b)** $847.61   **c)** $508.99   **d)** $6.62   **e)** $847.92

39. a) $2.60   **b)** $546.92   **c)** $382.68   **d)** $5.36   **e)** $549.68

40. a) $33,925   **b)** $1345.49   **c)** $515.68
   d) $832.35   **e)** Yes

41. a) $13,485   **b)** $756.73   **c)** $24.42
   d) $285,907.80   **e)** $196,007.80

42. a) 4.5%   **b)** $1797.73   **c)** 6.0%

43. $48,378.57   **44.** $3668.72

CHAPTER TEST, PAGE 642

1. a) $97.20   **d)** 2 years   **2.** $89.78

3. $1989.78   **4.** $2523.20   **5.** $123.20

6. a) $7961.99, $461.99   **b)** $3036.68, $536.68

7. $1997.50   **8.** $181.46   **9.** 8.5%

10. a) 4.5%   **b)** $64.02   **c)** $2836.28

11. a) $12.30   **b)** $1146.57   **c)** $765.67
   d) $10.72   **e)** $1144.99

12. $32,250   **13.** $3655

14. $2694.53   **15.** $1037.63   **16.** $1341.80   **17.** Yes

18. a) $409,451.80   **b)** $194,451.80

19. $25,858.69   **20.** $277.28

Chapter 11

SECTION 11.1, PAGE 654

1. Experiment   **3.** Event   **5.** Theoretical

7. a) 0   **b)** 1   **c)** 0, 1   **9.** 1

11. a) $\dfrac{7}{15}$ **b)** $\dfrac{1}{3}$ **c)** $\dfrac{1}{5}$ **13. a)** $\dfrac{3}{7}$ **b)** $\dfrac{8}{21}$ **c)** $\dfrac{1}{21}$

15. a) 0.3094   **b)** 0.1902   **c)** 0.1375

17. a) 0.1708   **b)** 0.8292   **c)** 0.0928   **d)** 0.9072

19. a) $\dfrac{24}{80} = 0.3$ **b)** $\dfrac{11}{22} = 0.5$ **c)** $\dfrac{8}{8} = 1$ **d)** $\dfrac{0}{90} = 0$

21. a) 0   **b)** $\dfrac{50}{250} = 0.2$ **c)** 1

23. a) $\dfrac{1}{5}$ **b)** $\dfrac{4}{5}$ **c)** $\dfrac{1}{3}$ **d)** $\dfrac{2}{3}$

25. $\dfrac{1}{4}$ **27.** $\dfrac{3}{4}$ **29.** $\dfrac{3}{13}$ **31.** $\dfrac{1}{52}$ **33.** $\dfrac{1}{26}$

35. a) $\dfrac{1}{4}$ **b)** $\dfrac{1}{2}$ **c)** $\dfrac{1}{4}$ **d)** $\dfrac{3}{4}$

37. a) $\dfrac{1}{2}$ **b)** $\dfrac{3}{8}$ **c)** $\dfrac{1}{8}$ **d)** $\dfrac{7}{8}$

39. $\dfrac{1}{5}$ **41.** $\dfrac{1}{10}$ **43.** $\dfrac{9}{10}$ **45.** $\dfrac{1}{12}$ **47.** $\dfrac{1}{6}$ **49.** $\dfrac{3}{4}$ **51.** $\dfrac{2}{9}$

53. $\dfrac{5}{9}$ **55.** 0   **57.** $\dfrac{4}{11}$ **59.** $\dfrac{7}{11}$ **61.** $\dfrac{3}{11}$ **63.** $\dfrac{5}{18}$

65. $\dfrac{13}{18}$ **67.** $\dfrac{11}{12}$ **69.** 0

71. No, it means that if a coin was flipped many times, about $\frac{1}{2}$ of the tosses would land heads up.

73. a) Roll a die many times and then determine the relative frequency of 5's to the total number of rolls.
   b) Answers will vary.   **c)** Answers will vary.

75. $\dfrac{13}{36}$ **77.** $\dfrac{23}{36}$ **79.** $\dfrac{1}{3}$ **81.** $\dfrac{23}{36}$

83. a) Answers will vary.   **b)** Answers will vary.
   c) Answers will vary.

85. a) Answers will vary.   **b)** Answers will vary.
   c) Answers will vary.

87. a) Answers will vary.   **b)** Answers will vary.
   c) Answers will vary.

88. 29 dots.

SECTION 11.2, PAGE 664

1. Against   **3.** 4 : 1   **5.** 2 : 1   **7.** $\dfrac{1}{4}$

9. a) $\dfrac{1}{100}$ **b)** $\dfrac{99}{100}$ **c)** 99 : 1   **d)** 1 : 99

11. a) $\dfrac{3}{5}$ **b)** $\dfrac{2}{5}$ **c)** 2 : 3   **d)** 3 : 2

13. 5 : 1   **15.** 2 : 4 or 1 : 2   **17.** 12 : 1, 1 : 12

19. 10 : 3, 3 : 10   **21.** 1 : 1   **23.** 5 : 3   **25.** 8 : 7   **27.** 7 : 8

29. 8 : 7   **31.** $\dfrac{1}{5}$ **33.** 1 : 4   **35.** 74 : 1

37. 0.36   **39.** 16 : 9   **41.** 4 : 21

43. a) $\dfrac{9}{11}$ **b)** $\dfrac{2}{11}$ **45. a)** $\dfrac{7}{11}$ **b)** $\dfrac{4}{11}$ **47. a)** $\dfrac{1}{101}$ **b)** $\dfrac{100}{101}$

49. a) $\dfrac{197}{199}$ **b)** $\dfrac{2}{199}$ **51. a)** 91 : 9   **b)** 9 : 91   **53.** 1 : 9

55. a) $\dfrac{1}{33}$ **b)** 32 : 1

57. Horse 2, $\dfrac{1}{16}$; Horse 3, $\dfrac{1}{2}$; Horse 4, $\dfrac{1}{4}$; Horse 5, $\dfrac{3}{16}$

59. 96.5 : 3.5 or 193 : 7

SECTION 11.3, PAGE 673

1. Expected   **3.** Positive   **5.** 52 hot dogs   **7.** $8800

9. a) 2.2 in.   **b)** 68.2 in.   **11. a)** $1.60   **b)** $43.40

13. a) −$0.50   **b)** $0.50   **15. a)** ≈ −$0.17   **b)** ≈ $0.17

17. a) Yes, because you have a positive expectation of $\dfrac{1}{5}$

   b) Yes, because you have a positive expectation of $\dfrac{1}{2}$

19. $7.50   **21.** −$1.75   **23.** $750   **25.** $416.67   **27.** $2.50

29. −$8.25   **31.** Gain of 290 employees   **33.** 56 employees

35. $167,000   **37.** ≈ 141.51 service calls   **39.** $13,600

41. a) $\dfrac{9}{16}, \dfrac{1}{4}, \dfrac{1}{8}, \dfrac{1}{16}$ **b)** $11.81   **43.** $5.50

45. a) −$2.00   **b)** $1.00   **47. a)** −$2.00   **b)** $1.00

49. a) $3.50   **b)** $5.50   **51. a)** $2.25   **b)** $4.25

53. An amount greater than $1200

55. −$0.053, or −5.3¢   **57. a)** $458.33   **b)** $308.33

SECTION 11.4, PAGE 683

1. Sample   **3.** 24   **5. a)** 676   **b)** 650

7. a) 576   **b)** 552   **9. a)** 4

9. b)

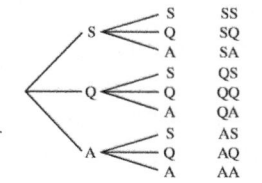

Sample Space

c) $\dfrac{1}{4}$ **d)** $\dfrac{1}{2}$ **e)** $\dfrac{1}{4}$ **f)** $\dfrac{3}{4}$

11. a) 9   **b)**

Sample Space

c) $\dfrac{1}{9}$ **d)** $\dfrac{4}{9}$ **e)** $\dfrac{5}{9}$

13. a) 12   **b)**

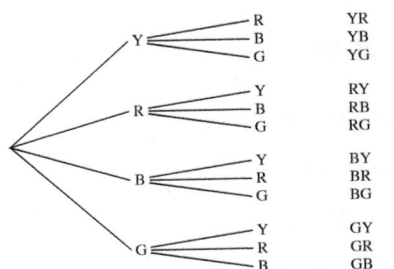

Sample Space

Y	YR
	YB
	YG
R	RY
	RB
	RG
B	BY
	BR
	BG
G	GY
	GR
	GB

c) $\frac{1}{2}$ **d)** 1   **e)** $\frac{1}{2}$

15. a) 9   **b)**

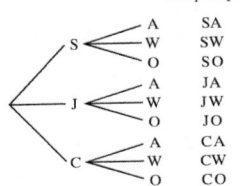

Sample Space

S → A, W, O : SA, SW, SO
J → A, W, O : JA, JW, JO
C → A, W, O : CA, CW, CO

c) $\frac{1}{3}$ **d)** $\frac{1}{9}$ **e)** $\frac{2}{3}$

17. a) 36   **b)**

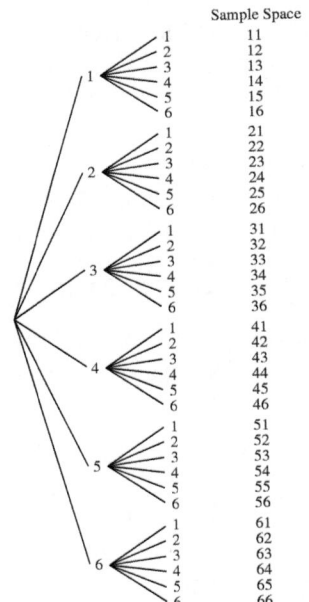

Sample Space

11 12 13 14 15 16
21 22 23 24 25 26
31 32 33 34 35 36
41 42 43 44 45 46
51 52 53 54 55 56
61 62 63 64 65 66

c) $\frac{1}{6}$ **d)** $\frac{5}{36}$ **e)** $\frac{1}{36}$ **f)** No

19. a) 12   **b)**

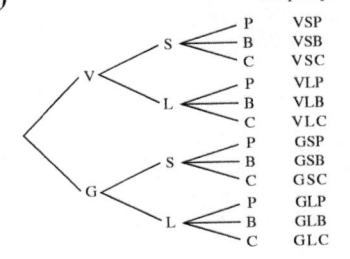

Sample Space

VSP VSB VSC
VLP VLB VLC
GSP GSB GSC
GLP GLB GLC

c) $\frac{1}{2}$ **d)** $\frac{1}{6}$ **e)** $\frac{2}{3}$

21. a) 16   **b)**

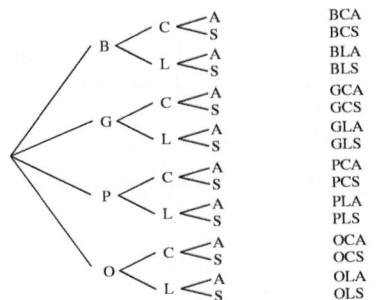

Sample Space

BCA BCS BLA BLS
GCA GCS GLA GLS
PCA PCS PLA PLS
OCA OCS OLA OLS

c) $\frac{1}{4}$ **d)** $\frac{1}{8}$ **e)** $\frac{3}{4}$

23. a) 27   **b)**

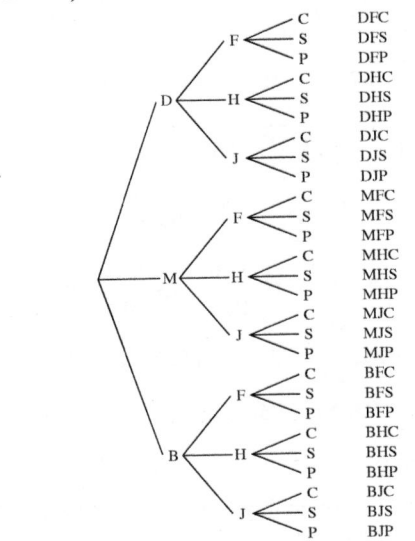

Sample Space

DFC DFS DFP
DHC DHS DHP
DJC DJS DJP
MFC MFS MFP
MHC MHS MHP
MJC MJS MJP
BFC BFS BFP
BHC BHS BHP
BJC BJS BJP

c) $\frac{1}{3}$ **d)** $\frac{1}{9}$ **e)** $\frac{2}{3}$

25. a) 18   **b)**

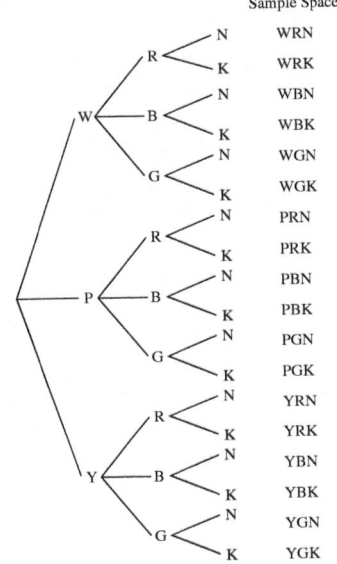

Sample Space

WRN WRK
WBN WBK
WGN WGK
PRN PRK
PBN PBK
PGN PGK
YRN YRK
YBN YBK
YGN YGK

c) $\frac{1}{18}$ **d)** $\frac{1}{3}$ **e)** $\frac{2}{3}$

27. a) $\frac{1}{3}$ b) $2:1$ c) $\frac{2}{3}$
   d) No, the probability of selecting a red chip is not the same as the probability of selecting a white chip.
   e) Answers will vary.
29. 3; 1 red, 1 blue, and 1 brown 30. 5 faces
31. a) 120 b) STORE 32. a) 720 b) BASKET

SECTION 11.5, PAGE 695
1. Compound 3. And 5. Independent 7. Independent
9. $P(A) + P(B) - P(A \text{ and } B)$ 11. $\frac{4}{5}$ 13. 0.6 15. 0.1
17. 0.4 19. $\frac{1}{3}$ 21. $\frac{2}{3}$ 23. $\frac{2}{13}$ 25. $\frac{4}{13}$ 27. $\frac{8}{13}$
29. a) $\frac{1}{16}$ b) $\frac{1}{19}$ 31. a) $\frac{1}{16}$ b) $\frac{5}{76}$ 33. a) $\frac{1}{40}$ b) $\frac{1}{38}$
35. a) $\frac{9}{25}$ b) $\frac{33}{95}$ 37. $\frac{11}{20}$ 39. $\frac{2}{5}$ 41. $\frac{1}{8}$ 43. $\frac{9}{64}$ 45. $\frac{3}{32}$
47. $\frac{15}{32}$ 49. $\frac{3}{7}$ 51. $\frac{16}{49}$ 53. $\frac{33}{49}$ 55. $\frac{1}{35}$ 57. $\frac{31}{35}$ 59. $\frac{1}{8}$
61. $\frac{7}{8}$ 63. a) $\frac{1}{32}$ b) $\frac{1}{2}$ 65. a) $\frac{1}{18}$ b) $\frac{2}{35}$
67. a) $\frac{1}{4}$ b) $\frac{17}{70}$ 69. $\frac{376}{925}$ 71. $\frac{549}{925}$ 73. $\frac{253}{580}$ 75. $\frac{1}{1015}$
77. $\frac{70}{253}$ 79. $\frac{7}{253}$ 81. $\frac{1}{24}$ 83. $\frac{5}{12}$ 85. 0.75 87. No
89. 0.36 91. 0.36 93. No 95. 0.00004
97. 0.000999 99. $\frac{14}{45}$
101. Favors dealer. The probability of at least one diamond is ≈ 0.44, which is less than 0.5.
103. $\frac{1}{9}$ 104. $\frac{1}{4}$ 105. $\frac{1}{2}$ 106. 1 107. Answers will vary.

SECTION 11.6, PAGE 703
1. Conditional 3. $\frac{3}{7}$ 5. a) $\frac{1}{4}$ b) $\frac{1}{2}$ 7. a) $\frac{2}{13}$ b) $\frac{2}{3}$
9. a) $\frac{1}{6}$ b) $\frac{1}{3}$ 11. a) $\frac{1}{2}$ b) $\frac{1}{3}$ 13. a) $\frac{1}{3}$ b) 0
15. $\frac{3}{4}$ 17. 1 19. $\frac{2}{3}$ 21. $\frac{3}{5}$ 23. $\frac{1}{2}$ 25. $\frac{3}{5}$ 27. $\frac{1}{7}$ 29. $\frac{1}{16}$
31. $\frac{1}{7}$ 33. $\frac{1}{18}$ 35. 0 37. $\frac{2}{3}$ 39. a) $\frac{23}{45}$ b) $\frac{3}{5}$
41. a) $\frac{5}{9}$ b) $\frac{15}{23}$ 43. a) 0.5941 b) 0.6251
45. a) 0.3425 b) 0.3605 47. $\frac{7}{13}$ 49. $\frac{4}{15}$ 51. $\frac{11}{15}$
53. $\frac{34}{55}$ 55. $\frac{59}{80}$ 57. $\frac{72}{131}$ 59. $\frac{10}{11}$ 61. $\frac{3}{19}$ 63. $\frac{44}{47}$
65. 0.5248 67. 0.6190 69. 0.1722 71. $\frac{93}{200}$ 73. $\frac{15}{52}$
75. a) 140 b) 120 c) $\frac{7}{10}$ d) $\frac{3}{5}$ e) $\frac{2}{3}$ f) $\frac{4}{7}$
   g) Because A and B are not independent events

77. a) 0.3 b) 0.5 c) Yes; $P(A \mid B) = P(A) \cdot P(B)$
78. $\frac{1}{3}$ 79. $\frac{1}{2}$ 80. $\frac{1}{3}$ 81. $\frac{1}{3}$ 82. 1 83. $\frac{1}{3}$

SECTION 11.7, PAGE 715
1. Permutation 3. $n!$ 5. $\dfrac{n!}{(n-r)!}$ 7. $_5P_3$ 9. 720 11. 1
13. 6 15. 1 17. 24 19. 1680 21. 10,000 23. 24
25. 280 27. a) 1,000,000 b) $\dfrac{1}{1,000,000} = 0.000001$
29. 1,000,000,000 31. a) 120 b) 120 c) 24 d) 6
33. a) 40,320 b) $\dfrac{1}{40,320}$ 35. a) 720 b) $\dfrac{1}{30}$
37. a) 676,000 b) 468,000 c) 104,000 d) 421,200
39. a) 362,880 b) 100,800 c) 60,480
41. 720 43. 15,120 45. 524,160
47. 362,880 49. 50,400 51. 22,680 53. 630
55. a) 3125 b) 128 c) 0.00032 57. 12,600 sec, or 3.5 hr
59. No. For example, $_5P_2 = 20$ but $_5P_3 = 60$
60. 56 61. 600 62. a) 360 b) CHOICE
63. a) 2520 b) STUDENT

SECTION 11.8, PAGE 723
1. Combination 3. Permutations
5. a) Permutation b) Combination 7. 10
9. a) 15 b) 360 11. a) 1 b) 1 13. a) 120 b) 720
15. $\frac{1}{26}$ 17. $\frac{4}{143}$ 19. $\frac{1}{30}$ 21. 120 23. 15 25. 56
27. 13,983,816 29. 495 31. 24 33. 378,378 35. 2352
37. 6930 39. 8820 41. 700 43. 600 45. a) 45 b) 56
47. a)

$$
\begin{array}{ccccccccc}
 & & & & 1 & & & & \\
 & & & 1 & & 1 & & & \\
 & & 1 & & 2 & & 1 & & \\
 & 1 & & 3 & & 3 & & 1 & \\
1 & & 4 & & 6 & & 4 & & 1
\end{array}
$$

   b) 1   5   10   10   5   1
49. a) 24 b) 24 51. 60,060
53. a) The order is important. Since the numbers may be repeated, it is not a true permutation lock.
   b) 64,000 c) 59,280

SECTION 11.9, PAGE 728
1. $\dfrac{_{12}C_4}{_{20}C_4}$ 3. $\dfrac{_4C_3}{_{52}C_3}$ 5. $\dfrac{_8C_5}{_{15}C_5}$ 7. $\dfrac{_{23}C_5}{_{120}C_5}$ 9. $\dfrac{1}{26}$ 11. $\dfrac{4}{143}$
13. $\dfrac{1}{30}$ 15. $\dfrac{40}{143}$ 17. $\dfrac{3}{10}$ 19. $\dfrac{7}{10}$ 21. 0.1434 23. 0.0244
25. $\dfrac{2}{33}$ 27. $\dfrac{4}{33}$ 29. $\dfrac{144}{81,719}$ 31. a) $\dfrac{1}{123,760}$ b) $\dfrac{1}{30,940}$
33. $\dfrac{2197}{33,320}$ 35. $\dfrac{44}{159}$ 37. a) $\dfrac{33}{54,145}$ b) $\dfrac{1}{2,598,960}$

39. a) $\dfrac{1}{2,162,160}$ **b)** $\dfrac{1}{6435}$

41. 1; Since there are more people than options, two or more people must have the same number of hairs on their head.

SECTION 11.10, PAGE 738

1. Trials   **3.** Success   **5.** 0.08192   **7.** 0.32922   **9.** 0.015625

11. a) $P(x) = (_nC_x)(0.14)^x(0.86)^{n-x}$
   b) $P(2) = (_{12}C_2)(0.14)^2(0.86)^{10}$ **c)** 0.28628

13. a) 0.00977   **b)** 0.24609   **c)** 0.00977

15. a) 0.4096   **b)** 0.5904   **17.** 0.22713

19. 0.27869   **21.** 0.06877   **23.** 0.36603

25. a) 0.01635   **b)** 0.00185   **c)** 0.99815

27. a) 0.11188   **b)** 0.29663   **29.** 0; it will be midnight.

REVIEW EXERCISES, PAGE 742

1. Answers will vary.   **2.** Answers will vary.

3. $\dfrac{8}{9}$ **4.** Answers will vary.   **5.** $\dfrac{8}{25}$ **6.** $\dfrac{1}{2}$ **7.** $\dfrac{4}{5}$

8. 1   **9.** $\dfrac{1}{5}$ **10.** $\dfrac{7}{20}$ **11.** $\dfrac{13}{60}$ **12.** $\dfrac{19}{30}$ **13.** $\dfrac{17}{20}$

14. a) 69 : 31   **b)** 31 : 69   **15.** 5 : 3   **16.** $\dfrac{3}{85}$ **17.** 3 : 1

18. a) −$1.20   **b)** −$3.60   **19. a)** −$0.23   **b)** $0.23
   c) Lose $23.08   **20.** 660 people   **21.** $4.00   **22.** $3.50

23. a)

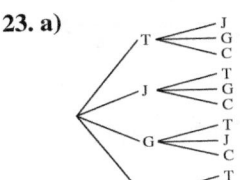

   b)   Sample Space   **c)** $\dfrac{1}{12}$
         TJ
         TG
         TC
         JT
         JG
         JC
         GT
         GJ
         GC
         CT
         CJ
         CG

24. a)

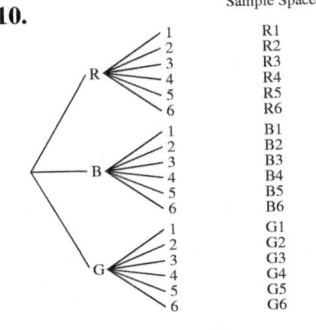

b)   Sample Space   **c)** $\dfrac{1}{4}$ **d)** $\dfrac{3}{4}$
      H1
      H2
      H3
      H4
      T1
      T2
      T3
      T4

25. $\dfrac{1}{4}$ **26.** $\dfrac{9}{64}$ **27.** $\dfrac{5}{16}$ **28.** $\dfrac{7}{8}$ **29.** 1   **30.** $\dfrac{3}{16}$

31. $\dfrac{1}{22}$ **32.** $\dfrac{14}{55}$ **33.** $\dfrac{41}{55}$ **34.** $\dfrac{1}{22}$ **35.** $\dfrac{5}{8}$

36. In favor, 3 : 5; against, 5 : 3   **37.** $3.75   **38.** $\dfrac{7}{8}$ **39.** $\dfrac{3}{20}$

40. $\dfrac{5}{6}$ **41.** $\dfrac{3}{22}$ **42.** $\dfrac{1}{2}$ **43.** $\dfrac{23}{40}$ **44.** $\dfrac{3}{17}$ **45.** $\dfrac{3}{4}$ **46.** $\dfrac{12}{17}$

47. a) 24   **b)** $4500   **48.** 30   **49.** 720   **50.** 504   **51.** 20

52. 3,628,800   **53. a)** $\dfrac{1}{12,103,014}$ **b)** $\dfrac{1}{302,575,350}$

54. 35,640   **55.** 560   **56.** $\dfrac{1}{221}$ **57.** $\dfrac{1}{12}$ **58.** $\dfrac{1}{18}$ **59.** $\dfrac{1}{24}$

60. $\dfrac{11}{12}$ **61.** $\dfrac{5}{182}$ **62.** $\dfrac{45}{364}$ **63.** $\dfrac{2}{13}$ **64.** $\dfrac{11}{13}$

65. a) $P(x) = (_nC_x)(0.6)^x(0.4)^{n-x}$
   b) $P(50) = (_{100}C_{50})(0.6)^{50}(0.4)^{50}$ **c)** 0.01034

66. 0.0512   **67. a)** 0.01024   **b)** 0.98976

CHAPTER TEST, PAGE 745

1. $\dfrac{11}{25}$ **2.** $\dfrac{4}{9}$ **3.** $\dfrac{5}{9}$ **4.** $\dfrac{2}{3}$ **5.** $\dfrac{1}{36}$

6. $\dfrac{5}{18}$ **7.** $\dfrac{5}{12}$ **8.** $\dfrac{8}{13}$ **9.** 18

10.

Sample Space

R — 1 R1 / 2 R2 / 3 R3 / 4 R4 / 5 R5 / 6 R6
B — 1 B1 / 2 B2 / 3 B3 / 4 B4 / 5 B5 / 6 B6
G — 1 G1 / 2 G2 / 3 G3 / 4 G4 / 5 G5 / 6 G6

11. $\dfrac{1}{18}$ **12.** $\dfrac{4}{9}$ **13.** $\dfrac{5}{6}$ **14.** 1,581,840   **15. a)** 5 : 4   **b)** 5 : 4

16. $0   **17. a)** $\dfrac{121}{228}$ **b)** $\dfrac{113}{228}$ **c)** $\dfrac{53}{113}$ **d)** $\dfrac{47}{107}$

18. 120   **19. a)** $\dfrac{3}{38}$ **b)** $\dfrac{35}{38}$ **20.** 0.2304

Chapter 12

SECTION 12.1, PAGE 754

1. Statistics   3. Descriptive   5. Sample   7. Random

9. Stratified   11. Stratified sample   13. Cluster sample

15. Systematic sample   17. Convenience sample

19. Random sample

21. The patients may have improved on their own without taking honey.

23. Half the students in a population are expected to be below average.

25. A recommended toothpaste may not be better than all other types of toothpaste.

27. Most driving is done close to home. Thus, one might expect more accidents close to home.

29. We don't know how many of each professor's students were surveyed. Perhaps more of Professor Fogal's students than Professor Bond's students were surveyed. Also, because more students prefer a teacher does not mean that he or she is a better teacher. For example, a particular teacher may be an easier grader and that may be why that teacher is preferred.

31. Just because they are more expensive does not mean that they will last longer.

33. There may be deep sections in the pond, so it may not be safe to go wading.

35. a)

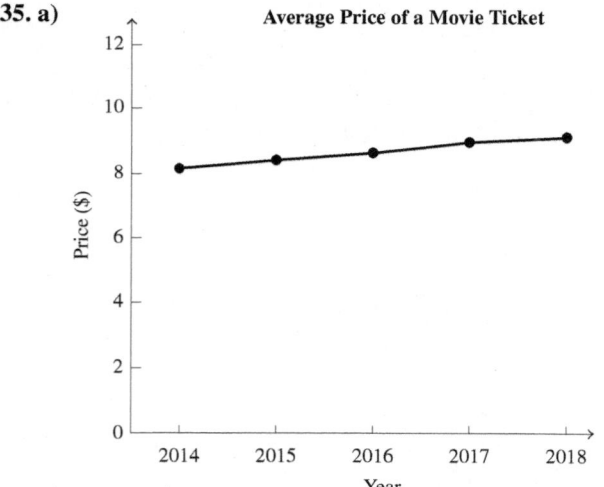

b)

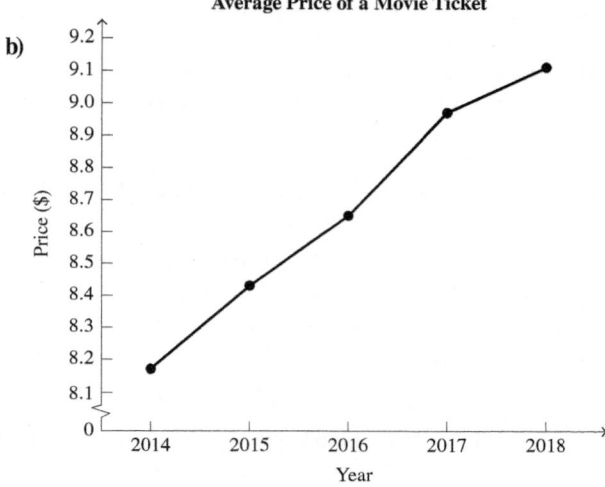

37. a)

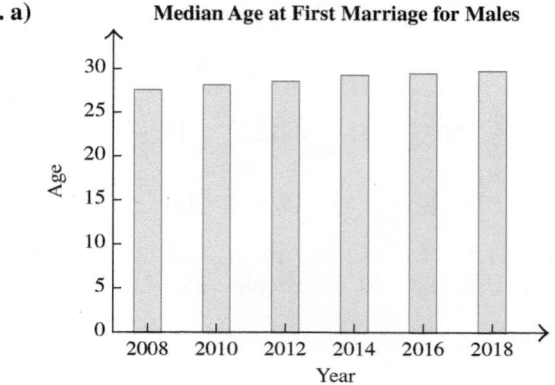

b)

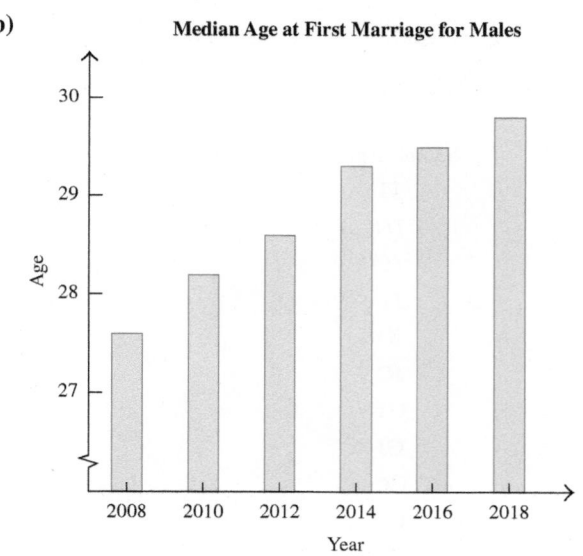

39. a)

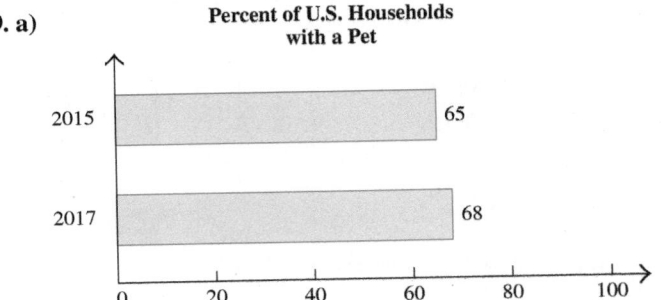

Percent of U.S. Households
with a Pet

2015 — 65

2017 — 68

0   20   40   60   80   100

b) Answers will vary.

41. Yes, the sum of its parts is 121%. The sum of the parts of a circle graph should be 100%. When the total percent of responses is more than 100%, a circle graph is not an appropriate graph to display the data. A bar graph is more appropriate in this situation.

43. Biased

SECTION 12.2, PAGE 766

1. Frequency **3.** Mark **5.** Histogram **7. a)** Stem **b)** Leaf

9. a) 20 **b)** 7 **c)** 19 **d)** 16–22 **e)** 51–57

11.

Number of Visits	Number of Students
0	3
1	8
2	3
3	5
4	2
5	7
6	2
7	3
8	4
9	1
10	2

13.

Copies Sold (millions)	Number of Books
0–99	21
100–199	7
200–299	1
300–399	0
400–499	0
500–599	1

15.

Copies Sold (millions)	Number of Books
50–99	21
100–149	6
150–199	1
200–249	1
250–299	0
300–349	0
350–399	0
400–499	0
450–449	0
500–549	1

17.

Population (millions)	Number of Cities
12.5–17.4	9
17.5–22.4	8
22.5–27.4	2
27.5–32.4	0
32.5–37.4	0
37.5–42.4	1

19.

Population (millions)	Number of Cities
12.0–17.9	10
18.0–23.9	8
24.0–29.9	1
30.0–35.9	0
36.0–41.9	1

21.

Cost of Living	Number of States
80.0–94.9	21
95.0–109.9	17
110.0–124.9	4
125.0–139.9	7
140.0–154.9	0
155.0–169.9	0
170.0–184.9	0
185.0–199.9	1

23.

Cost of Living	Number of States
85.7–100.6	30
100.7–115.6	8
115.7–130.6	6
130.7–145.6	5
145.7–160.6	0
160.7–175.6	0
175.7–190.6	1

25. 1|2 represents 12

```
0| 4  6  7  8
1| 2  2  3  5  6  7  8  9
2| 1  2  3  5  7
3| 3  4
4| 0
```

27. a)

Ticket Sales (millions $)	Number of Movies
877–996	7
997–1116	0
1117–1236	4
1237–1356	2
1357–1476	0
1477–1596	0
1597–1716	1
1717–1836	1

b) and **c)**

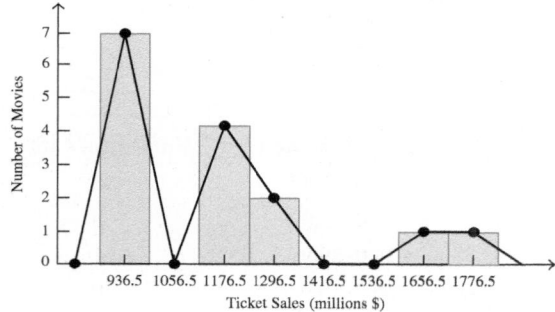

Top-15-Grossing Movies in the United States

29. a)

Age	Number of People
20–30	10
31–41	14
42–52	9
53–63	4
64–74	3

b) and c)

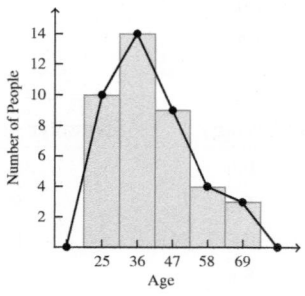

Ages of 40 People
Visiting an Art Museum

d) 2|3 represents 23

```
2| 0  0  3  3  6  6  9
3| 0  0  0  1  1  2  3  4  4  5  5  7  8  9
4| 0  0  0  2  5  7  9  9  9
5| 0  1  1  4  7
6| 2  3  6  9
7| 2
```

31. a) and **b)**

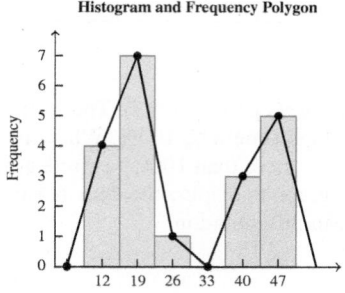

Histogram and Frequency Polygon

33. a) 28   **b)** 4   **c)** 2   **d)** 75

e)

Number of Televisions	Number of Homes
0	2
1	4
2	8
3	6
4	4
5	3
6	1

35. a) 7   **b)** 16   **c)** 36

d)

Number of Messages	Number of People	Number of Messages	Number of People
3	2	7	3
4	3	8	8
5	7	9	6
6	4	10	3

e)

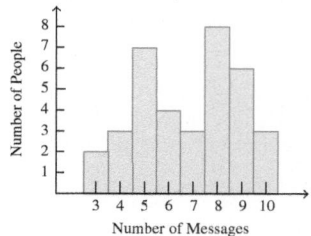

Number of E-mail Messages Sent

37.

Tuition:	$10,430.40
Room/board:	$9921.60
Books/supplies:	$763.20
Other	$4324.80

39. a) Answers will vary.   **b)** Answers will vary.
   c) Answers will vary.   **d)** Answers will vary.
   e) Answers will vary.

41. a) There are 6 F's.   **b)** Answers will vary.

SECTION 12.3, PAGE 778

1. Average   **3.** Mean   **5.** Mode   **7.** Quartiles

9. a) $\bar{x}$ **b)** μ

11. 17, 16, 16, 20   **13.** 64.3, 67, none, 61

15. 8,8, No mode, 8   **17.** 13.1, 11, 1, 18.5

19. 11.9, 12.5, 13, 11.5   **21.** 76.4, 73, none, 79

23. a) 4.9, 5, 5, 6   **b)** 5.3, 5, 5, 6   **c)** Only the mean
   d) The mean and the midrange

25. a) $42,100   **b)** $33,500   **c)** $32,000   **d)** $58,500
   e) The median because it is lower.   **f)** The mean because is
   it higher.

27. a) 4.8 million   **b)** 4.1 million   **c)** 3.5 million
   d) 7.2 million

29. a) $8.88   **b)** $8.25   **c)** no mode   **d)** $9.87

31. 405   **33. a)** Yes   **b)** No   **c)** No   **d)** Yes

35. 93 or greater   **37. a)** 25 or greater   **b)** Yes, 95 or greater
   c) 17 or greater   **d)** 77 or greater   **39.** $87.45

41. a) $91   **b)** $85   **c)** $99

43. a) No   **b)** Yes, Kendra was in better relative position.

45. a) $600   **b)** $610   **c)** 25%
   d) 25%   **e)** 17%   **f)** $62,000

47. Answers will vary. The National Center for Health uses the
   median for averages in this exercise.

49. One example is 1, 1, 2, 5, 6.

51. One example is 81, 82, 83, 85, 86, 87.

53. One example: 1, 2, 3, 3, 4, 5 changed to 1, 2, 3, 4, 4, 5.

55. The data must be ranked.

57. He is taller than approximately 35 percent of all
   kindergarten children.   **59.** Second quartile, median

61. a)

Ruth	Mantle
0.290	0.300
0.359	0.365
0.301	0.304
0.272	0.275
0.315	0.321

   b) Mantle's is greater in every case.
   c) Ruth: 0.316; Mantle: 0.311; Ruth's is greater.
   d) Answers will vary.

   e) Ruth: 0.307; Mantle: 0.313; Mantle's is greater.
   f) Answers will vary.   **g)** Answers will vary.

63. 90   **65. a)** Answers will vary.   **b)** Answers will vary.
   c) Answers will vary.

66. a) Answers will vary. One example is 2, 3, 5, 7, 7.
   b) Answers will vary.

SECTION 12.4, PAGE 788

1. Variability   **3.** Standard deviation   **5.** Sample
7. 11, $\sqrt{16.5} \approx 4.06$ **9.** 6, $\sqrt{4.67} \approx 2.16$
11. 11, $\sqrt{15.2} \approx 3.90$ **13.** 5, $\sqrt{3} \approx 1.73$
15. $60, $\sqrt{550} \approx 23.45 **17.** $190, $\sqrt{4725.25} \approx 68.74
19. a) $63, $\sqrt{631.6} \approx 25.13 **b)** Answers will vary.
   c) Answers remain the same, range: $63, standard
   deviation $\approx$ $25.13.

21. a) Answers will vary.   **b)** Answers will vary.
   c) Answers will vary.
   d) If each number in a set of data is multiplied by n, the
   mean and standard deviation of the new set of data will
   be n times that of the original set of data.
   e) 20, 10

23. Answers will vary.

25. The first set will have the greater standard deviation
   because the scores have a greater spread about the mean.

27. They would be the same, since the spread of data about
   each mean is the same.

29. a) The standard deviation increases. There is a greater
   spread from the mean as they get older.
   b) Mean: $\approx$ 100 lb; normal range: $\approx$ 60 to 140 lb
   c) Mean: $\approx$ 62 in.; normal range: $\approx$ 53 to 68 in.
   d) $\approx$ 140 lb   **e)** $\approx$ 40 lb   **f)** 5%

31. a)

East		West	
Number of Oil Changes Made	Number of Days	Number of Oil Changes Made	Number of Days
15–20	2	15–20	0
21–26	2	21–26	0
27–32	5	27–32	6
33–38	4	33–38	9
39–44	7	39–44	4
45–50	1	45–50	6
51–56	1	51–56	0
57–62	2	57–62	0
63–68	1	63–68	0

b)

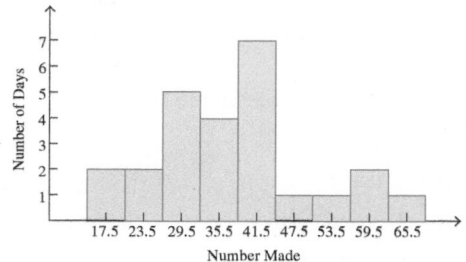

Number of Oil Changes Made Daily at East Store

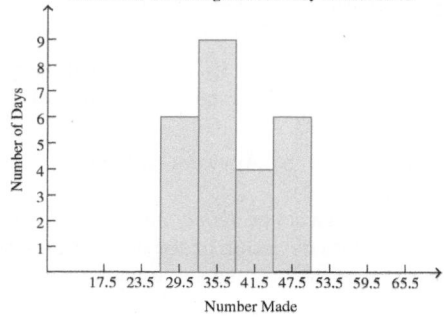

Number of Oil Changes Made Daily at West Store

c) They appear to have about the same mean, since they are both centered around 38.

d) The distribution for East is more spread out. Therefore, East has a greater standard deviation.

e) East: 38, West: 38   **f)** East: ≈ 12.64, West: ≈ 5.98

32. Answers will vary.   **33.** 6, 6, 6, 6, 6

SECTION 12.5, PAGE 804

1. Rectangular   **3. a)** Right   **b)** Left   **5.** Bell   **7.** 0

9. a) 68%   **b)** 95%   **c)** 99.7%

11. Answers will vary.   **13.** Answers will vary.

15. Normal   **17.** Skewed right   **19.** Bimodal

21. Rectangular   **23.** 0.5000   **25.** 0.8185   **27.** 0.9474

29. 0.0143   **31.** 0.9878   **33.** 0.0429   **35.** 79.67%

37. 97.13%   **39.** 93.24%   **41.** 0.88%   **43.** 21.96%

45. a) Luisa, Sarah, Eleanor   **b)** Jenny, Shenice
c) Sadaf, Heather, Kim-Liu

47. 50%   **49.** 50%   **51.** 81.85%   **53.** ≈ 53 adults

55. 69.15%   **57.** 24.17%   **59.** 2.28%   **61.** 93.32%

63. 59.87%   **65.** 50%   **67.** 11.51%   **69.** ≈ 23 cars

71. 25.1%   **73.** ≈ 187 females   **75.** 27.43%   **77.** 0.62%

79. ≈ 83 families   **81.** 1.79%

83. The standard deviation is too large.

85. a) B **b)** C **c)** A

87. The mean is the greatest value. The median is lower than the mean. The mode is the lowest value.

89. Answers will vary.

91. a) Katie: $z = 2.4$; Stella: $z = 1.7$
b) Katie. Her z-score is higher than Stella's z-score, which means her sales are further above the mean than Stella's sales.

93. -1.18 **94.** Answers will vary.   **95.** 2

SECTION 12.6, PAGE 819

1. Coefficient   **3. a)** 1   **b)** -1 **c)** 0   **5.** Positive

7. No correlation   **9.** Strong positive correlation

11. Yes   **13.** No   **15.** Yes   **17.** No

19. a)   **b)** 0.981   **c)** Yes   **d)** Yes

21. a)   **b)** 0.228   **c)** No   **d)** No

23. a)   **b)** 0.999   **c)** Yes   **d)** Yes

25. a)   **b)** -0.968 **c)** Yes   **d)** Yes

27. $y = 1.26x + 3.51$ **29.** $y = 0.18x + 24.82$

31. $y = 0.81x + 5.84$ **33.** $y = -0.08x + 9.50$

35. a) 0.987   **b)** Yes   **c)** $y = 1.58x + 2.91$

37. a) -0.979 **b)** Yes   **c)** $y = -1.07x + 83.30$
d) ≈ 48 cups

39. a) 0.961   **b)** Yes   **c)** $y = 0.20x - 1.05$ **d)** \$3.95

41. a) 0.993   **b)** Yes   **c)** $y = 3.90x + 1.94$
d) ≈ 21 kilocalories

43. a) Answers will vary.   **b)** Answers will vary.
c)

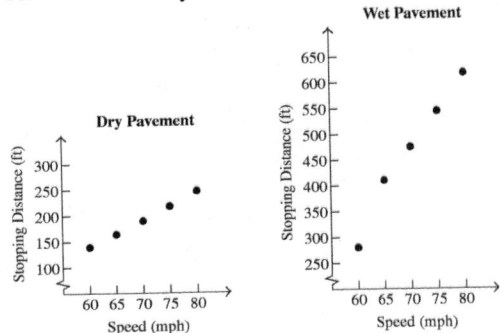

d) 0.999   **e)** 0.990   **f)** Answers will vary.
g) $y = 5.36x - 183.40$ **h)** $y = 16.22x - 669.80$
i) Dry, 229.3 ft; wet, 579.1 ft

45. a) Correlation will not change.   **b)** Answers will vary.

47. a) Answers will vary.   **b)** Answers will vary.
c) Answers will vary.   **d)** Answers will vary.
e) Answers will vary.   **f)** Answers will vary.
g) Answers will vary.

49. a) ≈ 0.922 **b)** Should be the same.

REVIEW EXERCISES, PAGE 825

1. a) A population consists of all items or people of interest.
b) A sample is a subset of the population.

2. a) A random sample is one where every item in the population has the same chance of being selected.
b) A systematic sample is obtained by selecting a random starting point and then selecting every nth item in a population.
c) A cluster sample consists of dividing the population into sections. Then randomly select sections to use and either select all the items in the selected sections or a random sample of items from the selected sections.
d) A stratified sample consists of dividing the population into strata and then taking a random sample from each strata.
e) A convenience sample uses data that are easily or readily obtained.

3. The candy bars may have lots of calories, or fat, or sodium. Therefore, it may not be healthy to eat them.

4. Sales may not necessarily be a good indicator of profit. Expenses must also be considered.

5. a)

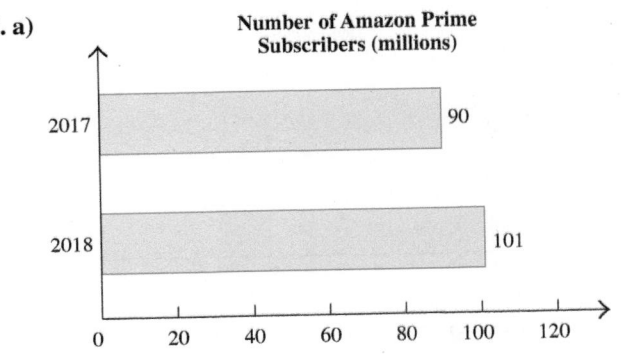

Number of Amazon Prime
Subscribers (millions)

b)

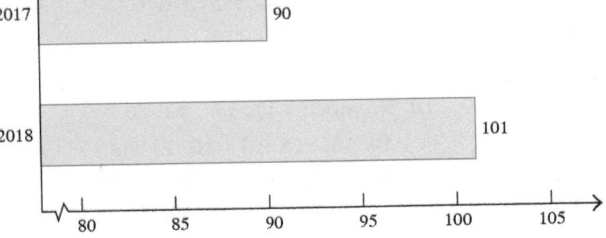

Number of Amazon Prime
Subscribers (millions)

6. a)

Class	Frequency
35	1
36	3
37	6
38	2
39	3
40	0
41	4
42	1
43	3
44	1
45	1

b) and c)

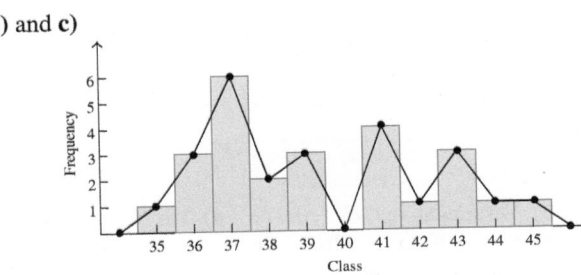

7. a)

High Temperature	Number of Cities
58–62	1
63–67	4
68–72	9
73–77	10
78–82	11
83–87	4
88–92	1

b) and c)

Average Monthly High Temperature
in July for Selected Cities

d) 5|8 represents 58

```
5|8
6|3 6 6 7 8 8 9
7|0 1 1 1 2 2 3 3 3 4 5 5 5 6 6 7 9 9 9
8|0 0 0 0 1 2 2 2 3 4 4 7
9|1
```

8. 78   **9.** 79   **10.** No mode   **11.** 78   **12.** 26
13. $\sqrt{80} \approx 8.94$ **14.** 18   **15.** 18   **16.** 12 and 17
17. 18.5   **18.** 19   **19.** $\sqrt{40} \approx 6.32$ **20.** 68.26%
21. 95.44%   **22.** 94.52%   **23.** 5.48%   **24.** 72.57%
25. 34.13%   **26.** 34.46%   **27.** 28.98%   **28.** 2.28%

29. a)

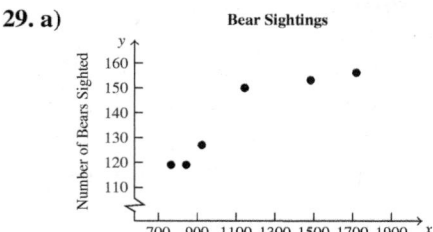

b) Yes; positive   **c)** 0.925   **d)** Yes
e) $y = 0.04x + 88.17$ **f)** ≈ 148 bears

30. a)

b) Yes; negative   **c)** -0.973 **d)** Yes
e) $y = -79.4x + 246.7$ **f)** ≈ 120 sold

31. 180 lb   **32.** 185 lb   **33.** 25%   **34.** 25%   **35.** 14%
36. 19,200 lb   **37.** 238 lb   **38.** 150.6 lb   **39.** 2
40. 0 and 2   **41.** 2   **42.** 3.5   **43.** 7   **44.** $\sqrt{3.79} \approx 1.95$
45.

Number of Children	Number of Families
0	5
1	4
2	5
3	3
4	1
5	0
6	1
7	1

46. and 47.

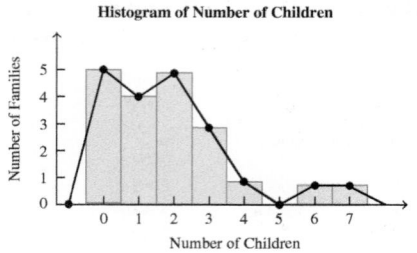

48. No, it is skewed to the right.   **49.** Answers will vary.
50. Answers will vary.

CHAPTER TEST, PAGE 828

1. 37   **2.** 38   **3.** 38   **4.** 34.5   **5.** 25   **6.** $\sqrt{84} \approx 9.17$
7.

Class	Frequency
25–30	7
31–36	5
37–42	1
43–48	7
49–54	5
55–60	3
61–66	2

8.

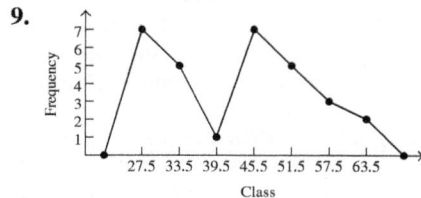

9.

10. $815   **11.** $790   **12.** 75%   **13.** 79%   **14.** $82,000
15. $860   **16.** 87.10%   **17.** 89.44%   **18.** 10.56%
19. 94.52%

20. a)   Time Spent Studying   **b)** Yes

c) 0.950   **d)** Yes   **e)** $y = 0.77x + 24.86$ **f)** ≈ 83

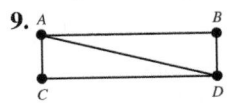

1. Graph   **3.** Edge   **5.** Path   **7.** Degree

9.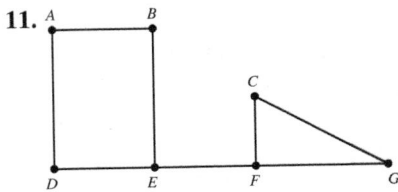

11.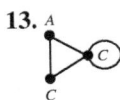

13.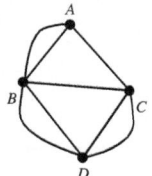

15. a) No, there is no edge connecting vertices *C* and *D*.
   b) Yes, there are edges connecting vertex *A* to vertex *B*, vertex *B* to vertex *C*, vertex *C* to vertex *E*, and vertex *E* to vertex *D*.
   c) No, the path does not begin and end with the same vertex.
   d) Yes, it is a path that begins and ends with the same vertex.

17. a) Yes. One example is *B, A, C, E, D, B, C*.   **b)** No   **c)** No

19. Yes. One example is *B, A, C, E, D, B*.

21.   **23.**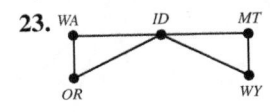

25.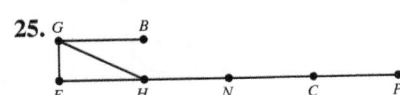

27. 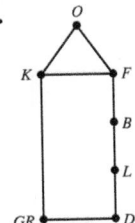   **29.**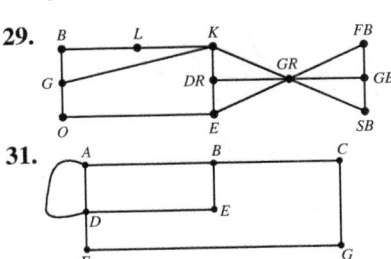

31. (see above)

33. Disconnected   **35.** Connected

37. a) Edge *BC*   **b)** Loop *FF*

39. a) Edge *BC*   **b)** No loops

41. Answers will vary.

43. a) Answers will vary.   **b)** Answers will vary.
   c) Answers will vary.
   d) The sum of the degrees is equal to twice the number of edges, which is true, since each edge must connect two vertices. Each edge then contributes two to the sum of the degrees.

44.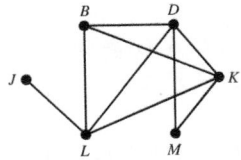

Other answers are possible.

45. a) Answers will vary.   **b)** Answers will vary.

1. Euler   **3.** No   **5.** Odd

7. a) Yes. One example is *A, B, D, E, C, A, D, C*.
   b) No. This graph has exactly two odd vertices, *A* and *C*. Each Euler path must begin at vertex *A* and end at vertex *C* or vice versa.

9. No. A graph with exactly two odd vertices has no Euler circuits.

11. No. A graph with more than two odd vertices has neither an Euler path nor an Euler circuit.

13. No. A graph with more than two odd vertices has neither an Euler path nor an Euler circuit.

15. *A, B, C, D, E, F, B, D, F, A*   **17.** *C, D, E, F, A, B, D, F, B, C*

19. *E, F, A, B, C, D, F, B, D, E*

21. a) Yes   **b)** They could start on either island and finish at the other.

23. a)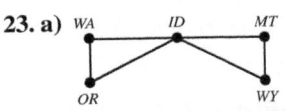
   b) Yes; *WA, ID, MT, WY, ID, OR, WA*

25. a)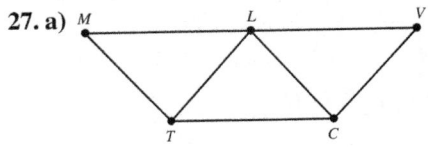
   b) Yes; *S, T, N, A, P, N, Q, J, S, Q, T*   **c)** No

27. a)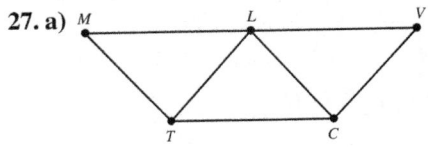
   b) Yes; *T, M, L, V, C, L, T, C*   **c)** No

29. a)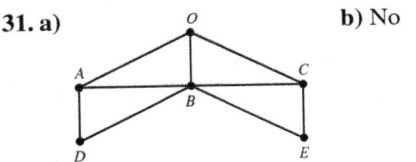
   b) Yes   **c)** *O, A, B, C, D, F, C, E, O, A*

31. a) 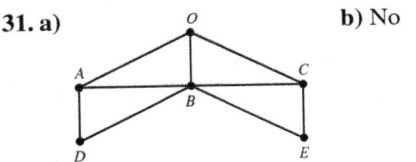   **b)** No

33. a) Yes
   b) The residents would need to start at the intersection of Maple Cir., Walnut St., and Willow St. or at the intersection of Walnut St. and Oak St.

35. A, B, E, D, C, A, D, B

37. A, B, H, E, B, C, E, F, H, G, F, D, C, A, G

39. A, B, C, D, I, H, G, B, E, H, C, E, G, F, A

41. A, C, D, G, H, F, C, F, E, B, A

43. A, B, C, E, B, D, E, F, I, E, H, D, G, H, I, J, F, C, A

45. UT, CO, NM, AZ, CA, NV, UT, AZ, NV

47. B, A, E, H, I, J, K, D, C, G, G, J, F, C, B, F, I, E, B

49. J, G, G, C, F, J, K, D, C, B, F, I, E, B, A, E, H, I, J

51. a) Yes. There are no odd vertices.
  b) Yes. There are no odd vertices.

53. a) No. There are more than two odd vertices.
  b) No. There is at least one odd vertex.

55. a) No
  b) California, Nevada, and Louisiana (and others) have an odd number of states bordering them. Since a graph of the United States would have more than two odd vertices, no Euler path and no Euler circuit exist.

57. a) 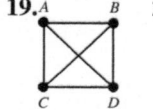   **b)**   **c)**

SECTION 13.3, PAGE 860

1. Salesman **3.** Hamilton **5.** Euler **7.** Force

9. B, A, E, F, G, H, D, C and H, D, C, G, F, E, A, B

11. A, B, C, D, G, F, E, H and E, H, F, G, D, C, A, B

13. A, G, J, D, E, B, C, F, I, H and J, D, A, G, F, I, H, E, B, C

15. A, B, D, G, E, H, F, C, A and D, B, A, C, F, H, E, G, D

17. A, B, C, F, I, E, H, G, D, A and A, E, B, C, F, I, H, G, D, A

19. 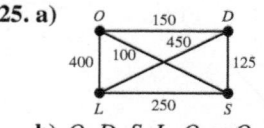   **21.** 5! = 120 ways

23. 12! = 479,001,600 ways

25. a)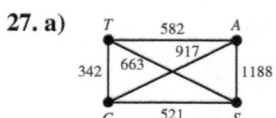
  b) O, D, S, L, O or O, L, S, D, O
  c) 925 feet

27. a)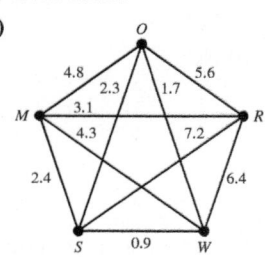
  d) T, A, S, C, T or T, C, S, A, T
  c) 2633 miles

29. a)   **b)** O, W, S, M, R, O; 13.7 miles
  c) Answers will vary.

31. a) 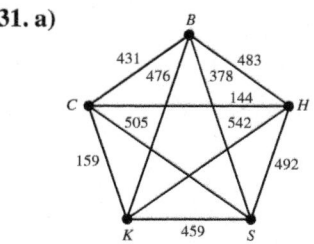   **b)** S, B, C, H, K, S for $1954
  c) Answers will vary.

33. a) Answers will vary. **b)** Answers will vary.
  c) Answers will vary. **d)** Answers will vary.

35. A, E, D, N, O, F, G, Q, P, T, M, L, C, B, J, K, S, R, I, H, A; other answers are possible.

SECTION 13.4, PAGE 871

1. Tree **3.** Circuits **5.** Minimum-cost

7.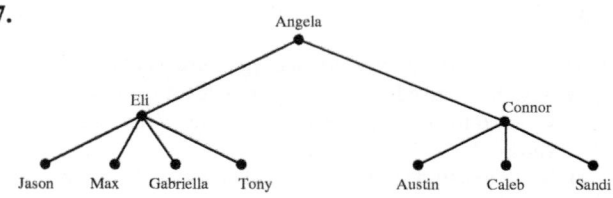

9.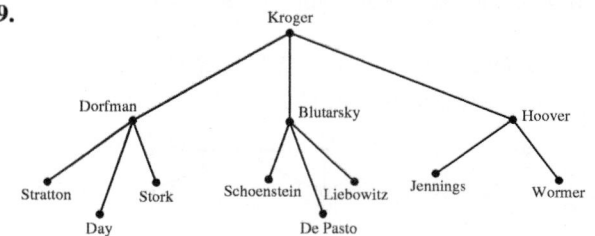

11.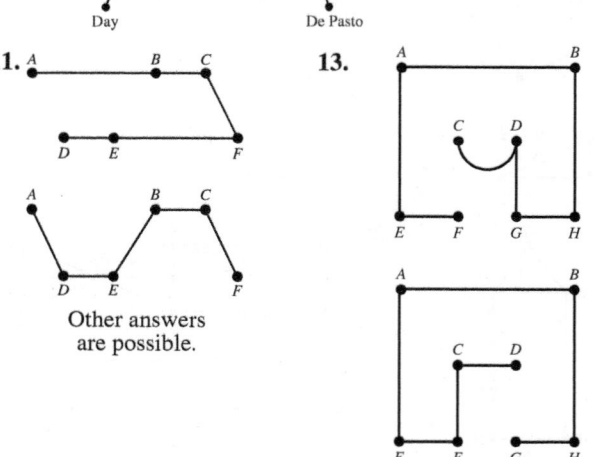
Other answers are possible.

13. Other answers are possible.

15.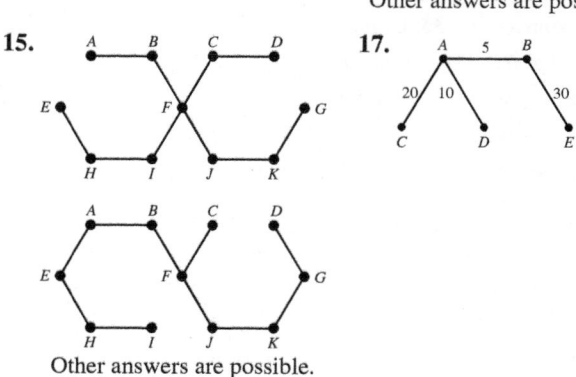
Other answers are possible.

17.

19.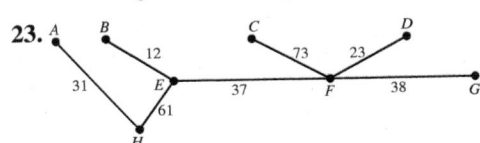

21.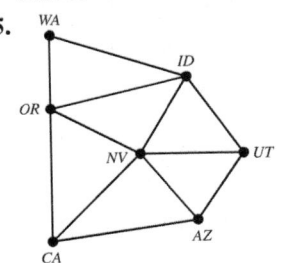

2.

Other answers are possible.

3. *E, C, D, F, E, G, A, B, H, G*

4. No. A path that includes each edge exactly one time would start at vertex *E* and end at vertex *G*, or vice versa.

5.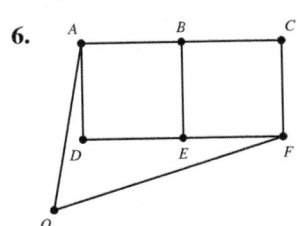

23.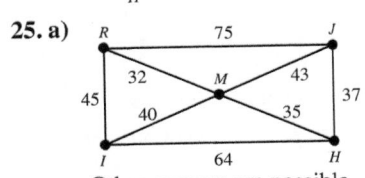

25. a)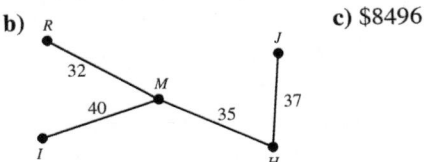

Other answers are possible.

b) 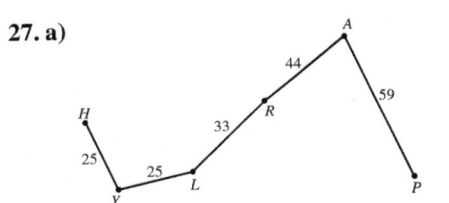   **c)** $8496

6.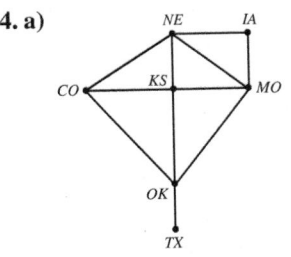

7. Connected   **8.** Disconnected   **9.** Edge *BD*

10. *C, A, B, F, H, G, C, D, E, D, E, F*; other answers are possible.

11. *F, E, D, E, D, C, G, H, F, B, A, C*; other answers are possible.

12. *A, B, E, G, F, D, C, A, D, E, A*

13. *E, D, C, A, E, B, A, D, F, G, E*

27. a)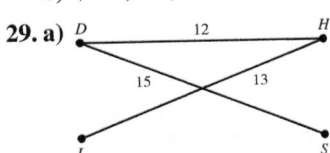

b) $241,800,000

29. a)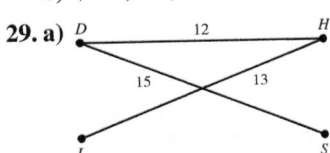

b) $460,000

14. a)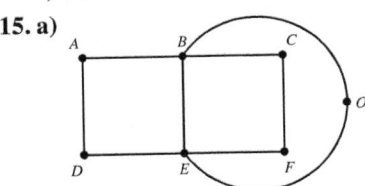

b) Yes. *CO, NE, IA, MO, NE, KS, MO, OK, CO, KS, OK, TX*; other answers are possible.

c) No

31. a)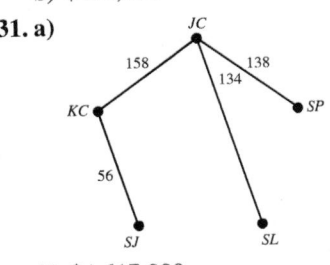

b) $4,617,000

33. Answers will vary.   **35.** Answers will vary.

REVIEW EXERCISES, PAGE 876

1.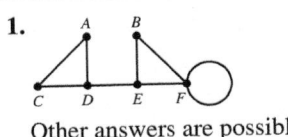

Other answers are possible.

15. a)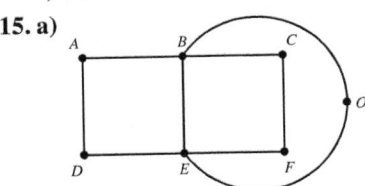

b) Yes

c) The person may start in any room or outside and will finish in the location from which he or she started.

16. a) Yes

b) The officer would have to start at either the intersection of Dayne St., Gibson Pl., and Alvarez Ave. or at the intersection of Chambers St., Fletcher Ct., and Alvarez Ave.

17. *C, A, B, G, F, A, D, C, F, D, B, E, D, G, E*; other answers are possible.

18. *A, B, D, E, I, J, O, N, L, K, G, H, L, M, I, H, D, C, G, F, A*; other answers are possible.

19. *C, A, E, K, I, F, G, J, L, H, B, D* and *D, B, H, L, J, G, F, C, A, E, K, I*; other answers are possible.

20. *A, B, E, F, J, I, L, K, H, G, C, D, A* and *I, J, F, E, B, A, D, C, G, H, K, L, I*; other answers are possible.

21. 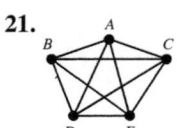   **22.** 4! = 24 ways

23. a) 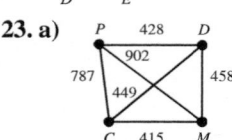   **b)** *P, D, M, C, P* or *P, C, M, D, P*
c) $2088

24. a)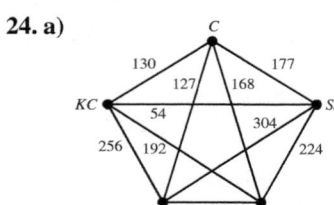

b) *SJ, KC, C, SL, Sp, SJ* traveling a total of 745 miles
c) *Sp, C, SL, KC, SJ, Sp* traveling a total of 829 miles

25.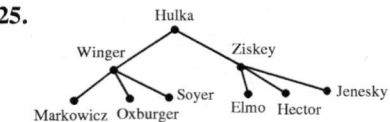

26.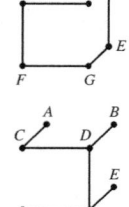

Other answers are possible.

27.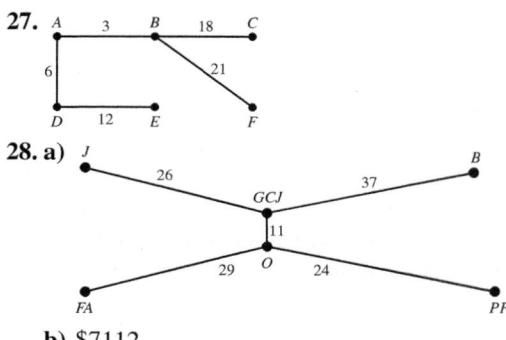

28. a)

b) $7112

1.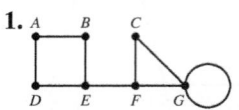
Other answers are possible.

2.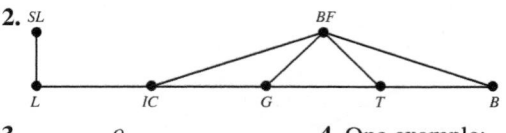

3. 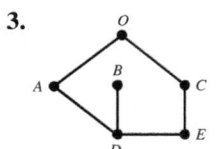   **4.** One example:

5. *A, B, E, D, C, A, D, B*; other answers are possible.

6. *A, B, D, G, H, F, C, B, E, D, F, E, C, A*; other answers are possible.

7. Yes. The person may start in room *A* and end at room *B*, or vice versa.

8. *A, D, E, A, F, E, H, F, I, G, F, B, G, C, B, A*; other answers are possible.

9. *B, A, D, E, F, C, G*; other answers are possible.

10. *A, B, C, G, E, D, H, I, K, J, F, A*; other answers are possible.

11. 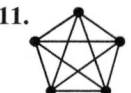   **12.** 5! = 120 ways

13.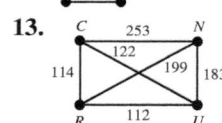

14. *C, R, N, U, C* or *C, U, N, R, C* for $618

15. *C, R, U, N, C* for $662

16. 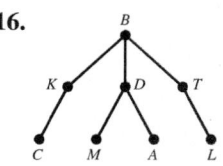   **17.**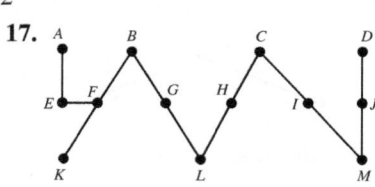

Other answers are possible.

18.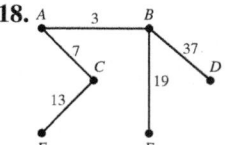

19.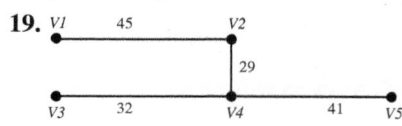

20. $183.75

Chapter 14

SECTION 14.1, PAGE 895

1. Majority **3.** $\dfrac{n(n-1)}{2}$ **5.** Plurality

7. Pairwise comparison **9. a)** Li **b)** No

11.

Number of Votes	3	1	2	2	1
First	B	D	H	H	D
Second	D	B	B	D	H
Third	H	H	D	B	B

13. 21 **15.** Guitar **17.** Keyboard **19.** Chicago

21. Three-way tie, no winner **23.** Portland **25.** No winner, a tie between Honolulu and Portland **27.** Brownstein

29. Marquez **31.** Lehigh Road **33.** Ontario Road

35. a) Bishara **b)** Bishara **c)** Bishara **d)** Bishara

37. a) Petunias **b)** Petunias **c)** Geraniums
  d) Tie between Petunias and Zinnias

39. Voters are able to provide more information, such as ranking choices, with the Borda count method. **41.** Brownstein

43. a) If there were only two columns, only two of the candidates were the first choice of the voters. If each of the 15 voters cast a ballot, one of the voters must have received a majority of votes because 15 cannot be split evenly.
  b) An odd number cannot be divided evenly, so one of the two first-choice candidates must receive more than half of the votes.

45. a) Wildcats, Rams and Tigers tied, Comets
  b) Rams, Tigers, Wildcats, Comets

47. a) 150 **b)** 50 **c)** Yes

49. One possible answer is:

Number of Votes	5	5	2
First	A	B	C
Second	B	A	B
Third	C	C	A

SECTION 14.2, PAGE 908

1. Majority **3.** Head-to-head **5.** Borda count

7. Plurality with elimination

9. New Orleans wins with the Borda count method, but Orlando has a majority of first-place votes.

11. a) Plan A is preferred
  c) b) No. C wins by the plurality method, but A is favored when compared to B and to C.

13. No. Lounge areas wins by the Borda count method, but parking is favored when compared to security and to lounge areas.

15. No. C wins by plurality with elimination, but B is favored over each of the other candidates using head-to-head comparisons.

17. No. A wins by the plurality method. If B drops out, C wins by the plurality method.

19. Yes. B wins by the Borda count method. If C drops out, B still wins.

21. No. C wins by plurality with elimination. If the voters change their preference, B wins by plurality with elimination.

23. No. D wins by pairwise comparison. If the voters change their preference, B wins by pairwise comparison.

25. No. B wins by pairwise comparison. If A, C, and E drop out, D wins by pairwise comparison.

27. No. B wins by the Borda count method, but A has a majority of first-place votes.

29. a) Savannah **b)** Savannah
  c) Savannah **d)** Savannah
  e) Savannah **f)** None of them

31. a) Steve Harvey **b)** Bradley Cooper **c)** Yes

33. A candidate who holds a plurality only gains strength and holds an even larger lead if more favorable votes are added.

35. Answers will vary.

37. Answers will vary.

SECTION 14.3, PAGE 927

1. Divisor **3.** Upper **5.** Quota **7.** Hamilton's

9. a) Webster's **b)** Adams' **c)** Jefferson's

11. a) 50,000 **b)** 29.40, 54.60, 17.14, 53.86

13. a) 24.75, 55.38, 17.38, 54.62 **b)** 24, 55, 17, 54

15. a) 24.06, 53.85, 16.90, 53.12 **b)** 25, 54, 17, 54

17. 24, 55, 17, 54 **19. a)** 27 **b)** 11.33, 7.93, 5.74

21. a) 11.86, 8.29, 6.01 **b)** 11, 8, 6

23. a) 10.55, 7.38, 5.34 **b)** 11, 8, 6 **25.** 11, 8, 6

27. a) 11 **b)** 7.82, 9.27, 11.82, 21.09 **c)** 8, 9, 12, 21

29. 8, 9, 12, 21 **31. a)** 52 **b)** 33.58, 136.44, 40.98, 18.02, 20.98 **c)** 34, 136, 41, 18, 21 **33.** 33, 137, 41, 18, 21

35. a) 90 **b)** 42.67, 32.07, 26.58, 18.69 **c)** 43, 32, 26, 19

37. 43, 32, 26, 19 **39. a)** 750 **b)** 12.09, 20.37, 17.08, 7.63, 33.67, 9.17 **c)** 12, 20, 17, 8, 34, 9 **41.** 12, 20, 17, 8, 34, 9

43. a) 12 **b)** 62.58, 81.67, 41.92, 13.83 **c)** 62, 82, 42, 14

45. 63, 82, 42, 13 **47. a)** 7, 2, 2, 2, 8, 14, 4, 5, 10, 10, 13, 2, 6, 2, 18 **b)** 7, 1, 2, 2, 8, 14, 4, 5, 10, 10, 13, 2, 6, 2, 19
  c) Benefited: Virginia; Disadvantage: Delaware

49. Answers will vary. One possible answer is A: 743, B: 367, C: 432, D: 491, E: 519, F: 388

SECTION 14.4, PAGE 936

1. Population **3.** Alabama **5.** Small

7. No. The new apportionment is 14, 11, 11, 12, 13. No clinic suffers a loss, so the Alabama paradox doesn't occur.

9. a) 6, 8, 16 **b)** Yes. When the number of seats increases, states B and C gain a seat and state A loses a seat.

11. a) 61, 40, 99 **b)** No

13. a) 7, 16, 9, 11, 11 **b)** Yes. Division B loses an internship to division A even though the population of division B grew faster than the population of division A.

15. a) 8, 40 **b)** Yes. The apportionment is now 9, 39, 7, and group B loses a manager.

17. a) 7, 26 **b)** No. The apportionment is 7, 26, 7.

REVIEW EXERCISES, PAGE 939

1. a) Comstock **b)** No **2. a)** Michelle **b)** Yes

3.

Number of Votes	3	2	1	1	3
First	B	A	D	D	C
Second	A	C	C	A	B
Third	C	D	A	B	A
Fourth	D	B	B	C	D

4.

Number of Votes	2	2	2	1
First	C	A	B	C
Second	A	C	A	B
Third	B	B	C	A

5. 11   **6.** Chipotle Mexican Grill   **7.** Domino's Pizza

8. Chipotle Mexican Grill   **9.** Three-way tie, no winner

10. Chipotle Mexican Grill   **11.** 110   **12.** Volleyball

13. Soccer   **14.** Volleyball   **15.** None, a three-way tie

16. Soccer   **17. a)** Yes, American Music   **b)** American Music
c) American Music   **d)** American Music   **e)** American Music

18. a) No   **b)** Seattle   **c)** Las Vegas   **d)** Las Vegas
e) Las Vegas

19. a) A tie between *Chronicles of Narnia* and *Where on Earth Is Carmen Sandiego*?
b) *Where on Earth Is Carmen Sandiego*?
c) A tie between *Chronicles of Narnia* and *Where on Earth Is Carmen Sandiego*?

20. No. A has a majority, but B wins by the Borda count method.

21. No. B wins by the Borda count method, but A is preferred over each of the other candidates using head-to-head comparisons.

22. a) C wins the election by plurality with elimination.
b) When the order is changed A wins. Therefore, the monotonicity criterion is not satisfied.

c) No. If B drops out, A is the winner by plurality with elimination. Therefore, the irrelevant alternatives criterion is not satisfied.

23. a) Ragu   **b)** Prego   **c)** Tie between Newman's Own and Barilla   **d)** Barilla   **e)** Ragu
f) Plurality method, Borda count method, and plurality with elimination method all violate the head-to-head criterion.

24. a) Yes, Fleetwood Mac   **b)** Boston   **c)** Fleetwood Mac
d) REO Speedwagon   **e)** Fleetwood Mac
f) The plurality method and the plurality with elimination method

25. The Borda count method

26. The plurality with elimination method

27. The pairwise comparison method and the Borda count method

28. 4, 3, 3   **29.** 5, 2, 3   **30.** 4, 3, 3   **31.** 5, 2, 3

32. Yes. The apportionment is 5, 2, 4. Region B loses one truck.

33. 11, 7, 5   **34.** 11, 7, 5   **35.** 10, 7, 6   **36.** 11, 7, 5

37. No. The apportionment with the additional population is still 11, 7, 5.   **38.** 4, 46   **39.** 4, 46   **40.** 5, 45   **41.** 4, 46

42. Yes. The new apportionment is 5, 45, 5. State B loses a seat.

CHAPTER TEST, PAGE 942

1. 12   **2.** No   **3.** Pizza   **4.** Pizza   **5.** Deli sandwiches

6. Deli sandwiches   **7.** 142   **8.** Salamander   **9.** Iguana

10. Hamster   **11.** Lemming

12. Plurality method, Borda count method, and plurality with elimination method

13. Yes, El Capitan has a majority of first-place votes but mule deer wins.

14. 6, 9, 15   **15.** 6, 9, 15   **16.** 6, 9, 15   **17.** 6, 9, 15

18. No. The apportionment is 6, 9, 16.

19. No. The apportionment is 6, 9, 15.

20. No. The apportionment is 6, 9, 15, 5.

Credits

PHOTO CREDITS

Chapter 1

Cover, Busà Photography/Getty Images; Page 1, ESB Professional/Shutterstock; Page 2, Tatiana Makotra/123RF; Page 4, National Aeronautics and Space Administration; Page 9T, sirtravelalot/Shutterstock; Page 9B, William Perugini/123RF; Page 11, lorcel/123RF; Page 14TL, Seksak Kerdkanno/123RF; Page 14TR, Allen R. Angel; Page 14B, Allen R. Angel; Page 18TL, Allen R. Angel; Page 18BL, Nitr/Fotolia; Page 18TR, Luciano Mortula-LGM/Fotolia; Page 18BR, Nick Stubbs/Shutterstock; Page 19, rawpixel/123RF; Page 20, rawpixel/123RF; Page 22, Associated Press; Page 25L, william87/Fotolia; Page 25R, Thomas Andreas/Shutterstock; Page 29, Romrodphoto/Shutterstock; Page 31, GagliardiPhotography/Shutterstock; Page 39, Sean Xu/Shutterstock

Chapter 2

Page 41, wavebreakmedia/Shutterstock; Page 42, Freeograph/Shutterstock; Page 45, Ganna Tokolova/Shutterstock; Page 47, National Aeronautics and Space Administration; Page 48, Olivier Le Queinec/Shutterstock; Page 50T, Popular graphic art print filing series (Library of Congress); Page 50B, Jon Osumi/Shutterstock; Page 52, Amelandfoto/Shutterstock; Page 54T, Nejron Photo/Shutterstock; Page 54B, nednapa/Shutterstock; Page 55L, EPG_Euro-PhotoGraphics/Shutterstock; Page 55R, Marzanna Syncerz/Fotolia; Page 56T, Allen R. Angel; Page 56B, Amble Design/Shutterstock; Page 64, Christopher Boswell/Shutterstock; Page 65, action sports/Shutterstock; Page 66, Ingram Publishing/Getty Images; Page 67, Andrey_Popov/Shutterstock; Page 72, ChameleonsEye/Shutterstock; Page 76, Robert Kneschke/Shutterstock; Page 80L, sirtravelalot/Shutterstock; Page 80R, Stokkete/Shutterstock; Page 81, Kzenon/Shutterstock; Page 82, Jupiterimages/Getty Images

Chapter 3

Page 91, William Perugini/123RF; Page 92, chutima kuanamon/Alamy Stock Photo; Page 93, lunamarina/Shutterstock; Page 94, Pictorial Press Ltd/Alamy Stock Photo; Page 97, wang Tom/123RF; Page 99, Jim/Fotolia; Page 100, ProStockStudio/Shutterstock; Page 101L, racorn/123RF GB Ltd; Page 101R, Jason Patrick Ross/Shutterstock; Page 102, Zety Akhzar/Shutterstock; Page 106, Jiri Vaclavek/Shutterstock; Page 112, BIGANDT.COM/Shutterstock; Page 113, Brocreative/Shutterstock; Page 114, Ioan Panaite/Shutterstock; Page 115, Samuel Borges Photography/Shutterstock; Page 116, Lemusique/Shutterstock; Page 118L, Georgejmclittle/Shutterstock; Page 118R, lunamarina/Shutterstock; Page 120, Bryan Pollard/Shutterstock; Page 124L, Nataly Studio/Shutterstock; Page 124R, J Main/Shutterstock; Page 126T, Linn Currie/Shutterstock; Page 126B, stockfour/Shutterstock; Page 131, Viorel Sima/Shutterstock; Page 132, Jaap Arriens/Alamy Stock Photo; Page 133, AF archive/Alamy Stock Photo; Page 134, sint/Shutterstock; Page 136, Nestor Rizhniak/Shutterstock; Page 137R, Diane Macdonald/Getty Images; Page 137L, ESB Professional/Shutterstock; Page 138, Oranzy Photography/Shutterstock; Page 139, Tropical studio/Shutterstock; Page 141, K.A.Willis/Shutterstock; Page 143, Onur ERSIN/Shutterstock; Page 145, PhotoProCorp/Shutterstock; Page 147L, littlenySTOCK/Shutterstock; Page 147R, gpointstudio/Shutterstock; Page 148, Iakov Filimonov/Shutterstock; Page 149, Hero Images Inc./Alamy Stock Photo; Page 153, Frank L Junior/Shutterstock; Page 154, ravl/Shutterstock; Page 155, Pressmaster/Shutterstock; Page 158, IanDagnall Computing/Alamy Stock Photo

Chapter 4

Page 166, Dmitry Kalinovsky/123RF; Page 167, verdateo/123RF; Page 168, Album/Alamy Stock Photo; Page 171T, Allen R. Angel; Page 171B, Allen R. Angel; Page 175, Nicholas Pitt/Getty Images; Page 176, Feel good studio/Shutterstock; Page 177, ekkapon/Shutterstock; Page 178, Wavebreak Media Ltd/123RF; Page 180, Ruslan Olinchuk/Fotolia; Page 183, djile/Shutterstock; Page 184, Lisa S./Shutterstock; Page 185, Ljupco Smokovski/Shutterstock; Page 186, foreverhappy/123RF; Page 189, TOLBERT PHOTO/Alamy Stock Photo; Page 190, Pressmaster/Shutterstock; Page 198T, Jupiterimages/Getty Images; Page 198B, Radu Bercan/Shutterstock; Page 199, Pressmaster/Shutterstock; Page 200, SSPL/Getty Images

Chapter 5

Page 207, Stuart Jenner/Shutterstock; Page 208, ronstik/Shutterstock; Page 214, Oleksiy Mark/Shutterstock; Page 216, Gary/Fotolia; Page 217, ALEXANDRE FABRO/Shutterstock; Page 219, Khamidulin Sergey/Shutterstock; Page 222, Vixit/Shutterstock; Page 227, Bart Sadowski/Shutterstock; Page 229, Duplass/Shutterstock; Page 237, Pavel L Photo and Video/Shutterstock; Page 239, Ecuadorpostales/Shutterstock; Page 241, Hurst Photo/Shutterstock; Page 244, Robert Nyholm/123RF; Page 245, Lanmas/Alamy Stock Photo; Page 250, LightField Studios/Shutterstock; Page 251L, SergiyN/Shutterstock; Page 251R, Alexandr Shevchenko/Shutterstock; Page 252, Alexander Raths/Shutterstock; Page 253, robertpinna/Fotolia; Page 258, Maria Sbytova/Shutterstock; Page 259, Alex Mit/Shutterstock; Page 262, Ron Tarver/Newscom; Page 264, Library of Congress; Page 266, pjmorley/Shutterstock; Page 267, Danita Delmont/Shutterstock; Page 268, Kateryna Kon/Shutterstock; Page 270, serpeblu/Shutterstock; Page 272, Cryber/Shutterstock; Page 274T, rudall30/Shutterstock; Page 274B, Terry Oakley/Alamy; Page 275, kzww/Shutterstock; Page 276, Eiji Ueda/Shutterstock; Page 277, Dan Breckwoldt/Shutterstock; Page 278L, Olga Drabovich/Shutterstock; Page 278R, Henry Bonn/Fotolia; Page 278B, akg-images/Superstock; Page 279, Caryl Bryer Fallert-Gentry

Chapter 6

Page 287, Lucky Business/Shutterstock; Page 288, ohrim/Shutterstock; Page 290, Alfonso Vicente/Alamy Stock Photo; Page 297, OSTILL is Franck Camhi/Shutterstock; Page 298, Brenda Carson/Shutterstock; Page 299, Library of Congress Prints and Photographs Division [LC-USZ62-60242]; Page 303, Minerva Studio/Shutterstock; Page 305, Elena Hramova/Shutterstock; Page 307, hadynyah/iStock/Getty Images; Page 308, Alan Stoddard/Shutterstock; Page 310, eurobanks/Shutterstock; Page 311L, Joe Seer/Shutterstock; Page 311R, Olesya Baron/Shutterstock; Page 312, Andrey_Popov/Shutterstock; Page 313, zhangyang13576997233/Shutterstock; Page 314, antoniodiaz/Shutterstock; Page 315, dwphotos/Shutterstock; Page 316, Featureflash Photo Agency/Shutterstock; Page 317, Christine Abbott; Page 318, Dinis Tolipov/123RF; Page 319L, Sergey Novikov/123RF; Page 319R, HONGQI ZHANG/123RF; Page 320, Lisa F. Young/Shutterstock; Page 326, wavebreakmedia/Shutterstock; Page 327T, sirtravelalot/Shutterstock; Page 327B,

Phovoir/Shutterstock; Page 328, Disney/Marvel/Kobal/Shutterstock; Page 331, OZGIOUN SAMPRI/Shutterstock; Page 333, CTK/Alamy Stock Photo; Page 336, Ollyy/Shutterstock; Page 339, sebra/Shutterstock; Page 341, Goncharov_Artem/Shutterstock; Page 348, Steve Mann/Shutterstock; Page 349T, Library of Congress (Photoduplication); Page 349B, wavebreakmedia/Shutterstock; Page 352, karamysh/Shutterstock; Page 353L, Andrey_Popov/Shutterstock; Page 353R, Iakov Filimonov/Shutterstock; Page 355, VDB Photos/Shutterstock; Page 361, Iakov Filimonov/Shutterstock; Page 363, Andrey_Popov/Shutterstock; Page 364, lzf/Shutterstock; Page 365T, Chuck Wagner/Shutterstock; Page 365B, wavebreakmedia/Shutterstock; Page 372, sciencephotos/Alamy Stock Photo; Page 375, Photo Melon/Shutterstock; Page 376, iofoto/Shutterstock; Page 378, SeventyFour/Shutterstock; Page 379, NASA Pictures/Alamy Stock Photo; Page 381, Mike Focus/Shutterstock; Page 383, Vanessa van Rensburg/Shutterstock; Page 385, phaitoon/123RF; Page 390, Okrasyuk/Shutterstock; Page 392, North Wind Picture Archives/Alamy Stock Photo

Chapter 7

Page 399, Greg Balfour Evans/Alamy Stock Photo; Page 400, Creatas Images/Getty Images; Page 401L, Allen R. Angel; Page 401R, Allen R. Angel; Page 401B, Evgeny Malkov/123RF; Page 402, National Aeronautics and Space Administration; Page 404, Allen R. Angel; Page 405, Allen R. Angel; Page 407, cyo bo/Shutterstock; Page 408R, Craig Hanson/Shutterstock; Page 408L, Oleg Znamenskiy/123RF; Page 409T, Allen R. Angel; Page 409B, Gareth Boden/Pearson Education Ltd; Page 410, Africa Studio/Fotolia; Page 411, AboutLife/Shutterstock; Page 413, DesignPrax/Shutterstock; Page 415, michaeljung/Fotolia; Page 416, Lotus_studio/Shutterstock; Page 417L, Borislav Marinic/123RF; Page 417R, sirchitvises/123RF; Page 418, jfergusonphotos/Fotolia; Page 419T, CleverPencil/Getty Images; Page 419B, Andrey Armyagov/Fotolia; Page 420, Alex Horvath/ZUMA Press, Inc./Alamy Stock Photo; Page 421, dbrus/Fotolia; Page 424, Kamuran Ağbaba/123RF; Page 425, Stephen VanHorn/Shutterstock; Page 426L, Steve Byland/Shutterstock; Page 426R, Michael Potter11/Shutterstock; Page 427L, Christopher Boswell/Shutterstock; Page 427R, Allen R. Angel; Page 429, Vaidas Bucys/123RF; Page 430, Agcuesta/Fotolia; Page 432, SkyBlodgett/Shutterstock; Page 433, National Aeronautics and Space Administration; Page 434, Allen R. Angel; Page 435, Shutterstock; Page 436, dirk ercken/123RF; Page 437, wavebreakmedia/Shutterstock; Page 438TR, Allen R. Angel; Page 438L, Ariel Bravy/Shutterstock; Page 438BR, Allen R. Angel

Chapter 8

Page 444, Petr David Josek/AP/Shutterstock; Page 445, BrunoWeltmann/Shutterstock; Page 447, Daniel M Ernst/Shutterstock; Page 457, LightField Studios/Shutterstock; Page 459, TCD/Prod.DB/Alamy Stock Photo; Page 460, boris64/iStock/Getty Images; Page 467, Dmitry Argunov/Shutterstock; Page 470, Pictorial Press Ltd / Alamy Stock Photo; Page 472T, Dmytro Zinkevych/123RF; Page 472B, Charles Rex Arbogast/AP Images; Page 477, Allen R. Angel; Page 478, racorn/123RF; Page 481, fabio formaggio/123RF; Page 487, Artens/Shutterstock; Page 490T, Georgiy Pashin/Fotolia; Page 490B, Irina Afonskaya/Shutterstock; Page 491, Stefan Dahl Langstrup/Alamy Stock Photo; Page 494, Science History Images/Alamy Stock Photo; Page 500T, Serp/Shutterstock; Page 500B, superjoseph/Fotolia; Page 512, Ian Allenden/123RF; Page 517, NASA images/Shutterstock; Page 518, bennymarty/123RF; Page 520L, Vladimir Caplinskij; Page 520R, KristinaSh/Shutterstock; Page 522, Andrew Hazelden

Chapter 9

Page 531, Tomasz Trojanowski/Shutterstock; Page 532, NikoNomad/Shutterstock; Page 536, dani shlom/Shutterstock; Page 540, CREATISTA/Shutterstock; Page 541, emmeci74/Fotolia; Page 542, Coprid/Fotolia; Page 550, Volt Collection/Shutterstock; Page 554, Allen R. Angel; Page 555, agsandrew/Fotolia; Page 557, Christoph Riddle/Shutterstock; Page 558, Cathy Yeulet/123RF; Page 559, Kzenon/Shutterstock; Page 560, Allen R. Angel; Page 561, Peter Sobolev/Shutterstock; Page 568, William Graham/Alamy Stock Photo

Chapter 10

Page 573, wavebreakmedia/Shutterstock; Page 574, hedgehog94/Shutterstock; Page 576, Aliaksandr Mazurkevich/123RF; Page 577, Cal Sport Media/Alamy Stock Photo; Page 578, Cara-Foto/Shutterstock; Page 580, Mikkel Bigandt/123RF; Page 581, Brent Hofacker/Shutterstock; Page 583, Jacek Chabraszewski/Shutterstock; Page 584, Nejron Photo/Shutterstock; Page 586, ValeStock/Shutterstock; Page 587, Ljupco Smokovski/Shutterstock; Page 589, Carlos Enrique Santa Maria/123RF; Page 591, Syda Productions/Shutterstock; Page 592B, Graham Oliver/123RF; Page 592T, Zoriana Zaitseva/Shutterstock; Page 593L, Medioimages/Photodisc/Getty Images; Page 593R, Iakov Filimonov/Shutterstock; Page 594, Lyudmyla Kharlamova/Shutterstock; Page 595, RuslanDashinsky/iStock/Getty Images; Page 596, ARENA Creative/Shutterstock; Page 597, Fotolia; Page 600, Library of Congress; Page 601, Shamleen/Shutterstock; Page 602, ESB Professional/Shutterstock; Page 603, Blend Images/Shutterstock; Page 604, YanLev/Shutterstock; Page 607T, Igor Kardasov/Shutterstock; Page 607B, Sallehudin Ahmad/Shutterstock; Page 609, nevodka/Shutterstock; Page 611, DGLimages/Shutterstock; Page 613, Elena Elisseeva/Shutterstock; Page 614, Tetra Images/Getty Images; Page 615L, Anna Kraynova/Shutterstock; Page 615R, hedgehog94/Shutterstock; Page 617T, Parkol/Shutterstock; Page 617B, ALPA PROD/Shutterstock; Page 618, Flamingo Images/Shutterstock; Page 619, Hurst Photo/Shutterstock; Page 620, Gabriela Beres/Shutterstock; Page 626, Susan Law Cain/Shutterstock; Page 627, Rene Frederick/Getty Images; Page 628, Andy Dean Photography/Shutterstock; Page 629, qingwa/Fotolia; Page 630, kzenon/123RF; Page 631, Neil Fraser/Alamy Stock Photo; Page 632, K. Geijer/Fotolia; Page 634, amriphoto/Getty Images; Page 635, Duncan Noakes/Fotolia; Page 637L, Spiroview Inc/Shutterstock; Page 637R, Martin Molcan/123RF

Chapter 11

Page 644, Comaniciu Dan/Shutterstock; Page 645, Andrei Kholmov/Shutterstock; Page 647, Cal Sport Media/Alamy Stock Photo; Page 649, Allen R. Angel; Page 652, James Pierce/123RF; Page 654, Andre Jenny/Alamy Stock Photo; Page 655, chingyunsong/Shutterstock; Page 657, pilipphoto/Shutterstock; Page 659, UPI/Alamy Stock Photo; Page 664, Andrey_Popov/Shutterstock; Page 665, Keith Brofsky/Getty Images; Page 666, PA Images/Alamy Stock Photo; Page 667, Shutterstock; Page 668, EQRoy/Shutterstock; Page 669, michaeljung/Shutterstock; Page 670, sirtravelalot/Shutterstock; Page 674L, Alexey Stiop/Corn; Page 674R, Dmytro Zinkevych/Shutterstock; Page 678, Antonio Guillem/Shutterstock; Page 680T, Piotr Zajac/Shutterstock; Page 680B, Sharkshock/Shutterstock; Page 684, Shutterstock; Page 685, Flashon Studio/Shutterstock; Page 687, Gareth Boden/Pearson Education Ltd; Page 688, ben bryant/Shutterstock; Page 689, Rocco Macri/123RF; Page 697, Kat72/Shutterstock; Page 690, Mikael Damkier/Shutterstock; Page 692, Casey Sykes/The Grand Rapids Press/AP Images; Page 693, Dmitrii Shironosov/123RF; Page 699, Artens/Shutterstock; Page 701, Anna Om/Shutterstock; Page 702, Cliff Lipson/CBS/Getty Images; Page 704, Sergey Nivens/Shutterstock; Page 705L, Antonio Guillem/123RF; Page 705R, Rob Wilson/Shutterstock;

Page 707, Frank11/Shutterstock; Page 709, Dusan Petkovic/Shutterstock; Page 710, TCD/Prod.DB/Alamy Stock Photo; Page 713T, Shutterstock; Page 713B, flairimages/Fotolia; Page 717, Shutterstock; Page 718, PA Images/Alamy Stock Photo; Page 719, VGstockstudio/Shutterstock; Page 720, dennizn/Shutterstock; Page 723L, sirtravelalot/Shutterstock; Page 723R, AA World Travel Library/Alamy Stock Photo; Page 724, Igor Bulgarin/Shutterstock; Page 726, vladimirfloyd/Fotolia; Page 727, Allen R. Angel; Page 728T, PictureLux/The Hollywood Archive/Alamy Stock Photo; Page 728B, actionsports/123RF; Page 729T, ermess/Shutterstock; Page 729B, Mark Herreid/Shutterstock; Page 730R, Paylessimages/Fotolia; Page 730L, B Christopher/Alamy Stock Photo; Page 731, Paul Vasarhelyi/Shutterstock; Page 736, Comstock/Getty Images; Page 739T, Philip Scalia/Alamy Stock Photo; Page 739B, JJ pixs/Shutterstock

Chapter 12

Page 747, Leremy/Shutterstock; Page 748, Soloviova Liudmyla/Shutterstock; Page 749, Andrey_Popov/Shutterstock; Page 750, SPL/Science Source; Page 751, Chris Tefme/Shutterstock; Page 754T, BlueSkyImage/Shutterstock; Page 754B, Wavebreak Media Ltd/123rf; Page 755L, Jenifoto/Fotolia; Page 755R, Tsomka/Shutterstock; Page 757, Shutterstock; Page 763, Katherine Wallman; Page 764, bjul/Shutterstock; Page 767, fuyu liu/Shutterstock; Page 768, Adam Gault/DigitalVision/Getty Images; Page 769, Ufuk Uyanik/Alamy Stock Photo; Page 771, Alessandro Landi/123RF; Page 772, 914 collection/Alamy Stock Photo; Page 773B, DenisTangneyJr/E+/Getty Images; Page 773T, DenisTangneyJr/iStock/Getty Images; Page 779L, Miranda Drummond, Syracuse University; Page 779R, Dave Allen Photography/Shutterstock; Page 782T, Pictorial Press Ltd/Alamy Stock Photo; Page 782B, Anonymous/AP Photo; Page 784T, Amble Design/Shutterstock; Page 784B, g-stockstudio/Shutterstock; Page 786, Arapov Sergey/Shutterstock; Page 789, Antonio Guillem/123rf; Page 792, Lisa F. Young/Shutterstock; Page 796, VDB Photos/Shutterstock; Page 803, Ahturner/Shutterstock; Page 805, Phoo Chan/Shutterstock; Page 806L, Darrin Henry/Shutterstock; Page 806R, Vasin Leenanuruksa/123RF; Page 807, Jules Selmes/Pearson Education Ltd; Page 808, Digital Vision/Photodisc/Getty Images; Page 809, Aaron Kohr/Shutterstock; Page 821, Ian Allenden/Shutterstock; Page 822, Antonio Diaz/Shutterstock; Page 823, Stockbyte/Getty Images

Chapter 13

Page 829, Country Gate Productions/Shutterstock; Page 830, David Grossman/Alamy Stock Photo; Page 831, Samuel Borges Photography/Shutterstock; Page 832, Nate Loper/Shutterstock; Page 833, melissamn/Shutterstock; Page 840, ND700/Shutterstock; Page 842, nobeastsofierce/Shutterstock; Page 843, Johnny Adolphson/Shutterstock; Page 847, ArcadeImages/Alamy Stock Photo; Page 852, Eric Buermeyer/Shutterstock; Page 853, matej_z/Shutterstock; Page 855, Everett Collection Inc/Alamy Stock Photo; Page 858, Kathy Hutchins/Shutterstock; Page 861L, Matt Smith Photographer/Shutterstock; Page 861R, Derren/Shutterstock; Page 862, Sean Pavone/Shutterstock; Page 863, Royal Irish Academy; Page 864, sangkhom sangkakam/Shutterstock; Page 867, FPG/Archive Photos/Getty Images; Page 870, Rodan/Shutterstock; Page 872TTL, Frank L Junior/Shutterstock; Page 872TTR, Juliet Photography/Fotolia; Page 872, Darya Prokapalo/Fotolia; Page 872TBL, Pjhpix/Fotolia; Page 872TBR, ANDRII GLUSHCHENKO/123RF; PageTC 872BTL, Anton Starikov/123RF; Page 872BTR, Dominator/Fotolia; Page 872BR, Dusty Cline/Shutterstock; Page 872BB, Gerasimenuk/Fotolia; Page 872BBL, DonNichols/iStock/Getty Images; Page 873, Ahturner/Shutterstock

Chapter 14

Page 881, David R. Frazier Photolibrary, Inc./Alamy Stock Photo; Page 882T, FamVeld/Shutterstock; Page 882B, Zhao jian kang/Shutterstock; Page 883, Christopher Meder/Shutterstock; Page 884, carlofranco/E+/Getty Images; Page 885, Burlingham/Shutterstock; Page 888, Nick Fox/Shutterstock; Page 891, neelsky/Shutterstock; Page 893, James Landy/Library of Congress Prints and Photographs Division [LC-USZ62-64279]; Page 894, Kathy Hutchins/Shutterstock; Page 895, viki2win/Shutterstock; Page 896R, Josemaria Toscano/123RF; Page 896L, Sean Pavone/123RF; Page 897, PictureLux/The Hollywood Archive/Alamy Stock Photo; Page 898, Chad McDermott/123RF; Page 899T, Tinseltown/Shutterstock; Page 899B, Fotolia; Page 901, Hero Images/ALamy Stock Photo; Page 905, sirtravelalot/Shutterstock; Page 908, WENN Ltd/Alamy Stock Photo; Page 909L, GTS Productions/Shutterstock; Page 909R, AVAVA/Shutterstock; Page 910, Svetlana Kolpakova/Shutterstock; Page 912L, James Kirkikis/123RF; Page 912R, Andre Luiz Moreira/Shutterstock; Page 914, Syda Productions/ Shutterstock; Page 915, Richard Ellis/Alamy Stock Photo; Page 916, Library of Congress Prints and Photographs Division; Page 918, W.H. Gallagher Co., N.Y/Library of Congress Prints and Photographs Division; Page 921, Brady-Handy Photograph Collection/Library of Congress Prints and Photographs Division; Page 923, Healy, G. P. A. (George Peter Alexander), 1813–1894, artist, J.C. Tichenor/Library of Congress Prints and Photographs Division; Page 928L, EpicStockMedia/Shutterstock; Page 928R, Nate Allred/Shutterstock; Page 928, georgerudy/123RF; Page 930, Adam Radosavljevic/Shutterstock; Page 934, Wavebreak Media Ltd/123RF; Page 936, Andrey Popov/Shutterstock; Page 937T, Dmitry Kalinovsky/Shutterstock; Page 937B, scanrail/123RF

Index of Applications

Index

Note: Figures and tables are indicated by *f* and *t*.